TABLE A.2 Cumulative normal distribution (continued)

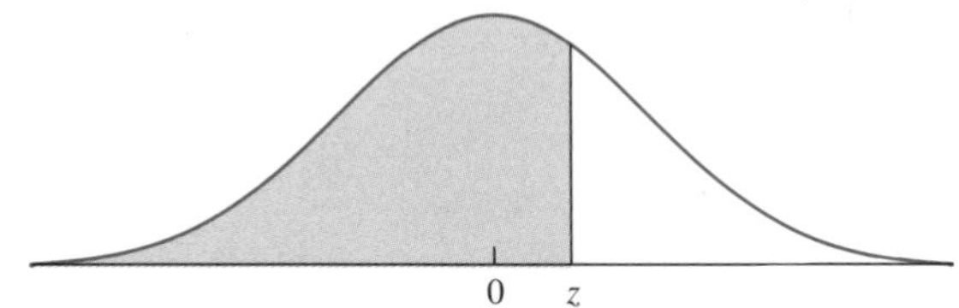

z	0.00	0.01	0.02	0.03	0.04	0.05	0.06	0.07	0.08	0.09
0.0	.5000	.5040	.5080	.5120	.5160	.5199	.5239	.5279	.5319	.5359
0.1	.5398	.5438	.5478	.5517	.5557	.5596	.5636	.5675	.5714	.5753
0.2	.5793	.5832	.5871	.5910	.5948	.5987	.6026	.6064	.6103	.6141
0.3	.6179	.6217	.6255	.6293	.6331	.6368	.6406	.6443	.6480	.6517
0.4	.6554	.6591	.6628	.6664	.6700	.6736	.6772	.6808	.6844	.6879
0.5	.6915	.6950	.6985	.7019	.7054	.7088	.7123	.7157	.7190	.7224
0.6	.7257	.7291	.7324	.7357	.7389	.7422	.7454	.7486	.7517	.7549
0.7	.7580	.7611	.7642	.7673	.7704	.7734	.7764	.7794	.7823	.7852
0.8	.7881	.7910	.7939	.7967	.7995	.8023	.8051	.8078	.8106	.8133
0.9	.8159	.8186	.8212	.8238	.8264	.8289	.8315	.8340	.8365	.8389
1.0	.8413	.8438	.8461	.8485	.8508	.8531	.8554	.8577	.8599	.8621
1.1	.8643	.8665	.8686	.8708	.8729	.8749	.8770	.8790	.8810	.8830
1.2	.8849	.8869	.8888	.8907	.8925	.8944	.8962	.8980	.8997	.9015
1.3	.9032	.9049	.9066	.9082	.9099	.9115	.9131	.9147	.9162	.9177
1.4	.9192	.9207	.9222	.9236	.9251	.9265	.9279	.9292	.9306	.9319
1.5	.9332	.9345	.9357	.9370	.9382	.9394	.9406	.9418	.9429	.9441
1.6	.9452	.9463	.9474	.9484	.9495	.9505	.9515	.9525	.9535	.9545
1.7	.9554	.9564	.9573	.9582	.9591	.9599	.9608	.9616	.9625	.9633
1.8	.9641	.9649	.9656	.9664	.9671	.9678	.9686	.9693	.9699	.9706
1.9	.9713	.9719	.9726	.9732	.9738	.9744	.9750	.9756	.9761	.9767
2.0	.9772	.9778	.9783	.9788	.9793	.9798	.9803	.9808	.9812	.9817
2.1	.9821	.9826	.9830	.9834	.9838	.9842	.9846	.9850	.9854	.9857
2.2	.9861	.9864	.9868	.9871	.9875	.9878	.9881	.9884	.9887	.9890
2.3	.9893	.9896	.9898	.9901	.9904	.9906	.9909	.9911	.9913	.9916
2.4	.9918	.9920	.9922	.9925	.9927	.9929	.9931	.9932	.9934	.9936
2.5	.9938	.9940	.9941	.9943	.9945	.9946	.9948	.9949	.9951	.9952
2.6	.9953	.9955	.9956	.9957	.9959	.9960	.9961	.9962	.9963	.9964
2.7	.9965	.9966	.9967	.9968	.9969	.9970	.9971	.9972	.9973	.9974
2.8	.9974	.9975	.9976	.9977	.9977	.9978	.9979	.9979	.9980	.9981
2.9	.9981	.9982	.9982	.9983	.9984	.9984	.9985	.9985	.9986	.9986
3.0	.9987	.9987	.9987	.9988	.9988	.9989	.9989	.9989	.9990	.9990
3.1	.9990	.9991	.9991	.9991	.9992	.9992	.9992	.9992	.9993	.9993
3.2	.9993	.9993	.9994	.9994	.9994	.9994	.9994	.9995	.9995	.9995
3.3	.9995	.9995	.9995	.9996	.9996	.9996	.9996	.9996	.9996	.9997
3.4	.9997	.9997	.9997	.9997	.9997	.9997	.9997	.9997	.9997	.9998
3.5	.9998	.9998	.9998	.9998	.9998	.9998	.9998	.9998	.9998	.9998
3.6	.9998	.9998	.9999	.9999	.9999	.9999	.9999	.9999	.9999	.9999

Statistics for Engineers and Scientists

William Navidi
Colorado School of Mines

Boston Burr Ridge, IL Dubuque, IA Madison, WI New York San Francisco St. Louis
Bangkok Bogotá Caracas Kuala Lumpur Lisbon London Madrid Mexico City
Milan Montreal New Delhi Santiago Seoul Singapore Sydney Taipei Toronto

STATISTICS FOR ENGINEERS AND SCIENTISTS

Published by McGraw-Hill, a business unit of The McGraw-Hill Companies, Inc., 1221 Avenue of the Americas, New York, NY 10020.

This book is printed on acid-free paper.

1 2 3 4 5 6 7 8 9 0 DOC/DOC 0 9 8 7 6 5 4

ISBN 0–07–255160–7

Senior Sponsoring Editor: *Suzanne Jeans*
Managing Developmental Editor: *Debra D. Matteson*
Senior Marketing Manager: *Mary K. Kittell*
Lead Project Manager: *Peggy J. Selle*
Senior Production Supervisor: *Sherry L. Kane*
Lead Media Project Manager: *Audrey A. Reiter*
Media Technology Producer: *Eric A. Weber*
Senior Designer: *David W. Hash*
Supplement Producer: *Brenda A. Ernzen*
Compositor: *Lachina Publishing Services*
Typeface: *10.5/12 Times*
Printer: *R. R. Donnelley Crawfordsville, IN*

Library of Congress Cataloging-in-Publication Data

Navidi, William Cyrus.
Statistics for engineers and scientists / William Navidi.—1st ed.
p. cm.
Includes index.
ISBN 0–07–255160–7
1. Mathematical statistics—Simulation methods. 2. Bootstrap (Statistics). 3. Linear models (Statistics). I. Title.

QA276.4.N38 2006
519.5—dc22

2004051329
CIP

www.mhhe.com

To Catherine, Sarah, and Thomas

ABOUT THE AUTHOR

William Navidi is Professor of Mathematical and Computer Sciences at the Colorado School of Mines. He received the B.A. degree in mathematics from New College, the M.A. in mathematics from Michigan State University, and the Ph.D. in statistics from the University of California at Berkeley. Professor Navidi has authored more than 50 research papers both in statistical theory and in a wide variety of applications including computer networks, epidemiology, molecular biology, chemical engineering, and geophysics.

BRIEF CONTENTS

CONTENTS

PREFACE

MOTIVATION

The idea for this book grew out of discussions between the statistics faculty and the engineering faculty at the Colorado School of Mines regarding our introductory statistics course for engineers. Our engineering faculty felt that the students needed substantial coverage of propagation of error, as well as more emphasis on model-fitting skills. The statistics faculty believed that students needed to become more aware of some important practical statistical issues such as the checking of model assumptions and the use of simulation.

My view is that an introductory statistics text for students in engineering and science should offer all these topics in some depth. In addition, it should be flexible enough to allow for a variety of choices to be made regarding coverage, because there are many different ways to design a successful introductory statistics course. Finally, it should provide examples that present important ideas in realistic settings. Accordingly, the book has the following features:

- The book is flexible in its presentation of probability, allowing instructors wide latitude in choosing the depth and extent of their coverage of this topic.
- The book contains many examples that feature real, contemporary data sets, both to motivate students and to show connections to industry and scientific research.
- The book contains many examples of computer output and exercises suitable for solving with computer software.
- The book provides extensive coverage of propagation of error.
- The book presents a solid introduction to simulation methods and the bootstrap, including applications to verifying normality assumptions, computing probabilities, estimating bias, computing confidence intervals, and testing hypotheses.
- The book provides more extensive coverage of linear model diagnostic procedures than is found in most introductory texts. This includes material on examination of residual plots, transformations of variables, and principles of variable selection in multivariate models.
- The book covers the standard introductory topics, including descriptive statistics, probability, confidence intervals, hypothesis tests, linear regression, factorial experiments, and statistical quality control.

MATHEMATICAL LEVEL

Most of the book will be mathematically accessible to those whose background includes one semester of calculus. The exceptions are multivariate propagation of error, which requires partial derivatives, and joint probability distributions, which require multiple integration. These topics may be skipped on first reading, if desired.

COMPUTER USE

Over the past 20 years, the development of fast and cheap computing has revolutionized statistical practice; indeed, this is one of the main reasons that statistical methods have been penetrating ever more deeply into scientific work. Scientists and engineers today must not only be adept with computer software packages, they must also have the skill to draw conclusions from computer output and to state those conclusions in words. Accordingly, the book contains exercises and examples that involve interpreting, as well as generating, computer output, especially in the chapters on linear models and factorial experiments.

The modern availability of computers and statistical software has produced an important educational benefit as well, by making simulation methods accessible to introductory students. Simulation makes the fundamental principles of statistics come alive. The material on simulation presented here is designed to reinforce some basic statistical ideas, and to introduce students to some of the uses of this powerful tool.

CONTENT

Chapter 1 covers sampling and descriptive statistics. The reason that statistical methods work is that samples, when properly drawn, are likely to resemble their populations. Therefore Chapter 1 begins by describing some ways to draw valid samples. The second part of the chapter discusses descriptive statistics.

Chapter 2 is about probability. There is a wide divergence in preferences of instructors regarding how much and how deeply to cover this subject. Accordingly, I have tried to make this chapter as flexible as possible. The major results are derived from axioms, with proofs given for most of them. This should enable instructors to take a mathematically rigorous approach. On the other hand, I have attempted to illustrate each result with an example or two, in a scientific context where possible, that is designed to present the intuition behind the result. Instructors who prefer a more informal approach may therefore focus on the examples rather than the proofs.

Chapter 3 covers propagation of error, which is sometimes called "error analysis" or, by statisticians, "the delta method." The coverage is more extensive than in most texts, but the topic is so important that I thought it was worthwhile. The presentation is designed to enable instructors to adjust the amount of coverage to fit the needs of of the course.

Chapter 4 presents many of the probability distribution functions commonly used in practice. Probability plots and the Central Limit Theorem are also covered. The final section introduces simulation methods to assess normality assumptions, compute probabilities, and estimate bias.

Chapters 5 and 6 cover confidence intervals and hypothesis testing, respectively. The *P*-value approach to hypothesis testing is emphasized, but fixed-level testing and power calculations are also covered. The multiple testing problem is covered in some depth. Simulation methods to compute confidence intervals and to test hypotheses are introduced as well.

Chapter 7 covers correlation and simple linear regression. I have worked hard to emphasize that linear models are appropriate only when the relationship between the variables is linear. This point is all the more important since it is often overlooked in practice by engineers and scientists (not to mention statisticians). It is not hard to find in the scientific literature straight-line fits and correlation coefficient summaries for plots that show obvious curvature or for which the slope of the line is determined by a few influential points. Therefore this chapter includes a lengthy section on checking model assumptions and transforming variables.

Chapter 8 covers multiple regression. Model selection methods are given particular emphasis, because choosing the variables to include in a model is an essential step in many real-life analyses. The topic of confounding is given careful treatment as well.

Chapter 9 discusses some commonly used experimental designs and the methods by which their data are analyzed. One-way and two-way analysis of variance methods, along with randomized complete block designs and 2^p factorial designs, are covered fairly extensively.

Chapter 10 presents the topic of statistical quality control, discussing control charts, CUSUM charts, and process capability; and concluding with a brief discussion of six-sigma quality.

RECOMMENDED COVERAGE

The book contains enough material for a year-long course. For a one-semester course, there are a number of options. In our three-hour course at the Colorado School of Mines, we cover all of the first four chapters, except for joint distributions and the exponential, gamma, and Weibull distributions. We then cover the material on confidence intervals and hypothesis testing in Chapters 5 and 6, going quickly over the two-sample methods and power calculations and omitting distribution-free methods and the chi-square and F tests. We finish by covering as much of the material on correlation and simple linear regression in Chapter 7 as time permits.

A course with a somewhat different emphasis can be fashioned by including more material on probability, spending more time on two-sample methods and power, and reducing coverage of propagation of error, simulation, or regression. Many other options are available; for example, one may choose to include material on factorial experiments in place of some of the preceding topics. Sample syllabi, emphasizing a variety of approaches and course lengths, can be found in the Instructor's Manual available at the Online Learning Center at www.mhhe.com/navidi.

ACKNOWLEDGMENTS

I am indebted to many people for contributions at every stage of development. My colleagues in the Engineering Division at the Colorado School of Mines have been patient and generous in helping me to appreciate the ways in which statistical ideas interact with engineering practice; Terry Parker deserves particular mention in this regard. Colleagues who taught from and students who studied from drafts of the manuscript found many errors and gave me many valuable suggestions. In particular, Barbara Moskal and Gus Greivel each taught several times from the evolving manuscript and were helpful and supportive throughout, and Melissa Laeser found many interesting data sets from published sources. Mike Colagrosso of the Colorado School of Mines developed some excellent applets, as did Chris Boisclair and the team at Link-Systems. Jessica Kohlschmidt, at The Ohio State University, developed PowerPoint slides to supplement the text, and Jackie Miller, also at The Ohio State University, read the entire manuscript, found many errors, and made many valuable suggestions for improvement.

The staff at McGraw-Hill has been extremely capable and supportive. Project Manager Peggy Selle was always patient and helpful. Copy editor Lucy Mullins deserves thanks as well. The guidance of developmental editors Maja Lorkovich, Kate Scheinman, Lisa Kalner-Williams, and Debra Matteson has considerably improved the final product. I deeply appreciate the patience and faith of my sponsoring editor, Suzanne Jeans, and publisher, Betsy Jones, in bringing this project to fruition.

William Navidi

ACKNOWLEDGMENTS OF REVIEWERS AND CONTRIBUTORS

This text reflects the generous contributions of well over one hundred statistics instructors and their students, who, through numerous reviews, surveys, and class tests, helped us understand how to meet their needs and how to make improvements when we fell short. The ideas of these instructors and students are woven throughout the book, from its content and organization to its supplements.

The author and the engineering team at McGraw-Hill are grateful to these colleagues for their thoughtful comments and contributions during the development of the text and its supplements and media resources.

Sabri Abou-Ward
University of Toronto, Canada

Jeff Arthur
Oregon State University

Georgiana Baker
University of South Carolina

Barb Barnet
University of Wisconsin, Platteville

Samantha Bates
Virgina Polytechnic Institute

Beno Benhabib
University of Toronto, Canada

L. Bernaards
University of California, Los Angeles

Mary Besterfield-Saore
University of Pittsburgh

J. D. Bhattacharyya
Iowa State University

Amy Biesterfeld
University of Colorado, Boulder

Rick Bilonick
Carnegie Mellon University

Peter Bloomfield
North Carolina State University

Saul Blumenthal
Ohio State University

Robert J. Boik
Montana State University

Tamara Bowcutt
California State University at Chico

Donald Boyd
Montana State University

Raymond Brown
Drexel University

Ronald B. Bucinell
Union College

Karen M. Bursic
University of Pittsburgh

John Butler
California Polytechnic State University, San Luis Obispo

M. Jaya Chandra
Pennsylvania State University

Huann-Shen Chen
Michigan Technological University

Nicolas Christou
University of California, Los Angeles

Dave Clark
Kettering University

Peyton Cook
University of Tulsa

William Cooper
University of Minnesota

Casey Cremins
University of Maryland

Dan Dalthorp
Cornell University

Valentin Deaconu
University of Nevada, Reno

Darinka Dentcheva
Stevens Institute of Technology

Art Diaz
San Jose State University

Don Edwards
University of Toronto, Canada

Judith Ekstrand
San Francisco State University

Timothy C. Elston
North Carolina State University

Randy Eubank
Texas A&M University

Thomas Z. Fahidy
University of Waterloo, Canada

Nasser Fard
Northeastern University

Steve From
University of Nebraska at Omaha

Gary Gadbury
University of Missouri, Rolla

Pierre Gharghouri
Ryerson University, Canada

Sampson Gholston
University of Alabama

Robert Gilbert
The University of Texas at Austin

Dave Goldsman
Georgia Institute of Technology

Pierre Goovaerts
University of Michigan

Canan Bilen Green
North Dakota State University

Trevor Hale
Ohio University

Jim Handley
Montana Tech of the University of Montana

James Higgins
Kansas State University

David Hoeppener
University of Utah

Carol O'Connor Holloman
University of Louisville

Joseph Horowitz
University of Massachusetts, Amherst

Wei-Min Huang
Lehigh University

Norma Hubele
Arizona State University

Floyd Hummel
Pennsylvania State University

Wafik Iskander
West Virginia University

Roger Johnson
South Dakota School of Mines and Technology

Scott Jordan
Arkansas Tech University

Kailash Kapur
University of Washington

David Kender
Wright State University

Kim Wee Kong
Multimedia University, Malaysia

Ravindra Khattree
Oakland University

Vadim Khayms
Stanford University

Claire Komives
San Jose State University

Thomas Koon
Binghamton University

Milo Koretsky
Oregon State University

Roger Korus
University of Idaho

Tomasz Kozubowski
University of Nevada, Reno

Gary Kretchik
University of Alaska, Anchorage

Hillel Kumin
University of Oklahoma

Samuel Labi
Purdue University

Robert Lacher
South Dakota State University

John Lawson
Brigham Young University

Stephen Lee
University of Idaho

Marvin Lentner
Virgina Polytechnic Institute

Liza Levina
University of Michigan

Quingchong John Liu
Oakland University

Graham Lord
Princeton University

Arthur Lubin
Illinois Institute of Technology

Zhen Luo
Pennsylvania State University

James Lynch
University of South Carolina

Peter MacDonald
McMaster University, Canada

James Maneval
Bucknell University

Munir Mahmood
Rochester Institute of Technology

Glen Marotz
University of Kansas

Timothy Matis
New Mexico State University

Laura McSweeney
Fairfield University

Megan Meece
University of Florida

Vivien Miller
Mississippi State University

Bradley Mingels
University of Massachusetts, Lowell

Satya Mishra
University of South Alabama

N. N. Murulidhar
National Institute of Technology, India

Steve Ng Hoi-Kow
Hong Kong Institute of Vocational Education, Hong Kong

Zulkifli Mohd Nopiah
Universiti Kebangsaan Malaysia, Malaysia

Bob O'Donnell
Oregon State University

Nancy Olmstead
Milwaukee School of Engineering

K. P. Patil
Veermeeta Jijabai Technological Institute, India

Arunkumar Pennathur
University of Texas at El Paso

Joseph Petrucelli
Worcester Polytechnic University

Elizabeth Podlaha
Louisiana State University

Michael Pore
The University of Texas at Austin

Jeffrey Proehl
Dartmouth College

Jorge Prozzi
The University of Texas at Austin

D. Ramachandran
California State University at Sacramento

Amelia Regan
University of California, Irvine

Larry Ringer
Texas A&M University

Iris V. Rivero
Texas Tech University

Timothy Robinson
University of Wyoming

Derrick Rollins
Iowa State University

Andrew Ross
Lehigh University

Manual Rossetti
University of Arkansas

V. A. Samaranayake
University of Missouri, Rolla

Doug Schmucker
Valaparaiso University

David Schrady
Naval Postgraduate School

Neil Schwertman
California State University, Chico

Nong Shang
Rensselaer Polytechnic Institute

Galit Shmueli
Carnegie Mellon University

Ruey-Lin Sheu
National Cheng Kung University, Taiwan

Tony Smith
University of Pennsylvania

Jery Stedinger
Cornell University

Frank E. Stratton
San Diego State University

John Ting
University of Massachusetts, Lowell

Gilberto Urroz
Utah State University

Margot Vigeant
Bucknell University

Joseph G. Voelkel
Rochester Institute of Technology

Natascha Vukasinovic
Utah State University

Amy Wagler
Oklahoma State University

Tse-Wei Wang
University of Tennessee, Knoxville

Bill Warde
Oklahoma State University

Simon Washington
University of Arizona

Daniel Weiner
Boston University

Alison Weir
University of Toronto at Mississauga, Canada

Bruce Westermo
San Diego State University

Grant Willson
The University of Texas at Austin

Jae Yoon
Old Dominion University

Ali Zargar
San Jose State University

Key Features

Content Overview

This book allows flexible coverage because there are many ways to design a successful introductory statistics course.

- **Flexible coverage of probability** addresses the needs of different courses. Allowing for a mathematically rigorous approach, the major results are derived from axioms, with proofs given for most of them. On the other hand, each result is illustrated with an example or two to promote intuitive understanding. Instructors who prefer a more informal approach may therefore focus on the examples rather than the proofs and skip the optional sections.

- **Extensive coverage of propagation of error,** sometimes called "error analysis" or "the delta method," is provided in a separate chapter. The coverage is more thorough than in most texts. The format is flexible so that the amount of coverage can be tailored to the needs of the course.

- **A solid introduction to simulation methods and the bootstrap** is presented in the final sections of Chapters 4, 5, and 6.

- **Extensive coverage of linear model diagnostic procedures** in Chapter 7 includes a lengthy section on checking model assumptions and transforming variables. The chapter emphasizes that linear models are appropriate only when the relationship between the variables is linear. This point is all the more important since it is often overlooked in practice by engineers and scientists (not to mention statisticians).

Key Features

Real-World Data Sets

With a fresh approach to the subject, the author uses contemporary real-world data sets to motivate students and show a direct connection to industry and research.

The article "Virgin Versus Recycled Wafers for Furnace Qualification: Is the Expense Justified?" (V. Czitrom and J. Reece, in *Statistical Case Studies for Industrial Process Improvement*, ASA and SIAM, 1997:87–104) describes a process for growing a thin silicon dioxide layer onto silicon wafers that are to be used in semiconductor manufacture. Table 1.6 presents thickness measurements, in angstroms (Å), of the oxide layer for 24 wafers. Nine measurements were made on each wafer. The wafers were produced in two separate runs, with 12 wafers in each run.

TABLE 1.6 Oxide layer thicknesses for silicon wafers

	Wafer	Thicknesses(Å)								
Run 1	1	90.0	92.2	94.9	92.7	91.6	88.2	92.0	98.2	96.0
	2	91.8	94.5	93.9	77.3	92.0	89.9	87.9	92.8	93.3
	3	90.3	91.1	93.3	93.5	87.2	88.1	90.1	91.9	94.5
	4	92.6	90.3	92.8	91.6	92.7	91.7	89.3	95.5	93.6
	5	91.1	89.8	91.5	91.5	90.6	93.1	88.9	92.5	92.4
	6	76.1	90.2	96.8	84.6	93.3	95.7	90.9	100.3	95.2
	7	92.4	91.7	91.6	91.1	88.0	92.4	88.7	92.9	92.6
	8	91.3	90.1	95.4	89.6	90.7	95.8	91.7	97.9	95.7
	9	96.7	93.7	93.9	87.9	90.4	92.0	90.5	95.2	94.3
	10	92.0	94.6	93.7	94.0	89.3	90.1	91.3	92.7	94.5
	11	94.1	91.5	95.3	92.8	93.4	92.2	89.4	94.5	95.4
	12	91.7	97.4	95.1	96.7	77.5	91.4	90.5	95.2	93.1
Run 2	1	93.0	89.9	93.6	89.0	93.6	90.9	89.8	92.4	93.0
	2	91.4	90.6	92.2	91.9	92.4	87.6	88.9	90.9	92.8
	3	91.9	91.8	92.8	96.4	93.8	86.5	92.7	90.9	92.8
	4	90.6	91.3	94.9	88.3	87.9	92.2	90.7	91.3	93.6
	5	93.1	91.8	94.6	88.9	90.0	97.9			
	6	90.8	91.5	91.5	91.5	94.0	91.0			
	7	88.0	91.8	90.5	90.4	90.3	91.5			
	8	88.3	96.0	92.8	93.7	89.6	89.6			
	9	94.2	92.2	95.8	92.5	91.0	91.4			
	10	101.5	103.1	103.2	103.5	96.1	102.5			
	11	92.8	90.8	92.2	91.7	89.0	88.5			
	12	92.1	93.4	94.0	94.7	90.8	92.1			

The 12 wafers in each run were of several different
several different furnace locations. The purpose in collect
whether the thickness of the oxide layer was affected eith
furnace location. This was therefore a factorial experimen
location as the factors, and oxide layer thickness as the o
designed so that there was not supposed to be any systemati
between one run and another. The first step in the analysis
the data in each run to help determine if this condition wa
of the observations should be deleted. The results are pres

22. The article "Seismic Hazard in Greece Based on Different Strong Ground Motion Parameters" (S. Koutrakis, G. Karakaisis, et al., *Journal of Earthquake Engineering*, 2002:75–109) presents a study of seismic events in Greece during the period 1978–1997. Of interest is the duration of "strong ground motion," which is the length of time that the acceleration of the ground exceeds a specified value. For each event, measurements of the duration of strong ground motion were made at one or more locations. Table SE22 presents, for each of 121 such measurements, the data for the duration of time y (in seconds) that the ground acceleration exceeded twice the acceleration due to gravity, the magnitude m of the earthquake, the distance d (in km) of the measurement from the epicenter, and two indicators of the soil type s_1 and s_2, defined as follows: $s_1 = 1$ if the soil consists of soft alluvial deposits, $s_1 = 0$ otherwise, and $s_2 = 1$ if the soil consists of tertiary or older rock, $s_2 = 0$ otherwise. Cases where both $s_1 = 0$ and $s_2 = 0$ correspond to intermediate soil conditions. The article presents repeated measurements at some locations, which we have not included here.

Use the data in Table SE22 to construct a linear model to predict duration y from some or all of the variables m, d, s_1, and s_2. Be sure to consider transformations of the variables, as well as powers of and interactions between the independent variables. Describe the steps taken to construct your model. Plot the residuals versus the fitted values to verify that your model satisfies the necessary assumptions. In addition, note that the data is presented in chronological order. Make a plot to determine whether time should be included as an independent variable.

TABLE SE22

y	m	d	s_1									d	s_1	s_2
8.82	6.4	30	1									15	0	0
4.08	5.2	7	0									128	1	0
15.90	6.9	105	1									13	0	0
6.04	5.8	15	0									19	1	0
0.15	4.9	16	1									68	1	0
5.06	6.2	75	1									10	0	0
0.01	6.6	119	0									45	0	1
4.13	5.1	10	1									18	1	0
0.02	5.3	22	0									14	0	1
2.14	4.5	12	0									15	0	0
4.41	5.2	17	0	0	5.43	5.2	49	0	1	6.18	5.2	13	0	0
17.19	5.9	9	0	0	4.78	5.5	1	0	0	4.56	5.5	1	0	0
5.14	5.5	10	1	0	2.82	5.5	20	0	1	0.94	5.0	6	0	1
0.05	4.9	14	1	0	3.51	5.7	22	0	0	2.85	4.6	21	1	0
20.00	5.8	16	1	0	13.92	5.8	34	1	0	4.21	4.7	20	1	0
12.04	6.1	31	0	0	3.96	6.1	44	0	0	1.93	5.7	39	1	0
0.87	5.0	65	1	0	6.91	5.4	16	0	0	1.56	5.0	44	1	0
0.62	4.8	11	1	0	5.63	5.3	6	1	0	5.03	5.1	2	1	0
8.10	5.4	12	1	0	0.10	5.2	21	1	0	0.51	4.9	14	1	0
1.30	5.8	34	1	0	5.10	4.8	16	1	0	13.14	5.6	5	1	0
11.92	5.6	5	0	0	16.52	5.5	15	1	0	8.16	5.5	12	1	0
3.93	5.7	65	1	0	19.84	5.7	50	1	0	10.04	5.1	28	1	0
2.00	5.4	27	0	1	1.65	5.4	27	1	0	0.79	5.4	35	0	0
0.43	5.4	31	0	1	1.75	5.4	30	0	1	0.02	5.4	32	1	0
14.22	6.5	20	0	1	6.37	6.5	90	1	0	0.10	6.5	61	0	1
0.06	6.5	72	0	1	2.78	4.9	8	0	0	5.43	5.2	9	0	0
1.48	5.2	27	0	0	2.14	5.2	22	0	0	0.81	4.6	9	0	0
3.27	5.1	12	0	0	0.92	5.2	29	0	0	0.73	5.2	22	0	0
6.36	5.2	14	0	0	3.18	4.8	15	0	0	11.18	5.0	8	0	0
0.18	5.0	19	0	0	1.20	5.0	19	0	0	2.54	4.5	6	0	0
0.31	4.5	12	0	0	4.37	4.7	5	0	0	1.55	4.7	13	0	1
1.90	4.7	12	0	0	1.02	5.0	14	0	0	0.01	4.5	17	0	0
0.29	4.7	5	1	0	0.71	4.8	4	1	0	0.21	4.8	5	0	1
6.26	6.3	9	1	0	4.27	6.3	9	0	1	0.04	4.5	3	1	0
3.44	5.4	4	1	0	3.25	5.4	4	0	1	0.01	4.5	1	1	0
2.32	5.4	5	1	0	0.90	4.7	4	1	0	1.19	4.7	3	1	0
1.49	5.0	4	1	0	0.37	5.0	4	0	1	2.66	5.4	1	1	0
2.85	5.4	1	0	1	21.07	6.4	78	0	1	7.47	6.4	104	0	0
0.01	6.4	86	0	1	0.04	6.4	105	0	1	30.45	6.6	51	1	0
9.34	6.6	116	0	1	15.30	6.6	82	0	1	12.78	6.6	65	1	0
10.47	6.6	117	0	0										

Learning Supplements for the Student

Student Resources CD-ROM

Packaged free with every new copy of the text, this CD provides all the data sets from the text as well as applets based on text content to reinforce a visual understanding of statistics.

- **All text data sets** are provided for download in various formats:
  - ASCII comma delimited
  - ASCII tab delimited
  - MINITAB
  - Excel
  - SAS
  - SPSS
  - TI-89

- **Java Applets,** created specifically for this calculus-based course, provide interactive exercises based on text content which allow students to alter variables and explore "What if?" scenarios. Also included in the applet suite are **Simulation Applets**, which reinforce the excellent text coverage of simulation methods. The applets allow students to see the text simulation examples in action and to alter the parameters for further exploration.

- **A Guide to Simulation in MINITAB,** prepared by the author, describes how the simulation examples in the text may be implemented in MINITAB.

- **Tools and Resources,** including a link to the book Online Learning Center, offers online instructor and student resources at *www.mhhe.com/navidi.*

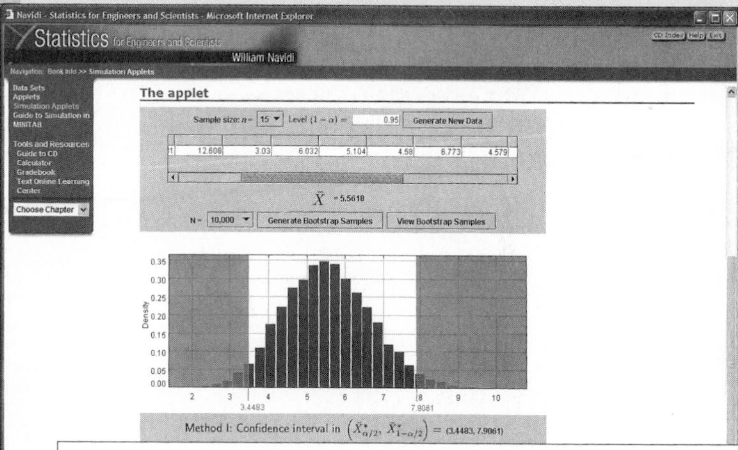

Central Limit Theorem Simulation

This applet evaluates the accuracy of the central limit theorem approximation by computing sample means of data drawn from a gamma distribution, with parameters r and λ with $\lambda = 1$.

How to use the applet

1. Choose a sample size, n, from the drop down box. Choose a sample size between 1 and 100.
2. Enter value for the parameter, r, of the gamma distribution into the cells at the top. *Note, this cell acts like a spreadsheet application; to change its entry, delete the current value, type a new one, and then press enter.* The value of r must be a positive integer. When you update the value of the parameter, the plot in lower-left titled "Gamma Density Function" is updated to reflect the new distribution.
3. Click the button that reads "Generate Data." 1000 samples of size n will be generated from $\Gamma(r, 1)$.
4. You can view the generated data and each sample mean, $\bar{X}$, by clicking the button that reads "View Data."
5. A histogram of the sample means is plotted in the center plot, and a normal curve is superimposed.
6. The lower-right plot is a normal probability plot of the sample means.

The applet

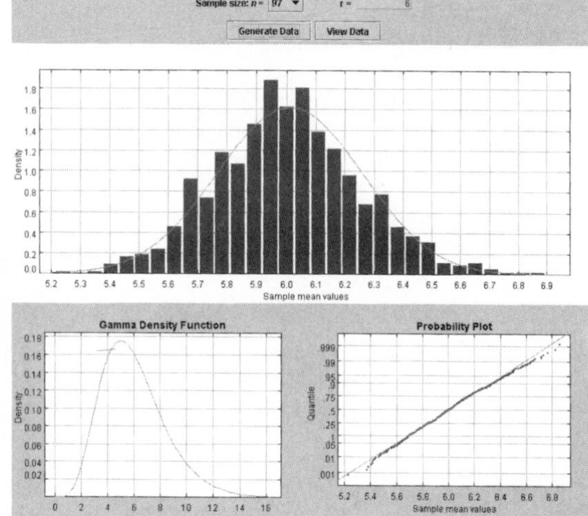

You must have Java to interact with the simulation applet. Use **the Java tester** to determine if you have Java. The Java applet has been tested with Java version 1.4. It works best with the latest version of Java directly from Sun. If you don't have Java, **here is a screenshot** of what you are missing.

Key Features

Computer Output

The book contains exercises and examples that involve interpreting, as well as generating, computer output.

12. The following MINITAB output presents the results of a hypothesis test for a population mean μ.

```
One-Sample Z: X

Test of mu = 73.5 vs not = 73.5
The assumed standard deviation = 2.3634

Variable    N     Mean   StDev  SE Mean         95% CI                Z      P
X         145  73.2461  2.3634   0.1963  (72.8614, 73.6308)   -1.29  0.196
```

a. Is this a one-tailed or two-tailed test?
b. What is the null hypothesis?
c. What is the P-value?
d. Use the output and an appropriate table to compute the P-value for the test of $H_0: \mu \geq 73.6$ versus $H_1: \mu < 73.6$.
e. Use the output and an appropriate table to compute a 99% confidence interval for μ.

7. In a study of the lung function of children, the volume of air exhaled under force in one second is called FEV_1. (FEV_1 stands for forced expiratory volume in one second.) Measurements were made on a group of children each year for two years. A linear model was fit to predict this year's FEV_1 as a function of last year's FEV_1 (in liters), the child's gender (0 = Male, 1 = Female), the child's height (in m), and the ambient atmospheric pressure (in mm). The following MINITAB output presents the results of fitting the model

$$\text{FEV}_1 = \beta_0 + \beta_1 \text{ Last FEV}_1 + \beta_2 \text{ Gender} + \beta_3 \text{ Height} + \beta_4 \text{ Pressure} + \varepsilon$$

```
The regression equation is
FEV1 = -0.219+0.779 Last FEV -0.108 Gender+1.354 Height -0.00134 Pressure

Predictor        Coef       StDev          T        P
Constant     -0.21947      0.4503      -0.49    0.627
Last FEV        0.779     0.04909      15.87    0.000
Gender       -0.10827      0.0352      -3.08    0.002
Height         1.3536      0.2880       4.70    0.000
Pressure   -0.0013431   0.0004722      -2.84    0.005

S = 0.22039        R-Sq = 93.5%          R-Sq(adj) = 93.3%

Analysis of Variance

Source            DF         SS          MS        F        P
Regression         4     111.31      27.826   572.89    0.000
Residual Error   160     7.7716    0.048572
Total            164     119.08
```

a. Predict the FEV_1 for a boy who is 1.4 m tall, if the measurement was taken at a pressure of 730 mm and last year's measurement was 2.113 L.
b. If two girls differ in height by 5 cm, by how much would you expect their FEV_1 measurements to differ, other things being equal?
c. The constant term β_0 is estimated to be negative. But FEV_1 must always be positive. Is something wrong? Explain.
d. The health physicist in charge of this experiment wants to redesign the algorithm that electronically records the measurement, in order to automatically adjust for the atmospheric pressure. A barometer will be attached to the device to input the pressure. Use the preceding MINITAB output to determine how to compute an adjusted FEV_1 value as a function of the measured FEV_1 value and the pressure.

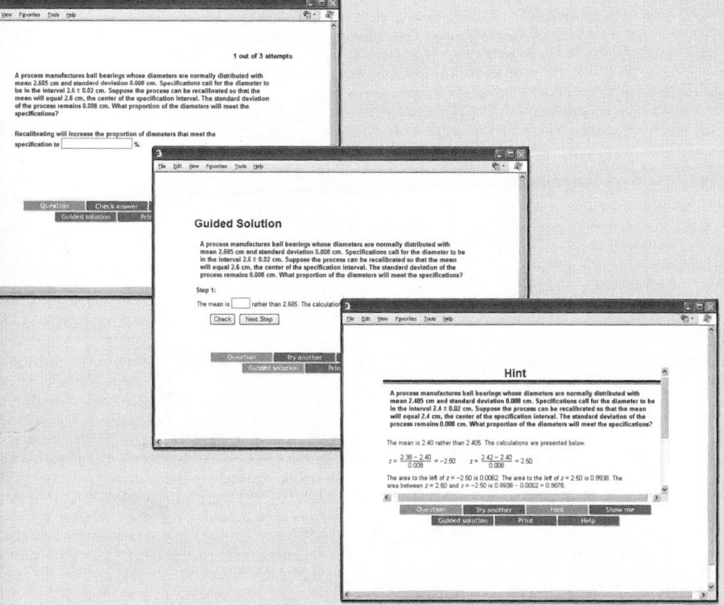

- **Practice Quizzes,** offering both chapter-level and cumulative testing, provide a virtually unlimited number of practice versions of text examples. Our algorithmic test generator offers the following options:
  —The **Guided Solution** button leads you step-by-step through the solution, prompting you to complete each step.
  —The **Hint** button produces a worked-out solution to a similar problem.

Student's Solutions Manual

ISBN: 0-07-310959-2

The *Student's Solutions Manual* for use with *Statistics for Engineers and Scientists* provides detailed solutions to all odd-numbered exercises in the book.

MINITAB 14 Student Version

The student version of this professional software is available packaged with this text. This package also provides a MINITAB manual with information on data and file management, conducting various statistical analyses, and creating presentation-style graphics.

Teaching Supplements for the Instructor

Instructor's Resource CD-ROM

Offering a wide range of class preparation and presentation resources, this CD helps instructors to easily transition to our text.

- **Electronic Solutions Manual** provides PDF files by chapter with detailed solutions to all text exercises
- **PowerPoint Lecture Notes** for each chapter of the text can be customized to fit your classroom presentation needs
- **Suggested Syllabi** provide useful roadmaps for many different versions of the course
- **Correlation Guides** match the organization and coverage in our text to other popular engineering statistics textbooks

Online delivery of these resources and more is available on our **Online Learning Center** at *www.mhhe.com/navidi*

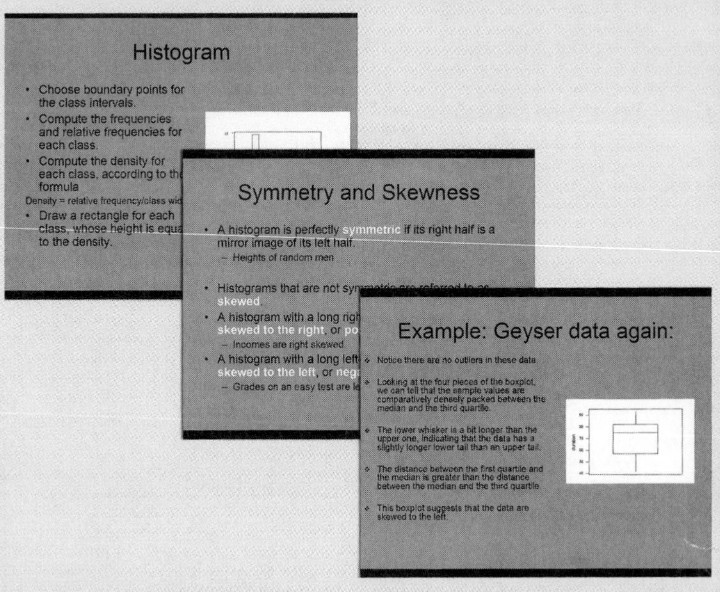

COSMOS CD-ROM

This CD, available to instructors only, provides electronic solutions delivered via our database management tool. McGraw-Hill's COSMOS is a powerful solutions manual tool to help instructors streamline the creation of assignments, quizzes, and tests by using problems and solutions from the texbook—as well as their own custom material.

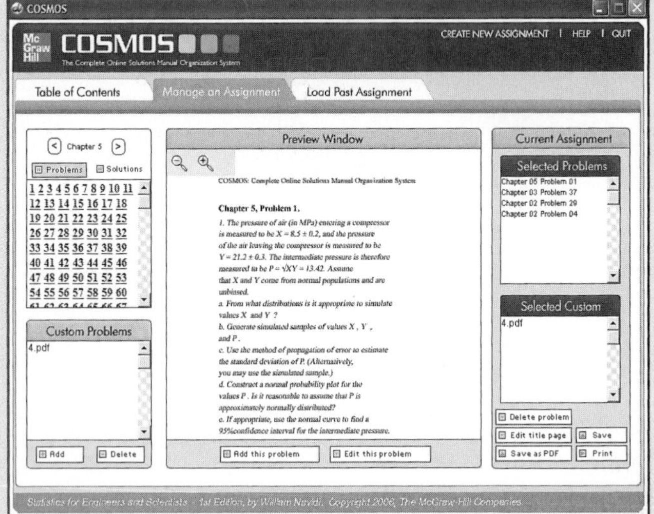

Chapter 1

Sampling and Descriptive Statistics

Introduction

The collection and analysis of data are fundamental to science and engineering. Scientists discover the principles that govern the physical world, and engineers learn how to design important new products and processes, by analyzing data collected in scientific experiments. A major difficulty with scientific data is that they are subject to random variation, or uncertainty. That is, when scientific measurements are repeated, they come out somewhat differently each time. This poses a problem: How can one draw conclusions from the results of an experiment when those results could have come out differently? To address this question, a knowledge of statistics is essential. Statistics is the field of study concerned with the collection, analysis, and interpretation of uncertain data. The methods of statistics allow scientists and engineers to design valid experiments and to draw reliable conclusions from the data they produce.

Although our emphasis in this book is on the applications of statistics to science and engineering, it is worth mentioning that the analysis and interpretation of data are playing an ever-increasing role in all aspects of modern life. For better or worse, huge amounts of data are collected about our opinions and our lifestyles, for purposes ranging from the creation of more effective marketing campaigns to the development of social policies designed to improve our way of life. On almost any given day, newspaper articles are published that purport to explain social or economic trends through the analysis of data. A basic knowledge of statistics is therefore necessary not only to be an effective scientist or engineer, but also to be a well-informed member of society.

The Basic Idea

The basic idea behind all statistical methods of data analysis is to make inferences about a population by studying a relatively small sample chosen from it. As an illustration, consider a machine that makes steel rods for use in optical storage devices. The specification for the diameter of the rods is 0.45 ± 0.02 cm. During the last hour, the machine has made 1000 rods. The quality engineer wants to know approximately how many of these rods meet the specification. He does not have time to measure all 1000 rods. So he draws a random sample of 50 rods, measures them, and finds that 46 of them (92%) meet the diameter specification. Now, it is unlikely that the sample of 50 rods represents the population of 1000 perfectly. The proportion of good rods in the population is likely to differ somewhat from the sample proportion of 92%. What the engineer needs to know is just how large that difference is likely to be. For example, is it plausible that the population percentage could be as high as 95%? 98%? As low as 90%? 85%?

Here are some specific questions that the engineer might need to answer on the basis of these sample data:

1. The engineer needs to compute a rough estimate of the likely size of the difference between the sample proportion and the population proportion. How large is a typical difference for this kind of sample?
2. The quality engineer needs to note in a logbook the percentage of acceptable rods manufactured in the last hour. Having observed that 92% of the sample rods were good, he will indicate the percentage of acceptable rods in the population as an interval of the form $92\% \pm x\%$, where x is a number calculated to provide reasonable certainty that the true population percentage is in the interval. How should x be calculated?
3. The engineer wants to be fairly certain that the percentage of good rods is at least 90%; otherwise he will shut down the process for recalibration. How certain can he be that at least 90% of the 1000 rods are good?

Much of this book is devoted to addressing questions like these. The first of these questions requires the computation of a **standard deviation**, which we will discuss in Chapters 2 and 4. The second question requires the construction of a **confidence interval**, which we will learn about in Chapter 5. The third calls for a **hypothesis test**, which we will study in Chapter 6.

The remaining chapters in the book cover other important topics. For example, the engineer in our example may want to know how the amount of carbon in the steel rods is related to their tensile strength. Issues like this can be addressed with the methods of **correlation** and **regression**, which are covered in Chapters 7 and 8. It may also be important to determine how to adjust the manufacturing process with regard to several factors, in order to produce optimal results. This requires the design of **factorial experiments**, which are discussed in Chapter 9. Finally, the engineer will need to develop a plan for monitoring the quality of the product manufactured by the process. Chapter 10 covers the topic of **statistical quality control**, in which statistical methods are used to maintain quality in an industrial setting.

The topics listed here concern methods of drawing conclusions from data. These methods form the field of **inferential statistics**. Before we discuss these topics, we must first learn more about methods of collecting data and of summarizing clearly the basic information they contain. These are the topics of **sampling** and **descriptive statistics**, and they are covered in the rest of this chapter.

1.1 Sampling

As mentioned, statistical methods are based on the idea of analyzing a **sample** drawn from a **population**. For this idea to work, the sample must be chosen in an appropriate way. For example, let us say that we wished to study the heights of students at the Colorado School of Mines by measuring a sample of 100 students. How should we choose the 100 students to measure? Some methods are obviously bad. For example, choosing the students from the rosters of the football and basketball teams would undoubtedly result in a sample that would fail to represent the height distribution of the population of students. You might think that it would be reasonable to use some conveniently obtained sample, for example, all students living in a certain dorm or all students enrolled in engineering statistics. After all, there is no reason to think that the heights of these students would tend to differ from the heights of students in general. Samples like this are not ideal, however, because they can turn out to be misleading in ways that are not anticipated. The best sampling methods involve **random sampling**. There are many different random sampling methods, the most basic of which is **simple random sampling**.

To understand the nature of a simple random sample, think of a lottery. Imagine that 10,000 lottery tickets have been sold and that 5 winners are to be chosen. What is the fairest way to choose the winners? The fairest way is to put the 10,000 tickets in a drum, mix them thoroughly, and then reach in and one by one draw 5 tickets out. These 5 winning tickets are a simple random sample from the population of 10,000 lottery tickets. Each ticket is equally likely to be one of the 5 tickets drawn. More importantly, each collection of 5 tickets that can be formed from the 10,000 is equally likely to comprise the group of 5 that is drawn. It is this idea that forms the basis for the definition of a simple random sample.

Summary

- A **population** is the entire collection of objects or outcomes about which information is sought.
- A **sample** is a subset of a population, containing the objects or outcomes that are actually observed.
- A **simple random sample** of size n is a sample chosen by a method in which each collection of n population items is equally likely to comprise the sample, just as in a lottery.

Since a simple random sample is analogous to a lottery, it can often be drawn by the same method now used in many lotteries: with a computer random number generator. Suppose there are N items in the population. One assigns to each item in the population an integer between 1 and N. Then one generates a list of random integers between 1 and N and chooses the corresponding population items to comprise the simple random sample.

Example 1.1

A physical education professor wants to study the physical fitness levels of students at her university. There are 20,000 students enrolled at the university, and she wants to draw a sample of size 100 to take a physical fitness test. She obtains a list of all 20,000 students, numbered from 1 to 20,000. She uses a computer random number generator to generate 100 random integers between 1 and 20,000 and then invites the 100 students corresponding to those numbers to participate in the study. Is this a simple random sample?

Solution

Yes, this is a simple random sample. Note that it is analogous to a lottery in which each student has a ticket and 100 tickets are drawn.

Example 1.2

A quality engineer wants to inspect rolls of wallpaper in order to obtain information on the rate at which flaws in the printing are occurring. She decides to draw a sample of 50 rolls of wallpaper from a day's production. Each hour for 5 hours, she takes the 10 most recently produced rolls and counts the number of flaws on each. Is this a simple random sample?

Solution

No. Not every subset of 50 rolls of wallpaper is equally likely to comprise the sample. To construct a simple random sample, the engineer would need to assign a number to each roll produced during the day and then generate random numbers to determine which rolls comprise the sample.

In some cases, it is difficult or impossible to draw a sample in a truly random way. In these cases, the best one can do is to sample items by some convenient method. For example, imagine that a construction engineer has just received a shipment of 1000 concrete blocks, each weighing approximately 50 pounds. The blocks have been delivered in a large pile. The engineer wishes to investigate the crushing strength of the blocks by measuring the strengths in a sample of 10 blocks. To draw a simple random sample would require removing blocks from the center and bottom of the pile, which might be quite difficult. For this reason, the engineer might construct a sample simply by taking 10 blocks off the top of the pile. A sample like this is called a **sample of convenience**.

Definition

A **sample of convenience** is a sample that is not drawn by a well-defined random method.

The big problem with samples of convenience is that they may differ systematically in some way from the population. For this reason samples of convenience should not be used, except in situations where it is not feasible to draw a random sample. When it is necessary to take a sample of convenience, it is important to think carefully about all the ways in which the sample might differ systematically from the population. If it is reasonable to believe that no important systematic difference exists, then it may be acceptable to treat the sample of convenience as if it were a simple random sample. With regard to the concrete blocks, if the engineer is confident that the blocks on the top of the pile do not differ systematically in any important way from the rest, then he may treat the sample of convenience as a simple random sample. If, however, it is possible that blocks in different parts of the pile may have been made from different batches of mix or may have different curing times or temperatures, a sample of convenience could give misleading results.

Some people think that a simple random sample is guaranteed to reflect its population perfectly. This is not true. Simple random samples always differ from their populations in some ways, and occasionally may be substantially different. Two different samples from the same population will differ from each other as well. This phenomenon is known as **sampling variation**. Sampling variation is one of the reasons that scientific experiments produce somewhat different results when repeated, even when the conditions appear to be identical.

A quality inspector samples 40 bolts from a large shipment and measures the length of each. He finds that 34 of them, or 85%, meet a length specification. He concludes that exactly 85% of the bolts in the shipment meet the specification. The inspector's supervisor concludes that the proportion of good bolts is likely to be close to, but not exactly equal to, 85%. Which conclusion is appropriate?

Solution

Because of sampling variation, simple random samples don't reflect the population perfectly. They are often fairly close, however. It is therefore appropriate to infer that the proportion of good bolts in the lot is likely to be close to the sample proportion, which is 85%. It is not likely that the population proportion is equal to 85%, however.

Example 1.4

Continuing Example 1.3, another inspector repeats the study with a different simple random sample of 40 bolts. She finds that 36 of them, or 90%, are good. The first inspector claims that she must have done something wrong, since his results showed that 85%, not 90%, of bolts are good. Is he right?

Solution

No, he is not right. This is sampling variation at work. Two different samples from the same population will differ from each other and from the population.

Since simple random samples don't reflect their populations perfectly, why is it important that sampling be done at random? The benefit of a simple random sample is that there is no systematic mechanism tending to make the sample unrepresentative.

The differences between the sample and its population are due entirely to random variation. Since the mathematical theory of random variation is well understood, we can use mathematical models to study the relationship between simple random samples and their populations. For a sample not chosen at random, there is generally no theory available to describe the mechanisms that caused the sample to differ from its population. Therefore, nonrandom samples are often difficult to analyze reliably.

In Examples 1.1 to 1.4, the populations consisted of actual physical objects—the students at a university, the concrete blocks in a pile, the bolts in a shipment. Such populations are called **tangible populations**. Tangible populations are always finite. After an item is sampled, the population size decreases by 1. In principle, one could in some cases return the sampled item to the population, with a chance to sample it again, but this is rarely done in practice.

Engineering data are often produced by measurements made in the course of a scientific experiment, rather than by sampling from a tangible population. To take a simple example, imagine that an engineer measures the length of a rod five times, being as careful as possible to take the measurements under identical conditions. No matter how carefully the measurements are made, they will differ somewhat from one another, because of variation in the measurement process that cannot be controlled or predicted. It turns out that it is often appropriate to consider data like these to be a simple random sample from a population. The population, in these cases, consists of all the values that might possibly have been observed. Such a population is called a **conceptual population**, since it does not consist of actual objects.

> A simple random sample may consist of values obtained from a process under identical experimental conditions. In this case, the sample comes from a population that consists of all the values that might possibly have been observed. Such a population is called a **conceptual population**.

Example 1.5 involves a conceptual population.

A geologist weighs a rock several times on a sensitive scale. Each time, the scale gives a slightly different reading. Under what conditions can these readings be thought of as a simple random sample? What is the population?

Solution

If the physical characteristics of the scale remain the same for each weighing, so that the measurements are made under identical conditions, then the readings may be considered to be a simple random sample. The population is conceptual. It consists of all the readings that the scale could in principle produce.

Note that in Example 1.5, it is the physical characteristics of the measurement process that determine whether the data are a simple random sample. In general, when deciding whether a set of data may be considered to be a simple random sample, it is very useful to have an understanding of the process that generated the data. Statistical methods can sometimes help, especially when the sample is large, but knowledge of the mechanism that produced the data is more important.

A new chemical process has been designed that is supposed to produce a higher yield of a certain chemical than does an old process. To study the yield of this process, we run it 50 times and record the 50 yields. Under what conditions might it be reasonable to treat this as a simple random sample? Describe some conditions under which it might not be appropriate to treat this as a simple random sample.

Solution

To answer this, we must first specify the population. The population is conceptual and consists of the set of all yields that will result from this process as many times as it will ever be run. What we have done is to sample the first 50 yields of the process. *If, and only if,* we are confident that the first 50 yields are generated under identical conditions, and that they do not differ in any systematic way from the yields of future runs, then we may treat them as a simple random sample.

Be cautious, however. There are many conditions under which the 50 yields could fail to be a simple random sample. For example, with chemical processes, it is sometimes the case that runs with higher yields tend to be followed by runs with lower yields, and vice versa. Sometimes yields tend to increase over time, as process engineers learn from experience how to run the process more efficiently. In these cases, the yields are not being generated under identical conditions and would not comprise a simple random sample.

Example 1.6 shows once again that a good knowledge of the nature of the process under consideration is important in deciding whether data may be considered to be a simple random sample. Statistical methods can sometimes be used to show that a given data set is *not* a simple random sample. For example, sometimes experimental conditions gradually change over time. A simple but effective method to detect this condition is to plot the observations in the order they were taken. A simple random sample should show no obvious pattern or trend.

Figure 1.1 (page 8) presents plots of three samples in the order they were taken. The plot in Figure 1.1a shows an oscillatory pattern. The plot in Figure 1.1b shows an increasing trend. Neither of these samples should be treated as a simple random sample. The plot in Figure 1.1c does not appear to show any obvious pattern or trend. It might be appropriate to treat these data as a simple random sample. However, before making that decision, it is still important to think about the process that produced the data, since there may be concerns that don't show up in the plot (see Example 1.7).

Sometimes the question as to whether a data set is a simple random sample depends on the population under consideration. This is one case in which a plot can look good, yet the data are not a simple random sample. Example 1.7 provides an illustration.

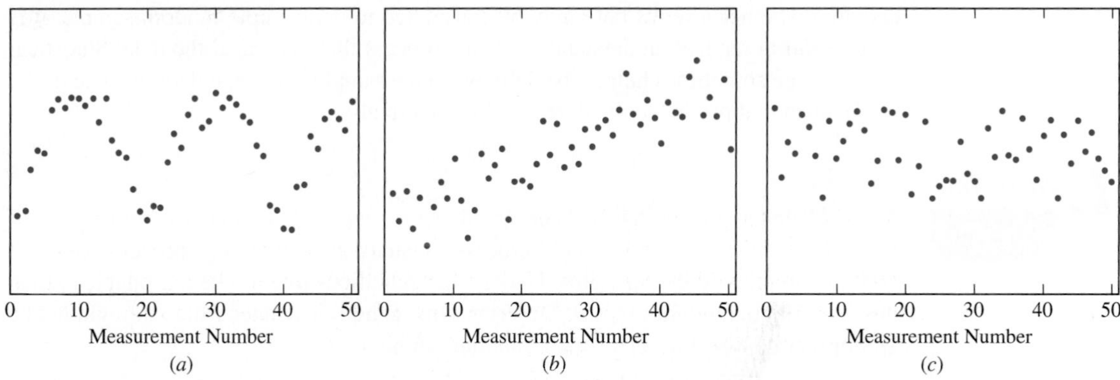

FIGURE 1.1 Three plots of observed values versus the order in which they were made. *(a)* The values show a definite pattern over time. This is not a simple random sample. *(b)* The values show a trend over time. This is not a simple random sample. *(c)* The values do not show a pattern or trend. It may be appropriate to treat these data as a simple random sample.

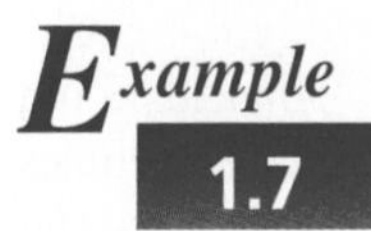

A new chemical process is run 10 times each morning for five consecutive mornings. A plot of yields in the order they are run does not exhibit any obvious pattern or trend. If the new process is put into production, it will be run 10 hours each day, from 7 A.M. until 5 P.M. Is it reasonable to consider the 50 yields to be a simple random sample? What if the process will always be run in the morning?

Solution

Since the intention is to run the new process in both the morning and the afternoon, the population consists of all the yields that would ever be observed, including both morning and afternoon runs. The sample is drawn from only that portion of the population that consists of morning runs, and thus it is not a simple random sample. There are many things that could go wrong if this is used as a simple random sample. For example, ambient temperatures may differ between morning and afternoon, which could affect yields.

If the process will only be run in the morning, then the population consists only of morning runs. Since the sample does not exhibit any obvious pattern or trend, it might well be appropriate to consider it to be a simple random sample.

Independence

The items in a sample are said to be **independent** if knowing the values of some of them does not help to predict the values of the others. With a finite, tangible population, the items in a simple random sample are not strictly independent, because as each item is drawn, the population changes. This change can be substantial when the population is small. However, when the population is very large, this change is negligible and the items can be treated as if they were independent.

To illustrate this idea, imagine that we draw a simple random sample of 2 items from the population

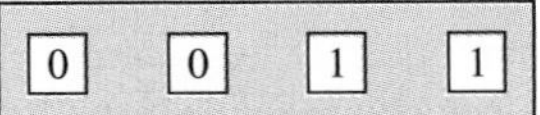

For the first draw, the numbers 0 and 1 are equally likely. But the value of the second item is clearly influenced by the first; if the first is 0, the second is more likely to be 1, and vice versa. Thus the sampled items are dependent. Now assume we draw a sample of size 2 from this population:

Again on the first draw, the numbers 0 and 1 are equally likely. But unlike the previous example, these two values remain almost equally likely the second draw as well, no matter what happens on the first draw. With the large population, the sample items are for all practical purposes independent.

It is reasonable to wonder how large a population must be in order that the items in a simple random sample may be treated as independent. A rule of thumb is that when sampling from a finite population, the items may be treated as independent so long as the sample comprises 5% or less of the population.

Interestingly, it is possible to make a population behave as though it were infinitely large, by replacing each item after it is sampled. This method is called **sampling with replacement**. With this method, the population is exactly the same on every draw and the sampled items are truly independent.

With a conceptual population, we require that the sample items be produced under identical experimental conditions. In particular, then, no sample value may influence the conditions under which the others are produced. Therefore, the items in a simple random sample from a conceptual population may be treated as independent. We may think of a conceptual population as being infinite, or equivalently, that the items are sampled with replacement.

Summary

- The items in a sample are **independent** if knowing the values of some of the items does not help to predict the values of the others.
- Items in a simple random sample may be treated as independent in many cases encountered in practice. The exception occurs when the population is finite and the sample comprises a substantial fraction (more than 5%) of the population.

Other Sampling Methods

In addition to simple random sampling there are other sampling methods that are useful in various situations. In **weighted sampling**, some items are given a greater chance of being selected than others, like a lottery in which some people have more tickets than others. In **stratified random sampling**, the population is divided up into subpopulations, called **strata**, and a simple random sample is drawn from each stratum. In **cluster sampling**, items are drawn from the population in groups, or clusters. Cluster sampling is useful when the population is too large and spread out for simple random sampling to be feasible. For example, many U.S. government agencies use cluster sampling to sample the U.S. population to measure sociological factors such as income and unemployment. A good source of information on sampling methods is Cochran (1977).

Simple random sampling is not the only valid method of random sampling. But it is the most fundamental, and we will focus most of our attention on this method. From now on, unless otherwise stated, the terms "sample" and "random sample" will be taken to mean "simple random sample."

Types of Experiments

There are many types of experiments that can be used to generate data. We briefly describe a few of them. In a **one-sample** experiment, there is only one population of interest, and a single sample is drawn from it. For example, imagine that a process is being designed to produce polyethylene that will be used to line pipes. An experiment in which several specimens of polyethylene are produced by this process, and the tensile strength of each is measured, is a one-sample experiment. The measured strengths are considered to be a simple random sample from a conceptual population of all the possible strengths that can be observed for specimens manufactured by this process. One-sample experiments can be used to determine whether a process meets a certain standard, for example, whether it provides sufficient strength for a given application.

In a **multisample** experiment, there are two or more populations of interest, and a sample is drawn from each population. For example, if several competing processes are being considered for the manufacture of polyethylene, and tensile strengths are measured on a sample of specimens from each process, this is a multisample experiment. Each process corresponds to a separate population, and the measurements made on the specimens from a particular process are considered to be a simple random sample from that population. The usual purpose of multisample experiments is to make comparisons among populations. In this example, the purpose might be to determine which process produced the greatest strength or to determine whether there is any difference in the strengths of polyethylene made by the different processes.

In many multisample experiments, the populations are distinguished from one another by the varying of one or more **factors** that may affect the outcome. Such experiments are called **factorial experiments**. For example, in his M.S. thesis at the Colorado School of Mines, G. Fredrickson measured the Charpy V-notch impact toughness for a large number of welds. Each weld was made with one of two types of base metals and had its toughness measured at one of several temperatures. This was a factorial experiment with two factors: base metal and temperature. The data consisted of several toughness measurements made at each combination of base metal and temperature. In a factorial experiment, each combination of the factors for which data are collected defines a population, and a simple random sample is drawn from each population. The purpose of a factorial experiment is to determine how varying the levels of the factors affects the outcome being measured. In his experiment Fredrickson found that for each type of base metal, the toughness remained unaffected by temperature unless the temperature was very low—below −100°C. As the temperature was decreased from −100°C to −200°C, the toughness dropped steadily.

Types of Data

When a numerical quantity is assigned to each item in a sample, the resulting set of values is called **numerical** or **quantitative**. In some cases, sample items are placed into categories. Then the data are **categorical** or **qualitative**. Example 1.8 provides an illustration.

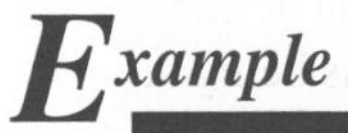

Example 1.8

The article "Hysteresis Behavior of CFT Column to H-Beam Connections with External T-Stiffeners and Penetrated Elements" (C. Kang, K. Shin, et al., *Engineering Structures*, 2001:1194–1201) reported the results of cyclic loading tests on concrete-filled tubular (CFT) column to H-beam welded connections. Several test specimens were loaded until failure. Some failures occurred at the welded joint; others occurred through buckling in the beam itself. For each specimen, the location of the failure was recorded, along with the torque applied at failure [in kilonewton-meters (kN · m)]. The results for the first five specimens were as follows:

Specimen	Torque (kN · m)	Failure Location
1	165	Weld
2	237	Beam
3	222	Beam
4	255	Beam
5	194	Weld

Which data are numerical, and which data are categorical?

Solution

The torques, in the middle column, are numerical data. The failure locations, in the rightmost column, are categorical data.

Exercises for Section 1.1

1. Each of the following processes involves sampling from a population. Define the population, and state whether it is tangible or conceptual.

a. A shipment of bolts is received from a vendor. To check whether the shipment is acceptable with regard to shear strength, an engineer reaches into the container and selects 10 bolts, one by one, to test.

b. The resistance of a certain resistor is measured five times with the same ohmmeter.

c. A graduate student majoring in environmental science is part of a study team that is assessing the risk posed to human health of a certain contaminant present in the tap water in their town. Part of the assessment process involves estimating the amount of time that people who live in that town are in contact with tap water. The student recruits residents of the town to keep diaries for a month, detailing day by day the amount of time they were in contact with tap water.

d. Eight welds are made with the same process, and the strength of each is measured.

e. A quality engineer needs to estimate the percentage of parts manufactured on a certain day that are defective. At 2:30 in the afternoon he samples the last 100 parts to be manufactured.

2. If you wanted to estimate the mean height of all the students at a university, which one of the following sampling strategies would be best? Why? Note that none of the methods are true simple random samples.

i. Measure the heights of 50 students found in the gym during basketball intramurals.

ii. Measure the heights of all engineering majors.

iii. Measure the heights of the students selected by choosing the first name on each page of the campus phone book.

3. True or false:

a. A simple random sample is guaranteed to reflect exactly the population from which it was drawn.

b. A simple random sample is free from any systematic tendency to differ from the population from which it was drawn.

4. A quality engineer draws a simple random sample of 50 O-rings from a lot of several thousand. She measures the thickness of each, and finds that 45 of them, or 90%, meet a certain specification. Which of the following statements is correct?

i. The proportion of O-rings in the entire lot that meet the specification is likely to be equal to 90%.

ii. The proportion of O-rings in the entire lot that meet the specification is likely to be close to 90% but is not likely to equal 90%.

5. A certain process for manufacturing plastic bottles has been in use for a long time, and it is known that 10% of the bottles it produces are defective. A new process that is supposed to reduce the proportion of defectives is being tested. In a simple random sample of 100 bottles produced by the new process, 10 were defective.

a. One of the engineers suggests that the test proves that the new process is no better than the old process, since the proportion of defectives is the same. Is this conclusion justified? Explain.

b. Assume that there had been only 9 defective bottles in the sample of 100. Would this have proven that the new process is better? Explain.

c. Which outcome represents stronger evidence that the new process is better: finding nine defective bottles in the sample, or finding two defective bottles in the sample?

6. Refer to Exercise 5. True or false:

a. If the proportion of defectives in the sample is less than 10%, it is safe to conclude that the new process is better.

b. If the proportion of defectives in the sample is only slightly less than 10%, the difference could well be due entirely to sampling variation, and it is not safe to conclude that the new process is better.

c. If the proportion of defectives in the sample is a lot less than 10%, it is very unlikely that the difference is due entirely to sampling variation, so it is safe to conclude that the new process is better.

d. No matter how few defectives are in the sample, the result could well be due entirely to sampling variation, so it is not safe to conclude that the new process is better.

7. To determine whether a sample should be treated as a simple random sample, which is more important: a good knowledge of statistics, or a good knowledge of the process that produced the data?

1.2 Summary Statistics

A sample is often a long list of numbers. To help make the important features of a sample stand out, we compute summary statistics. The two most commonly used summary statistics are the **sample mean** and the **sample standard deviation**. The mean gives an indication of the center of the data, and the standard deviation gives an indication of how spread out the data are.

The Sample Mean

The sample mean is also called the "arithmetic mean," or, more simply, the "average." It is the sum of the numbers in the sample, divided by how many there are.

Definition

Let $X_1, \ldots, X_n$ be a sample. The **sample mean** is

$$\overline{X} = \frac{1}{n}\sum_{i=1}^{n} X_i \tag{1.1}$$

Note that it is customary to use a letter with a bar over it (e.g., $\overline{X}$) to denote a sample mean. Note also that the sample mean has the same units as the sample values $X_1, \ldots, X_n$.

A simple random sample of five men is chosen from a large population of men, and their heights are measured. The five heights (in inches) are 65.51, 72.30, 68.31, 67.05, and 70.68. Find the sample mean.

Solution
We use Equation (1.1). The sample mean is

$$\overline{X} = \frac{1}{5}(65.51 + 72.30 + 68.31 + 67.05 + 70.68) = 68.77 \text{ in.}$$

The Standard Deviation

Here are two lists of numbers: 28, 29, 30, 31, 32 and 10, 20, 30, 40, 50. Both lists have the same mean of 30. But clearly the lists differ in an important way that is not captured by the mean: the second list is much more spread out than the first. The **standard deviation** is a quantity that measures the degree of spread in a sample.

Let $X_1, \ldots, X_n$ be a sample. The basic idea behind the standard deviation is that when the spread is large, the sample values will tend to be far from their mean, but when the spread is small, the values will tend to be close to their mean. So the first step in calculating the standard deviation is to compute the distances (also called deviations) from each sample value to the sample mean. The deviations are $(X_1 - \overline{X}), \ldots, (X_n - \overline{X})$. Now some of these deviations are positive and some are negative. Large negative deviations are just as indicative of spread as large positive deviations are. To make all the deviations positive we square them, obtaining the squared deviations $(X_1 - \overline{X})^2, \ldots, (X_n - \overline{X})^2$. From the squared deviations we can compute a measure of spread called the **sample variance**. The sample variance is the average of the squared deviations, except that we divide by $n - 1$ instead of n. It is customary to denote the sample variance with the letter s^2.

Definition

Let $X_1, \ldots, X_n$ be a sample. The **sample variance** is the quantity

$$s^2 = \frac{1}{n-1} \sum_{i=1}^{n} (X_i - \overline{X})^2 \tag{1.2}$$

An equivalent formula, which can be easier to compute, is

$$s^2 = \frac{1}{n-1} \left(\sum_{i=1}^{n} X_i^2 - n\overline{X}^2 \right) \tag{1.3}$$

While the sample variance is an important quantity, it has a serious drawback as a measure of spread. Its units are not the same as the units of the sample values; instead they are the squared units. To obtain a measure of spread whose units are the same as those of the sample values, we simply take the square root of the variance. This quantity is known as the **sample standard deviation**. It is customary to denote the sample standard deviation by the letter s (the square root of s^2).

Definition

Let $X_1, \ldots, X_n$ be a sample. The **sample standard deviation** is the quantity

$$s = \sqrt{\frac{1}{n-1} \sum_{i=1}^{n} (X_i - \overline{X})^2} \tag{1.4}$$

An equivalent formula, which can be easier to compute, is

$$s = \sqrt{\frac{1}{n-1} \left(\sum_{i=1}^{n} X_i^2 - n\overline{X}^2 \right)} \tag{1.5}$$

The sample standard deviation is the square root of the sample variance.

It is natural to wonder why the sum of the squared deviations is divided by $n - 1$ rather than n. The purpose in computing the sample standard deviation is to estimate the amount of spread in the population from which the sample was drawn. Ideally, therefore, we would compute deviations from the mean of all the items in the population, rather than the deviations from the sample mean. However, the population mean is in general unknown, so the sample mean is used in its place. It is a mathematical fact that the deviations around the sample mean tend to be a bit smaller than the deviations around the population mean and that dividing by $n - 1$ rather than n provides exactly the right correction.

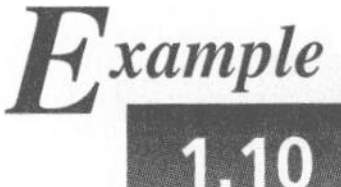

Example 1.10

Find the sample variance and the sample standard deviation for the height data in Example 1.9.

Solution

We'll first compute the sample variance by using Equation (1.2). The sample mean is $\overline{X} = 68.77$ (see Example 1.9). The sample variance is therefore

$$\begin{aligned} s^2 = \frac{1}{4}[&(65.51 - 68.77)^2 + (72.30 - 68.77)^2 + (68.31 - 68.77)^2 \\ &+ (67.05 - 68.77)^2 + (70.68 - 68.77)^2] = 7.47665 \end{aligned}$$

Alternatively, we can use Equation (1.3):

$$s^2 = \frac{1}{4}[65.51^2 + 72.30^2 + 68.31^2 + 67.05^2 + 70.68^2 - 5(68.77^2)] = 7.47665$$

The sample standard deviation is the square root of the sample variance:

$$s = \sqrt{7.47665} = 2.73$$

What would happen to the sample mean, variance, and standard deviation if the heights in Example 1.9 were measured in centimeters rather than inches? Let's denote the heights in inches by X_1, X_2, X_3, X_4, X_5, and the heights in centimeters by Y_1, Y_2, Y_3, Y_4, Y_5. The relationship between X_i and Y_i is then given by $Y_i = 2.54X_i$. If you go back to Example 1.9, convert to centimeters, and compute the sample mean, you will find that the sample means in centimeters and in inches are related by the equation $\overline{Y} = 2.54\overline{X}$. Thus if we multiply each sample item by a constant, the sample mean is multiplied by the same constant. As for the sample variance, you will find that the deviations are related by the equation $(Y_i - \overline{Y}) = 2.54(X_i - \overline{X})$. It follows that $s_Y^2 = 2.54^2 s_X^2$, and that $s_Y = 2.54 s_X$.

What if each man in the sample put on 2-inch heels? Then each sample height would increase by 2 inches and the sample mean would increase by 2 inches as well. In general, if a constant is added to each sample item, the sample mean increases (or decreases) by the same constant. The deviations, however, do not change, so the sample variance and standard deviation are unaffected.

Summary

- If $X_1, \ldots, X_n$ is a sample and $Y_i = a + bX_i$, where a and b are constants, then $\overline{Y} = a + b\overline{X}$.
- If $X_1, \ldots, X_n$ is a sample and $Y_i = a + bX_i$, where a and b are constants, then $s_Y^2 = b^2 s_X^2$, and $s_Y = |b| s_X$.

Outliers

Sometimes a sample may contain a few points that are much larger or smaller than the rest. Such points are called **outliers**. See Figure 1.2 for an example. Sometimes outliers result from data entry errors; for example, a misplaced decimal point can result in a value that is an order of magnitude different from the rest. Outliers should always be scrutinized, and any outlier that is found to result from an error should be corrected or deleted. Not all outliers are errors. Sometimes a population may contain a few values that are much different from the rest, and the outliers in the sample reflect this fact.

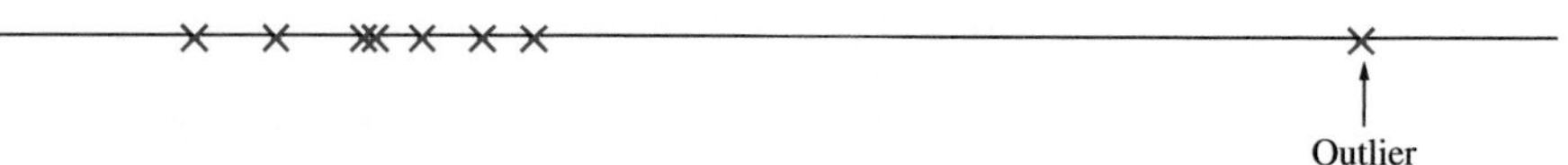

FIGURE 1.2 A data set that contains an outlier.

Outliers are a real problem for data analysts. For this reason, when people see outliers in their data, they sometimes try to find a reason, or an excuse, to delete them. An outlier should not be deleted, however, unless there is reasonable certainty that it results from an error. If a population truly contains outliers, but they are deleted from the sample, the sample will not characterize the population correctly.

The Sample Median

The **median**, like the mean, is a measure of center. To compute the median of a sample, order the values from smallest to largest. The sample median is the middle number. If the sample size is an even number, it is customary to take the sample median to be the average of the two middle numbers.

Definition

If n numbers are ordered from smallest to largest:

- If n is odd, the sample median is the number in position $\dfrac{n+1}{2}$.
- If n is even, the sample median is the average of the numbers in positions $\dfrac{n}{2}$ and $\dfrac{n}{2} + 1$.

Example 1.11

Find the sample median for the height data in Example 1.9.

Solution

The five heights, arranged in increasing order, are 65.51, 67.05, 68.31, 70.68, 72.30. The sample median is the middle number, which is 68.31.

The median is often used as a measure of center for samples that contain outliers. To see why, consider the sample consisting of the values 1, 2, 3, 4, and 20. The mean is 6, and the median is 3. It is reasonable to think that the median is more representative of the sample than the mean is. See Figure 1.3.

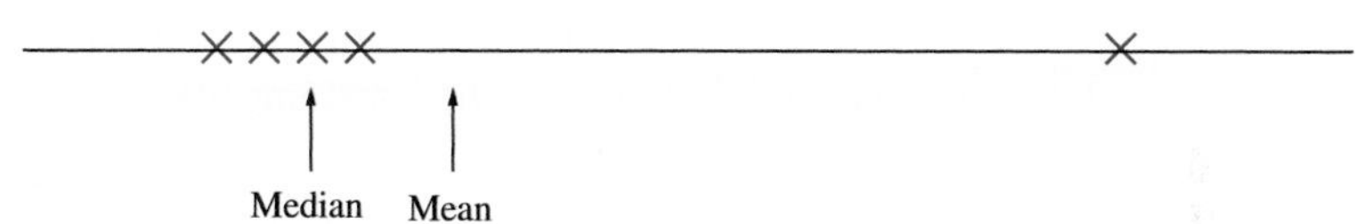

FIGURE 1.3 When a sample contains outliers, the median may be more representative of the sample than the mean is.

The Trimmed Mean

Like the median, the **trimmed mean** is a measure of center that is designed to be unaffected by outliers. The trimmed mean is computed by arranging the sample values in order, "trimming" an equal number of them from each end, and computing the mean of those remaining. If $p\%$ of the data are trimmed from each end, the resulting trimmed mean is called the "$p\%$ trimmed mean." There are no hard-and-fast rules on how many values to trim. The most commonly used trimmed means are the 5%, 10%, and 20% trimmed means. Note that the median can be thought of as an extreme form of trimmed mean, obtained by trimming away all but the middle one or two sample values.

Since the number of data points trimmed must be a whole number, it is impossible in many cases to trim the exact percentage of data that is called for. If the sample size is denoted by n, and a $p\%$ trimmed mean is desired, the number of data points to be trimmed is $np/100$. If this is not a whole number, the simplest thing to do when computing by hand is to round it to the nearest whole number and trim that amount.

Example 1.12

In the article "Evaluation of Low-Temperature Properties of HMA Mixtures" (P. Sebaaly, A. Lake, and J. Epps, *Journal of Transportation Engineering*, 2002:578–583), the following values of fracture stress (in megapascals) were measured for a sample of 24 mixtures of hot-mixed asphalt (HMA).

30	75	79	80	80	105	126	138	149	179	179	191
223	232	232	236	240	242	245	247	254	274	384	470

Compute the mean, median, and the 5%, 10%, and 20% trimmed means.

Solution

The mean is found by averaging together all 24 numbers, which produces a value of 195.42. The median is the average of the 12th and 13th numbers, which is $(191 + 223)/2 = 207.00$. To compute the 5% trimmed mean, we must drop 5% of the data from each end. This comes to $(0.05)(24) = 1.2$ observations. We round 1.2 to 1, and trim one observation off each end. The 5% trimmed mean is the average of the remaining 22 numbers:

$$\frac{75 + 79 + \cdots + 274 + 384}{22} = 190.45$$

To compute the 10% trimmed mean, round off $(0.1)(24) = 2.4$ to 2. Drop 2 observations from each end, and then average the remaining 20:

$$\frac{79 + 80 + \cdots + 254 + 274}{20} = 186.55$$

To compute the 20% trimmed mean, round off $(0.2)(24) = 4.8$ to 5. Drop 5 observations from each end, and then average the remaining 14:

$$\frac{105 + 126 + \cdots + 242 + 245}{14} = 194.07$$

The Mode and the Range

The **mode** and the **range** are summary statistics that are of limited use but are occasionally seen. The sample mode is the most frequently occurring value in a sample. If several values occur with equal frequency, each one is a mode. The range is the difference between the largest and smallest values in a sample. It is a measure of spread, but it is rarely used, because it depends only on the two extreme values and provides no information about the rest of the sample.

Find the modes and the range for the sample in Example 1.12.

Solution

There are three modes: 80, 179, and 232. Each of these values appears twice, and no other value appears more than once. The range is $470 - 30 = 440$.

Quartiles

The median divides the sample in half. **Quartiles** divide it as nearly as possible into quarters. A sample has three quartiles. There are several different ways to compute quartiles, but all of them give approximately the same result. The simplest method when computing by hand is as follows: Let n represent the sample size. Order the sample values from smallest to largest. To find the first quartile, compute the value $0.25(n + 1)$. If this is an integer, then the sample value in that position is the first quartile. If not, then take the average of the sample values on either side of this value. The third quartile is computed in the same way, except that the value $0.75(n + 1)$ is used. The second quartile uses the value $0.5(n + 1)$. The second quartile is identical to the median. We note that some computer packages use slightly different methods to compute quartiles, so their results may not be quite the same as the ones obtained by the method described here.

Find the first and third quartiles of the asphalt data in Example 1.12.

Solution

The sample size is $n = 24$. To find the first quartile, compute $(0.25)(25) = 6.25$. The first quartile is therefore found by averaging the 6th and 7th data points, when the sample is arranged in increasing order. This yields $(105 + 126)/2 = 115.5$. To find the third quartile, compute $(0.75)(25) = 18.75$. We average the 18th and 19th data points to obtain $(242 + 245)/2 = 243.5$.

Percentiles

The pth percentile of a sample, for a number p between 0 and 100, divides the sample so that as nearly as possible $p\%$ of the sample values are less than the pth percentile, and $(100 - p)\%$ are greater. There are many ways to compute percentiles; they all produce similar results. We describe here a method analogous to the method described for computing quartiles. Order the sample values from smallest to largest, and then compute the quantity $(p/100)(n + 1)$, where n is the sample size. If this quantity is an integer, the sample value in this position is the pth percentile. Otherwise average the two sample values on either side. Note that the first quartile is the 25th percentile, the median is the 50th percentile, and the third quartile is the 75th percentile. Some computer packages use slightly different methods to compute percentiles, so their results may differ slightly from the ones obtained by this method.

Percentiles are often used to interpret scores on standardized tests. For example, if a student is informed that her score on a college entrance exam is on the 64th percentile, this means that 64% of the students who took the exam got lower scores.

Find the 65th percentile of the asphalt data in Example 1.12.

Solution

The sample size is $n = 24$. To find the 65th percentile, compute $(0.65)(25) = 16.25$. The 65th percentile is therefore found by averaging the 16th and 17th data points, when the sample is arranged in increasing order. This yields $(236 + 240)/2 = 238$.

In practice, the summary statistics we have discussed are often calculated on a computer, using a statistical software package. The summary statistics are sometimes called **descriptive statistics** because they describe the data. We present an example of the calculation of summary statistics from the software package MINITAB. Then we will show how these statistics can be used to discover some important features of the data.

For a Ph.D. thesis that investigated factors affecting diesel vehicle emissions, J. Yanowitz of the Colorado School of Mines obtained data on emissions of particulate matter (PM) for a sample of 138 vehicles driven at low altitude (near sea level) and for a sample of 62 vehicles driven at high altitude (approximately one mile above sea level). All the vehicles were manufactured between 1991 and 1996. The samples contained roughly equal proportions of high- and low-mileage vehicles. The data, in units of grams of particulates per gallon of fuel consumed, are presented in Tables 1.1 and 1.2. At high altitude, the barometric pressure is lower, so the effective air/fuel ratio is lower as well. For this reason it was thought that PM emissions might be greater at higher altitude. We would like to compare the samples to determine whether the data support this assumption. It is difficult to do this simply by examining the raw data in the tables. Computing summary statistics makes the job much easier. Figure 1.4 presents summary statistics for both samples, as computed by MINITAB.

TABLE 1.1 Particulate matter (PM) emissions (in g/gal) for 138 vehicles driven at low altitude

1.50	0.87	1.12	1.25	3.46	1.11	1.12	0.88	1.29	0.94	0.64	1.31	2.49
1.48	1.06	1.11	2.15	0.86	1.81	1.47	1.24	1.63	2.14	6.64	4.04	2.48
2.98	7.39	2.66	11.00	4.57	4.38	0.87	1.10	1.11	0.61	1.46	0.97	0.90
1.40	1.37	1.81	1.14	1.63	3.67	0.55	2.67	2.63	3.03	1.23	1.04	1.63
3.12	2.37	2.12	2.68	1.17	3.34	3.79	1.28	2.10	6.55	1.18	3.06	0.48
0.25	0.53	3.36	3.47	2.74	1.88	5.94	4.24	3.52	3.59	3.10	3.33	4.58
6.73	7.82	4.59	5.12	5.67	4.07	4.01	2.72	3.24	5.79	3.59	3.48	2.96
5.30	3.93	3.52	2.96	3.12	1.07	5.30	5.16	7.74	5.41	3.40	4.97	11.23
9.30	6.50	4.62	5.45	4.93	6.05	5.82	10.19	3.62	2.67	2.75	8.92	9.93
6.96	5.78	9.14	10.63	8.23	6.83	5.60	5.41	6.70	5.93	4.51	9.04	7.71
7.21	4.67	4.49	4.63	2.80	2.16	2.97	3.90					

TABLE 1.2 Particulate matter (PM) emissions (in g/gal) for 62 vehicles driven at high altitude

7.59	6.28	6.07	5.23	5.54	3.46	2.44	3.01	13.63	13.02	23.38	9.24	3.22
2.06	4.04	17.11	12.26	19.91	8.50	7.81	7.18	6.95	18.64	7.10	6.04	5.66
8.86	4.40	3.57	4.35	3.84	2.37	3.81	5.32	5.84	2.89	4.68	1.85	9.14
8.67	9.52	2.68	10.14	9.20	7.31	2.09	6.32	6.53	6.32	2.01	5.91	5.60
5.61	1.50	6.46	5.29	5.64	2.07	1.11	3.32	1.83	7.56			

Descriptive Statistics: LowAltitude, HiAltitude

Variable	N	Mean	SE Mean	TrMean	StDev
LoAltitude	138	3.715	0.218	3.526	2.558
HiAltitude	62	6.596	0.574	6.118	4.519

Variable	Minimum	Q1	Median	Q3	Maximum
LoAltitude	0.250	1.468	3.180	5.300	11.230
HiAltitude	1.110	3.425	5.750	7.983	23.380

FIGURE 1.4 MINITAB output presenting descriptive statistics for the PM data in Tables 1.1 and 1.2.

In Figure 1.4, the quantity labeled "N" is the sample size. Following that is the sample mean. The next quantity (SE Mean) is the **standard error of the mean**. The standard error of the mean is equal to the standard deviation divided by the square root of the sample size. This is a quantity that is not used much as a descriptive statistic, although it is very important for applications such as constructing confidence intervals and hypothesis tests, which we will cover in Chapters 5 and 6. Following the standard error of the mean is the 5% trimmed mean (TrMean), and the standard deviation. Finally, the second line of the output provides the minimum, median, and maximum, as well as the first and third quartiles (Q1 and Q3). We note that the values of the quartiles produced by the computer package differ slightly from the values that would be computed by the methods we describe. This is not surprising, since there are several ways to compute these values. The differences are not large enough to have any practical importance.

The summary statistics tell a lot about the differences in PM emissions between high- and low-altitude vehicles. First, note that the mean is indeed larger for the high-altitude vehicles than for the low-altitude vehicles (6.60 vs. 3.71), which supports the hypothesis that emissions tend to be greater at high altitudes. Now note that the maximum value for the high-altitude vehicles (23.38) is much higher than the maximum for the low-altitude vehicles (11.23). This shows that there are one or more high-altitude vehicles whose emissions are much higher than the highest of the low-altitude vehicles. Could the difference in mean emissions be due entirely to these vehicles? To answer this, compare the medians, the first and third quartiles, and the trimmed means. These statistics are not affected much by a few large values, yet all of them are noticeably larger for the high-altitude vehicles. Therefore, we can conclude that the high-altitude vehicles not only contain a few very high emitters, they also have higher emissions than the low-altitude vehicles in general. Finally note that the standard deviation is larger for the high-altitude vehicles, which indicates that the values for the high-altitude vehicles are more spread out than those for the low-altitude vehicles. At least some of this difference in spread must be due to the one or more high-altitude vehicles with very high emissions.

Summary Statistics for Categorical Data

With categorical data, each sample item is assigned a category rather than a numerical value. But to work with categorical data, numerical summaries are needed. The two most commonly used ones are the **frequencies** and the **sample proportions** (sometimes called **relative frequencies**). The frequency for a given category is simply the number of sample items that fall into that category. The sample proportion is the frequency divided by the sample size.

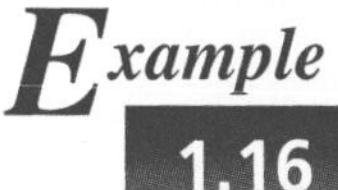

A process manufactures crankshaft journal bearings for an internal combustion engine. Bearings whose thicknesses are between 1.486 and 1.490 mm are classified as conforming, which means that they meet the specification. Bearings thicker than this are reground, and bearings thinner than this are scrapped. In a sample of 1000 bearings, 910 were conforming, 53 were reground, and 37 were scrapped. Find the frequencies and sample proportions.

Solution

The frequencies are 910, 53, and 37. The sample proportions are $910/1000 = 0.910$, $53/1000 = 0.053$, and $37/1000 = 0.037$.

Sample Statistics and Population Parameters

Each of the sample statistics we have discussed has a population counterpart. This is easy to see when the population is finite. For example, for a finite population of numerical values, the population mean is simply the average of all the values in the population; the population median is the middle value, or average of the two middle values; and so on. In fact, any numerical summary used for a sample can be used for a finite population, just by applying the methods of calculation to the population values rather than the sample values. One small exception occurs for the population variance, where we divide by n rather than $n - 1$. There is a difference in terminology for numerical summaries of populations as opposed to samples. Numerical summaries of a sample are called **statistics**, while numerical summaries of a population are called **parameters**. Of course, in practice, the entire population is never observed, so the population parameters cannot be calculated directly. Instead, the sample statistics are used to estimate the values of the population parameters.

The methods for computing sample statistics require that the sample be finite. Therefore, when a population contains an infinite number of values, the methods for computing sample statistics cannot be applied to compute population parameters. For infinite populations, parameters such as the mean and variance are computed by procedures that generalize the methods used to compute sample statistics, and which involve infinite sums or integrals. We will describe these procedures in Chapter 2.

Summary

- A numerical summary of a sample is called a **statistic**.
- A numerical summary of a population is called a **parameter**.
- Statistics are often used to estimate parameters.

Exercises for Section 1.2

1. True or false: For any list of numbers, half of them will be below the mean.

2. Is the sample mean always the most frequently occurring value? If so, explain why. If not, give an example.

3. Is the sample mean always equal to one of the values in the sample? If so, explain why. If not, give an example.

4. Is the sample median always equal to one of the values in the sample? If so, explain why. If not, give an example.

5. Find a sample size for which the median will always equal one of the values in the sample.

6. In a certain company, every worker received a \$50-per-week raise. How does this affect the mean salary? The standard deviation of the salaries?

7. In another company, every worker received a 5% raise. How does this affect the mean salary? The standard deviation of the salaries?

8. The Apgar score is used to rate reflexes and responses of newborn babies. Each baby is assigned a score by a medical professional, and the possible values are the integers from 0 to 10. A sample of 1000 babies born in a certain county is taken and the number with each score is as follows:

Score	0	1	2	3	4	5	6	7	8	9	10
Number of babies	1	3	2	4	25	35	198	367	216	131	18

   a. Find the sample mean Apgar score.
   b. Find the sample standard deviation of the Apgar scores.
   c. Find the sample median of the Apgar scores.
   d. What is the first quartile of the scores?
   e. What proportion of the scores is greater than the mean?
   f. What proportion of the scores is more than one standard deviation greater than the mean?
   g. What proportion of the scores is within one standard deviation of the mean?

9. A statistics class of 40 students took a quiz. The highest possible score was 4 points. Ten students scored 4 points, 12 students scored 3 points, 8 students scored 2 points, 6 students scored 1 point, and 4 students scored 0. Compute the mean, median, and standard deviation of the scores.

10. Another statistics class of 60 students took the same quiz. In this class, 15 students scored 4 points, 18 students scored 3 points, 12 students scored 2 points, 9 students scored 1 point, and 6 students scored 0. Compute the mean, median, and standard deviation of the scores.

11. In yet another statistics class, the total number of students is unknown. In this class, 25% of the students scored 4 points, 30% scored 3 points, 20% scored 2 points, 15% scored 1 point, and 10% scored 0.
   a. Is it possible to compute the mean score for this class? If so, compute it. If not, explain why not.
   b. Is it possible to compute the median score for this class? If so, compute it. If not, explain why not.

c. Is it possible to compute the sample standard deviation of the scores for this class? If so, compute it. If not, explain why not.

12. Each of 16 students measured the circumference of a tennis ball by four different methods, which were:

Method A: Estimate the circumference by eye.

Method B: Measure the diameter with a ruler, and then compute the circumference.

Method C: Measure the circumference with a ruler and string.

Method D: Measure the circumference by rolling the ball along a ruler.

The results (in cm) are as follows, in increasing order for each method:

Method A: 18.0, 18.0, 18.0, 20.0, 22.0, 22.0, 22.5, 23.0, 24.0, 24.0, 25.0, 25.0, 25.0, 25.0, 26.0, 26.4.

Method B: 18.8, 18.9, 18.9, 19.6, 20.1, 20.4, 20.4, 20.4, 20.4, 20.5, 21.2, 22.0, 22.0, 22.0, 22.0, 23.6.

Method C: 20.2, 20.5, 20.5, 20.7, 20.8, 20.9, 21.0, 21.0, 21.0, 21.0, 21.0, 21.5, 21.5, 21.5, 21.5, 21.6.

Method D: 20.0, 20.0, 20.0, 20.0, 20.2, 20.5, 20.5, 20.7, 20.7, 20.7, 21.0, 21.1, 21.5, 21.6, 22.1, 22.3.

a. Compute the mean measurement for each method.
b. Compute the median measurement for each method.
c. Compute the 20% trimmed mean measurement for each method.
d. Compute the first and third quartiles for each method.
e. Compute the standard deviation of the measurements for each method.
f. For which method is the standard deviation the largest? Why should one expect this method to have the largest standard deviation?
g. Other things being equal, is it better for a measurement method to have a smaller standard deviation or a larger standard deviation? Or doesn't it matter? Explain.

13. Refer to Exercise 12.

a. If the measurements for one of the methods were converted to inches (1 inch = 2.54 cm), how would this affect the mean? The median? The quartiles? The standard deviation?
b. If the students remeasured the ball, using a ruler marked in inches, would the effects on the mean, median, quartiles, and standard deviation be the same as in part (a)? Explain.

14. A list of 10 numbers has a mean of 20, a median of 18, and a standard deviation of 5. The largest number on the list is 39.27. By accident, this number is changed to 392.7.

a. What is the value of the mean after the change?
b. What is the value of the median after the change?
c. What is the value of the standard deviation after the change?

15. Why doesn't anyone talk about the fourth quartile? Or do they?

16. In each of the following data sets, tell whether the outlier seems certain to be due to an error, or whether it could conceivably be correct.

a. A rock is weighed five times. The readings in grams are 48.5, 47.2, 4.91, 49.5, 46.3.
b. A sociologist samples five families in a certain town and records their annual income. The incomes are \$34,000, \$57,000, \$13,000, \$1,200,000, \$62,000.

1.3 Graphical Summaries

Stem-and-Leaf Plots

The mean, median, and standard deviation are numerical summaries of a sample or of a population. Graphical summaries are used as well to help visualize a list of numbers. The graphical summary that we will discuss first is the **stem-and-leaf plot**. A stem-and-leaf plot is a simple way to summarize a data set.

As an example, the data in Table 1.3 concern the geyser Old Faithful in Yellowstone National Park. This geyser alternates periods of eruption, which typically last from 1.5 to 4 minutes, with periods of dormancy, which are considerably longer. Table 1.3 presents the durations, in minutes, of 60 dormant periods. The list has been sorted into numerical order.

TABLE 1.3 Durations (in minutes) of dormant periods of the geyser Old Faithful

42	45	49	50	51	51	51	51	53	53
55	55	56	56	57	58	60	66	67	67
68	69	70	71	72	73	73	74	75	75
75	75	76	76	76	76	76	79	79	80
80	80	80	81	82	82	82	83	83	84
84	84	85	86	86	86	88	90	91	93

Figure 1.5 presents a stem-and-leaf plot of the geyser data. Each item in the sample is divided into two parts: a **stem**, consisting of the leftmost one or two digits, and the **leaf**, which consists of the next significant digit. In the geyser data, the stem consists of the tens digit and the leaf consists of the ones digit. Each line of the stem-and-leaf plot contains all of the sample items with a given stem. The stem-and-leaf plot is a compact way to represent the data. It also gives some indication of its shape. For the geyser data, we can see that there are relatively few durations in the 60–69 minute interval, compared with the 50–59, 70–79, or 80–89 minute intervals.

Stem	Leaf
4	259
5	0111133556678
6	067789
7	01233455556666699
8	000012223344456668
9	013

FIGURE 1.5 Stem-and-leaf plot for the geyser data in Table 1.3.

When there are a great many sample items with the same stem, it is often necessary to assign more than one row to that stem. As an example, Figure 1.6 (page 26) presents a computer-generated stem-and-leaf plot, produced by MINITAB, for the PM data in Table 1.2 in Section 1.2. The middle column, consisting of 0s, 1s, and 2s, contains the stems, which are the tens digits. To the right of the stems are the leaves, consisting of the ones digits for each of the sample items. Since many numbers are less than 10, the 0

```
Stem-and-leaf of HiAltitude      N = 62
Leaf Unit = 1.0

   4    0    1111
  19    0    222222223333333
 (14)   0    44445555555555
  29    0    66666666777777
  15    0    8889999
   8    1    0
   7    1    233
   4    1
   4    1    7
   3    1    89
   1    2
   1    2    3
```

FIGURE 1.6 Stem-and-leaf plot of the PM data in Table 1.2 in Section 1.2 as produced by MINITAB.

stem must be assigned several lines, five in this case. Specifically, the first line contains the sample items whose ones digits are either 0 or 1, the next line contains the items whose ones digits are either 2 or 3, and so on. For consistency, all the stems are assigned several lines in the same way, even though there are few enough values for the 1 and 2 stems that they could have fit on fewer lines.

The output in Figure 1.6 contains a cumulative frequency column to the left of the stem-and-leaf plot. The upper part of this column provides a count of the number of items at or above the current line, and the lower part of the column provides a count of the number of items at or below the current line. Next to the line that contains the median is the count of items in that line, shown in parentheses.

A good feature of stem-and-leaf plots is that they display all the sample values. One can reconstruct the sample in its entirety from a stem-and-leaf plot—with one important exception: The order in which the items were sampled cannot be determined.

Dotplots

A **dotplot** is a graph that can be used to give a rough impression of the shape of a sample. It is useful when the sample size is not too large and when the sample contains some repeated values. Figure 1.7 presents a dotplot for the geyser data in Table 1.3. For each value in the sample a vertical column of dots is drawn, with the number of dots in the column equal to the number of times the value appears in the sample. The dotplot gives a good indication of where the sample values are concentrated and where the gaps are. For example, it is immediately apparent from Figure 1.7 that the sample contains no dormant periods between 61 and 65 minutes in length.

Stem-and-leaf plots and dotplots are good methods for informally examining a sample, and they can be drawn fairly quickly with pencil and paper. They are rarely used

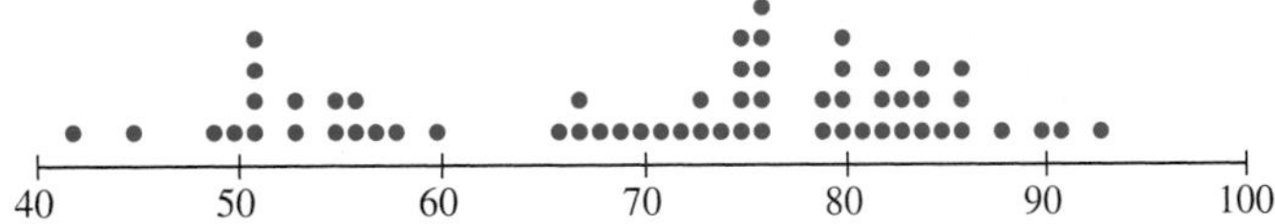

FIGURE 1.7 Dotplot for the geyser data in Table 1.3.

in formal presentations, however. Graphics more commonly used in formal presentations include the histogram and the boxplot, which we will now discuss.

Histograms

A **histogram** is a graphic that gives an idea of the "shape" of a sample, indicating regions where sample points are concentrated and regions where they are sparse. We will construct a histogram for the PM emissions of 62 vehicles driven at high altitude, as presented in Table 1.2 in Section 1.2. The sample values range from a low of 1.11 to a high of 23.38, in units of grams of emissions per gallon of fuel. The first step is to construct a **frequency table**, shown in Table 1.4.

TABLE 1.4 Frequency table for PM emissions of 62 vehicles driven at high altitude

Class Interval (g/gal)	Frequency	Relative Frequency	Density
1–< 3	12	0.194	0.0970
3–< 5	11	0.177	0.0885
5–< 7	18	0.290	0.1450
7–< 9	9	0.145	0.0725
9–< 11	5	0.081	0.0405
11–< 15	3	0.048	0.0120
15–< 25	4	0.065	0.0065

The intervals in the left-hand column are called **class intervals**. They divide the sample into groups. The notation 1–< 3, 3–< 5, and so on, indicates that a point on the boundary will go into the class on its right. For example, a sample value equal to 3 will go into the class 3–< 5, not 1–< 3.

There is no hard-and-fast rule as to how to choose the endpoints of the class intervals. In general, it is good to have more intervals rather than fewer, but it is also good to have large numbers of sample points in the intervals. Striking the proper balance is a matter of judgment and of trial and error. In many cases it is reasonable to take the number of class intervals roughly equal to the square root of the sample size. For the PM data, class intervals that are two units wide work well, except for the larger values (e.g., greater than 11), where the data thin out. We have therefore grouped the values between 11 and 15 into a class, and all the values greater than 15 into another class.

The column labeled "Frequency" in Table 1.4 presents the numbers of data points that fall into each of the class intervals. The column labeled "Relative Frequency" presents the frequencies divided by the total number of data points, which for these data is 62.

The relative frequency of a class interval is the proportion of data points that fall into the interval. Note that since every data point is in exactly one class interval, the relative frequencies must sum to 1. Finally, the column labeled "Density" presents the relative frequency divided by the class width. For example, in the first row, the relative frequency is 0.194 and the class width is 2 ($3 - 1 = 2$). The density is therefore $0.194/2 = 0.0970$. The last class has width 10 and relative frequency 0.065, so its density is $0.065/10 = 0.0065$. The purpose of the density is to adjust the relative frequency for the width of the class. Other things being equal, wider classes tend to contain more sample items than the narrower classes, and thus tend to have larger relative frequencies. Dividing the relative frequency by the class width adjusts for this tendency. The density can be thought of as the relative frequency per unit.

Figure 1.8 presents the histogram for Table 1.4. The units on the horizontal axis are the units of the data, in this case g/gal. Each class interval is represented by a rectangle. The height of each rectangle is the density of the sample in that class interval, which is given in the fourth column of Table 1.4. The *area* of each rectangle is thus the relative frequency of the class interval, which is given in the third column of Table 1.4. Since the relative frequencies sum to 1, the area under the entire histogram must be equal to 1.

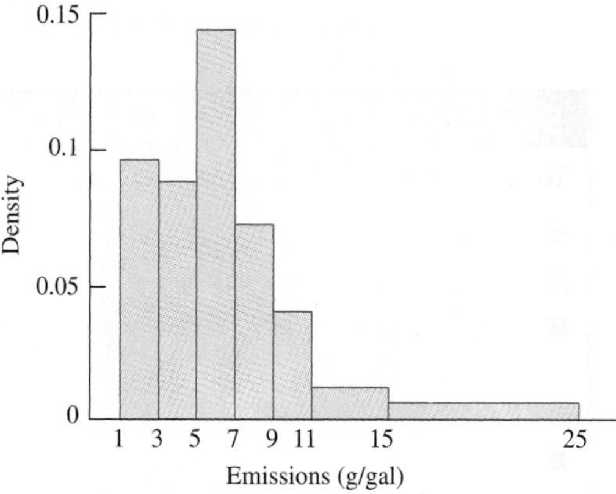

FIGURE 1.8 Histogram for the PM emissions for high-altitude vehicles. The frequency table is presented in Table 1.4.

Example 1.17

Use the histogram in Figure 1.8 to determine the proportion of the vehicles in the sample with emissions between 7 and 11 g/gal.

Solution

The proportion is the area under the histogram between 7 and 11. This is found by adding the areas of the rectangles for the two class intervals covered. The result is $(2)(0.0725) + (2)(0.0405) = 0.226$. Note that this result can also be obtained from the frequency table. The proportion of data points with values between 7 and 9 is 0.145, and the proportion between 9 and 11 is 0.081. The proportion between 7 and 11 is therefore equal to $0.145 + 0.081 = 0.226$.

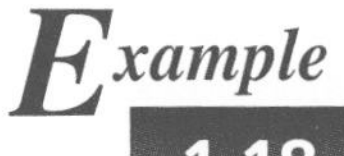

Use the histogram to estimate the proportion of vehicles in the sample with emissions between 6 and 10 g/gal.

Solution

The histogram does not give the exact answer, because the values 6 and 10 are not endpoints of class intervals. We estimate the proportion by computing the area under the histogram between the values 6 and 10. This consists of half the area of the rectangle over the interval 5 to 7, plus the whole rectangle over the interval 7 to 9, plus half the rectangle over the interval 9 to 11. The area is $(1)(0.1450)+(2)(0.0725)+(1)(0.0405) = 0.3305$.

What is the density of the sample at 6 g/gal?

Solution

The density is the height of the histogram at this point. This height is 0.1450.

Summary

To construct a histogram:

- Choose boundary points for the class intervals.
- Compute the frequencies and relative frequencies for each class.
- Compute the density for each class, according to the formula

$$\text{Density} = \frac{\text{relative frequency}}{\text{class width}}$$

- Draw a rectangle for each class, whose height is equal to the density.

Equal Class Widths

Most statistical software packages draw histograms and will give you the option of specifying the class intervals or of having the software choose them for you. When asked to choose class intervals, most software packages make them all the same width. For example, Figure 1.9 (page 30) presents a histogram of the data in Figure 1.8, with class intervals chosen by MINITAB. MINITAB chooses all the intervals to have a width of 2.

For many data sets, equal class widths are fine. But for data such as these, with several outliers stretching out to the right, equal class widths are less desirable. To understand why, realize that there are only 7 data points out of 62 with values greater than 11. In Figure 1.9, more than half of the class intervals are devoted to these 7 points, and they form a string of small rectangles of varying sizes. These rectangles contain a lot of visual

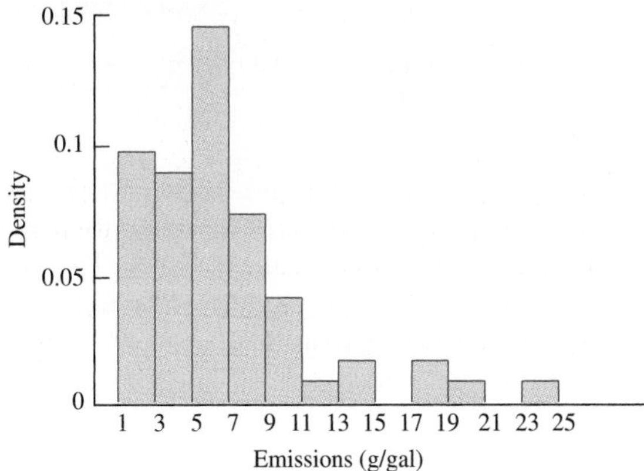

FIGURE 1.9 Histogram for the data in Figure 1.8, with equal class widths as chosen by MINITAB.

structure, which distracts from the more important structure in the bulk of the data to the left. In contrast, the histogram in Figure 1.8 groups these 7 points into two class intervals, which presents a smoother appearance and better enables the eye to appreciate the structure of the data set as a whole.

The Histogram and the Sample Mean and Variance

Both the sample mean and the sample variance have physical interpretations with regard to the histogram. Imagine that the histogram in Figure 1.8 is a thin plate standing on the horizontal axis. Assume that the mass of each rectangle is proportional to its area. The horizontal component of the center of mass is the point on the x axis where the histogram would balance if supported there. To find the horizontal component of the center of mass of the histogram, we would treat each rectangle as if its mass were concentrated at its midpoint. We would multiply the midpoint of each rectangle by its area and add the products to obtain the center of mass. The midpoints of the rectangles are the midpoints of the class intervals, and the areas are the relative frequencies (see Table 1.4). Thus the center of mass of the histogram in Figure 1.8 is given by

$$(2)(0.194) + (4)(0.177) + \cdots + (20)(0.065) = 6.730 \tag{1.6}$$

This value is fairly close to the sample mean, which is 6.596 as shown in the MINITAB output (Figure 1.4, in Section 1.2). To relate the center of mass to the sample mean, note that if each item in the sample had a value equal to the midpoint of its class interval, then expression (1.6) would equal the sample mean. Thus the center of mass of the histogram is an approximation to the sample mean. The narrower the rectangles are, the closer each sample item is to the center of its class interval, and the closer the center of mass of the histogram comes to the sample mean.

To develop a physical interpretation for the sample variance, imagine a solid rod inserted vertically into the histogram through the center of mass (sample mean). Now imagine grasping the rod and twirling the histogram around it. The more spread out the histogram, the more difficult it would be to twirl. The physical quantity that measures the difficulty in twirling is the moment of inertia. For each rectangle in the histogram, the moment of inertia around the center of mass is approximately given by the squared distance from the midpoint of the rectangle to the center of mass, multiplied by the area of the rectangle. The moment of inertia for the entire histogram is the sum of the moments of the rectangles, which is

$$(2 - 6.730)^2(0.194) + (4 - 6.730)^2(0.177) + \cdots + (20 - 6.730)^2(0.065) = 20.25 \tag{1.7}$$

This value is close to the sample variance, which is 20.42. [The sample variance can be found from the MINITAB output (Figure 1.4, in Section 1.2) by squaring the standard deviation, which is 4.519.] If the value of each item in the sample were exactly equal to the midpoint of its class interval, Equation (1.7) would give the sample variance exactly. As it is, the moment of inertia of the histogram around the center of mass is an approximation to the sample variance. The narrower the rectangles are, the closer the approximation is.

The fact that the sample mean and variance correspond to physical properties of the histogram is very useful. In Chapter 2, we will develop methods to compute the population mean and variance for an infinite population by representing the population with a curve and computing the center of mass and moment of inertia.

Symmetry and Skewness

A histogram is perfectly **symmetric** if its right half is a mirror image of its left half. Histograms that are not symmetric are referred to as **skewed**. In practice, no sample has a perfectly symmetric histogram; all exhibit some degree of skewness. In a skewed histogram, one side, or tail, is longer than the other. A histogram with a long right-hand tail is said to be **skewed to the right**, or **positively skewed**. A histogram with a long left-hand tail is said to be **skewed to the left**, or **negatively skewed**. While there is a formal mathematical method for measuring the skewness of a histogram, it is rarely used; instead people judge the degree of skewness informally by looking at the histogram. Figure 1.10 (page 32) presents some histograms for hypothetical samples. Note that for a histogram that is skewed to the right (Figure 1.10c), the mean is greater than the median, because more than half of the data will be to the left of the center of mass. In the same way, the mean is less than the median for a histogram that is skewed to the left (Figure 1.10a). The histogram for the PM data (Figure 1.8) is skewed to the right. The sample mean is 6.596, which is greater than the sample median of 5.75.

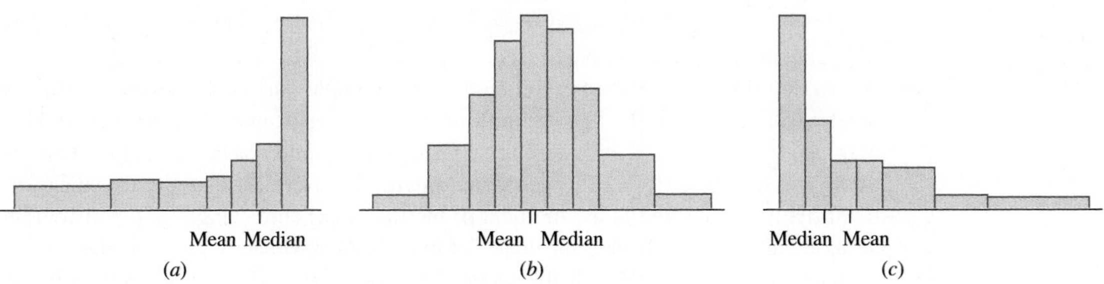

FIGURE 1.10 *(a)* A histogram skewed to the left. The mean is less than the median. *(b)* A nearly symmetric histogram. The mean and median are approximately equal. *(c)* A histogram skewed to the right. The mean is greater than the median.

Unimodal and Bimodal Histograms

We have used the term "mode" to refer to the most frequently occurring value in a sample. This term is also used in regard to histograms and other curves to refer to a peak, or local maximum. A histogram is **unimodal** if it has only one peak, or mode, and **bimodal** if it has two clearly distinct modes. In principle, a histogram can have more than two modes, but this does not happen often in practice. The histograms in Figure 1.10 are all unimodal. Figure 1.11 presents a bimodal histogram for a hypothetical sample.

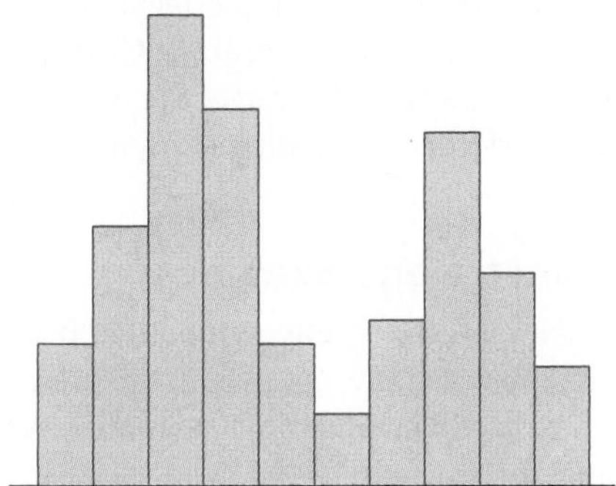

FIGURE 1.11 A bimodal histogram.

In some cases, a bimodal histogram indicates that the sample can be divided into two subsamples that differ from each other in some scientifically important way. Each sample corresponds to one of the modes. As an example, Table 1.5 presents the durations of 60 dormant periods of the geyser Old Faithful (originally presented in Table 1.3). Along with the durations of the dormant period, in minutes, the duration of the eruption immediately preceding the dormant period is classified either as short (less than 3 minutes) or long (more than 3 minutes).

TABLE 1.5 Durations of dormant periods (in minutes) and of the previous eruptions of the geyser Old Faithful

Dormant	Eruption	Dormant	Eruption	Dormant	Eruption	Dormant	Eruption
76	Long	90	Long	45	Short	84	Long
80	Long	42	Short	88	Long	70	Long
84	Long	91	Long	51	Short	79	Long
50	Short	51	Short	80	Long	60	Long
93	Long	79	Long	49	Short	86	Long
55	Short	53	Short	82	Long	71	Long
76	Long	82	Long	75	Long	67	Short
58	Short	51	Short	73	Long	81	Long
74	Long	76	Long	67	Long	76	Long
75	Long	82	Long	68	Long	83	Long
80	Long	84	Long	86	Long	76	Long
56	Short	53	Short	72	Long	55	Short
80	Long	86	Long	75	Long	73	Long
69	Long	51	Short	75	Long	56	Short
57	Long	85	Long	66	Short	83	Long

Figure 1.12a presents a histogram for all 60 durations. Figures 1.12b and c present histograms for the durations following short and long eruptions, respectively. The histogram for all the durations is clearly bimodal. The histograms for the durations following short or long eruptions are both unimodal, and their modes form the two modes of the histogram for the full sample.

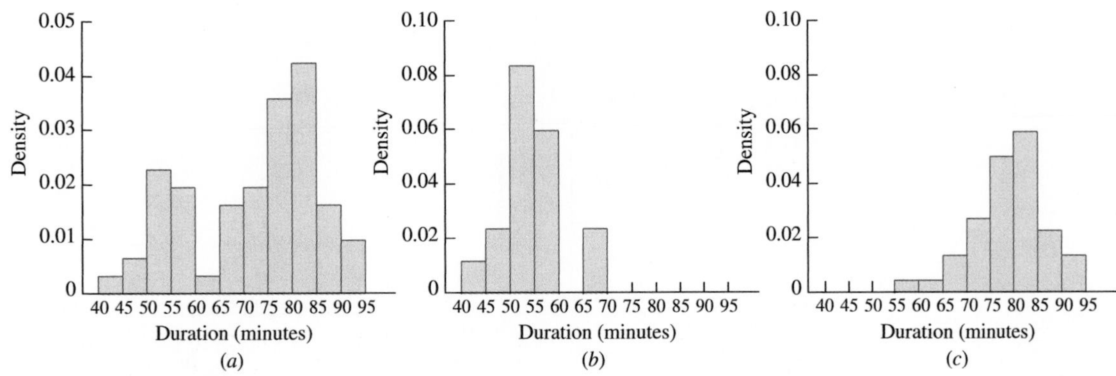

FIGURE 1.12 *(a)* Histogram for all 60 durations in Table 1.5. This histogram is bimodal. *(b)* Histogram for the durations in Table 1.5 that follow short eruptions. *(c)* Histogram for the durations in Table 1.5 that follow long eruptions. The histograms for the durations following short eruptions and for those following long eruptions are both unimodal, but the modes are in different places. When the two samples are combined, the histogram is bimodal.

Summary

Histograms have the following properties:

- The area of each rectangle represents the proportion of the sample that is in the corresponding class interval.
- The height of each rectangle represents the density of the sample in the corresponding class interval.
- The total area under the histogram is equal to 1.
- The sample mean is approximately equal to the center of mass of the histogram. The approximation gets closer as the rectangles get narrower.
- The sample variance is approximately equal to the moment of inertia of the histogram around its center of mass. The approximation gets closer as the rectangles get narrower.

Making the Heights Equal to the Frequencies

In this book, we use the term "histogram" to refer to a graphic in which the heights of the rectangles represent *densities* (so that the areas represent relative frequencies). However, some people draw histograms with the heights of the rectangles equal to the *frequencies*. In fact, this is the default method in many software packages. Setting the heights equal to the frequencies (or to the relative frequencies) can produce a distorted picture of the data. As an example, Figure 1.13 presents a histogram for the PM data in which the heights are equal to the frequencies, using the same class intervals as the histogram shown in Figure 1.8. This histogram visually exaggerates the proportion of vehicles in the largest

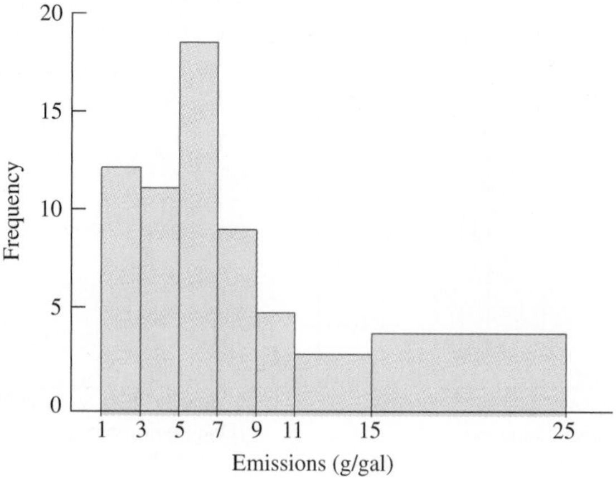

FIGURE 1.13 Histogram for the sample in Table 1.4, with the heights equal to the frequencies. The large rectangle over the interval 15–25 is misleading; in fact only 6.5% of the sample is in that interval. Compare with a correctly drawn histogram in Figure 1.8 on page 28.

two classes. The reason for this is that these class intervals are wider than the rest, and the histogram does not adjust for this.

If all the class intervals are the same width, then the histogram will have the same shape whether the heights represent densities or frequencies. Setting the heights equal to the frequencies is not misleading in this case. When it is desired to make class intervals of differing widths, however, it is important that the heights of the rectangles be equal to the *densities*, and not to the frequencies or relative frequencies.

Finally, we point out that when the heights are equal to the frequencies or to the relative frequencies, the total area of the rectangles is generally not equal to 1. If the class intervals all have the same width, then the sample mean will still be approximated by the center of mass; otherwise it generally will not be.

Boxplots

A **boxplot** is a graphic that presents the median, the first and third quartiles, and any outliers that are present in a sample. Boxplots are easy to understand, but there is a bit of terminology that goes with them. The **interquartile range** is the difference between the third quartile and the first quartile. Note that since 75% of the data is less than the third quartile, and 25% of the data is less than the first quartile, it follows that 50%, or half, of the data are between the first and third quartiles. The interquartile range is therefore the distance needed to span the middle half of the data.

We have defined outliers as points that are unusually large or small. If IQR represents the interquartile range, then for the purpose of drawing boxplots, any point that is more than 1.5 IQR above the third quartile, or more than 1.5 IQR below the first quartile, is considered an outlier. Some texts define a point that is more than 3 IQR from the first or third quartile as an **extreme outlier**. These definitions of outliers are just conventions for drawing boxplots and need not be used in other situations.

Figure 1.14 presents a boxplot for some hypothetical data. The plot consists of a box whose bottom side is the first quartile and whose top side is the third quartile. A horizontal line is drawn at the median. The "outliers" are plotted individually and are indicated by crosses in the figure. Extending from the top and bottom of the box are vertical lines called "whiskers." The whiskers end at the most extreme data point that is not an outlier.

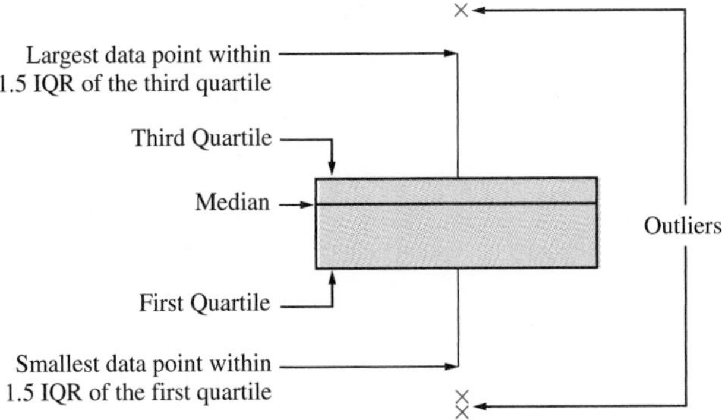

FIGURE 1.14 Anatomy of a boxplot.

Apart from any outliers, a boxplot can be thought of as having four pieces: the two parts of the box separated by the median line, and the two whiskers. Again apart from outliers, each of these four parts represents one-quarter of the data. The boxplot therefore indicates how large an interval is spanned by each quarter of the data, and in this way it can be used to determine the regions in which the sample values are more densely crowded and the regions in which they are more sparse.

Steps in the Construction of a Boxplot

- Compute the median and the first and third quartiles of the sample. Indicate these with horizontal lines. Draw vertical lines to complete the box.
- Find the largest sample value that is no more than 1.5 IQR above the third quartile, and the smallest sample value that is no more than 1.5 IQR below the first quartile. Extend vertical lines (whiskers) from the quartile lines to these points.
- Points more than 1.5 IQR above the third quartile, or more than 1.5 IQR below the first quartile, are designated as outliers. Plot each outlier individually.

Figure 1.15 presents a boxplot for the geyser data presented in Table 1.5. First note that there are no outliers in these data. Comparing the four pieces of the boxplot, we can tell that the sample values are comparatively densely packed between the median and the third quartile, and more sparse between the median and the first quartile. The lower whisker is a bit longer than the upper one, indicating that the data has a slightly longer lower tail than an upper tail. Since the distance between the median and the first quartile is greater than the distance between the median and the third quartile, and since the lower quarter of the data produces a longer whisker than the upper quarter, this boxplot suggests that the data are skewed to the left.

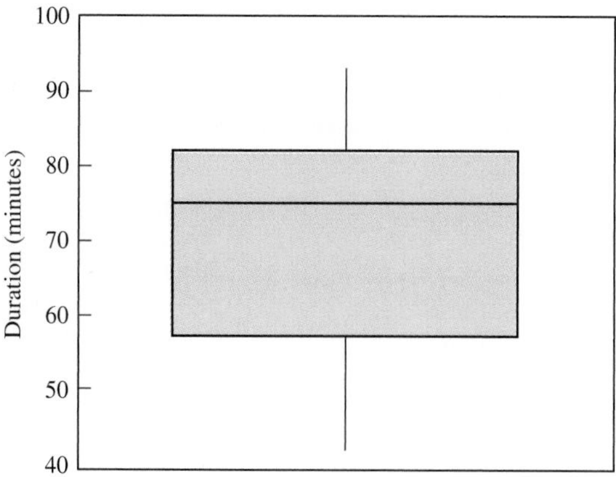

FIGURE 1.15 Boxplot for the Old Faithful dormant period data presented in Table 1.5.

A histogram for these data was presented in Figure 1.12a. The histogram presents a more general impression of the spread of the data. Importantly, the histogram indicates that the data are bimodal, which a boxplot cannot do.

Comparative Boxplots

A major advantage of boxplots is that several of them may be placed side by side, allowing for easy visual comparison of the features of several samples. Tables 1.1 and 1.2 (in Section 1.2) presented PM emissions data for vehicles driven at high and low altitudes. Figure 1.16 presents a side-by-side comparison of the boxplots for these two samples.

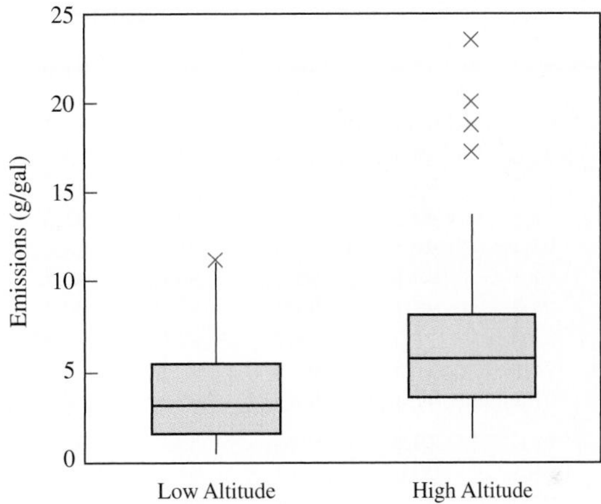

FIGURE 1.16 Comparative boxplots for PM emissions data for vehicles driven at high versus low altitudes.

The comparative boxplots in Figure 1.16 show that vehicles driven at low altitude tend to have lower emissions. In addition, there are several outliers among the data for high-altitude vehicles whose values are much higher than any of the values for the low-altitude vehicles (there is also one low-altitude value that barely qualifies as an outlier). We conclude that at high altitudes, vehicles have somewhat higher emissions in general, and that a few vehicles have much higher emissions. The box for the high-altitude vehicles is a bit taller, and the lower whisker a bit longer, than that for the low-altitude vehicles. We conclude that apart from the outliers, the spread in values is slightly larger for the high-altitude vehicles and is much larger when the outliers are considered.

In Figure 1.4 (in Section 1.2) we compared the values of some numerical descriptive statistics for these two samples, and reached some conclusions similar to the previous ones. The visual nature of the comparative boxplots in Figure 1.16 makes comparing the features of samples much easier.

We have mentioned that it is important to scrutinize outliers to determine whether they have resulted from errors, in which case they may be deleted. By identifying outliers, boxplots can be useful in this regard. The following example provides an illustration.

The article "Virgin Versus Recycled Wafers for Furnace Qualification: Is the Expense Justified?" (V. Czitrom and J. Reece, in *Statistical Case Studies for Industrial Process Improvement*, ASA and SIAM, 1997:87–104) describes a process for growing a thin silicon dioxide layer onto silicon wafers that are to be used in semiconductor manufacture. Table 1.6 presents thickness measurements, in angstroms (Å), of the oxide layer for 24 wafers. Nine measurements were made on each wafer. The wafers were produced in two separate runs, with 12 wafers in each run.

TABLE 1.6 Oxide layer thicknesses for silicon wafers

	Wafer	Thicknesses (Å)								
Run 1	1	90.0	92.2	94.9	92.7	91.6	88.2	92.0	98.2	96.0
	2	91.8	94.5	93.9	77.3	92.0	89.9	87.9	92.8	93.3
	3	90.3	91.1	93.3	93.5	87.2	88.1	90.1	91.9	94.5
	4	92.6	90.3	92.8	91.6	92.7	91.7	89.3	95.5	93.6
	5	91.1	89.8	91.5	91.5	90.6	93.1	88.9	92.5	92.4
	6	76.1	90.2	96.8	84.6	93.3	95.7	90.9	100.3	95.2
	7	92.4	91.7	91.6	91.1	88.0	92.4	88.7	92.9	92.6
	8	91.3	90.1	95.4	89.6	90.7	95.8	91.7	97.9	95.7
	9	96.7	93.7	93.9	87.9	90.4	92.0	90.5	95.2	94.3
	10	92.0	94.6	93.7	94.0	89.3	90.1	91.3	92.7	94.5
	11	94.1	91.5	95.3	92.8	93.4	92.2	89.4	94.5	95.4
	12	91.7	97.4	95.1	96.7	77.5	91.4	90.5	95.2	93.1
Run 2	1	93.0	89.9	93.6	89.0	93.6	90.9	89.8	92.4	93.0
	2	91.4	90.6	92.2	91.9	92.4	87.6	88.9	90.9	92.8
	3	91.9	91.8	92.8	96.4	93.8	86.5	92.7	90.9	92.8
	4	90.6	91.3	94.9	88.3	87.9	92.2	90.7	91.3	93.6
	5	93.1	91.8	94.6	88.9	90.0	97.9	92.1	91.6	98.4
	6	90.8	91.5	91.5	91.5	94.0	91.0	92.1	91.8	94.0
	7	88.0	91.8	90.5	90.4	90.3	91.5	89.4	93.2	93.9
	8	88.3	96.0	92.8	93.7	89.6	89.6	90.2	95.3	93.0
	9	94.2	92.2	95.8	92.5	91.0	91.4	92.8	93.6	91.0
	10	101.5	103.1	103.2	103.5	96.1	102.5	102.0	106.7	105.4
	11	92.8	90.8	92.2	91.7	89.0	88.5	87.5	93.8	91.4
	12	92.1	93.4	94.0	94.7	90.8	92.1	91.2	92.3	91.1

The 12 wafers in each run were of several different types and were processed in several different furnace locations. The purpose in collecting the data was to determine whether the thickness of the oxide layer was affected either by the type of wafer or the furnace location. This was therefore a factorial experiment, with wafer type and furnace location as the factors, and oxide layer thickness as the outcome. The experiment was designed so that there was not supposed to be any systematic difference in the thicknesses between one run and another. The first step in the analysis was to construct a boxplot for the data in each run to help determine if this condition was in fact met, and whether any of the observations should be deleted. The results are presented in Figure 1.17.

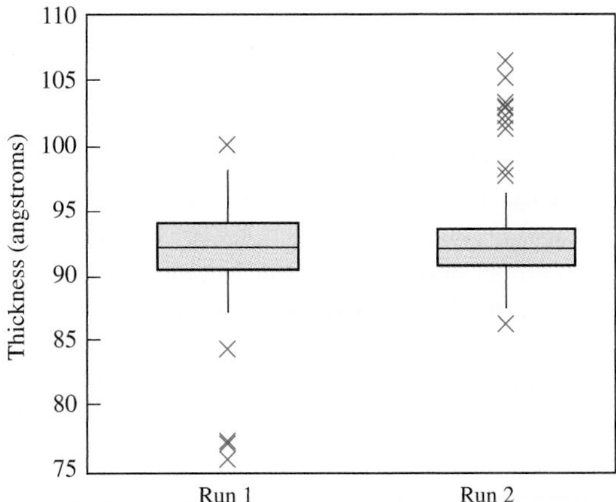

FIGURE 1.17 Comparative boxplots for oxide layer thickness data.

The boxplots show that there were several outliers in each run. Note that apart from these outliers, there are no striking differences between the samples, and therefore no evidence of any systematic difference between the runs. The next task is to inspect the outliers, to determine which, if any, should be deleted. By examining the data in Table 1.6, it can be seen that the eight largest measurements in run 2 occurred on a single wafer: number 10.

It was then determined that this wafer had been contaminated with a film residue, which caused the large thickness measurements. It would therefore be appropriate to delete these measurements. In the actual experiment, the engineers had data from several other runs available, and for technical reasons, decided to delete the entire run, rather than to analyze a run that was missing one wafer. In run 1, the three smallest measurements were found to have been caused by a malfunctioning gauge, and were therefore appropriately deleted. No cause could be determined for the remaining two outliers in run 1, so they were included in the analysis.

Multivariate Data

Sometimes the items in a population may have several values associated with them. For example, imagine choosing a random sample of days and determining the average temperature and humidity on each day. Each day in the population provides two values, temperature and humidity. The random sample therefore would consist of pairs of numbers. If the precipitation were measured on each day as well, the sample would consist of triplets. In principle, any number of quantities could be measured on each day, producing a sample in which each item is a list of numbers.

Data for which each item consists of more than one value is called **multivariate data**. When each item is a pair of values, the data are said to be **bivariate**. One of the most useful graphical summaries for numerical bivariate data is the **scatterplot**. If the data consist of ordered pairs $(x_1, y_1), \dots, (x_n, y_n)$, then a scatterplot is constructed simply by plotting each point on a two-dimensional coordinate system. Scatterplots can also be used to summarize multivariate data when each item consists of more than two values. One simply constructs separate scatterplots for each pair of values.

The following example illustrates the usefulness of scatterplots. The article "Advances in Oxygen Equivalence Equations for Predicting the Properties of Titanium Welds" (D. Harwig, W. Ittiwattana, and H. Castner, *The Welding Journal*, 2001:126s–136s) presents data concerning the chemical composition and strength characteristics of a number of titanium welds. Figure 1.18 presents two scatterplots. Figure 1.18a is a plot of the yield strength [in thousands of pounds per square inch (ksi)] versus carbon content (in %) for some of these welds. Figure 1.18b is a plot of the yield strength (in ksi) versus nitrogen content (in %) for the same welds.

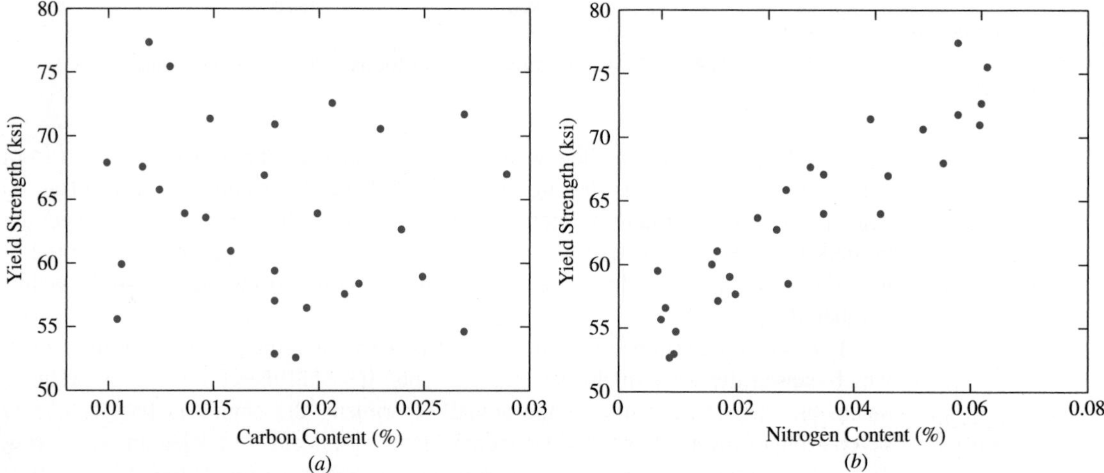

FIGURE 1.18 ***(a)*** A scatterplot showing that there is not much of a relationship between carbon content and yield strength for a certain group of welds. ***(b)*** A scatterplot showing that for these same welds, higher nitrogen content is associated with higher yield strength.

The plot of yield strength versus nitrogen content (Figure 1.18b) shows some clear structure—the points seem to be following a line from lower left to upper right. In this way, the plot illustrates a relationship between nitrogen content and yield strength: Welds with higher nitrogen content tend to have higher yield strength. This scatterplot might lead investigators to try to predict strength from nitrogen content or to try to increase nitrogen content to increase strength. (The fact that there is a relationship on a scatterplot does not guarantee that these attempts will be successful, as we will discuss in Section 7.1.) In contrast, there does not seem to be much structure to the scatterplot of yield strength versus carbon content, and thus there is no evidence of a relationship between these two quantities. This scatterplot would discourage investigators from trying to predict strength from carbon content.

Exercises for Section 1.3

1. As part of a quality-control study aimed at improving a production line, the weights (in ounces) of 50 bars of soap are measured. The results are as follows, sorted from smallest to largest.

11.6	12.6	12.7	12.8	13.1	13.3	13.6	13.7	13.8	14.1
14.3	14.3	14.6	14.8	15.1	15.2	15.6	15.6	15.7	15.8
15.8	15.9	15.9	16.1	16.2	16.2	16.3	16.4	16.5	16.5
16.5	16.6	17.0	17.1	17.3	17.3	17.4	17.4	17.4	17.6
17.7	18.1	18.3	18.3	18.3	18.5	18.5	18.8	19.2	20.3

   a. Construct a stem-and-leaf plot for these data.

   b. Construct a histogram for these data.

   c. Construct a dotplot for these data.

   d. Construct a boxplot for these data. Does the boxplot show any outliers?

2. Following is a list of the number of hazardous waste sites in each of the 50 states of the United States as of April 1995. The data are taken from *The World Almanac and Book of Facts 1996* (World Almanac Books, Mahwah, NJ, 1996). The list has been sorted into numerical order.

1	2	3	4	4	5	6	8	8	9
10	10	10	11	11	11	12	12	12	12
13	13	14	15	16	17	17	18	18	19
19	20	22	23	24	25	29	30	33	37
38	39	40	55	58	77	81	96	102	107

   a. Construct a stem-and-leaf plot for these data.

   b. Construct a histogram for these data.

   c. Construct a dotplot for these data.

   d. Construct a boxplot for these data. Does the boxplot show any outliers?

3. Refer to Table 1.2 (p. 20). Construct a stem-and-leaf plot with the ones digit as the stem (for values greater than or equal to 10 the stem will have two digits) and the tenths digit as the leaf. How many stems are there (be sure to include leafless stems)? What are some advantages and disadvantages of this plot, compared to the one in Figure 1.6 (p. 26)?

4. Two methods were studied for the recovery of protein. Thirteen runs were made using each method, and the fraction of protein recovered was recorded for each run. The results are as follows:

Method 1	Method 2
0.32	0.25
0.35	0.40
0.37	0.48
0.39	0.55
0.42	0.56
0.47	0.58
0.51	0.60
0.58	0.65
0.60	0.70
0.62	0.76
0.65	0.80
0.68	0.91
0.75	0.99

   a. Construct a histogram for the results of each method.

   b. Construct comparative boxplots for the two methods.

   c. Using the boxplots, what differences can be seen between the results of the two methods?

5. Each of 32 students, comprised of two lab sections of 16 students each, estimated the circumference of a tennis ball by eye. The results, in centimeters, are given here. (The results for the first group of students also appear in Exercise 12 of Section 1.2.)

Group 1	Group 2
18.0	15.0
18.0	18.0
18.0	18.0
20.0	19.0
22.0	19.0
22.0	19.0
22.5	19.0
23.0	19.5
24.0	20.0
24.0	20.0
25.0	20.0
25.0	20.0
25.0	20.0
25.0	22.0
26.0	24.0
26.4	25.0

a. Construct a histogram for each group.
b. Construct comparative boxplots for the two groups.
c. Using the boxplots, what differences can be seen between the results of the first and second group?

6. Sketch a histogram for which
a. The mean is greater than the median.
b. The mean is less than the median.
c. The mean is approximately equal to the median.

7. The following histogram presents the distribution of systolic blood pressure for a sample of women. Use it to answer the following questions.
a. Is the percentage of women with blood pressures above 130 mm closest to 25%, 50%, or 75%?
b. In which interval are there more women: 130–135 or 140–150 mm?
c. In the interval 125–130 mm, the height of the histogram is about 0.024. What percentage of the women had blood pressures in this interval?

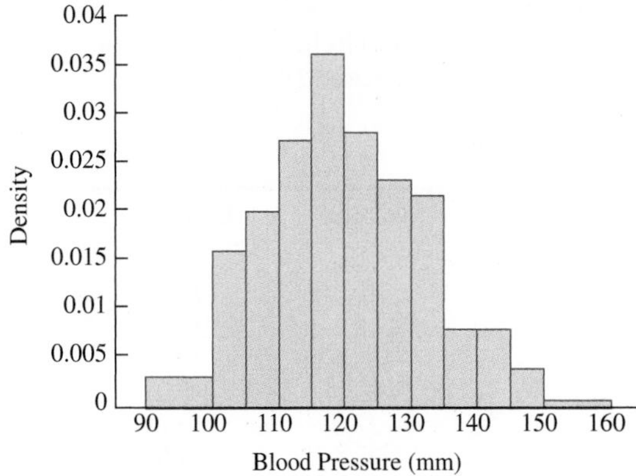

8. The following histogram presents the amounts of silver [in parts per million (ppm)] found in a sample of rocks. One rectangle from the histogram is missing. What is its height?

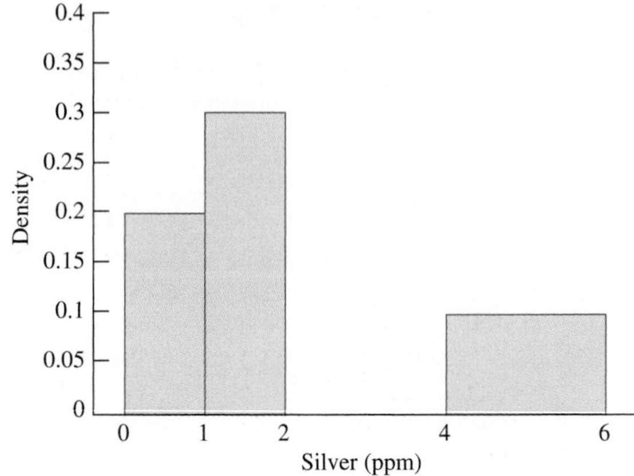

9. A sample of 100 men has average height 70 in. and standard deviation 2.5 in. A sample of 100 women has average height 64 in. and standard deviation 2.5 in. If both samples are combined, the standard deviation of all 200 heights will be ________
i. less than 2.5 in.
ii. greater than 2.5 in.
iii. equal to 2.5 in.
iv. can't tell from the information given.

(*Hint:* Don't do any calculations. Just try to sketch, very roughly, histograms for each sample separately, and then one for the combined sample.)

10. Following are boxplots comparing the charge [in coulombs per mole (C/mol) × 10^{-25}] at pH 4.0 and pH 4.5 for a collection of proteins (from the article "Optimal Synthesis of Protein Purification Processes," E. Vasquez-Alvarez, M. Leinqueo, and J. Pinto, *Biotechnology Progress* 2001:685–695). True or false:
a. The median charge for the pH of 4.0 is greater than the 75th percentile of charge for the pH of 4.5.
b. Approximately 25% of the charges for pH 4.5 are less than the smallest charge at pH 4.0.
c. About half the sample values for pH 4.0 are between 2 and 4.
d. There is a greater proportion of values outside the box for pH 4.0 than for pH 4.5.

e. Both samples are skewed to the right.

f. Both samples contain outliers.

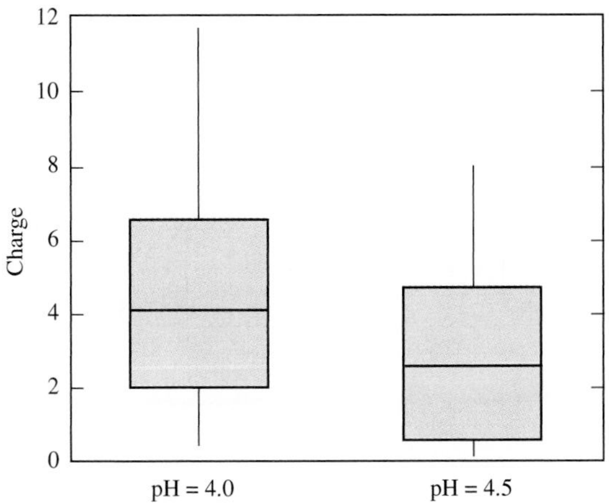

11. Following are summary statistics for two data sets, A and B.

	A	B
Minimum	0.066	−2.235
1st Quartile	1.42	5.27
Median	2.60	8.03
3rd Quartile	6.02	9.13
Maximum	10.08	10.51

a. Compute the interquartile ranges for both A and B.

b. Do the summary statistics for A provide enough information to construct a boxplot? If so, construct the boxplot. If not, explain why.

c. Do the summary statistics for B provide enough information to construct a boxplot? If so, construct the boxplot. If not, explain why.

12. Match each histogram to the boxplot that represents the same data set.

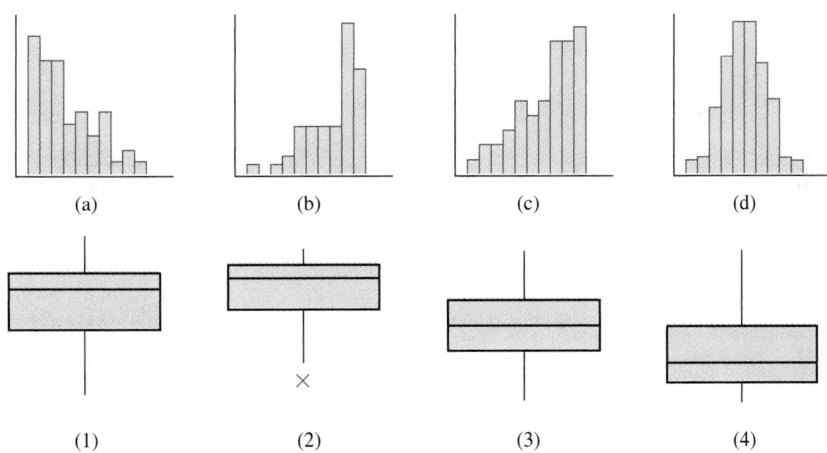

13. Refer to the asphalt data in Example 1.12 (p. 17).

a. Construct a boxplot for the asphalt data.

b. Which values, if any, are outliers?

c. Construct a dotplot for the asphalt data.

d. For purposes of constructing boxplots, an outlier is defined to be a point whose distance from the nearest quartile is more than 1.5 IQR. A more general, and less precise, definition is that an outlier is any point that is detached from the bulk of the data. Are there any points in the asphalt data set that are outliers under this more general definition, but not under the boxplot definition? If so, which are they?

14. Match each scatterplot to the statement that best describes it.

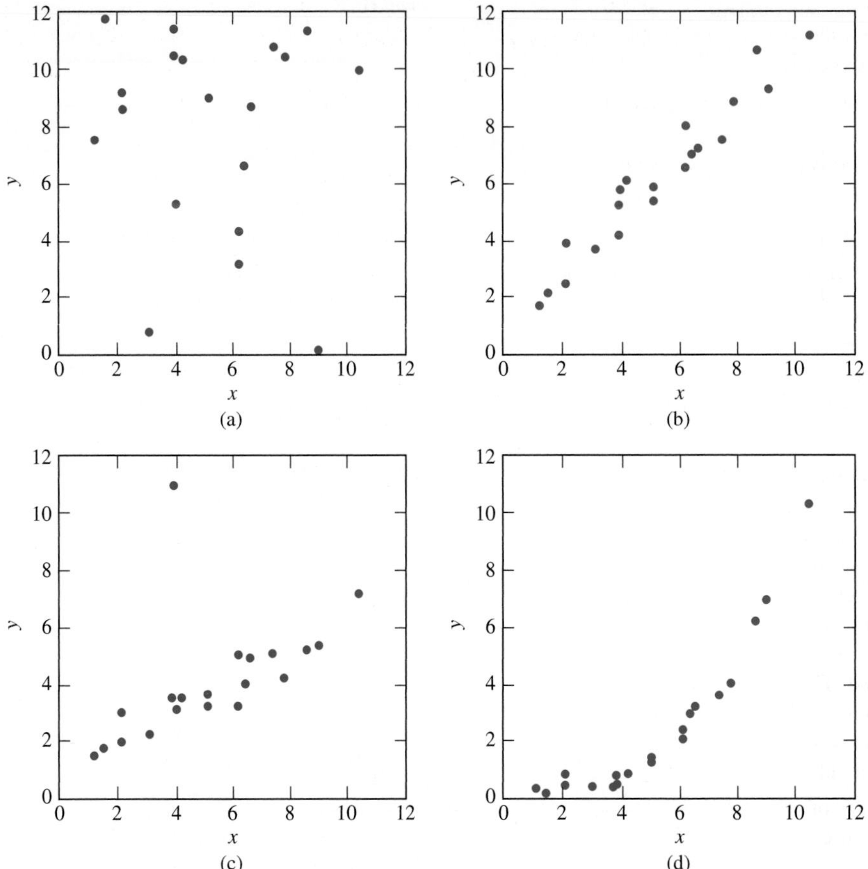

i. The relationship between x and y is approximately linear.
ii. The relationship between x and y is nonlinear.
iii. There isn't much of any relationship between x and y.
iv. The relationship between x and y is approximately linear, except for an outlier.

15. For the following data:

x	1.4	2.4	4.0	4.9	5.7	6.3	7.8	9.0	9.3	11.0
y	2.3	3.7	5.7	9.9	6.9	15.8	15.4	36.9	34.6	53.2

a. Make a scatterplot of y versus x. Is the relationship between x and y approximately linear, or is it nonlinear?
b. Compute the natural logarithm of each y value. This is known as *making a log transformation of* y. Make a scatterplot of $\ln y$ versus x. Is the relationship between x and $\ln y$ approximately linear, or is it nonlinear?
c. In general, it is easier to work with quantities that have an approximate linear relationship than with quantities that have a nonlinear relationship. For these data, do you think it would be easier to work with x and y or with x and $\ln y$? Explain.

Supplementary Exercises for Chapter 1

1. A vendor converts the weights on the packages she sends out from pounds to kilograms (1 kg $\approx$ 2.2 lb).
   a. How does this affect the mean weight of the packages?
   b. How does this affect the standard deviation of the weights?

2. Refer to Exercise 1. The vendor begins using heavier packaging, which increases the weight of each package by 50 g.
   a. How does this affect the mean weight of the packages?
   b. How does this affect the standard deviation of the weights?

3. Integrated circuits consist of electric channels that are etched onto silicon wafers. A certain proportion of circuits are defective because of "undercutting," which occurs when too much material is etched away so that the channels, which consist of the unetched portions of the wafers, are too narrow. A redesigned process, involving lower pressure in the etching chamber, is being investigated. The goal is to reduce the rate of undercutting to less than 5%. Out of the first 100 circuits manufactured by the new process, only 4 show evidence of undercutting. True or false:
   a. Since only 4% of the 100 circuits had undercutting, we can conclude that the goal has been reached.
   b. Although the sample percentage is under 5%, this may represent sampling variation, so the goal may not yet be reached.
   c. There is no use in testing the new process, because no matter what the result is, it could just be due to sampling variation.
   d. If we sample a large enough number of circuits, and if the percentage of defective circuits is far enough below 5%, then it is reasonable to conclude that the goal has been reached.

4. A coin is tossed twice and comes up heads both times. Someone says, "There's something wrong with this coin. A coin is supposed to come up heads only half the time, not every time."
   a. Is it reasonable to conclude that something is wrong with the coin? Explain.
   b. If the coin came up heads 100 times in a row, would it be reasonable to conclude that something is wrong with the coin? Explain.

5. The smallest number on a list is changed from 12.9 to 1.29.
   a. Is it possible to determine by how much the mean changes? If so, by how much does it change?
   b. Is it possible to determine by how much the median changes? If so, by how much does it change? What if the list consists of only two numbers?
   c. Is it possible to determine by how much the standard deviation changes? If so, by how much does it change?

6. There are 15 numbers on a list, and the smallest number is changed from 12.9 to 1.29.
   a. Is it possible to determine by how much the mean changes? If so, by how much does it change?
   b. Is it possible to determine the value of the mean after the change? If so, what is the value?
   c. Is it possible to determine by how much the median changes? If so, by how much does it change?
   d. Is it possible to determine by how much the standard deviation changes? If so, by how much does it change?

7. There are 15 numbers on a list, and the mean is 25. The smallest number on the list is changed from 12.9 to 1.29.
   a. Is it possible to determine by how much the mean changes? If so, by how much does it change?
   b. Is it possible to determine the value of the mean after the change? If so, what is the value?
   c. Is it possible to determine by how much the median changes? If so, by how much does it change?
   d. Is it possible to determine by how much the standard deviation changes? If so, by how much does it change?

8. The article "The Selection of Yeast Strains for the Production of Premium Quality South African Brandy Base Products" (C. Steger and M. Lambrechts, *Journal of Industrial Microbiology and Biotechnology*, 2000:431–440) presents detailed information on the volatile compound composition of base wines made

from each of 16 selected yeast strains. Following are the concentrations of total esters (in mg/L) in each of the wines.

284.34	173.01	229.55	312.95	215.34	188.72
144.39	172.79	139.38	197.81	303.28	256.02
658.38	105.14	295.24	170.41		

a. Compute the mean concentration.
b. Compute the median concentration.
c. Compute the first quartile of the concentrations.
d. Compute the third quartile of the concentrations.
e. Construct a boxplot for the concentrations. What features does it reveal?

9. Concerning the data represented in the following boxplot, which one of the following statements is true?

i. The mean is greater than the median.
ii. The mean is less than the median.
iii. The mean is approximately equal to the median.

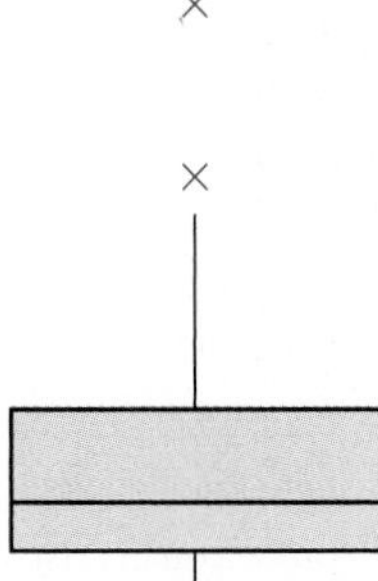

10. True or false: In any boxplot,

a. The length of the whiskers is equal to 1.5 IQR, where IQR is the interquartile range.
b. The length of the whiskers may be greater than 1.5 IQR, where IQR is the interquartile range.
c. The length of the whiskers may be less than 1.5 IQR, where IQR is the interquartile range.
d. The values at the ends of the whiskers are always values in the data set used to construct the boxplot.

11. For each of the following histograms, determine whether the vertical axis has been labeled correctly.

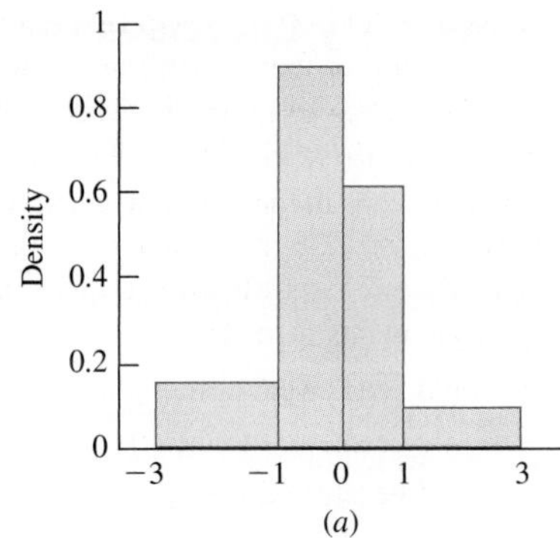

(*a*)

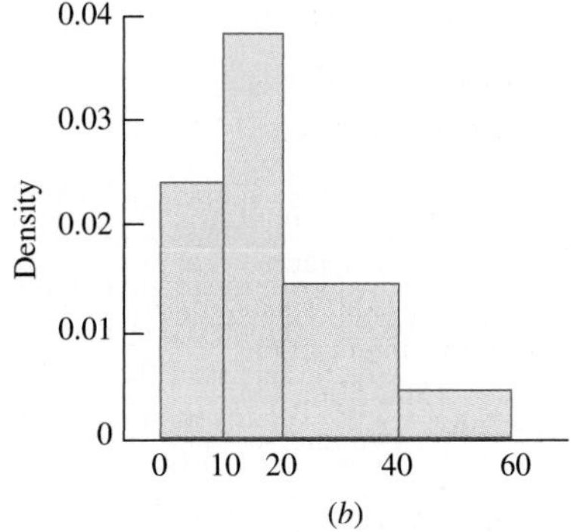

(*b*)

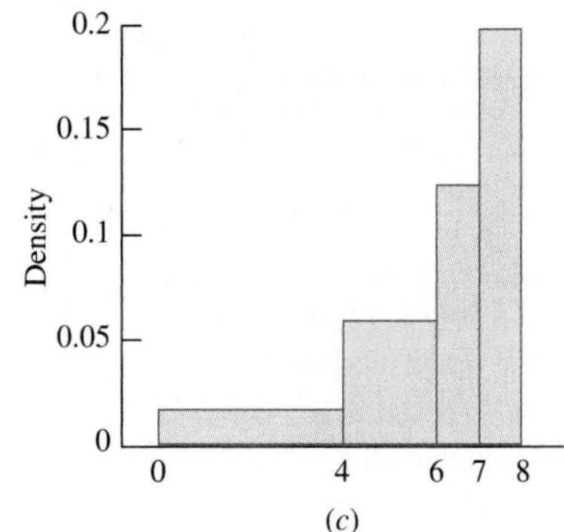

(*c*)

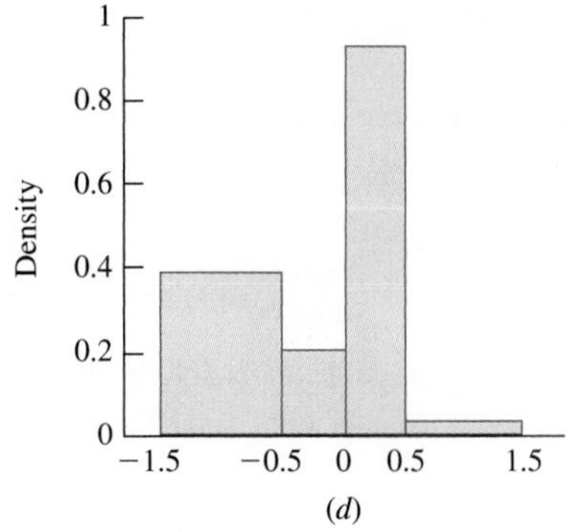

(*d*)

12. In the article "Occurrence and Distribution of Ammonium in Iowa Groundwater" (K. Schilling, *Water Environment Research*, 2002:177–186), ammonium concentrations (in mg/L) were measured at a total of 349 alluvial wells in the state of Iowa. The mean concentration was 0.27, the median was 0.10, and the standard deviation was 0.40. If a histogram of these 349 measurements were drawn,

i. it would be skewed to the right.
ii. it would be skewed to the left.
iii. it would be approximately symmetric.
iv. its shape could not be determined without knowing the relative frequencies.

13. The article "Vehicle-Arrival Characteristics at Urban Uncontrolled Intersections" (V. Rengaraju and V. Rao, *Journal of Transportation Engineering*, 1995:317–323) presents data on traffic characteristics at 10 intersections in Madras, India. One characteristic measured was the speeds of the vehicles traveling through the intersections. The accompanying table gives the 15th, 50th, and 85th percentiles of speed (in km/h) for two intersections.

	Percentile		
Intersection	**15th**	**50th**	**85th**
A	27.5	37.5	40.0
B	24.5	26.5	36.0

a. If a histogram for speeds of vehicles through intersection A were drawn, do you think it would be skewed to the left, skewed to the right, or approximately symmetric? Explain.

b. If a histogram for speeds of vehicles through intersection B were drawn, do you think it would be skewed to the left, skewed to the right, or approximately symmetric? Explain.

14. The *cumulative frequency* and the *cumulative relative frequency* for a given class interval are the sums of the frequencies and relative frequencies, respectively, over all classes up to and including the given class. For example, if there are five classes, with frequencies 11, 7, 3, 14, and 5, the cumulative frequencies would be 11, 18, 21, 35, and 40, and the cumulative relative frequencies would be 0.275, 0.450, 0.525, 0.875, and 1.000. Construct a table presenting frequencies, relative frequencies, cumulative frequencies, and cumulative relative frequencies, for the data in Exercise 1 of Section 1.3, using the class intervals 11–< 12, 12–< 13, ..., 20–< 21.

15. The article "Computing and Using Rural versus Urban Measures in Statistical Applications" (C. Goodall, K. Kafadar, and J. Tukey, *The American Statistician*, 1998:101–111) discusses methods to measure the degree to which U.S. counties are urban rather than rural. The following frequency table presents population frequencies of U.S. counties. Populations are on the $\log_2$ scale; thus the first interval contains counties whose populations are at least $2^6 = 64$ but less than $2^{12.4} = 5404$, and so on.

$\log_2$ Population	Number of Counties
6.0–< 12.4	305
12.4–< 13.1	294
13.1–< 13.6	331
13.6–< 14.0	286
14.0–< 14.4	306
14.4–< 14.8	273
14.8–< 15.3	334
15.3–< 16.0	326
16.0–< 17.0	290
17.0–< 23.0	323

a. Construct a histogram from the frequency table.

b. Estimate the proportion of counties whose populations are greater than 100,000.

c. Is the histogram skewed to the left, skewed to the right, or approximately symmetric?

d. Construct a histogram using the actual populations rather than their logs. Why do you think the article transformed the populations to the log scale?

16. The article "Hydrogeochemical Characteristics of Groundwater in a Mid-Western Coastal Aquifer System" (S. Jeen, J. Kim, et al., *Geosciences Journal*, 2001:339–348) presents measurements of various properties of shallow groundwater in a certain aquifer system in Korea. Following are measurements of electrical conductivity (in microsiemens per centimeter) for 23 water samples.

2099	528	2030	1350	1018	384	1499
1265	375	424	789	810	522	513
488	200	215	486	257	557	260
461	500					

a. Find the mean.

b. Find the standard deviation.

c. Find the median.

d. Construct a dotplot.

e. Find the 10% trimmed mean.

f. Find the first quartile.

g. Find the third quartile.

h. Find the interquartile range.

i. Construct a boxplot.

j. Which of the points, if any, are outliers?

k. If a histogram were constructed, would it be skewed to the left, skewed to the right, or approximately symmetric?

17. Water scarcity has traditionally been a major concern in the Canary Islands. Water rights are divided into shares, which are privately owned. The article "The Social Construction of Scarcity. The Case of Water in Tenerife (Canary Islands)" (F. Aguilera-Klink, E. Pérez-Moriana, and J. Sánchez-Garcia, *Ecological Economics*, 2000:233–245) discusses the extent to which many of the shares are concentrated among a few owners. The following table presents the number of owners who own various numbers of shares. (There were 15 owners who owned 50 shares or more; these are omitted.) Note that it is possible to own a fractional number of shares; for example, the interval 2–< 3 contains 112 individuals who owned at least 2 but less than 3 shares.

Number of Shares	Number of Owners
0–< 1	18
1–< 2	165
2–< 3	112
3–< 4	87
4–< 5	43
5–< 10	117
10–< 15	51
15–< 20	32
20–< 25	10
25–< 30	8
30–< 50	8

a. Construct a histogram for these data.

b. Approximate the median number of shares owned by finding the point for which the areas on either side are equal.

c. Approximate the first quartile of the number of shares owned by finding the point for which 25% of the area is to the left.

d. Approximate the third quartile of the number of shares owned by finding the point for which 75% of the area is to the left.

e. Approximate the mean number of shares owned by computing the center of mass of the histogram.

f. Approximate the variance of the number of shares owned by computing the moment of inertia around the mean, under the assumption that all the mass of a rectangle is concentrated at its midpoint.

18. The Editor's Report in the November 2003 issue of *Technometrics* provides the following information regarding the length of time taken to review articles that were submitted for publication during the year 2002. For computational purposes, interpret the last category (> 9) as 9–< 15.

Time (months)	Number of Articles
0–< 1	45
1–< 2	17
2–< 3	18
3–< 4	19
4–< 5	12
5–< 6	14
6–< 7	13
7–< 8	22
8–< 9	11
> 9	18

a. Construct a histogram for these data.

b. Approximate the median review time by finding the point for which the areas on either side are equal.

c. Approximate the first quartile of the review times by finding the point for which 25% of the area is to the left.

d. Approximate the third quartile of the review times by finding the point for which 75% of the area is to the left.

e. Approximate the mean review time by computing the center of mass of the histogram.

f. Approximate the variance of the review times by computing the moment of inertia around the mean, under the assumption that all the mass of a rectangle is concentrated at its midpoint.

19. The article "The Ball-on-Three-Ball Test for Tensile Strength: Refined Methodology and Results for Three Hohokam Ceramic Types" (M. Beck, *American Antiquity*, 2002:558–569) discusses the strength of ancient ceramics. Several specimens of each of three types of ceramic were tested. The loads (in kg) required to crack the specimens are as follows:

Ceramic Type	Loads (kg)
Sacaton	15, 30, 51, 20, 17, 19, 20, 34, 17, 15, 23, 19, 15, 18, 16, 22, 29, 15, 13, 15
Gila Plain	27, 18, 28, 25, 55, 21, 18, 34, 23, 30, 20, 30, 31, 25, 28, 26, 17, 19, 16, 24, 19, 9, 31, 19, 27, 20, 45, 15
Casa Grande	20, 16, 20, 36, 27, 35, 66, 15, 18, 24, 21, 30, 20, 24, 23, 21, 13, 21

a. Construct comparative boxplots for the three samples.

b. How many outliers does each sample contain?

c. Comment on the features of the three samples.

Chapter 2

Probability

Introduction

The development of the theory of probability was financed by seventeenth-century gamblers, who hired some of the leading mathematicians of the day to calculate the correct odds for certain games of chance. Later, people realized that scientific processes involve chance as well, and since then the methods of probability have been used to study the physical world.

Probability is now an extensive branch of mathematics. Many books are devoted to the subject, and many researchers have dedicated their professional careers to its further development. In this chapter we present an introduction to the ideas of probability that are most important to the study of statistics.

2.1 Basic Ideas

To make a systematic study of probability, we need some terminology. An **experiment** is a process that results in an outcome that cannot be predicted in advance with certainty. Tossing a coin, rolling a die, measuring the diameter of a bolt, weighing the contents of a box of cereal, and measuring the breaking strength of a length of fishing line are all examples of experiments. To discuss an experiment in probabilistic terms, we must specify its possible outcomes:

Definition

The set of all possible outcomes of an experiment is called the **sample space** for the experiment.

For tossing a coin, we can use the set {Heads, Tails} as the sample space. For rolling a six-sided die, we can use the set $\{1, 2, 3, 4, 5, 6\}$. These sample spaces are finite. Some experiments have sample spaces with an infinite number of outcomes. For example, imagine that a punch with diameter 10 mm punches holes in sheet metal. Because of variations in the angle of the punch and slight movements in the sheet metal, the diameters of the holes vary between 10.0 and 10.2 mm. For the experiment of punching a hole, then, a reasonable sample space is the interval (10.0, 10.2), or in set notation, $\{x \mid 10.0 < x < 10.2\}$. This set obviously contains an infinite number of outcomes.

For many experiments, there are several sample spaces to choose from. For example, assume that a process manufactures steel pins whose lengths vary between 5.20 and 5.25 cm. An obvious choice for the sample space for the length of a pin is the set $\{x \mid 5.20 < x < 5.25\}$. However, if the object were simply to determine whether the pin was too short, too long, or within specification limits, a good choice for the sample space might be {too short, too long, within specifications}.

When discussing experiments, we are often interested in a particular subset of outcomes. For example, we might be interested in the probability that a die comes up an even number. The sample space for the experiment is $\{1, 2, 3, 4, 5, 6\}$, and coming up even corresponds to the subset $\{2, 4, 6\}$. In the hole punch example, we might be interested in the probability that a hole has a diameter less than 10.1 mm. This corresponds to the subset $\{x \mid 10.0 < x < 10.1\}$. There is a special name for a subset of a sample space:

Definition

A subset of a sample space is called an **event**.

Note that for any sample space, the empty set $\emptyset$ is an event, as is the entire sample space. A given event is said to have occurred if the outcome of the experiment is one of the outcomes in the event. For example, if a die comes up 2, the events $\{2, 4, 6\}$ and $\{1, 2, 3\}$ have both occurred, along with every other event that contains the outcome "2."

Example 2.1

An electrical engineer has on hand two boxes of resistors, with four resistors in each box. The resistors in the first box are labeled 10 Ω (ohms), but in fact their resistances are 9, 10, 11, and 12 Ω. The resistors in the second box are labeled 20 Ω, but in fact their resistances are 18, 19, 20, and 21 Ω. The engineer chooses one resistor from each box and determines the resistance of each.

Let A be the event that the first resistor has a resistance greater than 10, let B be the event that the second resistor has a resistance less than 19, and let C be the event that the sum of the resistances is equal to 28. Find a sample space for this experiment, and specify the subsets corresponding to the events A, B, and C.

Solution

A good sample space is the set of ordered pairs in which the first component is the resistance of the first resistor and the second component is the resistance of the second resistor. We will denote this sample space by $\mathcal{S}$.

$$\mathcal{S} = \{(9, 18), (9, 19), (9, 20), (9, 21), (10, 18), (10, 19), (10, 20), (10, 21), (11, 18), (11, 19), (11, 20), (11, 21), (12, 18), (12, 19), (12, 20), (12, 21)\}$$

The events A, B, and C are given by

$A = \{(11, 18), (11, 19), (11, 20), (11, 21), (12, 18), (12, 19), (12, 20), (12, 21)\}$
$B = \{(9, 18), (10, 18), (11, 18), (12, 18)\}$
$C = \{(9, 19), (10, 18)\}$

Combining Events

We often construct events by combining simpler events. Because events are subsets of sample spaces, it is traditional to use the notation of sets to describe events constructed in this way. We review the necessary notation here.

- The **union** of two events A and B, denoted $A \cup B$, is the set of outcomes that belong either to A, to B, or to both. In words, $A \cup B$ means "A or B." Thus the event $A \cup B$ occurs whenever either A or B (or both) occurs.
- The **intersection** of two events A and B, denoted $A \cap B$, is the set of outcomes that belong both to A and to B. In words, $A \cap B$ means "A and B." Thus the event $A \cap B$ occurs whenever both A and B occur.
- The **complement** of an event A, denoted A^c, is the set of outcomes that do not belong to A. In words, A^c means "not A." Thus the event A^c occurs whenever A does *not* occur.

Events can be graphically illustrated with Venn diagrams. Figure 2.1 illustrates the events $A \cup B$, $A \cap B$, and $B \cap A^c$.

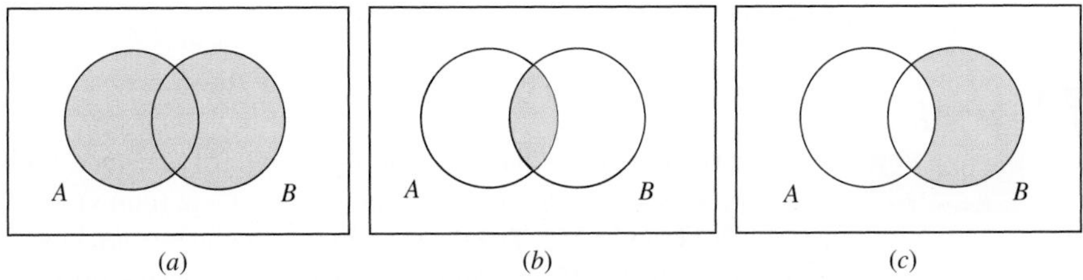

FIGURE 2.1 Venn diagrams illustrating various events: (a) $A \cup B$, (b) $A \cap B$, (c) $B \cap A^c$.

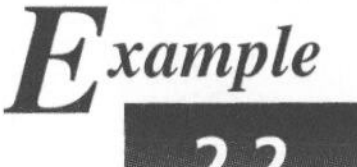

Refer to Example 2.1. Find $B \cup C$ and $A \cap B^c$.

Solution

The event $B \cup C$ contains all the outcomes that belong either to B or to C, or to both. Therefore

$$B \cup C = \{(9, 18), (10, 18), (11, 18), (12, 18), (9, 19)\}$$

The event B^c contains those outcomes in the sample space that do not belong to B. It follows that the event $A \cap B^c$ contains the outcomes that belong to A and do not belong to B. Therefore

$$A \cap B^c = \{(11, 19), (11, 20), (11, 21), (12, 19), (12, 20), (12, 21)\}$$

Mutually Exclusive Events

There are some events that can never occur together. For example, it is impossible that a coin can come up both heads and tails, and it is impossible that a steel pin can be both too long and too short. Events like this are said to be **mutually exclusive**.

Definition

- The events A and B are said to be **mutually exclusive** if they have no outcomes in common.
- More generally, a collection of events $A_1, A_2, \ldots, A_n$ is said to be mutually exclusive if no two of them have any outcomes in common.

The Venn diagram in Figure 2.2 illustrates mutually exclusive events.

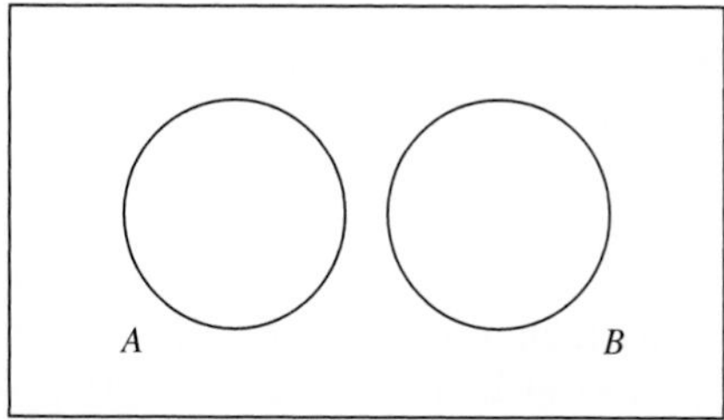

FIGURE 2.2 The events A and B are mutually exclusive.

Example 2.3

Refer to Example 2.1. If the experiment is performed, is it possible for events A and B both to occur? How about B and C? A and C? Which pair of events is mutually exclusive?

Solution

If the outcome is (11, 18) or (12, 18), then events A and B both occur. If the outcome is (10, 18), then both B and C occur. It is impossible for A and C both to occur, because these events are mutually exclusive, having no outcomes in common.

Probabilities

Each event in a sample space has a **probability** of occurring. Intuitively, the probability is a quantitative measure of how likely the event is to occur. Formally speaking, there are several interpretations of probability; the one we shall adopt is that the probability of an event is the proportion of times the event would occur in the long run, if the experiment were to be repeated over and over again.

We often use the letter P to stand for probability. Thus when tossing a coin, the notation "$P(\text{heads}) = 1/2$" means that the probability that the coin lands heads is equal to 1/2.

Summary

Given any experiment and any event A:

- The expression $P(A)$ denotes the probability that the event A occurs.
- $P(A)$ is the proportion of times that event A would occur in the long run, if the experiment were to be repeated over and over again.

In many situations, the only way to estimate the probability of an event is to repeat the experiment many times and determine the proportion of times that the event occurs. For example, if it is desired to estimate the probability that a printed circuit board manufactured by a certain process is defective, it is usually necessary to produce a number of boards and test them to determine the proportion that are defective. In some cases, probabilities can be determined through knowledge of the physical nature of an experiment. For example, if it is known that the shape of a die is nearly a perfect cube and that its mass is distributed nearly homogeneously, it may be assumed that each of the six faces is equally likely to land upward when the die is rolled.

Once the probabilities of some events have been found through scientific knowledge or experience, the probabilities of other events can be computed mathematically. For example, if it has been estimated through experimentation that the probability that a printed circuit board is defective is 0.10, an estimate of the probability that a board is not defective can be calculated to be 0.90. As another example, assume that steel pins manufactured by a certain process can fail to meet a length specification either by being too short or too long. By measuring a large number of pins, it is estimated that the probability that a pin is too short is 0.02, and the probability that a pin is too long is 0.03. It can then be estimated that the probability that a pin fails to meet the specification is 0.05.

In practice, scientists and engineers estimate the probabilities of some events on the basis of scientific understanding and experience and then use mathematical rules to compute estimates of the probabilities of other events. In the rest of this section and in Section 2.2, we will explain some of these rules and show how to use them.

Axioms of Probability

The subject of probability is based on three commonsense rules, known as axioms. They are:

The Axioms of Probability

1. Let $\mathcal{S}$ be a sample space. Then $P(\mathcal{S}) = 1$.
2. For any event A, $0 \leq P(A) \leq 1$.
3. If A and B are mutually exclusive events, then $P(A \cup B) = P(A) + P(B)$. More generally, if $A_1, A_2, \ldots$ are mutually exclusive events, then $P(A_1 \cup A_2 \cup \cdots) = P(A_1) + P(A_2) + \cdots$.

With a little thought, it is easy to see that the three axioms do indeed agree with common sense. The first axiom says that the outcome of an experiment is always in the sample space. This is obvious, because by definition the sample space contains all the possible outcomes of the experiment. The second axiom says that the long-run frequency of any event is always between 0 and 100%. For an example illustrating the third axiom, we previously discussed a process that manufactures steel pins, in which the probability that a pin is too short is 0.02 and the probability that a pin is too long is 0.03. The third axiom says that the probability that the pin is either too short or too long is $0.02 + 0.03 = 0.05$.

We now present two simple rules that are helpful in computing probabilities. These rules are intuitively obvious, and they can also be proved from the axioms. Proofs are provided at the end of the section.

For any event A,

$$P(A^c) = 1 - P(A) \qquad (2.1)$$

Let $\emptyset$ denote the empty set. Then

$$P(\emptyset) = 0 \qquad (2.2)$$

Equation (2.1) says that the probability that an event does not occur is equal to 1 minus the probability that it does occur. For example, if there is a 40% chance of rain, there is a 60% chance that it does not rain. Equation (2.2) says that it is impossible for an experiment to have no outcome.

Example 2.4

A target on a test firing range consists of a bull's-eye with two concentric rings around it. A projectile is fired at the target. The probability that it hits the bull's-eye is 0.10, the probability that it hits the inner ring is 0.25, and the probability that it hits the outer ring is 0.45. What is the probability that the projectile hits the target? What is the probability that it misses the target?

Solution

Hitting the bull's-eye, hitting the inner ring, and hitting the outer ring are mutually exclusive events, since it is impossible for more than one of these events to occur. Therefore, using Axiom 3,

$$
\begin{aligned}
P(\text{hits target}) &= P(\text{bull's-eye}) + P(\text{inner ring}) + P(\text{outer ring}) \\
&= 0.10 + 0.25 + 0.45 \\
&= 0.80
\end{aligned}
$$

We can now compute the probability that the projectile misses the target by using Equation (2.1):

$$
\begin{aligned}
P(\text{misses target}) &= 1 - P(\text{hits target}) \\
&= 1 - 0.80 \\
&= 0.20
\end{aligned}
$$

Example 2.5

The following table presents probabilities for the number of times that a certain computer system will crash in the course of a week. Let A be the event that there are more than two crashes during the week, and let B be the event that the system crashes at least once. Find a sample space. Then find the subsets of the sample space that correspond to the events A and B. Then find $P(A)$ and $P(B)$.

Number of Crashes	Probability
0	0.60
1	0.30
2	0.05
3	0.04
4	0.01

Solution

A sample space for the experiment is the set $\{0, 1, 2, 3, 4\}$. The events are $A = \{3, 4\}$ and $B = \{1, 2, 3, 4\}$. To find $P(A)$, notice that A is the event that either 3 crashes happen or 4 crashes happen. The events "3 crashes happen" and "4 crashes happen" are mutually exclusive. Therefore, using Axiom 3, we conclude that

$$
\begin{aligned}
P(A) &= P(\text{3 crashes happen or 4 crashes happen}) \\
&= P(\text{3 crashes happen}) + P(\text{4 crashes happen}) \\
&= 0.04 + 0.01 \\
&= 0.05
\end{aligned}
$$

We will compute $P(B)$ in two ways. First, note that B^c is the event that no crashes happen. Therefore, using Equation (2.1),

$$
\begin{aligned}
P(B) &= 1 - P(B^c) \\
&= 1 - P(\text{0 crashes happen}) \\
&= 1 - 0.60 \\
&= 0.40
\end{aligned}
$$

For a second way to compute $P(B)$, note that B is the event that 1 crash happens or 2 crashes happen or 3 crashes happen or 4 crashes happen. These events are mutually exclusive. Therefore, using Axiom 3, we conclude that

$$\begin{aligned} P(B) &= P(1 \text{ crash}) + P(2 \text{ crashes}) + P(3 \text{ crashes}) + P(4 \text{ crashes}) \\ &= 0.30 + 0.05 + 0.04 + 0.01 \\ &= 0.40 \end{aligned}$$

In Example 2.5, we computed the probabilities of the events $A = \{3, 4\}$ and $B = \{1, 2, 3, 4\}$ by summing the probabilities of the outcomes in each of the events: $P(A) = P(3) + P(4)$ and $P(B) = P(1) + P(2) + P(3) + P(4)$. This method works in general. Since any two outcomes in a sample space are mutually exclusive, the probability of any event that contains a finite number of outcomes can be found by summing the probabilities of the outcomes that comprise the event.

If A is an event, and $A = \{E_1, E_2, \ldots, E_n\}$, then

$$P(A) = P(E_1) + P(E_2) + \cdots + P(E_n) \quad (2.3)$$

Sample Spaces with Equally Likely Outcomes

For some experiments, a sample space can be constructed in which all the outcomes are equally likely. A simple example is the roll of a fair die, in which the sample space is $\{1, 2, 3, 4, 5, 6\}$ and each of these outcomes has probability 1/6. Another type of experiment that results in equally likely outcomes is the random selection of an item from a population of items. The items in the population can be thought of as the outcomes in a sample space, and each item is equally likely to be selected.

A population from which an item is sampled at random can be thought of as a sample space with equally likely outcomes.

If a sample space contains N equally likely outcomes, the probability of each outcome is $1/N$. This is so, because the probability of the whole sample space must be 1, and this probability is equally divided among the N outcomes. If A is an event that contains k outcomes, then $P(A)$ can be found by summing the probabilities of the k outcomes, so $P(A) = k/N$.

If $\mathcal{S}$ is a sample space containing N equally likely outcomes, and if A is an event containing k outcomes, then

$$P(A) = \frac{k}{N} \quad (2.4)$$

Example 2.6

An extrusion die is used to produce aluminum rods. Specifications are given for the length and the diameter of the rods. For each rod, the length is classified as too short, too long, or OK, and the diameter is classified as too thin, too thick, or OK. In a population of 1000 rods, the number of rods in each class is as follows:

	Diameter		
Length	**Too Thin**	**OK**	**Too Thick**
Too Short	10	3	5
OK	38	900	4
Too Long	2	25	13

A rod is sampled at random from this population. What is the probability that it is too short?

Solution

We can think of each of the 1000 rods as an outcome in a sample space. Each of the 1000 outcomes is equally likely. We'll solve the problem by counting the number of outcomes that correspond to the event. The number of rods that are too short is $10 + 3 + 5 = 18$. Since the total number of rods is 1000,

$$P(\text{too short}) = \frac{18}{1000}$$

The Addition Rule

If A and B are mutually exclusive events, then $P(A \cup B) = P(A) + P(B)$. This rule can be generalized to cover the case where A and B are not mutually exclusive. Example 2.7 illustrates the reasoning.

Example 2.7

Refer to Example 2.6. If a rod is sampled at random, what is the probability that it is either too short or too thick?

Solution

First we'll solve this problem by counting the number of outcomes that correspond to the event. In the following table the numbers of rods that are too thick are circled, and the numbers of rods that are too short have rectangles around them. Note that there are 5 rods that are both too short and too thick.

	Diameter		
Length	**Too Thin**	**OK**	**Too Thick**
Too Short	10	3	5
OK	38	900	4
Too Long	2	25	13

Of the 1000 outcomes, the number that are either too short or too thick is $10 + 3 + 5 + 4 + 13 = 35$. Therefore

$$P(\text{too short or too thick}) = \frac{35}{1000}$$

Now we will solve the problem in a way that leads to a more general method. In the sample space, there are $10+3+5 = 18$ rods that are too short and $5+4+13 = 22$ rods that are too thick. But if we try to find the number of rods that are either too short or too thick by adding $18+22$, we get too large a number (40 instead of 35). The reason is that there are five rods that are both too short and too thick, and these are counted twice. We can still solve the problem by adding 18 and 22, but we must then subtract 5 to correct for the double counting.

We restate this reasoning, using probabilities:

$$P(\text{too short}) = \frac{18}{1000}, \quad P(\text{too thick}) = \frac{22}{1000}, \quad P(\text{too short and too thick}) = \frac{5}{1000}$$

$$\begin{aligned} P(\text{too short or too thick}) &= P(\text{too short}) + P(\text{too thick}) - P(\text{too short and too thick}) \\ &= \frac{18}{1000} + \frac{22}{1000} - \frac{5}{1000} \\ &= \frac{35}{1000} \end{aligned}$$

The method of Example 2.7 holds for any two events in any sample space. In general, to find the probability that either of two events occurs, add the probabilities of the events and then subtract the probability that they both occur.

Summary

Let A and B be any events. Then

$$P(A \cup B) = P(A) + P(B) - P(A \cap B) \tag{2.5}$$

A proof of this result, based on the axioms, is provided at the end of this section. Note that if A and B are mutually exclusive, then $P(A \cap B) = 0$, so Equation (2.5) reduces to Axiom 3 in this case.

In a process that manufactures aluminum cans, the probability that a can has a flaw on its side is 0.02, the probability that a can has a flaw on the top is 0.03, and the probability that a can has a flaw on both the side and the top is 0.01. What is the probability that a randomly chosen can has a flaw? What is the probability that it has no flaw?

Solution

We are given that $P(\text{flaw on side}) = 0.02$, $P(\text{flaw on top}) = 0.03$, and $P(\text{flaw on side and flaw on top}) = 0.01$. Now $P(\text{flaw}) = P(\text{flaw on side or flaw on top})$. Using Equation (2.5),

$$\begin{aligned} P(\text{flaw on side or flaw on top}) &= P(\text{flaw on side}) + P(\text{flaw on top}) \\ &\quad - P(\text{flaw on side and flaw on top}) \\ &= 0.02 + 0.03 - 0.01 \\ &= 0.04 \end{aligned}$$

To find the probability that a can has no flaw, we compute

$$P(\text{no flaw}) = 1 - P(\text{flaw}) = 1 - 0.04 = 0.96$$

Proof that $P(A^c) = 1 - P(A)$

Let $\mathcal{S}$ be a sample space and let A be an event. Then A and A^c are mutually exclusive, so by Axiom 3,

$$P(A \cup A^c) = P(A) + P(A^c)$$

But $A \cup A^c = \mathcal{S}$, and by Axiom 1, $P(\mathcal{S}) = 1$. Therefore

$$P(A \cup A^c) = P(\mathcal{S}) = 1$$

It follows that $P(A) + P(A^c) = 1$, so $P(A^c) = 1 - P(A)$.

Proof that $P(\emptyset) = 0$

Let $\mathcal{S}$ be a sample space. Then $\emptyset = \mathcal{S}^c$. Therefore $P(\emptyset) = 1 - P(\mathcal{S}) = 1 - 1 = 0$.

Proof that $P(A \cup B) = P(A) + P(B) - P(A \cap B)$

Let A and B be any two events. The key to the proof is to write $A \cup B$ as the union of three mutually exclusive events: $A \cap B^c$, $A \cap B$, and $A^c \cap B$.

$$A \cup B = (A \cap B^c) \cup (A \cap B) \cup (A^c \cap B) \tag{2.6}$$

The following figure illustrates Equation (2.6).

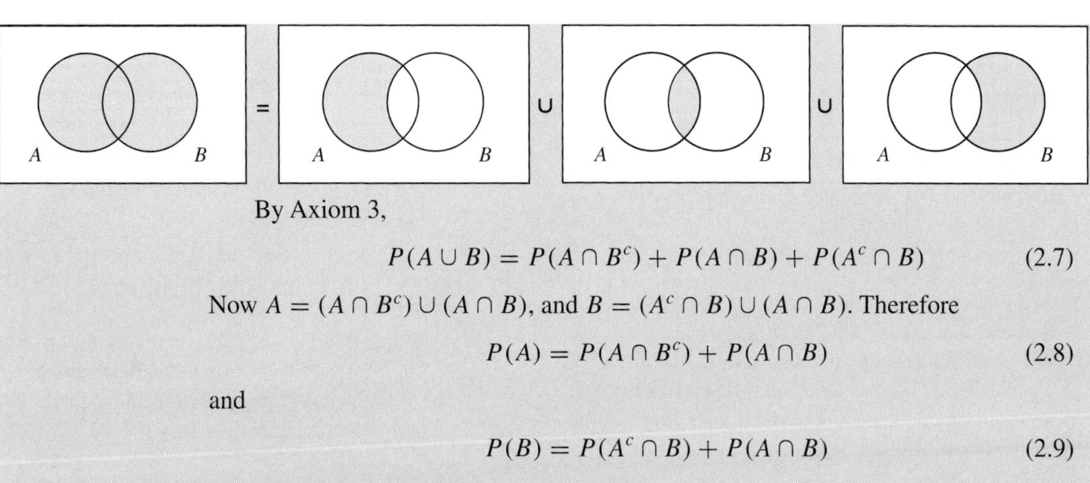

By Axiom 3,

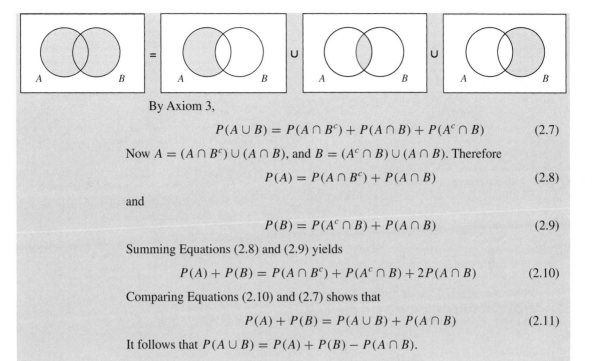

$$P(A \cup B) = P(A \cap B^c) + P(A \cap B) + P(A^c \cap B) \tag{2.7}$$

Now $A = (A \cap B^c) \cup (A \cap B)$, and $B = (A^c \cap B) \cup (A \cap B)$. Therefore

$$P(A) = P(A \cap B^c) + P(A \cap B) \tag{2.8}$$

and

$$P(B) = P(A^c \cap B) + P(A \cap B) \tag{2.9}$$

Summing Equations (2.8) and (2.9) yields

$$P(A) + P(B) = P(A \cap B^c) + P(A^c \cap B) + 2P(A \cap B) \tag{2.10}$$

Comparing Equations (2.10) and (2.7) shows that

$$P(A) + P(B) = P(A \cup B) + P(A \cap B) \tag{2.11}$$

It follows that $P(A \cup B) = P(A) + P(B) - P(A \cap B)$.

Exercises for Section 2.1

1. The probability that a microcircuit is defective is 0.08. What is the probability that it is not defective?

2. An octahedral die (eight faces) has the number 1 painted on two of its faces, the number 2 painted on three of its faces, the number 3 painted on two of its faces, and the number 4 painted on one face. The die is rolled. Assume that each face is equally likely to come up.

a. Find a sample space for this experiment.

b. Find P(even number).

c. If the die were loaded so that the face with the 4 on it were twice as likely to come up as each of the other seven faces, would this change the sample space? Explain.

d. If the die were loaded so that the face with the 4 on it were twice as likely to come up as each of the other seven faces, would this change the value of P(even number)? Explain.

3. Sixty percent of large purchases made at a certain computer retailer are personal computers, 30% are laptop computers, and 10% are peripheral devices such as printers. As part of an audit, one purchase record is sampled at random.

a. What is the probability that it is a personal computer?

b. What is the probability that it is either a personal computer or a laptop computer?

4. An item manufactured by a certain process has probability 0.10 of being defective. True or false:

a. If a sample of 100 items is drawn, exactly 10 of them will be defective.

b. If a sample of 100 items is drawn, the number of defectives is likely to be close to 10, but not exactly equal to 10.

c. As more and more items are sampled, the proportion of defective items will approach 10%.

5. A quality-control engineer samples 100 items manufactured by a certain process, and finds that 15 of them are defective. True or false:
   a. The probability that an item produced by this process is defective is 0.15.
   b. The probability that an item produced by this process is defective is likely to be close to 0.15, but not exactly equal to 0.15.

6. A system contains two components, A and B. The system will function so long as either A or B functions. The probability that A functions is 0.95, the probability that B functions is 0.90, and the probability that both function is 0.88. What is the probability that the system functions?

7. A system contains two components, A and B. The system will function only if both components function. The probability that A functions is 0.98, the probability that B functions is 0.95, and the probability that either A or B functions is 0.99. What is the probability that the system functions?

8. Human blood may contain either or both of two antigens, A and B. Blood that contains only the A antigen is called type A, blood that contains only the B antigen is called type B, blood that contains both antigens is called type AB, and blood that contains neither antigen is called type O. At a certain blood bank, 35% of the blood donors have type A blood, 10% have type B, and 5% have type AB.
   a. What is the probability that a randomly chosen blood donor is type O?
   b. A recipient with type A blood may safely receive blood from a donor whose blood does not contain the B antigen. What is the probability that a randomly chosen blood donor may donate to a recipient with type A blood?

9. True or false: If A and B are mutually exclusive,
   a. $P(A \cup B) = 0$
   b. $P(A \cap B) = 0$
   c. $P(A \cup B) = P(A \cap B)$
   d. $P(A \cup B) = P(A) + P(B)$

2.2 Counting Methods*

When computing probabilities, it is sometimes necessary to determine the number of outcomes in a sample space. In this section we will describe several methods for doing this. The basic rule, which we will call **the fundamental principle of counting**, is presented by means of Example 2.9.

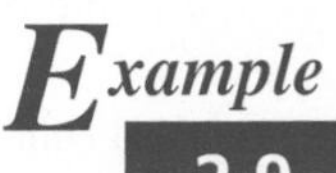

Example 2.9

A certain make of automobile is available in any of three colors: red, blue, or green, and comes with either a large or small engine. In how many ways can a buyer choose a car?

Solution

There are three choices of color and two choices of engine. A complete list of choices is written in the following 3×2 table. The total number of choices is $(3)(2) = 6$.

	Red	Blue	Green
Large	Red, Large	Blue, Large	Green, Large
Small	Red, Small	Blue, Small	Green, Small

*This section is optional.

To generalize Example 2.9, if there are n_1 choices of color and n_2 choices of engine, a complete list of choices can be written in an $n_1 \times n_2$ table, so the total number of choices is $n_1 n_2$.

> If an operation can be performed in n_1 ways, and if for each of these ways a second operation can be performed in n_2 ways, then the total number of ways to perform the two operations is $n_1 n_2$.

The fundamental principle of counting states that this reasoning can be extended to any number of operations.

> **The Fundamental Principle of Counting**
> Assume that k operations are to be performed. If there are n_1 ways to perform the first operation, and if for each of these ways there are n_2 ways to perform the second operation, and if for each choice of ways to perform the first two operations there are n_3 ways to perform the third operation, and so on, then the total number of ways to perform the sequence of k operations is $n_1 n_2 \cdots n_k$.

When ordering a certain type of computer, there are 3 choices of hard drive, 4 choices for the amount of memory, 2 choices of video card, and 3 choices of monitor. In how many ways can a computer be ordered?

Solution

The total number of ways to order a computer is $(3)(4)(2)(3) = 72$.

Permutations

A **permutation** is an ordering of a collection of objects. For example, there are six permutations of the letters A, B, C: ABC, ACB, BAC, BCA, CAB, and CBA. With only three objects, it is easy to determine the number of permutations just by listing them all. But with a large number of objects this would not be feasible. The fundamental principle of counting can be used to determine the number of permutations of any set of objects. For example, we can determine the number of permutations of a set of three objects as follows. There are 3 choices for the object to place first. After that choice is made, there are 2 choices remaining for the object to place second. Then there is 1 choice left for the object to place last. Therefore, the total number of ways to order three objects is $(3)(2)(1) = 6$. This reasoning can be generalized. The number of permutations of a collection of n objects is

$$n(n-1)(n-2)\cdots(3)(2)(1)$$

This is the product of the integers from 1 to n. This product can be written with the symbol $n!$, read "n factorial."

Definition

For any positive integer n, $n! = n(n-1)(n-2)\cdots(3)(2)(1)$.

Also, we define $0! = 1$.

The number of permutations of n objects is $n!$.

Five people stand in line at a movie theater. Into how many different orders can they be arranged?

Solution

The number of permutations of a collection of five people is $5! = (5)(4)(3)(2)(1) = 120$.

Sometimes we are interested in counting the number of permutations of subsets of a certain size chosen from a larger set. This is illustrated in Example 2.12.

Example 2.12

Five lifeguards are available for duty one Saturday afternoon. There are three lifeguard stations. In how many ways can three lifeguards be chosen and ordered among the stations?

Solution

We use the fundamental principle of counting. There are 5 ways to choose a lifeguard to occupy the first station, then 4 ways to choose a lifeguard to occupy the second station, and finally 3 ways to choose a lifeguard to occupy the third station. The total number of permutations of three lifeguards chosen from 5 is therefore $(5)(4)(3) = 60$.

The reasoning used to solve Example 2.12 can be generalized. The number of permutations of k objects chosen from a group of n objects is

$$(n)(n-1)\cdots(n-k+1)$$

This expression can be simplified by using factorial notation:

$$\begin{aligned}(n)(n-1)\cdots(n-k+1) &= \frac{n(n-1)\cdots(n-k+1)(n-k)(n-k-1)\cdots(3)(2)(1)}{(n-k)(n-k-1)\cdots(3)(2)(1)} \\ &= \frac{n!}{(n-k)!}\end{aligned}$$

Summary

The number of permutations of k objects chosen from a group of n objects is

$$\frac{n!}{(n-k)!}.$$

Combinations

In some cases, when choosing a set of objects from a larger set, we don't care about the ordering of the chosen objects; we care only which objects are chosen. For example, we may not care which lifeguard occupies which station; we might care only which three lifeguards are chosen. Each distinct group of objects that can be selected, without regard to order, is called a **combination**. We will now show how to determine the number of combinations of k objects chosen from a set of n objects. We will illustrate the reasoning with the result of Example 2.12. In that example, we showed that there are 60 permutations of three objects chosen from five. Denoting the objects A, B, C, D, E, Figure 2.3 presents a list of all 60 permutations.

ABC	ABD	ABE	ACD	ACE	ADE	BCD	BCE	BDE	CDE
ACB	ADB	AEB	ADC	AEC	AED	BDC	BEC	BED	CED
BAC	BAD	BAE	CAD	CAE	DAE	CBD	CBE	DBE	DCE
BCA	BDA	BEA	CDA	CEA	DEA	CDB	CEB	DEB	DEC
CAB	DAB	EAB	DAC	EAC	EAD	DBC	EBC	EBD	ECD
CBA	DBA	EBA	DCA	ECA	EDA	DCB	ECB	EDB	EDC

FIGURE 2.3 The 60 permutations of three objects chosen from five.

The 60 permutations in Figure 2.3 are arranged in 10 columns of 6 permutations each. Within each column, the three objects are the same, and the column contains the six different permutations of those three objects. Therefore, each column represents a distinct combination of three objects chosen from five, and there are 10 such combinations. Figure 2.3 thus shows that the number of combinations of three objects chosen from five can be found by dividing the number of permutations of three objects chosen from five, which is $5!/(5-3)!$, by the number of permutations of three objects, which is 3! In summary, the number of combinations of three objects chosen from five is $\frac{5!}{3!(5-3)!}$.

The number of combinations of k objects chosen from n is often denoted by the symbol $\binom{n}{k}$. The reasoning used to derive the number of combinations of three objects chosen from five can be generalized to derive an expression for $\binom{n}{k}$.

Summary

The number of combinations of k objects chosen from a group of n objects is

$$\binom{n}{k} = \frac{n!}{k!(n-k)!} \tag{2.12}$$

Example 2.13

At a certain event, 30 people attend, and 5 will be chosen at random to receive door prizes. The prizes are all the same, so the order in which the people are chosen does not matter. How many different groups of five people can be chosen?

Solution

Since the order of the five chosen people does not matter, we need to compute the number of combinations of 5 chosen from 30. This is

$$\begin{aligned}\binom{30}{5} &= \frac{30!}{5!25!} \\ &= \frac{(30)(29)(28)(27)(26)}{(5)(4)(3)(2)(1)} \\ &= 142{,}506\end{aligned}$$

Choosing a combination of k objects from a set of n divides the n objects into two subsets: the k that were chosen and the $n - k$ that were not chosen. Sometimes a set is to be divided up into more than two subsets. For example, assume that in a class of 12 students, a project is assigned in which the students will work in groups. Three groups are to be formed, consisting of five, four, and three students. We can calculate the number of ways in which the groups can be formed as follows. We consider the process of dividing the class into three groups as a sequence of two operations. The first operation is to select a combination of 5 students to comprise the group of 5. The second operation is to select a combination of 4 students from the remaining 7, to comprise the group of 4. The group of 3 will then automatically consist of the 3 students who are left.

The number of ways to perform the first operation is

$$\binom{12}{5} = \frac{12!}{5!7!}$$

After the first operation has been performed, the number of ways to perform the second operation is

$$\binom{7}{4} = \frac{7!}{4!3!}$$

The total number of ways to perform the sequence of two operations is therefore

$$\frac{12!}{5!7!}\frac{7!}{4!3!} = \frac{12!}{5!4!3!} = 27{,}720$$

Notice that the numerator in the final answer is the factorial of the total group size, while the denominator is the product of the factorials of the sizes of the groups chosen from it. This holds in general.

Summary

The number of ways of dividing a group of n objects into groups of $k_1, \ldots, k_r$ objects, where $k_1 + \cdots + k_r = n$, is

$$\frac{n!}{k_1! \cdots k_r!} \tag{2.13}$$

A die is rolled 20 times. Given that three of the rolls came up 1, five came up 2, four came up 3, two came up 4, three came up 5, and three came up 6, how many different arrangements of the outcomes are there?

Solution

There are 20 outcomes. They are divided into six groups, namely, the group of three outcomes that came up 1, the group of five outcomes that came up 2, and so on. The number of ways to divide the 20 outcomes into six groups of the specified sizes is

$$\frac{20!}{3!5!4!2!3!3!} = 1.955 \times 10^{12}$$

When a sample space consists of equally likely outcomes, the probability of an event can be found by dividing the number of outcomes in the event by the total number of outcomes in the sample space. Counting rules can sometimes be used to compute these numbers, as the following example illustrates:

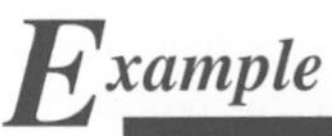

A box of bolts contains 8 thick bolts, 5 medium bolts, and 3 thin bolts. A box of nuts contains 6 that fit the thick bolts, 4 that fit the medium bolts, and 2 that fit the thin bolts. One bolt and one nut are chosen at random. What is the probability that the nut fits the bolt?

Solution

The sample space consists of all pairs of nuts and bolts, and each pair is equally likely to be chosen. The event that the nut fits the bolt corresponds to the set of all matching pairs of nuts and bolts. Therefore

$$P(\text{nut fits bolt}) = \frac{\text{number of matching pairs of nuts and bolts}}{\text{number of pairs of nuts and bolts}}$$

There are $6 + 4 + 2 = 12$ nuts, and $8 + 5 + 3 = 16$ bolts. Therefore

$$\text{Number of pairs of nuts and bolts} = (12)(16) = 192$$

The number of matching pairs is found by summing the number of pairs of thick nuts and bolts, the number of pairs of medium nuts and bolts, and the number of pairs of thin nuts and bolts. These numbers are

$$\text{Number of pairs of thick nuts and bolts} = (6)(8) = 48$$
$$\text{Number of pairs of medium nuts and bolts} = (4)(5) = 20$$
$$\text{Number of pairs of thin nuts and bolts} = (2)(3) = 6$$

Therefore

$$P(\text{nut fits bolt}) = \frac{48 + 20 + 6}{192} = 0.3854$$

Exercises for Section 2.2

1. DNA molecules consist of chemically linked sequences of the bases adenine, guanine, cytosine, and thymine, denoted A, G, C, and T. A sequence of three bases is called a *codon*.

a. How many different codons are there?

b. The bases A and G are *purines*, while C and T are *pyrimidines*. How many codons are there whose first and third bases are purines and whose second base is a pyrimidine?

c. How many codons consist of three different bases?

2. A chemical engineer is designing an experiment to determine the effect of temperature, stirring rate, and type of catalyst on the yield of a certain reaction. She wants to study five different reaction temperatures, two different stirring rates, and four different catalysts. If each run of the experiment involves a choice of one temperature, one stirring rate, and one catalyst, how many different runs are possible?

3. Ten engineers have applied for a management position in a large firm. Four of them will be selected as finalists for the position. In how many ways can this selection be made?

4. A committee of eight people must choose a president, a vice-president, and a secretary. In how many ways can this be done?

5. A test consists of 15 questions. Ten are true–false questions, and five are multiple-choice questions that have four choices each. A student must select an answer for each question. In how many ways can this be done?

6. In a certain state, license plates consist of three letters followed by three numbers.

a. How many different license plates can be made?

b. How many different license plates can be made in which no letter or number appears more than once?

c. A license plate is chosen at random. What is the probability that no letter or number appears more than once?

7. A computer password consists of eight characters.

a. How many different passwords are possible if each character may be any lowercase letter or digit?

b. How many different passwords are possible if each character may be any lowercase letter or digit, and at least one character must be a digit?

c. A computer system requires that passwords contain at least one digit. If eight characters are generated at random, and each is equally likely to be any of the 26 letters or 10 digits, what is the probability that a valid password will be generated?

8. A company has hired 15 new employees, and must assign 6 to the day shift, 5 to the graveyard shift, and 4 to the night shift. In how many ways can the assignment be made?

9. One drawer in a dresser contains 8 blue socks and 6 white socks. A second drawer contains 4 blue socks and 2 white socks. One sock is chosen from each drawer. What is the probability that they match?

10. A drawer contains 6 red socks, 4 green socks, and 2 black socks. Two socks are chosen at random. What is the probability that they match?

2.3 Conditional Probability and Independence

A sample space contains all the possible outcomes of an experiment. Sometimes we obtain some additional information about an experiment that tells us that the outcome comes from a certain part of the sample space. In this case, the probability of an event is based on the outcomes in that part of the sample space. A probability that is based on a part of a sample space is called a **conditional probability**. We explore this idea through some examples.

In Example 2.6 (in Section 2.1) we discussed a population of 1000 aluminum rods. For each rod, the length is classified as too short, too long, or OK, and the diameter is classified as too thin, too thick, or OK. These 1000 rods form a sample space in which each rod is equally likely to be sampled. The number of rods in each category is presented in Table 2.1. Of the 1000 rods, 928 meet the diameter specification. Therefore, if a rod is sampled, $P(\text{diameter OK}) = 928/1000 = 0.928$. This probability is called the **unconditional probability**, since it is based on the entire sample space. Now assume that a rod is sampled, and its length is measured and found to meet the specification. What is the probability that the diameter also meets the specification? The key to computing this probability is to realize that knowledge that the length meets the specification reduces the sample space from which the rod is drawn. Table 2.2 presents this idea. Once we know that the length specification is met, we know that the rod will be one of the 942 rods in the sample space presented in Table 2.2.

TABLE 2.1 Sample space containing 1000 aluminum rods

	Diameter		
Length	**Too Thin**	**OK**	**Too Thick**
Too Short	10	3	5
OK	38	900	4
Too Long	2	25	13

TABLE 2.2 Reduced sample space containing 942 aluminum rods that meet the length specification

	Diameter		
Length	**Too Thin**	**OK**	**Too Thick**
Too Short	—	—	—
OK	38	900	4
Too Long	—	—	—

Of the 942 rods in this sample space, 900 of them meet the diameter specification. Therefore, if we know that the rod meets the length specification, the probability that the rod meets the diameter specification is 900/942. We say that the **conditional probability** that the rod meets the diameter specification **given** that it meets the length specification is equal to 900/942, and we write $P(\text{diameter OK} \mid \text{length OK}) = 900/942 = 0.955$. Note that the conditional probability $P(\text{diameter OK} \mid \text{length OK})$ differs from the unconditional probability $P(\text{diameter OK})$, which was computed from the full sample space (Table 2.1) to be 0.928.

Compute the conditional probability $P(\text{diameter OK} \mid \text{length too long})$. Is this the same as the unconditional probability $P(\text{diameter OK})$?

Solution

The conditional probability $P(\text{diameter OK} \mid \text{length too long})$ is computed under the assumption that the rod is too long. This reduces the sample space to the 40 items indicated in boldface in the following table.

	Diameter		
Length	**Too Thin**	**OK**	**Too Thick**
Too Short	10	3	5
OK	38	900	4
Too Long	**2**	**25**	**13**

Of the 40 outcomes, 25 meet the diameter specification. Therefore

$$P(\text{diameter OK} \mid \text{length too long}) = \frac{25}{40} = 0.625$$

The unconditional probability $P(\text{diameter OK})$ is computed on the basis of all 1000 outcomes in the sample space and is equal to $928/1000 = 0.928$. In this case, the conditional probability differs from the unconditional probability.

Let's look at the solution to Example 2.16 more closely. We found that

$$P(\text{diameter OK} \mid \text{length too long}) = \frac{25}{40}$$

In the answer 25/40, the denominator, 40, represents the number of outcomes that satisfy the condition that the rod is too long, while the numerator, 25, represents the number of outcomes that satisfy both the condition that the rod is too long and that its diameter is OK. If we divide both the numerator and denominator of this answer by the number of outcomes in the full sample space, which is 1000, we obtain

$$P(\text{diameter OK} \mid \text{length too long}) = \frac{25/1000}{40/1000}$$

Now 40/1000 represents the *probability* of satisfying the condition that the rod is too long. That is,

$$P(\text{length too long}) = \frac{40}{1000}$$

The quantity 25/1000 represents the *probability* of satisfying both the condition that the rod is too long and that the diameter is OK. That is,

$$P(\text{diameter OK and length too long}) = \frac{25}{1000}$$

We can now express the conditional probability as

$$P(\text{diameter OK} \mid \text{length too long}) = \frac{P(\text{diameter OK and length too long})}{P(\text{length too long})}$$

This reasoning can be extended to construct a definition of conditional probability that holds for any sample space:

Definition

Let A and B be events with $P(B) \neq 0$. The conditional probability of A given B is

$$P(A|B) = \frac{P(A \cap B)}{P(B)} \tag{2.14}$$

Figure 2.4 presents Venn diagrams to illustrate the idea of conditional probability.

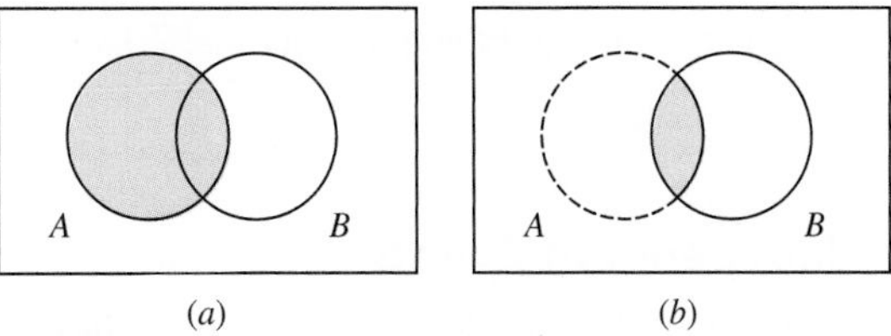

FIGURE 2.4 ***(a)*** The diagram represents the unconditional probability $P(A)$. $P(A)$ is illustrated by considering the event A in proportion to the entire sample space, which is represented by the rectangle. ***(b)*** The diagram represents the conditional probability $P(A|B)$. Since the event B is known to occur, the event B now becomes the sample space. For the event A to occur, the outcome must be in the intersection $A \cap B$. The conditional probability $P(A|B)$ is therefore illustrated by considering the intersection $A \cap B$ in proportion to the entire event B.

Refer to Example 2.8 in Section 2.1. What is the probability that a can will have a flaw on the side, given that it has a flaw on top?

Solution

We are given that $P(\text{flaw on top}) = 0.03$, and $P(\text{flaw on side and flaw on top}) = 0.01$. Using Equation (2.14),

$$\begin{aligned} P(\text{flaw on side} \mid \text{flaw on top}) &= \frac{P(\text{flaw on side and flaw on top})}{P(\text{flaw on top})} \\ &= \frac{0.01}{0.03} \\ &= 0.33 \end{aligned}$$

Refer to Example 2.8 (in Section 2.1). What is the probability that a can will have a flaw on the top, given that it has a flaw on the side?

Solution

We are given that $P(\text{flaw on side}) = 0.02$, and $P(\text{flaw on side and flaw on top}) = 0.01$. Using Equation (2.14),

$$\begin{aligned} P(\text{flaw on top} \mid \text{flaw on side}) &= \frac{P(\text{flaw on top and flaw on side})}{P(\text{flaw on side})} \\ &= \frac{0.01}{0.02} \\ &= 0.5 \end{aligned}$$

The results of Examples 2.17 and 2.18 show that in general, $P(A|B) \neq P(B|A)$.

Independent Events

Sometimes the knowledge that one event has occurred does not change the probability that another event occurs. In this case the conditional and unconditional probabilities are the same, and the events are said to be **independent**. We present an example.

If an aluminum rod is sampled from the sample space presented in Table 2.1, find $P(\text{too long})$ and $P(\text{too long} \mid \text{too thin})$. Are these probabilities different?

Solution

$$P(\text{too long}) = \frac{40}{1000} = 0.040$$

$$\begin{aligned} P(\text{too long} \mid \text{too thin}) &= \frac{P(\text{too long and too thin})}{P(\text{too thin})} \\ &= \frac{2/1000}{50/1000} \\ &= 0.040 \end{aligned}$$

The conditional probability and the unconditional probability are the same. The information that the rod is too thin does not change the probability that the rod is too long.

Example 2.19 shows that knowledge that an event occurs sometimes does not change the probability that another event occurs. In these cases, the two events are said to be **independent**. The event that a rod is too long and the event that a rod is too thin are independent. We now give a more precise definition of the term, both in words and in symbols.

Definition

Two events A and B are **independent** if the probability of each event remains the same whether or not the other occurs.

In symbols: If $P(A) \neq 0$ and $P(B) \neq 0$, then A and B are independent if

$$P(B|A) = P(B) \quad \text{or, equivalently,} \quad P(A|B) = P(A) \tag{2.15}$$

If either $P(A) = 0$ or $P(B) = 0$, then A and B are independent.

If A and B are independent, then the following pairs of events are also independent: A and B^c, A^c and B, and A^c and B^c. The proof of this fact is left as an exercise.

The concept of independence can be extended to more than two events:

Definition

Events $A_1, A_2, \ldots, A_n$ are independent if the probability of each remains the same no matter which of the others occur.

In symbols: Events $A_1, A_2, \ldots, A_n$ are independent if for each A_i, and each collection $A_{j1}, \ldots, A_{jm}$ of events with $P(A_{j1} \cap \cdots \cap A_{jm}) \neq 0$,

$$P(A_i|A_{j1} \cap \cdots \cap A_{jm}) = P(A_i) \tag{2.16}$$

The Multiplication Rule

Sometimes we know $P(A|B)$ and we wish to find $P(A \cap B)$. We can obtain a result that is useful for this purpose by multiplying both sides of Equation (2.14) by $P(B)$. This leads to the multiplication rule.

If A and B are two events with $P(B) \neq 0$, then

$$P(A \cap B) = P(B)P(A|B) \tag{2.17}$$

If A and B are two events with $P(A) \neq 0$, then

$$P(A \cap B) = P(A)P(B|A) \tag{2.18}$$

If $P(A) \neq 0$ and $P(B) \neq 0$, then Equations (2.17) and (2.18) both hold.

When two events are independent, then $P(A|B) = P(A)$ and $P(B|A) = P(B)$, so the multiplication rule simplifies:

If A and B are independent events, then

$$P(A \cap B) = P(A)P(B) \tag{2.19}$$

This result can be extended to any number of events. If $A_1, A_2, \ldots, A_n$ are independent events, then for each collection $A_{j1}, \ldots, A_{jm}$ of events

$$P(A_{j1} \cap A_{j2} \cap \cdots \cap A_{jm}) = P(A_{j1})P(A_{j2}) \cdots P(A_{jm}) \tag{2.20}$$

In particular,

$$P(A_1 \cap A_2 \cap \cdots \cap A_n) = P(A_1)P(A_2) \cdots P(A_n) \tag{2.21}$$

Example 2.20

A vehicle contains two engines, a main engine and a backup. The engine component fails only if both engines fail. The probability that the main engine fails is 0.05, and the probability that the backup engine fails is 0.10. Assume that the main and backup engines function independently. What is the probability that the engine component fails?

Solution

The probability that the engine component fails is the probability that both engines fail. Therefore

$$P(\text{engine component fails}) = P(\text{main engine fails and backup engine fails})$$

Since the engines function independently, we may use Equation (2.19):

$$
\begin{aligned}
P(\text{main engine fails and backup engine fails}) &= P(\text{main fails})P(\text{backup fails}) \\
&= (0.10)(0.05) \\
&= 0.005
\end{aligned}
$$

Example 2.21

A system contains two components, A and B. Both components must function for the system to work. The probability that component A fails is 0.08, and the probability that component B fails is 0.05. Assume the two components function independently. What is the probability that the system functions?

Solution

The probability that the system functions is the probability that both components function. Therefore

$$
P(\text{system functions}) = P(\text{A functions and B functions})
$$

Since the components function independently,

$$
\begin{aligned}
P(\text{A functions and B functions}) &= P(\text{A functions})P(\text{B functions}) \\
&= [1 - P(\text{A fails})][1 - P(\text{B fails})] \\
&= (1 - 0.08)(1 - 0.05) \\
&= 0.874
\end{aligned}
$$

Example 2.22

Of the microprocessors manufactured by a certain process, 20% are defective. Five microprocessors are chosen at random. Assume they function independently. What is the probability that they all work?

Solution

For $i = 1, \ldots, 5$, let A_i denote the event that the ith microprocessor works. Then

$$
\begin{aligned}
P(\text{all 5 work}) &= P(A_1 \cap A_2 \cap A_3 \cap A_4 \cap A_5) \\
&= P(A_1)P(A_2)P(A_3)P(A_4)P(A_5) \\
&= (1 - 0.20)^5 \\
&= 0.328
\end{aligned}
$$

Example 2.23

In Example 2.22, what is the probability that at least one of the microprocessors works?

Solution

The easiest way to solve this problem is to notice that

$$
P(\text{at least one works}) = 1 - P(\text{all are defective})
$$

Now, letting D_i denote the event that the ith microprocessor is defective,

$$\begin{aligned} P(\text{all are defective}) &= P(D_1 \cap D_2 \cap D_3 \cap D_4 \cap D_5) \\ &= P(D_1)P(D_2)P(D_3)P(D_4)P(D_5) \\ &= (0.20)^5 \\ &= 0.0003 \end{aligned}$$

Therefore $P(\text{at least one works}) = 1 - 0.0003 = 0.9997$.

Equations (2.19) and (2.20) tell us how to compute probabilities when we know that events are independent, but they are usually not much help when it comes to deciding whether two events really *are* independent. In most cases, the best way to determine whether events are independent is through an understanding of the process that produces the events. Here are a few illustrations:

- A die is rolled twice. It is reasonable to believe that the outcome of the second roll is not affected by the outcome of the first roll. Therefore, knowing the outcome of the first roll does not help to predict the outcome of the second roll. The two rolls are independent.
- A certain chemical reaction is run twice, using different equipment each time. It is reasonable to believe that the yield of one reaction will not affect the yield of the other. In this case the yields are independent.
- A chemical reaction is run twice in succession, using the same equipment. In this case, it might not be wise to assume that the yields are independent. For example, a low yield on the first run might indicate that there is more residue than usual left behind. This might tend to make the yield on the next run higher. Thus knowing the yield on the first run could help to predict the yield on the second run.
- The items in a simple random sample may be treated as independent, unless the population is finite and the sample comprises more than about 5% of the population (see the discussion of independence in Section 1.1).

The Law of Total Probability

The law of total probability is illustrated in Figure 2.5. A sample space contains the events A_1, A_2, A_3, and A_4. These events are mutually exclusive, since no two overlap. They are also **exhaustive**, which means that their union covers the whole sample space. Every outcome in the sample space belongs to one and only one of the events A_1, A_2, A_3, A_4.

The event B is any event. In Figure 2.5, each of the events A_i intersects B, forming the events $A_1 \cap B$, $A_2 \cap B$, $A_3 \cap B$, and $A_4 \cap B$. It is clear from Figure 2.5 that the events $A_1 \cap B$, $A_2 \cap B$, $A_3 \cap B$, and $A_4 \cap B$ are mutually exclusive and that they cover B. Every outcome in B belongs to one and only one of the events $A_1 \cap B$, $A_2 \cap B$, $A_3 \cap B$, $A_4 \cap B$. It follows that

$$B = (A_1 \cap B) \cup (A_2 \cap B) \cup (A_3 \cap B) \cup (A_4 \cap B)$$

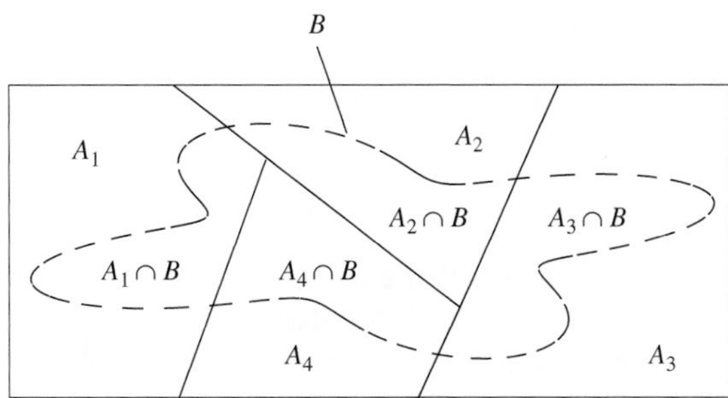

FIGURE 2.5 The mutually exclusive and exhaustive events A_1, A_2, A_3, A_4 divide the event B into mutually exclusive subsets.

which is a union of mutually exclusive events. Therefore

$$P(B) = P(A_1 \cap B) + P(A_2 \cap B) + P(A_3 \cap B) + P(A_4 \cap B)$$

Since $P(A_i \cap B) = P(B|A_i)P(A_i)$,

$$P(B) = P(B|A_1)P(A_1) + P(B|A_2)P(A_2) + P(B|A_3)P(A_3) + P(B|A_4)P(A_4) \tag{2.22}$$

Equation (2.22) is a special case of the law of total probability, restricted to the case where there are four mutually exclusive and exhaustive events. The intuition behind the law of total probability is quite simple. The events A_1, A_2, A_3, A_4 break the event B into pieces. The probability of B is found by adding up the probabilities of the pieces.

We could redraw Figure 2.5 to have any number of events A_i. This leads to the general case of the law of total probability.

Law of Total Probability

If $A_1, \ldots, A_n$ are mutually exclusive and exhaustive events, and B is any event, then

$$P(B) = P(A_1 \cap B) + \cdots + P(A_n \cap B) \tag{2.23}$$

Equivalently, if $P(A_i) \neq 0$ for each A_i,

$$P(B) = P(B|A_1)P(A_1) + \cdots + P(B|A_n)P(A_n) \tag{2.24}$$

Example 2.24

Customers who purchase a certain make of car can order an engine in any of three sizes. Of all cars sold, 45% have the smallest engine, 35% have the medium-size one, and 20% have the largest. Of cars with the smallest engine, 10% fail an emissions test within two years of purchase, while 12% of those with the medium size and 15% of those with the largest engine fail. What is the probability that a randomly chosen car will fail an emissions test within two years?

Solution

Let B denote the event that a car fails an emissions test within two years. Let A_1 denote the event that a car has a small engine, A_2 the event that a car has a medium-size engine, and A_3 the event that a car has a large engine. Then

$$P(A_1) = 0.45 \qquad P(A_2) = 0.35 \qquad P(A_3) = 0.20$$

The probability that a car will fail a test, given that it has a small engine, is 0.10. That is, $P(B|A_1) = 0.10$. Similarly, $P(B|A_2) = 0.12$, and $P(B|A_3) = 0.15$. By the law of total probability (Equation 2.24),

$$\begin{aligned} P(B) &= P(B|A_1)P(A_1) + P(B|A_2)P(A_2) + P(B|A_3)P(A_3) \\ &= (0.10)(0.45) + (0.12)(0.35) + (0.15)(0.20) \\ &= 0.117 \end{aligned}$$

Sometimes problems like Example 2.24 are solved with the use of tree diagrams. Figure 2.6 presents a tree diagram for Example 2.24. There are three primary branches on the tree, corresponding to the three engine sizes. The probabilities of the engine sizes are listed on their respective branches. At the end of each primary branch are two secondary branches, representing the events of failure and no failure. The conditional probabilities of failure and no failure, given engine size, are listed on the secondary branches. By multiplying along each of the branches corresponding to the event B = fail, we obtain the probabilities $P(B|A_i)P(A_i)$. Summing these probabilities yields $P(B)$, as desired.

Bayes' Rule

If A and B are two events, we have seen that in general $P(A|B) \neq P(B|A)$. Bayes' rule provides a formula that allows us to calculate one of the conditional probabilities if we know the other one. To see how it works, assume that we know $P(B|A)$ and we wish to calculate $P(A|B)$. Start with the definition of conditional probability (Equation 2.14):

$$P(A|B) = \frac{P(A \cap B)}{P(B)}$$

Now use Equation (2.18) to substitute $P(B|A)P(A)$ for $P(A \cap B)$:

$$P(A|B) = \frac{P(B|A)P(A)}{P(B)} \tag{2.25}$$

Equation (2.25) is essentially Bayes' rule. When Bayes' rule is written, the expression $P(B)$ in the denominator is usually replaced with a more complicated expression derived from the law of total probability. Specifically, since the events A and A^c are mutually exclusive and exhaustive, the law of total probability shows that

$$P(B) = P(B|A)P(A) + P(B|A^c)P(A^c) \tag{2.26}$$

Substituting the right-hand side of Equation (2.26) for $P(B)$ in Equation (2.25) yields Bayes' rule. A more general version of Bayes' rule can be derived as well, by considering a collection $A_1, \dots, A_n$ of mutually exclusive and exhaustive events and using the

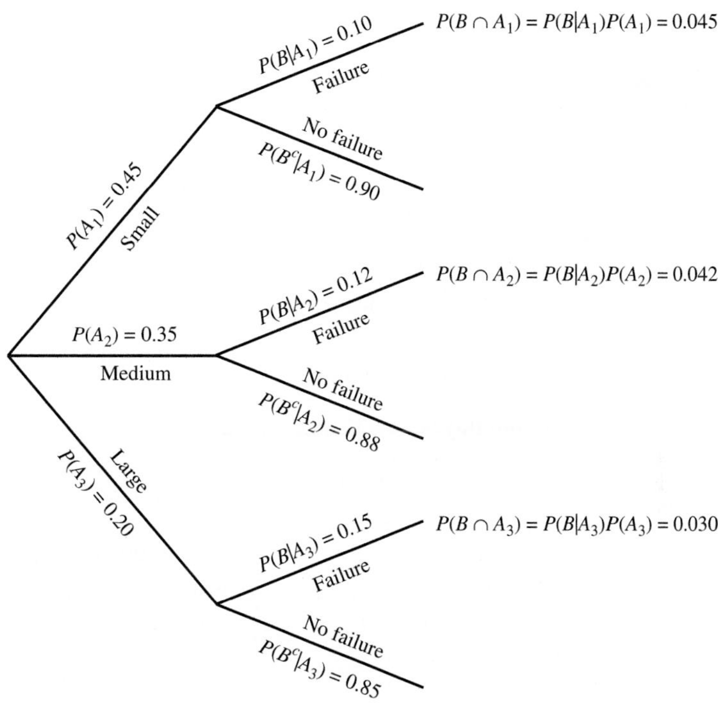

FIGURE 2.6 Tree diagram for the solution to Example 2.24.

law of total probability to replace $P(B)$ with the expression on the right-hand side of Equation (2.24).

Bayes' Rule

Special Case: Let A and B be events with $P(A) \neq 0$, $P(A^c) \neq 0$, and $P(B) \neq 0$. Then

$$P(A|B) = \frac{P(B|A)P(A)}{P(B|A)P(A) + P(B|A^c)P(A^c)} \tag{2.27}$$

General Case: Let $A_1, \ldots, A_n$ be mutually exclusive and exhaustive events with $P(A_i) \neq 0$ for each A_i. Let B be any event with $P(B) \neq 0$. Then

$$P(A_k|B) = \frac{P(B|A_k)P(A_k)}{\sum_{i=1}^{n} P(B|A_i)P(A_i)} \tag{2.28}$$

Example 2.25 shows how Bayes' rule can be used to discover an important and surprising result in the field of medical testing.

The proportion of people in a given community who have a certain disease is 0.005. A test is available to diagnose the disease. If a person has the disease, the probability that the test will produce a positive signal is 0.99. If a person does not have the disease, the probability that the test will produce a positive signal is 0.01. If a person tests positive, what is the probability that the person actually has the disease?

Solution

Let D represent the event that the person actually has the disease, and let $+$ represent the event that the test gives a positive signal. We wish to find $P(D|+)$. We are given the following probabilities:

$$P(D) = 0.005 \qquad P(+|D) = 0.99 \qquad P(+|D^c) = 0.01$$

Using Bayes' rule (Equation 2.27),

$$\begin{aligned} P(D|+) &= \frac{P(+|D)P(D)}{P(+|D)P(D) + P(+|D^c)P(D^c)} \\ &= \frac{(0.99)(0.005)}{(0.99)(0.005) + (0.01)(0.995)} \\ &= 0.332 \end{aligned}$$

In Example 2.25, only about a third of the people who test positive for the disease actually have the disease. Note that the test is fairly accurate; it correctly classifies 99% of both diseased and nondiseased individuals. The reason that a large proportion of those who test positive are actually disease-free is that the disease is rare—only 0.5% of the population has it. Because many diseases are rare, it is the case for many medical tests that most positives are false positives, even when the test is fairly accurate. For this reason, when a test comes out positive, a second test is usually given before a firm diagnosis is made.

Refer to Example 2.24. A record for a failed emissions test is chosen at random. What is the probability that it is for a car with a small engine?

Solution

Let B denote the event that a car failed an emissions test. Let A_1 denote the event that a car has a small engine, A_2 the event that a car has a medium-size engine, and A_3 the event that a car has a large engine. We wish to find $P(A_1|B)$. The following probabilities are given in Example 2.24:

$$\begin{array}{lll} P(A_1) = 0.45 & P(A_2) = 0.35 & P(A_3) = 0.20 \\ P(B|A_1) = 0.10 & P(B|A_2) = 0.12 & P(B|A_3) = 0.15 \end{array}$$

By Bayes' rule,

$$P(A_1|B) = \frac{P(B|A_1)P(A_1)}{P(B|A_1)P(A_1) + P(B|A_2)P(A_2) + P(B|A_3)P(A_3)}$$

$$= \frac{(0.10)(0.45)}{(0.10)(0.45) + (0.12)(0.35) + (0.15)(0.20)}$$

$$= 0.385$$

Application to Reliability Analysis

Reliability analysis is the branch of engineering concerned with estimating the failure rates of systems. While some problems in reliability analysis require advanced mathematical methods, there are many problems that can be solved with the methods we have learned so far. We begin with an example illustrating the computation of the reliability of a system consisting of two components connected *in series.*

Example 2.27

A system contains two components, A and B, connected in series as shown in the following diagram.

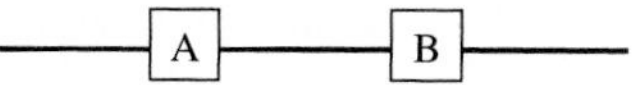

The system will function only if both components function. The probability that A functions is given by $P(A) = 0.98$, and the probability that B functions is given by $P(B) = 0.95$. Assume that A and B function independently. Find the probability that the system functions.

Solution

Since the system will function only if both components function, it follows that

$$\begin{aligned} P(\text{system functions}) &= P(A \cap B) \\ &= P(A)P(B) \text{ by the assumption of independence} \\ &= (0.98)(0.95) \\ &= 0.931 \end{aligned}$$

Example 2.28 illustrates the computation of the reliability of a system consisting of two components connected *in parallel.*

Example 2.28

A system contains two components, C and D, connected in parallel as shown in the following diagram.

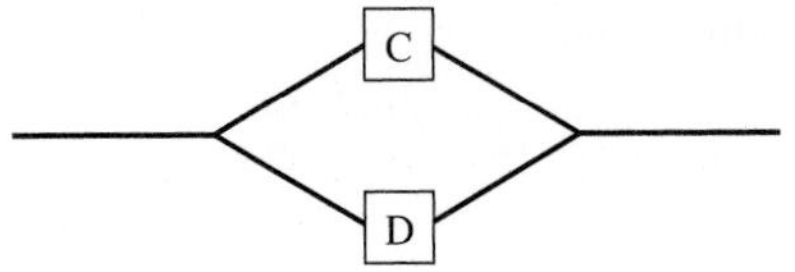

The system will function if either C or D functions. The probability that C functions is 0.90, and the probability that D functions is 0.85. Assume C and D function independently. Find the probability that the system functions.

Solution

Since the system will function so long as either of the two components functions, it follows that

$$\begin{aligned} P(\text{system functions}) &= P(C \cup D) \\ &= P(C) + P(D) - P(C \cap D) \\ &= P(C) + P(D) - P(C)P(D) \\ &\qquad \text{by the assumption of independence} \\ &= 0.90 + 0.85 - (0.90)(0.85) \\ &= 0.985 \end{aligned}$$

The reliability of more complex systems can often be determined by decomposing the system into a series of subsystems, each of which contains components connected either in series or in parallel. Example 2.29 illustrates the method.

Example 2.29

The thesis "Dynamic, Single-stage, Multiperiod, Capacitated Production Sequencing Problem with Multiple Parallel Resources" (D. Ott, M.S. thesis, Colorado School of Mines, 1998) describes a production method used in the manufacture of aluminum cans. The following schematic diagram, slightly simplified, depicts the process.

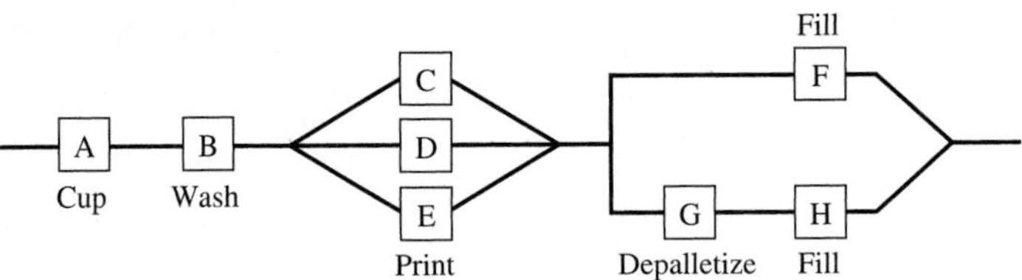

The initial input into the process consists of coiled aluminum sheets, approximately 0.25 mm thick. In a process known as "cupping," these sheets are uncoiled and shaped into can bodies, which are cylinders that are closed on the bottom and open on top. These can bodies are then washed and sent to the printer, which prints the label on the can. In practice there are several printers on a line; the diagram presents a line with three printers. The printer deposits the cans onto pallets, which are wooden structures that hold 7140 cans each. The cans next go to be filled. Some fill lines can accept cans directly from the pallets, but others can accept them only from cell bins, which are large containers holding approximately 100,000 cans each. To use these fill lines, the cans must be transported from the pallets to cell bins, in a process called depalletizing. In practice there are several fill lines; the diagram presents a case where there are two fill lines, one of which will accept cans from the pallets, and the other of which will not. In the filling process the cans are filled, and the can top is seamed on. The cans are then packaged and shipped to distributors.

It is desired to estimate the probability that the process will function for one day without failing. Assume that the cupping process has probability 0.995 of functioning successfully for one day. Since this component is denoted by "A" in the diagram, we will express this probability as $P(A) = 0.995$. Assume that the other process components have the following probabilities of functioning successfully during a one-day period: $P(B) = 0.99$, $P(C) = P(D) = P(E) = 0.95$, $P(F) = 0.90$, $P(G) = 0.90$, $P(H) = 0.98$. Assume the components function independently. Find the probability that the process functions successfully for one day.

Solution

We can solve this problem by noting that the entire process can be broken down into subsystems, each of which consists of simple series or parallel component systems. Specifically, subsystem 1 consists of the cupping and washing components, which are connected in series. Subsystem 2 consists of the printers, which are connected in parallel. Subsystem 3 consists of the fill lines, which are connected in parallel, with one of the two lines consisting of two components connected in series.

We compute the probabilities of successful functioning for each subsystem, denoting the probabilities p_1, p_2, and p_3.

$$\begin{aligned} P(\text{subsystem 1 functions}) = p_1 &= P(A \cap B) \\ &= P(A)P(B) \\ &= (0.995)(0.990) \\ &= 0.985050 \end{aligned}$$

$$\begin{aligned} P(\text{subsystem 2 functions}) = p_2 &= 1 - P(\text{subsystem 2 fails}) \\ &= 1 - P(C^c \cap D^c \cap E^c) \\ &= 1 - P(C^c)P(D^c)P(E^c) \\ &= 1 - (0.05)^3 \\ &= 0.999875 \end{aligned}$$

Subsystem 3 functions if F functions, or if both G and H function. Therefore

$$\begin{aligned} P(\text{subsystem 3 functions}) = p_3 &= P(F \cup (G \cap H)) \\ &= P(F) + P(G \cap H) - P(F \cap G \cap H) \\ &= P(F) + P(G)P(H) - P(F)P(G)P(H) \\ &= (0.90) + (0.90)(0.98) - (0.90)(0.90)(0.98) \\ &= 0.988200 \end{aligned}$$

The entire process consists of the three subsystems connected in series. Therefore, for the process to function, all three subsystems must function. We conclude that

$$\begin{aligned} P(\text{system functions}) &= P(\text{systems 1, 2, and 3 all function}) \\ &= p_1 p_2 p_3 \\ &= (0.985050)(0.999875)(0.988200) \\ &= 0.973 \end{aligned}$$

We remark that the assumption that the components function independently is crucial in the solutions of Examples 2.27, 2.28, and 2.29. When this assumption is not met, it can be very difficult to make accurate reliability estimates. If the assumption of independence is used without justification, reliability estimates may be misleading.

Exercises for Section 2.3

1. A box contains 10 fuses. Eight of them are rated at 10 amperes (A) and the other two are rated at 15 A. Two fuses are selected at random.

a. What is the probability that the first fuse is rated at 15 A?

b. What is the probability that the second fuse is rated at 15 A, given that the first fuse is rated at 10 A?

c. What is the probability that the second fuse is rated at 15 A, given that the first fuse is rated at 15 A?

2. Refer to Exercise 1. Fuses are randomly selected from the box, one by one, until a 15 A fuse is selected.

a. What is the probability that the first two fuses are both 10 A?

b. What is the probability that a total of two fuses are selected from the box?

c. What is the probability that more than three fuses are selected from the box?

3. On graduation day at a large university, one graduate is selected at random. Let A represent the event that the student is an engineering major, and let B represent the event that the student took a calculus course in college. Which probability is greater, $P(A|B)$ or $P(B|A)$? Explain.

4. The article "Integrating Risk Assessment and Life Cycle Assessment: A Case Study of Insulation" (Y. Nishioka, J. Levy, et al., *Risk Analysis*, 2002:1003–1017) estimates that 5.6% of a certain population has asthma, and that an asthmatic has probability 0.027 of suffering an asthma attack on a given day. A person is chosen at random from this population. What is the probability that this person has an asthma attack on that day?

5. Oil wells drilled in region A have probability 0.2 of producing. Wells drilled in region B have probability 0.09 of producing. One well is drilled in each region. Assume the wells produce independently.

a. What is the probability that both wells produce?

b. What is the probability that neither well produces?

c. What is the probability that at least one of the two produces?

6. Of all failures of a certain type of computer hard drive, it is determined that in 20% of them only the sector containing the file allocation table is damaged, in 70% of them only nonessential sectors are damaged, and in 10% of the cases both the allocation sector and one or more nonessential sectors are damaged. A failed drive is selected at random and examined.

a. What is the probability that the allocation sector is damaged?

b. What is the probability that a nonessential sector is damaged?

c. If the drive is found to have a damaged allocation sector, what is the probability that some nonessential sectors are damaged as well?

d. If the drive is found to have a damaged nonessential sector, what is the probability that the allocation sector is damaged as well?

e. If the drive is found to have a damaged allocation sector, what is the probability that no nonessential sectors are damaged?

f. If the drive is found to have a damaged nonessential sector, what is the probability that the allocation sector is not damaged?

7. In the process of producing engine valves, the valves are subjected to a first grind. Valves whose thicknesses are within the specification are ready for installation. Those valves whose thicknesses are above the specification are reground, while those whose thicknesses are below the specification are scrapped. Assume that after the first grind, 70% of the valves meet the specification, 20% are reground, and 10% are scrapped. Furthermore, assume that of those valves that are reground, 90% meet the specification, and 10% are scrapped.

a. Find the probability that a valve is ground only once.

b. Given that a valve is ground only once, what is the probability that it is scrapped?

c. Find the probability that a valve is scrapped.

d. Given that a valve is scrapped, what is the probability that it was ground twice?

e. Find the probability that the valve meets the specification (after either the first or second grind).

f. Given that a valve meets the specification (after either the first or second grind), what is the probability that it was ground twice?

g. Given that a valve meets the specification, what is the probability that it was ground once?

8. Sarah and Thomas each roll a die. Whoever gets the higher number wins; if they both roll the same number, neither wins.

a. What is the probability that Thomas wins?

b. If Sarah rolls a 3, what is the probability that she wins?

c. If Sarah rolls a 3, what is the probability that Thomas wins?

d. If Sarah wins, what is the probability that Thomas rolled a 3?

e. If Sarah wins, what is the probability that Sarah rolled a 3?

9. A particular automatic sprinkler system has two different types of activation devices for each sprinkler head. One type has a reliability of 0.9; that is, the probability that it will activate the sprinkler when it should is 0.9. The other type, which operates independently of the first type, has a reliability of 0.8. If either device is triggered, the sprinkler will activate. Suppose a fire starts near a sprinkler head.

a. What is the probability that the sprinkler head will be activated?

b. What is the probability that the sprinkler head will not be activated?

c. What is the probability that both activation devices will work properly?

d. What is the probability that only the device with reliability 0.9 will work properly?

10. A fast-food restaurant chain has 600 outlets in the United States. The following table categorizes cities by size and location, and presents the number of restaurants in cities of each category. A restaurant is to be chosen at random from the 600 to test market a new style of chicken.

	Region			
Population of City	**NE**	**SE**	**SW**	**NW**
Under 50,000	30	35	15	5
50,000–500,000	60	90	70	30
Over 500,000	150	25	30	60

a. If the restaurant is located in a city with a population over 500,000, what is the probability that it is in the Northeast?

b. If the restaurant is located in the Southeast, what is the probability that it is in a city with a population under 50,000?

c. If the restaurant is located in the Southwest, what is the probability that it is in a city with a population of 500,000 or less?

d. If the restaurant is located in a city with a population of 500,000 or less, what is the probability that is in the Southwest?

e. If the restaurant is located in the South (either SE or SW), what is the probability that it is in a city with a population of 50,000 or more?

11. Nuclear power plants have redundant components in important systems to reduce the chance of catastrophic failure. Assume that a plant has two gauges to measure the level of coolant in the reactor core and that each gauge has probability 0.01 of failing. Assume that one potential cause of gauge failure is that the electric cables leading from the core to the control room where the gauges are located may burn up in a fire. Someone wishes to estimate the probability that both gauges fail, and makes the following calculation:

$$
\begin{aligned}
P(\text{both gauges fail}) &= P(\text{first gauge fails}) \times \\
&\quad P(\text{second gauge fails}) \\
&= (0.01)(0.01) \\
&= 0.0001
\end{aligned}
$$

   a. What assumption is being made in this calculation?
   b. Explain why this assumption is probably not justified in the present case.
   c. Is the probability of 0.0001 likely to be too high or too low? Explain.

12. Refer to Exercise 11. Is it possible for the probability that both gauges fail to be greater than 0.01? Explain.

13. A lot of 10 components contains 3 that are defective. Two components are drawn at random and tested. Let A be the event that the first component drawn is defective, and let B be the event that the second component drawn is defective.
   a. Find $P(A)$.
   b. Find $P(B|A)$.
   c. Find $P(A \cap B)$.
   d. Find $P(A^c \cap B)$.
   e. Find $P(B)$.
   f. Are A and B independent? Explain.

14. A lot of 1000 components contains 300 that are defective. Two components are drawn at random and tested. Let A be the event that the first component drawn is defective, and let B be the event that the second component drawn is defective.
   a. Find $P(A)$.
   b. Find $P(B|A)$.
   c. Find $P(A \cap B)$.
   d. Find $P(A^c \cap B)$.
   e. Find $P(B)$.
   f. Find $P(A|B)$.
   g. Are A and B independent? Is it reasonable to treat A and B as though they were independent? Explain.

15. In a lot of n components, 30% are defective. Two components are drawn at random and tested. Let A be the event that the first component drawn is defective, and let B be the event that the second component drawn is defective. For which lot size n will A and B be more nearly independent: $n = 10$ or $n = 10{,}000$? Explain.

16. Items are inspected for flaws by two quality inspectors. If a flaw is present, it will be detected by the first inspector with probability 0.9, and by the second inspector with probability 0.7. Assume the inspectors function independently.
   a. If an item has a flaw, what is the probability that it will be found by both inspectors?
   b. If an item has a flaw, what is the probability that it will be found by at least one of the two inspectors?
   c. Assume that the second inspector examines only those items that have been passed by the first inspector. If an item has a flaw, what is the probability that the second inspector will find it?

17. Refer to Exercise 16. Assume that both inspectors inspect every item and that if an item has no flaw, then neither inspector will detect a flaw.
   a. Assume that the probability that an item has a flaw is 0.10. If an item is passed by the first inspector, what is the probability that it actually has a flaw?
   b. Assume that the probability that an item has a flaw is 0.10. If an item is passed by both inspectors, what is the probability that it actually has a flaw?

18. A quality-control program at a plastic bottle production line involves inspecting finished bottles for flaws such as microscopic holes. The proportion of bottles that actually have such a flaw is only 0.0002. If a bottle has a flaw, the probability is 0.995 that it will fail the inspection. If a bottle does not have a flaw, the probability is 0.99 that it will pass the inspection.
   a. If a bottle fails inspection, what is the probability that it has a flaw?
   b. Which of the following is the more correct interpretation of the answer to part (a)?

i. Most bottles that fail inspection do not have a flaw.

ii. Most bottles that pass inspection do have a flaw.

c. If a bottle passes inspection, what is the probability that it does not have a flaw?

d. Which of the following is the more correct interpretation of the answer to part (c)?

i. Most bottles that fail inspection do have a flaw.

ii. Most bottles that pass inspection do not have a flaw.

e. Explain why a small probability in part (a) is not a problem, so long as the probability in part (c) is large.

19. Refer to Example 2.25 (p. 80).

a. If a man tests negative, what is the probability that he actually has the disease?

b. For many medical tests, it is standard procedure to repeat the test when a positive signal is given. If repeated tests are independent, what is the probability that a man will test positive on two successive tests if he has the disease?

c. Assuming repeated tests are independent, what is the probability that a man tests positive on two successive tests if he does not have the disease?

d. If a man tests positive on two successive tests, what is the probability that he has the disease?

20. A system contains two components, A and B, connected in series, as shown in the diagram.

Assume A and B function independently. For the system to function, both components must function.

a. If the probability that A fails is 0.05, and the probability that B fails is 0.03, find the probability that the system functions.

b. If both A and B have probability p of failing, what must the value of p be so that the probability that the system functions is 0.90?

c. If three components are connected in series, and each has probability p of failing, what must the value of p be so that the probability that the system functions is 0.90?

21. A system contains two components, C and D, connected in parallel as shown in the diagram.

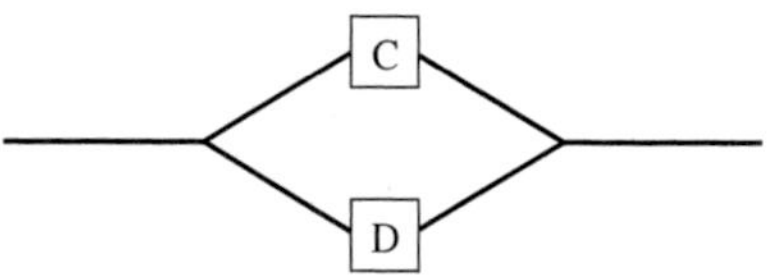

Assume C and D function independently. For the system to function, either C or D must function.

a. If the probability that C fails is 0.08 and the probability that D fails is 0.12, find the probability that the system functions.

b. If both C and D have probability p of failing, what must the value of p be so that the probability that the system functions is 0.99?

c. If three components are connected in parallel, function independently, and each has probability p of failing, what must the value of p be so that the probability that the system functions is 0.99?

d. If components function independently, and each component has probability 0.5 of failing, what is the minimum number of components that must be connected in parallel so that the probability that the system functions is at least 0.99?

22. A system consists of four components connected as shown in the following diagram.

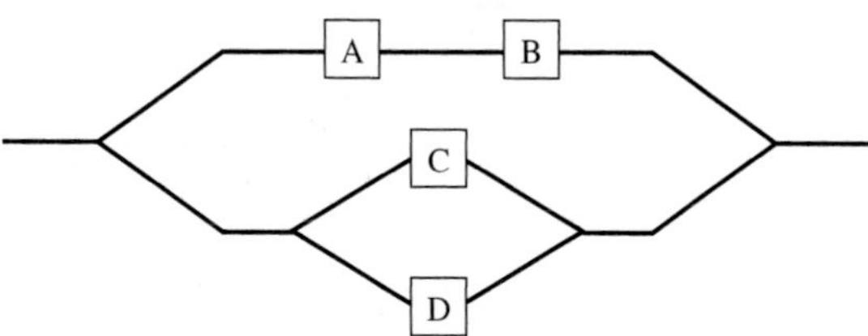

Assume A, B, C, and D function independently. If the probabilities that A, B, C, and D fail are 0.10, 0.05, 0.10, and 0.20, respectively, what is the probability that the system functions?

23. If A and B are independent events, prove that the following pairs of events are independent: A^c and B, A and B^c, and A^c and B^c.

2.4 Random Variables

In many situations, it is desirable to assign a numerical value to each outcome of an experiment. Such an assignment is called a **random variable**. To make the idea clear, we present an example. Suppose that an electrical engineer has on hand six resistors. Three of them are labeled 10 Ω and the other three are labeled 20 Ω. The engineer wants to connect a 10 Ω resistor and a 20 Ω resistor in series, to create a resistance of 30 Ω. Now suppose that in fact the three resistors labeled 10 Ω have actual resistances of 9, 10, and 11 Ω, and that the three resistors labeled 20 Ω have actual resistances of 19, 20, and 21 Ω. The process of selecting one resistor of each type is an experiment whose sample space consists of nine equally likely outcomes. The sample space is presented in the following table.

Outcome	Probability
(9, 19)	1/9
(9, 20)	1/9
(9, 21)	1/9
(10, 19)	1/9
(10, 20)	1/9
(10, 21)	1/9
(11, 19)	1/9
(11, 20)	1/9
(11, 21)	1/9

Now what is important to the engineer in this experiment is the sum of the two resistances, rather than their individual values. Therefore we assign to each outcome a number equal to the sum of the two resistances selected. This assignment, represented by the letter X, is presented in the following table.

Outcome	X	Probability
(9, 19)	28	1/9
(9, 20)	29	1/9
(9, 21)	30	1/9
(10, 19)	29	1/9
(10, 20)	30	1/9
(10, 21)	31	1/9
(11, 19)	30	1/9
(11, 20)	31	1/9
(11, 21)	32	1/9

The function X, which assigns a numerical value to each outcome in the sample space, is a random variable.

> A **random variable** assigns a numerical value to each outcome in a sample space.

It is customary to denote random variables with uppercase letters. The letters X, Y, and Z are most often used.

We can compute probabilities for random variables in an obvious way. In the example just presented, the event $X = 29$ corresponds to the event $\{(9, 20), (10, 19)\}$ of the sample space. Therefore $P(X = 29) = P(\{(9, 20), (10, 19)\}) = 2/9$.

Example 2.30

List the possible values of the random variable X, and find the probability of each of them.

Solution

The possible values are 28, 29, 30, 31, and 32. To find the probability of one of these values, we add the probabilities of the outcomes in the sample space that correspond to the value. The results are given in the following table.

x	$P(X = x)$
28	1/9
29	2/9
30	3/9
31	2/9
32	1/9

The table of probabilities in Example 2.30 contains all the information needed to compute any probability regarding the random variable X. Note that the outcomes of the sample space are not presented in the table. When the probabilities pertaining to a random variable are known, we usually do not think about the sample space; we just focus on the probabilities.

There are two important types of random variables, **discrete** and **continuous**. A discrete random variable is one whose possible values form a discrete set; in other words, the values can be ordered, and there are gaps between adjacent values. The random variable X, just described, is discrete. In contrast, the possible values of a continuous random variable always contain an interval, that is, all the points between some two numbers. We will provide precise definitions of these types of random variables later in this section.

We present some more examples of random variables.

Example 2.31

Computer chips often contain surface imperfections. For a certain type of computer chip, 9% contain no imperfections, 22% contain 1 imperfection, 26% contain 2 imperfections, 20% contain 3 imperfections, 12% contain 4 imperfections, and the remaining

11% contain 5 imperfections. Let Y represent the number of imperfections in a randomly chosen chip. What are the possible values for Y? Is Y discrete or continuous? Find $P(Y = y)$ for each possible value y.

Solution

The possible values for Y are the integers 0, 1, 2, 3, 4, and 5. The random variable Y is discrete, because it takes on only integer values. Nine percent of the outcomes in the sample space are assigned the value 0. Therefore $P(Y = 0) = 0.09$. Similarly $P(Y = 1) = 0.22$, $P(Y = 2) = 0.26$, $P(Y = 3) = 0.20$, $P(Y = 4) = 0.12$, and $P(Y = 5) = 0.11$.

Example 2.32

A certain type of magnetic disk must function in an environment where it is exposed to corrosive gases. It is known that 10% of all such disks have lifetimes less than or equal to 100 hours, 50% have lifetimes greater than 100 hours but less than or equal to 500 hours, and 40% have lifetimes greater than 500 hours. Let Z represent the number of hours in the lifetime of a randomly chosen disk. Is Z continuous or discrete? Find $P(Z \leq 500)$. Can we compute all the probabilities for Z? Explain.

Solution

The lifetime of a component is not limited to a list of discretely spaced values; Z is continuous. Of all the components, 60% have lifetimes less than or equal to 500 hours. Therefore $P(Z \leq 500) = 0.60$. We do not have enough information to compute all the probabilities for Z. We can compute some of them, for example, $P(Z \leq 100) = 0.10$, $P(100 < Z \leq 500) = 0.50$, and $P(Z > 500) = 0.40$. But we do not know, for example, the proportion of components that have lifetimes between 100 and 200 hours, or between 200 and 300 hours, so we cannot find the probability that the random variable Z falls into either of these intervals. To compute all the probabilities for Z, we would need to be able to compute the probability for every possible interval, for example, $P(200 < Z \leq 300)$, $P(200 < Z \leq 201)$, $P(200 < Z \leq 200.1)$, and so on. We will see how this can be done later in this section, when we discuss continuous random variables.

Random Variables and Populations

It is often useful to think of a value of a random variable as having been sampled from a population. For example, consider the random variable Y described in Example 2.31. Observing a value for this random variable is like sampling a value from a population consisting of the integers 0, 1, 2, 3, 4, and 5 in the following proportions: 0s, 9%; 1s, 22%; 2s, 26%; 3s, 20%; 4s, 12%; and 5s, 11%. For a continuous random variable, it is appropriate to imagine an infinite population containing all the possible values of the random variable. For example, for the random variable Z in Example 2.32 we would imagine a population containing all the positive numbers, with 10% of the population values less than or equal to 100, 50% greater than 100 but less than or equal to 500, and 40% greater than 500. The proportion of population values in any interval would be equal to the probability that the variable Z is in that interval.

Methods for working with random variables differ somewhat between discrete and continuous random variables. We begin with the discrete case.

Discrete Random Variables

We begin by reviewing the definition of a discrete random variable.

Definition

A random variable is **discrete** if its possible values form a discrete set. This means that if the possible values are arranged in order, there is a gap between each value and the next one. The set of possible values may be infinite; for example, the set of all integers or the set of all positive integers.

It is common for the possible values of a discrete random variable to be a set of integers. For any discrete random variable, if we specify the list of its possible values along with the probability that the random variable takes on each of these values, then we have completely described the population from which the random variable is sampled. We illustrate with an example.

The number of flaws in a 1-inch length of copper wire manufactured by a certain process varies from wire to wire. Overall, 48% of the wires produced have no flaws, 39% have one flaw, 12% have two flaws, and 1% have three flaws. Let X be the number of flaws in a randomly selected piece of wire. Then

$$P(X = 0) = 0.48 \qquad P(X = 1) = 0.39 \qquad P(X = 2) = 0.12 \qquad P(X = 3) = 0.01$$

The list of possible values 0, 1, 2, 3, along with the probabilities for each, provide a complete description of the population from which X is drawn. This description has a name—the **probability mass function**.

Definition

The **probability mass function** of a discrete random variable X is the function $p(x) = P(X = x)$. The probability mass function is sometimes called the **probability distribution**.

Thus for the random variable X representing the number of flaws in a length of wire, $p(0) = 0.48$, $p(1) = 0.39$, $p(2) = 0.12$, $p(3) = 0.01$, and $p(x) = 0$ for any value of x other than 0, 1, 2, or 3. Note that if the values of the probability mass function are added over all the possible values of X, the sum is equal to 1. This is true for any probability mass function. The reason is that summing the values of a probability mass function over all the possible values of the corresponding random variable produces the probability that the random variable is equal to one of its possible values, and this probability is always equal to 1.

The probability mass function can be represented by a graph in which a vertical line is drawn at each of the possible values of the random variable. The heights of the lines are equal to the probabilities of the corresponding values. The physical interpretation of this graph is that each line represents a mass equal to its height. Figure 2.7 presents a graph of the probability mass function of the random variable X.

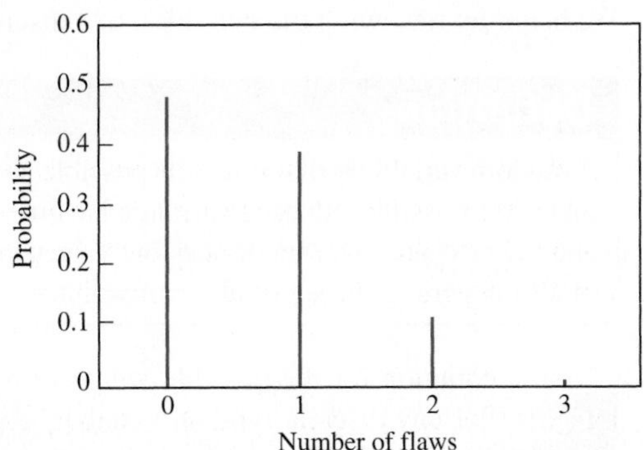

FIGURE 2.7 Probability mass function of X, the number of flaws in a randomly chosen piece of wire.

The Cumulative Distribution Function of a Discrete Random Variable

The probability mass function specifies the probability that a random variable is equal to a given value. A function called the **cumulative distribution function** specifies the probability that a random variable is less than or equal to a given value. The cumulative distribution function of the random variable X is the function $F(x) = P(X \leq x)$.

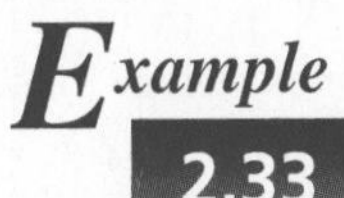

Let $F(x)$ denote the cumulative distribution function of the random variable X that represents the number of flaws in a randomly chosen wire. Find $F(2)$. Find $F(1.5)$.

Solution

Since $F(2) = P(X \leq 2)$, we need to find $P(X \leq 2)$. We do this by summing the probabilities for the values of X that are less than or equal to 2, namely, 0, 1, and 2. Thus

$$\begin{aligned} F(2) &= P(X \leq 2) \\ &= P(X = 0) + P(X = 1) + P(X = 2) \\ &= 0.48 + 0.39 + 0.12 \\ &= 0.99 \end{aligned}$$

Now $F(1.5) = P(X \leq 1.5)$. Therefore, to compute $F(1.5)$ we must sum the probabilities for the values of X that are less than or equal to 1.5, which are 0 and 1. Thus

$$\begin{aligned} F(1.5) &= P(X \le 1.5) \\ &= P(X = 0) + P(X = 1) \\ &= 0.48 + 0.39 \\ &= 0.87 \end{aligned}$$

In general, for any discrete random variable X, the cumulative distribution function $F(x)$ can be computed by summing the probabilities of all the possible values of X that are less than or equal to x. Note that $F(x)$ is defined for any number x, not just for the possible values of X.

Summary

Let X be a discrete random variable. Then

- The probability mass function of X is the function $p(x) = P(X = x)$.
- The cumulative distribution function of X is the function $F(x) = P(X \le x)$.
- $F(x) = \sum_{t \le x} p(t) = \sum_{t \le x} P(X = t)$.
- $\sum_x p(x) = \sum_x P(X = x) = 1$, where the sum is over all the possible values of X.

Plot the cumulative distribution function $F(x)$ of the random variable X that represents the number of flaws in a randomly chosen wire.

Solution

First we compute $F(x)$ for each of the possible values of X, which are 0, 1, 2, and 3.

$$\begin{aligned} F(0) &= P(X \le 0) = 0.48 \\ F(1) &= P(X \le 1) = 0.48 + 0.39 = 0.87 \\ F(2) &= P(X \le 2) = 0.48 + 0.39 + 0.12 = 0.99 \\ F(3) &= P(X \le 3) = 0.48 + 0.39 + 0.12 + 0.01 = 1 \end{aligned}$$

For any value x, we compute $F(x)$ by summing the probabilities of all the possible values of X that are less than or equal to x. For example, if $1 \le x < 2$, the possible values of X that are less than or equal to x are 0 and 1, so $F(x) = P(X = 0) + P(X = 1) = F(1) = 0.87$. Therefore

$$F(x) = \begin{cases} 0 & x < 0 \\ 0.48 & 0 \le x < 1 \\ 0.87 & 1 \le x < 2 \\ 0.99 & 2 \le x < 3 \\ 1 & x \ge 3 \end{cases}$$

A plot of $F(x)$ is presented in the following figure.

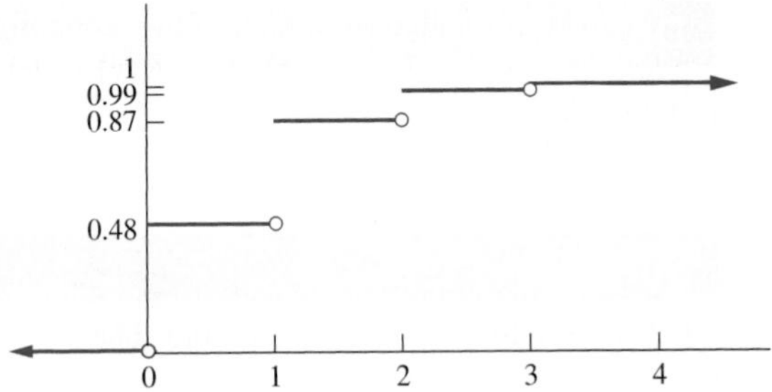

For a discrete random variable, the graph of $F(x)$ consists of a series of horizontal lines (called "steps") with jumps at each of the possible values of X. Note that the size of the jump at any point x is equal to the value of the probability mass function $p(x) = P(X = x)$.

Mean and Variance for Discrete Random Variables

We saw in Section 1.3 that the mean of a sample is approximately equal to the horizontal component of the center of mass of a sample histogram, which is the point on the x axis where the histogram would balance if supported there. By analogy, we define the **population mean** of a discrete random variable to be the horizontal component of the center of mass of the graph of its probability mass function. The population mean of a random variable X may also be called the **expectation**, or **expected value**, of X and can be denoted by μ_X, by $E(X)$, or simply by μ. Sometimes we will drop the word "population," and simply refer to the population mean as the mean.

Find the mean of the random variable X representing the number of flaws in a randomly chosen piece of wire.

Solution

The mean is the center of mass of the graph of the probability mass function (Figure 2.7). The center of mass is computed by multiplying the height of each line by its value on the horizontal axis, and then adding the products. The values are 0, 1, 2, and 3. The heights are $P(X = 0) = 0.48$, $P(X = 1) = 0.39$, $P(X = 2) = 0.12$, and $P(X = 3) = 0.01$. The mean is therefore

$$\begin{aligned}\mu_X &= 0 \cdot P(X = 0) + 1 \cdot P(X = 1) + 2 \cdot P(X = 2) + 3 \cdot P(X = 3) \\ &= (0)(0.48) + (1)(0.39) + (2)(0.12) + (3)(0.01) \\ &= 0.6600\end{aligned}$$

In general, the mean of a discrete random variable is found by multiplying each possible value of the random variable by its probability, and then summing.

Definition

Let X be a discrete random variable with probability mass function $p(x) = P(X = x)$.

The mean of X is given by

$$\mu_X = \sum_x x P(X = x) \tag{2.29}$$

where the sum is over all possible values of X.

The mean of X is sometimes called the expectation, or expected value, of X and may also be denoted by $E(X)$ or by μ.

In Section 1.3, we showed that the variance of a sample is approximately equal to the moment of inertia of the sample histogram around the sample mean. We define the **population variance** of a discrete random variable to be the moment of inertia of the graph of its probability mass function around the population mean μ. The population variance of a random variable X is often referred to simply as the **variance** of X. It can be denoted by σ_X^2, by $V(X)$, or simply by σ^2. To compute it, multiply the height of each line in the graph of the probability mass function by the square of its horizontal distance from the population mean, and then add the products. It's easier to understand this when it is presented in a formula:

$$\sigma_X^2 = \sum_x (x - \mu_X)^2 P(X = x)$$

By performing some algebra, an alternate formula can be obtained.

$$\sigma_X^2 = \sum_x x^2 P(X = x) - \mu_X^2$$

A derivation of this formula is given at the end of this section.

We also define the **population standard deviation** to be the square root of the population variance. We denote the population standard deviation of a random variable X by σ_X or simply by σ. As with the mean, we will sometimes drop the word "population," and simply refer to the population variance and population standard deviation as the variance and standard deviation, respectively.

Summary

Let X be a discrete random variable with probability mass function $p(x) = P(X = x)$. Then

- The variance of X is given by

$$\sigma_X^2 = \sum_x (x - \mu_X)^2 P(X = x) \tag{2.30}$$

- An alternate formula for the variance is given by

$$\sigma_X^2 = \sum_x x^2 P(X = x) - \mu_X^2 \tag{2.31}$$

- The variance of X may also be denoted by $V(X)$ or by σ^2.
- The standard deviation is the square root of the variance: $\sigma_X = \sqrt{\sigma_X^2}$.

Example 2.36

Find the variance and standard deviation for the random variable X representing the number of flaws in a randomly chosen piece of wire.

Solution

In Example 2.35, we computed the mean of X to be $\mu_X = 0.6600$. We compute the variance by using Equation (2.30):

$$\begin{aligned}\sigma_X^2 &= (0 - 0.6600)^2 P(X = 0) + (1 - 0.6600)^2 P(X = 1) + (2 - 0.6600)^2 P(X = 2) \\ &\quad + (3 - 0.6600)^2 P(X = 3) \\ &= (0.4356)(0.48) + (0.1156)(0.39) + (1.7956)(0.12) + (5.4756)(0.01) \\ &= 0.5244\end{aligned}$$

The standard deviation is $\sigma_X = \sqrt{0.5244} = 0.724$.

Example 2.37

Use the alternate formula, Equation (2.31), to compute the variance of X, the number of flaws in a randomly chosen wire.

Solution

In Example 2.35, the mean was computed to be $\mu_X = 0.6600$. The variance is therefore

$$\begin{aligned}\sigma_X^2 &= 0^2 P(X = 0) + 1^2 P(X = 1) + 2^2 P(X = 2) + 3^2 P(X = 3) - (0.6600)^2 \\ &= (0)(0.48) + (1)(0.39) + (4)(0.12) + (9)(0.01) - (0.6600)^2 \\ &= 0.5244\end{aligned}$$

Example 2.38

A resistor in a certain circuit is specified to have a resistance in the range 99 Ω–101 Ω. An engineer obtains two resistors. The probability that both of them meet the specification is 0.36, the probability that exactly one of them meets the specification is 0.48, and the probability that neither of them meets the specification is 0.16. Let X represent the number of resistors that meet the specification. Find the probability mass function, and the mean, variance, and standard deviation of X.

Solution

The probability mass function is $P(X = 0) = 0.16$, $P(X = 1) = 0.48$, $P(X = 2) = 0.36$, and $P(X = x) = 0$ for $x \neq 0, 1$, or 2. The mean is

$$\begin{aligned}\mu_X &= (0)(0.16) + (1)(0.48) + (2)(0.36) \\ &= 1.200\end{aligned}$$

The variance is

$$\begin{aligned}\sigma_X^2 &= (0 - 1.200)^2(0.16) + (1 - 1.200)^2(0.48) + (2 - 1.200)^2(0.36) \\ &= 0.4800\end{aligned}$$

The standard deviation is $\sigma_X = \sqrt{0.4800} = 0.693$.

The Probability Histogram

When the possible values of a discrete random variable are evenly spaced, the probability mass function can be represented by a histogram, with rectangles centered at the possible values of the random variable. The area of a rectangle centered at a value x is equal to $P(X = x)$. Such a histogram is called a **probability histogram**, because the areas represent probabilities. Figure 2.7 presented the graph of the probability mass function of a random variable X representing the number of flaws in a wire. Figure 2.8 (page 98) presents a probability histogram for this random variable.

The probability that the value of a random variable falls into a given interval is given by an area under the probability histogram. Example 2.39 illustrates the idea.

Example 2.39

Find the probability that a randomly chosen wire has more than one flaw. Indicate this probability as an area under the probability histogram.

Solution

We wish to find $P(X > 1)$. Since no wire has more than three flaws, the proportion of wires that have more than one flaw can be found by adding the proportion that have two flaws to the proportion that have three flaws. In symbols, $P(X > 1) = P(X = 2) + P(X = 3)$. The probability mass function specifies that $P(X = 2) = 0.12$ and $P(X = 3) = 0.01$. Therefore $P(X > 1) = 0.12 + 0.01 = 0.13$.

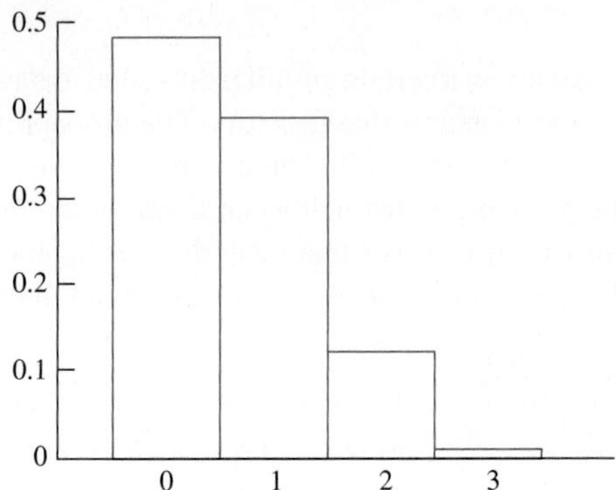

FIGURE 2.8 Probability histogram for X, the number of flaws in a randomly chosen piece of wire. Compare with Figure 2.7.

This probability is given by the area under the probability histogram corresponding to those rectangles centered at values greater than 1 (see Figure 2.9). There are two such rectangles; their areas are $P(X = 2) = 0.12$ and $P(X = 3) = 0.01$. This is another way to show that $P(X > 1) = 0.12 + 0.01 = 0.13$.

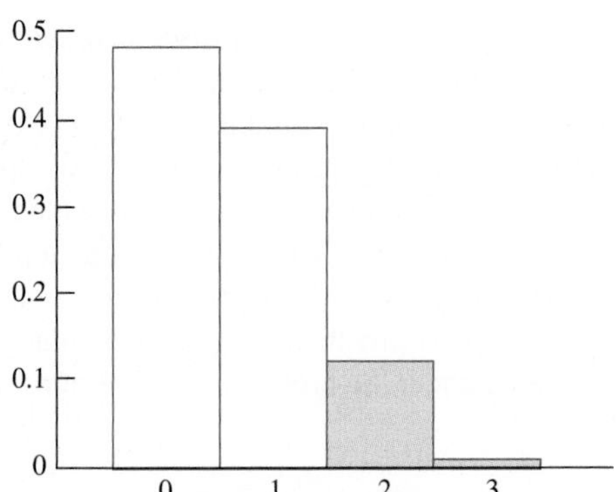

FIGURE 2.9 Probability histogram for X, the number of flaws in a randomly chosen piece of wire. The area corresponding to values of X greater than 1 is shaded. This area is equal to $P(X > 1)$.

In Chapter 4, we will see that probabilities for discrete random variables can sometimes be approximated by computing the area under a curve. Representing the discrete probabilities with a probability histogram will make it easier to understand how this is done.

Continuous Random Variables

Table 1.4 (in Section 1.3) presented class intervals for the emissions, in grams of particulates per gallon of fuel consumed, of a sample of 62 vehicles. Note that emissions is a continuous variable, because its possible values are not restricted to some discretely spaced set. The class intervals are chosen so that each interval contains a reasonably large number of vehicles. If the sample were larger, we could make the intervals narrower. In particular, if we had information on the entire population, containing millions of vehicles, we could make the intervals extremely narrow. The histogram would then look quite smooth and could be approximated with a curve, which might look like Figure 2.10.

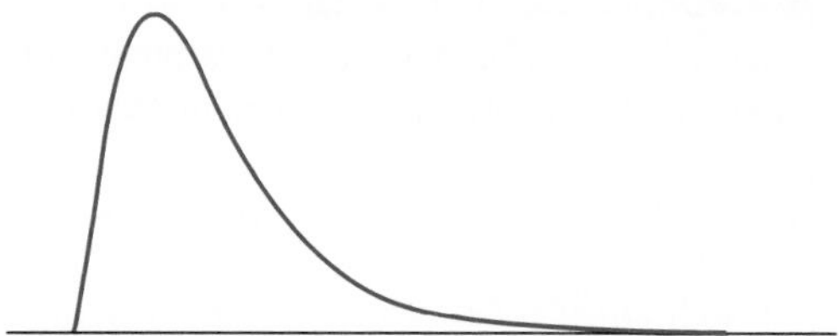

FIGURE 2.10 The histogram for a large continuous population could be drawn with extremely narrow rectangles and might look like this curve.

If a vehicle were chosen at random from this population to have its emissions measured, the emissions level X would be a random variable. The probability that X falls between any two values a and b is equal to the area under the histogram between a and b. Because the histogram in this case is represented by a curve, the probability would be found by computing an integral.

The random variable X described here is an example of a **continuous random variable**. A continuous random variable is defined to be a random variable whose probabilities are represented by areas under a curve. This curve is called the **probability density function**. Because the probability density function is a curve, the computations of probabilities involve integrals, rather than the sums that are used in the discrete case.

Definition

A random variable is **continuous** if its probabilities are given by areas under a curve. The curve is called a **probability density function** for the random variable.

The probability density function is sometimes called the **probability distribution**.

Computing Probabilities with the Probability Density Function

Let X be a continuous random variable. Let the function $f(x)$ be the probability density function of X. Let a and b be any two numbers, with $a < b$.

The proportion of the population whose values of X lie between a and b is given by $\int_a^b f(x)\,dx$, the area under the probability density function between a and b. This is the probability that the random variable X takes on a value between a and b. Note that the area under the curve does not depend on whether the endpoints a and b are included in the interval. Therefore probabilities involving X do not depend on whether endpoints are included.

Summary

Let X be a continuous random variable with probability density function $f(x)$. Let a and b be any two numbers, with $a < b$. Then

$$P(a \le X \le b) = P(a \le X < b) = P(a < X \le b) = P(a < X < b) = \int_a^b f(x)\,dx$$

In addition,

$$P(X \le a) = P(X < a) = \int_{-\infty}^{a} f(x)\,dx \tag{2.32}$$

$$P(X \ge a) = P(X > a) = \int_{a}^{\infty} f(x)\,dx \tag{2.33}$$

If $f(x)$ is the probability density function of a random variable X, then the area under the entire curve from $-\infty$ to ∞ is the probability that the value of X is between $-\infty$ and ∞. This probability must be equal to 1, because the value of X is always between $-\infty$ and ∞. Therefore the area under the entire curve $f(x)$ is equal to 1.

Summary

Let X be a continuous random variable with probability density function $f(x)$. Then

$$\int_{-\infty}^{\infty} f(x)\,dx = 1$$

Example 2.40

A hole is drilled in a sheet-metal component, and then a shaft is inserted through the hole. The shaft clearance is equal to the difference between the radius of the hole and the radius of the shaft. Let the random variable X denote the clearance, in millimeters. The probability density function of X is

$$f(x) = \begin{cases} 1.25(1 - x^4) & 0 < x < 1 \\ 0 & \text{otherwise} \end{cases}$$

Components with clearances larger than 0.8 mm must be scrapped. What proportion of components are scrapped?

Solution

Figure 2.11 presents the probability density function of X. Note that the density $f(x)$ is 0 for $x \leq 0$ and for $x \geq 1$. This indicates that the clearances are always between 0 and 1 mm. The proportion of components that must be scrapped is $P(X > 0.8)$, which is equal to the area under the probability density function to the right of 0.8.

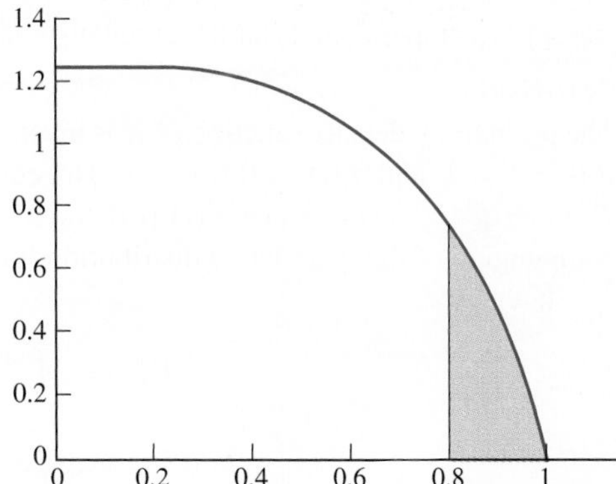

FIGURE 2.11 Graph of the probability density function of X, the clearance of a shaft. The area shaded is equal to $P(X > 0.8)$.

This area is given by

$$\begin{aligned} P(X > 0.8) &= \int_{0.8}^{\infty} f(x)\,dx \\ &= \int_{0.8}^{1} 1.25(1 - x^4)\,dx \\ &= 1.25\left(x - \frac{x^5}{5}\right)\Bigg|_{0.8}^{1} \\ &= 0.0819 \end{aligned}$$

The Cumulative Distribution Function of a Continuous Random Variable

The cumulative distribution function of a continuous random variable X is $F(x) = P(X \leq x)$, just as it is for a discrete random variable. For a discrete random variable, $F(x)$ can be found by summing values of the probability mass function. For a continuous random variable, the value of $F(x)$ is obtained by integrating the probability density function. Since $F(x) = P(X \leq x)$, it follows from Equation (2.32) that $F(x) = \int_{-\infty}^{x} f(t)\,dt$, where $f(t)$ is the probability density function.

Definition

Let X be a continuous random variable with probability density function $f(x)$. The cumulative distribution function of X is the function

$$F(x) = P(X \leq x) = \int_{-\infty}^{x} f(t)\,dt \tag{2.34}$$

Example 2.41

Refer to Example 2.40. Find the cumulative distribution function $F(x)$ and plot it.

Solution

The probability density function of X is given by $f(t) = 0$ if $t \leq 0$, $f(t) = 1.25(1 - t^4)$ if $0 < t < 1$, and $f(t) = 0$ if $t \geq 1$. The cumulative distribution function is given by $F(x) = \int_{-\infty}^{x} f(t)\,dt$. Since $f(t)$ is defined separately on three different intervals, the computation of the cumulative distribution function involves three separate cases.

If $x \leq 0$:

$$\begin{aligned} F(x) &= \int_{-\infty}^{x} f(t)\,dt \\ &= \int_{-\infty}^{x} 0\,dt \\ &= 0 \end{aligned}$$

If $0 < x < 1$:

$$\begin{aligned} F(x) &= \int_{-\infty}^{x} f(t)\,dt \\ &= \int_{-\infty}^{0} f(t)\,dt + \int_{0}^{x} f(t)\,dt \\ &= \int_{-\infty}^{0} 0\,dt + \int_{0}^{x} 1.25(1 - t^4)\,dt \\ &= 0 + 1.25\left(t - \frac{t^5}{5}\right)\Bigg|_{0}^{x} \\ &= 1.25\left(x - \frac{x^5}{5}\right) \end{aligned}$$

If $x > 1$:

$$\begin{aligned} F(x) &= \int_{-\infty}^{x} f(t)\,dt \\ &= \int_{-\infty}^{0} f(t)\,dt + \int_{0}^{1} f(t)\,dt + \int_{1}^{x} f(t)\,dt \end{aligned}$$

$$
\begin{aligned}
&= \int_{-\infty}^{0} 0\,dt + \int_{0}^{1} 1.25(1 - t^4)\,dt + \int_{1}^{x} 0\,dt \\
&= 0 + 1.25\left(t - \frac{t^5}{5}\right)\Bigg|_0^1 + 0 \\
&= 0 + 1 + 0 \\
&= 1
\end{aligned}
$$

Therefore

$$
F(x) = \begin{cases} 0 & x \le 0 \\ 1.25\left(x - \dfrac{x^5}{5}\right) & 0 < x < 1 \\ 1 & x \ge 1 \end{cases}
$$

A plot of $F(x)$ is presented here.

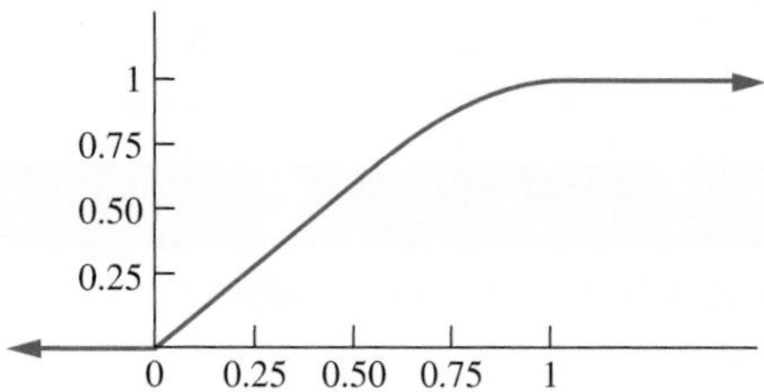

Note that the cumulative distribution function $F(x)$ in Example 2.41 is a continuous function—there are no jumps in its graph. This is characteristic of continuous random variables. The cumulative distribution function of a continuous random variable will always be continuous, while the cumulative distribution function of a noncontinuous random variable will never be continuous.

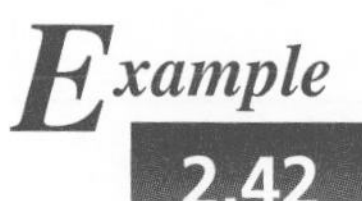

Refer to Example 2.40. Use the cumulative distribution function to find the probability that the shaft clearance is less than 0.5 mm.

Solution

Let X denote the shaft clearance. We need to find $P(X \le 0.5)$. This is equivalent to finding $F(0.5)$, where $F(x)$ is the cumulative distribution function. Using the results of Example 2.41, $F(0.5) = 1.25(0.5 - 0.5^5/5) = 0.617$.

Mean and Variance for Continuous Random Variables

The population mean and variance of a continuous random variable are defined in the same way as those of a discrete random variable, except that the probability density function is used instead of the probability mass function. Specifically, if X is a continuous random variable, its population mean is defined to be the center of mass of its probability density function, and its population variance is the moment of inertia around a vertical axis through the center of mass. The formulas are analogous to Equations (2.29) through (2.31), with the sums replaced by integrals.

As was the case with discrete random variables, we will sometimes drop the word "population" and refer to the population mean, population variance, and population standard deviation more simply as the mean, variance, and standard deviation, respectively.

Definition

Let X be a continuous random variable with probability density function $f(x)$. Then the mean of X is given by

$$\mu_X = \int_{-\infty}^{\infty} x f(x)\, dx \tag{2.35}$$

The mean of X is sometimes called the expectation, or expected value, of X and may also be denoted by $E(X)$ or by μ.

Definition

Let X be a continuous random variable with probability density function $f(x)$. Then

- The variance of X is given by

$$\sigma_X^2 = \int_{-\infty}^{\infty} (x - \mu_X)^2 f(x)\, dx \tag{2.36}$$

- An alternate formula for the variance is given by

$$\sigma_X^2 = \int_{-\infty}^{\infty} x^2 f(x)\, dx - \mu_X^2 \tag{2.37}$$

- The variance of X may also be denoted by $V(X)$ or by σ^2.
- The standard deviation is the square root of the variance: $\sigma_X = \sqrt{\sigma_X^2}$.

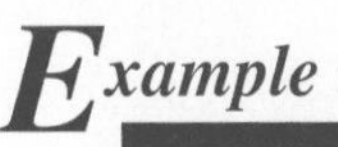

Example 2.43

Refer to Example 2.40. Find the mean clearance and the variance of the clearance.

Solution

Using Equation (2.35), the mean clearance is given by

$$\begin{aligned}\mu_X &= \int_{-\infty}^{\infty} x f(x)\, dx \\ &= \int_0^1 x[1.25(1 - x^4)]\, dx\end{aligned}$$

$$= 1.25\left(\frac{x^2}{2} - \frac{x^6}{6}\right)\Bigg|_0^1$$
$$= 0.4167$$

Having computed $\mu_X = 0.4167$, we can now compute σ_X^2. It is easiest to use the alternate formula, Equation (2.37):

$$\sigma_X^2 = \int_{-\infty}^{\infty} x^2 f(x)\,dx - \mu_X^2$$
$$= \int_0^1 x^2[1.25(1 - x^4)]\,dx - (0.4167)^2$$
$$= 1.25\left(\frac{x^3}{3} - \frac{x^7}{7}\right)\Bigg|_0^1 - (0.4167)^2$$
$$= 0.0645$$

The Population Median and Percentiles

In Section 1.2, we defined the median of a sample to be the middle number, or the average of the two middle numbers, when the sample values are arranged from smallest to largest. Intuitively, the sample median is the point that divides the sample in half. The population median is defined analogously. In terms of the probability density function, the median is the point at which half the area under the curve is to the left, and half the area is to the right. Thus if X is a continuous random variable with probability density function $f(x)$, the median of X is the point x_m that solves the equation $P(X \le x_m) = \int_{-\infty}^{x_m} f(x)\,dx = 0.5$.

The median is a special case of a **percentile**. Let $0 < p < 100$. The pth percentile of a population is the value x_p such that $p\%$ of the population values are less than or equal to x_p. Thus if X is a continuous random variable with probability density function $f(x)$, the pth percentile of X is the point x_p that solves the equation $P(X \le x_p) = \int_{-\infty}^{x_p} f(x)\,dx = p/100$. Note that the median is the 50th percentile. Figure 2.12 illustrates the median and the 90th percentile for a hypothetical population.

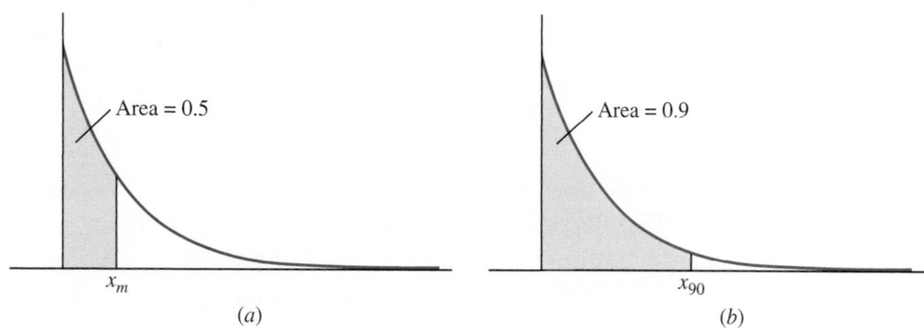

FIGURE 2.12 *(a)* Half of the population values are less than the median x_m. *(b)* Ninety percent of the population values are less than the 90th percentile x_{90}.

Definition

Let X be a continuous random variable with probability mass function $f(x)$ and cumulative distribution function $F(x)$.

- The median of X is the point x_m that solves the equation $F(x_m) = P(X \le x_m) = \int_{-\infty}^{x_m} f(x)\,dx = 0.5$.
- If p is any number between 0 and 100, the pth percentile is the point x_p that solves the equation $F(x_p) = P(X \le x_p) = \int_{-\infty}^{x_p} f(x)\,dx = p/100$.
- The median is the 50th percentile.

We note that it is possible to construct continuous random variables for which there is an interval of points that satisfy the definition of the median or a percentile. Such random variables are seldom found in practice.

A certain radioactive mass emits alpha particles from time to time. The time between emissions, in seconds, is random, with probability density function

$$f(x) = \begin{cases} 0.1e^{-0.1x} & x > 0 \\ 0 & x \le 0 \end{cases}$$

Find the median time between emissions. Find the 60th percentile of the times.

Solution

The median x_m is the solution to $\int_{-\infty}^{x_m} f(x)\,dx = 0.5$. We therefore must solve

$$\begin{aligned}
\int_0^{x_m} 0.1e^{-0.1x}\,dx &= 0.5 \\
-e^{-0.1x}\Big|_0^{x_m} &= 0.5 \\
1 - e^{-0.1x_m} &= 0.5 \\
e^{-0.1x_m} &= 0.5 \\
-0.1x_m &= \ln 0.5 \\
0.1x_m &= 0.6931 \\
x_m &= 6.931 \text{ s}
\end{aligned}$$

Half of the times between emissions are less than 6.931 s, and half are greater.

The 60th percentile x_{60} is the solution to $\int_{-\infty}^{x_{60}} f(x)\,dx = 0.6$. We proceed as before, substituting x_{60} for x_m, and 0.6 for 0.5. We obtain

$$\begin{aligned}
1 - e^{-0.1x_{60}} &= 0.6 \\
e^{-0.1x_{60}} &= 0.4 \\
-0.1x_{60} &= \ln 0.4
\end{aligned}$$

$$0.1x_{60} = 0.9163$$
$$x_{60} = 9.163 \text{ s}$$

Sixty percent of the times between emissions are less than 9.163 s, and 40% are greater.

Derivation of the Alternate Formula for the Variance

We derive Equation (2.31). We begin with Equation (2.30).

$$\sigma_X^2 = \sum_x (x - \mu_X)^2 P(X = x)$$

Multiplying out $(x - \mu_X)^2$, we obtain

$$\sigma_X^2 = \sum_x (x^2 - 2x\mu_X + \mu_X^2) P(X = x)$$

Distributing the term $P(X = x)$ over the terms in the parentheses yields

$$\sigma_X^2 = \sum_x [x^2 P(X = x) - 2x\mu_X P(X = x) + \mu_X^2 P(X = x)]$$

Summing the terms separately,

$$\sigma_X^2 = \sum_x x^2 P(X = x) - \sum_x 2x\mu_X P(X = x) + \sum_x \mu_X^2 P(X = x) \tag{2.38}$$

Now $\sum_x 2x\mu_X P(X = x) = 2\mu_X \sum_x x P(X = x) = 2\mu_X \mu_X = 2\mu_X^2$, and $\sum_x \mu_X^2 P(X = x) = \mu_X^2 \sum_x P(X = x) = \mu_X^2(1) = \mu_X^2$.

Substituting into Equation (2.38) yields

$$\sigma_X^2 = \sum_x x^2 P(X = x) - 2\mu_X^2 + \mu_X^2$$

We conclude that

$$\sigma_X^2 = \sum_x x^2 P(X = x) - \mu_X^2$$

To derive the alternate formula (2.37) for the variance of a continuous random variable from Equation (2.36), the same steps may be used; replacing $\sum_x$ with $\int_{-\infty}^{\infty}$, and $P(X = x)$ with $f(x)\,dx$.

Exercises for Section 2.4

1. Determine whether each of the following random variables is discrete or continuous.
   a. The number of heads in 100 tosses of a coin.
   b. The length of a rod randomly chosen from a day's production.
   c. The final exam score of a randomly chosen student from last semester's engineering statistics class.
   d. The age of a randomly chosen Colorado School of Mines student.

e. The age that a randomly chosen Colorado School of Mines student will be on his or her next birthday.

2. The following table presents the probability mass function of the number of defects X in a randomly chosen printed-circuit board.

x	0	1	2	3
$p(x)$	0.5	0.3	0.1	0.1

a. Find $P(X < 2)$.

b. Find $P(X \geq 1)$.

c. Find μ_X.

d. Find σ_X^2.

3. A chemical supply company ships a certain solvent in 10-gallon drums. Let X represent the number of drums ordered by a randomly chosen customer. Assume X has the following probability mass function:

x	1	2	3	4	5
$p(x)$	0.4	0.2	0.2	0.1	0.1

a. Find the mean number of drums ordered.

b. Find the variance of the number of drums ordered.

c. Find the standard deviation of the number of drums ordered.

d. Let Y be the number of gallons ordered. Find the probability mass function of Y.

e. Find the mean number of gallons ordered.

f. Find the variance of the number of gallons ordered.

g. Find the standard deviation of the number of gallons ordered.

4. An old car with a four-cylinder engine is brought in for a tune-up. Let X represent the number of cylinders with low compression.

a. Which of the three functions given in the following table is a possible probability mass function of X? Explain.

	x				
	0	**1**	**2**	**3**	**4**
$p_1(x)$	0.2	0.2	0.3	0.1	0.1
$p_2(x)$	0.1	0.3	0.3	0.2	0.2
$p_3(x)$	0.1	0.2	0.4	0.2	0.1

b. For the possible probability mass function, compute μ_X and σ_X^2.

5. A certain type of component is packaged in lots of four. Let X represent the number of properly functioning components in a randomly chosen lot. Assume that the probability that exactly x components function is proportional to x; in other words, assume that the probability mass function of X is given by

$$p(x) = \begin{cases} cx & x = 1, 2, 3, \text{ or } 4 \\ 0 & \text{otherwise} \end{cases}$$

where c is a constant.

a. Find the value of the constant c so that $p(x)$ is a probability mass function.

b. Find $P(X = 2)$.

c. Find the mean number of properly functioning components.

d. Find the variance of the number of properly functioning components.

e. Find the standard deviation of the number of properly functioning components.

6. The output of a chemical process is continually monitored to ensure that the concentration remains within acceptable limits. Whenever the concentration drifts outside the limits, the process is shut down and recalibrated. Let X be the number of times in a given week that the process is recalibrated. The following table presents values of the cumulative distribution function $F(x)$ of X.

x	*F(x)*
0	0.17
1	0.53
2	0.84
3	0.97
4	1.00

a. What is the probability that the process is recalibrated fewer than two times during a week?

b. What is the probability that the process is recalibrated more than three times during a week?

c. What is the probability that the process is recalibrated exactly once during a week?

d. What is the probability that the process is not recalibrated at all during a week?

e. What is the most probable number of recalibrations to occur during a week?

7. On 100 different days, a traffic engineer counts the number of cars that pass through a certain intersection between 5 P.M. and 5:05 P.M. The results are presented in the following table.

Number of Cars	Number of Days	Proportion of Days
0	36	0.36
1	28	0.28
2	15	0.15
3	10	0.10
4	7	0.07
5	4	0.04

a. Let X be the number of cars passing through the intersection between 5 P.M. and 5:05 P.M. on a randomly chosen day. Someone suggests that for any positive integer x, the probability mass function of X is $p_1(x) = (0.2)(0.8)^x$. Using this function, compute $P(X = x)$ for values of x from 0 through 5 inclusive.

b. Someone else suggests that for any positive integer x, the probability mass function is $p_2(x) = (0.4)(0.6)^x$. Using this function, compute $P(X = x)$ for values of x from 0 through 5 inclusive.

c. Compare the results of parts (a) and (b) to the data in the table. Which probability mass function appears to be the better model? Explain.

d. Someone says that neither of the functions is a good model since neither one agrees with the data exactly. Is this right? Explain.

8. Microprocessing chips are randomly sampled one by one from a large population, and tested to determine if they are acceptable for a certain application. Ninety percent of the chips in the population are acceptable.

a. What is the probability that the first chip chosen is acceptable?

b. What is the probability that the first chip is unacceptable, and the second is acceptable?

c. Let X represent the number of chips that are tested up to and including the first acceptable chip. Find $P(X = 3)$.

d. Find the probability mass function of X.

9. Refer to Exercise 8. Let Y be the number of chips tested up to and including the second acceptable chip.

a. What is the smallest possible value for Y?

b. What is the probability that Y takes on that value?

c. Let X represent the number of chips that are tested up to and including the first acceptable chip. Find $P(Y = 3|X = 1)$.

d. Find $P(Y = 3|X = 2)$.

e. Find $P(Y = 3)$.

10. Three components are randomly sampled, one at a time, from a large lot. As each component is selected, it is tested. If it passes the test, a success (S) occurs; if it fails the test, a failure (F) occurs. Assume that 80% of the components in the lot will succeed in passing the test. Let X represent the number of successes among the three sampled components.

a. What are the possible values for X?

b. Find $P(X = 3)$.

c. The event that the first component fails and the next two succeed is denoted by FSS. Find $P(\text{FSS})$.

d. Find $P(\text{SFS})$ and $P(\text{SSF})$.

e. Use the results of parts (c) and (d) to find $P(X = 2)$.

f. Find $P(X = 1)$.

g. Find $P(X = 0)$.

h. Find μ_X.

i. Find σ_X^2.

j. Let Y represent the number of successes if four components are sampled. Find $P(Y = 3)$.

11. The hydrogenation of benzene to cyclohexane is promoted with a finely divided porous nickel catalyst. The catalyst particles can be considered to be spheres of various sizes. All the particles have masses between 10 and 70 μg. Let X be the mass of a randomly chosen particle. The probability density function of X is given by

$$f(x) = \begin{cases} \dfrac{x - 10}{1800} & 10 < x < 70 \\ 0 & \text{otherwise} \end{cases}$$

a. What proportion of particles have masses less than 50 μg?

b. Find the mean mass of the particles.

c. Find the standard deviation of the particle masses.

d. Find the cumulative distribution function of the particle masses.

e. Find the median of the particle masses.

12. Specifications call for the thickness of aluminum sheets that are to be made into cans to be between 8 and 11 thousandths of an inch. Let X be the thickness of an aluminum sheet. Assume the probability density function of X is given by

$$f(x) = \begin{cases} \dfrac{x}{54} & 6 < x < 12 \\ 0 & \text{otherwise} \end{cases}$$

a. What proportion of sheets will meet the specification?

b. Find the mean thickness of a sheet.

c. Find the variance of the thickness of a sheet.

d. Find the standard deviation of the thickness of a sheet.

e. Find the cumulative distribution function of the thickness.

f. Find the median thickness.

g. Find the 10th percentile of the thickness.

h. A particular sheet is 10 thousandths of an inch thick. What proportion of sheets are thicker than this?

13. A radioactive mass emits particles from time to time. The time between two emissions is random. Let T represent the time in seconds between two emissions. Assume that the probability density function of T is given by

$$f(t) = \begin{cases} 0.2e^{-0.2t} & t > 0 \\ 0 & t \le 0 \end{cases}$$

a. Find the mean time between emissions.

b. Find the standard deviation of the time between emissions.

c. Find the cumulative distribution function of the time between emissions.

d. Find the probability that the time between emissions will be less than 10 seconds.

e. Find the median time between emissions.

f. Find the 90th percentile of the times between emissions.

14. A process that manufactures piston rings produces rings whose diameters (in centimeters) vary according to the probability density function

$$f(x) = \begin{cases} 3[1 - 16(x - 10)^2] & 9.75 < x < 10.25 \\ 0 & \text{otherwise} \end{cases}$$

a. Find the mean diameter of rings manufactured by this process.

b. Find the standard deviation of the diameters of rings manufactured by this process. (*Hint*: Equation 2.36 may be easier to use than Equation 2.37.)

c. Find the cumulative distribution function of piston ring diameters.

d. What proportion of piston rings have diameters less than 9.75 cm?

e. What proportion of piston rings have diameters between 9.75 and 10.25 cm?

15. Refer to Exercise 14. A competing process produces rings whose diameters (in centimeters) vary according to the probability density function

$$f(x) = \begin{cases} 15[1 - 25(x - 10.05)^2]/4 & 9.85 < x < 10.25 \\ 0 & \text{otherwise} \end{cases}$$

Specifications call for the diameter to be 10.0 ± 0.1 cm. Which process is better, this one or the one in Exercise 14? Explain.

16. Particles are a major component of air pollution in many areas. It is of interest to study the sizes of contaminating particles. Let X represent the diameter, in micrometers, of a randomly chosen particle. Assume that in a certain area, the probability density function of X is inversely proportional to the volume of the particle; that is, assume that

$$f(x) = \begin{cases} \dfrac{c}{x^3} & x \geq 1 \\ 0 & x < 1 \end{cases}$$

where c is a constant.

a. Find the value of c so that $f(x)$ is a probability density function.

b. Find the mean particle diameter.

c. Find the cumulative distribution function of the particle diameter.

d. Find the median particle diameter.

e. The term PM_{10} refers to particles 10 μm or less in diameter. What proportion of the contaminating particles are PM_{10}?

f. The term $PM_{2.5}$ refers to particles 2.5 μm or less in diameter. What proportion of the contaminating particles are $PM_{2.5}$?

g. What proportion of the PM_{10} particles are $PM_{2.5}$?

17. An environmental scientist is concerned with the rate at which a certain toxic solution is absorbed into the skin. Let X be the volume in microliters of solution absorbed by 1 in^2 of skin in 1 min. Assume that the probability density function of X is well approximated by the function $f(x) = (2\sqrt{2\pi})^{-1}e^{-(x-10)^2/8}$, defined for $-\infty < x < \infty$.

a. Find the mean volume absorbed in 1 min.

b. (Hard) Find the standard deviation of the volume absorbed in 1 min.

2.5 Linear Functions of Random Variables

In practice we often construct new random variables by performing arithmetic operations on other random variables. For example, we might add a constant to a random variable, multiply a random variable by a constant, or add two or more random variables together. In this section, we describe how to compute means and variances of random variables constructed in these ways, and we present some practical examples. The presentation in this section is intuitive. A more rigorous presentation is provided in Section 2.6. For those desiring such a presentation, Section 2.6 may be covered in addition to, or in place of, this section.

Adding a Constant

When a constant is added to a random variable, the mean is increased by the value of the constant, but the variance and standard deviation are unchanged. For example, assume that steel rods produced by a certain machine have a mean length of 5.0 in. and a variance of $\sigma^2 = 0.003$ in^2. Each rod is attached to a base that is exactly 1.0 in. long. The mean length of the assembly will be $5.0 + 1.0 = 6.0$ in. Since each length is increased by the same amount, the spread in the lengths does not change, so the variance remains the same. To put this in statistical terms, let X be the length of a randomly chosen rod, and let $Y = X + 1$ be the length of the assembly. Then $\mu_Y = \mu_{X+1} = \mu_X + 1$, and $\sigma_Y^2 = \sigma_{X+1}^2 = \sigma_X^2$. In general, when a constant is added to a random variable, the mean is shifted by that constant, and the variance is unchanged.

Summary

If X is a random variable and b is a constant, then

$$\mu_{X+b} = \mu_X + b \tag{2.39}$$

$$\sigma^2_{X+b} = \sigma^2_X \tag{2.40}$$

Multiplying by a Constant

Often we need to multiply a random variable by a constant. This might be done, for example, to convert to a more convenient set of units. We continue the example of steel rod production to show how multiplication by a constant affects the mean, variance, and standard deviation of a random variable.

If we measure the lengths of the rods described earlier in centimeters rather than inches, the mean length will be (2.54 cm/in.)(5.0 in.) = 12.7 cm. In statistical terms, let the random variable X be the length in inches of a randomly chosen rod, and let $Y = 2.54X$ be the length in centimeters. Then $\mu_Y = 2.54\mu_X$. In general, when a random variable is multiplied by a constant, its mean is multiplied by the same constant.

Summary

If X is a random variable and a is a constant, then

$$\mu_{aX} = a\mu_X \tag{2.41}$$

When the length X of a rod is measured in inches, the variance σ^2_X must have units of in^2. If $Y = 2.54X$ is the length in centimeters, then σ^2_Y must have units of cm^2. Therefore we obtain σ^2_Y by multiplying σ^2_X by 2.54^2, which is the conversion factor from in^2 to cm^2. In general, when a random variable is multiplied by a constant, its variance is multiplied by the *square* of the constant.

Summary

If X is a random variable and a is a constant, then

$$\sigma^2_{aX} = a^2\sigma^2_X \tag{2.42}$$

$$\sigma_{aX} = |a|\sigma_X \tag{2.43}$$

If a random variable is multiplied by a constant and then added to another constant, the effect on the mean and variance can be determined by combining Equations (2.39) and (2.41) and Equations (2.40) and (2.42). The results are presented in the following summary.

Summary

If X is a random variable, and a and b are constants, then

$$\mu_{aX+b} = a\mu_X + b \tag{2.44}$$

$$\sigma^2_{aX+b} = a^2\sigma^2_X \tag{2.45}$$

$$\sigma_{aX+b} = |a|\sigma_X \tag{2.46}$$

Note that Equations (2.44) through (2.46) are analogous to results for the sample mean and standard deviation presented in Section 1.2.

The molarity of a solute in solution is defined to be the number of moles of solute per liter of solution (1 mole = 6.02×10^{23} molecules). If the molarity of a stock solution of concentrated sulfuric acid (H_2SO_4) is X, and if one part of the solution is mixed with N parts water, the molarity Y of the dilute solution is given by $Y = X/(N + 1)$. Assume that the stock solution is manufactured by a process that produces a molarity with mean 18 and standard deviation 0.1. If 100 mL of stock solution is added to 300 mL of water, find the mean and standard deviation of the molarity of the dilute solution.

Solution

The molarity of the dilute solution is $Y = 0.25X$. The mean and standard deviation of X are $\mu_X = 18$ and $\sigma_X = 0.1$, respectively. Therefore

$$\begin{aligned}
\mu_Y &= \mu_{0.25X} \\
&= 0.25\mu_X \qquad \text{(using Equation 2.41)} \\
&= 0.25(18.0) \\
&= 4.5
\end{aligned}$$

Also,

$$\begin{aligned}
\sigma_Y &= \sigma_{0.25X} \\
&= 0.25\sigma_X \qquad \text{(using Equation 2.43)} \\
&= 0.25(0.1) \\
&= 0.025
\end{aligned}$$

Means of Linear Combinations of Random Variables

Consider the case of adding two random variables. For example, assume that there are two machines that fabricate a certain metal part. The mean daily production of machine

A is 100 parts, and the mean daily production of machine B is 150 parts. Clearly the mean daily production from the two machines together is 250 parts. Putting this in mathematical notation, let X be the number of parts produced on a given day by machine A, and let Y be the number of parts produced on the same day by machine B. The total number of parts is $X + Y$, and we have that $\mu_{X+Y} = \mu_X + \mu_Y$.

This idea extends to any number of random variables.

If $X_1, X_2, \ldots, X_n$ are random variables, then the mean of the sum $X_1 + X_2 + \cdots + X_n$ is given by

$$\mu_{X_1+X_2+\cdots+X_n} = \mu_{X_1} + \mu_{X_2} + \cdots + \mu_{X_n} \tag{2.47}$$

The sum $X_1 + X_2 + \cdots + X_n$ is a special case of a **linear combination**:

If $X_1, \ldots, X_n$ are random variables and $c_1, \ldots, c_n$ are constants, then the random variable

$$c_1X_1 + \cdots + c_nX_n$$

is called a **linear combination** of $X_1, \ldots, X_n$.

To find the mean of a linear combination of random variables, we can combine Equations (2.41) and (2.47):

If X and Y are random variables, and a and b are constants, then

$$\mu_{aX+bY} = \mu_{aX} + \mu_{bY} = a\mu_X + b\mu_Y \tag{2.48}$$

More generally, if $X_1, X_2, \ldots, X_n$ are random variables and $c_1, c_2, \ldots, c_n$ are constants, then the mean of the linear combination $c_1X_1 + c_2X_2 + \cdots + c_nX_n$ is given by

$$\mu_{c_1X_1+c_2X_2+\cdots+c_nX_n} = c_1\mu_{X_1} + c_2\mu_{X_2} + \cdots + c_n\mu_{X_n} \tag{2.49}$$

Independent Random Variables

The notion of independence for random variables is very much like the notion of independence for events. Two random variables are independent if knowledge regarding one of them does not affect the probabilities of the other. When two events are independent, the probability that both occur is found by multiplying the probabilities for each event (see Equations 2.19 and 2.20 in Section 2.3). There are analogous formulas for independent random variables. The notation for these formulas is as follows. Let X be a random variable and let S be a set of numbers. The notation "$X \in S$" means that the value of the random variable X is in the set S.

Definition

If X and Y are **independent** random variables, and S and T are sets of numbers, then

$$P(X \in S \text{ and } Y \in T) = P(X \in S)P(Y \in T) \quad (2.50)$$

More generally, if $X_1, \ldots, X_n$ are independent random variables, and $S_1, \ldots, S_n$ are sets, then

$$P(X_1 \in S_1 \text{ and } X_2 \in S_2 \text{ and} \cdots \text{ and } X_n \in S_n) = P(X_1 \in S_1)P(X_2 \in S_2) \cdots P(X_n \in S_n) \quad (2.51)$$

Rectangular plastic covers for a compact disc (CD) tray have specifications regarding length and width. Let X be the length and Y be the width, each measured to the nearest millimeter, of a randomly sampled cover. The probability mass function of X is given by $P(X = 129) = 0.2$, $P(X = 130) = 0.7$, and $P(X = 131) = 0.1$. The probability mass function of Y is given by $P(Y = 15) = 0.6$ and $P(Y = 16) = 0.4$. The area of a cover is given by $A = XY$. Assume X and Y are independent. Find the probability that the area is 1935 mm^2.

Solution

The area will be equal to 1935 if $X = 129$ and $Y = 15$. Therefore

$$\begin{aligned} P(A = 1935) &= P(X = 129 \text{ and } Y = 15) \\ &= P(X = 129)P(Y = 15) \qquad \text{since } X \text{ and } Y \text{ are independent} \\ &= (0.2)(0.6) \\ &= 0.12 \end{aligned}$$

Equations (2.50) and (2.51) tell how to compute probabilities for independent random variables, but they are not usually much help in determining whether random variables actually are independent. In general, the best way to determine whether random variables are independent is through an understanding of the process that generated them.

Variances of Linear Combinations of Independent Random Variables

We have seen that the mean of a sum of random variables is always equal to the sum of the means (Equation 2.47). In general, the formula for the variance of a sum of random variables is a little more complicated than this. But when random variables are *independent*, the result is simple: the variance of the sum is the sum of the variances.

If $X_1, X_2, \ldots, X_n$ are *independent* random variables, then the variance of the sum $X_1 + X_2 + \cdots + X_n$ is given by

$$\sigma^2_{X_1+X_2+\cdots+X_n} = \sigma^2_{X_1} + \sigma^2_{X_2} + \cdots + \sigma^2_{X_n} \tag{2.52}$$

To find the variance of a linear combination of random variables, we can combine Equations (2.52) and (2.42):

If $X_1, X_2, \ldots, X_n$ are *independent* random variables and $c_1, c_2, \ldots, c_n$ are constants, then the variance of the linear combination $c_1X_1 + c_2X_2 + \cdots + c_nX_n$ is given by

$$\sigma^2_{c_1X_1+c_2X_2+\cdots+c_nX_n} = c_1^2\sigma^2_{X_1} + c_2^2\sigma^2_{X_2} + \cdots + c_n^2\sigma^2_{X_n} \tag{2.53}$$

Two frequently encountered linear combinations are the sum and the difference of two random variables. Interestingly enough, when the random variables are independent, the variance of the sum is the same as the variance of the difference.

If X and Y are *independent* random variables with variances σ^2_X and σ^2_Y, then the variance of the sum $X + Y$ is

$$\sigma^2_{X+Y} = \sigma^2_X + \sigma^2_Y \tag{2.54}$$

The variance of the difference $X - Y$ is

$$\sigma^2_{X-Y} = \sigma^2_X + \sigma^2_Y \tag{2.55}$$

The fact that the variance of the difference is the *sum* of the variances may seem counterintuitive. However, it follows from Equation (2.53) by setting $c_1 = 1$ and $c_2 = -1$.

A piston is placed inside a cylinder. The clearance is the distance between the edge of the piston and the wall of the cylinder and is equal to one-half the difference between the cylinder diameter and the piston diameter. Assume the piston diameter has a mean of 80.85 cm with a standard deviation of 0.02 cm. Assume the cylinder diameter has a mean of 80.95 cm with a standard deviation of 0.03 cm. Find the mean clearance. Assuming that the piston and cylinder are chosen independently, find the standard deviation of the clearance.

Solution

Let X_1 represent the diameter of the cylinder and let X_2 the diameter of the piston. The clearance is given by $C = 0.5X_1 - 0.5X_2$. Using Equation (2.49), the mean perimeter is

$$\begin{aligned}\mu_C &= \mu_{0.5X_1-0.5X_2}\\ &= 0.5\mu_{X_1} - 0.5\mu_{X_2}\\ &= 0.5(80.95) - 0.5(80.85)\\ &= 0.050\end{aligned}$$

Since X_1 and X_2 are independent, we can use Equation (2.53) to find the standard deviation σ_C:

$$\begin{aligned}\sigma_C &= \sqrt{\sigma^2_{0.5X_1-0.5X_2}}\\ &= \sqrt{(0.5)^2\sigma^2_{X_1} + (-0.5)^2\sigma^2_{X_2}}\\ &= \sqrt{0.25(0.02)^2 + 0.25(0.03)^2}\\ &= 0.018\end{aligned}$$

Independence and Simple Random Samples

When a simple random sample of numerical values is drawn from a population, each item in the sample can be thought of as a random variable. The items in a simple random sample may be treated as independent, except when the sample is a large proportion (more than 5%) of a finite population (see the discussion of independence in Section 1.1). From here on, unless explicitly stated to the contrary, we will assume this exception has not occurred, so that the values in a simple random sample may be treated as independent random variables.

Summary

If $X_1, X_2, \ldots, X_n$ is a simple random sample, then $X_1, X_2, \ldots, X_n$ may be treated as independent random variables, all with the same distribution.

When $X_1, \ldots, X_n$ are independent random variables, all with the same distribution, it is sometimes said that $X_1, \ldots, X_n$ are **independent and identically distributed (i.i.d)**.

The Mean and Variance of a Sample Mean

The most frequently encountered linear combination is the sample mean. Specifically, if $X_1, \ldots, X_n$ is a simple random sample from a population with mean μ and variance σ^2, then the sample mean $\overline{X}$ is the linear combination

$$\overline{X} = \frac{1}{n}X_1 + \cdots + \frac{1}{n}X_n$$

From this fact we can compute the mean and variance of $\overline{X}$.

$$\begin{aligned}\mu_{\overline{X}} &= \mu_{\frac{1}{n}X_1+\cdots+\frac{1}{n}X_n}\\ &= \frac{1}{n}\mu_{X_1} + \cdots + \frac{1}{n}\mu_{X_n} \qquad \text{(using Equation 2.49)}\end{aligned}$$

$$= \frac{1}{n}\mu + \cdots + \frac{1}{n}\mu$$
$$= (n)\left(\frac{1}{n}\right)\mu$$
$$= \mu$$

As discussed previously, the items in a simple random sample may be treated as independent random variables. Therefore

$$\begin{aligned}
\sigma^2_{\overline{X}} &= \sigma^2_{\frac{1}{n}X_1+\cdots+\frac{1}{n}X_n} \\
&= \frac{1}{n^2}\sigma^2_{X_1} + \cdots + \frac{1}{n^2}\sigma^2_{X_n} \qquad \text{(using Equation 2.53)} \\
&= \frac{1}{n^2}\sigma^2 + \cdots + \frac{1}{n^2}\sigma^2 \\
&= (n)\left(\frac{1}{n^2}\right)\sigma^2 \\
&= \frac{\sigma^2}{n}
\end{aligned}$$

Summary

If $X_1, \ldots, X_n$ is a simple random sample from a population with mean μ and variance σ^2, then the sample mean $\overline{X}$ is a random variable with

$$\mu_{\overline{X}} = \mu \tag{2.56}$$

$$\sigma^2_{\overline{X}} = \frac{\sigma^2}{n} \tag{2.57}$$

The standard deviation of $\overline{X}$ is

$$\sigma_{\overline{X}} = \frac{\sigma}{\sqrt{n}} \tag{2.58}$$

Example 2.48

A process that fills plastic bottles with a beverage has a mean fill volume of 2.013 L and a standard deviation of 0.005 L. A case contains 24 bottles. Assuming that the bottles in a case are a simple random sample of bottles filled by this method, find the mean and standard deviation of the average volume per bottle in a case.

Solution

Let $V_1, \ldots, V_{24}$ represent the volumes in 24 bottles in a case. This is a simple random sample from a population with mean $\mu = 2.013$ and standard deviation $\sigma = 0.005$. The average volume is $\overline{V} = (V_1 + \cdots + V_{24})/24$. Using Equation (2.56),

$$\mu_{\overline{V}} = \mu = 2.013$$

Using Equation (2.58),

$$\sigma_{\overline{V}} = \frac{\sigma}{\sqrt{24}} = 0.001$$

Exercises for Section 2.5

1. If X and Y are independent random variables with means $\mu_X = 10.5$ and $\mu_Y = 5.7$, and standard deviations $\sigma_X = 0.5$ and $\sigma_Y = 0.3$, find the means and standard deviations of the following quantities:

a. $2X$

b. $X - Y$

c. $3X + 2Y$

2. If a resistor with resistance R ohms carries a current of I amperes, the potential difference across the resistor, in volts, is given by $V = IR$. The resistance of a randomly selected resistor that is labeled 100 Ω has mean 100 Ω and standard deviation 10 Ω. A current of 3 A is set up in a randomly selected resistor.

a. Find μ_V.

b. Find σ_V.

3. The lifetime of a certain lightbulb in a certain application has mean 700 hours and standard deviation 20 hours. As each bulb burns out, it is replaced with a new bulb. Find the mean and standard deviation of the length of time that five bulbs will last.

4. Two resistors, with resistances R_1 and R_2, are connected in series. The combined resistance R is given by $R = R_1 + R_2$. Assume that R_1 has mean 50 Ω and standard deviation 5 Ω, and that R_2 has mean 100 Ω and standard deviation 10 Ω.

a. Find μ_R.

b. Assuming R_1 and R_2 to be independent, find σ_R.

5. A piece of plywood is composed of five layers. The layers are a simple random sample from a population whose thickness has mean 0.125 in. and standard deviation 0.005 in.

a. Find the mean thickness of a piece of plywood.

b. Find the standard deviation of the thickness of a piece of plywood.

6. The manufacture of a certain item requires two machines operating consecutively. The time on the first machine has mean 10 min and standard deviation 2 min. The time on the second machine has mean 15 min and standard deviation 3 min. Assume the times spent on the two machines are independent.

a. Find the mean of the total time spent on the two machines.

b. Find the standard deviation of the total time spent on the two machines.

7. The molarity of a solute in solution is defined to be the number of moles of solute per liter of solution (1 mole = 6.02×10^{23} molecules). If X is the molarity of a solution of magnesium chloride ($MgCl_2$), and Y is the molarity of a solution of ferric chloride ($FeCl_3$), the molarity of chloride ion (Cl^-) in a solution made of equal parts of the solutions of $MgCl_2$ and $FeCl_3$ is given by $M = X + 1.5Y$. Assume that X has mean 0.125 and standard deviation 0.05, and that Y has mean 0.350 and standard deviation 0.10.

a. Find μ_M.

b. Assuming X and Y to be independent, find σ_M.

8. A machine that fills cardboard boxes with cereal has a fill weight whose mean is 12.02 oz, with a standard deviation of 0.03 oz. A case consists of 12 boxes randomly sampled from the output of the machine.

a. Find the mean of the total weight of the cereal in the case.

b. Find the standard deviation of the total weight of the cereal in the case.

c. Find the mean of the average weight per box of the cereal in the case.

d. Find the standard deviation of the average weight per box of the cereal in the case.

e. How many boxes must be included in a case for the standard deviation of the average weight per box to be 0.005 oz?

9. The four sides of a picture frame consist of two pieces selected from a population whose mean length is 30 cm with standard deviation 0.1 cm, and two pieces selected from a population whose mean length is 45 cm with standard deviation 0.3 cm.

a. Find the mean perimeter.

b. Assuming the four pieces are chosen independently, find the standard deviation of the perimeter.

10. A gas station earns \$1.60 in revenue for each gallon of regular gas it sells, \$1.75 for each gallon of midgrade gas, and \$1.90 for each gallon of premium gas. Let X_1, X_2, and X_3 denote the numbers of gallons of regular, midgrade, and premium gasoline sold in a day. Assume that X_1, X_2, and X_3 have means $\mu_1 = 1500$, $\mu_2 = 500$, and $\mu_3 = 300$, and standard deviations $\sigma_1 = 180$, $\sigma_2 = 90$, and $\sigma_3 = 40$, respectively.

a. Find the mean daily revenue.

b. Assuming X_1, X_2, and X_3 to be independent, find the standard deviation of the daily revenue.

11. In the article "An Investigation of the Ca–CO_3–CaF_2–K_2SiO_3–SiO_2–Fe Flux System Using the Submerged Arc Welding Process on HSLA-100 and AISI-1018 Steels" (G. Fredrickson, M.S. thesis, Colorado School of Mines, 1992), the carbon equivalent P of a weld metal is defined to be a linear combination of the weight percentages of carbon (C), manganese (Mn), copper (Cu), chromium (Cr), silicon (Si), nickel (Ni), molybdenum (Mo), vanadium (V), and boron (B). The carbon equivalent is given by

$$P = \mathrm{C} + \frac{\mathrm{Mn} + \mathrm{Cu} + \mathrm{Cr}}{20} + \frac{\mathrm{Si}}{30} + \frac{\mathrm{Ni}}{60} + \frac{\mathrm{Mo}}{15} + \frac{\mathrm{V}}{10} + 5\mathrm{B}$$

Means and standard deviations of the weight percents of these chemicals were estimated from measurements on 45 weld metals produced on HSLA-100 steel base metal. Assume the means and standard deviations (SD) are as given in the following table.

	Mean	SD
C	0.0695	0.0018
Mn	1.0477	0.0269
Cu	0.8649	0.0225
Cr	0.7356	0.0113
Si	0.2171	0.0185
Ni	2.8146	0.0284
Mo	0.5913	0.0031
V	0.0079	0.0006
B	0.0006	0.0002

a. Find the mean carbon equivalent of weld metals produced from HSLA-1000 steel base metal.

b. Assuming the weight percents to be independent, find the standard deviation of the carbon equivalent of weld metals produced from HSLA-1000 steel base metal.

2.6 Jointly Distributed Random Variables*

In this section, results concerning several random variables are presented in a more rigorous fashion than in Section 2.5. For those desiring such a presentation, this section may be covered in addition to, or in place of, Section 2.5.

We have said that observing a value of a random variable is like sampling a value from a population. In some cases, the items in the population may each have several random variables associated with them. For example, imagine choosing a student at random from a list of all the students registered at a university and measuring that student's height and weight. Each individual in the population of students corresponds to two random variables, height and weight. If we also determined the student's age,

*This section is optional.

each individual would correspond to three random variables. In principle, any number of random variables may be associated with each item in a population.

When two or more random variables are associated with each item in a population, the random variables are said to be **jointly distributed**. If all the random variables are discrete, they are said to be **jointly discrete**. If all the random variables are continuous, they are said to be **jointly continuous**. We will study these two cases separately.

Jointly Discrete Random Variables

Example 2.46 (in Section 2.5) discussed the lengths and widths of rectangular plastic covers for a CD tray that is installed in a personal computer. Measurements are rounded to the nearest millimeter. Let X denote the measured length and Y the measured width. The possible values of X are 129, 130, and 131, and the possible values for Y are 15 and 16. Both X and Y are discrete, so X and Y are **jointly discrete**. There are six possible values for the ordered pair (X, Y): $(129,\ 15)$, $(129,\ 16)$, $(130,\ 15)$, $(130,\ 16)$, $(131,\ 15)$, and $(131,\ 16)$. Assume that the probabilities of each of these ordered pairs are as given in the following table.

x	y	$P(X = x \text{ and } Y = y)$
129	15	0.12
129	16	0.08
130	15	0.42
130	16	0.28
131	15	0.06
131	16	0.04

The **joint probability mass function** is the function $p(x, y) = P(X = x \text{ and } Y = y)$. So, for example, we have $p(129,\ 15) = 0.12$, and $p(130,\ 16) = 0.28$.

Sometimes we are given a joint probability mass function of two random variables, but we are interested in only one of them. For example, we might be interested in the probability mass function of X, the length of the CD cover, but not interested in the width Y. We can obtain the probability mass function of either one of the variables X or Y separately by summing the appropriate values of the joint probability mass function. Examples 2.49 and 2.50 illustrate the method.

Example 2.49

Find the probability that a CD cover has a length of 129 mm.

Solution

It is clear from the previous table that 12% of the CD covers in the population have a length of 129 and a width of 15, and 8% have a length of 129 and a width of 16.

Therefore 20% of the items in the population have a length of 129. The probability that a CD cover has a length of 129 mm is 0.20. In symbols, we have

$$\begin{aligned}P(X = 129) &= P(X = 129 \text{ and } Y = 15) + P(X = 129 \text{ and } Y = 16)\\ &= 0.12 + 0.08\\ &= 0.20\end{aligned}$$

2.50

Find the probability that a CD cover has a width of 16 mm.

Solution

We need to find $P(Y = 16)$. We can find this quantity by summing the probabilities of all pairs (x, y) for which $y = 16$. We obtain

$$\begin{aligned}P(Y = 16) &= P(X = 129 \text{ and } Y = 16) + P(X = 130 \text{ and } Y = 16)\\ &\quad + P(X = 131 \text{ and } Y = 16)\\ &= 0.08 + 0.28 + 0.04\\ &= 0.40\end{aligned}$$

Examples 2.49 and 2.50 show that we can find the probability mass function of X (or Y) by summing the joint probability mass function over all values of Y (or X). Table 2.3 presents the joint probability mass function of X and Y. The probability mass function of X appears in the rightmost column and is obtained by summing along the rows. The probability mass function of Y appears in the bottom row and is obtained by summing down the columns. Note that the probability mass functions of X and of Y appear in the margins of the table. For this reason they are often referred to as **marginal** probability mass functions.

TABLE 2.3 Joint and marginal probability mass functions for the length X and width Y of a CD cover

	y		
x	**15**	**16**	$p_X(x)$
129	0.12	0.08	0.20
130	0.42	0.28	0.70
131	0.06	0.04	0.10
$p_Y(y)$	0.60	0.40	

Finally, if we sum the joint probability density function over all possible values of x and y, we obtain the probability that X and Y take values somewhere within their possible ranges, and this probability is equal to 1.

Summary

If X and Y are jointly discrete random variables:

- The joint probability mass function of X and Y is the function

$$p(x, y) = P(X = x \text{ and } Y = y)$$

- The marginal probability mass functions of X and of Y can be obtained from the joint probability mass function as follows:

$$p_X(x) = P(X = x) = \sum_y p(x, y) \qquad p_Y(y) = P(Y = y) = \sum_x p(x, y)$$

  where the sums are taken over all the possible values of Y and of X, respectively.

- The joint probability mass function has the property that

$$\sum_x \sum_y p(x, y) = 1$$

  where the sum is taken over all the possible values of X and Y.

Jointly Continuous Random Variables

We have seen that if X is a continuous random variable, its probabilities are found by integrating its probability density function. We say that the random variables X and Y are **jointly continuous** if their probabilities are found by integrating a function of two variables, called the **joint probability density function** of X and Y. To find the probability that X and Y take values in any region, we integrate the joint probability density function over that region, as Example 2.51 shows.

Example 2.51

Assume that for a certain type of washer, both the thickness and the hole diameter vary from item to item. Let X denote the thickness in millimeters and let Y denote the hole diameter in millimeters, for a randomly chosen washer. Assume that the joint probability density function of X and Y is given by

$$f(x, y) = \begin{cases} \frac{1}{6}(x + y) & \text{if } 1 \le x \le 2 \text{ and } 4 \le y \le 5 \\ 0 & \text{otherwise} \end{cases}$$

Find the probability that a randomly chosen washer has a thickness between 1.0 and 1.5 mm, and a hole diameter between 4.5 and 5 mm.

Solution

We need to find $P(1 \le X \le 1.5 \text{ and } 4.5 \le Y \le 5)$. The large rectangle in the figure indicates the region where the joint density is positive. The shaded rectangle indicates the region where $1 \le x \le 1.5$ and $4.5 \le y \le 5$, over which the joint density is to be integrated.

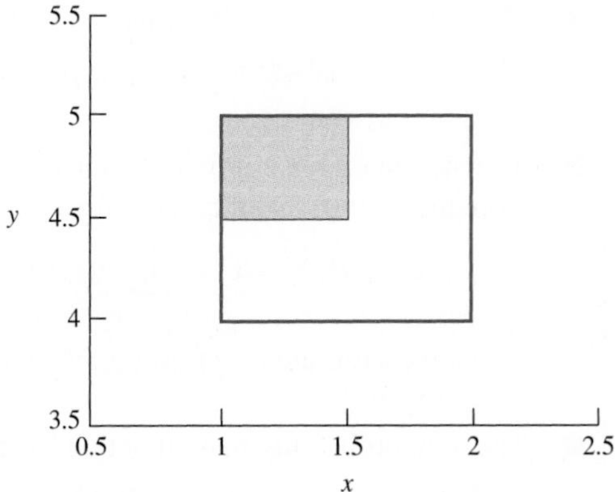

We integrate the joint probability density function over the indicated region:

$$\begin{aligned}
P(1 \le X \le 1.5 \text{ and } 4.5 \le Y \le 5) &= \int_1^{1.5} \int_{4.5}^{5} \frac{1}{6}(x+y)\, dy\, dx \\
&= \int_1^{1.5} \left\{ \frac{xy}{6} + \frac{y^2}{12} \Big|_{y=4.5}^{y=5} \right\} dx \\
&= \int_1^{1.5} \left(\frac{x}{12} + \frac{19}{48} \right) dx \\
&= \frac{1}{4}
\end{aligned}$$

Note that if a joint probability density function is integrated over the entire plane, that is, if the limits are $-\infty$ to ∞ for both x and y, we obtain the probability that both X and Y take values between $-\infty$ and ∞, which is equal to 1.

Summary

If X and Y are jointly continuous random variables, with joint probability density function $f(x,y)$, and $a < b$, $c < d$, then

$$P(a \le X \le b \text{ and } c \le Y \le d) = \int_a^b \int_c^d f(x,y)\, dy\, dx$$

The joint probability density function has the property that

$$\int_{-\infty}^{\infty} \int_{-\infty}^{\infty} f(x,y)\, dy\, dx = 1$$

We have seen that if X and Y are jointly discrete, the probability mass function of either variable may be found by summing the joint probability mass function over all the values of the other variable. When computed this way, the probability mass function is called the marginal probability mass function. By analogy, if X and Y are jointly continuous, the probability density function of either variable may be found by integrating the joint probability density function with respect to the other variable. When computed this way, the probability density function is called the **marginal probability density function**. Example 2.52 illustrates the idea.

Example 2.52

Refer to Example 2.51. Find the marginal probability density function of the thickness X of a washer. Find the marginal probability density function of the hole diameter Y of a washer.

Solution

Denote the marginal probability density function of X by $f_X(x)$, and the marginal probability density function of Y by $f_Y(y)$. Then

$$
\begin{aligned}
f_X(x) &= \int_{-\infty}^{\infty} f(x,y)\,dy \\
&= \int_4^5 \frac{1}{6}(x+y)\,dy \\
&= \frac{1}{6}\left(x + \frac{9}{2}\right) \qquad \text{for } 1 \le x \le 2
\end{aligned}
$$

and

$$
\begin{aligned}
f_Y(y) &= \int_{-\infty}^{\infty} f(x,y)\,dx \\
&= \int_1^2 \frac{1}{6}(x+y)\,dx \\
&= \frac{1}{6}\left(y + \frac{3}{2}\right) \qquad \text{for } 4 \le y \le 5
\end{aligned}
$$

Summary

If X and Y are jointly continuous with joint probability density function $f(x,y)$, then the marginal probability density functions of X and of Y are given, respectively, by

$$
f_X(x) = \int_{-\infty}^{\infty} f(x,y)\,dy \qquad f_Y(y) = \int_{-\infty}^{\infty} f(x,y)\,dx
$$

Example 2.53

The article "Performance Comparison of Two Location Based Routing Protocols for Ad Hoc Networks" (T. Camp, J. Boleng, et al., *Proceedings of the Twenty-first Annual Joint Conference of IEEE Computer and Communications Societies* 2002:1678–1687) describes a model for the movement of a mobile computer. Assume that a mobile computer moves within the region A bounded by the x axis, the line $x = 1$, and the line $y = x$ in such a way that if (X, Y) denotes the position of the computer at a given time, the joint density of X and Y is given by

$$f(x,y) = \begin{cases} 8xy & (x,y) \in A \\ 0 & (x,y) \notin A \end{cases}$$

Find $P(X > 0.5 \text{ and } Y < 0.5)$.

Solution

The region A is the triangle shown in Figure 2.13, with the region $X > 0.5$ and $Y < 0.5$ shaded in. To find $P(X > 0.5 \text{ and } Y < 0.5)$, we integrate the joint density over the shaded region.

$$\begin{aligned} P(X > 0.5 \text{ and } Y < 0.5) &= \int_{0.5}^{1} \int_{0}^{0.5} 8xy \, dy \, dx \\ &= \int_{0.5}^{1} \left\{ 4xy^2 \Big|_{y=0}^{y=0.5} \right\} dx \\ &= \int_{0.5}^{1} x \, dx \\ &= 0.375 \end{aligned}$$

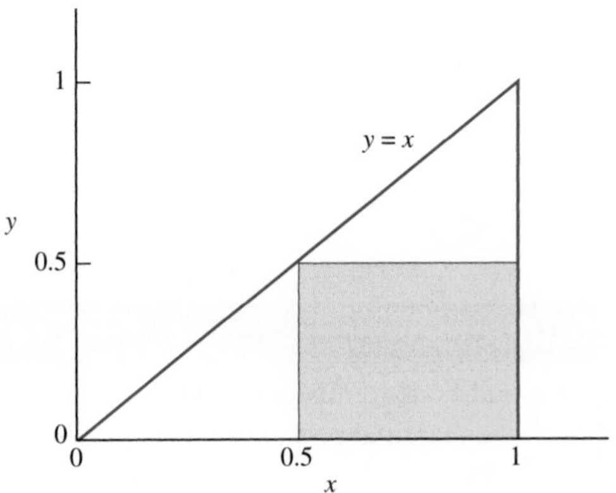

FIGURE 2.13 The triangle represents the region where the joint density of X and Y is positive. By integrating the joint density over the shaded square, we find the probability that the point (X, Y) lies in the shaded square.

Example 2.54

Refer to Example 2.53. Find the marginal densities of X and of Y.

Solution

To compute $f_X(x)$, the marginal density of X, we fix x and integrate the joint density along the vertical line through x, as shown in Figure 2.14. The integration is with respect to y, and the limits of integration are $y = 0$ to $y = x$.

$$\begin{aligned} f_X(x) &= \int_0^x 8xy\,dy \\ &= 4xy^2 \Big|_{y=0}^{y=x} \\ &= 4x^3 \qquad \text{for } 0 < x < 1 \end{aligned}$$

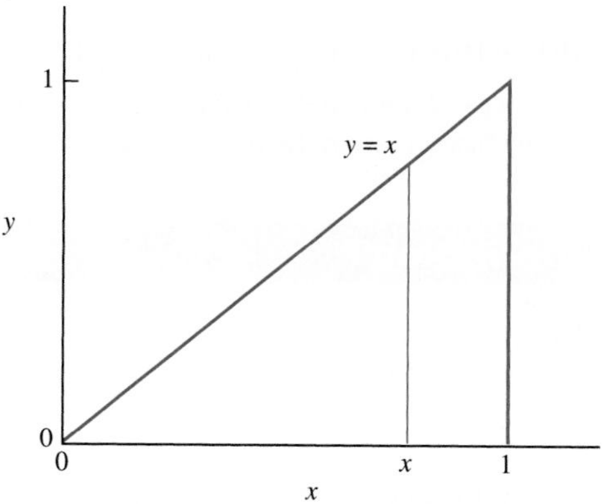

FIGURE 2.14 The marginal density $f_X(x)$ is computed by integrating the joint density along the vertical line through x.

To compute $f_Y(y)$, the marginal density of Y, we fix y and integrate the joint density along the horizontal line through y, as shown in Figure 2.15 (page 128). The integration is with respect to x, and the limits of integration are $x = y$ to $x = 1$.

$$\begin{aligned} f_Y(y) &= \int_y^1 8xy\,dy \\ &= 4x^2 y \Big|_{x=y}^{x=1} \\ &= 4y - 4y^3 \qquad \text{for } 0 < y < 1 \end{aligned}$$

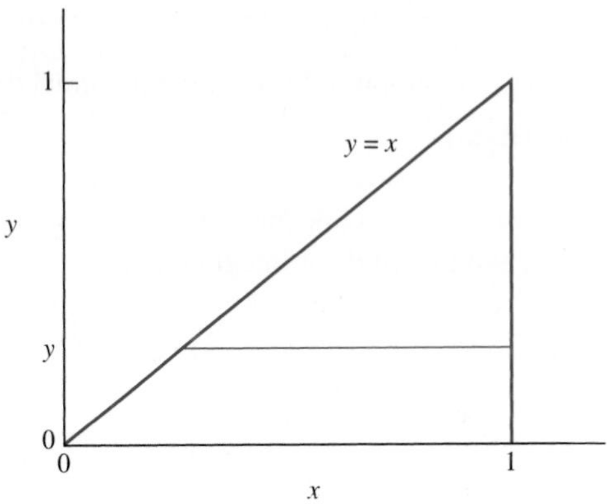

FIGURE 2.15 The marginal density $f_Y(y)$ is computed by integrating the joint density along the horizontal line through y.

More than Two Random Variables

The ideas of joint probability mass functions and joint probability density functions extend easily to more than two random variables. We present the definitions here.

Definition

- If the random variables $X_1, \ldots, X_n$ are jointly discrete, the joint probability mass function is

$$p(x_1, \ldots, x_n) = P(X_1 = x_1, \ \ldots, \ X_n = x_n)$$

- If the random variables $X_1, \ldots, X_n$ are jointly continuous, they have a joint probability density function $f(x_1, \ldots, x_n)$, where

$$P(a_1 \le X_1 \le b_1, \ \ldots, \ a_n \le X_n \le b_n) = \int_{a_n}^{b_n} \cdots \int_{a_1}^{b_1} f(x_1, \ldots, x_n)\, dx_1 \cdots dx_n$$

for any constants $a_1 \le b_1, \ldots, a_n \le b_n$.

Means of Functions of Random Variables

Sometimes we are given a random variable X and we need to work with a function of X. If X is a random variable, and $h(X)$ is a function of X, then $h(X)$ is a random variable as well. If we wish to compute the mean of $h(X)$, it can be done by using the probability mass function or probability density function of X. It is not necessary to know the probability mass function or probability density function of $h(X)$.

Let X be a random variable, and let $h(X)$ be a function of X. Then

- If X is discrete with probability mass function $p(x)$, the mean of $h(X)$ is given by

$$\mu_{h(X)} = \sum_x h(x)p(x) \tag{2.59}$$

where the sum is taken over all the possible values of X.

- If X is continuous with probability density function $f(x)$, the mean of $h(X)$ is given by

$$\mu_{h(X)} = \int_{-\infty}^{\infty} h(x)f(x)\,dx \tag{2.60}$$

Note that if we substitute $h(X) = (X - \mu_X)^2$ in either Equation (2.59) or (2.60), the right-hand side of the equation becomes an expression for the variance of X. It follows that $\sigma_X^2 = \mu_{(X-\mu_X)^2}$. We can obtain another expression for the variance of X by substituting $h(X) = X^2$ and subtracting μ_X^2 from both sides of the equation. We conclude that $\sigma_X^2 = \mu_{X^2} - \mu_X^2$.

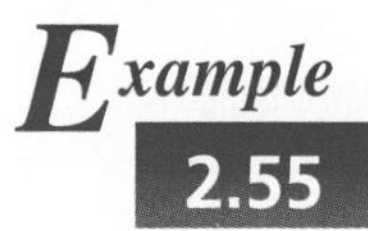

An internal combustion engine contains several cylinders bored into the engine block. Let X represent the bore diameter of a cylinder, in millimeters. Assume that the probability density function of X is

$$f(x) = \begin{cases} 10 & 80.5 < x < 80.6 \\ 0 & \text{otherwise} \end{cases}$$

Let $A = \pi X^2/4$ represent the area of the bore. Find the mean of A.

Solution

$$\begin{aligned} \mu_A &= \int_{-\infty}^{\infty} \frac{\pi x^2}{4} f(x)\,dx \\ &= \int_{80.5}^{80.6} \frac{\pi x^2}{4}(10)\,dx \\ &= 5096 \end{aligned}$$

The mean area is 5096 mm^2.

If $h(X) = aX + b$ is a linear function of X, then the mean μ_{aX+b} and the variance σ_{aX+b}^2 can be expressed in terms of μ_X and σ_X^2. These results were presented in Equations (2.44) through (2.46) in Section 2.5; we repeat them here.

If X is a random variable, and a and b are constants, then

$$\mu_{aX+b} = a\mu_X + b \tag{2.61}$$

$$\sigma^2_{aX+b} = a^2\sigma^2_X \tag{2.62}$$

$$\sigma_{aX+b} = |a|\sigma_X \tag{2.63}$$

Proofs of these results are presented at the end of this section.

If X and Y are jointly distributed random variables, and $h(X,Y)$ is a function of X and Y, then the mean of $h(X,Y)$ can be computed from the joint probability mass function or joint probability density function of X and Y.

If X and Y are jointly distributed random variables, and $h(X,Y)$ is a function of X and Y, then

- If X and Y are jointly discrete with joint probability mass function $p(x,y)$,

$$\mu_{h(X,Y)} = \sum_x \sum_y h(x,y)p(x,y) \tag{2.64}$$

where the sum is taken over all the possible values of X and Y.

- If X and Y are jointly continuous with joint probability density function $f(x,y)$,

$$\mu_{h(X,Y)} = \int_{-\infty}^{\infty}\int_{-\infty}^{\infty} h(x,y)f(x,y)\,dx\,dy \tag{2.65}$$

Example 2.56

The displacement of a piston in an internal combustion engine is defined to be the volume that the top of the piston moves through from the top to the bottom of its stroke. Let X represent the diameter of the cylinder bore, in millimeters, and let Y represent the length of the piston stroke in millimeters. The displacement is given by $D = \pi X^2 Y/4$. Assume X and Y are jointly distributed with joint probability mass function

$$f(x,y) = \begin{cases} 100 & 80.5 < x < 80.6 \text{ and } 65.1 < y < 65.2 \\ 0 & \text{otherwise} \end{cases}$$

Find the mean of D.

Solution

$$\mu_D = \int_{-\infty}^{\infty}\int_{-\infty}^{\infty} \frac{\pi x^2 y}{4} f(x,y)\,dx\,dy$$

$$= \int_{65.1}^{65.2} \int_{80.5}^{80.6} \frac{\pi x^2 y}{4} (100) \, dx \, dy$$

$$= 331{,}998$$

The mean displacement is 331,998 mm^3, or $\approx$ 332 mL.

Conditional Distributions

If X and Y are jointly distributed random variables, then knowing the value of X may change probabilities regarding the random variable Y. For example, let X represent the height in inches and Y represent the weight in pounds of a randomly chosen college student. Let's say that we are interested in the probability $P(Y \geq 200)$. If we know the joint density of X and Y, we can determine this probability by computing the marginal density of Y. Now let's say that we learn that the student's height is $X = 78$. Clearly, this knowledge changes the probability that $Y \geq 200$. To compute this new probability, the idea of a **conditional** distribution is needed.

We will first discuss the case where X and Y are jointly discrete. Let x be any value for which $P(X = x) > 0$. Then the conditional probability that $Y = y$ given $X = x$ is $P(Y = y \mid X = x)$. We will express this conditional probability in terms of the joint and marginal probability mass functions. Let $p(x, y)$ denote the joint probability mass function of X and Y, and let $p_X(x)$ denote the marginal probability mass function of X. Then the conditional probability

$$P(Y = y \mid X = x) = \frac{P(X = x \text{ and } Y = y)}{P(X = x)} = \frac{p(x, y)}{p_X(x)}$$

The **conditional probability mass function** of Y given $X = x$ is the conditional probability $P(Y = y \mid X = x)$, considered as a function of y and x.

Definition

Let X and Y be jointly discrete random variables, with joint probability mass function $p(x, y)$. Let $p_X(x)$ denote the marginal probability mass function of X and let x be any number for which $p_X(x) > 0$.

The conditional probability mass function of Y given $X = x$ is

$$p_{Y|X}(y \mid x) = \frac{p(x, y)}{p_X(x)} \tag{2.66}$$

Note that for any particular values of x and y, the value of $p_{Y|X}(y \mid x)$ is just the conditional probability $P(Y = y \mid X = x)$.

Example 2.57

Table 2.3 presents the joint probability mass function of the length X and width Y of a CD cover. Compute the conditional probability mass function $p_{Y|X}(y \mid 130)$.

Solution

The possible values for Y are $y = 15$ and $y = 16$. From Table 2.3, $P(Y = 15 \text{ and } X = 130) = 0.42$, and $P(X = 130) = 0.70$. Therefore,

$$\begin{aligned} p_{Y|X}(15 \mid 130) &= P(Y = 15 \mid X = 130) \\ &= \frac{P(Y = 15 \text{ and } X = 130)}{P(X = 130)} \\ &= \frac{0.42}{0.70} \\ &= 0.60 \end{aligned}$$

The value of $p_{Y|X}(16 \mid 130)$ can be computed with a similar calculation. Alternatively, note that $p_{Y|X}(16 \mid 130) = 1 - p_{Y|X}(15 \mid 130)$, since $y = 15$ and $y = 16$ are the only two possible values for Y. Therefore $p_{Y|X}(16 \mid 130) = 0.4$. The conditional probability mass function of Y given $X = 130$ is therefore $p_{Y|X}(15 \mid 130) = 0.60$, $p_{Y|X}(16 \mid 130) = 0.40$, and $p_{Y|X}(y \mid 130) = 0$ for any value of y other than 15 or 16.

The analog to the conditional probability mass function for jointly continuous random variables is the **conditional probability density function**. The definition of the conditional probability density function is just like that of the conditional probability mass function, with mass functions replaced by density functions.

Definition

Let X and Y be jointly continuous random variables, with joint probability density function $f(x, y)$. Let $f_X(x)$ denote the marginal probability density function of X and let x be any number for which $f_X(x) > 0$.

The conditional probability density function of Y given $X = x$ is

$$f_{Y|X}(y \mid x) = \frac{f(x, y)}{f_X(x)} \tag{2.67}$$

Example 2.58

(Continuing Example 2.51.) The joint probability density function of the thickness X and hole diameter Y (both in millimeters) of a randomly chosen washer is $f(x, y) = (1/6)(x + y)$ for $1 \le x \le 2$ and $4 \le y \le 5$. Find the conditional probability density function of Y given $X = 1.2$. Find the probability that the hole diameter is less than or equal to 4.8 mm given that the thickness is 1.2 mm.

Solution

In Example 2.52 we computed the marginal probability density functions

$$f_X(x) = \frac{1}{6}(x + 4.5) \quad \text{for } 1 \le x \le 2 \qquad f_Y(y) = \frac{1}{6}(y + 1.5) \quad \text{for } 4 \le y \le 5$$

The conditional probability density function of Y given $X = 1.2$ is

$$\begin{aligned} f_{Y|X}(y \mid 1.2) &= \frac{f(1.2,\, y)}{f_X(1.2)} \\ &= \begin{cases} \dfrac{(1/6)(1.2 + y)}{(1/6)(1.2 + 4.5)} & \text{if } 4 \le y \le 5 \\ 0 & \text{otherwise} \end{cases} \\ &= \begin{cases} \dfrac{1.2 + y}{5.7} & \text{if } 4 \le y \le 5 \\ 0 & \text{otherwise} \end{cases} \end{aligned}$$

The probability that the hole diameter is less than or equal to 4.8 mm given that the thickness is 1.2 mm is $P(Y \le 4.8 \mid X = 1.2)$. This is found by integrating $f_{Y|X}(y|1.2)$ over the region $y \le 4.8$:

$$\begin{aligned} P(Y \le 4.8 \mid X = 1.2) &= \int_{-\infty}^{4.8} f_{Y|X}(y \mid 1.2)\, dy \\ &= \int_{4}^{4.8} \frac{1.2 + y}{5.7}\, dy \\ &= 0.786 \end{aligned}$$

Conditional Expectation

Expectation is another term for mean. A **conditional expectation** is an expectation, or mean, calculated using a conditional probability mass function or conditional probability density function. The conditional expectation of Y given $X = x$ is denoted $E(Y \mid X = x)$ or $\mu_{Y|X}$. We illustrate with Examples 2.59 through 2.61.

Example 2.59

Table 2.3 presents the joint probability mass function of the length X and width Y of a CD cover. Compute the conditional expectation $E(Y \mid X = 130)$.

Solution

We computed the conditional probability mass function $p_{Y|X}(y \mid 130)$ in Example 2.57. The conditional expectation $E(Y \mid X = 130)$ is calculated using the definition of the mean of a discrete random variable and the conditional probability mass function. Specifically,

$$
\begin{aligned}
E(Y \mid X = 130) &= \sum_y y\, p_{Y|X}(y \mid 130) \\
&= 15\, p_{Y|X}(15 \mid 130) + 16\, p_{Y|X}(16 \mid 130) \\
&= 15(0.60) + 16(0.40) \\
&= 15.4
\end{aligned}
$$

Example 2.60

Refer to Example 2.58. Find the conditional expectation of Y given that $X = 1.2$.

Solution

Since X and Y are jointly continuous, we use the definition of the mean of a continuous random variable to compute the conditional expectation.

$$
\begin{aligned}
E(Y \mid X = 1.2) &= \int_{-\infty}^{\infty} y f_{Y|X}(y \mid 1.2)\, dy \\
&= \int_4^5 y \frac{1.2 + y}{5.7}\, dy \\
&= 4.5146
\end{aligned}
$$

Example 2.61

Refer to Example 2.58. Find the value μ_Y (which can be called the unconditional mean of Y). Does it differ from $E(Y \mid X = 1.2)$?

Solution

The value μ_Y is calculated using the marginal probability mass function of Y. Thus

$$
\begin{aligned}
\mu_Y &= \int_{-\infty}^{\infty} y f_Y(y)\, dy \\
&= \int_4^5 y \frac{1}{6}(y + 1.5)\, dy \\
&= 4.5139
\end{aligned}
$$

The conditional expectation in this case differs slightly from the unconditional expectation.

Independent Random Variables

The notion of independence for random variables is very much like the notion of independence for events. Two random variables are independent if knowledge regarding one of them does not affect the probabilities of the other. We present here a definition of independence of random variables in terms of their joint probability mass or joint probability density function. A different but logically equivalent definition was presented in Section 2.5.

Definition

Two random variables X and Y are independent, provided that

- If X and Y are jointly discrete, the joint probability mass function is equal to the product of the marginals:

$$p(x,y) = p_X(x)p_Y(y)$$

- If X and Y are jointly continuous, the joint probability density function is equal to the product of the marginals:

$$f(x,y) = f_X(x)f_Y(y)$$

Random variables $X_1, \ldots, X_n$ are independent, provided that

- If $X_1, \ldots, X_n$ are jointly discrete, the joint probability mass function is equal to the product of the marginals:

$$p(x_1, \ldots, x_n) = p_{X_1}(x_1) \cdots p_{X_n}(x_n)$$

- If $X_1, \ldots, X_n$ are jointly continuous, the joint probability density function is equal to the product of the marginals:

$$f(x_1, \ldots, x_n) = f_{X_1}(x_1) \cdots f_{X_n}(x_n)$$

Intuitively, when two random variables are independent, knowledge of the value of one of them does not affect the probability distribution of the other. In other words, the conditional distribution of Y given X is the same as the marginal distribution of Y.

If X and Y are independent random variables, then

- If X and Y are jointly discrete, and x is a value for which $p_X(x) > 0$, then

$$p_{Y|X}(y \mid x) = p_Y(y)$$

- If X and Y are jointly continuous, and x is a value for which $f_X(x) > 0$, then

$$f_{Y|X}(y \mid x) = f_Y(y)$$

Example 2.62

The joint probability mass function of the length X and thickness Y of a CD tray cover is given in Table 2.3. Are X and Y independent?

Solution

We must check to see if $P(X = x \text{ and } Y = y) = P(X = x)P(Y = y)$ for every value of x and y. We begin by checking $x = 129$, $y = 15$:

$$P(X = 129 \text{ and } Y = 15) = 0.12 = (0.20)(0.60) = P(X = 129)P(Y = 15)$$

Continuing in this way, we can verify that $P(X = x \text{ and } Y = y) = P(X = x)P(Y = y)$ for every value of x and y. Therefore X and Y are independent.

2.63

(Continuing Example 2.51.) The joint probability density function of the thickness X and hole diameter Y of a randomly chosen washer is $f(x,y) = (1/6)(x + y)$ for $1 \le x \le 2$ and $4 \le y \le 5$. Are X and Y independent?

Solution

In Example 2.52 we computed the marginal probability mass functions

$$f_X(x) = \frac{1}{6}\left(x + \frac{9}{2}\right) \qquad f_Y(y) = \frac{1}{6}\left(y + \frac{3}{2}\right)$$

Clearly $f(x,y) \ne f_X(x)f_Y(y)$. Therefore X and Y are not independent.

Covariance

When two random variables are not independent, it is useful to have a measure of the strength of the relationship between them. The **population covariance** is a measure of a certain type of relationship known as a *linear* relationship. We will usually drop the term "population," and refer simply to the covariance.

Definition

Let X and Y be random variables with means μ_X and μ_Y. The covariance of X and Y is

$$\text{Cov}(X,Y) = \mu_{(X-\mu_X)(Y-\mu_Y)} \tag{2.68}$$

An alternate formula is

$$\text{Cov}(X,Y) = \mu_{XY} - \mu_X\mu_Y \tag{2.69}$$

A proof of the equivalence of these two formulas is presented at the end of the section. It is important to note that the units of $\text{Cov}(X,Y)$ are the units of X multiplied by the units of Y.

How does the covariance measure the strength of the linear relationship between X and Y? The covariance is the mean of the product of the deviations $(X - \mu_X)(Y - \mu_Y)$. If a Cartesian coordinate system is constructed with the origin at (μ_X, μ_Y), this product will be positive in the first and third quadrants, and negative in the second and fourth quadrants (see Figure 2.16). It follows that if $\text{Cov}(X,Y)$ is strongly positive, then values of (X,Y) in the first and third quadrants will be observed much more often than values in the second and fourth quadrants. In a random sample of points, therefore, larger values of X would tend to be paired with larger values of Y, while smaller values of X would tend to be paired with smaller values of Y (see Figure 2.16a). Similarly, if $\text{Cov}(X,Y)$ is strongly negative, the points in a random sample would be more likely to lie in the second and fourth quadrants, so larger values of X would tend to be paired with smaller values of Y (see Figure 2.16b). Finally, if $\text{Cov}(X,Y)$ is near 0, there would be little tendency for larger values of X to be paired with either larger or smaller values of Y (see Figure 2.16c).

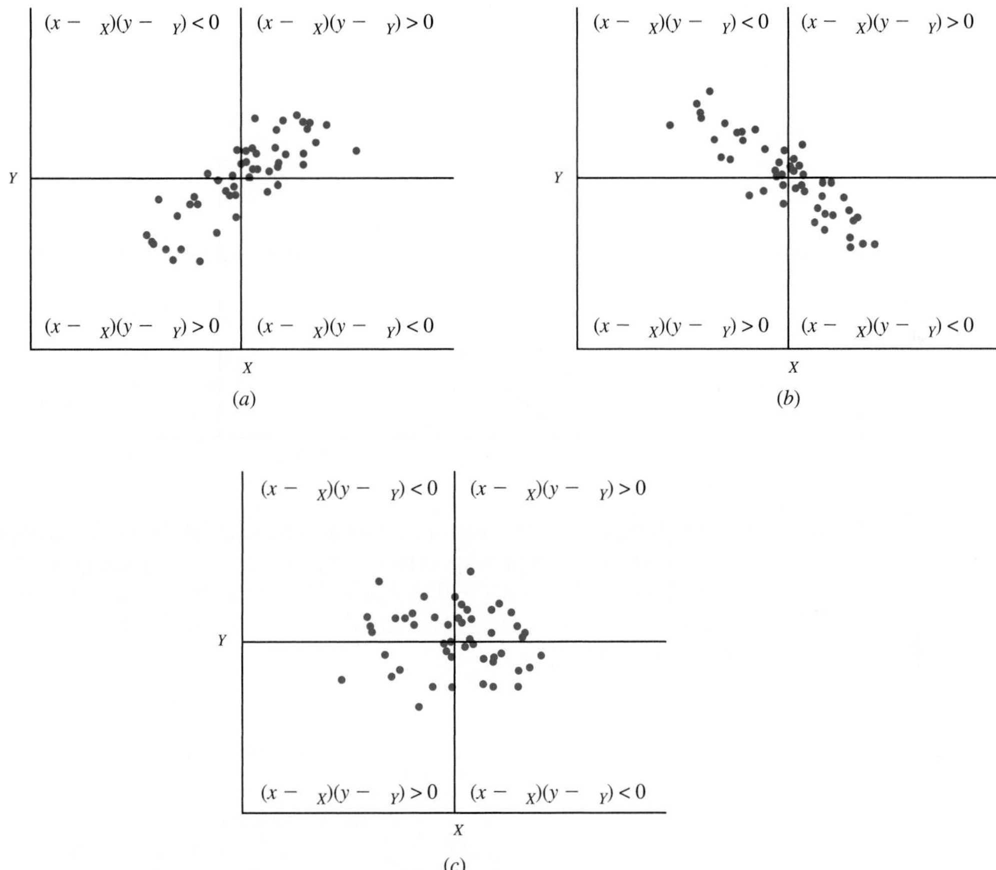

FIGURE 2.16 *(a)* A random sample of points from a population with positive covariance. *(b)* A random sample of points from a population with negative covariance. *(c)* A random sample of points from a population with covariance near 0.

Example 2.64

Continuing Example 2.53, a mobile computer is moving in the region A bounded by the x axis, the line $x = 1$, and the line $y = x$ (see Figure 2.13). If (X,Y) denotes the position of the computer at a given time, the joint density of X and Y is given by

$$f(x,y) = \begin{cases} 8xy & (x,y) \in A \\ 0 & (x,y) \notin A \end{cases}$$

Find $\text{Cov}(X,Y)$.

Solution

We will use the formula $\text{Cov}(X,Y) = \mu_{XY} - \mu_X\mu_Y$ (Equation 2.69). First we compute μ_{XY}:

$$\mu_{XY} = \int_{-\infty}^{\infty}\int_{-\infty}^{\infty} xyf(x,y)\,dy\,dx$$

Now the joint density is positive on the triangle shown.

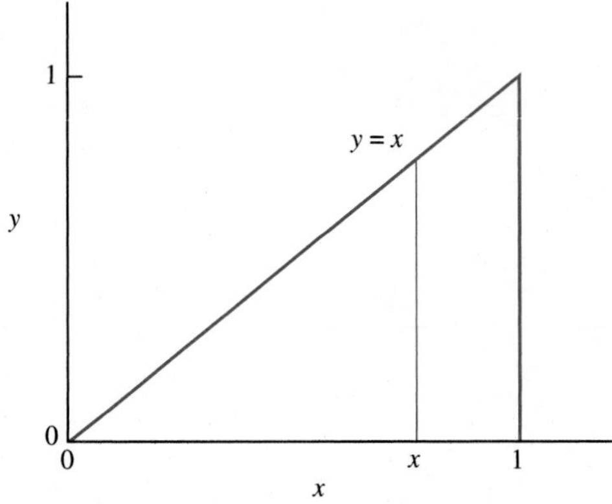

To compute the integral over this region, we fix a value of x, as shown. We compute the inner integral by integrating with respect to y along the vertical line through x. The limits of integration along this line are $y = 0$ to $y = x$. Then we compute the outer integral by integrating with respect to x over all possible values of x, so the limits of integration on the outer integral are $x = 0$ to $x = 1$.

Therefore

$$\begin{aligned}\mu_{XY} &= \int_0^1\int_0^x xy(8xy)\,dy\,dx \\ &= \int_0^1\left(\int_0^x 8x^2y^2\,dy\right)dx \\ &= \int_0^1 \frac{8x^5}{3}\,dx \\ &= \frac{4}{9}\end{aligned}$$

To find μ_X and μ_Y, we will use the marginal densities computed in Example 2.54. These are

$$f_X(x) = \begin{cases} 4x^3 & 0 < x < 1 \\ 0 & \text{otherwise}\end{cases}$$

$$f_Y(y) = \begin{cases} 4y - 4y^3 & 0 < y < 1 \\ 0 & \text{otherwise}\end{cases}$$

We now compute μ_X and μ_Y:

$$\begin{aligned}\mu_X &= \int_{-\infty}^{\infty} x f_X(x)\,dx \\ &= \int_0^1 4x^4\,dx \\ &= \frac{4}{5}\end{aligned}$$

$$\begin{aligned}\mu_Y &= \int_{-\infty}^{\infty} y f_Y(y)\,dy \\ &= \int_0^1 \left(4y^2 - 4y^4\right)\,dy \\ &= \frac{8}{15}\end{aligned}$$

Now $\text{Cov}(X,Y) = \frac{4}{9} - \left(\frac{4}{5}\right)\left(\frac{8}{15}\right) = \frac{4}{225} = 0.01778.$

Example 2.65

Quality-control checks on wood paneling involve counting the number of surface flaws on each panel. On a given 2×8 ft panel, let X be the number of surface flaws due to uneven application of the final coat of finishing material, and let Y be the number of surface flaws due to inclusions of foreign particles in the finish. The joint probability mass function $p(x,y)$ of X and Y is presented in the following table. The marginal probability mass functions are presented as well, in the margins of the table. Find the covariance of X and Y.

		y		
x	**0**	**1**	**2**	$p_X(x)$
0	0.05	0.10	0.20	0.35
1	0.05	0.15	0.05	0.25
2	0.25	0.10	0.05	0.40
$p_Y(y)$	0.35	0.35	0.30	

Solution

We will use the formula $\text{Cov}(X,Y) = \mu_{XY} - \mu_X\mu_Y$ (Equation 2.69). First we compute μ_{XY}:

$$\begin{aligned}\mu_{XY} &= \sum_{x=0}^{2}\sum_{y=0}^{2} xy\,p(x,y) \\ &= (1)(1)(0.15) + (1)(2)(0.05) + (2)(1)(0.10) + (2)(2)(0.05) \\ &= 0.65 \qquad \text{(omitting terms equal to 0)}\end{aligned}$$

We use the marginals to compute μ_X and μ_Y:

$$\mu_X = (0)(0.35) + (1)(0.25) + (2)(0.40) = 1.05$$

$$\mu_Y = (0)(0.35) + (1)(0.35) + (2)(0.30) = 0.95$$

It follows that $\text{Cov}(X,Y) = 0.65 - (1.05)(0.95) = -0.3475$.

Correlation

If X and Y are jointly distributed random variables, $\text{Cov}(X,Y)$ measures the strength of the linear relationship between them. As mentioned previously, the covariance has units, which are the units of X multiplied by the units of Y. This is a serious drawback in practice, because one cannot use the covariance to determine which of two pairs of random variables is more strongly related, since the two covariances will have different units. What is needed is a measure of the strength of a linear relationship that is a pure number. The **population correlation** is such a measure. We will usually drop the term "population," and refer simply to the correlation. We will denote the correlation between random variables X and Y by $\rho_{X,Y}$.

The correlation is a scaled version of the covariance. Specifically, to compute the correlation between X and Y, one first computes the covariance, and then gets rid of the units by dividing by the product of the standard deviations of X and Y. It can be proved by advanced methods that the correlation is always between -1 and 1.

Summary

Let X and Y be jointly distributed random variables with standard deviations σ_X and σ_Y. The correlation between X and Y is denoted $\rho_{X,Y}$ and is given by

$$\rho_{X,Y} = \frac{\text{Cov}(X,Y)}{\sigma_X \sigma_Y} \tag{2.70}$$

For any two random variables X and Y:

$$-1 \leq \rho_{X,Y} \leq 1$$

Example 2.66

Refer to Example 2.64. Find $\rho_{X,Y}$.

Solution

In Example 2.64, we computed $\text{Cov}(X,Y) = 0.01778$, $\mu_X = 4/5$, and $\mu_Y = 8/15$. We now must compute σ_X and σ_Y. To do this we use the marginal densities of X and of Y, which were computed in Example 2.54. These are $f_X(x) = 4x^3$ for $0 < x < 1$, and $f_Y(y) = 4y - 4y^3$ for $0 < y < 1$. We obtain

$$
\begin{aligned}
\sigma_X^2 &= \int_{-\infty}^{\infty} x^2 f_X(x)\,dx - \mu_X^2 \\
&= \int_0^1 4x^5\,dx - \left(\frac{4}{5}\right)^2 \\
&= 0.02667
\end{aligned}
$$

$$
\begin{aligned}
\sigma_Y^2 &= \int_{-\infty}^{\infty} y^2 f_Y(y)\,dy - \mu_Y^2 \\
&= \int_0^1 (4y^3 - 4y^5)\,dy - \left(\frac{8}{15}\right)^2 \\
&= 0.04889
\end{aligned}
$$

It follows that $\rho_{X,Y} = \dfrac{0.01778}{\sqrt{(0.02667)(0.04889)}} = 0.492.$

Example 2.67

Refer to Example 2.65. Find $\rho_{X,Y}$.

Solution

In Example 2.65, we computed $\mathrm{Cov}(X,Y) = -0.3475$, $\mu_X = 1.05$, and $\mu_Y = 0.95$. We now must compute σ_X and σ_Y. To do this we use the marginal densities of X and of Y, which were presented in the table in Example 2.65. We obtain

$$
\begin{aligned}
\sigma_X^2 &= \sum_{x=0}^{2} x^2 p_X(x) - \mu_X^2 \\
&= (0^2)(0.35) + (1^2)(0.25) + (2^2)(0.40) - 1.05^2 \\
&= 0.7475
\end{aligned}
$$

$$
\begin{aligned}
\sigma_Y^2 &= \sum_{y=0}^{2} y^2 p_Y(y) - \mu_Y^2 \\
&= (0^2)(0.35) + (1^2)(0.35) + (2^2)(0.30) - 0.95^2 \\
&= 0.6475
\end{aligned}
$$

It follows that

$$\rho_{X,Y} = \frac{-0.3475}{\sqrt{(0.7475)(0.6475)}} = -0.499$$

Covariance, Correlation, and Independence

When $\mathrm{Cov}(X,Y) = \rho_{X,Y} = 0$, there is no linear relationship between X and Y. In this case we say that X and Y are **uncorrelated**. Note that if $\mathrm{Cov}(X,Y) = 0$, then it is always the case that $\rho_{X,Y} = 0$, and vice versa. If X and Y are independent random variables, then X and Y are always uncorrelated, since there is no relationship, linear or otherwise, between them. It is mathematically possible to construct random variables that are uncorrelated but not independent. This phenomenon is rarely seen in practice, however.

Summary

- If $\text{Cov}(X,Y) = \rho_{X,Y} = 0$, then X and Y are said to be uncorrelated.
- If X and Y are independent, then X and Y are uncorrelated.
- It is mathematically possible for X and Y to be uncorrelated without being independent. This rarely occurs in practice.

A proof of the fact that independent random variables are always uncorrelated is presented at the end of this section. An example of random variables that are uncorrelated but not independent is presented in Exercise 26.

Linear Combinations of Random Variables

We discussed linear combinations of random variables in Section 2.5. We review the results here and include additional results on the variance of a linear combination of dependent random variables.

If $X_1, \ldots, X_n$ are random variables and $c_1, \ldots, c_n$ are constants, then the random variable

$$c_1 X_1 + \cdots + c_n X_n$$

is called a **linear combination** of $X_1, \ldots, X_n$.

If $X_1, \ldots, X_n$ are random variables and $c_1, \ldots, c_n$ are constants, then

$$\mu_{c_1 X_1 + \cdots + c_n X_n} = c_1 \mu_{X_1} + \cdots + c_n \mu_{X_n} \tag{2.71}$$

$$\sigma^2_{c_1 X_1 + \cdots + c_n X_n} = c_1^2 \sigma^2_{X_1} + \cdots + c_n^2 \sigma^2_{X_n} + 2 \sum_{i=1}^{n-1} \sum_{j=i+1}^{n} c_i c_j \, \text{Cov}(X_i, X_j) \tag{2.72}$$

Proofs of these results for the case $n = 2$ are presented at the end of this section. Equation (2.72) is the most general result regarding the variance of a linear combination of random variables. As a special case, note that if $X_1, \ldots, X_n$ are independent, then all the covariances are equal to 0, so the result simplifies:

If $X_1, \ldots, X_n$ are *independent* random variables and $c_1, \ldots, c_n$ are constants, then

$$\sigma^2_{c_1X_1+\cdots+c_nX_n} = c_1^2\sigma^2_{X_1} + \cdots + c_n^2\sigma^2_{X_n} \tag{2.73}$$

In particular,

$$\sigma^2_{X_1+\cdots+X_n} = \sigma^2_{X_1} + \cdots + \sigma^2_{X_n} \tag{2.74}$$

Finally, we present some special cases of Equations (2.72) and (2.74) in which there are only two random variables:

If X and Y are random variables, then

$$\sigma^2_{X+Y} = \sigma^2_X + \sigma^2_Y + 2\,\mathrm{Cov}(X,Y) \tag{2.75}$$

$$\sigma^2_{X-Y} = \sigma^2_X + \sigma^2_Y - 2\,\mathrm{Cov}(X,Y) \tag{2.76}$$

If X and Y are *independent* random variables, then

$$\sigma^2_{X+Y} = \sigma^2_X + \sigma^2_Y \tag{2.77}$$

$$\sigma^2_{X-Y} = \sigma^2_X + \sigma^2_Y \tag{2.78}$$

Note that the variance of the difference $X - Y$ of two independent random variables is the *sum* of the variances.

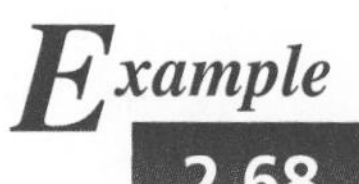

(Continuing Example 2.53.) Assume that the mobile computer moves from a random position (X, Y) vertically to the point $(X, 0)$, and then along the x axis to the origin. Find the mean and variance of the distance traveled.

Solution

The distance traveled is the sum $X + Y$. The means of X and of Y were computed in Example 2.64. They are $\mu_X = 4/5 = 0.800$, and $\mu_Y = 8/15 = 0.533$. We compute

$$\begin{aligned}\mu_{X+Y} &= \mu_X + \mu_Y \\ &= 0.800 + 0.533 \\ &= 1.333\end{aligned}$$

To compute σ^2_{X+Y}, we use Equation (2.75). In Example 2.64 we computed $\mathrm{Cov}(X,Y) = 0.01778$. In Example 2.66 we computed $\sigma^2_X = 0.02667$ and $\sigma^2_Y = 0.04889$. Therefore

$$\begin{aligned}\sigma^2_{X+Y} &= \sigma^2_X + \sigma^2_Y + 2\,\mathrm{Cov}(X,Y)\\ &= 0.02667 + 0.04889 + 2(0.01778)\\ &= 0.1111\end{aligned}$$

The Mean and Variance of a Sample Mean

We review the procedures for computing the mean and variance of a sample mean, which were presented in Section 2.5. When a simple random sample of numerical values is drawn from a population, each item in the sample can be thought of as a random variable. Unless the sample is a large proportion (more than 5%) of the population, the items in the sample may be treated as independent (see the discussion of independence in Section 1.1). From here on, unless stated to the contrary, we will assume that the values in a simple random sample may be treated as independent random variables.

> If $X_1, \ldots, X_n$ is a simple random sample, then $X_1, \ldots, X_n$ may be treated as independent random variables, all with the same distribution.

The most frequently encountered linear combination is the sample mean. Specifically, if $X_1, \ldots, X_n$ is a simple random sample, then $X_1, \ldots, X_n$ are independent, and the sample mean $\overline{X}$ is the linear combination

$$\overline{X} = \frac{1}{n}X_1 + \cdots + \frac{1}{n}X_n$$

Formulas for the mean and variance of $\overline{X}$ may therefore be derived from Equations (2.71) and (2.73), respectively, by setting $c_1 = c_2 = \cdots = c_n = 1/n$.

> If $X_1, \ldots, X_n$ is a simple random sample from a population with mean μ and variance σ^2, then the sample mean $\overline{X}$ is a random variable with
>
> $$\mu_{\overline{X}} = \mu \qquad (2.79)$$
>
> $$\sigma^2_{\overline{X}} = \frac{\sigma^2}{n} \qquad (2.80)$$
>
> The standard deviation of $\overline{X}$ is
>
> $$\sigma_{\overline{X}} = \frac{\sigma}{\sqrt{n}} \qquad (2.81)$$

The article "Water Price Influence on Apartment Complex Water Use" (D. Agthe and R. Billings, *Journal of Water Resources Planning and Management*, 2002:366–369) discusses the volume of water used in apartments in 308 complexes in Tucson, Arizona.

The volume used per apartment during the summer had mean 20.4 m^3 and standard deviation 11.1 m^3. Find the mean and standard deviation for the sample mean water use in a sample of 100 apartments. How many apartments must be sampled so that the sample mean water use will have a standard deviation equal to 0.5 m^3?

Solution

Let $X_1, \ldots, X_{100}$ be the amounts of water used in a sample of 100 apartments. Then $X_1, \ldots, X_{100}$ come from a population with mean $\mu = 20.4$ and standard deviation $\sigma = 11.1$. We conclude that the sample mean $\overline{X}$ has mean $\mu_{\overline{X}} = \mu = 20.4$, and standard deviation $\sigma_{\overline{X}} = \sigma/\sqrt{100} = 1.11$. Let n be the sample size required so that $\sigma_{\overline{X}} = 0.5$. Then $\sigma/\sqrt{n} = 11.1/\sqrt{n} = 0.5$. Solving for n, we obtain $n \approx 493$.

Application to Portfolio Management

Equation (2.72) and its variants play an important role in the field of finance. Assume that an investor has a fixed number of dollars to invest. She may choose from a variety of investments, for example, stocks, bonds, and real estate. After one year she will sell her investment; let X denote her profit (or loss). The value of X cannot be predicted with certainty, so economists treat it as a random variable. The mean μ_X indicates the amount that the investment can be expected to earn on the average. The standard deviation σ_X reflects the *volatility*, or *risk*, of the investment. If σ_X is very small, then it is nearly certain that the investment will earn close to its mean return μ_X, so the risk is low. If σ_X is large, the return can vary over a wide range, so the risk is high. In general, if two investments have the same mean return, the one with the smaller standard deviation is preferable, since it earns the same return on the average with lower risk.

An investor has \$200 to invest. He will invest \$100 in each of two investments. Let X and Y denote the returns on the two investments. Assume that $\mu_X = \mu_Y = \$5$, $\sigma_X = \sigma_Y = \$2$, and $\rho_{X,Y} = 0.5$. Find the mean and standard deviation of the total return on the two investments.

Solution

The total return is $X + Y$. The mean is

$$\begin{aligned}\mu_{X+Y} &= \mu_X + \mu_Y \\ &= \$5 + \$5 \\ &= \$10\end{aligned}$$

Using Equation (2.75), the standard deviation is $\sigma_{X+Y} = \sqrt{\sigma_X^2 + \sigma_Y^2 + 2\,\mathrm{Cov}(X,Y)}$. Now $\mathrm{Cov}(X,Y) = \rho_{X,Y}\sigma_X\sigma_Y = (0.5)(2)(2) = 2$. Therefore

$$\begin{aligned}\sigma_{X+Y} &= \sqrt{2^2 + 2^2 + 2(2)} \\ &= \$3.46\end{aligned}$$

It is instructive to compare the result of Example 2.70 with the result that would occur if the entire \$200 were invested in a single investment. Example 2.71 analyzes that possibility.

Example 2.71

If the investor in Example 2.70 invests the entire \$200 in one of the two investments, find the mean and standard deviation of the return.

Solution

Assume the investor invests in the investment whose return on \$100 is X (the result is the same if Y is chosen). Since \$200, rather than \$100, is invested, the return will be $2X$. The mean return is

$$\mu_{2X} = 2\mu_X = 2(5) = \$10$$

The standard deviation is

$$\sigma_{2X} = 2\sigma_X = 2(2) = \$4$$

Comparing the results of Examples 2.70 and 2.71 shows that the mean returns of the two investment strategies are the same, but the standard deviation (i.e., risk) is lower when the investment capital is divided between two investments. This is the principle of *diversification*. When two investments are available whose returns have the same mean and same risk, it is always advantageous to divide one's capital between them, rather than to invest in only one of them.

Proof that $\mu_{aX+b} = a\mu_X + b$

We assume that X is a continuous random variable with density function $f(x)$. Then

$$\begin{aligned}
\mu_{aX+b} &= \int_{-\infty}^{\infty} (ax+b) f(x)\,dx \quad \text{(Equation 2.60)} \\
&= \int_{-\infty}^{\infty} ax\, f(x)dx + \int_{-\infty}^{\infty} b\, f(x)\,dx \\
&= a\int_{-\infty}^{\infty} x\, f(x)dx + b\int_{-\infty}^{\infty} f(x)\,dx \\
&= a\mu_X + b(1) \\
&= a\mu_X + b
\end{aligned}$$

The proof in the case that X is a discrete random variable is similar, with the integrals replaced by sums.

Proof that $\mu_{aX+bY} = a\mu_X + b\mu_Y$

Let X and Y be jointly continuous with joint density $f(x, y)$ and marginal densities $f_X(x)$ and $f_Y(y)$. Let a and b be constants. Then

$$\begin{aligned}
\mu_{aX+bY} &= \int_{-\infty}^{\infty}\int_{-\infty}^{\infty}(ax+by)f(x,y)\,dx\,dy \\
&= \int_{-\infty}^{\infty}\int_{-\infty}^{\infty}axf(x,y)\,dx\,dy + \int_{-\infty}^{\infty}\int_{-\infty}^{\infty}byf(x,y)\,dx\,dy \\
&= a\int_{-\infty}^{\infty}\int_{-\infty}^{\infty}xf(x,y)\,dy\,dx + b\int_{-\infty}^{\infty}\int_{-\infty}^{\infty}yf(x,y)\,dx\,dy \\
&= a\int_{-\infty}^{\infty}x\left[\int_{-\infty}^{\infty}f(x,y)\,dy\right]dx + b\int_{-\infty}^{\infty}y\left[\int_{-\infty}^{\infty}f(x,y)\,dx\right]dy \\
&= a\int_{-\infty}^{\infty}xf_X(x)\,dx + b\int_{-\infty}^{\infty}yf_Y(y)\,dy \\
&= a\mu_X + b\mu_Y
\end{aligned}$$

The proof in the case that X and Y are jointly discrete is similar, with the integrals replaced by sums.

Proof that $\sigma^2_{aX+b} = a^2\sigma^2_X$

We will use the notation $E(X)$ interchangeably with μ_X, $E(Y)$ interchangeably with μ_Y, and so forth. Let $Y = aX + b$. Then

$$\begin{aligned}
\sigma^2_{aX+b} &= \sigma^2_Y \\
&= E(Y^2) - \mu^2_Y \\
&= E[(aX+b)^2] - \mu^2_{aX+b} \\
&= E(a^2X^2 + 2abX + b^2) - (a\mu_X + b)^2 \\
&= E(a^2X^2) + E(2abX) + E(b^2) - (a\mu_X + b)^2 \\
&= a^2E(X^2) + 2abE(X) + b^2 - a^2\mu^2_X - 2ab\mu_X - b^2 \\
&= a^2[E(X^2) - \mu^2_X] \\
&= a^2\sigma^2_X
\end{aligned}$$

Proof that $\sigma^2_{aX+bY} = a^2\sigma^2_X + b^2\sigma^2_Y + 2ab\,\mathrm{Cov}(X,Y)$

We will use the notation $E(X)$ interchangeably with μ_X, $E(Y)$ interchangeably with μ_Y, and so forth.

$$\begin{aligned}
\sigma^2_{aX+bY} &= E[(aX+bY)^2] - \mu^2_{aX+bY} \\
&= E(a^2X^2 + 2abXY + b^2Y^2) - \mu^2_{aX+bY} \\
&= E(a^2X^2) + E(2abXY) + E(b^2Y^2) - (a\mu_X + b\mu_Y)^2 \\
&= a^2E(X^2) + 2abE(XY) + b^2E(Y^2) - a^2\mu^2_X - 2ab\mu_X\mu_Y - b^2\mu^2_Y \\
&= a^2[E(X^2) - \mu^2_X] + b^2[E(Y^2) - \mu^2_Y] + 2ab[E(XY) - \mu_X\mu_Y] \\
&= a^2\sigma^2_X + b^2\sigma^2_Y + 2ab\,\mathrm{Cov}(X,Y)
\end{aligned}$$

Proof of the equivalence of Equations (2.68) and (2.69)

We will use the notation $E(X)$ interchangeably with μ_X, $E(Y)$ interchangeably with μ_Y, and so forth. We must show that

$$E[(X-\mu_X)(Y-\mu_Y)] = \mu_{XY} - \mu_X\mu_Y$$

Now

$$\begin{aligned}E[(X-\mu_X)(Y-\mu_Y)] &= E(XY - X\mu_Y - Y\mu_X + \mu_X\mu_Y)\\ &= E(XY) - E(X\mu_Y) - E(Y\mu_X) + E(\mu_X\mu_Y)\\ &= E(XY) - \mu_Y E(X) - \mu_X E(Y) + \mu_X\mu_Y\\ &= \mu_{XY} - \mu_Y\mu_X - \mu_X\mu_Y + \mu_X\mu_Y\\ &= \mu_{XY} - \mu_X\mu_Y\end{aligned}$$

Proof that if *X* and *Y* are independent then *X* and *Y* are uncorrelated

Let X and Y be independent random variables. We will show that $\mu_{XY} = \mu_X\mu_Y$, from which it will follow that $\mathrm{Cov}(X,Y) = \rho_{X,Y} = 0$. We will assume that X and Y are jointly continuous with joint density $f(x,y)$ and marginal densities $f_X(x)$ and $f_Y(y)$. The key to the proof is the fact that since X and Y are independent, $f(x,y) = f_X(x)f_Y(y)$.

$$\begin{aligned}\mu_{XY} &= \int_{-\infty}^{\infty}\int_{-\infty}^{\infty} xyf(x,y)\,dx\,dy\\ &= \int_{-\infty}^{\infty}\int_{-\infty}^{\infty} xy\ f_X(x)f_Y(y)\,dx\,dy\\ &= \int_{-\infty}^{\infty} x\ f_X(x)\,dx \int_{-\infty}^{\infty} y\ f_Y(y)\,dy\\ &= \mu_X\mu_Y\end{aligned}$$

The proof in the case that X and Y are jointly discrete is similar, with the integrals replaced by sums.

Exercises for Section 2.6

1. In a randomly chosen lot of 1000 bolts, let X be the number that fail to meet a length specification, and let Y be the number that fail to meet a diameter specification. Assume that the joint probability mass function of X and Y is given in the following table.

	y		
x	**0**	**1**	**2**
0	0.40	0.12	0.08
1	0.15	0.08	0.03
2	0.10	0.03	0.01

a. Find $P(X = 0 \text{ and } Y = 2)$.

b. Find $P(X > 0 \text{ and } Y \leq 1)$.

c. Find $P(X \leq 1)$.

d. Find $P(Y > 0)$.

e. Find the probability that all the bolts in the lot meet the length specification.

f. Find the probability that all the bolts in the lot meet the diameter specification.

g. Find the probability that all the bolts in the lot meet both specifications.

2. Refer to Exercise 1.

a. Find the marginal probability mass function $p_X(x)$.

b. Find the marginal probability mass function $p_Y(y)$.

c. Find μ_X.

d. Find μ_Y.

e. Find σ_X.

f. Find σ_Y.

g. Find $\text{Cov}(X,Y)$.

h. Find $\rho_{X,Y}$.

i. Are X and Y independent? Explain.

3. Refer to Exercise 1.

a. Find the conditional probability mass function $p_{Y|X}(y\,|\,1)$.

b. Find the conditional probability mass function $p_{X|Y}(x\,|\,1)$.

c. Find the conditional expectation $E(Y \mid X = 1)$.

d. Find the conditional expectation $E(X \mid Y = 1)$.

4. A computer program can make calls to two subroutines, A and B. In a randomly chosen run, let X be the number of calls to subroutine A and let Y denote the number of calls to subroutine B. The joint probability mass function of X and Y is given in the following table.

	y		
x	**1**	**2**	**3**
1	0.15	0.10	0.10
2	0.10	0.20	0.15
3	0.05	0.05	0.10

a. Find the marginal probability mass function of X.

b. Find the marginal probability mass function of Y.

c. Are X and Y independent? Explain.

d. Find μ_X and μ_Y.

e. Find σ_X and σ_Y.

f. Find $\text{Cov}(X,Y)$.

g. Find $\rho(X,Y)$.

5. Refer to Exercise 4. The total number of calls to the two subroutines is $X + Y$.

a. Find μ_{X+Y}.

b. Find σ_{X+Y}.

c. Find $P(X + Y = 4)$.

6. Refer to Exercise 4.

a. Find the conditional probability mass function $p_{Y|X}(y\,|\,2)$.

b. Find the conditional probability mass function $p_{X|Y}(x\,|\,3)$.

c. Find the conditional expectation $E(Y \mid X = 2)$.

d. Find the conditional expectation $E(X \mid Y = 3)$.

7. Refer to Exercise 4. Assume that each execution of subroutine A takes 100 ms, and that each execution of subroutine B takes 200 ms.

a. Express the number of milliseconds used in all the calls to the two subroutines in terms of X and Y.

b. Find the mean number of milliseconds used in all the calls to the two subroutines.

c. Find the standard deviation of the number of milliseconds used in all the calls to the two subroutines.

8. The number of customers in line at a supermarket express checkout counter is a random variable whose probability mass function is given in the following table.

x	0	1	2	3	4	5
$p(x)$	0.10	0.25	0.30	0.20	0.10	0.05

For each customer, the number of items to be purchased is a random variable with probability mass function

y	1	2	3	4	5	6
$p(y)$	0.05	0.15	0.25	0.30	0.15	0.10

Let X denote the number of customers in line, and let Y denote the total number of items purchased by all the customers in line. Assume the number of items purchased by one customer is independent of the number of items purchased by any other customer.

a. Find $P(X = 2 \text{ and } Y = 2)$.

b. Find $P(X = 2 \text{ and } Y = 6)$.

c. Find $P(Y = 2)$.

9. At a certain intersection on a four-lane highway, there are two left-turn-only lanes. Lane A is on the extreme left, and lane B is next to it. Let X represent the number of vehicles in lane A, and let Y represent the number of vehicles in lane B, when the signal changes to green. Assume that X and Y have the joint probability mass function $p(x,y)$ given in the following table.

	y				
x	**0**	**1**	**2**	**3**	**4**
0	0.05	0.04	0.01	0.00	0.00
1	0.05	0.10	0.03	0.02	0.00
2	0.03	0.05	0.15	0.05	0.02
3	0.00	0.02	0.08	0.10	0.05
4	0.00	0.00	0.02	0.05	0.08

a. Find the marginal probability mass function of X.

b. Find the marginal probability mass function of Y.

c. Are X and Y independent? Explain.

d. Find μ_X and μ_Y.

e. Find σ_X and σ_Y.

f. Find $\text{Cov}(X,Y)$.

g. Find $\rho(X,Y)$.

10. Refer to Exercise 9.

a. Find the mean of the total number of vehicles in the two lanes.

b. Find the variance of the total number of vehicles in the two lanes.

c. Find the probability that the total number of vehicles in the two lanes is exactly 6.

11. Refer to Exercise 9.

a. Find the conditional probability mass function $p_{Y|X}(y \mid 3)$.

b. Find the conditional probability mass function $p_{X|Y}(x \mid 4)$.

c. Find the conditional expectation $E(Y \mid X = 3)$.

d. Find the conditional expectation $E(X \mid Y = 4)$.

12. Let X represent the number of cars and Y the number of trucks that pass through a certain toll booth in a one-minute time interval. The joint probability mass function of X and Y is given in the following table.

	y		
x	**0**	**1**	**2**
0	0.10	0.05	0.05
1	0.10	0.10	0.05
2	0.05	0.20	0.10
3	0.05	0.05	0.10

a. Find the marginal probability mass function $p_X(x)$.

b. Find the marginal probability mass function $p_Y(y)$.

c. Find μ_X.

d. Find μ_Y.

e. Find σ_X.

f. Find σ_Y.

g. Find $\text{Cov}(X,Y)$.

h. Find $\rho_{X,Y}$.

13. Refer to Exercise 12. Let Z represent the total number of vehicles that pass through the toll booth in a one-minute time interval.

a. Find μ_Z.

b. Find σ_Z.

c. Find $P(Z = 2)$.

14. Refer to Exercise 12. Assume that the toll for cars is \$2, and the toll for trucks is \$5. Let T represent the total amount of toll paid by the vehicles passing through the toll booth in a one-minute time interval.

a. Find μ_T.

b. Find σ_T.

c. Find $P(T = 9)$.

15. Refer to Exercise 12.

a. Find the conditional probability mass function $p_{Y|X}(y \mid 3)$.

b. Find the conditional probability mass function $p_{X|Y}(x \mid 1)$.

c. Find the conditional expectation $E(Y \mid X = 3)$.

d. Find the conditional expectation $E(X \mid Y = 1)$.

16. For continuous random variables X and Y with joint probability density function

$$f(x,y) = \begin{cases} 4xy & 0 < x < 1 \text{ and } 0 < y < 1 \\ 0 & \text{otherwise} \end{cases}$$

a. Find $P(X < 0.5 \text{ and } Y > 0.75)$.

b. Find the marginal probability density functions $f_X(x)$ and $f_Y(y)$.

c. Are X and Y independent? Explain.

17. Refer to Example 2.51 (p. 123).

a. Find $\mathrm{Cov}(X,Y)$.

b. Find $\rho_{X,Y}$.

18. For continuous random variables X and Y with joint probability density function

$$f(x,y) = \begin{cases} x+y & 0 < x < 1 \text{ and } 0 < y < 1 \\ 0 & \text{otherwise} \end{cases}$$

a. Find $P(X \leq 0.25 \text{ and } Y \leq 0.85)$.

b. Find $\mathrm{Cov}(X,Y)$.

c. Find $\rho_{X,Y}$.

d. Are X and Y independent? Explain.

19. Refer to Exercise 18.

a. Find the marginal probability density functions $f_X(x)$ and $f_Y(y)$.

b. Find the conditional probability density function $f_{Y|X}(y\,|\,0.75)$.

c. Find the conditional expectation $E(Y \mid X = 0.75)$.

20. Let X denote the amount of shrinkage (in %) undergone by a randomly chosen fiber of a certain type when heated to a temperature of 120°C. Let Y represent the additional shrinkage (in %) when the fiber is heated to 140°C. Assume that the joint probability density function of X and Y is given by

$$f(x,y) = \begin{cases} \dfrac{48xy^2}{49} & 3 < x < 4 \text{ and } 0.5 < y < 1 \\ 0 & \text{otherwise} \end{cases}$$

a. Find $P(X < 3.25 \text{ and } Y > 0.8)$.

b. Find the marginal probability density functions $f_X(x)$ and $f_Y(y)$.

c. Are X and Y independent? Explain.

21. Measurements are made on the length and width (in centimeters) of a rectangular component. Because of measurement error, the measurements are random variables. Let X denote the length measurement and let Y denote the width measurement. Assume that the probability density function of X is

$$f(x) = \begin{cases} 10 & 9.95 < x < 10.05 \\ 0 & \text{otherwise} \end{cases}$$

and that the probability density function of Y is

$$g(y) = \begin{cases} 5 & 4.9 < y < 5.1 \\ 0 & \text{otherwise} \end{cases}$$

Assume that the measurements X and Y are independent.

a. Find $P(X < 9.98)$.

b. Find $P(Y > 5.01)$.

c. Find $P(X < 9.98 \text{ and } Y > 5.01)$.

d. Find μ_X.

e. Find μ_Y.

f. Let $A = XY$ be the area computed from the measurements X and Y. Find μ_A.

22. The thickness X of a wooden shim (in millimeters) has probability density function

$$f(x) = \begin{cases} \dfrac{3}{4} - \dfrac{3(x-5)^2}{4} & 4 \leq x \leq 6 \\ 0 & \text{otherwise} \end{cases}$$

a. Find μ_X.

b. Find σ_X^2.

c. Let Y denote the thickness of a shim in inches (1 mm = 0.0394 in.). Find μ_Y and σ_Y^2.

d. If three shims are selected independently and stacked one on top of another, find the mean and variance of the total thickness.

23. The lifetime of a certain component, in years, has probability density function

$$f(x) = \begin{cases} e^{-x} & x > 0 \\ 0 & x \leq 0 \end{cases}$$

Two such components, whose lifetimes are independent, are available. As soon as the first component fails, it is replaced with the second component. Let X denote the lifetime of the first component, and let Y denote the lifetime of the second component.

a. Find the joint probability density function of X and Y.

b. Find $P(X \leq 1 \text{ and } Y > 1)$.

c. Find μ_X.

d. Find μ_{X+Y}.

e. Find $P(X + Y \leq 2)$. (*Hint*: Sketch the region of the plane where $x + y \leq 2$, and then integrate the joint probability density function over that region.)

24. The number of bytes downloaded per second on an information channel has mean 10^5 and standard deviation 10^4. Among the factors influencing the rate is congestion, which produces alternating periods of faster and slower transmission. Let X represent the number of bytes downloaded in a randomly chosen five-second period.

a. Is it reasonable to assume that $\mu_X = 5 \times 10^5$? Explain.

b. Is it reasonable to assume that $\sigma_X = \sqrt{5} \times 10^4$? Explain.

25. The article "Abyssal Peridotites > 3800 Ma from Southern West Greenland: Field Relationships, Petrography, Geochronology, Whole-Rock and Mineral Chemistry of Dunite and Harzburgite Inclusions in the Itsaq Gneiss Complex" (C. Friend, V. Bennett, and A. Nutman, *Contributions to Mineralogy and Petrology*, 2002:71–92) describes the chemical compositions of certain minerals in the early Archaean mantle. For a certain type of olivine assembly, the silicon dioxide (SiO_2) content (in weight %) in a randomly chosen rock has mean 40.25 and standard deviation 0.36.

a. Find the mean and standard deviation of the sample mean SiO_2 content in a random sample of 10 rocks.

b. How many rocks must be sampled so that the standard deviation of the sample mean SiO_2 content is 0.05?

26. Here are two random variables that are uncorrelated but not independent. Let X and Y have the following joint probability mass function:

x	y	$p(x, y)$
−1	1	1/3
0	0	1/3
1	1	1/3

a. Use the definition of independence on p. 135 to show that X and Y are not independent (in fact $Y = |X|$, so Y is actually a function of X).

b. Show that X and Y are uncorrelated.

27. An investor has \$100 to invest, and two investments between which to divide it. If she invests the entire amount in the first investment, her return will be X, while if she invests the entire amount in the second investment, her return will be Y. Both X and Y have mean \$6 and standard deviation (risk) \$3. The correlation between X and Y is 0.3.

a. Express the return in terms of X and Y if she invests \$30 in the first investment and \$70 in the second.

b. Find the mean return and the risk if she invests \$30 in the first investment and \$70 in the second.

c. Find the mean return and the risk, in terms of K, if she invests \$$K$ in the first investment and \$$(100 - K)$ in the second.

d. Find the value of K that minimizes the risk in part (c).

e. Prove that the value of K that minimizes the risk in part (c) is the same for any correlation $\rho_{X,Y}$ with $\rho_{X,Y} \neq 1$.

28. If X is a random variable, prove that $\text{Cov}(X, X) = \sigma_X^2$.

29. Let X and Y be random variables, and a and b be constants.

a. Prove that $\text{Cov}(aX, bY) = ab\,\text{Cov}(X, Y)$.

b. Prove that if $a > 0$ and $b > 0$, then $\rho_{aX,bY} = \rho_{X,Y}$. Conclude that the correlation coefficient is unaffected by changes in units.

Supplementary Exercises for Chapter 2

1. A system consists of four components connected as shown.

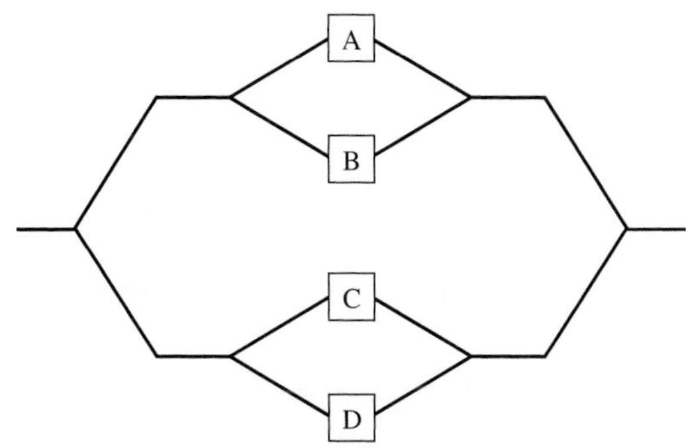

Assume A, B, C, and D function independently. If the probabilities that A, B, C, and D fail are 0.1, 0.2, 0.05, and 0.3, respectively, what is the probability that the system functions?

2. A die is rolled until a 6 appears. What is the probability that more than five rolls are necessary?

3. Silicon wafers are used in the manufacture of integrated circuits. Of the wafers manufactured by a certain process, 10% have resistances below specification and 5% have resistances above specification.

   a. What is the probability that the resistance of a randomly chosen wafer does not meet the specification?
   b. If a randomly chosen wafer has a resistance that does not meet the specification, what is the probability that it is too low?

4. Two production lines are used to pack sugar into 5 kg bags. Line 1 produces twice as many bags as does line 2. One percent of the bags from line 1 are defective in that they fail to meet a purity specification, while 3% of the bags from line 2 are defective. A bag is randomly chosen for inspection.

   a. What is the probability that it came from line 1?
   b. What is the probability that it is defective?
   c. If the bag is defective, what is the probability that it came from line 1?
   d. If the bag is not defective, what is the probability that it came from line 1?

5. A customer receives a shipment of 10,000 fuses. She randomly samples three of the fuses and tests them. If any of the three are defective, she will return the shipment. If in fact 1000 of the fuses in the shipment are defective, what is the probability that she will return the shipment?

6. In a certain type of automobile engine, the cylinder head is fastened to the block by 10 bolts, each of which should be torqued to 60 N·m. Assume that the torques of the bolts are independent.

   a. If each bolt is torqued correctly with probability 0.99, what is the probability that all the bolts on a cylinder head are torqued correctly?
   b. The goal is for 95% of the engines to have all their bolts torqued correctly. What must be the probability that a bolt is torqued correctly in order to reach this goal?

7. An electronic message consists of a string of bits (0s and 1s). The message must pass through two relays before being received. At each relay the probability is 0.1 that the bit will be reversed before being relayed (i.e., a 1 will be changed to a 0, or a 0 to a 1). Find the probability that the value of a bit received at its final destination is the same as the value of the bit that was sent.

8. The reading given by a thermometer calibrated in ice water (actual temperature 0°C) is a random variable with probability density function

$$f(x) = \begin{cases} k(1 - x^2) & -1 < x < 1 \\ 0 & \text{otherwise} \end{cases}$$

where k is a constant.

   a. Find the value of k.
   b. What is the probability that the thermometer reads above 0°C?
   c. What is the probability that the reading is within 0.25°C of the actual temperature?
   d. What is the mean reading?
   e. What is the median reading?
   f. What is the standard deviation?

9. Two dice are rolled. Given that two different numbers come up, what is the probability that one of the dice comes up 6?

10. In a lot of 10 components, 2 are sampled at random for inspection. Assume that in fact exactly 2 of the 10 components in the lot are defective. Let X be the number of sampled components that are defective.

a. Find $P(X = 0)$.

b. Find $P(X = 1)$.

c. Find $P(X = 2)$.

d. Find the probability mass function of X.

e. Find the mean of X.

f. Find the standard deviation of X.

11. There are two fuses in an electrical device. Let X denote the lifetime of the first fuse, and let Y denote the lifetime of the second fuse (both in years). Assume the joint probability density function of X and Y is

$$f(x,y) = \begin{cases} \frac{1}{6}e^{-x/2-y/3} & x > 0 \text{ and } y > 0 \\ 0 & \text{otherwise} \end{cases}$$

a. Find $P(X \leq 2 \text{ and } Y \leq 3)$.

b. Find the probability that both fuses last at least 3 years.

c. Find the marginal probability density function of X.

d. Find the marginal probability density function of Y.

e. Are X and Y independent? Explain.

12. Let A and B be events with $P(A) = 0.3$ and $P(A \cup B) = 0.7$.

a. For what value of $P(B)$ will A and B be mutually exclusive?

b. For what value of $P(B)$ will A and B be independent?

13. A snowboard manufacturer has three plants, one in the eastern United States, one in the western United States, and one in Canada. Production records show that the U.S. plants each produced 10,000 snowboards last month, while the Canadian plant produced 8000 boards. Of all the boards manufactured in Canada last month, 4% had a defect that caused the boards to delaminate prematurely. Records kept at the U.S. plants show that 3% of the boards manufactured in the eastern United States and 6% of the boards manufactured in the western United States had this defect as well.

a. What proportion of the boards manufactured last month were defective?

b. What is the probability that a snowboard is defective and was manufactured in Canada?

c. Given that a snowboard is defective, what is the probability that it was manufactured in the United States?

14. The article "Traps in Mineral Valuations — Proceed With Care" (W. Lonegan, *Journal of the Australasian Institute of Mining and Metallurgy*, 2001:18–22) models the value (in millions of dollars) of a mineral deposit yet to be mined as a random variable X with probability mass function $p(x)$ given by $p(10) = 0.40$, $p(60) = 0.50$, $p(80) = 0.10$, and $p(x) = 0$ for values of x other than 10, 60, or 80.

a. Is this article treating the value of a mineral deposit as a discrete or a continuous random variable?

b. Compute μ_X.

c. Compute σ_X.

d. The project will be profitable if the value is more than \$50 million. What is the probability that the project is profitable?

15. Six new graduates are hired by an engineering firm. Each is assigned at random to one of six cubicles arranged in a row in the back of the room that houses the engineering staff. Two of the graduates are Bill and Cathy. What is the probability that they are assigned adjacent cubicles?

16. A closet contains four pairs of shoes. If four shoes are chosen at random, what is the probability that the chosen shoes do not contain a pair?

17. Let X and Y be independent random variables with $\mu_X = 2$, $\sigma_X = 1$, $\mu_Y = 2$, and $\sigma_Y = 3$. Find the means and variances of the following quantities.

a. $3X$

b. $X + Y$

c. $X - Y$

d. $2X + 6Y$

18. Let X and Y be random variables with $\mu_X = 1$, $\sigma_X = 2$, $\mu_Y = 3$, $\sigma_Y = 1$, and $\rho_{X,Y} = 0.5$. Find the means and variances of the following quantities.

a. $X + Y$

b. $X - Y$

c. $3X + 2Y$

d. $5Y - 2X$

19. A steel manufacturer is testing a new additive for manufacturing an alloy of steel. The joint probability mass function of tensile strength (in thousands of pounds/in^2) and additive concentration is

	Tensile Strength		
Concentration of Additive	**100**	**150**	**200**
0.02	0.05	0.06	0.11
0.04	0.01	0.08	0.10
0.06	0.04	0.08	0.17
0.08	0.04	0.14	0.12

a. What are the marginal probability mass functions for X (additive concentration) and Y (tensile strength)?

b. Are X and Y independent? Explain.

c. Given that a specimen has an additive concentration of 0.04, what is the probability that its strength is 150 or more?

d. Given that a specimen has an additive concentration of 0.08, what is the probability that its tensile strength is greater than 125?

e. A certain application calls for the tensile strength to be 175 or more. What additive concentration should be used to make the probability of meeting this specification the greatest?

20. Refer to Exercise 19.

a. Find μ_X.

b. Find μ_Y.

c. Find σ_X.

d. Find σ_Y.

e. Find Cov(X,Y).

f. Find $\rho_{X,Y}$.

21. Refer to Exercise 19.

a. Compute the conditional mass function $p_{Y|X}(y \mid 0.06)$.

b. Compute the conditional mass function $p_{X|Y}(x \mid 100)$.

c. Compute the conditional expectation $E(Y \mid X = 0.06)$.

d. Compute the conditional expectation $E(X \mid Y = 100)$.

22. A certain plant runs three shifts per day. Of all the items produced by the plant, 50% of them are produced on the first shift, 30% on the second shift, and 20% on the third shift. Of all the items produced on the first shift, 1% are defective, while 2% of the items produced on the second shift and 3% of the items produced on the third shift are defective.

a. An item is sampled at random from the day's production, and it turns out to be defective. What is the probability that it was manufactured during the first shift?

b. An item is sampled at random from the day's production, and it turns out not to be defective. What is the probability that it was manufactured during the third shift?

23. The article "Uncertainty and Climate Change" (G. Heal and B. Kriström, *Environmental and Resource Economics*, 2002:3–39) considers three scenarios, labeled A, B, and C, for the impact of global warming on income. For each scenario, a probability mass function for the loss of income is specified. These are presented in the following table.

	Probability		
Loss (%)	**Scenario A**	**Scenario B**	**Scenario C**
0	0.65	0.65	0.65
2	0	0	0.24
5	0.2	0.24	0.1
10	0	0	0.01
15	0.1	0.1	0
20	0	0.01	0
25	0.05	0	0

a. Compute the mean and standard deviation of the loss under scenario A.

b. Compute the mean and standard deviation of the loss under scenario B.

c. Compute the mean and standard deviation of the loss under scenario C.

d. Under each scenario, compute the probability that the loss is less than 10%.

24. Refer to Exercise 23. Assume that the probabilities that each of the three scenarios occurs are $P(\text{A}) = 0.20$, $P(\text{B}) = 0.30$, and $P(\text{C}) = 0.50$.

a. Find the probability that scenario A occurs and that the loss is 5%.

b. Find the probability that the loss is 5%.

c. Find the probability that scenario A occurs given that the loss is 5%.

25. A box contains four 75 W lightbulbs, three 60 W lightbulbs, and three burned-out lightbulbs. Two bulbs are selected at random from the box. Let X represent the number of 75 W bulbs selected, and let Y represent the number of 60 W bulbs selected.

a. Find the joint probability mass function of X and Y.

b. Find μ_X.

c. Find μ_Y.

d. Find σ_X.

e. Find σ_Y.

f. Find $\mathrm{Cov}(X,Y)$.

g. Find $\rho_{X,Y}$.

26. A stock solution of hydrochloric acid (HCl) supplied by a certain vendor contains small amounts of several impurities, including copper and nickel. Let X denote the amount of copper and let Y denote the amount of nickel, in parts per ten million, in a randomly selected bottle of solution. Assume that the joint probability density function of X and Y is given by

$$f(x,y) = \begin{cases} c(x+y)^2 & 0 < x < 1 \text{ and } 0 < y < 1 \\ 0 & \text{otherwise} \end{cases}$$

a. Find the value of the constant c so that $f(x,y)$ is a joint density function.

b. Compute the marginal density function $f_X(x)$.

c. Compute the conditional density function $f_{Y|X}(y \mid x)$.

d. Compute the conditional expectation $E(Y \mid X = 0.4)$.

e. Are X and Y independent? Explain.

27. Refer to Exercise 26.

a. Find μ_X.

b. Find σ_X^2.

c. Find $\mathrm{Cov}(X,Y)$.

d. Find $\rho_{X,Y}$.

28. A fair coin is tossed five times. Which sequence is more likely, HTTHH or HHHHH? Or are they equally likely? Explain.

29. A penny and a nickel are tossed. The penny has probability 0.4 of coming up heads, and the nickel has probability 0.6 of coming up heads. Let $X = 1$ if the penny comes up heads, and let $X = 0$ if the penny comes up tails. Let $Y = 1$ if the nickel comes up heads, and let $Y = 0$ if the nickel comes up tails.

a. Find the probability mass function of X.

b. Find the probability mass function of Y.

c. Is it reasonable to assume that X and Y are independent? Why?

d. Find the joint probability mass function of X and Y.

30. Two fair dice are rolled. Let X represent the number on the first die, and let Y represent the number on the second die. Find μ_{XY}.

31. A box contains three cards, labeled 1, 2, and 3. Two cards are chosen at random, with the first card being replaced before the second card is drawn. Let X represent the number on the first card, and let Y represent the number on the second card.

a. Find the joint probability mass function of X and Y.

b. Find the marginal probability mass functions $p_X(x)$ and $p_Y(y)$.

c. Find μ_X and μ_Y.

d. Find μ_{XY}.

e. Find $\mathrm{Cov}(X,Y)$.

32. Refer to Exercise 31. Assume the first card is not replaced before the second card is drawn.

a. Find the joint probability mass function of X and Y.

b. Find the marginal probability mass functions $p_X(x)$ and $p_Y(y)$.

c. Find μ_X and μ_Y.

d. Find μ_{XY}.

e. Find $\mathrm{Cov}(X,Y)$.

Chapter 3

Propagation of Error

Introduction

Measurement is fundamental to scientific work. Scientists and engineers often perform calculations with measured quantities, for example, computing the density of an object by dividing a measurement of its mass by a measurement of its volume, or computing the area of a rectangle by multiplying measurements of its length and width.

Any measuring procedure contains error. As a result, measured values generally differ somewhat from the true values that are being measured. When a calculation is performed with measurements, the errors in the measurements produce an error in the calculated value. We say that the error is **propagated** from the measurements to the calculated value. If we have some knowledge concerning the sizes of the errors in measurements such as the length and width of a rectangle, there are methods for obtaining knowledge concerning the likely size of the error in a calculated quantity such as the area. The subject of propagation of error concerns these methods and is the topic of this chapter.

3.1 Measurement Error

A geologist is weighing a rock on a scale. She weighs the rock five times and obtains the following measurements (in grams):

$$251.3 \quad 252.5 \quad 250.8 \quad 251.1 \quad 250.4$$

The measurements are all different, and it is likely that none of them is equal to the true mass of the rock. The difference between a measured value and the true value is called the **error** in the measured value. Any measuring procedure contains many sources of error. For example, assume that the rock measurements were read from a dial. If the scale was not calibrated properly, this will pull each measurement away from the true value by some fixed amount. Thus imperfect calibration contributes errors of equal magnitude to each measurement. Interpolation between graduation marks on the dial is another source of error. The magnitude of the error due to interpolation is likely to vary from measurement to measurement and is likely to be positive for some measurements and negative for others. It may be reasonable to assume that interpolation errors average out to zero in the long run.

In general, we can think of the error in a measurement as being composed of two parts, the **systematic error**, or **bias**, and the **random error**. The bias is the part of the error that is the same for every measurement. The random error, on the other hand, varies from measurement to measurement, and averages out to zero in the long run. Some sources of error contribute both to bias and to random error. For example, consider parallax error. Parallax is the difference in the apparent position of the dial indicator when observed from different angles. The magnitude of the parallax error in any particular measurement depends on the position of the observer relative to the dial. Since the position will vary somewhat from reading to reading, parallax contributes to random error. If the observer tends to lean somewhat to one side rather than another, parallax will contribute to bias as well.

Any measurement can be considered to be the sum of the true value plus contributions from each of the two components of error:

$$\text{Measured value} = \text{true value} + \text{bias} + \text{random error} \tag{3.1}$$

Since part of the error is random, it is appropriate to use a statistical model to study measurement error. We model each measured value as a random variable, drawn from a population of possible measurements. The mean μ of the population represents that part of the measurement that is the same for every measurement. Therefore, μ is the sum of the true value and the bias. The standard deviation σ of the population is the standard deviation of the random error. It represents the variation that is due to the fact that each measurement has a different value for its random error. Intuitively, σ represents the size of a typical random error.

We are interested in two aspects of the measuring process. First is its **accuracy**. The accuracy is determined by the bias, which is the difference between the mean measurement μ and the true value being measured. The smaller the bias, the more accurate the measuring process. If the mean μ is equal to the true value, so that the bias is 0, the measuring process is said to be **unbiased**.

The other aspect of the measuring process that is of interest is the **precision**. Precision refers to the degree to which repeated measurements of the same quantity tend to agree with each other. If repeated measurements come out nearly the same every time, the precision is high. If they are widely spread out, the precision is low. The precision is therefore determined by the standard deviation σ of the measurement process. The smaller the value of σ, the more precise the measuring process. Engineers and scientists often refer to σ as the **random uncertainty** or **statistical uncertainty** in the measuring process. We will refer to σ simply as the **uncertainty**.

Summary

- A measured value is a random variable with mean μ and standard deviation σ.
- The bias in the measuring process is the difference between the mean measurement and the true value:

$$\text{Bias} = \mu - \text{true value}$$

- The uncertainty in the measuring process is the standard deviation σ.
- The smaller the bias, the more accurate the measuring process.
- The smaller the uncertainty, the more precise the measuring process.

When reporting a measured value, it is important to report an estimate of the bias and uncertainty along with it, in order to describe the accuracy and precision of the measurement. It is in general easier to estimate the uncertainty than the bias. Figures 3.1 and 3.2 illustrate the reason for this. Figure 3.1 illustrates a hypothetical experiment involving repeated measurements, under differing conditions regarding bias and uncertainty. The sets of measurements in Figure 3.1a and b are fairly close together, indicating that the uncertainty is small. The sets of measurements in Figure 3.1a and c are centered near the true value, indicating that the bias is small.

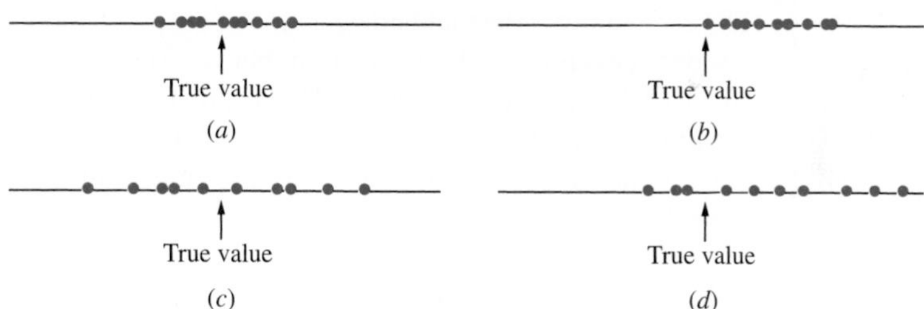

FIGURE 3.1 *(a)* Both bias and uncertainty are small. *(b)* Bias is large; uncertainty is small. *(c)* Bias is small; uncertainty is large. *(d)* Both bias and uncertainty are large.

In real life, of course, we do not know the true value being measured. Thus the plot of measurements shown in Figure 3.1 would look like Figure 3.2 (page 160). We can still determine that the sets of measurements in Figure 3.2a and b have smaller uncertainty. But without additional information about the true value, we cannot estimate the bias.

We conclude from Figures 3.1 and 3.2 that *uncertainty can be estimated from repeated measurements, but in order to estimate the bias, we must have additional information about the true value*. We might obtain this additional information, for example, by repeatedly measuring a standard quantity whose true value is known and estimating the bias to be the difference between the average of the measurements and the known true value. Another way to estimate the bias would be to compare the average of a large number of measurements to a measurement made with a more elaborate process for

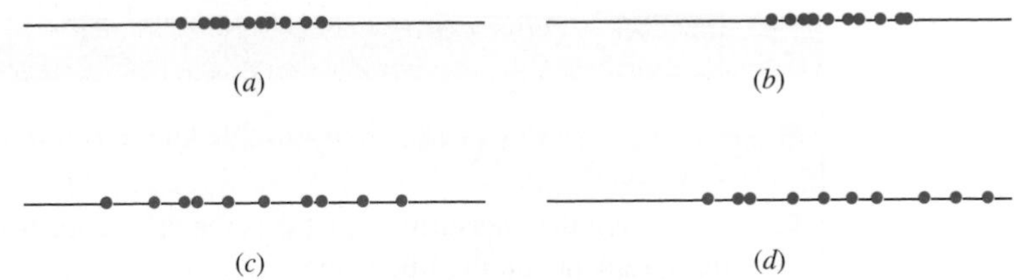

FIGURE 3.2 We can estimate the uncertainty from the set of repeated measurements, but without knowing the true value, we cannot estimate the bias.

which the bias is known to be negligible. Estimating the bias is essentially the process of calibration, for which information external to the measuring device is needed.

A laboratory sample of gas is known to have a carbon monoxide (CO) concentration of 50 parts per million (ppm). A spectrophotometer is used to take five independent measurements of this concentration. The five measurements, in ppm, are 51, 47, 53, 53, and 48. Estimate the bias and the uncertainty in a spectrophotometer measurement.

Solution

The five measurements are regarded as a random sample from the population of possible measurements. The bias is equal to the mean of this population minus the true value of 50. The uncertainty is the standard deviation of the population. We do not know the mean and standard deviation of the population, but we can approximate them with the mean and standard deviation of the sample. The mean of the five measurements is 50.4. Therefore we estimate the bias to be $50.4 - 50 = 0.4$ ppm. The standard deviation of the five measurements is 2.8 ppm. Therefore we estimate the uncertainty in each measurement to be 2.8 ppm.

A different spectrophotometer is now used to measure the CO concentration in another gas sample. The true concentration in this sample is unknown. Five measurements are made (in ppm). They are 62, 63, 61, 62, and 59. Estimate the uncertainty in a measurement from this spectrophotometer. Can we estimate the bias?

Solution

The uncertainty in a single measurement is estimated with the sample standard deviation, which is 1.5 ppm. The sample mean is 61.4 ppm, but to estimate the bias, we would have to subtract the true concentration from this. Since we do not know the true concentration, we cannot estimate the bias.

In practice, estimates of uncertainty are sometimes very crude. In Examples 3.1 and 3.2 we suggested estimating the uncertainty σ with the sample standard deviation of five measurements. Estimates based on small samples like this are sometimes widely off the mark. When possible, it is best to base estimates of uncertainty on large samples. However, an estimate from a small sample is better than none at all.

Summary

Let $X_1, \ldots, X_n$ be independent measurements, all made by the same process on the same quantity.

- The sample standard deviation s can be used to estimate the uncertainty.
- Estimates of uncertainty are often crude, especially when based on small samples.
- If the true value is known, the sample mean $\overline{X}$ can be used to estimate the bias: Bias $\approx \overline{X} -$ true value
- If the true value is unknown, the bias cannot be estimated from repeated measurements.

An important example of bias estimation is the calibration of scales in supermarkets and other commercial establishments to ensure that they do not systematically over- or underweigh goods sold to customers. This calibration procedure follows a chain of comparisons with external standards, beginning at the county level and ending up near Paris, France, where the world's ultimate standard for weight (technically mass) is located. This is the International Prototype Kilogram, a platinum–iridium cylinder whose mass is by definition exactly 1 kg. A replica of The Kilogram, located at the National Institute of Standards and Technology in Washington, serves as the standard for measures in the United States. Use of this replica, rather than The Kilogram, introduces a bias into every measure of weight in the United States. By comparing the U.S. replica to The Kilogram, this bias has been estimated to be -1.9×10^{-8} kg. That is, the U.S. replica appears to be lighter than The Kilogram by about 19 parts in a billion. For this reason, all weight measurements made at the National Institute of Standards and Technology are adjusted upward by 19 parts in a billion to compensate. Note that this adjustment factor could not have been estimated by repeated weighing of the replica; comparison with an external standard was required.

From here on we will assume, unless otherwise stated, that bias has been reduced to a negligible level. We will describe measurements in the form

$$\text{Measured value} \pm \sigma \tag{3.2}$$

where σ is the uncertainty in the process that produced the measured value.

Expression (3.2) has the form $a \pm b$, where a and b are numbers. It is important to realize that expressions containing the symbol $\pm$ can have many meanings. The meaning here is that a is a measured value and b is the uncertainty in a. Some people use $a \pm b$ to indicate that b is the maximum value for the error, or that b is a multiple of the uncertainty, typically two or three times the uncertainty. Yet another meaning will be presented in Chapter 5, where we will use the notation $a \pm b$ to denote a *confidence interval*, which is an interval computed in a way so as to be likely to contain the true value. Whenever you encounter the symbol $\pm$, you should be sure to understand the context in which it is used.

The spectrophotometer in Example 3.1 has been recalibrated, so we may assume that the bias is negligible. The spectrophotometer is now used to measure the CO concentration in another gas sample. The measurement is 55.1 ppm. How should this measurement be expressed?

Solution

From the repeated measurements in Example 3.1, the uncertainty in a measurement from this instrument was estimated to be 2.8 ppm. Therefore we report the CO concentration in this gas sample as 55.1 ± 2.8 ppm.

Exercises for Section 3.1

1. The boiling point of water is measured four times. The results are 110.01°C, 110.02°C, 109.99°C, and 110.01°C. Which of the following statements best describes this measuring process?

i. Accurate but not precise
ii. Precise but not accurate
iii. Neither accurate nor precise
iv. Both accurate and precise

2. Two apparatuses are used to measure the melting point of *p*-amino-benzene. Equal numbers of measurements are taken on each apparatus. The result from the first apparatus is 90 ± 1°C, and the result from the second apparatus is 90 ± 2°C.

a. Is it possible to tell which apparatus is more accurate? If so, state which one. If not, explain why.
b. Is it possible to tell which apparatus is more precise? If so, state which one. If not, explain why.

3. The length of an object is given as 3.21 ± 0.02 cm. True or false:

a. The length was measured to be 3.21 cm.
b. The true length of the object is 3.21 cm.
c. The bias in the measurement is 0.02 cm.
d. The uncertainty in the measurement is 0.02 cm.

4. For some measuring processes, the uncertainty is approximately proportional to the value of the measurement. For example, a certain scale is said to have an uncertainty of $\pm 2\%$. An object is weighed on this scale.

a. Given that the reading is 100 g, express the uncertainty in this measurement in grams.
b. Given that the reading is 50 g, express the uncertainty in this measurement in grams.

5. A person stands on a bathroom scale. The reading is 150 lb. After the person gets off the scale, the reading is 2 lb.

a. Is it possible to estimate the uncertainty in this measurement? If so, estimate it. If not, explain why not.
b. Is it possible to estimate the bias in this measurement? If so, estimate it. If not, explain why not.

6. A person gets on and off a bathroom scale four times. The four readings (in pounds) are 148, 151, 150, and 152. Each time after the person gets off the scale, the reading is 2 lb.

a. Is it possible to estimate the uncertainty in these measurements? If so, estimate it. If not, explain why not.
b. Is it possible to estimate the bias in these measurements? If so, estimate it. If not, explain why not.

7. In a hypothetical scenario, the National Institute of Standards and Technology has received a new replica of The Kilogram. It is weighed five times. The measurements are as follows, in units of micrograms above 1 kg: 114.3, 82.6, 136.4, 126.8, 100.7.

a. Is it possible to estimate the uncertainty in these measurements? If so, estimate it. If not, explain why not.

b. Is it possible to estimate the bias in these measurements? If so, estimate it. If not, explain why not.

8. The Kilogram is now weighed five times on a different scale. The measurements are as follows, in units of micrograms above 1 kg: 25.6, 26.8, 26.2, 26.8, 25.4.

   a. Is it possible to estimate the uncertainty in these measurements? If so, estimate it. If not, explain why not.

   b. Is it possible to estimate the bias in these measurements? If so, estimate it. If not, explain why not.

9. A new and unknown weight is weighed on the same scale that was used in Exercise 8, and the measurement is 127 μg above 1 kg. Using the information in Exercise 8, is it possible to come up with a more accurate measurement? If so, what is it? If not, explain why not.

10. The article "Calibration of an FTIR Spectrometer" (P. Pankratz, *Statistical Case Studies for Industrial and Process Improvement*, SIAM-ASA, 1997:19–38) describes the use of a spectrometer to make five measurements of the carbon content (in ppm) of a certain silicon wafer whose true carbon content was known to be 1.1447 ppm. The measurements were 1.0730, 1.0825, 1.0711, 1.0870, and 1.0979.

    a. Is it possible to estimate the uncertainty in these measurements? If so, estimate it. If not, explain why not.

    b. Is it possible to estimate the bias in these measurements? If so, estimate it. If not, explain why not.

11. The length of a rod was measured eight times. The measurements in centimeters, in the order they were taken, were 21.20, 21.22, 21.25, 21.26, 21.28, 21.30, 21.32, 21.35.

    a. Do these measurements appear to be a random sample from a population of possible measurements? Why or why not?

    b. Is it possible to estimate the uncertainty in these measurements? Explain.

3.2 Linear Combinations of Measurements

Often we add constants to measurements, multiply measurements by constants, or add two or more measurements together. This section describes how uncertainties are affected by these arithmetic operations. Since measurements are random variables, and uncertainties are the standard deviations of these random variables, the results used to compute standard deviations of linear combinations of random variables can be applied to compute uncertainties in linear combinations of measurements. Results for independent random variables were presented in Section 2.5; more general results were presented in Section 2.6. In this section we apply these results to independent measurements. We discuss dependent measurements as well, at the end of the section.

We begin by stating the basic results used to compute uncertainties in linear combinations of independent measurements, and then follow with some examples.

If X is a measurement and c is a constant, then

$$\sigma_{cX} = |c|\sigma_X \tag{3.3}$$

If $X_1, \ldots, X_n$ are independent measurements and $c_1, \ldots, c_n$ are constants, then

$$\sigma_{c_1X_1+\cdots+c_nX_n} = \sqrt{c_1^2\sigma_{X_1}^2 + \cdots + c_n^2\sigma_{X_n}^2} \tag{3.4}$$

The radius of a circle is measured to be 3.0 ± 0.1 cm. Estimate the circumference and find the uncertainty in the estimate.

Solution

Let R denote the radius of the circle. The measured value of R is 3.0 cm, and the uncertainty is the standard deviation of this measurement, which is $\sigma_R = 0.1$ cm. The circumference is given by $C = 2\pi R$. The uncertainty in C is σ_C, the standard deviation of C. Since 2π is a constant, we have

$$\begin{aligned} \sigma_C &= |2\pi|\sigma_R && \text{(using Equation 3.3)} \\ &= (6.28)(0.1 \text{ cm}) \\ &= 0.63 \text{ cm} \end{aligned}$$

The circumference is 18.84 ± 0.63 cm.

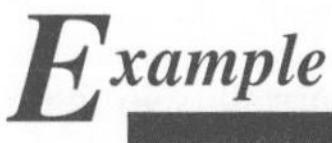

An item is formed by placing two components end to end. The lengths of the components are measured independently, by a process that yields a random measurement with uncertainty 0.1 cm. The length of the item is estimated by adding the two measured lengths. Assume that the measurements are 4.10 cm and 3.70 cm. Estimate the length of the item and find the uncertainty in the estimate.

Solution

Let X be the measured length of the first component, and let Y be the measured length of the second component. The estimated length is 7.80 cm. The uncertainty is

$$\begin{aligned} \sigma_{X+Y} &= \sqrt{\sigma_X^2 + \sigma_Y^2} && \text{(using Equation 3.4 with } c_1 = c_2 = 1\text{)} \\ &= \sqrt{(0.1)^2 + (0.1)^2} \\ &= 0.14 \text{ cm} \end{aligned}$$

The estimated length is 7.80 ± 0.14 cm.

Example 3.6

A surveyor is measuring the perimeter of a rectangular lot. He measures two adjacent sides to be 50.11 ± 0.05 m and 75.21 ± 0.08 m. These measurements are independent. Estimate the perimeter of the lot and find the uncertainty in the estimate.

Solution

Let $X = 50.11$ and $Y = 75.21$ be the two measurements. The perimeter is estimated by $P = 2X + 2Y = 250.64$ m, and the uncertainty in P is

$$\begin{aligned} \sigma_P &= \sigma_{2X+2Y} \\ &= \sqrt{4\sigma_X^2 + 4\sigma_Y^2} && \text{(using Equation 3.4)} \end{aligned}$$

$$= \sqrt{4(0.05)^2 + 4(0.08)^2}$$
$$= 0.19 \text{ m}$$

The perimeter is 250.64 ± 0.19 m.

In Example 3.6, the surveyor's assistant suggests computing the uncertainty in P by a different method. He reasons that since $P = X + X + Y + Y$, then

$$\begin{aligned}\sigma_P &= \sigma_{X+X+Y+Y}\\ &= \sqrt{\sigma_X^2 + \sigma_X^2 + \sigma_Y^2 + \sigma_Y^2}\\ &= \sqrt{(0.05)^2 + (0.05)^2 + (0.08)^2 + (0.08)^2}\\ &= 0.13 \text{ m}\end{aligned}$$

This disagrees with the value of 0.19 m calculated in Example 3.6. What went wrong?

Solution

What went wrong is that the four terms in the sum for P are not all independent. Specifically, $X + X$ is not the sum of independent quantities; neither is $Y + Y$. In order to use Equation (3.4) to compute the uncertainty in P, we must express P as the sum of independent quantities, that is, $P = 2X + 2Y$, as in Example 3.6.

Repeated Measurements

One of the best ways to reduce uncertainty is to take several independent measurements and average them. The measurements in this case are a simple random sample from a population, and their average is the sample mean. Methods for computing the mean and standard deviation of a sample mean were presented in Sections 2.5 and 2.6. These methods can be applied to compute the mean and uncertainty in the average of independent repeated measurements.

If $X_1, \ldots, X_n$ are n *independent* measurements, each with mean μ and uncertainty σ, then the sample mean $\overline{X}$ is a measurement with mean

$$\mu_{\overline{X}} = \mu \tag{3.5}$$

and with uncertainty

$$\sigma_{\overline{X}} = \frac{\sigma}{\sqrt{n}} \tag{3.6}$$

With a little thought, we can see how important these results are for applications. What these results say is that if we perform many independent measurements of the same quantity, then the average of these measurements has the same mean as each individual measurement, but the standard deviation is reduced by a factor equal to the square root of the sample size. In other words, the average of several repeated measurements has the same accuracy as, and is more precise than, any single measurement.

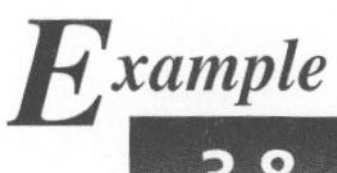

The length of a component is to be measured by a process whose uncertainty is 0.05 cm. If 25 independent measurements are made and the average of these is used to estimate the length, what will the uncertainty be? How much more precise is the average of 25 measurements than a single measurement?

Solution

The uncertainty in the average of 25 measurements is $0.05/\sqrt{25} = 0.01$ cm. The uncertainty in a single measurement is 0.05 cm. The uncertainty in the average of 25 independent measurements is therefore less than that of a single measurement by a factor of 5, which is the square root of the number of measurements that are averaged. Thus the average of 25 independent measurements is five times more precise than a single measurement.

Example 3.9

The mass of a rock is measured five times on a scale whose uncertainty is unknown. The five measurements (in grams) are 21.10, 21.05, 20.98, 21.12, and 21.05. Estimate the mass of the rock and find the uncertainty in the estimate.

Solution

Let $\overline{X}$ represent the average of the five measurements, and let s represent the sample standard deviation. We compute $\overline{X} = 21.06$ g and $s = 0.0543$ g. Using Equation (3.6), we would estimate the length of the component to be $\overline{X} \pm \sigma/\sqrt{5}$. We do not know σ, which is the uncertainty, or standard deviation, of the measurement process. However, we can approximate σ with s, the sample standard deviation of the five measurements. We therefore estimate the mass of the rock to be $21.06 \pm 0.0543/\sqrt{5}$, or 21.06 ± 0.02 g.

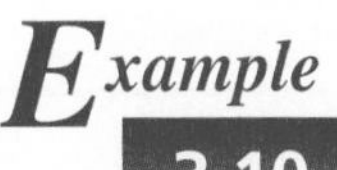

In Example 3.6 two adjacent sides of a rectangular lot were measured to be $X = 50.11 \pm 0.05$ m and $Y = 75.21 \pm 0.08$ m. Assume that the budget for this project is sufficient to allow 14 more measurements to be made. Each side has already been measured once. One engineer suggests allocating the new measurements equally to each side, so that each will be measured eight times. A second engineer suggests using all 14 measurements on the longer side, since that side is measured with greater uncertainty. Estimate the uncertainty in the perimeter under each plan. Which plan results in the smaller uncertainty?

Solution

Under the first plan, let $\overline{X}$ represent the average of eight measurements of the shorter side, and let $\overline{Y}$ represent the average of eight measurements of the longer side. The perimeter will be estimated by $2\overline{X} + 2\overline{Y}$. The uncertainty in the perimeter under the first plan is therefore

$$\begin{aligned}
\sigma_{2\overline{X}+2\overline{Y}} &= \sqrt{4\sigma_{\overline{X}}^2 + 4\sigma_{\overline{Y}}^2} && \text{(using Equation 3.4)} \\
&= \sqrt{4\left(\frac{\sigma_X}{\sqrt{8}}\right)^2 + 4\left(\frac{\sigma_Y}{\sqrt{8}}\right)^2} && \text{(using Equation 3.6)} \\
&= \sqrt{\frac{4(0.05)^2}{8} + \frac{4(0.08)^2}{8}} \\
&= 0.067 \text{ m}
\end{aligned}$$

Under the second plan, the perimeter will be estimated by $2X + 2\overline{Y}$, where X is a single measurement of the shorter side and $\overline{Y}$ is the average of 15 measurements of the longer side. The uncertainty in the perimeter under the second plan is therefore

$$\begin{aligned}
\sigma_{2X+2\overline{Y}} &= \sqrt{4\sigma_X^2 + 4\sigma_{\overline{Y}}^2} && \text{(using Equation 3.4)} \\
&= \sqrt{4\sigma_X^2 + 4\left(\frac{\sigma_Y}{\sqrt{15}}\right)^2} && \text{(using Equation 3.6)} \\
&= \sqrt{4(0.05)^2 + \frac{4(0.08)^2}{15}} \\
&= 0.11 \text{ m}
\end{aligned}$$

The first plan is better.

Repeated Measurements with Differing Uncertainties

Sometimes repeated measurements may have differing uncertainties. This can happen, for example, when the measurements are made with different instruments. It turns out that the best way to combine the measurements in this case is with a weighted average, rather than with the sample mean. Examples 3.11 and 3.12 explore this idea.

Example 3.11

An engineer measures the period of a pendulum (in seconds) to be 2.0 ± 0.2 s. Another independent measurement is made with a more precise clock, and the result is 2.2 ± 0.1 s. The average of these two measurements is 2.1 s. Find the uncertainty in this quantity.

Solution

Let X represent the measurement with the less precise clock, so $X = 2.0$ s, with uncertainty $\sigma_X = 0.2$ s. Let Y represent the measurement on the more precise clock, so $Y = 2.2$ s, with uncertainty $\sigma_Y = 0.1$ s. The average is $(1/2)X + (1/2)Y = 2.10$, and the uncertainty in this average is

$$\begin{aligned}
\sigma_{\text{avg}} &= \sqrt{\frac{1}{4}\sigma_X^2 + \frac{1}{4}\sigma_Y^2} \\
&= \sqrt{\frac{1}{4}(0.2)^2 + \frac{1}{4}(0.1)^2} \\
&= 0.11 \text{ s}
\end{aligned}$$

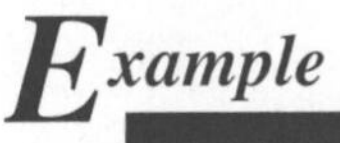

3.12

In Example 3.11, another engineer suggests that since Y is a more precise measurement than X, a weighted average in which Y is weighted more heavily than X might be more precise than the unweighted average. Specifically, the engineer suggests that by choosing an appropriate constant c between 0 and 1, the weighted average $cX + (1-c)Y$ might have a smaller uncertainty than the unweighted average $(1/2)X + (1/2)Y$ considered in Example 3.11. Express the uncertainty in the weighted average $cX + (1-c)Y$ in terms of c, and find the value of c that minimizes the uncertainty.

Solution

The uncertainty in the weighted average is

$$\begin{aligned}\sigma &= \sqrt{c^2\sigma_X^2 + (1-c)^2\sigma_Y^2} \\ &= \sqrt{0.04c^2 + 0.01(1-c)^2} \\ &= \sqrt{0.05c^2 - 0.02c + 0.01}\end{aligned}$$

We now must find the value of c minimizing σ. This is equivalent to finding the value of c minimizing σ^2. We take the derivative of $\sigma^2 = 0.05c^2 - 0.02c + 0.01$ with respect to c and set it equal to 0:

$$\frac{d\sigma^2}{dc} = 0.10c - 0.02 = 0$$

Solving for c, we obtain

$$c = 0.2$$

The most precise weighted average is therefore $0.2X + 0.8Y = 2.16$. The uncertainty in this estimate is

$$\sigma_{\text{best}} = \sqrt{(0.2)^2\sigma_X^2 + (0.8)^2\sigma_Y^2} = \sqrt{(0.2)^2(0.2)^2 + (0.8)^2(0.1)^2} = 0.09 \text{ s}$$

Note that this is less than the uncertainty of 0.11 s found for the unweighted average used in Example 3.11.

The ratio of the coefficients of X and Y in the best weighted average is equal to the ratio of the variances of Y and X: $\sigma_Y^2/\sigma_X^2 = 0.1^2/0.2^2 = 0.25 = 0.2/0.8 = c/(1-c)$. We can therefore express the coefficients in terms of the variances: $c = \sigma_Y^2/(\sigma_X^2 + \sigma_Y^2) = 0.2$ and $1 - c = \sigma_X^2/(\sigma_X^2 + \sigma_Y^2) = 0.8$. This relationship holds in general.

Summary

If X and Y are *independent* measurements of the same quantity, with uncertainties σ_X and σ_Y, respectively, then the weighted average of X and Y with the smallest uncertainty is given by $c_{\text{best}}X + (1 - c_{\text{best}})Y$, where

$$c_{\text{best}} = \frac{\sigma_Y^2}{\sigma_X^2 + \sigma_Y^2} \qquad 1 - c_{\text{best}} = \frac{\sigma_X^2}{\sigma_X^2 + \sigma_Y^2} \tag{3.7}$$

Linear Combinations of Dependent Measurements

Imagine that X and Y are measurements with uncertainties σ_X and σ_Y, and we wish to compute the uncertainty in the sum $X + Y$. If X and Y are *dependent*, the uncertainty in the sum may be either greater than or less than it would be in the independent case, and it cannot be determined from σ_X and σ_Y alone. For example, if positive random errors in X tend to occur alongside negative random errors in Y, and vice versa, the random errors will tend to cancel out when computing the sum $X + Y$, so the uncertainty in $X + Y$ will be smaller than in the independent case. On the other hand, if the random errors in X and Y tend to have the same sign, the uncertainty in $X + Y$ will be larger than in the independent case.

The quantity that measures the relationship between the random errors in X and Y is the covariance, which was discussed in Section 2.6. In general, if $X_1, \ldots, X_n$ are measurements, and if the covariance of each pair of measurements is known, Equation (2.72) (in Section 2.6) can be used to compute the uncertainty in a linear combination of the measurements.

In practice, when measurements are dependent, it is often the case that not enough is known about the dependence to quantify it. In these cases, an upper bound may be placed on the uncertainty in a linear combination of the measurements. The result is presented here; a proof is provided at the end of the section.

If $X_1, \ldots, X_n$ are measurements and $c_1, \ldots, c_n$ are constants, then

$$\sigma_{c_1X_1+\cdots+c_nX_n} \leq |c_1|\sigma_{X_1} + \cdots + |c_n|\sigma_{X_n} \tag{3.8}$$

The expression on the right-hand side of the inequality (3.8) is a conservative estimate of the uncertainty in $c_1X_1 + \cdots + c_nX_n$.

Example 3.13

A surveyor is measuring the perimeter of a rectangular lot. He measures two adjacent sides to be 50.11 ± 0.05 m and 75.21 ± 0.08 m. These measurements are not necessarily independent. Find a conservative estimate of the uncertainty in the perimeter of the lot.

Solution

Denote the two measurements by X_1 and X_2. The uncertainties are then $\sigma_{X_1} = 0.05$ and $\sigma_{X_2} = 0.08$, and the perimeter is given by $P = 2X_1 + 2X_2$. Using the inequality (3.8), we obtain

$$\begin{aligned}\sigma_P &= \sigma_{2X_1+2X_2} \\ &\leq 2\sigma_{X_1} + 2\sigma_{X_2} \\ &= 2(0.05) + 2(0.08) \\ &= 0.26 \text{ m}\end{aligned}$$

The uncertainty in the perimeter is no greater than 0.26 m. In Example 3.6, we computed the uncertainty to be 0.19 m when X and Y are independent.

Derivation of the Inequality $\sigma_{c_1X_1+\cdots+c_nX_n} \le |c_1|\sigma_{X_1} + \cdots + |c_n|\sigma_{X_n}$

This derivation requires material from Section 2.6. Let $X_1, \ldots, X_n$ be random variables, and $c_1, \ldots, c_n$ be constants. By Equation (2.72) (in Section 2.6),

$$\sigma^2_{c_1X_1+\cdots+c_nX_n} = c_1^2\sigma^2_{X_1} + \cdots + c_n^2\sigma^2_{X_n} + 2\sum_{i=1}^{n-1}\sum_{j=i+1}^{n} c_ic_j \operatorname{Cov}(X_i, X_j)$$

Now $\rho_{X_i,X_j} = \dfrac{\operatorname{Cov}(X_i, X_j)}{\sigma_{X_i}\sigma_{X_j}}$. Since $|\rho_{X_i,X_j}| \le 1$, it follows that

$$|\operatorname{Cov}(X_i, X_j)| \le \sigma_{X_i}\sigma_{X_j}$$

Since $c_ic_j \operatorname{Cov}(X_i, X_j) \le |c_i||c_j||\operatorname{Cov}(X_i, X_j)|$, it follows that

$$c_ic_j \operatorname{Cov}(X_i, X_j) \le |c_i||c_j|\sigma_{X_i}\sigma_{X_j}$$

Substituting, we obtain

$$\sigma^2_{c_1X_1+\cdots+c_nX_n} \le c_1^2\sigma^2_{X_1} + \cdots + c_n^2\sigma^2_{X_n} + 2\sum_{i=1}^{n-1}\sum_{j=i+1}^{n} |c_i||c_j|\sigma_{X_i}\sigma_{X_j} \tag{3.9}$$

Since $c_i^2 = |c_i|^2$, the right-hand side of inequality (3.9) can be factored:

$$c_1^2\sigma^2_{X_1} + \cdots + c_n^2\sigma^2_{X_n} + 2\sum_{i=1}^{n-1}\sum_{j=i+1}^{n} |c_i||c_j|\sigma_{X_i}\sigma_{X_j} = (|c_1|\sigma_{X_1} + \cdots + |c_n|\sigma_{X_n})^2$$

Substituting into inequality (3.9) and taking square roots, we obtain

$$\sigma_{c_1X_1+\cdots+c_nX_n} \le |c_1|\sigma_{X_1} + \cdots + |c_n|\sigma_{X_n}$$

Exercises for Section 3.2

1. Assume that X and Y are independent measurements with uncertainties $\sigma_X = 0.2$ and $\sigma_Y = 0.4$. Find the uncertainties in the following quantities:

a. $3X$

b. $X - Y$

c. $2X + 3Y$

2. A measurement of the diameter of a cylinder has an uncertainty of 2 mm. How many measurements must be made so that the diameter can be estimated with an uncertainty of only 0.5 mm?

3. In the article "The World's Longest Continued Series of Sea Level Observations" (M. Ekman, *Paleogeography*, 1988:73–77), the mean annual level of land uplift in Stockholm, Sweden, was estimated to be 4.93 ± 0.23 mm for the years 1774–1884 and to be 3.92 ± 0.19 mm for the years 1885–1984. Estimate the difference in the mean annual uplift between these two time periods, and find the uncertainty in the estimate.

4. A cylindrical hole is bored through a steel block, and a cylindrical piston is machined to fit into the hole. The diameter of the hole is 20.00 ± 0.01 cm, and the diameter of the piston is 19.90 ± 0.02 cm. The clearance is one-half the difference between the diameters. Estimate the clearance and find the uncertainty in the estimate.

5. The width and height of a 2×4 piece of lumber are actually 1.5×3.5 in. Assume that the uncertainty in these quantities is negligible. The length of a 2×4 is measured to be 72 ± 0.1 in. Estimate the volume of the piece of lumber, and find the uncertainty in the estimate.

6. The period T of a simple pendulum is given by $T = 2\pi\sqrt{L/g}$ where L is the length of the pendulum and g is the acceleration due to gravity. Thus if L and T are measured, we can estimate g with $g = 4\pi^2 L/T^2$. Assume that the period is known to be $T = 1.5$ s with negligible uncertainty, and that L is measured to be 0.559 ± 0.005 m. Estimate g, and find the uncertainty in the estimate.

7. The Beer–Lambert law relates the absorbance A of a solution to the concentration C of a species in solution by $A = MLC$, where L is the path length and M is the molar absorption coefficient. Assume that $C = 1.25$ mol/cm^3 and $L = 1$ cm, both with negligible uncertainty, and that $A = 1.30 \pm 0.05$. Estimate M, and find the uncertainty in the estimate.

8. In a Couette flow, two large flat plates lie one on top of another, separated by a thin layer of fluid. If a shear stress is applied to the top plate, the viscosity of the fluid produces motion in the bottom plate as well. The velocity V in the top plate relative to the bottom plate is given by $V = \tau h/\mu$, where τ is the shear stress applied to the top plate, h is the thickness of the fluid layer, and μ is the viscosity of the fluid. Assume that $\mu = 1.49$ Pa · s and $h = 10$ mm, both with negligible uncertainty.

a. Suppose that $\tau = 30.0 \pm 0.1$ Pa. Estimate V, and find the uncertainty in the estimate.

b. If it is desired to estimate V with an uncertainty of 0.2 mm/s, what must be the uncertainty in τ?

9. According to Newton's law of cooling, the temperature T of a body at time t is given by $T = T_a + (T_0 - T_a)e^{-kt}$, where T_a is the ambient temperature, T_0 is the initial temperature, and k is the cooling rate constant. For a certain type of beverage container, the value of k is known to be 0.025 min^{-1}.

a. Assume that $T_a = 36$°F exactly and that $T_0 = 72.0 \pm 0.5$°F. Estimate the temperature T at time $t = 10$ min, and find the uncertainty in the estimate.

b. Assume that $T_0 = 72$°F exactly and that $T_a = 36.0 \pm 0.5$°F. Estimate the temperature T at time $t = 10$ min, and find the uncertainty in the estimate.

10. In the article "Influence of Crack Width on Shear Behaviour of SIFCON" (C. Fritz and H. Reinhardt, *High Performance Fiber Reinforced Cement Composites: Proceedings of the International RILEM/ACI Workshop*, 1992), the maximum shear stress τ of a cracked concrete member is given to be $\tau = \tau_0(1 - kw)$, where τ_0 is the maximum shear stress for a crack width of zero, w is the crack width in mm, and k is a constant estimated from experimental data. Assume $k = 0.29 \pm 0.05$ mm^{-1}. Given that $\tau_0 = 50$ MPa and $w = 1.0$ mm, both with negligible uncertainty, estimate τ and find the uncertainty in the estimate.

11. Nine independent measurements are made of the length of a rod. The average of the nine measurements is $\overline{X} = 5.238$ cm, and the standard deviation is $s = 0.081$ cm.

a. Is the uncertainty in the value 5.238 cm closest to 0.009, 0.027, or 0.081 cm? Explain.

b. Another rod is measured once by the same process. The measurement is 5.423 cm. Is the uncertainty in this value closest to 0.009, 0.027, or 0.081 cm? Explain.

12. A certain scale has an uncertainty of 3 g and a bias of 2 g.

a. A single measurement is made on this scale. What are the bias and uncertainty in this measurement?

b. Four independent measurements are made on this scale. What are the bias and uncertainty in the average of these measurements?

c. Four hundred independent measurements are made on this scale. What are the bias and uncertainty in the average of these measurements?

d. As more measurements are made, does the uncertainty get smaller, get larger, or stay the same?

e. As more measurements are made, does the bias get smaller, get larger, or stay the same?

13. The volume of a rock is measured by placing the rock in a graduated cylinder partially filled with water and measuring the increase in volume. Eight independent measurements are made. The average of the measurements is 87.0 mL, and the standard deviation is 2.0 mL.

a. Estimate the volume of the rock, and find the uncertainty in the estimate.

b. Eight additional measurements are made, for a total of 16. What is the uncertainty, approximately, in the average of the 16 measurements?

c. Approximately how many measurements would be needed to reduce the uncertainty to 0.4 mL?

14. A student measures the spring constant k of a spring by loading it and measuring the extension. (According to Hooke's law, if l is the load and e is the extension, then $k = l/e$.) Assume five independent measurements are made, and the measured values of k (in N/m) are 36.4, 35.4, 38.6, 36.6, and 38.0.

a. Estimate the spring constant, and find the uncertainty in the estimate.

b. Find an approximate value for the uncertainty in the average of 10 measurements.

c. Approximately how many measurements must be made to reduce the uncertainty to 0.3 N/m?

d. A second spring, similar to the first, is measured once. The measured value for k is 39.3. Approximately how much is the uncertainty in this measurement?

15. A certain chemical process is run 10 times at a temperature of 65°C and 10 times at a temperature of 80°C. The yield at each run was measured as a percent of a theoretical maximum. The data are presented in the following table.

65°C	71.3	69.1	70.3	69.9	71.1	70.7	69.8	68.5	70.9	69.8
80°C	90.3	90.8	91.2	90.7	89.0	89.7	91.3	91.2	89.7	91.1

a. For each temperature, estimate the mean yield and find the uncertainty in the estimate.

b. Estimate the difference between the mean yields at the two temperatures, and find the uncertainty in the estimate.

16. An object is weighed four times, and the results, in milligrams, are 234, 236, 233, and 229. The object is then weighed four times on a different scale, and the results, in milligrams, are 236, 225, 245, and 240. The average of all eight measurements will be used to estimate the weight. Someone suggests estimating the uncertainty in this estimate as follows: Compute the standard deviation of all eight measurements. Call this quantity s. The uncertainty is then $s/\sqrt{8}$. Is this correct? Explain.

17. The length of a component is to be estimated through repeated measurement.

a. Ten independent measurements are made with an instrument whose uncertainty is 0.05 mm. Let $\overline{X}$ denote the average of these measurements. Find the uncertainty in $\overline{X}$.

b. A new measuring device, whose uncertainty is only 0.02 mm, becomes available. Five independent measurements are made with this device. Let $\overline{Y}$ denote the average of these measurements. Find the uncertainty in $\overline{Y}$.

c. In order to decrease the uncertainty still further, it is decided to combine the estimates $\overline{X}$ and $\overline{Y}$. One engineer suggests estimating the length with $(1/2)\overline{X} + (1/2)\overline{Y}$. A second engineer argues that since $\overline{X}$ is based on 10 measurements, while $\overline{Y}$ is based on only five, a better estimate is $(10/15)\overline{X} + (5/15)\overline{Y}$. Find the uncertainty in each of these estimates. Which is smaller?

d. Find the value c such that the weighted average $c\overline{X} + (1 - c)\overline{Y}$ has minimum uncertainty. Find the uncertainty in this weighted average.

18. The lengths of two components will be measured several times. The uncertainty in each measurement of the length of the first component is $\sigma_1 = 0.02$ cm, and the uncertainty in each measurement of the length of the second component is $\sigma_2 = 0.08$ cm. Let $\overline{X}$ denote the average of the measurements of the first component, and let $\overline{Y}$ denote the average of the measurements of the second component. The total length of the two components will be estimated with the quantity $\overline{X} + \overline{Y}$.

a. Find the uncertainty in the total length if the first component is measured 4 times and the second component is measured 12 times.

b. Find the uncertainty in the total length in terms of n if the first component is measured n times and the second component is measured $16 - n$ times.

c. Determine the best way to allocate 16 measurements between the components by determining the value of n that minimizes the uncertainty.

3.3 Uncertainties for Functions of One Measurement

The examples we have seen so far involve estimating uncertainties in linear functions of measurements. In many cases we wish to estimate the uncertainty in a nonlinear function of a measurement. For example, if the radius R of a circle is measured to be 5.00 ± 0.01 cm, what is the uncertainty in the area A? In statistical terms, we know that the standard deviation σ_R is 0.01 cm, and we must calculate the standard deviation of A, where A is the function of R given by $A = \pi R^2$.

The type of problem we wish to solve is this: Given a random variable X, with known standard deviation σ_X, and given a function $U = U(X)$, how do we compute the standard deviation σ_U? If U is a linear function, the methods of Section 3.2 apply. If U is not linear, we can still approximate σ_U, by multiplying σ_X by the absolute value of the derivative dU/dX. The approximation will be good so long as σ_X is small.

If X is a measurement whose uncertainty σ_X is small, and if U is a function of X, then

$$\sigma_U \approx \left| \frac{dU}{dX} \right| \sigma_X \tag{3.10}$$

In practice, we evaluate the derivative dU/dX at the observed measurement X.

Equation (3.10) is known as the **propagation of error** formula. Its derivation is given at the end of this section.

Propagation of Error Uncertainties Are Only Approximate

The uncertainties computed by using Equation (3.10) are often only rough approximations. For this reason, these uncertainties should be expressed with no more than two significant digits. Indeed, some authors suggest using only one significant digit.

Nonlinear Functions Are Biased

If X is an unbiased measurement of a true value μ_X, and if the function $U = U(X)$ is a nonlinear function of X, then in most cases U will be a *biased* estimate of the true value $U(\mu_X)$. In practice this bias is usually ignored. It can be shown by advanced methods that in general, the size of the bias depends mostly on the magnitudes of σ_X and of the second derivative d^2U/dX^2. Therefore, so long as the uncertainty σ_X is small, the bias in U will in general be small as well, except for some fairly unusual circumstances when the second derivative is quite large. Of course, if X is a measurement with non-negligible bias, then the bias in U may be large. These ideas are explored further in Supplementary Exercise 22 at the end of this chapter.

The radius R of a circle is measured to be 5.00 ± 0.01 cm. Estimate the area of the circle and find the uncertainty in this estimate.

Solution

The area A is given by $A = \pi R^2$. The estimate of A is $\pi(5.00 \text{ cm})^2 = 78.5 \text{ cm}^2$. Now $\sigma_R = 0.01$ cm, and $dA/dR = 2\pi R = 10\pi$ cm. We can now find the uncertainty in A:

$$\begin{aligned} \sigma_A &= \left|\frac{dA}{dR}\right| \sigma_R \\ &= (10\pi \text{ cm})(0.01 \text{ cm}) \\ &= 0.31 \text{ cm}^2 \end{aligned}$$

We estimate the area of the circle to be $78.5 \pm 0.3 \text{ cm}^2$.

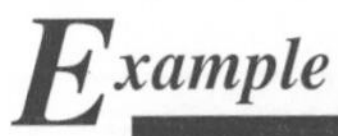

A rock identified as cobble-sized quartzite has a mass m of 674.0 g. Assume this measurement has negligible uncertainty. The volume V of the rock will be measured by placing it in a graduated cylinder partially filled with water and measuring the volume of water displaced. The density D of the rock will be computed as $D = m/V$. Assume the volume of displaced water is 261.0 ± 0.1 mL. Estimate the density of the rock and find the uncertainty in this estimate.

Solution

Substituting $V = 261.0$ mL, the estimate of the density D is $674.0/261.0 = 2.582$ g/mL. Treating $m = 674.0$ as a known constant, $dD/dV = -674.0/V^2 = -674.0/(261.0)^2 = -0.010 \text{ g/mL}^2$. We know that $\sigma_V = 0.1$ mL. The uncertainty in D is therefore

$$\begin{aligned} \sigma_D &= \left|\frac{dD}{dV}\right| \sigma_V \\ &= |-0.010|(0.1 \text{ g/mL}) \\ &= 0.001 \text{ g/mL} \end{aligned}$$

We estimate the density to be 2.582 ± 0.001 g/mL.

Relative Uncertainties for Functions of One Measurement

We have been referring to the standard deviation σ_U of a measurement U as the uncertainty in U. A more complete name for σ_U is the **absolute uncertainty**, because it is expressed in the same units as the measurement U. Sometimes we wish to express the uncertainty as a fraction of the true value, which (assuming no bias) is the mean measurement μ_U. This is called the **relative uncertainty** in U. The relative uncertainty can also be called the **coefficient of variation**. In practice, since μ_U is unknown, the measured value U is used in its place when computing the relative uncertainty.

Summary

If U is a measurement whose true value is μ_U, and whose uncertainty is σ_U, the relative uncertainty in U is the quantity σ_U/μ_U.

The relative uncertainty is a pure number, without units. It is frequently expressed as a percent. In practice μ_U is unknown, so if the bias is negligible, we estimate the relative uncertainty with σ_U/U.

There are two ways to compute the relative uncertainty in a quantity U. One is simply to use Equation (3.10) to compute the absolute uncertainty σ_U, and then divide by U. To develop the second method, we will compute the absolute uncertainty in $\ln U$:

$$\sigma_{\ln U} = \frac{d(\ln U)}{dU}\sigma_U = \frac{\sigma_U}{U}$$

This equation shows that the absolute uncertainty in $\ln U$ is equal to the relative uncertainty in U. Thus the second way to compute the relative uncertainty in U is to compute $\ln U$, and then use Equation (3.10) to compute the absolute uncertainty in $\ln U$.

Summary

There are two methods for approximating the relative uncertainty σ_U/U in a function $U = U(X)$:

1. Compute σ_U using Equation (3.10), and then divide by U.
2. Compute $\ln U$ and use Equation (3.10) to find $\sigma_{\ln U}$, which is equal to σ_U/U.

Both of the methods work in every instance, so one may use whichever is easiest for a given problem. This choice is usually dictated by whether it is easier to compute the derivative of U or of $\ln U$.

Example 3.16

The radius of a circle is measured to be 5.00 ± 0.01 cm. Estimate the area, and find the relative uncertainty in the estimate.

Solution

In Example 3.14 the area $A = \pi R^2$ was computed to be 78.5 ± 0.3 cm^2. The absolute uncertainty is therefore $\sigma_A = 0.3$ cm^2, and the relative uncertainty is $\sigma_A/A = 0.3/78.5 = 0.004$. We can therefore express the area as $A = 78.5$ cm^2 $\pm 0.4\%$.

If we had not already computed σ_A, it would be easier to compute the relative uncertainty by computing the absolute uncertainty in $\ln A$. Since $\ln A = \ln \pi + 2 \ln R$, $d \ln A/dR = 2/R = 0.4$. The relative uncertainty in A is therefore

$$\begin{aligned}
\frac{\sigma_A}{A} &= \sigma_{\ln A} \\
&= \left| \frac{d \ln A}{dR} \right| \sigma_R \\
&= 0.4\sigma_R \\
&= (0.4)(0.01) \\
&= 0.4\%
\end{aligned}$$

Example 3.17

The acceleration of a mass down a frictionless inclined plane is given by $a = g \sin\theta$, where g is the acceleration due to gravity and θ is the angle of inclination of the plane. Assume the uncertainty in g is negligible. If $\theta = 0.60 \pm 0.01$ rad, find the relative uncertainty in a.

Solution

The relative uncertainty in a is the absolute uncertainty in $\ln a$. Now $\ln a = \ln g + \ln(\sin\theta)$, where $\ln g$ is constant. Therefore $d \ln a/d\theta = d \ln(\sin\theta)/d\theta = \cos\theta/\sin\theta = \cot\theta = \cot(0.60) = 1.46$. The uncertainty in θ is $\sigma_\theta = 0.01$. The relative uncertainty in a is therefore

$$\begin{aligned}
\frac{\sigma_a}{a} &= \sigma_{\ln a} \\
&= \left| \frac{d \ln a}{d\theta} \right| \sigma_\theta \\
&= (1.46)(0.01) \\
&= 1.5\%
\end{aligned}$$

Note that the relative uncertainty in $a = g \sin\theta$ does not depend on the constant g.

Derivation of the Propagation of Error Formula

We derive the propagation of error formula for a nonlinear function U of a random variable X by approximating it with a linear function, and then using the methods of Section 3.2. To find a linear approximation to U, we use a first-order Taylor series approximation. This is known as **linearizing the problem**; it is a commonly used technique in science and engineering.

Let $U(X)$ be a differentiable function. Let μ_X be any point. Then if X is close to μ_X, the first-order Taylor series approximation for $U(X)$ is

$$U(X) - U(\mu_X) \approx \frac{dU}{dX}(X - \mu_X) \tag{3.11}$$

The derivative dU/dX is evaluated at μ_X.

Now let X be a measurement, and let $U(X)$ (which we will also refer to as U) be a quantity calculated from X. Let μ_X denote the mean of X. For any reasonably precise measurement, X will be close enough to μ_X for the Taylor series approximation to be valid.

Adding $U(\mu_X)$ to both sides of Equation (3.11) yields

$$U \approx U(\mu_X) + \frac{dU}{dX}(X - \mu_X)$$

Multiplying through by dU/dX and rearranging terms yields

$$U \approx \left(U(\mu_X) - \frac{dU}{dX}\mu_X\right) + \frac{dU}{dX}X$$

Now the quantity dU/dX is a constant, because it is evaluated at μ_X. Therefore the quantity $U(\mu_X) - (dU/dX)\,\mu_X$ is also constant. It follows from Equation (2.46) (in Section 2.5) that

$$\sigma_U = \left|\frac{dU}{dX}\right| \sigma_X$$

This is the **propagation of error** formula. When applying this formula, we evaluate the derivative dU/dX at the observed measurement X, since we do not know the value μ_X.

Exercises for Section 3.3

1. Find the uncertainty in Y, given that $X = 4.0 \pm 0.4$ and

a. $Y = X^2$

b. $Y = \sqrt{X}$

c. $Y = 1/X$

d. $Y = \ln X$

e. $Y = e^X$

f. $Y = \sin X$ (X is in units of radians)

2. Given that X and Y are related by the given equation, and that $X = 2.0 \pm 0.2$, estimate Y, and find the uncertainty in the estimate.

a. $XY = 1$

b. $X/Y = 2$

c. $X\sqrt{Y} = 3$

d. $Y\sqrt{X} = 4$

3. The acceleration g due to gravity is estimated by dropping an object and measuring the time it takes to travel a certain distance. Assume the distance s is known to be exactly 5 m, and that the time is measured to be $t = 1.01 \pm 0.02$ s. Estimate g, and find the uncertainty in the estimate. (Note that $g = 2s/t^2$.)

4. The velocity V of sound in air at temperature T is given by $V = 20.04\sqrt{T}$, where T is measured in kelvins (K) and V is in m/s. Assume that $T = 300 \pm 0.4$ K. Estimate V, and find the uncertainty in the estimate.

5. The period T of a simple pendulum is given by $T = 2\pi\sqrt{L/g}$ where L is the length of the pendulum and g is the acceleration due to gravity.

a. Assume $g = 9.80$ m/s^2 exactly, and that $L = 0.742 \pm 0.005$ m. Estimate T, and find the uncertainty in the estimate.

b. Assume $L = 0.742$ m exactly, and that $T = 1.73 \pm 0.01$ s. Estimate g, and find the uncertainty in the estimate.

6. The height h of capillary rise of water in a clean glass tube is given by $h = k/r$, where r is the radius of the tube and k is a constant that depends on the temperature of the water. Assume that at a temperature of 10°C, $k = 7.57$ mm^2. Assume that the radius of the tube is 2.0 ± 0.1 mm. Estimate h, and find the uncertainty in the estimate.

7. The friction velocity F of water flowing through a pipe is given by $F = \sqrt{gdh/4l}$, where g is the acceleration due to gravity, d is the diameter of the pipe, l is the length of the pipe, and h is the head loss. Estimate F, and find the uncertainty in the estimate, assuming that $g = 9.80$ m/s^2 exactly, and that

a. $d = 0.15$ m and $l = 30.0$ m, both with negligible uncertainty, and $h = 5.33 \pm 0.02$ m.

b. $h = 5.33$ m and $l = 30.0$ m, both with negligible uncertainty, and $d = 0.15 \pm 0.03$ m.

c. $d = 0.15$ m and $h = 5.33$ m, both with negligible uncertainty, and $l = 30.00 \pm 0.04$ m.

8. The refractive index n of a piece of glass is related to the critical angle θ by $n = 1/\sin\theta$. Assume that the critical angle is measured to be 0.70 ± 0.02 rad. Estimate the refractive index, and find the uncertainty in the estimate.

9. The density of a rock will be measured by placing it into a graduated cylinder partially filled with water, and then measuring the volume of water displaced. The density D is given by $D = m/(V_1 - V_0)$, where m is the mass of the rock, V_0 is the initial volume of water, and V_1 is the volume of water plus rock. Assume the mass of the rock is 750 g, with negligible uncertainty, and that $V_0 = 500.0 \pm 0.1$ mL and $V_1 = 813.2 \pm 0.1$ mL. Estimate the density of the rock, and find the uncertainty in the estimate.

10. The conversion of ammonium cyanide to urea is a second-order reaction. This means that the concentration C of ammonium cyanide at time t is given by $1/C = kt + 1/C_0$, where C_0 is the initial concentration and k is the rate constant. Assume the initial concentration is known to be 0.1 mol/L exactly. Assume that time can be measured with negligible uncertainty.

a. After 45 minutes, the concentration of ammonium cyanide is measured to be 0.0811 ± 0.0005 mol/L. Estimate the rate constant k, and find the uncertainty in the estimate.

b. Use the result in part (a) to estimate the time when the concentration of ammonium cyanide will be 0.0750 mol/L, and find the uncertainty in this estimate.

11. Convert the following absolute uncertainties to relative uncertainties.

a. 37.2 ± 0.1

b. 8.040 ± 0.003

c. 936 ± 37

d. 54.8 ± 0.3

12. Convert the following relative uncertainties to absolute uncertainties.

a. $48.41 \pm 0.3\%$

b. $991.7 \pm 0.6\%$

c. $0.011 \pm 9\%$

d. $7.86 \pm 1\%$

13. The acceleration g due to gravity is estimated by dropping an object and measuring the time it takes to travel a certain distance. Assume the distance s is known to be exactly 2.2 m. The time is measured to be $t = 0.67 \pm 0.02$ s. Estimate g, and find the relative uncertainty in the estimate. (Note that $g = 2s/t^2$.)

14. Refer to Exercise 4. Assume that $T = 298.4 \pm 0.2$ K. Estimate V, and find the relative uncertainty in the estimate.

15. Refer to Exercise 5.

a. Assume $g = 9.80$ m/s^2 exactly, and that $L = 0.855 \pm 0.005$ m. Estimate T, and find the relative uncertainty in the estimate.

b. Assume $L = 0.855$ m exactly, and that $T = 1.856 \pm 0.005$ s. Estimate g, and find the relative uncertainty in the estimate.

16. Refer to Exercise 6. Assume that the radius of the tube is $r = 2.5 \pm 0.2$ mm and that $k = 7.57$ mm^2 with negligible uncertainty. Estimate h, and find the relative uncertainty in the estimate.

17. Refer to Exercise 7. Estimate F, and find the relative uncertainty in the estimate, assuming that $g = 9.80$ m/s^2 exactly and that

a. $d = 0.20$ m and $l = 35.0$ m, both with negligible uncertainty, and $h = 4.51 \pm 0.03$ m.

b. $h = 4.51$ m and $l = 35.0$ m, both with negligible uncertainty, and $d = 0.20 \pm 0.008$ m.

c. $d = 0.20$ m and $h = 4.51$ m, both with negligible uncertainty, and $l = 35.00 \pm 0.4$ m.

18. Refer to Exercise 8. Assume the critical angle is measured to be 0.90 ± 0.01 rad. Estimate the refractive index and find the relative uncertainty the estimate.

19. Refer to Exercise 9. Assume that the mass of the rock is 288.2 g with negligible uncertainty, the initial volume of water in the cylinder is 400 ± 0.1 mL, and the volume of water plus rock is 516 ± 0.2 mL. Estimate the density of the rock, and find the relative uncertainty in the estimate.

20. In a chemical reaction run at a certain temperature, the concentration C of a certain reactant at time t is given by $1/C = kt + 1/C_0$, where C_0 is the initial concentration and k is the rate constant. Assume the initial concentration is known to be 0.04 mol/L exactly. Assume that time is measured with negligible uncertainty.

a. After 30 s, the concentration C is measured to be $0.0038 \pm 2.0 \times 10^{-4}$ mol/L. Estimate the rate constant k, and find the relative uncertainty in the estimate.

b. After 50 s, the concentration C is measured to be $0.0024 \pm 2.0 \times 10^{-4}$ mol/L. Estimate the rate constant k and find the relative uncertainty in the estimate.

c. Denote the estimates of the rate constant k in parts (a) and (b) by $\widehat{k}_1$ and $\widehat{k}_2$, respectively. The geometric mean $\sqrt{\widehat{k}_1 \widehat{k}_2}$ is used as an estimate of k. Find the relative uncertainty in this estimate.

3.4 Uncertainties for Functions of Several Measurements

Often we need to estimate a quantity as a function of several measurements. For example, we might measure the mass m and volume V of a rock and compute the density as $D = m/V$. In Example 3.15 we saw how to estimate the uncertainty in D when one of the quantities, in this case V, was measured with uncertainty while m was treated as a known constant. However, in practice we might need to estimate the uncertainty in D when both m and V are measured with uncertainty.

In this section we will learn how to estimate the uncertainty in a quantity that is a function of several *independent* uncertain measurements. The basic formula is given here.

If $X_1, X_2, \ldots, X_n$ are *independent* measurements whose uncertainties $\sigma_{X_1}, \sigma_{X_2}, \ldots, \sigma_{X_n}$ are small, and if $U = U(X_1, X_2, \ldots, X_n)$ is a function of $X_1, X_2, \ldots, X_n$, then

$$\sigma_U \approx \sqrt{\left(\frac{\partial U}{\partial X_1}\right)^2 \sigma_{X_1}^2 + \left(\frac{\partial U}{\partial X_2}\right)^2 \sigma_{X_2}^2 + \cdots + \left(\frac{\partial U}{\partial X_n}\right)^2 \sigma_{X_n}^2} \qquad (3.12)$$

In practice, we evaluate the partial derivatives at the point $(X_1, X_2, \ldots, X_n)$.

Equation (3.12) is the **multivariate propagation of error formula**. It is important to note that it is valid only when the measurements $X_1, X_2, \ldots, X_n$ are independent. A derivation of the formula is given at the end of the section. As in the case of one measurement, the uncertainties computed by the propagation of error formula are often only rough approximations.

Nonlinear functions of measurements are generally biased (see the discussion concerning functions of one measurement in Section 3.3). However, so long as the measurements $X_1, \ldots, X_n$ are unbiased, and the uncertainties $\sigma_{X_1}, \ldots, \sigma_{X_n}$ are all small, the bias in U will usually be small enough to ignore. Exceptions to this rule, which are fairly unusual, can occur when some of the second- or higher-order partial derivatives of U with respect to the X_i are quite large. Of course, if one or more of $X_1, \ldots, X_n$ are substantially biased, then U may be substantially biased as well. These ideas are explored further in Exercise 23 in the Supplementary Exercises at the end of this chapter.

We now present some examples that illustrate the use of multivariate propagation of error.

Example 3.18

Assume the mass of a rock is measured to be $m = 674.0 \pm 1.0$ g, and the volume of the rock is measured to be $V = 261.0 \pm 0.1$ mL. Estimate the density of the rock and find the uncertainty in the estimate.

Solution

Substituting $m = 674.0$ g and $V = 261.0$ mL, the estimate of the density D is $674.0/261.0 = 2.582$ g/mL. Since $D = m/V$, the partial derivatives of D are

$$\frac{\partial D}{\partial m} = \frac{1}{V} = 0.0038 \text{ mL}^{-1}$$

$$\frac{\partial D}{\partial V} = \frac{-m}{V^2} = -0.0099 \text{ g/mL}^2$$

The uncertainty in D is therefore

$$\begin{aligned}\sigma_D &= \sqrt{\left(\frac{\partial D}{\partial m}\right)^2 \sigma_m^2 + \left(\frac{\partial D}{\partial V}\right)^2 \sigma_V^2} \\ &= \sqrt{(0.0038)^2(1.0)^2 + (-0.0099)^2(0.1)^2} \\ &= 0.0040 \text{ g/mL}\end{aligned}$$

The density of the rock is 2.582 ± 0.004 g/mL.

One of the great benefits of the multivariate propagation of error formula is that it enables one to determine which measurements are most responsible for the uncertainty in the final result. Example 3.19 illustrates this.

Example 3.19

The density of the rock in Example 3.18 is to be estimated again with different equipment, in order to improve the precision. Which would improve the precision of the density estimate more: decreasing the uncertainty in the mass estimate to 0.5 g, or decreasing the uncertainty in the volume estimate to 0.05 mL?

Solution

From Example 3.18, $\sigma_D = \sqrt{(0.0038)^2\sigma_m^2 + (-0.0099)^2\sigma_V^2}$. We have a choice between having $\sigma_m = 0.5$ and $\sigma_V = 0.1$, or having $\sigma_m = 1.0$ and $\sigma_V = 0.05$. The first choice results in $\sigma_D = 0.002$ g/mL, while the second choice results in $\sigma_D = 0.004$ g/mL. It is better to reduce σ_m to 0.5 g.

Example 3.20

Two resistors with resistances R_1 and R_2 are connected in parallel. The combined resistance R is given by $R = (R_1 R_2)/(R_1 + R_2)$. If R_1 is measured to be $100 \pm 10\ \Omega$, and R_2 is measured to be $20 \pm 1\ \Omega$, estimate R and find the uncertainty in the estimate.

Solution

The estimate of R is $(100)(20)/(100 + 20) = 16.67\ \Omega$. To compute σ_R, we first compute the partial derivatives of R:

$$\frac{\partial R}{\partial R_1} = \left(\frac{R_2}{R_1 + R_2}\right)^2 = 0.0278$$

$$\frac{\partial R}{\partial R_2} = \left(\frac{R_1}{R_1 + R_2}\right)^2 = 0.694$$

Now $\sigma_{R_1} = 10\ \Omega$, and $\sigma_{R_2} = 1\ \Omega$. Therefore

$$\begin{aligned}\sigma_R &= \sqrt{\left(\frac{\partial R}{\partial R_1}\right)^2 \sigma_{R_1}^2 + \left(\frac{\partial R}{\partial R_2}\right)^2 \sigma_{R_2}^2} \\ &= \sqrt{(0.0278)^2(10)^2 + (0.694)^2(1)^2} \\ &= 0.75\ \Omega\end{aligned}$$

The combined resistance is $16.67 \pm 0.75\ \Omega$.

Example 3.21

In Example 3.20, the $100\pm10\ \Omega$ resistor can be replaced with a more expensive $100\pm1\ \Omega$ resistor. How much would this reduce the uncertainty in the combined resistance? Is it worthwhile to make the replacement?

Solution

Using the method of Example 3.20, the uncertainty in the combined resistance R with the new resistor would be

$$\sqrt{(0.0278)^2(1)^2 + (0.694)^2(1)^2} = 0.69\ \Omega$$

There is not much reduction from the uncertainty of $0.75\ \Omega$ using the old resistor. Almost all the uncertainty in the combined resistance is due to the uncertainty in the $20\ \Omega$ resistor. The uncertainty in the $100\ \Omega$ resistor can be neglected for most practical purposes. There is little benefit in replacing this resistor.

Note that in Example 3.20, one component (the $100\ \Omega$ resistor) had larger uncertainty, both in absolute terms and relative to the measured value, than the other. Even so, Example 3.21 showed that the uncertainty in the combined resistance was only slightly affected by the uncertainty in this component. The lesson is that one cannot predict the impact that the uncertainties in individual measurements will have on the uncertainty in the final calculation from the magnitudes of those uncertainties alone. One must use the propagation of error formula.

Uncertainties for Functions of Dependent Measurements

If $X_1, X_2, \ldots, X_n$ are not independent, the uncertainty in a function $U = U(X_1, X_2, \ldots, X_n)$ can be estimated if the covariance of each pair (X_i, X_j) is known. (Covariance is discussed in Section 2.6.) In many situations, the covariances are not known. In these cases, a conservative estimate of the uncertainty in U may be computed. We present this result here.

> If $X_1, X_2, \ldots, X_n$ are measurements whose uncertainties $\sigma_{X_1}, \sigma_{X_2}, \ldots, \sigma_{X_n}$ are small, and if $U = U(X_1, X_2, \ldots, X_n)$ is a function of $(X_1, X_2, \ldots, X_n)$, then a conservative estimate of σ_U is given by
>
> $$\sigma_U \le \left|\frac{\partial U}{\partial X_1}\right|\sigma_{X_1} + \left|\frac{\partial U}{\partial X_2}\right|\sigma_{X_2} + \cdots + \left|\frac{\partial U}{\partial X_n}\right|\sigma_{X_n} \tag{3.13}$$
>
> In practice, we evaluate the partial derivatives at the point $(X_1, X_2, \ldots, X_n)$.
>
> The inequality (3.13) is valid in almost all practical situations; in principle it can fail if some of the second partial derivatives of U are quite large.

Example 3.22

Refer to Example 3.20. Find a conservative estimate for the uncertainty in the total resistance R if R_1 and R_2 are not known to be independent.

Solution

We have $\sigma_{R_1} = 10\ \Omega$, $\sigma_{R_2} = 1\ \Omega$, $\partial R/\partial R_1 = 0.0278$, and $\partial R/\partial R_2 = 0.694$. Therefore

$$\begin{aligned}\sigma_R &\le \left|\frac{\partial R}{\partial R_1}\right|\sigma_{R_1} + \left|\frac{\partial R}{\partial R_2}\right|\sigma_{R_2} \\ &= (0.0278)(10) + (0.694)(1) \\ &= 0.97\ \Omega\end{aligned}$$

The uncertainty in the total resistance is conservatively estimated by 0.97 Ω. In Example 3.20, we computed the uncertainty to be 0.75 Ω when R_1 and R_2 are independent.

Relative Uncertainties for Functions of Several Measurements

In Section 3.3, we presented methods for calculating relative uncertainties for functions of one variable. The methods for calculating relative uncertainties for functions of several variables are similar.

There are two methods for approximating the relative uncertainty σ_U/U in a function $U = U(X_1, X_2, \ldots, X_n)$:

1. Compute σ_U using Equation (3.12), and then divide by U.
2. Compute $\ln U$, and then use Equation (3.12) to find $\sigma_{\ln U}$, which is equal to σ_U/U.

Both of the methods work in every instance, so one may use whichever is easiest for a given problem. This choice is usually dictated by whether it is easier to compute partial derivatives of U or of $\ln U$.

Example 3.23

Two perpendicular sides of a rectangle are measured to be $X = 2.0 \pm 0.1$ cm and $Y = 3.2 \pm 0.2$ cm. Find the relative uncertainty in the area $A = XY$.

Solution

This is easily computed by finding the absolute uncertainty in $\ln A = \ln X + \ln Y$. We begin by computing the partial derivatives of $\ln A$:

$$\frac{\partial \ln A}{\partial X} = \frac{1}{X} = 0.50 \qquad \frac{\partial \ln A}{\partial Y} = \frac{1}{Y} = 0.31$$

We are given that $\sigma_X = 0.1$ and $\sigma_Y = 0.2$. The relative uncertainty in A is

$$\begin{aligned}\frac{\sigma_A}{A} = \sigma_{\ln A} &= \sqrt{\left(\frac{\partial \ln A}{\partial X}\right)^2 \sigma_X^2 + \left(\frac{\partial \ln A}{\partial Y}\right)^2 \sigma_Y^2} \\ &= \sqrt{(0.50)^2(0.1)^2 + (0.31)^2(0.2)^2} \\ &= 0.080\end{aligned}$$

The relative uncertainty in A is 0.080, or 8%. The area of the rectangle is $6.4\text{ cm}^2 \pm 8\%$.

Example 3.24

An Atwood machine consists of two masses X and Y ($X > Y$) attached to the ends of a light string that passes over a light frictionless pulley. When the masses are released, the larger mass X accelerates down with acceleration

$$a = g\frac{X - Y}{X + Y}$$

Suppose that X and Y are measured as $X = 100 \pm 1$ g, and $Y = 50 \pm 1$ g. Assume that g, the acceleration due to gravity, is known with negligible uncertainty. Find the relative uncertainty in the acceleration a.

Solution

The relative uncertainty in a is equal to the absolute uncertainty in $\ln a = \ln g + \ln(X - Y) - \ln(X + Y)$. We treat g as a constant, since its uncertainty is negligible. The partial derivatives are

$$\frac{\partial \ln a}{\partial X} = \frac{1}{X - Y} - \frac{1}{X + Y} = 0.0133$$

$$\frac{\partial \ln a}{\partial Y} = -\frac{1}{X - Y} - \frac{1}{X + Y} = -0.0267$$

The uncertainties in X and Y are $\sigma_X = \sigma_Y = 1$. The relative uncertainty in a is

$$\begin{aligned} \frac{\sigma_a}{a} = \sigma_{\ln a} &= \sqrt{\left(\frac{\partial \ln a}{\partial X}\right)^2 \sigma_X^2 + \left(\frac{\partial \ln a}{\partial Y}\right)^2 \sigma_Y^2} \\ &= \sqrt{(0.0133)^2(1)^2 + (-0.0267)^2(1)^2} \\ &= 0.030 \end{aligned}$$

The relative uncertainty in a is 0.030, or 3%. Note that this value does not depend on g.

Derivation of the Multivariate Propagation of Error Formula

We derive the propagation of error formula for a nonlinear function U of a random variable X by approximating it with a multivariate linear function (i.e., linearizing the problem), and then using the methods of Section 3.2. To find a linear approximation to U, we use a first-order multivariate Taylor series approximation. Let $U = U(X_1, X_2, \ldots, X_n)$ be a function whose partial derivatives all exist. Let $(\mu_1, \mu_2, \ldots, \mu_n)$ be any point. Then if $X_1, X_2, \ldots, X_n$ are close to $\mu_1, \mu_2, \ldots, \mu_n$, respectively, the linearization of U is

$$\begin{aligned} U(X_1, X_2, \ldots, X_n) - U(\mu_1, \mu_2, \ldots, \mu_n) &\approx \frac{\partial U}{\partial X_1}(X_1 - \mu_1) \\ &+ \frac{\partial U}{\partial X_2}(X_2 - \mu_2) + \cdots + \frac{\partial U}{\partial X_n}(X_n - \mu_n) \end{aligned} \tag{3.14}$$

Each partial derivative is evaluated at the point $(\mu_1, \mu_2, \ldots, \mu_n)$.

If $X_1, X_2, \ldots, X_n$ are *independent* measurements, the linear approximation leads to a method for approximating the uncertainty in U, given the uncertainties in $X_1, X_2, \ldots, X_n$. The derivation is similar to the one-variable case presented at the end of Section 3.3. Let $\mu_1, \mu_2, \ldots, \mu_n$ be the means of $X_1, X_2, \ldots, X_n$, respectively. Then for any reasonably precise measurements, $X_1, X_2, \ldots, X_n$ will be close enough to $\mu_1, \mu_2, \ldots, \mu_n$ for the linearization to be valid.

We can rewrite Equation (3.14) as

$$U \approx \left(U(\mu_1, \mu_2, \ldots, \mu_n) - \frac{\partial U}{\partial X_1}\mu_1 - \frac{\partial U}{\partial X_2}\mu_2 - \cdots - \frac{\partial U}{\partial X_n}\mu_n \right) + \frac{\partial U}{\partial X_1}X_1 + \frac{\partial U}{\partial X_2}X_2 + \cdots + \frac{\partial U}{\partial X_n}X_n \qquad (3.15)$$

The quantities $\partial U/\partial X_1, \partial U/\partial X_2, \ldots, \partial U/\partial X_n$ are all constant, since they are evaluated at the point $(\mu_1, \mu_2, \ldots, \mu_n)$. Therefore the quantity

$$U(\mu_1, \mu_2, \ldots, \mu_n) - \frac{\partial U}{\partial X_1}\mu_1 - \frac{\partial U}{\partial X_2}\mu_2 - \cdots - \frac{\partial U}{\partial X_n}\mu_n$$

is constant as well. It follows from Equation (2.40) (in Section 2.5) and Equation (3.4) (in Section 3.2) that

$$\sigma_U \approx \sqrt{\left(\frac{\partial U}{\partial X_1}\right)^2 \sigma_{X_1}^2 + \left(\frac{\partial U}{\partial X_2}\right)^2 \sigma_{X_2}^2 + \cdots + \left(\frac{\partial U}{\partial X_n}\right)^2 \sigma_{X_n}^2}$$

Exercises for Section 3.4

1. Find the uncertainty in U, assuming that $X = 10.0 \pm 0.5$, $Y = 5.0 \pm 0.1$, and

a. $U = XY^2$

b. $U = X^2 + Y^2$

c. $U = (X + Y^2)/2$

2. Refer to Exercise 8 in Section 3.2. Assume that $\tau = 30.0 \pm 0.1$ Pa, $h = 10.0 \pm 0.2$ mm, and $\mu = 1.49$ Pa · s with negligible uncertainty.

a. Estimate V and find the uncertainty in the estimate.

b. Which would provide a greater reduction in the uncertainty in V: reducing the uncertainty in τ to 0.01 Pa or reducing the uncertainty in h to 0.1 mm?

3. When air enters a compressor at pressure P_1 and leaves at pressure P_2, the intermediate pressure is given by $P_3 = \sqrt{P_1 P_2}$. Assume that $P_1 = 10.1 \pm 0.3$ MPa and $P_2 = 20.1 \pm 0.4$ MPa.

a. Estimate P_3, and find the uncertainty in the estimate.

b. Which would provide a greater reduction in the uncertainty in P_3: reducing the uncertainty in P_1 to 0.2 MPa or reducing the uncertainty in P_2 to 0.2 MPa?

4. One way to measure the water content of a soil is to weigh the soil both before and after drying it in an oven. The water content is $W = (M_1 - M_2)/M_1$, where M_1 is the mass before drying and M_2 is the mass after drying. Assume that $M_1 = 1.32 \pm 0.01$ kg and $M_2 = 1.04 \pm 0.01$ kg.

a. Estimate W, and find the uncertainty in the estimate.

b. Which would provide a greater reduction in the uncertainty in W: reducing the uncertainty in M_1 to 0.005 kg or reducing the uncertainty in M_2 to 0.005 kg?

5. The lens equation says that if an object is placed at a distance p from a lens, and an image is formed at a distance q from the lens, then the focal length f satisfies the equation $1/f = 1/p + 1/q$. Assume that $p = 2.3 \pm 0.2$ cm and $q = 3.1 \pm 0.2$ cm.

a. Estimate f, and find the uncertainty in the estimate.

b. Which would provide a greater reduction in the uncertainty in f: reducing the uncertainty in p to 0.1 cm or reducing the uncertainty in q to 0.1 cm?

6. The pressure P, temperature T, and volume V of one mole of an ideal gas are related by the equation $PV = 8.31T$, when P is measured in kilopascals, T is measured in kelvins, and V is measured in liters.

a. Assume that $P = 242.52 \pm 0.03$ kPa and $V = 10.103 \pm 0.002$ L. Estimate T, and find the uncertainty in the estimate.

b. Assume that $P = 242.52 \pm 0.03$ kPa and $T = 290.11 \pm 0.02$ K. Estimate V, and find the uncertainty in the estimate.

c. Assume that $V = 10.103 \pm 0.002$ L and $T = 290.11 \pm 0.02$ K. Estimate P, and find the uncertainty in the estimate.

7. Refer to Exercise 7 in Section 3.3. Assume that $g = 9.80$ m/s^2 exactly, $d = 0.18 \pm 0.02$ m, $h = 4.86 \pm 0.06$ m, and $l = 32.04 \pm 0.01$ m. Estimate F, and find the uncertainty in the estimate.

8. In the article "Temperature-Dependent Optical Constants of Water Ice in the Near Infrared: New Results and Critical Review of the Available Measurements" (B. Rajaram, D. Glandorf, et al., *Applied Optics*, 2001:4449–4462), the imaginary index of refraction of water ice is presented for various frequencies and temperatures. At a frequency of 372.1 cm^{-1} and a temperature of 166 K, the index is estimated to be 0.00116. At the same frequency and at a temperature of 196 K, the index is estimated to be 0.00129. The uncertainty is reported to be 10^{-4} for each of these two estimated indices. The ratio of the indices is estimated to be $0.00116/0.00129 = 0.899$. Find the uncertainty in this ratio.

9. Refer to Exercise 10 in Section 3.2. Assume that $\tau_0 = 50 \pm 1$ MPa, $w = 1.2 \pm 0.1$ mm, and $k = 0.29 \pm 0.05$ mm^{-1}.

a. Estimate τ, and find the uncertainty in the estimate.

b. Which would provide the greatest reduction in the uncertainty in τ: reducing the uncertainty in τ_0 to 0.1 MPa, reducing the uncertainty in w to 0.01 mm, or reducing the uncertainty in k to 0.025 mm^{-1}?

c. A new, somewhat more expensive process would allow both τ_0 and w to be measured with negligible uncertainty. Is it worthwhile to implement the process? Explain.

10. According to Snell's law, the angle of refraction θ_2 of a light ray traveling in a medium of index of refraction n is related to the angle of incidence θ_1 of a ray traveling in a vacuum through the equation $\sin\theta_1 = n\sin\theta_2$. Assume that $\theta_1 = 0.3672 \pm 0.005$ rad and $\theta_2 = 0.2943 \pm 0.004$ rad. Estimate n, and find the uncertainty in the estimate.

11. Archaeologists studying meat storage methods employed by the Nunamiut in northern Alaska have developed a Meat Drying Index. Following is a slightly simplified version of the index given in the article "A Zooarchaeological Signature for Meat Storage: Rethinking the Drying Utility Index" (T. Friesen, *American Antiquity*, 2001:315–331). Let m represent the weight of meat, b the weight of bone, and g the gross weight of some part of a caribou. The Meat Drying Index y is given by $y = mb/g$. Assume that for a particular caribou rib, the following measurements are made (in grams): $g = 3867.4 \pm 0.3$, $b = 1037.0 \pm 0.2$, $m = 2650.4 \pm 0.1$.

a. Estimate y, and find the uncertainty in the estimate.

b. Which would provide the greatest reduction in the uncertainty in y: reducing the uncertainty in g to 0.1 g, reducing the uncertainty in b to 0.1 g, or reducing the uncertainty in m to 0?

12. The resistance R (in ohms) of a cylindrical conductor is given by $R = kl/d^2$, where l is the length, d is the diameter, and k is a constant of proportionality. Assume that $l = 14.0 \pm 0.1$ cm and $d = 4.4 \pm 0.1$ cm.

a. Estimate R, and find the uncertainty in the estimate. Your answer will be in terms of the proportionality constant k.

b. Which would provide the greater reduction in the uncertainty in R: reducing the uncertainty in l to 0.05 cm or reducing the uncertainty in d to 0.05 cm?

13. A cylindrical wire of radius R elongates when subjected to a tensile force F. Let L_0 represent the initial length of the wire and let L_1 represent the final length. Young's modulus for the material is given by

$$Y = \frac{FL_0}{\pi R^2(L_1 - L_0)}$$

Assume that $F = 800 \pm 1$ N, $R = 0.75 \pm 0.1$ mm, $L_0 = 25.0 \pm 0.1$ mm, and $L_1 = 30.0 \pm 0.1$ mm.

a. Estimate Y, and find the uncertainty in the estimate.

b. Of the uncertainties in F, R, L_0, and L_1, only one has a non-negligible effect on the uncertainty in Y. Which one is it?

14. According to Newton's law of cooling, the time t needed for an object at an initial temperature T_0 to cool to a temperature T in an environment with ambient temperature T_a is given by

$$t = \frac{\ln(T_0 - T_a)}{k} - \frac{\ln(T - T_a)}{k}$$

where k is a constant. Assume that for a certain type of container, $k = 0.025 \text{ min}^{-1}$. Let t be the number of minutes needed to cool the container to a temperature of 50°F. Assume that $T_0 = 70.1 \pm 0.2$°F and $T_a = 35.7 \pm 0.1$°F. Estimate t, and find the uncertainty in the estimate.

15. Refer to Exercise 14. In an experiment to determine the value of k, the temperature T at time $t = 10$ min is measured to be $T = 54.1 \pm 0.2$°F. Assume that $T_0 = 70.1 \pm 0.2$°F and $T_a = 35.7 \pm 0.1$°F. Estimate k, and find the uncertainty in the estimate.

16. The vertical displacement v of a cracked slurry infiltrated fiber concrete member at maximum shear stress is given by $v = a + bw$, where w is the crack width, and a and b are estimated from data to be $a = 2.5 \pm 0.1$ mm and $b = 0.05 \pm 0.01$. Assume that $w = 1.2 \pm 0.1$ mm.

a. Estimate v, and find the uncertainty in the estimate.

b. Of the uncertainties in w, a, and b, only one has a non-negligible effect on the uncertainty in v. Which one is it?

17. The shape of a bacterium can be approximated by a cylinder of radius r and height h capped on each end by a hemisphere. The volume and surface area of the bacterium are given by

$$V = \pi r^2(h + 4r/3)$$
$$S = 2\pi r(h + 2r)$$

It is known that the rate R at which a chemical is absorbed into the bacterium is $R = c(S/V)$, where c is a constant of proportionality. Assume that for a certain bacterium, $r = 0.9 \pm 0.1$ μm and $h = 1.7 \pm 0.1$ μm.

a. Are the computed values of S and V independent? Explain.

b. Assuming the measurements of r and h to be independent, estimate R and find the uncertainty in the estimate. Your answer will be in terms of c.

18. Estimate U, and find the relative uncertainty in the estimate, assuming that $X = 5.0 \pm 0.2$, $Y = 10.0 \pm 0.5$, and

a. $U = X\sqrt{Y}$

b. $U = 2Y/\sqrt{X}$

c. $U = X^2 + Y^2$

19. Refer to Exercise 8 in Section 3.2. Assume that $\tau = 35.2 \pm 0.1$ Pa, $h = 12.0 \pm 0.3$ mm, and $\mu = 1.49$ Pa · s with negligible uncertainty. Estimate V, and find the relative uncertainty in the estimate.

20. Refer to Exercise 3. Assume that $P_1 = 15.3 \pm 0.2$ MPa and $P_2 = 25.8 \pm 0.1$ MPa. Estimate P_3, and find the relative uncertainty in the estimate.

21. Refer to Exercise 5. Assume that $p = 4.3 \pm 0.1$ cm and $q = 2.1 \pm 0.2$ cm. Estimate f, and find the relative uncertainty in the estimate.

22. Refer to Exercise 6.

a. Assume that $P = 224.51 \pm 0.04$ kPa and $V = 11.237 \pm 0.002$ L. Estimate T, and find the relative uncertainty in the estimate.

b. Assume that $P = 224.51 \pm 0.04$ kPa and $T = 289.33 \pm 0.02$ K. Estimate V, and find the relative uncertainty in the estimate.

c. Assume that $V = 11.203 \pm 0.002$ L and $T = 289.33 \pm 0.02$ K. Estimate P, and find the relative uncertainty in the estimate.

23. Refer to Exercise 10. Estimate n, and find the relative uncertainty in the estimate, from the following measurements: $\theta_1 = 0.216 \pm 0.003$ rad and $\theta_2 = 0.456 \pm 0.005$ rad.

24. Refer to Exercise 12. Assume that $l = 10.0$ cm ± 0.5% and $d = 10.4$ cm ± 0.5%.

a. Estimate R, and find the relative uncertainty in the estimate. Does the relative uncertainty depend on k?

b. Assume that either l or d can be remeasured with relative uncertainty 0.2%. Which should be remeasured to provide the greater improvement in the relative uncertainty of the resistance?

25. Refer to Exercise 13. Assume that $F = 750 \pm 1$ N, $R = 0.65 \pm 0.09$ mm, $L_0 = 23.7 \pm 0.2$ mm, and $L_1 = 27.7 \pm 0.2$ mm. Estimate Y, and find the relative uncertainty in the estimate.

26. Refer to Exercise 14. Assume that $T_0 = 73.1 \pm 0.1$°F, $T_a = 37.5 \pm 0.2$°F, $k = 0.032 \text{ min}^{-1}$ with negligible uncertainty, and $T = 50$°F exactly. Estimate t, and find the relative uncertainty in the estimate.

27. Refer to Exercise 17. Assume that for a certain bacterium, $r = 0.8 \pm 0.1$ μm and $h = 1.9 \pm 0.1$ μm.

a. Estimate S, and find the relative uncertainty in the estimate.

b. Estimate V, and find the relative uncertainty in the estimate.

c. Estimate R, and find the relative uncertainty in the estimate.

d. Does the relative uncertainty in R depend on c?

28. Let X and Y be independent measurements, and let c, n, and m be constants. Show that the relative uncertainty in $U = X^n Y^m$ is

$$\frac{\sigma_U}{U} = \sqrt{\left(n\frac{\sigma_X}{X}\right)^2 + \left(m\frac{\sigma_Y}{Y}\right)^2}$$

Supplementary Exercises for Chapter 3

1. Assume that X, Y, and Z are independent measurements with $X = 25 \pm 1$, $Y = 5.0 \pm 0.3$, and $Z = 3.5 \pm 0.2$. Find the uncertainties in each of the following quantities:

a. $X + YZ$

b. $X/(Y - Z)$

c. $X\sqrt{Y + e^Z}$

d. $X \ln(Y^2 + Z)$

2. Assume that X, Y, and Z are independent measurements and that the relative uncertainty in X is 5%, the relative uncertainty in Y is 10%, and the relative uncertainty in Z is 15%. Find the relative uncertainty in each of the following quantities:

a. XY/Z

b. $X^{3/2}(YZ)^{1/3}$

c. $X^3\sqrt{Z/Y}$

3. An item is to be constructed by laying two components end to end. The length of each component will be measured.

a. If the uncertainty in measuring the length of each component is 0.1 mm, what is the uncertainty in the combined length of the two components?

b. If it is desired to estimate the length of the item with an uncertainty of 0.05 mm, what must be the uncertainty in the measurement of each individual component? Assume the uncertainties in the two measurements are the same.

4. For some genetic mutations, it is thought that the frequency of the mutant gene in men increases linearly with age. If m_1 is the frequency at age t_1, and m_2 is the frequency at age t_2, then the yearly rate of increase is estimated by $r = (m_2 - m_1)/(t_2 - t_1)$. In a polymerase chain reaction assay, the frequency in 20-year-old men was estimated to be 17.7 ± 1.7 per μg DNA, and the frequency in 40-year-old men was estimated to be 35.9 ± 5.8 per μg DNA. Assume that age is measured with negligible uncertainty.

a. Estimate the yearly rate of increase, and find the uncertainty in the estimate.

b. Find the relative uncertainty in the estimated rate of increase.

5. The Darcy–Weisbach equation states that the power-generating capacity in a hydroelectric system that is lost due to head loss is given by $P = \eta\gamma QH$, where η is the efficiency of the turbine, γ is the specific gravity of water, Q is the flow rate, and H is the head loss. Assume that $\eta = 0.85 \pm 0.02$, $H = 3.71 \pm 0.10$ m, $Q = 60 \pm 1$ m^3/s, and $\gamma = 9800$ N/m^3 with negligible uncertainty.

a. Estimate the power loss (the units will be in watts), and find the uncertainty in the estimate.

b. Find the relative uncertainty in the estimated power loss.

c. Which would provide the greatest reduction in the uncertainty in P: reducing the uncertainty in η to 0.01, reducing the uncertainty in H to 0.05, or reducing the uncertainty in Q to 0.5?

6. Let A and B represent two variants (alleles) of the DNA at a certain locus on the genome. Let p represent the proportion of alleles in a population that are of type A, and let q represent the proportion of alleles that are of type B. The Hardy–Weinberg equilibrium principle states that the proportion P_{AB} of organisms that are of type AB is equal to pq. In a population survey of a particular species, the proportion of alleles of type A is estimated to be 0.360 ± 0.048 and the proportion of alleles of type B is independently estimated to be 0.250 ± 0.043.

a. Estimate the proportion of organisms that are of type AB, and find the uncertainty in the estimate.
b. Find the relative uncertainty in the estimated proportion.
c. Which would provide a greater reduction in the uncertainty in the proportion: reducing the uncertainty in the type A proportion to 0.02 or reducing the uncertainty in the type B proportion to 0.02?

7. The heating capacity of a calorimeter is known to be 4 kJ/°C, with negligible uncertainty. The number of dietary calories (kilocalories) per gram of a substance is given by $C = cH(\Delta T)/m$, where C is the number of dietary calories, H is the heating capacity of the calorimeter, ΔT is the increase in temperature in °C caused by burning the substance in the calorimeter, m is the mass of the substance in grams, and $c = 0.2390$ cal/kJ is the conversion factor from kilojoules to dietary calories. An amount of mayonnaise with mass 0.40 ± 0.01 g is burned in a calorimeter. The temperature increase is 2.75 ± 0.02°C.

a. Estimate the number of dietary calories per gram of mayonnaise, and find the uncertainty in the estimate.
b. Find the relative uncertainty in the estimated number of dietary calories.
c. Which would provide a greater reduction in the uncertainty in C: reducing the uncertainty in the mass to 0.005 g or reducing the uncertainty in ΔT to 0.01°C?

8. Twenty-two independent measurements were made of the hardness of a weld, using the Rockwell A scale. The average was 65.52 and the standard deviation was 0.63.

a. Estimate the hardness of this weld, and find the uncertainty in the estimate.
b. A single measurement is made of the hardness of another weld from the same base metal. This measurement is 61.3. What is the uncertainty in this measurement?

9. The article "Insights into Present-Day Crustal Motion in the Central Mediterranean Area from GPS Surveys" (M. Anzidei, P. Baldi, et al., *Geophysical Journal International*, 2001:98–100) reports that the components of velocity of the earth's crust in Zimmerwald, Switzerland, are 22.10 ± 0.34 mm/year in a northerly direction and 14.3 ± 0.32 mm/year in an easterly direction.

a. Estimate the velocity of the earth's crust, and find the uncertainty in the estimate.
b. Using your answer to part (a), estimate the number of years it will take for the crust to move 100 mm, and find the uncertainty in the estimate.

10. If two gases have molar masses M_1 and M_2, Graham's law states that the ratio R of their rates of effusion through a small opening is given by $R = \sqrt{M_1/M_2}$. The effusion rate of an unknown gas through a small opening is measured to be 1.66 ± 0.03 times greater than the effusion rate of carbon dioxide. The molar mass of carbon dioxide may be taken to be 44 g/mol with negligible uncertainty.

a. Estimate the molar mass of the unknown gas, and find the uncertainty in the estimate.
b. Find the relative uncertainty in the estimated molar mass.

11. A laminated item is made up of six layers. The two outer layers each have a thickness of 1.25 ± 0.10 mm and the four inner layers each have a thickness of 0.80 ± 0.05 mm. Assume the thicknesses of the layers are independent. Estimate the thickness of the item, and find the uncertainty in the estimate.

12. The article "Effect of Varying Solids Concentration and Organic Loading on the Performance of Temperature Phased Anaerobic Digestion Process" (S. Vandenburgh and T. Ellis, *Water Environment Research*, 2002:142–148) discusses experiments to determine the effect of the solids concentration on the performance of treatment methods for wastewater sludge. In the first experiment, the concentration of solids (in g/L) was 43.94 ± 1.18. In the second experiment, which was independent of the first, the concentration was 48.66 ± 1.76. Estimate the difference in the concentration between the two experiments, and find the uncertainty in the estimate.

13. In the article "Measurements of the Thermal Conductivity and Thermal Diffusivity of Polymer Melts with the Short-Hot-Wire Method" (X. Zhang, W. Hendro, et al., *International Journal of Thermophysics*,

2002:1077–1090), the thermal diffusivity of a liquid measured by the transient short-hot-wire method is given by

$$\lambda = \frac{VIA}{\pi l a}$$

where λ is the thermal diffusivity; V and I are the voltage and current applied to the hot wire, respectively; l is the length of the wire; and A and a are quantities involving temperature whose values are estimated separately. In this article, the relative uncertainties of these quantities are given as follows: V, 0.01%; I, 0.01%; l, 1%; A, 0.1%; a, 1%.

a. Find the relative uncertainty in λ.

b. Which would reduce the relative uncertainty more: reducing the relative uncertainty in l to 0.5% or reducing the relative uncertainties in V, I, and A each to 0?

14. A cable is made up of several parallel strands of wire. The strength of the cable can be estimated from the strengths of the individual wires by either of two methods. In the *ductile wire* method, the strength of the cable is estimated to be the sum of the strengths of the wires. In the *brittle wire* method, the strength of the cable is estimated to be the strength of the weakest wire multiplied by the number of wires. A particular cable is composed of 12 wires. Four of them have strength 6000 ± 20 lb, four have strength 5700 ± 30 lb, and four have strength 6200 ± 40 lb.

a. Estimate the strength of the cable, and find the uncertainty in the estimate, using the ductile wire method.

b. Estimate the strength of the cable, and find the uncertainty in the estimate, using the brittle wire method.

15. Refer to Exercise 14. A cable is composed of 16 wires. The strength of each wire is 5000 ± 20 lb.

a. Will the estimated strength of the cable be the same under the ductile wire method as under the brittle wire method?

b. Will the uncertainty in the estimated strength of the cable be the same under the ductile wire method as under the brittle wire method? Explain why or why not.

16. The mean yield from process A is estimated to be 80 ± 5, where the units are percent of a theoretical maximum. The mean yield from process B is estimated to be 90 ± 3. The relative increase obtained from process B is therefore estimated to be $(90 - 80)/80 = 0.125$. Find the uncertainty in this estimate.

17. The flow rate of water through a cylindrical pipe is given by $Q = \pi r^2 v$, where r is the radius of the pipe and v is the flow velocity.

a. Assume that $r = 3.00 \pm 0.03$ m and $v = 4.0 \pm 0.2$ m/s. Estimate Q, and find the uncertainty in the estimate.

b. Assume that $r = 4.00 \pm 0.04$ m and $v = 2.0 \pm 0.1$ m/s. Estimate Q, and find the uncertainty in the estimate.

c. If r and v have not been measured, but it is known that the relative uncertainty in r is 1% and that the relative uncertainty in v is 5%, is it possible to compute the relative uncertainty in Q? If so, compute the relative uncertainty. If not, explain what additional information is needed.

18. The conversion of cyclobutane (C_4H_8) to ethylene (C_2H_4) is a first-order reaction. This means that the concentration of cyclobutane at time t is given by $\ln C = \ln C_0 - kt$, where C is the concentration at time t, C_0 is the initial concentration, t is the time since the reaction started, and k is the rate constant. Assume that $C_0 = 0.2$ mol/L with negligible uncertainty. After 300 seconds at a constant temperature, the concentration is measured to be $C = 0.174 \pm 0.005$ mol/L. Assume that time can be measured with negligible uncertainty.

a. Estimate the rate constant k, and find the uncertainty in the estimate. The units of k will be s^{-1}.

b. Find the relative uncertainty in k.

c. The half-life $t_{1/2}$ of the reaction is the time it takes for the concentration to be reduced to one-half its initial value. The half-life is related to the rate constant by $t_{1/2} = (\ln 2)/k$. Using the result found in part (a), find the uncertainty in the half-life.

d. Find the relative uncertainty in the half-life.

19. The decomposition of nitrogen dioxide (NO_2) into nitrogen monoxide (NO) and oxygen is a second-order reaction. This means that the concentration C of NO_2 at time t is given by $1/C = kt + 1/C_0$, where C_0 is the initial concentration and k is the rate constant. Assume the initial concentration is known to be 0.03 mol/L exactly. Assume that time can be measured with negligible uncertainty.

a. After 40 s, the concentration C is measured to be $0.0023 \pm 2.0 \times 10^{-4}$ mol/L. Estimate the rate constant k, and find the uncertainty in the estimate.

b. After 50 s, the concentration C is measured to be $0.0018 \pm 2.0 \times 10^{-4}$ mol/L. Estimate the rate constant k, and find the uncertainty in the estimate.

c. Denote the estimates of the rate constant k in parts (a) and (b) by $\widehat{k}_1$ and $\widehat{k}_2$, respectively. The average $(\widehat{k}_1 + \widehat{k}_2)/2$ is used as an estimate of k. Find the uncertainty in this estimate.

d. Find the value of c so that the weighted average $c\widehat{k}_1 + (1 - c)\widehat{k}_2$ has the smallest uncertainty.

20. Two students want to measure the acceleration a of a cart rolling down an inclined plane. The cart starts at rest and travels a distance s down the plane. The first student estimates the acceleration by measuring the instantaneous velocity v just as the cart has traveled s meters, and uses the formula $a = v^2/2s$. The second student estimates the acceleration by measuring the time, in seconds, that the cart takes to travel s meters, and uses the formula $a = 2s/t^2$. Assume that $s = 1$ m, and that there is negligible uncertainty in s. Assume that $v = 3.2 \pm 0.1$ m/s and that $t = 0.63 \pm 0.01$ s. Assume that the measurements of v and t are independent.

a. Compute the acceleration using the method of the first student. Call this estimate a_1. Find the uncertainty in a_1.

b. Compute the acceleration using the method of the second student. Call this estimate a_2. Find the uncertainty in a_2.

c. Find the weighted average of a_1 and a_2 that has the smallest uncertainty. Find the uncertainty in this weighted average.

21. A track has the shape of a square capped on two opposite sides by semicircles. The length of a side of the square is measured to be 181.2 ± 0.1 m.

a. Compute the area of the square and its uncertainty.

b. Compute the area of one of the semicircles and its uncertainty.

c. Let S denote the area of the square as computed in part (a), and let C denote the area of one of the semicircles as computed in part (b). The area enclosed by the track is $A = S + 2C$. Someone computes the uncertainty in A as $\sigma_A = \sqrt{\sigma_S^2 + 4\sigma_C^2}$. Is this correct? If so, explain why. If not, compute the uncertainty in A correctly.

22. If X is an unbiased measurement of a true value μ_X, and $U(X)$ is a nonlinear function of X, then in most cases U is a biased estimate of the true value $U(\mu_X)$. In most cases this bias is ignored. If it is important to reduce this bias, however, a *bias-corrected* estimate is $U(X) - (1/2)(d^2U/dX^2)\sigma_X^2$. In general the bias-corrected estimate is not unbiased, but has a smaller bias than $U(X)$.

Assume that the radius of a circle is measured to be $r = 3.0 \pm 0.1$ cm.

a. Estimate the area A, and find the uncertainty in the estimate, without bias correction.

b. Compute the bias-corrected estimate of A.

c. Compare the difference between the bias-corrected and non-bias-corrected estimates to the uncertainty in the non-bias-corrected estimate. Is bias correction important in this case? Explain.

23. If $X_1, X_2, \ldots, X_n$ are independent and unbiased measurements of true values $\mu_1, \mu_2, \ldots, \mu_n$, and $U(X_1, X_2, \ldots, X_n)$ is a nonlinear function of $X_1, X_2, \ldots, X_n$, then in general $U(X_1, X_2, \ldots, X_n)$ is a biased estimate of the true value $U(\mu_1, \mu_2, \ldots, \mu_n)$. A bias-corrected estimate is $U(X_1, X_2, \ldots, X_n) - (1/2)\sum_{i=1}^{n}(\partial^2 U/\partial X_i^2)\sigma_{X_i}^2$.

When air enters a compressor at pressure P_1 and leaves at pressure P_2, the intermediate pressure is given by $P_3 = \sqrt{P_1 P_2}$. Assume that $P_1 = 8.1 \pm 0.1$ MPa and $P_2 = 15.4 \pm 0.2$ MPa.

a. Estimate P_3, and find the uncertainty in the estimate, without bias correction.

b. Compute the bias-corrected estimate of P_3.

c. Compare the difference between the bias-corrected and non-bias-corrected estimates to the uncertainty in the non-bias-corrected estimate. Is bias correction important in this case? Explain.

Chapter 4

Commonly Used Distributions

Introduction

Statistical inference involves drawing a sample from a population and analyzing the sample data to learn about the population. In many situations, one has an approximate knowledge of the probability mass function or probability density function of the population. In these cases, the probability mass or density function can often be well approximated by one of several standard families of curves, or functions. In this chapter, we describe some of these standard functions, and for each one we describe some conditions under which it is appropriate.

4.1 The Bernoulli Distribution

Imagine an experiment that can result in one of two outcomes. One outcome is labeled "success," and the other outcome is labeled "failure." The probability of success is denoted by p. The probability of failure is therefore $1-p$. Such a trial is called a **Bernoulli trial** with success probability p. The simplest Bernoulli trial is the toss of a coin. The two outcomes are heads and tails. If we define heads to be the success outcome, then p is the probability that the coin comes up heads. For a fair coin, $p = 1/2$. Another example of a Bernoulli trial is the selection of a component from a population of components, some of which are defective. If we define "success" as a defective component, then p is the proportion of defective components in the population.

For any Bernoulli trial, we define a random variable X as follows: If the experiment results in success, then $X = 1$. Otherwise $X = 0$. It follows that X is a discrete random variable, with probability mass function $p(x)$ defined by

$$p(0) = P(X = 0) = 1 - p$$

$$p(1) = P(X = 1) = p$$

$$p(x) = 0 \text{ for any value of } x \text{ other than 0 or 1}$$

The random variable X is said to have the **Bernoulli distribution** with parameter p. The notation is $X \sim \text{Bernoulli}(p)$. Figure 4.1 presents probability histograms for the Bernoulli(0.5) and Bernoulli(0.8) probability mass functions.

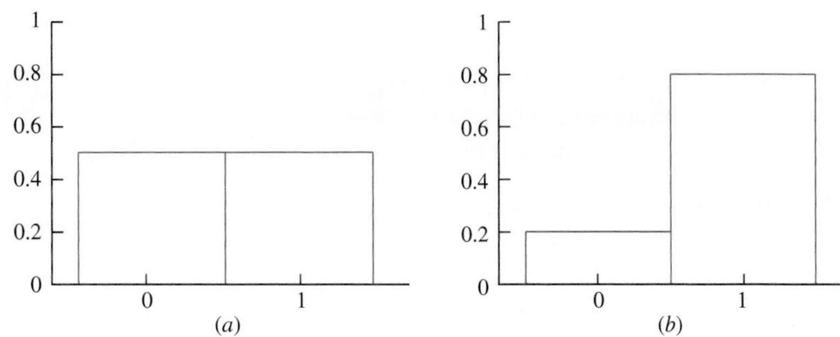

FIGURE 4.1 *(a)* The Bernoulli(0.5) probability histogram. *(b)* The Bernoulli(0.8) probability histogram.

Example 4.1

A coin has probability 0.5 of landing heads when tossed. Let $X = 1$ if the coin comes up heads, and $X = 0$ if the coin comes up tails. What is the distribution of X?

Solution

Since $X = 1$ when heads comes up, heads is the success outcome. The success probability, $P(X = 1)$, is equal to 0.5. Therefore $X \sim \text{Bernoulli}(0.5)$.

Example 4.2

A die has probability 1/6 of coming up 6 when rolled. Let $X = 1$ if the die comes up 6, and $X = 0$ otherwise. What is the distribution of X?

Solution

The success probability is $p = P(X = 1) = 1/6$. Therefore $X \sim \text{Bernoulli}(1/6)$.

Example 4.3

Ten percent of the components manufactured by a certain process are defective. A component is chosen at random. Let $X = 1$ if the component is defective, and $X = 0$ otherwise. What is the distribution of X?

Solution

The success probability is $p = P(X = 1) = 0.1$. Therefore $X \sim \text{Bernoulli}(0.1)$.

Mean and Variance of a Bernoulli Random Variable

It is easy to compute the mean and variance of a Bernoulli random variable. If $X \sim \text{Bernoulli}(p)$, then, using Equations (2.29) and (2.30) (in Section 2.4), we compute

$$\begin{aligned}\mu_X &= (0)(1-p) + (1)(p) \\ &= p\end{aligned}$$

$$\begin{aligned}\sigma_X^2 &= (0-p)^2(1-p) + (1-p)^2(p) \\ &= p(1-p)\end{aligned}$$

Summary

If $X \sim \text{Bernoulli}(p)$, then

$$\mu_X = p \quad (4.1)$$

$$\sigma_X^2 = p(1-p) \quad (4.2)$$

Refer to Example 4.3. Find μ_X and σ_X^2.

Solution

Since $X \sim \text{Bernoulli}(0.1)$, the success probability p is equal to 0.1. Using Equations (4.1) and (4.2), $\mu_X = 0.1$ and $\sigma_X^2 = 0.1(1 - 0.1) = 0.09$.

Exercises for Section 4.1

1. A basketball player is about to attempt a shot from the top of the key. The probability that he makes the shot is 0.55.

a. Let $X = 1$ if he makes the shot, $X = 0$ if he doesn't. Find the mean and variance of X.

b. If he makes the shot, his team scores 2 points; if he misses, his team scores 0 points. Let Y be the number of points scored. Does Y have a Bernoulli distribution? If so, find the success probability. If not, explain why not.

c. Find the mean and variance of Y.

2. At a certain fast-food restaurant, 25% of drink orders are for a small drink, 35% for a medium, and 40% for a large. Let $X = 1$ if a randomly chosen order is for a small, and let $X = 0$ otherwise. Let $Y = 1$ if the order is for a medium, and let $Y = 0$ otherwise. Let $Z = 1$ if it is for either a small or a medium, and let $Z = 0$ otherwise.

a. Let p_X denote the success probability for X. Find p_X.

b. Let p_Y denote the success probability for Y. Find p_Y.

c. Let p_Z denote the success probability for Z. Find p_Z.

d. Is it possible for both X and Y to equal 1?

e. Does $p_Z = p_X + p_Y$?

f. Does $Z = X + Y$? Explain.

3. When a certain glaze is applied to a ceramic surface, the probability is 5% that there will be discoloration, 20% that there will be a crack, and 23% that there will be either discoloration or a crack, or both. Let $X = 1$

if there is discoloration, and let $X = 0$ otherwise. Let $Y = 1$ if there is a crack, and let $Y = 0$ otherwise. Let $Z = 1$ if there is either discoloration or a crack, or both, and let $Z = 0$ otherwise.

a. Let p_X denote the success probability for X. Find p_X.
b. Let p_Y denote the success probability for Y. Find p_Y.
c. Let p_Z denote the success probability for Z. Find p_Z.
d. Is it possible for both X and Y to equal 1?
e. Does $p_Z = p_X + p_Y$?
f. Does $Z = X + Y$? Explain.

4. Let X and Y be Bernoulli random variables. Let $Z = X + Y$.

a. Show that if X and Y cannot both be equal to 1, then Z is a Bernoulli random variable.
b. Show that if X and Y cannot both be equal to 1, then $p_Z = p_X + p_Y$.
c. Show that if X and Y *can* both be equal to 1, then Z is not a Bernoulli random variable.

5. A penny and a nickel are tossed. Both are fair coins. Let $X = 1$ if the penny comes up heads, and let $X = 0$ otherwise. Let $Y = 1$ if the nickel comes up heads, and let $Y = 0$ otherwise. Let $Z = 1$ if both the penny and nickel come up heads, and let $Z = 0$ otherwise.

a. Let p_X denote the success probability for X. Find p_X.
b. Let p_Y denote the success probability for Y. Find p_Y.
c. Let p_Z denote the success probability for Z. Find p_Z.
d. Are X and Y independent?
e. Does $p_Z = p_X p_Y$?
f. Does $Z = XY$? Explain.

6. Two dice are rolled. Let $X = 1$ if the dice come up doubles and let $X = 0$ otherwise. Let $Y = 1$ if the sum is 6, and let $Y = 0$ otherwise. Let $Z = 1$ if the dice come up both doubles and with a sum of 6 (that is, double 3), and let $Z = 0$ otherwise.

a. Let p_X denote the success probability for X. Find p_X.
b. Let p_Y denote the success probability for Y. Find p_Y.
c. Let p_Z denote the success probability for Z. Find p_Z.
d. Are X and Y independent?
e. Does $p_Z = p_X p_Y$?
f. Does $Z = XY$? Explain.

7. Let X and Y be Bernoulli random variables. Let $Z = XY$.

a. Show that Z is a Bernoulli random variable.
b. Show that if X and Y are independent, then $p_Z = p_X p_Y$.

4.2 The Binomial Distribution

Sampling a single component from a lot and determining whether it is defective is an example of a Bernoulli trial. In practice, we might sample several components from a very large lot and count the number of defectives among them. This amounts to conducting several independent Bernoulli trials and counting the number of successes. The number of successes is a random variable, which is said to have a **binomial distribution**.

We now present a formal description of the binomial distribution. Assume that a series of n Bernoulli trials is conducted, each with the same success probability p. Assume further that the trials are *independent*, that is, that the outcome of one trial does not influence the outcomes of any of the other trials. Let the random variable X equal the number of successes in these n trials. Then X is said to have the **binomial distribution** with parameters n and p. The notation is $X \sim \text{Bin}(n, p)$. X is a discrete random variable, and its possible values are $0, 1, \ldots, n$.

Summary

If a total of n Bernoulli trials are conducted, and

- The trials are independent
- Each trial has the same success probability p
- X is the number of successes in the n trials

then X has the binomial distribution with parameters n and p, denoted $X \sim \text{Bin}(n, p)$.

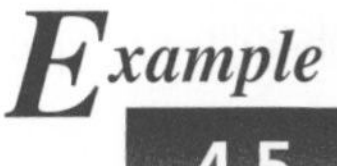

A fair coin is tossed 10 times. Let X be the number of heads that appear. What is the distribution of X?

Solution

There are 10 independent Bernoulli trials, each with success probability $p = 0.5$. The random variable X is equal to the number of successes in the 10 trials. Therefore $X \sim \text{Bin}(10, 0.5)$.

Recall from the discussion of independence in Section 1.1 that when drawing a sample from a finite, tangible population, the sample items may be treated as independent if the population is very large compared to the size of the sample. Otherwise the sample items are not independent. In some cases, the purpose of drawing a sample may be to classify each sample item into one of two categories. For example, we may sample a number of items from a lot and classify each one as defective or nondefective. In these cases, each sampled item is a Bernoulli trial, with one category counted as a success and the other counted as a failure. When the population of items is large compared to the number sampled, these Bernoulli trials are nearly independent, and the number of successes among them has, for all practical purposes, a binomial distribution. When the population size is not large compared to the sample, however, the Bernoulli trials are not independent, and the number of successes among them do not have a binomial distribution. A good rule of thumb is that if the sample size is 5% or less of the population, the binomial distribution may be used.

Summary

Assume that a finite population contains items of two types, successes and failures, and that a simple random sample is drawn from the population. Then if the sample size is no more than 5% of the population, the binomial distribution may be used to model the number of successes.

A lot contains several thousand components, 10% of which are defective. Seven components are sampled from the lot. Let X represent the number of defective components in the sample. What is the distribution of X?

Solution

Since the sample size is small compared to the population (i.e., less than 5%), the number of successes in the sample approximately follows a binomial distribution. Therefore we model X with the Bin(7, 0.1) distribution.

Probability Mass Function of a Binomial Random Variable

We now derive the probability mass function of a binomial random variable by considering an example. A biased coin has probability 0.6 of coming up heads. The coin is tossed three times. Let X be the number of heads. Then $X \sim \text{Bin}(3, 0.6)$. We will compute $P(X = 2)$.

There are three arrangements of two heads in three tosses of a coin, HHT, HTH, and THH. We first compute the probability of HHT. The event HHT is a sequence of independent events: H on the first toss, H on the second toss, T on the third toss. We know the probabilities of each of these events separately:

$$P(\text{H on the first toss}) = 0.6, \; P(\text{H on the second toss}) = 0.6, \; P(\text{T on the third toss}) = 0.4$$

Since the events are independent, the probability that they all occur is equal to the product of their probabilities (Equation 2.20 in Section 2.3). Thus

$$P(\text{HHT}) = (0.6)(0.6)(0.4) = (0.6)^2(0.4)^1$$

Similarly, $P(\text{HTH}) = (0.6)(0.4)(0.6) = (0.6)^2(0.4)^1$, and $P(\text{THH}) = (0.4)(0.6)(0.6) = (0.6)^2(0.4)^1$. It is easy to see that all the different arrangements of two heads and one tail have the same probability. Now

$$\begin{aligned} P(X = 2) &= P(\text{HHT or HTH or THH}) \\ &= P(\text{HHT}) + P(\text{HTH}) + P(\text{THH}) \\ &= (0.6)^2(0.4)^1 + (0.6)^2(0.4)^1 + (0.6)^2(0.4)^1 \\ &= 3(0.6)^2(0.4)^1 \end{aligned}$$

Examining this result, we see that the number 3 represents the number of arrangements of two successes (heads) and one failure (tails), 0.6 is the success probability p, the exponent 2 is the number of successes, 0.4 is the failure probability $1 - p$, and the exponent 1 is the number of failures.

We can now generalize this result to produce a formula for the probability of x successes in n independent Bernoulli trials with success probability p, in terms of x, n, and p. In other words, we can compute $P(X = x)$ where $X \sim \text{Bin}(n, p)$. We can see that

$$P(X = x) = (\text{number of arrangements of } x \text{ successes in } n \text{ trials}) \cdot p^x(1-p)^{n-x} \qquad (4.3)$$

All we need to do now is to provide an expression for the number of arrangements of x successes in n trials. To describe this number, we need factorial notation. For any positive integer n, the quantity $n!$ (read "n factorial") is the number

$$(n)(n-1)(n-2)\cdots(3)(2)(1)$$

We also define $0! = 1$. The number of arrangements of x successes in n trials is $n!/x!(n-x)!$. (A derivation of this result is presented in Section 2.2.) We can now define the probability mass function for a binomial random variable.

If $X \sim \text{Bin}(n, p)$, the probability mass function of X is

$$p(x) = P(X = x) = \begin{cases} \dfrac{n!}{x!(n-x)!}p^x(1-p)^{n-x} & x = 0, 1, \ldots, n \\ 0 & \text{otherwise} \end{cases} \qquad (4.4)$$

Figure 4.2 presents probability histograms for the Bin(10, 0.4) and Bin(20, 0.1) probability mass functions.

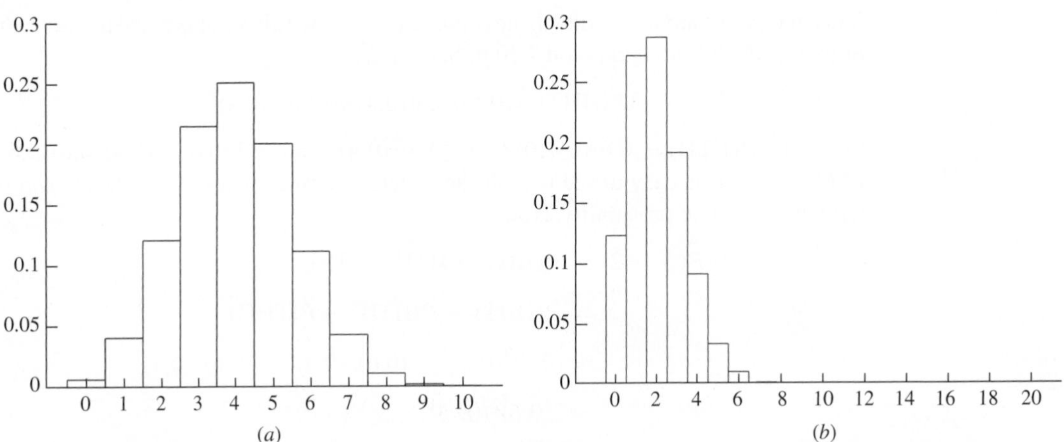

FIGURE 4.2 *(a)* The Bin(10, 0.4) probability histogram. *(b)* The Bin(20, 0.1) probability histogram.

Example 4.7

Find the probability mass function of the random variable X if $X \sim \text{Bin}(10, 0.4)$. Find $P(X = 5)$.

Solution

We use Equation (4.4) with $n = 10$ and $p = 0.4$. The probability mass function is

$$p(x) = \begin{cases} \dfrac{10!}{x!(10-x)!}(0.4)^x(0.6)^{10-x} & x = 0, 1, \ldots, 10 \\ 0 & \text{otherwise} \end{cases}$$

$$\begin{aligned} P(X = 5) = p(5) &= \frac{10!}{5!(10-5)!}(0.4)^5(0.6)^{10-5} \\ &= 0.2007 \end{aligned}$$

Example 4.8

A fair die is rolled eight times. Find the probability that no more than 2 sixes come up.

Solution

Each roll of the die is a Bernoulli trial with success probability 1/6. Let X denote the number of sixes in 8 rolls. Then $X \sim \text{Bin}(8, 1/6)$. We need to find $P(X \leq 2)$. Using the probability mass function,

$$\begin{aligned} P(X \leq 2) &= P(X = 0 \text{ or } X = 1 \text{ or } X = 2) \\ &= P(X = 0) + P(X = 1) + P(X = 2) \\ &= \frac{8!}{0!(8-0)!}\left(\frac{1}{6}\right)^0\left(\frac{5}{6}\right)^{8-0} + \frac{8!}{1!(8-1)!}\left(\frac{1}{6}\right)^1\left(\frac{5}{6}\right)^{8-1} \\ &\quad + \frac{8!}{2!(8-2)!}\left(\frac{1}{6}\right)^2\left(\frac{5}{6}\right)^{8-2} \\ &= 0.2326 + 0.3721 + 0.2605 \\ &= 0.8652 \end{aligned}$$

Table A.1 (in Appendix A) presents binomial probabilities of the form $P(X \leq x)$ for $n \leq 20$ and selected values of p. Examples 4.9 and 4.10 illustrate the use of this table.

Example 4.9

A large industrial firm allows a discount on any invoice that is paid within 30 days. Of all invoices, 10% receive the discount. In a company audit, 12 invoices are sampled at random. What is the probability that fewer than 4 of the 12 sampled invoices receive the discount?

Solution

Let X represent the number of invoices in the sample that receive discounts. Then $X \sim \text{Bin}(12, 0.1)$. The probability that fewer than four invoices receive discounts is $P(X \leq 3)$. We consult Table A.1 with $n = 12$, $p = 0.1$, and $x = 3$. We find that $P(X \leq 3) = 0.974$.

Sometimes the best way to compute the probability of an event is to compute the probability that the event does not occur, and then subtract from 1. Example 4.10 provides an illustration.

Refer to Example 4.9. What is the probability that more than 1 of the 12 sampled invoices receives a discount?

Solution

Let X represent the number of invoices in the sample that receive discounts. We wish to compute the probability $P(X > 1)$. Table A.1 presents probabilities of the form $P(X \leq x)$. Therefore we note that $P(X > 1) = 1 - P(X \leq 1)$. Consulting the table with $n = 12$, $p = 0.1$, $x = 1$, we find that $P(X \leq 1) = 0.659$. Therefore $P(X > 1) = 1 - 0.659 = 0.341$.

A Binomial Random Variable Is a Sum of Bernoulli Random Variables

Assume n independent Bernoulli trials are conducted, each with success probability p. Let $Y_1, \ldots, Y_n$ be defined as follows: $Y_i = 1$ if the ith trial results in success, and $Y_i = 0$ otherwise. Then each of the random variables Y_i has the Bernoulli(p) distribution. Now let X represent the number of successes among the n trials. Then $X \sim \text{Bin}(n, p)$. Since each Y_i is either 0 or 1, the sum $Y_1 + \cdots + Y_n$ is equal to the number of the Y_i that have the value 1, which is the number of successes among the n trials. Therefore $X = Y_1 + \cdots + Y_n$. This shows that a binomial random variable can be expressed as a sum of Bernoulli random variables. Put another way, sampling a single value from a Bin(n, p) population is equivalent to drawing a sample of size n from a Bernoulli(p) population, and then summing the sample values.

The Mean and Variance of a Binomial Random Variable

With a little thought, it is easy to see how to compute the mean of a binomial random variable. For example, if a fair coin is tossed 10 times, we expect on the average to see five heads. The number 5 comes from multiplying the success probability (0.5) by the number of trials (10). This method works in general. If we perform n Bernoulli trials, each with success probability p, the mean number of successes is np. Therefore, if $X \sim \text{Bin}(n, p)$, then $\mu_X = np$. We can verify this intuition by noting that X is the sum of n Bernoulli variables, each with mean p. The mean of X is therefore the sum of the means of the Bernoulli random variables that compose it, which is equal to np.

We can compute σ_X^2 by again noting that X is the sum of independent Bernoulli random variables and recalling that the variance of a Bernoulli random variable is $p(1 - p)$. The variance of X is therefore the sum of the variances of the Bernoulli random variables that compose it, which is equal to $np(1 - p)$.

Summary

If $X \sim \text{Bin}(n,\ p)$, then the mean and variance of X are given by

$$\mu_X = np \tag{4.5}$$

$$\sigma_X^2 = np(1-p) \tag{4.6}$$

Using the binomial probability mass function (Equation 4.4), we could in principle compute the mean and variance of a binomial random variable by using the definitions of mean and variance for a discrete random variable (Equations 2.29 and 2.30 in Section 2.4). These expressions involve sums that are tedious to evaluate. It is much easier to think of a binomial random variable as a sum of independent Bernoulli random variables.

Using a Sample Proportion to Estimate a Success Probability

In many cases we do not know the success probability p associated with a certain Bernoulli trial, and we wish to estimate its value. A natural way to do this is to conduct n independent trials and count the number X of successes. To estimate the success probability p we compute the sample proportion $\widehat{p}$.

$$\widehat{p} = \frac{\text{number of successes}}{\text{number of trials}} = \frac{X}{n}$$

This notation follows a pattern that is important to know. The success probability, which is unknown, is denoted by p. The sample proportion, which is known, is denoted $\widehat{p}$. The "hat" ($\widehat{\ }$) indicates that $\widehat{p}$ is used to estimate the unknown value p.

A quality engineer is testing the calibration of a machine that packs ice cream into containers. In a sample of 20 containers, 3 are underfilled. Estimate the probability p that the machine underfills a container.

Solution

The sample proportion of underfilled containers is $\widehat{p} = 3/20 = 0.15$. We estimate that the probability p that the machine underfills a container is 0.15 as well.

Uncertainty in the Sample Proportion

It is important to realize that the sample proportion $\widehat{p}$ is just an *estimate* of the success probability p, and in general, is *not equal* to p. If another sample were taken, the value of $\widehat{p}$ would probably come out differently. In other words, there is uncertainty in $\widehat{p}$. For $\widehat{p}$ to be a useful estimate, we must compute its bias and its uncertainty. We now do this. Let n denote the sample size, and let X denote the number of successes, where $X \sim \text{Bin}(n,\ p)$.

The bias is the difference $\mu_{\widehat{p}} - p$. Since $\widehat{p} = X/n$, it follows from Equation (2.41) (in Section 2.5) that

$$\begin{aligned}\mu_{\widehat{p}} = \mu_{X/n} &= \frac{\mu_X}{n} \\ &= \frac{np}{n} = p\end{aligned}$$

Since $\mu_{\widehat{p}} = p$, $\widehat{p}$ is unbiased; in other words, its bias is 0.

The uncertainty is the standard deviation $\sigma_{\widehat{p}}$. From Equation (4.6), the standard deviation of X is $\sigma_X = \sqrt{np(1-p)}$. Since $\widehat{p} = X/n$, it follows from Equation (2.43) (in Section 2.5) that

$$\begin{aligned}\sigma_{\widehat{p}} = \sigma_{X/n} &= \frac{\sigma_X}{n} \\ &= \frac{\sqrt{np(1-p)}}{n} = \sqrt{\frac{p(1-p)}{n}}\end{aligned}$$

In practice, when computing the uncertainty in $\widehat{p}$, we don't know the success probability p, so we approximate it with $\widehat{p}$.

Summary

If $X \sim \text{Bin}(n,\ p)$, then the sample proportion $\widehat{p} = X/n$ is used to estimate the success probability p.

- $\widehat{p}$ is unbiased.
- The uncertainty in $\widehat{p}$ is

$$\sigma_{\widehat{p}} = \sqrt{\frac{p(1-p)}{n}} \tag{4.7}$$

In practice, when computing $\sigma_{\widehat{p}}$, we substitute $\widehat{p}$ for p, since p is unknown.

Example 4.12

The safety commissioner in a large city wants to estimate the proportion of buildings in the city that are in violation of fire codes. A random sample of 40 buildings is chosen for inspection, and 4 of them are found to have fire code violations. Estimate the proportion of buildings in the city that have fire code violations, and find the uncertainty in the estimate.

Solution

Let p denote the proportion of buildings in the city that have fire code violations. The sample size (number of trials) is $n = 40$. The number of buildings with violations (successes) is $X = 4$. We estimate p with the sample proportion $\widehat{p}$:

$$\widehat{p} = \frac{X}{n} = \frac{4}{40} = 0.10$$

Using Equation (4.7), the uncertainty in $\widehat{p}$ is

$$\sigma_{\widehat{p}} = \sqrt{\frac{p(1-p)}{n}}$$

Substituting $\widehat{p} = 0.1$ for p and 40 for n, we obtain

$$\sigma_{\widehat{p}} = \sqrt{\frac{(0.10)(0.90)}{40}}$$
$$= 0.047$$

In Example 4.12, it turned out that the uncertainty in the sample proportion was rather large. We can reduce the uncertainty by increasing the sample size. Example 4.13 shows how to compute the approximate sample size needed to reduce the uncertainty to a specified amount.

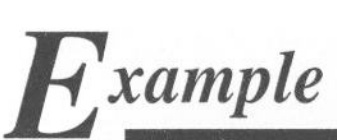

4.13

In Example 4.12, approximately how many additional buildings must be inspected so that the uncertainty in the sample proportion of buildings in violation will be only 0.02?

Solution

We need to find the value of n so that $\sigma_{\widehat{p}} = \sqrt{p(1-p)/n} = 0.02$. Approximating p with $\widehat{p} = 0.1$, we obtain

$$\sigma_{\widehat{p}} = \sqrt{\frac{(0.1)(0.9)}{n}} = 0.02$$

Solving for n yields $n = 225$. We have already sampled 40 buildings, so we need to sample 185 more.

Sometimes it is desired to estimate the value of a function $f(p)$ of a success probability p. In these cases we estimate $f(p)$ with $f(\widehat{p})$, where $\widehat{p}$ is the sample proportion. We can then use the propagation of error method (Equation 3.10 in Section 3.3) to find the uncertainty in $f(\widehat{p})$.

Example

4.14

In a sample of 100 newly manufactured automobile tires, 7 are found to have minor flaws in the tread. If four newly manufactured tires are selected at random and installed on a car, estimate the probability that none of the four tires have a flaw, and find the uncertainty in this estimate.

Solution

Let p represent the probability that a tire has no flaw. We begin by computing the sample proportion $\widehat{p}$ and finding its uncertainty. The sample proportion is $\widehat{p} = 93/100 = 0.93$. The uncertainty in $\widehat{p}$ is given by $\sigma_{\widehat{p}} = \sqrt{p(1-p)/n}$. We substitute $n = 100$ and $\widehat{p} = 0.93$ for p to obtain

$$\sigma_{\widehat{p}} = \sqrt{\frac{(0.93)(0.07)}{100}} = 0.0255$$

Now the probability that all four tires have no flaw is p^4. We estimate this probability with $\widehat{p}^4 = 0.93^4 = 0.7481$. We use Equation (3.10) to compute the uncertainty in $\widehat{p}^4$:

$$\begin{aligned}\sigma_{\widehat{p}^4} &\approx \left| \frac{d}{d\widehat{p}} \widehat{p}^4 \right| \sigma_{\widehat{p}} \\ &= 4\widehat{p}^3 \sigma_{\widehat{p}} \\ &= 4(0.93)^3(0.0255) \\ &= 0.082\end{aligned}$$

Exercises for Section 4.2

1. Let $X \sim \text{Bin}(8, 0.4)$. Find

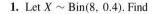

a. $P(X = 2)$

b. $P(X = 4)$

c. $P(X < 2)$

d. $P(X > 6)$

e. μ_X

f. σ_X^2

2. A sample of five items is drawn from a large lot in which 10% of the items are defective.

a. Find the probability that none of the sampled items are defective.

b. Find the probability that exactly one of the sampled items is defective.

c. Find the probability that one or more of the sampled items is defective.

d. Find the probability that fewer than two of the sampled items are defective.

3. A fair coin is tossed 10 times.

a. What is the probability of obtaining exactly three heads?

b. Find the mean number of heads obtained.

c. Find the variance of the number of heads obtained.

d. Find the standard deviation of the number of heads obtained.

4. In a large shipment of automobile tires, 5% have a certain flaw. Four tires are chosen at random to be installed on a car.

a. What is the probability that none of the tires have a flaw?

b. What is the probability that exactly one of the tires has a flaw?

c. What is the probability that one or more of the tires has a flaw?

5. In a random pattern of eight bits used to test a microcircuit, each bit is equally likely to be 0 or 1. Assume the values of the bits are independent.

a. What is the probability that all eight bits are 1?

b. What is the probability that exactly three of the bits are 1?

c. What is the probability that at least six of the bits are 1?

d. What is the probability that at least two of the bits are 1?

6. A quality engineer takes a random sample of 100 steel rods from a day's production, and finds that 92 of them meet specifications.

a. Estimate the proportion of that day's production that meets specifications, and find the uncertainty in the estimate.

b. Estimate the number of rods that must be sampled to reduce the uncertainty to 1%.

7. In a random sample of 100 parts ordered from vendor A, 12 were defective. In a random sample of 200 parts ordered from vendor B, 10 were defective.

a. Estimate the proportion of parts from vendor A that are defective, and find the uncertainty in the estimate.

b. Estimate the proportion of parts from vendor B that are defective, and find the uncertainty in the estimate.

c. Estimate the difference in the proportions, and find the uncertainty in the estimate.

8. Of the items manufactured by a certain process, 20% are defective. Of the defective items, 60% can be repaired.

a. Find the probability that a randomly chosen item is defective and cannot be repaired.

b. Find the probability that exactly 2 of 20 randomly chosen items are defective and cannot be repaired.

9. Of the bolts manufactured for a certain application, 90% meet the length specification and can be used immediately, 6% are too long and can be used after being cut, and 4% are too short and must be scrapped.

a. Find the probability that a randomly selected bolt can be used (either immediately or after being cut).

b. Find the probability that fewer than 9 out of a sample of 10 bolts can be used (either immediately or after being cut).

10. A distributor receives a large shipment of components. The distributor would like to accept the shipment if 10% or fewer of the components are defective and to return it if more than 10% of the components are defective. She decides to sample 10 components, and to return the shipment if more than 1 of the 10 is defective.

a. If the proportion of defectives in the batch is in fact 10%, what is the probability that she will return the shipment?

b. If the proportion of defectives in the batch is 20%, what is the probability that she will return the shipment?

c. If the proportion of defectives in the batch is 2%, what is the probability that she will return the shipment?

d. The distributor decides that she will accept the shipment only if none of the sampled items are defective. What is the minimum number of items she should sample if she wants to have a probability no greater than 0.01 of accepting the shipment if 20% of the components in the shipment are defective?

11. A certain large shipment comes with a guarantee that it contains no more than 15% defective items. If the proportion of defective items in the shipment is greater than 15%, the shipment may be returned. You draw a random sample of 10 items. Let X be the number of defective items in the sample.

a. If in fact 15% of the items in the shipment are defective (so that the shipment is good, but just barely), what is $P(X \geq 7)$?

b. Based on the answer to part (a), if 15% of the items in the shipment are defective, would 7 defectives in a sample of size 10 be an unusually large number?

c. If you found that 7 of the 10 sample items were defective, would this be convincing evidence that the shipment should be returned? Explain.

d. If in fact 15% of the items in the shipment are defective, what is $P(X \geq 2)$?

e. Based on the answer to part (d), if 15% of the items in the shipment are defective, would 2 defectives in a sample of size 10 be an unusually large number?

f. If you found that 2 of the 10 sample items were defective, would this be convincing evidence that the shipment should be returned? Explain.

12. An insurance company offers a discount to homeowners who install smoke detectors in their homes. A company representative claims that 80% or more of policyholders have smoke detectors. You draw a random sample of eight policyholders. Let X be the number of policyholders in the sample who have smoke detectors.

a. If exactly 80% of the policyholders have smoke detectors (so the representative's claim is true, but just barely), what is $P(X \leq 1)$?

b. Based on the answer to part (a), if 80% of the policyholders have smoke detectors, would one policyholder with a smoke detector in a sample of size 8 be an unusually small number?

c. If you found that one of the eight sample policyholders had a smoke detector, would this be convincing evidence that the claim is false? Explain.

d. If exactly 80% of the policyholders have smoke detectors, what is $P(X \leq 6)$?

e. Based on the answer to part (d), if 80% of the policyholders have smoke detectors, would six policyholders with smoke detectors in a sample of size 8 be an unusually small number?

f. If you found that six of the eight sample policyholders had smoke detectors, would this be convincing evidence that the claim is false? Explain.

13. Porcelain figurines are sold for \$10 if flawless, and for \$3 if there are minor cosmetic flaws. Of the figurines made by a certain company, 90% are flawless and 10% have minor cosmetic flaws. In a sample of 100 figurines that are sold, let Y be the revenue earned by selling them and let X be the number of them that are flawless.

a. Express Y as a function of X.

b. Find μ_Y.

c. Find σ_Y.

14. One design for a system requires the installation of two identical components. The system will work if at least one of the components works. An alternative design requires four of these components, and the system will work if at least two of the four components work. If the probability that a component works is 0.9, and if the components function independently, which design has the greater probability of functioning?

15. Refer to Example 4.14. Estimate the probability that exactly one of the four tires has a flaw, and find the uncertainty in the estimate.

16. If p is a success probability, the quantity $p/(1-p)$ is called the *odds*. Odds are commonly estimated in medical research. The article "A Study of Twelve Southern California Communities with Differing Levels and Types of Air Pollution" (J. Peters, E. Avol, et al., *The American Journal of Respiratory and Critical Care Medicine*, 1999:760–767) reports an assessment of respiratory health of southern California children. Assume that 88 boys in a sample of 612 reported being diagnosed with bronchitis during the last 12 months.

a. Estimate the proportion p of boys who have been diagnosed with bronchitis, and find the uncertainty in the estimate.

b. Estimate the odds, and find the uncertainty in the estimate.

4.3 The Poisson Distribution

The Poisson distribution arises frequently in scientific work. One way to think of the Poisson distribution is as an approximation to the binomial distribution when n is large and p is small. We illustrate with an example.

A mass contains 10,000 atoms of a radioactive substance. The probability that a given atom will decay in a one minute time period is 0.0002. Let X represent the number of atoms that decay in one minute. Now each atom can be thought of as a Bernoulli trial, where success occurs if the atom decays. Thus X is the number of successes in 10,000 independent Bernoulli trials, each with success probability 0.0002, so the distribution of X is Bin(10,000, 0.0002). The mean of X is $\mu_X = (10{,}000)(0.0002) = 2$.

Another mass contains 5000 atoms, and each of these atoms has probability 0.0004 of decaying in a one-minute time interval. Let Y represent the number of atoms that decay in one minute from this mass. By the reasoning in the previous paragraph, $Y \sim \text{Bin}(5000,\ 0.0004)$ and $\mu_Y = (5000)(0.0004) = 2$.

In each of these cases, the number of trials n and the success probability p are different, but the mean number of successes, which is equal to the product np, is the same. Now assume that we wanted to compute the probability that exactly three atoms decay in one minute for each of these masses. Using the binomial probability mass function, we would compute as follows:

$$P(X = 3) = \frac{10{,}000!}{3!\,9997!}(0.0002)^3(0.9998)^{9997} = 0.180465091$$

$$P(Y = 3) = \frac{5000!}{3!\,4997!}(0.0004)^3(0.9996)^{4997} = 0.180483143.$$

It turns out that these probabilities are very nearly equal to each other. Although it is not obvious from the formula for the binomial probability mass function, it is the case that when n is large and p is small the mass function depends almost entirely on the mean np, and very little on the specific values of n and p. We can therefore approximate the binomial mass function with a quantity that depends on the product np only. Specifically, if n is large and p is small, and we let $\lambda = np$, it can be shown by advanced methods that for all x,

$$\frac{n!}{x!(n-x)!}p^x(1-p)^{n-x} \approx e^{-\lambda}\frac{\lambda^x}{x!} \tag{4.8}$$

We are led to define a new probability mass function, called the Poisson probability mass function. The Poisson probability mass function is defined by

$$p(x) = P(X = x) = \begin{cases} e^{-\lambda}\dfrac{\lambda^x}{x!} & \text{if } x \text{ is a non-negative integer} \\ 0 & \text{otherwise} \end{cases} \tag{4.9}$$

If X is a random variable whose probability mass function is given by Equation (4.9), then X is said to have the **Poisson distribution** with parameter λ. The notation is $X \sim \text{Poisson}(\lambda)$.

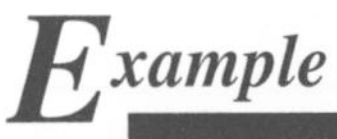

4.15

If $X \sim \text{Poisson}(3)$, compute $P(X = 2)$, $P(X = 10)$, $P(X = 0)$, $P(X = -1)$, and $P(X = 0.5)$.

Solution

Using the probability mass function (4.9), with $\lambda = 3$, we obtain

$$P(X = 2) = e^{-3}\frac{3^2}{2!} = 0.2240$$

$$P(X = 10) = e^{-3}\frac{3^{10}}{10!} = 0.0008$$

$$P(X = 0) = e^{-3}\frac{3^0}{0!} = 0.0498$$

$P(X = -1) = 0$ because -1 is not a non-negative integer

$P(X = 0.5) = 0$ because 0.5 is not a non-negative integer

Example 4.16

If $X \sim \text{Poisson}(4)$, compute $P(X \leq 2)$ and $P(X > 1)$.

Solution

$$\begin{aligned} P(X \leq 2) &= P(X = 0) + P(X = 1) + P(X = 2) \\ &= e^{-4}\frac{4^0}{0!} + e^{-4}\frac{4^1}{1!} + e^{-4}\frac{4^2}{2!} \\ &= 0.0183 + 0.0733 + 0.1465 \\ &= 0.2381 \end{aligned}$$

To find $P(X > 1)$, we might try to start by writing

$$P(X > 1) = P(X = 2) + P(X = 3) + \cdots$$

This leads to an infinite sum that is difficult to compute. Instead, we write

$$\begin{aligned} P(X > 1) &= 1 - P(X \leq 1) \\ &= 1 - [P(X = 0) + P(X = 1)] \\ &= 1 - \left(e^{-4}\frac{4^0}{0!} + e^{-4}\frac{4^1}{1!}\right) \\ &= 1 - (0.0183 + 0.0733) \\ &= 0.9084 \end{aligned}$$

For the radioactive masses described at the beginning of this section, we would use the Poisson mass function to approximate either $P(X = x)$ or $P(Y = x)$ by substituting $\lambda = 2$ into Equation (4.9). Table 4.1 shows that the approximation is excellent.

TABLE 4.1 An example of the Poisson approximation to the binomial probability mass function*

x	$P(X = x)$, $X \sim \text{Bin}(10{,}000, 0.0002)$	$P(Y = x)$, $Y \sim \text{Bin}(5000, 0.0004)$	Poisson Approximation, Poisson (2)
0	0.135308215	0.135281146	0.135335283
1	0.270670565	0.270670559	0.270670566
2	0.270697637	0.270724715	0.270670566
3	0.180465092	0.180483143	0.180447044
4	0.090223521	0.090223516	0.090223522
5	0.036082189	0.036074965	0.036089409
6	0.012023787	0.012017770	0.012029803
7	0.003433993	0.003430901	0.003437087
8	0.000858069	0.000856867	0.000859272
9	0.000190568	0.000190186	0.000190949

*When n is large and p is small, the Bin(n, p) probability mass function is well approximated by the Poisson (λ) probability mass function (Equation 4.9), with $\lambda = np$. Here $X \sim \text{Bin}(10{,}000, 0.0002)$ and $Y \sim \text{Bin}(5000, 0.0004)$, so $\lambda = np = 2$, and the Poisson approximation is Poisson(2).

Summary

If $X \sim \text{Poisson}(\lambda)$, then

- X is a discrete random variable whose possible values are the non-negative integers.
- The parameter λ is a positive constant.
- The probability mass function of X is

$$p(x) = P(X = x) = \begin{cases} e^{-\lambda}\dfrac{\lambda^x}{x!} & \text{if } x \text{ is a non-negative integer} \\ 0 & \text{otherwise} \end{cases}$$

- The Poisson probability mass function is very close to the binomial probability mass function when n is large, p is small, and $\lambda = np$.

The Mean and Variance of a Poisson Random Variable

To compute the mean and variance of a Poisson random variable, we can use the probability mass function along with the definitions given by Equations (2.29) and (2.30) (in Section 2.4). Rigorous derivations of the mean and variance using this method are given at the end of the section. We present an intuitive approach here. If $X \sim \text{Poisson}(\lambda)$, we can think of X as a binomial random variable with large n, small p, and $np = \lambda$. Since the mean of a binomial random variable is np, it follows that the mean of a Poisson random variable is λ. The variance of a binomial random variable is $np(1 - p)$. Since p is very small, we replace $1 - p$ with 1, and conclude that the variance of a Poisson random variable is $np = \lambda$. Note that the variance of a Poisson random variable is equal to its mean.

Summary

If $X \sim \text{Poisson}(\lambda)$, then the mean and variance of X are given by

$$\mu_X = \lambda \tag{4.10}$$

$$\sigma_X^2 = \lambda \tag{4.11}$$

Figure 4.3 (page 210) presents probability histograms for the Poisson(1) and Poisson(10) probability mass functions.

One of the earliest industrial uses of the Poisson distribution involved an application to the brewing of beer. A crucial step in the brewing process is the addition of yeast culture to prepare mash for fermentation. The living yeast cells are kept suspended in a liquid medium. Because the cells are alive, their concentration in the medium changes over time. Therefore, just before the yeast is added, it is necessary to estimate the concentration of yeast cells per unit volume of suspension, so as to be sure to add the right amount.

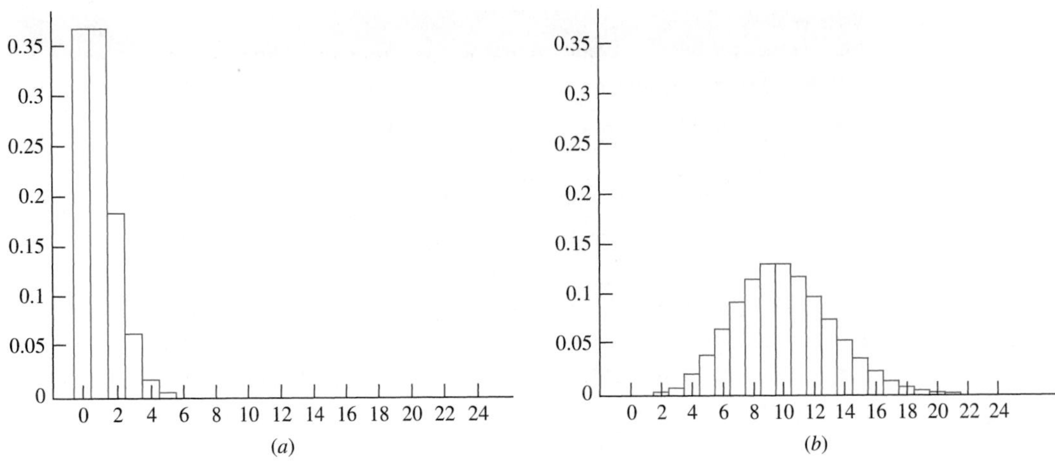

FIGURE 4.3 *(a)* The Poisson(1) probability histogram. *(b)* The Poisson(10) probability histogram.

Up until the early part of the twentieth century, this posed a problem for brewers. They estimated the concentration in the obvious way, by withdrawing a small volume of the suspension and counting the yeast cells in it under a microscope. Of course, the estimates determined this way were subject to uncertainty, but no one knew how to compute the uncertainty. Thus no one knew by how much the concentration in the sample was likely to differ from the actual concentration.

William Sealy Gosset, a young man in his mid-twenties who was employed by the Guinness Brewing Company of Dublin, Ireland, discovered in 1904 that the number of yeast cells in a sampled volume of suspension follows a Poisson distribution. He was then able to develop methods to compute the needed uncertainty. Gosset's discovery not only enabled Guinness to produce a more consistent product, it showed that the Poisson distribution could have important applications in many situations. Gosset wanted to publish his result, but his managers at Guinness considered his discovery to be proprietary information and forbade publication. Gosset published it anyway, but to hide this fact from his employers, he used the pseudonym "Student."

In Example 4.17, we will follow a train of thought that leads to Student's result. Before we get to it though, we will mention that four years after publishing this result, Student made another discovery that solved one of the most important outstanding problems in statistics, and which has profoundly influenced work in virtually all fields of science ever since. We will discuss this result in Section 5.3.

Particles (e.g., yeast cells) are suspended in a liquid medium at a concentration of 10 particles per mL. A large volume of the suspension is thoroughly agitated, and then 1 mL is withdrawn. What is the probability that exactly eight particles are withdrawn?

Solution

So long as the volume withdrawn is a small fraction of the total, the solution to this problem does not depend on the total volume of the suspension, but only on the concentration of particles in it. Let V be the total volume of the suspension, in mL. Then the total number of particles in the suspension is $10V$. Think of each of the $10V$ particles as a Bernoulli trial. A particle "succeeds" if it is withdrawn. Now 1 mL out of the total of V mL is to be withdrawn. Therefore the amount to be withdrawn is $1/V$ of the total, so it follows that each particle has probability $1/V$ of being withdrawn. Let X denote the number of particles withdrawn. Then X represents the number of successes in $10V$ Bernoulli trials, each with probability $1/V$ of success. Therefore $X \sim \text{Bin}(10V, 1/V)$. Since V is large, $10V$ is large and $1/V$ is small. Thus to a very close approximation, $X \sim \text{Poisson}(10)$. We compute $P(X = 8)$ with the Poisson probability mass function: $P(X = 8) = e^{-10}10^8/8! = 0.1126$.

In Example 4.17, λ had the value 10 because the mean number of particles in 1 mL of suspension (the volume withdrawn) was 10.

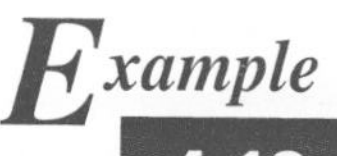

Particles are suspended in a liquid medium at a concentration of 6 particles per mL. A large volume of the suspension is thoroughly agitated, and then 3 mL are withdrawn. What is the probability that exactly 15 particles are withdrawn?

Solution

Let X represent the number of particles withdrawn. The mean number of particles in a 3 mL volume is 18. Then $X \sim \text{Poisson}(18)$. The probability that exactly 15 particles are withdrawn is

$$\begin{aligned} P(X = 15) &= e^{-18}\frac{18^{15}}{15!} \\ &= 0.0786 \end{aligned}$$

Note that for the solutions to Examples 4.17 and 4.18 to be correct, it is important that the amount of suspension withdrawn not be too large a fraction of the total. For example, if the total volume in Example 4.18 was 3 mL, so that the entire amount was withdrawn, it would be certain that all 18 particles would be withdrawn, so the probability of withdrawing 15 particles would be zero.

Grandma bakes chocolate chip cookies in batches of 100. She puts 300 chips into the dough. When the cookies are done, she gives you one. What is the probability that your cookie contains no chocolate chips?

Solution

This is another instance of particles in a suspension. Let X represent the number of chips in your cookie. The mean number of chips is 3 per cookie, so $X \sim \text{Poisson}(3)$. It follows that $P(X = 0) = e^{-3}3^0/0! = 0.0498$.

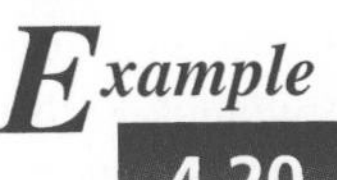

4.20

Grandma's grandchildren have been complaining that Grandma is too stingy with the chocolate chips. Grandma agrees to add enough chips to the dough so that only 1% of the cookies will contain no chips. How many chips must she include in a batch of 100 cookies to achieve this?

Solution

Let n be the number of chips to include in a batch of 100 cookies, and let X be the number of chips in your cookie. The mean number of chips is $0.01n$ per cookie, so $X \sim \text{Poisson}(0.01n)$. We must find the value of n for which $P(X = 0) = 0.01$. Using the Poisson$(0.01n)$ probability mass function,

$$\begin{aligned} P(X = 0) &= e^{-0.01n}\frac{(0.01n)^0}{0!} \\ &= e^{-0.01n} \end{aligned}$$

Setting $e^{-0.01n} = 0.01$, we obtain $n \approx 461$.

Examples 4.17 through 4.20 show that for particles distributed uniformly throughout a medium, the number of particles that happen to fall in a small portion of the medium follows a Poisson distribution. In these examples, the particles were actual particles and the medium was spatial in nature. There are many cases, however, when the "particles" represent events and the medium is time. We saw such an example previously, where the number of radioactive decay events in a fixed time interval turned out to follow a Poisson distribution. We now present another.

4.21

Assume that the number of hits on a certain website during a fixed time interval follows a Poisson distribution. Assume that the mean rate of hits is 5 per minute. Find the probability that there will be exactly 17 hits in the next three minutes.

Solution

Let X be the number of hits in three minutes. The mean number of hits in three minutes is $(5)(3) = 15$, so $X \sim \text{Poisson}(15)$. Using the Poisson$(15)$ probability mass function,

$$\begin{aligned} P(X = 17) &= e^{-15}\frac{15^{17}}{17!} \\ &= 0.0847 \end{aligned}$$

Example 4.22

In Example 4.21, let X be the number of hits in t minutes. Find the probability mass function of X, in terms of t.

Solution

The mean number of hits in t minutes is $5t$, so $X \sim \text{Poisson}(5t)$. The probability mass function of X is

$$p(x) = P(X = x) = \frac{e^{-5t}(5t)^x}{x!} \qquad x = 0, 1, 2, \ldots$$

Using the Poisson Distribution to Estimate a Rate

Often experiments are done to estimate a rate λ that represents the mean number of events that occur in one unit of time or space. In these experiments, the number of events X that occur in t units is counted, and the rate λ is estimated with the quantity $\widehat{\lambda} = X/t$. (Note that since the quantity X/t is used to estimate λ, it is denoted $\widehat{\lambda}$.) If the numbers of events in disjoint intervals of time or space are independent, and if events cannot occur simultaneously, then X follows a Poisson distribution. A process that produces such events is called a **Poisson process**. Since the mean number of events that occur in t units of time or space is equal to λt, $X \sim \text{Poisson}(\lambda t)$.

Summary

Let λ denote the mean number of events that occur in one unit of time or space. Let X denote the number of events that are observed to occur in t units of time or space. Then if $X \sim \text{Poisson}(\lambda t)$, λ is estimated with $\widehat{\lambda} = X/t$.

A suspension contains particles at an unknown concentration of λ per mL. The suspension is thoroughly agitated, and then 4 mL are withdrawn and 17 particles are counted. Estimate λ.

Solution

Let $X = 17$ represent the number of particles counted, and let $t = 4$ mL be the volume of suspension withdrawn. Then $\widehat{\lambda} = X/t = 17/4 = 4.25$ particles per mL.

Uncertainty in the Estimated Rate

It is important to realize that the estimated rate or concentration $\widehat{\lambda}$ is just an *estimate* of the true rate or concentration λ. In general $\widehat{\lambda}$ does not equal λ. If the experiment were repeated, the value of $\widehat{\lambda}$ would probably come out differently. In other words, there is uncertainty in $\widehat{\lambda}$. For $\widehat{\lambda}$ to be a useful estimate, we must compute its bias and its uncertainty. The calculations are similar to those for the sample proportion that were presented in Section 4.2. Let X be the number of events counted in t units of time or space, and assume that $X \sim \text{Poisson}(\lambda t)$.

The bias is the difference $\mu_{\widehat{\lambda}} - \lambda$. Since $\widehat{\lambda} = X/t$, it follows from Equation (2.41) (in Section 2.5) that

$$\mu_{\widehat{\lambda}} = \mu_{X/t} = \frac{\mu_X}{t}$$
$$= \frac{\lambda t}{t} = \lambda$$

Since $\mu_{\widehat{\lambda}} = \lambda$, $\widehat{\lambda}$ is unbiased.

The uncertainty is the standard deviation $\sigma_{\widehat{\lambda}}$. Since $\widehat{\lambda} = X/t$, it follows from Equation (2.43) (in Section 2.5) that $\sigma_{\widehat{\lambda}} = \sigma_X/t$. Since $X \sim \text{Poisson}(\lambda t)$, it follows from Equation (4.11) that $\sigma_X = \sqrt{\lambda t}$. Therefore

$$\sigma_{\widehat{\lambda}} = \frac{\sigma_X}{t} = \frac{\sqrt{\lambda t}}{t} = \sqrt{\frac{\lambda}{t}}$$

In practice, the value of λ is unknown, so we approximate it with $\widehat{\lambda}$.

Summary

If $X \sim \text{Poisson}(\lambda t)$, we estimate the rate λ with $\widehat{\lambda} = \dfrac{X}{t}$.

- $\widehat{\lambda}$ is unbiased.
- The uncertainty in $\widehat{\lambda}$ is

$$\sigma_{\widehat{\lambda}} = \sqrt{\frac{\lambda}{t}} \tag{4.12}$$

In practice, we substitute $\widehat{\lambda}$ for λ in Equation (4.12), since λ is unknown.

Example 4.24

A 5 mL sample of a suspension is withdrawn, and 47 particles are counted. Estimate the mean number of particles per mL, and find the uncertainty in the estimate.

Solution

The number of particles counted is $X = 47$. The volume withdrawn is $t = 5$ mL. The estimated mean number of particles per mL is

$$\widehat{\lambda} = \frac{47}{5} = 9.4$$

The uncertainty in the estimate is

$$\sigma_{\widehat{\lambda}} = \sqrt{\frac{\lambda}{t}}$$
$$= \sqrt{\frac{9.4}{5}} \qquad \text{approximating } \lambda \text{ with } \widehat{\lambda} = 9.4$$
$$= 1.4$$

Example 4.25

A certain mass of a radioactive substance emits alpha particles at a mean rate of λ particles per second. A physicist counts 1594 emissions in 100 seconds. Estimate λ, and find the uncertainty in the estimate.

Solution

The estimate of λ is $\widehat{\lambda} = 1594/100 = 15.94$ emissions per second. The uncertainty is

$$\begin{aligned}\sigma_{\widehat{\lambda}} &= \sqrt{\frac{\lambda}{t}} \\ &= \sqrt{\frac{15.94}{100}} \qquad \text{approximating } \lambda \text{ with } \widehat{\lambda} = 15.94 \\ &= 0.40\end{aligned}$$

Example 4.26

In Example 4.25, for how many seconds should emissions be counted to reduce the uncertainty to 0.3 emissions per second?

Solution

We want to find the time t for which $\sigma_{\widehat{\lambda}} = \sqrt{\lambda/t} = 0.3$. From Example 4.25, $\widehat{\lambda} = 15.94$. Substituting this value for λ, we obtain

$$\sigma_{\widehat{\lambda}} = \sqrt{\frac{15.94}{t}} = 0.3$$

Solving for t yields $t = 177$ seconds.

Sometimes it is desired to estimate a function $f(\lambda)$ of a Poisson rate λ. We estimate $f(\lambda)$ with $f(\widehat{\lambda})$. We can then use the propagation of error method (Equation 3.10 in Section 3.3) to find the uncertainty in $f(\widehat{\lambda})$.

Example 4.27

The number of flaws on a sheet of aluminum manufactured by a certain process follows a Poisson distribution. In a sample of 100 m^2 of aluminum, 200 flaws are counted. Estimate the probability that a given square meter of aluminum has no flaws, and find the uncertainty in the estimate.

Solution

Let λ represent the mean number of flaws per square meter. We will begin by computing $\widehat{\lambda}$ and its uncertainty. We have observed $X = 200$ flaws in $t = 100$ m^2 of aluminum. Therefore $\widehat{\lambda} = 200/100 = 2.00$. The uncertainty in $\widehat{\lambda}$ is

$$\sigma_{\widehat{\lambda}} = \sqrt{\frac{\lambda}{t}}$$

$$= \sqrt{\frac{2}{100}} \qquad \text{approximating } \lambda \text{ with } \widehat{\lambda} = 2$$
$$= 0.1414$$

What we want to estimate is the probability that a square meter of aluminum contains no flaws. We first express this probability as a function of λ. To do this, let Y represent the number of flaws in a 1 m^2 sheet of aluminum. Then $Y \sim \text{Poisson}(\lambda)$. We want to estimate $P(Y = 0)$. Using the Poisson probability mass function, this probability is given by

$$P(Y = 0) = \frac{e^{-\lambda}\lambda^0}{0!} = e^{-\lambda}$$

The probability that a square meter contains no flaws is therefore estimated with $e^{-\widehat{\lambda}} = e^{-2.00} = 0.1353$. To find the uncertainty in this estimate, we use the propagation of error method (Equation 3.10).

$$\sigma_{e^{-\widehat{\lambda}}} \approx \left| \frac{d}{d\widehat{\lambda}} e^{-\widehat{\lambda}} \right| \sigma_{\widehat{\lambda}}$$
$$= \left| -e^{-\widehat{\lambda}} \right| \sigma_{\widehat{\lambda}}$$
$$= e^{-2.00}(0.1414)$$
$$= 0.0191$$

In the case of particles in a suspension, or radioactive decay events, enough is known about the underlying physical principles governing these processes that we were able to argue from first principles to show that the distribution of the number of events is Poisson. There are many other cases where empirical evidence suggests that the Poisson distribution may be appropriate, but the laws governing the process are not sufficiently well understood to make a rigorous derivation possible. Examples include the number of hits on a website, the number of traffic accidents at an intersection, and the number of trees in a section of forest.

Derivation of the Mean and Variance of a Poisson Random Variable

Let $X \sim \text{Poisson}(\lambda)$. We will show that $\mu_X = \lambda$ and $\sigma_X^2 = \lambda$. Using the definition of population mean for a discrete random variable (Equation 2.29 in Section 2.4):

$$\mu_X = \sum_{x=0}^{\infty} xP(X = x)$$
$$= \sum_{x=0}^{\infty} xe^{-\lambda}\frac{\lambda^x}{x!}$$
$$= (0)(e^{-\lambda})\left(\frac{\lambda^0}{0!}\right) + \sum_{x=1}^{\infty} xe^{-\lambda}\frac{\lambda^x}{x!}$$

$$= 0 + \sum_{x=1}^{\infty} e^{-\lambda} \frac{\lambda^x}{(x-1)!}$$

$$= \sum_{x=1}^{\infty} e^{-\lambda} \frac{\lambda^x}{(x-1)!}$$

$$= \lambda \sum_{x=1}^{\infty} e^{-\lambda} \frac{\lambda^{x-1}}{(x-1)!}$$

$$= \lambda \sum_{x=0}^{\infty} e^{-\lambda} \frac{\lambda^x}{x!}$$

Now the sum $\sum_{x=0}^{\infty} e^{-\lambda}\lambda^x/x!$ is the sum of the Poisson(λ) probability mass function over all its possible values. Therefore $\sum_{x=0}^{\infty} e^{-\lambda}\lambda^x/x! = 1$, so

$$\mu_X = \lambda$$

We use Equation (2.31) (in Section 2.4) to show that $\sigma_X^2 = \lambda$.

$$\sigma_X^2 = \sum_{x=0}^{\infty} x^2 e^{-\lambda} \frac{\lambda^x}{x!} - \mu_X^2 \tag{4.13}$$

Substituting $x(x-1) + x$ for x^2 and λ for μ_X in Equation (4.13), we obtain

$$\sigma_X^2 = \sum_{x=0}^{\infty} x(x-1) e^{-\lambda} \frac{\lambda^x}{x!} + \sum_{x=0}^{\infty} x e^{-\lambda} \frac{\lambda^x}{x!} - \lambda^2 \tag{4.14}$$

Now $x(x-1) = 0$ if $x = 0$ or 1, and $\sum_{x=0}^{\infty} xe^{-\lambda}\lambda^x/x! = \mu_X = \lambda$. We may therefore begin the sum on the right-hand side of Equation (4.14) at $x = 2$, and substitute λ for $\sum_{x=0}^{\infty} xe^{-\lambda}\lambda^x/x!$. We obtain

$$\sigma_X^2 = \sum_{x=2}^{\infty} x(x-1) e^{-\lambda} \frac{\lambda^x}{x!} + \lambda - \lambda^2$$

$$= \sum_{x=2}^{\infty} e^{-\lambda} \frac{\lambda^x}{(x-2)!} + \lambda - \lambda^2$$

$$= \lambda^2 \sum_{x=2}^{\infty} e^{-\lambda} \frac{\lambda^{x-2}}{(x-2)!} + \lambda - \lambda^2$$

$$= \lambda^2 \sum_{x=0}^{\infty} e^{-\lambda} \frac{\lambda^x}{x!} + \lambda - \lambda^2$$

$$= \lambda^2(1) + \lambda - \lambda^2$$

$$= \lambda$$

Exercises for Section 4.3

1. Let $X \sim \text{Poisson}(4)$. Find

a. $P(X = 1)$
b. $P(X = 0)$
c. $P(X < 2)$
d. $P(X > 1)$
e. μ_X
f. σ_X

2. The concentration of particles in a suspension is 2 per mL. The suspension is thoroughly agitated, and then 3 mL is withdrawn. Let X represent the number of particles that are withdrawn. Find

a. $P(X = 5)$
b. $P(X \leq 2)$
c. $P(X > 1)$
d. μ_X
e. σ_X

3. Suppose that 0.03% of plastic containers manufactured by a certain process have small holes that render them unfit for use. Let X represent the number of containers in a random sample of 10,000 that have this defect. Find

a. $P(X = 3)$
b. $P(X \leq 2)$
c. $P(1 \leq X < 4)$
d. μ_X
e. σ_X

4. One out of every 5000 individuals in a population carries a certain defective gene. A random sample of 1000 individuals is studied.

a. What is the probability that exactly one of the sample individuals carries the gene?
b. What is the probability that none of the sample individuals carries the gene?
c. What is the probability that more than two of the sample individuals carry the gene?
d. What is the mean of the number of sample individuals that carry the gene?
e. What is the standard deviation of the number of sample individuals that carry the gene?

5. The number of messages received by a computer bulletin board is a Poisson random variable with a mean rate of 8 messages per hour.

a. What is the probability that five messages are received in a given hour?
b. What is the probability that 10 messages are received in 1.5 hours?
c. What is the probability that fewer than three messages are received in one-half hour?

6. A certain type of circuit board contains 300 diodes. Each diode has probability $p = 0.002$ of failing.

a. What is the probability that exactly two diodes fail?
b. What is the mean number of diodes that fail?
c. What is the standard deviation of the number of diodes that fail?
d. A board works if none of its diodes fail. What is the probability that a board works?
e. Five boards are shipped to a customer. What is the probability that four or more of them work?

7. A random variable X has a binomial distribution, and a random variable Y has a Poisson distribution. Both X and Y have means equal to 3. Is it possible to determine which random variable has the larger variance? Choose one of the following answers:

i. Yes, X has the larger variance.
ii. Yes, Y has the larger variance.
iii. No, we need to know the number of trials, n, for X.
iv. No, we need to know the success probability, p, for X.
v. No, we need to know the value of λ for Y.

8. A chemist wishes to estimate the concentration of particles in a certain suspension. She withdraws 3 mL of the suspension and counts 48 particles. Estimate the concentration in particles per mL and find the uncertainty in the estimate.

9. A microbiologist wants to estimate the concentration of a certain type of bacterium in a wastewater sample. She puts a 0.5 mL sample of the wastewater on a microscope slide and counts 39 bacteria. Estimate the concentration of bacteria, per mL, in this wastewater, and find the uncertainty in the estimate.

10. Grandma is trying out a new recipe for raisin bread. Each batch of bread dough makes three loaves, and each loaf contains 20 slices of bread.
    a. If she puts 100 raisins into a batch of dough, what is the probability that a randomly chosen slice of bread contains no raisins?
    b. If she puts 200 raisins into a batch of dough, what is the probability that a randomly chosen slice of bread contains 5 raisins?
    c. How many raisins must she put in so that the probability that a randomly chosen slice will have no raisins is 0.01?

11. Mom and Grandma are each baking chocolate chip cookies. Each gives you two cookies. One of Mom's cookies has 14 chips in it and the other has 11. Grandma's cookies have 6 and 8 chips.
    a. Estimate the mean number of chips in one of Mom's cookies.
    b. Estimate the mean number of chips in one of Grandma's cookies.
    c. Find the uncertainty in the estimate for Mom's cookies.
    d. Find the uncertainty in the estimate for Grandma's cookies.
    e. Estimate how many more chips there are on the average in one of Mom's cookies than in one of Grandma's. Find the uncertainty in this estimate.

12. You have received a radioactive mass that is claimed to have a mean decay rate of at least 1 particle per second. If the mean decay rate is less than 1 per second, you may return the product for a refund. Let X be the number of decay events counted in 10 seconds.
    a. If the mean decay rate is exactly 1 per second (so that the claim is true, but just barely), what is $P(X \leq 1)$?
    b. Based on the answer to part (a), if the mean decay rate is 1 particle per second, would one event in 10 seconds be an unusually small number?
    c. If you counted one decay event in 10 seconds, would this be convincing evidence that the product should be returned? Explain.
    d. If the mean decay rate is exactly 1 per second, what is $P(X \leq 8)$?
    e. Based on the answer to part (d), if the mean decay rate is 1 particle per second, would eight events in 10 seconds be an unusually small number?
    f. If you counted eight decay events in 10 seconds, would this be convincing evidence that the product should be returned? Explain.

13. Someone claims that a certain suspension contains at least seven particles per mL. You sample 1 mL of solution. Let X be the number of particles in the sample.
    a. If the mean number of particles is exactly seven per mL (so that the claim is true, but just barely), what is $P(X \leq 1)$?
    b. Based on the answer to part (a), if the suspension contains seven particles per mL, would one particle in a 1 mL sample be an unusually small number?
    c. If you counted one particle in the sample, would this be convincing evidence that the claim is false? Explain.
    d. If the mean number of particles is exactly 7 per mL, what is $P(X \leq 6)$?
    e. Based on the answer to part (d), if the suspension contains seven particles per mL, would six particles in a 1 mL sample be an unusually small number?
    f. If you counted six particles in the sample, would this be convincing evidence that the claim is false? Explain.

14. A physicist wants to estimate the rate of emissions of alpha particles from a certain source. He makes two counts. First, he measures the background rate by counting the number of particles in 100 seconds in the absence of the source. He counts 36 background emissions. Then, with the source present, he counts 324 emissions in 100 seconds. This represents the sum of source emissions plus background emissions.
    a. Estimate the background rate, in emissions per second, and find the uncertainty in the estimate.
    b. Estimate the sum of the source plus background rate, in emissions per second, and find the uncertainty in the estimate.
    c. Estimate the rate of source emissions in particles per second, and find the uncertainty in the estimate.

d. Which will provide the smaller uncertainty in estimating the rate of emissions from the source: (1) counting the background only for 150 seconds and the background plus the source for 150 seconds, or (2) counting the background for 100 seconds and the source plus the background for 200 seconds? Compute the uncertainty in each case.

e. Is it possible to reduce the uncertainty to 0.03 particles per second if the background rate is measured for only 100 seconds? If so, for how long must the source plus background be measured? If not, explain why not.

15. Refer to Example 4.27. Estimate the probability that a 1 m^2 sheet of aluminum has exactly one flaw, and find the uncertainty in this estimate.

4.4 Some Other Discrete Distributions

In this section we discuss several discrete distributions that are useful in various situations.

The Hypergeometric Distribution

When a finite population contains two types of items, which may be called successes and failures, and a simple random sample is drawn from the population, each item sampled constitutes a Bernoulli trial. As each item is selected, the proportion of successes in the remaining population decreases or increases, depending on whether the sampled item was a success or a failure. For this reason the trials are not independent, so the number of successes in the sample does not follow a binomial distribution. Instead, the distribution that properly describes the number of successes in this situation is called the **hypergeometric distribution**.

As an example, assume that a lot of 20 items contains 6 that are defective, and that 5 items are sampled from this lot at random. Let X be the number of defective items in the sample. We will compute $P(X = 2)$. To do this, we first count the total number of different groups of 5 items that can be sampled from the population of 20. (We refer to each group of 5 items as a combination.) The number of combinations of 5 items is the number of different samples that can be drawn, and each of them is equally likely. Then we determine how many of these combinations of 5 items contain exactly 2 defectives. The probability that a combination of 5 items contains exactly 2 defectives is the quotient

$$P(X = 2) = \frac{\text{number of combinations of 5 items that contain 2 defectives}}{\text{number of combinations of 5 items that can be chosen from 20}}$$

In general, the number of combinations of k items that can be chosen from a group of n items is denoted $\binom{n}{k}$ and is equal to (see Equation 2.12 in Section 2.2 for a derivation)

$$\binom{n}{k} = \frac{n!}{k!(n-k)!}$$

The number of combinations of 5 items that can be chosen from 20 is therefore

$$\binom{20}{5} = \frac{20!}{5!(20-5)!} = 15{,}504$$

To determine the number of combinations of 5 that contain exactly 2 defectives, we describe the construction of such a combination as a sequence of two operations. First, select 2 items from the 6 defective ones; second, select 3 items from the 14 nondefective ones. The number of combinations of 2 items chosen from 6 is

$$\binom{6}{2} = \frac{6!}{2!(6-2)!} = 15$$

and the number of combinations of 3 items chosen from 14 is

$$\binom{14}{3} = \frac{14!}{3!(14-3)!} = 364$$

The total number of combinations of 5 items that can be made up of 2 defectives and 3 nondefectives is therefore the product $\binom{6}{2}\binom{14}{3} = (15)(364) = 5460$ (this is an application of the fundamental principal of counting; see Section 2.2 for a more detailed discussion). We conclude that

$$\begin{aligned} P(X = 2) &= \frac{\binom{6}{2}\binom{14}{3}}{\binom{20}{5}} \\ &= \frac{5460}{15{,}504} \\ &= 0.3522 \end{aligned}$$

To compute $P(X = 2)$ in the preceding example, it was necessary to know the number of items in the population (20), the number of defective items in the population (6), and the number of items sampled (5). The probability mass function of the random variable X is determined by these three parameters. Specifically, X is said to have the hypergeometric distribution with parameters 20, 6, and 5, which we denote $X \sim \text{H}(20, 6, 5)$. We now generalize this idea.

Summary

Assume a finite population contains N items, of which R are classified as successes and $N - R$ are classified as failures. Assume that n items are sampled from this population, and let X represent the number of successes in the sample. Then X has the hypergeometric distribution with parameters N, R, and n, which can be denoted $X \sim \text{H}(N, R, n)$.

The probability mass function of X is

$$p(x) = P(X = x) = \begin{cases} \dfrac{\binom{R}{x}\binom{N-R}{n-x}}{\binom{N}{n}} & \max(0, R+n-N) \le x \le \min(n, R) \\ 0 & \text{otherwise} \end{cases} \tag{4.15}$$

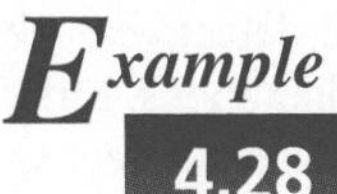

Example 4.28

Of 50 buildings in an industrial park, 12 have electrical code violations. If 10 buildings are selected at random for inspection, what is the probability that exactly 3 of the 10 have code violations?

Solution

Let X represent the number of sampled buildings that have code violations. Then $X \sim \text{H}(50, 12, 10)$. We must find $P(X = 3)$. Using Equation (4.15),

$$P(X = 3) = \frac{\binom{12}{3}\binom{38}{7}}{\binom{50}{10}}$$

$$= \frac{(220)(12{,}620{,}256)}{10{,}272{,}278{,}170}$$

$$= 0.2703$$

Mean and Variance of the Hypergeometric Distribution

The mean and variance of the hypergeometric distribution are presented in the following box. Their derivations are omitted.

> If $X \sim \text{H}(N, R, n)$, then
>
> $$\mu_X = \frac{nR}{N} \qquad (4.16)$$
>
> $$\sigma_X^2 = n\left(\frac{R}{N}\right)\left(1 - \frac{R}{N}\right)\left(\frac{N-n}{N-1}\right) \qquad (4.17)$$

Example 4.29

Refer to Example 4.28. Find the mean and variance of X.

Solution

$X \sim \text{H}(50, 12, 10)$, so

$$\mu_X = \frac{(10)(12)}{50}$$

$$= 2.4000$$

$$\sigma_X^2 = (10)\left(\frac{12}{50}\right)\left(1 - \frac{12}{50}\right)\left(\frac{50-10}{50-1}\right)$$

$$= 1.4890$$

Comparison with the Binomial Distribution

A population of size N contains R successes and $N - R$ failures. Imagine that a sample of n items is drawn from this population with replacement, that is, with each sampled item being returned to the population after it is drawn. Then the sampled items result from a sequence of independent Bernoulli trials, and the number of successes X in the sample has a binomial distribution with n trials and success probability $p = R/N$.

In practice, samples are seldom drawn with replacement, because there is no need to sample the same item twice. Instead, sampling is done without replacement, where each item is removed from the population after it is sampled. The sampled items then result from dependent Bernoulli trials, because the population changes as each item is sampled. For this reason the distribution of the number of successes, X, is $\text{H}(N, R, n)$ rather than $\text{Bin}(n, R/N)$.

When the sample size n is small compared to the population size N (i.e., no more than 5%), the difference between sampling with and without replacement is slight, and the binomial distribution $\text{Bin}(n, R/N)$ is a good approximation to the hypergeometric distribution $\text{H}(N, R, n)$. Note that the mean of $\text{H}(N, R, n)$ is nR/N, the same as that of $\text{Bin}(n, R/N)$. This indicates that whether the sampling is done with or without replacement, the proportion of successes in the sample is the same on the average as the proportion of successes in the population. The variance of $\text{Bin}(n, R/N)$ is $n(R/N)(1 - R/N)$, and the variance of $\text{H}(N, R, n)$ is obtained by multiplying this by the factor $(N - n)/(N - 1)$. Note that when n is small relative to N, this factor is close to 1.

The Geometric Distribution

Assume that a sequence of independent Bernoulli trials is conducted, each with the same success probability p. Let X represent the number of trials up to and including the first success. Then X is a discrete random variable, which is said to have the **geometric distribution** with parameter p. We write $X \sim \text{Geom}(p)$.

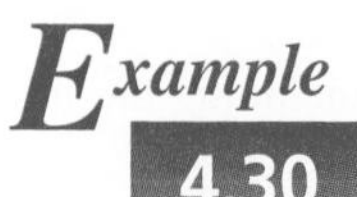

A test of weld strength involves loading welded joints until a fracture occurs. For a certain type of weld, 80% of the fractures occur in the weld itself, while the other 20% occur in the beam. A number of welds are tested. Let X be the number of tests up to and including the first test that results in a beam fracture. What is the distribution of X?

Solution

Each test is a Bernoulli trial, with success defined as a beam fracture. The success probability is therefore $p = 0.2$. The number of trials up to and including the first success has a geometric distribution with parameter $p = 0.2$. Therefore $X \sim \text{Geom}(0.2)$.

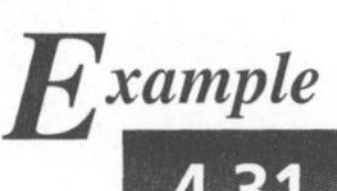

Refer to Example 4.30. Find $P(X = 3)$.

Solution

The event $X = 3$ occurs when the first two trials result in failure and the third trial results in success. It follows that

$$\begin{aligned} P(X = 3) &= P(\text{FFS}) \\ &= (0.8)(0.8)(0.2) \\ &= 0.128 \end{aligned}$$

The result of Example 4.31 can be generalized to produce the probability mass function of a geometric random variable.

> If $X \sim \text{Geom}(p)$, then the probability mass function of X is
>
> $$p(x) = P(X = x) = \begin{cases} p(1-p)^{x-1} & x = 1, 2, \ldots \\ 0 & \text{otherwise} \end{cases}$$

The Mean and Variance of a Geometric Distribution

The mean and variance of the geometric distribution are given in the following box. Their derivations require the manipulation of infinite series and are omitted.

> If $X \sim \text{Geom}(p)$, then
>
> $$\mu_X = \frac{1}{p} \tag{4.18}$$
>
> $$\sigma_X^2 = \frac{1-p}{p^2} \tag{4.19}$$

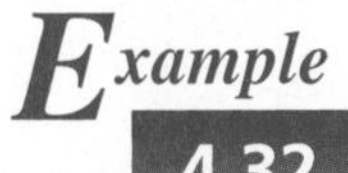

Refer to Example 4.30. Let X denote the number of tests up to and including the first beam fracture. Find the mean and variance of X.

Solution

Since $X \sim \text{Geom}(0.2)$, $\mu_X = 1/0.2 = 5$, and $\sigma_X^2 = (1 - 0.2)/(0.2^2) = 20$.

The Negative Binomial Distribution

The negative binomial distribution is an extension of the geometric distribution. Let r be a positive integer. Assume that independent Bernoulli trials, each with success probability

p, are conducted, and let X denote the number of trials up to and including the rth success. Then X has the **negative binomial distribution** with parameters r and p. We write $X \sim \text{NB}(r, p)$.

Example 4.33

(Continuing Example 4.30.) In a test of weld strength, 80% of tests result in a fracture in the weld, while the other 20% result in a fracture in the beam. Let X denote the number of tests up to and including the third beam fracture. What is the distribution of X? Find $P(X = 8)$.

Solution

Since X represents the number of trials up to and including the third success, and since the success probability is $p = 0.2$, $X \sim \text{NB}(3, 0.2)$. We will compute $P(X = 8)$, and the method of computation will lead to a derivation of the probability mass function of a negative binomial random variable. Since $X \sim \text{NB}(3, 0.2)$, the event $X = 8$ means that the third success occurred on the eighth trial. Another way to say this is that there were exactly two successes in the first 7 trials, and the eighth trial was a success. Since all the trials are independent, it follows that

$$P(X = 8) = P(\text{exactly 2 successes in first 7 trials})\,P(\text{success on eighth trial})$$

Now the number of successes in the first 7 trials has the Bin(7, 0.2) distribution, so

$$P(\text{exactly 2 successes in first 7 trials}) = \binom{7}{2}(0.2)^2(0.8)^5$$

The probability that the eighth trial (or any other trial) results in success is 0.2. Therefore

$$\begin{aligned} P(X = 8) &= \binom{7}{2}(0.2)^2(0.8)^5(0.2) \\ &= \binom{7}{2}(0.2)^3(0.8)^5 \\ &= 0.05505 \end{aligned}$$

We generalize the result of Example 4.33 to produce the probability mass function of a negative binomial random variable.

If $X \sim \text{NB}(r, p)$, then the probability mass function of X is

$$p(x) = P(X = x) = \begin{cases} \binom{x-1}{r-1} p^r (1-p)^{x-r} & x = r, r+1, \ldots \\ 0 & \text{otherwise} \end{cases}$$

Note that the smallest possible value for X is r, since it takes at least r trials to produce r successes. Note also that when $r = 1$, the negative binomial distribution is the same as the geometric distribution. In symbols, $\text{NB}(1, p) = \text{Geom}(p)$.

A Negative Binomial Random Variable Is a Sum of Geometric Random Variables

Assume that a sequence of 8 independent Bernoulli trials, each with success probability p, comes out as follows:

$$\text{F F S F S F F S}$$

If X is the number of trials up to and including the third success, then $X \sim \text{NB}(3, p)$, and for this sequence of trials, $X = 8$. Denote the number of trials up to and including the first success by Y_1. For this sequence, $Y_1 = 3$, but in general, $Y_1 \sim \text{Geom}(p)$. Now start counting, beginning from the first trial after the first success, up to and including the second success. Denote this number of trials by Y_2. For this sequence $Y_2 = 2$, but in general, $Y_2 \sim \text{Geom}(p)$. Finally, count the number of trials, beginning from the first trial after the second success, up to and including the third success. Denote this number of trials by Y_3. For this sequence $Y_3 = 3$, but again in general, $Y_3 \sim \text{Geom}(p)$. It is clear that $X = Y_1 + Y_2 + Y_3$. Furthermore, since the trials are independent, Y_1, Y_2, and Y_3 are independent. This shows that if $X \sim \text{NB}(3, p)$, then X is the sum of three independent $\text{Geom}(p)$ random variables. This result can be generalized to any positive integer r.

Summary

If $X \sim \text{NB}(r, p)$, then

$$X = Y_1 + \cdots + Y_r$$

where $Y_1, \ldots, Y_r$ are independent random variables, each with the $\text{Geom}(p)$ distribution.

The Mean and Variance of the Negative Binomial Distribution

If $X \sim \text{NB}(r, p)$, then $X = Y_1 + \cdots + Y_r$, where $Y_1, \ldots, Y_r$ are independent random variables, each with the $\text{Geom}(p)$ distribution. It follows that the mean of X is the sum of the means of the Ys, and the variance of X is the sum of the variances. Each Y_i has mean $1/p$ and variance $(1 - p)/p^2$. Therefore $\mu_X = r/p$ and $\sigma_X^2 = r(1 - p)/p^2$.

Summary

If $X \sim \text{NB}(r, p)$, then

$$\mu_X = \frac{r}{p} \tag{4.20}$$

$$\sigma_X^2 = \frac{r(1 - p)}{p^2} \tag{4.21}$$

Refer to Example 4.33. Find the mean and variance of X, where X represents the number of tests up to and including the third beam fracture.

Solution

Since $X \sim \text{NB}(3,\ 0.2)$, it follows that

$$\mu_X = \frac{3}{0.2} = 15$$

$$\sigma_X^2 = \frac{3(1-0.2)}{0.2^2} = 60$$

The Multinomial Distribution

A Bernoulli trial is a process that results in one of two possible outcomes. A generalization of the Bernoulli trial is the **multinomial trial**, which is a process that can result in any of k outcomes, where $k \geq 2$. For example, the rolling of a die is a multinomial trial, with the six possible outcomes 1, 2, 3, 4, 5, 6. Each outcome of a multinomial trial has a probability of occurrence. We denote the probabilities of the k outcomes by $p_1, \ldots, p_k$. For example, in the roll of a fair die, $p_1 = p_2 = \cdots = p_6 = 1/6$.

Now assume that n independent multinomial trials are conducted, each with the same k possible outcomes and with the same probabilities $p_1, \ldots, p_k$. Number the outcomes $1, 2, \ldots, k$. For each outcome i, let X_i denote the number of trials that result in that outcome. Then $X_1, \ldots, X_k$ are discrete random variables. The collection $X_1, \ldots, X_k$ is said to have the **multinomial distribution** with parameters $n, p_1, \ldots, p_k$. We write $X_1, \ldots, X_k \sim \text{MN}(n,\ p_1, \ldots, p_k)$. Note that it is the whole collection $X_1, \ldots, X_k$ that has the multinomial distribution, rather than any single X_i.

The items produced on an assembly line are inspected, and each is classified either as conforming (acceptable), downgraded, or rejected. Overall, 70% of the items are conforming, 20% are downgraded, and 10% are rejected. Assume that four items are chosen independently and at random. Let X_1, X_2, X_3 denote the numbers among the 4 that are conforming, downgraded, and rejected, respectively. What is the distribution of X_1, X_2, X_3?

Solution

Each item is a multinomial trial with three possible outcomes, conforming, downgraded, and rejected. The probabilities associated with the outcomes are $p_1 = 0.7$, $p_2 = 0.2$, and $p_3 = 0.1$. The random variables X_1, X_2, X_3 refer to the numbers of each outcome in 4 independent trials. Therefore $X_1, X_2, X_3 \sim \text{MN}(4,\ 0.7,\ 0.2,\ 0.1)$.

To show how to compute probabilities concerning multinomial random variables, we will compute $P(X_1 = 2, X_2 = 1, \text{and } X_3 = 1)$, where X_1, X_2, X_3 are defined in

Example 4.35. This will lead to a derivation of the multinomial probability mass function. We begin by noting that there are 12 arrangements of two conforming (C), one downgraded (D), and one rejected (R) among four trials. They are listed here.

CCDR	CCRD	CDCR	CDRC	CRCD	CRDC
DCCR	DCRC	DRCC	RCCD	RCDC	RDCC

Each of these 12 arrangements is equally probable. We compute the probability of CCDR. The event CCDR is a sequence of four outcomes: C on the first trial, C on the second trial, D on the third trial, and R on the fourth trial. Since the trials are independent, the probability of the sequence of outcomes is equal to the product of their individual probabilities.

$$P(\text{CCDR}) = (0.7)(0.7)(0.2)(0.1) = (0.7)^2(0.2)(0.1)$$

Since each of the 12 arrangements has the same probability,

$$P(X_1 = 2, X_2 = 1, X_3 = 1) = (12)(0.7)^2(0.2)(0.1) = 0.1176$$

In this calculation, the number of arrangements was small enough to count by listing them all. To compute probabilities like this in general, we need a formula. The formula is given in the following box. A derivation is presented in Section 2.2.

Assume n independent trials are performed, each of which results in one of k possible outcomes. Let $x_1, \ldots, x_k$ be the numbers of trials resulting in outcomes $1, 2, \ldots, k$, respectively. The number of arrangements of the outcomes among the n trials is

$$\frac{n!}{x_1!\, x_2! \cdots x_k!}$$

We can now specify the multinomial probability mass function.

If $X_1, \ldots, X_k \sim \text{MN}(n, p_1, \ldots, p_k)$, then the probability mass function of $X_1, \ldots, X_k$ is

$$p(x_1, \ldots, x_k) = P(X_1 = x_1, \ldots, X_k = x_k)$$
$$= \begin{cases} \dfrac{n!}{x_1!\, x_2! \cdots x_k!} p_1^{x_1} p_2^{x_2} \cdots p_k^{x_k} & x_i = 0, 1, 2, \ldots, n \\ & \text{and } \sum x_i = n \\ 0 & \text{otherwise} \end{cases} \tag{4.22}$$

Note that the multinomial distribution differs from the other distributions that we have studied in that it concerns several random variables simultaneously. We can express this fact by saying that $p(x_1, \ldots, x_k)$ is the **joint probability mass function** of $X_1, \ldots, X_k$. Joint probability mass functions are discussed more fully in Section 2.6.

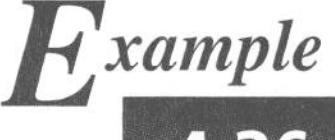

Example 4.36

Alkaptonuria is a genetic disease that results in the lack of an enzyme necessary to break down homogentisic acid. Some people are carriers of alkaptonuria, which means that they do not have the disease themselves, but they can potentially transmit it to their offspring. According to the laws of genetic inheritance, an offspring both of whose parents are carriers of alkaptonuria has probability 0.25 of being unaffected, 0.5 of being a carrier, and 0.25 of having the disease. In a sample of 10 offspring of carriers of alkaptonuria, what is the probability that 3 are unaffected, 5 are carriers, and 2 have the disease?

Solution

Let X_1, X_2, X_3 denote the numbers among the 10 offspring who are unaffected, carriers, and diseased, respectively. Then $X_1, X_2, X_3 \sim \text{MN}(10, 0.25, 0.50, 0.25)$. It follows from Equation (4.22) that

$$\begin{aligned} P(X_1 = 3, X_2 = 5, X_3 = 2) &= \frac{10!}{3!\,5!\,2!}(0.25)^3(0.50)^5(0.25)^2 \\ &= (2520)(0.015625)(0.03125)(0.0625) \\ &= 0.07690 \end{aligned}$$

Sometimes we want to focus on only one of the possible outcomes of a multinomial trial. In this situation, we can consider the outcome of interest a "success," and any other outcome a "failure." In this way it can be seen that the number of occurrences of any particular outcome has a binomial distribution.

> If $X_1, \ldots, X_k \sim \text{MN}(n, p_1, \ldots, p_k)$, then for each i
>
> $$X_i \sim \text{Bin}(n, p_i)$$

Example 4.37

Refer to Example 4.36. Find the probability that exactly 4 of 10 offspring are unaffected.

Solution

Let X represent the number of unaffected offspring in a sample of 10. Then $X \sim \text{Bin}(10, 0.25)$, so

$$\begin{aligned} P(X = 4) &= \frac{10!}{4!\,6!}(0.25)^4(0.75)^6 \\ &= 0.1460 \end{aligned}$$

Exercises for Section 4.4

1. Fifteen cars have been brought into a dealership for warranty work. Assume that 5 of the cars have serious engine problems, while the other 10 have minor problems. Six cars are chosen at random to be worked on. What is the probability that 2 of them have serious problems?

2. A shipment contains 40 items. Five items will be sampled at random and tested. If two or more are defective, the shipment will be returned.
   a. If the shipment in fact contains five defective items, what is the probability that it will be accepted (not returned)?
   b. If the shipment in fact contains 10 defective items, what is the probability that it will be returned?

3. The probability that a computer running a certain operating system crashes on any given day is 0.1. Find the probability that the computer crashes for the first time on the twelfth day after the operating system is installed.

4. A traffic light at a certain intersection is green 50% of the time, yellow 10% of the time, and red 40% of the time. A car approaches this intersection once each day. Let X represent the number of days that pass up to and including the first time the car encounters a red light. Assume that each day represents an independent trial.
   a. Find $P(X = 3)$.
   b. Find $P(X \leq 3)$.
   c. Find μ_X.
   d. Find σ_X^2.

5. Refer to Exercise 4. Let Y denote the number of days up to and including the third day on which a red light is encountered.
   a. Find $P(Y = 7)$.
   b. Find μ_Y.
   c. Find σ_Y^2.

6. Refer to Exercise 4. What is the probability that in a sequence of 10 days, four green lights, one yellow light, and five red lights are encountered?

7. If $X \sim \text{Geom}(p)$, what is the most probable value of X?
   i. 0
   ii. $1/p$
   iii. p
   iv. 1
   v. $(1 - p)/p^2$

8. A process that fills packages is stopped whenever a package is detected whose weight falls outside the specification. Assume that each package has probability 0.01 of falling outside the specification and that the weights of the packages are independent.
   a. Find the mean number of packages that will be filled before the process is stopped.
   b. Find the variance of the number of packages that will be filled before the process is stopped.
   c. Assume that the process will not be stopped until four packages whose weight falls outside the specification are detected. Find the mean and variance of the number of packages that will be filled before the process is stopped.

9. In a lot of 10 microcircuits, 3 are defective. Four microcircuits are chosen at random to be tested. Let X denote the number of tested circuits that are defective.
   a. Find $P(X = 2)$.
   b. Find μ_X.
   c. Find σ_X.

10. Of customers ordering a certain type of personal computer, 20% order an upgraded graphics card, 30% order extra memory, 15% order both the upgraded graphics card and extra memory, and 35% order neither. Fifteen orders are selected at random. Let X_1, X_2, X_3, X_4 denote the respective numbers of orders in the four given categories.
    a. Find $P(X_1 = 3, X_2 = 4, X_3 = 2, \text{ and } X_4 = 6)$.
    b. Find $P(X_1 = 3)$.

11. A certain make of car comes equipped with an engine in one of four sizes (in liters): 2.8, 3.0, 3.3, or 3.8. Ten percent of customers order the 2.8 liter engine, 40% order the 3.0, 30% order the 3.3, and 20% order the 3.8. A random sample of 20 orders is selected for audit.

a. What is the probability that the numbers of orders for the 2.8, 3.0, 3.3, and 3.8 liter engines are 3, 7, 6, and 4, respectively?

b. What is the probability that more than 10 orders are for 3.0 liter engines?

12. A thermocouple placed in a certain medium produces readings within 0.1°C of the true temperature 70% of the time, readings more than 0.1°C above the true temperature 10% of the time, and readings more than 0.1°C below the true temperature 20% of the time.

a. In a series of 10 independent readings, what is the probability that 5 are within 0.1°C of the true temperature, 2 are more than 0.1°C above, and 3 are more than 0.1°C below?

b. What is the probability that more than 8 of the readings are within 0.1°C of the true temperature?

4.5 The Normal Distribution

The **normal distribution** (also called the **Gaussian distribution**) is by far the most commonly used distribution in statistics. This distribution provides a good model for many, although not all, continuous populations. Part of the reason for this is the Central Limit Theorem, which we shall discuss in Section 4.10.

The normal distribution is continuous rather than discrete. The mean of a normal random variable may have any value, and the variance may have any positive value. The probability density function of a normal random variable with mean μ and variance σ^2 is given by

$$f(x) = \frac{1}{\sigma\sqrt{2\pi}} e^{-(x-\mu)^2/2\sigma^2} \tag{4.23}$$

At the end of the section, we derive the fact that μ and σ^2 are the mean and variance, respectively. If X is a random variable whose probability density function is normal with mean μ and variance σ^2, we write $X \sim N(\mu, \sigma^2)$.

Summary

If $X \sim N(\mu, \sigma^2)$, then the mean and variance of X are given by

$$\mu_X = \mu$$

$$\sigma_X^2 = \sigma^2$$

Figure 4.4 (page 232) presents a plot of the normal probability density function with mean μ and standard deviation σ. The normal probability density function is sometimes called the **normal curve.** Note that the normal curve is symmetric around μ, so that μ is the median as well as the mean. It is also the case that for any normal population

- About 68% of the population is in the interval $\mu \pm \sigma$.
- About 95% of the population is in the interval $\mu \pm 2\sigma$.
- About 99.7% of the population is in the interval $\mu \pm 3\sigma$.

The proportion of a normal population that is within a given number of standard deviations of the mean is the same for any normal population. For this reason, when dealing with normal populations, we often convert from the units in which the population items were originally measured to **standard units**. Standard units tell how many standard deviations an observation is from the population mean.

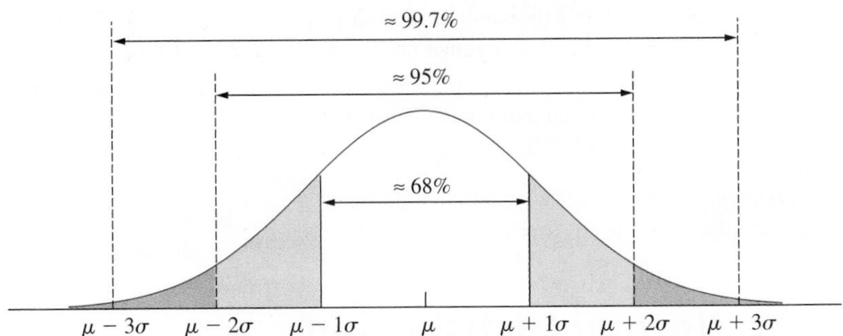

FIGURE 4.4 Probability density function of a normal random variable with mean μ and variance σ^2.

Example 4.38

Assume that the heights in a population of women follow the normal curve with mean $\mu = 64$ inches and standard deviation $\sigma = 3$ inches. The heights of two randomly chosen women are 67 inches and 62 inches. Convert these heights to standard units.

Solution

A height of 67 inches is 3 inches more than the mean of 64, and 3 inches is equal to one standard deviation. So 67 inches is one standard deviation above the mean and is thus equivalent to one standard unit. A height of 62 inches is 0.67 standard deviations below the mean, so 62 inches is equivalent to −0.67 standard units.

In general, we convert to standard units by subtracting the mean and dividing by the standard deviation. Thus, if x is an item sampled from a normal population with mean μ and variance σ^2, the standard unit equivalent of x is the number z, where

$$z = \frac{x - \mu}{\sigma} \tag{4.24}$$

The number z is sometimes called the "z-score" of x. The z-score is an item sampled from a normal population with mean 0 and standard deviation 1. This normal population is called the **standard normal population**.

Example 4.39

Aluminum sheets used to make beverage cans have thicknesses (in thousandths of an inch) that are normally distributed with mean 10 and standard deviation 1.3. A particular sheet is 10.8 thousandths of an inch thick. Find the z-score.

Solution

The quantity 10.8 is an observation from a normal population with mean $\mu = 10$ and standard deviation $\sigma = 1.3$. Therefore

$$z = \frac{10.8 - 10}{1.3}$$
$$= 0.62$$

Example 4.40

Refer to Example 4.39. The thickness of a certain sheet has a z-score of -1.7. Find the thickness of the sheet in the original units of thousandths of inches.

Solution

We use Equation (4.24), substituting -1.7 for z and solving for x. We obtain

$$-1.7 = \frac{x - 10}{1.3}$$

Solving for x yields $x = 7.8$. The sheet is 7.8 thousandths of an inch thick.

The proportion of a normal population that lies within a given interval is equal to the area under the normal probability density above that interval. This would suggest that we compute these proportions by integrating the normal probability density given in Equation (4.23). Interestingly enough, areas under this curve cannot be found by the method, taught in elementary calculus, of finding the antiderivative of the function and plugging in the limits of integration. This is because the antiderivative of this function is an infinite series and cannot be written down exactly. Instead, areas under this curve must be approximated numerically.

Areas under the standard normal curve (mean 0, variance 1) have been extensively tabulated. A typical such table, called a **standard normal table**, or **z table**, is given as Table A.2 (in Appendix A). To find areas under a normal curve with a different mean and variance, we convert to standard units and use the z table. Table A.2 provides areas in the left-hand tail of the curve for values of z. Other areas can be calculated by subtraction or by using the fact that the total area under the curve is equal to 1. We now present several examples to illustrate the use of the z table.

Example 4.41

Find the area under the normal curve to the left of $z = 0.47$.

Solution

From the z table, the area is 0.6808. See Figure 4.5.

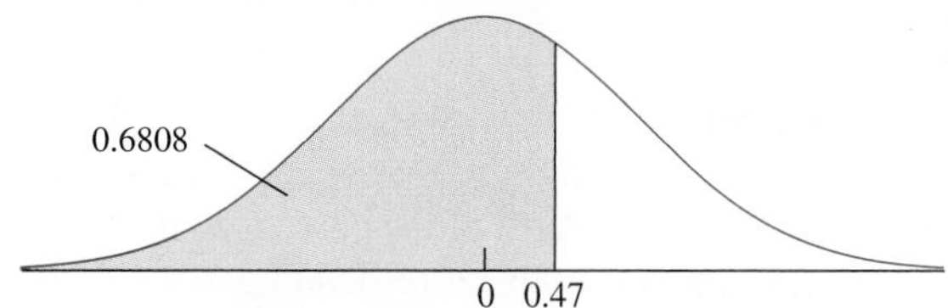

FIGURE 4.5 Solution to Example 4.41.

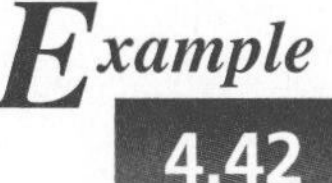

Example 4.42

Find the area under the normal curve to the right of $z = 1.38$.

Solution

From the z table, the area to the *left* of $z = 1.38$ is 0.9162. Therefore the area to the right is $1 - 0.9162 = 0.0838$. See Figure 4.6.

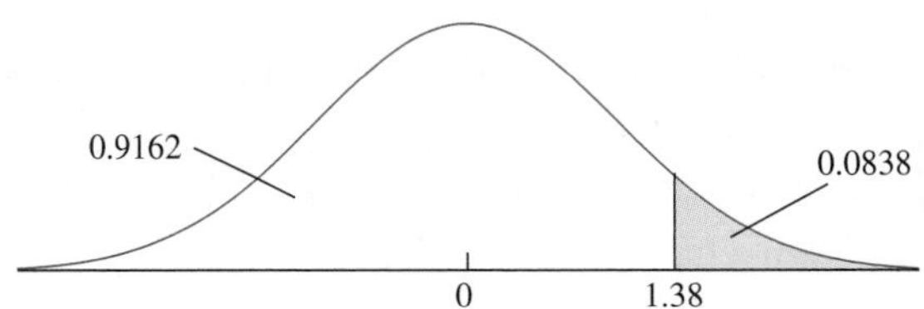

FIGURE 4.6 Solution to Example 4.42.

Example 4.43

Find the area under the normal curve between $z = 0.71$ and $z = 1.28$.

Solution

From the z table, the area to the left of $z = 1.28$ is 0.8997. The area to the left of $z = 0.71$ is 0.7611. The area between $z = 0.71$ and $z = 1.28$ is therefore $0.8997 - 0.7611 = 0.1386$. See Figure 4.7.

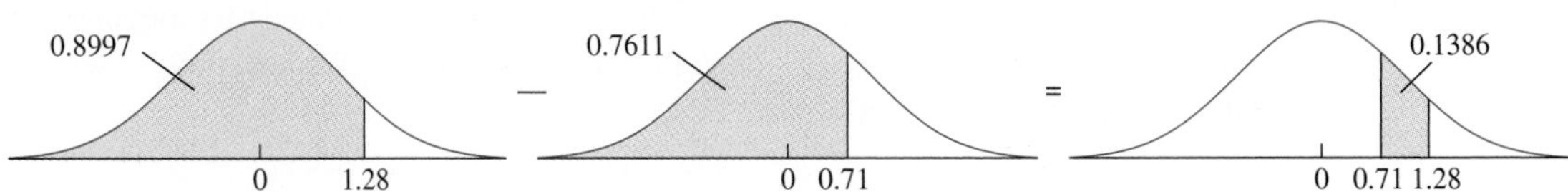

FIGURE 4.7 Solution to Example 4.43.

Example 4.44

What z-score corresponds to the 75th percentile of a normal curve? The 25th percentile? The median?

Solution

To answer this question, we use the z table in reverse. We need to find the z-score for which 75% of the area of the curve is to the left. From the body of the table, the closest area to 75% is 0.7486, corresponding to a z-score of 0.67. Therefore the 75th percentile is approximately 0.67. By the symmetry of the curve, the 25th percentile is $z = -0.67$ (this can also be looked up in the table directly). See Figure 4.8. The median is $z = 0$.

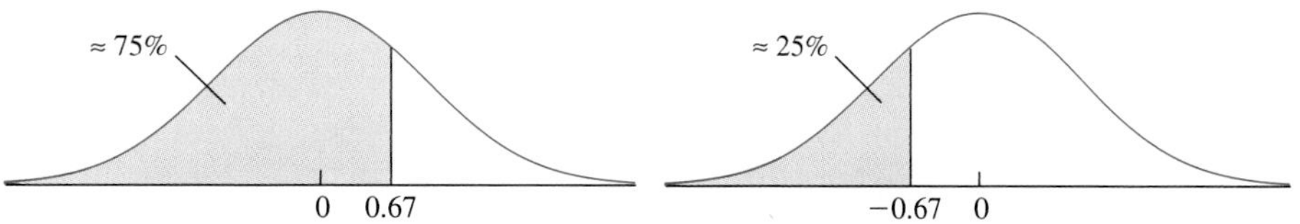

FIGURE 4.8 Solution to Example 4.44.

Example **4.45**

Lifetimes of batteries in a certain application are normally distributed with mean 50 hours and standard deviation 5 hours. Find the probability that a randomly chosen battery lasts between 42 and 52 hours.

Solution

Let X represent the lifetime of a randomly chosen battery. Then $X \sim N(50, 5^2)$. Figure 4.9 presents the probability density function of the $N(50, 5^2)$ population. The shaded area represents $P(42 < X < 52)$, the probability that a randomly chosen battery has a lifetime between 42 and 52 hours. To compute the area, we will use the z table. First we need to convert the quantities 42 and 52 to standard units. We have

$$z = \frac{42 - 50}{5} = -1.60 \qquad z = \frac{52 - 50}{5} = 0.40$$

From the z table, the area to the left of $z = -1.60$ is 0.0548, and the area to the left of $z = 0.40$ is 0.6554. The probability that a battery has a lifetime between 42 and 52 hours is $0.6554 - 0.0548 = 0.6006$.

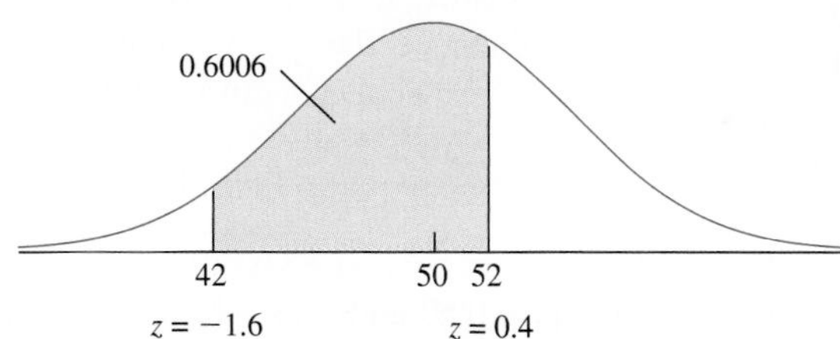

FIGURE 4.9 Solution to Example 4.45.

Example **4.46**

Refer to Example 4.45. Find the 40th percentile of battery lifetimes.

Solution

From the z table, the closest area to 0.4000 is 0.4013, corresponding to a z-score of -0.25. The population of lifetimes has mean 50 and standard deviation 5. The 40th percentile is the point 0.25 standard deviations below the mean. We find this value by converting the z-score to a raw score, using Equation (4.24):

$$-0.25 = \frac{x - 50}{5}$$

Solving for x yields $x = 48.75$. The 40th percentile of battery lifetimes is 48.75 hours. See Figure 4.10.

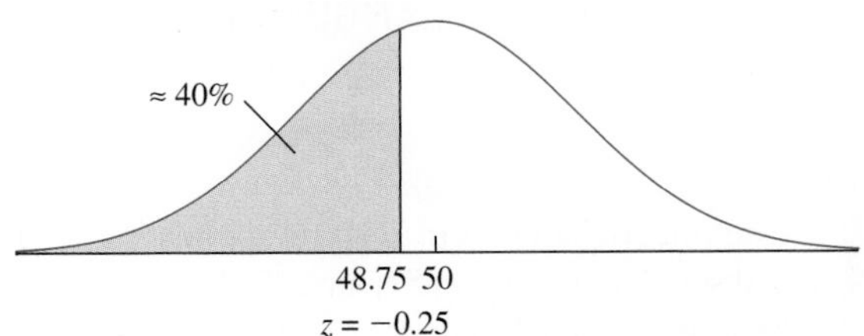

FIGURE 4.10 Solution to Example 4.46.

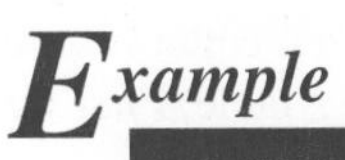

Example 4.47

A process manufactures ball bearings whose diameters are normally distributed with mean 2.505 cm and standard deviation 0.008 cm. Specifications call for the diameter to be in the interval 2.5 ± 0.01 cm. What proportion of the ball bearings will meet the specification?

Solution

Let X represent the diameter of a randomly chosen ball bearing. Then $X \sim N(2.505, 0.008^2)$. Figure 4.11 presents the probability density function of the $N(2.505, 0.008^2)$ population. The shaded area represents $P(2.49 < X < 2.51)$, which is the proportion of ball bearings that meet the specification.

We compute the z-scores of 2.49 and 2.51:

$$z = \frac{2.49 - 2.505}{0.008} = -1.88 \qquad z = \frac{2.51 - 2.505}{0.008} = 0.63$$

The area to the left of $z = -1.88$ is 0.0301. The area to the left of $z = 0.63$ is 0.7357. The area between $z = 0.63$ and $z = -1.88$ is $0.7357 - 0.0301 = 0.7056$. Approximately 70.56% of the diameters will meet the specification.

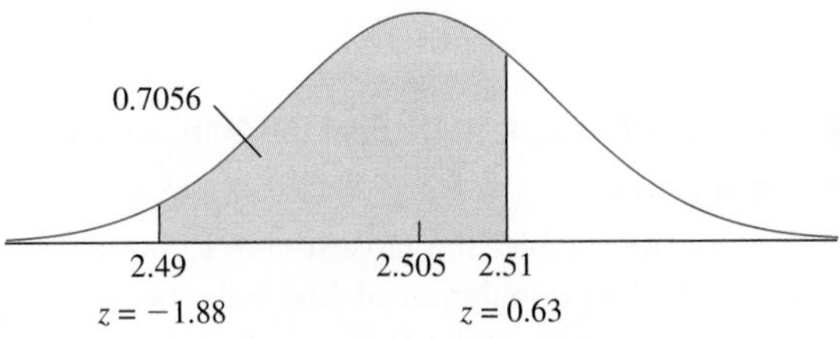

FIGURE 4.11 Solution to Example 4.47.

Refer to Example 4.47. The process can be recalibrated so that the mean will equal 2.5 cm, the center of the specification interval. The standard deviation of the process remains 0.008 cm. What proportion of the diameters will meet the specifications?

Solution

The method of solution is the same as in Example 4.47. The mean is 2.500 rather than 2.505. The calculations are as follows:

$$z = \frac{2.49 - 2.50}{0.008} = -1.25 \qquad z = \frac{2.51 - 2.50}{0.008} = 1.25$$

The area to the left of $z = -1.25$ is 0.1056. The area to the left of $z = 1.25$ is 0.8944. The area between $z = 1.25$ and $z = -1.25$ is $0.8944 - 0.1056 = 0.7888$. See Figure 4.12. Recalibrating will increase the proportion of diameters that meet the specification to 78.88%.

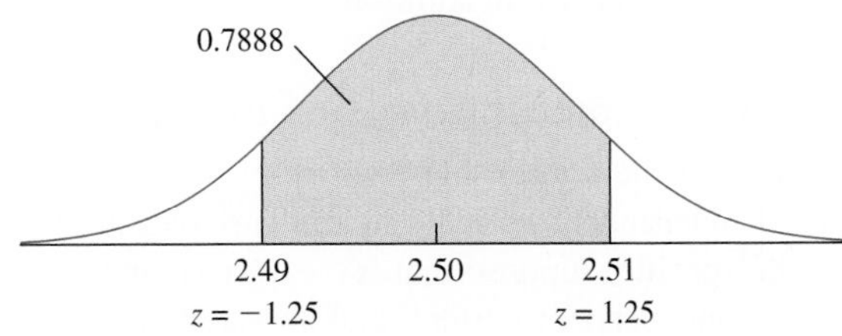

FIGURE 4.12 Solution to Example 4.48.

Refer to Examples 4.47 and 4.48. Assume that the process has been recalibrated so that the mean diameter is now 2.5 cm. To what value must the standard deviation be lowered so that 95% of the diameters will meet the specification?

Solution

The specification interval is 2.49–2.51 cm. We must find a value for σ so that this interval spans the middle 95% of the population of ball bearing diameters. See Figure 4.13 (page 238). The z-score that has 2.5% of the area to its left is $z = -1.96$. The z-score that has 2.5% of the area to its right is $z = 1.96$ (this follows from the symmetry of the curve). It follows that the lower specification limit, 2.49, has a z-score of -1.96, while the upper limit of 2.51 has a z-score of 1.96. Either of these facts may be used to find σ. From Equation (4.24),

$$1.96 = \frac{2.51 - 2.50}{\sigma}$$

Solving for σ yields $\sigma = 0.0051$ cm.

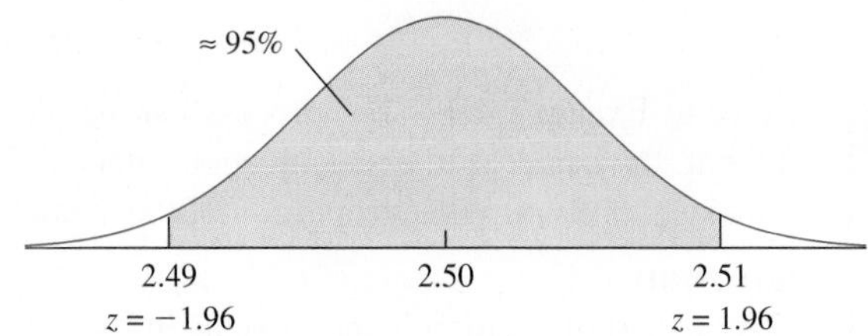

FIGURE 4.13 Solution to Example 4.49. If $\sigma = 0.0051$, then approximately 95% of the population will fall between 2.49 and 2.51.

Estimating the Parameters of a Normal Distribution

The parameters μ and σ of a normal distribution represent its mean and variance, respectively. Therefore, if $X_1, \ldots, X_n$ are a random sample from a $N(\mu, \sigma^2)$ distribution, μ is estimated with the sample mean $\overline{X}$ and σ^2 is estimated with the sample variance s^2. As with any sample mean, the uncertainty in $\overline{X}$ is $\sigma/\sqrt{n}$, which we replace with $s/\sqrt{n}$ if σ is unknown. In addition $\mu_{\overline{X}} = \mu$, so $\overline{X}$ is an unbiased estimate of μ.

Linear Combinations of Independent Normal Random Variables

One of the remarkable features of the normal distribution is that linear combinations of independent normal random variables are themselves normal random variables. To be specific, suppose that $X_1 \sim N(\mu_1, \sigma_1^2)$, $X_2 \sim N(\mu_2, \sigma_2^2)$, $\ldots$, $X_n \sim N(\mu_n, \sigma_n^2)$ are independent normal random variables. Note that the means and variances of these random variables can differ from one another. Let $c_1, c_2, \ldots, c_n$ be constants. Then the linear combination $c_1X_1 + c_2X_2 + \cdots + c_nX_n$ is a normally distributed random variable. The mean and variance of the linear combination are $c_1\mu_1 + c_2\mu_2 + \cdots + c_n\mu_n$ and $c_1^2\sigma_1^2 + c_2^2\sigma_2^2 + \cdots + c_n^2\sigma_n^2$, respectively (see Equations 2.49 and 2.53 in Section 2.5).

Summary

Let $X_1, X_2, \ldots, X_n$ be independent and normally distributed with means $\mu_1, \mu_2, \ldots, \mu_n$ and variances $\sigma_1^2, \sigma_2^2, \ldots, \sigma_n^2$. Let $c_1, c_2, \ldots, c_n$ be constants, and $c_1X_1 + c_2X_2 + \cdots + c_nX_n$ be a linear combination. Then

$$c_1X_1+c_2X_2+\cdots+c_nX_n \sim N(c_1\mu_1+c_2\mu_2+\cdots+c_n\mu_n,\ c_1^2\sigma_1^2+c_2^2\sigma_2^2+\cdots+c_n^2\sigma_n^2) \tag{4.25}$$

Example 4.50

In the article "Advances in Oxygen Equivalent Equations for Predicting the Properties of Titanium Welds" (D. Harwig, W. Ittiwattana, and H. Castner, *The Welding Journal*, 2001:126s–136s), the authors propose an oxygen equivalence equation to predict the strength, ductility, and hardness of welds made from nearly pure titanium. The equation is $E = 2C + 3.5N + O$, where E is the oxygen equivalence, and C, N, and O are the proportions by weight, in parts per million, of carbon, nitrogen, and oxygen, respectively (a constant term involving iron content has been omitted). Assume that for a particular grade of commercially pure titanium, the quantities C, N, and O are approximately independent and normally distributed with means $\mu_C = 150$, $\mu_N = 200$, $\mu_O = 1500$, and standard deviations $\sigma_C = 30$, $\sigma_N = 60$, $\sigma_O = 100$. Find the distribution of E. Find $P(E > 3000)$.

Solution

Since E is a linear combination of independent normal random variables, its distribution is normal. We must now find the mean and variance of E. Using Equation (4.25), we compute

$$\begin{aligned}\mu_E &= 2\mu_C + 3.5\mu_N + 1\mu_O\\ &= 2(150) + 3.5(200) + 1(1500)\\ &= 2500\\ \sigma_E^2 &= 2^2\sigma_C^2 + 3.5^2\sigma_N^2 + 1^2\sigma_O^2\\ &= 2^2(30^2) + 3.5^2(60^2) + 1^2(100^2)\\ &= 57{,}700\end{aligned}$$

We conclude that $E \sim N(2500,\ 57{,}700)$.

To find $P(E > 3000)$, we compute the z-score: $z = (3000 - 2500)/\sqrt{57{,}700} = 2.08$. The area to the right of $z = 2.08$ under the standard normal curve is 0.0188. So $P(E > 3000) = 0.0188$.

If $X_1, \ldots, X_n$ is a random sample from any population with mean μ and variance σ^2, then the sample mean $\overline{X}$ has mean $\mu_{\overline{X}} = \mu$ and variance $\sigma^2_{\overline{X}} = \sigma^2/n$. If the population is normal, then $\overline{X}$ is normal as well, because it is a linear combination of $X_1, \ldots, X_n$ with coefficients $c_1 = \cdots = c_n = 1/n$.

Summary

Let $X_1, \ldots, X_n$ be independent and normally distributed with mean μ and variance σ^2. Then

$$\overline{X} \sim N\left(\mu,\ \frac{\sigma^2}{n}\right) \tag{4.26}$$

Other important linear combinations are the sum and difference of two random variables. If X and Y are independent normal random variables, the sum $X + Y$ and the difference $X - Y$ are linear combinations. The distributions of $X + Y$ and $X - Y$ can be determined by using Equation (4.25) with $c_1 = 1$, $c_2 = 1$ for $X + Y$ and $c_1 = 1$, $c_2 = -1$ for $X - Y$.

Summary

Let X and Y be independent, with $X \sim N(\mu_X, \sigma_X^2)$ and $Y \sim N(\mu_Y, \sigma_Y^2)$. Then

$$X + Y \sim N(\mu_X + \mu_Y, \sigma_X^2 + \sigma_Y^2) \tag{4.27}$$

$$X - Y \sim N(\mu_X - \mu_Y, \sigma_X^2 + \sigma_Y^2) \tag{4.28}$$

How Can I Tell Whether My Data Come from a Normal Population?

In practice, we often have a sample from some population, and we must use the sample to decide whether the population distribution is close to normal. If the sample is reasonably large, the sample histogram may give a good indication. Large samples from normal populations have histograms that look something like the normal density function—peaked in the center, and decreasing more or less symmetrically on either side. Probability plots, which will be discussed in Section 4.9, provide another good way of determining whether a reasonably large sample comes from a population that is approximately normal. For small samples, it can be difficult to tell whether the normal distribution is appropriate. One important fact is this: *Samples from normal populations rarely contain outliers.* Therefore the normal distribution should generally not be used for data sets that contain outliers. This is especially true when the sample size is small. Unfortunately, for small data sets that do not contain outliers, it is difficult to determine whether the population is approximately normal. In general, some knowledge of the process that generated the data is needed.

Derivation of the Mean and Variance for a Normal Random Variable

Let $X \sim N(\mu, \sigma^2)$. We show that $\mu_X = \mu$ and $\sigma_X^2 = \sigma^2$. Using the definition of the population mean of a continuous random variable (Equation 2.35 in Section 2.4),

$$\mu_X = \int_{-\infty}^{\infty} \frac{1}{\sigma\sqrt{2\pi}} x e^{-(x-\mu)^2/2\sigma^2}\, dx$$

Make the substitution $z = (x - \mu)/\sigma$. Then $x = \sigma z + \mu$, and $dx = \sigma dz$. We obtain

$$\begin{aligned}\mu_X &= \int_{-\infty}^{\infty} \frac{1}{\sigma\sqrt{2\pi}} (\sigma z + \mu)\sigma e^{-z^2/2}\, dz \\ &= \sigma \int_{-\infty}^{\infty} \frac{1}{\sqrt{2\pi}} z e^{-z^2/2}\, dz + \mu \int_{-\infty}^{\infty} \frac{1}{\sqrt{2\pi}} e^{-z^2/2}\, dz\end{aligned}$$

Direct computation shows that

$$\int_{-\infty}^{\infty} \frac{1}{\sqrt{2\pi}} z e^{-z^2/2}\, dz = 0$$

Also,

$$\int_{-\infty}^{\infty} \frac{1}{\sqrt{2\pi}} e^{-z^2/2}\, dz = 1$$

since it is the integral of the $N(0,\ 1)$ probability density function over all its possible values.

Therefore

$$\mu_X = \sigma(0) + \mu(1) = \mu$$

To show that $\sigma_X^2 = \sigma^2$, we use Equation (2.36) (in Section 2.4):

$$\sigma_X^2 = \int_{-\infty}^{\infty} (x - \mu_X)^2 \frac{1}{\sigma\sqrt{2\pi}} e^{-(x-\mu)^2/2\sigma^2}\, dx$$

Make the substitution $z = (x - \mu)/\sigma$. Recall that $\mu_X = \mu$. Then

$$\begin{aligned}\sigma_X^2 &= \int_{-\infty}^{\infty} \sigma^2 z^2 \frac{1}{\sigma\sqrt{2\pi}} \sigma e^{-z^2/2}\, dz \\ &= \sigma^2 \int_{-\infty}^{\infty} \frac{1}{\sqrt{2\pi}} z^2 e^{-z^2/2}\, dz\end{aligned}$$

Integrating by parts twice shows that

$$\int_{-\infty}^{\infty} \frac{1}{\sqrt{2\pi}} z^2 e^{-z^2/2}\, dz = 1$$

Therefore $\sigma_X^2 = \sigma^2$.

Exercises for Section 4.5

1. Find the area under the normal curve

a. To the right of $z = -0.85$.

b. Between $z = 0.40$ and $z = 1.30$.

c. Between $z = -0.30$ and $z = 0.90$.

d. Outside $z = -1.50$ to $z = -0.45$.

2. Find the area under the normal curve

a. To the left of $z = 0.56$.

b. Between $z = -2.93$ and $z = -2.06$.

c. Between $z = -1.08$ and $z = 0.70$.

d. Outside $z = 0.96$ to $z = 1.62$.

3. Scores on a standardized test are approximately normally distributed with a mean of 480 and a standard deviation of 90.

a. What proportion of the scores are above 700?

b. What is the 25th percentile of the scores?

c. If someone's score is 600, what percentile is she on?

d. What proportion of the scores are between 420 and 520?

4. Assume heights of women in a population follow the normal curve with mean 64.3 in. and standard deviation 2.6 in.

a. What proportion of women have heights between 60 and 66 in.?

b. A certain woman has a height 0.5 standard deviations above the mean. What proportion of women are taller than she is?

c. How tall is a woman whose height is on the 90th percentile?

d. A woman is chosen at random from this population. What is the probability that she is more than 67 in. tall?

e. Five women are chosen at random from this population. What is the probability that exactly one of them is more than 67 in. tall?

5. The strength of an aluminum alloy is normally distributed with mean 10 gigapascals (GPa) and standard deviation 1.4 GPa.

a. What is the probability that a specimen of this alloy will have a strength greater than 12 GPa?

b. Find the first quartile of the strengths of this alloy.

c. Find the 95th percentile of the strengths of this alloy.

6. In a certain university, math SAT scores in the entering freshman class averaged 650 and had a standard deviation of 100. The maximum possible score is 800. Is it possible that the histogram for the scores of these freshmen follows the normal curve? Explain.

7. Penicillin is produced by the *Penicillium* fungus, which is grown in a broth whose sugar content must be carefully controlled. The optimum sugar concentration is 4.9 mg/mL. If the concentration exceeds 6.0 mg/mL, the fungus dies and the process must be shut down for the day.

a. If sugar concentration in batches of broth is normally distributed with mean 4.9 mg/mL and standard deviation 0.6 mg/mL, on what proportion of days will the process shut down?

b. The supplier offers to sell broth with a sugar content that is normally distributed with mean 5.2 mg/mL and standard deviation 0.4 mg/mL. Will this broth result in fewer days of production lost? Explain.

8. A chromatography method used to purify a protein also destroys some of it, in a process called denaturation. A particular method recovers a mean of 55% (0.55) of the protein and has a standard deviation of 0.15. The amount recovered is normally distributed.

a. In a certain industrial process, one cannot afford to have a recovery less than 0.30 more than 5% of the time. Does this process meet this requirement? Explain.

b. In another process, the recovery must be more than 0.50 at least 95% of the time. If the mean recovery is normally distributed with a mean of 0.60, what is the largest value the standard deviation can have that will meet the requirement?

9. A cylindrical hole is drilled in a block, and a cylindrical piston is placed in the hole. The clearance is equal to one-half the difference between the diameters of the hole and the piston. The diameter of the hole is normally distributed with mean 15 cm and standard deviation 0.025 cm, and the diameter of the piston is normally distributed with mean 14.88 cm and standard deviation 0.015 cm.

a. Find the mean clearance.

b. Find the standard deviation of the clearance.

c. What is the probability that the clearance is less than 0.05 cm?

d. Find the 25th percentile of the clearance.

e. Specifications call for the clearance to be between 0.05 and 0.09 cm. What is the probability that the clearance meets the specification?

f. It is possible to adjust the mean hole diameter. To what value should it be adjusted so as to maximize the probability that the clearance will be between 0.05 and 0.09 cm?

10. Shafts manufactured for use in optical storage devices have diameters that are normally distributed with mean $\mu = 0.652$ cm and standard deviation $\sigma = 0.003$ cm. The specification for the shaft diameter is 0.650 ± 0.005 cm.

a. What proportion of the shafts manufactured by this process meet the specifications?

b. The process mean can be adjusted through calibration. If the mean is set to 0.650 cm, what proportion of the shafts will meet specifications?

c. If the mean is set to 0.650 cm, what must the standard deviation be so that 99% of the shafts will meet specifications?

11. The fill volume of cans filled by a certain machine is normally distributed with mean 12.05 oz and standard deviation 0.03 oz.

a. What proportion of cans contain less than 12 oz?

b. The process mean can be adjusted through calibration. To what value should the mean be set so that 99% of the cans will contain 12 oz or more?

c. If the process mean remains at 12.05 oz, what must the standard deviation be so that 99% of the cans will contain 12 oz or more?

12. A film-coating process produces films whose thicknesses are normally distributed with a mean of 110 microns and a standard deviation of 10 microns. For a certain application, the minimum acceptable thickness is 90 microns.

a. What proportion of films will be too thin?

b. To what value should the mean be set so that only 1% of the films will be too thin?

c. If the mean remains at 110, what must the standard deviation be so that only 1% of the films will be too thin?

13. A fiber-spinning process currently produces a fiber whose strength is normally distributed with a mean of 75 N/m^2. The minimum acceptable strength is 65 N/m^2.

a. Ten percent of the fiber produced by the current method fails to meet the minimum specification. What is the standard deviation of fiber strengths in the current process?

b. If the mean remains at 75 N/m^2, what must the standard deviation be so that only 1% of the fiber will fail to meet the specification?

c. If the standard deviation is 5 N/m^2, to what value must the mean be set so that only 1% of the fiber will fail to meet the specification?

14. The quality-assurance program for a certain adhesive formulation process involves measuring how well the adhesive sticks a piece of plastic to a glass surface. When the process is functioning correctly, the adhesive strength X is normally distributed with a mean of 200 N and a standard deviation of 10 N. Each hour, you make one measurement of the adhesive strength. You are supposed to inform your supervisor if your measurement indicates that the process has strayed from its target distribution.

a. Find $P(X \leq 160)$, under the assumption that the process is functioning correctly.

b. Based on your answer to part (a), if the process is functioning correctly, would a strength of 160 N be unusually small? Explain.

c. If you observed an adhesive strength of 160 N, would this be convincing evidence that the process was no longer functioning correctly? Explain.

d. Find $P(X \geq 203)$, under the assumption that the process is functioning correctly.

e. Based on your answer to part (d), if the process is functioning correctly, would a strength of 203 N be unusually large? Explain.

f. If you observed an adhesive strength of 203 N, would this be convincing evidence that the process was no longer functioning correctly? Explain.

g. Find $P(X \leq 195)$, under the assumption that the process is functioning correctly.

h. Based on your answer to part (g), if the process is functioning correctly, would a strength of 195 N be unusually small? Explain.

i. If you observed an adhesive strength of 195 N, would this be convincing evidence that the process was no longer functioning correctly? Explain.

15. A light fixture holds two lightbulbs. Bulb A is of a type whose lifetime is normally distributed with mean 800 hours and standard deviation 100 hours. Bulb B has a lifetime that is normally distributed with mean 900 hours and standard deviation 150 hours. Assume that the lifetimes of the bulbs are independent.

a. What is the probability that bulb B lasts longer than bulb A?

b. What is the probability that bulb B lasts more than 200 hours longer than bulb A?

c. Another light fixture holds only one bulb. A bulb of type A is installed, and when it burns out, a bulb of type B is installed. What is the probability that the total lifetime of the two bulbs is more than 2000 hours?

16. The molarity of a solute in solution is defined to be the number of moles of solute per liter of solution (1 mole $= 6.02 \times 10^{23}$ molecules). If X is the molarity of a solution of sodium chloride (NaCl), and Y is the molarity of a solution of sodium carbonate (Na_2CO_3), the molarity of sodium ion (Na^+) in a solution made of equal parts NaCl and Na_2CO_3 is given by $M = 0.5X + Y$. Assume X and Y are independent and normally distributed, and that X has mean 0.450 and standard deviation 0.050, and Y has mean 0.250 and standard deviation 0.025.

a. What is the distribution of M?

b. Find $P(M > 0.5)$.

17. A company receives a large shipment of bolts. The bolts will be used in an application that requires a torque of 100 J. Before the shipment is accepted, a quality engineer will sample 12 bolts and measure the torque needed to break each of them. The shipment will be accepted if the engineer concludes that fewer than 1% of the bolts in the shipment have a breaking torque of less than 100 J.

a. If the 12 values are 107, 109, 111, 113, 113, 114, 114, 115, 117, 119, 122, 124, compute the sample mean and sample standard deviation.

b. Assume the 12 values are sampled from a normal population, and assume the the sample mean and standard deviation calculated in part (a) are actually the population mean and standard deviation. Compute the proportion of bolts whose breaking torque is less than 100 J. Will the shipment be accepted?

c. What if the 12 values had been 108, 110, 112, 114, 114, 115, 115, 116, 118, 120, 123, 140? Use the method outlined in parts (a) and (b) to determine whether the shipment would have been accepted.

d. Compare the sets of 12 values in parts (a) and (c). In which sample are the bolts stronger?

e. Is the method valid for both samples? Why or why not?

4.6 The Lognormal Distribution

For data that contain outliers, the normal distribution is generally not appropriate. The **lognormal** distribution, which is related to the normal distribution, is often a good choice for these data sets. The lognormal distribution is derived from the normal distribution as follows: If X is a normal random variable with mean μ and variance σ^2, then the random variable $Y = e^X$ is said to have the **lognormal distribution** with parameters μ and σ^2. Note that if Y has the lognormal distribution with parameters μ and σ^2, then $X = \ln Y$ has the normal distribution with mean μ and variance σ^2.

Summary

- If $X \sim N(\mu,\ \sigma^2)$, then the random variable $Y = e^X$ has the lognormal distribution with parameters μ and σ^2.
- If Y has the lognormal distribution with parameters μ and σ^2, then the random variable $X = \ln Y$ has the $N(\mu,\ \sigma^2)$ distribution.

The probability density function of a lognormal random variable with parameters μ and σ is

$$f(x) = \begin{cases} \dfrac{1}{\sigma x\sqrt{2\pi}} \exp\left[-\dfrac{1}{2\sigma^2}(\ln x - \mu)^2\right] & \text{if } x > 0 \\ 0 & \text{if } x \le 0 \end{cases} \tag{4.29}$$

Figure 4.14 presents a graph of the lognormal density function with parameters $\mu = 0$ and $\sigma = 1$. Note that the density function is highly skewed. This is the reason that

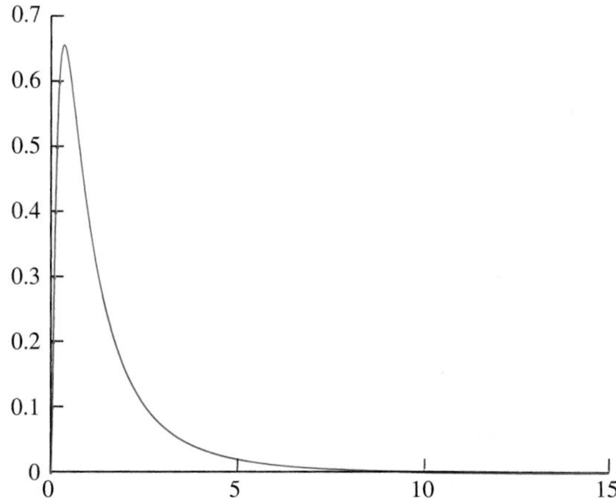

FIGURE 4.14 The probability density function of the lognormal distribution with parameters $\mu = 0$ and $\sigma = 1$.

the lognormal distribution is used to model processes that tend to produce occasional large values, or outliers.

It can be shown by advanced methods that if Y is a lognormal random variable with parameters μ and σ^2, then the mean $E(Y)$ and variance $V(Y)$ are given by

$$E(Y) = e^{\mu+\sigma^2/2} \qquad V(Y) = e^{2\mu+2\sigma^2} - e^{2\mu+\sigma^2} \tag{4.30}$$

Note that if Y has the lognormal distribution, the parameters μ and σ^2 *do not* refer to the mean and variance of Y. They refer instead to the mean and variance of the normal random variable $\ln Y$. In Equation (4.30) we used the notation $E(Y)$ instead of μ_Y, and $V(Y)$ instead of σ_Y^2, in order to avoid confusion with μ and σ.

Lifetimes of a certain component are lognormally distributed with parameters $\mu = 1$ day and $\sigma = 0.5$ days. Find the mean lifetime of these components. Find the standard deviation of the lifetimes.

Solution

Let Y represent the lifetime of a randomly chosen component. The mean of Y is found by Equation (4.30) to be $e^{1+0.5^2/2} = 3.08$ days. The variance is $e^{2(1)+2(0.5)^2} - e^{2(1)+0.5^2} = 2.6948$. The standard deviation is therefore $\sqrt{2.6948} = 1.64$ days.

To compute probabilities involving lognormal random variables, take the log and use the z table (Table A.2). Examples 4.52 and 4.53 illustrate the method.

Example 4.52

Refer to Example 4.51. Find the probability that a component lasts longer than four days.

Solution

Let Y represent the lifetime of a randomly chosen component. We need to find $P(Y > 4)$. We cannot use the z table for Y, because Y is not sampled from a normal population. However, $\ln Y$ is sampled from a normal population; specifically, $\ln Y \sim N(1, 0.5^2)$. We express $P(Y > 4)$ as a probability involving $\ln Y$:

$$P(Y > 4) = P(\ln Y > \ln 4) = P(\ln Y > 1.386)$$

The z-score of 1.386 is

$$\begin{aligned} z &= \frac{1.386 - 1.000}{0.5} \\ &= 0.77 \end{aligned}$$

From the z table, we find that $P(\ln Y > 1.386) = 0.2206$. (See Figure 4.15.) We conclude that approximately 22% of the components will last longer than four days.

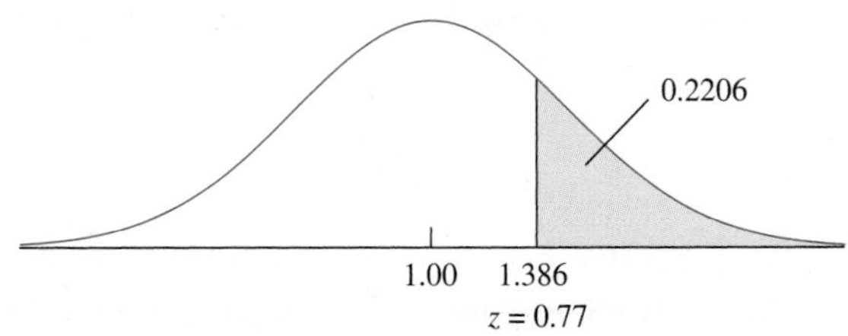

FIGURE 4.15 Solution to Example 4.52.

Example 4.53

Refer to Example 4.51. Find the median lifetime. Find the 80th percentile of the lifetimes.

Solution

Let Y represent the lifetime of a randomly chosen component. Let m denote the median lifetime. Then $P(Y \leq m) = 0.5$. Taking logs, we have $P(\ln Y \leq \ln m) = 0.5$. This means that $\ln m$ is the median of $\ln Y$. Now $\ln Y \sim N(1, 0.5^2)$. Therefore $\ln m = 1$, so $m = e^1 = 2.718$.

To find the 80th percentile, p_{80}, set $P(Y \leq p_{80}) = 0.80$. Then $P(\ln Y \leq \ln p_{80}) = 0.80$. This means that $\ln p_{80}$ is the 80th percentile of $\ln Y$. Now $\ln Y \sim N(1, 0.5^2)$. From the z table, the z-score of the 80th percentile is 0.84. Therefore $\ln p_{80} = 1 + (0.84)(0.5) = 1.42$, so $p_{80} = e^{1.42} = 4.14$.

Estimating the Parameters of a Lognormal Distribution

If Y is a random variable whose distribution is lognormal with parameters μ and σ^2, then μ and σ^2 are the mean and variance, respectively, of $\ln Y$. Therefore if $Y_1, \ldots, Y_n$ is a random sample from a lognormal population, we first transform to the log scale, defining $X_1 = \ln Y_1, \quad \ldots, \quad X_n = \ln Y_n$. Now $X_1, \ldots, X_n$ is a random sample from $N(\mu, \sigma^2)$. We estimate μ with $\overline{X}$ and σ^2 with the sample variance s_X^2. As with any sample mean, the uncertainty in $\overline{X}$ is $\sigma_{\overline{X}} = \sigma/\sqrt{n}$, and if σ is unknown, we estimate it with the sample standard deviation s_X.

The diameters (in mm) of seeds of a certain plant are lognormally distributed. A random sample of five seeds had diameters 1.52, 2.22, 2.64, 2.00, and 1.69. Estimate the parameters μ and σ.

Solution

To estimate μ and σ, we take logs of the five sample values, to obtain 0.419, 0.798, 0.971, 0.693, and 0.525. The sample mean is 0.681, and the sample standard deviation is 0.218. We therefore estimate $\widehat{\mu} = 0.681, \widehat{\sigma} = 0.218$.

How Can I Tell Whether My Data Come from a Lognormal Population?

As stated previously, samples from normal populations rarely contain outliers. In contrast, samples from lognormal populations often contain outliers in the right-hand tail. That is, the samples usually contain a few values that are much larger than the rest of the data. This of course reflects the long right-hand tail in the lognormal density function (Figure 4.14). For samples with outliers on the right, we *transform* the data, by taking the natural logarithm (or any logarithm) of each value. We then try to determine whether these logs come from a normal population, by plotting them on a histogram, or on a probability plot. Probability plots will be discussed in Section 4.9.

Note that the lognormal density has only one long tail, on the right. For this reason, samples from lognormal populations will have outliers on the right, but not on the left. The lognormal distribution should therefore not be used for samples with a few unusually small observations. In addition, lognormal populations contain only positive values, so the lognormal distribution should not be used for samples that contain zeros or negative values. Finally, it is important to note that the log transformation will not always produce a sample that is close to normal. One has to plot a histogram or probability plot (see Section 4.9) to check.

Figure 4.16 (page 248) presents two histograms. The first shows the monthly production for 255 gas wells, in units of thousand cubic feet. This histogram clearly has a long right-hand tail, so we conclude that the data do not come from a normal population. The second shows the natural logs of the monthly productions. This histogram is closer to the normal curve, although some departure from normality can be detected.

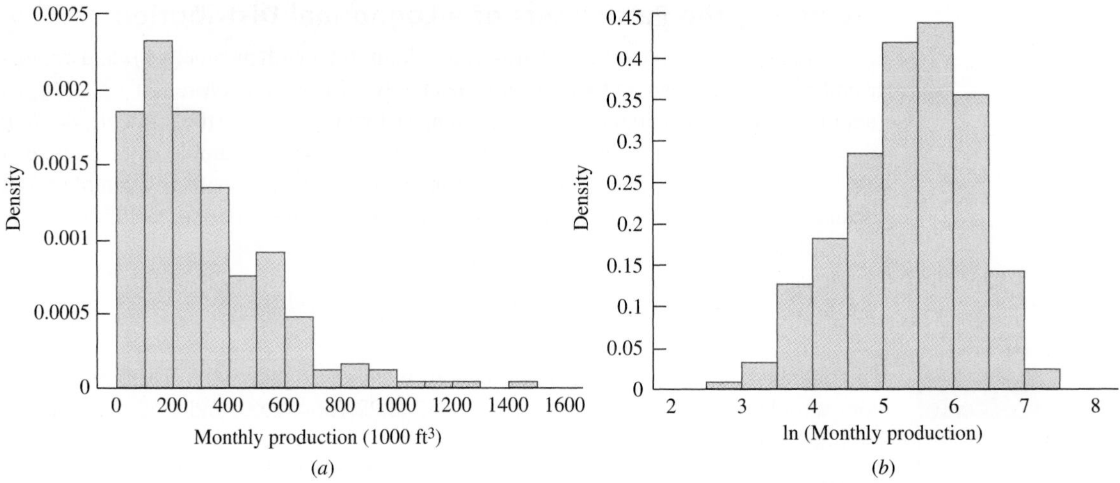

FIGURE 4.16 *(a)* A histogram showing monthly production for 255 gas wells. There is a long right-hand tail. *(b)* A histogram showing the natural logs of the monthly productions. The distribution of the logged data is much closer to normal.

Exercises for Section 4.6

1. The lifetime (in days) of a certain electronic component that operates in a high-temperature environment is lognormally distributed with $\mu = 1.2$ and $\sigma = 0.4$.
   a. Find the mean lifetime.
   b. Find the probability that a component lasts between three and six days.
   c. Find the median lifetime.
   d. Find the 90th percentile of the lifetimes.

2. When a pesticide comes into contact with skin, a certain percentage of it is absorbed. The percentage of the pesticide that will be absorbed during a given time period can be modeled with a lognormal distribution. Assume that for a given pesticide, the amount that is absorbed (in percent) in a two-hour time period is lognormally distributed with $\mu = 1.5$ and $\sigma = 0.5$.
   a. Find the mean percent absorbed.
   b. Find the median percent absorbed.
   c. Find the probability that the percent absorbed is more than 10.
   d. Find the probability that the percent absorbed is less than 5.
   e. Find the 75th percentile of the percent absorbed.
   f. Find the standard deviation of the percent absorbed.

3. The body mass index (BMI) of a person is defined to be the person's body mass divided by the square of the person's height. The article "Influences of Parameter Uncertainties within the ICRP 66 Respiratory Tract Model: Particle Deposition" (W. Bolch, E. Farfan, et al., *Health Physics*, 2001:378–394) states that body mass index (in kg/m^2) in men aged 25–34 is lognormally distributed with parameters $\mu = 3.215$ and $\sigma = 0.157$.
   a. Find the mean BMI for men aged 25–34.
   b. Find the standard deviation of BMI for men aged 25–34.
   c. Find the median BMI for men aged 25–34.
   d. What proportion of men aged 25–34 have a BMI less than 22?
   e. Find the 75th percentile of BMI for men aged 25–34.

4. The article "Stochastic Estimates of Exposure and Cancer Risk from Carbon Tetrachloride Released to the Air from the Rocky Flats Plant" (A. Rood,

P. McGavran, et al., *Risk Analysis*, 2001:675–695) models the increase in the risk of cancer due to exposure to carbon tetrachloride as lognormal with $\mu = -15.65$ and $\sigma = 0.79$.

a. Find the mean risk.
b. Find the median risk.
c. Find the standard deviation of the risk.
d. Find the 5th percentile.
e. Find the 95th percentile.

5. The article "Withdrawal Strength of Threaded Nails" (D. Rammer, S. Winistorfer, and D. Bender, *Journal of Structural Engineering* 2001:442–449) describes an experiment comparing the ultimate withdrawal strengths (in N/mm) for several types of nails. For an annularly threaded nail with shank diameter 3.76 mm driven into spruce-pine-fir lumber, the ultimate withdrawal strength was modeled as lognormal with $\mu = 3.82$ and $\sigma = 0.219$. For a helically threaded nail under the same conditions, the strength was modeled as lognormal with $\mu = 3.47$ and $\sigma = 0.272$.

a. What is the mean withdrawal strength for annularly threaded nails?
b. What is the mean withdrawal strength for helically threaded nails?
c. For which type of nail is it more probable that the withdrawal strength will be greater than 50 N/mm?
d. What is the probability that a helically threaded nail will have a greater withdrawal strength than the median for annularly threaded nails?
e. An experiment is performed in which withdrawal strengths are measured for several nails of both types. One nail is recorded as having a withdrawal strength of 20 N/mm, but its type is not given. Do you think it was an annularly threaded nail or a helically threaded nail? Why? How sure are you?

6. Choose the best answer, and explain. If X is a random variable with a lognormal distribution, then _______

i. the mean of X is always greater than the median.
ii. the mean of X is always less than the median.
iii. the mean may be greater than, less than, or equal to the median, depending on the value of σ.

7. The prices of stocks or other financial instruments are often modeled with a lognormal distribution. An investor is considering purchasing stock in one of two companies, A or B. The price of a share of stock today is \$1 for both companies. For company A, the value of the stock one year from now is modeled as lognormal with parameters $\mu = 0.05$ and $\sigma = 0.1$. For company B, the value of the stock one year from now is modeled as lognormal with parameters $\mu = 0.02$ and $\sigma = 0.2$.

a. Find the mean of the price of one share of company A one year from now.
b. Find the probability that the price of one share of company A one year from now will be greater than \$1.20.
c. Find the mean of the price of one share of company B one year from now.
d. Find the probability that the price of one share of company B one year from now will be greater than \$1.20.

8. A manufacturer claims that the tensile strength of a certain composite (in MPa) has the lognormal distribution with $\mu = 5$ and $\sigma = 0.5$. Let X be the strength of a randomly sampled specimen of this composite.

a. If the claim is true, what is $P(X < 20)$?
b. Based on the answer to part (a), if the claim is true, would a strength of 20 MPa be unusually small?
c. If you observed a tensile strength of 20 MPa, would this be convincing evidence that the claim is false? Explain.
d. If the claim is true, what is $P(X < 130)$?
e. Based on the answer to part (d), if the claim is true, would a strength of 130 MPa be unusually small?
f. If you observed a tensile strength of 130 MPa, would this be convincing evidence that the claim is false? Explain.

4.7 The Exponential Distribution

The **exponential distribution** is a continuous distribution that is sometimes used to model the time that elapses before an event occurs. Such a time is often called a **waiting time**. The exponential distribution is sometimes used to model the lifetime of a component. Also, there is a close connection between the exponential distribution and the Poisson distribution.

The probability density function of the exponential distribution involves a parameter, which is a positive constant λ whose value determines the density function's location and shape.

Definition

The probability density function of the exponential distribution with parameter $\lambda > 0$ is

$$f(x) = \begin{cases} \lambda e^{-\lambda x} & x > 0 \\ 0 & x \leq 0 \end{cases} \tag{4.31}$$

Figure 4.17 presents the probability density function of the exponential distribution for various values of λ. If X is a random variable whose distribution is exponential with parameter λ, we write $X \sim \text{Exp}(\lambda)$.

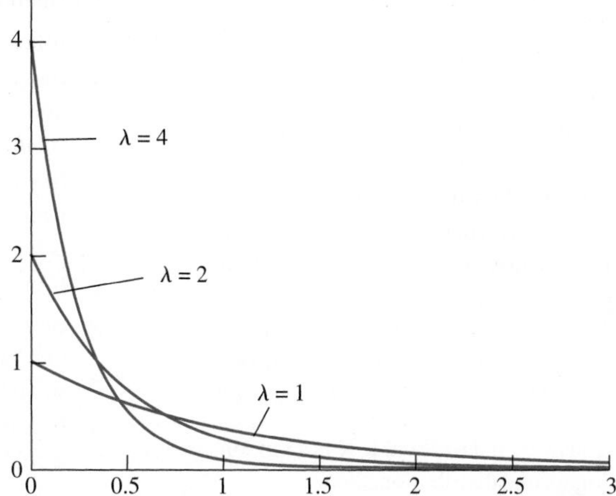

FIGURE 4.17 Plots of the exponential probability density function for various values of λ.

The cumulative distribution function of the exponential distribution is easy to compute. For $x \leq 0$, $F(x) = P(X \leq x) = 0$. For $x > 0$, the cumulative distribution function is

$$F(x) = P(X \leq x) = \int_0^x \lambda e^{-\lambda t}\, dt = 1 - e^{-\lambda x}$$

Summary

If $X \sim \text{Exp}(\lambda)$, the cumulative distribution function of X is

$$F(x) = P(X \leq x) = \begin{cases} 1 - e^{-\lambda x} & x > 0 \\ 0 & x \leq 0 \end{cases} \tag{4.32}$$

The mean and variance of an exponential random variable can be computed by using integration by parts. Derivations are provided at the end of the section.

If $X \sim \text{Exp}(\lambda)$, then

$$\mu_X = \frac{1}{\lambda} \tag{4.33}$$

$$\sigma_X^2 = \frac{1}{\lambda^2} \tag{4.34}$$

Example 4.55

If $X \sim \text{Exp}(2)$, find μ_X, σ_X^2, and $P(X \leq 1)$.

Solution

We compute μ_X and σ_X^2 from Equations (4.33) and (4.34), substituting $\lambda = 2$. We obtain $\mu_X = 0.5$, $\sigma_X^2 = 0.25$. Using Equation (4.32), we find that

$$P(X \leq 1) = 1 - e^{-2(1)} = 0.865$$

Example 4.56

Refer to Example 4.55. Find the median of X. Find the 30th percentile of X.

Solution

Let m denote the median of X. Then $P(X \leq m) = 0.5$. Using Equation (4.32), we find that $1 - e^{-2m} = 0.5$. Solving for m, we obtain $m = 0.3466$.

Let p_{30} denote the 30th percentile. Then $P(X \leq p_{30}) = 0.30$. Using Equation (4.32), we find that $1 - e^{-2p_{30}} = 0.30$. Solving for p_{30}, we obtain $p_{30} = 0.1783$.

The Exponential Distribution and the Poisson Process

We mentioned that the exponential distribution is sometimes used to model the waiting time to an event. It turns out that the exponential distribution is the correct model for waiting times whenever the events follow a Poisson process. Recall from Section 4.3 that events follow a Poisson process with rate parameter λ when the numbers of events in disjoint intervals are independent, and the number X of events that occur in any time interval of length t has a Poisson distribution with mean λt, in other words, when

$X \sim \text{Poisson}(\lambda t)$. The connection between the exponential distribution and the Poisson process is as follows:

> If events follow a Poisson process with rate parameter λ, and if T represents the waiting time from any starting point until the next event, then $T \sim \text{Exp}(\lambda)$.

A proof of this fact is given at the end of the section.

Example 4.57

A radioactive mass emits particles according to a Poisson process at a mean rate of 15 particles per minute. At some point, a clock is started. What is the probability that more than 5 seconds will elapse before the next emission? What is the mean waiting time until the next particle is emitted?

Solution

We will measure time in seconds. Let T denote the time in seconds that elapses before the next particle is emitted. The mean rate of emissions is 0.25 per second, so the rate parameter is $\lambda = 0.25$, and $T \sim \text{Exp}(0.25)$. The probability that more than 5 seconds will elapse before the next emission is equal to

$$\begin{aligned} P(T > 5) &= 1 - P(T \le 5) \\ &= 1 - (1 - e^{-0.25(5)}) \\ &= e^{-1.25} \\ &= 0.2865 \end{aligned}$$

The mean waiting time is $\mu_T = \dfrac{1}{0.25} = 4$.

Lack of Memory Property

The exponential distribution has a property known as the lack of memory property, which we illustrate with Examples 4.58 and 4.59.

Example 4.58

The lifetime of a particular integrated circuit has an exponential distribution with mean 2 years. Find the probability that the circuit lasts longer than three years.

Solution

Let T represent the lifetime of the circuit. Since $\mu_T = 2$, $\lambda = 0.5$. We need to find $P(T > 3)$.

$$\begin{aligned} P(T > 3) &= 1 - P(T \le 3) \\ &= 1 - (1 - e^{-0.5(3)}) \\ &= e^{-1.5} \\ &= 0.223 \end{aligned}$$

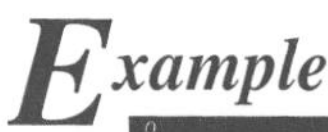

4.59

Refer to Example 4.58. Assume the circuit is now four years old and is still functioning. Find the probability that it functions for more than three additional years. Compare this probability with the probability that a new circuit functions for more than three years, which was calculated in Example 4.58.

Solution

We are given that the lifetime of the circuit will be more than four years, and we must compute the probability that the lifetime will be more than $4 + 3 = 7$ years. The probability is given by

$$P(T > 7 | T > 4) = \frac{P(T > 7 \text{ and } T > 4)}{P(T > 4)}$$

If $T > 7$, then $T > 4$ as well. Therefore $P(T > 7 \text{ and } T > 4) = P(T > 7)$. It follows that

$$\begin{aligned} P(T > 7 | T > 4) &= \frac{P(T > 7)}{P(T > 4)} \\ &= \frac{e^{-0.5(7)}}{e^{-0.5(4)}} \\ &= e^{-0.5(3)} \\ &= e^{-1.5} \\ &= 0.223 \end{aligned}$$

The probability that a 4-year-old circuit lasts 3 additional years is exactly the same as the probability that a new circuit lasts 3 years.

Examples 4.58 and 4.59 illustrate the lack of memory property. The probability that we must wait an additional t units, given that we have already waited s units, is the same as the probability that we must wait t units from the start. The exponential distribution does not "remember" how long we have been waiting. In particular, if the lifetime of a component follows the exponential distribution, then the probability that a component that is s time units old will last an additional t time units is the same as the probability that a new component will last t time units. In other words, a component whose lifetime follows an exponential distribution does not show any effects of age or wear.

The calculations in Examples 4.58 and 4.59 could be repeated for any values s and t in place of 4 and 3, and for any value of λ in place of 0.5. We now state the lack of memory property in its general form:

Lack of Memory Property

If $T \sim \text{Exp}(\lambda)$, and t and s are positive numbers, then

$$P(T > t + s \mid T > s) = P(T > t)$$

4.60

The number of hits on a website follows a Poisson process with a rate of 3 per minute. What is the probability that more than a minute goes by without a hit? If 2 minutes have gone by without a hit, what is the probability that a hit will occur in the next minute?

Solution

Let T denote the waiting time in minutes until the next hit. Then $T \sim \text{Exp}(3)$. The probability that one minute elapses with no hits is $P(T > 1) = e^{-3(1)} = 0.0498$. Because of the lack of memory property, the probability that one additional minute elapses without a hit, given that two minutes have gone by without a hit, is also equal to 0.0498. The probability that a hit does occur in the next minute is therefore equal to $1 - 0.0498 = 0.9502$.

Using the Exponential Distribution to Estimate a Rate

If $X \sim \text{Exp}(\lambda)$, then $\mu_X = 1/\lambda$, so $\lambda = 1/\mu_X$. It follows that if $X_1, \ldots, X_n$ is a random sample from $\text{Exp}(\lambda)$, a reasonable estimate of λ is $\widehat{\lambda} = 1/\overline{X}$.

We will discuss the bias in $\widehat{\lambda} = 1/\overline{X}$. As with any sample mean $\overline{X}$, $\mu_{\overline{X}} = \mu$, so $\overline{X}$ is an unbiased estimate of μ. However, $\mu_{1/\overline{X}} \neq 1/\mu$, because $1/\mu$ is not a linear function of μ. Therefore $\widehat{\lambda} = 1/\overline{X}$ is a biased estimate of $\lambda = 1/\mu$. Using advanced methods, it can be shown that $\mu_{\widehat{\lambda}} \approx \lambda + \lambda/n$, so the bias is approximately λ/n. Thus for a sufficiently large sample size n the bias is negligible, but it can be substantial when the sample size is small.

We can estimate the uncertainty in $\widehat{\lambda}$ by using the propagation of error method (Equation 3.10 in Section 3.3):

$$\sigma_{\widehat{\lambda}} \approx \left| \frac{d}{d\overline{X}} \frac{1}{\overline{X}} \right| \sigma_{\overline{X}}$$

For this expression to be useful, we need to know $\sigma_{\overline{X}}$. Now the standard deviation of an $\text{Exp}(\lambda)$ distribution is $\sigma = 1/\lambda$ (this follows from Equation 4.34; note that the standard deviation is the same as the mean). Therefore $\sigma_{\overline{X}} = \sigma/\sqrt{n} = 1/(\lambda\sqrt{n})$. We can replace λ with its estimate $1/\overline{X}$ to obtain

$$\sigma_{\overline{X}} \approx \frac{\overline{X}}{\sqrt{n}}$$

Now we can estimate the uncertainty $\sigma_{\widehat{\lambda}}$:

$$\begin{aligned}\sigma_{\widehat{\lambda}} &\approx \left| \frac{d}{d\overline{X}} \frac{1}{\overline{X}} \right| \sigma_{\overline{X}} \\ &= \frac{1}{\overline{X}^2} \frac{\overline{X}}{\sqrt{n}} \\ &= \frac{1}{\overline{X}\sqrt{n}}\end{aligned}$$

This propagation of error estimate is fairly good when the sample size is at least 20 or so. For smaller sample sizes it underestimates the uncertainty.

Summary

If $X_1, \ldots, X_n$ is a random sample from Exp(λ), then the parameter λ is estimated with

$$\widehat{\lambda} = \frac{1}{\overline{X}} \tag{4.35}$$

This estimator is biased. The bias is approximately equal to λ/n. The uncertainty in $\widehat{\lambda}$ is estimated with

$$\sigma_{\widehat{\lambda}} = \frac{1}{\overline{X}\sqrt{n}} \tag{4.36}$$

This uncertainty estimate is reasonably good when the sample size is more than 20.

Correcting the Bias

Since $\mu_{\widehat{\lambda}} = \mu_{1/\overline{X}} \approx \lambda + \lambda/n = (n+1)\lambda/n$, it follows that $\mu_{n/(n+1)\overline{X}} \approx \lambda$. In other words, the quantity $n/[(n+1)\overline{X}]$ is a less biased estimate of λ than is $1/\overline{X}$. This is known as a *bias-corrected* estimate.

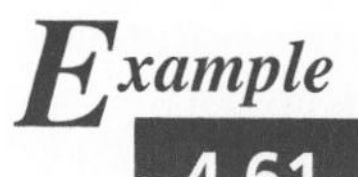
Example 4.61

A random sample of size 5 is taken from an Exp(λ) distribution. The values are 7.71, 1.32, 7.46, 6.53, and 0.44. Find a bias-corrected estimate of λ.

Solution

The sample mean is $\overline{X} = 4.6920$. The sample size is $n = 5$. The bias-corrected estimate of λ is 5/[6(4.6920)] = 0.178.

Derivation of the Mean and Variance of an Exponential Random Variable

To derive Equation (4.33), we begin with Equation (2.35) (in Section 2.4):

$$\mu_X = \int_{-\infty}^{\infty} x f(x)\, dx$$

Substituting the exponential probability density function (4.31) for $f(x)$, we obtain

$$\mu_X = \int_0^{\infty} \lambda x e^{-\lambda x}\, dx$$

Integrating by parts, setting $u = x$ and $dv = \lambda e^{-\lambda x}$ yields

$$\mu_X = -xe^{-\lambda x}\Big|_0^{\infty} + \int_0^{\infty} e^{-\lambda x}\, dx \tag{4.37}$$

We evaluate the first quantity on the right-hand side of Equation (4.37):

$$\begin{aligned}
-xe^{-\lambda x}\Big|_0^\infty &= \lim_{x\to\infty} -xe^{-\lambda x} - 0 \\
&= \lim_{x\to\infty} -\frac{x}{e^{\lambda x}} \\
&= \lim_{x\to\infty} -\frac{1}{\lambda e^{\lambda x}} \qquad \text{by L'Hospital's rule} \\
&= 0
\end{aligned}$$

Therefore

$$\begin{aligned}
\mu_X &= \int_0^\infty e^{-\lambda x}\,dx \\
&= \frac{1}{\lambda}
\end{aligned}$$

To derive Equation (4.34), we begin with Equation (2.37) (in Section 2.4):

$$\sigma_X^2 = \int_{-\infty}^{\infty} x^2 f(x)\,dx - \mu_X^2$$

Substituting the exponential probability density function (4.31) for $f(x)$, and $1/\lambda$ for μ_X, we obtain

$$\sigma_X^2 = \int_0^\infty \lambda x^2 e^{-\lambda x}\,dx - \frac{1}{\lambda^2} \tag{4.38}$$

We evaluate the integral $\int_0^\infty \lambda x^2 e^{-\lambda x}\,dx$, using integration by parts. Setting $u = x^2$ and $dv = \lambda e^{-\lambda x}$ we obtain

$$\int_0^\infty \lambda x^2 e^{-\lambda x}\,dx = -x^2 e^{-\lambda x}\Big|_0^\infty + \int_0^\infty 2xe^{-\lambda x}\,dx \tag{4.39}$$

We evaluate the first quantity on the right-hand side of Equation (4.39):

$$\begin{aligned}
-x^2 e^{\lambda x}\Big|_0^\infty &= \lim_{x\to\infty} -x^2 e^{-\lambda x} - 0 \\
&= \lim_{x\to\infty} -\frac{x^2}{e^{\lambda x}} \\
&= \lim_{x\to\infty} -\frac{2x}{\lambda e^{\lambda x}} \qquad \text{by L'Hospital's rule} \\
&= \lim_{x\to\infty} -\frac{2}{\lambda^2 e^{\lambda x}} \qquad \text{by L'Hospital's rule} \\
&= 0
\end{aligned}$$

Therefore

$$\int_0^\infty \lambda x^2 e^{-\lambda x}\,dx = \int_0^\infty 2xe^{-\lambda x}\,dx$$

$$= \frac{2}{\lambda}\int_0^{\infty} \lambda x e^{-\lambda x}\,dx$$
$$= \frac{2}{\lambda}\mu_X$$
$$= \left(\frac{2}{\lambda}\right)\left(\frac{1}{\lambda}\right)$$
$$= \frac{2}{\lambda^2}$$

Substituting into (4.38) yields

$$\sigma_X^2 = \frac{2}{\lambda^2} - \frac{1}{\lambda^2}$$
$$= \frac{1}{\lambda^2}$$

Derivation of the Relationship Between the Exponential Distribution and the Poisson Process

Let T represent the waiting time until the next event in a Poisson process with rate parameter λ. We show that $T \sim \text{Exp}(\lambda)$ by showing that the cumulative distribution function of T is $F(t) = 1 - e^{-\lambda t}$, which is the cumulative distribution function of $\text{Exp}(\lambda)$.

First, if $t \le 0$, then $F(t) = P(T \le t) = 0$. Now let $t > 0$. We begin by computing $P(T > t)$. The key is to realize that $T > t$ if and only if no events happen in the next t time units. Let X represent the number of events that happen in the next t time units. Now $T > t$ if and only if $X = 0$, so $P(T > t) = P(X = 0)$.

Since $X \sim \text{Poisson}(\lambda t)$,

$$P(X = 0) = e^{-\lambda t}\frac{\lambda^0}{0!}$$
$$= e^{-\lambda t}$$

Therefore $P(T > t) = e^{-\lambda t}$. The cumulative distribution function of T is $F(t) = 0$ for $t \le 0$, and for $t > 0$

$$F(t) = P(T \le t)$$
$$= 1 - P(T > t)$$
$$= 1 - e^{-\lambda t}$$

Since $F(t)$ is the cumulative distribution function of $\text{Exp}(\lambda)$, it follows that $T \sim \text{Exp}(\lambda)$.

Exercises for Section 4.7

1. Let $T \sim \text{Exp}(0.45)$. Find

a. μ_T

b. σ_T^2

c. $P(T > 3)$

d. The median of T

2. The lifetime of a fuse in a certain application has an exponential distribution with mean 2 years.

a. What is the value of the parameter λ?

b. What is the median lifetime of such a fuse?

c. What is the standard deviation?

d. What is the 60th percentile?

e. Find the probability that a fuse lasts longer than five years.

f. If a fuse is one year old and still functioning, what is the probability that it will function for more than two additional years?

3. A catalyst researcher states that the diameters, in microns, of the pores in a new product she has made have the exponential distribution with parameter $\lambda = 0.25$.

a. What is the mean pore diameter?

b. What is the standard deviation of the pore diameters?

c. What proportion of the pores are less than 3 microns in diameter?

d. What proportion of the pores are greater than 11 microns in diameter?

e. What is the median pore diameter?

f. What is the third quartile of the pore diameters?

g. What is the 99th percentile of the pore diameters?

4. Someone claims that the waiting time, in minutes, between hits at a certain website has the exponential distribution with parameter $\lambda = 1$.

a. Let X be the waiting time until the next hit. If the claim is true, what is $P(X \geq 5)$?

b. Based on the answer to part (a), if the claim is true, is five minutes an unusually long time to wait?

c. If you waited five minutes until the next hit occurred, would you still believe the claim? Explain.

5. A certain type of component can be purchased new or used. Fifty percent of all new components last more than five years, but only 30% of used components last more than five years. Is it possible that the lifetimes of new components are exponentially distributed? Explain.

6. A radioactive mass emits particles according to a Poisson process at a mean rate of 2 per second. Let T be the waiting time, in seconds, between emissions.

a. What is the mean waiting time?

b. What is the median waiting time?

c. Find $P(T > 2)$.

d. Find $P(T < 0.1)$.

e. Find $P(0.3 < T < 1.5)$.

f. If 3 seconds have elapsed with no emission, what is the probability that there will be an emission within the next second?

7. The number of traffic accidents at a certain intersection is thought to be well modeled by a Poisson process with a mean of 3 accidents per year.

a. Find the mean waiting time between accidents.

b. Find the standard deviation of the waiting times between accidents.

c. Find the probability that more than one year elapses between accidents.

d. Find the probability that less than one month elapses between accidents.

e. If no accidents have occurred within the last six months, what is the probability that an accident will occur within the next year?

8. The distance between consecutive flaws on a roll of sheet aluminum is exponentially distributed with mean distance 3 m. Let X be the distance, in meters, between flaws.

a. What is the mean number of flaws per meter?

b. What is the probability that a 5 m length of aluminum contains exactly two flaws?

9. A light fixture contains five lightbulbs. The lifetime of each bulb is exponentially distributed with mean 200 hours. Whenever a bulb burns out, it is replaced. Let T be the time of the first bulb replacement. Let X_i, $i = 1, \ldots, 5$, be the lifetimes of the five bulbs. Assume the lifetimes of the bulbs are independent.

a. Find $P(X_1 > 100)$.

b. Find $P(X_1 > 100$ and $X_2 > 100$ and $\cdots$ and $X_5 > 100)$.

c. Explain why the event $T > 100$ is the same as $\{X_1 > 100$ and $X_2 > 100$ and $\cdots$ and $X_5 > 100\}$.

d. Find $P(T \leq 100)$.

e. Let t be any positive number. Find $P(T \leq t)$, which is the cumulative distribution function of T.

f. Does T have an exponential distribution?

g. Find the mean of T.

h. If there were n lightbulbs, and the lifetime of each was exponentially distributed with parameter λ, what would be the distribution of T?

4.8 The Gamma and Weibull Distributions

The gamma and Weibull distributions are extensions of the exponential distribution. Both of them involve a certain integral known as the **gamma function**. We first define the gamma function and state some of its properties.

Definition

For $r > 0$, the gamma function is defined by

$$\Gamma(r) = \int_0^\infty t^{r-1} e^{-t}\, dt \tag{4.40}$$

The gamma function has the following properties:

1. If r is an integer, then $\Gamma(r) = (r-1)!$.
2. For any r, $\Gamma(r+1) = r\Gamma(r)$.
3. $\Gamma(1/2) = \sqrt{\pi}$.

The Gamma Distribution

The **gamma distribution** is a continuous distribution, one of whose purposes is to extend the usefulness of the exponential distribution in modeling waiting times. The gamma probability density function has two parameters, r and λ, both of which are positive constants.

Definition

The probability density function of the gamma distribution with parameters $r > 0$ and $\lambda > 0$ is

$$f(x) = \begin{cases} \dfrac{\lambda^r x^{r-1} e^{-\lambda x}}{\Gamma(r)} & x > 0 \\ 0 & x \leq 0 \end{cases} \tag{4.41}$$

If X is a random variable whose probability density function is gamma with parameters r and λ, we write $X \sim \Gamma(r, \lambda)$. Note that when $r = 1$, the gamma distribution is the same as the exponential. In symbols, $\Gamma(1, \lambda) = \text{Exp}(\lambda)$. Figure 4.18 presents plots of the gamma probability density function for various values of r and λ.

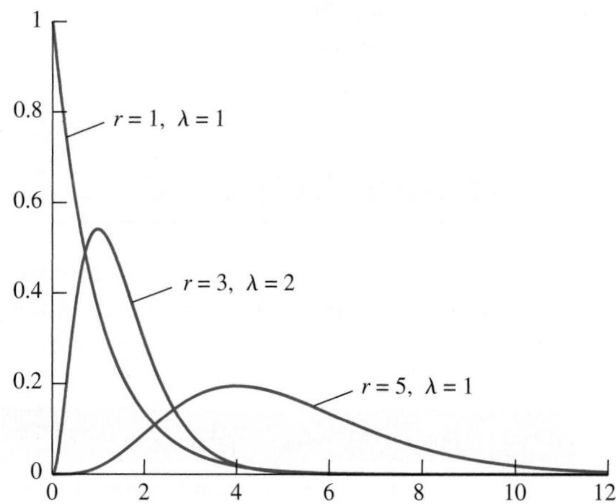

FIGURE 4.18 The gamma probability density function for various values of r and λ.

When the parameter r is an integer, the gamma distribution is a direct extension of the exponential distribution. To be specific, recall that if events follow a Poisson process with rate parameter λ, the waiting time until an event occurs is distributed as $\text{Exp}(\lambda)$. If r is any positive integer, then the waiting time until r events have occurred is distributed as $\Gamma(r, \lambda)$. We can say this another way. Let X_1 be the waiting time until the first event, and for $i > 1$ let X_i be the waiting time between events $i - 1$ and i. The waiting time until the rth event is the sum of the independent random variables $X_1 + \cdots + X_r$, each of which is distributed as $\text{Exp}(\lambda)$.

Summary

If $X_1, \ldots, X_r$ are independent random variables, each distributed as $\text{Exp}(\lambda)$, then the sum $X_1 + \cdots + X_r$ is distributed as $\Gamma(r, \lambda)$.

Since the mean and variance of an exponential random variable are given by $1/\lambda$ and $1/\lambda^2$, respectively, we can use the fact that a gamma random variable is the sum of independent exponential random variables to compute the mean and variance of a gamma random variable in the case when r is an integer. The results are presented in the following box, and in fact, they are valid for any values of r and λ.

If $X \sim \Gamma(r, \lambda)$, then

$$\mu_X = \frac{r}{\lambda} \tag{4.42}$$

$$\sigma_X^2 = \frac{r}{\lambda^2} \tag{4.43}$$

Example 4.62

Assume that arrival times at a drive-through window follow a Poisson process with mean rate $\lambda = 0.2$ arrivals per minute. Let T be the waiting time until the third arrival. Find the mean and variance of T. Find $P(T \leq 20)$.

Solution

The random variable T is distributed $\Gamma(3, 0.2)$. Using Equations (4.42) and (4.43) we compute $\mu_T = 3/0.2 = 15$ and $\sigma_T^2 = 3/(0.2^2) = 75$. To compute $P(T \leq 20)$ we reason as follows: $T \leq 20$ means that the third event occurs within 20 minutes. This is the same as saying that the number of events that occur within 20 minutes is greater than or equal to 3. Now let X denote the number of events that occur within 20 minutes. What we have said is that $P(T \leq 20) = P(X \geq 3)$. Now the mean of X is (20)(0.2) = 4, and X has a Poisson distribution, so $X \sim \text{Poisson}(4)$. It follows that

$$\begin{aligned}
P(T \leq 20) &= P(X \geq 3) \\
&= 1 - P(X \leq 2) \\
&= 1 - [P(X = 0) + P(X = 1) + P(X = 2)] \\
&= 1 - \left(e^{-4}\frac{4^0}{0!} + e^{-4}\frac{4^1}{1!} + e^{-4}\frac{4^2}{2!}\right) \\
&= 1 - (e^{-4} + 4e^{-4} + 8e^{-4}) \\
&= 0.7619
\end{aligned}$$

The method used in Example 4.62 to find $P(T \leq 20)$ can be used to find the cumulative distribution function $F(x) = P(T \leq x)$ when $T \sim \Gamma(r, \lambda)$ and r is a positive integer.

If $T \sim \Gamma(r, \lambda)$, and r is a positive integer, the cumulative distribution function of T is given by

$$F(x) = P(T \leq x) = \begin{cases} 1 - \displaystyle\sum_{j=0}^{r-1} e^{-\lambda x} \frac{(\lambda x)^j}{j!} & x > 0 \\ 0 & x \leq 0 \end{cases} \tag{4.44}$$

A gamma distribution for which the parameter r is a positive integer is sometimes called an **Erlang distribution**. If $r = k/2$ where k is a positive integer, the $\Gamma(r, 1/2)$ distribution is called a **chi-square distribution with k degrees of freedom**. The chi-square distribution is very important in statistical inference. We will discuss some of its uses in Section 6.10.

The Weibull Distribution

The Weibull distribution is a continuous distribution that is used in a variety of situations. A common application of the Weibull distribution is to model the lifetimes of components such as bearings, ceramics, capacitors, and dielectrics. The Weibull probability density function has two parameters, both positive constants, that determine its location and shape. We denote these parameters α and β. The probability density function of the Weibull distribution is

$$f(x) = \begin{cases} \alpha\beta^\alpha x^{\alpha-1} e^{-(\beta x)^\alpha} & x > 0 \\ 0 & x \le 0 \end{cases} \tag{4.45}$$

If X is a random variable whose probability density function is Weibull with parameters α and β, we write $X \sim \text{Weibull}(\alpha, \beta)$. Note that when $\alpha = 1$, the Weibull distribution is the same as the exponential distribution with parameter $\lambda = \beta$. In symbols, $\text{Weibull}(1, \beta) = \text{Exp}(\beta)$.

Figure 4.19 presents plots of the Weibull(α, β) probability density function for several choices of the parameters α and β. By varying the values of α and β, a wide variety of curves can be generated. Because of this, the Weibull distribution can be made to fit a wide variety of data sets. This is the main reason for the usefulness of the Weibull distribution.

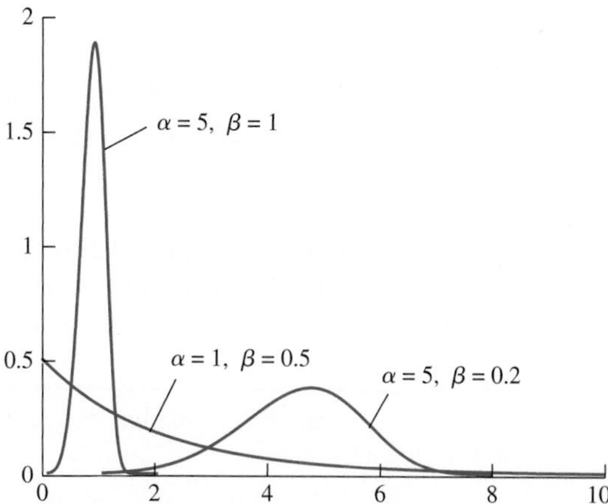

FIGURE 4.19 The Weibull probability density function for various choices of α and β.

The Weibull cumulative distribution function can be computed by integrating the probability density function:

$$F(x) = P(X \le x) = \begin{cases} \displaystyle\int_0^x \alpha\beta^\alpha t^{\alpha-1} e^{-(\beta t)^\alpha}\, dt = 1 - e^{-(\beta x)^\alpha} & x > 0 \\ 0 & x \le 0 \end{cases} \tag{4.46}$$

This integral is not as hard as it looks. Just substitute $u = (\beta t)^\alpha$, and $du = \alpha\beta^\alpha t^{\alpha-1}\, dt$.

The mean and variance of the Weibull distribution are expressed in terms of the gamma function.

If $X \sim \text{Weibull}(\alpha,\ \beta)$, then

$$\mu_X = \frac{1}{\beta}\Gamma\left(1 + \frac{1}{\alpha}\right) \tag{4.47}$$

$$\sigma_X^2 = \frac{1}{\beta^2}\left\{\Gamma\left(1 + \frac{2}{\alpha}\right) - \left[\Gamma\left(1 + \frac{1}{\alpha}\right)\right]^2\right\} \tag{4.48}$$

In the special case that $1/\alpha$ is an integer, then

$$\mu_X = \frac{1}{\beta}\left[\left(\frac{1}{\alpha}\right)!\right] \qquad \sigma_X^2 = \frac{1}{\beta^2}\left\{\left(\frac{2}{\alpha}\right)! - \left[\left(\frac{1}{\alpha}\right)!\right]^2\right\}$$

If the quantity $1/\alpha$ is an integer, then $1 + 1/\alpha$ and $1 + 2/\alpha$ are integers, so Property 1 of the gamma function can be used to compute μ_X and σ_X^2 exactly. If the quantity $1/\alpha$ is of the form $n/2$, where n is an integer, then in principle μ_X and σ_X^2 can be computed exactly through repeated applications of Properties 2 and 3 of the gamma function. For other values of α, μ_X and σ_X^2 must be approximated. Many computer packages can do this.

Example 4.63

In the article "Snapshot: A Plot Showing Program through a Device Development Laboratory" (D. Lambert, J. Landwehr, and M. Shyu, *Statistical Case Studies for Industrial Process Improvement*, ASA-SIAM 1997), the authors suggest using a Weibull distribution to model the duration of a bake step in the manufacture of a semiconductor. Let T represent the duration in hours of the bake step for a randomly chosen lot. If $T \sim \text{Weibull}(0.3,\ 0.1)$, what is the probability that the bake step takes longer than four hours? What is the probability that it takes between two and seven hours?

Solution

We use the cumulative distribution function, Equation (4.46). Substituting 0.3 for α and 0.1 for β, we have

$$P(T \le t) = 1 - e^{-(0.1t)^{0.3}}$$

Therefore

$$\begin{aligned} P(T > 4) &= 1 - P(T \le 4) \\ &= 1 - (1 - e^{-[(0.1)(4)]^{0.3}}) \end{aligned}$$

$$= e^{-(0.4)^{0.3}}$$
$$= e^{-0.7597}$$
$$= 0.468$$

The probability that the step takes between two and seven hours is

$$\begin{aligned} P(2 < T < 7) &= P(T \le 7) - P(T \le 2) \\ &= (1 - e^{-[(0.1)(7)]^{0.3}}) - (1 - e^{-[(0.1)(2)]^{0.3}}) \\ &= e^{-[(0.1)(2)]^{0.3}} - e^{-[(0.1)(7)]^{0.3}} \\ &= e^{-(0.2)^{0.3}} - e^{-(0.7)^{0.3}} \\ &= e^{-0.6170} - e^{-0.8985} \\ &= 0.132 \end{aligned}$$

Exercises for Section 4.8

1. Let $T \sim \Gamma(4, 0.5)$.

a. Find μ_T.

b. Find σ_T.

c. Find $P(T \le 1)$.

d. Find $P(T \ge 4)$.

2. The lifetime, in years, of a type of small electric motor operating under adverse conditions is exponentially distributed with $\lambda = 3.6$. Whenever a motor fails, it is replaced with another of the same type. Find the probability that fewer than six motors fail within one year.

3. Let $T \sim \text{Weibull}(0.5, 3)$.

a. Find μ_T.

b. Find σ_T.

c. Find $P(T < 1)$.

d. Find $P(T > 5)$.

e. Find $P(2 < T < 4)$.

4. If T is a continuous random variable that is always positive (such as a waiting time), with probability density function $f(t)$ and cumulative distribution function $F(t)$, then the **hazard function** is defined to be the function

$$h(t) = \frac{f(t)}{1 - F(t)}$$

The hazard function is the rate of failure per unit time, expressed as a proportion of the items that have not failed.

a. If $T \sim \text{Weibull}(\alpha, \beta)$, find $h(t)$.

b. For what values of α is the hazard rate increasing with time? For what values of α is it decreasing?

c. If T has an exponential distribution, show that the hazard function is constant.

5. In the article "Parameter Estimation with Only One Complete Failure Observation" (W. Pang, P. Leung, et al., *International Journal of Reliability, Quality, and Safety Engineering*, 2001:109–122), the lifetime, in hours, of a certain type of bearing is modeled with the Weibull distribution with parameters $\alpha = 2.25$ and $\beta = 4.474 \times 10^{-4}$.

a. Find the probability that a bearing lasts more than 1000 hours.

b. Find the probability that a bearing lasts less than 2000 hours.

c. Find the median lifetime of a bearing.

d. The hazard function is defined in Exercise 4. What is the hazard at $t = 2000$ hours?

6. The lifetime of a certain battery is modeled with the Weibull distribution with $\alpha = 2$ and $\beta = 0.1$.

a. What proportion of batteries will last longer than 10 hours?

b. What proportion of batteries will last less than 5 hours?

c. What proportion of batteries will last longer than 20 hours?

d. The hazard function is defined in Exercise 4. What is the hazard at $t = 10$ hours?

7. The lifetime of a cooling fan, in hours, that is used in a computer system has the Weibull distribution with $\alpha = 1.5$ and $\beta = 0.0001$.

   a. What is the probability that a fan lasts more than 10,000 hours?

   b. What is the probability that a fan lasts less than 5000 hours?

   c. What is the probability that a fan lasts between 3000 and 9000 hours?

8. Someone suggests that the lifetime T (in days) of a certain component can be modeled with the Weibull distribution with parameters $\alpha = 3$ and $\beta = 0.01$.

   a. If this model is correct, what is $P(T \leq 1)$?

   b. Based on the answer to part (a), if the model is correct, would one day be an unusually short lifetime? Explain.

   c. If you observed a component that lasted one day, would you find this model to be plausible? Explain.

   d. If this model is correct, what is $P(T \leq 90)$?

   e. Based on the answer to part (d), if the model is correct, would 90 days be an unusually short lifetime? An unusually long lifetime? Explain.

   f. If you observed a component that lasted 90 days, would you find this model to be plausible? Explain.

9. A system consists of two components connected in series. The system will fail when either of the two components fails. Let T be the time at which the system fails. Let X_1 and X_2 be the lifetimes of the two components. Assume that X_1 and X_2 are independent and that each has the Weibull distribution with $\alpha = 2$ and $\beta = 0.2$.

   a. Find $P(X_1 > 5)$.

   b. Find $P(X_1 > 5 \text{ and } X_2 > 5)$.

   c. Explain why the event $T > 5$ is the same as the event $\{X_1 > 5 \text{ and } X_2 > 5\}$.

   d. Find $P(T \leq 5)$.

   e. Let t be any positive number. Find $P(T \leq t)$, which is the cumulative distribution function of T.

   f. Does T have a Weibull distribution? If so, what are its parameters?

4.9 Probability Plots

Scientists and engineers frequently work with data that can be thought of as a random sample from some population. In many such cases, it is important to determine a probability distribution that approximately describes that population. In some cases, knowledge of the process that generated the data can guide the decision. More often, though, the only way to determine an appropriate distribution is to examine the sample to find a probability distribution that fits.

Probability plots provide a good way to do this. Given a random sample $X_1, \ldots, X_n$, a probability plot can determine whether the sample might have plausibly come from some specified population. We will present the idea behind probability plots with a simple example. A random sample of size 5 is drawn, and we want to determine whether the population from which it came might have been normal. The sample, arranged in increasing order, is

$$3.01,\ 3.35,\ 4.79,\ 5.96,\ 7.89$$

Denote the values, in increasing order, by $X_1, \ldots, X_n$ ($n = 5$ in this case). The first thing to do is to assign increasing, evenly spaced values between 0 and 1 to the X_i. There are several acceptable ways to do this; perhaps the simplest is to assign the value $(i - 0.5)/n$ to X_i. The following table shows the assignment for the given sample.

i	X_i	$(i - 0.5)/5$
1	3.01	0.1
2	3.35	0.3
3	4.79	0.5
4	5.96	0.7
5	7.89	0.9

The value $(i - 0.5)/n$ is chosen to reflect the position of X_i in the ordered sample. There are $i - 1$ values less than X_i, and i values less than or equal to X_i. The quantity $(i - 0.5)/n$ is a compromise between the proportions $(i - 1)/n$ and i/n.

The goal is to determine whether the sample might have come from a normal population. The most plausible normal distribution is the one whose mean and standard deviation are the same as the sample mean and standard deviation. The sample mean is $\overline{X} = 5.00$, and the sample standard deviation is $s = 2.00$. We will therefore determine whether this sample might have come from a $N(5, 2^2)$ distribution. Figure 4.20 is a plot of the five points $(X_i,\ (i - 0.5)/5)$. The curve is the cumulative distribution function (cdf) $F(x)$ of the $N(5, 2^2)$ distribution. Recall that $F(x) = P(X \le x)$ where $X \sim N(5, 2^2)$.

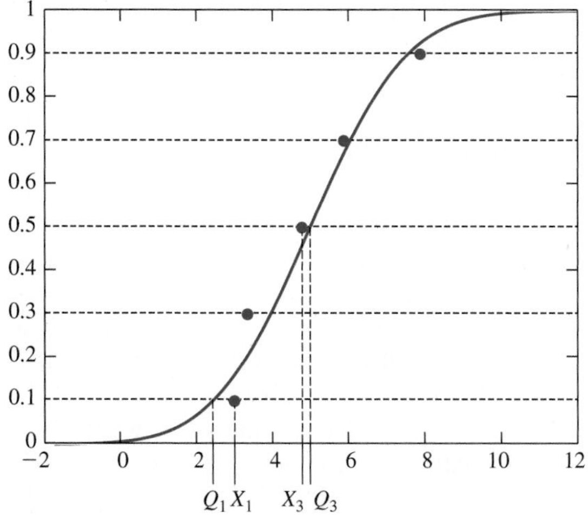

FIGURE 4.20 The curve is the cdf of $N(5, 2^2)$. If the sample points $X_1, \ldots, X_5$ came from this distribution, they are likely to lie close to the curve.

We have drawn horizontal lines through the sample points. We denote the x values of the points on the cdf that are crossed by the line, in increasing order, by $Q_1, \ldots, Q_5$. Now the horizontal line through $(X_1, 0.1)$ intersects the cdf at the point $(Q_1, 0.1)$. This means that the proportion of values in the $N(5, 2^2)$ population that are less than or equal to Q_1 is 0.1. Another way to say this is that Q_1 is the 10th percentile of the $N(5, 2^2)$ distribution. If the sample $X_1, \ldots, X_5$ truly came from a $N(5, 2^2)$ distribution, then it is reasonable to believe that the lowest value in the sample, X_1, would be fairly close to the 10th percentile of the population, Q_1. Intuitively, the reason for this is that we would expect that the lowest of five points would be likely to come from the lowest fifth, or 20%, of the population, and the 10th percentile is in the middle of that lowest 20%. Applying similar reasoning to the remaining points, we would expect each Q_i to be close to its corresponding X_i.

The **probability plot** consists of the points (X_i, Q_i). Since the distribution that generated the Q_i was a normal distribution, this is called a **normal probability plot**. If $X_1, \ldots, X_n$ do in fact come from the distribution that generated the Q_i, the points should lie close to a straight line. To construct the plot, we must compute the Q_i. These are the $100(i - 0.5)/n$ percentiles of the distribution that is suspected of generating the sample. In this example the Q_i are the 10th, 30th, 50th, 70th, and 90th percentiles of the $N(5, 2^2)$ distribution. We could approximate these values by looking up the z-scores corresponding to these percentiles, and then converting to raw scores. In practice, the Q_i are invariably calculated by a computer software package. The following table presents the X_i and the Q_i for this example.

i	X_i	Q_i
1	3.01	2.44
2	3.35	3.95
3	4.79	5.00
4	5.96	6.05
5	7.89	7.56

Figure 4.21 presents a normal probability plot for the sample $X_1, \ldots, X_5$. A straight line is superimposed on the plot, to make it easier to judge whether the points lie close

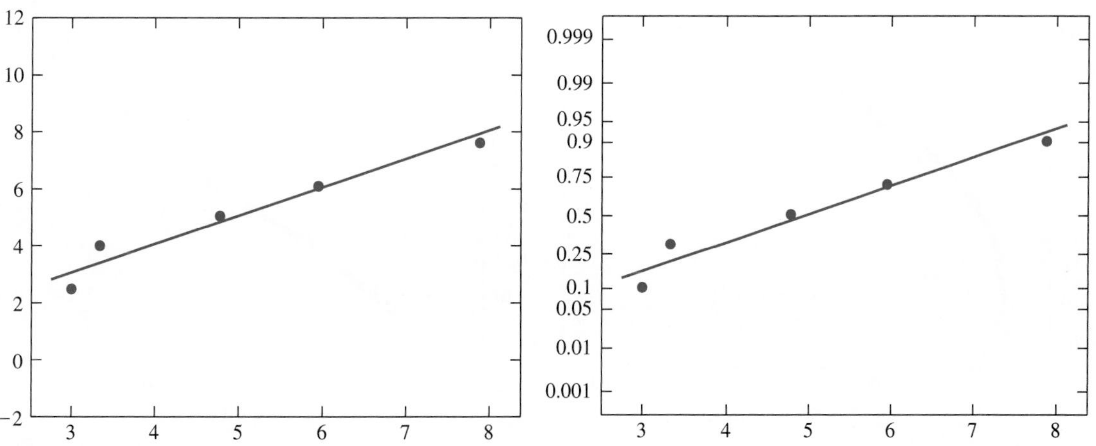

FIGURE 4.21 Normal probability plots for the sample $X_1, \ldots, X_5$. The plots are identical, except for the scaling on the vertical axis. The sample points lie approximately on a straight line, so it is plausible that they came from a normal population.

to a straight line or not. Two versions of the plot are presented; they are identical except for the scaling on the vertical axis. In the plot on the left, the values on the vertical axis represent the Q_i. In the plot on the right, the values on the vertical axis represent the percentiles (as decimals, so 0.1 is the 10th percentile) of the Q_i. For example, the 10th percentile of $N(5, 2^2)$ is 2.44, so the value 0.1 on the right-hand plot corresponds to the value 2.44 on the left-hand plot. The 50th percentile, or median, is 5, so the value 0.5 on the right-hand plot corresponds to the value 5 on the left-hand plot. Computer packages often scale the vertical axis like the plot on the right. In Figure 4.21, the sample points are close to the line, so it is quite plausible that the sample came from a normal distribution.

We remark that the points $Q_1, \ldots, Q_n$ are called **quantiles** of the distribution from which they are generated. Sometimes the sample points $X_1, \ldots, X_n$ are called **empirical quantiles**. For this reason the probability plot is sometimes called a quantile–quantile plot, or QQ plot.

In this example, we used a sample of only five points to make the calculations clear. In practice, probability plots work better with larger samples. A good rule of thumb is to require at least 30 points before relying on a probability plot. Probability plots can still be used for smaller samples, but they will detect only fairly large departures from normality.

Figure 4.22 shows two normal probability plots. The plot in Figure 4.22a is of the monthly productions of 255 gas wells. These data do not lie close to a straight line, and thus do not come from a population that is close to normal. The plot in Figure 4.22b is of the natural logs of the monthly productions. These data lie much closer to a straight line, although some departure from normality can be detected. (Figure 4.16 in Section 4.6 presents histograms of these data.)

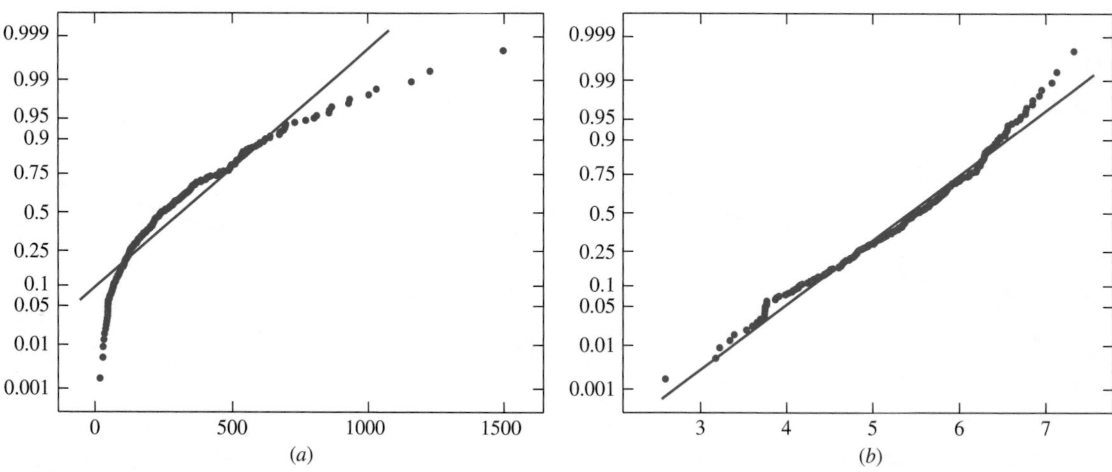

FIGURE 4.22 Two normal probability plots. *(a)* A plot of the monthly productions of 255 gas wells. These data do not lie close to a straight line, and thus do not come from a population that is close to normal. *(b)* A plot of the natural logs of the monthly productions. These data lie much closer to a straight line, although some departure from normality can be detected. See Figure 4.16 on page 248 for histograms of these data.

Interpreting Probability Plots

It's best not to use hard-and-fast rules when interpreting a probability plot. Judge the straightness of the plot by eye. When deciding whether the points on a probability plot lie close to a straight line or not, do not pay too much attention to the points at the very ends (high or low) of the sample, unless they are quite far from the line. It is common for a few points at either end to stray from the line somewhat. However, a point that is very far from the line when most other points are close is an outlier, and deserves attention.

Exercises for Section 4.9

1. Each of three samples has been plotted on a normal probability plot. For each, say whether the sample appears to have come from an approximately normal population.

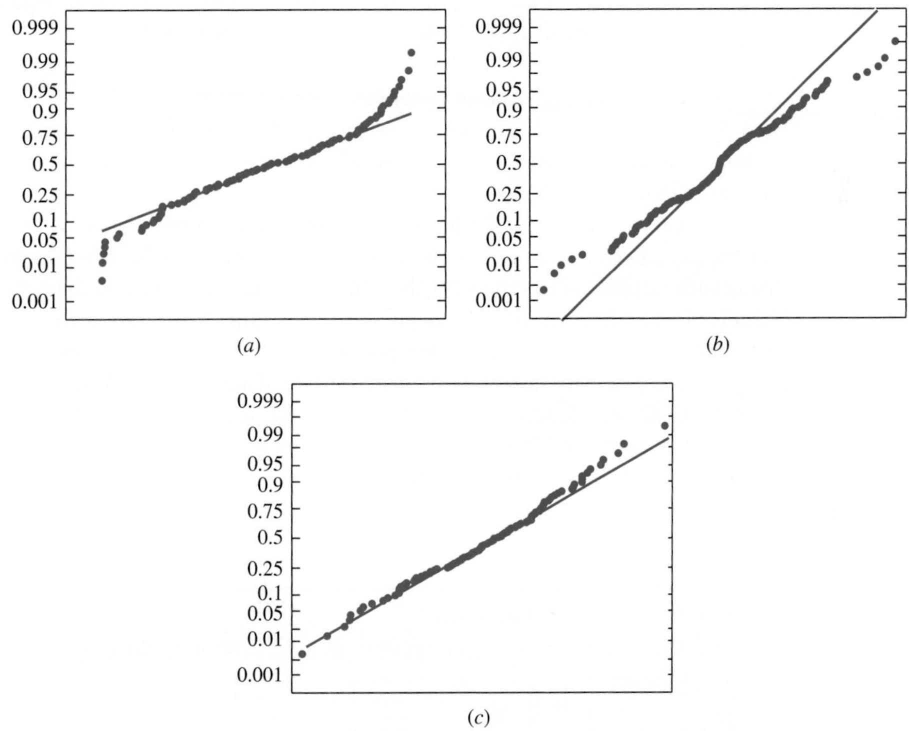

(a) (b) (c)

2. Construct a normal probability plot for the soapbars data in Exercise 1 in Section 1.3. Do these data appear to come from an approximately normal distribution?

3. Below are the durations (in minutes) of 40 eruptions of the geyser Old Faithful in Yellowstone National Park.

4.1	1.8	3.2	1.9	4.6	2.0	4.5	3.9	4.3	2.3
3.8	1.9	4.6	1.8	4.7	1.8	4.6	1.9	3.5	4.0
3.7	3.7	4.3	3.6	3.8	3.8	3.8	2.5	4.5	4.1
3.7	3.8	3.4	4.0	2.3	4.4	4.1	4.3	3.3	2.0

Construct a normal probability plot for these data. Do the data appear to come from an approximately normal distribution?

4. Below are the durations (in minutes) of 40 time intervals between eruptions of the geyser Old Faithful in Yellowstone National Park.

91	51	79	53	82	51	76	82	84	53
86	51	85	45	88	51	80	49	82	75
73	67	68	86	72	75	75	66	84	70
79	60	86	71	67	81	76	83	76	55

Construct a normal probability plot for these data. Do they appear to come from an approximately normal distribution?

5. Construct a normal probability plot for the PM data in Table 1.2 (p. 20). Do the PM data appear to come from a normal population?

6. Construct a normal probability plot for the logs of the PM data in Table 1.2. Do the logs of the PM data appear to come from a normal population?

7. Can the plot in Exercise 6 be used to determine whether the PM data appear to come from a lognormal population? Explain.

4.10 The Central Limit Theorem

The **Central Limit Theorem** is by far the most important result in statistics. Many commonly used statistical methods rely on this theorem for their validity. The Central Limit Theorem says that if we draw a large enough sample from a population, then the distribution of the sample mean is approximately normal, no matter what population it was drawn from. This allows us to compute probabilities for sample means using the z table, even though the population from which the sample was drawn is not normal. We now explain this more fully.

Let $X_1, \ldots, X_n$ be a simple random sample from a population with mean μ and variance σ^2. Let $\overline{X} = (X_1 + \cdots + X_n)/n$ be the sample mean. Now imagine drawing many such samples and computing their sample means. If one could draw every possible sample of size n from the original population, and compute the sample mean for each one, the resulting collection would be the population of sample means. One could construct the probability density function of this population. One might think that the shape of this probability density function would depend on the shape of the population from which the sample was drawn. The surprising thing is that if the sample size is sufficiently large, this is not so. If the sample size is large enough, the distribution of the sample mean is approximately normal, no matter what the distribution of the population from which the sample was drawn.

The Central Limit Theorem

Let $X_1, \ldots, X_n$ be a simple random sample from a population with mean μ and variance σ^2.

Let $\overline{X} = \dfrac{X_1 + \cdots + X_n}{n}$ be the sample mean.

Let $S_n = X_1 + \cdots + X_n$ be the sum of the sample observations.

Then if n is sufficiently large,

$$\overline{X} \sim N\left(\mu, \frac{\sigma^2}{n}\right) \quad \text{approximately} \qquad (4.49)$$

and

$$S_n \sim N(n\mu, n\sigma^2) \quad \text{approximately} \qquad (4.50)$$

Note that the Central Limit Theorem specifies that $\mu_{\overline{X}} = \mu$ and $\sigma^2_{\overline{X}} = \sigma^2/n$, which hold for any sample mean. The sum of the sample items is equal to the mean multiplied by the sample size, that is, $S_n = n\overline{X}$. It follows that $\mu_{S_n} = n\mu$ and $\sigma^2_{S_n} = n^2\sigma^2/n = n\sigma^2$ (see Equations 2.41 and 2.42 in Section 2.5).

The Central Limit Theorem says that $\overline{X}$ and S_n are approximately normally distributed, if the sample size n is large enough. The natural question to ask is: How large is large enough? The answer depends on the shape of the underlying population. If the sample is drawn from a nearly symmetric distribution, the normal approximation can be good even for a fairly small value of n. However, if the population is heavily skewed, a fairly large n may be necessary. Empirical evidence suggests that for most populations, a sample size of 30 or more is large enough for the normal approximation to be adequate (see Figure 4.23 on page 272).

For most populations, if the sample size is greater than 30, the Central Limit Theorem approximation is good.

Let X denote the number of flaws in a 1 in. length of copper wire. The probability mass function of X is presented in the following table.

x	$P(X = x)$
0	0.48
1	0.39
2	0.12
3	0.01

One hundred wires are sampled from this population. What is the probability that the average number of flaws per wire in this sample is less than 0.5?

Solution

The population mean number of flaws is $\mu = 0.66$, and the population variance is $\sigma^2 = 0.5244$. See Examples 2.35 and 2.36 (in Section 2.4) for the calculation of these quantities. Let $X_1, \ldots, X_{100}$ denote the number of flaws in the 100 wires sampled from this population. We need to find $P(\overline{X} < 0.5)$. Now the sample size is $n = 100$, which is a large sample. It follows from the Central Limit Theorem (expression 4.49) that $\overline{X} \sim N(0.66, 0.005244)$. The z-score of 0.5 is therefore

$$z = \frac{0.5 - 0.66}{\sqrt{0.005244}} = -2.21$$

From the z table, the area to the left of -2.21 is 0.0136. Therefore $P(\overline{X} < 0.5) = 0.0136$, so only 1.36% of samples of size 100 will have fewer than 0.5 flaws per wire. See Figure 4.24 (page 273).

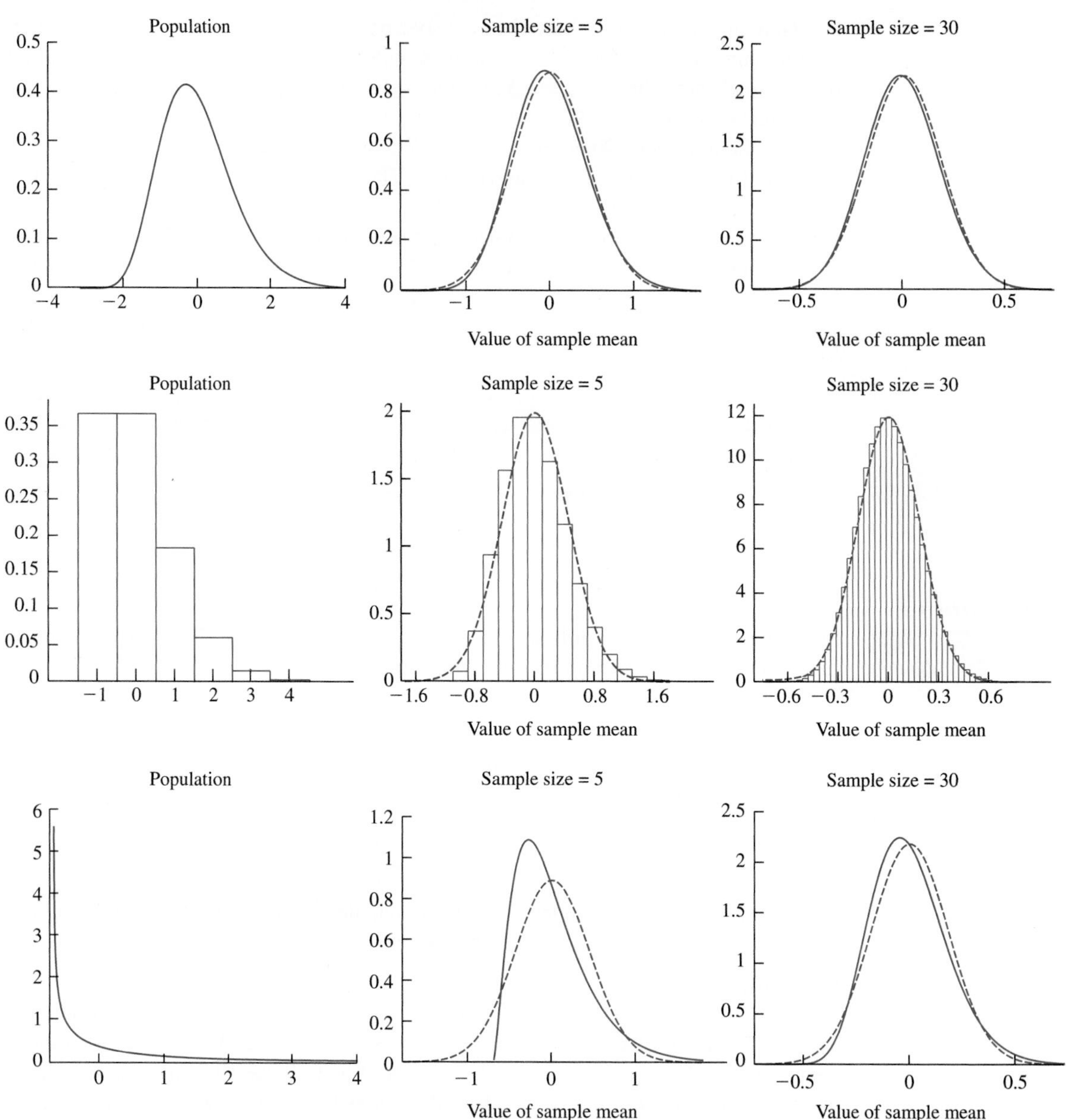

FIGURE 4.23 The leftmost plot in each row is the distribution (probability density function or probability mass function) of a random variable. The two plots to its right are the distributions of the sample mean (solid line) for samples of sizes 5 and 30, respectively, with the normal curve (dashed line) superimposed. **Top row:** Since the original distribution is nearly symmetric, the normal approximation is good even for a sample size as small as 5. **Middle row:** The original distribution is somewhat skewed. Even so, the normal approximation is reasonably close even for a sample of size 5, and very good for a sample of size 30. **Bottom row:** The original distribution is highly skewed. The normal approximation is not good for a sample size of 5, but is reasonably good for a sample of size 30. Note that two of the original distributions are continuous, and one is discrete. The Central Limit Theorem holds for both continuous and discrete distributions.

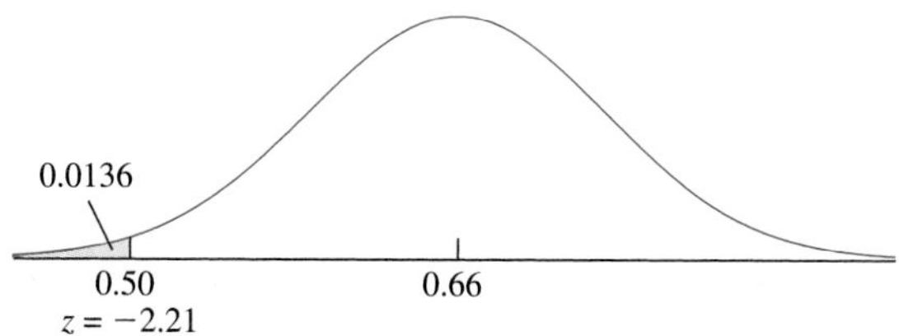

FIGURE 4.24 Solution to Example 4.64.

Note that in Example 4.64 we needed to know only the mean and variance of the population, not the probability mass function.

Example 4.65

At a large university, the mean age of the students is 22.3 years, and the standard deviation is 4 years. A random sample of 64 students is drawn. What is the probability that the average age of these students is greater than 23 years?

Solution

Let $X_1, \ldots, X_{64}$ be the ages of the 64 students in the sample. We wish to find $P(\overline{X} > 23)$. Now the population from which the sample was drawn has mean $\mu = 22.3$ and variance $\sigma^2 = 16$. The sample size is $n = 64$. It follows from the Central Limit Theorem (expression 4.49) that $\overline{X} \sim N(22.3,\ 0.25)$. The z-score for 23 is

$$z = \frac{23 - 22.3}{\sqrt{0.25}} = 1.40$$

From the z table, the area to the right of 1.40 is 0.0808. Therefore $P(\overline{X} > 23) = 0.0808$. See Figure 4.25.

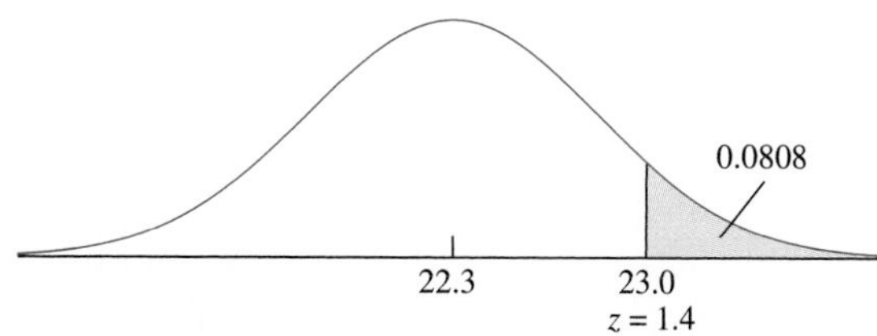

FIGURE 4.25 Solution to Example 4.65.

Normal Approximation to the Binomial

Recall from Section 4.2, that if $X \sim \text{Bin}(n,\ p)$, then $X = Y_1 + \cdots + Y_n$, where $Y_1, \ldots, Y_n$ is a sample from a Bernoulli(p) population. Therefore X is the sum of the sample observations. The sample proportion is

$$\widehat{p} = \frac{X}{n} = \frac{Y_1 + \cdots + Y_n}{n}$$

which is also the sample mean $\overline{Y}$. The Bernoulli(p) population has mean $\mu = p$ and variance $\sigma^2 = p(1-p)$. It follows from the Central Limit Theorem that if the number of trials n is large, then $X \sim N(np,\ np(1-p))$, and $\widehat{p} \sim N(p,\ p(1-p)/n)$.

Again the question arises, how large a sample is large enough? In the binomial case, the accuracy of the normal approximation depends on the mean number of successes np and on the mean number of failures $n(1-p)$. The larger the values of np and $n(1-p)$, the better the approximation. A common rule of thumb is to use the normal approximation whenever $np > 5$ and $n(1-p) > 5$. A better and more conservative rule is to use the normal approximation whenever $np > 10$ and $n(1-p) > 10$.

Summary

If $X \sim \text{Bin}(n,\ p)$, and if $np > 10$ and $n(1-p) > 10$, then

$$X \sim N(np,\ np(1-p)) \quad \text{approximately} \tag{4.51}$$

$$\widehat{p} \sim N\left(p,\ \frac{p(1-p)}{n}\right) \quad \text{approximately} \tag{4.52}$$

To illustrate the accuracy of the normal approximation to the binomial, Figure 4.26 presents the Bin(100, 0.2) probability histogram with the $N(20,\ 16)$ probability density function superimposed. While a slight degree of skewness can be detected in the binomial distribution, the normal approximation is quite close.

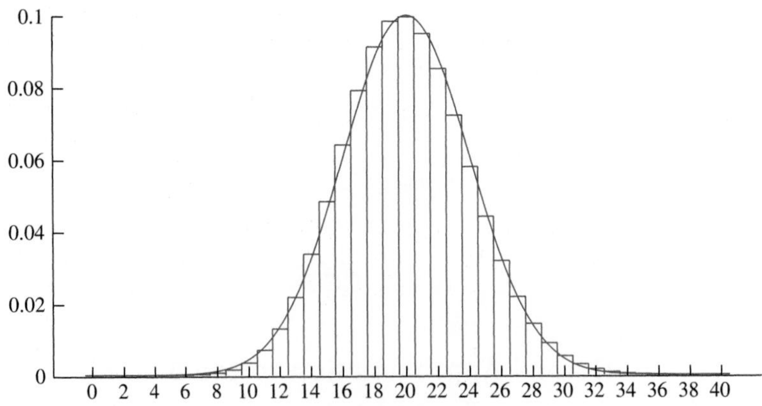

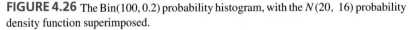

FIGURE 4.26 The Bin(100, 0.2) probability histogram, with the $N(20,\ 16)$ probability density function superimposed.

The Continuity Correction

The binomial distribution is discrete, while the normal distribution is continuous. The **continuity correction** is an adjustment, made when approximating a discrete distribution with a continuous one, that can improve the accuracy of the approximation. To see how it works, imagine that a fair coin is tossed 100 times. Let X represent the number of heads. Then $X \sim \text{Bin}(100, 0.5)$. Imagine that we wish to compute the probability that X is between 45 and 55. This probability will differ depending on whether the endpoints, 45 and 55, are included or excluded. Figure 4.27 illustrates the case where the endpoints are included, that is, where we wish to compute $P(45 \leq X \leq 55)$. The exact probability is given by the total area of the rectangles of the binomial probability histogram corresponding to the integers 45 to 55 inclusive. The approximating normal curve is superimposed. To get the best approximation, we should compute the area under the normal curve between 44.5 and 55.5. In contrast, Figure 4.28 (page 276) illustrates the case where we wish to compute $P(45 < X < 55)$. Here the endpoints are excluded. The exact probability is given by the total area of the rectangles of the binomial probability histogram corresponding to the integers 46 to 54. The best normal approximation is found by computing the area under the normal curve between 45.5 and 54.5.

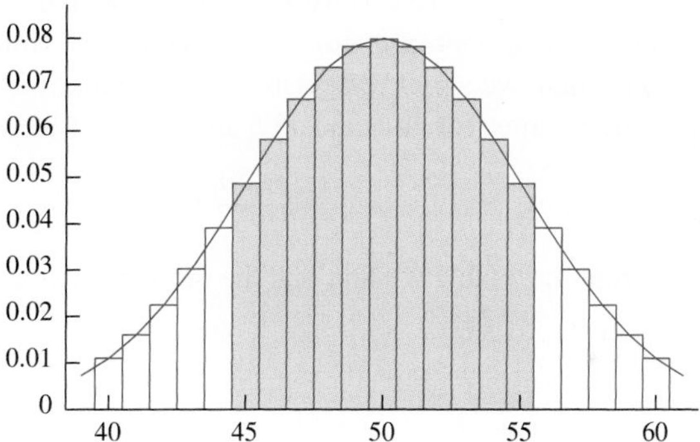

FIGURE 4.27 To compute $P(45 \leq X \leq 55)$, the areas of the rectangles corresponding to 45 and to 55 should be included. To approximate this probability with the normal curve, compute the area under the curve between 44.5 and 55.5.

In summary, to apply the continuity correction, determine precisely which rectangles of the discrete probability histogram you wish to include, and then compute the area under the normal curve corresponding to those rectangles.

Example 4.66

If a fair coin is tossed 100 times, use the normal curve to approximate the probability that the number of heads is between 45 and 55 *inclusive*.

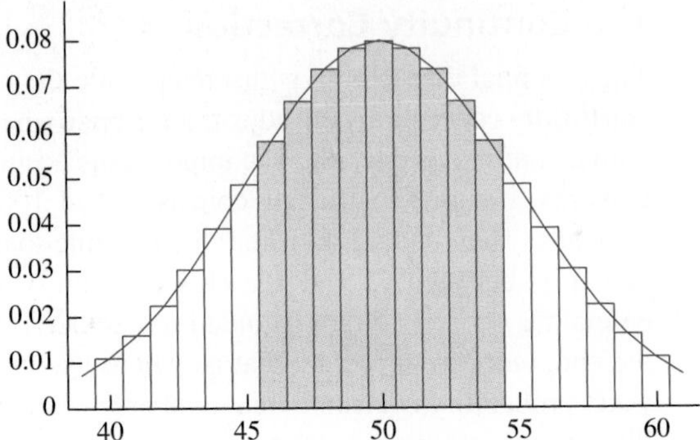

FIGURE 4.28 To compute $P(45 < X < 55)$, the areas of the rectangles corresponding to 45 and to 55 should be excluded. To approximate this probability with the normal curve, compute the area under the curve between 45.5 and 54.5.

Solution

This situation is illustrated in Figure 4.27. Let X be the number of heads obtained. Then $X \sim \text{Bin}(100,\ 0.5)$. Substituting $n = 100$ and $p = 0.5$ into Equation (4.51), we obtain the normal approximation $X \sim N(50, 25)$. Since the endpoints 45 and 55 are to be included, we should compute the area under the normal curve between 44.5 and 55.5. The z-scores for 44.5 and 55.5 are

$$z = \frac{44.5 - 50}{5} = -1.1, \qquad z = \frac{55.5 - 50}{5} = 1.1$$

From the z table we find that the probability is 0.7286. See Figure 4.29.

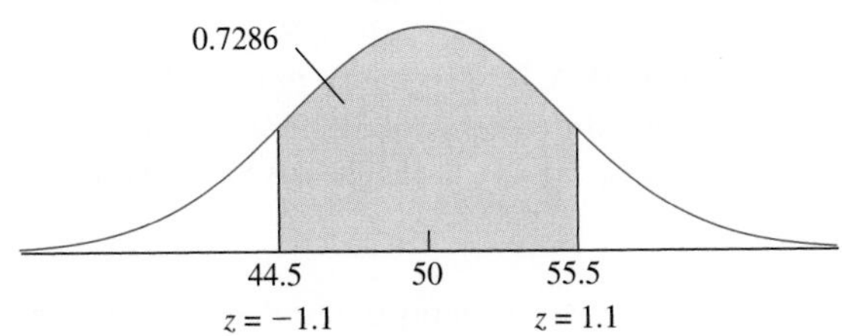

FIGURE 4.29 Solution to Example 4.66.

***Example* 4.67**

If a fair coin is tossed 100 times, use the normal curve to approximate the probability that the number of heads is between 45 and 55 *exclusive*.

Solution

This situation is illustrated in Figure 4.28. Let X be the number of heads obtained. As in Example 4.66, $X \sim \text{Bin}(100,\ 0.5)$, and the normal approximation is $X \sim N(50, 25)$. Since the endpoints 45 and 55 are to be excluded, we should compute the area under the normal curve between 45.5 and 54.5. The z-scores for 45.5 and 54.5 are

$$z = \frac{45.5 - 50}{5} = -0.9, \qquad z = \frac{54.5 - 50}{5} = 0.9$$

From the z table we find that the probability is 0.6318. See Figure 4.30.

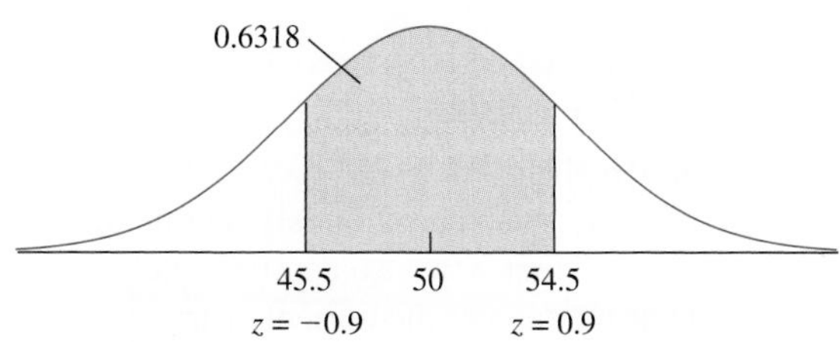

FIGURE 4.30 Solution to Example 4.67.

Example 4.68

In a certain large university, 25% of the students are over 21 years of age. In a sample of 400 students, what is the probability that more than 110 of them are over 21?

Solution

Let X represent the number of students who are over 21. Then $X \sim \text{Bin}(400,\ 0.25)$. We can use the normal approximation, which is $X \sim N(100,\ 75)$. Since we want to find the probability that the number of students is *more than* 110, the value 110 is excluded. We therefore find $P(X > 110.5)$. We compute the z-score for 110.5, which is

$$z = \frac{110.5 - 100}{\sqrt{75}} = 1.21$$

Using the z table, we find that $P(X > 110.5) = 0.1131$. See Figure 4.31.

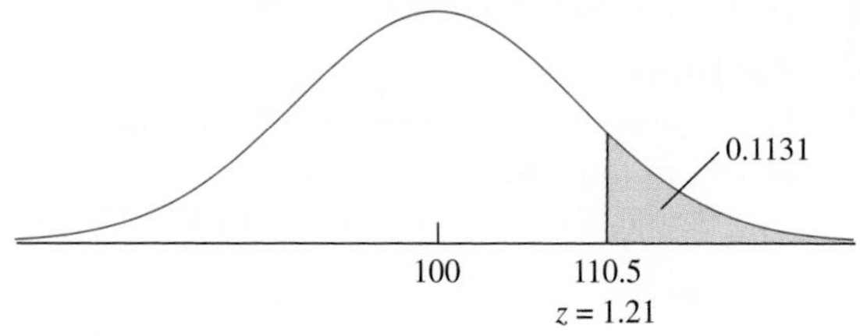

FIGURE 4.31 Solution to Example 4.68.

Accuracy of the Continuity Correction

The continuity correction improves the accuracy of the normal approximation to the binomial distribution in most cases. For binomial distributions with large n and small p, however, when computing a probability that corresponds to an area in the tail of the distribution, the continuity correction can in some cases reduce the accuracy of the normal approximation somewhat. This results from the fact that the normal approximation is not perfect; it fails to account for a small degree of skewness in these distributions. In summary, use of the continuity correction makes the normal approximation to the binomial distribution better in most cases, but not all.

Normal Approximation to the Poisson

Recall that if $X \sim \text{Poisson}(\lambda)$, then X is approximately binomial with n large and $np = \lambda$. Recall also that $\mu_X = \lambda$ and $\sigma_X^2 = \lambda$. It follows that if λ is sufficiently large, i.e., $\lambda > 10$, then X is approximately binomial, with $np > 10$. It follows from the central limit theorem that X is also approximately normal, with mean and variance both equal to λ. Thus we can use the normal distribution to approximate the Poisson.

Summary

If $X \sim \text{Poisson}(\lambda)$, where $\lambda > 10$, then

$$X \sim N(\lambda, \lambda) \quad \text{approximately} \tag{4.53}$$

Continuity Correction for the Poisson Distribution

Since the Poisson distribution is discrete, the continuity correction can in principle be applied when using the normal approximation. For areas that include the central part of the curve, the continuity correction generally improves the normal approximation, but for areas in the tails the continuity correction sometimes makes the approximation worse. We will not use the continuity correction for the Poisson distribution.

Example 4.69

The number of hits on a website follows a Poisson distribution, with a mean of 27 hits per hour. Find the probability that there will be 90 or more hits in three hours.

Solution

Let X denote the number of hits on the website in three hours. The mean number of hits in three hours is 81, so $X \sim \text{Poisson}(81)$. Using the normal approximation, $X \sim N(81, 81)$. We wish to find $P(X \geq 90)$. We compute the z-score of 90, which is

$$z = \frac{90 - 81}{\sqrt{81}} = 1.00$$

Using the z table, we find that $P(X \geq 90) = 0.1587$. See Figure 4.32.

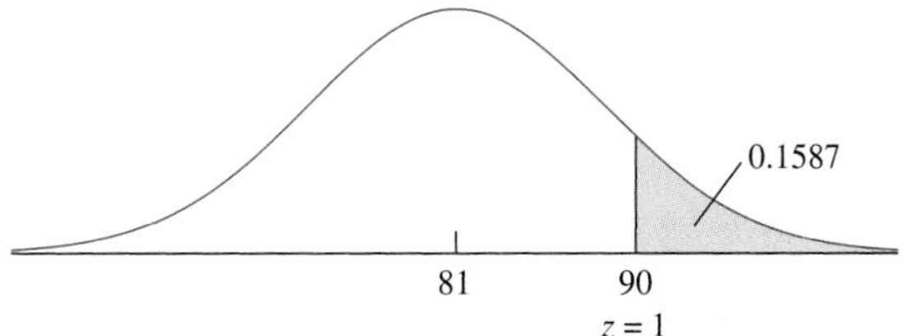

FIGURE 4.32 Solution to Example 4.69.

Exercises for Section 4.10

1. Bags of concrete mix labeled as containing 100 lb have a population mean weight of 100 lb and a population standard deviation of 0.5 lb.
   a. What is the probability that the mean weight of a random sample of 50 bags is less than 99.9 lb?
   b. If the population mean weight is increased to 100.15 lb, what is the probability that the mean weight of a sample of size 50 will be less than 100 lb?

2. A 500-page book contains 250 sheets of paper. The thickness of the paper used to manufacture the book has mean 0.08 mm and standard deviation 0.01 mm.
   a. What is the probability that a randomly chosen book is more than 20.2 mm thick (not including the covers)?
   b. What is the 10th percentile of book thicknesses?
   c. Someone wants to know the probability that a randomly chosen page is more than 0.1 mm thick. Is enough information given to compute this probability? If so, compute the probability. If not, explain why not.

3. A simple random sample of 100 men is chosen from a population with mean height 70 in. and standard deviation 2.5 in. What is the probability that the average height of the sample men is greater than 69.5 in?

4. Among all the income-tax forms filed in a certain year, the mean tax paid was \$2000 and the standard deviation was \$500. In addition, for 10% of the forms, the tax paid was greater than \$3000. A random sample of 625 tax forms is drawn.
   a. What is the probability that the average tax paid on the sample forms is greater than \$1980?
   b. What is the probability that more than 60 of the sampled forms have a tax of greater than \$3000?

5. A sample of 225 wires is drawn from the population of wires described in Example 4.64 (p. 271). Find the probability that fewer than 110 of these wires have no flaws.

6. Drums labeled 30 L are filled with a solution from a large vat. The amount of solution put into each drum is random with mean 30.01 L and standard deviation 0.1 L.
   a. What is the probability that the total amount of solution contained in 50 drums is more than 1500 L?
   b. If the total amount of solution in the vat is 2401 L, what is the probability that 80 drums can be filled without running out?
   c. How much solution should the vat contain so that the probability is 0.9 that 80 drums can be filled without running out?

7. A certain electronic component manufacturing process produces parts, 20% of which are defective. Parts are shipped in units of 400. Shipments containing more than 90 defective parts may be returned. You may assume that each shipment constitutes a simple random sample of parts.
   a. What is the probability that a given shipment will be returned?
   b. On a particular day, 500 shipments are made. What is the probability that 60 or more of these shipments are returned?
   c. A new manufacturing process is introduced that is supposed to reduce the percentage of defective parts. The goal of the company is to reduce the

probability that a shipment is returned to 0.01. What must the percentage of defective parts be in order to reach this goal?

8. True or false:

a. With a large sample, the sample histogram will closely resemble the normal curve.

b. With a large sample, the probability density function of the sample mean will closely resemble the normal curve.

9. The density of particles in a suspension is 50 per mL. A 5 mL volume of the suspension is withdrawn.

a. What is the probability that the number of particles withdrawn will be between 235 and 265?

b. What is the probability that the average number of particles per mL in the withdrawn sample is between 48 and 52?

c. If a 10 mL sample is withdrawn, what is the probability that the average number per mL of particles in the withdrawn sample is between 48 and 52?

d. How large a sample must be withdrawn so that the average number of particles per mL in the sample is between 48 and 52 with probability 95%?

10. A battery manufacturer claims that the lifetime of a certain type of battery has a population mean of 40 hours and a standard deviation of 5 hours. Let $\overline{X}$ represent the mean lifetime of the batteries in a simple random sample of size 100.

a. If the claim is true, what is $P(\overline{X} \leq 36.7)$?

b. Based on the answer to part (a), if the claim is true, is a sample mean lifetime of 36.7 hours unusually short?

c. If the sample mean lifetime of the 100 batteries were 36.7 hours, would you find the manufacturer's claim to be plausible? Explain.

d. If the claim is true, what is $P(\overline{X} \leq 39.8)$?

e. Based on the answer to part (d), if the claim is true, is a sample mean lifetime of 39.8 hours unusually short?

f. If the sample mean lifetime of the 100 batteries were 39.8 hours, would you find the manufacturer's claim to be plausible? Explain.

11. A new process has been designed to make ceramic tiles. The goal is to have no more than 5% of the tiles be nonconforming due to surface defects. A random sample of 1000 tiles is inspected. Let X be the number of nonconforming tiles in the sample.

a. If 5% of the tiles produced are nonconforming, what is $P(X \geq 75)$?

b. Based on the answer to part (a), if 5% of the tiles are nonconforming, is 75 nonconforming tiles out of 1000 an unusually large number?

c. If 75 of the sample tiles were nonconforming, would it be plausible that the goal had been reached? Explain.

d. If 5% of the tiles produced are nonconforming, what is $P(X \geq 53)$?

e. Based on the answer to part (d), if 5% of the tiles are nonconforming, is 53 nonconforming tiles out of 1000 an unusually large number?

f. If 53 of the sample tiles were nonconforming, would it be plausible that the goal had been reached? Explain.

12. *Radiocarbon dating:* Carbon-14 is a radioactive isotope of carbon that decays by emitting a beta particle. In the earth's atmosphere, approximately one carbon atom in 10^{12} is carbon-14. Living organisms exchange carbon with the atmosphere, so this same ratio holds for living tissue. After an organism dies, it stops exchanging carbon with its environment, and its carbon-14 ratio decreases exponentially with time. The rate at which beta particles are emitted from a given mass of carbon is proportional to the carbon-14 ratio, so this rate decreases exponentially with time as well. By measuring the rate of beta emissions in a sample of tissue, the time since the death of the organism can be estimated. Specifically, it is known that t years after death, the number of beta particle emissions occurring in any given time interval from 1 g of carbon follows a Poisson distribution with rate $\lambda = 15.3e^{-0.0001210t}$ events per minute. The number of years t since the death of an organism can therefore be expressed in terms of λ:

$$t = \frac{\ln 15.3 - \ln \lambda}{0.0001210}$$

An archaeologist finds a small piece of charcoal from an ancient campsite. The charcoal contains 1 g of carbon.

a. Unknown to the archaeologist, the charcoal is 11,000 years old. What is the true value of the emission rate λ?

b. The archaeologist plans to count the number X of emissions in a 25 minute interval. Find the mean and standard deviation of X.

c. The archaeologist then plans to estimate λ with $\widehat{\lambda} = X/25$. What is the mean and standard deviation of $\widehat{\lambda}$?

d. What value for $\widehat{\lambda}$ would result in an age estimate of 10,000 years?

e. What value for $\widehat{\lambda}$ would result in an age estimate of 12,000 years?

f. What is the probability that the age estimate is correct to within ± 1000 years?

4.11 Simulation

When fraternal (nonidentical) twins are born, they may be both boys, both girls, or one of each. Assume that each twin is equally likely to be a boy or a girl, and assume that the sexes of the twins are determined independently. What is the probability that both twins are boys? This probability is easy to compute, using the multiplication rule for independent events. The answer is $(0.5)(0.5) = 0.25$. But let's say that you did not know the multiplication rule. Is there another way that you could estimate this probability? You could do a scientific experiment, or study. You could obtain records of twin births from hospitals, and count the number in which both were boys. If you obtained a large enough number of records, the proportion in which both twins were boys would likely be close to 0.25, and you would have a good estimate of the probability.

Here is an easier way. There are two equally likely outcomes for the birth of a twin: boy and girl. There are also two equally likely outcomes for the toss of a coin: heads and tails. Therefore the number of heads in the toss of two coins has the same distribution as the number of boys in a twin birth (both are binomial with $n = 2$ trials and success probability $p = 0.5$). Rather than go to the trouble of monitoring actual births, you could toss two coins a large number of times. The proportion of tosses in which both coins landed heads could be used to estimate the proportion of births in which both twins are boys.

Estimating the probability that twins are both boys by estimating the probability that two coins both land heads is an example of a **simulation** experiment. If the sides of the coin are labeled "0" and "1," then the toss of a coin is an example of a **random number generator**. A random number generator is a procedure that produces a value that has the same statistical properties as a random quantity sampled from some specified distribution. The random number generated by the toss of a coin comes from a Bernoulli distribution with success probability $p = 0.5$.

Nowadays, computers can generate thousands of random numbers in a fraction of a second, and virtually every statistical software package contains routines that will generate random samples from a wide variety of distributions. When a scientific experiment is too costly, or physically difficult or impossible to perform, and when the probability distribution of the data that would be generated by the experiment is approximately known, computer-generated random numbers from the appropriate distribution can be used in place of actual experimental data. Such computer-generated numbers are called **simulated** or **synthetic** data.

Summary

Simulation refers to the process of generating random numbers and treating them as if they were data generated by an actual scientific experiment. The data so generated are called **simulated** or **synthetic** data.

Simulation methods have many uses, including estimating probabilities, estimating means and variances, verifying an assumption of normality, and estimating bias. We describe some of these methods in the rest of this section.

Using Simulation to Estimate a Probability

Simulation is often used to estimate probabilities that are difficult to calculate directly. Here is an example. An electrical engineer will connect two resistors, labeled 100 Ω and 25 Ω, in parallel. The actual resistances may differ from the labeled values. Denote the actual resistances of the resistors that are chosen by X and Y. The total resistance R of the assembly is given by $R = XY/(X + Y)$. Assume that $X \sim N(100, 10^2)$ and $Y \sim N(25, 2.5^2)$ and that the resistors are chosen independently. Assume that the specification for the resistance of the assembly is $19 < R < 21$. What is the probability that the assembly will meet the specification? In other words, what is $P(19 < R < 21)$?

We will estimate this probability with a simulation. The idea is to generate simulated data that is as close as possible in its distribution to data that would be generated in an actual experiment. In an actual experiment we would take a sample of N resistors labeled 100 Ω, whose actual resistances were $X_1, \ldots, X_N$, and then independently take an equal size sample of resistors labeled 25 Ω, whose actual resistances were $Y_1, \ldots, Y_N$. We would then construct N assemblies with resistances $R_1 = X_1Y_1/(X_1 + Y_1), \ldots, R_N = X_NY_N/(X_N + Y_N)$. The values $R_1, \ldots, R_N$ would be a random sample from the population of all possible values of the total resistance. The proportion of the sample values $R_1, \ldots, R_N$ that fell between 19 and 21 would be an estimate of $P(19 < R < 21)$.

In an actual experiment, $X_1, \ldots, X_N$ would be a random sample from $N(100, 10^2)$ and $Y_1, \ldots, Y_N$ would be a random sample from $N(25, 2.5^2)$. Therefore, in the simulated experiment, we will generate a random sample $X_1^*, \ldots, X_N^*$ from $N(100, 10^2)$ and, independently, a random sample $Y_1^*, \ldots, Y_N^*$ from $N(25, 2.5^2)$. We will then compute simulated total resistances $R_1^* = X_1^*Y_1^*/(X_1^* + Y_1^*), \ldots, R_N^* = X_N^*Y_N^*/(X_N^* + Y_N^*)$. We use the notation X_i^*, Y_i^*, and R_i^* to indicate that these are simulated values from a random number generator rather than actual data from a real experiment. Since the sample $X_1^*, \ldots, X_N^*$ comes from the same distribution as would an actual sample $X_1, \ldots, X_N$, and since $Y_1^*, \ldots, Y_N^*$ comes from the same distribution as would an actual sample $Y_1, \ldots, Y_N$, it follows that the sample $R_1^*, \ldots, R_N^*$ comes from the same distribution as would an actual sample of total resistances $R_1, \ldots, R_N$. Therefore we can treat $R_1^*, \ldots, R_N^*$ as if it were in fact a sample of actual resistances, even though it is really a sample of random numbers generated by a computer.

The results from a simulation with sample size $N = 100$ are given in Table 4.2 (page 284). This is a smaller sample than one would use in practice. In practice, samples of 1000, 10,000, or more are commonly used. Samples this size pose no problem for modern computers and their software, and the larger the sample, the more precise the results.

To make the calculations more transparent, we arrange the 100 values of R_i^* found in Table 4.2 in increasing order:

15.37	15.48	15.58	16.66	16.94	17.18	17.44	17.54	17.68	17.69
17.91	17.95	18.01	18.06	18.21	18.31	18.49	18.58	18.60	18.65
18.71	18.80	18.81	18.85	18.91	18.92	18.93	18.99	18.99	19.01
19.02	19.03	19.06	19.11	19.13	19.14	19.20	19.22	19.24	19.30
19.47	19.52	19.56	19.58	19.60	19.60	19.65	19.71	19.77	19.81
19.84	19.90	19.91	19.95	19.97	19.98	20.03	20.14	20.16	20.17
20.17	20.49	20.52	20.54	20.55	20.55	20.58	20.60	20.60	20.64
20.69	20.75	20.76	20.78	20.81	20.90	20.96	21.06	21.13	21.24
21.41	21.49	21.52	21.54	21.58	21.79	21.84	21.87	21.93	21.93
22.02	22.06	22.11	22.13	22.36	22.42	23.19	23.40	23.71	24.01

To estimate $P(19 < R < 21)$ we determine that 48 values out of the sample of 100 are in this range. We therefore estimate $P(19 < R < 21) = 0.48$. We note that with a larger sample we might use computer software to make this count.

Note the importance of the assumption that the resistance X of the first resistor and the resistance Y of the second resistor were independent. Because of this assumption, we could simulate the experiment by generating independent samples X^* and Y^*. If X and Y had been dependent, we would have had to generate X^* and Y^* to have the same *joint distribution* as X and Y. (Joint distributions are discussed in Section 2.6.) Fortunately, many real problems involve independent samples.

We now present another example of a probability estimated with a simulation.

Example 4.70

An engineer has to choose between two types of cooling fans to install in a computer. The lifetimes, in months, of fans of type A are exponentially distributed with mean 50 months, and the lifetimes of fans of type B are exponentially distributed with mean 30 months. Since type A fans are more expensive, the engineer decides that she will choose type A fans if the probability that a type A fan will last more than twice as long as a type B fan is greater than 0.5. Estimate this probability.

Solution

Let A represent the lifetime, in months, of a randomly chosen type A fan, and let B represent the lifetime, in months, of a randomly chosen type B fan. We need to compute $P(A > 2B)$. We perform a simulation experiment, using samples of size 1000. We generated a random sample $A_1^*, \ldots, A_{1000}^*$ from an exponential distribution with mean 50 ($\lambda = 0.02$) and a random sample $B_1^*, \ldots, B_{1000}^*$ from an exponential distribution with mean 30 ($\lambda = 0.033$). We then count the number of times that $A_i^* > 2B_i^*$. Table 4.3 (page 285) presents the first 10 values, and the last value. The column labeled "$A^* > 2B^*$" contains a "1" if $A_i^* > 2B_i^*$ and a "0" if $A_i^* \leq 2B_i^*$.

TABLE 4.2 Simulated data for resistances in a parallel circuit

	X^*	Y^*	R^*		X^*	Y^*	R^*		X^*	Y^*	R^*		X^*	Y^*	R^*
1	102.63	24.30	19.65	26	115.94	24.93	20.52	51	94.20	23.68	18.92	76	113.32	22.54	18.80
2	96.83	21.42	17.54	27	100.65	28.36	22.13	52	82.62	27.82	20.81	77	90.82	23.79	18.85
3	96.46	26.34	20.69	28	89.71	23.00	18.31	53	119.49	22.88	19.20	78	102.88	25.99	20.75
4	88.39	22.10	17.68	29	104.93	24.10	19.60	54	99.43	28.03	21.87	79	93.59	23.04	18.49
5	113.07	29.17	23.19	30	93.74	23.68	18.91	55	108.03	21.69	18.06	80	89.19	27.05	20.76
6	117.66	27.09	22.02	31	104.20	24.02	19.52	56	95.32	20.60	16.94	81	95.04	23.76	19.01
7	108.04	18.20	15.58	32	123.43	26.66	21.93	57	80.70	30.36	22.06	82	109.72	30.25	23.71
8	101.13	28.30	22.11	33	101.38	22.19	18.21	58	91.13	20.38	16.66	83	107.35	27.01	21.58
9	105.43	23.51	19.22	34	88.52	26.85	20.60	59	111.35	27.09	21.79	84	89.59	18.55	15.37
10	103.41	24.64	19.90	35	101.23	26.88	21.24	60	118.75	23.92	19.91	85	101.72	24.65	19.84
11	104.89	22.59	18.58	36	86.96	25.66	19.81	61	103.33	23.99	19.47	86	101.24	25.92	20.64
12	94.91	27.86	21.54	37	95.92	26.16	20.55	62	107.77	18.08	15.48	87	109.67	26.61	21.41
13	92.91	27.06	20.96	38	95.97	26.05	20.49	63	104.86	24.64	19.95	88	100.74	26.18	20.78
14	95.86	24.82	19.71	39	93.76	24.71	19.56	64	84.39	25.52	19.60	89	98.44	23.63	19.06
15	100.06	23.65	19.13	40	113.89	22.79	18.99	65	94.26	25.61	20.14	90	101.05	28.81	22.42
16	90.34	23.75	18.81	41	109.37	26.19	21.13	66	82.16	27.49	20.60	91	88.13	28.43	21.49
17	116.74	24.38	20.17	42	91.13	24.93	19.58	67	108.37	27.35	21.84	92	113.94	29.45	23.40
18	90.45	25.30	19.77	43	101.60	28.66	22.36	68	86.16	21.46	17.18	93	97.42	23.78	19.11
19	97.58	23.05	18.65	44	102.69	21.37	17.69	69	105.97	23.59	19.30	94	109.05	23.04	19.02
20	101.19	23.60	19.14	45	108.50	25.34	20.54	70	92.69	23.88	18.99	95	100.65	26.63	21.06
21	101.77	31.42	24.01	46	80.86	27.55	20.55	71	97.48	25.43	20.17	96	105.64	21.57	17.91
22	100.53	24.93	19.98	47	85.80	24.80	19.24	72	110.45	20.70	17.44	97	78.82	23.25	17.95
23	98.00	27.57	21.52	48	105.96	23.20	19.03	73	89.92	27.23	20.90	98	112.31	22.77	18.93
24	108.10	27.51	21.93	49	103.98	21.78	18.01	74	103.78	25.67	20.58	99	100.14	24.95	19.97
25	91.07	23.38	18.60	50	97.97	23.13	18.71	75	95.53	25.55	20.16	100	88.78	25.87	20.03

TABLE 4.3 Simulated data for Example 4.70

	A^*	B^*	$A^* > 2B^*$
1	25.554	12.083	1
2	66.711	11.384	1
3	61.189	15.191	1
4	9.153	119.150	0
5	98.794	45.258	1
6	14.577	139.149	0
7	65.126	9.877	1
8	13.205	12.106	0
9	20.535	21.613	0
10	62.278	13.289	1
⋮	⋮	⋮	⋮
1000	19.705	12.873	0

Among the first 10 pairs (A_i^*, B_i^*), there are 6 for which $A_i^* > 2B_i^*$. Therefore, if we were to base our results on the first 10 values, we would estimate $P(A > 2B) = 6/10 = 0.6$. Of course, 10 simulated pairs are not nearly enough to compute a reliable estimate. Among the 1000 simulated pairs, there were 460 for which $A_i^* > 2B_i^*$. We therefore estimate $P(A > 2B) = 0.460$. The engineer chooses type B. We note that this probability can be computed exactly with a multiple integral. The exact probability is $5/11 = 0.4545$. The simulation approximation is quite good.

A properly simulated sample from a given probability distribution is in fact a simple random sample from that distribution. Therefore the mean and variance of the simulated sample can be used to estimate the mean and variance of the distribution, and a probability plot may be used to determine whether the probability distribution is well approximated by a standard density function, such as the normal curve. We now present some examples.

Estimating Means and Variances

Example 4.71 shows how simulated values can be used to estimate a population mean and standard deviation.

4.71

Use the simulated values R_i^* in Table 4.2 to estimate the mean μ_R and standard deviation σ_R of the total resistance R.

Solution

We may treat the values $R_1^*, \ldots, R_{100}^*$ as if they were a random sample of actual resistances. Therefore we estimate μ_R with the sample mean $\overline{R}^*$ and σ_R with the sample standard deviation s_{R^*}. The sample mean and standard deviation of $R_1^*, \ldots, R_{100}^*$ are $\overline{R}^* = 19.856$ and $s_{R^*} = 1.6926$, respectively. These are the estimates of μ_R and σ_R, respectively.

Comparison with Propagation of Error

In Example 4.71, we used simulation to approximate the mean and standard deviation of a function of random variables $R = XY/(X + Y)$. The method of propagation of error, discussed in Section 3.4, can be used for this purpose as well (see Example 3.20). Of course, simulation can do many things that propagation of error cannot do, such as estimate probabilities and determine whether a given function of random variables is normally distributed. But if what is needed is to estimate the standard deviation of a function of random variables, it is natural to ask whether simulation or propagation of error is the better technique. The answer is that each method has advantages and disadvantages.

To make our discussion of this question concrete, let $X_1, \ldots, X_n$ be independent random variables, and let $U = U(X_1, \ldots, X_n)$ be a function. We wish to estimate σ_U. The first thing that needs to be said is that in many cases, both methods work well and give similar results, so it is just a matter of convenience which is used. Simulation has one advantage, which is that it does not require that the standard deviations of $X_1, \ldots, X_n$ be small, as propagation of error does. Propagation of error has two big advantages, however. First, it is not necessary to know the distributions of $X_1, \ldots, X_n$, as it is for simulation. Second, propagation of error can pinpoint which of the Xs contributes most to the uncertainty in U, which simulation cannot easily do.

Using Simulation to Determine Whether a Population Is Approximately Normal

One of the most frequently arising issues in data analysis is whether a population is approximately normally distributed. When a simulated sample is available from a population, this issue can be addressed.

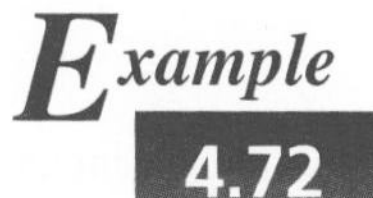

Example 4.72

Construct a histogram of the simulated values of R^* presented in Table 4.2. Construct a normal probability plot to determine whether the density of the total resistance R is approximately normal.

Solution

The histogram and probability plot are shown in the following figure. The histogram is approximately symmetric and has one mode. This is consistent with normality. The normal probability plot suggests a slight departure from normality, especially in the tails. It is fair to say that the distribution appears to be approximately normal. In practice, a sample size of 1000 or more would provide a more precise histogram. A sample of 100 is adequate for the probability plot, although it is no more trouble to generate a larger sample.

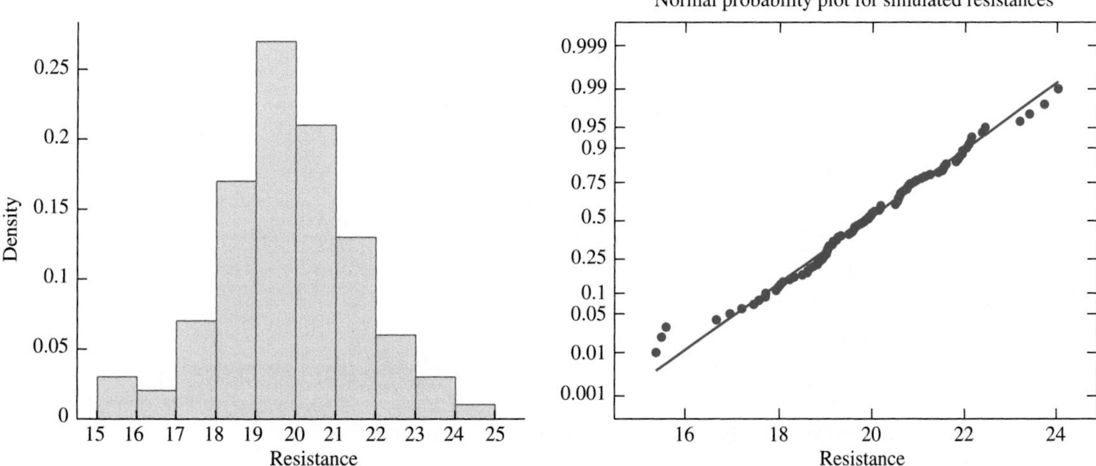

Example 4.73 shows how simulation can be used to determine whether a sample size is large enough for the Central Limit Theorem to hold, if the distribution from which the sample is drawn is known.

Example 4.73

The article "Dermal Absorption from Pesticide Residues" (M. Reddy and A. Bunge, *The Practical Applicability of Toxicokinetic Models in the Risk Assessment of Chemicals*, 2002:55–79) models the amount of pesticide absorbed into the system as a lognormal random variable whose mean is proportional to the dose. Assume that for a certain dose, the amount absorbed is lognormally distributed with parameters $\mu = 1$ and $\sigma = 0.5$. An experiment will be performed in which this dose will be applied in each of five independent trials, and the amount absorbed will be determined each time. Will the average amount absorbed be approximately normally distributed?

Solution

Let $X_1, \ldots, X_5$ be a random sample from a lognormal distribution with parameters $\mu = 1$ and $\sigma = 0.5$. The question asked is whether the sample mean $\overline{X}$ is approximately normally distributed. We will answer this question by generating 1000 simulated random samples of size 5 from this lognormal distribution, computing the sample mean of each of them, and then constructing a normal probability plot for the 1000 sample means. Table 4.4 (page 288) presents the first three and last three of the samples. The first five columns in each row of Table 4.4 constitute a simple random sample $X^*_{1i}, \ldots, X^*_{5i}$ from a lognormal distribution with parameters $\mu = 1$ and $\sigma = 0.5$. The sixth column is the sample mean $\overline{X}^*_i$. The 1000 entries in the sixth column are therefore a random sample of sample means. By constructing a normal probability plot of these values, we can determine whether the sample mean is approximately normally distributed.

TABLE 4.4 Simulated data for Example 4.73

	X_1^*	X_2^*	X_3^*	X_4^*	X_5^*	$\overline{X}^*$
1	2.3220	1.5087	1.2144	2.5092	3.3408	2.1790
2	3.3379	2.8557	1.0023	3.8088	2.3320	2.6673
3	2.9338	3.0364	3.1488	2.0380	4.7030	3.1720
⋮	⋮	⋮	⋮	⋮	⋮	⋮
998	4.7993	3.7609	1.5751	3.6382	2.0254	3.1598
999	3.7929	2.9527	6.3663	1.8057	10.4450	5.0725
1000	3.7680	4.5899	2.8609	2.1659	5.0658	3.6901

Following is a histogram and a normal probability plot of the 1000 values of $\overline{X}^*$. The histogram shows that the distribution is skewed to the right. The probability plot confirms that the distribution is far from normal.

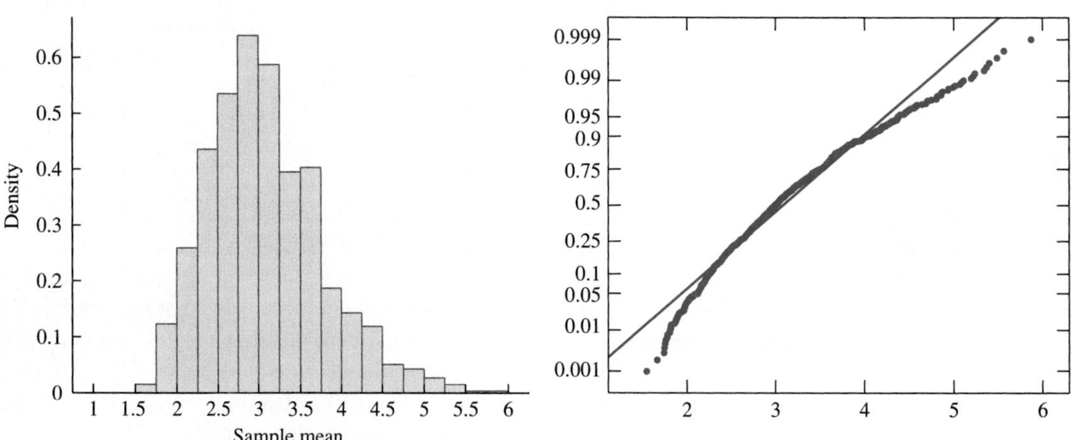

Using Simulation in Reliability Analysis

A system is made up of components, each of which has a lifetime that is random. The lifetime of the system is therefore also random. Reliability engineers often know, at least approximately, the probability distributions of the lifetimes of the components and wish to determine the probability distribution of the system. In practice, it can be very difficult to calculate the distribution of the system lifetime directly from the distributions of the component lifetimes. However, if the lifetimes of the components are independent, it can often be done easily with simulation. Following is a simple example.

Example 4.74

A system consists of components A and B connected in parallel as shown in the following schematic illustration. The lifetime in months of component A is distributed Exp(1), and the lifetime in months of component B is distributed Exp(0.5). The system will function until both A and B fail. Estimate the mean lifetime of the system, the probability that the system functions for less than 1 month, and the 10th percentile of the system lifetimes.

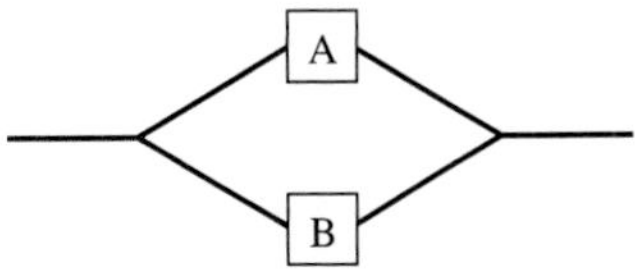

Solution

We generate a sample $A_1^*, \ldots, A_{1000}^*$ of simulated lifetimes of component A from an Exp(1) distribution. Then we generate a sample $B_1^*, \ldots, B_{1000}^*$ of simulated lifetimes of component B from an Exp(0.5) distribution. Note that the mean lifetime for component A is 1 month and the mean lifetime for component B is 1/0.5 = 2 months. The lifetime of the ith simulated system is $L_i^* = \max(A_i^*, B_i^*)$. Table 4.5 presents results for the first 10 samples and for the last sample.

TABLE 4.5 Simulated data for Example 4.74

	A^*	B^*	L^*
1	0.0245	0.5747	0.5747
2	0.3623	0.3998	0.3998
3	0.8858	1.7028	1.7028
4	0.1106	14.2252	14.2252
5	0.1903	0.4665	0.4665
6	2.2259	1.4138	2.2259
7	0.8881	0.9120	0.9120
8	3.3471	3.2134	3.3471
9	2.5475	1.3240	2.5475
10	0.3614	0.8383	0.8383
⋮	⋮	⋮	⋮
1000	0.3619	1.8799	1.8799

The sample mean of the first 10 values of L_i^* is 2.724. Five of them are less than 1. The 10th percentile of these 10 values is $(0.3998 + 0.4665)/2 = 0.43315$. So if we were to base our estimates on the first 10 samples, we would estimate the mean system lifetime to be 2.724 months, the probability that the system fails within a month to be 0.5, and the 10th percentile of system lifetimes to be 0.43315. Of course, 10 samples is not nearly enough for a reliable estimate. Based on all 1000 samples, the estimate of the mean lifetime was 2.29 months, the estimate of the probability of failure within a month was 0.278, and the 10th percentile was 0.516 months.

Using Simulation to Estimate Bias

Simulation can be used to estimate bias. Example 4.75 shows how.

If $X_1, \ldots, X_n$ is a random sample, then the sample standard deviation s is used to estimate the population standard deviation σ. However, s is a biased estimate. If $X_1, \ldots, X_6$ is a simple random sample from a $N(0,1)$ distribution, use simulation to estimate the bias in s. Estimate the standard deviation σ_s of s as well.

Solution

We will generate 1000 random samples $X^*_{1i}, \ldots, X^*_{6i}$ of size 6 from $N(0, 1)$, and for each one compute the sample standard deviation s^*_i. Table 4.6 presents the results for the first 10 samples and for the last sample.

TABLE 4.6 Simulated data for Example 4.75

	X^*_1	X^*_2	X^*_3	X^*_4	X^*_5	X^*_6	s^*
1	−0.4326	0.7160	−0.6028	0.8304	−0.1342	−0.3560	0.6160
2	−1.6656	1.5986	−0.9934	−0.0938	0.2873	−1.8924	1.3206
3	0.1253	−2.0647	1.1889	−0.4598	0.3694	0.4906	1.1190
4	−1.7580	0.1575	−0.8496	0.3291	−1.5780	−1.1100	0.8733
5	1.6867	0.3784	0.3809	0.4870	0.9454	−0.4602	0.7111
6	1.3626	0.7469	−2.1102	2.6734	−0.5311	1.1611	1.6629
7	−2.2424	−0.5719	−1.9659	0.1269	−0.2642	0.3721	1.0955
8	1.3765	−0.4187	−0.5014	1.9869	−0.0532	−0.7086	1.1228
9	−1.8045	0.5361	−0.9121	1.4059	−1.2156	−0.9619	1.2085
10	0.3165	0.6007	−0.5363	−0.2300	0.2626	0.0523	0.4092
⋮	⋮	⋮	⋮	⋮	⋮	⋮	⋮
1000	0.3274	0.1787	0.2006	−1.1602	1.1020	0.3173	0.7328

The values $s^*_1, \ldots, s^*_{1000}$ are a random sample from the population of all possible values of s that can be calculated from a normal sample of size 6. The sample mean $\overline{s}^*$ is therefore an estimate of the population mean μ_s. Now the true standard deviation of the distribution from which the simulated data were generated is $\sigma = 1$, so the bias in s is $\mu_s - 1$. We estimate the bias with $\overline{s}^* - 1$.

The sample mean of the first 10 values of s^*_i is 1.0139. Therefore if we were to base our results on the first 10 values, we would estimate the bias to be $1.0139 - 1 = 0.0139$.

Of course, 10 values is not enough to construct a reliable estimate. The sample mean of the 1000 values s_i^* is $\overline{s}^* = 0.9601$. We estimate the bias to be $0.9601 - 1 = -0.0399$.

The sample standard deviation of the 1000 values is 0.3156. This is the estimate of σ_s.

The Bootstrap

In the examples discussed so far in this section, the distribution from which to generate the simulated data has been specified. In some cases, this distribution must be determined from data. Simulation methods in which the distribution to be sampled from is determined from data are called **bootstrap** methods. To illustrate, we present a variation on Example 4.75 in which the distribution sampled from is determined from data.

A sample of size 6 is taken from a normal distribution whose mean and variance are unknown. The sample values are 5.23, 1.93, 5.66, 3.28, 5.93, and 6.21. The sample mean is $\overline{X} = 4.7067$, and the sample standard deviation is $s = 1.7137$. The value of s will be used to estimate the unknown population standard deviation σ. Estimate the bias in s.

Solution

If we knew the population mean μ and standard deviation σ of the normal distribution from which the sample came, we could use the method of Example 4.75, simulating from a $N(\mu, \sigma)$ distribution. Since we don't know these values, we will estimate them with the sample values $\overline{X} = 4.7067$ and $s = 1.7137$. We will proceed exactly as in Example 4.75, except that we will sample from a $N(4.7067, 1.7137^2)$ distribution. Since this distribution was determined from the data, this is a bootstrap method.

We will generate 1000 random samples $X_{1i}^*, \ldots, X_{6i}^*$ of size 6 from $N(4.7067, 1.7137^2)$, and for each one compute the sample standard deviation s_i^*. Table 4.7 presents the results for the first 10 samples and for the last sample.

TABLE 4.7 Simulated data for Example 4.76

	X_1^*	X_2^*	X_3^*	X_4^*	X_5^*	X_6^*	s^*
1	2.3995	4.8961	3.6221	6.9787	4.4311	4.5367	1.5157
2	2.6197	4.3102	3.2350	6.2619	4.4233	3.5903	1.2663
3	3.0114	5.2492	7.6990	6.0439	6.5965	3.7505	1.7652
4	3.9375	5.2217	1.9737	4.5434	3.0304	3.8632	1.1415
5	5.8829	5.3084	4.6003	2.6439	2.3589	2.3055	1.6054
6	7.8915	3.9731	5.1229	5.1749	3.5255	3.3330	1.6884
7	4.2737	5.5189	2.3314	5.1512	5.7752	4.0205	1.2705
8	5.8602	5.3280	5.5860	6.8256	7.5063	3.9393	1.2400
9	5.7813	4.9364	2.5893	3.7633	0.9065	3.8372	1.7260
10	3.3690	1.8618	2.7627	3.2837	3.9863	6.0382	1.4110
⋮	⋮	⋮	⋮	⋮	⋮	⋮	⋮
1000	2.0496	6.3385	6.2414	5.1580	3.7213	8.4576	2.2364

The values $s_1^*, \ldots, s_{1000}^*$ are a random sample from the population of all possible values of s that can be calculated from a normal sample of size 6. The sample mean $\overline{s}^*$ is therefore an estimate of the population mean μ_s. Now the standard deviation of the population from which the simulated data were generated is $\sigma^* = 1.7137$. We estimate the bias with $\overline{s}^* - 1.7137$.

The sample mean of the first 10 values of s_i^* is 1.4630. Therefore if we were to base our results on the first 10 values, we would estimate the bias to be $1.4630 - 1.7137 = -0.2507$. Of course, 10 values are not enough to construct a reliable estimate. The sample mean of all the 1000 values of s_i^* is 1.6188. We estimate the bias to be $1.6188 - 1.7137 = -0.0949$.

Bootstrap results can sometimes be used to adjust estimates to make them more accurate. Example 4.77 shows how this can be done with the sample standard deviation.

In Example 4.76, a sample of size 6 was taken from an $N(\mu, \sigma^2)$ population. The sample standard deviation $s = 1.7137$ is an estimate of the unknown population standard deviation σ. Use the results of the bootstrap in Example 4.76 to reduce the bias in this estimate.

Solution

We estimated the bias in s to be -0.0949. This means that on the average, the sample standard deviation computed from this $N(\mu, \sigma^2)$ population will be less than the true standard deviation σ by about -0.0949. We therefore adjust for the bias by adding 0.0949 to the estimate. The bias-corrected estimate of the population standard deviation is $1.7137 + 0.0949 = 1.81$.

Parametric and Nonparametric Bootstrap

In Example 4.76, we knew that the sample came from a normal distribution, but we didn't know the mean and variance. We therefore used the data to estimate the parameters μ and σ. This procedure is called the **parametric bootstrap**, because the data are used to estimate parameters. What if we hadn't known that the distribution was normal? Then we would have used the **nonparametric bootstrap**. In the nonparametric bootstrap, we simulate by sampling from the data itself. The nonparametric bootstrap is useful in constructing confidence intervals and in performing hypothesis tests. We will briefly describe the nonparametric bootstrap, and then present some applications in Sections 5.8 and 6.15.

If we had a sample $X_1, \ldots, X_n$ from an unknown distribution, we would simulate samples $X_{1i}^*, \ldots, X_{ni}^*$ as follows. Imagine placing the values $X_1, \ldots, X_n$ in a box, and drawing out one value at random. Then replace the value and draw again. The second draw is also a draw from the sample $X_1, \ldots, X_n$. Continue until n draws have been made. This is the first simulated sample, called a bootstrap sample: $X_{11}^*, \ldots, X_{n1}^*$. Note that since the sampling is done with replacement, the bootstrap sample will probably contain some of the original sample items more than once, and others not at all. Now draw more bootstrap samples; as many as one would draw in any simulation, perhaps 1000 or more. Then proceed just as in any other simulation.

For more information about the bootstrap and other simulation procedures, Efron and Tibshirani (1993) is an excellent source of information.

Exercises for Section 4.11

1. Vendor A supplies parts, each of which has probability 0.03 of being defective. Vendor B also supplies parts, each of which has probability 0.05 of being defective. You receive a shipment of 100 parts from each vendor.

a. Let X be the number of defective parts in the shipment from vendor A and let Y be the number of defective parts in the shipment from vendor B. What are the distributions of X and Y?

b. Generate simulated samples of size 1000 from the distributions of X and Y.

c. Use the samples to estimate the probability that the total number of defective parts is less than 10.

d. Use the samples to estimate the probability that the shipment from vendor A contains more defective parts than the shipment from vendor B.

e. Construct a normal probability plot for the total number of defective parts. Is this quantity approximately normally distributed?

2. There are two competing designs for a certain semiconductor circuit. The lifetime of the first (in hours) is exponentially distributed with $\lambda = 10^{-4}$, and the lifetime of the second is lognormally distributed with $\mu = 6$ and $\sigma^2 = 5.4$.

a. Use a simulated sample of size 1000 to estimate the probability that a circuit with the first design lasts longer than one with the second design.

b. Estimate the probability that a circuit with the first design lasts more than twice as long as one with the second design.

3. Rectangular plates are manufactured whose lengths are distributed $N(2.0, 0.1^2)$ and whose widths are distributed $N(3.0, 0.2^2)$. Assume the lengths and widths are independent. The area of a plate is given by $A = XY$.

a. Use a simulated sample of size 1000 to estimate the mean and variance of A.

b. Estimate the probability that $P(5.9 < A < 6.1)$.

c. Construct a normal probability plot for the areas. Is the area of a plate approximately normally distributed?

4. A cable is made up of four wires. The breaking strength of each wire is a normally distributed random variable with mean 10 kN and standard deviation 1 kN. The strength of the cable, using the brittle wire method, is estimated to be the strength of the weakest wire multiplied by the number of wires.

a. Use simulated samples of size 1000 to estimate the mean strength of this type of cable.

b. Estimate the median cable strength.

c. Estimate the standard deviation of the cable strength.

d. To be acceptable for a certain application, the probability that the cable breaks under a load of 28 kN must be less than 0.01. Does the cable appear to be acceptable? Explain.

5. The lifetime of a laser (in hours) is lognormally distributed with $\mu = 8$ and $\sigma^2 = 2.4$. Two such lasers are operating independently.

a. Use a simulated sample of size 1000 to estimate the probability that the sum of the two lifetimes is greater than 20,000 hours.

b. Estimate the probability that both lasers last more than 3000 hours.

c. Estimate the probability that both lasers fail before 10,000 hours.

6. *Estimating the value of π.* The following figure suggests how to estimate the value of π with a simulation.

In the figure, a circle with area equal to $\pi/4$ is inscribed in a square whose area is equal to 1. One hundred points have been randomly chosen from within the square. The probability that a point is inside the circle is equal to the fraction of the area of the square that is taken up by the circle, which is $\pi/4$. We can therefore estimate the value of $\pi/4$ by counting the number of points inside the circle, which is 79, and dividing by the total number of points, which is 100, to obtain the estimate $\pi/4 \approx 0.79$. From this we conclude that $\pi \approx 4(0.79) = 3.16$. This exercise presents a simulation experiment that is designed to estimate the value of π by generating 1000 points in the unit square.

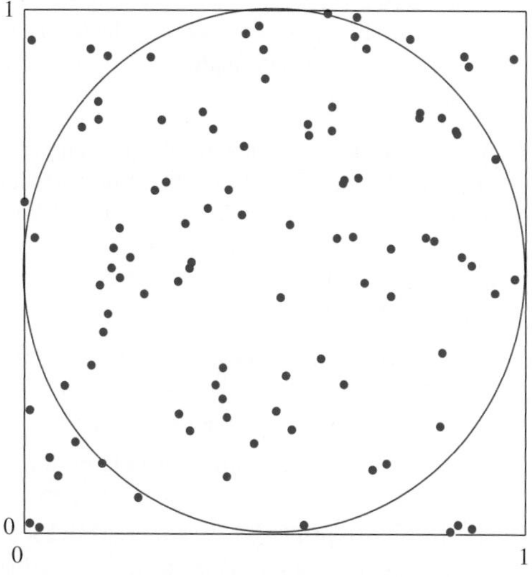

a. Generate 1000 x coordinates $X_1^*, \ldots, X_{1000}^*$. Use the **uniform distribution** with minimum value 0 and maximum value 1. The uniform distribution generates random values that are equally likely to come from any part of the interval (0,1).

b. Generate 1000 y coordinates $Y_1^*, \ldots, Y_{1000}^*$, again using the **uniform distribution** with minimum value 0 and maximum value 1.

c. Each point (X_i^*, Y_i^*) is inside the circle if its distance from the center (0.5,0.5) is less than 0.5. For each pair (X_i^*, Y_i^*), determine whether its distance from the center is less than 0.5. This can be done by computing the value $(X_i^* - 0.5)^2 + (Y_i^* - 0.5)^2$, which is the squared distance, and determining whether it is less than 0.25.

d. How many of the points are inside the circle? What is your estimate of π? (*Note*: With only 1000 points, it is not unlikely for your estimate to be off by as much as 0.05 or more. A simulation with 10,000 or 100,000 points is much more likely to provide an estimate that is very close to the true value.)

7. *Application to mobile computer networks.* Computer scientists often model the movement of a mobile computer as a random path within a rectangle. That is, two points are chosen at random within the rectangle, and the computer moves on a straight line from the first point to the second. In the study of mobile computer networks, it is important to know the mean length of a path (see the article "Stationary Distributions for Random Waypoint Models," W. Navidi and T. Camp, *IEEE Transactions on Mobile Computing*, 2004:99–108). It is very difficult to compute this mean directly, but it is easy to estimate it with a simulation. If the endpoints of a path are denoted (X_1, Y_1), and (X_2, Y_2), then the length of the path is $\sqrt{(X_2 - X_1)^2 + (Y_2 - Y_1)^2}$. The mean length is estimated by generating endpoints (X_1^*, Y_1^*), and (X_2^*, Y_2^*) for many paths, computing the length of each, and taking the mean. This exercise presents a simulation experiment that is designed to estimate the mean distance between two points randomly chosen within a square of side 1.

a. Generate 1000 sets of endpoints (X_{1i}^*, Y_{1i}^*), and (X_{2i}^*, Y_{2i}^*). Use the uniform distribution with minimum value 0 and maximum value 1 for each coordinate of each point. The uniform distribution generates values that are equally likely to come from any part of the interval (0,1).

b. Compute the 1000 path lengths $L_i^* = \sqrt{(X_{2i}^* - X_{1i}^*)^2 + (Y_{2i}^* - Y_{1i}^*)^2}$.

c. Compute the sample mean path length $\overline{L}^*$. The true mean, to six significant digits, is 0.521405. How close did you come?

d. Estimate the probability that a path is more than 1 unit long.

8. Refer to Example 4.74 (p. 289). In order to increase the lifetime of the system, the engineers have a choice between replacing component A with one whose lifetime is distributed Exp(1/2), or replacing component B with one whose lifetime is distributed Exp(1/3).

a. Generate, by simulation, a large number (at least 1000) of system lifetimes under the assumption that component A is replaced.

b. Generate, by simulation, a large number (at least 1000) of system lifetimes under the assumption that component B is replaced.

c. If the goal is to maximize the mean system lifetime, which is the better choice? Explain.

d. If the goal is to minimize the probability that the system fails within a month, which is the better choice? Explain.

e. If the goal is to maximize the 10th percentile of the system lifetimes, which is the better choice? Explain.

9. A system consists of components A and B connected in series, as shown in the following schematic illustration. The lifetime in months of component A is lognormally distributed with $\mu = 1$ and $\sigma = 0.5$, and the lifetime in months of component B is lognormally distributed with $\mu = 2$ and $\sigma = 1$. The system will function only so long as A and B both function.

a. Generate, by simulation, a large number (at least 1000) of system lifetimes.

b. Estimate the mean system lifetime.

c. Estimate the probability that the system fails within 2 months.

d. Estimate the 20th percentile of system lifetimes.

e. Construct a normal probability plot of system lifetimes. Is the system lifetime approximately normally distributed?

f. Construct a histogram of the system lifetimes. Is it skewed to the left, skewed to the right, or approximately symmetric?

10. A system consists of two subsystems connected in series, as shown in the following schematic illustration. Each subsystem consists of two components connected in parallel. The AB subsystem fails when both A and B have failed. The CD subsystem fails when both C and D have failed. The system fails as soon as one of the two subsystems fails. Assume that the lifetimes of the components, in months, have the following distributions: A: Exp(1), B: Exp(0.1), C: Exp(0.2), D: Exp(0.2).

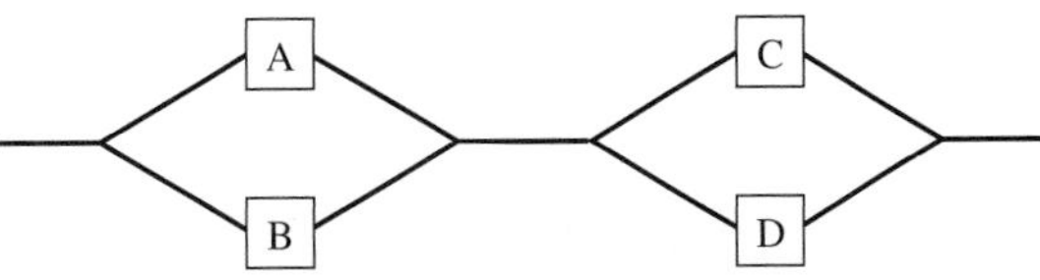

a. Generate, by simulation, a large number (at least 1000) of system lifetimes.

b. Estimate the mean system lifetime.

c. Estimate the median system lifetime.

d. Estimate the probability that the system functions for more than 6 months.

e. Estimate the 90th percentile of system lifetimes.

f. Estimate the probability that the AB subsystem fails before the CD subsystem does.

11. (Continues Exercise 12 on p. 280). The age of an ancient piece of organic matter can be estimated from the rate at which it emits beta particles as a result of carbon-14 decay. For example, if X is the number of particles emitted in 10 minutes by a 10,000-year-old bone fragment that contains 1 g of carbon, then X has a Poisson distribution with mean $\lambda = 45.62$. An archaeologist has found a small bone fragment that contains exactly 1 g of carbon. If t is the unknown age of the bone, in years, the archaeologist will count the number X of particles emitted in 10 minutes and compute an estimated age $\widehat{t}$ with the formula

$$\widehat{t} = \frac{\ln 15.3 - \ln(X/10)}{0.0001210}$$

Unknown to the archaeologist, the bone is exactly 10,000 years old, so X has a Poisson distribution with $\lambda = 45.62$.

a. Generate a simulated sample of 1000 values of X, and their corresponding values of $\widehat{t}$.

b. Estimate the mean of $\widehat{t}$.

c. Estimate the standard deviation of $\widehat{t}$.

d. Estimate the probability that $\widehat{t}$ will be within 1000 years of the actual age of 10,000 years.

e. Estimate the probability that $\widehat{t}$ will be more than 2000 years from the actual age of 10,000 years.

f. Construct a normal probability plot for $\widehat{t}$. Is $\widehat{t}$ approximately normally distributed?

12. A random sample will be drawn from a normal distribution, for the purpose of estimating the population mean μ. Since μ is the median as well as the mean, it seems that both the sample median m and the sample mean $\overline{X}$ are reasonable estimators. This exercise is designed to determine which of these estimators has the smaller uncertainty.

a. Generate a large number (at least 1000) samples of size 5 from a $N(0, 1)$ distribution.

b. Compute the sample medians $m_1^*, \ldots, m_{1000}^*$ for the 1000 samples.

c. Compute the mean $\overline{m}^*$ and the standard deviation s_{m^*} of $m_1^*, \ldots, m_{1000}^*$.

d. Compute the sample means $\overline{X}_1^*, \ldots, \overline{X}_{1000}^*$ for the 1000 samples.

e. Compute the mean and standard deviation $s_{\overline{X}^*}$ of $\overline{X}_1^*, \ldots, \overline{X}_{1000}^*$.

f. The true value of μ is 0. Estimate the bias and uncertainty (σ_m) in m. (*Note*: In fact, the median is unbiased, so your bias estimate should be close to 0.)

g. Estimate the bias and uncertainty ($\sigma_{\overline{X}}$) in $\overline{X}$. Is your bias estimate close to 0? Explain why it should be. Is your uncertainty estimate close to $1/\sqrt{5}$? Explain why it should be.

13. A random sample of size 8 is taken from a Exp(λ) distribution, where λ is unknown. The sample values are 2.74, 6.41, 4.96, 1.65, 6.38, 0.19, 0.52, and 8.38. This exercise shows how to use the bootstrap to estimate the bias and uncertainty ($\sigma_{\widehat{\lambda}}$) in the estimate $\widehat{\lambda} = 1/\overline{X}$.

a. Compute $\widehat{\lambda} = 1/\overline{X}$ for the given sample.

b. Generate 1000 bootstrap samples of size 8 from an Exp($\widehat{\lambda}$) distribution.

c. Compute the values $\widehat{\lambda}_i^* = 1/\overline{X}_i^*$ for each of the 1000 bootstrap samples.

d. Compute the sample mean $\overline{\widehat{\lambda}^*}$ and the sample standard deviation $s_{\widehat{\lambda}^*}$ of $\widehat{\lambda}_1^*, \ldots, \widehat{\lambda}_{1000}^*$.

e. Estimate the bias and uncertainty ($\sigma_{\widehat{\lambda}}$) in $\widehat{\lambda}$.

Supplementary Exercises for Chapter 4

1. An airplane has 100 seats for passengers. Assume that the probability that a person holding a ticket appears for the flight is 0.90. If the airline sells 105 tickets, what is the probability that everyone who appears for the flight will get a seat?

2. The number of large cracks in a length of pavement along a certain street has a Poisson distribution with a mean of 1 crack per 100 m.

a. What is the probability that there will be exactly 8 cracks in a 500 m length of pavement?

b. What is the probability that there will be no cracks in a 100 m length of pavement?

c. Let T be the distance in meters between two successive cracks. What is the probability density function of T?

d. What is the probability that the distance between two successive cracks will be more than 50 m?

3. Pea plants contain two genes for seed color, each of which may be Y (for yellow seeds) or G (for green seeds). Plants that contain one of each type of gene are called heterozygous. According to the Mendelian theory of genetics, if two heterozygous plants are crossed, each of their offspring will have probability 0.75 of having yellow seeds and probability 0.25 of having green seeds.

a. Out of 10 offspring of heterozygous plants, what is the probability that exactly 3 have green seeds?

b. Out of 10 offspring of heterozygous plants, what is the probability that more than 2 have green seeds?

c. Out of 100 offspring of heterozygous plants, what is the probability that more than 30 have green seeds?

d. Out of 100 offspring of heterozygous plants, what is the probability that between 30 and 35 inclusive have green seeds?

e. Out of 100 offspring of heterozygous plants, what is the probability that fewer than 80 have yellow seeds?

4. A simple random sample $X_1, \dots, X_n$ is drawn from a population, and the quantities $\ln X_1, \dots, \ln X_n$ are plotted on a normal probability plot. The points approximately follow a straight line. True or false:

   a. $X_1, \dots, X_n$ come from a population that is approximately lognormal.

   b. $X_1, \dots, X_n$ come from a population that is approximately normal.

   c. $\ln X_1, \dots, \ln X_n$ come from a population that is approximately lognormal.

   d. $\ln X_1, \dots, \ln X_n$ come from a population that is approximately normal.

5. The Environmental Protection Agency (EPA) has contracted with your company for equipment to monitor water quality for several lakes in your water district. A total of 10 devices will be used. Assume that each device has a probability of 0.01 of failure during the course of the monitoring period.

   a. What is the probability that none of the devices fail?

   b. What is the probability that two or more devices fail?

   c. If the EPA requires the probability that none of the devices fail to be at least 0.95, what is the largest individual failure probability allowable?

6. In the article "Occurrence and Distribution of Ammonium in Iowa Groundwater" (K. Schilling, *Water Environment Research*, 2002:177–186), ammonium concentrations (in mg/L) were measured at a large number of wells in the state of Iowa. The mean concentration was 0.71, the median was 0.22, and the standard deviation was 1.09. Is it possible to determine whether these concentrations are approximately normally distributed? If so, say whether they are normally distributed, and explain how you know. If not, describe the additional information you would need to determine whether they are normally distributed.

7. Medication used to treat a certain condition is administered by syringe. The target dose in a particular application is μ. Because of the variations in the syringe, in reading the scale, and in mixing the fluid suspension, the actual dose administered is normally distributed with mean μ and variance σ^2.

   a. What is the probability that the dose administered differs from the mean μ by less than σ?

   b. If X represents the dose administered, find the value of z so that $P(X < \mu + z\sigma) = 0.90$.

   c. If the mean dose is 10 mg, the variance is 2.6 mg^2, and a clinical overdose is defined as a dose larger than 15 mg, what is the probability that a patient will receive an overdose?

8. You have a large box of resistors whose resistances are normally distributed with mean 10 Ω and standard deviation 1 Ω.

   a. What proportion of the resistors have resistances between 9.3 and 10.7 Ω?

   b. If you sample 100 resistors, what is the probability that 50 or more of them will have resistances between 9.3 and 10.7 Ω?

   c. How many resistors must you sample so that the probability is 0.99 that 50 or more of the sampled resistors will have resistances between 9.3 and 10.7 Ω?

9. The intake valve clearances on new engines of a certain type are normally distributed with mean 200 μm and standard deviation 10 μm.

   a. What is the probability that the clearance is greater than 215 μm?

   b. What is the probability that the clearance is between 180 and 205 μm?

   c. An engine has six intake valves. What is the probability that exactly two of them have clearances greater than 215 μm?

10. The stiffness of a certain type of steel beam used in building construction has mean 30 kN/mm and standard deviation 2 kN/mm.

    a. Is it possible to compute the probability that the stiffness of a randomly chosen beam is greater than 32 kN/mm? If so, compute the probability. If not, explain why not.

    b. In a sample of 100 beams, is it possible to compute the probability that the sample mean stiffness of the beams is greater than 30.2 kN/mm? If so, compute the probability. If not, explain why not.

11. In a certain process, the probability of producing an oversize assembly is 0.05.

a. In a sample of 300 randomly chosen assemblies, what is the probability that fewer than 20 of them are oversize?

b. In a sample of 10 randomly chosen assemblies, what is the probability that one or more of them is oversize?

c. To what value must the probability of an oversize assembly be reduced so that only 1% of lots of 300 assemblies contain 20 or more that are oversize?

12. A process for manufacturing plate glass leaves an average of three small bubbles per 10 m^2 of glass. The number of bubbles on a piece of plate glass follows a Poisson distribution.

a. What is the probability that a piece of glass 3×5 m will contain more than two bubbles?

b. What is the probability that a piece of glass 4×6 m will contain no bubbles?

c. What is the probability that 50 pieces of glass, each 3×6 m, will contain more than 300 bubbles in total?

13. Yeast cells are suspended in a liquid medium. A 2 mL sample of the suspension is withdrawn. A total of 56 yeast cells are counted.

a. Estimate the concentration of yeast cells per mL of suspension, and find the uncertainty.

b. What volume of suspension must be withdrawn to reduce the uncertainty to 1 cell per mL?

14. A plate is attached to its base by 10 bolts. Each bolt is inspected before installation, and the probability of passing the inspection is 0.9. Only bolts that pass the inspection are installed. Let X denote the number of bolts that are inspected in order to attach one plate.

a. Find $P(X = 12)$.

b. Find μ_X.

c. Find σ_X.

15. Thicknesses of shims are normally distributed with mean 1.5 mm and standard deviation 0.2 mm. Three shims are stacked, one atop another.

a. Find the probability that the stack is more than 5 mm thick.

b. Find the 80th percentile of the stack thickness.

c. What is the minimum number of shims to be stacked so that the probability that the stack is more than 5 mm thick is at least 0.99?

16. The lifetime of a microprocessor is exponentially distributed with mean 3000 hours.

a. What proportion of microprocessors will fail within 300 hours?

b. What proportion of microprocessors will function for more than 6000 hours?

c. A new microprocessor is installed alongside one that has been functioning for 1000 hours. Assume the two microprocessors function independently. What is the probability that the new one fails before the old one?

17. The lifetime of a bearing (in years) follows the Weibull distribution with parameters $\alpha = 1.5$ and $\beta = 0.8$.

a. What is the probability that a bearing lasts more than 1 year?

b. What is the probability that a bearing lasts less than 2 years?

18. The length of time to perform an oil change at a certain shop is normally distributed with mean 29.5 minutes and standard deviation 3 minutes. What is the probability that a mechanic can complete 16 oil changes in an eight hour day?

19. A surveyor will estimate the height of a cliff by taking n independent measurements and averaging them. Each measurement is unbiased and has standard deviation $\sigma = 1$ m. How many measurements should be taken for the probability to be 0.95 that the average is within ± 0.25 m of the true value?

20. Aluminum cans received by a beverage company are tested for conformance to a strength specification. From a very large lot of cans, 100 are sampled at random and tested. The shipment is rejected if 12 or more of the cans fail the test. Assume that 20% of the cans in the lot do not meet the specification. What is the probability that the lot will be rejected?

21. A cereal manufacturer claims that the gross weight (including packaging) of a box of cereal labeled as weighing 12 oz has a mean of 12.2 oz and a standard deviation of 0.1 oz. You gather 75 boxes and weigh them all together. Let S denote the total weight of the 75 boxes of cereal.

a. If the claim is true, what is $P(S \leq 914.8)$?

b. Based on the answer to part (a), if the claim is true, is 914.8 oz an unusually small total weight for a sample of 75 boxes?

c. If the total weight of the boxes were 914.8 oz, would you be convinced that the claim was false? Explain.

d. If the claim is true, what is $P(S \leq 910.3)$?

e. Based on the answer to part (d), if the claim is true, is 910.3 oz an unusually small total weight for a sample of 75 boxes?

f. If the total weight of the boxes were 910.3 oz, would you be convinced that the claim was false? Explain.

22. Someone claims that the number of hits on his website has a Poisson distribution with mean 20 per hour. Let X be the number of hits in five hours.

a. If the claim is true, what is $P(X \leq 95)$?

b. Based on the answer to part (a), if the claim is true, is 95 hits in a five-hour time period an unusually small number?

c. If you observed 95 hits in a five-hour time period, would this be convincing evidence that the claim is false? Explain.

d. If the claim is true, what is $P(X \leq 65)$?

e. Based on the answer to part (d), if the claim is true, is 65 hits in a five hour time period an unusually small number?

f. If you observed 65 hits in a five hour time period, would this be convincing evidence that the claim is false? Explain.

23. Let $X \sim \text{Geom}(p)$. Let $s \geq 0$ be an integer.

a. Show that $P(X > s) = (1 - p)^s$. (*Hint:* The probability that more than s trials are needed to obtain the first success is equal to the probability that the first s trials are all failures.)

b. Let $t \geq 0$ be an integer. Show that $P(X > s + t \mid X > s) = P(X > t)$. This is called the *lack of memory property*. [*Hint:* $P(X > s + t$ and $X > s) = P(X > s + t)$.]

c. A penny and a nickel are both fair coins. The penny is tossed three times and comes up tails each time. Now both coins will be tossed twice each, so that the penny will be tossed a total of five times and the nickel will be tossed twice. Use the lack of memory property to compute the conditional probability that all five tosses of the penny will be tails, given that the first three tosses were tails. Then compute the probability that both tosses of the nickel will be tails. Are both probabilities the same?

24. Let $X \sim \text{Bin}(n, p)$.

a. Show that if x is an integer between 1 and n inclusive, then

$$\frac{P(X = x)}{P(X = x - 1)} = \left(\frac{n - x + 1}{x}\right)\left(\frac{p}{1 - p}\right)$$

b. Show that if $X \sim \text{Bin}(n, p)$, the most probable value for X is the greatest integer less than or equal to $np + p$. [*Hint:* Use part (a) to show that $P(X = x) \geq P(X = x - 1)$ if and only if $x \leq np + p$.]

25. Let $X \sim \text{Poisson}(\lambda)$.

a. Show that if x is a positive integer, then

$$\frac{P(X = x)}{P(X = x - 1)} = \frac{\lambda}{x}$$

b. Show that if $X \sim \text{Poisson}(\lambda)$, the most probable value for X is the greatest integer less than or equal to λ. [*Hint:* Use part (a) to show that $P(X = x) \geq P(X = x - 1)$ if and only if $x \leq \lambda$.]

Chapter 5

Confidence Intervals

Introduction

In Chapter 4 we discussed estimates for various parameters: for example, $\widehat{p}$ as an estimate of a success probability p, and $\overline{X}$ as an estimate of a population mean μ. These estimates are called **point estimates**, because they are single numbers, or points. An important thing to remember about point estimates is that they are almost never exactly equal to the true values they are estimating. They are almost always off—sometimes by a little, sometimes by a lot. In order for a point estimate to be useful, it is necessary to describe just how far off the true value it is likely to be. One way to do this is by reporting an estimate of the standard deviation, or uncertainty, in the point estimate. In this chapter, we will show that when the estimate is normally distributed, we can obtain more information about its precision by computing a confidence interval. The following example presents the basic idea.

Assume that a large number of independent measurements, all using the same procedure, are made on the diameter of a piston. The sample mean of the measurements is 14.0 cm, and the uncertainty in this quantity, which is the standard deviation of the sample mean, is 0.1 cm. Assume that the measurements are unbiased. The value 14.0 comes from a normal distribution, because it is the average of a large number of measurements. Now the true diameter of the piston will certainly not be exactly equal to the sample mean of 14.0 cm. However, because the sample mean comes from a normal distribution, we can use its standard deviation to determine how close it is likely to be to the true diameter. For example, it is very unlikely that the sample mean will differ from the true diameter by more than three standard deviations. Therefore we have a high level of confidence that the true diameter is in the interval (13.7, 14.3). On the other hand, it is not too unlikely for the sample mean to differ from the true value by more than one standard deviation. Therefore we have a lower level of confidence that the true diameter is in the interval (13.9, 14.1).

The intervals (13.7, 14.3) and (13.9, 14.1) are **confidence intervals** for the true diameter of the piston. We will see in this chapter how to compute a quantitative measure of the level of confidence that we may have in these intervals, and in other intervals we may construct. Specifically, the results of Section 5.1 will show us that we may be 99.7%

confident that the true diameter of the piston is in the interval (13.7, 14.3), but only 68% confident that the true value is in the interval (13.9, 14.1).

5.1 Large-Sample Confidence Intervals for a Population Mean

We begin with an example. A quality-control engineer wants to estimate the mean fill weight of boxes that have been filled with cereal by a certain machine on a certain day. He draws a simple random sample of 100 boxes from the population of boxes that have been filled by that machine on that day. He computes the sample mean fill weight to be $\overline{X} = 12.05$ oz, and the sample standard deviation to be $s = 0.1$ oz.

Since the population mean will not be exactly equal to the sample mean of 12.05, it is best to construct a **confidence interval** around 12.05 that is likely to cover the population mean. We can then quantify our level of confidence that the population mean is actually covered by the interval. To see how to construct a confidence interval in this example, let μ represent the unknown population mean and let σ^2 represent the unknown population variance. Let $X_1, \ldots, X_{100}$ be the 100 fill weights of the sample boxes. The observed value of the sample mean is $\overline{X} = 12.05$. Since $\overline{X}$ is the mean of a large sample, the Central Limit Theorem specifies that it comes from a normal distribution whose mean is μ and whose standard deviation is $\sigma_{\overline{X}} = \sigma/\sqrt{100}$.

Figure 5.1 presents a normal curve, which represents the distribution of $\overline{X}$. The middle 95% of the curve, extending a distance $1.96\sigma_{\overline{X}}$ on either side of population mean μ, is indicated. The observed value $\overline{X} = 12.05$ is a single draw from this distribution. We have no way to know from what part of the curve this particular value of $\overline{X}$ was drawn. Figure 5.1 presents one possibility, which is that the sample mean $\overline{X}$ lies within

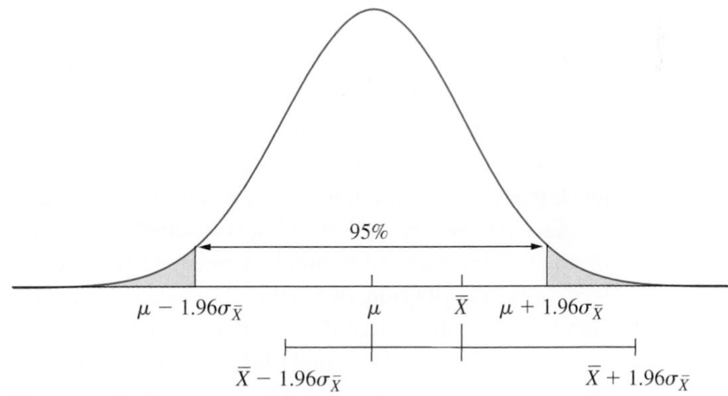

FIGURE 5.1 The sample mean $\overline{X}$ is drawn from a normal distribution with mean μ and standard deviation $\sigma_{\overline{X}} = \sigma/\sqrt{n}$. For this particular sample, $\overline{X}$ comes from the middle 95% of the distribution, so the 95% confidence interval $\overline{X} \pm 1.96\sigma_{\overline{X}}$ succeeds in covering the population mean μ.

the middle 95% of the distribution. Ninety-five percent of all the samples that could have been drawn fall into this category. The horizontal line below the curve in Figure 5.1 is an interval around $\overline{X}$ that is exactly the same length as the middle 95% of the distribution, namely, the interval $\overline{X} \pm 1.96\sigma_{\overline{X}}$. This interval is a **95% confidence interval** for the population mean μ. It is clear that this interval covers the population mean μ.

In contrast, Figure 5.2 represents a sample whose mean $\overline{X}$ lies outside the middle 95% of the curve. Only 5% of all the samples that could have been drawn fall into this category. For these more unusual samples the 95% confidence interval $\overline{X} \pm 1.96\sigma_{\overline{X}}$ fails to cover the population mean μ.

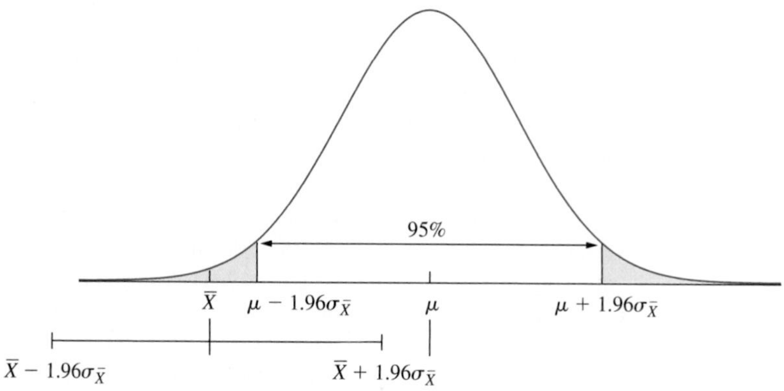

FIGURE 5.2 The sample mean $\overline{X}$ is drawn from a normal distribution with mean μ and standard deviation $\sigma_{\overline{X}} = \sigma/\sqrt{n}$. For this particular sample, $\overline{X}$ comes from the outer 5% of the distribution, so the 95% confidence interval $\overline{X} \pm 1.96\sigma_{\overline{X}}$ fails to cover the population mean μ.

We will now compute a 95% confidence interval $\overline{X} \pm 1.96\sigma_{\overline{X}}$ for the mean fill weight. The value of $\overline{X}$ is 12.05. The population standard deviation σ and thus $\sigma_{\overline{X}} = \sigma/\sqrt{100}$ are unknown. However, in this example, since the sample size is large, we may approximate σ with the sample standard deviation $s = 0.1$. We therefore compute a 95% confidence interval for the population mean fill weight μ to be $12.05 \pm (1.96)(0.01)$, or $(12.0304,\ 12.0696)$. We can say that we are **95% confident**, or **confident at the 95% level**, that the population mean fill weight lies between 12.0304 and 12.0696.

Does this 95% confidence interval actually cover the population mean μ? It depends on whether this particular sample happened to be one whose mean came from the middle 95% of the distribution, or whether it was a sample whose mean was unusually large or small, in the outer 5% of the distribution. There is no way to know for sure into which category this particular sample falls. But imagine that the engineer were to repeat this procedure every day, drawing a large sample and computing the 95% confidence interval $\overline{X} \pm 1.96\sigma_{\overline{X}}$. In the long run, 95% of the samples he draws will have means in the middle 95% of the distribution, so 95% of the confidence intervals he computes will cover the population mean. To put it another way, a 95% confidence interval is computed by a procedure that succeeds in covering the population mean 95% of the time.

We can use this same reasoning to compute intervals with various confidence levels. For example, we can construct a 68% confidence interval as follows. We know that the middle 68% of the normal curve corresponds to the interval extending a distance $1.0\sigma_{\overline{X}}$ on either side of the population mean μ. It follows that an interval of the same length around $\overline{X}$, specifically $\overline{X} \pm \sigma_{\overline{X}}$, will cover the population mean for 68% of the samples that could possibly be drawn. Therefore a 68% confidence interval for the mean fill weight of the boxes is $12.05 \pm (1.0)(0.01)$, or $(12.04, 12.06)$.

Note that the 95% confidence interval is wider than the 68% confidence interval. This is intuitively plausible. In order to increase our confidence that we have covered the true population mean, the interval must be made wider, to provide a wider margin for error. To take two extreme cases, we have 100% confidence that the true population mean is in the infinitely wide interval $(-\infty, \infty)$, and 0% confidence that the true population mean is in the zero-width interval $[12.05, 12.05]$ that contains the sample mean and no other point.

We now illustrate how to find a confidence interval with any desired level of confidence. Specifically, let α be a number between 0 and 1, and let $100(1 - \alpha)\%$ denote the required confidence level. Figure 5.3 presents a normal curve representing the distribution of $\overline{X}$. Define $z_{\alpha/2}$ to be the z-score that cuts off an area of $\alpha/2$ in the right-hand tail. For example, the z table (Table A.2) indicates that $z_{.025} = 1.96$, since 2.5% of the area under the standard normal curve is to the right of 1.96. Similarly, the quantity $-z_{\alpha/2}$ cuts off an area of $\alpha/2$ in the left-hand tail. The middle $1 - \alpha$ of the area under the curve corresponds to the interval $\mu \pm z_{\alpha/2}\sigma_{\overline{X}}$. By the reasoning shown in Figures 5.1 and 5.2, it follows that the interval $\overline{X} \pm z_{\alpha/2}\sigma_{\overline{X}}$ will cover the population mean μ for a proportion $1 - \alpha$ of all the samples that could possibly be drawn. Therefore a **level 100(1 − α)% confidence interval** for μ is $\overline{X} \pm z_{\alpha/2}\sigma_{\overline{X}}$, or $\overline{X} \pm z_{\alpha/2}\sigma/\sqrt{n}$.

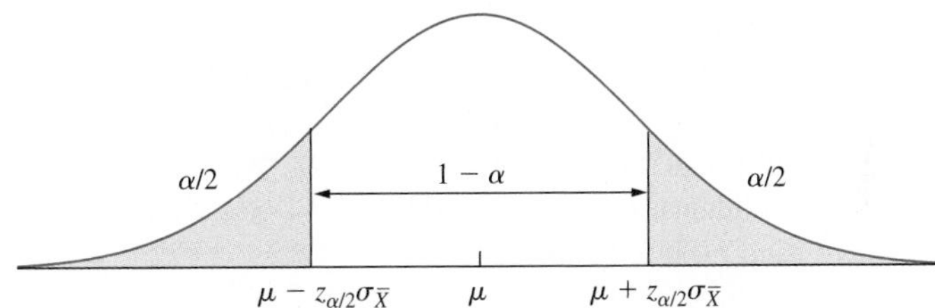

FIGURE 5.3 The sample mean $\overline{X}$ is drawn from a normal distribution with mean μ and standard deviation $\sigma_{\overline{X}} = \sigma/\sqrt{n}$. The quantity $z_{\alpha/2}$ is the z-score that cuts off an area of $\alpha/2$ in the right-hand tail. The quantity $-z_{\alpha/2}$ is the z-score that cuts off an area of $\alpha/2$ in the left-hand tail. The interval $\overline{X} \pm z_{\alpha/2}\sigma_{\overline{X}}$ will cover the population mean μ for a proportion $1 - \alpha$ of all samples that could possibly be drawn. Therefore $\overline{X} \pm z_{\alpha/2}\sigma_{\overline{X}}$ is a level $100(1 - \alpha)\%$ confidence interval for μ.

We note that even for large samples, the distribution of $\overline{X}$ is only *approximately* normal, rather than exactly normal. Therefore the levels stated for confidence intervals are approximate. When the sample size is large enough for the Central Limit Theorem to be used, the distinction between approximate and exact levels is generally ignored in practice.

Summary

Let $X_1, \ldots, X_n$ be a *large* ($n > 30$) random sample from a population with mean μ and standard deviation σ, so that $\overline{X}$ is approximately normal. Then a level $100(1-\alpha)\%$ confidence interval for μ is

$$\overline{X} \pm z_{\alpha/2}\sigma_{\overline{X}} \tag{5.1}$$

where $\sigma_{\overline{X}} = \sigma/\sqrt{n}$. When the value of σ is unknown, it can be replaced with the sample standard deviation s.

In particular,

- $\overline{X} \pm \dfrac{s}{\sqrt{n}}$ is a 68% confidence interval for μ.
- $\overline{X} \pm 1.645\dfrac{s}{\sqrt{n}}$ is a 90% confidence interval for μ.
- $\overline{X} \pm 1.96\dfrac{s}{\sqrt{n}}$ is a 95% confidence interval for μ.
- $\overline{X} \pm 2.58\dfrac{s}{\sqrt{n}}$ is a 99% confidence interval for μ.
- $\overline{X} \pm 3\dfrac{s}{\sqrt{n}}$ is a 99.7% confidence interval for μ.

Example 5.1

The sample mean and standard deviation for the fill weights of 100 boxes are $\overline{X} = 12.05$ and $s = 0.1$. Find an 85% confidence interval for the mean fill weight of the boxes.

Solution

To find an 85% confidence interval, set $1-\alpha = 0.85$ to obtain $\alpha = 0.15$ and $\alpha/2 = 0.075$. We then look in the table for $z_{.075}$, the z-score that cuts off 7.5% of the area in the right-hand tail. We find $z_{.075} = 1.44$. We approximate $\sigma_{\overline{X}} \approx s/\sqrt{n} = 0.01$. So the 85% confidence interval is $12.05 \pm (1.44)(0.01)$. This can be written as 12.05 ± 0.0144, or as $(12.0356, 12.0644)$.

Example 5.2

The article "Study on the Life Distribution of Microdrills" (Z. Yang, Y. Chen, and Y. Yang, *Journal of Engineering Manufacture*, 2002:301–305) reports that in a sample of 50 microdrills drilling a low-carbon alloy steel, the average lifetime (expressed as the number of holes drilled before failure) was 12.68 with a standard deviation of 6.83. Find a 95% confidence interval for the mean lifetime of microdrills under these conditions.

Solution

First let's translate the problem into statistical language. We have a simple random sample $X_1, \ldots, X_{50}$ of lifetimes. The sample mean and standard deviation are $\overline{X} = 12.68$ and $s = 6.83$. The population mean is unknown, and denoted by μ.

The confidence interval has the form $\overline{X} \pm z_{\alpha/2}\sigma_{\overline{X}}$, as specified in expression (5.1). Since we want a 95% confidence interval, the confidence level $1 - \alpha$ is equal to 0.95. Thus $\alpha = 0.05$, and $z_{\alpha/2} = z_{.025} = 1.96$. We approximate σ with $s = 6.83$, and obtain $\sigma_{\overline{X}} \approx 6.83/\sqrt{50} = 0.9659$. Thus the 95% confidence interval is $12.68 \pm (1.96)(0.9659)$. This can be written as 12.68 ± 1.89, or as (10.79, 14.57).

The following computer output (from MINITAB) presents the 95% confidence interval calculated in Example 5.2.

```
One-Sample Z

The assumed standard deviation = 6.830000

 N        Mean   SE Mean            95% CI
50   12.680000  0.965908  (10.786821, 14.573179)
```

Most of the output is self-explanatory. The quantity labeled "SE Mean" is the standard deviation of the sample mean $\sigma_{\overline{X}}$, approximated by $s/\sqrt{n}$. ("SE Mean" stands for standard error of the mean, which is another term for the standard deviation of the sample mean.)

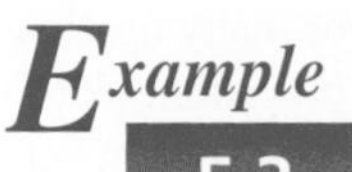

Example 5.3

In Example 5.2, find an 80% confidence interval.

Solution

To find an 80% confidence interval, set $1-\alpha = 0.80$ to obtain $\alpha = 0.20$. Then look in the table for $z_{.10}$, the z-score that cuts off 10% of the area in the right-hand tail. The value is $z_{.10} = 1.28$. So the 80% confidence interval is $12.68 \pm (1.28)(0.9659)$. This can be written as 12.68 ± 1.24, or as (11.44, 13.92).

We have seen how to compute a confidence interval with a given confidence level. It is also possible to compute the level of a given confidence interval. Example 5.4 illustrates the method.

Example 5.4

Based on the microdrill lifetime data presented in Example 5.2, an engineer reported a confidence interval of (11.09, 14.27) but neglected to specify the level. What is the level of this confidence interval?

Solution

The confidence interval has the form $\overline{X} \pm z_{\alpha/2}s/\sqrt{n}$. We will solve for $z_{\alpha/2}$, and then consult the z table to determine the value of α. Now $\overline{X} = 12.68$, $s = 6.83$, and $n = 50$. The upper confidence limit of 14.27 therefore satisfies the equation $14.27 = 12.68 + z_{\alpha/2}(6.83/\sqrt{50})$. Therefore $z_{\alpha/2} = 1.646$. From the z table, we determine that $\alpha/2$, the area to the right of 1.646, is approximately 0.05. The level is $100(1-\alpha)\%$, or 90%.

More About Confidence Levels

The confidence level of an interval measures the reliability of the method used to compute the interval. A level $100(1 - \alpha)\%$ confidence interval is one computed by a method that in the long run will succeed in covering the population mean a proportion $1 - \alpha$ of all the times that it is used. In practice, when one computes a confidence interval, one must decide what level of confidence to use. This decision involves a trade-off, because intervals with greater confidence levels are less precise. For example, a 68% confidence interval specifies the population mean to within $\pm 1.0\sigma_{\overline{X}}$, while a 95% confidence interval specifies it only to within $\pm 1.96\sigma_{\overline{X}}$, and therefore has only about half the precision of the 68% confidence interval. Figure 5.4 illustrates the trade-off between confidence and precision. One hundred samples were drawn from a population with mean μ. Figure 5.4b presents one hundred 95% confidence intervals, each based on one of these samples. The confidence intervals are all different, because each sample has a different mean $\overline{X}$. (They also have different values of s with which to approximate σ, but this has a much smaller effect.) About 95% of these intervals cover the population mean μ. Figure 5.4a presents 68% confidence intervals based on the same samples. These intervals are more precise (narrower), but many of them fail to cover the population mean. Figure 5.4c presents 99.7% confidence intervals. These intervals are very reliable. In the long run, only 3 in 1000 of these intervals will fail to cover the population mean. However, they are less precise (wider), and thus do not convey as much information.

The most common level of confidence used in practice is 95%. For many applications, this level provides a good compromise between reliability and precision. Confidence levels below 90% are rarely used. For some quality-assurance applications, where product reliability is extremely important, intervals with very high confidence levels, such as 99.7%, are used.

Probability versus Confidence

In the fill weight example discussed at the beginning of this section, a 95% confidence interval for the population mean μ was computed to be (12.304, 12.696). It is tempting to say that the probability is 95% that μ is between 12.304 and 12.696. This, however, is not correct. The term *probability* refers to random events, which can come out differently when experiments are repeated. The numbers 12.304 and 12.696 are fixed, not random. The population mean is also fixed. The mean fill weight is either in the interval 12.304 to 12.696, or it is not. There is no randomness involved. Therefore we say that we have 95% *confidence* (not probability) that the population mean is in this interval.

On the other hand, let's say that we are discussing a *method* used to compute a 95% confidence interval. The method will succeed in covering the population mean 95% of the time, and fail the other 5% of the time. In this case, whether the population mean is covered or not is a random event, because it can vary from experiment to experiment. Therefore it *is* correct to say that a *method* for computing a 95% confidence interval has probability 95% of covering the population mean.

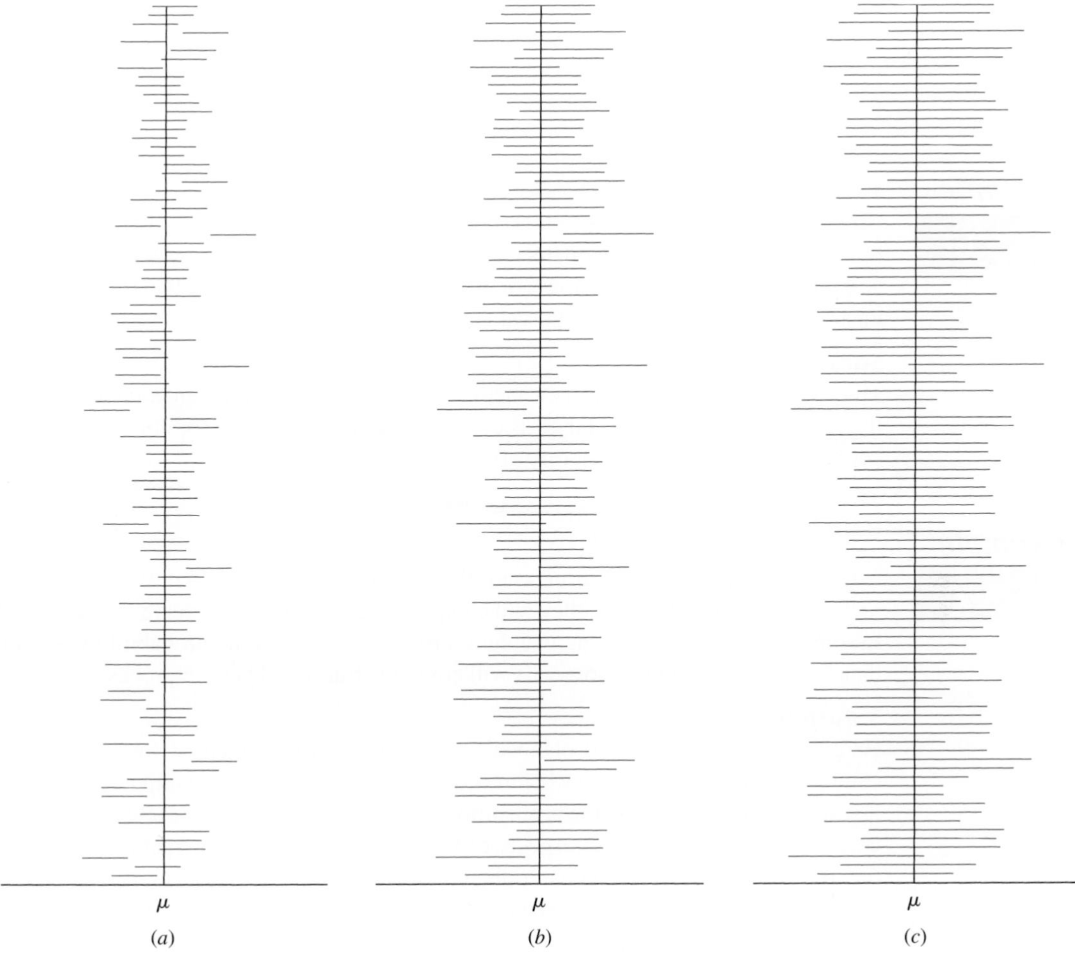

FIGURE 5.4 *(a)* One hundred 68% confidence intervals for a population mean, each computed from a different sample. Although precise, they fail to cover the population mean 32% of the time. This high failure rate makes the 68% confidence interval unacceptable for practical purposes. *(b)* One hundred 95% confidence intervals computed from these samples. This represents a good compromise between reliability and precision for many purposes. *(c)* One hundred 99.7% confidence intervals computed from these samples. These intervals fail to cover the population mean only three times in 1000. They are extremely reliable, but imprecise.

Example 5.5

A 90% confidence interval for the mean diameter (in cm) of steel rods manufactured on a certain extrusion machine is computed to be (14.73, 14.91). True or false: The probability that the mean diameter of rods manufactured by this process is between 14.73 and 14.91 is 90%.

Solution

False. A specific confidence interval is given. The mean is either in the interval or it isn't. We are 90% confident that the population mean is between 14.73 and 14.91. The term *probability* is inappropriate.

An engineer plans to compute a 90% confidence interval for the mean diameter of steel rods. She will measure the diameters of a large sample of rods, compute $\overline{X}$ and s, and then compute the interval $\overline{X} \pm 1.645s/\sqrt{n}$. True or false: The probability that the population mean diameter will be in this interval is 90%.

Solution

True. What is described here is a method for computing a confidence interval, rather than a specific numerical value. It is correct to say that a method for computing a 90% confidence interval has probability 90% of covering the population mean.

A team of geologists plans to measure the weights of 250 rocks. After weighing each rock a large number of times, they will compute a 95% confidence interval for its weight. Assume there is no bias in the weighing procedure. What is the probability that more than 240 of the confidence intervals will cover the true weights of the rocks?

Solution

Here we are discussing 250 planned implementations of a method for computing a confidence interval, not 250 specific intervals that have already been computed. Therefore it is appropriate to compute the probability that a specified number of these intervals will cover the true weights of their respective rocks. Since the weighing procedure is unbiased, the true weight of a rock is equal to the population mean of its measurements. We may think of each of the 250 confidence intervals as a Bernoulli trial, with success occurring if the confidence interval covers its population mean. Since a 95% confidence interval is computed by a process that covers the population mean 95% of the time, the success probability for each Bernoulli trial is 0.95. Let Y represent the number of confidence intervals that cover the true weight. Then $Y \sim \text{Bin}(250,\ 0.95) \approx N(237.5,\ 11.875)$. The standard deviation of Y is $\sigma = \sqrt{11.875} = 3.45$. Using the normal curve, the probability that $Y > 240$ is 0.1922. See Figure 5.5. Note that the continuity correction (see Section 4.10) has been used.

Determining the Sample Size Needed for a Confidence Interval of Specified Width

In Example 5.2, a 95% confidence interval was given by 12.68 ± 1.89, or $(10.79,\ 14.57)$. This interval specifies the mean to within ± 1.89. Now assume that this interval is too wide to be useful. Assume that it is desirable to produce a 95% confidence interval that specifies the mean to within ± 0.50. To do this, the sample size must be increased. We show how to calculate the sample size needed to obtain a confidence interval of any specified width.

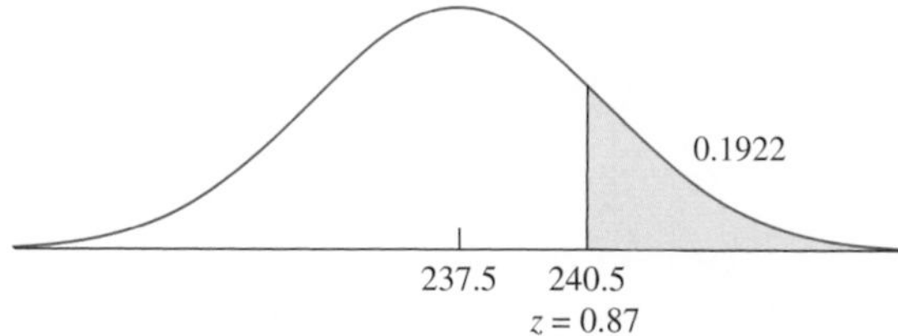

FIGURE 5.5 Solution to Example 5.7.

It follows from expression (5.1) that the width of a confidence interval for a population mean based on a sample of size n drawn from a population with standard deviation σ is $\pm z_{\alpha/2}\sigma/\sqrt{n}$. If the confidence level $100(1-\alpha)\%$ is specified, we can look up the value $z_{\alpha/2}$. If the population standard deviation σ is also specified, we can then compute the value of n needed to produce a specified width. In Example 5.2, the confidence level is 95% and the standard deviation is estimated to be 6.83. We look up $z_{\alpha/2} = z_{.025} = 1.96$. The sample size necessary to produce a 95% confidence interval with width ± 0.50 is found by solving the equation $(1.96)(6.83)/\sqrt{n} = 0.50$ for n. We obtain $n = 716.83$, which we round up to $n = 717$.

Example 5.8

In the fill weight example discussed earlier in this section, the sample standard deviation of weights from 100 boxes was $s = 0.1$ oz. How many boxes must be sampled to obtain a 99% confidence interval of width ± 0.012 oz?

Solution

The level is 99%, so $1-\alpha = 0.99$. Therefore $\alpha = 0.01$ and $z_{\alpha/2} = 2.58$. The value of σ is estimated with $s = 0.1$. The necessary sample size is found by solving $(2.58)(0.1)/\sqrt{n} = 0.012$. We obtain $n \approx 463$.

One-Sided Confidence Intervals

The confidence intervals discussed so far have been **two-sided**, in that they specify both a lower and an upper confidence bound. Occasionally we are interested only in one of these bounds. In these cases, one-sided confidence intervals are appropriate. For example, assume that a reliability engineer wants to estimate the mean crushing strength of a certain type of concrete block, to determine the sorts of applications for which it will be appropriate. The engineer will probably be interested only in a lower bound for the strength, since specifications for various applications will generally specify only a minimum strength.

Assume that a large sample has sample mean $\overline{X}$ and standard deviation $\sigma_{\overline{X}}$. Figure 5.6 (page 310) shows how the idea behind the two-sided confidence interval can be adapted to produce a one-sided confidence interval for the population mean μ. The normal curve represents the distribution of $\overline{X}$. For 95% of all the samples that could be drawn, $\overline{X} < \mu + 1.645\sigma_{\overline{X}}$, and therefore the interval $(\overline{X} - 1.645\sigma_{\overline{X}}, \infty)$ covers μ. This interval will fail to cover μ only if the sample mean is in the upper 5% of its distribution. The interval $(\overline{X} - 1.645\sigma_{\overline{X}}, \infty)$ is a 95% one-sided confidence interval for μ, and the quantity $\overline{X} - 1.645\sigma_{\overline{X}}$ is a 95% lower confidence bound for μ.

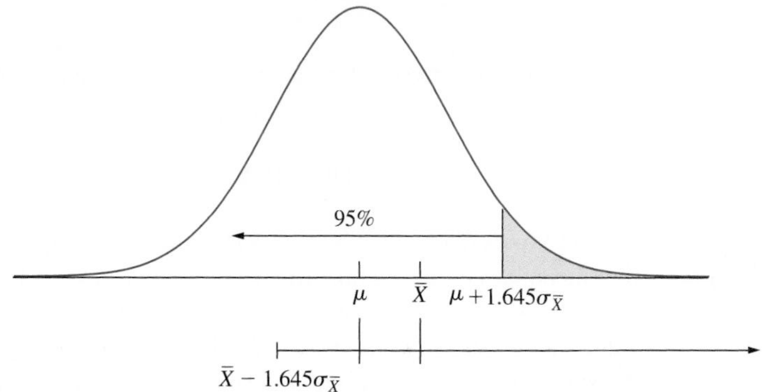

FIGURE 5.6 The sample mean $\overline{X}$ is drawn from a normal distribution with mean μ and standard deviation $\sigma_{\overline{X}} = \sigma/\sqrt{n}$. For this particular sample, $\overline{X}$ comes from the lower 95% of the distribution, so the 95% one-sided confidence interval $(\overline{X} - 1.645\sigma_{\overline{X}}, \infty)$ succeeds in covering the population mean μ.

By constructing a figure like Figure 5.6 with the lower 5% tail shaded, it can be seen that the quantity $\overline{X} + 1.645\sigma_{\overline{X}}$ is a 95% upper confidence bound for μ. We now generalize the method to produce one-sided confidence intervals of any desired level. Define z_α to be the z-score that cuts off an area α in the right-hand tail of the normal curve. For example, $z_{.05} = 1.645$. By the reasoning used to obtain a 95% confidence interval, it can be seen that a level $100(1 - \alpha)$% lower confidence bound for μ is given by $\overline{X} - z_\alpha\sigma_{\overline{X}}$, and a level $1 - \alpha$ upper confidence bound for μ is given by $\overline{X} + z_\alpha\sigma_{\overline{X}}$.

Summary

Let $X_1, \dots, X_n$ be a *large* ($n > 30$) random sample from a population with mean μ and standard deviation σ, so that $\overline{X}$ is approximately normal. Then level $100(1 - \alpha)$% lower confidence bound for μ is

$$\overline{X} - z_\alpha\sigma_{\overline{X}} \tag{5.2}$$

and level $100(1 - \alpha)$% upper confidence bound for μ is

$$\overline{X} + z_\alpha\sigma_{\overline{X}} \tag{5.3}$$

where $\sigma_{\overline{X}} = \sigma/\sqrt{n}$. When the value of σ is unknown, it can be replaced with the sample standard deviation s.

In particular,

- $\overline{X} + 1.28\dfrac{s}{\sqrt{n}}$ is a 90% upper confidence bound for μ.
- $\overline{X} + 1.645\dfrac{s}{\sqrt{n}}$ is a 95% upper confidence bound for μ.
- $\overline{X} + 2.33\dfrac{s}{\sqrt{n}}$ is a 99% upper confidence bound for μ.

The corresponding lower bounds are found by replacing the "+" with "−."

Refer to Example 5.2. Find both a 95% lower confidence bound and a 99% upper confidence bound for the mean lifetime of the microdrills.

Solution

The sample mean and standard deviation are $\overline{X} = 12.68$ and $s = 6.83$, respectively. The sample size is $n = 50$. We estimate $\sigma_{\overline{X}} \approx s/\sqrt{n} = 0.9659$. The 95% lower confidence bound is $\overline{X} - 1.645\sigma_{\overline{X}} = 11.09$, and the 99% upper confidence bound is $\overline{X} + 2.33\sigma_{\overline{X}} = 14.93$.

In Example 5.2, the 95% two-sided confidence interval was computed to be (10.79, 14.57). The 95% lower confidence bound of 11.09, computed in Example 5.9, is greater than the lower bound of the two-sided confidence interval. The reason for this is that the two-sided interval can fail in two ways—the value of μ may be too high or too low. The two-sided 95% confidence interval is designed to fail 2.5% of the time on the high side and 2.5% of the time on the low side. In contrast, the 95% lower confidence bound never fails on the high side. It is therefore designed to fail 5% of the time on the low side, so its lower limit is greater than that of the two-sided interval.

Confidence Intervals Must Be Based on Random Samples

The methods described in this section require that the data be a random sample from a population. When used for other samples, the results may not be meaningful. Following are two examples in which the assumption of random sampling is violated.

Example 5.10

A chemical engineer wishes to estimate the mean yield of a new process. The process is run 100 times over a period of several days. Figure 5.7 (page 312) presents the 100 yields plotted against time. Would it be appropriate to compute a confidence interval for the mean yield by calculating $\overline{X}$ and s for the yields, and then using expression (5.1)?

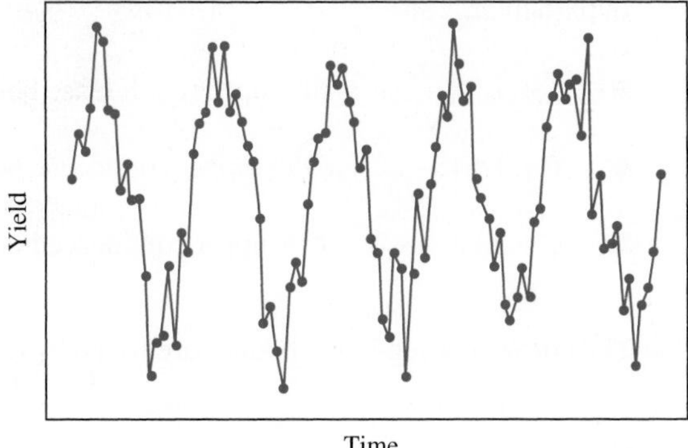

FIGURE 5.7 Yields from 100 runs of a chemical process, plotted against time. There is a clear pattern, indicating that the data do not constitute a random sample.

Solution

No. Expression (5.1) is valid only when the data are a random sample from a population. Figure 5.7 shows a cyclic pattern. This might indicate that the yield on each run is influenced by the yield on the previous run, which would violate the assumption of independence. Another possibility is that the yield is influenced by ambient conditions that fluctuate in a regular fashion. In either case, the data do not satisfy the conditions of a random sample, and expression (5.1) should not be used.

Example 5.11

The engineer in Example 5.10 is studying the yield of another process. Figure 5.8 presents yields from 100 runs of this process, plotted against time. Should expression (5.1) be used to compute a confidence interval for the mean yield of this process?

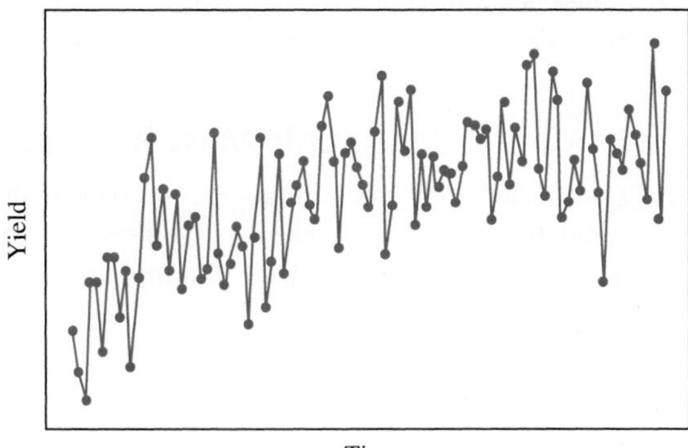

FIGURE 5.8 Yields from 100 runs of a chemical process, plotted against time. There is an increasing trend with time, at least in the initial part of the plot, indicating that the data do not constitute a random sample.

Solution

No. As in Example 5.10, there is a pattern in time. In this case, the yields tend to increase with time, at least in the initial part of the plot. This might indicate a "learning effect"; as an operator becomes more experienced in running a process, the results get better. A more thorough analysis of the data might indicate a point in time where the increase appears to stop, in which case the succeeding portion of the data might be used to form a confidence interval.

Exercises for Section 5.1

1. Find the value of $z_{\alpha/2}$ to use in expression (5.1) to construct a confidence interval with level

a. 90%

b. 83%

c. 99.5%

d. 75%

2. Find the levels of the confidence intervals that have the following values of $z_{\alpha/2}$:

a. $z_{\alpha/2} = 1.96$

b. $z_{\alpha/2} = 2.17$

c. $z_{\alpha/2} = 1.28$

d. $z_{\alpha/2} = 3.28$

3. As the confidence level goes up, the reliability goes ________ and the precision goes ________. *Options:* up, down.

4. Interpolation methods are used to estimate heights above sea level for locations where direct measurements are unavailable. In the article "Transformation of Ellipsoid Heights to Local Leveling Heights" (M. Yanalak and O. Baykal, *Journal of Surveying Engineering*, 2001:90–103), a second-order polynomial method of interpolation for estimating heights from GPS measurements is evaluated. In a sample of 74 locations, the errors made by the method averaged 3.8 cm, with a standard deviation of 4.8 cm.

a. Find a 95% confidence interval for the mean error made by this method.

b. Find a 98% confidence interval for the mean error made by this method.

c. A surveyor claims that the mean error is between 3.2 and 4.4 cm. With what level of confidence can this statement be made?

d. Approximately how many locations must be sampled so that a 95% confidence interval will specify the mean to within ± 0.7 cm?

e. Approximately how many locations must be sampled so that a 98% confidence interval will specify the mean to within ± 0.7 cm?

5. In a random sample of 100 batteries produced by a certain method, the average lifetime was 150 hours and the standard deviation was 25 hours.

a. Find a 95% confidence interval for the mean lifetime of batteries produced by this method.

b. Find a 99% confidence interval for the mean lifetime of batteries produced by this method.

c. An engineer claims that the mean lifetime is between 147 and 153 hours. With what level of confidence can this statement be made?

d. Approximately how many batteries must be sampled so that a 95% confidence interval will specify the mean to within ± 2 hours?

e. Approximately how many batteries must be sampled so that a 99% confidence interval will specify the mean to within ± 2 hours?

6. In a random sample of 53 concrete specimens, the average porosity (in %) was 21.6 and the standard deviation was 3.2.

a. Find a 90% confidence interval for the mean porosity of specimens of this type of concrete.

b. Find a 95% confidence interval for the mean porosity of specimens of this type of concrete.

c. What is the confidence level of the interval (21.0, 22.2)?

d. How many specimens must be sampled so that a 90% confidence interval specifies the mean to within ± 0.3?

e. How many specimens must be sampled so that a 95% confidence interval specifies the mean to within ± 0.3?

7. In a sample of 80 ten-penny nails, the average weight was 1.56 g and the standard deviation was 0.1 g.

a. Find a 95% confidence interval for the mean weight of this type of nail.

b. Find a 98% confidence interval for the mean weight of this type of nail.

c. What is the confidence level of the interval $(1.54,\ 1.58)$?

d. How many nails must be sampled so that a 95% confidence interval specifies the mean to within ± 0.01 g?

e. Approximately how many nails must be sampled so that a 98% confidence interval will specify the mean to within ± 0.01 g?

8. One step in the manufacture of a certain metal clamp involves the drilling of four holes. In a sample of 150 clamps, the average time needed to complete this step was 72 seconds and the standard deviation was 10 seconds.

a. Find a 95% confidence interval for the mean time needed to complete the step.

b. Find a 99.5% confidence interval for the mean time needed to complete the step.

c. What is the confidence level of the interval $(71,\ 73)$?

d. How many clamps must be sampled so that a 95% confidence interval specifies the mean to within ± 1.5 seconds?

e. How many clamps must be sampled so that a 99.5% confidence interval specifies the mean to within ± 1.5 seconds?

9. A supplier sells synthetic fibers to a manufacturing company. A simple random sample of 81 fibers is selected from a shipment. The average breaking strength of these is 29 lb, and the standard deviation is 9 lb.

a. Find a 95% confidence interval for the mean breaking strength of all the fibers in the shipment.

b. Find a 99% confidence interval for the mean breaking strength of all the fibers in the shipment.

c. What is the confidence level of the interval $(27.5,\ 30.5)$?

d. How many fibers must be sampled so that a 95% confidence interval specifies the mean to within ± 1 lb?

e. How many fibers must be sampled so that a 99% confidence interval specifies the mean to within ± 1 lb?

10. Refer to Exercise 5.

a. Find a 95% lower confidence bound for the mean lifetime of this type of battery.

b. An engineer claims that the mean lifetime is greater than 148 hours. With what level of confidence can this statement be made?

11. Refer to Exercise 6.

a. Find a 99% upper confidence bound for the mean porosity.

b. The claim is made that the mean porosity is less than 22.7%. With what level of confidence can this statement be made?

12. Refer to Exercise 7.

a. Find a 90% upper confidence bound for the mean weight.

b. Someone says that the mean weight is less than 1.585 g. With what level of confidence can this statement be made?

13. Refer to Exercise 8.

a. Find a 98% lower confidence bound for the mean time to complete the step.

b. An efficiency expert says that the mean time is greater than 70 seconds. With what level of confidence can this statement be made?

14. Refer to Exercise 9.

a. Find a 95% upper confidence bound for the mean breaking strength.

b. The supplier claims that the mean breaking strength is greater than 28 lb. With what level of confidence can this statement be made?

15. An investigator computes a 95% confidence interval for a population mean on the basis of a sample of size 70. If she wishes to compute a 95% confidence interval that is half as wide, how large a sample does she need?

16. A 95% confidence interval for a population mean is computed from a sample of size 50. Another 95%

confidence interval will be computed from a sample of size 200 drawn from the same population. Choose the best answer to fill in the blank: The interval from the sample of size 50 will be approximately ______ as the interval from the sample of size 200.

i. One-eighth as wide
ii. One-fourth as wide
iii. One-half as wide
iv. The same width
v. Twice as wide
vi. Four times as wide
vii. Eight times as wide

17. Based on tests conducted on a large sample of welded joints, a 90% confidence interval for the mean Rockwell B hardness of a certain type of weld was computed to be (83.2, 84.1). Find a 95% confidence interval for the mean Rockwell B hardness of this type of weld.

18. Sixty-four independent measurements were made of the speed of light. They averaged 299,795 km/s and had a standard deviation of 8 km/s. True or false:

a. A 95% confidence interval for the speed of light is $299{,}795 \pm 1.96$ km/s.
b. The probability is 95% that the speed of light is in the interval $299{,}795 \pm 1.96$.
c. If a 65th measurement is made, the probability is 95% that it would fall in the interval $299{,}795 \pm 1.96$.

19. A large box contains 10,000 ball bearings. A random sample of 120 is chosen. The sample mean diameter is 10 mm, and the standard deviation is 0.24 mm. True or false:

a. A 95% confidence interval for the mean diameter of the 120 bearings in the sample is $10 \pm (1.96)(0.24)/\sqrt{120}$.
b. A 95% confidence interval for the mean diameter of the 10,000 bearings in the box is $10 \pm (1.96)(0.24)/\sqrt{120}$.
c. A 95% confidence interval for the mean diameter of the 10,000 bearings in the box is $10 \pm (1.96)(0.24)/\sqrt{10{,}000}$.

20. Each day a quality engineer selects a random sample of 100 bolts from the day's production, measures their lengths, and computes a 95% confidence interval for the mean length of all the bolts manufactured that day. What is the probability that more than 15 of the confidence intervals constructed in the next 250 days will fail to cover the true mean?

21. Based on a sample of repair records, an engineer calculates a 95% confidence interval for the mean cost to repair a fiber-optic component to be ($140, $160). A supervisor summarizes this result in a report, saying, "We are 95% confident that the mean cost of repairs is less than $160." Is the supervisor underestimating the confidence, overestimating it, or getting it right? Explain.

22. A meteorologist measures the temperature in downtown Denver at noon on each day for one year. The 365 readings average 57°F and have standard deviation 20°F. The meteorologist computes a 95% confidence interval for the mean temperature at noon to be $57^\circ \pm (1.96)(20)/\sqrt{365}$. Is this correct? Why or why not?

5.2 Confidence Intervals for Proportions

The methods of Section 5.1, in particular expression (5.1), can be used to find confidence intervals for the mean of any population from which a large sample has been drawn. When the population has a Bernoulli distribution, this expression takes on a special form. We illustrate this with an example.

In Example 5.2 (in Section 5.1), a confidence interval was constructed for the mean lifetime of a certain type of microdrill when drilling a low-carbon alloy steel. Now assume that a specification has been set that a drill should have a minimum lifetime of 10 holes drilled before failure. A sample of 144 microdrills is tested, and 120, or 83.3%, meet this specification. Let p represent the proportion of microdrills in the population that will meet the specification. We wish to find a 95% confidence interval for p.

We begin by constructing an estimate for p. Let X represent the number of drills in the sample that meet the specification. Then $X \sim \text{Bin}(n, p)$, where $n = 144$ is the sample size. The estimate for p is $\widehat{p} = X/n$. In this example, $X = 120$, so $\widehat{p} = 120/144 = 0.833$. The uncertainty, or standard deviation of $\widehat{p}$, is $\sigma_{\widehat{p}} = \sqrt{p(1-p)/n}$. Since the sample size is large, it follows from the Central Limit Theorem (Equation 4.52 in Section 4.10) that

$$\widehat{p} \sim N\left(p, \frac{p(1-p)}{n}\right)$$

The reasoning illustrated by Figures 5.1 and 5.2 (in Section 5.1) shows that for 95% of all possible samples, the population proportion p satisfies the following inequality:

$$\widehat{p} - 1.96\sqrt{\frac{p(1-p)}{n}} < p < \widehat{p} + 1.96\sqrt{\frac{p(1-p)}{n}} \tag{5.4}$$

At first glance, expression (5.4) looks like a 95% confidence interval for p. However, the limits $\widehat{p} \pm 1.96\sqrt{p(1-p)/n}$ contain the unknown p, and so cannot be computed. The traditional approach is to replace p with $\widehat{p}$, obtaining the confidence interval $\widehat{p} \pm 1.96\sqrt{\widehat{p}(1-\widehat{p})/n}$. Recent research shows that this interval can be improved by modifying both n and $\widehat{p}$ slightly. Specifically, one should add 4 to the number of trials, and 2 to the number of successes. So in place of n we use $\widetilde{n} = n + 4$, and in place of $\widehat{p}$ we use $\widetilde{p} = (X + 2)/\widetilde{n}$. A 95% confidence interval for p is thus given by $\widetilde{p} \pm 1.96\sqrt{\widetilde{p}(1-\widetilde{p})/\widetilde{n}}$. In this example, $\widetilde{n} = 148$ and $\widetilde{p} = 122/148 = 0.8243$, so the 95% confidence interval is 0.8243 ± 0.0613, or $(0.763, 0.886)$.

We justified this confidence interval on the basis of the Central Limit Theorem, which requires n to be large. However, this method of computing confidence intervals is appropriate for any sample size n. When used with small samples, it may occasionally happen that the lower limit is less than 0 or that the upper limit is greater than 1. Since $0 < p < 1$, a lower limit less than 0 should be replaced with 0, and an upper limit greater than 1 should be replaced with 1.

Summary

Let X be the number of successes in n independent Bernoulli trials with success probability p, so that $X \sim \text{Bin}(n, p)$.

Define $\widetilde{n} = n + 4$, and $\widetilde{p} = \dfrac{X+2}{\widetilde{n}}$. Then a level $100(1-\alpha)\%$ confidence interval for p is

$$\widetilde{p} \pm z_{\alpha/2}\sqrt{\frac{\widetilde{p}(1-\widetilde{p})}{\widetilde{n}}} \tag{5.5}$$

If the lower limit is less than 0, replace it with 0. If the upper limit is greater than 1, replace it with 1.

The confidence interval given by expression (5.5) is sometimes called the *Agresti–Coull* interval, after Alan Agresti and Brent Coull, who developed it. For more information on this confidence interval, consult the article "Approximate Is Better Than 'Exact' for Interval Estimation of Binomial Proportions" (A. Agresti and B. Coull, *The American Statistician*, 1998:119–126).

5.12

Interpolation methods are used to estimate heights above sea level for locations where direct measurements are unavailable. In the article "Transformation of Ellipsoid Heights to Local Leveling Heights" (M. Yanalak and O. Baykal, *Journal of Surveying Engineering*, 2001:90–103), a weighted-average method of interpolation for estimating heights from GPS measurements is evaluated. The method made "large" errors (errors whose magnitude was above a commonly accepted threshold) at 26 of the 74 sample test locations. Find a 90% confidence interval for the proportion of locations at which this method will make large errors.

Solution

The number of successes is $X = 26$, and the number of trials is $n = 74$. We therefore compute $\widetilde{n} = 74 + 4 = 78$, $\widetilde{p} = (26 + 2)/78 = 0.3590$, and $\sqrt{\widetilde{p}(1 - \widetilde{p})/\widetilde{n}} = \sqrt{(0.3590)(0.6410)/78} = 0.0543$. For a 90% confidence interval, the value of $\alpha/2$ is 0.05, so $z_{\alpha/2} = 1.645$. The 90% confidence interval is therefore $0.3590 \pm (1.645)(0.0543)$, or (0.270, 0.448).

One-sided confidence intervals can be computed for proportions as well. They are analogous to the one-sided intervals for a population mean (Equations 5.2 and 5.3 in Section 5.1). The levels for one-sided confidence intervals are only roughly approximate for small samples.

Summary

Let X be the number of successes in n independent Bernoulli trials with success probability p, so that $X \sim \text{Bin}(n, p)$.

Define $\widetilde{n} = n + 4$, and $\widetilde{p} = \dfrac{X + 2}{\widetilde{n}}$. Then a level $100(1 - \alpha)\%$ lower confidence bound for p is

$$\widetilde{p} - z_{\alpha}\sqrt{\frac{\widetilde{p}(1 - \widetilde{p})}{\widetilde{n}}} \tag{5.6}$$

and level $100(1 - \alpha)\%$ upper confidence bound for p is

$$\widetilde{p} + z_{\alpha}\sqrt{\frac{\widetilde{p}(1 - \widetilde{p})}{\widetilde{n}}} \tag{5.7}$$

If the lower bound is less than 0, replace it with 0. If the upper bound is greater than 1, replace it with 1.

Example 5.13 shows how to compute the sample size necessary for a confidence interval to have a specified width when a preliminary value of $\widetilde{p}$ is known.

5.13

In Example 5.12, what sample size is needed to obtain a 95% confidence interval with width ± 0.08?

Solution

A 95% confidence interval has width $\pm 1.96\sqrt{\widetilde{p}(1-\widetilde{p})/\widetilde{n}}$, where $\widetilde{n} = n + 4$. Therefore we determine the sample size n by solving the equation $1.96\sqrt{\widetilde{p}(1-\widetilde{p})/(n+4)} = 0.08$. From the data in Example 5.12, $\widetilde{p} = 0.3590$. Substituting this value for $\widetilde{p}$ and solving, we obtain $n \approx 135$.

Sometimes we may wish to compute a necessary sample size without having a reliable estimate $\widetilde{p}$ available. The quantity $\widetilde{p}(1-\widetilde{p})$, which determines the width of the confidence interval, is maximized for $\widetilde{p} = 0.5$. Since the width is greatest when $\widetilde{p}(1-\widetilde{p})$ is greatest, we can compute a conservative sample size estimate by substituting $\widetilde{p} = 0.5$ and proceeding as in Example 5.13.

5.14

In Example 5.12, how large a sample is needed to guarantee that the width of the 95% confidence interval will be no greater than ± 0.08, if no preliminary sample has been taken?

Solution

A 95% confidence interval has width $\pm 1.96\sqrt{\widetilde{p}(1-\widetilde{p})/(n+4)}$. The widest the confidence interval could be, for a sample of size n, is $\pm 1.96\sqrt{(0.5)(1-0.5)/(n+4)}$, or $\pm 0.98/\sqrt{n+4}$. Solving the equation $0.98/\sqrt{n+4} = 0.08$ for n, we obtain $n \approx 147$. Note that this estimate is somewhat larger than the one obtained in Example 5.13.

The Traditional Method

The method we have described was developed quite recently (although it was created by simplifying a much older method). Many people still use a more traditional method. The traditional method uses the actual sample size n in place of $\widetilde{n}$, and the actual sample proportion $\widehat{p}$ in place of $\widetilde{p}$. Although this method is still widely used, it fails to achieve its stated coverage probability even for some fairly large values of n. This means that $100(1-\alpha)\%$ confidence intervals computed by the traditional method will cover the true proportion less than $100(1-\alpha)\%$ of the time. The traditional method cannot be used for small samples at all; one rule of thumb regarding the sample size is that both $n\widehat{p}$ (the number of successes) and $n(1-\widehat{p})$ (the number of failures) should be greater than 10.

Since the traditional method is still widely used, we summarize it in the following box. For very large sample sizes, the results of the traditional method are almost identical to those of the modern method. For small or moderately large sample sizes, the modern approach is better.

Summary

The Traditional Method for Computing Confidence Intervals for a Proportion (widely used but not recommended)

Let $\widehat{p}$ be the proportion of successes in a *large* number n of independent Bernoulli trials with success probability p. Then the traditional level $100(1-\alpha)\%$ confidence interval for p is

$$\widehat{p} \pm z_{\alpha/2}\sqrt{\frac{\widehat{p}(1-\widehat{p})}{n}} \qquad (5.8)$$

The method cannot be used unless the sample contains at least 10 successes and 10 failures.

Exercises for Section 5.2

1. Concentrations of atmospheric pollutants such as carbon monoxide (CO) can be measured with a spectrophotometer. In a calibration test, 50 measurements were taken of a laboratory gas sample that is known to have a CO concentration of 70 parts per million (ppm). A measurement is considered to be satisfactory if it is within 5 ppm of the true concentration. Of the 50 measurements, 37 were satisfactory.

a. What proportion of the sample measurements were satisfactory?

b. Find a 95% confidence interval for the proportion of measurements made by this instrument that will be satisfactory.

c. How many measurements must be taken to specify the proportion of satisfactory measurements to within ± 0.10 with 95% confidence?

d. Find a 99% confidence interval for the proportion of measurements made by this instrument that will be satisfactory.

e. How many measurements must be taken to specify the proportion of satisfactory measurements to within ± 0.10 with 99% confidence?

2. On a certain day, a large number of fuses were manufactured, each rated at 15 A. A sample of 75 fuses is drawn from the day's production, and 17 of them were found to have burnout amperages greater than 15 A.

a. Find a 95% confidence interval for the proportion of fuses manufactured that day whose burnout amperage is greater than 15 A.

b. Find a 98% confidence interval for the proportion of fuses manufactured that day whose burnout amperage is greater than 15 A.

c. Find the sample size needed for a 95% confidence interval to specify the proportion to within ± 0.05.

d. Find the sample size needed for a 98% confidence interval to specify the proportion to within ± 0.05.

e. If a 95% confidence interval is computed each day for 200 days, what is the probability that more than 192 of the confidence intervals cover the true proportions?

3. A soft-drink manufacturer purchases aluminum cans from an outside vendor. A random sample of 70 cans is selected from a large shipment, and each is tested for strength by applying an increasing load to the side of the can until it punctures. Of the 70 cans, 52 meet the specification for puncture resistance.

a. Find a 95% confidence interval for the proportion of cans in the shipment that meet the specification.

b. Find a 90% confidence interval for the proportion of cans in the shipment that meet the specification.

c. Find the sample size needed for a 95% confidence interval to specify the proportion to within ± 0.05.

d. Find the sample size needed for a 90% confidence interval to specify the proportion to within ± 0.05.

e. If a 90% confidence interval is computed each day for 300 days, what is the probability that more than 280 of the confidence intervals cover the true proportions?

4. Refer to Exercise 1. Find a 95% lower confidence bound for the proportion of satisfactory measurements.

5. Refer to Exercise 2. Find a 98% upper confidence bound for the proportion of fuses with burnout amperages greater than 15 A.

6. Refer to Exercise 3. Find a 99% lower confidence bound for the proportion of cans that meet the specification.

7. A random sample of 400 electronic components manufactured by a certain process are tested, and 30 are found to be defective.

a. Let p represent the proportion of components manufactured by this process that are defective. Find a 95% confidence interval for p.

b. How many components must be sampled so that the 95% confidence interval will specify the proportion defective to within ± 0.02?

c. (Hard) The company ships the components in lots of 200. Lots containing more than 20 defective components may be returned. Find a 95% confidence interval for the proportion of lots that will be returned.

8. Refer to Exercise 7. A device will be manufactured in which two of the components in Exercise 7 will be connected in series. The components function independently, and the device will function only if both components function. Let q be the probability that a device functions. Find a 95% confidence interval for q. (*Hint:* Express q in terms of p, and then use the result of Exercise 7a.)

9. The article "Leachate from Land Disposed Residential Construction Waste" (W. Weber, Y. Jang, et al., *Journal of Environmental Engineering*, 2002: 237–245) presents a study of contamination at landfills containing construction and demolition waste. Specimens of leachate were taken from a test site. Out of 42 specimens, 26 contained detectable levels of lead, 41 contained detectable levels of arsenic, and 32 contained detectable levels of chromium.

a. Find a 90% confidence interval for the probability that a specimen will contain a detectable level of lead.

b. Find a 95% confidence interval for the probability that a specimen will contain a detectable level of arsenic.

c. Find a 99% confidence interval for the probability that a specimen will contain a detectable level of chromium.

10. Stainless steels can be susceptible to stress corrosion cracking under certain conditions. A materials engineer is interested in determining the proportion of steel alloy failures that are due to stress corrosion cracking.

a. In the absence of preliminary data, how large a sample must be taken so as to be sure that the 95% confidence interval will specify the proportion to within ± 0.05?

b. In a sample of 100 failures, 20 of them were caused by stress corrosion cracking. Find a 95% confidence interval for the proportion of failures caused by stress corrosion cracking.

c. Based on the data in part (b), estimate the sample size needed so that the 95% confidence interval will specify the proportion to within ± 0.05.

11. For major environmental remediation projects to be successful, they must have public support. The article "Modelling the Non-Market Environmental Costs and Benefits of Biodiversity Using Contingent Value Data" (D. Macmillan, E. Duff, and D. Elston, *Environmental and Resource Economics*, 2001:391–410) reported the results of a survey in which Scottish voters were asked whether they would be willing to pay additional taxes in order to restore the Affric forest. Out of 189 who responded, 61 said they would be willing to pay.

a. Assuming that the 189 voters who responded constitute a random sample, find a 90% confidence interval for the proportion of voters who would be willing to pay to restore the Affric forest.

b. How many voters should be sampled to specify the proportion to within ± 0.03 with 90% confidence?

c. Another survey is planned, in which voters will be asked whether they would be willing to pay in order to restore the Strathspey forest. At this point, no estimate of this proportion is available. Find a conservative estimate of the sample size needed so that the proportion will be specified to within ± 0.03 with 90% confidence.

12. A stock market analyst notices that in a certain year, the price of IBM stock increased on 131 out of 252 trading days. Can these data be used to find a 95% confidence interval for the proportion of days that IBM stock increases? Explain.

5.3 Small-Sample Confidence Intervals for a Population Mean

The methods described in Section 5.1 for computing confidence intervals for a population mean require that the sample size be large. When the sample size is small, there are no good general methods for finding confidence intervals. However, when the population is approximately normal, a probability distribution called the Student's t distribution can be used to compute confidence intervals for a population mean. In this section, we describe this distribution and show how to use it.

The Student's *t* Distribution

If $\overline{X}$ is the mean of a large sample of size n from a population with mean μ and variance σ^2, then the Central Limit Theorem specifies that $\overline{X} \sim N(\mu, \sigma^2/n)$. The quantity $(\overline{X} - \mu)/(\sigma/\sqrt{n})$ then has a normal distribution with mean 0 and variance 1. In addition, the sample standard deviation s will almost certainly be close to the population standard deviation σ. For this reason the quantity $(\overline{X} - \mu)/(s/\sqrt{n})$ is approximately normal with mean 0 and variance 1, so we can look up probabilities pertaining to this quantity in the standard normal table (z table). This enables us to compute confidence intervals of various levels for the population mean μ.

What can we do if $\overline{X}$ is the mean of a *small* sample? If the sample size is small, s may not be close to σ, and $\overline{X}$ may not be approximately normal. If we know nothing about the population from which the small sample was drawn, there are no easy methods for computing confidence intervals. However, if the population is approximately normal, $\overline{X}$ will be approximately normal even when the sample size is small. It turns out that we can still use the quantity $(\overline{X} - \mu)/(s/\sqrt{n})$, but since s is not necessarily close to σ, this quantity will not have a normal distribution. Instead, it has the **Student's t distribution** with $n - 1$ degrees of freedom, which we denote t_{n-1}. The number of degrees of freedom for the t distribution is one less than the sample size.

The Student's t distribution was discovered in 1908 by William Sealy Gossett, a statistician who worked for the Guinness Brewing Company in Dublin, Ireland. The management at Guinness considered the discovery to be proprietary information, and forbade Gossett to publish it. He published it anyway, using the pseudonym "Student." Gossett had done this before; see Section 4.3.

Summary

Let $X_1, \ldots, X_n$ be a *small* (e.g., $n < 30$) sample from a *normal* population with mean μ. Then the quantity

$$\frac{\overline{X} - \mu}{s/\sqrt{n}}$$

has a Student's t distribution with $n - 1$ degrees of freedom, denoted t_{n-1}.

When n is large, the distribution of the quantity $(\overline{X} - \mu)/(s/\sqrt{n})$ is very close to normal, so the normal curve can be used, rather than the Student's t.

The probability density function of the Student's t distribution is different for different degrees of freedom. Figure 5.9 presents plots of the probability density function for several choices of degrees of freedom. The curves all have a shape similar to that of the normal, or z, curve with mean 0 and standard deviation 1. The t curves are more spread out, however. For example, the t curve with one degree of freedom corresponds to a sample size of 2. When drawing samples of size 2, it will frequently happen that the sample standard deviation s is much smaller than σ, which makes the value of $(\overline{X} - \mu)/(s/\sqrt{n})$ quite large (either positive or negative). For this reason, the t curve with one degree of freedom has a lot of area in the tails. For larger sample sizes, the value of s is less likely to be far from σ, and the t curve is closer to the normal curve. With 10 degrees of freedom (corresponding to a sample size of 11), the difference between the t curve and the normal curve is not great. If a t curve with 30 degrees of freedom were plotted in Figure 5.9, it would be indistinguishable from the normal curve.

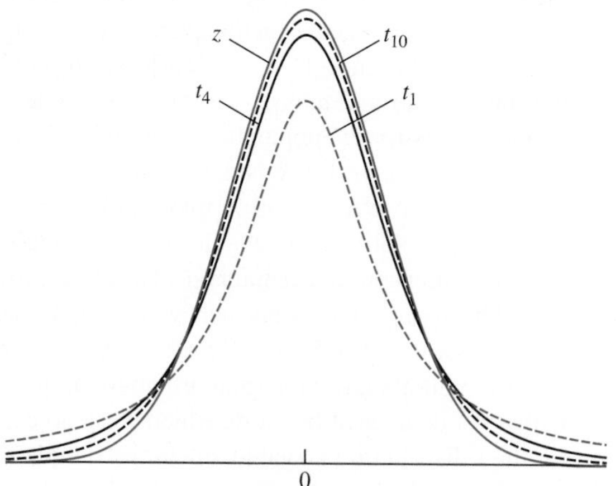

FIGURE 5.9 Plots of the probability density function of the Student's t curve for various degrees of freedom. The normal curve with mean 0 and variance 1 (z curve) is plotted for comparison. The t curves are more spread out than the normal, but the amount of extra spread decreases as the number of degrees of freedom increases.

Table A.3 (in Appendix A), called a ***t* table**, provides probabilities associated with the Student's t distribution. We present some examples to show how to use the table.

Example 5.15

A random sample of size 10 is to be drawn from a normal distribution with mean 4. The Student's t statistic $t = (\overline{X} - 4)/(s/\sqrt{10})$ is to be computed. What is the probability that $t > 1.833$?

Solution

This t statistic has $10-1 = 9$ degrees of freedom. From the t table, $P(t > 1.833) = 0.05$. See Figure 5.10.

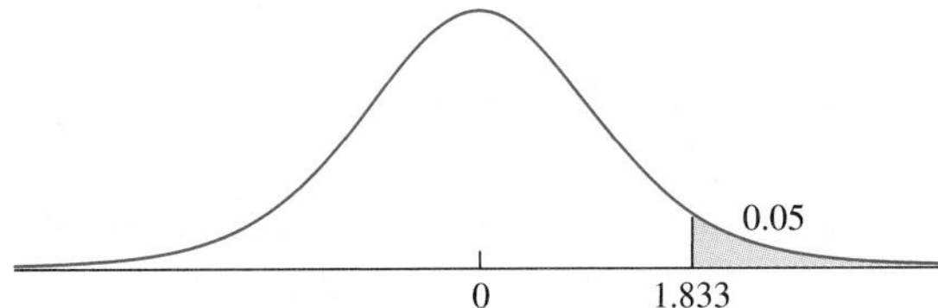

FIGURE 5.10 Solution to Example 5.15.

Example 5.16

Refer to Example 5.15. Find $P(t > 1.5)$.

Solution

Looking across the row corresponding to 9 degrees of freedom, we see that the t table does not list the value 1.5. We find that $P(t > 1.383) = 0.10$ and $P(t > 1.833) = 0.05$. We conclude that $0.05 < P(t > 1.5) < 0.10$. See Figure 5.11. A result more precise than this inequality can be obtained by linear interpolation:

$$P(t > 1.5) \approx 0.10 - \frac{1.5 - 1.383}{1.833 - 1.383}(0.10 - 0.05) = 0.0870$$

A computer package gives the answer correct to three significant digits as 0.0839.

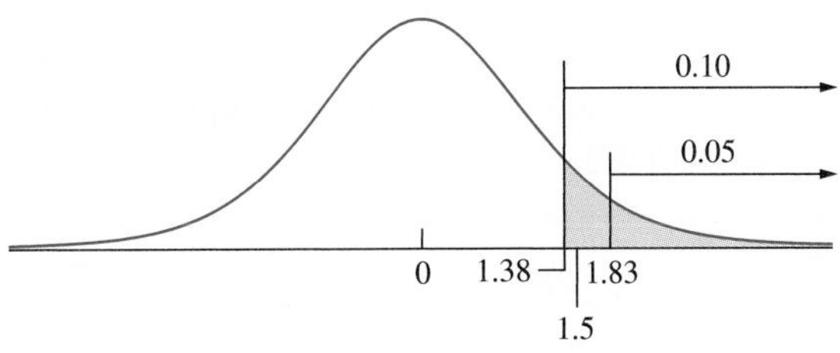

FIGURE 5.11 Solution to Example 5.16.

Example 5.17

Find the value for the t_{12} distribution whose upper-tail probability is 0.025.

Solution

Look down the column headed "0.025" to the row corresponding to 12 degrees of freedom. The value for t_{12} is 2.179.

Find the value for the t_{14} distribution whose lower-tail probability is 0.01.

Solution

Look down the column headed "0.01" to the row corresponding to 14 degrees of freedom. The value for t_{14} is 2.624. This value cuts off an area, or probability, of 1% in the upper tail. The value whose lower-tail probability is 1% is −2.624.

Don't Use the Student's *t* Statistic If the Sample Contains Outliers

For the Student's t statistic to be valid, the sample must come from a population that is approximately normal. Such samples rarely contain outliers. Therefore, methods involving the Student's t statistic should *not* be used for samples that contain outliers.

Confidence Intervals Using the Student's *t* Distribution

When the sample size is small, and the population is approximately normal, we can use the Student's t distribution to compute confidence intervals. We illustrate this with an example.

A metallurgist is studying a new welding process. He manufactures five welded joints and measures the yield strength of each. The five values (in ksi) are 56.3, 65.4, 58.7, 70.1, and 63.9. Assume that these values are a random sample from an approximately normal population. The task is to find a confidence interval for the mean strength of welds made by this process.

When the sample size is large, we don't need to worry much about the nature of the population, because the Central Limit Theorem guarantees that the quantity $\overline{X}$ will be approximately normally distributed. When the sample is small, however, the distribution of the population must be approximately normal.

The confidence interval in this situation is constructed much like the ones in Section 5.1, except that the z-score is replaced with a value from the Student's t distribution. The quantity

$$\frac{\overline{X} - \mu}{s/\sqrt{n}}$$

has a Student's t distribution with $n - 1$ degrees of freedom. Figure 5.12 shows the t_4 distribution. From the Student's t table, we find that 95% of the area under the curve is contained between the values $t = -2.776$ and $t = 2.776$. It follows that for 95% of all the samples that might have been chosen,

$$-2.776 < \frac{\overline{X} - \mu}{s/\sqrt{n}} < 2.776$$

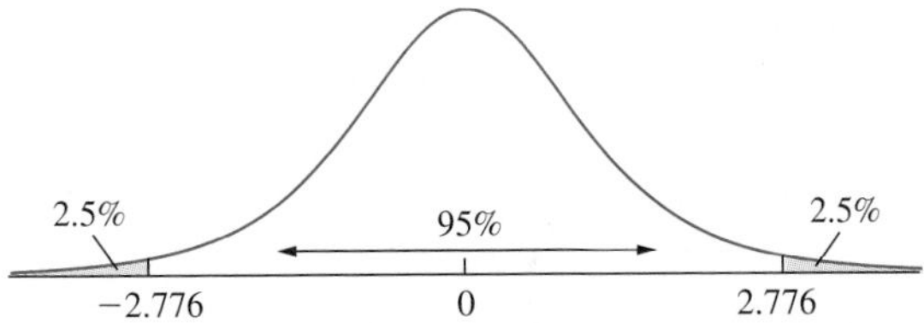

FIGURE 5.12 The Student's t distribution with four degrees of freedom. Ninety-five percent of the area falls between $t = -2.776$ and $t = 2.776$.

Put another way, for 95% of all the samples that might have been chosen, it is the case that

$$-2.776\frac{s}{\sqrt{n}} < \overline{X} - \mu < 2.776\frac{s}{\sqrt{n}}$$

Multiplying by -1 and adding $\overline{X}$ across the inequality, we obtain a 95% confidence interval for μ:

$$\overline{X} - 2.776\frac{s}{\sqrt{n}} < \mu < \overline{X} + 2.776\frac{s}{\sqrt{n}}$$

In this example, the sample mean is $\overline{X} = 62.88$, and the sample standard deviation $s = 5.4838$. The sample size is $n = 5$. Substituting values for $\overline{X}$, s, and n, we find that a 95% confidence interval for μ is $62.88 - 6.81 < \mu < 62.88 + 6.81$, or $(56.07,\ 69.69)$.

In general, to produce a level $100(1-\alpha)\%$ confidence interval, let $t_{n-1,\alpha/2}$ be the $1-\alpha/2$ quantile of the Student's t distribution with $n-1$ degrees of freedom, that is, the value which cuts off an area of $\alpha/2$ in the right-hand tail. For example, earlier we found that $t_{4,.025} = 2.776$. Then a level $100(1-\alpha)\%$ confidence interval for the population mean μ is $\overline{X} - t_{n-1,\alpha/2}(s/\sqrt{n}) < \mu < \overline{X} + t_{n-1,\alpha/2}(s/\sqrt{n})$, or $\overline{X} \pm t_{n-1,\alpha/2}(s/\sqrt{n})$.

Summary

Let $X_1, \ldots, X_n$ be a *small* random sample from a *normal* population with mean μ. Then a level $100(1-\alpha)\%$ confidence interval for μ is

$$\overline{X} \pm t_{n-1,\alpha/2}\frac{s}{\sqrt{n}} \tag{5.9}$$

How Do I Determine Whether the Student's *t* Distribution Is Appropriate?

The Student's t distribution is appropriate whenever the sample comes from a population that is approximately normal. Sometimes one knows from past experience whether a process produces data that are approximately normally distributed. In many cases, however, one must decide whether a population is approximately normal by examining the sample. Unfortunately, when the sample size is small, departures from normality may be hard to detect. A reasonable way to proceed is to construct a boxplot or dotplot of the sample. If these plots do not reveal a strong asymmetry or any outliers, then in most cases the Student's t distribution will be reliable. In principle, one can also determine whether a population is approximately normal by constructing a probability plot. With small samples, however, boxplots and dotplots are easier to draw, especially by hand.

Example 5.19

The article "Direct Strut-and-Tie Model for Prestressed Deep Beams" (K. Tan, K. Tong, and C. Tang, *Journal of Structural Engineering*, 2001:1076–1084) presents measurements of the nominal shear strength (in kN) for a sample of 15 prestressed concrete beams. The results are

580	400	428	825	850	875	920	550
575	750	636	360	590	735	950	

Is it appropriate to use the Student's t statistic to construct a 99% confidence interval for the mean shear strength? If so, construct the confidence interval. If not, explain why not.

Solution

To determine whether the Student's t statistic is appropriate, we will make a boxplot and a dotplot of the sample. These are shown in the following figure.

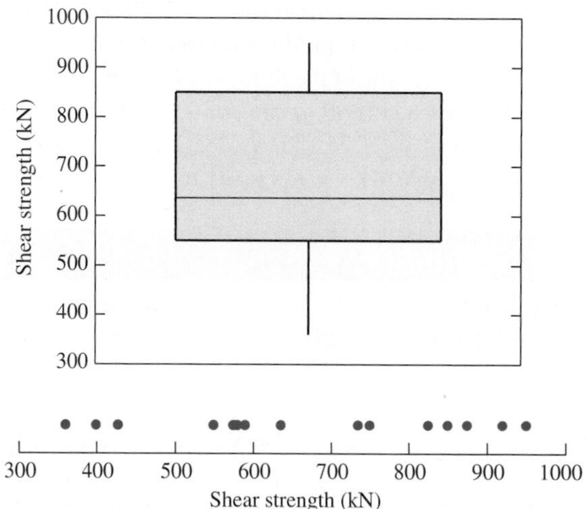

There is no evidence of a major departure from normality; in particular the plots are not strongly asymmetric, and there are no outliers. The Student's t method is appropriate. We therefore compute $\overline{X} = 668.27$ and $s = 192.089$. We use expression (5.9) with $n = 15$ and $\alpha/2 = 0.005$. From the t table with 14 degrees of freedom, we find $t_{14,.005} = 2.977$. The 99% confidence interval is $668.27 \pm (2.977)(192.089)/\sqrt{15}$, or $(520.62,\ 815.92)$.

The following computer output (from MINITAB) presents the confidence interval calculated in Example 5.19.

```
One-Sample T: Strength

Test of mu = 0 vs not = 0

Variable   N      Mean     StDev   SE Mean           99% CI
Strength  15  668.2667  192.0891  49.59718  (520.6159, 815.9175)
```

The output is self-explanatory. The quantity labeled "SE Mean" is the estimated standard deviation of the sample mean, $s/\sqrt{n}$.

In the article referred to in Example 5.19, cylindrical compressive strength (in MPa) was measured for 11 beams. The results were

38.43 38.43 38.39 38.83 38.45 38.35 38.43 38.31 38.32 38.48 38.50

Is it appropriate to use the Student's t statistic to construct a 95% confidence interval for the mean cylindrical compressive strength? If so, construct the confidence interval. If not, explain why not.

Solution

As in Example 5.19, we will make a boxplot and a dotplot of the sample. These are shown in the following figure.

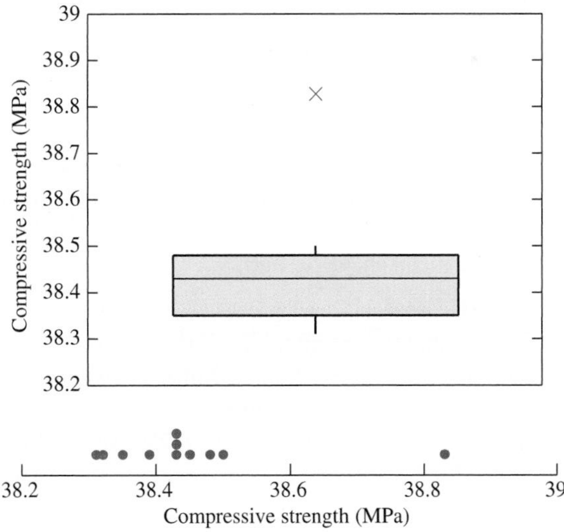

There is an outlier in this sample. The Student's t statistic should not be used.

An engineer reads a report that states that a sample of eleven concrete beams had an average compressive strength of 38.45 MPa with standard deviation 0.14 MPa. Should the t curve be used to find a confidence interval for the mean compressive strength?

Solution

No. The problem is that there is no way of knowing whether the measurements came from a normal population. For example, if the measurements contained an outlier (as in Example 5.20), the confidence interval would be invalid.

The Student's t distribution can be used to compute one-sided confidence intervals. The formulas are analogous to those used with large samples.

Let $X_1, \ldots, X_n$ be a *small* random sample from a *normal* population with mean μ. Then a level $100(1-\alpha)\%$ upper confidence bound for μ is

$$\overline{X} + t_{n-1,\alpha}\frac{s}{\sqrt{n}} \tag{5.10}$$

and a level $100(1-\alpha)\%$ lower confidence bound for μ is

$$\overline{X} - t_{n-1,\alpha}\frac{s}{\sqrt{n}} \tag{5.11}$$

Use z, Not t, If σ Is Known

Occasionally a small sample may be taken from a normal population whose standard deviation σ is known. In these cases, we do not use the Student's t curve, because we are not approximating σ with s. Instead, we use the z table. Example 5.22 illustrates the method.

Refer to Example 5.19. Assume that on the basis of a very large number of previous measurements of other beams, the population of shear strengths is known to be approximately normal, with standard deviation $\sigma = 180.0$ kN. Find a 99% confidence interval for the mean shear strength.

Solution

We compute $\overline{X} = 668.27$. We do not need to compute s, because we know the population standard deviation σ. Since we want a 99% confidence interval, $\alpha/2 = 0.005$. Because we know σ, we use $z_{\alpha/2} = z_{.005}$, rather than a Student's t value, to compute the confidence interval. From the z table, we obtain $z_{.005} = 2.58$. The confidence interval is $668.27 \pm (2.58)(180.0)/\sqrt{15}$, or $(548.36, 788.18)$.

It is important to remember that when the sample size is small, the population must be approximately normal, whether or not the standard deviation is known.

Summary

Let $X_1, \ldots, X_n$ be a random sample (of any size) from a *normal* population with mean μ. If the standard deviation σ is known, then a level $100(1-\alpha)\%$ confidence interval for μ is

$$\overline{X} \pm z_{\alpha/2}\frac{\sigma}{\sqrt{n}} \tag{5.12}$$

Occasionally one has a single value that is sampled from a normal population with known standard deviation. In these cases a confidence interval for μ can be derived as a special case of expression (5.12) by setting $n = 1$.

Summary

Let X be a single value sampled from a *normal* population with mean μ. If the standard deviation σ is known, then a level $100(1-\alpha)\%$ confidence interval for μ is

$$X \pm z_{\alpha/2}\sigma \tag{5.13}$$

Exercises for Section 5.3

1. Find the value of $t_{n-1,\alpha/2}$ needed to construct a two-sided confidence interval of the given level with the given sample size:

a. Level 90%, sample size 9.

b. Level 95%, sample size 5.

c. Level 99%, sample size 29.

d. Level 95%, sample size 2.

2. Find the value of $t_{n-1,\alpha}$ needed to construct an upper or lower confidence bound in each of the situations in Exercise 1.

3. Find the level of a two-sided confidence interval that is based on the given value of $t_{n-1,\alpha/2}$ and the given sample size.

a. $t = 2.179$, sample size 13.

b. $t = 3.365$, sample size 6.

c. $t = 1.729$, sample size 20.

d. $t = 3.707$, sample size 7.

e. $t = 3.707$, sample size 27.

4. True or false: The Student's t distribution may be used to construct a confidence interval for the mean of any population, so long as the sample size is small.

5. The article "Ozone for Removal of Acute Toxicity from Logyard Run-off" (M. Zenaitis and S. Duff, *Ozone Science and Engineering*, 2002:83–90) presents chemical analyses of runoff water from sawmills in British Columbia. Included were measurements of pH for six water specimens: 5.9, 5.0, 6.5, 5.6, 5.9, 6.5. Assuming these to be a random sample of water specimens from an approximately normal population, find a 95% confidence interval for the mean pH.

6. The following are summary statistics for a data set. Would it be appropriate to use the Student's t distribution to construct a confidence interval from these data? Explain.

```
 N       Mean     Median    TrMean     StDev    SE Mean
10      8.905      6.105     6.077     9.690      3.064

Minimum   Maximum       Q1        Q3
  0.512    39.920    1.967     8.103
```

7. The article "An Automatic Visual System for Marble Tile Classification" (L. Carrino, W. Polini, and S. Turchetta, *Journal of Engineering Manufacture*, 2002:1095–1108) describes a measure for the shade of marble tile in which the amount of light reflected by the tile is measured on a scale of 0–255. A perfectly black tile would reflect no light and measure 0, and a perfectly white tile would measure 255. A sample of nine Mezza Perla tiles were measured, with the following results:

204.999	206.149	202.102	207.048	203.496
206.343	203.496	206.676	205.831	

Is it appropriate to use the Student's t statistic to construct a 95% confidence interval for the mean shade of Mezza Perla tile? If so, construct the confidence interval. If not, explain why not.

8. A chemist made eight independent measurements of the melting point of tungsten. She obtained a sample mean of 3410.14°C and a sample standard deviation of 1.018°C.

   a. Find a 95% confidence interval for the melting point of tungsten.

   b. Find a 98% confidence interval for the melting point of tungsten.

   c. If the eight measurements had been 3409.76, 3409.80, 3412.66, 3409.79, 3409.76, 3409.77, 3409.80, and 3409.78, would the confidence intervals found in parts (a) and (b) be valid? Explain.

9. Eight independent measurements are taken of the diameter of a piston. The measurements (in inches) are 3.236, 3.223, 3.242, 3.244, 3.228, 3.253, 3.253, and 3.230.

   a. Make a dotplot of the eight values.

   b. Should the t curve be used to find a 99% confidence interval for the diameter of this piston? If so, find the confidence interval. If not, explain why not.

   c. Eight independent measurements are taken of the diameter of another piston. The measurements this time are 3.295, 3.232, 3.261, 3.248, 3.289, 3.245, 3.576, and 3.201. Make a dotplot of these values.

   d. Should the t curve be used to find a 95% confidence interval for the diameter of this piston? If so, find the confidence interval. If not, explain why not.

10. Five measurements are taken of the octane rating for a particular type of gasoline. The results (in %) are 87.0, 86.0, 86.5, 88.0, 85.3. Find a 99% confidence interval for the mean octane rating for this type of gasoline.

11. A model of heat transfer from a cylinder immersed in a liquid predicts that the heat-transfer coefficient for the cylinder will become constant at very low flow rates of the fluid. A sample of 10 measurements is taken. The results, in W/m^2 K, are

13.7	12.0	13.1	14.1	13.1
14.1	14.4	12.2	11.9	11.8

Find a 95% confidence interval for the heat-transfer coefficient.

12. Surfactants are chemical agents, such as detergents, that lower the surface tension of a liquid. Surfactants play an important role in the cleaning of contaminated soils. In an experiment to determine the effectiveness of a certain method for removing toluene from sand, the sand was washed with a surfactant, and then rinsed with deionized water. Of interest was the amount of toluene that came out in the rinse. In five such experiments, the amounts of toluene removed in the rinse cycle, expressed as a percentage of the total amount originally present, were 5.0, 4.8, 9.0, 10.0, and 7.3. Find a 95% confidence interval for the percentage of toluene removed in the rinse. (This exercise is based on the article "Laboratory Evaluation of the Use of Surfactants for Ground Water Remediation and the Potential for Recycling Them," D. Lee, R. Cody, and B. Hoyle, *Ground Water Monitoring and Remediation*, 2001:49–57.)

13. In an experiment to measure the rate of absorption of pesticides through skin, 500 μg of uniconazole was applied to the skin of four rats. After 10 hours, the amounts absorbed (in μg) were 0.5, 2.0, 1.4, and 1.1. Find a 90% confidence interval for the mean amount absorbed.

14. The following MINITAB output presents a confidence interval for a population mean.

```
One-Sample T: X

Variable     N       Mean      StDev    SE Mean           95% CI
X           10    6.59635    0.11213    0.03546    (6.51613, 6.67656)
```

a. How many degrees of freedom does the Student's t distribution have?

b. Use the information in the output, along with the t table, to compute a 99% confidence interval.

15. The following MINITAB output presents a confidence interval for a population mean, but some of the numbers got smudged and are now illegible. Fill in the missing numbers for (a), (b), and (c).

```
One-Sample T: X
Variable     N       Mean     StDev    SE Mean         99% CI
X           20    2.39374      (a)    0.52640      ( (b), (c) )
```

16. The concentration of carbon monoxide (CO) in a gas sample is measured by a spectrophotometer and found to be 85 ppm. Through long experience with this instrument, it is believed that its measurements are unbiased and normally distributed, with an uncertainty (standard deviation) of 8 ppm. Find a 95% confidence interval for the concentration of CO in this sample.

5.4 Confidence Intervals for the Difference Between Two Means

We now investigate examples in which we wish to estimate the difference between the means of two populations. The data will consist of two samples, one from each population. The basic idea is quite simple. We will compute the difference of the sample means and the standard deviation of that difference. Then a simple modification of expression (5.1) (in Section 5.1) will provide the confidence interval. The method we describe is based on the results concerning the sum and difference of two independent normal random variables that were presented in Section 4.5. We review these results here:

Let X and Y be independent, with $X \sim N(\mu_X, \sigma_X^2)$ and $Y \sim N(\mu_Y, \sigma_Y^2)$. Then

$$X + Y \sim N(\mu_X + \mu_Y, \sigma_X^2 + \sigma_Y^2) \tag{5.14}$$

$$X - Y \sim N(\mu_X - \mu_Y, \sigma_X^2 + \sigma_Y^2) \tag{5.15}$$

We will now see how to construct a confidence interval for the difference between two population means. As an example, assume that a new design of lightbulb has been developed that is thought to last longer than an old design. A simple random sample of 144 new lightbulbs has an average lifetime of 578 hours and a standard deviation of 22 hours. A simple random sample of 64 old lightbulbs has an average lifetime of 551 hours and a standard deviation of 33 hours. The samples are independent, in that the lifetimes for one sample do not influence the lifetimes for the other. We wish to find a 95% confidence interval for the difference between the mean lifetimes of lightbulbs of the two designs.

We begin by translating the problem into statistical language. We have a simple random sample $X_1, \ldots, X_{144}$ of lifetimes of new lightbulbs. The sample mean is $\overline{X} = 578$ and the sample standard deviation is $s_X = 22$. We have another simple random sample $Y_1, \ldots, Y_{64}$ of lifetimes of old lightbulbs. This sample has mean $\overline{Y} = 551$ and standard deviation $s_Y = 33$. The population means and standard deviations are unknown. Denote the mean of the population of lifetimes of new lightbulbs by μ_X, and the mean of the population of old lightbulbs by μ_Y. Denote the corresponding standard deviations by σ_X and σ_Y. We are interested in the difference $\mu_X - \mu_Y$.

We can construct the confidence interval for $\mu_X - \mu_Y$ by determining the distribution of $\overline{X} - \overline{Y}$. By the Central Limit Theorem, $\overline{X}$ comes from a normal distribution with mean μ_X and standard deviation $\sigma_X/\sqrt{144}$, and $\overline{Y}$ comes from a normal distribution with mean μ_Y and standard deviation $\sigma_Y/\sqrt{64}$. Since the samples are independent, it follows from expression (5.15) that the difference $\overline{X} - \overline{Y}$ comes from a normal distribution with mean $\mu_X - \mu_Y$ and variance $\sigma^2_{\overline{X}-\overline{Y}} = \sigma^2_X/144 + \sigma^2_Y/64$. Figure 5.13 illustrates the distribution of $\overline{X} - \overline{Y}$ and indicates that the middle 95% of the curve has width $\pm 1.96\sigma_{\overline{X}-\overline{Y}}$.

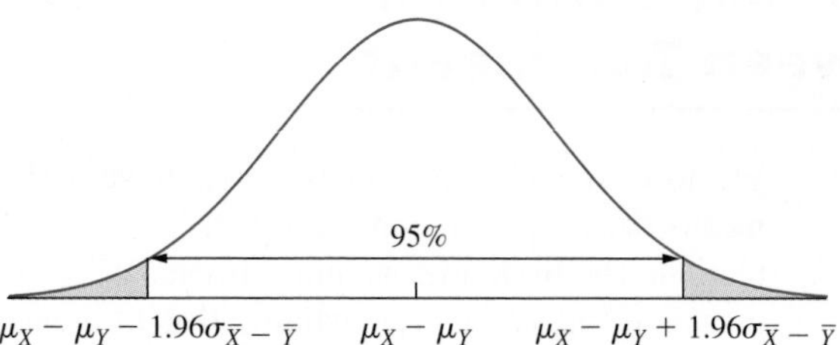

FIGURE 5.13 The observed difference $\overline{X} - \overline{Y} = 27$ is drawn from a normal distribution with mean $\mu_X - \mu_Y$ and standard deviation $\sigma_{\overline{X}-\overline{Y}} = \sqrt{\sigma^2_X/144 + \sigma^2_Y/64}$.

Estimating the population standard deviations σ_X and σ_Y with the sample standard deviations $s_X = 22$ and $s_Y = 33$, respectively, we estimate $\sigma_{\overline{X}-\overline{Y}} \approx \sqrt{22^2/144 + 33^2/64} = 4.514$. The 95% confidence interval for $\mu_X - \mu_Y$ is therefore $578 - 551 \pm 1.96(4.514)$, or 27 ± 8.85.

Summary

Let $X_1, \ldots, X_{n_X}$ be a *large* random sample of size n_X from a population with mean μ_X and standard deviation σ_X, and let $Y_1, \ldots, Y_{n_Y}$ be a *large* random sample of size n_Y from a population with mean μ_Y and standard deviation σ_Y. If the two samples are independent, then a level $100(1 - \alpha)\%$ confidence interval for $\mu_X - \mu_Y$ is

$$\overline{X} - \overline{Y} \pm z_{\alpha/2}\sqrt{\frac{\sigma_X^2}{n_X} + \frac{\sigma_Y^2}{n_Y}} \tag{5.16}$$

When the values of σ_X and σ_Y are unknown, they can be replaced with the sample standard deviations s_X and s_Y.

Example 5.23

The chemical composition of soil varies with depth. The article "Sampling Soil Water in Sandy Soils: Comparative Analysis of Some Common Methods" (M. Ahmed, M. Sharma, et al., *Communications in Soil Science and Plant Analysis*, 2001: 1677–1686) describes chemical analyses of soil taken from a farm in Western Australia. Fifty specimens were each taken at depths 50 and 250 cm. At a depth of 50 cm, the average NO_3 concentration (in mg/L) was 88.5 with a standard deviation of 49.4. At a depth of 250 cm, the average concentration was 110.6 with a standard deviation of 51.5. Find a 95% confidence interval for the difference between the NO_3 concentrations at the two depths.

Solution

Let $X_1, \ldots, X_{50}$ represent the concentrations of the 50 specimens taken at 50 cm, and let $Y_1, \ldots, Y_{50}$ represent the concentrations of the 50 specimens taken at 250 cm. Then $\overline{X} = 88.5$, $\overline{Y} = 110.6$, $s_X = 49.4$, and $s_Y = 51.5$. The sample sizes are $n_X = n_Y = 50$. Both samples are large, so we can use expression (5.16). Since we want a 95% confidence interval, $z_{\alpha/2} = 1.96$. The 95% confidence interval for the difference $\mu_Y - \mu_X$ is $110.6 - 88.5 \pm 1.96\sqrt{49.4^2/50 + 51.5^2/50}$, or 22.1 ± 19.8.

Exercises for Section 5.4

1. The melting points of two alloys are being compared. Thirty-five specimens of alloy 1 were melted. The average melting temperature was 517.0°F and the standard deviation was 2.4°F. Forty-seven specimens of alloy 2 were melted. The average melting temperature was 510.1°F and the standard deviation was 2.1°F. Find a 99% confidence interval for the difference between the melting points.

2. In an experiment to determine the effect of temperature on the deposition rate of tungsten on silicon wafers, 64 wafers were processed at 400°C and 88 wafers were processed at 425°C. The deposition rate for the wafers processed at 400°C averaged 1840 Å/min, with a standard deviation of 244 Å/min. The wafers processed at 425°C

averaged 2475 Å/min, with a standard deviation of 760 Å/min. Find a 95% confidence interval for the difference between the mean deposition rates.

3. The article "Vehicle-Arrival Characteristics at Urban Uncontrolled Intersections" (V. Rengaraju and V. Rao, *Journal of Transportation Engineering*, 1995: 317–323) presents data on traffic characteristics at 10 intersections in Madras, India. At one particular intersection, the average speed for a sample of 39 cars was 26.50 km/h, with a standard deviation of 2.37 km/h. The average speed for a sample of 142 motorcycles was 37.14 km/h, with a standard deviation of 3.66 km/h. Find a 95% confidence interval for the difference between the mean speeds of motorcycles and cars.

4. A stress analysis was conducted on random samples of epoxy-bonded joints from two species of wood. A random sample of 120 joints from species A had a mean shear stress of 1250 psi and a standard deviation of 350 psi, and a random sample of 90 joints from species B had a mean shear stress of 1400 psi and a standard deviation of 250 psi. Find a 98% confidence interval for the difference in mean shear stress between the two species.

5. In a study to compare two different corrosion inhibitors, specimens of stainless steel were immersed for four hours in a solution containing sulfuric acid and a corrosion inhibitor. Forty-seven specimens in the presence of inhibitor A had a mean weight loss of 242 mg and a standard deviation of 20 mg, and 42 specimens in the presence of inhibitor B had a mean weight loss of 220 mg and a standard deviation of 31 mg. Find a 95% confidence interval for the difference in mean weight loss between the two inhibitors.

6. An electrical engineer wishes to compare the mean lifetimes of two types of transistors in an application involving high-temperature performance. A sample of 60 transistors of type A were tested and were found to have a mean lifetime of 1827 hours and a standard deviation of 168 hours. A sample of 180 transistors of type B were tested and were found to have a mean lifetime of 1658 hours and a standard deviation of 225 hours. Find a 95% confidence interval for the difference between the mean lifetimes of the two types of transistors.

7. In a study of the effect of cooling rate on the hardness of welded joints, 50 welds cooled at a rate of 10°C/s had an average Rockwell (B) hardness of 91.1 and a standard deviation of 6.23, and 40 welds cooled at a rate of 30°C/s had an average hardness of 90.7 and a standard deviation of 4.34.

a. Find a 95% confidence interval for the difference in hardness between welds cooled at the different rates.

b. Someone says that the cooling rate has no effect on the hardness. Do these data contradict this claim? Explain.

8. Refer to Exercise 7. Ten more welds will be made in order to increase the precision of the confidence interval. Which would increase the precision the most, cooling all 10 welds at the rate of 10°C/s, cooling all 10 welds at the rate of 30°C/s, or cooling 5 welds at 10°C/s and 5 at 30°C/s? Explain.

9. The article "The Prevalence of Daytime Napping and Its Relationship to Nighttime Sleep" (J. Pilcher, K. Michalkowski, and R. Carrigan), *Behavioral Medicine*, 2001:71–76) presents results of a study of sleep habits in a large number of subjects. In a sample of 87 young adults, the average time per day spent in bed (either awake or asleep) was 7.70 hours, with a standard deviation of 1.02 hours, and the average time spent in bed asleep was 7.06 hours, with a standard deviation of 1.11 hours. The mean time spent in bed awake was estimated to be $7.70 - 7.06 = 0.64$ hours. Is it possible to compute a 95% confidence interval for the mean time spent in bed awake? If so, construct the confidence interval. If not possible, explain why not.

10. The article "Occurrence and Distribution of Ammonium in Iowa Groundwater" (K. Schilling, *Water Environment Research*, 2002:177–186) describes measurements of ammonium concentrations (in mg/L) at a large number of wells in the state of Iowa. These included 349 alluvial wells and 143 quaternary wells. The concentrations at the alluvial wells averaged 0.27 with a standard deviation of 0.40, and those at the quaternary wells averaged 1.62 with a standard deviation of 1.70. Find a 95% confidence interval for the difference in mean concentrations between alluvial and quaternary wells.

5.5 Confidence Intervals for the Difference Between Two Proportions

In a Bernoulli population, the mean is equal to the success probability p, which is the proportion of successes in the population. When independent trials are performed from each of two Bernoulli populations, we can use methods similar to those presented in Section 5.4 to find a confidence interval for the difference between the two success probabilities. We present an example to illustrate.

Eighteen of 60 light trucks produced on assembly line A had a defect in the steering mechanism, which needed to be repaired before shipment. Only 16 of 90 trucks produced on assembly line B had this defect. Assume that these trucks can be considered to be two independent simple random samples from the trucks manufactured on the two assembly lines. We wish to find a 95% confidence interval for the difference between the proportions of trucks with this defect on the two assembly lines.

This is a situation in which we would have to be careful in practice to make sure that it is reasonable to consider the data to be simple random samples. Choosing trucks sequentially off the line might not be a good idea, for example, if there are systematic fluctuations in quality over time. We will assume that the sampling has been done by some well-thought-out and appropriate procedure.

The construction of the confidence interval here proceeds in a manner similar to that in Section 5.4, with means replaced by proportions. Let p_X represent the proportion of trucks in the population from line A that had the defect, and let p_Y represent the corresponding proportion from line B. The values of p_X and p_Y are unknown. We wish to find a 95% confidence interval for $p_X - p_Y$.

Let X represent the number of trucks in the sample from line A that had defects, and let Y represent the corresponding number from line B. Then X is a binomial random variable with $n_X = 60$ trials and success probability p_X, and Y is a binomial random variable with $n_Y = 90$ trials and success probability p_Y. The sample proportions are $\widehat{p}_X$ and $\widehat{p}_Y$. In this example the observed values are $X = 18$, $Y = 16$, $\widehat{p}_X = 18/60$, and $\widehat{p}_Y = 16/90$. Since the sample sizes are large, it follows from the central limit theorem that $\widehat{p}_X$ and $\widehat{p}_Y$ are both approximately normally distributed with means p_X and p_Y and standard deviations $\sigma_{\widehat{p}_X} = \sqrt{p_X(1-p_X)/n_X}$ and $\sigma_{\widehat{p}_Y} = \sqrt{p_Y(1-p_Y)/n_Y}$. It follows that the difference $\widehat{p}_X - \widehat{p}_Y$ has a normal distribution with mean $p_X - p_Y$ and standard deviation $\sqrt{p_X(1-p_X)/n_X + p_Y(1-p_Y)/n_Y}$. We conclude that for 95% of all possible samples, the difference $p_X - p_Y$ satisfies the following inequality:

$$\widehat{p}_X - \widehat{p}_Y - 1.96\sqrt{\frac{p_X(1-p_X)}{n_X} + \frac{p_Y(1-p_Y)}{n_Y}}$$
$$< p_X - p_Y <$$
$$\widehat{p}_X - \widehat{p}_Y + 1.96\sqrt{\frac{p_X(1-p_X)}{n_X} + \frac{p_Y(1-p_Y)}{n_Y}} \tag{5.17}$$

Expression 5.17 is not a confidence interval, because the quantities $\sqrt{p_X(1-p_X)/n_X + p_Y(1-p_Y)/n_Y}$ depend on the unknown true values p_X and p_Y.

The traditional approach is to replace p_X and p_Y with $\widehat{p}_X$ and $\widehat{p}_Y$, producing the confidence interval $\widehat{p}_X - \widehat{p}_Y \pm z_{\alpha/2}\sqrt{\widehat{p}_X(1-\widehat{p}_X)/n_X + \widehat{p}_Y(1-\widehat{p}_Y)/n_Y}$. Recent research has shown that this interval can be improved by slightly modifying n_X, n_Y, $\widehat{p}_X$, and $\widehat{p}_Y$. Simply add 1 to each of the numbers of successes X and Y, and add 2 to each of the numbers of trials n_X and n_Y. Thus we define $\widetilde{n}_X = n_X + 2$, $\widetilde{n}_Y = n_Y + 2$, $\widetilde{p}_X = (X+1)/\widetilde{n}_X$, and $\widetilde{p}_Y = (Y+1)/\widetilde{n}_Y$. The 95% confidence interval is $\widetilde{p}_X - \widetilde{p}_Y \pm z_{\alpha/2}\sqrt{\widetilde{p}_X(1-\widetilde{p}_X)/\widetilde{n}_X + \widetilde{p}_Y(1-\widetilde{p}_Y)/\widetilde{n}_Y}$. In this example, $\widetilde{n}_X = 62$, $\widetilde{n}_Y = 92$, $\widetilde{p}_X = 19/62 = 0.3065$, and $\widetilde{p}_Y = 17/92 = 0.1848$. We thus obtain $0.3065 - 0.1848 \pm 0.1395$, or $(-0.0178,\ 0.2612)$.

To obtain a level $100(1-\alpha)$ confidence interval, replace 1.96 with $z_{\alpha/2}$. Although we justified this confidence interval by using the Central Limit Theorem, which assumes that n_X and n_Y are large, this method has been found to give good results for almost all sample sizes.

Summary

Let X be the number of successes in n_X independent Bernoulli trials with success probability p_X, and let Y be the number of successes in n_Y independent Bernoulli trials with success probability p_Y, so that $X \sim \text{Bin}(n_X,\ p_X)$ and $Y \sim \text{Bin}(n_Y,\ p_Y)$. Define $\widetilde{n}_X = n_X + 2$, $\widetilde{n}_Y = n_Y + 2$, $\widetilde{p}_X = (X+1)/\widetilde{n}_X$, and $\widetilde{p}_Y = (Y+1)/\widetilde{n}_Y$.

Then a level $100(1-\alpha)\%$ confidence interval for the difference $p_X - p_Y$ is

$$\widetilde{p}_X - \widetilde{p}_Y \pm z_{\alpha/2}\sqrt{\frac{\widetilde{p}_X(1-\widetilde{p}_X)}{\widetilde{n}_X} + \frac{\widetilde{p}_Y(1-\widetilde{p}_Y)}{\widetilde{n}_Y}} \tag{5.18}$$

If the lower limit of the confidence interval is less than -1, replace it with -1. If the upper limit of the confidence interval is greater than 1, replace it with 1.

The adjustment described here for the two-sample confidence interval is similar to the one described in Section 5.2 for the one-sample confidence interval. In both cases, a total of two successes and four trials are added. In the two-sample case, these are divided between the samples, so that one success and two trials are added to each sample. In the one-sample case, two successes and four trials are added to the one sample. The confidence interval given by expression (5.18) can be called the *Agresti–Caffo* interval, after Alan Agresti and Brian Caffo, who developed it. For more information about this confidence interval, consult the article "Simple and Effective Confidence Intervals for Proportions and Differences of Proportions Result from Adding Two Successes and Two Failures" (A. Agresti and B. Caffo, *The American Statistician*, 2000:280–288).

Methods for estimating strength and stiffness requirements should be conservative, in that they should overestimate rather than underestimate. The success rate of such a method can be measured by the probability of an overestimate. The article "Discrete Bracing Analysis for Light-Frame Wood-Truss Compression Webs" (M. Waltz, T. McLain, et al.,

Journal of Structural Engineering, 2000:1086–1093) presents the results of an experiment that evaluated a standard method (Plaut's method) for estimating the brace force for a compression web brace. In a sample of 380 short test columns (4 to 6 ft in length), the method overestimated the force for 304 of them, and in a sample of 394 long test columns (8 to 10 ft in length), the method overestimated the force for 360 of them. Find a 95% confidence interval for the difference between the success rates for long columns and short columns.

Solution

The number of successes in the sample of short columns is $X = 304$, and the number of successes in the sample of long columns is $Y = 360$. The numbers of trials are $n_X = 380$ and $n_Y = 394$. We compute $\tilde{n}_X = 382$, $\tilde{n}_Y = 396$, $\tilde{p}_X = (304 + 1)/382 = 0.7984$, and $\tilde{p}_Y = (360 + 1)/396 = 0.9116$. The value of $z_{\alpha/2}$ is 1.96. The 95% confidence interval is $0.9116 - 0.7984 \pm 1.96\sqrt{(0.7984)(0.2016)/382 + (0.9116)(0.0884)/396}$, or 0.1132 ± 0.0490.

The Traditional Method

Many people use the traditional method for computing confidence intervals for the difference between proportions. This method uses the sample proportions $\widehat{p}_X$ and $\widehat{p}_Y$ and the actual sample sizes n_X and n_Y. The traditional method gives results very similar to those of the modern method previously described for large or moderately large sample sizes. For small sample sizes, however, the traditional confidence interval fails to achieve its coverage probability; in other words, level $100(1-\alpha)\%$ confidence intervals computed by the traditional method cover the true value less than $100(1-\alpha)\%$ of the time.

Summary

The Traditional Method for Computing Confidence Intervals for the Difference Between Proportions (widely used but not recommended)

Let $\widehat{p}_X$ be the proportion of successes in a *large* number n_X of independent Bernoulli trials with success probability p_X, and let $\widehat{p}_Y$ be the proportion of successes in a *large* number n_Y of independent Bernoulli trials with success probability p_Y. Then the traditional level $100(1-\alpha)\%$ confidence interval for $p_X - p_Y$ is

$$\widehat{p}_X - \widehat{p}_Y \pm z_{\alpha/2}\sqrt{\frac{\widehat{p}_X(1-\widehat{p}_X)}{n_X} + \frac{\widehat{p}_Y(1-\widehat{p}_Y)}{n_Y}} \tag{5.19}$$

This method cannot be used unless both samples contain at least 10 successes and 10 failures.

Exercises for Section 5.5

1. In a test of the effect of dampness on electric connections, 100 electric connections were tested under damp conditions and 150 were tested under dry conditions. Twenty of the damp connections failed and only 10 of the dry ones failed. Find a 90% confidence interval for the difference between the proportions of connections that fail when damp as opposed to dry.

2. The specification for the pull strength of a wire that connects an integrated circuit to its frame is 10 g or more. In a sample of 85 units made with gold wire, 68 met the specification, and in a sample of 120 units made with aluminum wire, 105 met the specification. Find a 95% confidence interval for the difference in the proportions of units that meet the specification between units with gold wire and those with aluminum wire.

3. In a random sample of 340 cars driven at low altitudes, 46 of them produced more than 10 g of particulate pollution per gallon of fuel consumed. In a random sample of 85 cars driven at high altitudes, 21 of them produced more than 10 g of particulate pollution per gallon of fuel consumed. Find a 98% confidence interval for the difference between the proportions for high-altitude and low-altitude vehicles.

4. Out of 1200 pieces of gravel from one plant, 110 pieces are classified as "large." Out of 900 pieces from another plant, 95 are classified as large. Find a 99% confidence interval for the difference between the proportions of large gravel pieces produced at the two plants.

5. Two processes for manufacturing a certain microchip are being compared. A sample of 400 chips was selected from a less expensive process, and 62 chips were found to be defective. A sample of 100 chips was selected from a more expensive process, and 12 were found to be defective.
   a. Find a 95% confidence interval for the difference between the proportions of defective chips produced by the two processes.
   b. In order to increase the precision of the confidence interval, additional chips will be sampled. Three sampling plans of equal cost are being considered. In the first plan, 100 additional chips from the less expensive process will be sampled. In the second plan 50 additional chips from the more expensive process will be sampled. In the third plan, 50 chips from the less expensive and 25 chips from the more expensive process will be sampled. Which plan is most likely to provide the greatest increase in the precision of the confidence interval? Explain.

6. The article "Occurrence and Distribution of Ammonium in Iowa Groundwater" (K. Schilling, *Water Environment Research*, 2002:177–186) describes measurements of ammonium concentrations (in mg/L) at a large number of wells in the state of Iowa. These included 349 alluvial wells and 143 quaternary wells. Of the alluvial wells, 182 had concentrations above 0.1, and 112 of the quaternary wells had concentrations above 0.1. Find a 95% confidence interval for the difference between the proportions of the two types of wells with concentrations above 0.1.

7. The article referred to in Exercise 9 in Section 5.2 describes an experiment in which a total of 42 leachate specimens were tested for the presence of several contaminants. Of the 42 specimens, 26 contained detectable levels of lead, and 32 contained detectable levels of chromium. Is it possible, using the methods of this section, to find a 95% confidence interval for the difference between the probability that a specimen will contain a detectable amount of lead and the probability that it will contain a detectable amount of chromium? If so, find the confidence interval. If not, explain why not.

8. The article "Case Study Based Instruction of DOE and SPC" (J. Brady and T. Allen, *The American Statistician*, 2002:312–315) describes an effort by an engineering team to reduce the defect rate in the manufacture of a certain printed circuit board. The team decided to reconfigure the transistor heat sink. A total of 1500 boards were produced the week before the reconfiguration was implemented, and 345 of these were defective. A total of 1500 boards were produced the week after reconfiguration, and 195 of these were defective. Find a 95% confidence interval for the decrease in the defective rate after the reconfiguration.

9. Repeat business is a good measure of customer satisfaction. At the end of year, a vendor of computer supplies drew a sample of 120 accounts and found

that 56 of them had ordered more than once. Then he drew a sample of 80 accounts from the previous year and found that 30 of them had ordered more than once. Find a 95% confidence interval for the difference between the two proportions of customers who ordered more than once.

10. The article "Accidents on Suburban Highways—Tennessee's Experience" (R. Margiotta and A. Chatterjee, *Journal of Transportation Engineering*, 1995:255–261) compares rates of traffic accidents at intersections with raised medians with rates at intersections with two-way left-turn lanes. Out of 4644 accidents at intersections with raised medians, 2280 were rear-end accidents, and out of 4584 accidents at two-way left-turn lanes, 1982 were rear-end accidents. Assuming these to be random samples of accidents from two types of intersections, find a 90% confidence interval for the difference between the proportions of accidents that are of the rear-end type at the two types of intersections.

11. In a certain year, there were 80 days with measurable snowfall in Denver, and 63 days with measurable snowfall in Chicago. A meteorologist computes $(80 + 1)/(365 + 2) = 0.22$, $(63 + 1)/(365 + 2) = 0.17$, and proposes to compute a 95% confidence interval for the difference between the proportions of snowy days in the two cities as follows:

$$0.22 - 0.17 \pm 1.96\sqrt{\frac{(0.22)(0.78)}{367} + \frac{(0.17)(0.83)}{367}}$$

Is this a valid confidence interval? Explain.

5.6 Small-Sample Confidence Intervals for the Difference Between Two Means

The Student's t distribution can be used in some cases where samples are small and, thus, where the Central Limit Theorem does not apply. We present an example.

A sample of six welds of one type had an average ultimate testing strength (in ksi) of 83.2 and a standard deviation of 5.2, and a sample of 10 welds of another type had an average strength of 71.3 and a standard deviation of 3.1. Assume that both sets of welds are random samples from normal populations. We wish to find a 95% confidence interval for the difference between the mean strengths of the two types of welds.

Both sample sizes are small, so the Central Limit Theorem does not apply. If both populations are normal, the Student's t distribution can be used to compute a confidence interval for the difference between the two population means. The method is similar to that presented in Section 5.4 for the case where the samples are large, except that the z-score is replaced with a value from the Student's t distribution.

If $X_1, \ldots, X_{n_X}$ is a sample of size n_X from a *normal* population with mean μ_X and $Y_1, \ldots, Y_{n_Y}$ is a sample of size n_Y from a *normal* population with mean μ_Y, then the quantity

$$\frac{(\overline{X} - \overline{Y}) - (\mu_X - \mu_Y)}{\sqrt{s_X^2/n_X + s_Y^2/n_Y}}$$

has an approximate Student's t distribution.

The number of degrees of freedom to use for this distribution is given by

$$\nu = \frac{\left(\dfrac{s_X^2}{n_X} + \dfrac{s_Y^2}{n_Y}\right)^2}{\dfrac{(s_X^2/n_X)^2}{n_X - 1} + \dfrac{(s_Y^2/n_Y)^2}{n_Y - 1}} \quad \text{rounded down to the nearest integer.} \tag{5.20}$$

In our example, let $X_1, \ldots, X_6$ be the 6 welds of the first type, and let $Y_1, \ldots, Y_{10}$ be the 10 welds of the second type. Substituting $s_X = 5.2$, $s_Y = 3.1$, $n_X = 6$, $n_Y = 10$ into Equation (5.20) yields

$$\nu = \frac{\left(\dfrac{5.2^2}{6} + \dfrac{3.1^2}{10}\right)^2}{\dfrac{(5.2^2/6)^2}{5} + \dfrac{(3.1^2/10)^2}{9}} = 7.18 \approx 7$$

If both populations are normal, then the quantity

$$[(\overline{X} - \overline{Y}) - (\mu_X - \mu_Y)]/\sqrt{s_X^2/6 + s_Y^2/10}$$

has an approximate Student's t distribution with seven degrees of freedom. Figure 5.14 presents this distribution. Ninety-five percent of the area under the curve is contained between the values $t = -2.365$ and $t = 2.365$. It follows that for 95% of all the samples that might have been chosen,

$$-2.365 < \frac{(\overline{X} - \overline{Y}) - (\mu_X - \mu_Y)}{\sqrt{s_X^2/6 + s_Y^2/10}} < 2.365$$

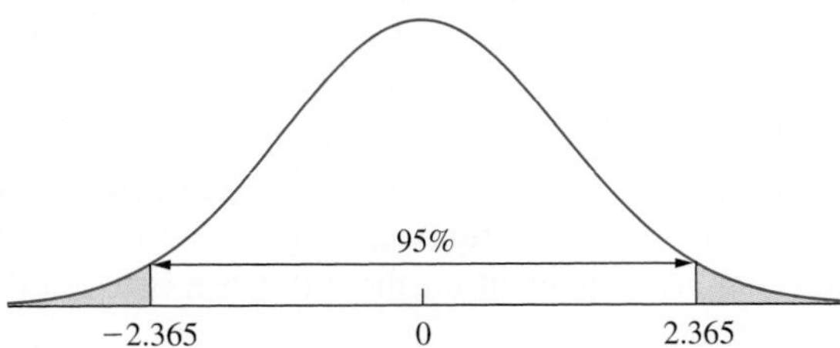

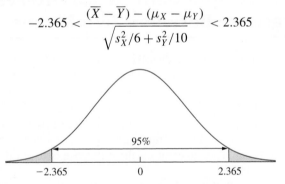

FIGURE 5.14 The Student's t distribution with seven degrees of freedom. Ninety-five percent of the area lies between $t = -2.365$ and $t = 2.365$.

By the reasoning used in Section 5.3, a 95% confidence interval for the difference $\mu_X - \mu_Y$ is $\overline{X} - \overline{Y} \pm 2.365\sqrt{s_X^2/6 + s_Y^2/10}$. Substituting $\overline{X} = 83.2$, $\overline{Y} = 71.3$, $s_X = 5.2$, and $s_Y = 3.1$, we find that a 95% confidence interval for $\mu_X - \mu_Y$ is 11.9 ± 5.53, or $(6.37, 17.43)$.

In general, to produce a level $100(1 - \alpha)\%$ confidence interval, let $t_{\nu,\alpha/2}$ be the $1 - \alpha/2$ quantile of the Student's t distribution with ν degrees of freedom, that is, the value that cuts off an area of $\alpha/2$ in the right-hand tail. For example, previously we found that $t_{7,.025} = 2.365$. Then a level $100(1 - \alpha)\%$ confidence interval for the difference between population means $\mu_X - \mu_Y$, when the sample sizes are n_X and n_Y, respectively, is $\overline{X} - \overline{Y} \pm t_{\nu,\alpha/2}\sqrt{s_X^2/n_X + s_Y^2/n_Y}$.

Summary

Let $X_1, \ldots, X_{n_X}$ be a random sample of size n_X from a *normal* population with mean μ_X, and let $Y_1, \ldots, Y_{n_Y}$ be a random sample of size n_Y from a *normal* population with mean μ_Y. Assume the two samples are independent.

If the populations do not necessarily have the same variance, a level $100(1-\alpha)\%$ confidence interval for $\mu_X - \mu_Y$ is

$$\overline{X} - \overline{Y} \pm t_{\nu,\alpha/2}\sqrt{\frac{s_X^2}{n_X} + \frac{s_Y^2}{n_Y}} \tag{5.21}$$

The number of degrees of freedom, ν, is given by

$$\nu = \frac{\left(\dfrac{s_X^2}{n_X} + \dfrac{s_Y^2}{n_Y}\right)^2}{\dfrac{(s_X^2/n_X)^2}{n_X - 1} + \dfrac{(s_Y^2/n_Y)^2}{n_Y - 1}}$$ rounded down to the nearest integer.

Example 5.25

Resin-based composites are used in restorative dentistry. The article "Reduction of Polymerization Shrinkage Stress and Marginal Leakage Using Soft-Start Polymerization" (C. Ernst, N. Brand, et al., *Journal of Esthetic and Restorative Dentistry*, 2003:93–104) presents a comparison of the surface hardness of specimens cured for 40 seconds with constant power with that of specimens cured for 40 seconds with exponentially increasing power. Fifteen specimens were cured with each method. Those cured with constant power had an average surface hardness (in N/mm^2) of 400.9 with a standard deviation of 10.6. Those cured with exponentially increasing power had an average surface hardness of 367.2 with a standard deviation of 6.1. Find a 98% confidence interval for the difference in mean hardness between specimens cured by the two methods.

Solution

We have $\overline{X} = 400.9$, $s_X = 10.6$, $n_X = 15$, $\overline{Y} = 367.2$, $s_Y = 6.1$, and $n_Y = 15$. The number of degrees of freedom is given by Equation (5.20) to be

$$\nu = \frac{\left(\dfrac{10.6^2}{15} + \dfrac{6.1^2}{15}\right)^2}{\dfrac{(10.6^2/15)^2}{15-1} + \dfrac{(6.1^2/15)^2}{15-1}} = 22.36 \approx 22$$

From the t table (Table A.3 in Appendix A), we find that $t_{22,.01} = 2.508$. We use expression (5.21) to find that the 98% confidence interval is

$$400.9 - 367.2 \pm 2.508\sqrt{10.6^2/15 + 6.1^2/15}, \text{ or } 33.7 \pm 7.9.$$

When the Populations Have Equal Variances

Occasionally a situation arises in which two populations are known to have nearly equal variances. In these cases a method that exploits this fact may be used to produce confidence intervals that are usually narrower than those produced by the more general method described before. As with the more general method, both populations must be approximately normal. We present an example.

Two standard weights, each labeled 100 g, are each weighed several times on the same scale. The first weight is weighed 8 times, and the mean scale reading is 18.2 μg above 100 g, with a standard deviation of 2.0 μg. The second weight is weighed 18 times, and the mean reading is 16.4 g above 100 g, with a standard deviation of 1.8 μg. Assume that each set of readings is a sample from an approximately normal population. Since the same scale is used for all measurements, and since the true weights are nearly equal, it is reasonable to assume that the population standard deviations of the readings are the same for both weights. Assume the measurements are unbiased (it is actually enough to assume that the bias is the same for both weights). We wish to find a 95% confidence interval for the difference between the true weights.

Let $X_1, \ldots, X_8$ represent the readings for the first weight, and let $Y_1, \ldots, Y_{18}$ represent the readings for the second weight. Let μ_X and μ_Y be the true weights, which are the means of the populations from which these samples were drawn. By assumption, both populations follow normal distributions with the same variance σ^2. Therefore $\overline{X}$ has a normal distribution with mean μ_X and variance $\sigma^2/8$, and $\overline{Y}$ has a normal distribution with mean μ_Y and variance $\sigma^2/18$. The difference $\overline{X} - \overline{Y}$ therefore has a normal distribution with mean $\mu_X - \mu_Y$ and variance $\sigma^2(1/8 + 1/18)$. Since σ^2 is unknown, we must estimate it. We could estimate this quantity using either of the sample variances $s_X^2 = \sum_{i=1}^{8}(X_i - \overline{X})^2/(8-1)$ or $s_Y^2 = \sum_{i=1}^{18}(Y_i - \overline{Y})^2/(18-1)$. But the best estimate is obtained by combining the information in both samples. The best estimate is the **pooled variance** $s_p^2 = (7s_X^2 + 17s_Y^2)/(7 + 17)$. The pooled variance s_p^2 is a weighted average of the two sample variances. The weights are equal to the sample sizes minus one. It is logical to use a weighted average so that the sample variance based on the larger sample counts more. Substituting the given values for s_X and s_Y, the value of the pooled variance is $s_p^2 = [7(2.0^2) + 17(1.8^2)]/(7 + 17) = 3.4617$, so $s_p = 1.8606$.

The quantity $[(\overline{X} - \overline{Y}) - (\mu_X - \mu_Y)]/\left(s_p\sqrt{1/8 + 1/18}\right)$ has the Student's t distribution with $8 + 18 - 2 = 24$ degrees of freedom. From the t table, we find that $t_{24,.025} = 2.064$. It follows that for 95% of all the samples that might have been chosen,

$$-2.064 < \frac{(\overline{X} - \overline{Y}) - (\mu_X - \mu_Y)}{s_p\sqrt{1/8 + 1/18}} < 2.064$$

By the reasoning used in Section 5.3, a 95% confidence interval for $\mu_X - \mu_Y$ is $\overline{X} - \overline{Y} \pm 2.064 s_p\sqrt{1/8 + 1/18}$. Substituting $\overline{X} = 18.2$, $\overline{Y} = 16.4$, and $s_p = 1.8606$, we find that a 95% confidence interval for $\mu_X - \mu_Y$ is 1.8 ± 1.6318, or $(0.1682, 3.4318)$.

Summary

Let $X_1, \ldots, X_{n_X}$ be a random sample of size n_X from a *normal* population with mean μ_X, and let $Y_1, \ldots, Y_{n_Y}$ be a random sample of size n_Y from a *normal* population with mean μ_Y. Assume the two samples are independent.

If the populations are known to have nearly the same variance, a level $100(1-\alpha)\%$ confidence interval for $\mu_X - \mu_Y$ is

$$\overline{X} - \overline{Y} \pm t_{n_X+n_Y-2,\alpha/2} \cdot s_p \sqrt{\frac{1}{n_X} + \frac{1}{n_Y}} \tag{5.22}$$

The quantity s_p^2 is the pooled variance, given by

$$s_p^2 = \frac{(n_X - 1)s_X^2 + (n_Y - 1)s_Y^2}{n_X + n_Y - 2} \tag{5.23}$$

A machine is used to fill plastic bottles with bleach. A sample of 18 bottles had a mean fill volume of 2.007 L and a standard deviation of 0.010 L. The machine was then moved to another location. A sample of 10 bottles filled at the new location had a mean fill volume of 2.001 L and a standard deviation of 0.012 L. It is believed that moving the machine may have changed the mean fill volume, but is unlikely to have changed the standard deviation. Assume that both samples come from approximately normal populations. Find a 99% confidence interval for the difference between the mean fill volumes at the two locations.

Solution

We have $\overline{X} = 2.007$, $s_X = 0.010$, $n_X = 18$, $\overline{Y} = 2.001$, $s_Y = 0.012$, and $n_Y = 10$. Since we believe that the population standard deviations are equal, we estimate their common value with the pooled standard deviation, using Equation (5.23). We obtain

$$s_p = \sqrt{\frac{(18-1)(0.010^2) + (10-1)(0.012^2)}{18+10-2}} = 0.0107$$

The number of degrees of freedom is $18 + 10 - 2 = 26$. We use expression (5.22) to find the 99% confidence interval. Consulting the t table with 26 degrees of freedom, we find that $t_{26,\,.005} = 2.779$. The 99% confidence interval is therefore

$$2.007 - 2.001 \pm 2.779(0.0107)\sqrt{1/18 + 1/10}, \text{ or } 0.006 \pm 0.012.$$

Don't Assume the Population Variances Are Equal Just Because the Sample Variances Are Close

The confidence interval given by expression (5.22) requires that the population variances be equal, or nearly so. In situations where the *sample* variances are nearly equal, it is tempting to assume that the population variances are nearly equal as well. However, when the sample sizes are small, the sample variances are not necessarily good approximations to the population variances. Thus it is possible for the sample variances to be close even when the population variances are fairly far apart. In general, population variances should be assumed to be equal only when there is knowledge about the processes that produced the data that justifies this assumption.

The confidence interval given by expression (5.21) produces good results in almost all cases, whether the population variances are equal or not. (Exceptions can occur when the samples are of very different sizes.) Therefore, when in doubt about whether the population variances are equal, use expression (5.21).

Exercises for Section 5.6

1. The carbon content (in parts per million) was measured five times for each of two different silicon wafers. The measurements were as follows:

Wafer A:	1.10	1.15	1.16	1.10	1.14
Wafer B:	1.20	1.18	1.16	1.18	1.15

Find a 99% confidence interval for the difference in carbon content between the two wafers.

2. In a study of the rate at which the pesticide hexaconazole is absorbed through skin, skin specimens were exposed to 24 μg of hexaconazole. Four specimens were exposed for 30 minutes and four others were exposed for 60 minutes. The amounts (in μg) that were absorbed were

30 minutes:	3.1	3.3	3.4	3.0
60 minutes:	3.7	3.6	3.7	3.4

Find a 95% confidence interval for the mean amount absorbed in the time interval between 30 and 60 minutes after exposure.

3. The article "Differences in Susceptibilities of Different Cell Lines to Bilirubin Damage" (K. Ngai, C. Yeung, and C. Leung, *Journal of Paediatric Child Health*, 2000:36–45) reports an investigation into the toxicity of bilirubin on several cell lines. Ten sets of human liver cells and 10 sets of mouse fibroblast cells were placed into solutions of bilirubin in albumin with a 1.4 bilirubin/albumin molar ratio for 24 hours. In the 10 sets of human liver cells, the average percentage of cells surviving was 53.9 with a standard deviation of 10.7. In the 10 sets of mouse fibroblast cells, the average percentage of cells surviving was 73.1 with a standard deviation of 9.1. Find a 98% confidence interval for the difference in survival percentages between the two cell lines.

4. A medical geneticist is studying the frequency of a certain genetic mutation in men of various ages. For 10 men aged 20–29, the mean number of mutant sequences per μg DNA was 47.2 and the standard deviation was 15.1. For 12 men aged 60–69 the mean was 109.5 and the standard deviation was 31.2. Find a 99% confidence interval for the difference in the mean mutant frequency between men in their twenties and men in their sixties.

5. During the spring of 1999, many fuel storage facilities in Serbia were destroyed by bombing. As a result, significant quantities of oil products were spilled and burned, resulting in soil pollution. The article "Mobility of Heavy Metals Originating from Bombing of Industrial Sites" (B. Škrbić, J. Novaković, and N. Miljević, *Journal of Environmental Science and Health*, 2002:7–16) reports measurements of heavy metal concentrations at several industrial sites in June 1999, just after the bombing, and again in March of 2000. At the Smederevo site, on the banks of the Danube river, eight soil specimens taken in 1999 had an average lead concentration (in mg/kg) of 10.7 with a standard deviation of 3.3. Four specimens taken in 2000 had an average lead concentration of 33.8 with a standard deviation of 0.50. Find a 95% confidence interval for the increase in lead concentration between June 1999 and March 2000.

6. The article "Quality of the Fire Clay Coal Bed, Southeastern Kentucky" (J. Hower, W. Andrews, et al., *Journal of Coal Quality*, 1994:13–26) contains measurements on samples of coal from several counties in Kentucky. In units of percent ash, five samples from Knott County had an average aluminum dioxide (AlO_2) content of 32.17 and a standard deviation of 2.23. Six samples from Leslie County had an average AlO_2 content of 26.48 and a standard deviation of 2.02. Find a 98% confidence interval for the difference in AlO_2 content between coal samples from the two counties.

7. The article "The Frequency Distribution of Daily Global Irradiation at Kumasi" (F. Akuffo and A. Brew-Hammond, *Solar Energy*, 1993:145–154) defines the daily clearness index for a location to be the ratio of daily global irradiation to extraterrestrial irradiation. Measurements were taken in the city of Ibadan, Nigeria, over a five-year period. For five months of May, the clearness index averaged 0.498 and had standard deviation 0.036. For five months of July, the average was 0.389 and the standard deviation was 0.049. Find a 95% confidence interval for the difference in mean clearness index between May and July.

8. Several specimens of coal were sampled from each of two mines, and the heat capacity (in kilocalories per pound) was measured for each specimen. The results are as follows:

Mine 1:	4167	4268	4159	4285	4229
	4386	4103			
Mine 2:	3924	3988	4096	4026	
	4235	4178			

Find a 90% confidence interval for the difference in heat capacity between coal from mine 1 and coal from mine 2.

9. The breaking strength of hockey stick shafts made of two different graphite-Kevlar composites yield the following results (in newtons):

Composite A:	487.3	444.5	467.7	456.3	449.7
	459.2	478.9	461.5	477.2	
Composite B:	488.5	501.2	475.3	467.2	462.5
	499.7	470.0	469.5	481.5	485.2
	509.3	479.3	478.3	491.5	

Find a 98% confidence interval for the difference between the mean breaking strengths of hockey stick shafts made of the two materials.

10. The article "Permeability, Diffusion and Solubility of Gases" (B. Flaconnèche, et al., *Oil and Gas Science and Technology*, 2001:262–278) reported on a study of the effect of temperature and other factors on gas transport coefficients in semicrystalline polymers. The permeability coefficient (in 10^{-6} cm^3 (STP) /cm · s · MPa) of CO_2 was measured for extruded medium-density polyethelene at both 60°C and 61°C. The results are as follows:

60°C:	54	51	61	67	57	69	60
	60	63	62				
61°C:	58	60	66	66	68	61	60

Find a 95% confidence interval for the difference in the permeability coefficent between 60°C and 61°C.

11. A computer system administrator notices that computers running a particular operating system seem to freeze up more often as the installation of the operating system ages. She measures the time (in minutes) before freeze-up for seven computers one month after installation, and for nine computers seven months after installation. The results are as follows:

One month after install:	207.4	233.1	215.9
	235.1	225.6	244.4
	245.3		
Seven months after install:	84.3	53.2	127.3
	201.3	174.2	246.2
	149.4	156.4	103.3

Find a 95% confidence interval for the mean difference in time to freeze-up between the first month and the seventh.

12. In the article "Bactericidal Properties of Flat Surfaces and Nanoparticles Derivatized with Alkylated Polyethylenimines" (J. Lin, S. Qiu, et al., *Biotechnology Progress*, 2002:1082–1086), experiments were described in which alkylated polyethylenimines were attached to surfaces and to nanoparticles to make them bactericidal. In one series of experiments, the bactericidal efficiency against the bacterium *E. coli* was compared for a methylated versus a nonmethylated polymer. The mean percentage of bacterial cells killed with the methylated polymer was 95 with a standard deviation of 1, and the mean percentage of bacterial cells killed with the nonmethylated polymer was 70 with a standard deviation of 6. Assume that five independent measurements were made on each type of polymer.

Find a 95% confidence interval for the increase in bactericidal efficiency of the methylated polymer.

13. A sample of 8 room air conditioners of a certain model had a mean sound pressure of 52 decibels (dB) and a standard deviation of 5 dB, and a sample of 12 air conditioners of a different model had a mean sound pressure of 46 dB and a standard deviation of 2 dB. Find a 98% confidence interval for the difference in mean sound pressure between the two models.

5.7 Confidence Intervals with Paired Data

The methods discussed so far for finding confidence intervals on the basis of two samples have required that the samples be independent. In some cases, it is better to design an experiment so that each item in one sample is paired with an item in the other. Following is an example.

A tire manufacturer wishes to compare the tread wear of tires made of a new material with that of tires made of a conventional material. One tire of each type is placed on each front wheel of each of 10 front-wheel-drive automobiles. The choice as to which type of tire goes on the right wheel and which goes on the left is made with the flip of a coin. Each car is driven for 40,000 miles, then the tires are removed, and the depth of the tread on each is measured. The results are presented in Figure 5.15.

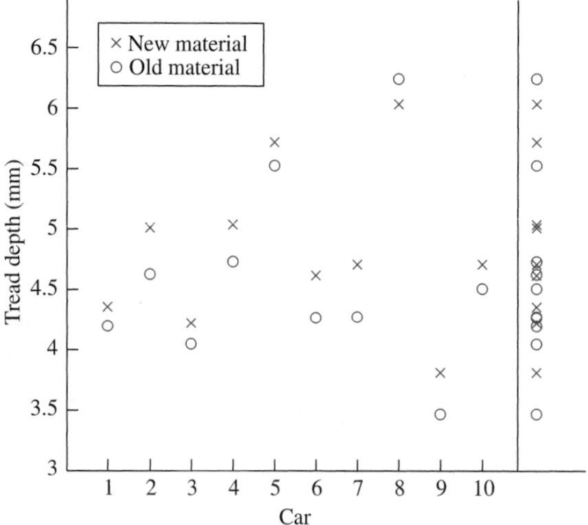

FIGURE 5.15 Tread depth for 10 pairs of tires.

The column on the right-hand side of Figure 5.15 presents the results for all 20 tires. There is considerable overlap in tread wear for the two samples. It is difficult to tell from the column whether there is a difference between the old and the new types of tire. However, when the data are examined in pairs, it is clear that the tires of the new type generally have more tread than those of the old type. The reason that analyzing the pairs presents a clearer picture of the result is that the cars vary greatly in the amount of wear they produce. Heavier cars, and those whose driving patterns involve many starts and stops, will generally produce more wear than others. The aggregated data in the column on the right-hand side of the figure includes this variability between cars as well as the variability in wear between tires. When the data are considered in pairs, the variability between the cars disappears, because both tires in a pair come from the same car.

Table 5.1 presents, for each car, the depths of tread for both the tires as well as the difference between them. We wish to find a 95% confidence interval for the mean difference in tread wear between old and new materials in a way that takes advantage of the reduced variability produced by the paired design. The way to do this is to think of a population of *pairs* of values, in which each pair consists of measurements from an old type tire and a new type tire on the same car. For each pair in the population, there is a difference (New − Old); thus there is a population of differences. The data are then a random sample from the population of pairs, and their differences are a random sample from the population of differences.

TABLE 5.1 Depths of tread, in mm, for tires made of new and old material

	Car									
	1	**2**	**3**	**4**	**5**	**6**	**7**	**8**	**9**	**10**
New material	4.35	5.00	4.21	5.03	5.71	4.61	4.70	6.03	3.80	4.70
Old material	4.19	4.62	4.04	4.72	5.52	4.26	4.27	6.24	3.46	4.50
Difference	0.16	0.38	0.17	0.31	0.19	0.35	0.43	−0.21	0.34	0.20

To put this into statistical notation, let $(X_1, Y_1), \ldots, (X_{10}, Y_{10})$ be the 10 observed pairs, with X_i representing the tread on the tire made from the new material on the ith car and Y_i representing the tread on the tire made from the old material on the ith car. Let $D_i = X_i - Y_i$ represent the difference between the treads for the tires on the ith car. Let μ_X and μ_Y represent the population means for X and Y, respectively. We wish to find a 95% confidence interval for the difference $\mu_X - \mu_Y$. Let μ_D represent the population mean of the differences. Then $\mu_D = \mu_X - \mu_Y$. It follows that a confidence interval for μ_D will also be a confidence interval for $\mu_X - \mu_Y$.

Since the sample $D_1, \ldots, D_{10}$ is a random sample from a population with mean μ_D, we can use one-sample methods to find confidence intervals for μ_D. In this example, since the sample size is small, we use the Student's t method of Section 5.3. The observed values of the sample mean and sample standard deviation are

$$\overline{D} = 0.232 \qquad s_D = 0.183$$

The sample size is 10, so there are nine degrees of freedom. The appropriate t value is $t_{9,.025} = 2.262$. The confidence interval using expression (5.9) (in Section 5.3) is therefore $0.232 \pm (2.262)(0.183)/\sqrt{10}$, or $(0.101,\ 0.363)$. When the number of pairs is large, the large-sample methods of Section 5.1, specifically expression (5.1), can be used.

Summary

Let $D_1, \ldots, D_n$ be a *small* random sample ($n \leq 30$) of differences of pairs. If the population of differences is approximately normal, then a level $100(1-\alpha)\%$ confidence interval for the mean difference μ_D is given by

$$\overline{D} \pm t_{n-1,\alpha/2}\frac{s_D}{\sqrt{n}} \tag{5.24}$$

Note that this interval is the same as that given by expression (5.9).

If the sample size is large, a level $100(1-\alpha)\%$ confidence interval for the mean difference μ_D is given by

$$\overline{D} \pm z_{\alpha/2}\sigma_{\overline{D}} \tag{5.25}$$

In practice $\sigma_{\overline{D}}$ is approximated with $s_D/\sqrt{n}$. Note that this interval is the same as that given by expression (5.1).

Exercises for Section 5.7

1. The article "Simulation of the Hot Carbonate Process for Removal of CO_2 and H_2S from Medium Btu Gas" (K. Park and T. Edgar, *Energy Progress*, 1984: 174–180) presents an equation used to estimate the equilibrium vapor pressure of CO_2 in a potassium carbonate solution. The actual equilibrium pressure (in kPa) was measured in nine different reactions and compared with the value estimated from the equation. The results are presented in the following table:

Reaction	Estimated	Experimental	Difference
1	45.10	42.95	2.15
2	85.77	79.98	5.79
3	151.84	146.17	5.67
4	244.30	228.22	16.08
5	257.67	240.63	17.04
6	44.32	41.99	2.33
7	84.41	82.05	2.36
8	150.47	149.62	0.85
9	253.81	245.45	8.36

Find a 95% confidence interval for the mean difference between the estimated and actual pressures.

2. The article "Effect of Refrigeration on the Potassium Bitartrate Stability and Composition of Italian Wines" (A. Versari, D. Barbanti, et al., *Italian Journal of Food Science* 2002:45–52) reports a study in which eight types of white wine had their tartaric acid concentration (in g/L) measured both before and after a cold stabilization process. The results are presented in the following table:

Wine Type	Before	After	Difference
1	2.86	2.59	0.27
2	2.85	2.47	0.38
3	1.84	1.58	0.26
4	1.60	1.56	0.04
5	0.80	0.78	0.02
6	0.89	0.66	0.23
7	2.03	1.87	0.16
8	1.90	1.71	0.19

Find a 95% confidence interval for the mean difference between the tartaric acid concentrations before and after the cold stabilization process.

3. In an experiment to determine whether there is a systematic difference between the weights obtained with two different scales, 10 rock specimens were weighed, in grams, on each scale. The following data were obtained:

Specimen	Weight on scale 1	Weight on scale 2
1	11.23	11.27
2	14.36	14.41
3	8.33	8.35
4	10.50	10.52
5	23.42	23.41
6	9.15	9.17
7	13.47	13.52
8	6.47	6.46
9	12.40	12.45
10	19.38	19.35

Assume that the difference between the scales, if any, does not depend on the object being weighed. Find a 98% confidence interval for this difference.

4. In a certain type of engine, the flywheel is fastened to the crankshaft flange block by eight bolts, which must be tightened in sequence. An engineer wants to determine how much difference there is between the torque of the first and the last bolts tightened. In a sample of seven engines, the torque on each of these bolts is measured. Following are the values, in N·m, for bolts 1 and 8.

Engine	Bolt 1	Bolt 8
1	105.28	105.38
2	105.03	105.96
3	104.53	105.19
4	104.68	105.61
5	104.98	105.42
6	105.21	105.20
7	105.30	105.98

Find a 90% confidence interval for the difference between the torques.

5. In a study of the dissolve time of various confections, nine people each dissolved one piece of chocolate and one piece of butterscotch. The dissolve times, in seconds, are presented in the following table:

Person	Chocolate	Butterscotch
1	51	53
2	47	40
3	90	155
4	65	90
5	27	33
6	105	68
7	90	72
8	54	52
9	93	77

Find a 98% confidence interval for the difference between the mean dissolve times of chocolate and butterscotch.

6. A sample of 10 diesel trucks were run both hot and cold to estimate the difference in fuel economy. The results, in mi/gal, are presented in the following table. (From "In-use Emissions from Heavy-Duty Diesel Vehicles," J. Yanowitz, Ph.D. thesis, Colorado School of Mines, 2001.)

Truck	Hot	Cold
1	4.56	4.26
2	4.46	4.08
3	6.49	5.83
4	5.37	4.96
5	6.25	5.87
6	5.90	5.32
7	4.12	3.92
8	3.85	3.69
9	4.15	3.74
10	4.69	4.19

Find a 98% confidence interval for the difference in mean fuel mileage between hot and cold engines.

7. For a sample of nine automobiles, the mileage (in 1000s of miles) at which the original front brake pads were worn to 10% of their original thickness was measured, as was the mileage at which the original rear brake pads were worn to 10% of their original thickness. The results are given in the following table.

Automobile	Front	Rear
1	32.8	41.2
2	26.6	35.2
3	35.6	46.1
4	36.4	46.0
5	29.2	39.9
6	40.9	51.7
7	40.9	51.6
8	34.8	46.1
9	36.6	47.3

Find a 95% confidence interval for the difference in mean lifetime between the front and rear brake pads.

8. Refer to Exercise 7. Someone suggests that the paired design be replaced with a design in which 18 cars are sampled, the lifetime of the front brakes is measured on 9 of them, and the lifetime of the rear brakes is measured on the other 9. A confidence interval for the difference between the means would then be constructed by using expression (5.21) (in Section 5.6). He claims that this design will produce a more precise confidence interval, since 18 cars will be used instead of 9.

a. Will the new design produce a valid confidence interval? Explain.

b. Is it likely that the confidence interval produced by the new design will be more precise than, less precise than, or about equally precise as the confidence interval produced by the paired design? Explain. (*Hint:* Look at Figure 5.15.)

9. A tire manufacturer is interested in testing the fuel economy for two different tread patterns. Tires of each tread type are driven for 1000 miles on each of 18 different cars. The mileages, in mi/gal, are presented in the following table.

Car	Tread A	Tread B
1	24.1	20.3
2	22.3	19.7
3	24.5	22.5
4	26.1	23.2
5	22.6	20.4
6	23.3	23.5
7	22.4	21.9
8	19.9	18.6
9	27.1	25.8
10	23.5	21.4
11	25.4	20.6
12	24.9	23.4
13	23.7	20.3
14	23.9	22.5
15	24.6	23.5
16	26.4	24.5
17	21.5	22.4
18	24.6	24.9

a. Find a 99% confidence interval for the mean difference in fuel economy.

b. A confidence interval based on the data in the table has width ± 0.5 mi/gal. Is the level of this confidence interval closest to 80%, 90%, or 95%?

10. Refer to Exercise 9. In a separate experiment, 18 cars were outfitted with tires with tread type A, and another 18 were outfitted with tires with tread type B. Each car was driven 1000 miles. The cars with tread type A averaged 23.93 mi/gal, with a standard deviation of 1.79 mi/gal. The cars with tread type B averaged 22.19 mi/gal, with a standard deviation of 1.95 mi/gal.

a. Which method should be used to find a confidence interval for the difference between the mean mileages of the two tread types: expression (5.24) (in this section) or expression (5.21) (in Section 5.6)?

b. Using the appropriate method, find a 99% confidence interval for the difference between the mean mileages of the two tread types.

c. Is the confidence interval found in part (b) wider than the one found in Exercise 9? Why is this so?

5.8 Using Simulation to Construct Confidence Intervals

If $X_1, \ldots, X_n$ are independent random variables with known standard deviations $\sigma_1, \ldots, \sigma_n$, and $U = U(X_1, \ldots, X_n)$ is a function of $X_1, \ldots, X_n$, then the method of propagation of error (see Chapter 3) can be used to estimate the standard deviation, or uncertainty, in U. If in addition the random variables $X_1, \ldots, X_n$ are approximately normally distributed, it will often (not always) be the case that U is approximately normally distributed as well. In these cases expression (5.13) (in Section 5.3) can be used to compute a confidence interval for the mean μ_U of U. To determine whether U is approximately normally distributed, simulation can be used.

To provide a concrete example, assume that a process manufactures steel washers, whose radii are normally distributed with unknown mean μ_R and known standard deviation $\sigma_R = 0.1$ cm. A single washer, selected at random, is observed to have a radius of $R = 2.5$ cm. Since R comes from a normal population with known standard deviation, expression (5.13) can be used to find a confidence interval for the mean radius μ_R. A 95% confidence interval for μ_R is $R \pm 1.96\sigma_R = 2.5 \pm 0.196$. Now let's consider the area of the washer. The area of the sampled washer is given by $A = \pi R^2 = 3.14(2.5^2) = 19.63\ \text{cm}^2$. We can use the method of propagation of error (Equation 3.10 in Section 3.3), to estimate the standard deviation, or uncertainty, σ_A, by

$$\sigma_A = \left|\frac{dA}{dR}\right| \sigma_R = 2\pi R \sigma_R = 2(3.14)(2.5)(0.10) = 1.57\ \text{cm}^2$$

We now have a single sampled value, $A = 19.63$, from the population of all possible areas, and we have an estimate of the standard deviation of this population, $\sigma_A = 1.57$. Can we find a confidence interval for the mean area μ_A? If the distribution of areas were normally distributed, we could find a confidence interval for μ_A by the same method used for μ_R, obtaining $A \pm 1.96\sigma_A = 19.63 \pm 1.96(1.57)$. The assumption of normality is necessary to justify this method. How can we determine whether the distribution of areas is normal? If we had a large sample of areas, we could construct a normal probability plot. We don't have a large sample of areas, but we can simulate one as follows (see Section 4.11 for a discussion of the basic principles of simulation).

We start by generating a large sample of simulated radii R^*. We want the distribution of the population from which this sample is drawn to be as close as possible to the distribution of the population from which the observation $R = 2.5$ was drawn. We know that the value $R = 2.5$ was drawn from a normal population with $\sigma_R = 0.1$. We don't know μ_R, but we can use the observed value $R = 2.5$ as an approximation to μ_R for the purposes of simulation. Thus we will generate a large sample of simulated radii $R_1^*, \ldots, R_N^*$ from a $N(2.5, 0.1^2)$ distribution. (The notation R_i^* indicates that this is a simulated value, rather than a value observed in an actual experiment.)

To understand exactly what we can do with the simulated values, let's imagine that we had a large sample of actual washers, and that their radii $R_1, \ldots, R_n$ had been determined. What are the similarities and differences between the actual sample $R_1, \ldots, R_n$ and the simulated sample $R_1^*, \ldots, R_N^*$? The actual sample comes from a population that is normally distributed, whose standard deviation is known to be 0.1, and whose mean μ_R is unknown. The simulated sample comes from a population that is also normally distributed, whose standard deviation is also equal to 0.1, and whose mean has been set to 2.5 (the value of the one actual observation of R). Thus the simulated population has the same shape (normal) and spread (standard deviation) as the actual population. The simulated and actual populations have different means (2.5 for the simulated population, and the unknown value μ_R for the actual population).

Now for each R_i^* we compute a simulated area $A_i^* = \pi R_i^{*2}$. Imagine we had an actual sample of radii $R_1, \ldots, R_n$, and computed an actual sample $A_1, \ldots, A_n$ of areas from it. Because the simulated sample of radii $R_1^*, \ldots, R_n^*$ comes from a population whose shape and spread are the same as the actual population of radii, it is reasonable to assume that the simulated sample of areas $A_1^*, \ldots, A_n^*$ comes from a population whose shape and spread are very similar to the actual population of areas. In other words, it is reasonable to assume that the sample standard deviation of the simulated sample $A_1^*, \ldots, A_n^*$ is close to the actual population standard deviation σ_A, and it is reasonable to assume that if the simulated sample $A_1^*, \ldots, A_n^*$ comes from a population that is approximately normal, the actual population of areas is approximately normal as well. The mean of the simulated population of areas will differ from that of the actual population of areas, however. The reason for this is that the means of the simulated and actual populations of radii differ. The mean of the simulated population of areas will be close to the one actual observed value of A, which is 19.63. The mean of the actual population of areas is the unknown value μ_A.

We construct a normal probability plot for $A_1^*, \ldots, A_n^*$. If it shows that the population of areas is approximately normal, then we can assume that the actual observed A came from an approximately normal population, and we can find a confidence interval for μ_A. Figure 5.16 presents a normal probability plot for a sample of 1000 areas. With the exception of a few points at either end, the normality assumption seems well satisfied.

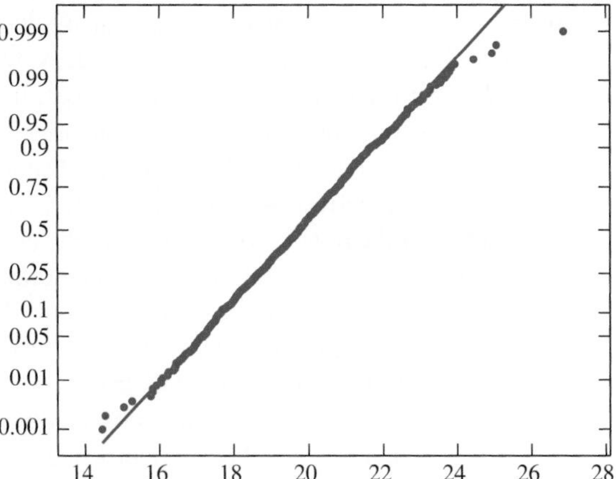

FIGURE 5.16 Normal probability plot for 1000 simulated areas. The assumption of normality is justified.

The value of A actually observed is $A = 19.63$. The standard deviation of the population from which A was drawn, as estimated by the method of propagation of error, is $\sigma_A = 1.57$. Since this population is approximately normal, a 95% confidence interval for the mean area μ_A is $19.63 \pm 1.96(1.57)$, or (16.55, 22.71).

In the confidence interval just calculated, the standard deviation σ_A was obtained by the method of propagation of error. It is also acceptable to use the sample standard deviation of the simulated values $A_1^*, \ldots, A_n^*$ to estimate σ_A. In the 1000 values we simulated, the sample standard deviation was 1.5891, which is close to the value 1.57 estimated by propagation of error. This close agreement is typical.

It is important to note that the center of the confidence interval is the actual observed value A, and not the mean of the simulated values $\overline{A}^*$. The reason for this is that we are finding a confidence interval for the mean of the actual population of areas μ_A, and the observed value A has been sampled from this population. The simulated values have been sampled from a population whose mean is different from that of the actual population. Therefore $\overline{A}^*$ is not an appropriate choice for the center of the confidence interval.

The method just described can be very useful when making measurements whose measurement errors are normally distributed. We present some examples.

Example 5.27

The length and width of a rectangle are measured as $X = 3.0 \pm 0.1$ and $Y = 3.5 \pm 0.2$ cm, respectively. Assume that the measurements come from normal populations and are unbiased. Assume the standard deviations $\sigma_X = 0.1$ and $\sigma_Y = 0.2$ are known. Find a 95% confidence interval for the area of the rectangle.

Solution

Let $A = XY$ denote the measured area of the rectangle. The observed value of A is $A = (3.0)(3.5) = 10.5$. We will use the method of propagation of error to estimate σ_A. Then we will use simulation to check that the distribution of A is approximately normal. To estimate σ_A, we first compute the partial derivatives of A with respect to X and Y.

$$\frac{\partial A}{\partial X} = Y = 3.5 \qquad \frac{\partial A}{\partial Y} = X = 3.0$$

The standard deviation σ_A is then estimated with

$$\begin{aligned} \sigma_A &= \sqrt{\left(\frac{\partial A}{\partial X}\right)^2 \sigma_X^2 + \left(\frac{\partial A}{\partial Y}\right)^2 \sigma_Y^2} \\ &= \sqrt{(3.5)^2(0.1)^2 + (3.0)^2(0.2)^2} \\ &= 0.6946 \end{aligned}$$

To check normality, we generated 1000 simulated values $X_1^*, \ldots, X_{1000}^*$ from a $N(3.0,\ 0.1^2)$ distribution, and 1000 simulated values $Y_1^*, \ldots, Y_{1000}^*$ from a $N(3.5,\ 0.2^2)$ distribution. Note that we used the observed values 3.0 and 3.5 to approximate the unknown means μ_X and μ_Y (which are the true length and width, respectively) for the purposes of simulation. We then computed 1000 simulated areas $A_1^*, \ldots, A_{1000}^*$. A normal probability plot appears in the following figure. The normality assumption is justified. A 95% confidence interval for the area of the rectangle is $10.5 \pm 1.96(0.6946)$.

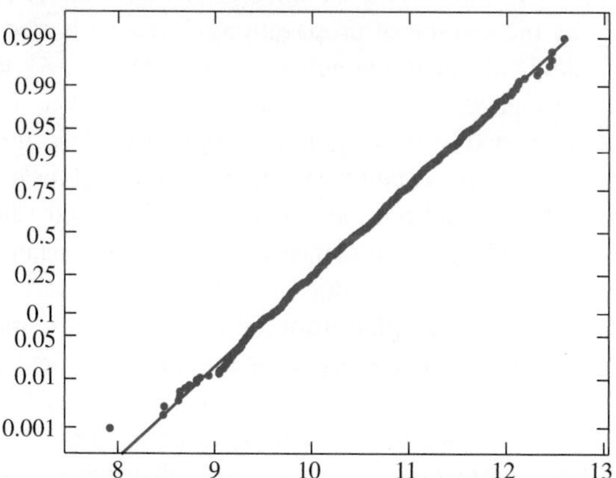

Technical note: In Example 5.27 we are actually finding a confidence interval for the mean μ_A of the measured area. The true area of the rectangle is the product of the true length and the true width, which is $\mu_X \mu_Y$. The value μ_A differs slightly from the product $\mu_X \mu_Y$, but the difference is negligible for practical purposes.

Example 5.28

Two resistors whose resistances are measured to be X and Y are connected in parallel. The total resistance is estimated with $R = (XY)/(X+Y)$. Assume that $X = 10.0 \pm 1.0\ \Omega$, $Y = 20.0 \pm 2.0\ \Omega$, and that X and Y come from normal populations and are unbiased. Find a 95% confidence interval for the total resistance.

Solution

The observed value of R is $(10)(20)/(10+20) = 6.667\ \Omega$. To compute σ_R, we first compute the partial derivatives of R:

$$\frac{\partial R}{\partial X} = \left(\frac{Y}{X+Y}\right)^2 = 0.4444$$

$$\frac{\partial R}{\partial Y} = \left(\frac{X}{X+Y}\right)^2 = 0.1111$$

Now $\sigma_X = 1\ \Omega$ and $\sigma_Y = 2\ \Omega$. Therefore

$$\begin{aligned} \sigma_R &= \sqrt{\left(\frac{\partial R}{\partial X}\right)^2 \sigma_X^2 + \left(\frac{\partial R}{\partial Y}\right)^2 \sigma_Y^2} \\ &= \sqrt{(0.4444)^2(1)^2 + (0.1111)^2(2)^2} \\ &= 0.497\ \Omega \end{aligned}$$

To check normality, we generated 1000 simulated values $X_1^*, \ldots, X_{1000}^*$ from a $N(10, 1.0^2)$ distribution, and 1000 simulated values $Y_1^*, \ldots, Y_{1000}^*$ from a $N(20, 2.0^2)$ distribution. Note that we use the observed values 10 and 20 to approximate the means μ_X and μ_Y for the purposes of simulation. We then computed 1000 simulated values $R_1^*, \ldots, R_{1000}^*$. A normal probability plot appears in the following figure. The normality assumption is justified. A 95% confidence interval for the total resistance is $6.667 \pm 1.96(0.497)$.

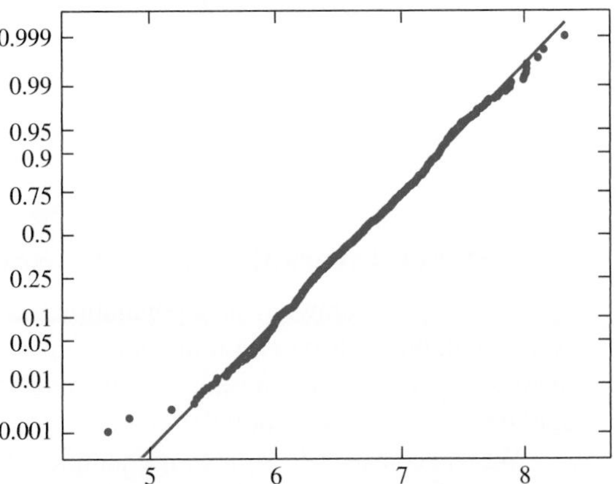

The technical note appearing after Example 5.27 applies to Example 5.28 as well. We are actually finding a confidence interval for the mean μ_R of the measured total resistance. This mean μ_R is slightly different from the true total resistance $\mu_X \mu_Y/(\mu_X + \mu_Y)$, but the difference is negligible for practical purposes.

In some cases the distribution of a function $U(X_1, \ldots, X_n)$ is not normal even when $X_1, \ldots, X_n$ are normal. For this reason it is important to check normality with a simulation. Example 5.29 provides an illustration.

Example 5.29

The mass of a rock is measured to be $M = 10 \pm 0.4$ g, and its volume is measured to be $V = 1.0 \pm 0.2$ mL. The density is estimated to be $D = M/V$. Assume M and V come from normal populations and are unbiased. Is D normally distributed? Can the method described in Examples 5.27 and 5.28, which is based on the normal curve, be used to find a 95% confidence interval for the density of the rock?

Solution

We generated 1000 simulated values $M_1^*, \ldots, M_{1000}^*$ from a $N(10, 0.4^2)$ distribution, and 1000 simulated values $V_1^*, \ldots, V_{1000}^*$ from a $N(1.0, 0.2^2)$ distribution. We then computed values $D_i^* = M_i^*/V_i^*$. A normal probability plot of the D_i^* appears in the following figure. The normality assumption is not justified. The method based on the normal curve cannot be used to find a confidence interval for the density of the rock.

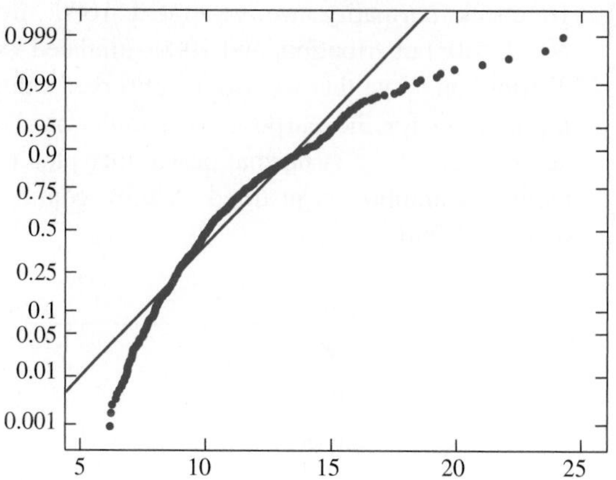

Confidence Intervals Using the Bootstrap

When a sample is drawn from a population that is not normal, and a confidence interval is required, methods based on the bootstrap (see Section 4.11) may be used. There are many such methods; we present a simple one here and show how to use it to construct a confidence interval for a population mean.

We will consider the following example. The article "In-use Emissions from Heavy-Duty Diesel Vehicles" (J. Yanowitz, Ph.D. thesis, Colorado School of Mines, 2001) presents fuel efficiency measurements (in mi/gal) on a sample of 7 trucks. The data are as follows:

7.69 4.97 4.56 6.49 4.34 6.24 4.45

Assume this is a random sample from a population of trucks, and assume that we wish to construct a 95% confidence interval for the mean fuel efficiency μ of this population. A look at the sample suggests that there is a gap near the middle of the distribution, since there are no trucks in the sample with values between 5 and 6. Therefore one might not wish to assume that the data were normal. The bootstrap provides a method for constructing a confidence interval whose level will be approximately 95% (or any other value that one might specify).

To construct a bootstrap confidence interval, we must draw *bootstrap samples* from the data. A bootstrap sample is a sample of the same size as the data, drawn with replacement. To describe this in detail, denote the values in a random sample by $X_1, \ldots, X_n$. Imagine putting these values in a box and drawing one out at random. This would be the first value in the bootstrap sample; call it X_1^*. Then replace X_1^* in the box, and draw another value, X_2^*. Continue in this way until n values $X_1^*, \ldots, X_n^*$ have been drawn. This is a bootstrap sample. Note that each value in the bootstrap sample is drawn from the complete data sample, so that it is likely that some values will appear more than once while others will not appear at all.

We drew 1000 bootstrap samples from the given mileage data. The first 10 and the last one of them are presented in Table 5.2. The sample mean is computed for each bootstrap sample.

TABLE 5.2 Bootstrap samples from the mileage data

Sample	Sample Values							Sample Mean
1	4.97	6.49	7.69	4.97	7.69	4.56	4.45	5.8314
2	6.24	4.97	4.56	4.97	4.56	6.24	7.69	5.6043
3	4.34	4.45	4.56	4.45	6.24	4.97	4.34	4.7643
4	4.45	6.49	7.69	6.24	4.97	4.45	4.34	5.5186
5	6.24	4.34	4.45	7.69	4.56	4.34	4.45	5.1529
6	4.34	4.97	7.69	4.97	6.24	6.24	6.24	5.8129
7	4.45	6.49	6.24	4.97	4.34	7.69	4.34	5.5029
8	6.49	7.69	4.97	6.49	6.49	4.34	4.56	5.8614
9	7.69	4.45	4.45	4.45	4.45	4.56	4.56	4.9443
10	6.24	4.56	4.97	6.49	4.45	4.97	6.24	5.4171
⋮				⋮				⋮
1000	4.34	7.69	4.45	4.56	7.69	4.45	7.69	5.8386

To construct a bootstrap confidence interval, many bootstrap samples (minimum 1000) must be drawn. Since we want a confidence interval for the population mean μ, we compute the sample mean for each bootstrap sample. Let $\overline{X}_i^*$ denote the mean of the ith bootstrap sample. Since we want the level of the confidence interval to be as close to 95% as possible, we find the interval that spans the middle 95% of the bootstrap sample means. The endpoints of this interval are the 2.5 percentile and the 97.5 percentile of the list of bootstrap sample means. Denote these percentiles by $\overline{X}^*_{.025}$ and $\overline{X}^*_{.975}$, respectively.

We will compute these percentiles for the mileage data. Following are the smallest 26 and largest 26 of the 1000 bootstrap sample means $\overline{X}_i^*$.

Smallest 26:	4.4929	4.4971	4.5357	4.5400	4.5514	4.5557	4.5557	4.5829
	4.5986	4.6143	4.6429	4.6457	4.6729	4.6729	4.6900	4.6943
	4.7014	4.7157	4.7257	4.7257	4.7329	4.7371	4.7414	4.7486
	4.7643	4.7643						
Largest 26:	6.4757	6.4757	6.4800	6.4900	6.4986	6.5214	6.5443	6.5543
	6.5929	6.5929	6.6257	6.6257	6.6471	6.6671	6.6900	6.6929
	6.7057	6.7129	6.7514	6.7971	6.7971	6.8486	6.9329	6.9686
	7.0714	7.1043						

Using the method of percentile calculation presented in Chapter 1, the 2.5 percentile is the average of the 25th and 26th values in the ordered sample of 1000, and the 97.5 percentile is the average of the 975th and 976th values. Therefore in this case $\overline{X}^*_{.025} = 4.7643$ and $\overline{X}^*_{.975} = 6.4757$.

There are now two methods available to construct the confidence interval; which is better is a matter of some controversy. In method 1, the confidence interval is $(\overline{X}^*_{.025}, \overline{X}^*_{.975})$. Method 2 uses the mean $\overline{X}$ of the original sample in addition to the percentiles; the method 2 confidence interval is $(2\overline{X} - \overline{X}^*_{.975},\ 2\overline{X} - \overline{X}^*_{.025})$. For the mileage data, the 95% confidence interval computed by method 1 is (4.7643, 6.4757).

The sample mean for the mileage data is $\overline{X} = 5.5343$. Therefore the 95% confidence interval computed by method 2 is

$$\left(2(5.5343) - 6.4757,\ 2(5.5343) - 4.7643\right) = \left(4.5929,\ 6.3043\right)$$

The confidence intervals from the two methods are similar in this case.

Summary

Given a random sample $X_1, \ldots, X_n$ from a population with mean μ, a bootstrap confidence interval for μ with level approximately $100(1-\alpha)\%$ can be computed as follows:

- Draw a large number m ($m \geq 1000$) of bootstrap samples of size n with replacement from $X_1, \ldots, X_n$.
- Compute the mean of each bootstrap sample. Denote these bootstrap means by $\overline{X}_1^*, \ldots, \overline{X}_m^*$.
- Compute the $100\alpha/2$ and the $100(1 - \alpha/2)$ percentiles of the bootstrap means. Denote these values $\overline{X}_{\alpha/2}^*, \overline{X}_{1-\alpha/2}^*$.
- There are two methods for computing the confidence interval.
  Method 1: $(\overline{X}_{\alpha/2}^*,\ \overline{X}_{1-\alpha/2}^*)$ Method 2: $(2\overline{X} - \overline{X}_{1-\alpha/2}^*,\ 2\overline{X} - \overline{X}_{\alpha/2}^*)$

Although it is not obvious at first, there is a connection between the bootstrap method presented here for computing confidence intervals for a population mean and the large-sample method based on the normal curve. In both cases the width of the confidence interval should ideally equal the width of the middle 95% of the distribution of the sample mean $\overline{X}$. When the sample size is large, the distribution of $\overline{X}$ approximately follows the normal curve, so the width of the 95% confidence interval is made to equal the width of the middle 95% of the normal distribution (see Figure 5.1 in Section 5.1). The bootstrap is used when the distribution of $\overline{X}$ is not necessarily normal. The collection of bootstrap sample means $\overline{X}_i^*$ approximates a random sample from the distribution of $\overline{X}$, so this collection, rather than the normal curve, forms the basis for the confidence interval. The width of the bootstrap confidence interval is made to equal the width of the middle 95% of the bootstrap sample means in order to approximate the width of the unknown distribution of $\overline{X}$.

There are many different methods for computing bootstrap confidence intervals. The simple methods presented here work well when the population from which the bootstrap sample is drawn is approximately symmetric, but not so well when it is highly skewed. More sophisticated methods have been developed that produce good results under more general conditions. Additional information on this topic may be found in Efron and Tibshirani (1993).

Using Simulation to Evaluate Confidence Intervals

A level $100(1 - \alpha)\%$ confidence interval is one that is computed by a procedure that succeeds in covering the true value for $100(1-\alpha)\%$ of all the samples that could possibly be drawn. When the assumptions governing the use of a method are violated, this success rate (also called the **coverage probability**) may be lower. In practice, assumptions often do not hold precisely. Some methods are very sensitive to their assumptions, in that the coverage probability can become much less than $100(1-\alpha)\%$ even when the assumptions are only slightly violated. Other methods are "robust," which means that the coverage probability does not go much below $100(1-\alpha)\%$ so long as their assumptions are approximately satisfied. The advantage of a robust method is that it is useful over a wide range of conditions and requires less concern about assumptions. Simulation experiments provide a good way to evaluate the robustness of a statistical procedure. We present an experiment that will be instructive regarding the robustness of the Student's t method for constructing confidence intervals for a population mean (expression 5.9 in Section 5.3).

The Student's t distribution can be used to construct confidence intervals for a population mean, provided the sample comes from a population that is "approximately" normal. We will perform a simulation experiment to gain some insight into how rough this approximation can be. The following figure shows the probability density function for the $\Gamma(2.5, 0.5)$ distribution (this is also known as the chi-square distribution with five degrees of freedom). It is fairly skewed and does not look too much like the normal curve. The mean of this population is $\mu = 5$. If the Student's t method is applied to samples of size 5 from this population, what proportion of the time will a 95% confidence interval cover the true mean?

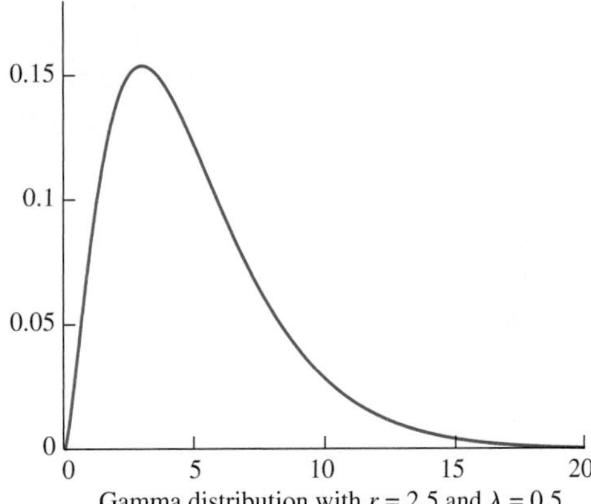

Gamma distribution with $r = 2.5$ and $\lambda = 0.5$

To address this question, we generated 10,000 samples of size 5 from the $\Gamma(2.5, 0.5)$ distribution. Denote the ith sample by $X_{1i}^*, X_{2i}^*, X_{3i}^*, X_{4i}^*, X_{5i}^*$; denote its sample mean by $\overline{X}_i^*$ and its sample standard deviation by s_i^*. For each sample, we computed a confidence interval using the formula for a 95% confidence interval based on the Student's t distribution (expression 5.9 in Section 5.3). The lower confidence limit is $L_i^* = \overline{X}_i^* - 2.776 s_i^*/\sqrt{5}$ (note that $t_{4,.025} = 2.776$). The upper confidence limit is $U_i^* = \overline{X}_i^* + 2.776 s_i^*/\sqrt{5}$. Table 5.3 (page 360) presents the results for the first 10 samples and for the last one. The rightmost column contains a "1" if $L_i^* < 5 < U_i^*$, in other words, if the ith confidence interval covers the true mean of 5.

TABLE 5.3 Simulated data from the Γ(2.5, 0.5) distribution

i	X_1^*	X_2^*	X_3^*	X_4^*	X_5^*	$\overline{X}^*$	s^*	L^*	U^*	$L^* < 5 < U^*$
1	2.58	6.54	3.02	3.40	1.23	3.36	1.96	0.92	5.79	1
2	1.28	1.44	1.45	10.22	4.17	3.71	3.83	−1.05	8.47	1
3	7.26	3.28	2.85	8.94	12.09	6.88	3.89	2.05	11.72	1
4	6.11	3.81	7.06	11.89	3.01	6.38	3.49	2.04	10.72	1
5	4.46	9.70	5.14	2.45	4.99	5.35	2.66	2.05	8.65	1
6	2.20	1.46	9.30	2.00	4.80	3.95	3.26	−0.09	7.99	1
7	7.17	13.33	6.19	10.31	8.49	9.10	2.83	5.59	12.61	1
8	1.97	1.81	4.13	1.28	5.16	2.87	1.68	0.78	4.95	0
9	3.65	1.98	8.19	7.20	3.81	4.97	2.61	1.72	8.21	1
10	3.39	2.31	1.86	5.97	5.28	3.76	1.80	1.52	6.00	1
⋮	⋮	⋮	⋮	⋮	⋮	⋮	⋮	⋮	⋮	⋮
10,000	7.30	7.21	1.64	3.54	3.41	4.62	2.52	1.49	7.75	1

Nine of the first 10 confidence intervals cover the true mean. So if we were to base our results on the first 10 samples, we would estimate the coverage probability of the confidence interval to be 0.90. Of course, 10 samples is not nearly enough. Out of all 10,000 samples, the true mean was covered in 9205 of them. We therefore estimate the coverage probability to be 0.9205. While this is less than 95%, it is not dramatically less. This simulation suggests that the Student's t procedure is fairly robust, in other words, that confidence intervals based on the Student's t distribution cover the true mean almost as often as they should, even when the population is somewhat different from normal.

If the population deviates substantially from normal, the Student's t method will not perform well. See Exercise 8.

Exercises for Section 5.8

1. The pressure of air (in MPa) entering a compressor is measured to be $X = 8.5 \pm 0.2$, and the pressure of the air leaving the compressor is measured to be $Y = 21.2 \pm 0.3$. The intermediate pressure is therefore measured to be $P = \sqrt{XY} = 13.42$. Assume that X and Y come from normal populations and are unbiased.

a. From what distributions is it appropriate to simulate values X^* and Y^*?

b. Generate simulated samples of values X^*, Y^*, and P^*.

c. Use the method of propagation of error to estimate the standard deviation of P. (Alternatively, you may use the simulated sample.)

d. Construct a normal probability plot for the values P^*. Is it reasonable to assume that P is approximately normally distributed?

e. If appropriate, use the normal curve to find a 95% confidence interval for the intermediate pressure.

2. The mass (in kg) of a soil specimen is measured to be $X = 1.18 \pm 0.02$. After the sample is dried in an oven, the mass of the dried soil is measured to be $Y = 0.85 \pm 0.02$. Assume that X and Y come from normal populations and are unbiased. The water content of the soil is measured to be

$$W = \frac{X - Y}{X}$$

a. From what distributions is it appropriate to simulate values X^* and Y^*?

b. Generate simulated samples of values X^*, Y^*, and W^*.

c. Use the method of propagation of error to estimate the standard deviation of W. (Alternatively, you may use the simulated sample.)

d. Construct a normal probability plot for the values W^*. Is it reasonable to assume that W is approximately normally distributed?

e. If appropriate, use the normal curve to find a 95% confidence interval for the water content.

3. A student measures the acceleration A of a cart moving down an inclined plane by measuring the time T that it takes the cart to travel 1 m and using the formula $A = 2/T^2$. Assume that $T = 0.55 \pm 0.01$ s, and that the measurement T comes from a normal population and is unbiased.

a. Generate an appropriate simulated sample of values A^*. Is it reasonable to assume that A is normally distributed?

b. Use the method of propagation of error to estimate the standard deviation of A. (Alternatively, you may use the simulated sample.)

c. If appropriate, use the normal curve to find a 95% confidence interval for the acceleration of the cart.

4. The initial temperature of a certain container is measured to be $T_0 = 20°\text{C}$. The ambient temperature is measured to be $T_a = 4°\text{C}$. An engineer uses Newton's law of cooling to compute the time needed to cool the container to a temperature of 10°C. Taking into account the physical properties of the container, this time (in minutes) is computed to be

$$T = 40 \ln\left(\frac{T_0 - T_a}{10 - T_a}\right)$$

Assume that the temperature measurements T_0 and T_a are unbiased and come from normal populations with standard deviation 0.1°C.

a. Generate an appropriate simulated sample of values T^*. Is it reasonable to assume that T is normally distributed?

b. Use the method of propagation of error to estimate the standard deviation of T. (Alternatively, you may use the simulated sample.)

c. If appropriate, use the normal curve to find a 95% confidence interval for the time needed to cool the container to a temperature of 10°C.

5. In the article "Occurrence and Distribution of Ammonium in Iowa Groundwater" (K. Schilling, *Water Environment Research*, 2002:177–186), ammonium concentrations (in mg/L) were measured at a large number of wells in the state of Iowa. These included 349 alluvial wells and 143 quaternary wells. The concentrations at the alluvial wells averaged 0.27 with a standard deviation of 0.40, and those at the quaternary wells averaged 1.62 with a standard deviation of 1.70. Since these standard deviations are based on large samples, assume they are negligibly different from the population standard deviations. An estimate for the ratio of the mean concentration in quaternary wells to the mean concentration in alluvial wells is $R = 1.62/0.27 = 6.00$.

a. Since the sample means 1.62 and 0.27 are based on large samples, it is reasonable to assume that they come from normal populations. Which distribution approximates the distribution of the sample mean concentration of alluvial wells, $N(0.27, 0.40^2)$ or $N(0.27, 0.40^2/349)$? Which distribution approximates the distribution of the sample mean concentration of quaternary wells, $N(1.62, 1.70^2)$ or $N(1.62, 1.70^2/143)$? Explain.

b. Generate a simulated sample of sample means and of ratios of sample means. Is it reasonable to assume that the ratio R is approximately normally distributed?

c. Use the method of propagation of error to estimate the standard deviation of R. (Alternatively, you may use the simulated sample.)

d. If appropriate, use the normal curve to find a 95% confidence interval for the ratio of the mean concentrations.

6. In Example 5.20 (in Section 5.3) the following measurements were given for the cylindrical compressive strength (in MPa) for 11 beams:

38.43	38.43	38.39	38.83	38.45	38.35
38.43	38.31	38.32	38.48	38.50	

One thousand bootstrap samples were generated from these data, and the bootstrap sample means were arranged in order. Refer to the smallest value as Y_1, the second smallest as Y_2, and so on, with the largest being Y_{1000}. Assume that $Y_{25} = 38.3818$, $Y_{26} = 38.3818$, $Y_{50} = 38.3909$, $Y_{51} = 38.3918$, $Y_{950} = 38.5218$, $Y_{951} = 38.5236$, $Y_{975} = 38.5382$, and $Y_{976} = 38.5391$.

a. Compute a 95% bootstrap confidence interval for the mean compressive strength, using method 1 as described on p. 358.
b. Compute a 95% bootstrap confidence interval for the mean compressive strength, using method 2 as described on p. 358.
c. Compute a 90% bootstrap confidence interval for the mean compressive strength, using method 1 as described on p. 358.
d. Compute a 90% bootstrap confidence interval for the mean compressive strength, using method 2 as described on p. 358.

7. Refer to Exercise 6.

a. Generate 1000 bootstrap samples from these data. Find the 2.5 and 97.5 percentiles.
b. Compute a 95% bootstrap confidence interval for the mean compressive strength, using method 1 as described on p. 358.
c. Compute a 95% bootstrap confidence interval for the mean compressive strength, using method 2 as described on p. 358.

8. This exercise continues the study of the robustness of the Student's t method for constructing confidence intervals. The following figure shows graphs of probability density functions for the $N(0, 1)$ distribution, the lognormal distribution with $\mu = 1$ and $\sigma^2 = 0.25$, and the gamma distribution with $r = 0.5$ and $\lambda = 0.5$ (this is also known as the chi-square distribution with one degree of freedom). For each of these distributions, generate 10,000 samples of size 5, and for each sample compute the upper and lower limits of a 95% confidence interval using the Student's t method. [If necessary, it is possible to compute the lognormal and gamma random values from normal random values. Specifically, to compute a value X from a lognormal distribution with $\mu = 1$ and $\sigma^2 = 0.25$, generate $Y \sim N(1, 0.25)$ and compute $X = e^Y$. To generate a value X from a gamma distribution with $r = 0.5$ and $\lambda = 0.5$, generate $Y \sim N(0, 1)$ and compute $X = Y^2$.]

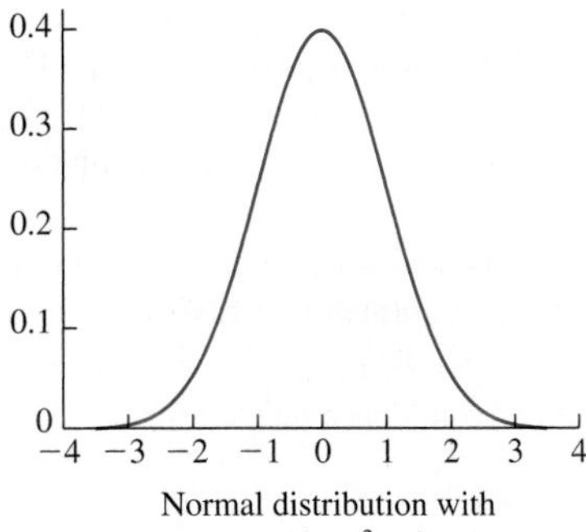

Normal distribution with $\mu = 0,\ \sigma^2 = 1$

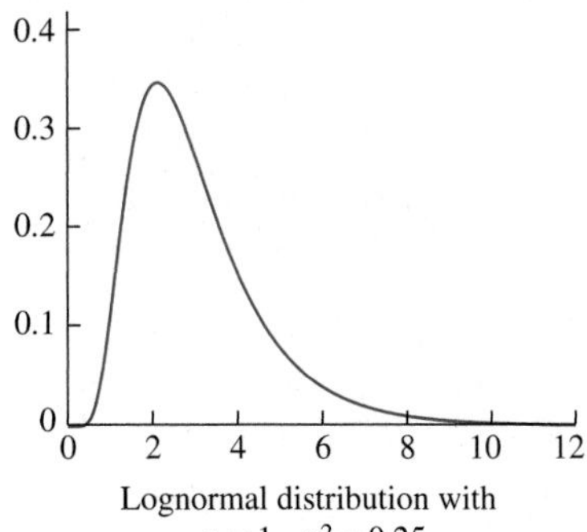

Lognormal distribution with $\mu = 1,\ \sigma^2 = 0.25$

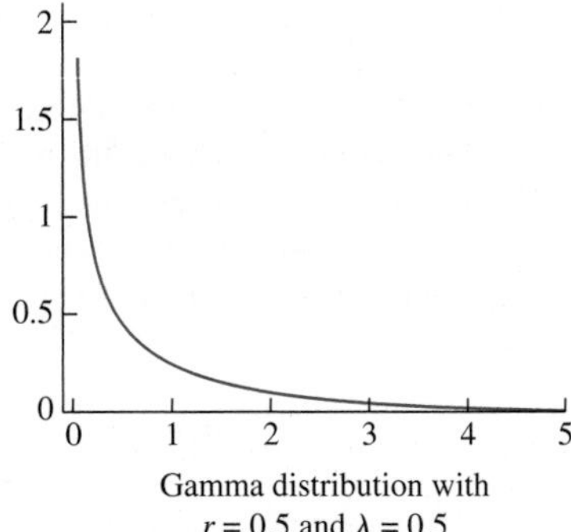

Gamma distribution with $r = 0.5$ and $\lambda = 0.5$

a. The true mean of the $N(0, 1)$ distribution is 0. Based on the simulation results, estimate the coverage probability (proportion of samples for which the confidence interval covers the true mean) for samples of size 5 from the $N(0, 1)$ distribution. (Since the assumptions underlying the Student's t

method are satisfied here, your answer should be very close to 95%.)

b. The true mean of the lognormal distribution with $\mu = 1$ and $\sigma^2 = 0.25$ is 3.0802. Based on the simulation results, estimate the coverage probability (proportion of samples for which the confidence interval covers the true mean) for samples of size 5 from the lognormal distribution with $\mu = 1$ and $\sigma^2 = 0.25$.

c. The true mean of the gamma distribution with $r = 0.5$ and $\lambda = 0.5$ is 1. Based on the simulation results, estimate the coverage probability (proportion of samples for which the confidence interval covers the true mean) for samples of size 5 from the gamma distribution with $r = 0.5$ and $\lambda = 0.5$.

9. This exercise is designed to compare the performance of the Agresti–Coull confidence interval for a proportion (expression 5.5 on p. 316) with that of the traditional confidence interval (expression 5.8 on p. 319). We will use sample sizes of $n = 10$, $n = 17$, and $n = 40$, with $p = 0.5$.

a. Generate 10,000 observations X_i^*, each from a binomial distribution with $n = 10$ and $p = 0.5$. For each observation, compute the upper and lower limits for both the Agresti–Coull 95% confidence interval and the traditional one. For each confidence interval, compute its length (upper limit − lower limit). Use the simulated data to estimate the coverage probability and mean width for both the Agresti-Coull and the traditional confidence interval.

b. Repeat part (a), using $n = 17$.

c. Repeat part (a), using $n = 40$.

d. The performance of the traditional confidence interval does not improve steadily as the sample size increases; instead it oscillates, so that the coverage probability can be better for a smaller sample than for a larger one. Compare the coverage probabilities for the traditional method for sample sizes of 17 and of 40. Do your results confirm this fact?

e. For which sample sizes does the Agresti–Coull interval have greater coverage probability than does the traditional one? For which sample size are the coverage probabilities nearly equal?

f. Other things being equal, a narrower confidence interval is better than a wider one. Which method produces confidence intervals with the narrower mean width?

10. A general method for finding a confidence interval for the difference between two means of normal populations is given by expression (5.21) on p. 341. A pooled method that can be used when the variances of the populations are known to be equal is given by expression (5.22) on p. 343. This exercise is designed to compare the coverage probabilities of these methods under a variety of conditions.

a. Let $n_X = 10$, $n_Y = 10$, $\sigma_X = 1$, and $\sigma_Y = 1$. Generate 10,000 pairs of samples: $X_{1i}^*, \ldots, X_{1n_X}^*$ from a $N(0, \sigma_X^2)$ distribution, and $Y_{1i}^*, \ldots, Y_{1n_Y}^*$ from a $N(0, \sigma_Y^2)$ distribution. For each pair of samples, compute a 95% confidence interval using the general method, and a 95% confidence interval using the pooled method. Note that each population has mean 0, so the true difference between the means is 0. Estimate the coverage probability for each method by computing the proportion of confidence intervals that cover the true value 0.

b. Repeat part (a), using $n_X = 10$, $n_Y = 10$, $\sigma_X = 1$, and $\sigma_Y = 5$.

c. Repeat part (a), using $n_X = 5$, $n_Y = 20$, $\sigma_X = 1$, and $\sigma_Y = 5$.

d. Repeat part (a), using $n_X = 20$, $n_Y = 5$, $\sigma_X = 1$, and $\sigma_Y = 5$.

e. Does the coverage probability for the general method differ substantially from 95% under any of the conditions in parts (a) through (d)? (It shouldn't.)

f. Based on the results in parts (a) through (d), under which conditions does the pooled method perform most poorly?

   i. When the sample sizes are equal and the variances differ.

   ii. When both the sample sizes and the variances differ, and the larger sample comes from the population with the larger variance.

   iii. When both the sample sizes and the variances differ, and the smaller sample comes from the population with the larger variance.

Supplementary Exercises for Chapter 5

1. A molecular biologist is studying the effectiveness of a particular enzyme to digest a certain sequence of DNA nucleotides. He divides six DNA samples into two parts, treats one part with the enzyme, and leaves the other part untreated. He then uses a polymerase chain reaction assay to count the number of DNA fragments that contain the given sequence. The results are as follows:

	Sample					
	1	**2**	**3**	**4**	**5**	**6**
Enzyme present	22	16	11	14	12	30
Enzyme absent	43	34	16	27	10	40

Find a 95% confidence interval for the difference between the mean numbers of fragments.

2. Refer to Exercise 1. Another molecular biologist repeats the study with a different design. She makes up 12 DNA samples, and then chooses 6 at random to be treated with the enzyme and 6 to remain untreated. The results are as follows:

Enzyme present:	12	15	14	22	22	20
Enzyme absent:	23	39	37	18	26	24

Find a 95% confidence interval for the difference between the mean numbers of fragments.

3. The article "Genetically Based Tolerance to Endosulfan, Chromium (VI) and Fluoranthene in the Grass Shrimp *Palaemonetes pugio*" (R. Harper-Arabie, Ph.D. Thesis, Colorado School of Mines, 2002) reported that out of 1985 eggs produced by shrimp at the Diesel Creek site in Charleston, South Carolina, 1919 hatched, and at the Shipyard Creed site, also in Charleston, 4561 out of 4988 eggs hatched. Find a 99% confidence interval for the difference between the proportions of eggs that hatch at the two sites.

4. Resistance measurements were made on a sample of 81 wires of a certain type. The sample mean resistance was 17.3 mΩ and the standard deviation was 1.2 mΩ.

a. Find a 98% confidence interval for the mean resistance of this type of wire.

b. What is the level of the confidence interval (17.1, 17.5)?

c. How many wires must be sampled so that a 98% confidence interval will specify the mean to within ±0.1 mΩ?

5. A sample of 125 pieces of yarn had mean breaking strength 6.1 N and standard deviation 0.7 N. A new batch of yarn was made, using new raw materials from a different vendor. In a sample of 75 pieces of yarn from the new batch, the mean breaking strength was 5.8 N and the standard deviation was 1.0 N. Find a 90% confidence interval for the difference in mean breaking strength between the two types of yarn.

6. Refer to Exercise 5. Additional pieces of yarn will be sampled in order to improve the precision of the confidence interval. Which would increase the precision the most: sampling 50 additional pieces of yarn from the old batch, 50 additional pieces from the new batch, or 25 additional pieces from each batch?

7. Leakage from underground fuel tanks has been a source of water pollution. In a random sample of 87 gasoline stations, 13 were found to have at least one leaking underground tank.

a. Find a 95% confidence interval for the proportion of gasoline stations with at least one leaking underground tank.

b. How many stations must be sampled so that a 95% confidence interval specifies the proportion to within ±0.03?

8. A new catalyst is being investigated for use in the production of a plastic chemical. Ten batches of the chemical are produced. The mean yield of the 10 batches is 72.5% and the standard deviation is 5.8%. Assume the yields are independent and approximately normally distributed. Find a 99% confidence interval for the mean yield when the new catalyst is used.

9. Three confidence intervals for the mean shear strength (in ksi) of anchor bolts of a certain type are computed, all from the same sample. The intervals are (4.01, 6.02), (4.20, 5.83), and (3.57, 6.46).

The levels of the intervals are 90%, 95%, and 99%. Which interval has which level?

10. A survey is to be conducted in which a random sample of residents in a certain city will be asked whether they favor or oppose the building of a new parking structure downtown. How many residents should be polled to be sure that a 95% confidence interval for the proportion who favor the construction specifies that proportion to within ± 0.05?

11. In the article "Groundwater Electromagnetic Imaging in Complex Geological and Topographical Regions: A Case Study of a Tectonic Boundary in the French Alps" (S. Houtot, P. Tarits, et al., *Geophysics*, 2002:1048–1060), the pH was measured for several water samples in various locations near Gittaz Lake in the French Alps. The results for 11 locations on the northern side of the lake and for 6 locations on the southern side are as follows:

Northern side:	8.1	8.2	8.1	8.2	8.2	7.4
	7.3	7.4	8.1	8.1	7.9	
Southern side:	7.8	8.2	7.9	7.9	8.1	8.1

Find a 98% confidence interval for the difference in pH between the northern and southern side.

12. Polychlorinated biphenyls (PCBs) are a group of synthetic oil-like chemicals that were at one time widely used as insulation in electrical equipment and were discharged into rivers. They were discovered to be a health hazard and were banned in the 1970s. Since then, much effort has gone into monitoring PCB concentrations in waterways. Assume that water samples are being drawn from a waterway in order to estimate the PCB concentration.

a. Suppose that a random sample of size 80 has a sample mean of 1.69 ppb and a sample standard deviation of 0.25 ppb. Find a 95% confidence interval for the PCB concentration.

b. Estimate the sample size needed so that a 95% confidence interval will specify the population mean to within ± 0.02 ppb.

13. A 90% confidence interval for a population mean based on 144 observations is computed to be (2.7, 3.4). How many observations must be made so that a 90% confidence interval will specify the mean to within ± 0.2?

14. A sample of 100 components is drawn, and a 95% confidence interval for the proportion defective specifies this proportion to within ± 0.06. To get a more precise estimate of the number defective, the sample size will be increased to 400, and the confidence interval will be recomputed. What will be the approximate width of the new confidence interval? Choose the best answer:

i. ± 0.015
ii. ± 0.03
iii. ± 0.06
iv. ± 0.12
v. ± 0.24

15. A metallurgist makes several measurements of the melting temperature of a certain alloy and computes a 95% confidence interval to be $2038 \pm 2°$C. Assume the measuring process for temperature is unbiased. True or false:

a. There is 95% probability that the true melting temperature is in the interval $2038 \pm 2°$C.

b. If the experiment were repeated, the probability is 95% that the mean measurement from that experiment would be in the interval $2038 \pm 2°$C.

c. If the experiment were repeated, and a 95% confidence interval computed, there is 95% probability that the confidence interval would cover the true melting point.

d. If one more measurement were made, the probability is 95% that it would be in the interval $2038 \pm 2°$C.

16. In a study of the lifetimes of electronic components, a random sample of 400 components are tested until they fail to function. The sample mean lifetime was 370 hours and the standard deviation was 650 hours. True or false:

a. An approximate 95% confidence interval for the mean lifetime of this type of component is from 306.3 to 433.7 hours.

b. About 95% of the sample components had lifetimes between 306.3 and 433.7 hours.

c. If someone takes a random sample of 400 components, divides the sample standard deviation of their lifetimes by 20, and then adds and subtracts that quantity from the sample mean, there is about a 68% chance that the interval so constructed will cover the mean lifetime of this type of component.

d. The z table can't be used to construct confidence intervals here, because the lifetimes of the components don't follow the normal curve.

e. About 68% of the components had lifetimes in the interval 370 ± 650 hours.

17. The temperature of a certain solution is estimated by taking a large number of independent measurements and averaging them. The estimate is 37°C, and the uncertainty (standard deviation) in this estimate is 0.1°C.

a. Find a 95% confidence interval for the temperature.

b. What is the confidence level of the interval 37 ± 0.1°C?

c. If only a small number of independent measurements had been made, what additional assumption would be necessary in order to compute a confidence interval?

d. Making the additional assumption, compute a 95% confidence interval for the temperature if 10 measurements were made.

18. Boxes of nails contain 100 nails each. A sample of 10 boxes is drawn, and each of the boxes is weighed. The average weight is 1500 g and the standard deviation is 5 g. Assume the weight of the box itself is negligible, so that all the weight is due to the nails in the box.

a. Let μ_{box} denote the mean weight of a box of nails. Find a 95% confidence interval for μ_{box}.

b. Let μ_{nail} denote the mean weight of a nail. Express μ_{nail} in terms of μ_{box}.

c. Find a 95% confidence interval for μ_{nail}.

19. Let X represent the number of events that are observed to occur in n units of time or space, and assume $X \sim \text{Poisson}(n\lambda)$, where λ is the mean number of events that occur in one unit of time or space. Assume X is large, so that $X \sim N(n\lambda, n\lambda)$. Follow steps (a) through (d) to derive a level $100(1-\alpha)$% confidence interval for λ. Then in part (e), you are asked to apply the result found in part (d).

a. Show that for a proportion $1-\alpha$ of all possible samples, $X - z_{\alpha/2}\sigma_X < n\lambda < X + z_{\alpha/2}\sigma_X$.

b. Let $\widehat{\lambda} = X/n$. Show that $\sigma_{\widehat{\lambda}} = \sigma_X/n$.

c. Conclude that for a proportion $1-\alpha$ of all possible samples, $\widehat{\lambda} - z_{\alpha/2}\sigma_{\widehat{\lambda}} < \lambda < \widehat{\lambda} + z_{\alpha/2}\sigma_{\widehat{\lambda}}$.

d. Use the fact that $\sigma_{\widehat{\lambda}} \approx \sqrt{\widehat{\lambda}/n}$ to derive an expression for the level $100(1-\alpha)$% confidence interval for λ.

e. A 5 mL sample of a certain suspension is found to contain 300 particles. Let λ represent the mean number of particles per mL in the suspension. Find a 95% confidence interval for λ.

20. The answer to Exercise 19 part (d) is needed for this exercise. A geologist counts 64 emitted particles in one minute from a certain radioactive rock.

a. Find a 95% confidence interval for the rate of emissions in units of particles per minute.

b. After four minutes, 256 particles are counted. Find a 95% confidence interval for the rate of emissions in units of particles per minute.

c. For how many minutes should errors be counted in order that the 95% confidence interval specifies the rate to within ± 1 particle per minute?

21. In a Couette flow, two large flat plates lie one atop another, separated by a thin layer of fluid. If a shear stress is applied to the top plate, the viscosity of the fluid produces motion in the bottom plate as well. The velocity V in the top plate relative to the bottom plate is given by $V = \tau h/\mu$, where τ is the shear stress applied to the top plate, h is the thickness of the fluid layer, and μ is the viscosity of the fluid.

Assume that μ, h, and τ are measured independently and that the measurements are unbiased and normally distributed. The measured values are $\mu = 1.6$ Pa·s, $h = 15$ mm, and $\tau = 25$ Pa. The uncertainties (standard deviations) of these measurements are $\sigma_\mu = 0.05$, $\sigma_h = 1.0$, and $\sigma_\tau = 1.0$.

a. Use the method of propagation of error to estimate V and its uncertainty σ_V.

b. Assuming the estimate of V to be normally distributed, find a 95% confidence interval for V.

c. Perform a simulation to determine whether or not the confidence interval found in part (b) is valid.

22. A sample of seven concrete blocks had their crushing strength measured in MPa. The results were

1367.6 1411.5 1318.7 1193.6 1406.2
1425.7 1572.4

Ten thousand bootstrap samples were generated from these data, and the bootstrap sample means were arranged in order. Refer to the smallest mean as Y_1, the second smallest as Y_2, and so on, with the largest being $Y_{10,000}$. Assume that $Y_{50} = 1283.4$, $Y_{51} = 1283.4$, $Y_{100} = 1291.5$, $Y_{101} = 1291.5$, $Y_{250} = 1305.5$, $Y_{251} = 1305.5$, $Y_{500} = 1318.5$, $Y_{501} = 1318.5$, $Y_{9500} = 1449.7$, $Y_{9501} = 1449.7$, $Y_{9750} = 1462.1$, $Y_{9751} = 1462.1$, $Y_{9900} = 1476.2$, $Y_{9901} = 1476.2$, $Y_{9950} = 1483.8$, and $Y_{9951} = 1483.8$.

a. Compute a 95% bootstrap confidence interval for the mean compressive strength, using method 1 as described on p. 358.

b. Compute a 95% bootstrap confidence interval for the mean compressive strength, using method 2 as described on p. 358.

c. Compute a 99% bootstrap confidence interval for the mean compressive strength, using method 1 as described on p. 358.

d. Compute a 99% bootstrap confidence interval for the mean compressive strength, using method 2 as described on p. 358.

23. Refer to Exercise 22.

a. Generate 10,000 bootstrap samples from the data in Exercise 22. Find the bootstrap sample mean percentiles that are used to compute a 99% confidence interval.

b. Compute a 99% bootstrap confidence interval for the mean compressive strength, using method 1 as described on p. 358.

c. Compute a 99% bootstrap confidence interval for the mean compressive strength, using method 2 as described on p. 358.

Chapter 6

Hypothesis Testing

Introduction

In Example 5.2 (in Section 5.1) a sample of 50 microdrills had an average lifetime of $\overline{X} = 12.68$ holes drilled and a standard deviation of $s = 6.83$. Let us assume that the main question is whether or not the population mean lifetime μ is greater than 11. We address this question by examining the value of the sample mean $\overline{X}$. We see that $\overline{X} > 11$, but because of the uncertainty in $\overline{X}$, this does not guarantee that $\mu > 11$. We would like to know just how certain we can be that $\mu > 11$. A confidence interval is not quite what we need. In Example 5.2, a 95% confidence interval for the population mean μ was computed to be (10.79, 14.57). This tells us that we are 95% confident that μ is between 10.79 and 14.57. But the confidence interval does not directly tell us how confident we can be that $\mu > 11$.

The statement "$\mu > 11$" is a **hypothesis** about the population mean μ. To determine just how certain we can be that a hypothesis such as this is true, we must perform a **hypothesis test**. A hypothesis test produces a number between 0 and 1 that measures the degree of certainty we may have in the truth of a hypothesis about a quantity such as a population mean or proportion. It turns out that hypothesis tests are closely related to confidence intervals. In general, whenever a confidence interval can be computed, a hypothesis test can also be performed, and vice versa.

6.1 Large-Sample Tests for a Population Mean

We begin with an example. A certain type of automobile engine emits a mean of 100 mg of oxides of nitrogen (NO_x) per second at 100 horsepower. A modification to the engine design has been proposed that may reduce NO_x emissions. The new design will be put into production if it can be demonstrated that its mean emission rate is less than 100 mg/s.

A sample of 50 modified engines are built and tested. The sample mean NO_x emission is 92 mg/s, and the sample standard deviation is 21 mg/s.

The population in this case consists of the emission rates from the engines that would be built if this modified design is put into production. If there were no uncertainty in the sample mean, then we could conclude that the modification would reduce emissions—from 100 to 92 mg/s. Of course, there is uncertainty in the sample mean. The population mean will actually be somewhat higher or lower than 92.

The manufacturers are concerned that the modified engines might not reduce emissions at all, that is, that the population mean might be 100 or more. They want to know whether this concern is justified. The question, therefore, is this: Is it plausible that this sample, with its mean of 92, could have come from a population whose mean is 100 or more?

This is the sort of question that hypothesis tests are designed to answer, and we will now construct a hypothesis test to answer this question. We have observed a sample with mean 92. There are two possible interpretations of this observation:

1. The population mean is actually greater than or equal to 100, and the sample mean is lower than this only because of random variation from the population mean. Thus emissions will not go down if the new design is put into production, and the sample is misleading.
2. The population mean is actually less than 100, and the sample mean reflects this fact. Thus the sample represents a real difference that can be expected if the new design is put into production.

These two explanations have standard names. The first is called the **null hypothesis**. In most situations, the null hypothesis says that the effect indicated by the sample is due only to random variation between the sample and the population. The second explanation is called the **alternate hypothesis**. The alternate hypothesis says that the effect indicated by the sample is real, in that it accurately represents the whole population.

In our example, the engine manufacturers are concerned that the null hypothesis might be true. A hypothesis test assigns a quantitative measure to the plausibility of the null hypothesis. After performing a hypothesis test, we will be able to tell the manufacturers, in numerical terms, precisely how valid their concern is.

To make things more precise, we express everything in symbols. The null hypothesis is denoted H_0. The alternate hypothesis is denoted H_1. As usual, the population mean is denoted μ. We have, therefore,

$$H_0: \mu \geq 100 \quad \text{versus} \quad H_1: \mu < 100$$

In performing a hypothesis test, we essentially put the null hypothesis on trial. We begin by assuming that H_0 is true, just as we begin a trial by assuming a defendant to be innocent. The random sample provides the evidence. The hypothesis test involves measuring the strength of the disagreement between the sample and H_0 to produce a number between 0 and 1, called a ***P*-value.** The P-value measures the plausibility of H_0. The smaller the P-value, the stronger the evidence is against H_0. If the P-value is sufficiently small, we may be willing to abandon our assumption that H_0 is true and believe H_1 instead. This is referred to as **rejecting** the null hypothesis.

In this example, let $X_1, \ldots, X_{50}$ be the emissions rates measured from the 50 sample engines. The observed value of the sample mean is $\overline{X} = 92$. We will also need to know the sample standard deviation, which is $s = 21$. We must assess the plausibility of H_0, which says that the population mean is 100 or more, given that we have observed a sample from this population whose mean is only 92. We will do this in two steps, as follows:

1. We will compute the distribution of $\overline{X}$ under the assumption that H_0 is true. This distribution is called the **null distribution** of $\overline{X}$.
2. We will compute the P-value. This is the probability, under the assumption that H_0 is true, of observing a value of $\overline{X}$ whose disagreement with H_0 is as least as great as that of the observed value of 92.

To perform step 1, note that $\overline{X}$ is the mean of a large sample, so the Central Limit Theorem specifies that it comes from a normal distribution whose mean is μ and whose variance is $\sigma^2/50$, where σ^2 is the population variance and 50 is the sample size. We must specify values for μ and for σ in order to determine the null distribution. Since we are assuming that H_0 is true, we assume that $\mu \geq 100$. This does not provide a specific value for μ. We take as the assumed value for μ the value closest to the alternate hypothesis H_1, for reasons that will be explained later in this section. Thus we assume $\mu = 100$. We do not know the population standard deviation σ. However, since the sample is large, we may approximate σ with the sample standard deviation $s = 21$. Thus we have determined that under H_0, $\overline{X}$ has a normal distribution with mean 100 and standard deviation $21/\sqrt{50} = 2.97$. The null distribution is $\overline{X} \sim N(100, 2.97^2)$.

We are now ready for step 2. Figure 6.1 illustrates the null distribution. The number 92 indicates the point on the distribution corresponding to the observed value of $\overline{X}$. How plausible is it that a number sampled from this distribution would be as small as 92? This is measured by the P-value. The P-value is the probability that a number drawn from the null distribution would disagree with H_0 at least as strongly as the observed value of $\overline{X}$, which is 92. Since H_0 specifies that the mean of $\overline{X}$ is greater than or equal to 100, values less than 92 are in greater disagreement with H_0. The P-value, therefore, is the probability that a number drawn from an $N(100, 2.97^2)$ distribution is less than or equal to 92. This probability is determined by computing the z-score:

$$z = \frac{92 - 100}{2.97} = -2.69$$

From the z table, the probability that a standard normal random variable z is less than or equal to -2.69 is 0.0036. The P-value for this test is 0.0036.

The P-value, as promised, provides a quantitative measure of the plausibility of H_0. But how do we interpret this quantity? The proper interpretation is rather subtle. The P-value tells us that if H_0 were true, the probability of drawing a sample whose mean was as far from H_0 as the observed value of 92 is only 0.0036. Therefore, one of the following two conclusions is possible:

- H_0 is false.
- H_0 is true, which implies that of all the samples that might have been drawn, only 0.36% of them have a mean as small as or smaller than that of the sample actually drawn. In other words, our sample mean lies in the most extreme 0.36% of its distribution.

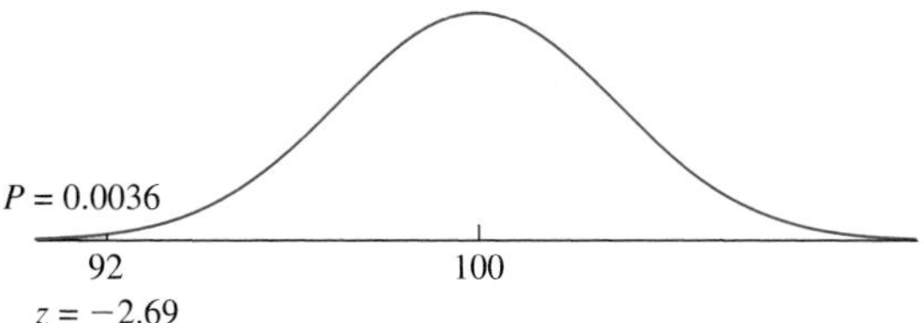

FIGURE 6.1 The null distribution of $\overline{X}$ is $N(100,\ 2.97^2)$. Thus if H_0 is true, the probability that $\overline{X}$ takes on a value as extreme as or more extreme than the observed value of 92 is 0.0036. This is the P-value.

In practice, events in the most extreme 0.36% of their distributions very seldom occur. Therefore we reject H_0 and conclude that the new engines will lower emissions.

The null hypothesis in this case specified only that $\mu \geq 100$. In assuming H_0 to be true, why did we choose the value $\mu = 100$, which is closest to H_1? To give H_0 a fair test, we must test it in its most plausible form. The most plausible value for μ, assuming H_0 to be true, is the value closest to H_1. To see this, look at Figure 6.1. Assume we had chosen a value for μ greater than 100 to represent H_0. Then the null distribution would have been shifted to the right. This would cause the sample mean of 92 to be even farther out in the tail. Then the P-value would have been even smaller. Therefore, among the values of μ consistent with H_0, the one closest to H_1 has the largest P-value, and thus is the most plausible. For this reason, when assuming H_0 to be true, we always use the value of the parameter closest to H_1 when performing a hypothesis test.

It is natural to ask how small the P-value should be in order to reject H_0. Some people use the "5% rule"; they reject H_0 if $P \leq 0.05$. However, there is no scientific justification for this or any other rule. We discuss this issue in more detail in Section 6.2.

Note that the method we have just described uses the Central Limit Theorem. It follows that for this method to be valid, the sample size must be reasonably large, say 30 or more. Hypothesis tests that are sometimes valid for small samples are presented in Section 6.4.

Finally, note that the calculation of the P-value was done by computing a z-score. For this reason, the z-score is called a **test statistic**. A test that uses a z-score as a test statistic is called a z test.

There are many kinds of hypothesis tests. All of them follow a basic series of steps, which are outlined in the following box.

Steps in Performing a Hypothesis Test

1. Define H_0 and H_1.
2. Assume H_0 to be true.
3. Compute a **test statistic**. A test statistic is a statistic that is used to assess the strength of the evidence against H_0.
4. Compute the P-value of the test statistic. The P-value is the probability, assuming H_0 to be true, that the test statistic would have a value whose disagreement with H_0 is as great as or greater than that actually observed. The P-value is also called the **observed significance level**.

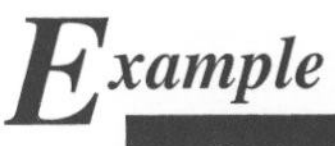

The article "Wear in Boundary Lubrication" (S. Hsu, R. Munro, and M. Shen, *Journal of Engineering Tribology*, 2002:427–441) discusses several experiments involving various lubricants. In one experiment, 45 steel balls lubricated with purified paraffin were subjected to a 40 kg load at 600 rpm for 60 minutes. The average wear, measured by the reduction in diameter, was 673.2 μm, and the standard deviation was 14.9 μm. Assume that the specification for a lubricant is that the mean wear be less than 675 μm. Find the P-value for testing $H_0 : \mu \geq 675$ versus $H_1 : \mu < 675$.

Solution

First let's translate the problem into statistical language. We have a simple random sample $X_1, \ldots, X_{45}$ of wear diameters. The sample mean and standard deviation are $\overline{X} = 673.2$ and $s = 14.9$. The population mean is unknown and denoted by μ. Before getting into the construction of the test, we'll point out again that the basic issue is the uncertainty in the sample mean. If there were no uncertainty in the sample mean, we could conclude that the lubricant would meet the specification, since $673.2 < 675$. The question is whether the uncertainty in the sample mean is large enough so that the population mean could plausibly be as high as 675.

To perform the hypothesis test, we follow the steps given earlier. The null hypothesis is that the lubricant does not meet the specification, and that the difference between the sample mean of 673.2 and 675 is due to chance. The alternate hypothesis is that the lubricant does indeed meet the specification.

We assume H_0 is true, so that the sample was drawn from a population with mean $\mu = 675$ (the value closest to H_1). We estimate the population standard deviation σ with the sample standard deviation $s = 14.9$. The test is based on $\overline{X}$. Under H_0, $\overline{X}$ comes from a normal population with mean 675 and standard deviation $14.9/\sqrt{45} = 2.22$. The P-value is the probability of observing a sample mean less than or equal to 673.2. The test statistic is the z-score, which is

$$z = \frac{673.2 - 675}{2.22} = -0.81$$

The P-value is 0.209 (see Figure 6.2). Therefore if H_0 is true, there is a 20.9% chance to observe a sample whose disagreement with H_0 is as least as great as that which was actually observed. Since 0.209 is not a very small probability, we do not reject H_0. Instead, we conclude that H_0 is plausible. The data do not show conclusively that the lubricant meets the specification. Note that we are *not* concluding that H_0 is *true*, only that it is *plausible*. We will discuss this distinction further in Section 6.2.

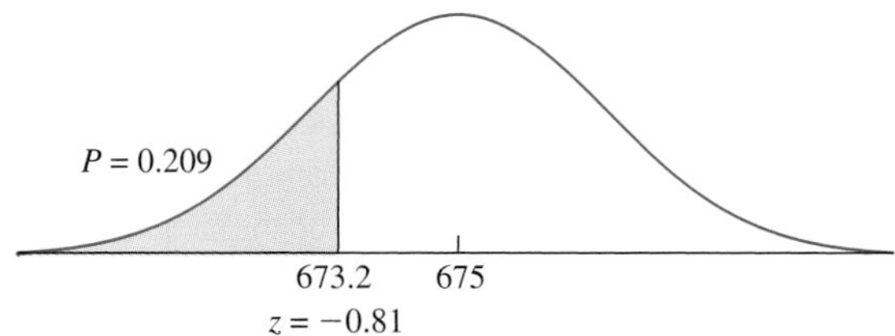

FIGURE 6.2 The null distribution of $\overline{X}$ is $N(675,\ 2.22^2)$. Thus if H_0 is true, the probability that $\overline{X}$ takes on a value as extreme as or more extreme than the observed value of 673.2 is 0.209. This is the P-value.

The following computer output (from MINITAB) presents the results of Example 6.1.

```
One-Sample Z: Wear

Test of mu = 675 vs < 675
The assumed standard deviation = 14.9

                                                95%
                                              Upper
Variable    N     Mean   StDev   SE Mean      Bound       Z       P
Wear       45  673.200    14.9     2.221    676.853   -0.81   0.209
```

The output states the null hypothesis as $\mu = 675$ rather than $\mu \geq 675$. This reflects the fact that the value $\mu = 675$ is used to construct the null distribution. The quantity "SE Mean" is the standard deviation of $\overline{X}$, estimated by $s/\sqrt{n}$. The output also provides a 95% upper confidence bound for μ.

In the examples shown so far, the null hypothesis specified that the population mean was less than or equal to something, or greater than or equal to something. In some cases, a null hypothesis specifies that the population mean is equal to a specific value. Example 6.2 provides an illustration.

Example 6.2

A scale is to be calibrated by weighing a 1000 g test weight 60 times. The 60 scale readings have mean 1000.6 g and standard deviation 2 g. Find the P-value for testing $H_0: \mu = 1000$ versus $H_1: \mu \neq 1000$.

Solution

Let μ denote the population mean reading. The null hypothesis says that the scale is in calibration, so that the population mean μ is equal to the true weight of 1000 g, and the difference between the sample mean reading and the true weight is due entirely to chance. The alternate hypothesis says that the scale is out of calibration.

In this example, the null hypothesis specifies that μ is *equal* to a specific value, rather than greater than or equal to or less than or equal to. For this reason, values of $\overline{X}$ that are *either* much larger *or* much smaller than μ will provide evidence against H_0. In the previous examples, only the values of $\overline{X}$ on one side of μ provided evidence against H_0.

We assume H_0 is true, and that therefore the sample readings were drawn from a population with mean $\mu = 1000$. We approximate the population standard deviation σ with $s = 2$. The null distribution of $\overline{X}$ is normal with mean 1000 and standard deviation $2/\sqrt{60} = 0.258$. The z-score of the observed value $\overline{X} = 1000.6$ is

$$z = \frac{1000.6 - 1000}{0.258} = 2.32$$

Since H_0 specifies $\mu = 1000$, regions in both tails of the curve are in greater disagreement with H_0 than the observed value of 1000.6. The P-value is the sum of the areas in both of these tails, which is 0.0204 (see Figure 6.3). Therefore, if H_0 is true, the probability of a result as extreme as or more extreme than that observed is only 0.0204. The evidence against H_0 is pretty strong. It would be prudent to reject H_0 and to recalibrate the scale.

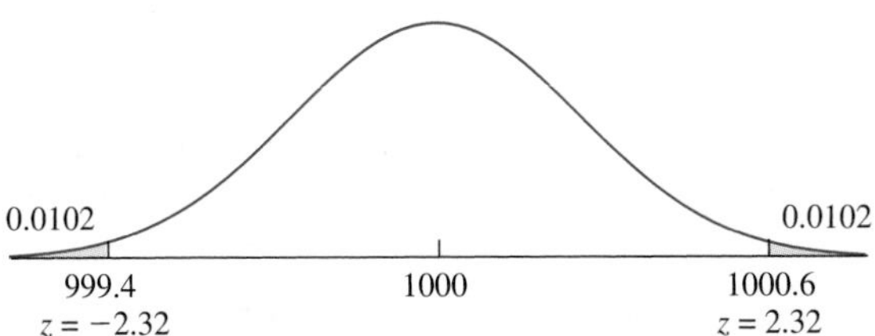

FIGURE 6.3 The null distribution of $\overline{X}$ is $N(1000,\ 0.258^2)$. Thus if H_0 is true, the probability that $\overline{X}$ takes on a value as extreme as or more extreme than the observed value of 1000.6 is 0.0204. This is the P-value.

When H_0 specifies a single value for μ, as in Example 6.2, both tails contribute to the P-value, and the test is said to be a **two-sided** or **two-tailed** test. When H_0 specifies only that μ is greater than or equal to, or less than or equal to a value, only one tail contributes to the P-value, and the test is called a **one-sided** or **one-tailed** test.

We conclude this section by summarizing the procedure used to perform a large-sample hypothesis test for a population mean.

Summary

Let $X_1, \ldots, X_n$ be a *large* (e.g., $n > 30$) sample from a population with mean μ and standard deviation σ.

To test a null hypothesis of the form $H_0: \mu \leq \mu_0$, $H_0: \mu \geq \mu_0$, or $H_0: \mu = \mu_0$:

- Compute the z-score: $z = \dfrac{\overline{X} - \mu_0}{\sigma/\sqrt{n}}$.
  If σ is unknown it may be approximated with s.
- Compute the P-value. The P-value is an area under the normal curve, which depends on the alternate hypothesis as follows:

Alternate Hypothesis	***P*-value**
$H_1: \mu > \mu_0$	Area to the right of z
$H_1: \mu < \mu_0$	Area to the left of z
$H_1: \mu \neq \mu_0$	Sum of the areas in the tails cut off by z and $-z$

Exercises for Section 6.1

1. Recently many companies have been experimenting with "telecommuting," allowing employees to work at home on their computers. Among other things, telecommuting is supposed to reduce the number of sick days taken. Suppose that at one firm, it is known that over the past few years employees have taken a mean of 5.4 sick days. This year, the firm introduces telecommuting. Management chooses a simple random sample of 80 employees to follow in detail, and, at the end of the year, these employees average 4.5 sick days with a standard deviation of 2.7 days. Let μ represent the mean number of sick days for all employees of the firm.

a. Find the P-value for testing $H_0: \mu \geq 5.4$ versus $H_1: \mu < 5.4$.

b. Either the mean number of sick days has declined since the introduction of telecommuting, or the sample is in the most extreme ________% of its distribution.

2. A simple random sample consists of 65 lengths of piano wire that were tested for the amount of extension under a load of 30 N. The average extension for the 65 lines was 1.102 mm and the standard deviation was 0.020 mm. Let μ represent the mean extension for all specimens of this type of piano wire.

a. Find the P-value for testing $H_0: \mu \leq 1.1$ versus $H_1: \mu > 1.1$.

b. Either the mean extension for this type of wire is greater than 1.1 mm, or the sample is in the most extreme ________% of its distribution.

3. The article "Evaluation of Mobile Mapping Systems for Roadway Data Collection" (H. Karimi, A. Khattak, and J. Hummer, *Journal of Computing in Civil Engineering*, 2000:168–173) describes a system for remotely measuring roadway elements such as the width of lanes and the heights of traffic signs. For a sample of 160 such elements, the average error (in percent) in the measurements was 1.90, with a standard deviation of 21.20. Let μ represent the mean error in this type of measurement.

a. Find the P-value for testing $H_0: \mu = 0$ versus $H_1: \mu \neq 0$.

b. Either the mean error for this type of measurement is nonzero, or the sample is in the most extreme ________% of its distribution.

4. An inspector measured the fill volume of a simple random sample of 100 cans of juice that were labeled as containing 12 oz. The sample had mean volume 11.98 oz and standard deviation 0.19 oz. Let μ represent the mean fill volume for all cans of juice recently filled by this machine. The inspector will test $H_0: \mu = 12$ versus $H_1: \mu \neq 12$.

a. Find the P-value.

b. Do you believe it is plausible that the mean fill volume is 12 oz, or are you convinced that the mean differs from 12 oz? Explain your reasoning.

5. When it is operating properly, a chemical plant has a mean daily production of at least 740 tons. The output is measured on a simple random sample of 60 days. The sample had a mean of 715 tons/day and a standard deviation of 24 tons/day. Let μ represent the mean daily output of the plant. An engineer tests $H_0: \mu \geq 740$ versus $H_1: \mu < 740$.

a. Find the P-value.

b. Do you believe it is plausible that the plant is operating properly or are you convinced that the plant is not operating properly? Explain your reasoning.

6. In a process that manufactures tungsten-coated silicon wafers, the target resistance for a wafer is 85 mΩ. In a simple random sample of 50 wafers, the sample mean resistance was 84.8 mΩ and the standard deviation was 0.5 mΩ. Let μ represent the mean resistance of the wafers manufactured by this process. A quality engineer tests $H_0: \mu = 85$ versus $H_1: \mu \neq 85$.

a. Find the P-value.

b. Do you believe it is plausible that the mean is on target, or are you convinced that the mean is not on target? Explain your reasoning.

7. An ultralow particulate air filter is used to maintain uniform airflow in production areas in a clean room. In one such room, the mean air velocity should be at least 40 cm/s. A simple random sample of 58 filters from a certain vendor were tested. The sample mean velocity was 39.6 cm/s, with a standard deviation of 7 cm/s. Let μ represent the mean air velocity obtained from filters supplied by this vendor. A test is made of $H_0: \mu \geq 40$ versus $H_1: \mu < 40$.

a. Find the P-value.

b. Do you believe it is plausible that the mean velocity is at least 40 cm/sec, or are you convinced that the mean is less than 40 cm/sec? Explain your reasoning.

8. A new concrete mix is being designed to provide adequate compressive strength for concrete blocks. The specification for a particular application calls for the blocks to have a mean compressive strength μ greater than 1350 kPa. A sample of 100 blocks is produced and tested. Their mean compressive strength is 1356 kPa and their standard deviation is 70 kPa. A test is made of $H_0: \mu \leq 1350$ versus $H_1: \mu > 1350$.

a. Find the P-value.

b. Do you believe it is plausible that the blocks do not meet the specification, or are you convinced that they do? Explain your reasoning.

9. Fill in the blank: If the null hypothesis is $H_0: \mu \leq 5$, then the mean of $\overline{X}$ under the null distribution is ________.

i. 0

ii. 5

iii. Any number less than or equal to 5.

iv. We can't tell unless we know H_1.

10. Fill in the blank: In a test of $H_0: \mu \geq 10$ versus $H_1: \mu < 10$, the sample mean was $\overline{X} = 8$ and the P-value was 0.04. This means that if $\mu = 10$, and the experiment were repeated 100 times, we would expect to obtain a value of $\overline{X}$ of 8 or less approximately ________ times.

i. 8

ii. 0.8

iii. 4

iv. 0.04

v. 80

11. An engineer takes a large number of independent measurements of the length of a component and obtains $\overline{X} = 5.2$ mm and $\sigma_{\overline{X}} = 0.1$ mm. Use this information to find the P-value for testing $H_0: \mu = 5.0$ versus $H_1: \mu \neq 5.0$.

12. The following MINITAB output presents the results of a hypothesis test for a population mean μ.

```
One-Sample Z: X

Test of mu = 73.5 vs not = 73.5
The assumed standard deviation = 2.3634

Variable    N     Mean    StDev  SE Mean          95% CI              Z      P
X         145  73.2461   2.3634   0.1963  (72.8614, 73.6308)    -1.29  0.196
```

a. Is this a one-tailed or two-tailed test?
b. What is the null hypothesis?
c. What is the P-value?
d. Use the output and an appropriate table to compute the P-value for the test of $H_0: \mu \geq 73.6$ versus $H_1: \mu < 73.6$.
e. Use the output and an appropriate table to compute a 99% confidence interval for μ.

13. The following MINITAB output presents the results of a hypothesis test for a population mean μ. Some of the numbers are missing. Fill in the numbers for (a) through (c).

```
One-Sample Z: X

Test of mu = 3.5 vs > 3.5
The assumed standard deviation = 2.00819

                                                       95%
                                                     Lower
Variable    N      Mean     StDev   SE Mean          Bound     Z     P
X          87   4.07114   2.00819       (a)        3.71700   (b)   (c)
```

6.2 Drawing Conclusions from the Results of Hypothesis Tests

Let's take a closer look at the conclusions reached in Examples 6.1 and 6.2 (in Section 6.1). In Example 6.2, we rejected H_0; in other words, we concluded that H_0 was false. In Example 6.1, we did not reject H_0. However, we did not conclude that H_0 was true. We could only conclude that H_0 was plausible.

In fact, the only two conclusions that can be reached in a hypothesis test are that H_0 is false or that H_0 is plausible. In particular, one can never conclude that H_0 is true. To understand why, think of Example 6.1 again. The sample mean was $\overline{X} = 673.2$, and the null hypothesis was $\mu \geq 675$. The conclusion was that 673.2 is close enough to 675 so that the null hypothesis *might* be true. But a sample mean of 673.2 obviously could not lead us to conclude that $\mu \geq 675$ *is* true, since 673.2 is less than 675. This is typical of many situations of interest. The test statistic is consistent with the alternate hypothesis and disagrees somewhat with the null. The only issue is whether the level of disagreement, measured with the P-value, is great enough to render the null hypothesis implausible.

How do we know when to reject H_0? The smaller the P-value, the less plausible H_0 becomes. A common rule of thumb is to draw the line at 5%. According to this rule of thumb, if $P \leq 0.05$, H_0 is rejected; otherwise H_0 is not rejected. In fact, there is no sharp dividing line between conclusive evidence against H_0 and inconclusive evidence, just as there is no sharp dividing line between hot and cold weather. So while this rule of thumb is convenient, it has no real scientific justification.

Summary

- The smaller the P-value, the more certain we can be that H_0 is false.
- The larger the P-value, the more plausible H_0 becomes, but we can never be certain that H_0 is true.
- A rule of thumb suggests to reject H_0 whenever $P \leq 0.05$. While this rule is convenient, it has no scientific basis.

Statistical Significance

Whenever the P-value is less than a particular threshold, the result is said to be "statistically significant" at that level. So, for example, if $P \leq 0.05$, the result is statistically significant at the 5% level; if $P \leq 0.01$, the result is statistically significant at the 1% level, and so on. If a result is statistically significant at the $100\alpha\%$ level, we can also say that the null hypothesis is "rejected at level $100\alpha\%$."

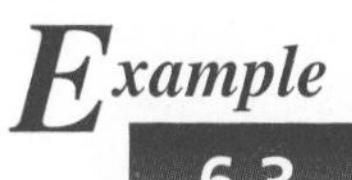

Example 6.3

A hypothesis test is performed of the null hypothesis $H_0 \colon \mu = 0$. The P-value turns out to be 0.03. Is the result statistically significant at the 10% level? The 5% level? The 1% level? Is the null hypothesis rejected at the 10% level? The 5% level? The 1% level?

Solution

The result is statistically significant at any level greater than or equal to 3%. Thus it is statistically significant at the 10% and 5% levels, but not at the 1% level. Similarly, we can reject the null hypothesis at any level greater than or equal to 3%, so H_0 is rejected at the 10% and 5% levels, but not at the 1% level.

Sometimes people report only that a test result was statistically significant at a certain level, without giving the P-value. It is common, for example, to read that a result was "statistically significant at the 5% level" or "statistically significant ($P < 0.05$)." This is a poor practice, for three reasons. First, it provides no way to tell whether the P-value was just barely less than 0.05, or whether it was a lot less. Second, reporting that a result was statistically significant at the 5% level implies that there is a big difference between a P-value just under 0.05 and one just above 0.05, when in fact there is little difference. Third, a report like this does not allow readers to decide for themselves whether the P-value is small enough to reject the null hypothesis. If a reader believes that the null hypothesis should not be rejected unless $P < 0.01$, then reporting only that $P < 0.05$ does not allow that reader to decide whether or not to reject H_0.

Reporting the P-value gives more information about the strength of the evidence against the null hypothesis and allows each reader to decide for himself or herself whether to reject. Software packages always output P-values; these should be included whenever the results of a hypothesis test are reported.

Summary

Let α be any value between 0 and 1. Then, if $P \leq \alpha$,

- The result of the test is said to be statistically significant at the $100\alpha\%$ level.
- The null hypothesis is rejected at the $100\alpha\%$ level.
- When reporting the result of a hypothesis test, report the P-value, rather than just comparing it to 5% or 1%.

The *P*-value Is Not the Probability That H_0 Is True

Since the P-value is a probability, and since small P-values indicate that H_0 is unlikely to be true, it is tempting to think that the P-value represents the probability that H_0 is true. This is emphatically not the case. The concept of probability discussed here is useful only when applied to outcomes that can turn out in different ways when experiments are repeated. It makes sense to define the P-value as the probability of observing an extreme value of a statistic such as $\overline{X}$, since the value of $\overline{X}$ could come out differently if the experiment were repeated. The null hypothesis, on the other hand, either is true or is not true. The truth or falsehood of H_0 cannot be changed by repeating the experiment. It is therefore not correct to discuss the "probability" that H_0 is true.

At this point we must mention that there is a notion of probability, different from that which we discuss in this book, in which one can compute a probability that a statement such as a null hypothesis is true. This kind of probability, called **subjective** probability, plays an important role in the theory of **Bayesian statistics**. The kind of probability we discuss in this book is called **frequentist** probability. A good reference for Bayesian statistics is Lee (1997).

Choose H_0 to Answer the Right Question

When performing a hypothesis test, it is important to choose H_0 and H_1 appropriately so that the result of the test can be useful in forming a conclusion. Examples 6.4 and 6.5 illustrate this.

Specifications for a water pipe call for a mean breaking strength μ of more than 2000 lb per linear foot. Engineers will perform a hypothesis test to decide whether or not to use a certain kind of pipe. They will select a random sample of 1 ft sections of pipe, measure their breaking strengths, and perform a hypothesis test. The pipe will not be used unless the engineers can conclude that $\mu > 2000$. Assume they test $H_0: \mu \leq 2000$ versus $H_1: \mu > 2000$. Will the engineers decide to use the pipe if H_0 is rejected? What if H_0 is not rejected?

Solution

If H_0 is rejected, the engineers will conclude that $\mu > 2000$, and they will use the pipe. If H_0 is not rejected, the engineers will conclude that μ *might* be less than or equal to 2000, and they will not use the pipe.

In Example 6.4, the engineers' action with regard to using the pipe will differ depending on whether H_0 is rejected or not. This is therefore a useful test to perform, and H_0 and H_1 have been specified correctly.

In Example 6.4, assume the engineers test $H_0: \mu \geq 2000$ versus $H_1: \mu < 2000$. Will the engineers decide to use the pipe if H_0 is rejected? What if H_0 is not rejected?

Solution

If H_0 is rejected, the engineers will conclude that $\mu < 2000$, and they will not use the pipe. If H_0 is not rejected, the engineers will conclude that μ *might* be greater than or equal to 2000, but that it also might not be. So again, they won't use the pipe.

In Example 6.5, the engineers' action with regard to using the pipe will be the same—they won't use it—whether or not H_0 is rejected. There is no point in performing this test. The hypotheses H_0 and H_1 have not been specified correctly.

Final note: In a one-tailed test, the equality always goes with the null hypothesis. Thus if μ_0 is the point that divides H_0 from H_1, we may have $H_0: \mu \leq \mu_0$ or $H_0: \mu \geq \mu_0$, but never $H_0: \mu < \mu_0$ or $H_0: \mu > \mu_0$. The reason for this is that when defining the null distribution, we represent H_0 with the value of μ closest to H_1. Without the equality, there is no value specified by H_0 that is the closest to H_1. Therefore the equality must go with H_0.

Statistical Significance Is Not the Same as Practical Significance

When a result has a small P-value, we say that it is "statistically significant." In common usage, the word *significant* means "important." It is therefore tempting to think that statistically significant results must always be important. This is not the case. Sometimes statistically significant results do not have any scientific or practical importance. We will illustrate this with an example. Assume that a process used to manufacture synthetic fibers is known to produce fibers with a mean breaking strength of 50 N. A new process, which would require considerable retooling to implement, has been developed. In a sample of 1000 fibers produced by this new method, the average breaking strength was 50.1 N, and the standard deviation was 1 N. Can we conclude that the new process produces fibers with greater mean breaking strength?

To answer this question, let μ be the mean breaking strength of fibers produced by the new process. We need to test $H_0\colon \mu \leq 50$ versus $H_1\colon \mu > 50$. In this way, if we reject H_0, we will conclude that the new process is better. Under H_0, the sample mean $\overline{X}$ has a normal distribution with mean 50 and standard deviation $1/\sqrt{1000} = 0.0316$. The z-score is

$$z = \frac{50.1 - 50}{0.0316} = 3.16$$

The P-value is 0.0008. This is very strong evidence against H_0. The new process produces fibers with a greater mean breaking strength.

What practical conclusion should be drawn from this result? On the basis of the hypothesis test, we are quite sure that the new process is better. Would it be worthwhile to implement the new process? Probably not. The reason is that the difference between the old and new processes, although highly statistically significant, amounts to only 0.1 N. It is unlikely that this is difference is large enough to matter.

The lesson here is that a result can be statistically significant without being large enough to be of practical importance. How can this happen? A difference is statistically significant when it is large compared to its standard deviation. In the example, a difference of 0.1 N was statistically significant because the standard deviation was only 0.0316 N. When the standard deviation is very small, even a small difference can be statistically significant.

The P-value does not measure practical significance. What it does measure is the degree of confidence we can have that the true value is really different from the value specified by the null hypothesis. When the P-value is small, then we can be confident that the true value is really different. This does not necessarily imply that the difference is large enough to be of practical importance.

The Relationship Between Hypothesis Tests and Confidence Intervals

Both confidence intervals and hypothesis tests are concerned with determining plausible values for a quantity such as a population mean μ. In a hypothesis test for a population mean μ, we specify a particular value of μ (the null hypothesis) and determine whether that value is plausible. In contrast, a confidence interval for a population mean μ can be thought of as the collection of all values for μ that meet a certain criterion of plausibility, specified by the confidence level $100(1 - \alpha)\%$. In fact, the relationship between confidence intervals and hypothesis tests is very close.

To be specific, the values contained within a two-sided level $100(1-\alpha)\%$ confidence interval are precisely those values for which the P-value of a two-tailed hypothesis test will be greater than α. To illustrate this, consider the following example (presented as Example 5.2 in Section 5.1). The sample mean lifetime of 50 microdrills was $\overline{X} = 12.68$ holes drilled and the standard deviation was $s = 6.83$. Setting α to 0.05 (5%), the 95% confidence interval for the population mean lifetime μ was computed to be (10.79, 14.57). Suppose we wanted to test the hypothesis that μ was equal to one of the endpoints of the confidence interval. For example, consider testing $H_0: \mu = 10.79$ versus $H_1: \mu \neq 10.79$. Under H_0, the observed value $\overline{X} = 12.68$ comes from a normal distribution with mean 10.79 and standard deviation $6.83/\sqrt{50} = 0.9659$. The z-score is $(12.68 - 10.79)/0.9659 = 1.96$. Since H_0 specifies that μ is *equal* to 10.79, both tails contribute to the P-value, which is 0.05, and thus equal to α (see Figure 6.4).

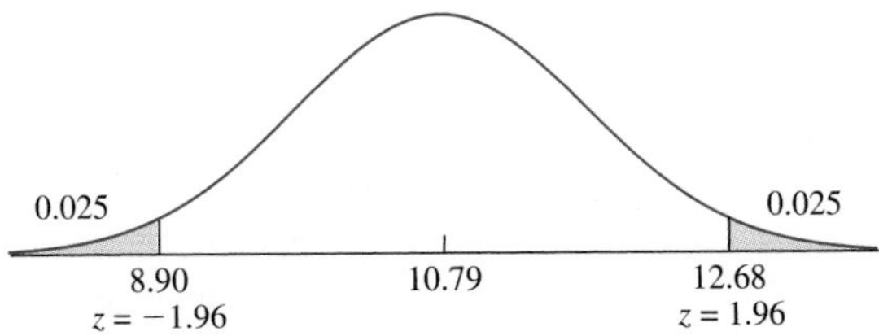

FIGURE 6.4 The sample mean $\overline{X}$ is equal to 12.68. Since 10.79 is an endpoint of a 95% confidence interval based on $\overline{X} = 12.68$, the P-value for testing $H_0: \mu = 10.79$ is equal to 0.05.

Now consider testing the hypothesis $H_0: \mu = 14.57$ versus $H_1: \mu \neq 14.57$, where 14.57 is the other endpoint of the confidence interval. This time we will obtain $z = (12.68 - 14.57)/0.9659 = -1.96$, and again the P-value is 0.05. It is easy to check that if we choose any value μ_0 in the interval (10.79, 14.57) and test $H_0: \mu = \mu_0$ versus $H_1: \mu \neq \mu_0$, the P-value will be greater than 0.05. On the other hand, if we choose $\mu_0 < 10.79$ or $\mu_0 > 14.57$, the P-value will be less than 0.05. Thus the 95% confidence interval consists of precisely those values of μ whose P-values are greater than 0.05 in a hypothesis test. In this sense, the confidence interval contains all the values that are plausible for the population mean μ.

It is easy to check that a one-sided level $100(1-\alpha)\%$ confidence interval consists of all the values for which the P-value in a one-tailed test would be greater than α. For example, with $\overline{X} = 12.68$, $s = 6.83$, and $n = 50$, the 95% lower confidence bound for the lifetime of the drills is 11.09. If $\mu_0 > 11.09$, then the P-value for testing $H_0: \mu \leq \mu_0$ will be greater than 0.05. Similarly, the 95% upper confidence bound for the lifetimes of the drills is 14.27. If $\mu_0 < 14.27$, then the P-value for testing $H_0: \mu \geq \mu_0$ will be greater than 0.05.

Exercises for Section 6.2

1. For which P-value is the null hypothesis more plausible: $P = 0.5$ or $P = 0.05$?

2. True or false:

a. If we reject H_0, then we conclude that H_0 is false.
b. If we do not reject H_0, then we conclude that H_0 is true.
c. If we reject H_0, then we conclude that H_1 is true.
d. If we do not reject H_0, then we conclude that H_1 is false.

3. If $P = 0.01$, which is the best conclusion?

i. H_0 is definitely false.
ii. H_0 is definitely true.
iii. There is a 1% probability that H_0 is true.
iv. H_0 might be true, but it's unlikely.
v. H_0 might be false, but it's unlikely.
vi. H_0 is plausible.

4. If $P = 0.50$, which is the best conclusion?

i. H_0 is definitely false.
ii. H_0 is definitely true.
iii. There is a 50% probability that H_0 is true.
iv. H_0 is plausible, and H_1 is false.
v. Both H_0 and H_1 are plausible.

5. True or false: If $P = 0.02$, then

a. The result is statistically significant at the 5% level.
b. The result is statistically significant at the 1% level.
c. The null hypothesis is rejected at the 5% level.
d. The null hypothesis is rejected at the 1% level.

6. A null hypothesis is rejected at the 5% level. True or false:

a. The P-value is greater than 5%.
b. The P-value is less than or equal to 5%.
c. The result is statistically significant at the 5% level.
d. The result is statistically significant at the 10% level.

---

7. The following MINITAB output (first shown in Exercise 12 in Section 6.1) presents the results of a hypothesis test for a population mean μ.

```
One-Sample Z: X

Test of mu = 73.5 vs not = 73.5
The assumed standard deviation = 2.3634

Variable    N     Mean    StDev  SE Mean          95% CI              Z      P
X         145  73.2461   2.3634   0.1963  (72.8614, 73.6308)  -1.29  0.196
```

a. Can H_0 be rejected at the 5% level? How can you tell?
b. Someone asks you whether the null hypothesis $H_0: \mu = 73$ versus $H_1: \mu \neq 73$ can be rejected at the 5% level. Can you answer without doing any calculations? How?

---

8. Let μ be the radiation level to which a radiation worker is exposed during the course of a year. The Environmental Protection Agency has set the maximum safe level of exposure at 5 rem per year. If a hypothesis test is to be performed to determine whether a workplace is safe, which is the most appropriate null hypothesis: $H_0: \mu \leq 5$, $H_0: \mu \geq 5$, or $H_0: \mu = 5$? Explain.

9. In each of the following situations, state the most appropriate null hypothesis regarding the population mean μ.

a. A new type of battery will be installed in heart pacemakers if it can be shown to have a mean lifetime greater than eight years.

b. A new material for manufacturing tires will be used if it can be shown that the mean lifetime of tires will be at least 60,000 miles.

c. A quality control inspector will recalibrate a flowmeter if the mean flow rate differs from 10 mL/s.

10. The installation of a radon abatement device is recommended in any home where the mean radon concentration is 4.0 picocuries per liter (pCi/L) or more, because it is thought that long-term exposure to sufficiently high doses of radon can increase the risk of cancer. Seventy-five measurements are made in a particular home. The mean concentration was 3.72 pCi/L and the standard deviation was 1.93 pCi/L.

a. The home inspector who performed the test says that since the mean measurement is less than 4.0, radon abatement is not necessary. Explain why this reasoning is incorrect.

b. Because of health concerns, radon abatement is recommended whenever it is plausible that the mean radon concentration may be 4.0 pCi/L or more. State the appropriate null and alternate hypotheses for determining whether radon abatement is appropriate.

c. Compute the P-value. Would you recommend radon abatement? Explain.

11. It is desired to check the calibration of a scale by weighing a standard 10 g weight 100 times. Let μ be the population mean reading on the scale, so that the scale is in calibration if $\mu = 10$. A test is made of the hypotheses $H_0: \mu = 10$ versus $H_1: \mu \neq 10$. Consider three possible conclusions: (i) The scale is in calibration. (ii) The scale is out of calibration. (iii) The scale might be in calibration.

a. Which of the three conclusions is best if H_0 is rejected?

b. Which of the three conclusions is best if H_0 is not rejected?

c. Is it possible to perform a hypothesis test in a way that makes it possible to demonstrate conclusively that the scale is in calibration? Explain.

12. A machine that fills cereal boxes is supposed to be calibrated so that the mean fill weight is 12 oz. Let μ denote the true mean fill weight. Assume that in a test of the hypotheses $H_0: \mu = 12$ versus $H_1: \mu \neq 12$, the P-value is 0.30.

a. Should H_0 be rejected on the basis of this test? Explain.

b. Can you conclude that the machine is calibrated to provide a mean fill weight of 12 oz? Explain.

13. A method of applying zinc plating to steel is supposed to produce a coating whose mean thickness is no greater than 7 microns. A quality inspector measures the thickness of 36 coated specimens and tests $H_0: \mu \leq 7$ versus $H_1: \mu > 7$. She obtains a P-value of 0.40. Since $P > 0.05$, she concludes that the mean thickness is within the specification. Is this conclusion correct? Explain.

14. Fill in the blank: A 95% confidence interval for μ is $(1.2, 2.0)$. Based on the data from which the confidence interval was constructed, someone wants to test $H_0: \mu = 1.4$ versus $H_1: \mu \neq 1.4$. The P-value will be ________.

i. Greater than 0.05
ii. Less than 0.05
iii. Equal to 0.05

15. Refer to Exercise 14. For which null hypothesis will $P = 0.05$?

i. $H_0: \mu = 1.2$
ii. $H_0: \mu \leq 1.2$
iii. $H_0: \mu \geq 1.2$

16. A scientist computes a 90% confidence interval to be $(4.38, 6.02)$. Using the same data, she also computes a 95% confidence interval to be $(4.22, 6.18)$, and a 99% confidence interval to be $(3.91, 6.49)$. Now she wants to test $H_0: \mu = 4$ versus $H_1: \mu \neq 4$. Regarding the P-value, which one of the following statements is true?

i. $P > 0.10$
ii. $0.05 < P < 0.10$
iii. $0.01 < P < 0.05$
iv. $P < 0.01$

17. The strength of a certain type of rubber is tested by subjecting pieces of the rubber to an abrasion test. For the rubber to be acceptable, the mean weight loss μ must be less than 3.5 mg. A large number of pieces of rubber that were cured in a certain way were subject to the abrasion test. A 95% upper confidence bound for the mean weight loss was computed from these data to be 3.45 mg. Someone suggests using these data to test $H_0: \mu \geq 3.5$ versus $H_1: \mu < 3.5$.

a. Is it possible to determine from the confidence bound whether $P < 0.05$? Explain.

b. Is it possible to determine from the confidence bound whether $P < 0.01$? Explain.

18. A shipment of fibers is not acceptable if the mean breaking strength of the fibers is less than 50 N. A large sample of fibers from this shipment was tested, and a 98% lower confidence bound for the mean breaking strength was computed to be 50.1 N. Someone suggests using these data to test the hypotheses $H_0: \mu \leq 50$ versus $H_1: \mu > 50$.

a. Is it possible to determine from the confidence bound whether $P < 0.01$? Explain.

b. Is it possible to determine from the confidence bound whether $P < 0.05$? Explain.

19. Refer to Exercise 17. It is discovered that the mean of the sample used to compute the confidence bound is $\overline{X} = 3.40$. Is it possible to determine whether $P < 0.01$? Explain.

20. Refer to Exercise 18. It is discovered that the standard deviation of the sample used to compute the confidence interval is 5 N. Is it possible to determine whether $P < 0.01$? Explain.

6.3 Tests for a Population Proportion

A population proportion is simply a population mean for a population of 0s and 1s: a Bernoulli population. For this reason, hypothesis tests for proportions are similar to those discussed in Section 6.1 for population means. Here is an example.

A supplier of semiconductor wafers claims that of all the wafers that he supplies, no more than 10% are defective. A sample of 400 wafers is tested, and 50 of them, or 12.5%, are defective. Can we conclude that the claim is false?

The hypothesis test here proceeds much like those in Section 6.1. What makes this problem distinct is that the sample consists of successes and failures, with "success" indicating a defective wafer. If the population proportion of defective wafers is denoted by p, then the supplier's claim is that $p \leq 0.1$. Since our hypothesis concerns a population proportion, it is natural to base the test on the sample proportion $\widehat{p}$. Making the reasonable assumption that the wafers are sampled independently, it follows from the Central Limit Theorem, since the sample size is large, that

$$\widehat{p} \sim N\left(p, \frac{p(1-p)}{n}\right) \tag{6.1}$$

where n is the sample size, equal to 400.

We must define the null hypothesis. The question asked is whether the data allow us to conclude that the supplier's claim is false. Therefore, the supplier's claim, which is that $p \leq 0.1$, must be H_0. Otherwise it would be impossible to prove the claim false, no matter what the data showed.

The null and alternate hypotheses are

$$H_0: p \leq 0.1 \quad \text{versus} \quad H_1: p > 0.1$$

To perform the hypothesis test, we assume H_0 to be true and take $p = 0.1$. Substituting $p = 0.1$ and $n = 400$ in expression (6.1) yields the null distribution of $\widehat{p}$:

$$\widehat{p} \sim N(0.1,\ 2.25 \times 10^{-4})$$

The standard deviation of $\widehat{p}$ is $\sigma_{\widehat{p}} = \sqrt{2.25 \times 10^{-4}} = 0.015$. The observed value of $\widehat{p}$ is $50/400 = 0.125$. The z-score of $\widehat{p}$ is

$$z = \frac{0.125 - 0.100}{0.015} = 1.67$$

The z table indicates that the probability that a standard normal random variable has a value greater than 1.67 is approximately 0.0475. The P-value is therefore 0.0475 (see Figure 6.5).

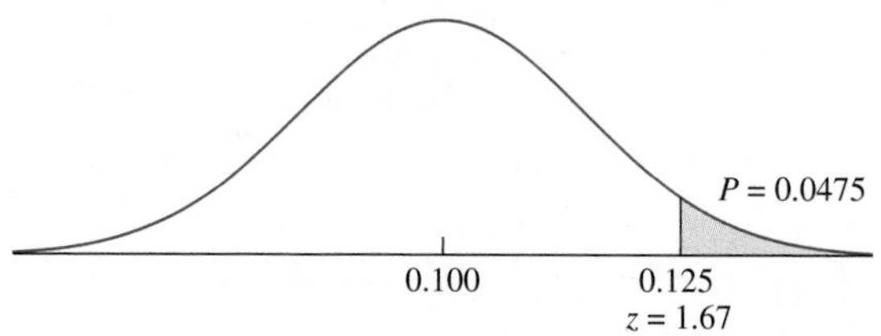

FIGURE 6.5 The null distribution of $\widehat{p}$ is $N(0.1,\ 0.015^2)$. Thus if H_0 is true, the probability that $\widehat{p}$ takes on a value as extreme as or more extreme than the observed value of 0.125 is 0.0475. This is the P-value.

What do we conclude about H_0? Either the supplier's claim is false, or we have observed a sample that is as extreme as all but 4.75% of the samples we might have drawn. Such a sample would be unusual, but not fantastically unlikely. There is every reason to be quite skeptical of the claim, but we probably shouldn't convict the supplier quite yet. If possible, it would be a good idea to sample more wafers.

Note that under the commonly used rule of thumb, we would reject H_0 and condemn the supplier, because P is less than 0.05. This example illustrates the weakness of this rule. If you do the calculations, you will find that if only 49 of the sample wafers had been defective rather than 50, the P-value would have risen to 0.0668, and the supplier would be off the hook. Thus the fate of the supplier hangs on the outcome of one single wafer out of 400. It doesn't make sense to draw such a sharp line. It's better just to report the P-value and wait for more evidence before reaching a firm conclusion.

The Sample Size Must Be Large

The test just described requires that the sample proportion be approximately normally distributed. This assumption will be justified whenever both $np_0 > 10$ and $n(1 - p_0) > 10$, where p_0 is the population proportion specified in the null distribution. Then the z-score can be used as the test statistic, making this a z-test.

Example 6.6

The article "Refinement of Gravimetric Geoid Using GPS and Leveling Data" (W. Thurston, *Journal of Surveying Engineering*, 2000:27–56) presents a method for measuring orthometric heights above sea level. For a sample of 1225 baselines, 926 gave results that were within the class C spirit leveling tolerance limits. Can we conclude that this method produces results within the tolerance limits more than 75% of the time?

Solution

Let p denote the probability that the method produces a result within the tolerance limits. The null and alternate hypotheses are

$$H_0: p \leq 0.75 \quad \text{versus} \quad H_1: p > 0.75$$

The sample proportion is $\widehat{p} = 926/1225 = 0.7559$. Under the null hypothesis, $\widehat{p}$ is normally distributed with mean 0.75 and standard deviation $\sqrt{(0.75)(1-0.75)/1225} = 0.0124$. The z-score is

$$z = \frac{0.7559 - 0.7500}{0.0124} = 0.48$$

The P-value is 0.3156 (see Figure 6.6). We cannot conclude that the method produces good results more than 75% of the time.

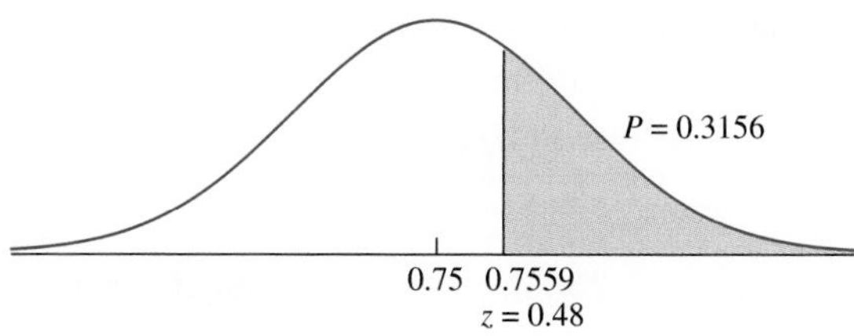

FIGURE 6.6 The null distribution of $\widehat{p}$ is $N(0.75,\ 0.0124^2)$. Thus if H_0 is true, the probability that $\widehat{p}$ takes on a value as extreme as or more extreme than the observed value of 0.7559 is 0.3156. This is the P-value.

The following computer output (from MINITAB) presents the results from Example 6.6.

```
Test and CI for One Proportion: GPS

Test of p = 0.75 vs p > 0.75
                                              95%
                                            Lower
Variable      X      N   Sample p           Bound   Z-Value   P-Value
GPS         926   1225   0.755918        0.735732      0.48     0.316
```

The output contains a 95% lower confidence bound as well as the P-value. Note that this lower bound is computed by the traditional method (expression 5.8 in Section 5.2 presents the two-sided version of this method).

Relationship with Confidence Intervals for a Proportion

A level $100(1-\alpha)\%$ confidence interval contains those values for a parameter for which the P-value of a hypothesis test will be greater than α. For the confidence intervals for a proportion presented in Section 5.2 and the hypothesis test presented here, this statement is only approximately true. The reason for this is that the methods presented in Section 5.2 are slight modifications (that are much easier to compute) of a more complicated confidence interval method for which the statement is exactly true.

Summary

Let X be the number of successes in n independent Bernoulli trials, each with success probability p; in other words, let $X \sim \text{Bin}(n, p)$.

To test a null hypothesis of the form $H_0: p \le p_0$, $H_0: p \ge p_0$, or $H_0: p = p_0$, assuming that both np_0 and $n(1-p_0)$ are greater than 10:

- Compute the z-score: $z = \dfrac{\widehat{p} - p_0}{\sqrt{p_0(1-p_0)/n}}$.
- Compute the P-value. The P-value is an area under the normal curve, which depends on the alternate hypothesis as follows:

Alternate Hypothesis	***P*-value**
$H_1: p > p_0$	Area to the right of z
$H_1: p < p_0$	Area to the left of z
$H_1: p \ne p_0$	Sum of the areas in the tails cut off by z and $-z$

Exercises for Section 6.3

1. A random sample of 300 electronic components manufactured by a certain process are tested, and 25 are found to be defective. Let p represent the proportion of components manufactured by this process that are defective. The process engineer claims that $p \le 0.05$. Does the sample provide enough evidence to reject the claim?

2. A random sample of 100 bolts is sampled from a day's production, and 2 of them are found to have diameters below specification. It is claimed that the proportion of defective bolts among those manufactured that day is less than 0.05. Is it appropriate to use the methods of this section to determine whether we can reject this claim? If so, state the appropriate null and alternate hypotheses and compute the P-value. If not, explain why not.

3. A telecommunications company provided its cable TV subscribers with free access to a new sports channel for a period of one month. It then chose a sample of 400 television viewers and asked them whether they would be willing to pay an extra $10 per month to continue to access the channel. Only 25 replied that they would be willing to pay. Can the company conclude that more than 5% of their subscribers would pay for the channel?

4. Incinerators can be a source of hazardous emissions into the atmosphere. Stack gas samples were collected from a sample of 50 incinerators in a major city. Of the 50 samples, only 18 met an environmental standard for the concentration of a hazardous compound. Can it be concluded that fewer than half of the incinerators in the city meet the standard?

5. Gravel pieces are classified as small, medium, or large. A vendor claims that at least 10% of the gravel pieces from her plant are large. In a random sample of 1600 pieces, 150 pieces were classified as large. Is this enough evidence to reject the claim?

6. A grinding machine will be qualified for a particular task if it can be shown to produce less than 8% defective parts. In a random sample of 300 parts, 12 were defective. On the basis of these data, can the machine be qualified?

7. A manufacturer of computer workstations is testing a new automated assembly process. The current process has a defect rate of 5%. In a sample of 400 workstations assembled by the new process, 15 had defects. Can it be concluded that new process has a lower defect rate?

8. Refer to Exercise 1 in Section 5.2. Can it be concluded that more than 60% of the measurements made by the instrument will be satisfactory?

9. Refer to Exercise 2 in Section 5.2. Can it be concluded that fewer than 40% of the fuses manufactured that day have burnout amperages greater than 15 A?

10. The following MINITAB output presents the results of a hypothesis test for a population proportion p.

```
Test and CI for One Proportion: X

Test of p = 0.4 vs p < 0.4

                                         95%
                                       Upper
Variable    X    N  Sample p           Bound  Z-Value  P-Value
X          73  240  0.304167        0.353013    -3.03    0.001
```

a. Is this a one-tailed or two-tailed test?
b. What is the null hypothesis?
c. Can H_0 be rejected at the 2% level? How can you tell?
d. Someone asks you whether the null hypothesis $H_0: p \geq 0.45$ versus $H_1: p < 0.45$ can be rejected at the 2% level. Can you answer without doing any calculations? How?
e. Use the output and an appropriate table to compute the P-value for the test of $H_0: p \leq 0.25$ versus $H_1: p > 0.25$.
f. Use the output and an appropriate table to compute a 90% confidence interval for p.

11. The following MINITAB output presents the results of a hypothesis test for a population proportion p. Some of the numbers are missing. Fill in the numbers for (a) through (c).

```
Test and CI for One Proportion: X

Test of p = 0.7 vs p < 0.7

                                          95%
                                        Upper
Variable     X    N  Sample p           Bound  Z-Value  P-Value
X          345  500       (a)        0.724021      (b)      (c)
```

6.4 Small-Sample Tests for a Population Mean

In Section 6.1, we described a method for testing a hypothesis about a population mean, based on a large sample. A key step in the method is to approximate the population standard deviation σ with the sample standard deviation s. The normal curve is then used to find the P-value. When the sample size is small, s may not be close to σ, which invalidates this large-sample method. However, when the population is approximately normal, the Student's t distribution can be used. We illustrate with an example.

Spacer collars for a transmission countershaft have a thickness specification of 38.98–39.02 mm. The process that manufactures the collars is supposed to be calibrated so that the mean thickness is 39.00 mm in the center of the specification window. A sample of six collars is drawn and measured for thickness. The six thicknesses are 39.030, 38.997, 39.012, 39.008, 39.019, and 39.002. Assume that the population of thicknesses of the collars is approximately normal. Can we conclude that the process needs recalibration?

Denoting the population mean by μ, the null and alternate hypotheses are

$$H_0: \mu = 39.00 \quad \text{versus} \quad H_1: \mu \neq 39.00$$

Note that H_0 specifies a single value for μ, since calibration requires that the mean be equal to the correct value. To construct the test statistic, note that since the population is assumed to follow a normal distribution, the quantity

$$t = \frac{\overline{X} - \mu}{s/\sqrt{n}}$$

has a Student's t distribution with $n - 1 = 5$ degrees of freedom. This is the test statistic.

In this example the observed values of the sample mean and standard deviation are $\overline{X} = 39.01133$ and $s = 0.011928$. The sample size is $n = 6$. The null hypothesis specifies that $\mu = 39$. The value of the test statistic is therefore

$$t = \frac{39.01133 - 39.00}{0.011928/\sqrt{6}} = 2.327$$

The P-value is the probability of observing a value of the test statistic whose disagreement with H_0 is as great as or greater than that actually observed. Since H_0 specifies that $\mu = 39.00$, this is a two-tailed test, so values both above and below 39.00 disagree with H_0. Therefore the P-value is the sum of the areas under the curve corresponding to $t > 2.327$ and $t < -2.327$.

Figure 6.7 illustrates the null distribution and indicates the location of the test statistic. From the t table (Table A.3 in Appendix A) the row corresponding to 5 degrees of freedom indicates that the value $t = \pm 2.015$ cuts off an area of 0.05 in each tail, for a total of 0.10, and that the value $t = \pm 2.571$ cuts off an area of 0.025 in each tail, for a total of 0.05. Thus the P-value is between 0.05 and 0.10. While we cannot conclusively state that the process is out of calibration, it doesn't look too good. It would be prudent to recalibrate.

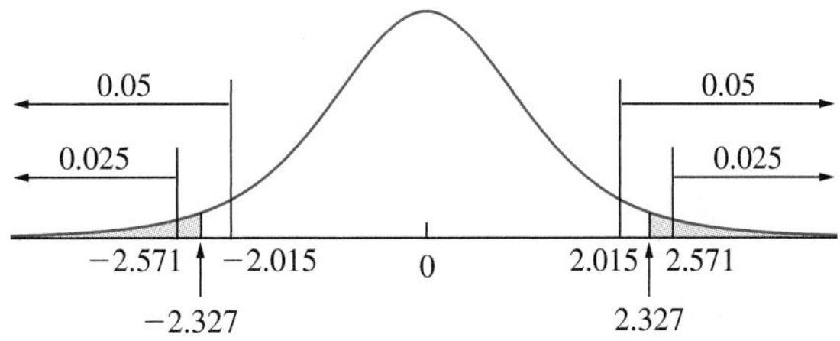

FIGURE 6.7 The null distribution of $t = (\overline{X} - 39.00)/(s/\sqrt{6})$ is Student's t with five degrees of freedom. The observed value of t, corresponding to the observed values $\overline{X} = 39.01133$ and $s = 0.011928$, is 2.327. If H_0 is true, the probability that t takes on a value as extreme as or more extreme than that observed is between 0.05 and 0.10. Because H_0 specified that μ was *equal* to a specific value, both tails of the curve contribute to the P-value.

In this example, the test statistic was a t statistic rather than a z-score. For this reason, this test is referred to as a t test.

Example 6.7

Before a substance can be deemed safe for landfilling, its chemical properties must be characterized. The article "Landfilling Ash/Sludge Mixtures" (J. Benoît, T. Eighmy, and B. Crannell, *Journal of Geotechnical and Geoenvironmental Engineering* 1999: 877–888) reports that in a sample of six replicates of sludge from a New Hampshire wastewater treatment plant, the mean pH was 6.68 with a standard deviation of 0.20. Can we conclude that the mean pH is less than 7.0?

Solution

Let μ denote the mean pH for this type of sludge. The null and alternate hypotheses are

$$H_0 : \mu \geq 7.0 \quad \text{versus} \quad H_1 : \mu < 7.0$$

Under H_0, the test statistic

$$t = \frac{\overline{X} - 7.0}{s/\sqrt{n}}$$

has a Student's t distribution with five degrees of freedom. Substituting $\overline{X} = 6.68$, $s = 0.20$, and $n = 6$, the value of the test statistic is

$$t = \frac{6.68 - 7.00}{0.20/\sqrt{6}} = -3.919$$

Consulting the t table, we find that the value $t = -3.365$ cuts off an area of 0.01 in the left-hand tail, and the value $t = -4.033$ cuts off an area of 0.005 (see Figure 6.8 page 392). We conclude that the P-value is between 0.005 and 0.01. There is strong evidence that the mean pH is less than 7.0.

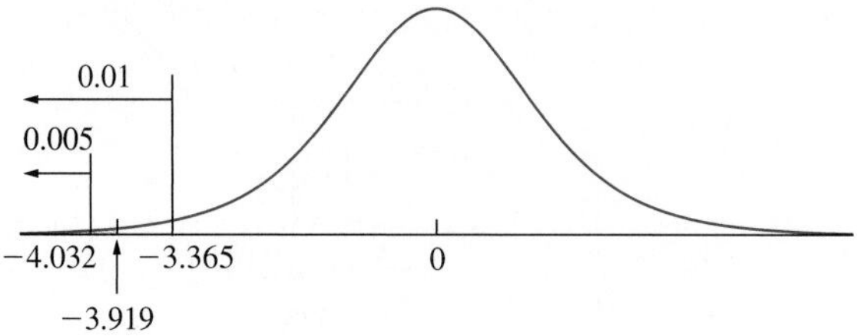

FIGURE 6.8 Solution to Example 6.7. The null distribution is Student's t with five degrees of freedom. The observed value of t is -3.919. If H_0 is true, the probability that t takes on a value as extreme as or more extreme than that observed is between 0.005 and 0.01.

The following computer output (from MINITAB) presents the results from Example 6.7.

```
One-Sample T: pH

Test of mu = 7 vs < 7
                                                 95%
                                               Upper
Variable  N    Mean   StDev   SE Mean          Bound       T       P
pH        6   6.680   0.200  0.081665        6.84453   -3.92   0.006
```

Note that the upper 95% confidence bound provided in the output is consistent with the alternate hypothesis. This indicates that the P-value is less than 5%.

Use z, Not t, If σ Is Known

Occasionally a small sample may be taken from a normal population whose standard deviation σ is known. In these cases, we do not use the Student's t curve, because we are not approximating σ with s. Instead, we use the z table and perform a z test. Example 6.8 illustrates the method.

Example 6.8

At the beginning of this section, we described a sample of six spacer collars, whose thicknesses (in mm) were 39.030, 38.997, 39.012, 39.008, 39.019, and 39.002. We denoted the population mean thickness by μ and tested the hypotheses

$$H_0 : \mu = 39.00 \quad \text{versus} \quad H_1 : \mu \neq 39.00$$

Now assume that these six spacer collars were manufactured just after the machine that produces them had been moved to a new location. Assume that on the basis of a very large number of collars manufactured before the move, the population of collar thicknesses is known to be very close to normal, with standard deviation $\sigma = 0.010$ mm, and that it is reasonable to assume that the move has not changed this. On the basis of the given data, can we reject H_0?

We compute $\overline{X} = 39.01133$. We do not need the value of s, because we know that $\sigma = 0.010$. Since the population is normal, $\overline{X}$ is normal even though the sample size is small. The null distribution is therefore

$$\overline{X} \sim N(39.00,\ 0.010^2)$$

The z-score is

$$z = \frac{39.01133 - 39.000}{0.010/\sqrt{6}} = 2.78$$

The P-value is 0.0054, so H_0 can be safely rejected.

Summary

Let $X_1, \ldots, X_n$ be a sample from a *normal* population with mean μ and standard deviation σ, where σ is unknown.

To test a null hypothesis of the form $H_0: \mu \le \mu_0$, $H_0: \mu \ge \mu_0$, or $H_0:\ \mu = \mu_0$:

- Compute the test statistic $t = \dfrac{\overline{X} - \mu_0}{s/\sqrt{n}}$.
- Compute the P-value. The P-value is an area under the Student's t curve with $n - 1$ degrees of freedom, which depends on the alternate hypothesis as follows:

Alternate Hypothesis	***P*-value**
$H_1: \mu > \mu_0$	Area to the right of t
$H_1: \mu < \mu_0$	Area to the left of t
$H_1: \mu \ne \mu_0$	Sum of the areas in the tails cut off by t and $-t$

- If σ is known, the test statistic is $z = \dfrac{\overline{X} - \mu_0}{\sigma/\sqrt{n}}$, and a z test should be performed.

Exercises for Section 6.4

1. Each of the following hypothetical data sets represents some repeated weighings of a standard weight that is known to have a mass of 100 g. Assume that the readings are a random sample from a population that follows the normal curve. Perform a t test to see whether the scale is properly calibrated, if possible. If impossible, explain why.

a. 100.02, 99.98, 100.03

b. 100.01

2. A geologist is making repeated measurements (in grams) on the mass of a rock. It is not known whether the measurements are a random sample from an approximately normal population. Below are three sets of replicate measurements, listed in the order they were made. For each set of readings, state whether the assumptions necessary for the validity of the t test appear to be met. If the assumptions are not met, explain why.

a. 213.03 212.95 213.04 213.00 212.99
213.01 221.03 213.05

b. 213.05 213.00 212.94 213.09 212.98
213.02 213.06 212.99

c. 212.92 212.95 212.97 213.00 213.01 213.04 213.05 213.06

3. A new process for synthesizing methanol from methane is under study to evaluate its technical feasibility. Design simulations indicate that the reactor must have a mean methane conversion greater than 35% for the process to be feasible. In an initial study, six runs were made. The average conversion was 39% and the standard deviation was 4%. If it can be concluded that the mean conversion μ is greater than 35%, further evaluation of the process will be conducted.

a. State the appropriate null and alternate hypotheses.

b. Find the P-value.

c. Should further evaluation of the process be conducted? Explain.

4. A particular type of gasoline is supposed to have a mean octane rating greater than 90%. Five measurements are taken of the octane rating, as follows:

90.1 88.8 89.5 91.0 92.1

Can you conclude that the mean octane rating is greater than 90%?

5. Specifications call for the wall thickness of two-liter polycarbonate bottles to average 4.0 mils. A quality-control engineer samples 7 two-liter polycarbonate bottles from a large batch and measures the wall thickness (in mils) in each. The following results are obtained.

4.065 3.967 4.028 4.008 4.195 4.057 4.010

Can it be concluded that the mean wall thickness differs from 4.0 mils?

6. As part of the quality-control program for a catalyst manufacturing line, the raw materials (alumina and a binder) are tested for purity. The process requires that the purity of the alumina be greater than 85%. A random sample from a recent shipment of alumina yielded the following results (in %):

93.2 87.0 92.1 90.1 87.3 93.6

A hypothesis test will be done to determine whether or not to accept the shipment.

a. State the appropriate null and alternate hypotheses.

b. Compute the P-value.

c. Should the shipment be accepted? Explain.

7. A sample of 18 pieces of laminate had a mean warpage of 1.88 mm and a standard deviation of 0.21 mm. Can it be concluded that the mean warpage for this type of laminate is less than 2 mm?

8. Refer to Exercise 12 in Section 5.3. Can you conclude that the mean amount of toluene removed in the rinse is less than 8%?

9. Refer to Exercise 13 in Section 5.3. Can you conclude that the mean amount of uniconazole absorbed is less than 2.5 μg?

10. The following MINITAB output presents the results of a hypothesis test for a population mean μ.

```
One-Sample T: X

Test of mu = 5.5 vs > 5.5

                                                 95%
                                                Lower
Variable  N     Mean     StDev   SE Mean       Bound      T      P
X         5  5.92563   0.15755   0.07046     5.77542   6.04  0.002
```

a. Is this a one-tailed or two-tailed test?

b. What is the null hypothesis?

c. Can H_0 be rejected at the 1% level? How can you tell?

d. Use the output and an appropriate table to compute the P-value for the test of $H_0: \mu \geq 6.5$ versus $H_1: \mu < 6.5$.

e. Use the output and an appropriate table to compute a 99% confidence interval for μ.

11. The following MINITAB output presents the results of a hypothesis test for a population mean μ. Some of the numbers are missing. Fill them in.

```
One-Sample T: X

Test of mu = 16 vs not = 16

Variable    N      Mean    StDev    SE Mean        95% CI           T        P
X          11   13.2874     (a)      1.8389    ( (b), (c) )      (d)    0.171
```

6.5 Large-Sample Tests for the Difference Between Two Means

We now investigate examples in which we wish to determine whether the means of two populations are equal. The data will consist of two samples, one from each population. The basic idea is quite simple. We will compute the difference of the sample means. If the difference is far from 0, we will conclude that the population means are different. If the difference is close to 0, we will conclude that the population means might be the same.

As an example, suppose that a production manager for a manufacturer of industrial machinery is concerned that ball bearings produced in environments with low ambient temperatures may have smaller diameters than those produced under higher temperatures. To investigate this concern, she samples 120 ball bearings that were manufactured early in the morning, before the shop was fully heated, and finds their mean diameter to be 5.068 mm and their standard deviation to be 0.011 mm. She independently samples 65 ball bearings manufactured during the afternoon and finds their mean diameter to be 5.072 mm and their standard deviation to be 0.007 mm. Can she conclude that ball bearings manufactured in the morning have smaller diameters, on average, than ball bearings manufactured in the afternoon?

We begin by translating the problem into statistical language. We have a simple random sample $X_1, \ldots, X_{120}$ of diameters of ball bearings manufactured in the morning, and another simple random sample $Y_1, \ldots, Y_{65}$ of diameters of ball bearings manufactured in the afternoon. Denote the population mean of diameters of bearings manufactured in the morning by μ_X, and the population mean of diameters of bearings manufactured in the afternoon by μ_Y. Denote the corresponding standard deviations by σ_X and σ_Y. These population means and standard deviations are unknown. The sample sizes are $n_X = 120$ and $n_Y = 65$. We are interested in the difference $\mu_X - \mu_Y$.

We must now determine the null and alternate hypotheses. The question asked is whether we can conclude that the population mean for the morning bearings is less than that for the afternoon bearings. Therefore the null and alternate hypotheses are

$$H_0: \mu_X - \mu_Y \geq 0 \quad \text{versus} \quad H_1: \mu_X - \mu_Y < 0$$

The test is based on $\overline{X} - \overline{Y}$. Since both sample sizes are large, $\overline{X}$ and $\overline{Y}$ are both approximately normally distributed. Since the samples are independent, it follows that the null distribution of $\overline{X} - \overline{Y}$ is

$$\overline{X} - \overline{Y} \sim N(\mu_X - \mu_Y,\ \sigma_{\overline{X}}^2 + \sigma_{\overline{Y}}^2) = N\left(\mu_X - \mu_Y,\ \frac{\sigma_X^2}{n_X} + \frac{\sigma_Y^2}{n_Y}\right) \tag{6.2}$$

The observed values are $\overline{X} = 5.068$ and $\overline{Y} = 5.072$ for the sample means, and $s_X = 0.011$ and $s_Y = 0.007$ for the sample standard deviations. Under H_0, $\mu_X - \mu_Y = 0$ (the value closest to H_1). We approximate the population variances σ_X^2 and σ_Y^2 with the sample variances $s_X^2 = 0.011^2$ and $s_Y^2 = 0.007^2$, respectively, and substitute $n_X = 120$ and $n_Y = 65$, to compute the standard deviation of the null distribution, obtaining $\sqrt{0.011^2/120 + 0.007^2/65} = 0.001327$. The null distribution of $\overline{X} - \overline{Y}$ is therefore

$$\overline{X} - \overline{Y} \sim N(0,\ 0.001327^2)$$

The observed value of $\overline{X} - \overline{Y}$ is $5.068 - 5.072 = -0.004$. The z-score is

$$z = \frac{-0.004 - 0}{0.001327} = -3.01$$

Figure 6.9 shows the null distribution and the location of the test statistic. The P-value is 0.0013. The manager's suspicion is correct. The bearings manufactured in the morning have a smaller mean diameter.

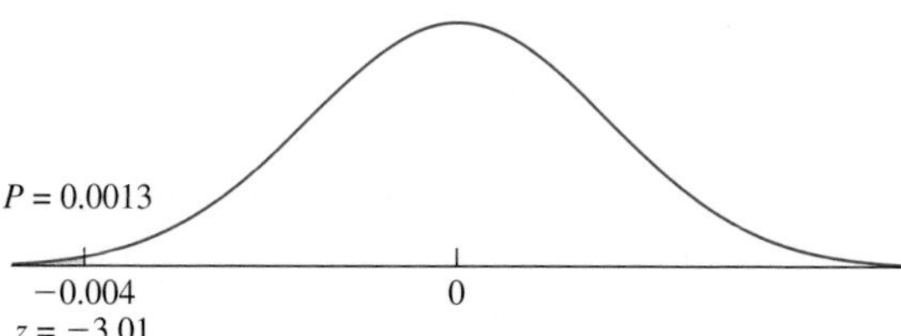

FIGURE 6.9 The null distribution of $\overline{X} - \overline{Y}$ is $N(0,\ 0.001327^2)$. Thus if H_0 is true, the probability that $\overline{X} - \overline{Y}$ takes on a value as extreme as or more extreme than the observed value of -0.004 is 0.0013. This is the P-value.

Note that we used the assumption that the samples were independent when computing the variance of $\overline{X} - \overline{Y}$. This is one condition that is usually easy to achieve in practice. Unless there is some fairly obvious connection between the items in the two samples, it is usually safe to assume they are independent.

Example 6.9

The article "Effect of Welding Procedure on Flux Cored Steel Wire Deposits" (N. Ramini de Rissone, I. de S. Bott, et al., *Science and Technology of Welding and Joining*, 2003:113–122) compares properties of welds made using carbon dioxide as a shielding gas with those of welds made using a mixture of argon and carbon dioxide.

One property studied was the diameter of inclusions, which are particles embedded in the weld. A sample of 544 inclusions in welds made using argon shielding averaged 0.37 μm in diameter, with a standard deviation of 0.25 μm. A sample of 581 inclusions in welds made using carbon dioxide shielding averaged 0.40 μm in diameter, with a standard deviation of 0.26 μm. (Standard deviations were estimated from a graph.) Can you conclude that the mean diameters of inclusions differ between the two shielding gases?

Solution

Let $\overline{X} = 0.37$ denote the sample mean diameter for argon welds. Then $s_X = 0.25$ and the sample size is $n_X = 544$. Let $\overline{Y} = 0.40$ denote the sample mean diameter for carbon dioxide welds. Then $s_Y = 0.26$ and the sample size is $n_Y = 581$. Let μ_X denote the population mean diameter for argon welds, and let μ_Y denote the population mean diameter for carbon dioxide welds. The null and alternate hypotheses are

$$H_0: \mu_X - \mu_Y = 0 \quad \text{versus} \quad H_1: \mu_X - \mu_Y \neq 0$$

We have observed $\overline{X} - \overline{Y} = 0.37 - 0.40 = -0.03$. This value was drawn from a normal population with mean $\mu_X - \mu_Y$, and variance approximated by $s_X^2/n_X + s_Y^2/n_Y$. Under H_0, we take $\mu_X - \mu_Y = 0$. Substituting values of s_X, s_Y, n_X, and n_Y, the standard deviation is $\sqrt{0.25^2/544 + 0.26^2/581} = 0.01521$. The null distribution of $\overline{X} - \overline{Y}$ is therefore

$$\overline{X} - \overline{Y} \sim N(0,\ 0.01521^2)$$

The z-score is

$$z = \frac{-0.03 - 0}{0.01521} = -1.97$$

This is a two-tailed test, and the P-value is 0.0488 (see Figure 6.10). A follower of the 5% rule would reject the null hypothesis. It is certainly reasonable to be skeptical about the truth of H_0.

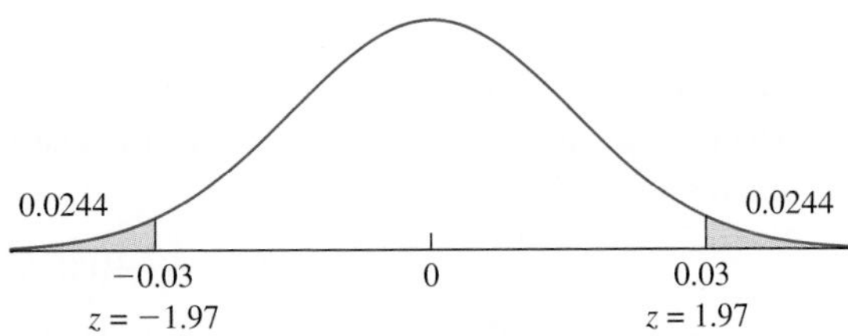

FIGURE 6.10 Solution to Example 6.9.

The following computer output (from MINITAB) presents the results of Example 6.9.

```
Two-sample T for Argon vs CO2

            N      Mean      StDev       SE Mean
Argon     544      0.37       0.25      0.010719
CO2       581      0.40       0.26      0.010787

Difference = mu (Argon) − mu (CO2)
Estimate for difference: 0.030000
95% confidence bound for difference:
(−0.0598366, −0.000163)
T-Test of difference = 0 (vs not = 0):
T-Value = −1.97 P-Value = 0.049 DF = 1122
```

Note that the computer uses the t statistic rather than the z statistic for this test. Many computer packages use the t statistic whenever a sample standard deviation is used to estimate a population standard deviation. When the sample size is large, the difference between t and z is negligible for practical purposes. When using tables rather than a computer, the z-score has the advantage that the P-value can be determined with greater precision with a z table than with a t table.

The methods described in this section can be used to test a hypothesis that two population means differ by a specified constant. Example 6.10 shows how.

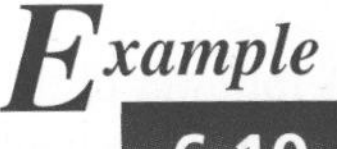

6.10

Refer to Example 6.9. Can you conclude that the mean diameter for carbon dioxide welds (μ_Y) exceeds that for argon welds (μ_X) by more than 0.015 μm?

Solution

The null and alternate hypotheses are

$$H_0: \mu_X - \mu_Y \geq -0.015 \quad \text{versus} \quad H_1: \mu_X - \mu_Y < -0.015$$

We observe $\overline{X} = 0.37$, $\overline{Y} = 0.40$, $s_X = 0.25$, $s_Y = 0.26$, $n_X = 544$, and $n_Y = 581$. Under H_0, we take $\mu_X - \mu_Y = -0.015$. The null distribution of $\overline{X} - \overline{Y}$ is given by expression (6.2) to be

$$\overline{X} - \overline{Y} \sim N(-0.015,\ 0.01521^2)$$

We observe $\overline{X} - \overline{Y} = 0.37 - 0.40 = -0.03$. The z-score is

$$z = \frac{-0.03 - (-0.015)}{0.01521} = -0.99$$

This is a one-tailed test. The P-value is 0.1611. We cannot conclude that the mean diameter of inclusions from carbon dioxide welds exceeds that of argon welds by more than 0.015 μm.

Summary

Let $X_1, \ldots, X_{n_X}$ and $Y_1, \ldots, Y_{n_Y}$ be *large* (e.g., $n_X > 30$ and $n_Y > 30$) samples from populations with means μ_X and μ_Y and standard deviations σ_X and σ_Y, respectively. Assume the samples are drawn independently of each other.

To test a null hypothesis of the form $H_0: \mu_X - \mu_Y \leq \Delta_0$, $H_0: \mu_X - \mu_Y \geq \Delta_0$, or $H_0: \mu_X - \mu_Y = \Delta_0$:

- Compute the z-score: $z = \dfrac{(\overline{X} - \overline{Y}) - \Delta_0}{\sqrt{\sigma_X^2/n_X + \sigma_Y^2/n_Y}}$. If σ_X and σ_Y are unknown they may be approximated with s_X and s_Y, respectively.
- Compute the P-value. The P-value is an area under the normal curve, which depends on the alternate hypothesis as follows:

Alternate Hypothesis	***P*-value**
$H_1: \mu_X - \mu_Y > \Delta_0$	Area to the right of z
$H_1: \mu_X - \mu_Y < \Delta_0$	Area to the left of z
$H_1: \mu_X - \mu_Y \neq \Delta_0$	Sum of the areas in the tails cut off by z and $-z$

Exercises for Section 6.5

1. The article "Measurement of Complex Permittivity of Asphalt Paving Materials" (J. Shang, J. Umana, et al., *Journal of Transportation Engineering*, 1999:347–356) compared the dielectric constants between two types of asphalt, HL3 and HL8, commonly used in pavements. For 42 specimens of HL3 asphalt the average dielectric constant was 5.92 with a standard deviation of 0.15, and for 37 specimens of HL8 asphalt the average dielectric constant was 6.05 with a standard deviation of 0.16. Can you conclude that the mean dielectric constant differs between the two types of asphalt?

2. To determine the effect of fuel grade on fuel efficiency, 80 new cars of the same make, with identical engines, were each driven for 1000 miles. Forty of the cars ran on regular fuel and the other 40 received premium grade fuel. The cars with the regular fuel averaged 27.2 mi/gal, with a standard deviation of 1.2 mi/gal. The cars with the premium fuel averaged 28.1 mi/gal and had a standard deviation of 2.0 mi/gal. Can you conclude that this type of car gets better mileage with premium fuel?

3. Two methods used to purify a protein are being compared. In 50 runs of method A, the mean recovery was 60% and the standard deviation was 15%, while in 60 runs of method B, the mean recovery was 65% and the standard deviation was 20%. Can you conclude that there is a difference in the two recovery rates?

4. Two machines used to fill soft drink containers are being compared. The number of containers filled each minute is counted for 60 minutes for each machine. During the 60 minutes, machine 1 filled an average of 73.8 cans per minute with a standard deviation of 5.2 cans per minute, and machine 2 filled an average of 76.1 cans per minute with a standard deviation of 4.1 cans per minute.
   a. If the counts are made each minute for 60 consecutive minutes, what assumption necessary to the validity of a hypothesis test may be violated?
   b. Assuming that all necessary assumptions are met, perform a hypothesis test. Can you conclude that machine 2 is faster than machine 1?

5. A statistics instructor who teaches a lecture section of 160 students wants to determine whether students have more difficulty with one-tailed hypothesis tests or with two-tailed hypothesis tests. On the next exam, 80 of the students, chosen at random, get a version of the exam with a 10-point question that requires a one-tailed test. The other 80 students get a question that is identical except that it requires a two-tailed test.

The one-tailed students average 7.79 points, and their standard deviation is 1.06 points. The two-tailed students average 7.64 points, and their standard deviation is 1.31 points.

a. Can you conclude that the mean score μ_1 on one-tailed hypothesis test questions is higher than the mean score μ_2 on two-tailed hypothesis test questions? State the appropriate null and alternate hypotheses, and then compute the P-value.

b. Can you conclude that the mean score μ_1 on one-tailed hypothesis test questions differs from the mean score μ_2 on two-tailed hypothesis test questions? State the appropriate null and alternate hypotheses, and then compute the P-value.

6. Fifty specimens of a new computer chip were tested for speed in a certain application, along with 50 specimens of chips with the old design. The average speed, in MHz, for the new chips was 495.6, and the standard deviation was 19.4. The average speed for the old chips was 481.2, and the standard deviation was 14.3.

a. Can you conclude that the mean speed for the new chips is greater than that of the old chips? State the appropriate null and alternate hypotheses, and then find the P-value.

b. A sample of 60 even older chips had an average speed of 391.2 MHz with a standard deviation of 17.2 MHz. Someone claims that the new chips average more than 100 MHz faster than these very old ones. Do the data provide convincing evidence for this claim? State the appropriate null and alternate hypotheses, and then find the P-value.

7. Two methods are being considered for a paint manufacturing process, in order to increase production. In a random sample of 100 days, the mean daily production using the first method was 625 tons and the standard deviation was 40 tons. In a random sample of 64 days, the mean daily production using the second method was 640 tons and the standard deviation was 50 tons. Assume the samples are independent.

a. Can you conclude that the second method yields the greater mean daily production?

b. Can you conclude that the mean daily production for the second method exceeds that of the first method by more than 10 tons?

8. Refer to Exercise 7 in Section 5.4. Can you conclude that the mean hardness of welds cooled at a rate of 10°C/s is greater than that of welds cooled at a rate of 30°C/s?

9. The Ungrounded Electric Corporation (UEC) claims that its power supplies for home computers last longer than its competitor's, Zircon Appliance Products (ZAP). Independent random samples of 75 of each of the two manufacturers' power supplies are chosen, and the sample means and standard deviations are computed:

UEC:	$\overline{X}_1 = 4387$ h	$s_1 = 252$ h
ZAP:	$\overline{X}_2 = 4260$ h	$s_2 = 231$ h

Can you conclude that the UEC power supplies outlast ZAP power supplies? What is the P-value for this test?

10. The following MINITAB output presents the results of a hypothesis test for the difference $\mu_X - \mu_Y$ between two population means:

```
Two-sample T for X vs Y

      N     Mean    StDev   SE Mean
X   135     3.94     2.65      0.23
Y   180     4.43     2.38      0.18

Difference = mu (X) - mu (Y)
Estimate for difference: -0.484442
95% upper bound for difference: -0.007380
T-Test of difference = 0 (vs <): T-Value = -1.68   P-Value = 0.047   DF = 270
```

a. Is this a one-tailed or two-tailed test?
b. What is the null hypothesis?
c. Can H_0 be rejected at the 5% level? How can you tell?
d. The output presents a Student's t test. Compute the P-value using a z test. Are the two results similar?
e. Use the output and an appropriate table to compute a 99% confidence interval for $\mu_X - \mu_Y$ based on the z statistic.

11. The following MINITAB output presents the results of a hypothesis test for the difference $\mu_X - \mu_Y$ between two population means. Some of the numbers are missing.

```
Two-sample T for X vs Y

     N      Mean     StDev     SE Mean
X   78      23.3      (i)        1.26
Y   63     20.63     3.02        (ii)

Difference = mu (X) - mu (Y)
Estimate for difference: 2.670
95% CI for difference: (0.05472, 5.2853)
T-Test of difference = 0 (vs not =): T-Value = 2.03  P-Value = 0.045  DF = 90
```

a. Fill in the missing numbers for (i) and (ii).
b. The output presents a Student's t test. Compute the P-value using a z test. Are the two results similar?
c. Use the output and an appropriate table to compute a 98% confidence interval for $\mu_X - \mu_Y$ based on the z statistic.

6.6 Tests for the Difference Between Two Proportions

The procedure for testing the difference between two proportions is similar to the procedure for testing the difference between two means. We illustrate with an example.

A mobile computer network consists of computers that maintain wireless communication with one another as they move about a given area. A routing protocol is an algorithm that determines how messages will be relayed from machine to machine along the network, so as to have the greatest chance of reaching their destination. The article "Performance Comparison of Two Location Based Routing Protocols" (T. Camp, J. Boleng, et al., *Proceedings of the IEEE International Conference on Communications*, 2002:3318–3324) compares the effectiveness of two routing protocols over a variety of metrics, including the rate of successful deliveries. Assume that using protocol A, 200 messages were sent, and 170 of them, or 85%, were successfully received. Using protocol B, 150 messages were sent, and 123 of them, or 82%, were successfully received. Can we conclude that protocol A has the higher success rate?

In this example, the samples consist of successes and failures. Let X represent the number of messages successfully sent using protocol A, and let Y represent the number of messages successfully sent using protocol B. The observed values in this example are $X = 170$ and $Y = 123$. Let p_X represent the proportion of messages that are successfully sent using protocol A, and let p_Y represent the corresponding proportion from protocol B. The values of p_X and p_Y are unknown.

The random variables X and Y have binomial distributions, with $n_X = 200$ and $n_Y = 150$ trials, respectively. The success probabilities are p_X and p_Y. The observed values of the sample proportions are $\widehat{p}_X = 170/200 = 0.85$ and $\widehat{p}_Y = 123/150 = 0.82$.

The null and alternate hypotheses are

$$H_0\colon p_X - p_Y \leq 0 \quad \text{versus} \quad H_1\colon p_X - p_Y > 0$$

The test is based on the statistic $\widehat{p}_X - \widehat{p}_Y$. We must determine the null distribution of this statistic. By the Central Limit Theorem, since n_X and n_Y are both large,

$$\widehat{p}_X \sim N\left(p_X, \frac{p_X(1-p_X)}{n_X}\right) \qquad \widehat{p}_Y \sim N\left(p_Y, \frac{p_Y(1-p_Y)}{n_Y}\right)$$

Therefore

$$\widehat{p}_X - \widehat{p}_Y \sim N\left(p_X - p_Y, \frac{p_X(1-p_X)}{n_X} + \frac{p_Y(1-p_Y)}{n_Y}\right) \tag{6.3}$$

To obtain the null distribution, we must substitute values for the mean $p_X - p_Y$ and the variance $p_X(1-p_X)/n_X + p_Y(1-p_Y)/n_Y$. The mean is easy. The null hypothesis specifies that $p_X - p_Y \leq 0$, so we take $p_X - p_Y = 0$. The variance is a bit trickier. At first glance, it might seem reasonable to approximate the standard deviation by substituting the sample proportions $\widehat{p}_X$ and $\widehat{p}_Y$ for the population proportions p_X and p_Y. However, the null hypothesis H_0 specifies that the population proportions are equal. Therefore we must estimate them both with a common value. The appropriate value is the **pooled proportion**, obtained by dividing the total number of successes in both samples by the total sample size. This value is

$$\widehat{p} = \frac{X+Y}{n_X+n_Y}$$

The null distribution of $\widehat{p}_X - \widehat{p}_Y$ is therefore estimated by substituting the pooled proportion $\widehat{p}$ for both p_X and p_Y into expression (6.3). This yields

$$\widehat{p}_X - \widehat{p}_Y \sim N\left(0, \widehat{p}(1-\widehat{p})\left(\frac{1}{n_X} + \frac{1}{n_Y}\right)\right) \tag{6.4}$$

In this example, $\widehat{p} = (170+123)/(200+150) = 0.837$. Under H_0, we take $p_X - p_Y = 0$. The null distribution of $\widehat{p}_X - \widehat{p}_Y$ is therefore normal with mean 0 and standard deviation $\sqrt{0.837(1-0.837)(1/200+1/150)} = 0.0399$. The observed value of $\widehat{p}_X - \widehat{p}_Y$ is $0.85 - 0.82 = 0.03$. The z-score is therefore

$$z = \frac{0.03-0}{0.0399} = 0.75$$

The P-value is 0.2266. Figure 6.11 illustrates the null distribution, and indicates the location of the test statistic. On the basis of this P-value, we cannot conclude that protocol B has the greater success rate. Note that for the Central Limit Theorem to be valid, both samples must be reasonably large. A good rule of thumb is that there should be at least 10 successes and 10 failures in each sample.

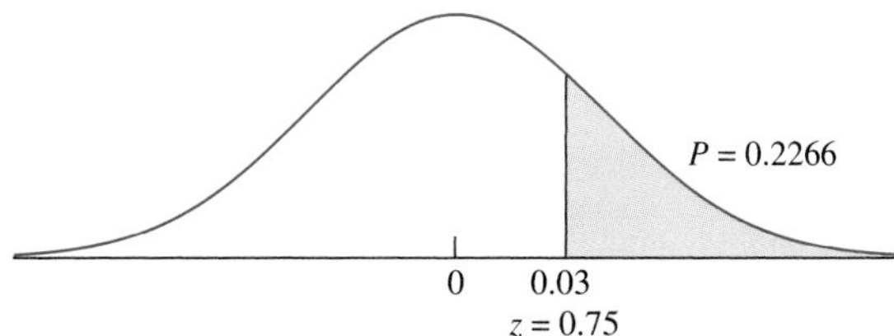

FIGURE 6.11 The null distribution of $\widehat{p}_X - \widehat{p}_Y$ is $N(0,\ 0.0399^2)$. Thus if H_0 is true, the probability that $\widehat{p}_X - \widehat{p}_Y$ takes on a value as extreme as or more extreme than the observed value of 0.03 is 0.2266. This is the P-value.

Example 6.11

Industrial firms often employ methods of "risk transfer," such as insurance or indemnity clauses in contracts, as a technique of risk management. The article "Survey of Risk Management in Major U.K. Companies" (S. Baker, K. Ponniah, and S. Smith, *Journal of Professional Issues in Engineering Education and Practice*, 1999:94–102) reports the results of a survey in which managers were asked which methods played a major role in the risk management strategy of their firms. In a sample of 43 oil companies, 22 indicated that risk transfer played a major role, while in a sample of 93 construction companies, 55 reported that risk transfer played a major role. (These figures were read from a graph.) Can we conclude that the proportion of oil companies that employ the method of risk transfer is less than the proportion of construction companies that do?

Solution

Let $\widehat{p}_X = 22/43 = 0.5116$ be the sample proportion of oil companies employing risk transfer methods, and let $\widehat{p}_Y = 55/93 = 0.5914$ be the corresponding sample proportion of industrial firms. The sample sizes are $n_X = 43$ and $n_Y = 93$. Let p_X and p_Y denote the population proportions for oil and industrial companies, respectively. The null and alternate hypotheses are

$$H_0\colon p_X - p_Y \geq 0 \quad \text{versus} \quad H_1\colon p_X - p_Y < 0$$

The test is based on $\widehat{p}_X - \widehat{p}_Y$. Since both samples are large, the null distribution of $\widehat{p}_X - \widehat{p}_Y$ is given by expression (6.4). The pooled proportion is

$$\widehat{p} = \frac{22 + 55}{43 + 93} = 0.5662$$

The null distribution is normal with mean 0 and standard deviation $\sqrt{0.5662(1 - 0.5662)(1/43 + 1/93)} = 0.0914$. The observed value of $\widehat{p}_X - \widehat{p}_Y$ is $0.5116 - 0.5914 = -0.0798$. The z-score is

$$z = \frac{-0.0798 - 0}{0.0914} = -0.87$$

The P-value is 0.1922 (see Figure 6.12 on page 404). We cannot conclude that the proportion of oil companies employing risk transfer methods is less than the proportion of industrial firms that do.

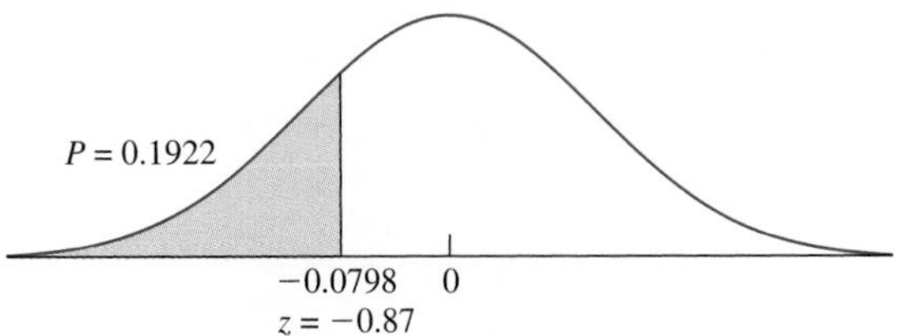

FIGURE 6.12 Solution to Example 6.11.

The following computer output (from MINITAB) presents the results of Example 6.11.

```
Test and CI for Two Proportions: Oil, Indus.

Variable     X    N   Sample p
Oil         22   43   0.511628
Indus.      55   93   0.591398

Difference = p (Oil) − p (Indus.)
Estimate for difference: −0.079770
95% Upper Bound for difference: 0.071079
Test for difference = 0 (vs < 0):  Z = −0.87  P-Value = 0.192
```

The output is self-explanatory. Note that the 95% upper confidence bound is computed by the traditional method (expression 5.19 in Section 5.5 presents the two-sided version of this method).

Summary

Let $X \sim \text{Bin}(n_X, p_X)$ and let $Y \sim \text{Bin}(n_Y, p_Y)$. Assume both n_X and n_Y are large, and that X and Y are independent.

To test a null hypothesis of the form $H_0: p_X - p_Y \le 0$, $H_0: p_X - p_Y \ge 0$, or $H_0: p_X - p_Y = 0$:

- Compute $\widehat{p}_X = \dfrac{X}{n_X}$, $\widehat{p}_Y = \dfrac{Y}{n_Y}$, and $\widehat{p} = \dfrac{X+Y}{n_X + n_Y}$.
- Compute the z-score: $z = \dfrac{\widehat{p}_X - \widehat{p}_Y}{\sqrt{\widehat{p}(1-\widehat{p})(1/n_X + 1/n_Y)}}$.
- Compute the P-value. The P-value is an area under the normal curve, which depends on the alternate hypothesis as follows:

Alternate Hypothesis	***P*-value**
$H_1: p_X - p_Y > 0$	Area to the right of z
$H_1: p_X - p_Y < 0$	Area to the left of z
$H_1: p_X - p_Y \ne 0$	Sum of the areas in the tails cut off by z and $-z$

Exercises for Section 6.6

1. Two extrusion machines that manufacture steel rods are being compared. In a sample of 1000 rods taken from machine 1, 960 met specifications regarding length and diameter. In a sample of 600 rods taken from machine 2, 582 met the specifications. Machine 2 is more expensive to run, so it is decided that machine 1 will be used unless it can be clearly shown that machine 2 produces a larger proportion of rods meeting specifications.
   a. State the appropriate null and alternate hypotheses for making the decision as to which machine to use.
   b. Compute the P-value.
   c. Which machine should be used?

2. Resistors labeled as 100 Ω are purchased from two different vendors. The specification for this type of resistor is that its actual resistance be within 5% of its labeled resistance. In a sample of 180 resistors from vendor A, 150 of them met the specification. In a sample of 270 resistors purchased from vendor B, 233 of them met the specification. Vendor A is the current supplier, but if the data demonstrate convincingly that a greater proportion of the resistors from vendor B meet the specification, a change will be made.
   a. State the appropriate null and alternate hypotheses.
   b. Find the P-value.
   c. Should a change be made?

3. The article "Strategic Management in Engineering Organizations" (P. Chinowsky, *Journal of Management in Engineering*, 2001:60–68) presents results of a survey on management styles that was administered both to private construction firms and to public agencies. Out of a total of 400 private firms contacted, 133 responded by completing the survey, while out of a total of 100 public agencies contacted, 50 responded. Can you conclude that the response rate differs between private firms and public agencies?

4. The article "Training Artificial Neural Networks with the Aid of Fuzzy Sets" (C. Juang, S. Ni, and C. Lu, *Computer-Aided Civil and Infrastructure Engineering*, 1999:407–415) describes the development of artificial neural networks designed to predict the collapsibility of soils. A model with one hidden layer made a successful prediction in 48 out of 60 cases, while a model with two hidden layers made a successful prediction in 44 out of 60 cases. Assuming these samples to be independent, can you conclude that the model with one hidden layer has a greater success rate?

5. In a survey of 100 randomly chosen holders of a certain credit card, 57 said that they were aware that use of the credit card could earn them frequent flier miles on a certain airline. After an advertising campaign to build awareness of this benefit, an independent survey of 200 credit card holders was made, and 135 said that they were aware of the benefit. Can you conclude that awareness of the benefit increased after the advertising campaign?

6. The article "Modeling the Inactivation of Particle-Associated Coliform Bacteria" (R. Emerick, F. Loge, et al., *Water Environment Research*, 2000:432–438) presents counts of the numbers of particles of various sizes in wastewater samples that contained coliform bacteria. Out of 161 particles 75–80 μm in diameter, 19 contained coliform bacteria, and out of 95 particles 90–95 μm in diameter, 22 contained coliform bacteria. Can you conclude that the larger particles are more likely to contain coliform bacteria?

7. To test the effectiveness of protective packaging, a firm shipped out 1200 orders in regular light packaging and 1500 orders in heavy-duty packaging. Of the orders shipped in light packaging, 20 arrived in damaged condition, while of the orders shipped in heavy-duty packaging, 15 arrived in damaged condition. Can you conclude that heavy-duty packaging reduces the proportion of damaged shipments?

8. In a sample of 100 lots of a chemical purchased from vendor A, 70 of them met a purity specification. In a sample of 70 lots purchased from vendor B, 61 met the specification. Can you conclude that a greater proportion of the lots from vendor B meet the specification?

9. In the article "Nitrate Contamination of Alluvial Groundwaters in the Nakdong River Basin, Korea" (J. Min, S. Yun, et al., *Geosciences Journal*, 2002: 35–46), 41 water samples taken from wells in the Daesan area were described, and 22 of these were found to meet quality standards for drinking water. Thirty-one samples were taken from the Yongdang area, and 18

of them were found to meet the standards. Can you conclude that the proportion of wells that meet the standards differs between the two areas?

10. In a clinical trial to compare the effectiveness of two pain relievers, a sample of 100 patients was given drug A, and an independent sample of 200 patients was given drug B. Of the patients on drug A, 76 reported substantial relief, while of the patients on drug B, 128 reported substantial relief. Can you conclude that drug A is more effective than drug B?

11. In order to determine whether to pitch a new advertising campaign more toward men or women, an advertiser provided each couple in a random sample of 500 married couples with a new type of TV remote control that is supposed to be easier to find when needed. Of the 500 husbands, 62% said that the new remote was easier to find than their old one. Of the 500 wives, only 54% said the new remote was easier to find. Let p_1 be the population proportion of married men who think that the new remote is easier to find, and let p_2 be the corresponding proportion of married women. Can the statistic $\widehat{p}_1 - \widehat{p}_2 = 0.62 - 0.54$ be used to test $H_0: p_1 - p_2 = 0$ versus $H_1: p_1 - p_2 \neq 0$? If so, perform the test and compute the P-value. If not, explain why not.

12. The following MINITAB output presents the results of a hypothesis test for the difference $p_1 - p_2$ between two population proportions.

```
Test and CI for Two Proportions

Sample      X     N   Sample p
1          41    97   0.422680
2          37    61   0.606557

Difference = p (1) - p (2)
Estimate for difference: -0.183877
95% CI for difference: (-0.341016, -0.026738)
Test for difference = 0 (vs not = 0):  Z = -2.25   P-Value = 0.024
```

a. Is this a one-tailed or two-tailed test?
b. What is the null hypothesis?
c. Can H_0 be rejected at the 5% level? How can you tell?

13. The following MINITAB output presents the results of a hypothesis test for the difference $p_1 - p_2$ between two population proportions. Some of the numbers are missing. Fill in the numbers for (a) through (d).

```
Test and CI for Two Proportions

Sample      X      N    Sample p
1         101    153      (a)
2         (b)     90    0.544444

Difference = p (1) - p (2)
Estimate for difference: 0.115686
95% CI for difference: (-0.0116695, 0.243042)
Test for difference = 0 (vs not = 0):  Z = (c)  P-Value = (d)
```

6.7 Small-Sample Tests for the Difference Between Two Means

The t test can be used in some cases where samples are small, and thus where the Central Limit Theorem does not apply. We present an example.

The article "The Achondroplasia Paternal Age Effect Is Not Explained By an Increase in Mutant Frequency" (I. Tiemann-Boege, W. Navidi, et al., *Proceedings of the National Academy of Sciences*, 2002:14952–14957) describes an experiment in which a number of DNA molecules is counted, and it needs to be determined whether these molecules contain a certain sequence of nucleotides. This is done by repeating the experiment with an added enzyme that digests the sequence of interest. If the mean count is lower with the enzyme present, then it can be concluded that the molecules being counted do indeed contain the sequence.

Assume that in six identically prepared specimens without the enzyme present, the numbers of molecules counted are 33, 30, 26, 22, 37, and 34. Assume that in four identically prepared specimens with the enzyme present, the counts were 22, 29, 25, and 23. Can we conclude that the counts are lower when the enzyme is present?

We have only a few observations for each process, so the Central Limit Theorem does not apply. If both populations are approximately normal, the Student's t distribution can be used to construct a hypothesis test.

Let $X_1, \ldots, X_6$ represent the counts obtained without the enzyme, and let $Y_1, \ldots, Y_4$ represent the counts obtained with the enzyme. Let μ_X and μ_Y be the means of the populations from which these samples are drawn; let n_X and n_Y denote the sample sizes. The null and alternate hypotheses are

$$H_0\colon \mu_X - \mu_Y \le 0 \quad \text{versus} \quad H_1\colon \mu_X - \mu_Y > 0$$

By assumption, both populations follow normal distributions. Therefore (as discussed in Section 5.6) the quantity

$$\frac{(\overline{X} - \overline{Y}) - (\mu_X - \mu_Y)}{\sqrt{s_X^2/n_X + s_Y^2/n_Y}} \tag{6.5}$$

has an approximate Student's t distribution with ν degrees of freedom, where

$$\nu = \frac{\left(\dfrac{s_X^2}{n_X} + \dfrac{s_Y^2}{n_Y}\right)^2}{\dfrac{(s_X^2/n_X)^2}{n_X - 1} + \dfrac{(s_Y^2/n_Y)^2}{n_Y - 1}} \quad \text{rounded down to the nearest integer.}$$

The observed values for the sample means and standard deviations are $\overline{X} = 30.333$, $\overline{Y} = 24.750$, $s_X = 5.538$, and $s_Y = 3.096$. The sample sizes are $n_X = 6$ and $n_Y = 4$. Substituting the values for the sample standard deviations and sample sizes, we compute $\nu = 7.89$, which we round down to 7. Under H_0, $\mu_X - \mu_Y = 0$. The test statistic is therefore

$$t = \frac{(\overline{X} - \overline{Y}) - 0}{\sqrt{s_X^2/n_X + s_Y^2/n_Y}}$$

Under H_0, the test statistic has a Student's t distribution with seven degrees of freedom. Substituting values for $\overline{X}, \overline{Y}, s_X, s_Y, n_X$, and n_Y, we compute the value of the test statistic to be

$$t = \frac{5.583 - 0}{2.740} = 2.038$$

Consulting the t table with seven degrees of freedom, we find that the value cutting off 5% in the right-hand tail is 1.895, and the value cutting off 2.5% is 2.365. The P-value is therefore between 0.025 and 0.05 (see Figure 6.13). We conclude that the mean count is lower when the enzyme is present.

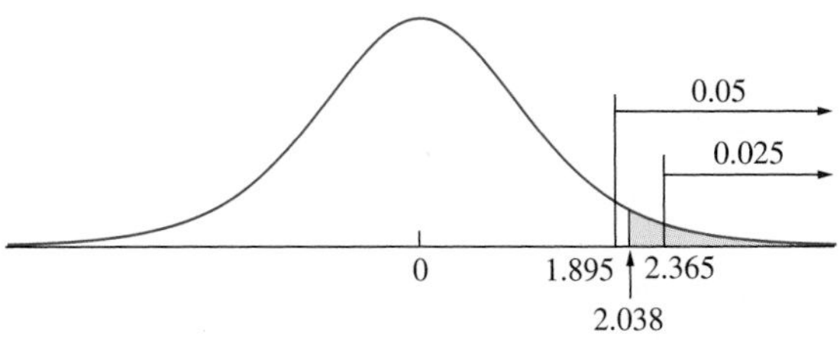

FIGURE 6.13 The null distribution is Student's t with seven degrees of freedom. The observed value of the test statistic is 2.038. If H_0 is true, the probability that t takes on a value as extreme as or more extreme than that observed is between 2.5% and 5%.

Good website design can make Web navigation easier. The article "The Implications of Visualization Ability and Structure Preview Design for Web Information Search Tasks" (H. Zhang and G. Salvendy, *International Journal of Human-Computer Interaction* 2001:75–95) presents a comparison of item recognition between two designs. A sample of 10 users using a conventional Web design averaged 32.3 items identified, with a standard deviation of 8.56. A sample of 10 users using a new structured Web design averaged 44.1 items identified, with a standard deviation of 10.09. Can we conclude that the mean number of items identified is greater with the new structured design?

Solution

Let $\overline{X} = 44.1$ be the sample mean for the structured Web design. Then $s_X = 10.09$ and $n_X = 10$. Let $\overline{Y} = 32.3$ be the sample mean for the conventional Web design. Then $s_Y = 8.56$ and $n_Y = 10$. Let μ_X and μ_Y denote the population mean measurements made by the structured and conventional methods, respectively. The null and alternate hypotheses are

$$H_0: \mu_X - \mu_Y \leq 0 \quad \text{versus} \quad H_1: \mu_X - \mu_Y > 0$$

The test statistic is

$$t = \frac{(\overline{X} - \overline{Y}) - 0}{\sqrt{s_X^2/n_X + s_Y^2/n_Y}}$$

Substituting values for $\overline{X}, \overline{Y}, s_X, S_Y, n_X$, and n_Y, we compute the value of the test statistic to be $t = 2.820$. Under H_0, this statistic has an approximate Student's t distribution, with the number of degrees of freedom given by

$$\nu = \frac{\left(\frac{10.09^2}{10} + \frac{8.56^2}{10}\right)^2}{\frac{(10.09^2/10)^2}{9} + \frac{(8.56^2/10)^2}{9}} = 17.53 \approx 17$$

Consulting the t table with 17 degrees of freedom, we find that the value cutting off 1% in the right-hand tail is 2.567, and the value cutting off 0.5% in the right-hand tail is 2.898. Therefore the area in the right-hand tail corresponding to values as extreme as or more extreme than the observed value of 2.820 is between 0.005 and 0.010. Therefore $0.005 < P < 0.01$ (see Figure 6.14). There is strong evidence that the mean number of items identified is greater for the new design.

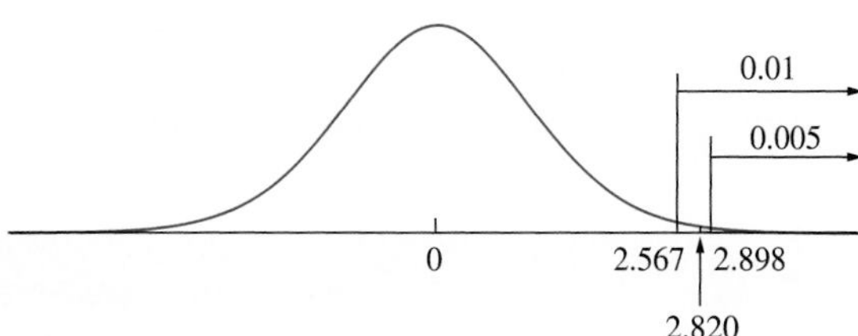

FIGURE 6.14 Solution to Example 6.12. The P-value is the area in the right-hand tail, which is between 0.005 and 0.01.

The following computer output (from MINITAB) presents the results from Example 6.12.

```
Two-Sample T-Test and CI: Struct, Conven

Two-sample T for C1 vs C2

           N      Mean     StDev    SE Mean
Struct    10     44.10     10.09    3.19074
Conven    10     32.30      8.56    2.70691

Difference = mu (Struct) - mu (Conven)
Estimate for difference: 11.8000
95% lower bound for difference: 4.52100
T-Test of difference = 0 (vs >):
T-Value = 2.82    P-Value = 0.006    DF = 17
```

Note that the 95% lower confidence bound is consistent with the alternate hypothesis. This indicates that the P-value is less than 5%.

The methods described in this section can be used to test a hypothesis that two population means differ by a specified constant. Example 6.13 shows how.

Example 6.13

Refer to Example 6.12. Can you conclude that the mean number of items identified with the new structured design exceeds that of the conventional design by more than 2?

Solution

The null and alternate hypotheses are

$$H_0: \mu_X - \mu_Y \leq 2 \quad \text{versus} \quad H_1: \mu_X - \mu_Y > 2$$

We observe $\overline{X} = 44.1$, $\overline{Y} = 32.3$, $s_X = 10.09$, $s_Y = 8.56$, $n_X = 10$, and $n_Y = 10$. Under H_0, we take $\mu_X - \mu_Y = 2$. The test statistic is given by expression (6.5) to be

$$t = \frac{(\overline{X} - \overline{Y}) - 2}{\sqrt{s_X^2/n_X + s_Y^2/n_Y}}$$

Under H_0, the test statistic has a Student's t distribution with 17 degrees of freedom. Note that the number of degrees of freedom is calculated in the same way as in Example 6.12. The value of the test statistic is $t = 2.342$. This is a one-tailed test. The P-value is between 0.01 and 0.025. We conclude that the mean number of items identified with the new structured design exceeds that of the conventional design by more than 2.

Summary

Let $X_1, \ldots, X_{n_X}$ and $Y_1, \ldots, Y_{n_Y}$ be samples from *normal* populations with means μ_X and μ_Y and standard deviations σ_X and σ_Y, respectively. Assume the samples are drawn independently of each other.

If σ_X and σ_Y are not known to be equal, then, to test a null hypothesis of the form $H_0: \mu_X - \mu_Y \leq \Delta_0$, $H_0: \mu_X - \mu_Y \geq \Delta_0$, or $H_0: \mu_X - \mu_Y = \Delta_0$:

- Compute $\nu = \dfrac{[(s_X^2/n_X) + (s_Y^2/n_Y)]^2}{[(s_X^2/n_X)^2/(n_X - 1)] + [(s_Y^2/n_Y)^2/(n_Y - 1)]}$, rounded down to the nearest integer.
- Compute the test statistic $t = \dfrac{(\overline{X} - \overline{Y}) - \Delta_0}{\sqrt{s_X^2/n_X + s_Y^2/n_Y}}$.
- Compute the P-value. The P-value is an area under the Student's t curve with ν degrees of freedom, which depends on the alternate hypothesis as follows:

Alternate Hypothesis	P-value
$H_1: \mu_X - \mu_Y > \Delta_0$	Area to the right of t
$H_1: \mu_X - \mu_Y < \Delta_0$	Area to the left of t
$H_1: \mu_X - \mu_Y \neq \Delta_0$	Sum of the areas in the tails cut off by t and $-t$

When the Populations Have Equal Variances

When the population variances are known to be nearly equal, the pooled variance (see Section 5.6) may be used. The pooled variance is given by

$$s_p^2 = \frac{(n_X - 1)s_X^2 + (n_Y - 1)s_Y^2}{n_X + n_Y - 2}$$

The test statistic for testing any of the null hypotheses $H_0: \mu_X - \mu_Y = 0$, $H_0: \mu_X - \mu_Y \leq 0$, or $H_0: \mu_X - \mu_Y \geq 0$ is

$$t = \frac{\overline{X} - \overline{Y}}{s_p\sqrt{1/n_X + 1/n_Y}}$$

Under H_0, the test statistic has a Student's t distribution with $n_X + n_Y - 2$ degrees of freedom.

Two methods have been developed to determine the nickel content of steel. In a sample of five replications of the first method on a certain kind of steel, the average measurement (in percent) was $\overline{X} = 3.16$ and the standard deviation was $s_X = 0.042$. The average of seven replications of the second method was $\overline{Y} = 3.24$ and the standard deviation was $s_Y = 0.048$. Assume that it is known that the population variances are nearly equal. Can we conclude that there is a difference in the mean measurements between the two methods?

Solution

Substituting the sample sizes $n_X = 5$ and $n_Y = 7$ along with the sample standard deviations $s_X = 0.042$ and $s_Y = 0.048$, we compute the pooled standard deviation, obtaining $s_p = 0.0457$.

The value of the test statistic is therefore

$$t = \frac{3.16 - 3.24}{0.0457\sqrt{1/5 + 1/7}} = -2.990$$

Under H_0, the test statistic has the Student's t distribution with 10 degrees of freedom. Consulting the Student's t table, we find that the area under the curve in each tail is between 0.01 and 0.005. Since the null hypothesis stated that the means are equal, this is a two-tailed test, so the P-value is the sum of the areas in both tails. We conclude that $0.01 < P < 0.02$ (see Figure 6.15 on page 412). There does appear to be a difference in the mean measurements between the two methods.

Don't Assume the Population Variances Are Equal Just Because the Sample Variances Are Close

It is tempting to assume that population variances are equal whenever the sample variances are approximately the same. This assumption is not justified, however, because it is possible for the sample variances to be nearly equal even when the population variances are quite different. The assumption that the population variances are equal

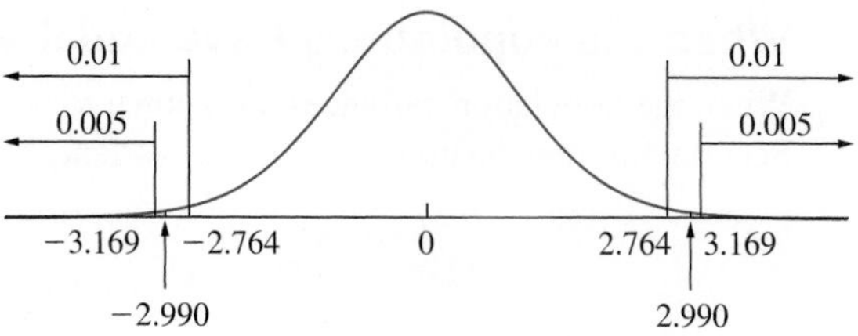

FIGURE 6.15 Solution to Example 6.14. The P-value is the sum of the areas in both tails, which is between 0.01 and 0.02.

should be made only when there is knowledge about the processes that produced the data that justifies this assumption. See the discussion on p. 343.

Summary

Let $X_1, \ldots, X_{n_X}$ and $Y_1, \ldots, Y_{n_Y}$ be samples from *normal* populations with means μ_X and μ_Y and standard deviations σ_X and σ_Y, respectively. Assume the samples are drawn independently of each other.

If σ_X and σ_Y are known to be equal, then, to test a null hypothesis of the form $H_0: \mu_X - \mu_Y \le \Delta_0$, $H_0: \mu_X - \mu_Y \ge \Delta_0$, or $H_0: \mu_X - \mu_Y = \Delta_0$:

- Compute $s_p = \sqrt{\dfrac{(n_X - 1)s_X^2 + (n_Y - 1)s_Y^2}{n_X + n_Y - 2}}$.
- Compute the test statistic $t = \dfrac{(\overline{X} - \overline{Y}) - \Delta_0}{s_p\sqrt{1/n_X + 1/n_Y}}$.
- Compute the P-value. The P-value is an area under the Student's t curve with $n_X + n_Y - 2$ degrees of freedom, which depends on the alternate hypothesis as follows:

Alternate Hypothesis	P-value
$H_1: \mu_X - \mu_Y > \Delta_0$	Area to the right of t
$H_1: \mu_X - \mu_Y < \Delta_0$	Area to the left of t
$H_1: \mu_X - \mu_Y \ne \Delta_0$	Sum of the areas in the tails cut off by t and $-t$

Exercises for Section 6.7

1. A crayon manufacturer is comparing the effects of two kinds of yellow dye on the brittleness of crayons. Dye B is more expensive than dye A, but it is thought that it might produce a stronger crayon. Four crayons are tested with each kind of dye, and the impact strength (in joules) is measured for each. The results are as follows:

Dye A:	1.0	2.0	1.2	3.0
Dye B:	3.0	3.2	2.6	3.4

a. Can you conclude that the mean strength of crayons made with dye B is greater than that of crayons made with dye A?

b. Can you conclude that the mean strength of crayons made with dye B exceeds that of crayons made with dye A by more than 1 J?

2. A study is conducted to determine whether semi-sweet chocolate dissolves more quickly than milk chocolate. Eight people each dissolved one piece of semi-sweet chocolate, while seven other people each dissolved one piece of milk chocolate. The dissolve times, in seconds, are as follows:

Semi-sweet:	30	55	50	22	46	45	30	44
Milk:	45	58	23	64	105	93	28	

Can you conclude that the mean dissolve time of milk chocolate differs from that of semi-sweet chocolate?

3. The article "Modeling Resilient Modulus and Temperature Correction for Saudi Roads" (H. Wahhab, I. Asi, and R. Ramadhan, *Journal of Materials in Civil Engineering*, 2001:298–305) describes a study designed to predict the resilient modulus of pavement from physical properties. One of the questions addressed was whether the modulus differs between rutted and nonrutted pavement. Measurements of the resilient modulus at 40°C (in 10^6 kPa) are presented below for 7 sections of rutted pavement and 12 sections of nonrutted pavement.

Rutted:	1.48	1.88	1.90	1.29	3.53	2.43	1.00
Nonrutted:	3.06	2.58	1.70	5.76	2.44	2.03	1.76
	4.63	2.86	2.82	1.04	5.92		

Perform a hypothesis test to determine whether it is plausible that the mean resilient modulus is the same for rutted and nonrutted pavement. Compute the P-value. What do you conclude?

4. The article "Time Series Analysis for Construction Productivity Experiments" (T. Abdelhamid and J. Everett, *Journal of Construction Engineering and Management* 1999:87–95) presents a study comparing the effectiveness of a video system that allows a crane operator to see the lifting point while operating the crane with the old system in which the operator relies on hand signals from a tagman. Three different lifts, A, B, and C, were studied. Lift A was of little difficulty, lift B was of moderate difficulty, and lift C was of high difficulty. Each lift was performed several times, both with the new video system and with the old tagman system. The time (in seconds) required to perform each lift was recorded. The following tables present the means, standard deviations, and sample sizes.

	Low Difficulty		
	Mean	**Standard Deviation**	**Sample Size**
Tagman	47.79	2.19	14
Video	47.15	2.65	40

	Moderate Difficulty		
	Mean	**Standard Deviation**	**Sample Size**
Tagman	69.33	6.26	12
Video	58.50	5.59	24

	High Difficulty		
	Mean	**Standard Deviation**	**Sample Size**
Tagman	109.71	17.02	17
Video	84.52	13.51	29

a. Can you conclude that the mean time to perform a lift of low difficulty is less when using the video system than when using the tagman system? Explain.

b. Can you conclude that the mean time to perform a lift of moderate difficulty is less when using the video system than when using the tagman system? Explain.

c. Can you conclude that the mean time to perform a lift of high difficulty is less when using the video system than when using the tagman system? Explain.

5. The article "Calibration of an FTIR Spectrometer" (P. Pankratz, *Statistical Case Studies for Industrial and Process Improvement*, SIAM-ASA, 1997:19–38) describes the use of a spectrometer to make five measurements of the carbon content (in ppm) of a certain silicon wafer on each of two successive days. The results were as follows:

Day 1:	2.1321	2.1385	2.0985	2.0941	2.0680
Day 2:	2.0853	2.1476	2.0733	2.1194	2.0717

Can you conclude that the calibration of the spectrometer has changed from the first day to the second day?

6. Two weights, each labeled as weighing 100 g, are each weighed several times on the same scale. The results, in units of μg above 100 g, are as follows:

First weight:	53	88	89	62	39	66
Second weight:	23	39	28	2	49	

Since the same scale was used for both weights, and since both weights are similar, it is reasonable to assume that the variance of the weighing does not depend on the object being weighed. Can you conclude that the weights differ?

7. The article "Mechanical Grading of Oak Timbers" (D. Kretschmann and D. Green, *Journal of Materials in Civil Engineering*, 1999:91–97) presents measurements of the ultimate compressive stress, in MPa, for green mixed oak 7 by 9 timbers from West Virginia and Pennsylvania. For 11 specimens of no. 1 grade lumber, the average compressive stress was 22.1 with a standard deviation of 4.09. For 7 specimens of no. 2 grade lumber, the average compressive stress was 20.4 with a standard deviation of 3.08. Can you conclude that the mean compressive stress is greater for no. 1 grade lumber than for no. 2 grade?

8. Two methods for measuring the molar heat of fusion of water are being compared. Ten measurements made by method A have a mean of 6.02 kilojoules per mole (kJ/mol) with a standard deviation of 0.02 (kJ/mol). Five measurements made by method B have a mean of 6.00 kJ/mol and a standard deviation of 0.01 kJ/mol. Can you conclude that the mean measurements differ between the two methods?

9. Refer to Exercise 8 in Section 5.6. Can you conclude that the heat capacities of coal from the two mines are different?

10. Refer to Exercise 2 in Section 5.6. Can you conclude that more than 0.1 μg is absorbed between 30 and 60 minutes after exposure?

11. Refer to Exercise 4 in Section 5.6.

    a. Can you conclude that the mean mutation frequency for men in their sixties is greater than that for men in their twenties?

    b. Can you conclude that the mean mutation frequency for men in their sixties is more than 25 sequences per μg greater than that for men in their twenties?

12. Refer to Exercise 9 in Section 5.6. Can you conclude that the mean breaking strength is greater for hockey sticks made from composite B?

13. Refer to Exercise 10 in Section 5.6. Can you conclude that the mean permeability coefficient at 60°C differs from that at 61°C?

14. The following MINITAB output presents the results of a hypothesis test for the difference $\mu_X - \mu_Y$ between two population means.

```
Two-sample T for X vs Y

      N     Mean    StDev    SE Mean
X    10    39.31     8.71        2.8
Y    10    29.12     4.79        1.5

Difference = mu (X) — mu (Y)
Estimate for difference: 10.1974
95% lower bound for difference: 4.6333
T-Test of difference = 0 (vs >): T-Value = 3.25   P-Value = 0.003   DF = 13
```

a. Is this a one-tailed or two-tailed test?
b. What is the null hypothesis?
c. Can H_0 be rejected at the 1% level? How can you tell?

15. The following MINITAB output presents the results of a hypothesis test for the difference $\mu_X - \mu_Y$ between two population means. Some of the numbers are missing. Fill in the numbers for (a) through (d).

```
Two-sample T for X vs Y

      N      Mean     StDev     SE Mean
X     6     1.755     0.482       (a)
Y    13     3.239      (b)      0.094
Difference = mu (X) − mu (Y)
Estimate for difference: (c)
95% CI for difference: (−1.99996, −0.96791)
T-Test of difference = 0 (vs not =): T-Value = (d)   P-Value = 0.000   DF = 7
```

6.8 Tests with Paired Data

We saw in Section 5.7 that it is sometimes better to design a two-sample experiment so that each item in one sample is paired with an item in the other. In this section, we present a method for testing hypotheses involving the difference between two population means on the basis of such paired data. We begin with an example.

Particulate matter (PM) emissions from automobiles are a serious environmental concern. Eight vehicles were chosen at random from a fleet, and their emissions were measured under both highway driving and stop-and-go driving conditions. The differences (stop-and-go emission − highway emission) were computed as well. The results, in milligrams of particulates per gallon of fuel, were as follows:

	Vehicle							
	1	2	3	4	5	6	7	8
Stop-and-go	1500	870	1120	1250	3460	1110	1120	880
Highway	941	456	893	1060	3107	1339	1346	644
Difference	559	414	227	190	353	−229	−226	236

Can we conclude that the mean level of emissions is less for highway driving than for stop-and-go driving?

The basic idea behind the construction of the hypothesis test in this example is the same as the idea behind the construction of confidence intervals for paired data presented in Section 5.7. We treat the collection of differences as a single random sample from a population of differences. Denote the population mean of the differences by μ_D and the standard deviation by σ_D. There are only eight differences, which is a small sample. If we assume that the population of differences is approximately normal, we can use the Student's t test, as presented in Section 6.4.

The observed value of the sample mean of the differences is $\overline{D} = 190.5$. The sample standard deviation is $s_D = 284.1$. The null and alternate hypotheses are

$$H_0: \mu_D \leq 0 \quad \text{versus} \quad H_1: \mu_D > 0$$

The test statistic is

$$t = \frac{\overline{D} - 0}{s_D/\sqrt{n}} = \frac{190.5 - 0}{284.1/\sqrt{8}} = 1.897$$

The null distribution of the test statistic is Student's t with seven degrees of freedom. Figure 6.16 presents the null distribution and indicates the location of the test statistic. This is a one-tailed test. The t table indicates that 5% of the area in the tail is cut off by a t value of 1.895, very close to the observed value of 1.897. The P-value is approximately 0.05. The following computer output (from MINITAB) presents this result.

```
Paired T-Test and CI: StopGo, Highway

Paired T for StopGo - Highway

              N      Mean      StDev    SE Mean
StopGo        8   1413.75    850.780    300.796
Highway       8   1223.25    820.850    290.214
Difference    8    190.50    284.104    100.446

95% lower bound for mean difference: 0.197215
T-Test of mean difference = 0 (vs > 0):
T-Value = 1.90   P-Value = 0.050
```

Note that the 95% lower bound is just barely consistent with the alternate hypothesis. This indicates that the P-value is just barely less than 0.05 (although it is given by 0.050 to two significant digits).

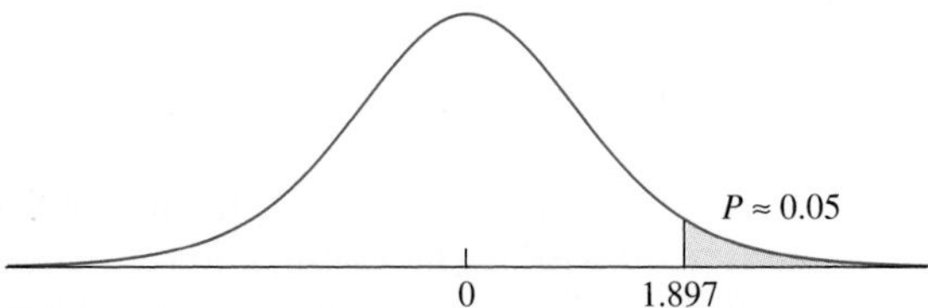

FIGURE 6.16 The null distribution of $t = (\overline{D} - 0)/(s_D/\sqrt{8})$ is t_7. The observed value of t, corresponding to the observed values $\overline{D} = 190.5$ and $s_p = 284.1$, is 1.897. If H_0 is true, the probability that t takes on a value as extreme as or more extreme than that observed is very close to 0.05.

Summary

Let $(X_1, Y_1), \ldots, (X_n, Y_n)$ be a sample of ordered pairs whose differences $D_1, \ldots, D_n$ are a sample from a *normal* population with mean μ_D.

To test a null hypothesis of the form $H_0: \mu_D \leq \mu_0$, $H_0: \mu_D \geq \mu_0$, or $H_0: \mu_D = \mu_0$:

- Compute the test statistic $t = \dfrac{\overline{D} - \mu_0}{s_D/\sqrt{n}}$.
- Compute the P-value. The P-value is an area under the Student's t curve with $n - 1$ degrees of freedom, which depends on the alternate hypothesis as follows:

Alternate Hypothesis	***P*-value**
$H_1: \mu_D > \mu_0$	Area to the right of t
$H_1: \mu_D < \mu_0$	Area to the left of t
$H_1: \mu_D \neq \mu_0$	Sum of the areas in the tails cut off by t and $-t$

- If the sample is large, the D_i need not be normally distributed, the test statistic is $z = \dfrac{\overline{D} - \mu_0}{s_D/\sqrt{n}}$, and a z-test should be performed.

Exercises for Section 6.8

1. Muscles flex when stimulated through electric impulses either to motor points (points on the muscle) or to nerves. The article "Force Assessment of the Stimulated Arm Flexors: Quantification of Contractile Properties" (J. Hong and P. Iaizzo, *Journal of Medical Engineering and Technology* 2002:28–35) reports a study in which both methods were applied to the upper-arm regions of each of several subjects. The latency time (delay between stimulus and contraction) was measured (in milliseconds) for each subject. The results for seven subjects are presented in the following table (one outlier has been deleted).

	Subject						
	1	**2**	**3**	**4**	**5**	**6**	**7**
Nerve	59	57	58	38	53	47	51
Motor point	56	52	56	32	47	42	48
Difference	3	5	2	6	6	5	3

Can you conclude that there is a difference in latency period between motor point and nerve stimulation?

2. The Valsalva maneuver involves blowing into a closed tube in order to create pressure in respiratory airways. Impedance cardiography is used during this maneuver to assess cardiac function. The article "Impedance Cardiographic Measurement of the Physiological Response to the Valsalva Manoeuvre" (R. Patterson and J. Zhang, *Medical and Biological Engineering and Computing*, 2003:40–43) presents a study in which the impedance ratio was measured for each of 11 subjects in both a standing and a reclining position. The results at an airway pressure of 10 mmHg are presented in the following table.

Subject	Standing	Reclining	Difference
1	1.45	0.98	0.47
2	1.71	1.42	0.29
3	1.81	0.70	1.11
4	1.01	1.10	−0.09
5	0.96	0.78	0.18
6	0.83	0.54	0.29
7	1.23	1.34	−0.11
8	1.00	0.72	0.28
9	0.80	0.75	0.05
10	1.03	0.82	0.21
11	1.39	0.60	0.79

Can you conclude that there is a difference between the mean impedance ratio measured in the standing position and that measured in the sitting position?

3. A dry etch process is used to etch silicon dioxide (SiO_2) off of silicon wafers. An engineer wishes to study the uniformity of the etching across the surface of the wafer. A total of 10 wafers are sampled after etching, and the etch rates (in Å/min) are measured at two different sites, one near the center of the wafer, and one near the edge. The results are presented in the following table.

Wafer	Center	Edge
1	586	582
2	568	569
3	587	587
4	550	543
5	543	540
6	552	548
7	562	563
8	577	572
9	558	559
10	571	566

Can you conclude that the etch rates differ between the center and the edge?

4. Two microprocessors are compared on a sample of six benchmark codes to determine whether there is a difference in speed. The times (in seconds) used by each processor on each code are given in the following table.

	Code					
	1	2	3	4	5	6
Processor A	27.2	18.1	27.2	19.7	24.5	22.1
Processor B	24.1	19.3	26.8	20.1	27.6	29.8

Can you conclude that the mean speeds of the two processors differ?

5. The compressive strength, in kilopascals, was measured for each of five concrete blocks both three and six days after pouring. The data are presented in the following table.

	Block				
	1	2	3	4	5
After 3 days	1341	1316	1352	1355	1327
After 6 days	1376	1373	1366	1384	1358

Can you conclude that the mean strength after six days is greater than the mean strength after three days?

6. Refer to Exercise 3 in Section 5.7. Can you conclude that scale 2 reads heavier, on the average, than scale 1?

7. Refer to Exercise 4 in Section 5.7. Can you conclude that there is a difference in tightness between bolts 1 and 8?

8. Refer to Exercise 7 in Section 5.7.

   a. Can you conclude that the mean lifetime of rear brake pads is greater than that of front brake pads?

   b. Can you conclude that the mean lifetime of rear brake pads exceeds that of front brake pads by more than 10,000 miles?

9. The management of a taxi cab company is trying to decide if they should switch from bias tires to radial tires to improve fuel economy. Each of 10 taxis was equipped with one of the two tire types and driven on a test course. Without changing drivers, tires were then switched to the other tire type and the test course

was repeated. The fuel economy (in mi/gal) for the 10 cars is as follows:

Car	Radial	Bias
1	32.1	27.1
2	36.1	31.5
3	32.3	30.4
4	29.5	26.9
5	34.3	29.9
6	31.9	28.7
7	33.4	30.2
8	34.6	31.8
9	35.2	33.6
10	32.7	29.9

a. Because switching tires on the taxi fleet is expensive, management does not want to switch unless a hypothesis test provides strong evidence that the mileage will be improved. State the appropriate null and alternate hypotheses, and find the P-value.

b. A cost-benefit analysis shows that it will be profitable to switch to radial tires if the mean mileage improvement is greater than 2 mi/gal. State the appropriate null and alternate hypotheses, and find the P-value, for a hypothesis test that is designed to form the basis for the decision whether to switch.

10. The following MINITAB output presents the results of a hypothesis test for the difference $\mu_X - \mu_Y$ between two population means.

```
Paired T for X - Y

              N       Mean      StDev    SE Mean
X            12    134.233     68.376     19.739
Y            12    100.601     94.583     27.304
Difference   12    33.6316    59.5113    17.1794

95% lower bound for mean difference: 2.7793
T-Test of mean difference = 0 (vs > 0):    T-Value = 1.96  P-Value = 0.038
```

a. Is this a one-tailed or two-tailed test?
b. What is the null hypothesis?
c. Can H_0 be rejected at the 1% level? How can you tell?
d. Use the output and an appropriate table to compute a 98% confidence interval for $\mu_X - \mu_Y$.

11. The following MINITAB output presents the results of a hypothesis test for the difference $\mu_X - \mu_Y$ between two population means. Some of the numbers are missing. Fill in the numbers for (a) through (d).

```
Paired T for X - Y

              N      Mean      StDev    SE Mean
X             7   12.4141     2.9235      (a)
Y             7    8.3476       (b)     1.0764
Difference    7      (c)     3.16758    1.19723

95% lower bound for mean difference: 1.74006
T-Test of mean difference = 0 (vs > 0):    T-Value = (d)  P-Value = 0.007
```

6.9 Distribution-Free Tests

The Student's t tests described in Sections 6.4 and 6.7 formally require that samples come from normal populations. Distribution-free tests get their name from the fact that the samples are not required to come from any specific distribution. While distribution-free tests do require assumptions for their validity, these assumptions are somewhat less restrictive than the assumptions needed for the t test. Distribution-free tests are sometimes called **nonparametric tests**.

We discuss two distribution-free tests in this section. The first, called the **Wilcoxon signed-rank test**, is a test for a population mean, analogous to the one-sample t test discussed in Section 6.4. The second, called the **Wilcoxon rank-sum test**, or the **Mann–Whitney test**, is analogous to the two-sample t test discussed in Section 6.7.

The Wilcoxon Signed-Rank Test

We illustrate this test with an example. The nickel content, in parts per thousand by weight, is measured for six welds. The results are 9.3, 0.9, 9.0, 21.7, 11.5, and 13.9. Let μ represent the mean nickel content for this type of weld. It is desired to test $H_0: \mu \geq 12$ versus $H_1: \mu < 12$. The Student's t test is not appropriate, because there are two outliers, 0.9 and 21.7, which indicate that the population is not normal. The Wilcoxon signed-rank test can be used in this situation. This test does not require the population to be normal. It does, however, require that the population be continuous (rather than discrete), and that the probability density function be symmetric. (The normal is a special case of a continuous symmetric population.) The given sample clearly comes from a continuous population, and the presence of outliers on either side make it reasonable to assume that the population is approximately symmetric as well. We therefore proceed as follows.

Under H_0, the population mean is $\mu = 12$. Since the population is assumed to be symmetric, the population median is 12 as well. To compute the rank-sum statistic, we begin by subtracting 12 from each sample observation to obtain differences. The difference closest to 0, ignoring sign, is assigned a rank of 1. The difference next closest to 0, again ignoring sign, is assigned a rank of 2, and so on. Finally, the ranks corresponding to negative differences are given negative signs. The following table shows the results.

x	$x - 12$	Signed Rank
11.5	−0.5	−1
13.9	1.9	2
9.3	−2.7	−3
9.0	−3.0	−4
21.7	9.7	5
0.9	−11.1	−6

Denote the sum of the positive ranks S_+ and the sum of the absolute values of the negative ranks S_-. Either S_+ or S_- may be used as a test statistic; we shall use S_+. In this example $S_+ = 2 + 5 = 7$, and $S_- = 1 + 3 + 4 + 6 = 14$. Note that since the sample size is 6, by necessity $S_+ + S_- = 1 + 2 + 3 + 4 + 5 + 6 = 21$. For any sample, it is the case that $S_+ + S_- = 1 + 2 + \cdots + n = n(n+1)/2$. In some cases, where there are many more positive ranks than negative ranks, it is easiest to first compute S_- by summing the negative ranks and then computing $S_+ = n(n+1)/2 - S_-$.

Figures 6.17 and 6.18 show how S_+ can be used as a test statistic. In Figure 6.17, $\mu > 12$. For this distribution, positive differences are more probable than negative differences and tend to be larger in magnitude as well. Therefore it is likely that the positive ranks will be greater both in number and in magnitude than the negative ranks, so S_+ is likely to situation is reversed. Here the positive ranks are likely to be fewer in number and smaller in magnitude, so S_+ is likely to be small.

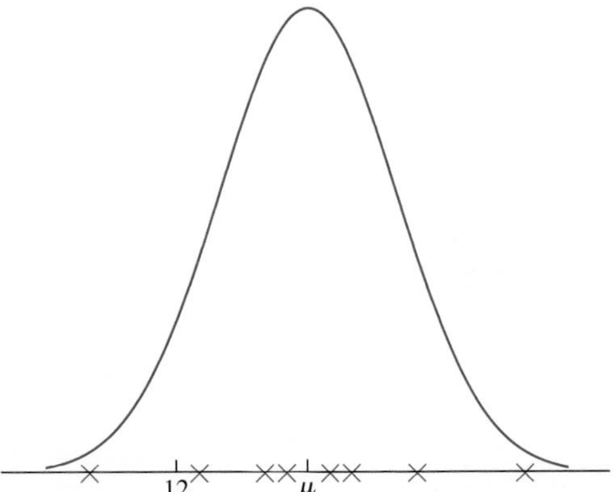

FIGURE 6.17 The true median is greater than 12. Sample observations are more likely to be above 12 than below 12. Furthermore, the observations above 12 will tend to have larger differences from 12 than the observations below 12. Therefore S_+ is likely to be large.

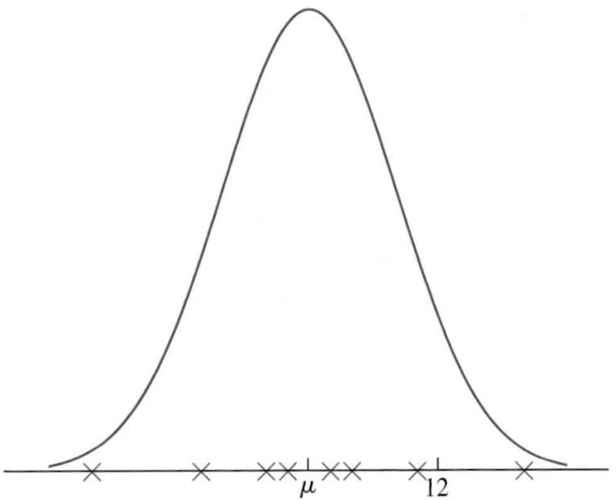

FIGURE 6.18 The true median is less than 12. Sample observations are more likely to be below 12 than above 12. Furthermore, the observations below 12 will tend to have larger differences from 12 than the observations above 12. Therefore S_+ is likely to be small.

We see that in general, large values of S_+ will provide evidence against a null hypothesis of the form $H_0: \mu \leq \mu_0$, while small values of S_+ will provide evidence against a null hypothesis of the form $H_0: \mu \geq \mu_0$.

In this example, the null hypothesis is $H_0: \mu \geq 12$, so a small value of S_+ will provide evidence against H_0. We observe $S_+ = 7$. The P-value is the probability of observing a value of S_+ that is less than or equal to 7 when H_0 is true. Table A.4 (in Appendix A) presents certain probabilities for the null distribution of S_+. Consulting this table with the sample size $n = 6$, we find that the probability of observing a value of 4 or less is 0.1094. The probability of observing a value of 7 or less must be greater than this, so we conclude that $P > 0.1094$, and thus do not reject H_0.

In the example discussed previously, the nickel content for six welds was measured to be 9.3, 0.9, 9.0, 21.7, 11.5, and 13.9. Use these data to test $H_0: \mu \leq 5$ versus $H_1: \mu > 5$.

Solution

The table of differences and signed ranks is as follows:

x	$x - 5$	Signed Rank
9.0	4.0	1
0.9	−4.1	−2
9.3	4.3	3
11.5	6.5	4
13.9	8.9	5
21.7	16.7	6

The observed value of the test statistic is $S_+ = 19$. Since the null hypothesis is of the form $H_0: \mu \leq \mu_0$, large values of S_+ provide evidence against H_0. Therefore the P-value is the area in the right-hand tail of the null distribution, corresponding to values greater than or equal to 19. Consulting Table A.4 shows that the P-value is 0.0469.

Use the data in Example 6.15 to test $H_0: \mu = 16$ versus $H_1: \mu \neq 16$.

Solution

The table of differences and signed ranks is as follows:

x	x − 16	Signed Rank
13.9	−2.1	−1
11.5	−4.5	−2
21.7	5.7	3
9.3	−6.7	−4
9.0	−7.0	−5
0.9	−15.1	−6

Since the null hypothesis is of the form $H_0 : \mu = \mu_0$, this is a two-tailed test. The observed value of the test statistic is $S_+ = 3$. Consulting Table A.4, we find that the area in the left-hand tail, corresponding to values less than or equal to 3, is 0.0781. The P-value is twice this amount, since it is the sum of areas in two equal tails. Thus the P-value is $2(0.0781) = 0.1562$.

Ties

Sometimes two or more of the quantities to be ranked have exactly the same value. Such quantities are said to be tied. The standard method for dealing with ties is to assign to each tied observation the average of the ranks they would have received if they had differed slightly. For example, the quantities 3, 4, 4, 5, 7 would receive the ranks 1, 2.5, 2.5, 4, 5, and the quantities 12, 15, 16, 16, 16, 20 would receive the ranks 1, 2, 4, 4, 4, 6.

Differences of Zero

If the mean under H_0 is μ_0, and one of the observations is equal to μ_0, then its difference is 0, which is neither positive nor negative. An observation that is equal to μ_0 cannot receive a signed rank. The appropriate procedure is to drop such observations from the sample altogether, and to consider the sample size to be reduced by the number of these observations. Example 6.17 serves to illustrate this point.

Example 6.17

Use the data in Example 6.15 to test $H_0 : \mu = 9$ versus $H_1 : \mu \neq 9$.

Solution

The table of differences and signed ranks is as follows:

x	x − 9	Signed Rank
9.0	0.0	—
9.3	0.3	1
11.5	2.5	2
13.9	4.9	3
0.9	−8.1	−4
21.7	12.7	5

The value of the test statistic is $S_+ = 11$. The sample size for the purposes of the test is 5, since the value 9.0 is not ranked. Entering Table A.4 with sample size 5, we find that if $S_+ = 12$, the P-value would be 2(0.1562) = 0.3124. We conclude that for $S_+ = 11$, $P > 0.3124$.

Large-Sample Approximation

When the sample size n is large, the test statistic S_+ is approximately normally distributed. A rule of thumb is that the normal approximation is good if $n > 20$. It can be shown by advanced methods that under H_0, S_+ has mean $n(n+1)/4$ and variance $n(n+1)(2n+1)/24$. The Wilcoxon signed-rank test is performed by computing the z-score of S_+, and then using the normal table to find the P-value. The z-score is

$$z = \frac{S_+ - n(n+1)/4}{\sqrt{n(n+1)(2n+1)/24}}$$

Example 6.18 illustrates the method.

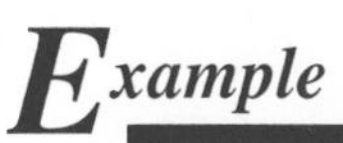

6.18

The article "Exact Evaluation of Batch-Ordering Inventory Policies in Two-Echelon Supply Chains with Periodic Review" (G. Chacon, *Operations Research*, 2001:79–98) presents an evaluation of a reorder point policy. Costs for 32 scenarios are estimated. Let μ represent the mean cost. Test $H_0: \mu \geq 70$ versus $H_1: \mu < 70$. The data, along with the differences and signed ranks, are presented in Table 6.1.

TABLE 6.1 Data for Example 6.18

x	$x - 70$	Signed Rank	x	$x - 70$	Signed Rank	x	$x - 70$	Signed Rank
79.26	9.26	1	30.27	–39.73	–12	11.48	–58.52	–23
80.79	10.79	2	22.39	–47.61	–13	11.28	–58.72	–24
82.07	12.07	3	118.39	48.39	14	10.08	–59.92	–25
82.14	12.14	4	118.46	48.46	15	7.28	–62.72	–26
57.19	–12.81	–5	20.32	–49.68	–16	6.87	–63.13	–27
55.86	–14.14	–6	16.69	–53.31	–17	6.23	–63.77	–28
42.08	–27.92	–7	16.50	–53.50	–18	4.57	–65.43	–29
41.78	–28.22	–8	15.95	–54.05	–19	4.09	–65.91	–30
100.01	30.01	9	15.16	–54.84	–20	140.09	70.09	31
100.36	30.36	10	14.22	–55.78	–21	140.77	70.77	32
30.46	–39.54	–11	11.64	–58.36	–22			

Solution

The sample size is $n = 32$, so the mean is $n(n+1)/4 = 264$ and the variance is $n(n+1)(2n+1)/24 = 2860$. The sum of the positive ranks is $S_+ = 121$. We compute

$$z = \frac{121 - 264}{\sqrt{2860}} = -2.67$$

Since the null hypothesis is of the form $H_0: \mu \geq \mu_0$, small values of S_+ provide evidence against H_0. Thus the P-value is the area under the normal curve to the left of $z = -2.67$. This area, and thus the P-value, is 0.0038.

The Wilcoxon Rank-Sum Test

The Wilcoxon rank-sum test, also called the Mann–Whitney test, can be used to test the difference in population means in certain cases where the populations are not normal. Two assumptions are necessary. First the populations must be continuous. Second, their probability density functions must be identical in shape and size; the only possible difference between them being their location. To describe the test, let $X_1, \ldots, X_m$ be a random sample from one population and let $Y_1, \ldots, Y_n$ be a random sample from the other. We adopt the notational convention that when the sample sizes are unequal, the smaller sample will be denoted $X_1, \ldots, X_m$. Thus the sample sizes are m and n, with $m \leq n$. Denote the population means by μ_X and μ_Y, respectively.

The test is performed by ordering the $m + n$ values obtained by combining the two samples, and assigning ranks $1, 2, \ldots, m + n$ to them. The test statistic, denoted by W, is the sum of the ranks corresponding to $X_1, \ldots, X_m$. Since the populations are identical with the possible exception of location, it follows that if $\mu_X < \mu_Y$, the values in the X sample will tend to be smaller than those in the Y sample, so the rank sum W will tend to be smaller as well. By similar reasoning, if $\mu_X > \mu_Y$, W will tend to be larger. We illustrate the test in Example 6.19.

Resistances, in mΩ, are measured for five wires of one type and six wires of another type. The results are as follows:

X: 36 28 29 20 38
Y: 34 41 35 47 49 46

Use the Wilcoxon rank-sum test to test $H_0: \mu_X \geq \mu_Y$ versus $H_1: \mu_X < \mu_Y$.

Solution

We order the 11 values and assign the ranks.

Value	Rank	Sample	Value	Rank	Sample
20	1	X	38	7	X
28	2	X	41	8	Y
29	3	X	46	9	Y
34	4	Y	47	10	Y
35	5	Y	49	11	Y
36	6	X			

The test statistic W is the sum of the ranks corresponding to the X values, so $W = 1+2+3+6+7 = 19$. To determine the P-value, we consult Table A.5 (in Appendix A). We note that small values of W provide evidence against $H_0: \mu_X \geq \mu_Y$, so the P-value is the area in the left-hand tail of the null distribution. Entering the table with $m = 5$ and $n = 6$ we find that the area to the left of $W = 19$ is 0.0260. This is the P-value.

Large-Sample Approximation

When both sample sizes m and n are greater than 8, it can be shown by advanced methods that the null distribution of the test statistic W is approximately normal with mean $m(m+n+1)/2$ and variance $mn(m+n+1)/12$. In these cases the test is performed by computing the z-score of W, and then using the normal table to find the P-value. The z-score is

$$z = \frac{W - m(m+n+1)/2}{\sqrt{mn(m+n+1)/12}}$$

Example 6.20 illustrates the method.

The article "Cost Analysis Between SABER and Design Bid Build Contracting Methods" (E. Henry and H. Brothers, *Journal of Construction Engineering and Management*, 2001:359–366) presents data on construction costs for 10 jobs bid by the traditional method (denoted X) and 19 jobs bid by an experimental system (denoted Y). The data, in units of dollars per square meter, and their ranks, are presented in Table 6.2. Test $H_0: \mu_X \le \mu_Y$ versus $H_1: \mu_X > \mu_Y$.

TABLE 6.2 Data for Example 6.20

Value	Rank	Sample	Value	Rank	Sample
57	1	X	613	16	X
95	2	Y	622	17	Y
101	3	Y	708	18	X
118	4	Y	726	19	Y
149	5	Y	843	20	Y
196	6	Y	908	21	Y
200	7	Y	926	22	X
233	8	Y	943	23	Y
243	9	Y	1048	24	Y
341	10	Y	1165	25	X
419	11	Y	1293	26	X
457	12	X	1593	27	X
584	13	X	1952	28	X
592	14	Y	2424	29	Y
594	15	Y			

Solution

The sum of the X ranks is $W = 1+12+13+16+18+22+25+26+27+28 = 188$. The sample sizes are $m = 10$ and $n = 19$. We use the normal approximation and compute

$$z = \frac{188 - 10(10+19+1)/2}{\sqrt{10(19)(10+19+1)/12}}$$
$$= 1.74$$

Large values of W provide evidence against the null hypothesis. Therefore the P-value is the area under the normal curve to the right of $z = 1.74$. From the z table we find that the P-value is 0.0409.

Distribution-Free Methods Are Not Assumption-Free

We have pointed out that the distribution-free methods presented here require certain assumptions for their validity. Unfortunately, this is sometimes forgotten in practice. It is tempting to turn automatically to a distribution-free procedure in any situation in which the Student's t test does not appear to be justified, and to assume that the results will always be valid. This is not the case. The necessary assumptions of symmetry for the signed-rank test and of identical shapes and spreads for the rank-sum test are actually rather restrictive. While these tests perform reasonably well under moderate violations of these assumptions, they are not universally applicable.

Exercises for Section 6.9

1. The article "Wastewater Treatment Sludge as a Raw Material for the Production of *Bacillus thuringiensis* Based Biopesticides" (M. Tirado Montiel, R. Tyagi, and J. Valero, *Water Research* 2001: 3807–3816) presents measurements of total solids, in g/L, for seven sludge specimens. The results (rounded to the nearest gram) are 20, 5, 25, 43, 24, 21, and 32. Assume the distribution of total solids is approximately symmetric.

a. Can you conclude that the mean concentration of total solids is greater than 14 g/L? Compute the appropriate test statistic and find the P-value.

b. Can you conclude that the mean concentration of total solids is less than 30 g/L? Compute the appropriate test statistic and find the P-value.

c. An environmental engineer claims that the mean concentration of total solids is equal to 18 g/L? Do the given data disprove this claim?

2. The thicknesses of eight pads designed for use in aircraft engine mounts are measured. The results, in mm, are 41.83, 41.01, 42.68, 41.37, 41.83, 40.50, 41.70, and 41.42. Assume that the thicknesses are a sample from an approximately symmetric distribution.

a. Can you conclude that the mean thickness is greater than 41 mm? Compute the appropriate test statistic and find the P-value.

b. Can you conclude that the mean thickness is less than 41.8 mm? Compute the appropriate test statistic and find the P-value.

c. The target thickness is 42 mm. Can you conclude that the mean thickness differs from the target value? Compute the appropriate test statistic and find the P-value.

3. The article "Reaction Modeling and Optimization Using Neural Networks and Genetic Algorithms: Case Study Involving TS-1-Catalyzed Hydroxylation of Benzene" (S. Nandi, P. Mukherjee, et al., *Industrial and Engineering Chemistry Research*, 2002:2159–2169) presents benzene conversions (in mole percent) for 24 different benzene-hydroxylation reactions. The results are

52.3	41.1	28.8	67.8	78.6	72.3	9.1	19.0
30.3	41.0	63.0	80.8	26.8	37.3	38.1	33.6
14.3	30.1	33.4	36.2	34.6	40.0	81.2	59.4.

a. Can you conclude that the mean conversion is less than 45? Compute the appropriate test statistic and find the P-value.

b. Can you conclude that the mean conversion is greater than 30? Compute the appropriate test statistic and find the P-value.

c. Can you conclude that the mean conversion differs from 55? Compute the appropriate test statistic and find the P-value.

4. The article "Abyssal Peridotites > 3,800 Ma from Southern West Greenland: Field Relationships, Petrography, Geochronology, Whole-Rock and Mineral Chemistry of Dunite and Harzburgite Inclusions in the Itsaq Gneiss Complex" (C. Friend, V. Bennett, and A. Nutman, *Contributions to Mineral Petrology*, 2002:71–92) presents silicon dioxide (SiO_2) concentrations (in weight percent) for 10 dunites. The results are

40.57 41.48 40.76 39.68 43.68 43.53
43.76 44.86 43.06 46.14.

   a. Can you conclude that the mean concentration is greater than 41? Compute the appropriate test statistic and find the P-value.
   b. Can you conclude that the mean concentration is less than 43? Compute the appropriate test statistic and find the P-value.
   c. Can you conclude that the mean concentration differs from 44? Compute the appropriate test statistic and find the P-value.

5. This exercise shows that the signed-rank test can be used with paired data. Two gauges that measure tire tread depth are being compared. Ten different locations on a tire are measured once by each gauge. The results, in mm, are presented in the following table.

Location	Gauge 1	Gauge 2	Difference
1	3.95	3.80	0.15
2	3.23	3.30	−0.07
3	3.60	3.59	0.01
4	3.48	3.61	−0.13
5	3.89	3.88	0.01
6	3.76	3.73	0.03
7	3.45	3.56	−0.11
8	3.01	3.02	−0.01
9	3.82	3.77	0.05
10	3.44	3.49	−0.05

Assume the differences are a sample from an approximately symmetric population with mean μ. Use the Wilcoxon signed-rank test to test $H_0: \mu = 0$ versus $H_1: \mu \neq 0$.

6. The article "n-Nonane Hydroconversion on Ni and Pt Containing HMFI, HMOR and HBEA" (G. Kinger and H. Vinek, *Applied Catalysis A: General*, 2002:139–149) presents hydroconversion rates (in μmol/g · s) of n-nonane over both HMFI and HBEA catalysts. The results are as follows:

HMFI: 0.43 0.93 1.91 2.56 3.72 6.19 11.00
HBEA: 0.73 1.12 1.24 2.93

Can you conclude that the mean rate differs between the two catalysts?

7. A new postsurgical treatment is being compared with a standard treatment. Seven subjects receive the new treatment, while seven others (the controls) receive the standard treatment. The recovery times, in days, are as follows:

Treatment (X):	12	13	15	19	20	21	27
Control (Y):	18	23	24	30	32	35	40

Can you conclude that the mean rate differs between the treatment and control?

8. In an experiment to determine the effect of curing time on compressive strength of concrete blocks, two samples of 15 blocks each were prepared identically except for curing time. The blocks in one sample were cured for two days, while the blocks in the other were cured for six days. The compressive strengths of the blocks, in MPa, are as follows:

Cured 2 days (X):	1326	1302	1314	1270
	1287	1328	1318	1296
	1306	1329	1255	1310
	1255	1291	1280	
Cured 6 days (Y):	1387	1301	1376	1397
	1399	1378	1343	1349
	1321	1364	1332	1396
	1372	1341	1374	

Can you conclude that the mean strength is greater for blocks cured for six days?

9. In a comparison of the effectiveness of distance learning with traditional classroom instruction, 12 students took a business administration course online, while 14 students took it in a classroom. The final exam scores were as follows.

Online:	66	75	85	64	88	77	74
	91	72	69	77	83		
Classroom:	80	83	64	81	75	80	86
	81	51	64	59	85	74	77

Can you conclude that the mean score differs between the two types of course?

10. A woman who has moved into a new house is trying to determine which of two routes to work has the shorter average driving time. Times in minutes for six trips on route A and five trips on route B are follows:

A:	16.0	15.7	16.4	15.9	16.2	16.3
B:	17.2	16.9	16.1	19.8	16.7	

Can you conclude that the mean time is less for route A?

6.10 The Chi-Square Test

In Section 4.1 we studied the Bernoulli trial, which is a process that results in one of two possible outcomes, labeled "success" and "failure." If a number of Bernoulli trials is conducted, and the number of successes is counted, we can test the null hypothesis that the success probability p is equal to a prespecified value p_0. This was covered in Section 6.3. If two sets of Bernoulli trials are conducted, with success probability p_1 for the first set and p_2 for the second set, we can test the null hypothesis that $p_1 = p_2$. This was covered in Section 6.6.

A generalization of the Bernoulli trial is the **multinomial trial** (see Section 4.4), which is an experiment that can result in any one of k outcomes, where $k \geq 2$. The probabilities of the k outcomes are denoted $p_1, \ldots, p_k$. For example, the roll of a fair die is a multinomial trial with six outcomes 1, 2, 3, 4, 5, 6; and probabilities $p_1 = p_2 = p_3 = p_4 = p_5 = p_6 = 1/6$. In this section, we generalize the tests for a Bernoulli probability to multinomial trials. We begin with an example in which we test the null hypothesis that the multinomial probabilities $p_1, p_2, \ldots, p_k$ are equal to a prespecified set of values $p_{01}, p_{02}, \ldots, p_{0k}$, so that the null hypothesis has the form $H_0: p_1 = p_{01}, p_2 = p_{02}, \ldots, p_k = p_{0k}$.

Imagine that a gambler wants to test a die to see whether it deviates from fairness. Let p_i be the probability that the number i comes up. The null hypothesis will state that the die is fair, so the probabilities specified under the null hypothesis are $p_{01} = \cdots = p_{06} = 1/6$. The null hypothesis is $H_0: p_1 = \cdots = p_6 = 1/6$.

The gambler rolls the die 600 times and obtains the results shown in Table 6.3 (page 430), in the column labeled "Observed." The results obtained are called the **observed values**. To test the null hypothesis, we construct a second column, labeled "Expected." This column contains the **expected values**. The expected value for a given outcome is the mean number of trials that would result in that outcome if H_0 were true. To compute the expected values, let N be the total number of trials. (In the die example, $N = 600$.) When H_0 is true, the probability that a trial results in outcome i is p_{0i}, so the expected number of trials resulting in outcome i is Np_{0i}. In the die example, the expected number of trials for each outcome is 100.

TABLE 6.3 Observed and expected values for 600 rolls of a die

Category	Observed	Expected
1	115	100
2	97	100
3	91	100
4	101	100
5	110	100
6	86	100
Total	600	600

The idea behind the hypothesis test is that if H_0 is true, then the observed and expected values are likely to be close to each other. Therefore we will construct a test statistic that measures the closeness of the observed to the expected values. The statistic is called the **chi-square statistic**. To define it, let k be the number of outcomes ($k = 6$ in the die example), and let O_i and E_i be the observed and expected numbers of trials, respectively, that result in outcome i. The chi-square statistic is

$$\chi^2 = \sum_{i=1}^{k} \frac{(O_i - E_i)^2}{E_i} \tag{6.6}$$

The larger the value of χ^2, the stronger the evidence against H_0. To determine the P-value for the test, we must know the null distribution of this test statistic. In general, we cannot determine the null distribution exactly. However, when the expected values are all sufficiently large, a good approximation is available. It is called the **chi-square distribution** with $k - 1$ degrees of freedom, denoted χ^2_{k-1}. Note that the number of degrees of freedom is one less than the number of categories. Use of the chi-square distribution is appropriate whenever all the expected values are greater than or equal to 5.

A table for the chi-square distribution (Table A.6) is provided in Appendix A. The table provides values for certain quantiles, or upper percentage points, for a large number of choices of degrees of freedom. As an example, Figure 6.19 presents the probability density function of the χ^2_{10} distribution. The upper 5% of the distribution is shaded. To find the upper 5% point in the table, look under $\alpha = 0.05$ and degrees of freedom $\nu = 10$. The value is 18.307.

We now compute the value of the chi-square statistic for the data in Table 6.3. The number of degrees of freedom is 5 (one less than the number of outcomes). Using Equation (6.6), the value of the statistic is

$$\begin{aligned}\chi^2 &= \frac{(115 - 100)^2}{100} + \cdots + \frac{(86 - 100)^2}{100} \\ &= 2.25 + \cdots + 1.96 \\ &= 6.12\end{aligned}$$

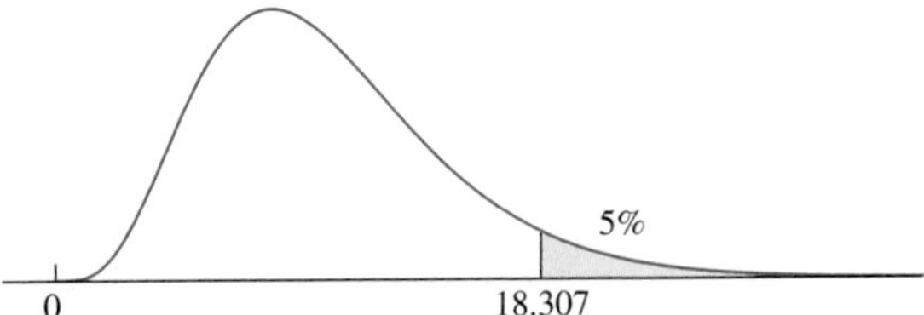

FIGURE 6.19 Probability density function of the χ^2_{10} distribution. The upper 5% point is 18.307. [See the chi-square table (Table A.6) in Appendix A.]

To determine the P-value for the test statistic, we first note that all the expected values are greater than or equal to 5, so use of the chi-square distribution is appropriate. We consult the chi-square table under five degrees of freedom. The upper 10% point is 9.236. We conclude that $P > 0.10$. (See Figure 6.20.) There is no evidence to suggest that the die is not fair.

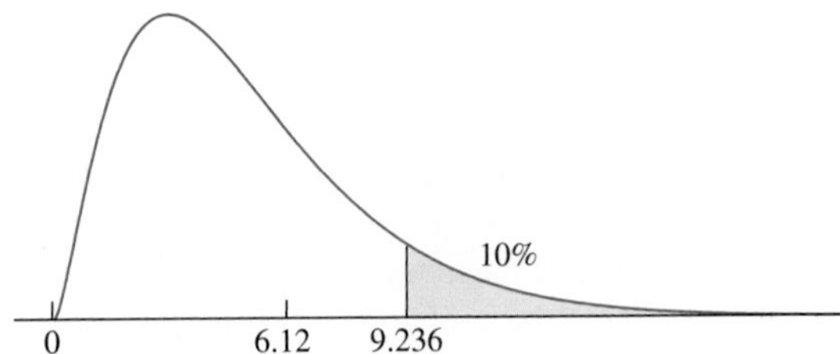

FIGURE 6.20 Probability density function of the χ^2_5 distribution. The observed value of the test statistic is 6.12. The upper 10% point is 9.236. Therefore the P-value is greater than 0.10.

The Chi-Square Test for Homogeneity

In the previous example, we tested the null hypothesis that the probabilities of the outcomes for a multinomial trial were equal to a prespecified set of values. Sometimes several multinomial trials are conducted, each with the same set of possible outcomes. The null hypothesis is that the probabilities of the outcomes are the same for each experiment. We present an example.

Four machines manufacture cylindrical steel pins. The pins are subject to a diameter specification. A pin may meet the specification, or it may be too thin or too thick. Pins are sampled from each machine, and the number of pins in each category is counted. Table 6.4 (page 432) presents the results.

Table 6.4 is an example of a **contingency table**. Each row specifies a category regarding one criterion (machine, in this case), and each column specifies a category regarding another criterion (thickness, in this case). Each intersection of row and column is called a **cell**, so there are 12 cells in Table 6.4.

TABLE 6.4 Observed numbers of pins in various categories with regard to a diameter specification

	Too Thin	OK	Too Thick	Total
Machine 1	10	102	8	120
Machine 2	34	161	5	200
Machine 3	12	79	9	100
Machine 4	10	60	10	80
Total	66	402	32	500

The number in the cell at the intersection of row i and column j is the number of trials whose outcome was observed to fall into row category i and into column category j. This number is called the **observed value** for cell ij. Note that we have included the totals of the observed values for each row and column. These are called the **marginal totals**.

The null hypothesis is that the proportion of pins that are too thin, OK, or too thick is the same for all machines. More generally, the null hypothesis says that no matter which row is chosen, the probabilities of the outcomes associated with the columns are the same. We will develop some notation with which to express H_0 and to define the test statistic.

Let I denote the number of rows in the table, and let J denote the number of columns. Let p_{ij} denote the probability that the outcome of a trial falls into column j given that it is in row i. Then the null hypothesis is

$$H_0\text{: For each column } j,\ p_{1j} = \cdots = p_{Ij} \tag{6.7}$$

Let O_{ij} denote the observed value in cell ij. Let $O_{i.}$ denote the sum of the observed values in row i, let $O_{.j}$ denote the sum of the observed values in column j, and let $O_{..}$ denote the sum of the observed values in all the cells (see Table 6.5).

TABLE 6.5 Notation for observed values

	Column 1	Column 2	...	Column *J*	Total
Row 1	O_{11}	O_{12}	...	O_{1J}	$O_{1.}$
Row 2	O_{21}	O_{22}	...	O_{2J}	$O_{2.}$
⋮	⋮	⋮	⋮	⋮	⋮
Row *I*	O_{I1}	O_{I2}	...	O_{IJ}	$O_{I.}$
Total	$O_{.1}$	$O_{.2}$	...	$O_{.J}$	$O_{..}$

To define a test statistic, we must compute an expected value for each cell in the table. Under H_0, the probability that the outcome of a trial falls into column j is the same for each row i. The best estimate of this probability is the proportion of trials whose outcome falls into column j. This proportion is $O_{.j}/O_{..}$. We need to compute the expected *number* of trials whose outcome falls into cell ij. We denote this expected value by E_{ij}. It is equal to the proportion of trials whose outcome falls into column j, multiplied by the number $O_{i.}$ of trials in row i. That is,

$$E_{ij} = \frac{O_{i.} O_{.j}}{O_{..}} \tag{6.8}$$

The test statistic is based on the differences between the observed and expected values:

$$\chi^2 = \sum_{i=1}^{I} \sum_{j=1}^{J} \frac{(O_{ij} - E_{ij})^2}{E_{ij}} \tag{6.9}$$

Under H_0, this test statistic has a chi-square distribution with $(I - 1)(J - 1)$ degrees of freedom. Use of the chi-square distribution is appropriate whenever the expected values are all greater than or equal to 5.

Example 6.21

Use the data in Table 6.4 to test the null hypothesis that the proportions of pins that are too thin, OK, or too thick are the same for all the machines.

Solution

We begin by using Equation (6.8) to compute the expected values E_{ij}. We show the calculations of E_{11} and E_{23} in detail:

$$E_{11} = \frac{(120)(66)}{500} = 15.84$$

$$E_{23} = \frac{(200)(32)}{500} = 12.80$$

The complete table of expected values is as follows:

Expected values for Table 6.4

	Too Thin	OK	Too Thick	Total
Machine 1	15.84	96.48	7.68	120.00
Machine 2	26.40	160.80	12.80	200.00
Machine 3	13.20	80.40	6.40	100.00
Machine 4	10.56	64.32	5.12	80.00
Total	66.00	402.00	32.00	500.00

We note that all the expected values are greater than 5. Therefore the chi-square test is appropriate. We use Equation (6.9) to compute the value of the chi-square statistic:

$$\begin{aligned}\chi^2 &= \frac{(10 - 15.84)^2}{15.84} + \cdots + \frac{(10 - 5.12)^2}{5.12} \\ &= \frac{34.1056}{15.84} + \cdots + \frac{23.8144}{5.12} \\ &= 15.5844\end{aligned}$$

Since there are four rows and three columns, the number of degrees of freedom is $(4 - 1)(3 - 1) = 6$. To obtain the P-value, we consult the chi-square table (Table A.6). Looking under six degrees of freedom, we find that the upper 2.5% point is 14.449, and the upper 1% point is 16.812. Therefore $0.01 < P < 0.025$. It is reasonable to conclude that the machines differ in the proportions of pins that are too thin, OK, or too thick.

Note that the observed row and column totals are identical to the expected row and column totals. This is always the case.

The following computer output (from MINITAB) presents the results of this hypothesis test.

```
Chi-Square Test: Thin, OK, Thick

Expected counts are printed below observed counts
Chi-Square contributions are printed below expected counts

            Thin        OK     Thick     Total
    1         10       102         8       120
           15.84     96.48      7.68
           2.153     0.316     0.013

    2         34       161         5       200
           26.40    160.80     12.80
           2.188     0.000     4.753

    3         12        79         9       100
           13.20     80.40      6.40
           0.109     0.024     1.056

    4         10        60        10        80
           10.56     64.32      5.12
           0.030     0.290     4.651

Total         66       402        32       500

Chi-Sq = 15.584 DF = 6, P-Value = 0.016
```

In the output, each cell (intersection of row and column) contains three numbers. The top number is the observed value, the middle number is the expected value, and the bottom number is the contribution $(O_{ij} - E_{ij})^2/E_{ij}$ made to the chi-square statistic from that cell.

The Chi-Square Test for Independence

In Example 6.21 the column totals were random, while the row totals were presumably fixed in advance, since they represented numbers of items sampled from various machines. In some cases, both row and column totals are random. In either case, we can test the null hypothesis that the probabilities of the column outcomes are the same for each row outcome, and the test is exactly the same in both cases. We present an example where both row and column totals are random.

The cylindrical steel pins in Example 6.21 are subject to a length specification as well as a diameter specification. With respect to the length, a pin may meet the specification, or it may be too short or too long. A total of 1000 pins are sampled and categorized with respect to both length and diameter specification. The results are presented in the following table. Test the null hypothesis that the proportions of pins that are too thin, OK, or too thick with respect to the diameter specification do not depend on the classification with respect to the length specification.

Observed Values for 1000 Steel Pins

	Diameter			
Length	**Too Thin**	**OK**	**Too Thick**	**Total**
Too Short	13	117	4	134
OK	62	664	80	806
Too Long	5	68	8	81
Total	80	849	92	1021

Solution

We begin by using Equation (6.8) to compute the expected values. The expected values are given in the following table.

Expected Values for 1000 Steel Pins

	Diameter			
Length	**Too Thin**	**OK**	**Too Thick**	**Total**
Too Short	10.50	111.43	12.07	134.0
OK	63.15	670.22	72.63	806.0
Too Long	6.35	67.36	7.30	81.0
Total	80.0	849.0	92.0	1021.0

We note that all the expected values are greater than or equal to 5. (One of the observed values is not; this doesn't matter.) Therefore the chi-square test is appropriate. We use Equation (6.9) to compute the value of the chi-square statistic:

$$\begin{aligned}\chi^2 &= \frac{(13-10.50)^2}{10.50} + \cdots + \frac{(8-7.30)^2}{7.30} \\ &= \frac{6.25}{10.50} + \cdots + \frac{0.49}{7.30} \\ &= 7.46\end{aligned}$$

Since there are three rows and three columns, the number of degrees of freedom is $(3-1)(3-1) = 4$. To obtain the P-value, we consult the chi-square table (Table A.6). Looking under four degrees of freedom, we find that the upper 10% point is 7.779. We conclude that $P > 0.10$. There is no evidence that the length and thickness are related.

Exercises for Section 6.10

1. Rivets are manufactured for a certain purpose. The length specification is 1.20–1.30 cm. It is thought that 90% of the rivets manufactured meet the specification, while 5% are too short and 5% are too long. In a random sample of 1000 rivets, 860 met the specification, 60 were too short, and 80 were too long. Can you conclude that the true percentages differ from 90%, 5%, and 5%?
   a. State the appropriate null hypothesis.
   b. Compute the expected values under the null hypothesis.
   c. Compute the value of the chi-square statistic.
   d. Find the P-value. What do you conclude?

2. Specifications for the diameter of a roller are 2.10–2.11 cm. Rollers that are too thick can be reground, while those that are too thin must be scrapped. Three machinists grind these rollers. Samples of rollers were collected from each machinist, and their diameters were measured. The results are as follows:

Machinist	Good	Regrind	Scrap	Total
A	328	58	14	400
B	231	48	21	300
C	409	73	18	500
Total	968	179	53	1200

   Can you conclude that the proportions of rollers in the three categories differ among the machinists?
   a. State the appropriate null hypothesis.
   b. Compute the expected values under the null hypothesis.
   c. Compute the value of the chi-square statistic.
   d. Find the P-value. What do you conclude?

3. The article "An Investment Tax Credit for Investing in New Technology: A Survey of California Firms" (R. Pope, *The Engineering Economist*, 1997: 269–287) examines the potential impact of a tax credit on capital investment. A number of firms were categorized by size ($>$ 100 employees vs. $\le$ 100 employees) and net excess capacity. The numbers of firms in each of the categories are presented in the following table:

Net Excess Capacity	Small	Large
$< 0\%$	66	115
0–10%	52	47
11–20%	13	18
21–30%	6	5
$> 30\%$	36	25

Can you conclude that the distribution of net excess capacity differs between small and large firms? Compute the relevant test statistic and P-value.

4. The article referred to in Exercise 3 categorized firms by size and percentage of full-operating-capacity labor force currently employed. The numbers of firms in each of the categories are presented in the following table.

Percent of Full-Operating-Capacity Labor Force Currently Employed	Small	Large
> 100%	6	8
95–100%	29	45
90–94%	12	28
85–89%	20	21
80–84%	17	22
75–79%	15	21
70–74%	33	29
< 70%	39	34

Can you conclude that the distribution of labor force currently employed differs between small and large firms? Compute the relevant test statistic and P-value.

5. For the given table of observed values,

   a. Construct the corresponding table of expected values.

   b. If appropriate, perform the chi-square test for the null hypothesis that the row and column outcomes are independent. If not appropriate, explain why.

	Observed Values		
	1	2	3
A	15	10	12
B	3	11	11
C	9	14	12

6. For the given table of observed values,

   a. Construct the corresponding table of expected values.

   b. If appropriate, perform the chi-square test for the null hypothesis that the row and column outcomes are independent. If not appropriate, explain why.

	Observed Values		
	1	2	3
A	25	4	11
B	3	3	4
C	42	3	5

7. Fill in the blank: For observed and expected values, ________

   i. The row totals in the observed table must be the same as the row totals in the expected table, but the column totals need not be the same.

   ii. The column totals in the observed table must be the same as the column totals in the expected table, but the row totals need not be the same.

   iii. Both the row and the column totals in the observed table must be the same as the row and the column totals, respectively, in the expected table.

   iv. Neither the row nor the column totals in the observed table need be the same as the row or the column totals in the expected table.

8. Because of printer failure, none of the observed values in the following table were printed, but some of the marginal totals were. Is it possible to construct the corresponding table of expected values from the information given? If so, construct it. If not, describe the additional information you would need.

	Observed Values			
	1	2	3	Total
A	—	—	—	25
B	—	—	—	—
C	—	—	—	40
D	—	—	—	75
Total	50	20	—	150

9. A random number generator is supposed to produce the digits 0 through 9 with equal probability. A sample of 200 digits was generated, with the following frequency for each of the digits.

Digit	0	1	2	3	4	5	6	7	8	9
Frequency	21	17	20	18	25	16	28	19	22	14

Do these data provide evidence that the random number generator is not working properly? Explain.

10. At an assembly plant for light trucks, routine monitoring of the quality of welds yields the following data:

	Number of Welds		
	High Quality	**Moderate Quality**	**Low Quality**
Day shift	470	191	42
Evening shift	445	171	28
Night shift	257	129	17

Can you conclude that the quality varies among shifts?

11. The article "Analysis of Unwanted Fire Alarm: Case Study" (W. Chow, N. Fong, and C. Ho, *Journal of Architectural Engineering*, 1999:62–65) presents a count of the number of false alarms at several sites. The numbers of false alarms each month, divided into those with known causes and those with unknown causes, are given in the following table. Can you conclude that the proportion of false alarms whose cause is known differs from month to month?

	Month											
	1	**2**	**3**	**4**	**5**	**6**	**7**	**8**	**9**	**10**	**11**	**12**
Known	20	13	21	26	23	18	14	10	20	20	18	14
Unknown	12	2	16	12	22	30	32	32	14	16	10	12

12. At a certain genetic locus on a chromosome, each individual has one of three different DNA sequences (alleles). The three alleles are denoted A, B, C. At another genetic locus on the same chromosome, each organism has one of three alleles, denoted 1, 2, 3. Each individual therefore has one of nine possible allele pairs: A1, A2, A3, B1, B2, B3, C1, C2, or C3. These allele pairs are called *haplotypes*. The loci are said to be in *linkage equilibrium* if the two alleles in an individual's haplotype are independent. Haplotypes were determined for 316 individuals. The following MINITAB output presents the results of a chi-square test for independence.

```
Chi-Square Test: A, B, C

Expected counts are printed below
observed counts
Chi-Square contributions are printed
below expected counts

            A        B        C    Total
1          66       44       34      144
        61.06    47.39    35.54
        0.399    0.243    0.067

2          36       38       20       94
        39.86    30.94    23.20
        0.374    1.613    0.442

3          32       22       24       78
        33.08    25.67    19.25
        0.035    0.525    1.170

Total     134      104       78      316

Chi-Sq = 4.868, DF = 4,
P-Value = 0.301
```

a. How many individuals were observed to have the haplotype B3?

b. What is the expected number of individuals with the haplotype A2?

c. Which of the nine haplotypes was least frequently observed?

d. Which of the nine haplotypes has the smallest expected count?

e. Can you conclude that the loci are not in linkage equilibrium (i.e., not independent)? Explain.

f. Can you conclude that the loci are in linkage equilibrium (i.e., independent)? Explain.

6.11 The *F* Test for Equality of Variance

The tests we have studied so far have involved means or proportions. Sometimes it is desirable to test a null hypothesis that two populations have equal variances. In general there is no good way to do this. In the special case where both populations are normal, however, a method is available.

Let $X_1, \ldots, X_m$ be a simple random sample from a $N(\mu_1, \sigma_1^2)$ population, and let $Y_1, \ldots, Y_n$ be a simple random sample from a $N(\mu_2, \sigma_2^2)$ population. Assume that the samples are chosen independently. The values of the means, μ_1 and μ_2, are irrelevant here; we are concerned only with the variances σ_1^2 and σ_2^2. Note that the sample sizes, m and n, may be different. Let s_1^2 and s_2^2 be the sample variances. That is,

$$s_1^2 = \frac{1}{m-1}\sum_{i=1}^{m}(X_i - \overline{X})^2 \qquad s_2^2 = \frac{1}{n-1}\sum_{i=1}^{n}(Y_i - \overline{Y})^2$$

Any of three null hypotheses may be tested. They are

$$H_0\colon \frac{\sigma_1^2}{\sigma_2^2} \le 1 \quad \text{or equivalently,} \quad \sigma_1^2 \le \sigma_2^2$$

$$H_0\colon \frac{\sigma_1^2}{\sigma_2^2} \ge 1 \quad \text{or equivalently,} \quad \sigma_1^2 \ge \sigma_2^2$$

$$H_0\colon \frac{\sigma_1^2}{\sigma_2^2} = 1 \quad \text{or equivalently,} \quad \sigma_1^2 = \sigma_2^2$$

The procedures for testing these hypotheses are similar, but not identical. We will describe the procedure for testing the null hypothesis $H_0\colon \sigma_1^2/\sigma_2^2 \le 1$ versus $H_1\colon \sigma_1^2/\sigma_2^2 > 1$, and then discuss how the procedure may be modified to test the other two hypotheses.

The test statistic is the ratio of the two sample variances:

$$F = \frac{s_1^2}{s_2^2} \tag{6.10}$$

When H_0 is true, we assume that $\sigma_1^2/\sigma_2^2 = 1$ (the value closest to H_1), or equivalently, that $\sigma_1^2 = \sigma_2^2$. When H_0 is true, s_1^2 and s_2^2 are, on average, the same size, so F is likely to be near 1. When H_0 is false, $\sigma_1^2 > \sigma_2^2$, so s_1^2 is likely to be larger than s_2^2, and F is likely to be greater than 1. In order to use F as a test statistic, we must know its null distribution. The null distribution is called an F distribution, which we now describe.

The *F* Distribution

Statistics that have an F distribution are ratios of quantities, such as the ratio of the two sample variances in Equation (6.10). The F distribution therefore has two values for the degrees of freedom: one associated with the numerator, and one associated with the denominator. The degrees of freedom are indicated with subscripts under the letter F. For example, the symbol $F_{3,16}$ denotes the F distribution with 3 degrees of freedom for the numerator and 16 degrees of freedom for the denominator. Note that the degrees of freedom for the numerator are always listed first.

A table for the F distribution is provided (Table A.7 in Appendix A). The table provides values for certain quantiles, or upper percentage points, for a large number of choices for the degrees of freedom. As an example, Figure 6.21 presents the probability density function of the $F_{3,16}$ distribution. The upper 5% of the distribution is shaded. To find the upper 5% point in the table, look under $\alpha = 0.050$, and degrees of freedom $\nu_1 = 3$, $\nu_2 = 16$. The value is 3.24.

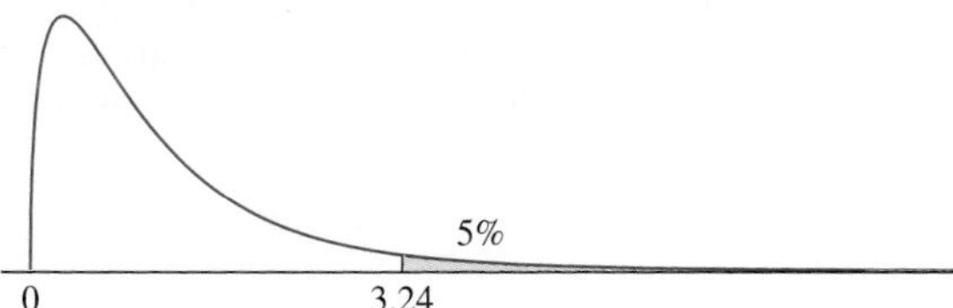

FIGURE 6.21 Probability density function of the $F_{3,16}$ distribution. The upper 5% point is 3.24. [See the F table (Table A.7) in Appendix A.]

The *F* Statistic for Testing Equality of Variance

The null distribution of the test statistic $F = s_1^2/s_2^2$ is $F_{m-1,\,n-1}$. The number of degrees of freedom for the numerator is one less than the sample size used to compute s_1^2, and the number of degrees of freedom for the denominator is one less than the sample size used to compute s_2^2. We illustrate the F test with an example.

Example 6.23

In a series of experiments to determine the absorption rate of certain pesticides into skin, measured amounts of two pesticides were applied to several skin specimens. After a time, the amounts absorbed (in μg) were measured. For pesticide A, the variance of the amounts absorbed in 6 specimens was 2.3, while for pesticide B, the variance of the amounts absorbed in 10 specimens was 0.6. Assume that for each pesticide, the amounts absorbed are a simple random sample from a normal population. Can we conclude that the variance in the amount absorbed is greater for pesticide A than for pesticide B?

Solution

Let σ_1^2 be the population variance for pesticide A, and let σ_2^2 be the population variance for pesticide B. The null hypothesis is

$$H_0: \frac{\sigma_1^2}{\sigma_2^2} \le 1$$

The sample variances are $s_1^2 = 2.3$ and $s_2^2 = 0.6$. The value of the test statistic is

$$F = \frac{2.3}{0.6} = 3.83$$

The null distribution of the test statistic is $F_{5,9}$. If H_0 is true, then s_1^2 will on the average be smaller than s_2^2. It follows that the larger the value of F, the stronger the evidence against H_0. Consulting the F table with five and nine degrees of freedom, we find that the upper 5% point is 3.48, while the upper 1% point is 6.06. We conclude that $0.01 < P < 0.05$. There is reasonably strong evidence against the null hypothesis. See Figure 6.22.

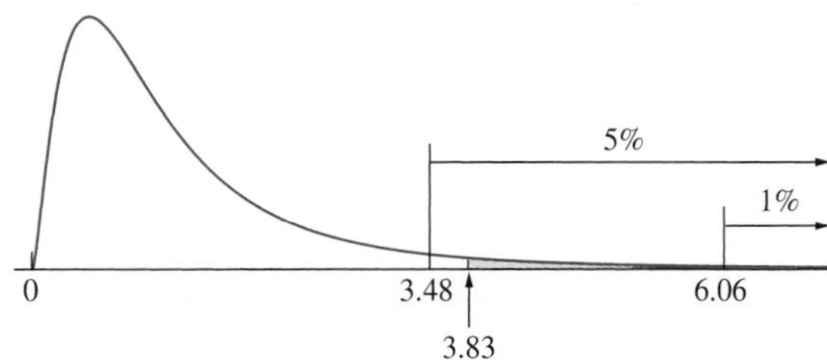

FIGURE 6.22 The observed value of the test statistic is 3.83. The upper 5% point of the $F_{5,9}$ distribution is 3.48; the upper 1% point is 6.06. Therefore the P-value is between 0.01 and 0.05.

We now describe the modifications to the procedure shown in Example 6.23 that are necessary to test the other null hypotheses. To test

$$H_0: \frac{\sigma_1^2}{\sigma_2^2} \geq 1$$

one could in principle use the test statistic s_1^2/s_2^2, with *small* values of the statistic providing evidence against H_0. However, since the F table contains only large values (i.e., greater than 1) for the F statistic, it is easier to use the statistic s_2^2/s_1^2. Under H_0, the distribution of s_2^2/s_1^2 is $F_{n-1,\,m-1}$.

Finally, we describe the method for testing the two-tailed hypothesis

$$H_0: \frac{\sigma_1^2}{\sigma_2^2} = 1$$

For this hypothesis, both large and small values of the statistic s_1^2/s_2^2 provide evidence against H_0. The procedure is to use either s_1^2/s_2^2 or s_2^2/s_1^2, whichever is greater than 1. The P-value for the two-tailed test is twice the P-value for the one-tailed test. In other words, the P-value of the two-tailed test is twice the upper tail area of the F distribution. We illustrate with an example.

In Example 6.23, $s_1^2 = 2.3$ with a sample size of 6, and $s_2^2 = 0.6$ with a sample size of 10. Test the null hypothesis

$$H_0: \sigma_1^2 = \sigma_2^2$$

Solution

The null hypothesis $\sigma_1^2 = \sigma_2^2$ is equivalent to $\sigma_1^2/\sigma_2^2 = 1$. Since $s_1^2 > s_2^2$, we use the test statistic s_1^2/s_2^2. In Example 6.23, we found that for the one-tailed test, $0.01 < P < 0.05$. Therefore for the two-tailed test, $0.02 < P < 0.10$.

The following computer output (from MINITAB) presents the solution to Example 6.24.

```
Test for Equal Variances

F-Test (normal distribution)
Test statistic = 3.83, p-value = 0.078
```

The *F* Test Is Sensitive to Departures from Normality

The *F* test, like the *t* test, requires that the samples come from normal populations. Unlike the *t* test, the *F* test for comparing variances is fairly sensitive to this assumption. If the shapes of the populations differ much from the normal curve, the *F* test may give misleading results. For this reason, the *F* test for comparing variances must be used with caution.

In Chapters 8 and 9, we will use the *F* distribution to perform certain hypothesis tests in the context of linear regression and analysis of variance. In these settings, the *F* test is less sensitive to violations of the normality assumption.

The *F* Test Cannot Prove That Two Variances Are Equal

In Section 6.7, we presented two versions of the *t* test for the difference between two means. One version is generally applicable, while the second version, which uses the pooled variance, is appropriate only when the population variances are equal. When deciding whether it is appropriate to assume population variances to be equal, it is tempting to perform an *F* test and assume the variances to be equal if the null hypothesis of equality is not rejected. Unfortunately this procedure is unreliable, for the basic reason that failure to reject the null hypothesis does not justify the assumption that the null hypothesis is true. In general, an assumption that population variances are equal cannot be justified by a hypothesis test.

Exercises for Section 6.11

1. Find the upper 5% point of $F_{7,20}$.

2. Find the upper 1% point of $F_{2,5}$.

3. An F test with five degrees of freedom in the numerator and seven degrees of freedom in the denominator produced a test statistic whose value was 7.46.

   a. What is the P-value if the test is one-tailed?

   b. What is the P-value if the test is two-tailed?

4. A broth used to manufacture a pharmaceutical product has its sugar content, in mg/mL, measured several times on each of three successive days.

Day 1:	5.0	4.8	5.1	5.1	4.8	5.1	4.8
	4.8	5.0	5.2	4.9	4.9	5.0	
Day 2:	5.8	4.7	4.7	4.9	5.1	4.9	5.4
	5.3	5.3	4.8	5.7	5.1	5.7	
Day 3:	6.3	4.7	5.1	5.9	5.1	5.9	4.7
	6.0	5.3	4.9	5.7	5.3	5.6	

   a. Can you conclude that the variability of the process is greater on the second day than on the first day?

   b. Can you conclude that the variability of the process is greater on the third day than on the second day?

5. Refer to Exercise 9 in Section 5.6. Can you conclude that the variance of the breaking strengths differs between the two composites?

6. Refer to Exercise 11 in Section 5.6. Can you conclude that the time to freeze-up is more variable in the seventh month than in the first month after installation?

6.12 Fixed-Level Testing

Critical Points and Rejection Regions

A hypothesis test measures the plausibility of the null hypothesis by producing a P-value. The smaller the P-value, the less plausible the null. We have pointed out that there is no scientifically valid dividing line between plausibility and implausibility, so it is impossible to specify a "correct" P-value below which we should reject H_0. When possible, it is best simply to report the P-value, and not to make a firm decision whether or not to reject. Sometimes, however, a decision has to be made. For example, if items are sampled from an assembly line to test whether the mean diameter is within tolerance, a decision must be made whether to recalibrate the process. If a sample of parts is drawn from a shipment and checked for defects, a decision must be made whether to accept or to return the shipment. If a decision is going to be made on the basis of a hypothesis test, there is no choice but to pick a cutoff point for the P-value. When this is done, the test is referred to as a **fixed-level** test.

Fixed-level testing is just like the hypothesis testing we have been discussing so far, except that a firm rule is set ahead of time for rejecting the null hypothesis. A value α, where $0 < \alpha < 1$, is chosen. Then the P-value is computed. If $P \leq \alpha$, the null hypothesis is rejected and the alternate hypothesis is taken as truth. If $P > \alpha$, then the null hypothesis is considered to be plausible. The value of α is called the **significance**

level, or, more simply, the **level**, of the test. Recall from Section 6.2 that if a test results in a P-value less than or equal to α, we say that the null hypothesis is rejected at level α (or $100\alpha\%$), or that the result is statistically significant at level α (or $100\alpha\%$). As we have mentioned, a common choice for α is 0.05.

Summary

To conduct a fixed-level test:

- Choose a number α, where $0 < \alpha < 1$. This is called the significance level, or the level, of the test.
- Compute the P-value in the usual way.
- If $P \leq \alpha$, reject H_0. If $P > \alpha$, do not reject H_0.

Refer to Example 6.1 in Section 6.1. The mean wear in a sample of 45 steel balls was $\overline{X} = 673.2\ \mu$m, and the standard deviation was $s = 14.9\ \mu$m. Let μ denote the population mean wear. A test of $H_0 : \mu \geq 675$ versus $H_1 : \mu < 675$ yielded a P-value of 0.209. Can we reject H_0 at the 25% level? Can we reject H_0 at the 5% level?

Solution

The P-value of 0.209 is less than 0.25, so if we had chosen a significance level of $\alpha = 0.25$, we would reject H_0. Thus we reject H_0 at the 25% level. Since $0.209 > 0.05$, we do not reject H_0 at the 5% level.

In a fixed-level test, a **critical point** is a value of the test statistic that produces a P-value exactly equal to α. A critical point is a dividing line for the test statistic just as the significance level is a dividing line for the P-value. If the test statistic is on one side of the critical point, the P-value will be less than α, and H_0 will be rejected. If the test statistic is on the other side of the critical point, the P-value will be greater than α, and H_0 will not be rejected. The region on the side of the critical point that leads to rejection is called the **rejection region**. The critical point itself is also in the rejection region.

A new concrete mix is being evaluated. The plan is to sample 100 concrete blocks made with the new mix, compute the sample mean compressive strength $\overline{X}$, and then test $H_0 : \mu \leq 1350$ versus $H_1 : \mu > 1350$, where the units are MPa. It is assumed from previous tests of this sort that the population standard deviation σ will be close to 70 MPa. Find the critical point and the rejection region if the test will be conducted at a significance level of 5%.

Solution

We will reject H_0 if the P-value is less than or equal to 0.05. The P-value for this test will be the area to the right of the value of $\overline{X}$. Therefore the P-value will be less than 0.05, and H_0 will be rejected, if the value of $\overline{X}$ is in the upper 5% of the null distribution (see Figure 6.23). The rejection region therefore consists of the upper 5% of the null distribution. The critical point is the boundary of the upper 5%. The null distribution is normal, and from the z table we find that the z-score of the point that cuts off the upper 5% of the normal curve is $z_{.05} = 1.645$. Therefore we can express the critical point as $z = 1.645$, and the rejection region as $z \geq 1.645$. It is often more convenient to express the critical point and rejection region in terms of $\overline{X}$, by converting the z-score to the original units. The null distribution has mean $\mu = 1350$ and standard deviation $\sigma_{\overline{X}} = \sigma/\sqrt{n} \approx 70/\sqrt{100} = 7$. Therefore the critical point can be expressed as $\overline{X} = 1350 + (1.645)(7) = 1361.5$. The rejection region consists of all values of $\overline{X}$ greater than or equal to 1361.5.

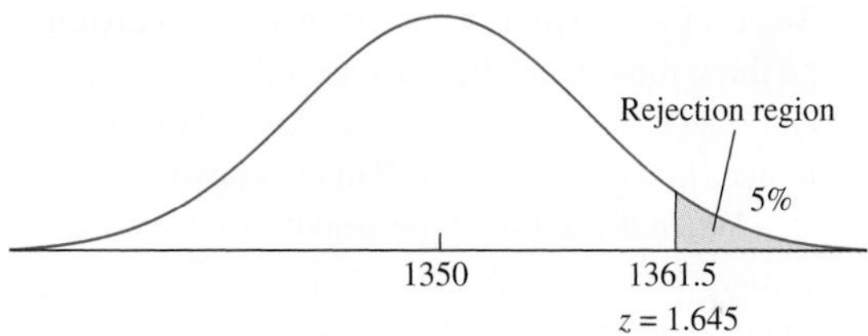

FIGURE 6.23 The rejection region for this one-tailed test consists of the upper 5% of the null distribution. The critical point is 1361.5, on the boundary of the rejection region.

Example 6.27

In a hypothesis test to determine whether a scale is in calibration, the null hypothesis is $H_0: \mu = 1000$ and the null distribution of $\overline{X}$ is $N(1000, 0.26^2)$. (This situation was presented in Example 6.2 in Section 6.1.) Find the rejection region if the test will be conducted at a significance level of 5%.

Solution

Since this is a two-tailed test, the rejection region is contained in both tails of the null distribution. Specifically, H_0 will be rejected if $\overline{X}$ is in either the upper or the lower 2.5%

of the null distribution (see Figure 6.24). The z-scores that cut off the upper and lower 2.5% of the distribution are $z = \pm 1.96$. Therefore the rejection region consists of all values of $\overline{X}$ greater than or equal to $1000 + (1.96)(0.26) = 1000.51$, along with all the values less than or equal to $1000 - (1.96)(0.26) = 999.49$. Note that there are two critical points, 999.49 and 1000.51.

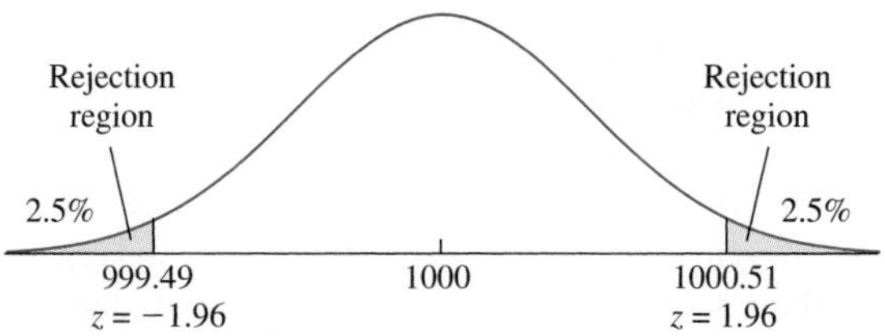

FIGURE 6.24 The rejection region for this two-tailed test consists of both the lower and the upper 2.5% of the null distribution. There are two critical points, 999.49 and 1000.51.

Type I and Type II Errors

Since a fixed-level test results in a firm decision, there is a chance that the decision could be the wrong one. There are exactly two ways in which the decision can be wrong. One can reject H_0 when it is in fact true. This is known as a type I error. Or, one can fail to reject H_0 when it is false. This is known as a type II error.

When designing experiments whose data will be analyzed with a fixed-level test, it is important to try to make the probabilities of type I and type II errors reasonably small. There is no use in conducting an experiment that has a large probability of leading to an incorrect decision. It turns out that it is easy to control the probability of a type I error, as shown by the following result.

> If α is the significance level that has been chosen for the test, then the probability of a type I error is never greater than α.

We illustrate this fact with the following example. Let $X_1, \ldots, X_n$ be a large random sample from a population with mean μ and variance σ^2. Then $\overline{X}$ is normally distributed with mean μ and variance σ^2/n. Assume that we are to test $H_0: \mu \leq 0$ versus $H_1: \mu > 0$ at the fixed level $\alpha = 0.05$. That is, we will reject H_0 if $P \leq 0.05$. The null distribution, shown in Figure 6.25, is normal with mean 0 and variance $\sigma_{\overline{X}}^2 = \sigma^2/n$. Assume the null hypothesis is true. We will compute the probability of a type I error and show that it is no greater than 0.05.

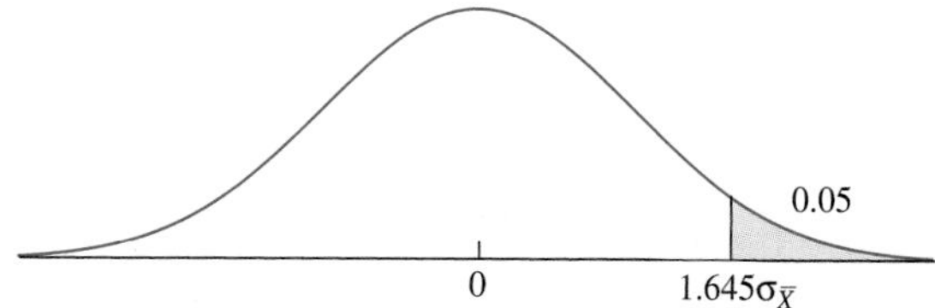

FIGURE 6.25 The null distribution with the rejection region for $H_0: \mu \leq 0$.

A type I error will occur if we reject H_0, which will occur if $P \leq 0.05$, which in turn will occur if $\overline{X} \geq 1.645\sigma_{\overline{X}}$. Therefore the rejection region is the region $\overline{X} \geq 1.645\sigma_{\overline{X}}$. Now since H_0 is true, $\mu \leq 0$. First, we'll consider the case where $\mu = 0$. Then the distribution of $\overline{X}$ is given by Figure 6.25. In this case, $P(\overline{X} \geq 1.645\sigma_{\overline{X}}) = 0.05$, so the probability of rejecting H_0 and making a type I error is equal to 0.05. Next, consider the case where $\mu < 0$. Then the distribution of $\overline{X}$ is obtained by shifting the curve in Figure 6.25 to the left, so $P(\overline{X} \geq 1.645\sigma_{\overline{X}}) < 0.05$, and the probability of a type I error is less than 0.05. We could repeat this illustration using any number α in place of 0.05. We conclude that if H_0 is true, the probability of a type I error is never greater than α. Furthermore, note that if μ is on the boundary of H_0 ($\mu = 0$ in this case), then the probability of a type I error is equal to α.

We can therefore make the probability of a type I error as small as we please, because it is never greater than the significance level α that we choose. Unfortunately, as we will see in Section 6.13, the smaller we make the probability of a type I error, the larger the probability of a type II error becomes. The usual strategy is to begin by choosing a value for α so that the probability of a type I error will be reasonably small. As we have mentioned, a conventional choice for α is 0.05. Then one computes the probability of a type II error and hopes that it is not too large. If it is large, it can be reduced only by redesigning the experiment—for example by increasing the sample size. Calculating and controlling the size of the type II error is somewhat more difficult than calculating and controlling the size of the type I error. We will discuss this in Section 6.13.

Summary

When conducting a fixed-level test at significance level α, there are two types of errors that can be made. These are

- Type I error: Reject H_0 when it is true.
- Type II error: Fail to reject H_0 when it is false.

The probability of a type I error is never greater than α.

Exercises for Section 6.12

1. A hypothesis test is performed, and the P-value is 0.03. True or false:

a. H_0 is rejected at the 5% level.

b. H_0 is rejected at the 2% level.

c. H_0 is not rejected at the 10% level.

2. A wastewater treatment program is designed to produce treated water with a pH of 7. Let μ represent the mean pH of water treated by this process. The pH of 60 water specimens will be measured, and a test of the hypotheses $H_0: \mu = 7$ versus $H_1: \mu \neq 7$ will be made. Assume it is known from previous experiments that the standard deviation of the pH of water specimens is approximately 0.5.

a. If the test is made at the 5% level, what is the rejection region?

b. If the sample mean pH is 6.87, will H_0 be rejected at the 10% level?

c. If the sample mean pH is 6.87, will H_0 be rejected at the 1% level?

d. If the value 7.2 is a critical point, what is the level of the test?

3. A new braking system is being evaluated for a certain type of car. The braking system will be installed if it can be conclusively demonstrated that the stopping distance under certain controlled conditions at a speed of 30 mi/h is less than 90 ft. It is known that under these conditions the standard deviation of stopping distance is approximately 5 ft. A sample of 150 stops will be made from a speed of 30 mi/h. Let μ represent the mean stopping distance for the new braking system.

a. State the appropriate null and alternate hypotheses.

b. Find the rejection region if the test is to be conducted at the 5% level.

c. Someone suggests rejecting H_0 if $\overline{X} \geq 89.4$ ft. Is this an appropriate rejection region, or is something wrong? If this is an appropriate rejection region, find the level of the test. Otherwise explain what is wrong.

d. Someone suggests rejecting H_0 if $\overline{X} \leq 89.4$ ft. Is this an appropriate rejection region, or is something wrong? If this is an appropriate rejection region, find the level of the test. Otherwise explain what is wrong.

e. Someone suggests rejecting H_0 if $\overline{X} \leq 89.4$ ft or if $\overline{X} \geq 90.6$ ft. Is this an appropriate rejection region, or is something wrong? If this is an appropriate rejection region, find the level of the test. Otherwise explain what is wrong.

4. A test is made of the hypotheses $H_0: \mu \leq 10$ versus $H_1: \mu > 10$. For each of the following situations, determine whether the decision was correct, a type I error occurred, or a type II error occurred.

a. $\mu = 8$, H_0 is rejected.

b. $\mu = 10$, H_0 is not rejected.

c. $\mu = 14$, H_0 is not rejected.

d. $\mu = 12$, H_0 is rejected.

5. A vendor claims that no more than 10% of the parts she supplies are defective. Let p denote the actual proportion of parts that are defective. A test is made of the hypotheses $H_0: p \leq 0.10$ versus $H_1: p > 0.10$. For each of the following situations, determine whether the decision was correct, a type I error occurred, or a type II error occurred.

a. The claim is true, and H_0 is rejected.

b. The claim is false, and H_0 is rejected.

c. The claim is true, and H_0 is not rejected.

d. The claim is false, and H_0 is not rejected.

6. A hypothesis test is to be performed, and it is decided to reject the null hypothesis if $P < 0.10$. If H_0 is in fact true, what is the maximum probability that it will be rejected?

7. A new process is being considered for the liquefaction of coal. The old process yielded a mean of 15 kg of distillate fuel per kilogram of hydrogen consumed in the process. Let μ represent the mean of the new process. A test of $H_0: \mu \leq 15$ versus $H_1: \mu > 15$ will be performed. The new process will be put into production if H_0 is rejected. Putting the new process into production is very expensive. Therefore it would be a costly error to put the new process into production if in fact it is no better than the old process. Which procedure provides a smaller probability for this error, to test at the 5% level or to test at the 1% level?

6.13 Power

A hypothesis test results in a type II error if H_0 is not rejected when it is false. The **power** of a test is the probability of *rejecting* H_0 when it is false. Therefore

$$\text{Power} = 1 - \text{P(type II error)}$$

To be useful, a test must have reasonably small probabilities of both type I and type II errors. The type I error is kept small by choosing a small value of α as the significance level. Then the power of the test is calculated. If the power is large, then the probability of a type II error is small as well, and the test is a useful one. Note that power calculations are generally done before data are collected. The purpose of a power calculation is to determine whether or not a hypothesis test, when performed, is likely to reject H_0 in the event that H_0 is false.

As an example of a power calculation, assume that a new chemical process has been developed that may increase the yield over that of the current process. The current process is known to have a mean yield of 80 and a standard deviation of 5, where the units are the percentage of a theoretical maximum. If the mean yield of the new process is shown to be greater than 80, the new process will be put into production. Let μ denote the mean yield of the new process. It is proposed to run the new process 50 times and then to test the hypothesis

$$H_0 : \mu \leq 80 \quad \text{versus} \quad H_1 : \mu > 80$$

at a significance level of 5%. If H_0 is rejected, it will be concluded that $\mu > 80$, and the new process will be put into production. Let us assume that if the new process had a mean yield of 81, then it would be a substantial benefit to put this process into production. If it is in fact the case that $\mu = 81$, what is the power of the test, that is, the probability that H_0 will be rejected?

Before presenting the solution, we note that in order to compute the power, it is necessary to specify a particular value of μ, in this case $\mu = 81$, for the alternate hypothesis. The reason for this is that the power is different for different values of μ. We will see that if μ is close to H_0, the power will be small, while if μ is far from H_0, the power will be large.

Computing the power involves two steps:

1. Compute the rejection region.
2. Compute the probability that the test statistic falls in the rejection region if the alternate hypothesis is true. This is the power.

We'll begin to find the power of the test by computing the rejection region, using the method illustrated in Example 6.26 in Section 6.12. We must first find the null distribution. We know that the statistic $\overline{X}$ has a normal distribution with mean μ and standard deviation $\sigma_{\overline{X}} = \sigma/\sqrt{n}$, where $n = 50$ is the sample size. Under H_0, we take $\mu = 80$. We must now find an approximation for σ. In practice this can be a difficult problem, because the sample has not yet been drawn, so there is no sample standard deviation s. There are several ways in which it may be possible to approximate σ. Sometimes a small preliminary sample has been drawn, for example in a feasibility study, and the standard deviation of this sample may be a satisfactory approximation for σ. In other cases, a sample from a similar population may exist, whose standard deviation may be used. In this example, there is a long history of a currently used process, whose standard deviation is 5. Let's say that it is reasonable to assume that the standard deviation of the new process will be similar to that of the current process. We will therefore assume that the population standard deviation for the new process is $\sigma = 5$ and that $\sigma_{\overline{X}} = 5/\sqrt{50} = 0.707$.

Figure 6.26 presents the null distribution of $\overline{X}$. Since H_0 specifies that $\mu \leq 80$, large values of $\overline{X}$ disagree with H_0, so the P-value will be the area to the right of the observed value of $\overline{X}$. The P-value will be less than or equal to 0.05 if $\overline{X}$ falls into the upper 5% of the null distribution. This upper 5% is the rejection region. The critical point has a z-score of 1.645, so its value is $80 + (1.645)(0.707) = 81.16$. We will reject H_0 if $\overline{X} \geq 81.16$. This is the rejection region.

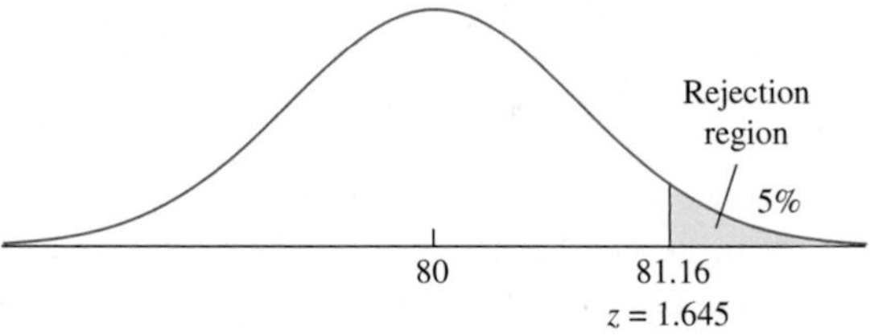

FIGURE 6.26 The hypothesis test will be conducted at a significance level of 5%. The rejection region for this test is the region where the P-value will be less than 0.05.

We are now ready to compute the power, which is the probability that $\overline{X}$ will fall into the rejection region if the alternate hypothesis $\mu = 81$ is true. Under this alternate hypothesis, the distribution of $\overline{X}$ is normal with mean 81 and standard deviation 0.707. Figure 6.27 presents the alternate distribution and the null distribution on the same

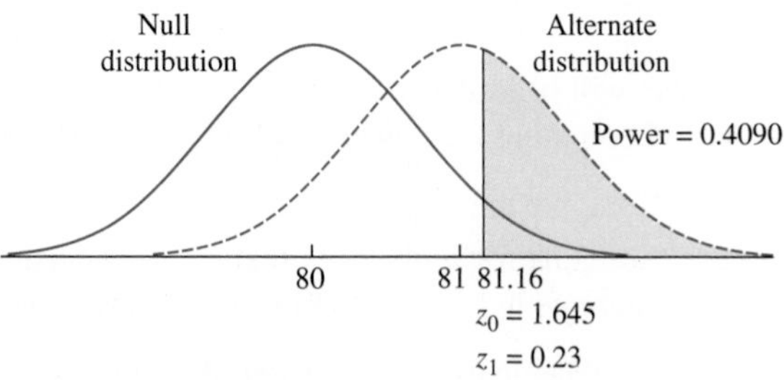

FIGURE 6.27 The rejection region, consisting of the upper 5% of the null distribution, is shaded. The z-score of the critical point is $z_0 = 1.645$ under the null distribution and $z_1 = 0.23$ under the alternate. The power is the area of the rejection region under the alternate distribution, which is 0.4090.

plot. Note that the alternate distribution is obtained by shifting the null distribution so that the mean becomes the alternate mean of 81 rather than the null mean of 80. Because the alternate distribution is shifted over, the probability that the test statistic falls into the rejection region is greater than it is under H_0. To be specific, the z-score under H_1 for the critical point 81.16 is $z = (81.16 - 81)/0.707 = 0.23$. The area to the right of $z = 0.23$ is 0.4090. This is the power of the test.

A power of 0.4090 is very low. It means that if the mean yield of new process is actually equal to 81, there is only a 41% chance that the proposed experiment will detect the improvement over the old process and allow the new process to be put into production. It would be unwise to invest time and money to run this experiment, since it has a large chance to fail.

It is natural to wonder how large the power must be for a test to be worthwhile to perform. As with P-values, there is no scientifically valid dividing line between sufficient and insufficient power. In general, tests with power greater than 0.80 or perhaps 0.90 are considered acceptable, but there are no well-established rules of thumb.

We have mentioned that the power depends on the value of μ chosen to represent the alternate hypothesis and is larger when the value is far from the null mean. Example 6.28 illustrates this.

Example 6.28

Find the power of the 5% level test of $H_0: \mu \leq 80$ versus $H_1: \mu > 80$ for the mean yield of the new process under the alternative $\mu = 82$, assuming $n = 50$ and $\sigma = 5$.

Solution

We have already completed the first step of the solution, which is to compute the rejection region. We will reject H_0 if $\overline{X} \geq 81.16$. Figure 6.28 presents the alternate and null distributions on the same plot. The z-score for the critical point of 81.16 under the alternate hypothesis is $z = (81.16 - 82)/0.707 = -1.19$. The area to the right of $z = -1.19$ is 0.8830. This is the power.

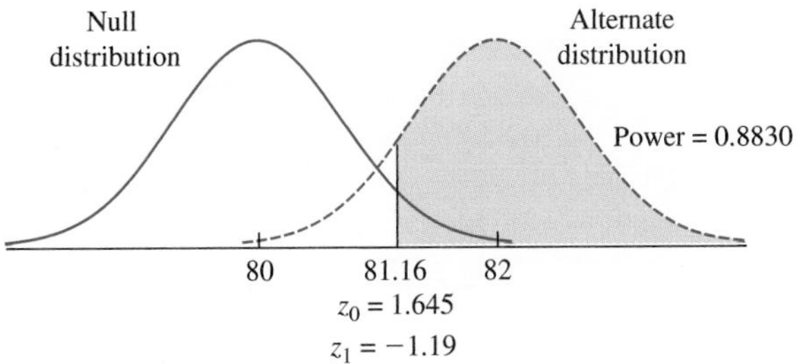

FIGURE 6.28 The rejection region, consisting of the upper 5% of the null distribution, is shaded. The z-score of the critical point is $z_0 = 1.645$ under the null distribution and $z_1 = -1.19$ under the alternate. The power is the area of the rejection region under the alternate distribution, which is 0.8830.

Since the alternate distribution is obtained by shifting the null distribution, the power depends on which alternate value is chosen for μ, and can range from barely greater than the significance level α all the way up to 1. If the alternate mean is chosen very close to the null mean, the alternate curve will be almost identical with the null, and the power will be very close to α. If the alternate mean is far from the null, almost all the area under the alternate curve will lie in the rejection region, and the power will be close to 1.

When power is not large enough, it can be increased by increasing the sample size. When planning an experiment, one can determine the sample size necessary to achieve a desired power. Example 6.29 illustrates this.

Example 6.29

In testing the hypothesis $H_0: \mu \leq 80$ versus $H_1: \mu > 80$ regarding the mean yield of the new process, how many times must the new process be run so that a test conducted at a significance level of 5% will have power 0.90 against the alternative $\mu = 81$, if it is assumed that $\sigma = 5$?

Solution

Let n represent the necessary sample size. We first use the null distribution to express the critical point for the test in terms of n. The null distribution of $\overline{X}$ is normal with mean 80 and standard deviation $5/\sqrt{n}$. Therefore the critical point is $80 + 1.645(5/\sqrt{n})$. Now, we use the alternate distribution to obtain a different expression for the critical point in terms of n. Refer to Figure 6.29. The power of the test is the area of the rejection region under the alternate curve. This area must be 0.90. Therefore the z-score for the critical point, under the alternate hypothesis, is $z = -1.28$. The critical point is therefore $81 - 1.28(5/\sqrt{n})$. We now have two different expressions for the critical point. Since there is only one critical point, these two expressions are equal. We therefore set them equal and solve for n.

$$80 + 1.645\left(\frac{5}{\sqrt{n}}\right) = 81 - 1.28\left(\frac{5}{\sqrt{n}}\right)$$

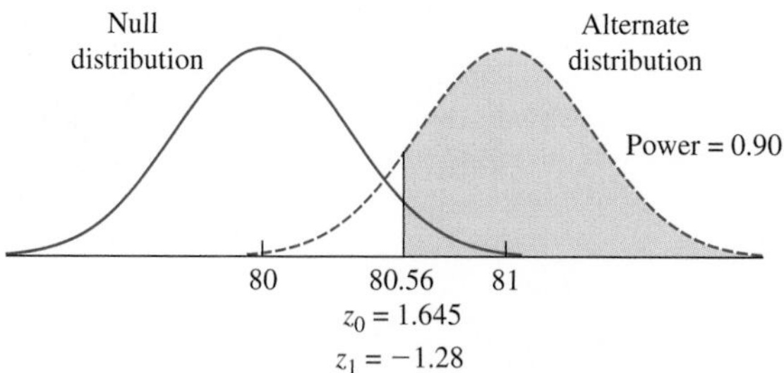

FIGURE 6.29 To achieve a power of 0.90 with a significance level of 0.05, the z-score for the critical point must be $z_0 = 1.645$ under the null distribution and $z_1 = -1.28$ under the alternate distribution.

Solving for n yields $n \approx 214$. The critical point can be computed by substituting this value for n into either side of the previous equation. The critical point is 80.56.

Using a Computer to Calculate Power

We have presented a method for calculating the power, and the sample size needed to attain a specified power, for a one-tailed large-sample test of a population mean. It is reasonably straightforward to extend this method to compute power and needed sample sizes for two-tailed tests and for tests for proportions. It is more difficult to compute power for a t test, F test, or chi-square test. Computer packages, however, can compute power and needed sample sizes for all these tests. We present some examples.

A pollster will conduct a survey of a random sample of voters in a community to estimate the proportion who support a measure on school bonds. Let p be the proportion of the population who support the measure. The pollster will test $H_0: p = 0.50$ versus $H_1: p \neq 0.50$ at the 5% level. If 200 voters are sampled, what is the power of the test if the true value of p is 0.55?

Solution

The following computer output (from MINITAB) presents the solution:

```
Power and Sample Size

Test for One Proportion

Testing proportion = 0.5 (versus not = 0.5)
Alpha = 0.05

Alternative    Sample
 Proportion      Size        Power
       0.55       200     0.292022
```

The first two lines of output state that this is a power calculation for a test for a single population proportion p. The next two lines state the null and alternate hypotheses, and the significance level of the test. Note that we have specified a two-tailed test with significance level $\alpha = 0.05$. Next is the alternative proportion, which is the value of p (0.55) that we are assuming to be true when the power is calculated. The sample size has been specified to be 200, and the power is computed to be 0.292.

Refer to Example 6.30. How many voters must be sampled so that the power will be 0.8 when the true value of $p = 0.55$?

Solution

The following computer output (from MINITAB) presents the solution:

```
Power and Sample Size

Test for One Proportion

Testing proportion = 0.5 (versus not = 0.5)
Alpha = 0.05

Alternative    Sample    Target
 Proportion      Size     Power    Actual Power
       0.55       783       0.8        0.800239
```

The needed sample size is 783. Note that the actual power is slightly higher than 0.80. Because the sample size is discrete, it is not possible to find a sample size that provides exactly the power requested (the target power). So MINITAB calculates the smallest sample size for which the power is greater than that requested.

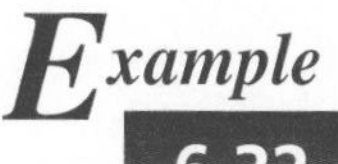

Example 6.32

Shipments of coffee beans are checked for moisture content. High moisture content indicates possible water contamination, leading to rejection of the shipment. Let μ represent the mean moisture content (in percent by weight) in a shipment. Five moisture measurements will be made on beans chosen at random from the shipment. A test of the hypothesis $H_0: \mu \leq 10$ versus $H_1: \mu > 10$ will be made at the 5% level, using the Student's t test. What is the power of the test if the true moisture content is 12% and the standard deviation is $\sigma = 1.5\%$?

Solution

The following computer output (from MINITAB) presents the solution:

```
Power and Sample Size

1-Sample t Test

Testing mean = null (versus > null)
Calculating power for mean = null + difference
Alpha = 0.05 Assumed standard deviation = 1.5

              Sample
Difference      Size      Power
         2         5   0.786485
```

The power depends only on the difference between the true mean and the null mean, which is $12 - 10 = 2$, and not on the means themselves. The power is 0.786. Note that the output specifies that this is the power for a one-tailed test.

6.33

Refer to Example 6.32. Find the sample size needed so that the power will be at least 0.9.

Solution

The following computer output (from MINITAB) presents the solution:

```
Power and Sample Size

1-Sample t Test

Testing mean = null (versus > null)
Calculating power for mean = null + difference
Alpha = 0.05 Assumed standard deviation = 1.5

                Sample    Target
Difference        Size     Power    Actual Power
         2           7       0.9        0.926750
```

The smallest sample size for which the power is 0.9 or more is 7. The actual power is 0.927.

To summarize, power calculations are important to ensure that experiments have the potential to provide useful conclusions. Many agencies that provide funding for scientific research require that power calculations be provided with every proposal in which hypothesis tests are to be performed.

Exercises for Section 6.13

1. A test has power 0.90 when $\mu = 15$. True or false:

a. The probability of rejecting H_0 when $\mu = 15$ is 0.90.

b. The probability of making a correct decision when $\mu = 15$ is 0.90.

c. The probability of making a correct decision when $\mu = 15$ is 0.10.

d. The probability that H_0 is true when $\mu = 15$ is 0.10.

2. A test has power 0.80 when $\mu = 3.5$. True or false:

a. The probability of rejecting H_0 when $\mu = 3.5$ is 0.80.

b. The probability of making a type I error when $\mu = 3.5$ is 0.80.

c. The probability of making a type I error when $\mu = 3.5$ is 0.20.

d. The probability of making a type II error when $\mu = 3.5$ is 0.80.

e. The probability of making a type II error when $\mu = 3.5$ is 0.20.

f. The probability that H_0 is false when $\mu = 3.5$ is 0.80.

3. If the sample size remains the same, and the level α increases, then the power will ________. *Options:* increase, decrease.

4. If the level α remains the same, and the sample size increases, then the power will ________. *Options:* increase, decrease.

5. A tire company claims that the lifetimes of its tires average 50,000 miles. The standard deviation of tire lifetimes is known to be 5000 miles. You sample 100 tires and will test the hypothesis that the mean tire lifetime is at least 50,000 miles against the alternative that it is less. Assume, in fact, that the true mean lifetime is 49,500 miles.

a. State the null and alternate hypotheses. Which hypothesis is true?

b. It is decided to reject H_0 if the sample mean is less than 49,400. Find the level and power of this test.

c. If the test is made at the 5% level, what is the power?

d. At what level should the test be conducted so that the power is 0.80?

e. You are given the opportunity to sample more tires. How many tires should be sampled in total so that the power is 0.80 if the test is made at the 5% level?

6. The mean drying time of a certain paint in a certain application is 12 minutes. A new additive will be tested to see if it reduces the drying time. One hundred specimens will be painted, and the sample mean drying time $\overline{X}$ will be computed. Assume the population standard deviation of drying times is $\sigma = 2$ minutes. Let μ be the mean drying time for the new paint. The null hypothesis $H_0: \mu \geq 12$ will be tested against the alternate $H_1: \mu < 12$. Assume that unknown to the investigators, the true mean drying time of the new paint is 11.5 minutes.

a. It is decided to reject H_0 if $\overline{X} \leq 11.7$. Find the level and power of this test.

b. For what values of $\overline{X}$ should H_0 be rejected so that the power of the test will be 0.90? What will the level then be?

c. For what values of $\overline{X}$ should H_0 be rejected so that the level of the test will be 5%? What will the power then be?

d. How large a sample is needed so that a 5% level test has power 0.90?

7. A power calculation has shown that if $\mu = 8$, the power of a test of $H_0: \mu \geq 10$ versus $H_1: \mu < 10$ is 0.90. If instead $\mu = 7$, which one of the following statements is true?

i. The power of the test will be less than 0.90.

ii. The power of the test will be greater than 0.90.

iii. We cannot determine the power of the test without knowing the population standard deviation σ.

8. A new process for producing silicon wafers for integrated circuits is supposed to reduce the proportion of defectives to 10%. A sample of 250 wafers will be tested. Let X represent the number of defectives in the sample. Let p represent the population proportion of defectives produced by the new process. A test will be made of $H_0: p \geq 0.10$ versus $H_1: p < 0.10$. Assume the true value of p is actually 0.06.

a. It is decided to reject H_0 if $X \leq 18$. Find the level of this test.

b. It is decided to reject H_0 if $X \leq 18$. Find the power of this test.

c. Should you use the same standard deviation for X to compute both the power and the level? Explain.

d. How many wafers should be sampled so that the power is 0.90 if the test is made at the 5% level?

9. The following MINITAB output presents the results of a power calculation for a test concerning a population proportion p.

```
Power and Sample Size

Test for One Proportion

Testing proportion = 0.5
(versus not = 0.5)
Alpha = 0.05

Alternative  Sample
 Proportion    Size      Power
        0.4     150   0.691332
```

a. Is the power calculated for a one-tailed or two-tailed test?

b. What is the null hypothesis for which the power is calculated?

c. For what alternative value of p is the power calculated?

d. If the sample size were 100, would the power be less than 0.7, greater than 0.7, or is it impossible to tell from the output? Explain.

e. If the sample size were 200, would the power be less than 0.6, greater than 0.6, or is it impossible to tell from the output? Explain.

f. For a sample size of 150, is the power against the alternative $p = 0.3$ less than 0.65, greater than 0.65, or is it impossible to tell from the output? Explain.

g. For a sample size of 150, is the power against the alternative $p = 0.45$ less than 0.65, greater than 0.65, or is it impossible to tell from the output? Explain.

10. The following MINITAB output presents the results of a power calculation for a test concerning a population mean μ.

```
Power and Sample Size

1-Sample t Test

Testing mean = null (versus > null)
Calculating power for mean = null + difference
Alpha = 0.05 Assumed standard deviation = 1.5

                  Sample    Target
Difference          Size     Power    Actual Power
         1            18      0.85        0.857299
```

a. Is the power calculated for a one-tailed or two-tailed test?
b. Assume that the value of μ used for the null hypothesis is $\mu = 3$. For what alternate value of μ is the power calculated?
c. If the sample size were 25, would the power be less than 0.85, greater than 0.85, or is it impossible to tell from the output? Explain.
d. If the difference were 0.5, would the power be less than 0.90, greater than 0.90, or is it impossible to tell from the output? Explain.
e. If the sample size were 17, would the power be less than 0.85, greater than 0.85, or is it impossible to tell from the output? Explain.

11. The following MINITAB output presents the results of a power calculation for a test of the difference between two means $\mu_1 - \mu_2$.

```
Power and Sample Size

2-Sample t Test

Testing mean 1 = mean 2 (versus not =)
Calculating power for mean 1 = mean 2 + difference
Alpha = 0.05 Assumed standard deviation = 5

                  Sample    Target
Difference          Size     Power       Actual Power
         3            60       0.9           0.903115
The sample size is for each group.
```

a. Is the power calculated for a one-tailed or two-tailed test?
b. If the sample sizes were 50 in each group, would the power be less than 0.9, greater than 0.9, or is it impossible to tell from the output? Explain.
c. If the difference were 4, would the power be less than 0.9, greater than 0.9, or is it impossible to tell from the output? Explain.

6.14 Multiple Tests

Sometimes a situation occurs in which it is necessary to perform many hypothesis tests. The basic rule governing this situation is that as more tests are performed, the confidence that we can place in our results decreases. In this section, we present an example to illustrate this point.

It is thought that applying a hard coating containing very small particles of tungsten carbide may reduce the wear on cam gears in a certain industrial application. There are many possible formulations for the coating, varying in the size and concentration of the tungsten carbide particles. Twenty different formulations were manufactured. Each one was tested by applying it to a large number of gears, and then measuring the wear on the gears after a certain period of time had elapsed. It is known on the basis of long experience that the mean wear for uncoated gears over this period of time is 100 μm. For each formulation, a test was made of the null hypothesis $H_0: \mu \geq 100$ μm. H_0 says that the formulation does not reduce wear. For 19 of the 20 formulations, the P-value was greater than 0.05, so H_0 was not rejected. For one formulation, H_0 was rejected. It might seem natural to conclude that this formulation really does reduce wear. Examples 6.34 through 6.37 will show that this conclusion is premature.

Example 6.34

If only one formulation were tested, and it in fact had no effect on wear, what is the probability that H_0 would be rejected, leading to a wrong conclusion?

Solution

If the formulation has no effect on wear, then $\mu = 100$ μm, so H_0 is true. Rejecting H_0 is then a type I error. The question is therefore asking for the probability of a type I error. In general, this probability is always less than or equal to the significance level of the test, which in this case is 5%. Since $\mu = 100$ is on the boundary of H_0, the probability of a type I error is equal to the significance level. The probability is 0.05 that H_0 will be rejected.

Example 6.35

Given that H_0 was rejected for one of the 20 formulations, is it plausible that this formulation actually has no effect on wear?

Solution

Yes. It is plausible that none of the formulations, including the one for which H_0 was rejected, have any effect on wear. There were 20 hypothesis tests made. For each test there was a 5% chance (i.e., 1 chance in 20) of a type I error. Therefore we expect on the average that out of every 20 true null hypotheses, one will be rejected. So rejecting H_0 in one out of the 20 tests is exactly what one would expect in the case that none of the formulations made any difference.

If in fact none of the 20 formulations have any effect on wear, what is the probability that H_0 will be rejected for one or more of them?

Solution

We first find the probability that the right conclusion (not rejecting H_0) is made for all the formulations. For each formulation, the probability that H_0 is not rejected is $1 - 0.05 = 0.95$, so the probability that H_0 is not rejected for any of the 20 formulations is $(0.95)^{20} = 0.36$. Therefore the probability is $1 - 0.36 = 0.64$ that we incorrectly reject H_0 for one or more of the formulations.

The experiment is repeated. This time, the operator forgets to apply the coatings, so each of the 20 wear measurements is actually made on uncoated gears. Is it likely that one or more of the formulations will appear to reduce wear, in that H_0 will be rejected?

Solution

Yes. Example 6.36 shows that the probability is 0.64 that one or more of the coatings will appear to reduce wear, even if they are not actually applied.

Examples 6.34 through 6.37 illustrate a phenomenon known as the **multiple testing problem**. Put simply, the multiple testing problem is this: When H_0 is rejected, we have strong evidence that it is false. But strong evidence is not certainty. Occasionally a true null hypothesis will be rejected. When many tests are performed, it is more likely that some true null hypotheses will be rejected. Thus when many tests are performed, it is difficult to tell which of the rejected null hypotheses are really false and which correspond to type I errors.

The Bonferroni Method

The Bonferroni method provides a way to adjust P-values upward when several hypothesis tests are performed. If a P-value remains small after the adjustment, the null hypothesis may be rejected. To make the Bonferroni adjustment, simply multiply the P-value by the number of tests performed. Here are two examples.

Four different coating formulations are tested to see if they reduce the wear on cam gears to a value below 100 μm. The null hypothesis $H_0\colon \mu \geq 100$ μm is tested for each formulation, and the results are

Formulation A:	$P = 0.37$
Formulation B:	$P = 0.41$
Formulation C:	$P = 0.005$
Formulation D:	$P = 0.21$

The operator suspects that formulation C may be effective, but he knows that the P-value of 0.005 is unreliable, because several tests have been performed. Use the Bonferroni adjustment to produce a reliable P-value.

Solution

Four tests were performed, so the Bonferroni adjustment yields $P = (4)(0.005) = 0.02$ for formulation C. So the evidence is reasonably strong that formulation C is in fact effective.

In Example 6.38, assume the P-value for formulation C had been 0.03 instead of 0.005. What conclusion would you reach then?

Solution

The Bonferroni adjustment would yield $P = (4)(0.03) = 0.12$. This is probably not strong enough evidence to conclude that formulation C is in fact effective. Since the original P-value was small, however, it is likely that one would not want to give up on formulation C quite yet.

The Bonferroni adjustment is conservative; in other words, the P-value it produces is never smaller than the true P-value. So when the Bonferroni-adjusted P-value is small, the null hypothesis can be rejected safely. Unfortunately, as Example 6.39 shows, there are many occasions in which the original P-value is small enough to arouse a strong suspicion that a null hypothesis may be false, but the Bonferroni adjustment does not allow the hypothesis to be rejected.

When the Bonferroni-adjusted P-value is too large to reject a null hypothesis, yet the original P-value is small enough to lead one to suspect that the hypothesis is in fact false, often the best thing to do is to retest the hypothesis that appears to be false, using data from a new experiment. If the P-value is again small, this time without multiple tests, this provides real evidence against the null hypothesis.

Real industrial processes are monitored frequently by sampling and testing process output to see whether it meets specifications. Every so often, the output appears to be outside the specifications. But in these cases, how do we know whether the process is really malfunctioning (out of control) or whether the result is a type I error? This is a version of the multiple testing problem that has received much attention. The subject of statistical quality control (see Chapter 10) is dedicated in large part to finding ways to overcome the multiple testing problem.

Exercises for Section 6.14

1. Six different settings are tried on a machine to see if any of them will reduce the proportion of defective parts. For each setting, an appropriate null hypothesis is tested to see if the proportion of defective parts has been reduced. The six P-values are 0.34, 0.27, 0.002, 0.45, 0.03, and 0.19.

   a. Find the Bonferroni-adjusted P-value for the setting whose P-value is 0.002. Can you conclude that this setting reduces the proportion of defective parts? Explain.

b. Find the Bonferroni-adjusted P-value for the setting whose P-value is 0.03. Can you conclude that this setting reduces the proportion of defective parts? Explain.

2. Five different variations of a bolt-making process are run to see if any of them can increase the mean breaking strength of the bolts over that of the current process. The P-values are 0.13, 0.34, 0.03, 0.28, and 0.38. Of the following choices, which is the best thing to do next?

i. Implement the process whose P-value was 0.03, since it performed the best.

ii. Since none of the processes had Bonferroni-adjusted P-values less than 0.05, we should stick with the current process.

iii. Rerun the process whose P-value was 0.03 to see if it remains small in the absence of multiple testing.

iv. Rerun all the five variations again, to see if any of them produce a small P-value the second time around.

3. Twenty formulations of a coating are being tested to see if any of them reduce gear wear. For the Bonferroni-adjusted P-value for a formulation to be 0.05, what must the original P-value be?

4. Five new paint additives have been tested to see if any of them can reduce the mean drying time from the current value of 12 minutes. Ten specimens have been painted with each of the new types of paint, and the drying times (in minutes) have been measured. The results are as follows:

	Additive				
	A	B	C	D	E
1	14.573	10.393	15.497	10.350	11.263
2	12.012	10.435	9.162	7.324	10.848
3	13.449	11.440	11.394	10.338	11.499
4	13.928	9.719	10.766	11.600	10.493
5	13.123	11.045	11.025	10.725	13.409
6	13.254	11.707	10.636	12.240	10.219
7	12.772	11.141	15.066	10.249	10.997
8	10.948	9.852	11.991	9.326	13.196
9	13.702	13.694	13.395	10.774	12.259
10	11.616	9.474	8.276	11.803	11.056

For each additive, perform a hypothesis test of the null hypothesis $H_0: \mu \geq 12$ against the alternate $H_1: \mu < 12$. You may assume that each population is approximately normal.

a. What are the P-values for the five tests?

b. On the basis of the results, which of the three following conclusions seems most appropriate? Explain your answer.

i. At least one of the new additives results in an improvement.

ii. None of the new additives result in an improvement.

iii. Some of the new additives may result in improvement, but the evidence is inconclusive.

5. Each day for 200 days, a quality engineer samples 144 fuses rated at 15 A and measures the amperage at which they burn out. He performs a hypothesis test of $H_0: \mu = 15$ versus $H_1: \mu \neq 15$, where μ is the mean burnout amperage of the fuses manufactured that day.

a. On 10 of the 200 days, H_0 is rejected at the 5% level. Does this provide conclusive evidence that the mean burnout amperage was different from 15 A on at least one of the 200 days? Explain.

b. Would the answer to part (a) be different if H_0 had been rejected on 20 of the 200 days? Explain.

6.15 Using Simulation to Perform Hypothesis Tests

If $X_1, \ldots, X_n$ are independent random variables, with known standard deviations $\sigma_1, \ldots, \sigma_n$, and $U = U(X_1, \ldots, X_n)$ is a function of $X_1, \ldots, X_n$, then the method of propagation of error (see Chapter 3) can be used to compute the standard deviation, or uncertainty, in U. If in addition the random variables $X_1, \ldots, X_n$ are approximately normally distributed, it will often, but not always, be the case that U is approximately normally distributed as well. In these cases we can perform hypothesis tests on the mean μ_U of U. To determine whether U is approximately normally distributed, simulation can be used. The method is similar to that described in Section 5.8.

We illustrate with an example. Let R represent the measurement of the radius of a cylinder, and let H represent the measurement of the height. Assume that these measurements are both unbiased and normally distributed. Let $V = \pi R^2 H$ denote the measurement of the volume of the cylinder that is computed from R and H. Now assume that $R = 4.8$ cm, $H = 10.1$ cm, and the uncertainties (standard deviations) are $\sigma_R = 0.1$ cm and $\sigma_H = 0.2$ cm. The measured volume is $V = \pi(4.8^2)(10.1) = 731.06$ cm^3. Suppose we wish to determine whether we can conclude that the true volume of the cylinder is greater than 700 cm^3. Let μ_V denote the mean of V. Since R and H are unbiased, with fairly small uncertainties, V is nearly unbiased (see the discussion on p. 173), so μ_V is close to the true volume of the cylinder. Therefore we can address the question concerning the true volume by performing a test of the hypotheses $H_0: \mu_V \leq 700$ versus $H_1: \mu_V > 700$.

We begin by using the method of propagation of error to compute the uncertainty in V:

$$\begin{aligned}
\sigma_V &= \sqrt{\left(\frac{\partial V}{\partial R}\right)^2 \sigma_R^2 + \left(\frac{\partial V}{\partial H}\right)^2 \sigma_H^2} \\
&= \sqrt{(2\pi RH)^2 \sigma_R^2 + (\pi R^2)^2 \sigma_H^2} \\
&= \sqrt{[2\pi(4.8)(10.1)]^2(0.1)^2 + [\pi(4.8^2)]^2(0.2)^2} \\
&= 33.73
\end{aligned}$$

Now if the measured volume V is normally distributed, we can proceed as follows: Under H_0, $V \sim N(700,\ 33.73^2)$. We observed the value $V = 731.06$. The P-value for $H_0: \mu_V \leq 700$ is $P(V \geq 731.06)$ where the probability is computed under the assumption that $V \sim N(700,\ 33.73^2)$. The z-score is $(731.06 - 700)/33.73 = 0.92$, and the P-value is 0.18.

The validity of the test just performed depends on the assumption that V is normally distributed. We check this assumption with a simulation. We first generate a large number N of values $R_1^*, \ldots, R_N^*$ for the radius measurement. We know that the radius measurement is normally distributed with standard deviation $\sigma_R = 0.1$. We do not know the mean radius measurement, which is equal to the true radius, but we can approximate it with the observed value 4.8. Thus we generate $R_1^*, \ldots, R_N^*$ from the distribution $N(4.8,\ 0.1^2)$. Similarly, we generate $H_1^*, \ldots, H_N^*$ from the distribution $N(10.1,\ 0.2^2)$.

Then we compute simulated volume measurements $V_i^* = \pi(R_i^*)^2 H_i^*$. A normal probability plot of the V_i^* can then be used to determine whether V is approximately normal.

Figure 6.30 presents a normal probability plot for a sample of 1000 values of V_i^*. The normality assumption is satisfied. The P-value of 0.18 is valid.

We remark finally that if the normality assumption is satisfied, the sample standard deviation of the V_i^* can be used in place of the value 33.73 computed by propagation of error. In the sample of 1000 that we generated, the sample standard deviation was 31.67, which is reasonably close to the value computed by propagation of error.

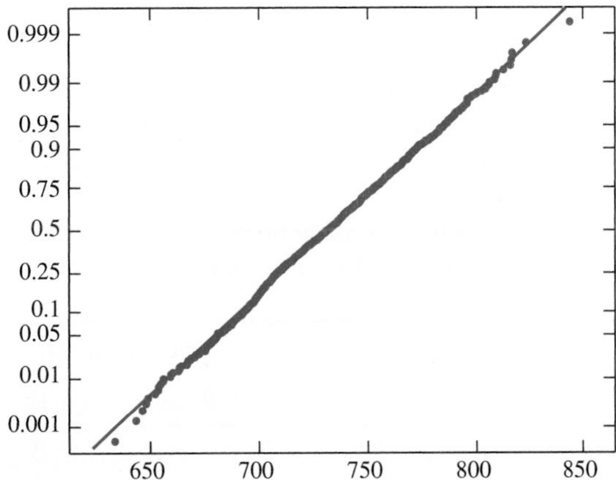

FIGURE 6.30 Normal probability plot for 1000 simulated volumes. The assumption of normality appears to be justified.

Testing Hypotheses with Bootstrap Confidence Intervals

Given a level $100(1-\alpha)\%$ confidence interval for a parameter, such as a population mean μ, we can reject at level $100\alpha\%$ the null hypothesis that the parameter is equal to any given value outside the interval, and cannot reject at level $100\alpha\%$ the null hypothesis that the parameter is equal to any given value inside the interval (see the discussion beginning on p. 381). This idea can be applied to a bootstrap confidence interval to construct a fixed-level hypothesis test. We present an example.

In Section 5.8, an approximate 95% confidence interval for the mean mileage, in mi/gal, of a population of trucks was found by a bootstrap method to be (4.7643, 6.4757). Can we conclude at the 5% level that the population mean mileage differs from 5 mi/gal? From 7 mi/gal?

Solution

A 95% confidence interval, whether computed by the bootstrap or by other means, contains the values that are not rejected at the 5% level. Therefore we conclude at the 5% level that the population mean differs from 7 mi/gal, but we cannot conclude at the 5% level that it differs from 5 mi/gal.

Randomization Tests

Randomization tests, also called permutation tests, were among the earliest methods developed to test the difference between two population means. While they require virtually no assumptions about the distribution of the data, they involve a great deal of computation, and so did not become truly feasible until recently. We present an example.

A crop scientist wants to determine whether the yield of lettuce will be increased by using a fertilizer with a higher nitrogen content. She conducts an experiment involving 20 plots. Ten plots are chosen at random to be treated with fertilizer A, which has a low nitrogen content. The other 10 plots are treated with fertilizer B, which has a higher nitrogen content.

The following table presents, for each plot, the treatment applied (A or B), and the yield, in number of heads of lettuce harvested.

	Plot Number																			
	1	**2**	**3**	**4**	**5**	**6**	**7**	**8**	**9**	**10**	**11**	**12**	**13**	**14**	**15**	**16**	**17**	**18**	**19**	**20**
Treatment	A	A	B	B	A	A	A	B	B	A	A	B	A	A	B	B	B	B	A	B
Yield	145	142	144	141	142	155	143	157	152	143	103	151	150	148	150	162	149	158	144	151

The null hypothesis is that there is no difference between the fertilizers with regard to yield; in other words, the yield for each plot would have been the same no matter which fertilizer it had received. For example the yield in plot 1 would have been 145 whether fertilizer A or B was used. If H_0 is true, then the 20 observed yields are constants, and the yields associated with fertilizer A are a simple random sample of 10 of these 20 constant yields. Denote the mean of the 10 yields associated with fertilizer A by $\overline{A}$, and the mean of the 10 yields associated with fertilizer B by $\overline{B}$. Since the main interest in the experiment is to determine whether fertilizer B increases the yield, a reasonable test statistic is the difference $\overline{B} - \overline{A}$. The observed value of this statistic is $151.5 - 141.5 = 10.0$. The larger the value of the test statistic, the stronger the evidence against H_0. The strength of the evidence is measured by the P-value. We now discuss how to calculate the P-value.

The experiment involves a random choice of 10 plots out of 20 to receive fertilizer A. In general, the number of different choices of k items to be selected from a group of n items is denoted $\binom{n}{k}$ and is given by (see Equation 2.12 in Section 2.2 for a derivation)

$$\binom{n}{k} = \frac{n!}{k!(n-k)!}$$

Therefore the number of possible choices for these 10 plots is $\binom{20}{10} = 184{,}756$. This means that there are 184,756 ways that the experiment could have come out; the actual experiment consists of observing one of these ways chosen at random. The choice that was actually made provided a value of $\overline{B} - \overline{A} = 10$ for the test statistic. Since, under H_0, the yields do not depend on which fertilizer was used, we could in principle compute the value of the test statistic $\overline{B} - \overline{A}$ for each of the 184,756 possible outcomes of the experiment. The P-value is the probability, under H_0, that the test statistic has a value greater than or equal to 10. This probability is equal to the proportion of the 184,756 possible outcomes of the experiment for which $\overline{B} - \overline{A} \geq 10$. Table 6.6 presents a partial listing of the possible outcomes of the experiment.

TABLE 6.6 Possible outcomes of the randomized experiment

Outcome	Yields Assigned to A	Yields Assigned to B	$\overline{A}$	$\overline{B}$	$\overline{B} - \overline{A}$
1	103 141 142 142 143 143 144 144 145 148	149 150 150 151 151 152 155 157 158 162	139.5	153.5	14.0
2	103 141 142 142 143 143 144 144 145 149	148 150 150 151 151 152 155 157 158 162	139.6	153.4	13.8
⋮	⋮	⋮	⋮	⋮	⋮
184,755	148 150 150 151 151 152 155 157 158 162	103 141 142 142 143 143 144 144 145 149	153.4	139.6	−13.8
184,756	149 150 150 151 151 152 155 157 158 162	103 141 142 142 143 143 144 144 145 148	153.5	139.5	−14.0

The exact P-value can be found by completing Table 6.6 and then determining the proportion of outcomes for which $\overline{B} - \overline{A} \geq 10$. This procedure is called a **randomization test**, or **permutation test**. Computing the exact P-value is an intensive task, even for a computer. An easier method, which is just as good in practice, is to work with a randomly generated collection of outcomes instead. This is done by generating a large number (1000 or more) randomly chosen subsets of 10 yields to assign to treatment A. Each chosen subset corresponds to one of the possible outcomes of the experiment, and for each subset, the value of the test statistic is computed. The P-value is approximated with the proportion of the randomly chosen outcomes for which the value of the test statistic is greater than or equal to the observed value of 10.

Table 6.7 (page 466) presents the first 5 and the last of 1000 randomly chosen outcomes for the experiment. Of the first five outcomes, none of them have values of $\overline{B} - \overline{A}$ greater than or equal to 10, so the estimate of the P-value based on these five is $0/5 = 0$. Of course, five outcomes is not enough upon which to base a reliable conclusion. Out of the full set of 1000 outcomes, only 9 had values of $\overline{B} - \overline{A}$ greater than or equal to 10. We therefore estimate the P-value to be 0.009, which is small enough to conclusively reject the null hypothesis that there is no difference between the fertilizers. It seems reasonable to conclude that fertilizer B tends to increase the yield.

Randomized experiments like the one just described play a major role in scientific investigations and are discussed more fully in Chapter 9. When no outliers are present,

TABLE 6.7 One thousand simulated outcomes of the randomized experiment

Outcome	Yields Assigned to A	Yields Assigned to B	$\overline{A}$	$\overline{B}$	$\overline{B} - \overline{A}$
1	157 151 144 150 142 150 155 144 143 141	145 148 142 143 103 152 158 149 162 151	147.70	145.30	−2.40
2	143 103 158 151 142 151 155 150 148 141	142 144 149 144 143 162 157 150 152 145	144.20	148.80	4.60
3	162 158 144 141 148 155 103 143 144 157	143 150 142 152 145 150 142 149 151 151	145.50	147.50	2.00
4	145 151 143 141 150 142 162 148 149 158	144 155 157 103 152 150 144 151 143 142	148.90	144.10	−4.80
5	152 148 144 142 157 155 162 103 150 151	145 150 158 149 144 143 141 143 151 142	146.40	146.60	0.20
⋮	⋮	⋮	⋮	⋮	⋮
1000	144 152 143 155 142 148 143 145 158 151	144 103 149 142 150 162 150 141 151 157	148.10	144.90	−3.20

it has been shown that the Student's t test for the difference between means (see Section 6.7) provides a good approximation to the randomization test when two treatments are being compared. Data from randomized experiments can generally be treated as though they consist of random samples from different populations; this is the approach we will take in Chapter 9. Freedman, Pisani, and Purves (1998) contains a good discussion of this topic. Rank tests (see Section 6.9) are sometimes used for these experiments as well.

Randomization tests can be used in some cases when the data consist of two samples from two populations, which is the situation discussed in Section 6.7. Thus randomization tests can be an alternative to the t test for the difference between means when outliers are present.

More information on randomization tests can be found in Efron and Tibshirani (1993).

Using Simulation to Estimate Power

For some tests, it is difficult to calculate power with a formula. In these cases, simulation can often be used to estimate the power. Following is an example.

Example 6.41

A new type of weld is being developed. If the mean fracture toughness of this weld is conclusively shown to be greater than 20 ft · lb, the weld will be used in a certain application. Assume that toughness is normally distributed with standard deviation equal to 4 ft · lb. Six welds will be made, and the fracture toughness of each will be measured.

A Student's t test will be made of the null hypothesis $H_0 : \mu \leq 20$ versus $H_1 : \mu > 20$. If the test is conducted at a significance level of 5%, what is the power of the test if the true mean toughness is 25 ft · lb?

Solution

Let $X_1, \ldots, X_6$ represent the six sample toughnesses, and let s represent their sample standard deviation. This is a sample from a $N(25, 16)$ distribution. The test statistic is $T = (\overline{X} - 20)/(s/\sqrt{6})$. Under H_0, this statistic has a Student's t distribution with five degrees of freedom. The null hypothesis will be rejected if the value of the test statistic is greater than $t_{5,.05} = 2.015$. The power, therefore, is equal to $P(T > 2.015)$. It is not easy to calculate this probability directly, since in fact the null hypothesis is false, so T does not in fact have a Student's t distribution. We can, however, estimate it with a simulation experiment.

We will generate 10,000 samples $X^*_{1i}, \ldots, X^*_{6i}$, each from a $N(25, 16)$ distribution. For each sample, we will compute the sample mean $\overline{X}^*_i$, the sample standard deviation s^*_i, and the test statistic $T^*_i = (\overline{X}^*_i - 20)/(s^*_i/\sqrt{6})$. Since each simulated sample is drawn from the same distribution as are the actual toughnesses of the welds, each simulated sample is statistically equivalent to a sample of actual welds. We can therefore estimate the power simply by computing the proportion of simulated samples for which the null hypothesis is rejected, that is, for which the value of the test statistic exceeds 2.015. Table 6.8 presents the results for the first 10 samples and for the last one. The rightmost column contains a "1" if the value of the test statistic exceeds 2.015 and a "0" otherwise.

The null hypothesis is rejected for 9 of the first 10 samples. If we were to base our results on the first 10 samples, we would estimate the power to be 0.9. Of course, 10 samples is not enough. Out of all 10,000 samples, the null hypothesis was rejected for 8366 of them. The estimate of the power is therefore 0.8366.

TABLE 6.8 Simulated data for Example 6.41

i	X^*_1	X^*_2	X^*_3	X^*_4	X^*_5	X^*_6	$\overline{X}^*$	s^*	T^*	$T^* > 2.015$
1	23.24	23.78	15.65	25.67	24.08	25.88	23.05	3.776	1.978	0
2	26.51	19.89	20.53	25.03	28.35	28.01	24.72	3.693	3.131	1
3	28.61	28.19	29.48	20.06	30.00	21.19	26.26	4.423	3.465	1
4	22.84	28.69	23.93	27.37	19.51	30.28	25.44	4.046	3.291	1
5	22.36	21.26	26.37	23.61	34.45	29.97	26.34	5.061	3.067	1
6	26.54	28.63	24.79	20.63	25.44	26.69	25.45	2.703	4.940	1
7	24.05	24.42	20.32	23.74	24.14	24.66	23.56	1.615	5.394	1
8	28.38	29.51	23.80	29.05	26.39	23.76	26.81	2.579	6.472	1
9	23.55	21.73	19.57	25.04	22.34	29.71	23.66	3.484	2.570	1
10	29.98	34.65	21.17	28.43	23.43	34.44	28.68	5.559	3.825	1
⋮	⋮	⋮	⋮	⋮	⋮	⋮	⋮	⋮	⋮	⋮
10,000	30.75	19.99	26.20	22.41	31.53	21.78	25.45	4.862	2.744	1

Exercises for Section 6.15

1. This exercise continues Exercise 9 in the Supplementary Exercises for Chapter 3. The article "Insights into Present-Day Crustal Motion in the Central Mediterranean Area from GPS Surveys" (M. Anzidei, P. Baldi, et al., *Geophysical Journal International*, 2001:98–100) reports measurements of the velocity of the earth's crust in Zimmerwald, Switzerland. The component of velocity in a northerly direction was measured to be $X = 22.10$, and the component in an easterly direction was measured to be $Y = 14.30$, where the units are mm/year. The uncertainties in the measurements were given as $\sigma_X = 0.34$ and $\sigma_Y = 0.32$.

a. Compute the estimated velocity V of the earth's crust, based on these measurements. Use the method of propagation of error to estimate its uncertainty.

b. Assuming the estimated velocity to be normally distributed, find the P-value for the hypothesis $H_0: \mu_V \leq 25$.

c. Assuming that the components of velocity in the northerly and easterly directions are independent and normally distributed, generate an appropriate simulated sample of values V^*. Is it reasonable to assume that V is approximately normally distributed?

2. A population geneticist is studying the genes found at two different locations on the genome. He estimates the proportion p_1 of organisms who have gene A at the first locus to be $\widehat{p}_1 = 0.42$, with uncertainty $\sigma_1 = 0.049$. He estimates the proportion of organisms that have gene B at a second locus to be $\widehat{p}_2 = 0.23$, with uncertainty $\sigma_2 = 0.043$. Under assumptions usually made in population genetics (Hardy–Weinberg equilibrium), $\widehat{p}_1$ and $\widehat{p}_2$ are independent and normally distributed, and the proportion p of organisms that have both genes A and B is estimated with $\widehat{p} = \widehat{p}_1\widehat{p}_2$.

a. Compute $\widehat{p}$ and use propagation of error to estimate its uncertainty.

b. Assuming $\widehat{p}$ to be normally distributed, find the P-value for testing $H_0: p \geq 0.10$.

c. Generate an appropriate simulated sample of values $\widehat{p}^*$. Is it reasonable to assume that $\widehat{p}$ is normally distributed?

3. Refer to Exercise 6 in Section 5.8. Let μ represent the population mean compressive strength, in MPa. Consider the following null hypotheses:

i. $H_0: \mu = 38.53$
ii. $H_0: \mu = 38.35$
iii. $H_0: \mu = 38.45$
iv. $H_0: \mu = 38.55$

a. Using the bootstrap data presented in Exercise 6 in Section 5.8, which of these null hypotheses can be rejected at the 5% level if a confidence interval is formed using method 1 on p. 358?

b. Using the bootstrap data presented in Exercise 6 in Section 5.8, which of these null hypotheses can be rejected at the 10% level if a confidence interval is formed using method 1 on p. 358?

4. Refer to Exercise 6 in Section 5.8. Let μ represent the population mean compressive strength, in MPa. Generate 1000 bootstrap samples.

a. Using the bootstrap data you generated, which of these null hypotheses can be rejected at the 5% level, using method 1 on p. 358?

b. Using the bootstrap data you generated, which of these null hypotheses can be rejected at the 10% level, using method 1 on p. 358?

c. If a bootstrap experiment is performed twice on the same data, is it necessary that the results will agree? Explain.

5. In the lettuce yield example presented on p. 464, would it be a good idea to use the t test described in Section 6.7 to determine whether the fertilizers differ in their effects on yield? Why or why not?

6. It is suspected that using premium gasoline rather than regular will increase the mileage for automobiles with a particular engine design. Sixteen cars are used in a randomized experiment. Eight are randomly chosen to be tested with regular gasoline, while the other eight are tested with premium gasoline. The results, in mi/gal, as follows:

Regular:	29.1	27.1	30.8	17.3	27.6	16.3
	28.4	30.2				
Premium:	28.3	32.0	27.4	35.3	29.9	35.6
	30.9	29.7				

a. Under the null hypothesis that each car would get the same mileage with either type of gasoline, how many different outcomes are possible for this experiment?

b. Let $\overline{R}$ and $\overline{P}$ denote the sample mean mileages for the regular and premium groups, respectively. Compute $\overline{R}$ and $\overline{P}$.

c. Perform a randomization test to determine whether it can be concluded that premium gasoline tends to increase mileage. Use the test statistic $\overline{P} - \overline{R}$. Generate at least 1000 random outcomes, and estimate the P-value.

d. Use the Student's t test described in Section 6.7 to test the null hypothesis that the mean mileage using regular is greater than or equal to the mean mileage for premium. Is this result reliable? Explain?

7. For the lettuce yield data, (p. 464) it is thought that the yields from fertilizer A might have a larger variance than the yields from fertilizer B.

a. Compute the sample variances s_A^2 and s_B^2 of the yields assigned to A and B, respectively, and the quotient s_A^2/s_B^2.

b. Someone suggests using the F test in Section 6.11 for this problem. Is this a good idea? Why or why not?

c. Perform a randomization test of $H_0: s_A^2 \le s_B^2$ versus $H_1: s_A^2 > s_B^2$, using the test statistic s_A^2/s_B^2, and a minimum of 1000 random outcomes.
(*Hint:* Proceed just as in the example in the text, but for each outcome compute s_A^2, s_B^2, and s_A^2/s_B^2 rather than $\overline{A}$, $\overline{B}$, and $\overline{B} - \overline{A}$.)

8. Refer to Exercise 6. Perform a randomization test to determine whether the mileage using regular gasoline has a greater variance than the mileage using premium gasoline. Generate at least 1000 random outcomes.

9. A certain wastewater treatment method is supposed to neutralize the wastewater so that the mean pH is 7. Measurements of pH will be made on seven specimens of treated wastewater, and a test of the hypotheses $H_0: \mu = 7$ versus $H_1: \mu \ne 7$ will be made using the Student's t test (Section 6.4). Assume that the true mean is $\mu = 6.5$, the pH measurements are normally distributed with mean μ and standard deviation 0.5, and the test is made at the 5% level.

a. Let $X_1, \ldots, X_7$ denote the pH measurements, let $\overline{X}$ denote their mean, and let s denote their sample standard deviation. What is the test statistic? For what values of the test statistic will H_0 be rejected?

b. Generate 10,000 samples $X_1^*, \ldots, X_7^*$ from the true distribution of pH measurements. For each sample, compute the test statistic and determine whether H_0 is rejected. Estimate the power of the test.

10. This exercise requires ideas from Section 2.6. In a two-sample experiment, when each item in one sample is paired with an item in the other, the paired t test (Section 6.8) can be used to test hypotheses regarding the difference between two population means. If one ignores the fact that the data are paired, one can use the two-sample t test (Section 6.7) as well. The question arises as to which test has the greater power. The following simulation experiment is designed to address this question.

Let $(X_1, Y_1), \ldots, (X_8, Y_8)$ be a random sample of eight pairs, with $X_1, \ldots, X_8$ drawn from an $N(0, 1)$ population and $Y_1, \ldots, Y_8$ drawn from an $N(1, 1)$ population. It is desired to test $H_0: \mu_X - \mu_Y = 0$ versus $H_1: \mu_X - \mu_Y \ne 0$. Note that $\mu_X = 0$ and $\mu_Y = 1$, so the true difference between the means is 1. Also note that the population variances are equal. If a test is to be made at the 5% significance level, which test has the greater power?

Let $D_i = X_i - Y_i$ for $i = 1, \ldots, 10$. The test statistic for the paired t test is $\overline{D}/(s_D/\sqrt{8})$, where s_D is the standard deviation of the D_i (see Section 6.8). Its null distribution is Student's t with seven degrees of freedom. Therefore the paired t test will reject H_0 if $|\overline{D}/(s_D/\sqrt{8})| > t_{7,.025} = 2.365$, so the power is $P(|\overline{D}/(s_D/\sqrt{8})| > 2.365)$.

For the two-sample t test when the population variances are equal, the test statistic is $\overline{D}/(s_p\sqrt{1/8 + 1/8}) = \overline{D}/(s_p/2)$, where s_p is the pooled standard deviation, which is equal in this case to $\sqrt{(s_X^2 + s_Y^2)/2}$. (See p. 411. Note that $\overline{D} = \overline{X} - \overline{Y}$.) The null distribution is Student's t with 14 degrees of freedom. Therefore the two-sample t test will reject H_0 if $|\overline{D}/(s_p\sqrt{1/8 + 1/8})| > t_{14,.025} = 2.145$, and the power is $P(|\overline{D}/(s_p\sqrt{1/8 + 1/8})| > 2.145)$.

The power of these tests depends on the correlation between X_i and Y_i.

a. Generate 10,000 samples $X_{1i}^*, \ldots, X_{8i}^*$ from an $N(0, 1)$ population and 10,000 samples $Y_{1i}^*, \ldots, Y_{8i}^*$ from an $N(1, 1)$ population. The random variables X_{ki}^* and Y_{ki}^* are independent in this experiment, so their correlation is 0. For each sample, compute the test statistics $\overline{D}^*/(s_D^*/\sqrt{8})$ and $\overline{D^*}/(s_p^*/2)$. Estimate the power of each test by computing the proportion of samples for which the test statistics exceeds its critical point (2.365 for the paired test, 2.145 for the two-sample test). Which test has greater power?

b. As in part (a), generate 10,000 samples $X_{1i}^*, \ldots, X_{8i}^*$ from an $N(0, 1)$ population. This time, instead of generating the values Y^* independently, generate them so the correlation between X_{ki}^* and Y_{ki}^* is 0.8. This can be done as follows: Generate 10,000 samples $Z_{1i}^*, \ldots, Z_{8i}$ from an $N(0, 1)$ population, independent of the X^* values. Then compute $Y_{ki} = 1 + 0.8X_{ki}^* + 0.6Z_{ki}^*$. The sample $Y_{1i}^*, \ldots, Y_{8i}^*$ will come from an $N(1, 1)$ population, and the correlation between X_{ki}^* and Y_{ki}^* will be 0.8, which means that large values of X_{ki}^* will tend to be paired with large values of Y_{ki}^*, and vice versa. Compute the test statistics and estimate the power of both tests, as in part (a). Which test has greater power?

Supplementary Exercises for Chapter 6

Exercises 1 to 4 describe experiments that require a hypothesis test. For each experiment, describe the appropriate test. State the appropriate null and alternate hypotheses, describe the test statistic, and specify which table should be used to find the P-value. If relevant, state the number of degrees of freedom for the test statistic.

1. A fleet of 100 taxis is divided into two groups of 50 cars each to see whether premium gasoline reduces maintenance costs. Premium unleaded fuel is used in group A, while regular unleaded fuel is used in group B. The total maintenance cost for each vehicle during a one-year period is recorded. Premium fuel will be used if it is shown to reduce maintenance costs.

2. A group of 15 swimmers is chosen to participate in an experiment to see if a new breathing style will improve their stamina. Each swimmer's pulse recovery rate is measured after a 20 minute workout using the old breathing style. The swimmers practice the new style for two weeks, and then measure their pulse recovery rates after a 20 minute workout using the new style. They will continue to use the new breathing style if it is shown to reduce pulse recovery time.

3. A new quality-inspection program is being tested to see if it will reduce the proportion of parts shipped out that are defective. Under the old program, the proportion of defective parts was 0.10. Two hundred parts that passed inspection under the new program will be sampled, and the number of defectives will be counted. The new program will be implemented if it is shown that the proportion of defectives is less than 0.10.

4. A new material is being tested for use in the manufacture of electrical conduit, to determine whether it will reduce the variance in crushing strength over the old material. Crushing strengths are measured for a sample of size 16 of the old material and a sample of size 20 of the new material. If it is shown that the crushing strength with the new material has smaller variance, the new material will be used.

5. Suppose you have purchased a filling machine for candy bags that is supposed to fill each bag with 16 oz of candy. Assume that the weights of filled bags are approximately normally distributed. A random sample of 10 bags yields the following data (in oz):

15.87	16.02	15.78	15.83	15.69	15.81
16.04	15.81	15.92	16.10		

On the basis of these data, can you conclude that the mean fill weight is actually less than 16 oz?

a. State the appropriate null and alternate hypotheses.

b. Compute the value of the test statistic.

c. Find the P-value and state your conclusion.

6. Are answer keys to multiple-choice tests generated randomly, or are they constructed to make it less likely for the same answer to occur twice in a row? This question was addressed in the article "Seek Whence: Answer Sequences and Their Consequences in Key-Balanced Multiple-Choice Tests" (M. Bar-Hillel and Y. Attali, *The American Statistician*, 2002:299–303). They studied 1280 questions on 10 real Scholastic Assessment Tests (SATs). Assume that all the questions had five choices (in fact 150 of them had only four choices). They found that for 192 of the questions, the correct choice (A, B, C, D, or E) was the same as the correct choice for the question immediately preceding. If the choices were generated at random, then the probability that a question would have the same correct choice as the one immediately preceding would be 0.20. Can you conclude that the choices for the SAT are not generated at random?

a. State the appropriate null and alternate hypotheses.

b. Compute the value of the test statistic.

c. Find the P-value and state your conclusion.

7. An automobile manufacturer wishes to compare the lifetimes of two brands of tire. She obtains samples of six tires of each brand. On each of six cars, she mounts one tire of each brand on each front wheel. The cars are driven until only 20% of the original tread remains. The distances, in miles, for each tire are presented in the following table. Can you conclude that there is a difference between the mean lifetimes of the two brands of tire?

Car	Brand 1	Brand 2
1	36,925	34,318
2	45,300	42,280
3	36,240	35,500
4	32,100	31,950
5	37,210	38,015
6	48,360	47,800
7	38,200	33,215

a. State the appropriate null and alternate hypotheses.

b. Compute the value of the test statistic.

c. Find the P-value and state your conclusion.

8. Twenty-one independent measurements were taken of the hardness (on the Rockwell C scale) of HSLA-100 steel base metal, and another 21 independent measurements were made of the hardness of a weld produced on this base metal. The standard deviation of the measurements made on the base metal was 3.06, and the standard deviation of the measurements made on the weld was 1.41. Assume that the measurements are independent random samples from normal populations. Can you conclude that measurements made on the base metal are more variable than measurements made on the weld?

9. There is concern that increased industrialization may be increasing the mineral content of river water. Ten years ago, the silicon content of the water in a certain river was 5 mg/L. Eighty-five water samples taken recently from the river have mean silicon content 5.6 mg/L and standard deviation 1.2 mg/L. Can you conclude that the silicon content of the water is greater today than it was 10 years ago?

10. The article "Modeling of Urban Area Stop-and-Go Traffic Noise" (P. Pamanikabud and C. Tharasawatipipat, *Journal of Transportation Engineering* 1999:152–159) presents measurements of traffic noise, in dBA, from 10 locations in Bangkok, Thailand. Measurements, presented in the following table, were made at each location, in both the acceleration and deceleration lanes.

Location	Acceleration	Deceleration
1	78.1	78.6
2	78.1	80.0
3	79.6	79.3
4	81.0	79.1
5	78.7	78.2
6	78.1	78.0
7	78.6	78.6
8	78.5	78.8
9	78.4	78.0
10	79.6	78.4

Can you conclude that there is a difference in the mean noise levels between acceleration and deceleration lanes?

11. A machine that grinds valves is set to produce valves whose lengths have mean 100 mm and standard deviation 0.1 mm. The machine is moved to a new location. It is thought that the move may have upset the calibration for the mean length, but that it is unlikely to have changed the standard deviation. Let μ represent the mean length of valves produced after the move. To test the calibration, a sample of 100 valves will be ground, their lengths will be measured, and a test will be made of the hypotheses $H_0: \mu = 100$ versus $H_1: \mu \neq 100$.

a. Find the rejection region if the test is made at the 5% level.

b. Find the rejection region if the test is made at the 10% level.

c. If the sample mean length is 99.97 mm, will H_0 be rejected at the 5% level?

d. If the sample mean length is 100.01 mm, will H_0 be rejected at the 10% level?

e. A critical point is 100.015 mm. What is the level of the test?

12. A process that manufactures glass sheets is supposed to be calibrated so that the mean thickness μ of the sheets is more than 4 mm. The standard deviation of the sheet thicknesses is known to be well approximated by $\sigma = 0.20$ mm. Thicknesses of each sheet in a sample of sheets will be measured, and a test of the hypothesis $H_0: \mu \leq 4$ versus $H_1: \mu > 4$ will be performed. Assume that, in fact, the true mean thickness is 4.04 mm.

a. If 100 sheets are sampled, what is the power of a test made at the 5% level?

b. How many sheets must be sampled so that a 5% level test has power 0.95?

c. If 100 sheets are sampled, at what level must the test be made so that the power is 0.90?

d. If 100 sheets are sampled, and the rejection region is $\overline{X} \geq 4.02$, what is the power of the test?

13. A machine manufactures bolts that are supposed to be 3 inches in length. Each day a quality engineer selects a random sample of 50 bolts from the day's production, measures their lengths, and performs a hypothesis test of $H_0: \mu = 3$ versus $H_1: \mu \neq 3$, where μ is the mean length of all the bolts manufactured that day. Assume that the population standard deviation for bolt lengths is 0.1 in. If H_0 is rejected at the 5% level, the machine is shut down and recalibrated.

a. Assume that on a given day, the true mean length of bolts is 3 in. What is the probability that the machine will be shut down? (This is called the **false alarm rate.**)

b. If the true mean bolt length on a given day is 3.01 in., find the probability that the equipment will be recalibrated.

14. Electric motors are assembled on four different production lines. Random samples of motors are taken from each line and inspected. The number that pass and that fail the inspection are counted for each line, with the following results:

	Line			
	1	**2**	**3**	**4**
Pass	482	467	458	404
Fail	57	59	37	47

Can you conclude that the failure rates differ among the four lines?

15. Refer to Exercise 14. The process engineer notices that the sample from line 3 has the lowest proportion of failures. Use the Bonferroni adjustment to determine whether she can conclude that the population proportion of failures on line 3 is less than 0.10.

Exercises 16 and 17 illustrate that distribution-free methods can produce misleading results when their assumptions are seriously violated.

16. Consider the following two samples:

X: 0 2 3 4 10 20 40 100 1000

Y: −738 162 222 242 252 258 259 260 262

a. Show that both samples have the same mean and variance.

b. Use the Wilcoxon rank-sum test to test the null hypothesis that the population means are equal. What do you conclude?

c. Do the assumptions of the rank-sum test appear to be satisfied? Explain why or why not.

17. The rank-sum test is sometimes thought of as a test for population medians. Under the assumptions of equal spread and shape, the means of two populations will differ if and only if the medians differ; therefore tests for equality of population means are also tests for equality of population medians. This exercise illustrates that when these assumptions are seriously violated, the rank-sum test can give misleading results concerning the equality of population medians. Consider the following two samples:

X: 1 2 3 4 5 6 7
20 40 50 60 70 80 90 100

Y: −10 −9 −8 −7 −6 −5 −4
20 21 22 23 24 25 26 27

a. Show that both samples have the same median.

b. Compute the P-value for a two-tailed rank-sum test. If small P-values provide evidence against the null hypothesis that the population medians are equal, would you conclude that the population medians are different?

c. Do the assumptions of the rank-sum test appear to be satisfied? Explain why or why not.

18. A new production process is being contemplated for the manufacture of stainless steel bearings. Measurements of the diameters of random samples of bearings from the old and the new processes produced the following data:

Old:	16.3	15.9	15.8	16.2	16.1	16.0
	15.7	15.8	15.9	16.1	16.3	16.1
	15.8	15.7	15.8	15.7		
New:	15.9	16.2	16.0	15.8	16.1	16.1
	15.8	16.0	16.2	15.9	15.7	16.2
	15.8	15.8	16.2	16.3		

a. Can you conclude at the 5% level that one process yields a different mean size bearing than the other?

b. Can you conclude at the 5% level that the variance of the new procedure is lower than the older procedure?

19. Two different chemical formulations of rocket fuel are considered for the peak thrust they deliver in a particular design for a rocket engine. The thrust/weight ratios (in kilograms force per gram) for each of the two fuels are measured several times. The results are as follows:

Fuel A:	54.3	52.9	57.9	58.2	53.4	51.4
	56.8	55.9	57.9	56.8	58.4	52.9
	55.5	51.3	51.8	53.3		
Fuel B:	55.1	55.5	53.1	50.5	49.7	50.1
	52.4	54.4	54.1	55.6	56.1	54.8
	48.4	48.3	55.5	54.7		

a. Assume the fuel processing plant is presently configured to produce fuel B and changeover costs are high. Since an increased thrust/weight ratio for rocket fuel is beneficial, how should the null and alternate hypotheses be stated for a test on which to base a decision whether to switch to fuel A?

b. Can you conclude at the 5% level that the switch to fuel A should be made?

20. Suppose the Environmental Protection Agency is in the process of monitoring the water quality in a large estuary in the eastern United States, in order to measure the PCB concentration (in parts per billion).

a. Suppose that a random sample of size 80 has a sample mean of 1.59 ppb and a sample standard deviation of 0.25 ppb. Test the hypothesis, at the 5% level, that the mean PCB concentration in the estuary is less than or equal to 1.50 ppb against the alternative that it is higher. Is H_0 rejected?

b. If the population mean is 1.6 ppb and the population standard deviation is 0.33 ppb, what is the probability that the null hypothesis $H_0: \mu \leq 1.50$ is rejected at the 5% level, if the sample size is 80?

c. If the population mean is 1.6 ppb and the population standard deviation is 0.33 ppb, what sample size is needed so that the probability is 0.99 that $H_0: \mu \leq 1.50$ is rejected at the 5% level?

21. Two machines are used to package laundry detergent. It is known that weights of boxes are normally distributed. Four boxes from each machine have their contents carefully weighed, with the following results (in grams):

Machine 1:	1752	1757	1751	1754
Machine 2:	1756	1750	1752	1746

An engineer wishes to test the null hypothesis that the mean weights of boxes from the two machines are equal. He decides to assume that the population variances are equal, reasoning as follows:

> The sample variances are $s_1^2 = 7.00$ for machine 1 and $s_2^2 = 17.33$ for machine 2. The F statistic for testing for equality of population variances is $F_{3,3} = s_2^2/s_1^2 = 2.48$. The upper 10% point of the $F_{3,3}$ distribution is 5.39. Since the null hypothesis specifies that the variances are equal, I determine that the P-value is greater than $2(0.10) = 0.20$. Therefore I do not reject the null hypothesis, and I conclude that the variances are equal.

a. Has the F test been done correctly?

b. Is the conclusion justified? Explain.

22. The article "Valuing Watershed Quality Improvements Using Conjoint Analysis" (S. Farber and B. Griner, *Ecological Economics*, 2000:63–76) presents the results of a mail survey designed to assess opinions on the value of improvement efforts in an acid-mine degraded watershed in Western Pennsylvania. Of the 510 respondents to the survey, 347 were male. Census data show that 48% of the target population is male. Can you conclude that the survey method employed in this study tends to oversample males? Explain.

23. Anthropologists can estimate the birthrate of an ancient society by studying the age distribution of skeletons found in ancient cemeteries. The numbers of skeletons found at two such sites, as reported in the article "Paleoanthropological Traces of a Neolithic Demographic Transition" (J. Bocquet-Appel, *Current Anthropology*, 2002:637–650) are given in the following table:

	Ages of Skeletons		
Site	**0–4 years**	**5–19 years**	**20 years or more**
Casa da Moura	27	61	126
Wandersleben	38	60	118

Do these data provide convincing evidence that the age distributions differ between the two sites?

24. Deforestation is a serious problem throughout much of India. The article "Factors Influencing People's Participation in Forest Management in India" (W. Lise, *Ecological Economics*, 2000:379–392) discusses the social forces that influence forest management policies in three Indian states: Haryana, Bihar, and Uttar Pradesh. The forest quality in Haryana is somewhat degraded, in Bihar it is very degraded, and in Uttar Pradesh it is well-stocked. In order to study the relationship between educational levels and attitudes toward forest management, random samples of adults in each of these states were surveyed and their educational levels were ascertained. The numbers of adults at each of several educational levels were recorded. The data are presented in the following table.

	Years of Education					
State	**0**	**1–4**	**5–6**	**7–9**	**10–11**	**12 or more**
Haryana	48	6	16	26	24	7
Bihar	34	24	7	32	16	10
Uttar Pradesh	20	9	25	30	17	34

Can you conclude that the educational levels differ among the three states? Explain.

Chapter 7

Correlation and Simple Linear Regression

Introduction

Scientists and engineers often collect data in order to determine the nature of a relationship between two quantities. For example, a chemical engineer may run a chemical process several times in order to study the relationship between the concentration of a certain catalyst and the yield of the process. Each time the process is run, the concentration x and the yield y are recorded. The experiment thus generates **bivariate** data; a collection of ordered pairs $(x_1, y_1), \ldots, (x_n, y_n)$. In many cases, ordered pairs generated in a scientific experiment will fall approximately along a straight line when plotted. In these situations the data can be used to compute an equation for the line. This equation can be used for many purposes; for example, in the catalyst versus yield experiment just described, it could be used to predict the yield y that will be obtained the next time the process is run with a specific catalyst concentration x.

The methods of correlation and simple linear regression, which are the subject of this chapter, are used to analyze bivariate data in order to determine whether a straight-line fit is appropriate, to compute the equation of the line if appropriate, and to use that equation to draw inferences about the relationship between the two quantities.

7.1 Correlation

One of the earliest applications of statistics was to study the variation in physical characteristics in human populations. To this end, statisticians invented a quantity called the **correlation coefficient** as a way of describing how closely related two physical characteristics were. The first published correlation coefficient was due to the English statistician Sir Francis Galton, who in 1888 measured the heights and forearm lengths of 348 adult men. (Actually, he measured the distance from the elbow to the tip of the middle finger, which is called a cubit.) If we denote the height of the ith man by x_i, and the length of his forearm by y_i, then Galton's data consist of 348 ordered pairs (x_i, y_i). Figure 7.1 presents a simulated re-creation of these data, based on a table constructed by Galton.

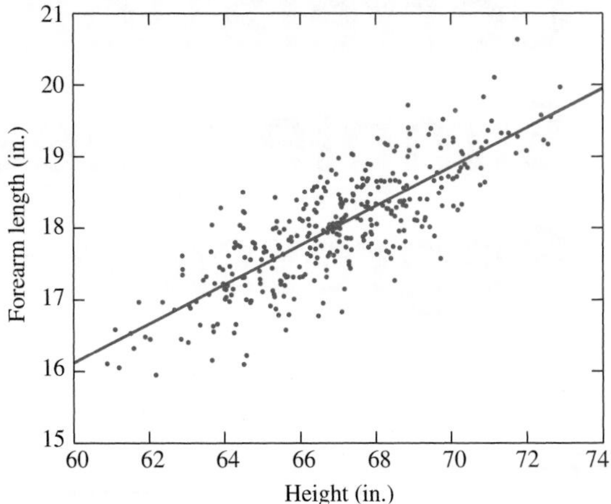

FIGURE 7.1 Heights and forearm lengths of 348 men.

The points tend to slope upward and to the right, indicating that taller men tend to have longer forearms. We say that there is a **positive association** between height and forearm length. The slope is roughly constant throughout the plot, indicating that the points are clustered around a straight line. The line superimposed on the plot is a special line known as the **least-squares line**. It is the line that fits the data best, in a sense to be described in Section 7.2. We will also learn how to compute the least-squares line in Section 7.2.

Figure 7.2 presents the results of a study of the relationship between the mean daily temperature and the mean daily humidity at a site near Riverside, California, during a recent winter. Again the points are clustered around the least-squares line. The line has a negative slope, indicating that days with higher humidity tend to have lower temperatures.

The degree to which the points in a scatterplot tend to cluster around a line reflects the strength of the linear relationship between x and y. The visual impression of a scatterplot can be misleading in this regard, because changing the scale of the axes can make the clustering appear tighter or looser. For this reason, we define the **correlation coefficient**, which is a numerical measure of the strength of the linear relationship between two variables. The correlation coefficient is usually denoted by the letter r.

Let $(x_1, y_1), \ldots, (x_n, y_n)$ represent n points on a scatterplot. To compute the correlation, first compute the means and standard deviations of the xs and ys, that is, $\overline{x}$, $\overline{y}$, s_x, and s_y. Then convert each x and y to standard units, or, in other words, compute

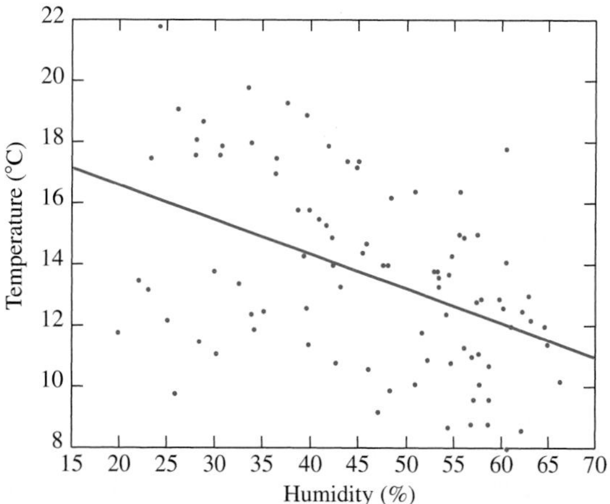

FIGURE 7.2 Humidity (in %) and temperature (in °C) for days in a recent winter in Riverside, California.

the z-scores: $(x_i - \overline{x})/s_x$, $(y_i - \overline{y})/s_y$. The correlation coefficient is the average of the products of the z-scores, except that we divide by $n - 1$ instead of n:

$$r = \frac{1}{n-1} \sum_{i=1}^{n} \left(\frac{x_i - \overline{x}}{s_x} \right) \left(\frac{y_i - \overline{y}}{s_y} \right) \tag{7.1}$$

We can rewrite Equation 7.1 in a way that is sometimes useful. By substituting $\sqrt{\sum_{i=1}^{n}(x_i - \overline{x})^2/(n-1)}$ for s_x and $\sqrt{\sum_{i=1}^{n}(y_i - \overline{y})^2/(n-1)}$ for s_y, we obtain

$$r = \frac{\sum_{i=1}^{n}(x_i - \overline{x})(y_i - \overline{y})}{\sqrt{\sum_{i=1}^{n}(x_i - \overline{x})^2}\sqrt{\sum_{i=1}^{n}(y_i - \overline{y})^2}} \tag{7.2}$$

In principle, the correlation coefficient can be calculated for any set of points. In many cases, the points constitute a random sample from a population of points. In these cases the correlation coefficient is often called the **sample correlation**, and it is an estimate of the population correlation. (Population correlation was discussed formally in Section 2.6; intuitively, you may imagine the population to consist of a large finite collection of points, and the population correlation to be the quantity computed using Equation 7.2 on the whole population, with sample means replaced by population means.) The sample correlation can be used to construct confidence intervals and perform hypothesis tests on the population correlation; these will be discussed later in this section. We point out that the correlation coefficient can also be used to measure the strength of a linear relationship in many cases where the points are not a random sample from a population; see the discussion of the coefficient of determination in Section 7.2.

It is a mathematical fact that the correlation coefficient is always between -1 and 1. Positive values of the correlation coefficient indicate that the least-squares line has a positive slope, which means that greater values of one variable are associated with greater values of the other. Negative values of the correlation coefficient indicate that the least-squares line has a negative slope, which means that greater values of one variable are associated with lesser values of the other. Values of the correlation coefficient close to 1 or to -1 indicate a strong linear relationship; values close to 0 indicate a weak linear relationship. The correlation coefficient is equal to 1 (or to -1) only when the points in the scatterplot lie exactly on a straight line of positive (or negative) slope, in other words, when there is a perfect linear relationship. As a technical note, if the points lie exactly on a horizontal or a vertical line, the correlation coefficient is undefined, because one of the standard deviations is equal to zero. Finally, a bit of terminology: Whenever $r \neq 0$, x and y are said to be **correlated**. If $r = 0$, x and y are said to be **uncorrelated**.

The correlation between height and forearm length in Figure 7.1 is 0.80. The correlation between temperature and humidity in Figure 7.2 is -0.46. Figures 7.3 and 7.4 on pages 479 and 480 present some examples of scatterplots with various correlations. In each plot, both x and y have mean 0 and standard deviation 1. All plots are drawn to the same scale.

How the Correlation Coefficient Works

Why does the formula (Equation 7.1) for the correlation coefficient r measure the strength of the linear association between two variables? Figure 7.5 on page 481 illustrates how the correlation coefficient works. In this scatterplot, the origin is placed at the point of averages $(\overline{x}, \overline{y})$. Therefore, in the first quadrant, the z-scores $(x_i - \overline{x})/s_x$ and $(y_i - \overline{y})/s_y$ are both positive, so their product is positive as well. Thus each point in the first quadrant contributes a positive amount to the sum in Equation (7.1). In the second quadrant, the z-scores for the x coordinates of the points are negative, while the z-scores for the y coordinates are positive. Therefore the products of the z-scores are negative, so each point in the second quadrant contributes a negative amount to the sum in Equation (7.1). Similarly, points in the third quadrant contribute positive amounts, and points in the fourth quadrant contribute negative amounts. Clearly, in Figure 7.5 there are more points in the first and third quadrants than in the second and fourth, so the correlation will be positive. If the plot had a negative slope, there would be more points in the second and fourth quadrants, and the correlation coefficient would be negative.

The Correlation Coefficient Is a Pure Number

In any sample $x_1, \ldots, x_n$, the mean $\overline{x}$ and the standard deviation s_x have the same units as $x_1, \ldots, x_n$. For this reason the z-scores $(x_i - \overline{x})/s_x$ are pure numbers. Since the correlation coefficient r is the average of products of z-scores, it too is a pure number, without units. This fact is crucial to the usefulness of r. For example, the units for the x and y coordinates in Figure 7.1 are both inches, while the corresponding units in Figure 7.2 are percent and degrees Celsius. If the correlation coefficients for the two plots had different units, it would be impossible to compare their values to determine which plot exhibited the stronger linear relationship. But since the correlation coefficients are pure numbers, they are directly comparable, and we can conclude that the relationship between heights of men and their forearm lengths in Figure 7.1 is more strongly linear than the relationship between temperature and humidity in Figure 7.2.

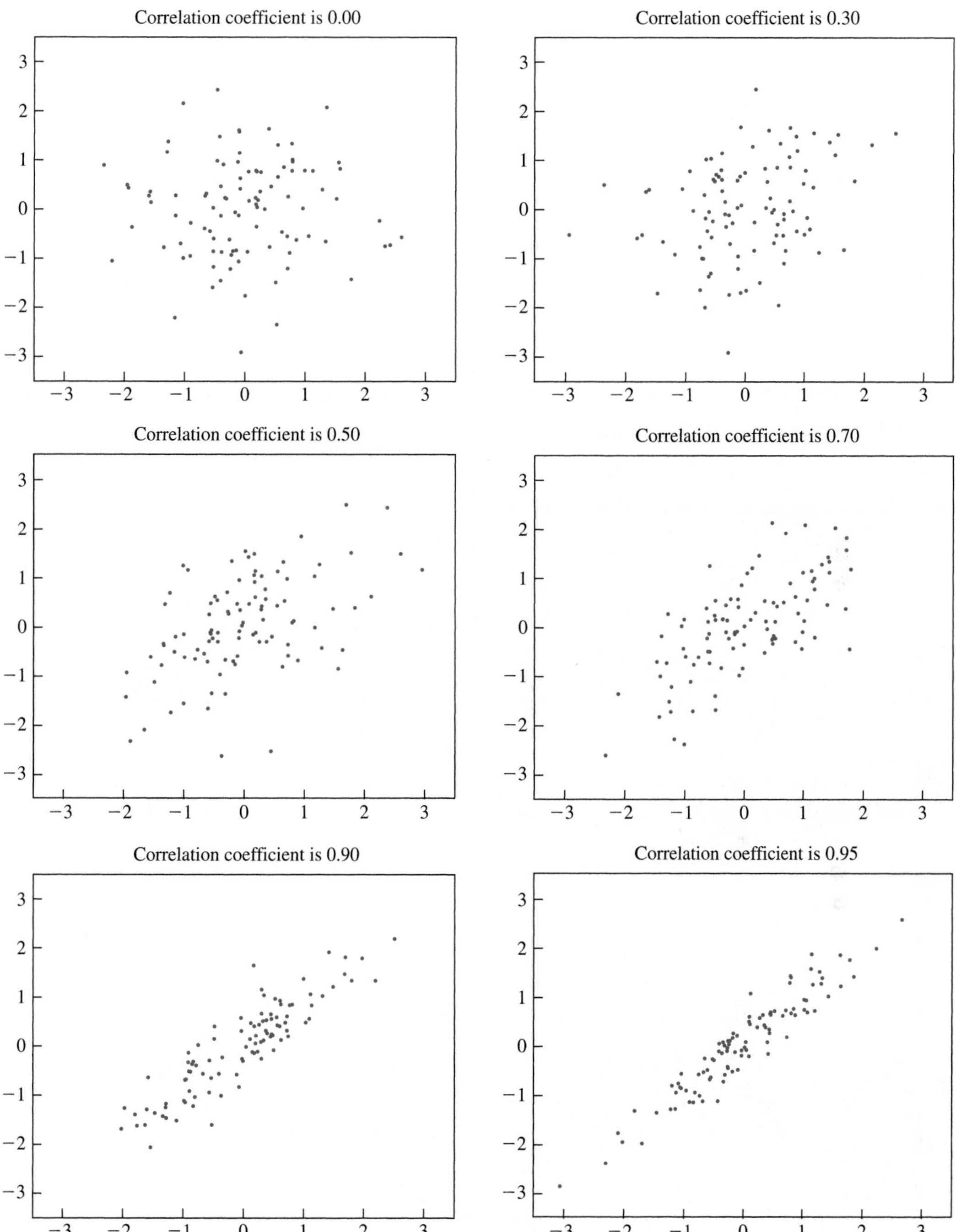

FIGURE 7.3 Examples of various levels of positive correlation.

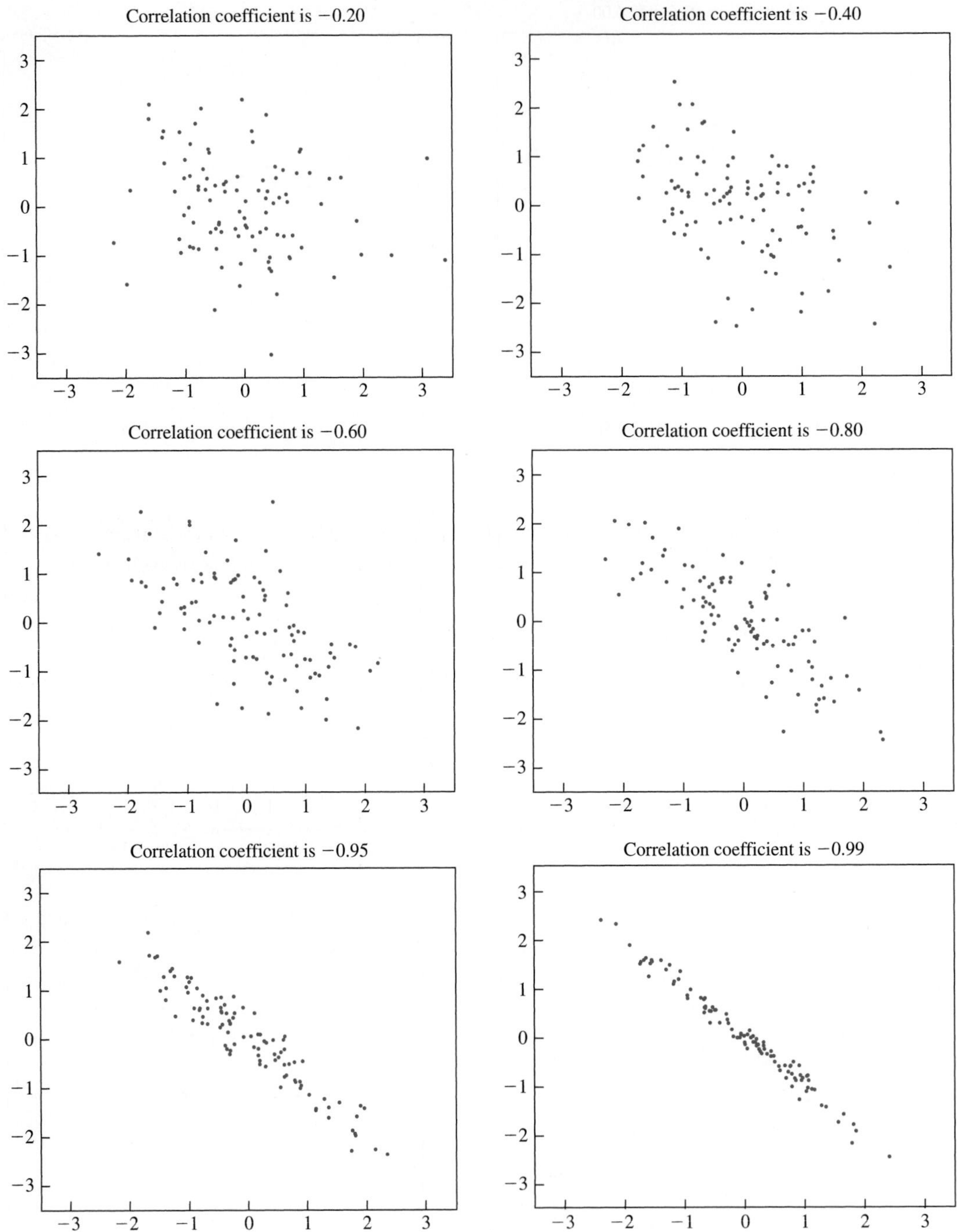

FIGURE 7.4 Examples of various levels of negative correlation.

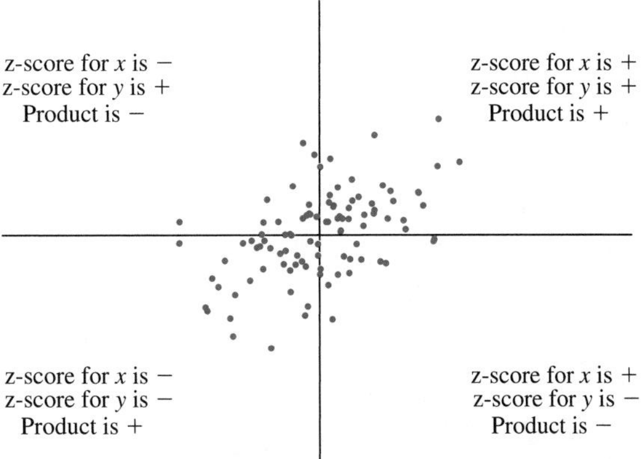

FIGURE 7.5 How the correlation coefficient works.

Another crucial property of the correlation coefficient is that it is unaffected by the units in which the measurements are made. For example, imagine that in Figure 7.1 the heights of the men were measured in centimeters rather than inches. Then each x_i would be multiplied by 2.54. But this would cause $\overline{x}$ and s_x to be multiplied by 2.54 as well, so the z-scores $(x_i - \overline{x})/s_x$ would be unchanged, and r would be unchanged as well. In a more fanciful example, imagine that each man stood on a platform 2 inches high while being measured. This would increase each x_i by 2, but the value of $\overline{x}$ would be increased by 2 as well. Thus the z-scores would be unchanged, so the correlation coefficient would be unchanged as well. Finally, imagine that we interchanged the values of x and y, using x to represent the forearm lengths, and y to represent the heights. Since the correlation coefficient is determined by the product of the z-scores, it does not matter which variable is represented by x and which by y.

Summary

The correlation coefficient remains unchanged under each of the following operations:

- Multiplying each value of a variable by a positive constant.
- Adding a constant to each value of a variable.
- Interchanging the values of x and y.

Figure 7.6 on page 482 presents plots of mean temperatures for the months of April and October for several U.S. cities. Whether the temperatures are measured in °C or °F, the correlation is the same. This is because converting from °C to °F involves multiplying by 1.8 and adding 32.

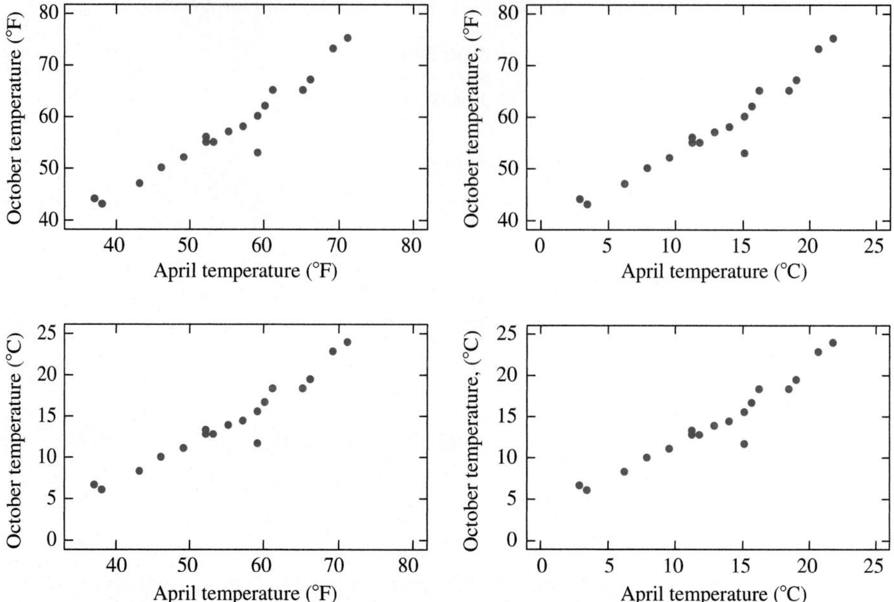

FIGURE 7.6 Mean April and October temperatures for several U.S. cities. The correlation coefficient is 0.96 for each plot; the choice of units does not matter.

The Correlation Coefficient Measures Only *Linear* Association

An object is fired upward from the ground with an initial velocity of 64 ft/s. At each of several times $x_1, \dots, x_n$, the heights $y_1, \dots, y_n$ of the object above the surface of the earth are measured. In the absence of friction, and assuming that there is no measurement error, the scatterplot of the points $(x_1, y_1), \dots, (x_n, y_n)$ will look like Figure 7.7. There is obviously a strong relationship between x and y; in fact the value of y is determined by x through the function $y = 64x - 16x^2$. Yet the correlation between x and y is equal to 0. Is something wrong? No. The value of 0 for the correlation indicates that there is no *linear* relationship between x and y, which is true. The relationship is purely *quadratic*. The lesson of this example is that the correlation coefficient should only be used when the relationship between the x and y is linear. Otherwise the results can be misleading.

Outliers

In Figure 7.8, the point (0, 3) is an outlier, a point that is detached from the main body of the data. The correlation for this scatterplot is $r = 0.26$, which indicates a weak linear relationship. Yet 10 of the 11 points have a perfect linear relationship. Outliers can greatly distort the correlation coefficient, especially in small data sets, and present a serious problem for data analysts. Some outliers are caused by data-recording errors or by failure to follow experimental protocol. These outliers can appropriately be corrected or deleted. Sometimes people delete outliers from a plot without cause, to give it a more

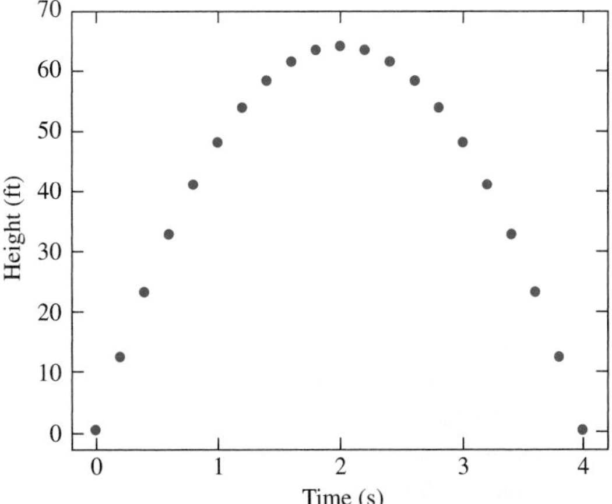

FIGURE 7.7 The relationship between the height of a free-falling object with a positive initial velocity and the time in free fall is quadratic. The correlation is equal to 0.

pleasing appearance. This is not appropriate, as it results in an underestimation of the variability of the process that generated the data. Interpreting data that contain outliers can be difficult, because there are few easy rules to follow.

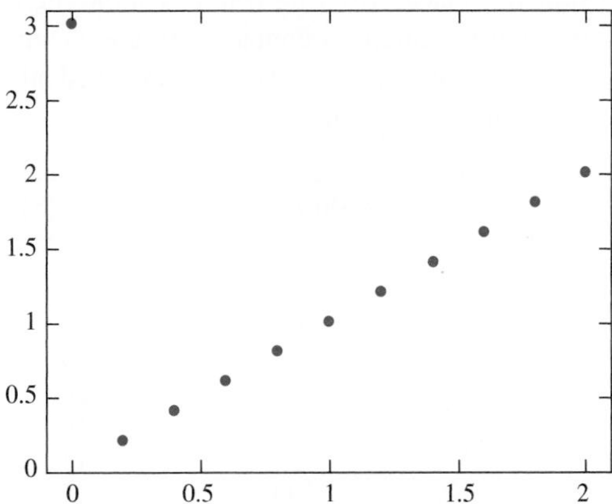

FIGURE 7.8 The correlation is 0.26. Because of the outlier, the correlation coefficient is misleading.

Correlation Is Not Causation

For children, vocabulary size is strongly correlated with shoe size. However, learning new words does not cause feet to grow, nor do growing feet cause one's vocabulary to increase. There is a third factor, namely age, that is correlated with both shoe size and vocabulary.

Older children tend to have both larger shoe sizes and larger vocabularies, and this results in a positive correlation between vocabulary and shoe size. This phenomenon is known as **confounding**. Confounding occurs when there is a third variable that is correlated with both of the variables of interest, resulting in a correlation between them.

To restate this example in more detail: Individuals with larger ages tend to have larger shoe sizes. Individuals with larger ages also tend to have larger vocabularies. It follows that individuals with larger shoe sizes will tend to have larger vocabularies. In other words, because both shoe size and vocabulary are positively correlated with age, they are positively correlated with each other.

In this example, the confounding was easy to spot. In many cases it is not so easy. The example shows that simply because two variables are correlated with each other, we cannot assume that a change in one will tend to cause a change in the other. Before we can conclude that two variables have a causal relationship, we must rule out the possibility of confounding.

Sometimes multiple regression (see Chapter 8) can be used to detect confounding. Sometimes experiments can be designed so as to reduce the possibility of confounding. The topic of **experimental design** (see Chapter 9) is largely concerned with this topic. Here is a simple example.

An environmental scientist is studying the rate of absorption of a certain chemical into skin. She places differing volumes of the chemical on different pieces of skin and allows the skin to remain in contact with the chemical for varying lengths of time. She then measures the volume of chemical absorbed into each piece of skin. She obtains the results shown in the following table.

Volume (mL)	Time (h)	Percent Absorbed
0.05	2	48.3
0.05	2	51.0
0.05	2	54.7
2.00	10	63.2
2.00	10	67.8
2.00	10	66.2
5.00	24	83.6
5.00	24	85.1
5.00	24	87.8

The scientist plots the percent absorbed against both volume and time, as shown in the following figure. She calculates the correlation between volume and absorption and obtains $r = 0.988$. She concludes that increasing the volume of the chemical causes the percentage absorbed to increase. She then calculates the correlation between time and absorption, obtaining $r = 0.987$. She concludes that increasing the time that the skin is in contact with the chemical causes the percentage absorbed to increase as well. Are these conclusions justified?

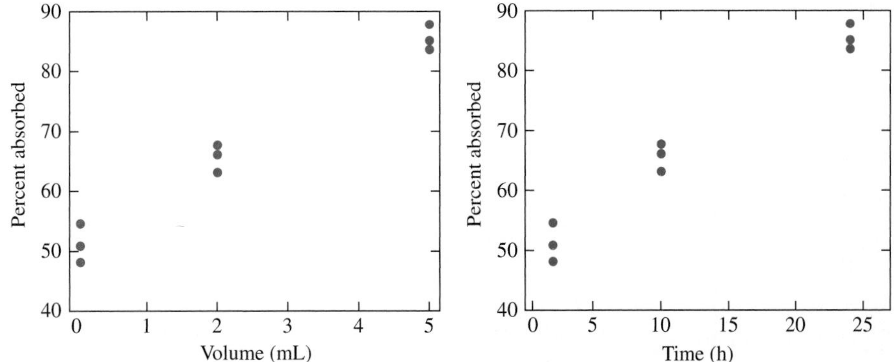

Solution

No. The scientist should look at the plot of time versus volume, presented in the following figure. The correlation between time and volume is $r = 0.999$, so these two variables are almost completely confounded. If *either* time or volume affects the percentage absorbed, *both* will appear to do so, because they are highly correlated with each other. For this reason, it is impossible to determine whether it is the time or the volume that is having an effect. This relationship between time and volume resulted from the design of the experiment and should have been avoided.

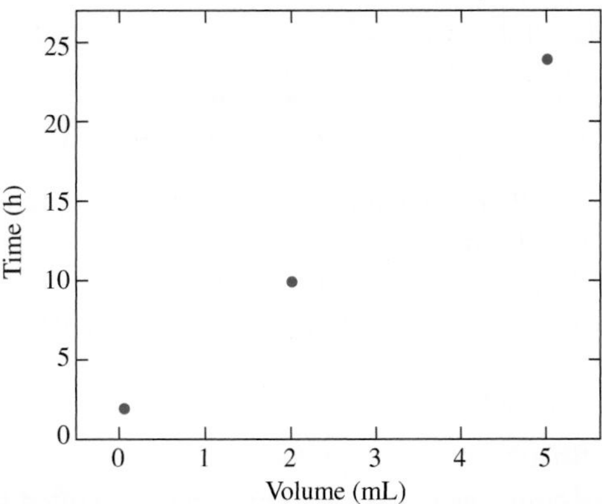

The scientist in Example 7.1 has repeated the experiment, this time with a new design. The results are presented in the following table.

Volume (mL)	Time (h)	Percent Absorbed
0.05	2	49.2
0.05	10	51.0
0.05	24	84.3
2.00	2	54.1
2.00	10	68.7
2.00	24	87.2
5.00	2	47.7
5.00	10	65.1
5.00	24	88.4

The scientist plots the percent absorbed against both volume and time, as shown in the following figure.

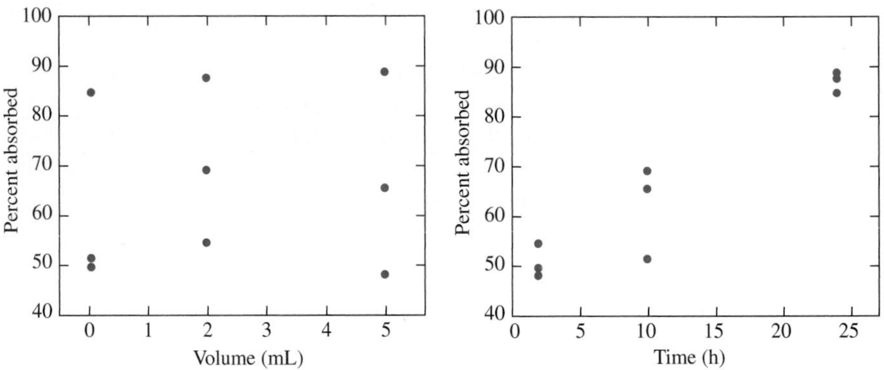

She then calculates the correlation between volume and absorption and obtains $r = 0.121$. She concludes that increasing the volume of the chemical has little or no effect on the percentage absorbed. She then calculates the correlation between time and absorption and obtains $r = 0.952$. She concludes that increasing the time that the skin is in contact with the chemical will cause the percentage absorbed to increase. Are these conclusions justified?

Solution

These conclusions are much better justified than the ones in Example 7.1. To see why, look at the plot of time versus volume in the following figure. This experiment has been designed so that time and volume are uncorrelated. It now appears that the time, but not the volume, has an effect on the percentage absorbed. Before making a final conclusion that increasing the time actually causes the percentage absorbed to increase, the scientist must make sure that there are no other potential confounders around. For example, if the ambient temperature varied with each replication of the experiment, and was highly correlated with time, then it might be the case that the temperature, rather than the time, was causing the percentage absorbed to vary.

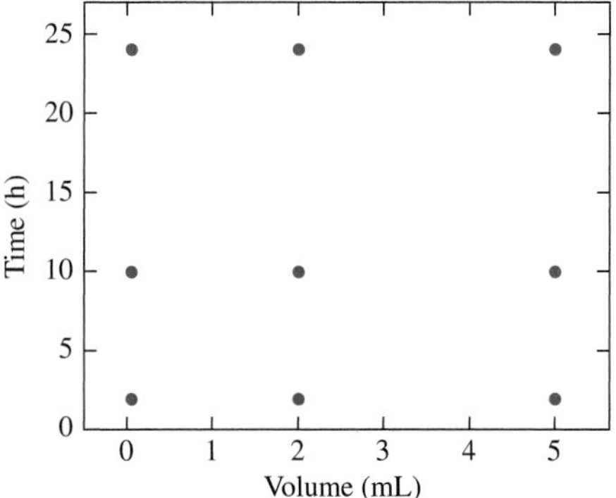

Inference on the Population Correlation

The rest of this section uses some ideas from Section 2.6. When the points (x_i, y_i) are a random sample from a population of ordered pairs, then each point can be thought of as an observation of an ordered pair of random variables (X, Y). The correlation coefficient, or sample correlation, r is then an estimate of the population correlation $\rho_{X,Y}$.

If the random variables X and Y have a certain joint distribution called a **bivariate normal distribution**, then the sample correlation r can be used to construct confidence intervals and perform hypothesis tests on the population correlation. In practice, if X and Y are both normally distributed, then it is a virtual certainty that X and Y will be bivariate normal, so the confidence intervals and tests described subsequently will be valid. (While it is mathematically possible to construct two normal random variables that jointly are not bivariate normal, the conditions under which this occurs are almost never seen in practice.)

Confidence intervals, and most tests, on $\rho_{X,Y}$ are based on the following result:

Let X and Y be random variables with the bivariate normal distribution.
Let ρ denote the population correlation between X and Y.
Let $(x_1, y_1), \ldots, (x_n, y_n)$ be a random sample from the joint distribution of X and Y.
Let r be the sample correlation of the n points.
Then the quantity

$$W = \frac{1}{2} \ln \frac{1+r}{1-r} \tag{7.3}$$

is approximately normally distributed, with mean given by

$$\mu_W = \frac{1}{2} \ln \frac{1+\rho}{1-\rho} \tag{7.4}$$

and variance given by

$$\sigma_W^2 = \frac{1}{n-3} \tag{7.5}$$

Note that μ_W is a function of the population correlation ρ. To construct confidence intervals, we will need to solve Equation (7.4) for ρ. We obtain

$$\rho = \frac{e^{2\mu_W} - 1}{e^{2\mu_W} + 1} \tag{7.6}$$

Example 7.3

In a study of reaction times, the time to respond to a visual stimulus (x) and the time to respond to an auditory stimulus (y) were recorded for each of 10 subjects. Times were measured in ms. The results are presented in the following table.

x	161	203	235	176	201	188	228	211	191	178
y	159	206	241	163	197	193	209	189	169	201

Find a 95% confidence interval for the correlation between the two reaction times.

Solution

Using Equation (7.1), we compute the sample correlation, obtaining $r = 0.8159$. Next we use Equation (7.3) to compute the quantity W:

$$\begin{aligned} W &= \frac{1}{2}\ln\frac{1+r}{1-r} \\ &= \frac{1}{2}\ln\frac{1+0.8159}{1-0.8159} \\ &= 1.1444 \end{aligned}$$

Since W is normally distributed with known standard deviation $\sigma = \sqrt{1/(10-3)} = 0.3780$ (Equation 7.5), a 95% confidence interval for μ_W is given by

$$\begin{gathered} 1.1444 - 1.96(0.3780) < \mu_W < 1.1444 + 1.96(0.3780) \\ 0.4036 < \mu_W < 1.8852 \end{gathered}$$

To obtain a 95% confidence interval for ρ we transform the inequality using Equation (7.6), obtaining

$$\frac{e^{2(0.4036)} - 1}{e^{2(0.4036)} + 1} < \frac{e^{2\mu_W} - 1}{e^{2\mu_W} + 1} < \frac{e^{2(1.8852)} - 1}{e^{2(1.8852)} + 1}$$

$$0.383 < \rho < 0.955$$

For testing null hypotheses of the form $\rho = \rho_0$, $\rho \leq \rho_0$, and $\rho \geq \rho_0$, where ρ_0 is a constant *not equal to 0*, the quantity W forms the basis of a test. Following is an example.

Example 7.4

Refer to Example 7.3. Find the P-value for testing $H_0: \rho \leq 0.3$ versus $H_1: \rho > 0.3$.

Solution

Under H_0 we take $\rho = 0.3$, so, using Equation (7.4),

$$\mu_W = \frac{1}{2}\ln\frac{1+0.3}{1-0.3}$$

$$= 0.3095$$

The standard deviation of W is $\sigma = \sqrt{1/(10-3)} = 0.3780$. It follows that under H_0, $W \sim N(0.3095, 0.3780^2)$. The observed value of W is $W = 1.1444$. The z-score is therefore

$$z = \frac{1.1444 - 0.3095}{0.3780} = 2.21$$

The P-value is 0.0136. We conclude that $\rho > 0.3$.

For testing null hypotheses of the form $\rho = 0$, $\rho \leq 0$, or $\rho \geq 0$, a somewhat simpler procedure is available. When $\rho = 0$, the quantity

$$U = \frac{r\sqrt{n-2}}{\sqrt{1-r^2}}$$

has a Student's t distribution with $n - 2$ degrees of freedom. Example 7.5 shows how to use U as a test statistic.

Example 7.5

Refer to Example 7.3. Test the hypothesis $H_0: \rho \leq 0$ versus $H_1: \rho > 0$.

Solution

Under H_0 we take $\rho = 0$, so the test statistic U has a Student's t distribution with $n - 2 = 8$ degrees of freedom. The sample correlation is $r = 0.8159$, so the value of U is

$$U = \frac{r\sqrt{n-2}}{\sqrt{1-r^2}}$$

$$= \frac{0.8159\sqrt{10-2}}{\sqrt{1-0.8159^2}}$$

$$= 3.991$$

Consulting the Student's t table with eight degrees of freedom, we find that the P-value is between 0.001 and 0.005. It is safe to conclude that $\rho > 0$.

Exercises for Section 7.1

1. Compute the correlation coefficient for the following data set.

x	1	2	3	4	5	6	7
y	2	1	4	3	7	5	6

2. For each of the following data sets, explain why the correlation coefficient is the same as for the data set in Exercise 1.

a.

x	1	2	3	4	5	6	7
y	5	4	7	6	10	8	9

b.

x	11	21	31	41	51	61	71
y	5	4	7	6	10	8	9

c.

x	53	43	73	63	103	83	93
y	4	6	8	10	12	14	16

3. For each of the following scatterplots, state whether the correlation coefficient is an appropriate summary, and explain briefly.

a.

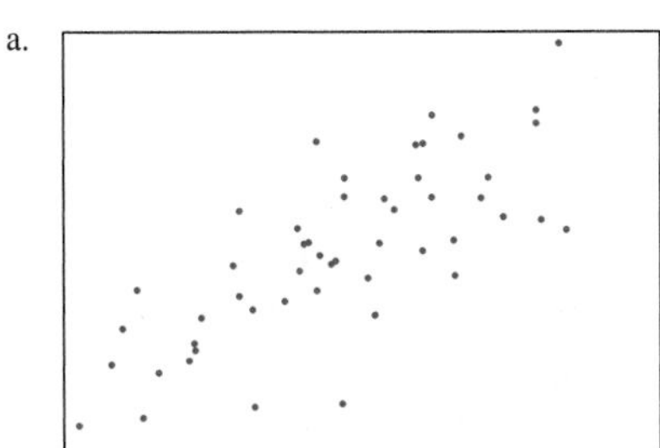

b.

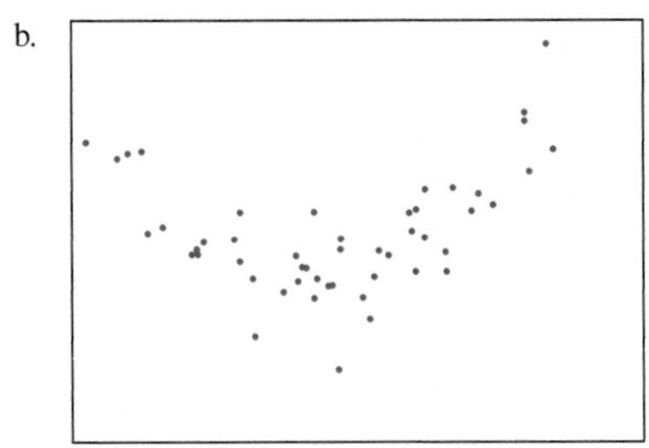

c.

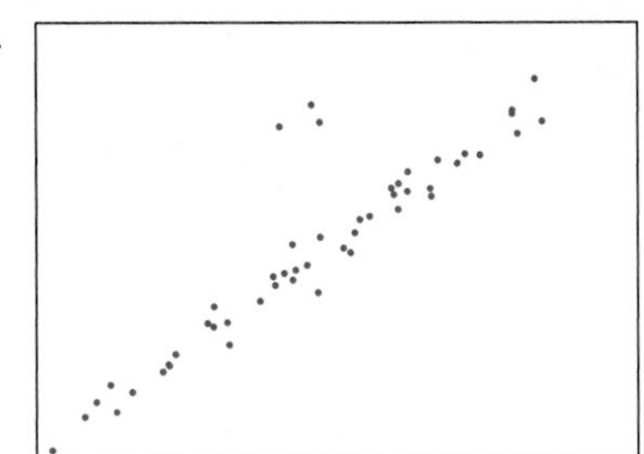

4. True or false, and explain briefly:
   a. If the correlation coefficient is positive, then above-average values of one variable are associated with above-average values of the other.
   b. If the correlation coefficient is negative, then below-average values of one variable are associated with below-average values of the other.
   c. If y is usually less than x, then the correlation between y and x will be negative.

5. An investigator collected data on heights and weights of college students. The correlation between height and weight for men was about 0.6, and for women it was about the same. If men and women are taken together, will the correlation between height and weight be more than 0.6, less than 0.6, or about equal to 0.6? It might be helpful to make a rough scatterplot.

6. In a study of ground motion caused by earthquakes, the peak velocity (in m/s) and peak acceleration (in m/s^2) were recorded for five earthquakes. The results are presented in the following table.

Velocity	1.54	1.60	0.95	1.30	2.92
Acceleration	7.64	8.04	8.04	6.37	5.00

   a. Compute the correlation coefficient between peak velocity and peak acceleration.
   b. Construct a scatterplot for these data.
   c. Is the correlation coefficient an appropriate summary for these data? Explain why or why not.
   d. Someone suggests converting the units from meters to centimeters and from seconds to minutes. What effect would this have on the correlation?

7. A chemical engineer is studying the effect of temperature and stirring rate on the yield of a certain product. The process is run 16 times, at the settings indicated in the following table. The units for yield are percent of a theoretical maximum.

Temperature (°C)	Stirring Rate (rpm)	Yield (%)
110	30	70.27
110	32	72.29
111	34	72.57
111	36	74.69
112	38	76.09
112	40	73.14
114	42	75.61
114	44	69.56
117	46	74.41
117	48	73.49
122	50	79.18
122	52	75.44
130	54	81.71
130	56	83.03
143	58	76.98
143	60	80.99

a. Compute the correlation between temperature and yield, between stirring rate and yield, and between temperature and stirring rate.

b. Do these data provide good evidence that increasing the temperature causes the yield to increase, within the range of the data? Or might the result be due to confounding? Explain.

c. Do these data provide good evidence that increasing the stirring rate causes the yield to increase, within the range of the data? Or might the result be due to confounding? Explain.

8. Another chemical engineer is studying the same process as in Exercise 7, and uses the following experimental matrix.

Temperature (°C)	Stirring Rate (rpm)	Yield (%)
110	30	70.27
110	40	74.95
110	50	77.91
110	60	82.69
121	30	73.43
121	40	73.14
121	50	78.27
121	60	74.89
132	30	69.07
132	40	70.83
132	50	79.18
132	60	78.10
143	30	73.71
143	40	77.70
143	50	74.31
143	60	80.99

a. Compute the correlation between temperature and yield, between stirring rate and yield, and between temperature and stirring rate.

b. Do these data provide good evidence that the yield is unaffected by temperature, within the range of the data? Or might the result be due to confounding? Explain.

c. Do these data provide good evidence that increasing the stirring rate causes the yield to increase, within the range of the data? Or might the result be due to confounding? Explain.

d. Which experimental design is better, this one or the one in Exercise 7? Explain.

9. Following are measurements of tensile strength in ksi (x) and Brinell hardness (y) for 10 specimens of cold drawn copper. Assume that tensile strength and Brinell hardness follow a bivariate normal distribution.

x	y
106.2	35.0
106.3	37.2
105.3	39.8
106.1	35.8
105.4	41.3
106.3	40.7
104.7	38.7
105.4	40.2
105.5	38.1
105.1	41.6

a. Find a 95% confidence interval for ρ, the population correlation between tensile strength and Brinell hardness.

b. Can you conclude that $\rho < 0.3$?

c. Can you conclude that $\rho \neq 0$?

10. In a sample of 400 ball bearings, the correlation coefficient between eccentricity and smoothness was $r = 0.10$.

a. Find the P-value for testing $H_0 : \rho \leq 0$ versus $H_1 : \rho > 0$. Can you conclude that $\rho > 0$?

b. Does the result in part (a) allow you to conclude that there is a strong correlation between eccentricity and smoothness? Explain.

11. A scatterplot contains four points: $(-2,-2)$, $(-1,-1)$, $(0,0)$, and $(1,1)$. A fifth point, $(2,y)$, is to be added to the plot. Let r represent the correlation between x and y.

a. Find the value of y so that $r = 1$.

b. Find the value of y so that $r = 0$.

c. Find the value of y so that $r = 0.5$.

d. Find the value of y so that $r = -0.5$.

e. Give a geometric argument to show that there is no value of y for which $r = -1$.

7.2 The Least-Squares Line

When two variables have a linear relationship, the scatterplot tends to be clustered around a line known as the least-squares line (see Figures 7.1 and 7.2 in Section 7.1). In this section we will learn how to compute the least-squares line and how it can be used to draw conclusions from data.

We begin by describing a hypothetical experiment. Springs are used in applications for their ability to extend (stretch) under load. The stiffness of a spring is measured by the "spring constant," which is the length that the spring will be extended by one unit of force or load.[1] To make sure that a given spring functions appropriately, it is necessary to estimate its spring constant with good accuracy and precision.

In our hypothetical experiment, a spring is hung vertically with the top end fixed, and weights are hung one at a time from the other end. After each weight is hung, the length of the spring is measured. Let $x_1, \ldots, x_n$ represent the weights, and let l_i represent the length of the spring under the load x_i. Hooke's law states that

$$l_i = \beta_0 + \beta_1 x_i \tag{7.7}$$

where β_0 is the length of the spring when unloaded and β_1 is the spring constant.

Let y_i be the *measured* length of the spring under load x_i. Because of measurement error, y_i will differ from the true length l_i. We write

$$y_i = l_i + \varepsilon_i \tag{7.8}$$

where ε_i is the error in the ith measurement. Combining (7.7) and (7.8), we obtain

$$y_i = \beta_0 + \beta_1 x_i + \varepsilon_i \tag{7.9}$$

In Equation (7.9) y_i is called the **dependent variable**, x_i is called the **independent variable**, β_0 and β_1 are the **regression coefficients**, and ε_i is called the **error**. Equation (7.9) is called a **linear model**.

Table 7.1 presents the results of the hypothetical experiment, and Figure 7.9 presents the scatterplot of y versus x. We wish to use these data to estimate the spring constant β_1 and the unloaded length β_0. If there were no measurement error, the points would lie on a straight line with slope β_1 and intercept β_0, and these quantities would be easy to determine. Because of measurement error, β_0 and β_1 cannot be determined exactly, but they can be estimated by calculating the least-squares line.

[1] The more traditional definition of the spring constant is the reciprocal of this quantity, namely, the force required to extend the spring one unit of length.

TABLE 7.1 Measured lengths of a spring under various loads

Weight (lb) x	Measured Length (in.) y	Weight (lb) x	Measured Length (in.) y
0.0	5.06	2.0	5.40
0.2	5.01	2.2	5.57
0.4	5.12	2.4	5.47
0.6	5.13	2.6	5.53
0.8	5.14	2.8	5.61
1.0	5.16	3.0	5.59
1.2	5.25	3.2	5.61
1.4	5.19	3.4	5.75
1.6	5.24	3.6	5.68
1.8	5.46	3.8	5.80

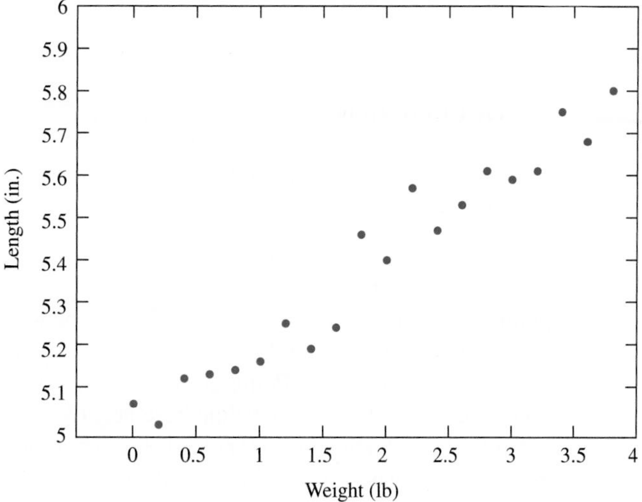

FIGURE 7.9 Plot of measured lengths of a spring versus load.

Figure 7.10 on page 494 presents the scatterplot of y versus x with the least-squares line superimposed. We write the equation of the line as

$$y = \widehat{\beta}_0 + \widehat{\beta}_1 x \tag{7.10}$$

The quantities $\widehat{\beta}_0$ and $\widehat{\beta}_1$ are called the **least-squares coefficients**. The coefficient $\widehat{\beta}_1$, the slope of the least-squares line, is an estimate of the true spring constant β_1, and the coefficient $\widehat{\beta}_0$, the intercept of the least-squares line, is an estimate of the true unloaded length β_0.

The least-squares line is the line that fits the data "best." We now define what we mean by "best." For each data point (x_i, y_i), the vertical distance to the point $(x_i, \widehat{y}_i)$ on the least-squares line is $e_i = y_i - \widehat{y}_i$ (see Figure 7.10). The quantity $\widehat{y}_i = \widehat{\beta}_0 + \widehat{\beta}_1 x_i$ is called the **fitted value**, and the quantity e_i is called the **residual** associated with the

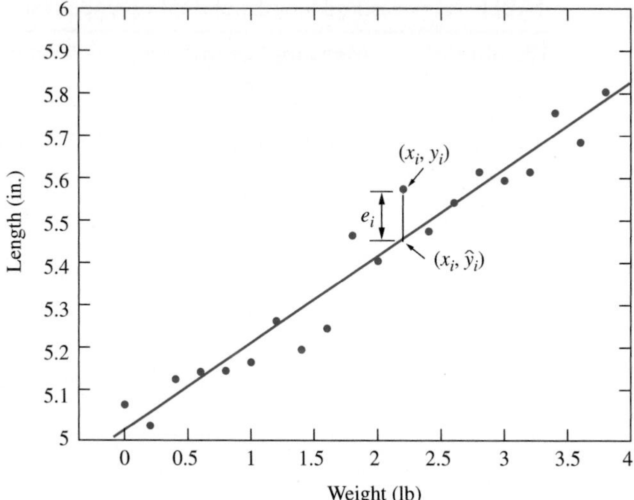

FIGURE 7.10 Plot of measured lengths of a spring versus load. The least-squares line $y = \widehat{\beta}_0 + \widehat{\beta}_1 x$ is superimposed. The vertical distance from a data point (x_i, y_i) to the point $(x_i, \widehat{y}_i)$ on the line is the ith residual e_i. The least-squares line is the line that minimizes the sum of the squared residuals.

point (x_i, y_i). The residual e_i is the difference between the value y_i observed in the data and the fitted value $\widehat{y}_i$ predicted by the least-squares line. This is the vertical distance from the point to the line. Points above the least-squares line have positive residuals, and points below the least-squares line have negative residuals. The closer the residuals are to 0, the closer the fitted values are to the observations and the better the line fits the data. We define the least-squares line to be the line for which the sum of the squared residuals $\sum_{i=1}^{n} e_i^2$ is minimized. In this sense, the least-squares line fits the data better than any other line.

In the Hooke's law example, there is only one independent variable (weight), since it is reasonable to assume that the only variable affecting the length of the spring is the amount of weight hung from it. In other cases, we may need to use several independent variables. For example, to predict the yield of a certain crop, we might need to know the amount of fertilizer used, the amount of water applied, and various measurements of chemical properties of the soil. Linear models like Hooke's law, with only one independent variable, are known as **simple linear regression** models. Linear models with more than one independent variable are called **multiple regression** models. This chapter covers simple linear regression. Multiple regression is covered in Chapter 8.

Computing the Equation of the Least-Squares Line

To compute the equation of the least-squares line, we must determine the values for the slope $\widehat{\beta}_1$ and the intercept $\widehat{\beta}_0$ that minimize the sum of the squared residuals $\sum_{i=1}^{n} e_i^2$.

To do this, we first express e_i in terms of $\widehat{\beta}_0$ and $\widehat{\beta}_1$:

$$e_i = y_i - \widehat{y}_i = y_i - \widehat{\beta}_0 - \widehat{\beta}_1 x_i \tag{7.11}$$

Therefore $\widehat{\beta}_0$ and $\widehat{\beta}_1$ are the quantities that minimize the sum

$$S = \sum_{i=1}^{n} e_i^2 = \sum_{i=1}^{n} (y_i - \widehat{\beta}_0 - \widehat{\beta}_1 x_i)^2 \tag{7.12}$$

These quantities are

$$\widehat{\beta}_1 = \frac{\sum_{i=1}^{n}(x_i - \overline{x})(y_i - \overline{y})}{\sum_{i=1}^{n}(x_i - \overline{x})^2} \tag{7.13}$$

$$\widehat{\beta}_0 = \overline{y} - \widehat{\beta}_1 \overline{x} \tag{7.14}$$

Derivations of these results are provided at the end of this section.

Computing Formulas

The quantities $\sum_{i=1}^{n}(x_i - \overline{x})^2$ and $\sum_{i=1}^{n}(x_i - \overline{x})(y_i - \overline{y})$ need to be computed in order to determine the equation of the least-squares line, and as we will soon see, the quantity $\sum_{i=1}^{n}(y_i - \overline{y})^2$ needs to be computed in order to determine how well the line fits the data. When computing these quantities by hand, there are alternate formulas that are often easier to use. They are given in the following box.

Computing Formulas

The expressions on the right are equivalent to those on the left, and are often easier to compute:

$$\sum_{i=1}^{n}(x_i - \overline{x})^2 = \sum_{i=1}^{n} x_i^2 - n\overline{x}^2 \tag{7.15}$$

$$\sum_{i=1}^{n}(y_i - \overline{y})^2 = \sum_{i=1}^{n} y_i^2 - n\overline{y}^2 \tag{7.16}$$

$$\sum_{i=1}^{n}(x_i - \overline{x})(y_i - \overline{y}) = \sum_{i=1}^{n} x_i y_i - n\overline{x}\,\overline{y} \tag{7.17}$$

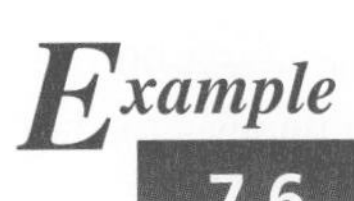

Using the Hooke's law data in Table 7.1, compute the least-squares estimates of the spring constant and the unloaded length of the spring. Write the equation of the least-squares line.

Solution

The estimate of the spring constant is $\widehat{\beta}_1$, and the estimate of the unloaded length is $\widehat{\beta}_0$. From Table 7.1 we compute:

$$\overline{x} = 1.9000 \qquad \overline{y} = 5.3885$$

$$\sum_{i=1}^{n}(x_i - \overline{x})^2 = \sum_{i=1}^{n} x_i^2 - n\overline{x}^2 = 26.6000$$

$$\sum_{i=1}^{n}(x_i - \overline{x})(y_i - \overline{y}) = \sum_{i=1}^{n} x_i y_i - n\overline{x}\,\overline{y} = 5.4430$$

Using Equations (7.13) and (7.14), we compute

$$\widehat{\beta}_1 = \frac{5.4430}{26.6000} = 0.2046$$

$$\widehat{\beta}_0 = 5.3885 - (0.2046)(1.9000) = 4.9997$$

The equation of the least-squares line is $y = \widehat{\beta}_0 + \widehat{\beta}_1 x$. Substituting the computed values for $\widehat{\beta}_0$ and $\widehat{\beta}_1$, we obtain

$$y = 4.9997 + 0.2046x$$

Using the equation of the least-squares line, we can compute the fitted values $\widehat{y}_i = \widehat{\beta}_0 + \widehat{\beta}_1 x_i$ and the residuals $e_i = y_i - \widehat{y}_i$ for each point (x_i, y_i) in the Hooke's law data set. The results are presented in Table 7.2. The point whose residual is shown in Figure 7.10 is the one where $x = 2.2$.

In the Hooke's law example, the quantity $\beta_0 + \beta_1 x$ represents the true length of the spring under a load x. Since $\widehat{\beta}_0$ and $\widehat{\beta}_1$ are estimates of the true values β_0 and β_1, the quantity $\widehat{y} = \widehat{\beta}_0 + \widehat{\beta}_1 x$ is an estimate of $\beta_0 + \beta_1 x$. Examples 7.7 and 7.8 illustrate this.

Using the Hooke's law data, estimate the length of the spring under a load of 1.3 lb.

Solution

In Example 7.6, the equation of the least-squares line was computed to be $y = 4.9997 + 0.2046x$. Using the value $x = 1.3$, we estimate the length of the spring under a load of 1.3 lb to be

$$\widehat{y} = 4.9997 + (0.2046)(1.3) = 5.27 \text{ in.}$$

TABLE 7.2 Measured lengths of a spring under various loads, with fitted values and residuals

Weight x	Measured Length y	Fitted Value $\widehat{y}$	Residual e	Weight x	Measured Length y	Fitted Value $\widehat{y}$	Residual e
0.0	5.06	5.00	0.06	2.0	5.40	5.41	−0.01
0.2	5.01	5.04	−0.03	2.2	5.57	5.45	0.12
0.4	5.12	5.08	0.04	2.4	5.47	5.49	−0.02
0.6	5.13	5.12	0.01	2.6	5.53	5.53	−0.00
0.8	5.14	5.16	−0.02	2.8	5.61	5.57	0.04
1.0	5.16	5.20	−0.04	3.0	5.59	5.61	−0.02
1.2	5.25	5.25	0.00	3.2	5.61	5.65	−0.04
1.4	5.19	5.29	−0.10	3.4	5.75	5.70	0.05
1.6	5.24	5.33	−0.09	3.6	5.68	5.74	−0.06
1.8	5.46	5.37	0.09	3.8	5.80	5.78	0.02

Example 7.8

Using the Hooke's law data, estimate the length of the spring under a load of 1.4 lb.

Solution

The estimate is $\widehat{y} = 4.9997 + (0.2046)(1.4) = 5.29$ in.

In Example 7.8, note that the measured length at a load of 1.4 was 5.19 in. (see Table 7.2). But the least-squares estimate of 5.29 in. is based on all the data and is more precise (has smaller uncertainty). We will learn how to compute uncertainties for the estimates $\widehat{y}$ in Section 7.3.

The Estimates Are Not the Same as the True Values

It is important to understand the difference between the least-squares *estimates* $\widehat{\beta}_0$ and $\widehat{\beta}_1$, and the *true values* β_0 and β_1. The true values are constants whose values are unknown. The estimates are quantities that are computed from the data. We may use the estimates as approximations for the true values.

In principle, an experiment such as the Hooke's law experiment could be repeated many times. The true values β_0 and β_1 would remain constant over the replications of the experiment. But each replication would produce different data, and thus different values of the estimates $\widehat{\beta}_0$ and $\widehat{\beta}_1$. The estimates $\widehat{\beta}_0$ and $\widehat{\beta}_1$ are thus *random variables*, since their values vary from experiment to experiment. To make full use of these estimates, we will need to be able to compute their standard deviations. We will discuss this topic in Section 7.3.

The Residuals Are Not the Same as the Errors

A collection of points $(x_1, y_1), \ldots, (x_n, y_n)$ follows a linear model if the x and y coordinates are related through the equation $y_i = \beta_0 + \beta_1 x_i + \varepsilon_i$. It is important to understand the difference between the residuals e_i and the errors ε_i. Each residual e_i is the difference

$y_i - \widehat{y}_i$ between an observed, or measured, value y_i and the fitted value $\widehat{y}_i = \widehat{\beta}_0 + \widehat{\beta}_1 x_i$ estimated from the least-squares line. Since the values y_i are known and the values $\widehat{y}_i$ can be computed from the data, the residuals can be computed. In contrast, the errors ε_i are the differences between the y_i and the values $\beta_0 + \beta_1 x_i$. Since the true values β_0 and β_1 are unknown, the errors are unknown as well. Another way to think of the distinction is that the residuals are the vertical distances from the observed values y_i to the least-squares line $\widehat{y} = \widehat{\beta}_0 + \widehat{\beta}_1 x$, and the errors are the distances from the y_i to the true line $y = \beta_0 + \beta_1 x$.

Summary

Given points $(x_1, y_1), \ldots, (x_n, y_n)$:

- The least-squares line is $\widehat{y} = \widehat{\beta}_0 + \widehat{\beta}_1 x$.
- $\widehat{\beta}_1 = \dfrac{\sum_{i=1}^{n}(x_i - \overline{x})(y_i - \overline{y})}{\sum_{i=1}^{n}(x_i - \overline{x})^2}$
- $\widehat{\beta}_0 = \overline{y} - \widehat{\beta}_1 \overline{x}$
- The quantities $\widehat{\beta}_0$ and $\widehat{\beta}_1$ can be thought of as estimates of a true slope β_1 and a true intercept β_0.
- For any x, $\widehat{y} = \widehat{\beta}_0 + \widehat{\beta}_1 x$ is an estimate of the quantity $\beta_0 + \beta_1 x$.

Don't Extrapolate Outside the Range of the Data

What if we wanted to estimate the length of the spring under a load of 100 lb? The least-squares estimate is $4.9997 + (0.2046)(100) = 25.46$ in. Should we believe this? No. None of the weights in the data set were this large. It is likely that the spring would be stretched out of shape, so Hooke's law would not hold. For many variables, linear relationships hold within a certain range, but not outside it. If we extrapolate a least-squares line outside the range of the data, therefore, there is no guarantee that it will properly describe the relationship. If we want to know how the spring will respond to a load of 100 lb, we must include weights of 100 lb or more in the data set.

Summary

Do not extrapolate a fitted line (such as the least-squares line) outside the range of the data. The linear relationship may not hold there.

Don't Use the Least-Squares Line When the Data Aren't Linear

In Section 7.1, we learned that the correlation coefficient should be used only when the relationship between x and y is linear. The same holds true for the least-squares line. When the scatterplot follows a curved pattern, it does not make sense to summarize it with a straight line. To illustrate this, Figure 7.11 presents a plot of the relationship

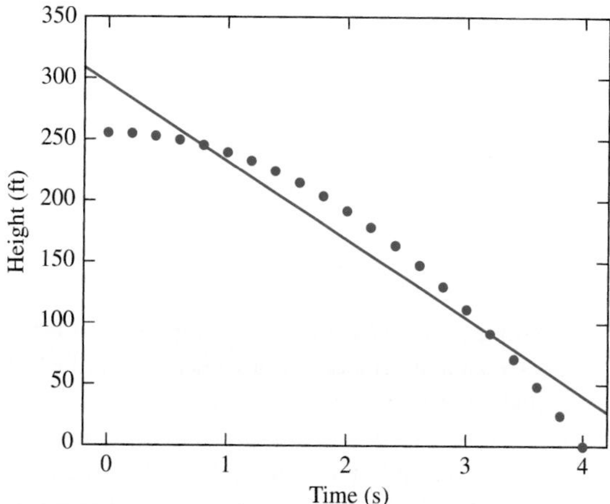

FIGURE 7.11 The relationship between the height of a free-falling object with a positive initial velocity and the time in free fall is not linear. The least-squares line does not fit the data well and should not be used to predict the height of the object at a given time.

between the height y of an object released from a height of 256 ft and the time x since its release. The relationship between x and y is nonlinear. The least-squares line does not fit the data well.

In some cases the least-squares line can be used for nonlinear data, after a process known as variable transformation has been applied. This topic is discussed in Section 7.4.

Another Look at the Least-Squares Line

The expression (7.13) for $\widehat{\beta}_1$ can be rewritten in a way that provides a useful interpretation. Starting with the definition of the correlation coefficient r (Equation 7.2 in Section 7.1) and multiplying both sides by $\sqrt{\sum_{i=1}^n (y_i - \overline{y})^2}/\sqrt{\sum_{i=1}^n (x_i - \overline{x})^2} = s_y/s_x$ yields the result

$$\widehat{\beta}_1 = r\frac{s_y}{s_x} \tag{7.18}$$

Equation (7.18) allows us to interpret the slope of the least-squares line in terms of the correlation coefficient. The units of $\widehat{\beta}_1$, the slope of the least-squares line, must be units of y/x. The correlation coefficient r is a pure number that measures the strength of the linear relationship between x and y. Equation (7.18) shows that the slope $\widehat{\beta}_1$ is proportional to the correlation coefficient, where the constant of proportionality is the quantity s_y/s_x that adjusts for the units in which x and y are measured.

Using Equation (7.13), we can write the equation of the least-squares line in point-slope form: Substituting $\overline{y} - \widehat{\beta}_1\overline{x}$ for $\widehat{\beta}_0$ in the equation for the least-squares line $\widehat{y} = \widehat{\beta}_0 + \widehat{\beta}_1 x$ and rearranging terms yields

$$\widehat{y} - \overline{y} = \widehat{\beta}_1(x - \overline{x}) \tag{7.19}$$

Combining Equations (7.18) and (7.19) yields

$$\widehat{y} - \overline{y} = r\frac{s_y}{s_x}(x - \overline{x}) \tag{7.20}$$

Thus the least-squares line is the line that passes through the center of mass of the scatterplot $(\overline{x}, \overline{y})$, with slope $\widehat{\beta}_1 = r(s_y/s_x)$.

Measuring Goodness-of-Fit

A goodness-of-fit statistic is a quantity that measures how well a model explains a given set of data. A linear model fits well if there is a strong linear relationship between x and y. We mentioned in Section 7.1 that the correlation coefficient r measures the strength of the linear relationship between x and y. Therefore r is a goodness-of-fit statistic for the linear model. We will now describe how r measures the goodness-of-fit. Figure 7.12 presents Galton's data on forearm lengths versus heights. The points on the scatterplot are (x_i, y_i) where x_i is the height of the ith man and y_i is the length of his forearm. Both the least-squares line and the horizontal line $y = \overline{y}$ are superimposed. Now imagine that we must predict the length of one of the forearms. If we have no knowledge of the man's height, we must predict his forearm length to be the average $\overline{y}$. Our prediction error is $y_i - \overline{y}$. If we predict the length of each forearm this way, the

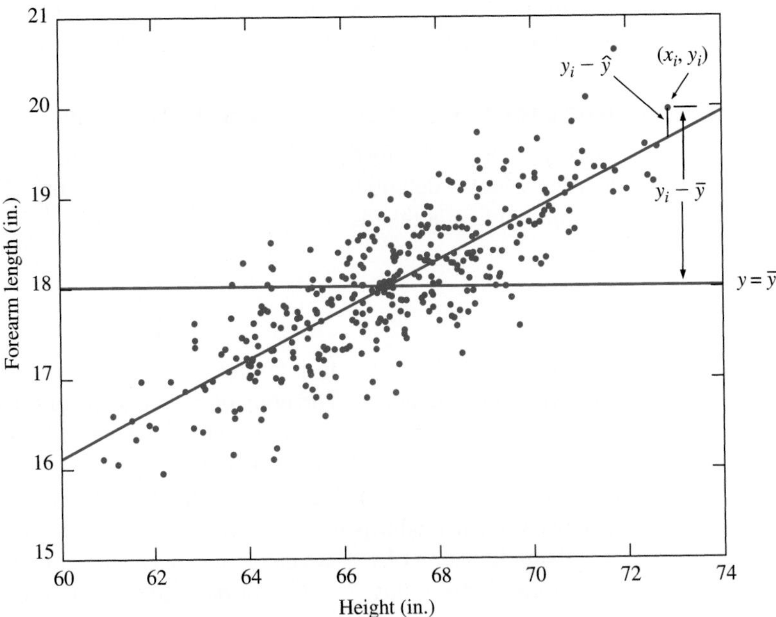

FIGURE 7.12 Heights and forearm lengths of men. The least-squares line and the horizontal line $y = \overline{y}$ are superimposed.

sum of squared prediction errors will be $\sum_{i=1}^{n}(y_i - \overline{y})^2$. If, on the other hand, we know the height of each man before predicting the length of his forearm, we can use the least-squares line, and we will predict the ith forearm length to be $\widehat{y}_i$. The prediction error will be the residual $y_i - \widehat{y}_i$, and the sum of squared prediction errors is $\sum_{i=1}^{n}(y_i - \widehat{y}_i)^2$. The strength of the linear relationship can be measured by computing the reduction in sum of squared prediction errors obtained by using $\widehat{y}_i$ rather than $\overline{y}$. This is the difference $\sum_{i=1}^{n}(y_i - \overline{y})^2 - \sum_{i=1}^{n}(y_i - \widehat{y}_i)^2$. The bigger this difference is, the more tightly clustered the points are around the least-squares line and the stronger the linear relationship is between x and y. Thus $\sum_{i=1}^{n}(y_i - \overline{y})^2 - \sum_{i=1}^{n}(y_i - \widehat{y}_i)^2$ is a goodness-of-fit statistic.

There is a problem with using $\sum_{i=1}^{n}(y_i - \overline{y})^2 - \sum_{i=1}^{n}(y_i - \widehat{y}_i)^2$ as a goodness-of-fit statistic, however. This quantity has units, namely the squared units of y. We could not use this statistic to compare the goodness-of-fit of two models fit to different data sets, since the units would be different. We need to use a goodness-of-fit statistic that is a pure number, so that we can measure goodness-of-fit on an absolute scale.

This is where the correlation coefficient r comes in. It is shown at the end of this section that

$$r^2 = \frac{\sum_{i=1}^{n}(y_i - \overline{y})^2 - \sum_{i=1}^{n}(y_i - \widehat{y}_i)^2}{\sum_{i=1}^{n}(y_i - \overline{y})^2}. \tag{7.21}$$

The quantity r^2, the square of the correlation coefficient, is called the **coefficient of determination**. It is the reduction in the sum of the squared prediction errors obtained by using $\widehat{y}_i$ rather than $\overline{y}$, expressed as a fraction of the sum of squared prediction errors $\sum_{i=1}^{n}(y_i - \overline{y})^2$, obtained by using $\overline{y}$. This interpretation of r^2 is important to know. In Chapter 8, we will see how it can be generalized to provide a measure of the goodness-of-fit of linear relationships involving several variables.

For a visual interpretation of r^2, look at Figure 7.12. For each point (x_i, y_i) on the scatterplot, the quantity $y_i - \overline{y}$ is the vertical distance from the point to the horizontal line $y = \overline{y}$. The quantity $y_i - \widehat{y}_i$ is the vertical distance from the point to the least-squares line. Thus the quantity $\sum_{i=1}^{n}(y_i - \overline{y})^2$ measures the overall spread of the points around the the line $y = \overline{y}$ and the quantity $\sum_{i=1}^{n}(y_i - \widehat{y}_i)^2$ measures the overall spread of the points around the least-squares line. The quantity $\sum_{i=1}^{n}(y_i - \overline{y})^2 - \sum_{i=1}^{n}(y_i - \widehat{y}_i)^2$ therefore measures the reduction in the spread of the points obtained by using the least-squares line rather than $y = \overline{y}$. The coefficient of determination r^2 expresses this reduction as a proportion of the spread around $y = \overline{y}$.

The sums of squares appearing in this discussion are used so often that statisticians have given them names. They call $\sum_{i=1}^{n}(y_i - \widehat{y}_i)^2$ the **error sum of squares** and $\sum_{i=1}^{n}(y_i - \overline{y})^2$ the **total sum of squares**. Their difference $\sum_{i=1}^{n}(y_i - \overline{y})^2 - \sum_{i=1}^{n}(y_i - \widehat{y}_i)^2$ is called the **regression sum of squares**. Clearly, the following relationship holds:

$$\text{Total sum of squares} = \text{Regression sum of squares} + \text{Error sum of squares}$$

Using the preceding terminology, we can write Equation (7.21) as

$$r^2 = \frac{\text{Regression sum of squares}}{\text{Total sum of squares}}$$

Since the total sum of squares is just the sample variance of the y_i without dividing by $n - 1$, statisticians (and others) often refer to r^2 as the **proportion of the variance in y explained by regression**.

Derivation of the Least-Squares Coefficients $\widehat{\beta}_0$ and $\widehat{\beta}_1$

We derive Equations (7.13) and (7.14). The least-squares coefficients $\widehat{\beta}_0$ and $\widehat{\beta}_1$ are the quantities that minimize the sum

$$S = \sum_{i=1}^{n}(y_i - \widehat{\beta}_0 - \widehat{\beta}_1 x_i)^2$$

We compute these values by taking partial derivatives of S with respect to $\widehat{\beta}_0$ and $\widehat{\beta}_1$ and setting them equal to 0. Thus $\widehat{\beta}_0$ and $\widehat{\beta}_1$ are the quantities that solve the simultaneous equations

$$\frac{\partial S}{\partial \widehat{\beta}_0} = -\sum_{i=1}^{n} 2(y_i - \widehat{\beta}_0 - \widehat{\beta}_1 x_i) = 0 \qquad (7.22)$$

$$\frac{\partial S}{\partial \widehat{\beta}_1} = -\sum_{i=1}^{n} 2x_i(y_i - \widehat{\beta}_0 - \widehat{\beta}_1 x_i) = 0 \qquad (7.23)$$

These equations can be written as a system of two linear equations in two unknowns:

$$n\widehat{\beta}_0 + \left(\sum_{i=1}^{n} x_i\right)\widehat{\beta}_1 = \sum_{i=1}^{n} y_i \qquad (7.24)$$

$$\left(\sum_{i=1}^{n} x_i\right)\widehat{\beta}_0 + \left(\sum_{i=1}^{n} x_i^2\right)\widehat{\beta}_1 = \sum_{i=1}^{n} x_i y_i \qquad (7.25)$$

We solve Equation (7.24) for $\widehat{\beta}_0$, obtaining

$$\begin{aligned}\widehat{\beta}_0 &= \frac{\sum_{i=1}^{n} y_i}{n} - \widehat{\beta}_1 \frac{\sum_{i=1}^{n} x_i}{n} \\ &= \overline{y} - \widehat{\beta}_1 \overline{x}\end{aligned}$$

This establishes Equation (7.14). Now substitute $\overline{y} - \widehat{\beta}_1 \overline{x}$ for $\widehat{\beta}_0$ in Equation (7.25) to obtain

$$\left(\sum_{i=1}^{n} x_i\right)(\overline{y} - \widehat{\beta}_1 \overline{x}) + \left(\sum_{i=1}^{n} x_i^2\right)\widehat{\beta}_1 = \sum_{i=1}^{n} x_i y_i \qquad (7.26)$$

Solving Equation (7.26) for $\widehat{\beta}_1$, we obtain

$$\widehat{\beta}_1 = \frac{\sum_{i=1}^{n} x_i y_i - n\overline{x}\,\overline{y}}{\sum_{i=1}^{n} x_i^2 - n\overline{x}^2}$$

To establish Equation (7.13), we must show that $\sum_{i=1}^{n}(x_i - \overline{x})^2 = \sum_{i=1}^{n} x_i^2 - n\overline{x}^2$ and that $\sum_{i=1}^{n}(x_i - \overline{x})(y_i - \overline{y}) = \sum_{i=1}^{n} x_i y_i - n\,\overline{x}\,\overline{y}$. (These are Equations 7.15 and 7.17.)

Now

$$\begin{aligned}
\sum_{i=1}^{n}(x_i - \overline{x})(y_i - \overline{y}) &= \sum_{i=1}^{n}(x_i y_i - \overline{x} y_i - \overline{y} x_i + \overline{x}\,\overline{y}) \\
&= \sum_{i=1}^{n} x_i y_i - \overline{x}\sum_{i=1}^{n} y_i - \overline{y}\sum_{i=1}^{n} x_i + \sum_{i=1}^{n}\overline{x}\,\overline{y} \\
&= \sum_{i=1}^{n} x_i y_i - n\,\overline{x}\,\overline{y} - n\,\overline{y}\,\overline{x} + n\,\overline{x}\,\overline{y} \\
&= \sum_{i=1}^{n} x_i y_i - n\,\overline{x}\,\overline{y}
\end{aligned}$$

Also

$$\begin{aligned}
\sum_{i=1}^{n}(x_i - \overline{x})^2 &= \sum_{i=1}^{n}(x_i^2 - 2\overline{x}x_i + \overline{x}^2) \\
&= \sum_{i=1}^{n} x_i^2 - 2\overline{x}\sum_{i=1}^{n} x_i + \sum_{i=1}^{n}\overline{x}^2 \\
&= \sum_{i=1}^{n} x_i^2 - 2n\overline{x}^2 + n\overline{x}^2 \\
&= \sum_{i=1}^{n} x_i^2 - n\overline{x}^2
\end{aligned}$$

Derivation of Equation (7.21)

We first show that

$$\sum_{i=1}^{n}(y_i - \overline{y})^2 = \sum_{i=1}^{n}(y_i - \widehat{y}_i)^2 + \sum_{i=1}^{n}(\widehat{y}_i - \overline{y})^2 \tag{7.27}$$

This result is known as the **analysis of variance identity**. To derive it, we begin by adding and subtracting $\widehat{y}_i$ from the left-hand side:

$$\begin{aligned}
\textstyle\sum_{i=1}^{n}(y_i - \overline{y})^2 &= \textstyle\sum_{i=1}^{n}[(y_i - \widehat{y}_i) + (\widehat{y}_i - \overline{y})]^2 \\
&= \textstyle\sum_{i=1}^{n}(y_i - \widehat{y}_i)^2 + \sum_{i=1}^{n}(\widehat{y}_i - \overline{y})^2 + 2\sum_{i=1}^{n}(y_i - \widehat{y}_i)(\widehat{y}_i - \overline{y})
\end{aligned}$$

Now we need only to show that $\sum_{i=1}^{n}(y_i - \widehat{y}_i)(\widehat{y}_i - \overline{y}) = 0$. Since $\widehat{y}_i = \widehat{\beta}_0 + \widehat{\beta}_1 x_i$ and $\widehat{\beta}_0 = \overline{y} - \widehat{\beta}_1\overline{x}$,

$$\widehat{y}_i = \overline{y} + \widehat{\beta}_1(x_i - \overline{x}) \tag{7.28}$$

Therefore

$$\sum_{i=1}^{n}(y_i - \widehat{y}_i)(\widehat{y}_i - \overline{y}) = \sum_{i=1}^{n}[(y_i - \overline{y}) - \widehat{\beta}_1(x_i - \overline{x})][\widehat{\beta}_1(x_i - \overline{x})]$$

$$= \widehat{\beta}_1 \sum_{i=1}^{n}(x_i - \overline{x})(y_i - \overline{y}) - \widehat{\beta}_1^2 \sum_{i=1}^{n}(x_i - \overline{x})^2 \quad (7.29)$$

Now $\widehat{\beta}_1 = \dfrac{\sum_{i=1}^{n}(x_i - \overline{x})(y_i - \overline{y})}{\sum_{i=1}^{n}(x_i - \overline{x})^2}$, so

$$\sum_{i=1}^{n}(x_i - \overline{x})(y_i - \overline{y}) = \widehat{\beta}_1 \sum_{i=1}^{n}(x_i - \overline{x})^2$$

Substituting into Equation (7.29), we obtain

$$\sum_{i=1}^{n}(y_i - \widehat{y}_i)(\widehat{y}_i - \overline{y}) = \widehat{\beta}_1^2 \sum_{i=1}^{n}(x_i - \overline{x})^2 - \widehat{\beta}_1^2 \sum_{i=1}^{n}(x_i - \overline{x})^2 = 0$$

This establishes the analysis of variance identity.

To derive Equation (7.21), Equation (7.28) implies that

$$\widehat{y}_i - \overline{y} = \widehat{\beta}_1(x_i - \overline{x}) \quad (7.30)$$

Square both sides of Equation (7.30) and sum, obtaining

$$\sum_{i=1}^{n}(\widehat{y}_i - \overline{y})^2 = \widehat{\beta}_1^2 \sum_{i=1}^{n}(x_i - \overline{x})^2$$

Now $\widehat{\beta}_1 = r\dfrac{s_y}{s_x}$ (Equation 7.18), so

$$\widehat{\beta}_1^2 = r^2 \frac{\sum_{i=1}^{n}(y_i - \overline{y})^2}{\sum_{i=1}^{n}(x_i - \overline{x})^2}$$

Substituting and canceling, we obtain

$$\sum_{i=1}^{n}(\widehat{y}_i - \overline{y})^2 = r^2 \sum_{i=1}^{n}(y_i - \overline{y})^2$$

so

$$r^2 = \frac{\sum_{i=1}^{n}(\widehat{y}_i - \overline{y})^2}{\sum_{i=1}^{n}(y_i - \overline{y})^2}$$

By the analysis of variance identity, $\sum_{i=1}^{n}(\widehat{y}_i - \overline{y})^2 = \sum_{i=1}^{n}(y_i - \overline{y})^2 - \sum_{i=1}^{n}(y_i - \widehat{y}_i)^2$. Therefore

$$r^2 = \frac{\sum_{i=1}^{n}(y_i - \overline{y})^2 - \sum_{i=1}^{n}(y_i - \widehat{y}_i)^2}{\sum_{i=1}^{n}(y_i - \overline{y})^2}$$

Exercises for Section 7.2

1. Each month for several months, the average temperature in °C (x) and the number of pounds of steam (y) consumed by a certain chemical plant were measured. The least-squares line computed from the resulting data is $y = 245.82 + 1.13x$.

a. Predict the number of pounds of steam consumed in a month where the average temperature is 65°C.

b. If two months differ in their average temperatures by 5°C, by how much do you predict the number of pounds of steam consumed to differ?

2. In a study of the relationship between the Brinell hardness (x) and tensile strength in ksi (y) of specimens of cold drawn copper, the least-squares line was $y = -196.32 + 2.42x$.

a. Predict the tensile strength of a specimen whose Brinell hardness is 102.7.

b. If two specimens differ in their Brinell hardness by 3, by how much do you predict their tensile strengths to differ?

3. A least-squares line is fit to a set of points. If the total sum of squares is $\sum(y_i - \overline{y})^2 = 9615$, and the error sum of squares is $\sum(y_i - \widehat{y}_i)^2 = 1450$, compute the coefficient of determination r^2.

4. A least-squares line is fit to a set of points. If the total sum of squares is $\sum(y_i - \overline{y})^2 = 181.2$, and the error sum of squares is $\sum(y_i - \widehat{y}_i)^2 = 33.9$, compute the coefficient of determination r^2.

5. In Galton's height data (Figure 7.1, in Section 7.1), the least-squares line for predicting forearm length (y) from height (x) is $y = -0.2967 + 0.2738x$.

a. Predict the forearm length of a man whose height is 70 in.

b. How tall must a man be so that we would predict his forearm length to be 19 in.?

c. All the men in a certain group have heights greater than the height computed in part (b). Can you conclude that all their forearms will be at least 19 in. long? Explain.

6. In a study relating the degree of warping, in mm, of a copper plate (y) to temperature in °C (x), the following summary statistics were calculated: $n = 40$, $\sum_{i=1}^{n}(x_i - \overline{x})^2 = 98{,}775$, $\sum_{i=1}^{n}(y_i - \overline{y})^2 = 19.10$, $\overline{x} = 26.36$, $\overline{y} = 0.5188$, $\sum_{i=1}^{n}(x_i - \overline{x})(y_i - \overline{y}) = 826.94$.

a. Compute the correlation r between the degree of warping and the temperature.

b. Compute the error sum of squares, the regression sum of squares, and the total sum of squares.

c. Compute the least-squares line for predicting warping from temperature.

d. Predict the warping at a temperature of 40°C.

e. At what temperature will we predict the warping to be 0.5 mm?

f. Assume it is important that the warping not exceed 0.5 mm. An engineer suggests that if the temperature is kept below the level computed in part (e), we can be sure that the warping will not exceed 0.5 mm. Is this a correct conclusion? Explain.

7. Inertial weight (in tons) and fuel economy (in mi/gal) were measured for a sample of seven diesel trucks. The results are presented in the following table. (From "In-Use Emissions from Heavy-Duty Diesel Vehicles," J. Yanowitz, Ph.D. thesis, Colorado School of Mines, 2001.)

Weight	Mileage
8.00	7.69
24.50	4.97
27.00	4.56
14.50	6.49
28.50	4.34
12.75	6.24
21.25	4.45

a. Construct a scatterplot of mileage (y) versus weight (x). Verify that a linear model is appropriate.

b. Compute the least-squares line for predicting mileage from weight.

c. If two trucks differ in weight by 5 tons, by how much would you predict their mileages to differ?

d. Predict the mileage for trucks with a weight of 15 tons.

e. What are the units of the estimated slope $\widehat{\beta}_1$?

f. What are the units of the estimated intercept $\widehat{\beta}_0$?

8. The processing of raw coal involves "washing," in which coal ash (nonorganic, incombustible material) is removed. The article "Quantifying Sampling Precision for Coal Ash Using Gy's Discrete Model of the Fundamental Error" (*Journal of Coal Quality*, 1989, 33–39) provides data relating the percentage of ash to the density of a coal particle. The average percentage ash for five densities of coal particles was measured. The data are presented in the following table:

Density (g/cm³)	Percent ash
1.25	1.93
1.325	4.63
1.375	8.95
1.45	15.05
1.55	23.31

a. Construct a scatterplot of percent ash (y) versus density (x). Verify that a linear model is appropriate.

b. Compute the least-squares line for predicting percent ash from density.

c. If two coal particles differed in density by 0.1 g/cm^3, by how much would you predict their percent ash to differ?

d. Predict the percent ash for particles with density 1.40 g/cm^3.

e. Compute the fitted values.

f. Compute the residuals. Which point has the residual with the largest magnitude?

g. Compute the correlation between density and percent ash.

h. Compute the regression sum of squares, the error sum of squares, and the total sum of squares.

i. Divide the regression sum of squares by the total sum of squares. What is the relationship between this quantity and the correlation coefficient?

9. In tests designed to measure the effect of a certain additive on the drying time of paint, the following data were obtained.

Concentration of Additive (%)	Drying Time (h)
4.0	8.7
4.2	8.8
4.4	8.3
4.6	8.7
4.8	8.1
5.0	8.0
5.2	8.1
5.4	7.7
5.6	7.5
5.8	7.2

a. Construct a scatterplot of drying time (y) versus additive concentration (x). Verify that a linear model is appropriate.

b. Compute the least-squares line for predicting drying time from additive concentration.

c. Compute the fitted value and the residual for each point.

d. If the concentration of the additive is increased by 0.1%, by how much would you predict the drying time to increase or decrease?

e. Predict the drying time for a concentration of 4.4%.

f. Can the least-squares line be used to predict the drying time for a concentration of 7%? If so, predict the drying time. If not, explain why not.

g. For what concentration would you predict a drying time of 8.2 hours?

h. The goal of this project is to reduce the drying time to 6 hours. Based on the given data, can you specify a concentration that is likely to produce this result? If so, specify the concentration. If not, explain why not.

10. Curing times in days (x) and compressive strengths in MPa (y) were recorded for several concrete specimens. The means and standard deviations of the x and y values were $\overline{x} = 5$, $s_x = 2$, $\overline{y} = 1350$, $s_y = 100$. The correlation between curing time and compressive strength was computed to be $r = 0.7$. Find the equation of the least-squares line to predict compressive strength from curing time.

11. Varying amounts of pectin were added to canned jellies, to study the relationship between pectin concentration in % (x) and a firmness index (y). The means and standard deviations of the x and y values were $\overline{x} = 3$, $s_x = 0.5$, $\overline{y} = 50$, $s_y = 10$. The correlation between curing time and firmness was computed to be $r = 0.5$. Find the equation of the least-squares line to predict firmness from pectin concentration.

12. An engineer wants to predict the value for y when $x = 4.5$, using the following data set.

x	y	$z = \ln y$	x	y	$z = \ln y$
1	0.2	−1.61	6	2.3	0.83
2	0.3	−1.20	7	2.9	1.06
3	0.5	−0.69	8	4.5	1.50
4	0.5	−0.69	9	8.7	2.16
5	1.3	0.26	10	12.0	2.48

a. Construct a scatterplot of the points (x, y).

b. Should the least-squares line be used to predict the value of y when $x = 4.5$? If so, compute the least-squares line and the predicted value. If not, explain why not.

c. Construct a scatterplot of the points (x, z), where $z = \ln y$.

d. Use the least-squares line to predict the value of z when $x = 4.5$. Is this an appropriate method of prediction? Explain why or why not.

e. Let $\widehat{z}$ denote the predicted value of z computed in part (d). Let $\widehat{y} = e^{\widehat{z}}$. Explain why $\widehat{y}$ is a reasonable predictor of the value of y when $x = 4.5$.

13. A simple random sample of 100 men aged 25–34 averaged 70 inches in height, and had a standard deviation of 3 inches. Their incomes averaged \$34,900 and had a standard deviation of \$17,200. Fill in the blank: From the least-squares line, we would predict that the income of a man 70 inches tall would be ________.

i. less than \$34,900.

ii. greater than \$34,900.

iii. equal to \$34,900.

iv. We cannot tell unless we know the correlation.

14. A mixture of sucrose and water was heated on a hot plate, and the temperature (in °C) was recorded each minute for 20 minutes by three thermocouples. The results are shown in the following table.

Time	T_1	T_2	T_3
0	20	18	21
1	18	22	11
2	29	22	26
3	32	25	35
4	37	37	33
5	36	46	35
6	46	45	44
7	46	44	43
8	56	54	63
9	58	64	68
10	64	69	62
11	72	65	65
12	79	80	80
13	84	74	75
14	82	87	78
15	87	93	88
16	98	90	91
17	103	100	103
18	101	98	109
19	103	103	107
20	102	103	104

a. Compute the least-squares line for estimating the temperature as a function of time, using T_1 as the value for temperature.

b. Compute the least-squares line for estimating the temperature as a function of time, using T_2 as the value for temperature.

c. Compute the least-squares line for estimating the temperature as a function of time, using T_3 as the value for temperature.

d. It is desired to compute a single line to estimate temperature as a function of time. One person suggests averaging the three slope estimates to obtain a single slope estimate, and averaging the three intercept estimates to obtain a single intercept estimate. Find the equation of the line that results from this method.

e. Someone else suggests averaging the three temperature measurements at each time to obtain $\overline{T} = (T_1 + T_2 + T_3)/3$. Compute the least-squares line using $\overline{T}$ as the value for temperature.

f. Are the results of parts (d) and (e) different?

7.3 Uncertainties in the Least-Squares Coefficients

In Section 7.2, the linear model was presented (Equation 7.9):

$$y_i = \beta_0 + \beta_1 x_i + \varepsilon_i$$

Here ε_i is the error in the ith observation y_i. In practice, ε_i represents the accumulation of error from many sources. For example, in the Hooke's law data, ε_i can be affected by errors in measuring the length of the spring, errors in measuring the weights of the loads placed on the spring, variations in the elasticity of the spring due to changes in ambient temperature or metal fatigue, and so on. If there were no error, the points would lie exactly on the least-squares line, and the slope $\widehat{\beta}_1$ and intercept $\widehat{\beta}_0$ of the least-squares line would equal the true values β_0 and β_1. Because of error, the points are scattered around the line, and the quantities $\widehat{\beta}_0$ and $\widehat{\beta}_1$ do not equal the true values. Each time the process is repeated, the values of ε_i, and thus the values of $\widehat{\beta}_0$ and $\widehat{\beta}_1$, will be different. In other words, ε_i, $\widehat{\beta}_0$, and $\widehat{\beta}_1$ are random variables. To be more specific, the errors ε_i create *uncertainty* in the estimates $\widehat{\beta}_0$ and $\widehat{\beta}_1$. It is intuitively clear that if the ε_i tend to be small in magnitude, the points will be tightly clustered around the line, and the uncertainty in the least-squares estimates $\widehat{\beta}_0$ and $\widehat{\beta}_1$ will be small. On the other hand, if the ε_i tend to be large in magnitude, the points will be widely scattered around the line, and the uncertainties (standard deviations) in the least-squares estimates $\widehat{\beta}_0$ and $\widehat{\beta}_1$ will be larger.

Assume we have n data points $(x_1, y_1), \ldots, (x_n, y_n)$, and we plan to fit the least-squares line. In order for the estimates $\widehat{\beta}_1$ and $\widehat{\beta}_0$ to be useful, we need to estimate just how large their uncertainties are. In order to make such an estimate, we need to know something about the nature of the errors ε_i. We will begin by studying the simplest situation, in which four important assumptions are satisfied. These are given in the following box.

Assumptions for Errors in Linear Models

In the simplest situation, the following assumptions are satisfied:

1. The errors $\varepsilon_1, \ldots, \varepsilon_n$ are random and independent. In particular, the magnitude of any error ε_i does not influence the value of the next error ε_{i+1}.
2. The errors $\varepsilon_1, \ldots, \varepsilon_n$ all have mean 0.
3. The errors $\varepsilon_1, \ldots, \varepsilon_n$ all have the same variance, which we denote by σ^2.
4. The errors $\varepsilon_1, \ldots, \varepsilon_n$ are normally distributed.

These assumptions are restrictive, so it is worthwhile to discuss briefly the degree to which it is acceptable to violate them in practice. When the sample size is large, the normality assumption (4) becomes less important. Mild violations of the assumption of constant variance (3) do not matter too much, but severe violations should be corrected. In Section 7.4, we discuss methods to correct certain violations of these assumptions.

Under these assumptions, the effect of the ε_i is largely governed by the magnitude of the variance σ^2, since it is this variance that determines how large the errors are likely to be. Therefore, in order to estimate the uncertainties in $\widehat{\beta}_0$ and $\widehat{\beta}_1$, we must first estimate the error variance σ^2. Since the magnitude of the variance is reflected in the degree of spread of the points around the least-squares line, it follows that by measuring this spread, we can estimate the variance. Specifically, the vertical distance from each data point (x_i, y_i) to the least-squares line is given by the residual e_i (see Figure 7.10 in Section 7.2). The spread of the points around the line can be measured by the sum of the squared residuals $\sum_{i=1}^n e_i^2$. The estimate of the error variance σ^2 is the quantity s^2 given by

$$s^2 = \frac{\sum_{i=1}^n e_i^2}{n-2} = \frac{\sum_{i=1}^n (y_i - \widehat{y}_i)^2}{n-2} \tag{7.31}$$

The estimate of the error variance is thus the average of the squared residuals, except that we divide by $n - 2$ rather than n. The reason for this is that since the least-squares line minimizes the sum $\sum_{i=1}^n e_i^2$, the residuals tend to be a little smaller than the errors ε_i. It turns out that dividing by $n - 2$ rather than n appropriately compensates for this.

There is an equivalent formula for s^2, involving the correlation coefficient r, that is often easier to calculate.

$$s^2 = \frac{(1-r^2)\sum_{i=1}^n (y_i - \overline{y})^2}{n-2} \tag{7.32}$$

We present a brief derivation of this result. Equation (7.21) (in Section 7.2) shows that $1-r^2 = \sum_{i=1}^n (y_i-\widehat{y}_i)^2 / \sum_{i=1}^n (y_i-\overline{y})^2$. Then $\sum_{i=1}^n (y_i-\widehat{y}_i)^2 = (1-r^2)\sum_{i=1}^n (y_i-\overline{y})^2$, and it follows that

$$s^2 = \frac{\sum_{i=1}^n (y_i - \widehat{y}_i)^2}{n-2} = \frac{(1-r^2)\sum_{i=1}^n (y_i - \overline{y})^2}{n-2}$$

Under assumptions 1 through 4, the observations y_i are also random variables. In fact, since $y_i = \beta_0 + \beta_1 x_i + \varepsilon_i$, it follows that y_i has a normal distribution with mean $\beta_0 + \beta_1 x_i$ and variance σ^2. In particular, β_1 represents the change in the mean of y associated with an increase of one unit in the value of x.

Summary

In the linear model $y_i = \beta_0 + \beta_1 x_i + \varepsilon_i$, under assumptions 1 through 4, the observations $y_1, \ldots, y_n$ are independent random variables that follow the normal distribution. The mean and variance of y_i are given by

$$\mu_{y_i} = \beta_0 + \beta_1 x_i$$

$$\sigma_{y_i}^2 = \sigma^2$$

The slope β_1 represents the change in the mean of y associated with an increase of one unit in the value of x.

We can now calculate the means and standard deviations of $\widehat{\beta}_0$ and $\widehat{\beta}_1$. The standard deviations are of course the uncertainties. Both $\widehat{\beta}_0$ and $\widehat{\beta}_1$ can be expressed as linear combinations of the y_i, so their means can be found using Equation (2.49) and their standard deviations can be found using Equation (2.53) (both equations in Section 2.5). Specifically, algebraic manipulation of Equations (7.13) and (7.14) (in Section 7.2) yields

$$\widehat{\beta}_1 = \sum_{i=1}^{n} \left[\frac{(x_i - \overline{x})}{\sum_{i=1}^{n}(x_i - \overline{x})^2} \right] y_i \tag{7.33}$$

$$\widehat{\beta}_0 = \sum_{i=1}^{n} \left[\frac{1}{n} - \frac{\overline{x}(x_i - \overline{x})}{\sum_{i=1}^{n}(x_i - \overline{x})^2} \right] y_i \tag{7.34}$$

Using the fact that each of the y_i has mean $\beta_0 + \beta_1 x_i$ and variance σ^2, Equations (2.49) and (2.53) yield the following results, after further manipulation:

$$\mu_{\widehat{\beta}_0} = \beta_0 \qquad \mu_{\widehat{\beta}_1} = \beta_1$$

$$\sigma_{\widehat{\beta}_0} = \sigma \sqrt{\frac{1}{n} + \frac{\overline{x}^2}{\sum_{i=1}^{n}(x_i - \overline{x})^2}} \qquad \sigma_{\widehat{\beta}_1} = \frac{\sigma}{\sqrt{\sum_{i=1}^{n}(x_i - \overline{x})^2}}$$

The estimates $\widehat{\beta}_0$ and $\widehat{\beta}_1$ are unbiased, since their means are equal to the true values. They are also normally distributed, because they are linear combinations of the independent normal random variables y_i. In practice, when computing the standard deviations, we usually don't know the value of σ, so we approximate it with s.

Summary

Under assumptions 1 through 4 (p. 508),

- The quantities $\widehat{\beta}_0$ and $\widehat{\beta}_1$ are normally distributed random variables.
- The means of $\widehat{\beta}_0$ and $\widehat{\beta}_1$ are the true values β_0 and β_1, respectively.
- The standard deviations of $\widehat{\beta}_0$ and $\widehat{\beta}_1$ are estimated with

$$s_{\widehat{\beta}_0} = s \sqrt{\frac{1}{n} + \frac{\overline{x}^2}{\sum_{i=1}^{n}(x_i - \overline{x})^2}} \tag{7.35}$$

and

$$s_{\widehat{\beta}_1} = \frac{s}{\sqrt{\sum_{i=1}^{n}(x_i - \overline{x})^2}} \tag{7.36}$$

where $s = \sqrt{\dfrac{(1 - r^2)\sum_{i=1}^{n}(y_i - \overline{y})^2}{n - 2}}$ is an estimate of the error standard deviation σ.

7.9

For the Hooke's law data, compute s, $s_{\widehat{\beta}_1}$, and $s_{\widehat{\beta}_0}$. Estimate the spring constant and the unloaded length, and find their uncertainties.

Solution

In Example 7.6 (in Section 7.2) we computed $\overline{x} = 1.9000$, $\overline{y} = 5.3885$, $\sum_{i=1}^{n}(x_i - \overline{x})^2 = 26.6000$, and $\sum_{i=1}^{n}(x_i - \overline{x})(y_i - \overline{y}) = 5.4430$. Now compute $\sum_{i=1}^{n}(y_i - \overline{y})^2 = 1.1733$. The correlation is $r = 5.4430/\sqrt{(26.6000)(1.1733)} = 0.9743$.

Using Equation (7.32), $s = \sqrt{\dfrac{(1 - 0.9743^2)(1.1733)}{18}} = 0.0575$.

Using Equation (7.35), $s_{\widehat{\beta}_0} = 0.0575\sqrt{\dfrac{1}{20} + \dfrac{1.9000^2}{26.6000}} = 0.0248$.

Using Equation (7.36), $s_{\widehat{\beta}_1} = \dfrac{0.0575}{\sqrt{26.6000}} = 0.0111$.

The More Spread in the *x* Values, the Better (Within Reason)

In the expressions for both of the uncertainties $s_{\widehat{\beta}_0}$ and $s_{\widehat{\beta}_1}$ in Equations (7.35) and (7.36), the quantity $\sum_{i=1}^{n}(x_i - \overline{x})^2$ appears in a denominator. This quantity measures the spread in the x values; when divided by the constant $n - 1$, it is just the sample variance of the x values. It follows that other things being equal, an experiment performed with more widely spread out x values will result in smaller uncertainties for $\widehat{\beta}_0$ and $\widehat{\beta}_1$, and thus more precise estimation of the true values β_0 and β_1. Of course, it is important not to use x values so large or so small that they are outside the range for which the linear model holds.

Summary

When one is able to choose the x values, it is best to spread them out widely. The more spread out the x values, the smaller the uncertainties in $\widehat{\beta}_0$ and $\widehat{\beta}_1$.

Specifically, the uncertainty $\sigma_{\widehat{\beta}_1}$ in $\widehat{\beta}_1$ is inversely proportional to $\sqrt{\sum_{i=1}^{n}(x_i - \overline{x})^2}$, or equivalently, to the sample standard deviation of $x_1, x_2, \ldots, x_n$.

Caution: If the range of x values extends beyond the range where the linear model holds, the results will not be valid.

There are two other ways to improve the accuracy of the estimated regression line. First, one can increase the size of the sum $\sum_{i=1}^{n}(x_i - \overline{x})^2$ by taking more observations, thus adding more terms to the sum. And second, one can decrease the size of the error variance σ^2, for example, by measuring more precisely. These two methods usually add to the cost of a project, however, while simply choosing more widely spread x values often does not.

Example 7.10

Two engineers are conducting independent experiments to estimate a spring constant for a particular spring. The first engineer suggests measuring the length of the spring with no load, and then applying loads of 1, 2, 3, and 4 lb. The second engineer suggests using loads of 0, 2, 4, 6, and 8 lb. Which result will be more precise? By what factor?

Solution

The sample standard deviation of the numbers 0, 2, 4, 6, 8 is twice as great as the sample standard deviation of the numbers 0, 1, 2, 3, 4. Therefore the uncertainty $\sigma_{\widehat{\beta}_1}$ for the first engineer is twice as large as for the second engineer, so the second engineer's estimate is twice as precise.

We have made two assumptions in the solution to this example. First, we assumed that the error variance σ^2 is the same for both engineers. If they are both using the same apparatus and the same measurement procedure, this could be a safe assumption. But if one engineer is able to measure more precisely, this needs to be taken into account. Second, we have assumed that a load of 8 lb is within the elastic zone of the spring, so that the linear model applies throughout the range of the data.

Inferences on the Slope and Intercept

Given a scatterplot with points $(x_1, y_1), \ldots, (x_n, y_n)$, we can compute the slope $\widehat{\beta}_1$ and intercept $\widehat{\beta}_0$ of the least-squares line. We consider these to be estimates of a true slope β_1 and intercept β_0. We will now explain how to use these estimates to find confidence intervals for, and to test hypotheses about, the true values β_1 and β_0. It turns out that the methods for a population mean, based on the Student's t distribution, can be easily adapted for this purpose.

We have seen that under assumptions 1 through 4, $\widehat{\beta}_0$ and $\widehat{\beta}_1$ are normally distributed with means β_0 and β_1, and standard deviations that are estimated by $s_{\widehat{\beta}_0}$ and $s_{\widehat{\beta}_1}$. The quantities $(\widehat{\beta}_0 - \beta_0)/s_{\widehat{\beta}_0}$ and $(\widehat{\beta}_1 - \beta_1)/s_{\widehat{\beta}_1}$ have Student's t distributions with $n - 2$ degrees of freedom. The number of degrees of freedom is $n-2$ because in the computation of $s_{\widehat{\beta}_0}$ and $s_{\widehat{\beta}_1}$ we divide the sum of squared residuals by $n - 2$. When the sample size n is large enough, the normal distribution is nearly indistinguishable from the Student's t and may be used instead. However, most software packages use the Student's t distribution regardless of sample size.

Summary

Under assumptions 1 through 4, the quantities $\dfrac{\widehat{\beta}_0 - \beta_0}{s_{\widehat{\beta}_0}}$ and $\dfrac{\widehat{\beta}_1 - \beta_1}{s_{\widehat{\beta}_1}}$ have Student's t distributions with $n - 2$ degrees of freedom.

Confidence intervals for β_0 and β_1 can be derived in exactly the same way as the Student's t based confidence interval for a population mean. Let $t_{n-2,\alpha/2}$ denote the point on the Student's t curve with $n - 2$ degrees of freedom that cuts off an area of $\alpha/2$ in the right-hand tail.

> Level $100(1 - \alpha)\%$ confidence intervals for β_0 and β_1 are given by
>
> $$\widehat{\beta}_0 \pm t_{n-2,\alpha/2} \cdot s_{\widehat{\beta}_0} \qquad \widehat{\beta}_1 \pm t_{n-2,\alpha/2} \cdot s_{\widehat{\beta}_1} \tag{7.37}$$
>
> where
>
> $$s_{\widehat{\beta}_0} = s\sqrt{\frac{1}{n} + \frac{\overline{x}^2}{\sum_{i=1}^n (x_i - \overline{x})^2}} \qquad s_{\widehat{\beta}_1} = \frac{s}{\sqrt{\sum_{i=1}^n (x_i - \overline{x})^2}}$$

We illustrate the preceding method with some examples.

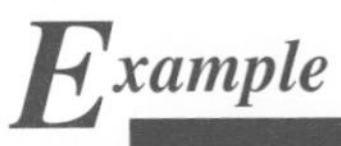

Example 7.11

Find a 95% confidence interval for the spring constant in the Hooke's law data.

Solution

The spring constant is β_1. We have previously computed $\widehat{\beta}_1 = 0.2046$ (Example 7.6 in Section 7.2) and $s_{\widehat{\beta}_1} = 0.0111$ (Example 7.9).

The number of degrees of freedom is $n - 2 = 20 - 2 = 18$, so the t value for a 95% confidence interval is $t_{18,.025} = 2.101$. The confidence interval for β_1 is therefore

$$0.2046 \pm (2.101)(0.0111) = 0.2046 \pm 0.0233 = (0.181,\ 0.228)$$

We are 95% confident that the increase in the length of the spring that will result from an increase of 1 lb in the load is between 0.181 and 0.228 in. Of course, this confidence interval is valid only within the range of the data (0 to 3.8 lb).

Example 7.12

In the Hooke's law data, find a 99% confidence interval for the unloaded length of the spring.

Solution

The unloaded length of the spring is β_0. We have previously computed $\widehat{\beta}_0 = 4.9997$ (Example 7.6) and $s_{\widehat{\beta}_0} = 0.0248$ (Example 7.9).

The number of degrees of freedom is $n - 2 = 20 - 2 = 18$, so the t value for a 99% confidence interval is $t_{18,.005} = 2.878$. The confidence interval for β_0 is therefore

$$4.9997 \pm (2.878)(0.0248) = 4.9997 \pm 0.0714 = (4.928,\ 5.071)$$

We are 99% confident that the unloaded length of the spring is between 4.928 and 5.071 in.

We can perform hypothesis tests on β_0 and β_1 as well. We present some examples.

Example 7.13

The manufacturer of the spring in the Hooke's law data claims that the spring constant β_1 is at least 0.23 in./lb. We have estimated the spring constant to be $\widehat{\beta}_1 = 0.2046$ in./lb. Can we conclude that the manufacturer's claim is false?

Solution

This calls for a hypothesis test. The null and alternate hypotheses are

$$H_0\colon \beta_1 \geq 0.23 \quad \text{versus} \quad H_1\colon \beta_1 < 0.23$$

The quantity

$$\frac{\widehat{\beta}_1 - \beta_1}{s_{\widehat{\beta}_1}}$$

has a Student's t distribution with $n - 2 = 20 - 2 = 18$ degrees of freedom. Under H_0, we take $\beta_1 = 0.23$. The test statistic is therefore

$$\frac{\widehat{\beta}_1 - 0.23}{s_{\widehat{\beta}_1}}$$

We have previously computed $\widehat{\beta}_1 = 0.2046$ and $s_{\widehat{\beta}_1} = 0.0248$. The value of the test statistic is therefore

$$\frac{0.2046 - 0.23}{0.0248} = -1.024$$

Consulting the Student's t table, we find that the P-value is between 0.10 and 0.25. We cannot reject the manufacturer's claim on the basis of these data.

The most commonly tested null hypothesis is $H_0\colon \beta_1 = 0$. If this hypothesis is true, then there is no tendency for y either to increase or decrease as x increases. This implies that x and y have no linear relationship. In general, if the hypothesis that $\beta_1 = 0$ is not rejected, the linear model should not be used to predict y from x.

Example 7.14

The ability of a welded joint to elongate under stress is affected by the chemical composition of the weld metal. In an experiment to determine the effect of carbon content (x) on elongation (y), 39 welds were stressed until fracture, and both carbon content (in parts per thousand) and elongation (in %) were measured. The following summary statistics were calculated:

$$\sum_{i=1}^{n}(x_i - \overline{x})^2 = 0.6561 \qquad \sum_{i=1}^{n}(x_i - \overline{x})(y_i - \overline{y}) = -3.9097 \qquad s = 4.3319$$

Assuming that x and y follow a linear model, compute the estimated change in elongation due to an increase of one part per thousand in carbon content. Should we use the linear model to predict elongation from carbon content?

Solution

The linear model is $y = \beta_0 + \beta_1 x + \varepsilon$, and the change in elongation (y) due to a one part per thousand increase in carbon content (x) is β_1. The null and alternate hypotheses are

$$H_0\colon \beta_1 = 0 \quad \text{versus} \quad H_1\colon \beta_1 \neq 0$$

The null hypothesis says that increasing the carbon content does not affect the elongation, while the alternate hypothesis says that is does. The quantity

$$\frac{\widehat{\beta}_1 - \beta_1}{s_{\widehat{\beta}_1}}$$

has a Student's t distribution with $n - 2 = 39 - 2 = 37$ degrees of freedom. Under H_0, $\beta_1 = 0$. The test statistic is therefore

$$\frac{\widehat{\beta}_1 - 0}{s_{\widehat{\beta}_1}}$$

We compute $\widehat{\beta}_1$ and $s_{\widehat{\beta}_1}$:

$$\widehat{\beta}_1 = \frac{\sum_{i=1}^{n}(x_i - \overline{x})(y_i - \overline{y})}{\sum_{i=1}^{n}(x_i - \overline{x})^2} = \frac{-3.9097}{0.6561} = -5.959$$

$$s_{\widehat{\beta}_1} = \frac{s}{\sqrt{\sum_{i=1}^{n}(x_i - \overline{x})^2}} = 5.348$$

The value of the test statistic is

$$\frac{-5.959 - 0}{5.348} = -1.114$$

The t table shows that the P-value is greater than 0.20. We cannot conclude that the linear model is useful for predicting elongation from carbon content.

Inferences on the Mean Response

In Example 7.8 (Section 7.2), we estimated the length of a spring under a load of 1.4 lb to be 5.29 in. Since this estimate was based on measurements that were subject to uncertainty, the estimate itself is subject to uncertainty. For the estimate to be more useful, we should construct a confidence interval around it to reflect its uncertainty. We now describe how to do this, for the general case where the load on the spring is x lb.

If a measurement y were taken of the length of the spring under a load of x lb, the mean of y would be the true length (or "mean response") $\beta_0 + \beta_1 x$, where β_1 is the true spring constant and β_0 is the true unloaded length of the spring. We estimate this length with $\widehat{y} = \widehat{\beta}_0 + \widehat{\beta}_1 x$. Since $\widehat{\beta}_0$ and $\widehat{\beta}_1$ are normally distributed with means β_0 and β_1, respectively, it follows that $\widehat{y}$ is normally distributed with mean $\beta_0 + \beta_1 x$.

To use $\widehat{y}$ to find a confidence interval, we must know its standard deviation. The standard deviation can be derived by expressing $\widehat{y}$ as a linear combination of the y_i and using Equation (2.53) (in Section 2.5). Equations (7.33) and (7.34) express $\widehat{\beta}_1$ and $\widehat{\beta}_0$ as linear combinations of the y_i. Since $\widehat{y} = \widehat{\beta}_0 + \widehat{\beta}_1 x$, these equations, after some algebraic manipulation, yield

$$\widehat{y} = \sum_{i=1}^{n} \left[\frac{1}{n} + (x - \overline{x}) \frac{x_i - \overline{x}}{\sum_{i=1}^{n}(x_i - \overline{x})^2} \right] y_i \tag{7.38}$$

Equation (2.53) now can be used to derive an expression for the standard deviation of $\widehat{y}$. The standard deviation depends on the error variance σ^2. Since in practice we don't usually know the value of σ, we approximate it with s. The standard deviation of $\widehat{y}$ is approximated by

$$s_{\widehat{y}} = s\sqrt{\frac{1}{n} + \frac{(x - \overline{x})^2}{\sum_{i=1}^{n}(x_i - \overline{x})^2}} \tag{7.39}$$

The quantity $[\widehat{y} - (\beta_0 + \beta_1 x)]/s_{\widehat{y}}$ has a Student's t distribution with $n - 2$ degrees of freedom. We can now provide the expression for a confidence interval for the mean response.

A level $100(1 - \alpha)\%$ confidence interval for the quantity $\beta_0 + \beta_1 x$ is given by

$$\widehat{\beta}_0 + \widehat{\beta}_1 x \pm t_{n-2,\alpha/2} \cdot s_{\widehat{y}} \tag{7.40}$$

where $s_{\widehat{y}} = s\sqrt{\frac{1}{n} + \frac{(x - \overline{x})^2}{\sum_{i=1}^{n}(x_i - \overline{x})^2}}$.

Example 7.15

Using the Hooke's law data, compute a 95% confidence interval for the length of a spring under a load of 1.4 lb.

Solution

We will calculate $\widehat{y}$, $s_{\widehat{y}}$, $\widehat{\beta}_0$, and $\widehat{\beta}_1$, and use expression (7.40). The number of points is $n = 20$. In Example 7.9, we computed $s = 0.0575$. In Example 7.6 (in Section 7.2), we computed $\overline{x} = 1.9$, $\sum_{i=1}^{n}(x_i - \overline{x})^2 = 26.6$, $\widehat{\beta}_1 = 0.2046$, and $\widehat{\beta}_0 = 4.9997$. Using $x = 1.4$, we now compute

$$\widehat{y} = \widehat{\beta}_0 + \widehat{\beta}_1 x = 4.9997 + (0.2046)(1.4) = 5.286$$

Using Equation (7.39) with $x = 1.4$, we obtain

$$s_{\widehat{y}} = 0.0575\sqrt{\frac{1}{20} + \frac{(1.4 - 1.9)^2}{26.6}} = 0.0140$$

The number of degrees of freedom is $n - 2 = 20 - 2 = 18$. We find that the t value is $t_{18,.025} = 2.101$. Substituting into (7.40) we determine the 95% confidence interval for the length $\beta_0 + \beta_1(1.4)$ to be

$$5.286 \pm (2.101)(0.0140) = 5.286 \pm 0.0294 = (5.26,\ 5.32)$$

Example 7.16

In a study of the relationship between oxygen content (x) and ultimate testing strength (y) of welds, the data presented in the following table were obtained for 29 welds. Here oxygen content is measured in parts per thousand, and strength is measured in ksi. Using a linear model, find a 95% confidence interval for the mean strength for welds with oxygen content 1.7 parts per thousand. (From the article "Advances in Oxygen Equivalence Equations for Predicting the Properties of Titanium Welds," D. Harwig, W. Ittiwattana, and H. Castner, *The Welding Journal*, 2001:126s–136s.)

Oxygen Content	Strength	Oxygen Content	Strength	Oxygen Content	Strength
1.08	63.00	1.16	68.00	1.17	73.00
1.19	76.00	1.32	79.67	1.40	81.00
1.57	66.33	1.61	71.00	1.69	75.00
1.72	79.67	1.70	81.00	1.71	75.33
1.80	72.50	1.69	68.65	1.63	73.70
1.65	78.40	1.78	84.40	1.70	91.20
1.50	72.00	1.50	75.05	1.60	79.55
1.60	83.20	1.70	84.45	1.60	73.95
1.20	71.85	1.30	70.25	1.30	66.05
1.80	87.15	1.40	68.05		

Solution

We calculate the following quantities (the computing formulas on p. 495 may be used):

$$\overline{x} = 1.51966 \quad \overline{y} = 75.4966 \quad \sum_{i=1}^{n}(x_i - \overline{x})^2 = 1.33770 \quad \sum_{i=1}^{n}(y_i - \overline{y})^2 = 1304.23$$

$$\sum_{i=1}^{n}(x_i - \overline{x})(y_i - \overline{y}) = 22.6377 \qquad \widehat{\beta}_0 = 49.7796 \qquad \widehat{\beta}_1 = 16.9229 \qquad s = 5.84090$$

The estimate of the mean strength for welds with an oxygen content of 1.7 is

$$\widehat{y} = \widehat{\beta}_0 + \widehat{\beta}_1(1.7) = 49.7796 + (16.9229)(1.7) = 78.5485$$

The standard deviation of $\widehat{y}$ is estimated to be

$$\begin{aligned} s_{\widehat{y}} &= s\sqrt{\frac{1}{n} + \frac{(x - \overline{x})^2}{\sum_{i=1}^{n}(x_i - \overline{x})^2}} \\ &= 5.84090\sqrt{\frac{1}{29} + \frac{(1.7 - 1.51966)^2}{1.33770}} \\ &= 1.4163 \end{aligned}$$

There are $n - 2 = 29 - 2 = 27$ degrees of freedom. The t value is therefore $t_{27,.025} = 2.052$. The 95% confidence interval is

$$78.5485 \pm (2.052)(1.4163) = 78.5485 \pm 2.9062 = (75.64,\ 81.45)$$

Hypothesis tests on the mean response can be conducted, using a Student's t distribution. Following is an example.

Refer to Example 7.15. Let μ_0 represent the true length of the spring under a load of 1.6 lb. Test the hypothesis $H_0: \mu_0 \leq 5.3$ versus $H_1: \mu_0 > 5.3$.

Solution

Since μ_0 is the true length of the spring under a load of 1.6 lb, $\mu_0 = \beta_0 + \beta_1(1.6)$. Now let $\widehat{y} = \widehat{\beta}_0 + \widehat{\beta}_1(1.6)$. The quantity

$$\frac{\widehat{y} - [\beta_0 + \beta_1(1.6)]}{s_{\widehat{y}}} = \frac{\widehat{y} - \mu_0}{s_{\widehat{y}}}$$

has a Student's t distribution with $n - 2 = 18$ degrees of freedom. Under H_0, we take $\mu_0 = 5.3$. The test statistic is therefore

$$\frac{\widehat{y} - 5.3}{s_{\widehat{y}}}$$

We compute $\widehat{y}$ and $s_{\widehat{y}}$:

$$\widehat{y} = \widehat{\beta}_0 + \widehat{\beta}_1(1.6) = 4.9997 + (0.2046)(1.6) = 5.3271$$

$$s_{\widehat{y}} = 0.0575\sqrt{\frac{1}{20} + \frac{(1.6 - 1.9)^2}{26.6000}} = 0.0133$$

The value of the test statistic is

$$\frac{5.3271 - 5.3}{0.0133} = 2.04$$

The P-value is between 0.025 and 0.05. It is reasonable to conclude that the true length is greater than 5.3 in.

Prediction Intervals for Future Observations

In Example 7.16 we found a confidence interval for the mean strength of welds with an oxygen content of 1.7 parts per thousand. Here is a somewhat different scenario: Assume we wish to predict the strength of a particular weld whose oxygen content is 1.7, rather than the mean strength of all such welds.

Using values calculated in Example 7.16, we predict this weld's strength to be $\widehat{y} = \widehat{\beta}_0 + \widehat{\beta}_1(1.7) = 49.7796 + (16.9229)(1.7) = 78.5485$. This prediction is the same as the estimate of the mean strength for all welds with an oxygen content of 1.7. Now we wish to put an interval around this prediction, to indicate its uncertainty. To compute this **prediction interval**, we must determine the uncertainty in the prediction.

The mean strength of welds with an oxygen content of 1.7 is $\beta_0 + \beta_1(1.7)$. The actual strength of a particular weld is equal to $\beta_0 + \beta_1(1.7) + \varepsilon$, where ε represents the random difference between the strength of the particular weld and the mean strength of all welds whose oxygen content is 1.7. The error in predicting the strength of the particular weld with $\widehat{y}$ is the prediction error

$$\widehat{y} - [\beta_0 + \beta_1(1.7)] - \varepsilon \tag{7.41}$$

The uncertainty in the prediction of the strength of the particular weld is the standard deviation of this prediction error. We briefly show how to compute this standard deviation. The quantity $\beta_0 + \beta_1(1.7)$ is constant and does not affect the standard deviation. The quantities $\widehat{y}$ and ε are independent, since $\widehat{y}$ is calculated from the data in Example 7.16, while ε applies to a weld that is not part of that data set. It follows that the standard deviation of the prediction error (7.41) is approximated by

$$s_{\text{pred}} = \sqrt{s_{\widehat{y}}^2 + s^2}$$

Using Equation (7.39) to substitute for $s_{\widehat{y}}$ yields

$$s_{\text{pred}} = s\sqrt{1 + \frac{1}{n} + \frac{(x - \overline{x})^2}{\sum_{i=1}^{n}(x_i - \overline{x})^2}} \tag{7.42}$$

The appropriate expression for the prediction interval can now be determined.

A level $100(1 - \alpha)\%$ prediction interval for the quantity $\beta_0 + \beta_1 x$ is given by

$$\widehat{\beta}_0 + \widehat{\beta}_1 x \pm t_{n-2,\alpha/2} \cdot s_{\text{pred}} \tag{7.43}$$

where $s_{\text{pred}} = s\sqrt{1 + \frac{1}{n} + \frac{(x - \overline{x})^2}{\sum_{i=1}^{n}(x_i - \overline{x})^2}}$.

Note that the prediction interval is wider than the confidence interval, because the value 1 is added to the quantity under the square root to account for the additional uncertainty.

Example 7.18

For the weld data in Example 7.16, find a 95% prediction interval for the strength of a particular weld whose oxygen content is 1.7 parts per thousand.

Solution

The predicted strength is $\widehat{y} = \widehat{\beta}_0 + \widehat{\beta}_1(1.7)$, which we have calculated in Example 7.16 to be 78.5485.

Using the quantities presented in Example 7.16, we compute the value of s_{pred} to be

$$s_{\text{pred}} = 5.84090\sqrt{1 + \frac{1}{29} + \frac{(1.7 - 1.51966)^2}{1.33770}} = 6.0102$$

There are $n - 2 = 29 - 2 = 27$ degrees of freedom. The t value is therefore $t_{27,.025} = 2.052$. The 95% prediction interval is therefore

$$78.5485 \pm (2.052)(6.0102) = 78.5485 \pm 12.3329 = (66.22,\ 90.88)$$

Both the confidence intervals and the prediction intervals we have described are specific to a given x value. In Examples 7.16 and 7.18, we took $x = 1.7$. By computing the intervals for many values of x and connecting the points with a smooth curve, we obtain **confidence bands** or **prediction bands**, respectively. Figure 7.13 illustrates 95% confidence bands and prediction bands for the weld data presented in Example 7.16. For any given oxygen content, the 95% confidence or prediction bands can be read off the figure.

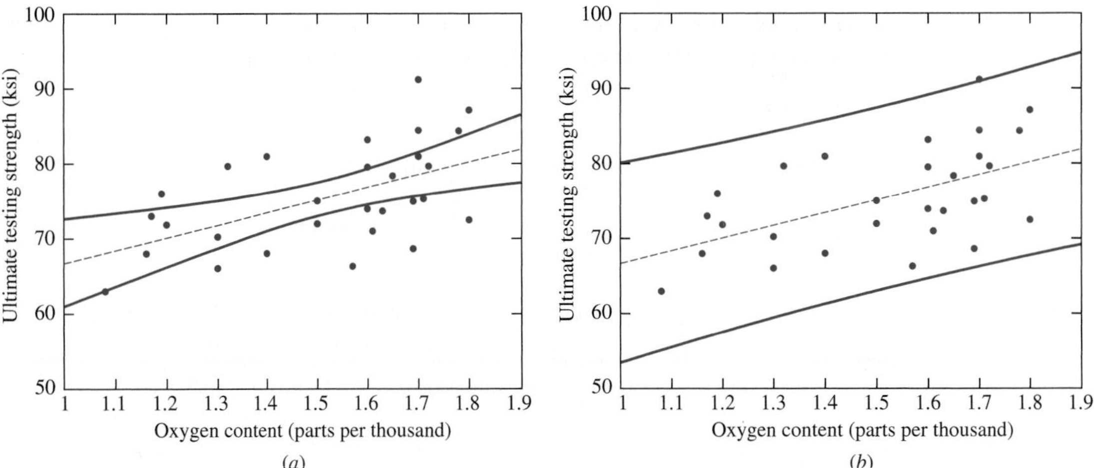

FIGURE 7.13 Oxygen content versus ultimate testing strength for 29 welds. In both plots, the dotted line is the least-squares line. *(a)* The two solid curves are the 95% confidence bands. Given any oxygen content, we are 95% confident that the mean strength for welds with that oxygen content lies between the upper and lower confidence limits. *(b)* The two solid curves are the 95% prediction bands. Given any specific weld, we are 95% confident that the strength for that particular weld lies between the upper and lower prediction limits corresponding to the oxygen content of that weld.

Confidence and prediction bands provide a nice visual presentation of the way in which the uncertainty depends on the value of the independent variable. Note that both the confidence interval and the prediction interval are narrowest when $x = \overline{x}$, and increases in width as x moves away from $\overline{x}$. This is due to the term $(x - \overline{x})^2$ appearing in a numerator in the expressions for $s_{\hat{y}}$ and s_{pred}. We conclude that *predictions based on the least-squares line are more precise near the center of the scatterplot and are less precise near the edges.*

Note that the confidence bands indicate confidence intervals for individual values of x. They do not provide a confidence region for the true line $y = \beta_0 + \beta_1 x$. In other words, we cannot be 95% confident that the true line lies between the 95% confidence bands.

Interpreting Computer Output

Nowadays, least-squares calculations are usually done on a computer, using a software package such as MINITAB. The following MINITAB output is for the Hooke's law data.

```
Regression Analysis: Length versus Weight

The regression equation is
Length = 5.00 + 0.205 Weight (1)

Predictor        Coef (2)    SE Coef (3)        T (4)       P (5)
Constant      4.99971        0.02477        201.81       0.000
Weight        0.20462        0.01115         18.36       0.000

S = 0.05749 (6)  R-Sq = 94.9% (7)          R-Sq(adj) = 94.6%

Analysis of Variance (8)

Source              DF          SS          MS          F          P
Regression           1      1.1138      1.1138     337.02      0.000
Residual Error      18      0.0595      0.0033
Total               19      1.1733

Unusual Observations (9)

Obs   Weight    Length       Fit    SE Fit   Residual   St Resid
 12     2.20    5.5700    5.4499    0.0133     0.1201      2.15R

R denotes an observation with a large standardized residual

Predicted Values for New Observations (10)

New Obs       Fit   SE Fit          95.0% CI              95.0% PI
1          5.2453   0.0150   ( 5.2137, 5.2769)   ( 5.1204, 5.3701)

Values of Predictors for New Observations (11)

New Obs   Weight
1           1.20
```

We will now explain the labeled quantities in the output:

(1) This is the equation of the least-squares line.

(2) **Coef:** The coefficients $\widehat{\beta}_0 = 4.99971$ and $\widehat{\beta}_1 = 0.20462$.

(3) **SE Coef:** The standard deviations $s_{\widehat{\beta}_0}$ and $s_{\widehat{\beta}_1}$. ("SE" stands for standard error, another term for standard deviation.)

(4) **T:** The values of the Student's t statistics for testing the hypotheses $\beta_0 = 0$ and $\beta_1 = 0$. The t statistic is equal to the coefficient divided by its standard deviation.

(5) **P:** The P-values for the tests of the hypotheses $\beta_0 = 0$ and $\beta_1 = 0$. The more important P-value is that for β_1. If this P-value is not small enough to reject the hypothesis that $\beta_1 = 0$, the linear model is not useful for predicting y from x. In this example, the P-values are extremely small, indicating that neither β_0 nor β_1 is equal to 0.

(6) **S:** The estimate s of the error standard deviation.

(7) **R-Sq:** This is r^2, the square of the correlation coefficient r, also called the coefficient of determination.

(8) **Analysis of Variance:** This table is not so important in simple linear regression, when there is only one independent variable. It is more important in multiple regression, where there are several independent variables. However, it is worth noting that the three numbers in the column labeled "SS" are the regression sum of squares $\sum_{i=1}^{n}(\widehat{y}_i - \overline{y})^2$, the error sum of squares $\sum_{i=1}^{n}(y_i - \widehat{y}_i)^2$, and their sum, the total sum of squares $\sum_{i=1}^{n}(y_i - \overline{y})^2$.

(9) **Unusual Observations:** Here MINITAB tries to alert you to data points that may violate some of the assumptions 1 through 4 previously discussed. MINITAB is conservative and will often list several such points even when the data are well described by a linear model. In Section 7.4, we will learn some graphical methods for checking the assumptions of the linear model.

(10) **Predicted Values for New Observations:** These are confidence intervals and prediction intervals for values of x that are specified by the user. Here we specified $x = 1.2$ for the weight. The "Fit" is the fitted value $\widehat{y} = \widehat{\beta}_0 + \widehat{\beta}_1 x$, and "SE Fit" is the standard deviation $s_{\widehat{y}}$. Then come the 95% confidence and prediction intervals, respectively.

(11) **Values of Predictors for New Observations:** This is simply a list of the x values for which confidence and prediction intervals have been calculated. It shows that these intervals refer to a weight of $x = 1.2$.

Exercises for Section 7.3

1. A new spring is being tested. Twenty-five weights x_i are hung, and the spring length y_i is measured for each. The weights are measured in lb, and the length is measured in inches. The following summary statistics are recorded:

$$\overline{x} = 2.40 \qquad \overline{y} = 12.18 \qquad \sum_{i=1}^{25}(x_i - \overline{x})^2 = 52.00$$

$$\sum_{i=1}^{25}(y_i - \overline{y})^2 = 498.96 \qquad \sum_{i=1}^{25}(x_i - \overline{x})(y_i - \overline{y}) = 160.27$$

Let β_0 represent the length of the spring at rest, and let β_1 represent the increase in length caused by a load of 1 lb. Assume that assumptions 1 through 4 on p. 508 hold.

a. Compute the least-squares estimates $\widehat{\beta}_0$ and $\widehat{\beta}_1$.
b. Compute the error variance estimate s^2.
c. Find 95% confidence intervals for β_0 and β_1.
d. The manufacturer of the spring claims that the spring constant is no more than 3 in./lb. Do the data provide sufficient evidence to disprove this claim?
e. It is also claimed that the unloaded length of the spring is at least 5.5 in. Do the data provide sufficient evidence to disprove this claim?
f. Find a 99% confidence interval for the length of the spring under a load of 1.5 lb.

g. Find a 99% prediction interval for the measured length of the spring under a load of 1.5 lb.
h. Which is more useful in this case, the confidence interval or the prediction interval? Explain.

2. For many chemicals, the amount that will dissolve in a given volume of water depends on the temperature. The following MINITAB output describes the fit of a linear model $y = \beta_0 + \beta_1 x + \varepsilon$ that expresses the number of grams of a certain chemical dissolved per liter of water (y) in terms of the temperature in °C (x). There are $n = 6$ observations.

```
Predictor          Coef      SE Coef        T         P
Constant         1.4381      0.62459     2.30     0.083
Temperature     0.30714      0.02063     14.9     0.000
```

a. There are $n = 6$ observations. How many degrees of freedom are there for the Student's t statistics?
b. Find a 95% confidence interval for β_1.
c. Find a 95% confidence interval for β_0.
d. Someone claims that if the temperature of the water is increased by 1°C, that the mean number of grams dissolved will increase by exactly 0.40. Use the given output to perform a hypothesis test to determine whether this claim can be disproved.
e. Someone claims that the mean number of grams that can be dissolved in a liter of water at 0°C is less than 1.0. Use the given output to perform a hypothesis test to determine whether this claim can be disproved.

3. Oxides of nitrogen (NO_x) form a major component of the air pollution produced by motor vehicles. In a study to determine the relationship between the load on an engine and NO_x production, a vehicle was run in a test lab at various speeds. Periodic measurements were taken of the horsepower (x) and NO_x emissions (y). NO_x emissions are measured in mg/s. The following MINITAB output describes the fit of a linear model to these data. Assume that assumptions 1 through 4 on p.508 hold.

```
The regression equation is
Chassis NOx = 44.5 + 0.845 Chassis HP

Predictor          Coef       SE Coef          T           P
Constant         44.534         2.704      16.47       0.000
Chassis HP      0.84451       0.03267      25.85       0.000

S = 24.62            R-Sq = 84.7%        R-Sq(adj) = 84.5%

Predicted Values for New Observations

New Obs       Fit     SE Fit          95.0% CI              95.0% PI
1           78.31       2.23    ( 73.89, 82.73)     ( 29.37, 127.26)

Values of Predictors for New Observations

New Obs     Chassis HP
1                 40.0
```

a. What are the slope and intercept of the least-squares line?
b. Is the linear model useful for predicting NO_x emission rates from horsepower? Explain.
c. Predict the NO_x emission rate if the engine is putting out 10 hp.
d. What is the correlation between horsepower and the NO_x emission rate?
e. The output provides a 95% confidence interval for the mean NO_x emission rate at 40 hp. There are $n = 123$ observations in this data set. Using the value "SE Fit," find a 90% confidence interval.

f. Someone plans to run the engine at 40 hp, and predicts that the NO_x emission rate will be 180 mg/s. Is this a reasonable prediction? If so, explain why. If not, give a reasonable range of predicted values.

4. In an experiment similar to that in Exercise 3, the engine is removed from the chassis and run. Measurements are made of horsepower and NO_x emission rate. The MINITAB output follows. Assume that assumptions 1 through 4 on p. 508 hold.

```
The regression equation is
Engine NOx = 33.0 + 0.753 Engine HP

Predictor        Coef     SE Coef        T        P
Constant       33.042       2.371    13.94    0.000
Engine HP     0.75269     0.02912    25.85    0.000

S = 22.79         R-Sq = 84.7%      R-Sq(adj) = 84.5%
```

a. What is the slope of the least-squares line?
b. There are 123 points in the data set. Find a 95% confidence interval for the slope.
c. Perform a test of the null hypothesis that the slope is greater than or equal to 0.8. What is the P-value?

5. Refer to Exercises 3 and 4. The engineer notices that the slope of the least-squares line in the experiment described in Exercise 4 is less than that in the experiment described in Exercise 3. He wishes to test the hypothesis that the effect of horsepower on the NO_x emission rate differs depending on whether the engine is in or out of the chassis. Let β_C denote the increase in emission rate due to an increase of 1 hp with the engine in the chassis, and β_E denote the corresponding increase with the engine out of the chassis.

a. Express the null hypothesis to be tested in terms of β_C and β_E.
b. Let $\widehat{\beta}_C$ and $\widehat{\beta}_E$ denote the slopes of the least-squares lines. Assume these slopes are independent. There are 123 observations in each data set. Test the null hypothesis in part (a). Can you conclude that the effect of horsepower differs in the two cases?

6. The article "Withdrawal Strength of Threaded Nails" (D. Rammer, S. Winistorfer, and D. Bender, *Journal of Structural Engineering* 2001:442–449) describes an experiment to investigate the relationship between the diameter of a nail (x) and its ultimate withdrawal strength (y). Annularly threaded nails were driven into Douglas fir lumber, and then their withdrawal strengths were measured in N/mm. The following results for 10 different diameters (in mm) were obtained.

x	2.52	2.87	3.05	3.43	3.68	3.76	3.76	4.50	4.50	5.26
y	54.74	59.01	72.92	50.85	54.99	60.56	69.08	77.03	69.97	90.70

a. Compute the least-squares line for predicting strength from diameter.
b. Compute the error standard deviation estimate s.
c. Compute a 95% confidence interval for the slope.
d. Find a 95% confidence interval for the mean withdrawal strength of nails 4 mm in diameter.
e. Can you conclude that the mean withdrawal strength of nails 4 mm in diameter is greater than 60 N/mm? Perform a hypothesis test and report the P-value.
f. Find a 95% prediction interval for the withdrawal strength of a particular nail whose diameter is 4 mm.

7. In a study to determine the relationship between the permeability (in cm/ms) and electrical resistance (in $k\Omega \cdot cm^2$) of human skin, the data in the following table were collected. Assume that assumptions 1 through 4 on p. 508 hold.

Perm.	Res.	Perm.	Res.	Perm.	Res.	Perm.	Res.
1.39	0.90	1.08	1.35	1.53	0.94	1.57	0.72
1.21	1.21	1.79	0.40	1.79	0.41	1.49	0.66
1.58	0.65	1.54	0.80	1.45	0.82	1.68	0.67
1.35	0.97	2.04	0.17	1.46	0.83	1.52	0.79
1.76	0.40	1.51	0.94	1.47	0.79	1.29	0.96
1.57	0.65	1.16	1.13	1.67	0.59	1.90	0.27
1.32	0.95	1.71	0.57	1.81	0.47	1.10	1.28
1.13	1.33	1.64	0.65	1.40	0.90	1.67	0.57
1.70	0.44	1.67	0.62	1.82	0.37	1.66	0.54
1.79	0.37	1.19	1.22	1.49	0.78	1.46	0.82
1.09	1.34	1.56	0.78	1.23	1.01	1.85	0.40
1.29	0.99	1.63	0.64				

a. Compute the least-squares line for predicting permeability (y) from resistance (x).
b. Compute 95% confidence intervals for β_0 and β_1.
c. Predict the permeability for skin whose resistance is 1.7 kΩ.
d. Find a 95% confidence interval for the mean permeability for all skin samples with a resistance of 1.7 kΩ.
e. Find a 95% prediction interval for the permeability of skin whose resistance is 1.7 kΩ.

8. Three engineers are independently estimating the spring constant of a spring, using the linear model specified by Hooke's law. Engineer A measures the length of the spring under loads of 0, 1, 3, 4, and 6 lb, for a total of five measurements. Engineer B uses the same loads, but repeats the experiment twice, for a total of 10 independent measurements. Engineer C uses loads of 0, 2, 6, 8, and 12 lb, measuring once for each load. The engineers all use the same measurement apparatus and procedure. Each engineer computes a 95% confidence interval for the spring constant.

a. If the width of the interval of engineer A is divided by the width of the interval of engineer B, the quotient will be approximately ________.
b. If the width of the interval of engineer A is divided by the width of the interval of engineer C, the quotient will be approximately ______________.
c. Each engineer computes a 95% confidence interval for the length of the spring under a load of 2.5 lb. Which interval is most likely to be the shortest? Which interval is most likely to be the longest?

9. In the weld data (Example 7.16) imagine that 95% confidence intervals are computed for the mean strength of welds with oxygen contents of 1.3, 1.5, and 1.8 parts per thousand. Which of the confidence intervals would be the shortest? Which would be the longest?

10. Refer to Exercise 1. If 95% confidence intervals are constructed for the length of the spring under loads of 2.15, 2.57, and 2.45 lb, which confidence interval would be the shortest? Which would be the longest?

11. In a study of copper bars, the relationship between shear stress in ksi (x) and shear strain in % (y) was summarized by the least-squares line $y = -20.00 + 2.56x$. There were a total of $n = 17$ observations, and the coefficient of determination was $r^2 = 0.9111$. If the total sum of squares was $\sum(y_i - \overline{y})^2 = 234.19$, compute the estimated error variance s^2.

12. In the manufacture of synthetic fiber, the fiber is often "set" by subjecting it to high temperatures. The object is to improve the shrinkage properties of the fiber. In a test of 25 yarn specimens, the relationship between temperature in °C (x) and shrinkage in % (y) was summarized by the least-squares line $y = -12.789 + 0.133x$. The total sum of squares was $\sum(y_i - \overline{y})^2 = 57.313$, and the estimated error variance was $s^2 = 0.0670$. Compute the coefficient of determination r^2.

13. In the following MINITAB output, some of the numbers have been accidentally erased. Recompute them, using the numbers still available. There are $n = 25$ points in the data set.

```
The regression equation is
Y = 1.71 + 4.27 X

Predictor        Coef     SE Coef        T       P
Constant      1.71348     6.69327      (a)     (b)
X             4.27473         (c)    3.768     (d)

S = 0.05749   R-Sq = 38.2%
```

14. In the following MINITAB output, some of the numbers have been accidentally erased. Recompute them, using the numbers still available. There are $n = 20$ points in the data set.

```
Predictor        Coef      SE Coef        T       P
Constant          (a)      0.43309    0.688     (b)
X             0.18917     0.065729      (c)     (d)

S = 0.67580   R-Sq = 31.0%
```

15. In order to increase the production of gas wells, a procedure known as "fracture treatment" is often used. Fracture fluid, which consists of fluid mixed with sand, is pumped into the well. The following figure presents a scatterplot of the monthly production versus the volume of fracture fluid pumped for 255 gas wells. Both production and fluid are expressed in units of volume per foot of depth of the well. The least-squares line is superimposed. The equation of the least-squares line is $y = 106.11 + 0.1119x$.

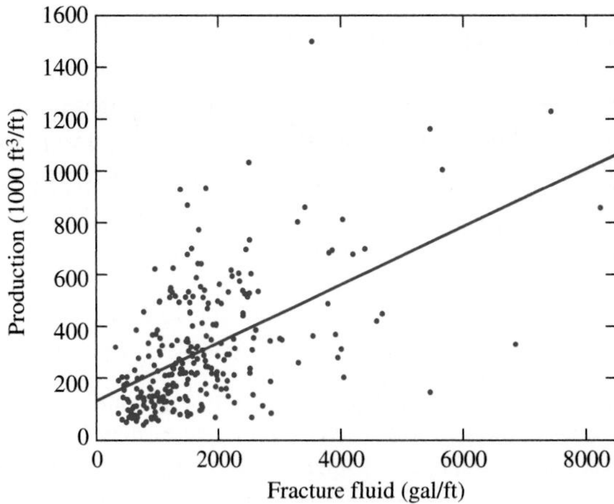

a. From the least-squares line, estimate the production for a well into which 4000 gal/ft are pumped.
b. From the least-squares line, estimate the production for a well into which 500 gal/ft are pumped.
c. A new well is dug, and 500 gal/ft of fracture fluid are pumped in. Based on the scatterplot, is it more likely that the production of this well will fall above or below the least-squares estimate?
d. What feature of the scatterplot indicates that assumption 3 on p. 508 is violated?

7.4 Checking Assumptions and Transforming Data

The methods discussed so far are valid under the assumption that the relationship between the variables x and y satisfies the linear model $y_i = \beta_0 + \beta_1 x_i + \varepsilon_i$, where the errors ε_i satisfy assumptions 1 through 4. We repeat these assumptions here.

Assumptions for Errors in Linear Models

1. The errors $\varepsilon_1, \ldots, \varepsilon_n$ are random and independent. In particular, the magnitude of any error ε_i does not influence the value of the next error ε_{i+1}.
2. The errors $\varepsilon_1, \ldots, \varepsilon_n$ all have mean 0.
3. The errors $\varepsilon_1, \ldots, \varepsilon_n$ all have the same variance, which we denote by σ^2.
4. The errors $\varepsilon_1, \ldots, \varepsilon_n$ are normally distributed.

As mentioned earlier, the normality assumption (4) is less important when the sample size is large. While mild violations of the assumption of constant variance (3) do not matter too much, severe violations should be corrected.

We need ways to check these assumptions to assure ourselves that our methods are appropriate. There have been innumerable diagnostic tools proposed for this purpose. Many books have been written on the topic. We will restrict ourselves here to a few of the most basic procedures.

The Plot of Residuals versus Fitted Values

The single best diagnostic for least-squares regression is a plot of residuals e_i versus fitted values $\widehat{y}_i$, sometimes called a **residual plot**. Figure 7.14 on page 528 presents such a plot for Galton's height versus forearm data (see Figure 7.1 in Section 7.1 for the original data). By mathematical necessity, the residuals have mean 0, and the correlation between the residuals and fitted values is 0 as well. The least-squares line is therefore horizontal, passing through 0 on the vertical axis. When the linear model is valid, and assumptions 1 through 4 are satisfied, the plot will show no substantial pattern. There should be no curve to the plot, and the vertical spread of the points should not vary too much over the horizontal range of the plot, except perhaps near the edges. These conditions are reasonably well satisfied for Galton's data. We have no reason to doubt the assumptions of the linear model.

A bit of terminology: When the vertical spread in a scatterplot doesn't vary too much, the scatterplot is said to be **homoscedastic**. The opposite of homoscedastic is **heteroscedastic**.

A good-looking residual plot does not by itself prove that the linear model is appropriate, because the assumptions of the linear model can fail in other ways. On the other hand, a residual plot with a serious defect does clearly indicate that the linear model is inappropriate.

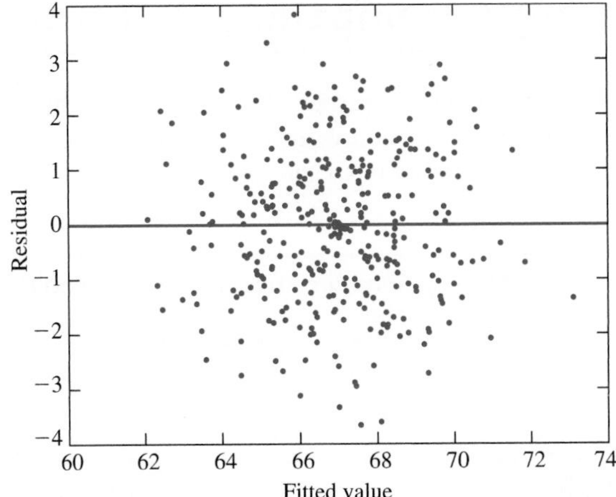

FIGURE 7.14 Plot of residuals (e_i) versus fitted values ($\widehat{y}_i$) for the Galton's height versus forearm data. There is no substantial pattern to the plot, and the vertical spread does not vary too much, except perhaps near the edges. This is consistent with the assumptions of the linear model.

Summary

If the plot of residuals versus fitted values

- Shows no substantial trend or curve, and
- Is **homoscedastic**, that is, the vertical spread does not vary too much along the horizontal length of plot, except perhaps near the edges,

then it is *likely*, but not *certain*, that the assumptions of the linear model hold.

However, if the residual plot *does* show a substantial trend or curve, or is **heteroscedastic**, it is certain that the assumptions of the linear model do *not* hold.

In many cases, the residual plot will exhibit curvature or heteroscedasticity, which reveal violations of assumptions. We will present three examples. Then we will present a method called **transforming the variables**, which can sometimes fix violations of assumptions and allow the linear model to be used.

Figure 7.15 presents a plot of atmospheric ozone concentrations versus NO_x concentrations measured on 359 days in a recent year at a site near Riverside, California. (NO_x stands for oxides of nitrogen, and refers to the sum of NO and NO_2.) Both concentrations are measured in parts per billion (ppb). Next to this plot is a residual plot. The plot is clearly **heteroscedastic**; that is, the vertical spread varies considerably with the fitted value. Specifically, when the fitted value (estimated ozone concentration) is large, the

residuals tend to be farther from 0. Since the magnitude of the spread in the residuals depends on the error variance σ^2, we conclude that the error variance is larger on days where the fitted value is larger. This is a violation of assumption 3, which states that the variance σ^2 is the same for all observations. The plot also contains an outlier (where the ozone concentration is near 100). The residual plot indicates that we should not use this linear model to predict the ozone concentration from the NO_x concentration.

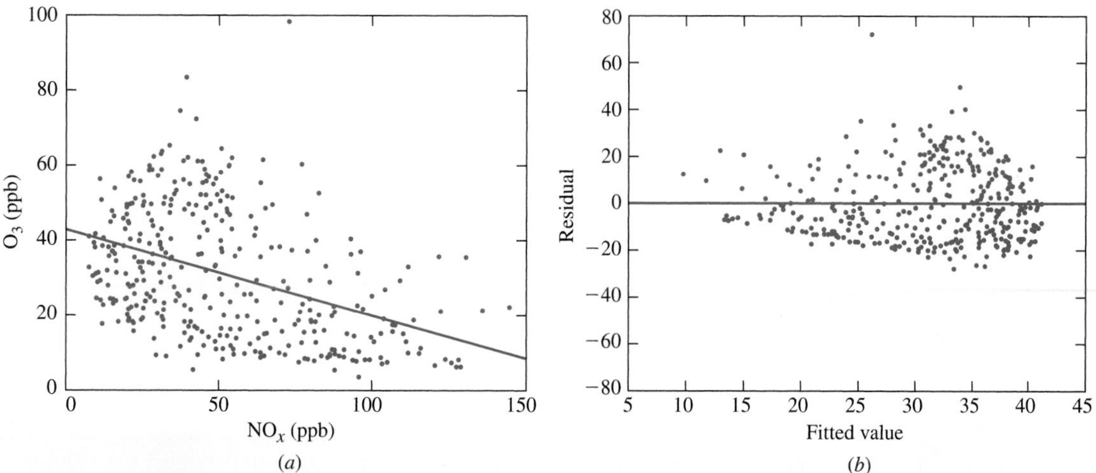

FIGURE 7.15 *(a)* Plot of ozone concentration versus NO_x concentration. The least-squares line is superimposed. *(b)* Plot of residuals (e_i) versus fitted values ($\widehat{y}_i$) for these data. The vertical spread clearly increases with the fitted value. This indicates a violation of the assumption of constant error variance.

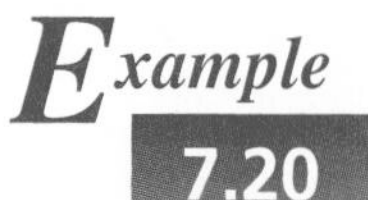

(Based on the article "Advances in Oxygen Equivalence Equations for Predicting the Properties of Titanium Welds," D. Harwig, W. Ittiwattana, and H. Castner, *The Welding Journal*, 2001:126s–136s.) The physical properties of a weld are influenced by the chemical composition of the weld material. One measure of the chemical composition is the Ogden–Jaffe number, which is a weighted sum of the percentages of carbon, oxygen, and nitrogen in the weld. In a study of 63 welds, the hardness of the weld (measured on the Rockwell B scale) was plotted against the Ogden–Jaffe number. The plot is presented in Figure 7.16 on page 530, along with a residual plot. The residual plot shows a pattern, with positive residuals concentrated in the middle of the plot, and negative residuals at either end. Technically, this indicates that the errors ε_i don't all have a mean of 0. This generally happens for one of two reasons: Either the relationship between the variables is nonlinear, or there are other variables that need to be included in the model. We conclude that we should not use this model to predict weld hardness from the Ogden–Jaffe number.

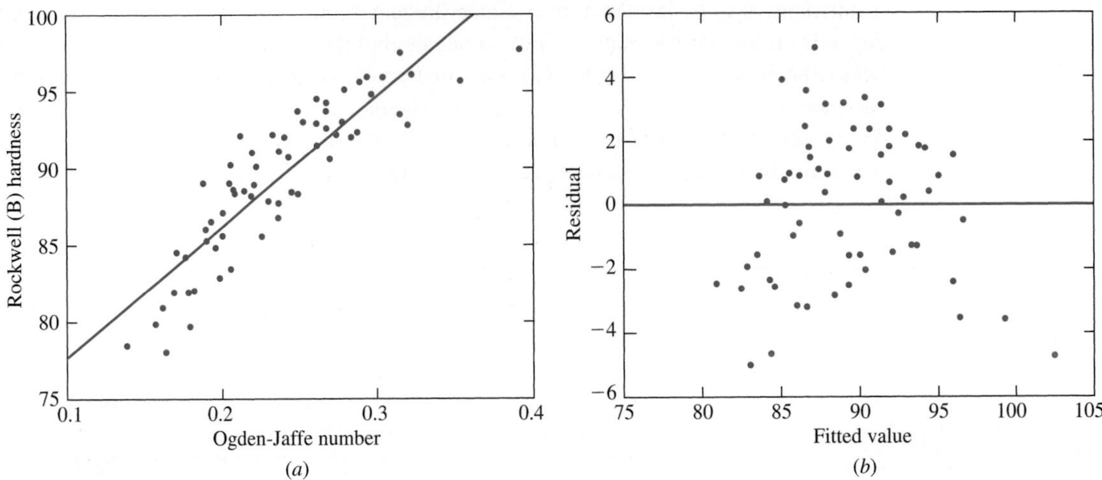

FIGURE 7.16 *(a)* Plot of Rockwell (B) hardness versus Ogden–Jaffe number. The least-squares line is superimposed. *(b)* Plot of residuals (e_i) versus fitted values ($\widehat{y}_i$) for these data. The residuals plot shows a trend, with positive residuals in the middle and negative residuals at either end.

Example 7.21

These data were presented in Exercise 15 in Section 7.3. For a group of 255 gas wells, the monthly production per foot of depth of the well is plotted against the volume of fracture fluid pumped into the well. This plot, along with the residual plot, is presented in Figure 7.17. The residual plot is strongly heteroscedastic, indicating that the error

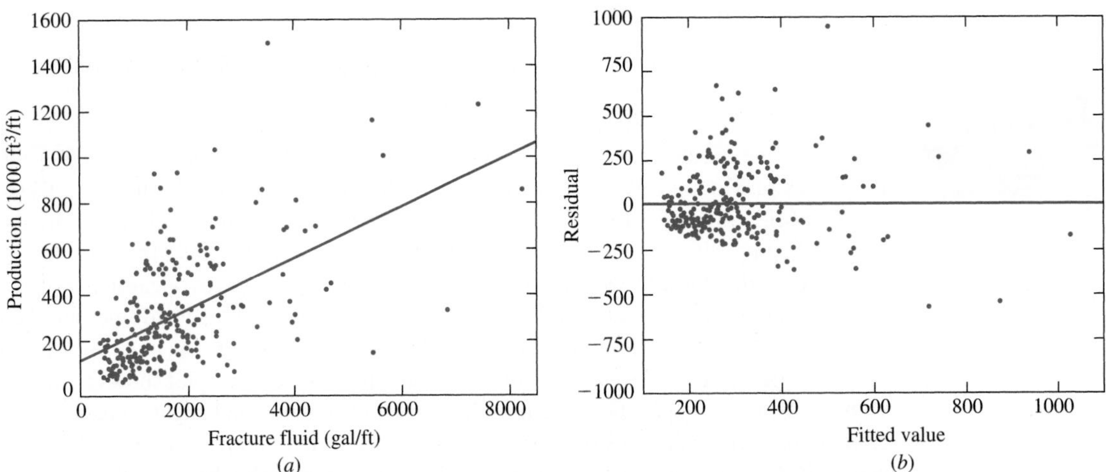

FIGURE 7.17 *(a)* Plot of monthly production versus volume of fracture fluid for 255 gas wells. *(b)* Plot of residuals (e_i) versus fitted values ($\widehat{y}_i$) for the gas well data. The vertical spread clearly increases with the fitted value. This indicates a violation of the assumption of constant error variance.

variance is larger for gas wells whose estimated production is larger. These of course are the wells into which more fracture fluid has been pumped. We conclude that we should not use this model to predict well production from the amount of fracture fluid pumped.

Transforming the Variables

If we fit the linear model $y = \beta_0 + \beta_1 x + \varepsilon$ and find that the residual plot is heteroscedastic, or exhibits a trend or pattern, we can sometimes fix the problem by raising x, y, or both to a power. It may be the case that a model of the form $y^a = \beta_0 + \beta_1 x^b + \varepsilon$ fits the data well. In general, replacing a variable with a function of itself is called **transforming** the variable. Specifically, raising a variable to a power is called a **power transformation**. Taking the logarithm of a variable is also considered to be a power transformation, even though the logarithm is not a power.

Here is a simple example that shows how a power transformation works. The following table presents values for hypothetical variables x, y, and y^2.

x	y	y²	x	y	y²
1.0	2.2	4.84	11.0	31.5	992.25
2.0	9.0	81.00	12.0	32.7	1069.29
3.0	13.5	182.25	13.0	34.9	1218.01
4.0	17.0	289.00	14.0	36.3	1317.69
5.0	20.5	420.25	15.0	37.7	1421.29
6.0	23.3	542.89	16.0	38.7	1497.69
7.0	25.2	635.04	17.0	40.0	1600.00
8.0	26.4	696.96	18.0	41.3	1705.69
9.0	27.6	761.76	19.0	42.5	1806.25
10.0	30.2	912.04	20.0	43.7	1909.69

The scatterplot of y versus x is presented in Figure 7.18, along with the residual plot. Clearly the linear model is inappropriate.

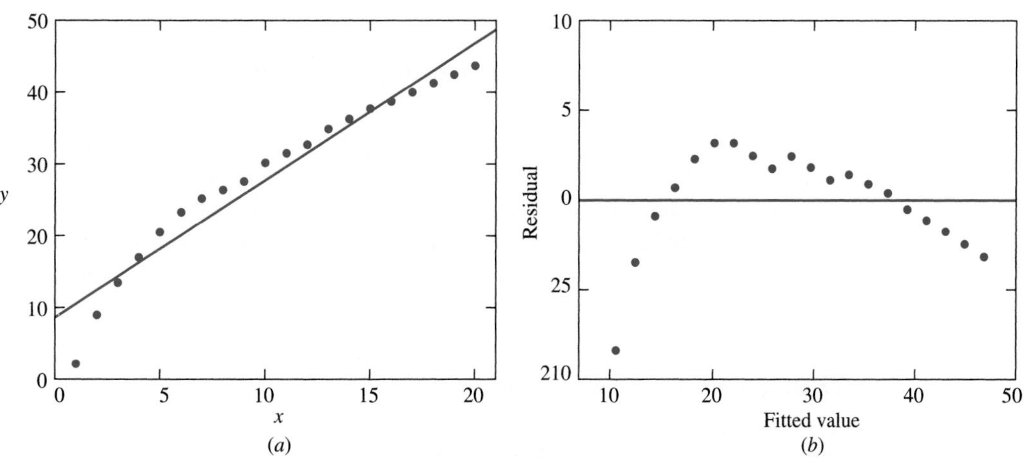

FIGURE 7.18 *(a)* Plot of y versus x with the least-squares line superimposed. *(b)* Plot of residuals versus fitted values. There is a strong pattern to the residual plot, indicating that the linear model is inappropriate.

The model $y = \beta_0 + \beta_1 x + \varepsilon$ does not fit the data. However, we can solve this problem by using y^2 in place of y. Figure 7.19 presents the scatterplot of y^2 versus x, along with the residual plot. The residual plot is homoscedastic, with no discernible trend or pattern.

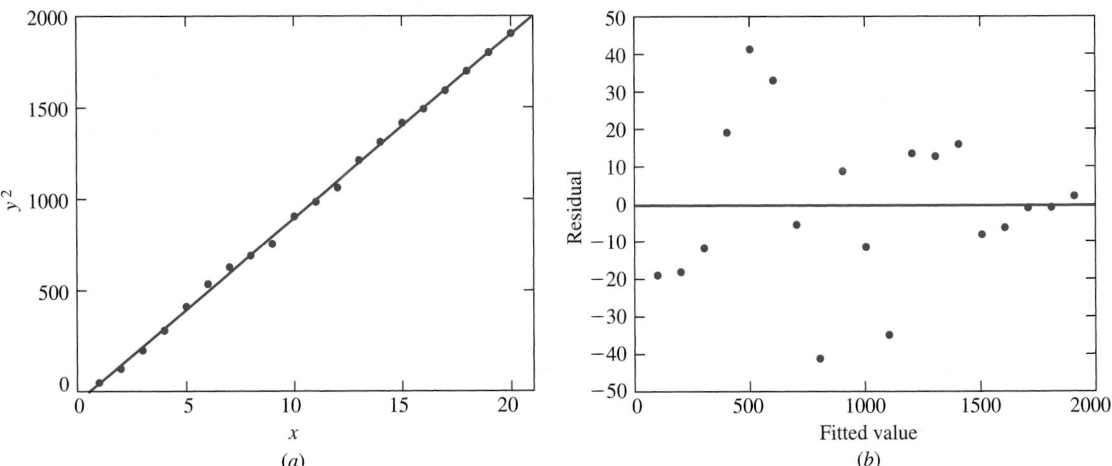

FIGURE 7.19 *(a)* Plot of y^2 versus x with the least-squares line superimposed. *(b)* Plot of residuals versus fitted values. The residual plot is approximately homoscedastic, with no discernible trend or pattern.

We conclude that the model $y^2 = \beta_0 + \beta_1 x + \varepsilon$ is a plausible model for these data. In this example, we transformed y, but did not need to transform x. In other cases, we may transform only x, or both x and y.

Determining Which Transformation to Apply

It is possible with experience to look at a scatterplot, or a residual plot, and make an educated guess as to how to transform the variables. Mathematical methods are also available to determine a good transformation. However, it is perfectly satisfactory to proceed by trial and error. Try various powers on both x and y (including $\ln x$ and $\ln y$), look at the residual plots, and hope to find one that is homoscedastic, with no discernible pattern. A more advanced discussion of transformation selection can be found in Draper and Smith (1998).

Transformations Don't Always Work

It is important to remember that power transformations don't always work. Sometimes, none of the residual plots look good, no matter what transformations are tried. In these cases, other methods must be used. One of these is multiple regression, discussed in Chapter 8. Some others are briefly mentioned at the end of this section.

Residual Plots with Only a Few Points Can Be Hard to Interpret

When there are only a few points in a residual plot, it can be hard to determine whether the assumptions of the linear model are met. Sometimes such a plot will at first glance appear to be heteroscedastic, or to exhibit a pattern, but upon closer inspection it turns out that this visual impression is caused by the placement of just one or two points. It is sometimes even difficult to determine whether such a plot contains an outlier. When one is faced with a sparse residual plot that is hard to interpret, a reasonable thing to do is to fit a linear model, but to consider the results tentative, with the understanding that the appropriateness of the model has not been established. If and when more data become available, a more informed decision can be made. Of course, not all sparse residual plots are hard to interpret. Sometimes there is a clear pattern, which cannot be changed just by shifting one or two points. In these cases, the linear model should not be used.

Refer to Example 7.19. Figure 7.15 presented a plot of ozone versus NO_x concentrations. It turns out that transforming ozone to its natural logarithm, ln Ozone, produces a satisfactory linear plot. Figure 7.20 presents the scatterplot of ln Ozone versus NO_x, and the corresponding residual plot. The residual plot is homoscedastic, with no discernible pattern. The outlier that was present in the original data is less prominent. The linear model looks good.

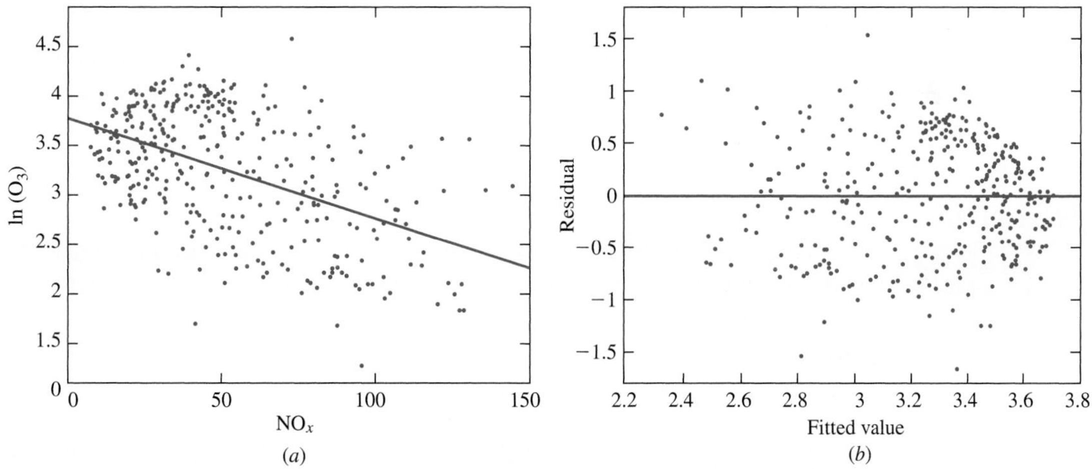

FIGURE 7.20 ***(a)*** Plot of the natural logarithm of ozone concentration versus NO_x concentration. The least-squares line is superimposed. ***(b)*** Plot of residuals (e_i) versus fitted values ($\widehat{y}_i$) for these data. The linear model looks good.

The following MINITAB output is for the transformed data.

```
Regression Analysis: LN OZONE versus NOx

The regression equation is
LN OZONE = 3.78 - 0.0101 NOx

Predictor          Coef     SE Coef        T       P
Constant        3.78238     0.05682    66.57   0.000
NOx          -0.0100976   0.0009497   -10.63   0.000

S = 0.5475          R-Sq = 24.1%      R-Sq(adj) = 23.8%

Analysis of Variance

Source            DF        SS        MS        F        P
Regression         1    33.882    33.882   113.05    0.000
Residual Error   357   106.996     0.300
Total            358   140.878

Predicted Values for New Observations

New Obs     Fit   SE Fit        95.0% CI              95.0% PI
1        3.2775   0.0289   ( 3.2206, 3.3344)   ( 2.1994, 4.3556)

Values of Predictors for New Observations

New Obs          NOx
1               50.0
```

The analysis of the transformed data gives results for the natural log of the ozone concentration. For some purposes, we can transform back to the original units. For example, we can use the transformed data to find prediction intervals for ozone values given a particular NO_x value. To do this, we simply use the methods of Section 7.3 to find the interval for ln Ozone, and then transform this interval back into the original units. Example 7.23 shows how.

Using the preceding MINITAB output, predict the ozone level when the NO_x level is 50 ppb, and find a 95% prediction interval for the ozone level on a day when the NO_x level is 50 ppb.

Solution

Let y represent the ozone level on a day when the NO_x level is 50 ppb, and let $\widehat{y}$ represent the predicted value for that level. We first compute the value $\ln \widehat{y}$, which is the predicted value for ln Ozone, using the coefficient estimates in the MINITAB output. For a NO_x value of 50, the prediction is

$$\ln \widehat{y} = 3.78238 - 0.0100976(50) = 3.2775$$

The predicted ozone value is therefore

$$\widehat{y} = e^{3.2775} = 26.51$$

To compute a 95% prediction interval for the ozone level y, we read off the prediction interval for $\ln y$ from the MINITAB output:

$$2.1994 < \ln y < 4.3556$$

Exponentiating across the inequality yields the 95% prediction interval for the ozone level:

$$e^{2.1994} < y < e^{4.3556}$$

$$9.02 < y < 77.91$$

It is important to note that the method used in Example 7.23 works only for the prediction interval. It does not work for the confidence interval for the mean response. When the dependent variable has been transformed, the confidence interval for the mean response cannot be obtained in the original units.

Example 7.24

Refer to Example 7.20. Figure 7.16 presented a plot of Rockwell (B) hardness versus the Ogden–Jaffe number for a group of welds. In this case, taking the reciprocal of the Ogden–Jaffe number (raising to the -1 power) produces an approximately linear relationship. Figure 7.21 presents the results. Note that in this case, we transformed the

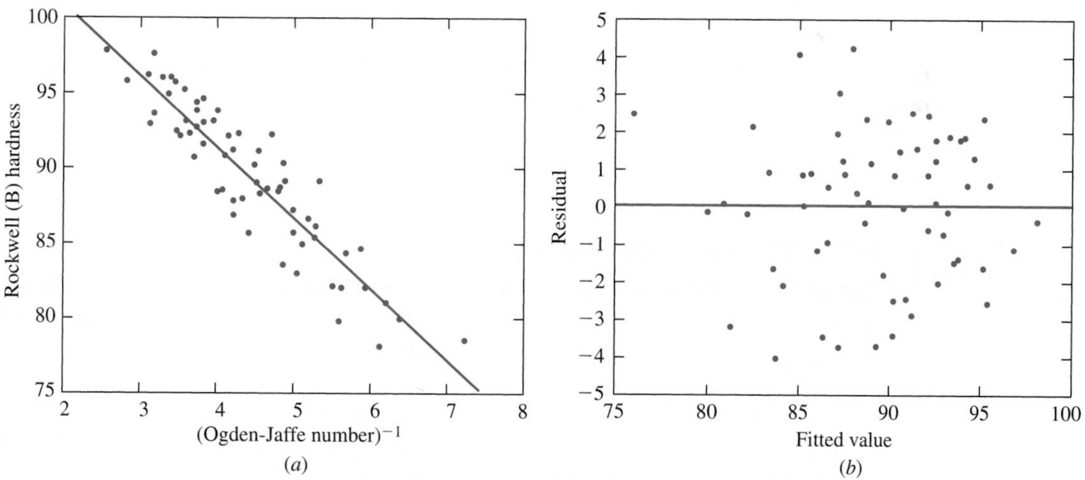

FIGURE 7.21 *(a)* Plot of hardness versus (Ogden–Jaffe number)$^{-1}$. The least-squares line is superimposed. *(b)* Plot of residuals (e_i) versus fitted values ($\widehat{y}_i$) for these data. The linear model looks good.

independent variable (x), while in Example 7.22 we transformed the dependent variable (y).

Refer to Example 7.21. Figure 7.17 presented a plot of production versus volume of fracture fluid for 255 gas wells. It turns out that an approximately linear relationship holds between the logarithm of production and the logarithm of the volume of fracture fluid. Figure 7.22 presents the results. Note that in this case both variables were transformed.

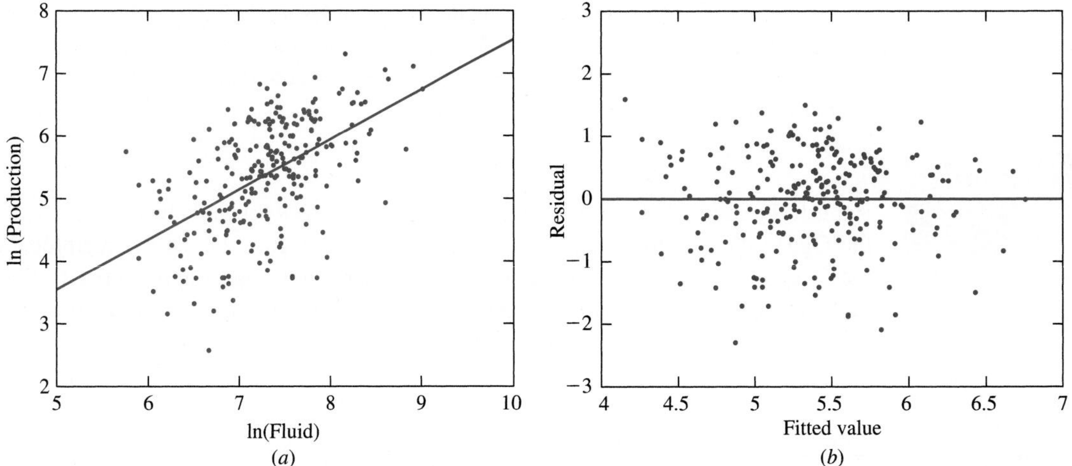

FIGURE 7.22 *(a)* Plot of the log of production versus the log of the volume of fracture fluid for 255 gas wells, with the least-squares line superimposed. *(b)* Plot of residuals versus fitted values. There is no substantial pattern to the residuals. The linear model looks good.

Outliers and Influential Points

Outliers are points that are detached from the bulk of the data. Both the scatterplot and the residual plot should be examined for outliers. The first thing to do with an outlier is to try to determine why it is different from the rest of the points. Sometimes outliers are caused by data-recording errors or equipment malfunction. In these cases, the outliers can be deleted from the data set. But many times the cause for an outlier cannot be determined with certainty. Deleting the outlier is then unwise, because it results in underestimating the variability of the process that generated the data.

Outliers can often be identified by visual inspection. Many software packages list points that have unusually large residuals; such a list will contain most of the outliers (and sometimes some innocuous points as well). Sometimes transforming the variables will eliminate outliers by moving them nearer to the bulk of the data. When transformations don't help, and when there is no justification for deleting the outliers, one approach is first to fit the line to the whole data set, and then to remove each outlier in turn, fitting the line to the data set with the one outlier deleted. If none of the outliers upon removal make a noticeable difference to the least-squares line or to the estimated standard deviations of the slope and intercept, then use the fit with the outliers included. If one or more of the outliers does make a difference when removed, then the range of values for the least-squares coefficients should be reported. In these cases computing confidence intervals or prediction intervals, or performing hypothesis tests, should be avoided.

An outlier that makes a considerable difference to the least-squares line when removed is called an **influential point**. Figure 7.23 presents an example of an influential outlier, along with one that is not influential. In general, outliers with unusual x values are more likely to be influential than those with unusual y values, but every outlier should be checked. Many software packages identify potentially influential points. Further information on treatment of outliers and influential points can be found in Draper and Smith (1998), Belsey, Kuh, and Welsch (1980), and Cook and Weisberg (1994).

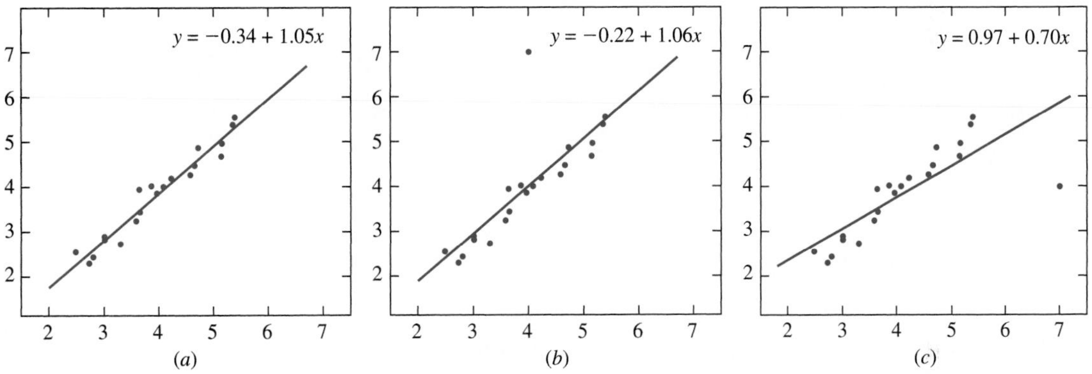

FIGURE 7.23 *(a)* Scatterplot with no outliers. *(b)* An outlier is added to the plot. There is little change in the least-squares line, so this point is not influential. *(c)* An outlier is added to the plot. There is a considerable change in the least-squares line, so this point is influential.

Finally, we remark that some authors restrict the definition of outliers to points that have unusually large residuals. Under this definition, a point that is far from the bulk of the data, yet near the least-squares line, is not an outlier. Such a point might or might not be influential.

The following example features a data set that contains two outliers. In a study to determine whether the frequency of a certain mutant gene increases with age, the number of mutant genes in a microgram of DNA was counted for each of 30 men. Two of the men had extremely large counts; their points are outliers. The least-squares line was fit to all 30 points, to each set of 29 points obtained by deleting an outlier, and then to the 28 points that remained after removing both outliers. Figure 7.24 presents scatterplots of frequency versus age for the full data set and for the sets with one and with both outliers deleted. The least-squares lines are superimposed. With the outliers included, the equation of the least-squares line is $y = -137.76 + 4.54x$. With the outliers removed, the equation of the least-squares line is $y = 31.86 + 1.23x$. These results are sufficiently different that both should be reported.

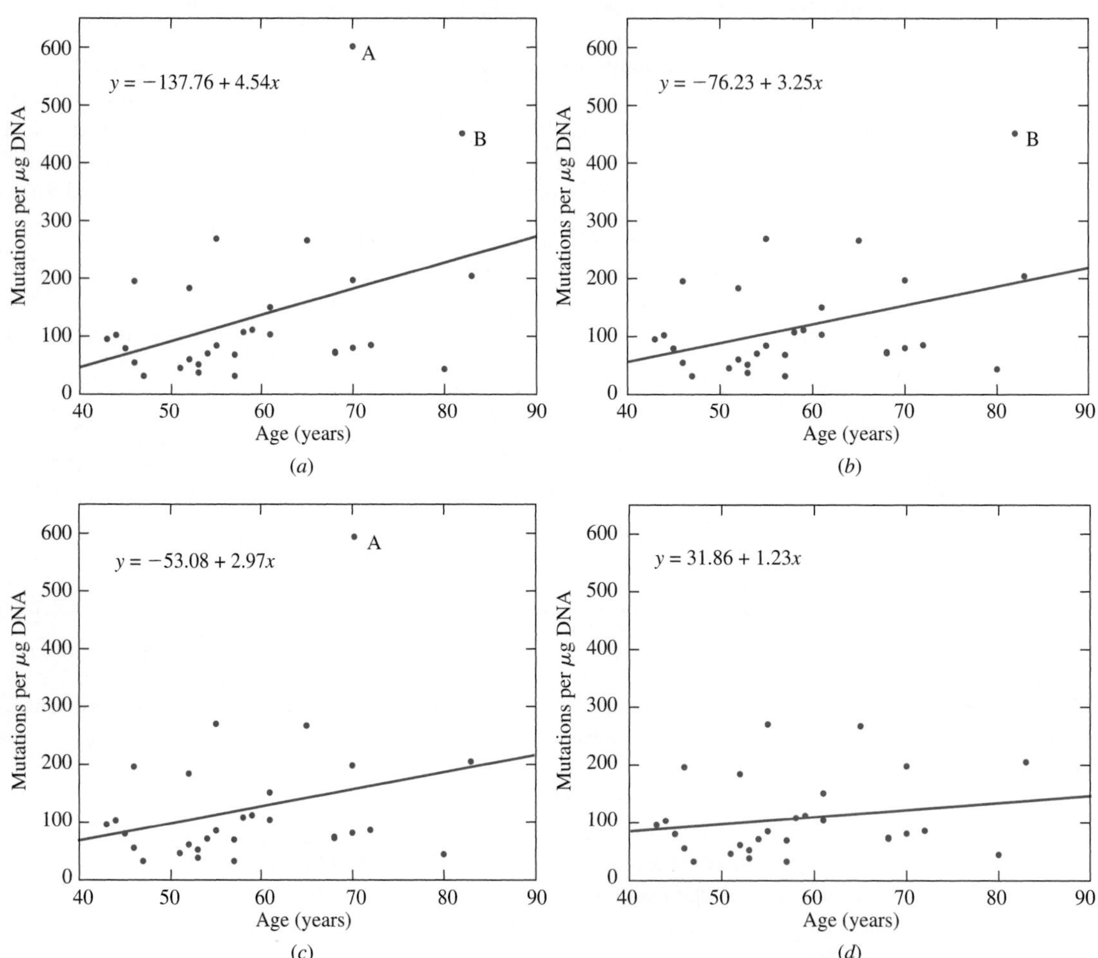

FIGURE 7.24 Mutation frequency versus age. *(a)* The plot contains two outliers, A and B. *(b)* Outlier A is deleted. The change in the least-squares line is noticeable although not extreme; this point is somewhat influential. *(c)* Outlier B is deleted. The change in the least-squares line is again noticeable but not extreme; this point is somewhat influential as well. Note that outlier B is a bit more influential than outlier A, even though B is visually somewhat closer to the bulk of the data. *(d)* Both outliers are deleted. The combined effect on the least-squares line is substantial.

Methods Other Than Transforming Variables

Transforming the variables is not the only method for analyzing data when the residual plot indicates a problem. When the residual plot is heteroscedastic, a technique called **weighted least-squares** is sometimes used. In this method, the x and y coordinates of each point are multiplied by a quantity known as a **weight**. Points in regions where the vertical spread is large are multiplied by smaller weights, while points in regions with less vertical spread are multiplied by larger weights. The effect is to make the points whose error variance is smaller have greater influence in the computation of the least-squares line.

When the residual plot shows a trend, this sometimes indicates that more than one independent variable is needed to explain the variation in the dependent variable. In these cases, more independent variables are added to the model, and multiple regression is used. Finally, some relationships are inherently nonlinear. For these, a method called **nonlinear regression** can be applied. Multiple regression is covered in Chapter 8. The other two methods are beyond the scope of this book. A good reference on these topics is Draper and Smith (1998).

To summarize, we present some generic examples of residual plots in Figure 7.25. For each one, we present a diagnosis and a prescription.

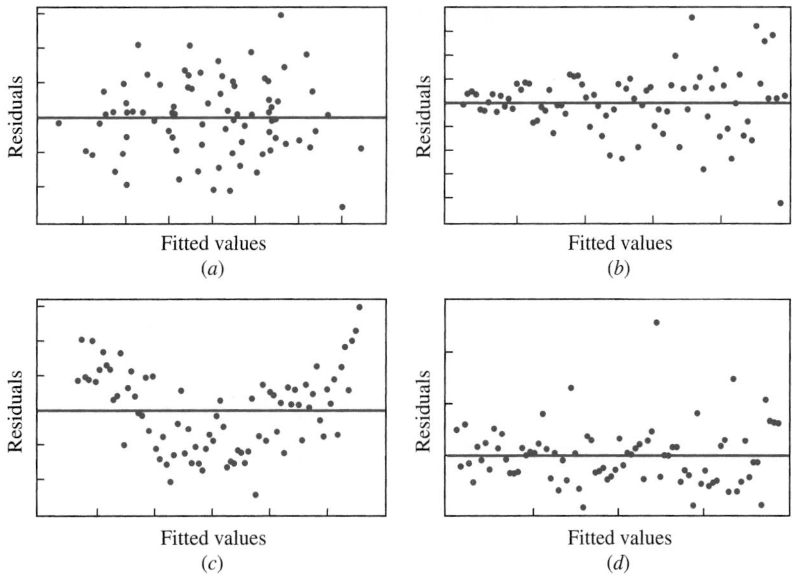

FIGURE 7.25 *(a)* No substantial pattern, plot is homoscedastic. Linear model is OK. *(b)* Heteroscedastic. Try a power transformation. *(c)* Discernible trend to residuals. Try a power transformation, or use multiple regression. *(d)* Outlier. Examine the offending data point to see if it is an error. If not, compute the least-squares line both with and without the outlier to see if it makes a noticeable difference.

Checking Independence and Normality

If the plot of residuals versus fitted values looks good, it may be advisable to perform additional diagnostics to further check the fit of the linear model. In particular, when the observations occur in a definite time order, it is desirable to plot the residuals against the order in which the observations were made. If there are trends in the plot, it indicates that the relationship between x and y may be varying with time. In these cases a variable representing time, or other variables related to time, should be included in the model as additional independent variables, and a multiple regression should be performed.

In the air pollution data in Example 7.19, with y representing ozone concentration and x representing NO_x concentration, the residual plot (Figure 7.15) for the model $\ln y = \beta_0 + \beta_1 x + \varepsilon$ is homoscedastic, with no discernible pattern or trend. These data were collected over the course of 359 days during a particular year. Figure 7.26 presents the plot of residuals versus time for these data. There is a clear pattern. The residuals are positive in the middle of the data, corresponding to the summer, and negative at the ends of the data, corresponding to the winter. Each residual is equal to the log of the observed ozone concentration on that day, minus the log of the value predicted by the model. We conclude that the values predicted by the model are too low in the summer and too high in the winter. It is clear that knowing the time of the year can improve our prediction of the ozone concentration over that provided by the model with NO_x concentration as the only independent variable. We therefore would fit a multiple regression model containing both time and NO_x as independent variables. Depending on the results of that fit, we might make further adjustments to the model. These ideas will be pursued further in Chapter 8.

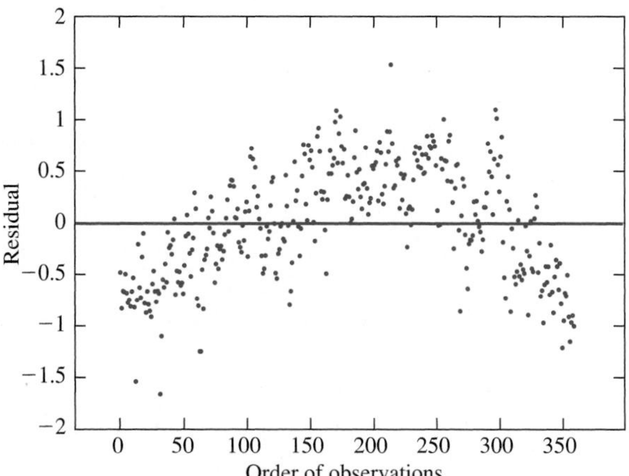

FIGURE 7.26 Plot of residuals versus order of the observations (time) for the ozone versus NO_x data. The model $\ln \text{Ozone} = \beta_0 + \beta_1 NO_x + \varepsilon$ was fit. The residuals show a clear pattern with time, indicating that a multiple regression model should be fit, with time as an additional variable.

Sometimes a plot of residuals versus time shows that the residuals oscillate with time. This indicates that the value of each error is influenced by the errors in previous observations, so therefore the errors are not independent. When this feature is severe, linear regression should not be used, and the methods of time series analysis should be used instead. A good reference on time series analysis is Brockwell and Davis (2002).

To check that the errors are normally distributed, a normal probability plot of the residuals can be made. If the probability plot has roughly the appearance of a straight line, the residuals are approximately normally distributed. It can be a good idea to make a probability plot when variables are transformed, since one sign of a good transformation is that the residuals are approximately normally distributed. As previously mentioned, the assumption of normality is not so important when the number of data points is large. Unfortunately, when the number of data points is small, it can be difficult to detect departures from normality.

Empirical Models and Physical Laws

How do we know whether the relationship between two variables is linear? In some cases, physical laws, such as Hooke's law, give us assurance that a linear model is correct. In other cases, such as the relationship between the log of the volume of fracture fluid pumped into a gas well and the log of its monthly production, there is no known physical law. In these cases, we use a linear model simply because it appears to fit the data well. A model that is chosen because it appears to fit the data, in the absence of physical theory, is called an **empirical model**. In real life, most data analysis is based on empirical models. It is less often that a known physical law applies. Of course, many physical laws started out as empirical models. If an empirical model is tested on many different occasions, under a wide variety of circumstances, and is found to hold without exception, it can gain the status of a physical law.

There is an important difference between the interpretation of results based on physical laws and the interpretation of results based on empirical models. A physical law may be regarded as *true*, whereas the best we can hope for from an empirical model is that it is *useful*. For example, in the Hooke's law data, we can be sure that the relationship between the load on the spring and its length is truly linear. We are sure that when we place another weight on the spring, the length of the spring can be accurately predicted from the linear model. For the gas well data, on the other hand, while the linear relationship describes the data well, we cannot be sure that it captures the true relationship between fracture fluid volume and production.

Here is a simple example that illustrates the point. Figure 7.27 on page 542 presents 20 triangles of varying shapes. Assume that we do not know the formula for the area of a triangle. We notice, however, that triangles with larger perimeters seem to have larger areas, so we fit a linear model:

$$\text{Area} = \beta_0 + \beta_1 \text{ (Perimeter)} + \varepsilon$$

The scatterplot of area versus perimeter, with the least-squares line superimposed, is shown to the right in Figure 7.27. The equation of the least-squares line is

$$\text{Area} = -1.232 + 1.373\ (\text{Perimeter})$$

The units in this equation are arbitrary. The correlation between area and perimeter is $r = 0.88$, which is strongly positive. The linear model appears to fit well. We could use this model to predict, for example, that a triangle with perimeter equal to 5 will have an area of 5.633.

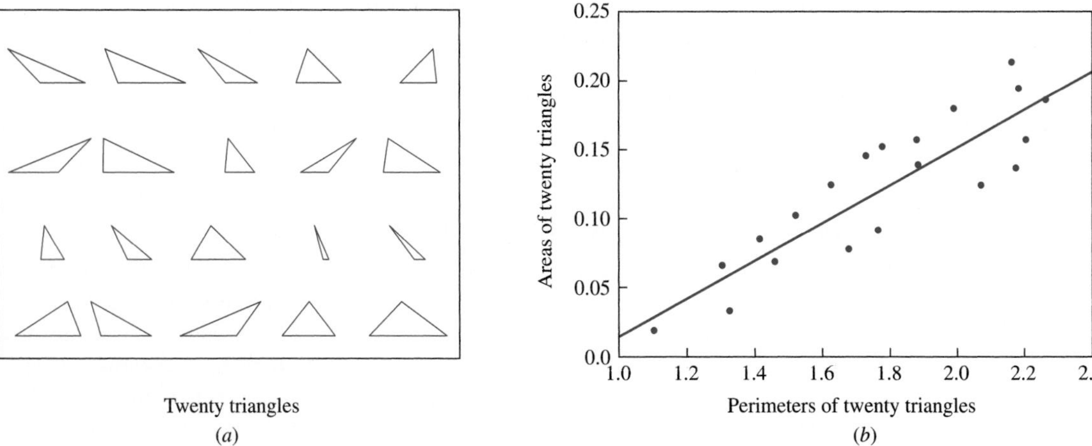

FIGURE 7.27 *(a)* Twenty triangles. *(b)* Area versus perimeter for 20 triangles. The correlation between perimeter and area is 0.88.

Now while this linear model may be useful, it is not true. The linear model correctly shows that there is a strong tendency for triangles with larger perimeters to have larger areas. In the absence of a better method, it may be of some use in estimating the areas of triangles. But it does not help to reveal the true mechanism behind the determination of area. The true mechanism, of course, is given by the law

$$\text{Area} = 0.5 \times \text{base} \times \text{height}$$

The results predicted by an empirical model may not hold up under replication. For example, a collection of triangles could be designed in such a way that the ones with the larger perimeters had smaller areas. In another collection, the area might appear to be proportional to the square of the perimeter, or to its logarithm. We cannot determine by statistical analysis of the triangle data how well the empirical model will apply to a triangle not yet observed. Deciding whether it is appropriate to apply the results of an empirical model to future observations is a matter of scientific judgment rather than statistics.

Summary

- Physical laws are applicable to all future observations.
- An empirical model is valid only for the data to which it is fit. It may or may not be useful in predicting outcomes for subsequent observations.
- Determining whether to apply an empirical model to a future observation requires scientific judgment rather than statistical analysis.

Exercises for Section 7.4

1. The following MINITAB output is for the least-squares fit of the model $\ln y = \beta_0 + \beta_1 \ln x + \varepsilon$, where y represents the monthly production of a gas well and x represents the volume of fracture fluid pumped in. (A scatterplot of these data is presented in Figure 7.22.)

```
Regression Analysis: LN PROD versus LN FLUID

The regression equation is
LN PROD = - 0.444 + 0.798 LN FLUID

Predictor        Coef     SE Coef        T        P
Constant      -0.4442      0.5853    -0.76    0.449
LN FLUID      0.79833     0.08010     9.97    0.000

S = 0.7459     R-Sq = 28.2%        R-Sq(adj) = 27.9%

Analysis of Variance

Source            DF        SS        MS        F        P
Regression         1    55.268    55.268    99.34    0.000
Residual Error   253   140.756     0.556
Total            254   196.024

Predicted Values for New Observations

New Obs       Fit    SE Fit       95.0% CI              95.0% PI
1          5.4457    0.0473   ( 5.3526, 5.5389)   ( 3.9738, 6.9176)

Values of Predictors for New Observations

New Obs    LN FLUID
1          7.3778
```

a. What is the equation of the least-squares line for predicting $\ln y$ from $\ln x$?
b. Predict the production of a well into which 2500 gal/ft of fluid have been pumped.
c. Predict the production of a well into which 1600 gal/ft of fluid have been pumped.
d. Find a 95% prediction interval for the production of a well into which 1600 gal/ft of fluid have been pumped. (*Note*: $\ln 1600 = 7.3778$.)

2. The processing of raw coal involves "washing," in which coal ash (nonorganic, incombustible material) is removed. The article "Quantifying Sampling Precision for Coal Ash Using Gy's Discrete Model of the Fundamental Error" (*Journal of Coal Quality*, 1989:33–39) provides data relating the percentage of ash to the volume of a coal particle. The average percentage of ash for six volumes of coal particles was measured. The data are as follows:

Volume (cm^3)	0.01	0.06	0.58	2.24	15.55	276.02
Percent ash	3.32	4.05	5.69	7.06	8.17	9.36

a. Compute the least-squares line for predicting percent ash (y) from volume (x). Plot the residuals versus the fitted values. Does the linear model seem appropriate? Explain.
b. Compute the least-squares line for predicting percent ash from ln volume. Plot the residuals versus the fitted values. Does the linear model seem appropriate? Explain.
c. Compute the least-squares line for predicting percent ash from $\sqrt{\text{volume}}$. Plot the residuals versus the fitted values. Does the linear model seem appropriate? Explain.
d. Using the most appropriate model, predict the percent ash for particles with a volume of 50 m^3.
e. Using the most appropriate model, construct a 95% confidence interval for the mean percent ash for particles with a volume of 50 m^3.

3. To determine the effect of temperature on the yield of a certain chemical process, the process is run 24 times at various temperatures. The temperature (in °C) and the yield (expressed as a percentage of a theoretical maximum) for each run are given in the following table. The results are presented in the order in which they were run, from earliest to latest.

Order	Temp	Yield	Order	Temp	Yield	Order	Temp	Yield
1	30	49.2	9	25	59.3	17	34	65.9
2	32	55.3	10	38	64.5	18	43	75.2
3	35	53.4	11	39	68.2	19	34	69.5
4	39	59.9	12	30	53.0	20	41	80.8
5	31	51.4	13	30	58.3	21	36	78.6
6	27	52.1	14	39	64.3	22	37	77.2
7	33	60.2	15	40	71.6	23	42	80.3
8	34	60.5	16	44	73.0	24	28	69.5

a. Compute the least-squares line for predicting yield (y) from temperature (x).
b. Plot the residuals versus the fitted values. Does the linear model seem appropriate? Explain.
c. Plot the residuals versus the order in which the observations were made. Is there a trend in the residuals over time? Does the linear model seem appropriate? Explain.

4. In rock blasting, the peak particle velocity (PPV) depends both on the distance from the blast and on the amount of charge. The article "Prediction of Particle Velocity Caused by Blasting for an Infrastructure Excavation Covering Granite Bedrock" (A. Kahriman, *Mineral Resources Engineering*, 2001:205–218) suggests predicting PPV (y) from the scaled distance (x), which is equal to the distance divided by the square root of the charge. The results for 15 blasts are presented in the following table.

PPV (mm/s)	Scaled Distance (m/kg$^{0.5}$)
1.4	47.33
15.7	9.6
2.54	15.8
1.14	24.3
0.889	23.0
1.65	12.7
1.4	39.3
26.8	8.0
1.02	29.94
4.57	10.9
6.6	8.63
1.02	28.64
3.94	18.21
1.4	33.0
1.4	34.0

a. Plot PPV versus scaled distance. Does the relationship appear to be linear?
b. Compute the least-squares line for the model ln PPV $= \beta_0 + \beta_1$ ln scaled distance $+ \varepsilon$. Plot the residuals versus fitted values. Does this linear model seem appropriate?
c. Use the least-squares line computed in part (b) to predict the PPV when the scaled distance is 20. Find a 95% prediction interval.

5. Good forecasting and control of preconstruction activities leads to more efficient use of time and resources in highway construction projects. Data on construction costs (in $1000s) and person-hours of labor required on several projects are presented in the following table and are taken from the article "Forecasting Engineering Manpower Requirements for Highway Preconstruction Activities" (K. Persad, J. O'Connor, and K. Varghese, *Journal of Management Engineering*, 1995:41–47). Each value represents an average of several projects, and two outliers have been deleted.

Person-Hours (x)	Cost (y)	Person-Hours (x)	Cost (y)
939	251	1069	355
5796	4690	6945	5253
289	124	4159	1177
283	294	1266	802
138	138	1481	945
2698	1385	4716	2327
663	345		

a. Compute the least-squares line for predicting y from x.
b. Plot the residuals versus the fitted values. Does the model seem appropriate?
c. Compute the least-squares line for predicting ln y from ln x.
d. Plot the residuals versus the fitted values. Does the model seem appropriate?
e. Using the more appropriate model, construct a 95% prediction interval for the cost of a project that requires 1000 person-hours of labor.

6. The article "Oxidation State and Activities of Chromium Oxides in CaO-SiO$_2$-CrO$_x$ Slag System" (Y. Xiao, L. Holappa, and M. Reuter, *Metallurgical and Materials Transactions B*, 2002:595–603) presents the amount x

(in mole percent) and activity coefficient y of $CrO_{1.5}$ for several specimens. The data, extracted from a larger table, are presented in the following table.

x	y	x	y	x	y
10.20	2.6	7.13	5.8	5.33	13.1
5.03	19.9	3.40	29.4	16.70	0.6
8.84	0.8	5.57	2.2	9.75	2.2
6.62	5.3	7.23	5.5	2.74	16.9
2.89	20.3	2.12	33.1	2.58	35.5
2.31	39.4	1.67	44.2	1.50	48.0

a. Compute the least-squares line for predicting y from x.
b. Plot the residuals versus the fitted values.
c. Compute the least-squares line for predicting y from $1/x$.
d. Plot the residuals versus the fitted values.
e. Using the better fitting line, find a 95% confidence interval for the mean value of y when $x = 5.0$.

7. A windmill is used to generate direct current. Data are collected on 45 different days to determine the relationship between wind speed in mi/h (x) and current in kA (y). The data are presented in the following table.

Day	Wind Speed	Current	Day	Wind Speed	Current	Day	Wind Speed	Current
1	4.2	1.9	16	3.7	2.1	31	2.6	1.4
2	1.4	0.7	17	5.9	2.2	32	7.7	2.8
3	6.6	2.2	18	6.0	2.6	33	6.1	2.4
4	4.7	2.0	19	10.7	3.2	34	5.5	2.2
5	2.6	1.1	20	5.3	2.3	35	4.7	2.3
6	5.8	2.6	21	5.1	1.9	36	4.0	2.0
7	1.8	0.3	22	4.9	2.3	37	2.3	1.2
8	5.8	2.3	23	8.3	3.1	38	11.9	3.0
9	7.3	2.6	24	7.1	2.3	39	8.6	2.5
10	7.1	2.7	25	9.2	2.9	40	5.6	2.1
11	6.4	2.4	26	4.4	1.8	41	4.2	1.7
12	4.6	2.2	27	8.0	2.6	42	6.2	2.3
13	1.6	1.1	28	10.5	3.0	43	7.7	2.6
14	2.3	1.5	29	5.1	2.1	44	6.6	2.9
15	4.2	1.5	30	5.8	2.5	45	6.9	2.6

a. Compute the least-squares line for predicting y from x. Make a plot of residuals versus fitted values.
b. Compute the least-squares line for predicting y from $\ln x$. Make a plot of residuals versus fitted values.
c. Compute the least-squares line for predicting $\ln y$ from x. Make a plot of residuals versus fitted values.
d. Compute the least-squares line for predicting $\sqrt{y}$ from x. Make a plot of residuals versus fitted values.
e. Which of the four models (a) through (d) fits best? Explain.
f. For the model that fits best, plot the residuals versus the order in which the observations were made. Do the residuals seem to vary with time?
g. Using the best model, predict the current when wind speed is 5.0 mi/h.
h. Using the best model, find a 95% prediction interval for the current on a given day when the wind speed is 5.0 mi/h.

8. Two radon detectors were placed in different locations in the basement of a home. Each provided an hourly measurement of the radon concentration, in units of pCi/L. The data are presented in the following table.

R_1	R_2	R_1	R_2	R_1	R_2	R_1	R_2
1.2	1.2	3.4	2.0	4.0	2.6	5.5	3.6
1.3	1.5	3.5	2.0	4.0	2.7	5.8	3.6
1.3	1.6	3.6	2.1	4.3	2.7	5.9	3.9
1.3	1.7	3.6	2.1	4.3	2.8	6.0	4.0
1.5	1.7	3.7	2.1	4.4	2.9	6.0	4.2
1.5	1.7	3.8	2.2	4.4	3.0	6.1	4.4
1.6	1.8	3.8	2.2	4.7	3.1	6.2	4.4
2.0	1.8	3.8	2.3	4.7	3.2	6.5	4.4
2.0	1.9	3.9	2.3	4.8	3.2	6.6	4.4
2.4	1.9	3.9	2.4	4.8	3.5	6.9	4.7
2.9	1.9	3.9	2.4	4.9	3.5	7.0	4.8
3.0	2.0	3.9	2.4	5.4	3.5		

a. Compute the least-squares line for predicting the radon concentration at location 2 from the concentration at location 1.
b. Plot the residuals versus the fitted values. Does the linear model seem appropriate?
c. Divide the data into two groups: points where $R_1 < 4$ in one group, points where $R_1 \geq 4$ in the other. Compute the least-squares line and the residual plot for each group. Does the line describe either group well? Which one?
d. Explain why it might be a good idea to fit a linear model to part of these data, and a nonlinear model to the other.

9. The article "The Equilibrium Partitioning of Titanium Between Ti^{3+} and Ti^{4+} Valency States in $CaO\text{-}SiO_2\text{-}TiO_x$ Slags" (G. Tranell, O. Ostrovski, and S. Jahanshahi, *Metallurgical and Materials Transactions B*, 2002:61–66) discusses the relationship between the redox ratio Ti^{3+}/Ti^{4+} and oxygen partial pressure p_{O_2} in $CaO\text{-}SiO_2\text{-}TiO_x$ melts. Several independent measurements of the redox ratio were made at each of five different partial pressures: 10^{-7}, 10^{-8}, 10^{-9}, 10^{-10}, and 10^{-12} atmospheres. The results for the runs at 14 mass percent TiO_x are presented in the following table.

Oxygen Partial Pressure	Redox Ratio Measurements
10^{-7}	0.011, 0.017, 0.034, 0.039
10^{-8}	0.018, 0.011, 0.026, 0.050, 0.034, 0.068, 0.061
10^{-9}	0.027, 0.038, 0.076, 0.088
10^{-10}	0.047, 0.069, 0.123, 0.162
10^{-12}	0.160, 0.220, 0.399, 0.469

a. Denoting the redox ratio by y and the partial pressure by x, theory states that y should be proportional to x^{β} for some β. Express this theoretical relationship as a linear model.
b. Compute the least-squares line for this linear model. Plot the residuals versus the fitted values. Does the linear model hold?
c. Further theoretical considerations suggest that under the conditions of this experiment, y should be proportional to $x^{-1/4}$. Are the data in the preceding table consistent with this theory? Explain.

10. The article "The Selection of Yeast Strains for the Production of Premium Quality South African Brandy Base Products" (C. Steger and M. Lambrechts, *Journal of Industrial Microbiology and Biotechnology*, 2000:431–440) presents detailed information on the volatile compound composition of base wines made from each of 16 selected yeast strains. Below are the concentrations of total esters and total volatile acids (in mg/L) in each of the wines.

Esters	Acids	Esters	Acids	Esters	Acids	Esters	Acids
284.34	445.70	173.01	265.43	229.55	210.58	312.95	203.62
215.34	332.59	188.72	166.73	144.39	254.82	172.79	342.21
139.38	356.88	197.81	291.72	303.28	215.83	256.02	152.38
658.38	192.59	105.14	412.42	295.24	442.55	170.41	391.30

a. Construct a scatterplot of acid concentration versus ester concentration. Indicate the outlier.
b. Compute the coefficients of the least-squares line for predicting acid level (y) from ester level (x), along with their estimated standard deviations.
c. Compute the P-value of the test of the null hypothesis $H_0: \beta_1 = 0$.
d. Delete the outlier, and recompute the coefficients of the least-squares line, along with their estimated standard deviations.
e. Compute the P-value of the test of the null hypothesis $H_0: \beta_1 = 0$ for the data with the outlier deleted.
f. Does a linear model appear to be useful for predicting acid concentration from ester concentration? Explain.

11. The article "Mathematical Modeling of the Argon-Oxygen Decarburization Refining Process of Stainless Steel: Part II. Application of the Model to Industrial Practice" (J. Wei and D. Zhu, *Metallurgical and Materials Transactions B*, 2001:212–217) presents the carbon content (in mass %) and bath temperature (in K) for 32 heats of austenitic stainless steel. These data are shown in the following table.

Carbon %	Temp.	Carbon %	Temp.	Carbon %	Temp.	Carbon %	Temp.
19	1975	17	1984	18	1962	17	1983
23	1947	20	1991	19	1985	20	1966
22	1954	19	1965	19	1946	21	1972
16	1992	22	1963	15	1986	17	1989
17	1965	18	1949	20	1946	18	1984
18	1971	22	1960	22	1950	23	1967
12	2046	20	1960	15	1979	13	1954
24	1945	19	1953	15	1989	15	1977

a. Compute the least-squares line for predicting bath temperature (y) from carbon content (x).
b. Identify two outliers. Compute the two least-squares lines that result from the deletion of each outlier individually, and the least-squares line that results from the deletion of both outliers.
c. Are the least-squares lines computed in parts (a) and (b) similar? If so, report the line that was fit to the full data set, along with 95% confidence intervals for the slope and intercept. If not, report the range of slopes, without a confidence interval.

12. The article "Mechanistic-Empirical Design of Bituminous Roads: An Indian Perspective" (A. Das and B. Pandey, *Journal of Transportation Engineering*, 1999:463–471) presents an equation of the form $y = a(1/x_1)^b(1/x_2)^c$ for predicting the number of repetitions for laboratory fatigue failure (y) in terms of the tensile strain at the bottom of the bituminous beam (x_1) and the resilient modulus (x_2). Transform this equation into a linear model, and express the linear model coefficients in terms of a, b, and c.

13. An engineer wants to determine the spring constant for a particular spring. She hangs various weights on one end of the spring and measures the length of the spring each time. A scatterplot of length (y) versus load (x) is depicted in the following figure.

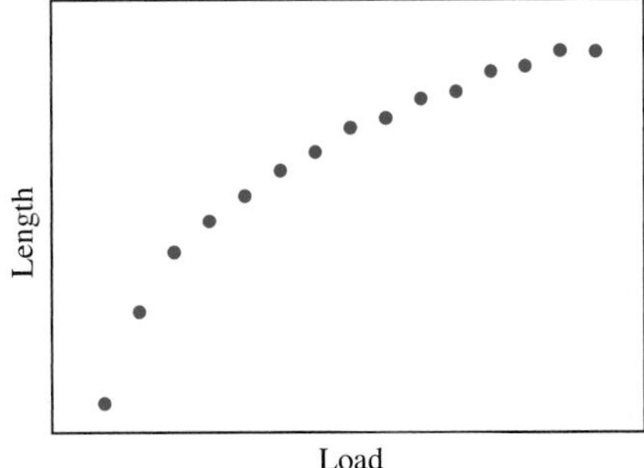

a. Is the model $y = \beta_0 + \beta_1 x$ an empirical model or a physical law?
b. Should she transform the variables to try to make the relationship more linear, or would it be better to redo the experiment? Explain.

Supplementary Exercises for Chapter 7

1. The Beer–Lambert law relates the absorbance A of a solution to the concentration C of a species in solution by $A = MLC$, where L is the path length and M is the molar absorption coefficient. Assume that $L = 1$ cm. Measurements of A are made at various concentrations. The data are presented in the following table.

Concentration (mol/cm^3)	1.00	1.20	1.50	1.70	2.00
Absorbance	0.99	1.13	1.52	1.73	1.96

a. Let $A = \widehat{\beta}_0 + \widehat{\beta}_1 C$ be the equation of the least-squares line for predicting absorbance (A) from concentration (C). Compute the values of $\widehat{\beta}_0$ and $\widehat{\beta}_1$.
b. What value does the Beer–Lambert law assign to β_0?
c. What physical quantity does $\widehat{\beta}_1$ estimate?
d. Test the hypothesis $H_0 : \beta_0 = 0$. Is the result consistent with the Beer–Lambert law?

2. In a test of military ordnance, a large number of bombs were dropped on a target from various heights. The initial velocity of the bombs in the direction of the ground was 0. Let y be the height in meters from which a bomb is dropped, let x be the time in seconds for the bomb to strike the ground, let $w = x^2$, and let $v = \sqrt{y}$. The relationship between x and y is given by $y = 4.9x^2$. For each of the following pairs of variables, state whether the correlation coefficient is an appropriate summary.

a. x and y
b. w and y
c. x and v
d. w and v
e. $\ln x$ and $\ln y$

3. Eruptions of the Old Faithful geyser in Yellowstone National Park typically last from 1.5 to 5 minutes. Between eruptions are dormant periods, which typically last from 50 to 100 minutes. A dormant period can also be thought of as the waiting time between eruptions. The durations in minutes for 60 consecutive dormant periods are given in the following table. It is desired to predict the length of a dormant period from the length of the dormant period immediately preceding it. To express this in symbols, denote the sequence of dormant periods $T_1, \ldots, T_{60}$. It is desired to predict T_{i+1} from T_i.

i	T_i	i	T_i	i	T_i	i	T_i	i	T_i	i	T_i
1	80	11	56	21	82	31	88	41	72	51	67
2	84	12	80	22	51	32	51	42	75	52	81
3	50	13	69	23	76	33	80	43	75	53	76
4	93	14	57	24	82	34	49	44	66	54	83
5	55	15	90	25	84	35	82	45	84	55	76
6	76	16	42	26	53	36	75	46	70	56	55
7	58	17	91	27	86	37	73	47	79	57	73
8	74	18	51	28	51	38	67	48	60	58	56
9	75	19	79	29	85	39	68	49	86	59	83
10	80	20	53	30	45	40	86	50	71	60	57

a. Construct a scatterplot of the points (T_i, T_{i+1}), for $i = 1, \ldots, 59$.

b. Compute the least-squares line for predicting T_{i+1} from T_i. (*Hint:* The values of the independent variable (x) are $T_1, \ldots, T_{59}$, and the values of the dependent variable (y) are $T_2, \ldots, T_{60}$.)

c. Find a 95% confidence interval for the slope β_1.

d. If the waiting time before the last eruption was 70 minutes, what is the predicted waiting time before the next eruption?

e. Find a 98% confidence interval for the mean waiting time before the next eruption when the time before the last eruption was 70 minutes.

f. Find a 99% prediction interval for the waiting time before the next eruption, if the time before the last eruption was 70 minutes.

4. Refer to Exercise 3.

a. Plot the residuals versus the fitted values. Does the plot indicate any serious violations of the standard assumptions?

b. Plot the residuals versus the order of the data. Does the plot indicate any serious violations of the standard assumptions?

5. A chemist is calibrating a spectrophotometer that will be used to measure the concentration of carbon monoxide (CO) in atmospheric samples. To check the calibration, samples of known concentration are measured. The true concentrations (x) and the measured concentrations (y) are given in the following table. Because of random error, repeated measurements on the same sample will vary. The machine is considered to be in calibration if its mean response is equal to the true concentration.

True concentration (ppm)	Measured concentration (ppm)
0	1
10	11
20	21
30	28
40	37
50	48
60	56
70	68
80	75
90	86
100	96

To check the calibration, the linear model $y = \beta_0 + \beta_1 x + \varepsilon$ is fit. Ideally, the value of β_0 should be 0 and the value of β_1 should be 1.

a. Compute the least-squares estimates $\widehat{\beta}_0$ and $\widehat{\beta}_1$.
b. Can you reject the null hypothesis $H_0 : \beta_0 = 0$?
c. Can you reject the null hypothesis $H_0 : \beta_1 = 1$?
d. Do the data provide sufficient evidence to conclude that the machine is out of calibration?
e. Compute a 95% confidence interval for the mean measurement when the true concentration is 20 ppm.
f. Compute a 95% confidence interval for the mean measurement when the true concentration is 80 ppm.
g. Someone claims that the machine is in calibration for concentrations near 20 ppm. Do these data provide sufficient evidence to disprove this claim? Explain.

6. The article "Experimental Measurement of Radiative Heat Transfer in Gas-Solid Suspension Flow System" (G. Han, K. Tuzla, and J. Chen, *AIChe Journal*, 2002:1910–1916) discusses the calibration of a radiometer. Several measurements were made on the electromotive force readings of the radiometer (in volts) and the radiation flux (in kilowatts per square meter). The results (read from a graph) are presented in the following table.

Heat flux (y)	15	31	51	55	67	89
Signal output (x)	1.08	2.42	4.17	4.46	5.17	6.92

a. Compute the least-squares line for predicting heat flux from the signal output.
b. If the radiometer reads 3.00 V, predict the heat flux.
c. If the radiometer reads 8.00 V, can the heat flux be predicted? If so, predict it. If not, explain why.

7. The article "Effect of Temperature on the Marine Immersion Corrosion of Carbon Steels" (R. Melchers, *Corrosion*, 2002:768–781) presents measurements of corrosion loss (in mm) for copper-bearing steel specimens immersed in seawater in 14 different locations. For each location, the average corrosion loss (in mm) was recorded, along with the mean water temperature (in °C). The results, after one year of immersion, are presented in the following table.

Corrosion	Mean Temp.	Corrosion	Mean Temp.
0.2655	23.5	0.2200	26.5
0.1680	18.5	0.0845	15.0
0.1130	23.5	0.1860	18.0
0.1060	21.0	0.1075	9.0
0.2390	17.5	0.1295	11.0
0.1410	20.0	0.0900	11.0
0.3505	26.0	0.2515	13.5

a. Compute the least-squares line for predicting corrosion loss (y) from mean temperature (x).
b. Find a 95% confidence interval for the slope.
c. Find a 95% confidence interval for the mean corrosion loss at a mean temperature of 20°C.
d. Find a 90% prediction interval for the corrosion loss of a specimen immersed at a mean temperature of 20°C.

8. The article "Measurements of the Thermal Conductivity and Thermal Diffusivity of Polymer Melts with the Short-Hot-Wire Method" (X. Zhang, W. Hendro, et al., *International Journal of Thermophysics*, 2002:1077–1090) presents measurements of the thermal conductivity and diffusivity of several polymers at several temperatures. The results for the thermal diffusivity of polycarbonate (in 10^{-7} m^{2}s^{-1}) are given in the following table.

Diff.	Temp. (°C)	Diff.	Temp. (°C)	Diff.	Temp. (°C)	Diff.	Temp. (°C)
1.43	28	1.36	107	1.05	159	1.26	215
1.53	38	1.34	119	1.13	169	0.86	225
1.43	61	1.29	130	1.03	181	1.01	237
1.34	83	1.36	146	1.06	204	0.98	248

a. Compute the least-squares line for predicting diffusivity (y) from temperature (x).
b. Find a 95% confidence interval for the slope.
c. Find a 95% confidence interval for the diffusivity of polycarbonate at a temperature of 100°C.
d. Find a 95% prediction interval for the diffusivity of polycarbonate at a temperature of 100°C.
e. Which is likely to be the more useful, the confidence interval or the prediction interval? Explain.

9. The article "Copper Oxide Mounted on Activated Carbon as Catalyst for Wet Air Oxidation of Aqueous Phenol. 1. Kinetic and Mechanistic Approaches" (P. Alvarez, D. McLurgh, and P. Plucinski, *Industrial Engineering and Chemistry Research*, 2002: 2147–2152) reports the results of experiments to describe the mechanism of the catalytic wet air oxidation of aqueous phenol. In one set of experiments, the initial oxidation rate (in kilograms of phenol per kilogram of catalyst per hour) and the oxygen concentration (in mol/m^3) were measured. The results (read from a graph) are presented in the following table.

Rate (y)	0.44	0.49	0.60	0.64	0.72
O_2 concentration (x)	3.84	4.76	6.08	7.06	8.28

a. It is known that x and y are related by an equation of the form $y = kx^r$, where r is the oxygen reaction order. Make appropriate transformations to express this as a linear equation.
b. Estimate the values of k and r by computing the least-squares line.
c. Based on these data, is it plausible that the oxygen reaction order is equal to 0.5? Explain.

10. A materials scientist is experimenting with a new material with which to make beverage cans. She fills cans with liquid at room temperature, and then refrigerates them to see how fast they cool. According to Newton's law of cooling, if t is the time refrigerated and y is the temperature drop at time t, then y is related to t by an equation of the form

$$\ln y = \beta_0 + \beta_1 t,$$

where β_0 is a constant that depends on the initial temperature of the can and the ambient temperature of the refrigerator, and β_1 is a constant that depends on the physical properties of the can. The scientist measures the temperature at regular intervals, and then fits this model to the data. The results are shown in the following figure. A scatterplot, with the least-squares line superimposed, is on the left, and the residual plot is on the right.

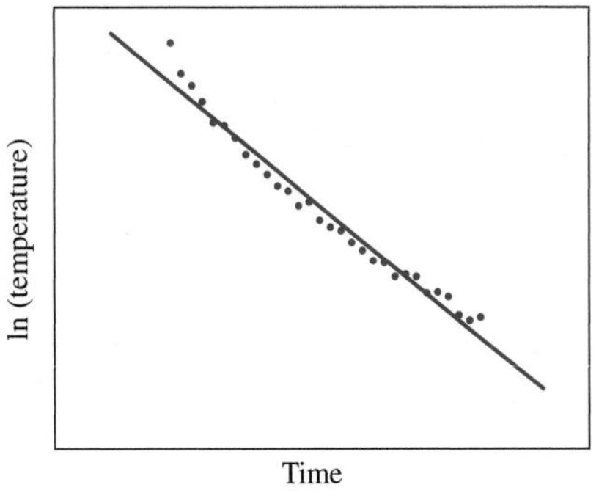

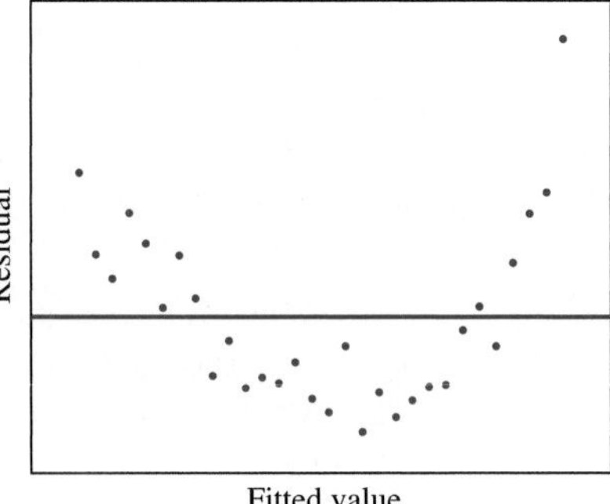

What should the scientist do next?

i. Try to find a transformation that makes the relationship more linear.
ii. Use the model as is, because Newton's law of cooling is a physical law.
iii. Use the model as is, because it fits well enough.
iv. Carefully examine the experimental setup to see what might have gone wrong.

11. Monitoring the yield of a particular chemical reaction at various reaction vessel temperatures produces the results shown in the following table.

Temp. (°C)	Yield (%)	Temp. (°C)	Yield (%)
150	77.4	250	88.9
150	76.7	250	89.2
150	78.2	250	89.7
200	84.1	300	94.8
200	84.5	300	94.7
200	83.7	300	95.9

a. Find the least-squares estimates for β_0, β_1, and σ^2 for the simple linear model Yield $= \beta_0 + \beta_1$ Temp $+ \varepsilon$.
b. Can you conclude that β_0 is not equal to 0?
c. Can you conclude that β_1 is not equal to 0?
d. Make a residual plot. Does the linear model seem appropriate?
e. Find a 95% confidence interval for the slope.
f. Find a 95% confidence interval for the mean yield at a temperature of 225°C.
g. Find a 95% prediction interval for a yield at a temperature of 225°C.

12. The article "Approach to Confidence Interval Estimation for Curve Numbers" (R. McCuen, *Journal of Hydrologic Engineering*, 2002:43–48) discusses the relationship between rainfall depth and runoff depth at several locations. At one particular location, rainfall depth and runoff depth were recorded for 13 rainstorms. Following is MINITAB output for a fit of the least-squares line to predict runoff depth from rainfall depth (both measured in inches).

```
The regression equation is
Runoff = -0.23 + 0.73 Rainfall

Predictor          Coef     SE Coef         T        P
Constant       -0.23429     0.23996     -0.98    0.350
Rainfall        0.72868     0.06353     11.47    0.000

S = 0.40229         R-Sq = 92.3%          R-Sq(adj) = 91.6%

Analysis of Variance

Source           DF         SS          MS          F        P
Regression        1     21.290      21.290     131.55    0.000
Residual Error   11      1.780     0.16184
Total            12     23.070
```

a. Predict the runoff for a storm with 2.5 in. of rainfall.
b. Someone claims that if two storms differ in their rainfall by 1 in., then their runoffs will differ, on the average, by 1 in. as well. Is this a plausible claim? Explain.
c. It is a fact that if the rainfall is 0, the runoff is 0. Is the least-squares line consistent with this fact? Explain.

13. Refer to Exercise 12. Someone wants to compute a 95% confidence interval for the mean runoff when the rainfall is 3 in. Can this be computed from the information in the MINITAB output shown in Exercise 12? Or is more information needed? Choose the best answer.

i. Yes, it can be computed from the MINITAB output.
ii. No, we also need to know the rainfall values that were used to compute the least-squares line.
iii. No, we also need to know the runoff values that were used to compute the least-squares line.
iv. No, we also need to know both the rainfall and the runoff values that were used to compute the least-squares line.

14. During the production of boiler plate, test pieces are subjected to a load, and their elongations are measured. In one particular experiment, five tests will be made, at loads (in MPa) of 11, 37, 54, 70, and 93. The least-squares line will be computed to predict elongation from load. Confidence intervals for the mean elongation will be computed for several different loads. Which of the following intervals will be the widest? Which will be the narrowest?

i. The 95% confidence interval for the mean elongation under a load of 53 MPa.
ii. The 95% confidence interval for the mean elongation under a load of 72 MPa.
iii. The 95% confidence interval for the mean elongation under a load of 35 MPa.

15. The article "Low-Temperature Heat Capacity and Thermodynamic Properties of 1,1,1-trifluoro-2,2-dichloroethane" (R. Varushchenko and A. Druzhinina, *Fluid Phase Equilibria*, 2002:109–119) describes an experiment in which samples of Freon R-123 were melted in a calorimeter. Various quantities of energy were supplied to the calorimeter for melting. The equilibrium melting temperatures (t) and fractions melted (f) were measured. The least-squares line was fit to the model $t = \beta_0 + \beta_1(1/f) + \varepsilon$, where $1/f$ is the reciprocal fraction. The results of the fit are as follows.

```
The regression equation is
Temperature = 145.74 - 0.052 Reciprocal Frac

Predictor        Coef      SE Coef         T        P
Constant      145.736      0.00848   17190.1    0.000
Recip Frac   -0.05180      0.00226   -22.906    0.000

S = 0.019516        R-Sq = 97.6%        R-Sq(adj) = 97.4%

Analysis of Variance

Source            DF         SS          MS         F        P
Regression         1      0.200       0.200    524.70    0.000
Residual Error    13    0.00495    0.000381
Total             14      0.205
```

a. Estimate the temperature at which half of the sample will melt (i.e., $f = 1/2$).
b. Can you determine the correlation coefficient between equilibrium temperature and reciprocal of the fraction melted from this output? If so, determine it. If not, explain what additional information is needed.
c. The triple-point temperature is the lowest temperature at which the whole sample will melt (i.e., $f = 1$). Estimate the triple-point temperature.

16. The article "Polyhedral Distortions in Tourmaline" (A. Ertl, J. Hughes, et al., *The Canadian Mineralogist*, 2002: 153–162) presents a model for calculating bond-length distortion in vanadium-bearing tourmaline. To check the accuracy of the model, several calculated values (x) were compared with directly observed values (y). The results (read from a graph) are presented in the following table.

Observed Value	Calculated Value	Observed Value	Calculated Value
0.33	0.36	0.74	0.78
0.36	0.36	0.79	0.86
0.54	0.58	0.97	0.97
0.56	0.64	1.03	1.11
0.66	0.64	1.10	1.06
0.66	0.67	1.13	1.08
0.74	0.58	1.14	1.17

a. Assume that the observed value y is an unbiased measurement of the true value. Show that if the calculated value x is accurate (i.e., equal to the true value), then $y = x + \varepsilon$, where ε is measurement error.
b. Compute the least-squares line $y = \widehat{\beta}_0 + \widehat{\beta}_1 x$.
c. Show that if the calculated value is accurate, then the true coefficients are $\beta_0 = 0$ and $\beta_1 = 1$.
d. Test the null hypotheses $\beta_0 = 0$ and $\beta_1 = 1$.
e. Is it plausible that the calculated value is accurate? Or can you conclude that it is not? Explain.

17. Consider the model $y = \beta x + \varepsilon$, where the intercept of the line is known to be zero. Assume that values $(x_1, y_1), \ldots,$ (x_n, y_n) are observed, and the least-squares estimate $\widehat{\beta}$ of β is to be computed.

a. Derive the least-squares estimate $\widehat{\beta}$ in terms of x_i and y_i.
b. Let σ^2 denote the variance of ε (which is also the variance of y). Derive the variance $\sigma^2_{\widehat{\beta}}$ of the least-squares estimate, in terms of σ^2 and the x_i.

18. Use Equation (7.33) (p. 510) to show that $\mu_{\widehat{\beta}_1} = \beta_1$.

19. Use Equation (7.34) (p. 510) to show that $\mu_{\widehat{\beta}_0} = \beta_0$.

20. Use Equation (7.33) (p. 510) to derive the formula $\sigma^2_{\widehat{\beta}_1} = \dfrac{\sigma^2}{\sum_{i=1}^n (x_i - \overline{x})^2}$.

21. Use Equation (7.34) (p. 510) to derive the formula $\sigma^2_{\widehat{\beta}_0} = \sigma^2 \left(\dfrac{1}{n} + \dfrac{\overline{x}^2}{\sum_{i=1}^n (x_i - \overline{x})^2} \right)$.

Chapter 8

Multiple Regression

Introduction

The methods of simple linear regression, discussed in Chapter 7, apply when we wish to fit a linear model relating the value of an independent variable y to the value of a single dependent variable x. There are many situations, however, in which a single independent variable is not enough. For example, the degree of wear on a lubricated bearing in a machine may depend both on the load on the bearing and on the physical properties of the lubricant. An equation that expressed wear as a function of load alone or of lubricant properties alone would fail as a predictor. In situations like this, there are several independent variables, $x_1, x_2, \ldots, x_p$, that are related to a dependent variable y. If the relationship between the dependent and independent variables is linear, the technique of **multiple regression** can be used.

8.1 The Multiple Regression Model

We describe the multiple regression model. Assume that we have a sample of n items, and that on each item we have measured a dependent variable y and p independent variables $x_1, \ldots, x_p$. The ith sample item thus gives rise to the ordered set $(y_i, x_{1i}, \ldots, x_{pi})$. We can then fit the **multiple regression model**

$$y_i = \beta_0 + \beta_1 x_{1i} + \cdots + \beta_p x_{pi} + \varepsilon_i \tag{8.1}$$

There are several special cases of the multiple regression model (8.1) that are often used in practice. One is the **polynomial regression model**, in which the independent variables are all powers of a single variable. The polynomial regression model of degree p is

$$y_i = \beta_0 + \beta_1 x_i + \beta_2 x_i^2 + \cdots + \beta_p x_i^p + \varepsilon_i \tag{8.2}$$

Multiple regression models can also be made up of powers of several variables. For example, a polynomial regression model of degree 2, also called a **quadratic model**, in two variables x_1 and x_2 is given by

$$y_i = \beta_0 + \beta_1 x_{1i} + \beta_2 x_{2i} + \beta_3 x_{1i} x_{2i} + \beta_4 x_{1i}^2 + \beta_5 x_{2i}^2 + \varepsilon_i \tag{8.3}$$

A variable that is the product of two other variables is called an **interaction**. In model (8.3), the variable $x_{1i}x_{2i}$ is the **interaction** between x_1 and x_2.

Models (8.2) and (8.3) are considered to be linear models, even though they contain nonlinear terms in the independent variables. The reason they are still linear models is that *they are linear in the coefficients* β_i.

Estimating the Coefficients

In any multiple regression model, the estimates $\widehat{\beta}_0, \widehat{\beta}_1, \ldots, \widehat{\beta}_p$ are computed by least-squares, just as in simple linear regression. The equation

$$\widehat{y} = \widehat{\beta}_0 + \widehat{\beta}_1 x_1 + \cdots + \widehat{\beta}_p x_p \tag{8.4}$$

is called the **least-squares equation** or **fitted regression equation**. Now define $\widehat{y}_i$ to be the y coordinate of the least-squares equation corresponding to the x values $(x_{1i}, \ldots, x_{pi})$. The residuals are the quantities $e_i = y_i - \widehat{y}_i$, which are the differences between the observed y values and the y values given by the equation. We want to compute $\widehat{\beta}_0, \widehat{\beta}_1, \ldots, \widehat{\beta}_p$ so as to minimize the sum of the squared residuals $\sum_{i=1}^n e_i^2$. To do this, we express e_i in terms of $\widehat{\beta}_0, \widehat{\beta}_1, \ldots, \widehat{\beta}_p$:

$$e_i = y_i - \widehat{\beta}_0 - \widehat{\beta}_1 x_{1i} - \cdots - \widehat{\beta}_p x_{pi} \tag{8.5}$$

Thus we wish to minimize the sum

$$\sum_{i=1}^{n} (y_i - \widehat{\beta}_0 - \widehat{\beta}_1 x_{1i} - \cdots - \widehat{\beta}_p x_{pi})^2 \tag{8.6}$$

To do this, we can take partial derivatives of (8.6) with respect to $\widehat{\beta}_0, \widehat{\beta}_1, \ldots, \widehat{\beta}_p$, set them equal to 0, and solve the resulting $p + 1$ equations in $p + 1$ unknowns. The expressions obtained for $\widehat{\beta}_0, \widehat{\beta}_1, \ldots, \widehat{\beta}_p$ are complicated. Fortunately, they have been coded into many software packages, so that you can calculate them on the computer. For each estimated coefficient $\widehat{\beta}_i$, there is an estimated standard deviation $s_{\widehat{\beta}_i}$. Expressions for these quantities are complicated as well, so nowadays people rely on computers to calculate them.

Sums of Squares

Much of the analysis in multiple regression is based on three fundamental quantities. They are the **regression sum of squares** (SSR), the **error sum of squares** (SSE), and the

total sum of squares (SST). We defined these quantities in Section 7.2, in our discussion of simple linear regression. The definitions hold for multiple regression as well. We repeat them here.

Definition: Sums of Squares

In the multiple regression model

$$y_i = \beta_0 + \beta_1 x_{1i} + \cdots + \beta_p x_{pi} + \varepsilon_i,$$

the following sums of squares are defined:

- Regression sum of squares: $\text{SSR} = \sum_{i=1}^{n}(\widehat{y}_i - \overline{y})^2$
- Error sum of squares: $\text{SSE} = \sum_{i=1}^{n}(y_i - \widehat{y}_i)^2$
- Total sum of squares: $\text{SST} = \sum_{i=1}^{n}(y_i - \overline{y})^2$

It can be shown that

$$\text{SST} = \text{SSR} + \text{SSE} \tag{8.7}$$

Equation (8.7) is called the **analysis of variance identity**. This identity is derived for simple linear regression at the end of Section 7.2.

We will now see how these sums of squares are used to derive the statistics used in multiple regression. As we did for simple linear regression, we will restrict our discussion to the simplest case, in which four assumptions about the errors ε_i are satisfied. We repeat these assumptions here.

Assumptions for Errors in Linear Models

In the simplest situation, the following assumptions are satisfied:

1. The errors $\varepsilon_1, \ldots, \varepsilon_n$ are random and independent. In particular, the magnitude of any error ε_i does not influence the value of the next error ε_{i+1}.
2. The errors $\varepsilon_1, \ldots, \varepsilon_n$ all have mean 0.
3. The errors $\varepsilon_1, \ldots, \varepsilon_n$ all have the same variance, which we denote by σ^2.
4. The errors $\varepsilon_1, \ldots, \varepsilon_n$ are normally distributed.

Just as in simple linear regression, these assumptions imply that the observations y_i are independent random variables. To be specific, each y_i has a normal distribution with mean $\beta_0 + \beta_1 x_{1i} + \cdots + \beta_p x_{pi}$ and variance σ^2. Each coefficient β_i represents the change in the mean of y associated with an increase of one unit in the value of x_i, when the other x variables are held constant.

> **Summary**
>
> In the multiple regression model $y_i = \beta_0 + \beta_1 x_{1i} + \cdots + \beta_p x_{pi} + \varepsilon_i$, under assumptions 1 through 4, the observations $y_1, \ldots, y_n$ are independent random variables that follow the normal distribution. The mean and variance of y_i are given by
>
> $$\mu_{y_i} = \beta_0 + \beta_1 x_{1i} + \cdots + \beta_p x_{pi}$$
>
> $$\sigma^2_{y_i} = \sigma^2$$
>
> Each coefficient β_i represents the change in the mean of y associated with an increase of one unit in the value of x_i, when the other x variables are held constant.

The Statistics s^2, R^2, and F

The three statistics most often used in multiple regression are the estimated error variance s^2, the coefficient of determination R^2, and the F statistic. Each of these has an analog in simple linear regression. We discuss them in turn.

In simple linear regression, the estimated error variance is $\sum_{i=1}^n (y_i - \widehat{y}_i)^2/(n-2)$. We divide by $n-2$ rather than n because the residuals ($e_i = y_i - \widehat{y}_i$) tend to be a little smaller than the errors ε_i. The reason that the residuals are a little smaller is that the two coefficients ($\widehat{\beta}_0$ and $\widehat{\beta}_1$) have been chosen to minimize $\sum_{i=1}^n (y_i - \widehat{y}_i)^2$. Now in the case of multiple regression, we are estimating $p+1$ coefficients rather than just two. Thus the residuals tend to be smaller still, so we must divide $\sum_{i=1}^n (y_i - \widehat{y}_i)^2$ by a still smaller denominator. It turns out that the appropriate denominator is equal to the number of observations (n) minus the number of parameters in the model ($p+1$). Therefore the estimated error variance is given by

$$s^2 = \frac{\sum_{i=1}^n (y_i - \widehat{y}_i)^2}{n-p-1} = \frac{\text{SSE}}{n-p-1} \tag{8.8}$$

The estimated variance $s^2_{\widehat{\beta}_i}$ of each least-squares coefficient $\widehat{\beta}_i$ is computed by multiplying s^2 by a rather complicated function of the variables x_{ij}. In practice, the values of $s^2_{\widehat{\beta}_i}$ are calculated on a computer. When assumptions 1 through 4 are satisfied, the quantity

$$\frac{\widehat{\beta}_i - \beta_i}{s_{\widehat{\beta}_i}}$$

has a Student's t distribution with $n-p-1$ degrees of freedom. The number of degrees of freedom is equal to the denominator used to compute the estimated error variance s^2 (Equation 8.8). This statistic is used to compute confidence intervals and to perform hypothesis tests on the values β_i, just as in simple linear regression.

In simple linear regression, the coefficient of determination, r^2, measures the goodness of fit of the linear model. The goodness-of-fit statistic in multiple regression is a quantity denoted R^2, which is also called the coefficient of determination. The value of R^2 is calculated in the same way as is r^2 in simple linear regression (Equation 7.21, in Section 7.2). That is,

$$R^2 = \frac{\sum_{i=1}^{n}(y_i - \overline{y})^2 - \sum_{i=1}^{n}(y_i - \widehat{y}_i)^2}{\sum_{i=1}^{n}(y_i - \overline{y})^2} = \frac{\text{SST} - \text{SSE}}{\text{SST}} = \frac{\text{SSR}}{\text{SST}} \tag{8.9}$$

In simple linear regression, a test of the null hypothesis $\beta_1 = 0$ is almost always made. If this hypothesis is not rejected, then the linear model may not be useful. The analogous null hypothesis in multiple regression is $H_0 : \beta_1 = \beta_2 = \cdots = \beta_p = 0$. This is a very strong hypothesis. It says that none of the independent variables has any linear relationship with the dependent variable. In practice, the data usually provide sufficient evidence to reject this hypothesis. The test statistic for this hypothesis is

$$F = \frac{\left[\sum_{i=1}^{n}(y_i - \overline{y})^2 - \sum_{i=1}^{n}(y_i - \widehat{y}_i)^2\right]/p}{\left[\sum_{i=1}^{n}(y_i - \widehat{y}_i)^2\right]/(n-p-1)} = \frac{[\text{SST} - \text{SSE}]/p}{\text{SSE}/(n-p-1)} = \frac{\text{SSR}/p}{\text{SSE}/(n-p-1)} \tag{8.10}$$

This is an F statistic; its null distribution is $F_{p,\,n-p-1}$. Note that the denominator of the F statistic is s^2 (Equation 8.8). The subscripts p and $n - p - 1$ are the **degrees of freedom** for the F statistic.

Slightly different versions of the F statistic can be used to test weaker null hypotheses. In particular, given a model with independent variables $x_1, \ldots, x_p$, we sometimes want to test the null hypothesis that some of them (say $x_{k+1}, \ldots, x_p$) are not linearly related to the dependent variable. To do this, a version of the F statistic can be constructed that will test the null hypothesis $H_0 : \beta_{k+1} = \cdots = \beta_p = 0$. We will discuss this further in Section 8.3.

An Example

Let us now look at an example of multiple regression. We first describe the data. A mobile ad hoc computer network consists of several computers (nodes) that move within a network area. Often messages are sent from one node to another. When the receiving node is out of range, the message must be sent to a nearby node, which then forwards it along a routing path toward its destination. The routing path is determined by a routine known as a routing protocol. The percentage of messages that are successfully delivered, which is called the goodput, is affected by the average node speed and by the length of time that the nodes pause at each destination. Table 8.1 presents average node speed, average pause time, and goodput for 25 simulated mobile ad hoc networks. These data were generated for a study described in the article "Metrics to Enable Adaptive Protocols for Mobile Ad Hoc Networks" (J. Boleng, W. Navidi, and T. Camp, *Proceedings of the 2002 International Conference on Wireless Networks*, 2002:293–298).

TABLE 8.1 Average node speed, pause time, and goodput for computer networks

Speed (m/s)	Pause Time (s)	Goodput (%)	Speed (m/s)	Pause Time (s)	Goodput (%)
5	10	95.111	20	40	87.800
5	20	94.577	20	50	89.941
5	30	94.734	30	10	62.963
5	40	94.317	30	20	76.126
5	50	94.644	30	30	84.855
10	10	90.800	30	40	87.694
10	20	90.183	30	50	90.556
10	30	91.341	40	10	55.298
10	40	91.321	40	20	78.262
10	50	92.104	40	30	84.624
20	10	72.422	40	40	87.078
20	20	82.089	40	50	90.101
20	30	84.937			

The following MINITAB output presents the results of fitting the model

$$\text{Goodput} = \beta_0 + \beta_1 \text{ Speed} + \beta_2 \text{ Pause} + \beta_3 \text{ Speed} \cdot \text{Pause} + \beta_4 \text{ Speed}^2 + \beta_5 \text{ Pause}^2 + \varepsilon$$

```
The regression equation is
Goodput = 96.0 - 1.82 Speed + 0.565 Pause
         + 0.0247 Speed*Pause + 0.0140 Speed^2
         -0.0118 Pause^2

Predictor        Coef      SE Coef        T       P
Constant       96.024        3.946    24.34   0.000
Speed         -1.8245       0.2376    -7.68   0.000
Pause          0.5652       0.2256     2.51   0.022
Speed*Pa     0.024731     0.003249     7.61   0.000
Speed^2      0.014020     0.004745     2.95   0.008
Pause^2     -0.011793     0.003516    -3.35   0.003

S = 2.942         R-Sq = 93.2%         R-Sq(adj) = 91.4%

Analysis of Variance

Source           DF        SS        MS       F       P
Regression        5   2240.49    448.10   51.77   0.000
Residual Error   19    164.46      8.66
Total            24   2404.95
```

```
Predicted Values for New Observations
New
Obs      Fit   SE Fit          95% CI               95% PI
  1   74.272    1.175   (71.812, 76.732)  (67.641, 80.903)

Values of Predictors for New Observations
New
Obs     Speed      Pause   Speed*Pause     Speed^2   Pause^2
  1      25.0       15.0           375         625       225
```

Much of the output is analogous to that of simple linear regression. The fitted regression equation is presented near the top of the output. Below that, the coefficient estimates $\widehat{\beta}_i$ and their estimated standard deviations $s_{\widehat{\beta}_i}$ are shown. Next to each standard deviation is the Student's t statistic for testing the null hypothesis that the true value of the coefficient is equal to 0. This statistic is equal to the quotient of the coefficient estimate and its standard deviation. Since there are $n = 25$ observations and $p = 5$ independent variables, the number of degrees of freedom for the Student's t statistic is $25 - 5 - 1 = 19$. The P-values for the tests are given in the next column. All the P-values are small, so it would be reasonable to conclude that each of the independent variables in the model is useful in predicting the goodput.

The quantity "S" is s, the estimated error standard deviation, and "R-sq" is the coefficient of determination R^2. The adjusted R^2, "R-sq(adj)," is primarily used in model selection. We will discuss this statistic in Section 8.3.

The analysis of variance table is analogous to the one found in simple linear regression. We'll go through it column by column. In the degrees of freedom column "DF," the degrees of freedom for regression is equal to the number of independent variables (5). Note that Speed^2, Pause^2, and Speed $\cdot$ Pause each count as separate independent variables, even though they can be computed from Speed and Pause. In the next row down, labeled "Residual Error," the number of degrees of freedom is 19, which represents the number of observations (25) minus the number of parameters estimated (6: the intercept, and coefficients for the five independent variables). Finally, the "Total" degrees of freedom is one less than the sample size of 25. Note that the total degrees of freedom is the sum of the degrees of freedom for regression and the degrees of freedom for error. Going down the column "SS," we find the regression sum of squares SSR, the error sum of squares SSE, and the total sum of squares SST. Note that SST = SSR + SSE. The column "MS" presents the **mean squares**, which are the sums of squares divided by their respective degrees of freedom. Note that the mean square for error is equal to s^2, the estimate for the error variance: ($s^2 = \mathrm{S}^2 = 2.942^2 = 8.66$). The column labeled "F" presents the mean square for regression divided by the mean square for error ($448.10/8.66 = 51.77$, allowing for roundoff error). This is the F statistic shown in Equation (8.10), and it is used to test the null hypothesis that none of the independent variables are linearly related to the dependent variable. The P-value for this test is approximately 0.

The output under the heading "Predicted Values for New Observations" presents confidence intervals on the mean response and predicted intervals for values of the dependent variables specified by the user. The values of the dependent variables that have been specified are listed under the heading "Values of Predictors for New Observations." The values of the independent variables in this output are Speed = 25 and Pause = 15. The quantity 74.242, labeled "Fit," is the value of $\widehat{y}$ obtained by substituting these values into the fitted regression equation. The quantity labeled "SE Fit" is the estimated standard deviation of $\widehat{y}$, which is used to compute the 95% confidence interval, labeled "95% CI." The quantity labeled "95% PI" is the 95% prediction interval for a future observation of the dependent variable when the independent variables are set to the given values. Like the confidence interval, this interval is centered at $\widehat{y}$, but it is wider, just as in simple linear regression.

Example 8.1

Use the multiple regression model to predict the goodput for a network with speed 12 m/s and pause time 25 s.

Solution

From the MINITAB output, the fitted model is

$$\begin{aligned}\text{Goodput} = 96.0 - 1.82\,\text{Speed} + 0.565\,\text{Pause} + 0.0247\,\text{Speed} \cdot \text{Pause} \\ + 0.0140\,\text{Speed}^2 - 0.0118\,\text{Pause}^2\end{aligned}$$

Substituting 12 for Speed and 25 for Pause, we find that the predicted goodput is 90.336.

Example 8.2

For the goodput data, find the residual for the point Speed = 20, Pause = 30.

Solution

The observed value of goodput (Table 8.1) is $y = 84.937$. The predicted value $\widehat{y}$ is found by substituting Speed = 20 and Pause = 30 into the fitted model presented in the solution to Example 8.1. This yields a predicted value for goodput of $\widehat{y} = 86.350$. The residual is given by $y - \widehat{y} = 84.937 - 86.350 = -1.413$.

It is straightforward to compute confidence intervals and to test hypotheses regarding the least-squares coefficients, by using the computer output. Examples 8.3 through 8.5 provide illustrations.

Example 8.3

Find a 95% confidence interval for the coefficient of Speed in the multiple regression model.

Solution

From the output, the estimated coefficient is -1.8245, with a standard deviation of 0.2376. To find a confidence interval, we use the Student's t distribution with 19 degrees of freedom. The degrees of freedom for the t statistic is equal to the degrees of freedom for error. The t value for a 95% confidence interval is $t_{19,.025} = 2.093$. The 95% confidence interval is

$$-1.8245 \pm (2.093)(0.2376) = -1.8245 \pm 0.4973 = (-2.3218,\ -1.3272)$$

Test the null hypothesis that the coefficient of Pause is less than or equal to 0.3.

Solution

The estimated coefficient of Pause is $\widehat{\beta}_2 = 0.5652$, with standard deviation $s_{\widehat{\beta}_2} = 0.2256$. The null hypothesis is $\beta_2 \le 0.3$. Under H_0, we take $\beta_2 = 0.3$, so the quantity

$$t = \frac{\widehat{\beta}_2 - 0.3}{0.2256}$$

has a Student's t distribution with 19 degrees of freedom. Note that the degrees of freedom for the t statistic is equal to the degrees of freedom for error. The value of the t statistic is $(0.5652 - 0.3)/0.2256 = 1.1755$. The P-value is between 0.10 and 0.25. It is plausible that $\beta_2 \le 0.3$.

Find a 95% confidence interval for the mean response μ_{y_i}, and a 95% prediction interval for a future observation when Speed = 25 and Pause = 15.

Solution

From the output, under the heading "Predicted Values for New Observations," the 95% confidence interval is (71.812, 76.732) and the 95% prediction interval is (67.641, 80.903).

Checking Assumptions in Multiple Regression

In multiple regression, as in simple linear regression, it is important to test the validity of the assumptions for errors in linear models (presented at the beginning of this section). The diagnostics for these assumptions used in the case of simple linear regression can be used in multiple regression as well. These are plots of residuals versus fitted values, normal probability plots of residuals, and plots of residuals versus the order in which the observations were made. It is also a good idea to make plots of the residuals versus each of the independent variables. If the residual plots indicate a violation of assumptions, transformations of the variables may be tried to cure the problem, as in simple linear regression.

Figure 8.1 presents a plot of the residuals versus the fitted values for the goodput data. Figure 8.2 and Figure 8.3 on page 566 present plots of the residuals versus speed and pause, respectively. The plot of residuals versus fitted values gives some impression of curvature, which is caused primarily by a few points at either end. The plots of residuals versus independent variables do not indicate any serious violations of the model assumptions. In practice, one might accept this model as fitting well enough, or one might use model selection techniques (discussed in Section 8.3) to explore alternative models.

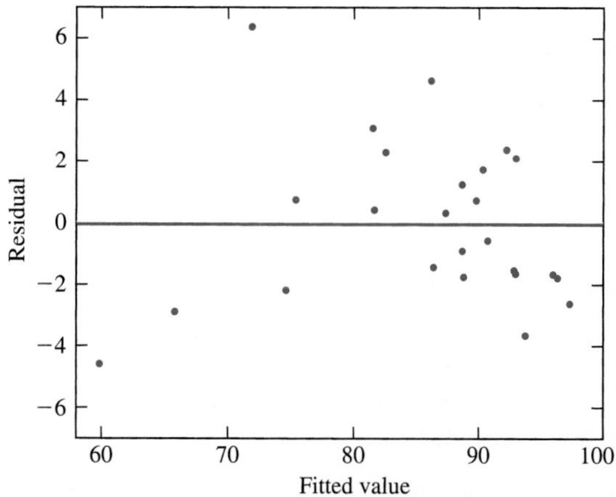

FIGURE 8.1 Plot of residuals versus fitted values for the Goodput data.

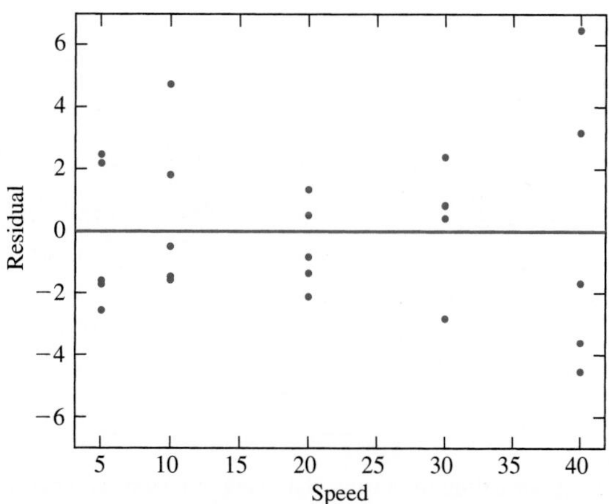

FIGURE 8.2 Plot of residuals versus Speed for the Goodput data.

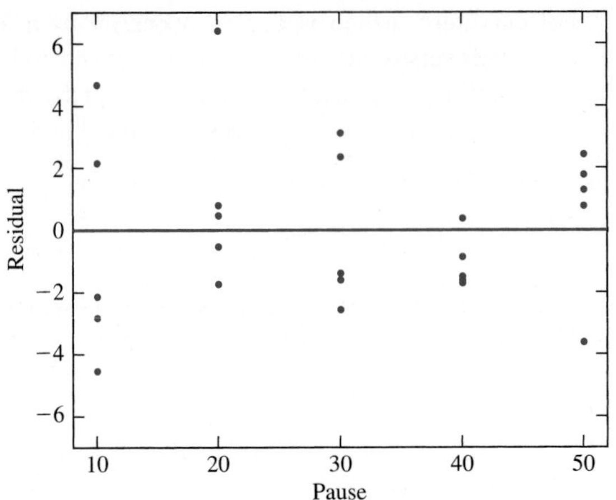

FIGURE 8.3 Plot of residuals versus Pause for the Goodput data.

Exercises for Section 8.1

1. In an experiment to determine the factors affecting fuel economy in trucks, the fuel consumption (in mi/gal), the weight (in tons), and the odometer reading (in 1000s of miles) were measured for 15 trucks. The following MINITAB output presents the results of fitting the model Miles per gallon $= \beta_0 + \beta_1$ Weight $+ \beta_2$ Odometer.

```
The regression equation is
MPG = 8.24 - 0.108 Weight - 0.00392 Odometer

Predictor          Coef      SE Coef        T        P
Constant         8.2407       0.2871    28.70    0.000
Weight         -0.10826      0.01194    -9.06    0.000
Odometer     -0.0039249     0.001406    -2.79    0.016

S = 0.3182          R-Sq = 87.8%      R-Sq(adj) = 85.7%

Analysis of Variance

Source            DF          SS        MS        F        P
Regression         2       8.720     4.360    43.05    0.000
Residual Error    12       1.215     0.101
Total             14       9.935
```

a. Predict the miles per gallon for a truck that weighs 10 tons and has an odometer reading of 50,000 miles.
b. If two trucks have the same weight, and one has 10,000 more miles on the odometer, by how much would you predict their miles per gallon would differ?

c. If two trucks have the same odometer reading, and one weighs 5 tons more than the other, by how much would you predict their miles per gallon to differ?

2. Refer to Exercise 1.

a. Find a 95% confidence interval for the coefficient of Weight.
b. Find a 99% confidence interval for the coefficient of Odometer.
c. Can you conclude that $\beta_1 < -0.05$? Perform the appropriate hypothesis test.
d. Can you conclude that $\beta_2 > -0.005$? Perform the appropriate hypothesis test.

3. The data used to fit the model in Exercise 1 are presented in the following table, along with the residuals and the fitted values. Plot the residuals versus the fitted values. Does the plot indicate that the linear model is reasonable? Explain.

MPG	Weight	Odometer	Residual	Fitted Value
7.28	10.5	15	0.235	7.045
5.63	23.0	71	0.158	5.472
5.26	27.5	36	0.138	5.122
6.58	14.5	113	0.353	6.227
5.01	30.5	39	0.224	4.786
6.73	14.0	97	0.386	6.344
5.37	21.0	195	0.168	5.202
7.28	8.5	8	−0.009	7.289
4.85	26.0	84	−0.246	5.096
5.08	26.5	25	−0.194	5.274
5.51	15.0	124	−0.620	6.130
4.75	30.0	25	−0.145	4.895
6.03	15.0	75	−0.292	6.322
5.26	22.5	192	0.209	5.051
5.60	16.0	139	−0.363	5.963

4. The article "Application of Analysis of Variance to Wet Clutch Engagement" (M. Mansouri, M. Khonsari, et al., *Proceedings of the Institution of Mechanical Engineers*, 2002:117–125) presents the following fitted model for predicting clutch engagement time in seconds (y) from engagement starting speed in m/s (x_1), maximum drive torque in N · m (x_2), system inertia in kg · m^2 (x_3), and applied force rate in kN/s (x_4):

$$y = -0.83 + 0.017x_1 + 0.0895x_2 + 42.771x_3 + 0.027x_4 - 0.0043x_2x_4$$

The sum of squares for regression was SSR = 1.08613 and the sum of squares for error was SSE = 0.036310. There were 44 degrees of freedom for error.

a. Predict the clutch engagement time when the starting speed is 20 m/s, the maximum drive torque is 17 N · m, the system inertia is 0.006 kg · m^2, and the applied force rate is 10 kN/s.
b. Is it possible to predict the change in engagement time associated with an increase of 2 m/s in starting speed? If so, find the predicted change. If not, explain why not.
c. Is it possible to predict the change in engagement time associated with an increase of 2 N · m in maximum drive torque? If so, find the predicted change. If not, explain why not.

d. Compute the coefficient of determination R^2.
e. Compute the F statistic for testing the null hypothesis that all the coefficients are equal to 0. Can this hypothesis be rejected?

5. In the article "Application of Statistical Design in the Leaching Study of Low-Grade Manganese Ore Using Aqueous Sulfur Dioxide" (P. Naik, L. Sukla, and S. Das, *Separation Science and Technology*, 2002:1375–1389), a fitted model for predicting the extraction of manganese in % (y) from particle size in mm (x_1), the amount of sulfur dioxide in multiples of the stoichiometric quantity needed for the dissolution of manganese (x_2), and the duration of leaching in minutes (x_3) is given as

$$
\begin{aligned}
y = {} & 56.145 - 9.046x_1 - 33.421x_2 + 0.243x_3 - 0.5963x_1x_2 - 0.0394x_1x_3 + 0.6022x_2x_3 \\
& + 0.6901x_1^2 + 11.7244x_2^2 - 0.0097x_3^2
\end{aligned}
$$

There were a total of $n = 27$ observations, with SSE $= 209.55$ and SST $= 6777.5$.

a. Predict the extraction percent when the particle size is 3 mm, the amount of sulfur dioxide is 1.5, and the duration of leaching is 20 minutes.
b. Is it possible to predict the change in extraction percent when the duration of leaching increases by one minute? If so, find the predicted change. If not, explain why not.
c. Compute the coefficient of determination R^2.
d. Compute the F statistic for testing the null hypothesis that all the coefficients are equal to 0. Can this hypothesis be rejected?

6. The article "Earthmoving Productivity Estimation Using Linear Regression Techniques" (S. Smith, *Journal of Construction Engineering and Management*, 1999:133–141) presents the following linear model to predict earthmoving productivity (in m^3 moved per hour):

$$
\begin{aligned}
\text{Productivity} = {} & -297.877 + 84.787x_1 + 36.806x_2 + 151.680x_3 - 0.081x_4 - 110.517x_5 \\
& - 0.267x_6 - 0.016x_1x_4 + 0.107x_4x_5 + 0.0009448x_4x_6 - 0.244x_5x_6
\end{aligned}
$$

where x_1 = number of trucks,
x_2 = number of buckets per load,
x_3 = bucket volume, in m^3
x_4 = haul length, in m
x_5 = match factor (ratio of hauling capacity to loading capacity),
x_6 = truck travel time, in s

a. If the bucket volume increases by 1 m^3, while other independent variables are unchanged, can you determine the change in the predicted productivity? If so, determine it. If not, state what other information you would need to determine it.
b. If the haul length increases by 1 m, can you determine the change in the predicted productivity? If so, determine it. If not, state what other information you would need to determine it.

7. In a study of the lung function of children, the volume of air exhaled under force in one second is called FEV_1. (FEV_1 stands for forced expiratory volume in one second.) Measurements were made on a group of children each year for two years. A linear model was fit to predict this year's FEV_1 as a function of last year's FEV_1 (in liters), the child's gender (0 = Male, 1 = Female), the child's height (in m), and the ambient atmospheric pressure (in mm). The following MINITAB output presents the results of fitting the model

$$\text{FEV}_1 = \beta_0 + \beta_1 \text{ Last FEV}_1 + \beta_2 \text{ Gender} + \beta_3 \text{ Height} + \beta_4 \text{ Pressure} + \varepsilon$$

```
The regression equation is
FEV1 = -0.219 + 0.779 Last FEV - 0.108 Gender + 1.354 Height - 0.00134 Pressure

Predictor        Coef         SE Coef          T         P
Constant      -0.21947        0.4503        -0.49     0.627
Last FEV         0.779       0.04909        15.87     0.000
Gender        -0.10827        0.0352        -3.08     0.002
Height          1.3536        0.2880         4.70     0.000
Pressure    -0.0013431     0.0004722        -2.84     0.005

S = 0.22039          R-Sq = 93.5%          R-Sq(adj) = 93.3%

Analysis of Variance

Source            DF          SS          MS          F         P
Regression         4      111.31      27.826     572.89     0.000
Residual Error   160      7.7716    0.048572
Total            164      119.08
```

a. Predict the FEV_1 for a boy who is 1.4 m tall, if the measurement was taken at a pressure of 730 mm and last year's measurement was 2.113 L.

b. If two girls differ in height by 5 cm, by how much would you expect their FEV_1 measurements to differ, other things being equal?

c. The constant term β_0 is estimated to be negative. But FEV_1 must always be positive. Is something wrong? Explain.

d. The health physicist in charge of this experiment wants to redesign the algorithm that electronically records the measurement, in order to automatically adjust for the atmospheric pressure. A barometer will be attached to the device to input the pressure. Use the preceding MINITAB output to determine how to compute an adjusted FEV_1 value as a function of the measured FEV_1 value and the pressure.

8. Refer to Exercise 7.

a. Find a 95% confidence interval for the coefficient of Last FEV.

b. Find a 98% confidence interval for the coefficient of Height.

c. Can you conclude that $\beta_2 < -0.08$? Perform the appropriate hypothesis test.

d. Can you conclude that $\beta_3 > 0.5$? Perform the appropriate hypothesis test.

9. The article "Drying of Pulps in Sprouted Bed: Effect of Composition on Dryer Performance" (M. Medeiros, S. Rocha, et al., *Drying Technology*, 2002:865–881) presents measurements of pH, viscosity (in kg/m · s), density (in g/cm^3), and BRIX (in %). The following MINITAB output presents the results of fitting the model

$$\text{pH} = \beta_0 + \beta_1 \text{ Viscosity} + \beta_2 \text{ Density} + \beta_3 \text{ BRIX} + \varepsilon$$

```
The regression equation is
pH = -1.79 + 0.000266 Viscosity + 9.82 Density - 0.300 BRIX

Predictor           Coef       SE Coef       T       P
Constant         -1.7914        6.2339   -0.29   0.778
Viscosity     0.00026626    0.00011517    2.31   0.034
Density           9.8184        5.7173    1.72   0.105
BRIX            -0.29982      0.099039   -3.03   0.008

S = 0.379578          R-Sq = 50.0%        R-Sq(adj) = 40.6%

Predicted Values for New Observations

New
Obs      Fit    SE Fit          95% CI                95% PI
  1   3.0875    0.1351   (2.8010, 3.3740)   (2.2333, 3.9416)
  2   3.7351    0.1483   (3.4207, 4.0496)   (2.8712, 4.5990)
  3   2.8576    0.2510   (2.3255, 3.3896)   (1.8929, 3.8222)

Values of Predictors for New Observations

New
Obs   Viscosity   Density   BRIX
  1        1000      1.05   19.0
  2        1200      1.08   18.0
  3        2000      1.03   20.0
```

a. Predict the pH for a pulp with a viscosity of 1500 kg/m · s, a density of 1.04 g/cm^3, and a BRIX of 17.5%.

b. If two pulps differ in density by 0.01 g/cm^3, by how much would you expect them to differ in pH, other things being equal?

c. The constant term β_0 is estimated to be negative. But pulp pH must always be positive. Is something wrong? Explain.

d. Find a 95% confidence interval for the mean pH of pulps with viscosity 1200 kg/m · s, density 1.08 g/cm^3, and BRIX 18.0%.

e. Find a 95% prediction interval for the pH of a pulp with viscosity 1000 kg/m · s, density 1.05 g/cm^3, and BRIX 19.0%.

f. Pulp A has viscosity 2000, density 1.03, and BRIX 20.0. Pulp B has viscosity 1000, density 1.05, and BRIX 19.0. Which pulp will have its pH predicted with greater precision? Explain.

10. A scientist has measured quantities y, x_1, and x_2. She believes that y is related to x_1 and x_2 through the equation $y = \alpha e^{\beta_1 x_1 + \beta_2 x_2}\delta$, where δ is a random error that is always positive. Find a transformation of the data that will enable her to use a linear model to estimate β_1 and β_2.

11. The following MINITAB output is for a multiple regression. Something went wrong with the printer, so some of the numbers are missing. Fill in the missing numbers.

Predictor	Coef	SE Coef	T	P
Constant	-0.58762	0.2873	(a)	0.086
X1	1.5102	(b)	4.30	0.005
X2	(c)	0.3944	-0.62	0.560
X3	1.8233	0.3867	(d)	0.003

S = 0.869 R-Sq = 90.2% R-Sq(adj) = 85.3%

Analysis of Variance

Source	DF	SS	MS	F	P
Regression	3	41.76	(e)	(f)	0.000
Residual Error	6	(g)	0.76		
Total	(h)	46.30			

12. The following MINITAB output is for a multiple regression. Some of the numbers got smudged and are illegible. Fill in the missing numbers.

Predictor	Coef	SE Coef	T	P
Constant	(a)	1.4553	5.91	0.000
X1	1.2127	(b)	1.71	0.118
X2	7.8369	3.2109	(c)	0.035
X3	(d)	0.8943	-3.56	0.005

S = 0.82936 R-Sq = 78.0% R-Sq(adj) = 71.4%

Source	DF	SS	MS	F	P
Regression	(e)	(f)	8.1292	11.818	0.001
Residual Error	10	6.8784	(g)		
Total	13	(h)			

13. The article "Evaluating Vent Manifold Inerting Requirements: Flash Point Modeling for Organic Acid-Water Mixtures" (R. Garland and M. Malcolm, *Process Safety Progress*, 2002:254–260) presents a model to predict the flash point (in °F) of a mixture of water, acetic acid, propionic acid, and butyric acid from the concentrations (in weight %) of the three acids. The results are as follows. The variable "Butyric Acid ∗ Acetic Acid" is the interaction between butyric acid concentration and acetic acid concentration.

Predictor	Coef	SE Coef	T	P
Constant	267.53	11.306	23.66	0.000
Acetic Acid	-1.5926	0.1295	-12.30	0.000
Propionic Acid	-1.3897	0.1260	-11.03	0.000
Butyric Acid	-1.0934	0.1164	-9.39	0.000
Butyric Acid*Acetic Acid	-0.002658	0.001145	-2.32	0.034

a. Predict the flash point for a mixture that is 30% acetic acid, 35% propionic acid, and 30% butyric acid. (*Note*: In the model, 30% is represented by 30, not by 0.30.)
b. Someone asks by how much the predicted flash point will change if the concentration of acetic acid is increased by 10% while the other concentrations are kept constant. Is it possible to answer this question? If so, answer it. If not, explain why not.
c. Someone asks by how much the predicted flash point will change if the concentration of propionic acid is increased by 10% while the other concentrations are kept constant. Is it possible to answer this question? If so, answer it. If not, explain why not.

14. In the article "Low-Temperature Heat Capacity and Thermodynamic Properties of 1,1,1-trifluoro-2, 2-dichloroethane" (R. Varushchenko and A. Druzhinina, *Fluid Phase Equilibria*, 2002:109–119), the relationship between vapor pressure (p) and heat capacity (t) is given as $p = t^{\beta_3} \cdot e^{\beta_0+\beta_1 t+\beta_2/t}\delta$, where δ is a random error that is always positive. Express this relationship as a linear model by using an appropriate transformation.

15. The following data were collected in an experiment to study the relationship between shear strength in kPa (y) and curing temperature in °C (x).

x	138	140	146	148	152	153
y	5390	5610	5670	5140	4480	4130

The least-squares quadratic model is $y = -291{,}576.77 + 4168.6479x - 14.6133933x^2$.

a. Using this equation, compute the residuals.
b. Compute the error sum of squares SSE and the total sum of squares SST.
c. Compute the error variance estimate s^2.
d. Compute the coefficient of determination R^2.
e. Compute the value of the F statistic for the hypothesis $H_0: \beta_1 = \beta_2 = 0$. How many degrees of freedom does this statistic have?
f. Can the hypothesis $H_0: \beta_1 = \beta_2 = 0$ be rejected at the 5% level? Explain.

16. The following data were collected in an experiment to study the relationship between the number of pounds of fertilizer (x) and the yield of tomatoes in bushels (y).

x	5	10	15	20	25
y	16	21	27	25	21

The least-squares quadratic model is $y = 4.8000 + 2.508571x - 0.07428571x^2$.

a. Using this equation, compute the residuals.
b. Compute the error sum of squares SSE and the total sum of squares SST.
c. Compute the error variance estimate s^2.
d. Compute the coefficient of determination R^2.
e. Compute the value of the F statistic for the hypothesis $H_0: \beta_1 = \beta_2 = 0$. How many degrees of freedom does this statistic have?
f. Can the hypothesis $H_0: \beta_1 = \beta_2 = 0$ be rejected at the 5% level? Explain.

17. The November 24, 2001, issue of *The Economist* published economic data for 15 industrialized nations. Included were the percent changes in gross domestic product (GDP), industrial production (IP), consumer prices (CP), and producer prices (PP) from Fall 2000 to Fall 2001, and the unemployment rate in Fall 2001 (UNEMP). An economist wants to construct a model to predict GDP from the other variables. A fit of the model

$$\text{GDP} = \beta_0 + \beta_1\text{IP} + \beta_2\text{UNEMP} + \beta_3\text{CP} + \beta_4\text{PP} + \varepsilon$$

yields the following output:

```
The regression equation is
GDP = 1.19 + 0.17 IP + 0.18 UNEMP + 0.18 CP - 0.18 PP

Predictor        Coef      SE Coef        T        P
Constant      1.18957      0.42180     2.82    0.018
IP            0.17326     0.041962     4.13    0.002
UNEMP         0.17918     0.045895     3.90    0.003
CP            0.17591      0.11365     1.55    0.153
PP           -0.18393     0.068808    -2.67    0.023
```

a. Predict the percent change in GDP for a country with IP = 0.5, UNEMP = 5.7, CP = 3.0, and PP = 4.1.
b. If two countries differ in unemployment rate by 1%, by how much would you predict their percent changes in GDP to differ, other things being equal?
c. CP and PP are both measures of the inflation rate. Which one is more useful in predicting GDP? Explain.
d. The producer price index for Sweden in September 2000 was 4.0, and for Austria it was 6.0. Other things being equal, for which country would you expect the percent change in GDP to be larger? Explain.

18. The article "Multiple Linear Regression for Lake Ice and Lake Temperature Characteristics" (S. Gao and H. Stefan, *Journal of Cold Regions Engineering*, 1999:59–77) presents data on maximum ice thickness in mm (y), average number of days per year of ice cover (x_1), average number of days the bottom temperature is lower than 8°C (x_2), and the average snow depth in mm (x_3) for 13 lakes in Minnesota. The data are presented in the following table.

y	x_1	x_2	x_3	y	x_1	x_2	x_3
730	152	198	91	730	157	204	90
760	173	201	81	650	136	172	47
850	166	202	69	850	142	218	59
840	161	202	72	740	151	207	88
720	152	198	91	720	145	209	60
730	153	205	91	710	147	190	63
840	166	204	70				

a. Fit the model $y = \beta_0 + \beta_1 x_1 + \beta_2 x_2 + \beta_3 x_3 + \varepsilon$. For each coefficient, find the P-value for testing the null hypothesis that the coefficient is equal to 0.
b. If two lakes differ by 2 in the average number of days per year of ice cover, with other variables being equal, by how much would you expect their maximum ice thicknesses to differ?
c. Do lakes with greater average snow depth tend to have greater or lesser maximum ice thickness? Explain.

19. In an experiment to estimate the acceleration of an object down an inclined plane, the object is released and its distance in meters (y) from the top of the plane is measured every 0.1 second from time $t = 0.1$ to $t = 1.0$. The data are presented in the following table.

t	y
0.1	0.03
0.2	0.1
0.3	0.27
0.4	0.47
0.5	0.73
0.6	1.07
0.7	1.46
0.8	1.89
0.9	2.39
1.0	2.95

The data follow the quadratic model $y = \beta_0 + \beta_1 t + \beta_2 t^2 + \varepsilon$, where β_0 represents the initial position of the object, β_1 represents the initial velocity of the object, and $\beta_2 = a/2$, where a is the acceleration of the object, assumed to be constant. In a perfect experiment, both the position and velocity of the object would be zero at time 0. However, due to experimental error, it is possible that the position and velocity at $t = 0$ are nonzero.

a. Fit the quadratic model $y = \beta_0 + \beta_1 t + \beta_2 t^2 + \varepsilon$.
b. Find a 95% confidence interval for β_2.
c. Find a 95% confidence interval for the acceleration a.
d. Compute the P-value for each coefficient.
e. Can you conclude that the initial position was not zero? Explain.
f. Can you conclude that the initial velocity was not zero? Explain.

8.2 Confounding and Collinearity

The subtitle of this section is: *Fitting separate models to each variable is not the same as fitting the multivariate model.* To illustrate what we are talking about, we review the gas well data, first described in Exercise 15 in Section 7.3. A total of 255 gas wells received "fracture treatment" in order to increase production. In this treatment, fracture fluid, which consists of fluid mixed with sand, is pumped into the well. The sand holds open the cracks in the rock, thus increasing the flow of gas. The main questions are these: Does increasing the volume of fluid pumped increase the production of the well? Does increasing the volume of sand increase the production of the well?

Other things being equal, deeper wells produce more gas, because they provide more surface through which the gas can permeate. For this reason, it is appropriate to express all variables in units per foot of depth of the well. Thus production is measured in units of ft^3 of gas per ft of depth, fluid is measured in units of gal/ft, and sand is measured in units of lb/ft.

We showed in Figure 7.17 (in Section 7.4) that a log transformation was needed to obtain homoscedasticity in the plot of production versus fluid. It turns out that a log transform is also required for the sand variable as well. Figure 8.4 shows the scatterplots of ln Production versus ln Fluid and ln Production versus ln Sand. Both fluid and sand appear to be strongly related to production.

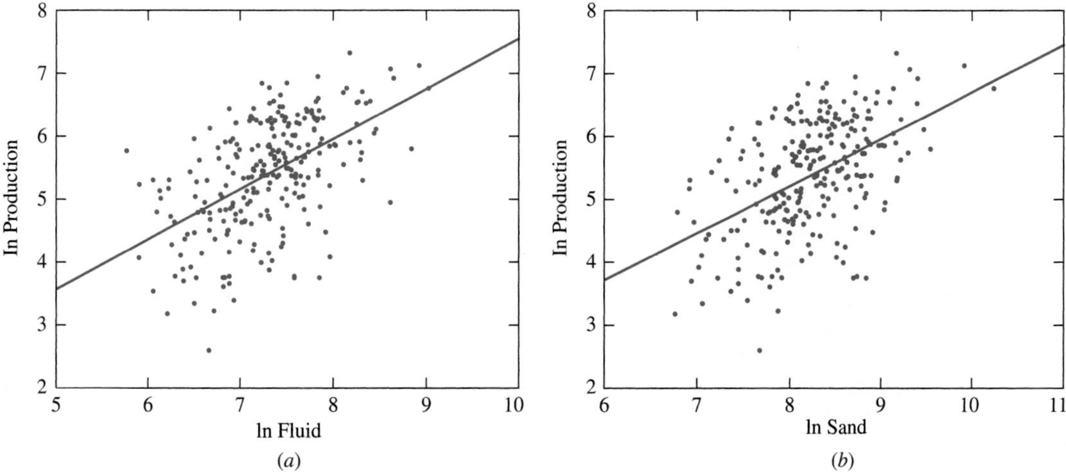

FIGURE 8.4 *(a)* A plot of the log of production versus the log of volume of fracture fluid for 255 gas wells, with the least-squares line superimposed. *(b)* A plot of the log of production versus the log of weight of sand for the same 255 wells. There appear to be strong linear relationships between the log of production and both the log of fluid and the log of sand.

To confirm the result that is apparent from the plots in Figure 8.4, we fit two simple linear regression models:

$$\ln \text{Production} = \beta_0 + \beta_1 \ln \text{Fluid} + \varepsilon$$
$$\ln \text{Production} = \beta_0 + \beta_1 \ln \text{Sand} + \varepsilon$$

The MINITAB output for these models is as follows:

```
The regression equation is
ln Prod = -0.444 + 0.798 ln Fluid

Predictor        Coef    SE Coef        T       P
Constant      -0.4442     0.5853    -0.76   0.449
ln Fluid      0.79833    0.08010     9.97   0.000

S = 0.7459          R-Sq = 28.2%     R-Sq(adj) = 27.9%

The regression equation is
ln Prod = -0.778 + 0.748 ln Sand

Predictor        Coef    SE Coef        T       P
Constant      -0.7784     0.6912    -1.13   0.261
ln Sand       0.74751    0.08381     8.92   0.000

S = 0.7678          R-Sq = 23.9%     R-Sq(adj) = 23.6%
```

Both ln Fluid and ln Sand have coefficients that are definitely different from 0 (the P-values for both are ≈ 0). We therefore might be tempted to conclude immediately that increasing either the volume of fluid or the volume of sand pumped into a well will increase production. But first we must consider the possibility of confounding.

The issue of confounding arises this way. Fluid and sand are pumped in together in a single mixture. It is logical to expect that wells that get more fluid also tend to get more sand. If this is true, then confounding is a possibility. Figure 8.5 presents the scatterplot of ln Fluid versus ln Sand. Sure enough, the amount of fluid pumped into a well is highly correlated with the amount of sand pumped in. It is quite possible, therefore, that either of the two univariate results previously presented may represent confounding rather than a real relationship. If production depends only on the volume of fluid, there will still be a relationship in the data between production and sand. If production depends only on the volume of sand, there will still be a relationship in the data between production and fluid.

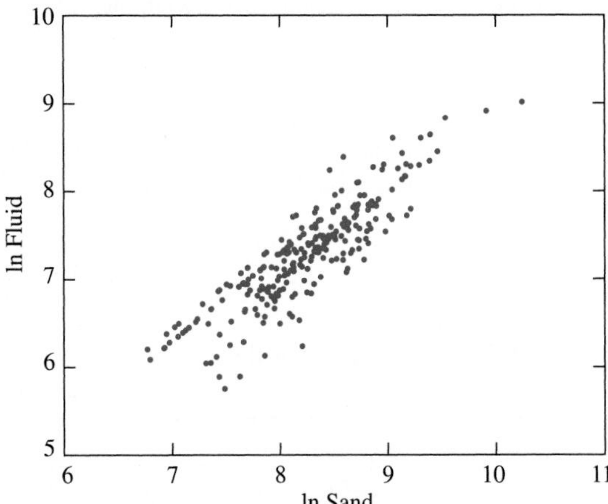

FIGURE 8.5 Scatterplot of ln Fluid versus ln Sand for 255 gas wells. There is clearly a strong linear relationship. Therefore apparent relationships between either fluid or sand and production may represent a confounding rather than a causal relationship.

Multiple regression provides a way to resolve the issue. The following MINITAB output is for the model

$$\ln \text{Production} = \beta_0 + \beta_1 \ln \text{Fluid} + \beta_2 \ln \text{Sand} + \varepsilon \tag{8.11}$$

```
The regression equation is
ln Prod = -0.729 + 0.670 ln Fluid + 0.148 ln Sand

Predictor        Coef      SE Coef        T        P
Constant      -0.7288       0.6719    -1.08    0.279
ln Fluid       0.6701       0.1687     3.97    0.000
ln Sand        0.1481       0.1714     0.86    0.389

S = 0.7463         R-Sq = 28.4%      R-Sq(adj) = 27.8%
```

We can see that the coefficient of ln Fluid is significantly different from 0, but the coefficient of ln Sand is not. If we assume that there is no other confounding going on (e.g., with the location of the wells), we can conclude that increasing the amount of fluid tends to increase production, but it is not clear that increasing the amount of sand has an effect. Therefore, one might increase the amount of fluid, but it might not be necessary to add more sand to it.

A final observation: None of the models have a particularly high value of R^2. This indicates that there are other important factors affecting production that have not been included in the models. In a more complete analysis, one would attempt to identify and measure some of these factors in order to build a model with greater predictive power.

Collinearity

When two independent variables are *very* strongly correlated, multiple regression may not be able to determine which is the important one. In this case, the variables are said to be **collinear**. The word *collinear* means to lie on the same line, and when two variables are highly correlated, their scatterplot is approximately a straight line. The word *multicollinear* is sometimes used as well. When collinearity is present, the set of independent variables is sometimes said to be **ill-conditioned**. Table 8.2 on page 578 presents some hypothetical data that illustrate the phenomenon of collinearity.

First we fit the simple linear models

$$y = \beta_0 + \beta_1 x_1 + \varepsilon$$
$$y = \beta_0 + \beta_1 x_2 + \varepsilon$$

The following MINITAB output shows that both x_1 and x_2 have a strong linear relationship with y. The values of r^2 are both around 0.96, so the correlations r between x_1 and y and between x_2 and y are both around 0.98.

TABLE 8.2 Collinear data

x_1	x_2	y
0.1	0.3	3.6
0.2	0.2	0.3
0.6	1.4	6.0
1.4	3.4	10.6
2.0	5.2	8.4
2.0	5.5	11.8
2.1	5.5	12.7
2.1	5.3	6.8
2.8	7.4	9.9
3.6	9.4	16.7
4.2	10.3	16.3
4.5	11.4	19.9
4.7	11.3	20.2
5.3	13.6	22.9
6.1	15.3	26.6
6.8	17.4	28.1
7.5	18.5	31.0
8.2	20.4	28.8
8.5	21.3	32.4
9.4	23.3	35.0

```
The regression equation is
Y = 2.90 + 3.53 X1

Predictor      Coef    SE Coef       T       P
Constant     2.8988     0.8224    3.52   0.002
X1           3.5326     0.1652   21.38   0.000

S = 2.080        R-Sq = 96.2%    R-Sq(adj) = 96.0%

The regression equation is
Y = 2.74 + 1.42 X2

Predictor      Coef    SE Coef       T       P
Constant     2.7431     0.8090    3.39   0.003
X2          1.42024    0.06485   21.90   0.000

S = 2.033        R-Sq = 96.4%    R-Sq(adj) = 96.2%
```

Figure 8.6 on page 579 presents the scatterplot of x_2 versus x_1. There is clearly a strong linear relationship, so we suspect that y may really have a relationship with only one of these variables, with the other being a confounder.

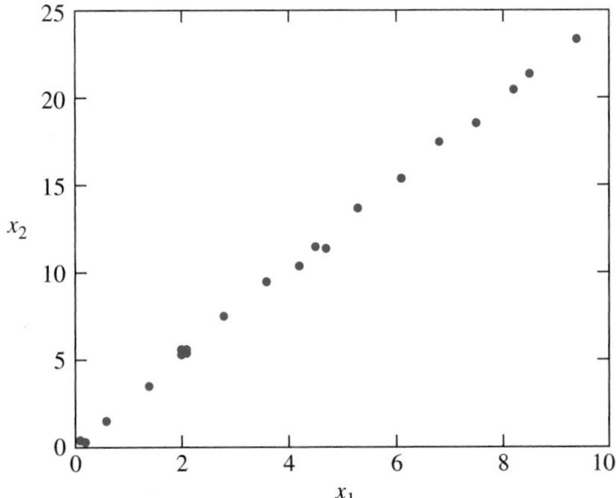

FIGURE 8.6 The independent variables x_1 and x_2 are *collinear*, because they have a strong linear relationship.

We therefore fit the multiple regression model

$$y = \beta_0 + \beta_1 x_1 + \beta_2 x_2 + \varepsilon$$

The MINITAB output is as follows.

```
The regression equation is
Y = 2.72 - 0.49 X1 + 1.62 X2

Predictor        Coef     SE Coef        T        P
Constant       2.7248      0.8488     3.21    0.005
X1             -0.490       4.460    -0.11    0.914
X2              1.617       1.791     0.90    0.379

S = 2.091           R-Sq = 96.4%    R-Sq(adj) = 96.0%
```

Surprisingly, the output appears to indicate that neither x_1 nor x_2 is linearly related to y, since both have large P-values. What is happening is that the linear relationship between x_1 and x_2 is so strong that it is simply impossible to determine which of the two is responsible for the linear relationship with y. Seen in this light, the large P-values make sense. It is plausible that the coefficient of x_1 is 0 and that only x_2 has a real relationship with y. Therefore the P-value for x_1 must be large. Likewise, it is plausible that the coefficient of x_2 is 0 and that only x_1 has a real relationship with y. Therefore the P-value for x_2 must be large as well.

In general, there is not much that can be done when variables are collinear. The only good way to fix the situation is to collect more data, including some values for the independent variables that are not on the same straight line. Then multiple regression will be able to determine which of the variables are really important.

Exercises for Section 8.2

1. In an experiment to determine factors related to weld toughness, the Charpy V-notch impact toughness in ft · lb (y) was measured for 22 welds at 0°C, along with the lateral expansion at the notch in % (x_1), and the brittle fracture surface in % (x_2). The data are presented in the following table.

y	x_1	x_2	y	x_1	x_2	y	x_1	x_2
32	20.0	28	27	16.0	29	25	14.6	36
39	23.0	28	43	26.2	27	25	10.4	29
20	12.8	32	22	9.6	32	20	11.6	30
21	16.0	29	22	15.2	32	20	12.6	31
25	10.2	31	18	8.8	43	24	16.2	36
20	11.6	28	32	20.4	24	18	9.2	34
32	17.6	25	22	12.2	36	28	16.8	30
29	17.8	28						

a. Fit the model $y = \beta_0 + \beta_1 x_1 + \varepsilon$. For each coefficient, test the null hypothesis that it is equal to 0.
b. Fit the model $y = \beta_0 + \beta_1 x_2 + \varepsilon$. For each coefficient, test the null hypothesis that it is equal to 0.
c. Fit the model $y = \beta_0 + \beta_1 x_1 + \beta_2 x_2 + \varepsilon$. For each coefficient, test the null hypothesis that it is equal to 0.
d. Which of the models in parts (a) through (c) is the best of the three? Why do you think so?

2. In a laboratory test of a new engine design, the emissions rate (in mg/s of oxides of nitrogen, NO_x) was measured as a function of engine speed (in rpm), engine torque (in ft · lb), and total horsepower. (From "In-Use Emissions from Heavy-Duty Diesel Vehicles," J. Yanowitz, Ph.D. thesis, Colorado School of Mines, 2001.) MINITAB output is presented for the following three models:

$$\text{NO}_x = \beta_0 + \beta_1 \text{ Speed} + \beta_2 \text{ Torque} + \varepsilon$$
$$\text{NO}_x = \beta_0 + \beta_1 \text{ Speed} + \beta_2 \text{ HP} + \varepsilon$$
$$\text{NO}_x = \beta_0 + \beta_1 \text{ Speed} + \beta_2 \text{ Torque} + \beta_3 \text{ HP} + \varepsilon$$

```
The regression equation is
NOx = -321 + 0.378 Speed - 0.160 Torque

Predictor        Coef     SE Coef        T        P
Constant      -320.59       98.14    -3.27    0.003
Speed         0.37820     0.06861     5.51    0.000
Torque       -0.16047     0.06082    -2.64    0.013

S = 67.13          R-Sq = 51.6%    R-Sq(adj) = 48.3%
```

```
The regression equation is
NOx = -380 + 0.416 Speed - 0.520 HP

Predictor        Coef     SE Coef        T        P
Constant       -380.1       104.8    -3.63    0.001
Speed         0.41641     0.07510     5.54    0.000
HP            -0.5198      0.1980    -2.63    0.014

S = 67.19         R-Sq = 51.5%     R-Sq(adj) = 48.2%

The regression equation is
NOx = -302 + 0.366 Speed - 0.211 Torque + 0.16 HP

Predictor        Coef     SE Coef        T        P
Constant       -301.8       347.3    -0.87    0.392
Speed          0.3660      0.2257     1.62    0.116
Torque        -0.2106      0.8884    -0.24    0.814
HP              0.164       2.889     0.06    0.955

S = 68.31         R-Sq = 51.6%     R-Sq(adj) = 46.4%
```

Of the variables Speed, Torque, and HP, which two are most nearly collinear? How can you tell?

3. Two chemical engineers, A and B, are working independently to develop a model to predict the viscosity of a product (y) from the pH (x_1) and the concentration of a certain catalyst (x_2). Each engineer has fit the linear model

$$y = \beta_0 + \beta_1 x_1 + \beta_2 x_2 + \varepsilon$$

The engineers have sent you MINITAB output summarizing their results:

```
Engineer A

Predictor        Coef     SE Coef        T        P
Constant        199.2      0.5047    394.7    0.000
pH             -1.569      0.4558    -3.44    0.007
Concent.       -4.730      0.5857    -8.08    0.000
```

```
Engineer B

Predictor        Coef     SE Coef        T        P
Constant        199.0       0.548    363.1    0.000
pH             -1.256       1.983    -0.63    0.544
Concent.       -3.636       1.952    -1.86    0.112
```

The engineers have also sent you the following scatterplots of pH versus concentration, but forgot to put their names on them.

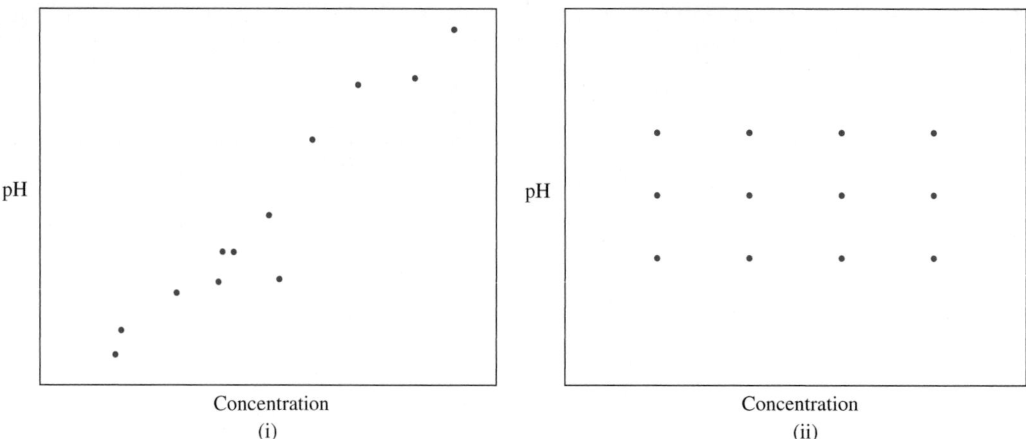

a. Which plot came from which engineer? How do you know?
b. Which engineer's experiment produced the more reliable results? Explain.

4. The article "Influence of Freezing Temperature on Hydraulic Conductivity of Silty Clay" (J. Konrad and M. Samson, *Journal of Geotechnical and Geoenvironmental Engineering*, 2000:180–187) describes a study of factors affecting hydraulic conductivity of soils. The measurements of hydraulic conductivity in units of 10^{-8} cm/s (y), initial void ratio (x_1), and thawed void ratio (x_2) for 12 specimens of silty clay are presented in the following table.

y	1.01	1.12	1.04	1.30	1.01	1.04	0.955	1.15	1.23	1.28	1.23	1.30
x_1	0.84	0.88	0.85	0.95	0.88	0.86	0.85	0.89	0.90	0.94	0.88	0.90
x_2	0.81	0.85	0.87	0.92	0.84	0.85	0.85	0.86	0.85	0.92	0.88	0.92

a. Fit the model $y = \beta_0 + \beta_1 x_1 + \varepsilon$. For each coefficient, test the null hypothesis that it is equal to 0.
b. Fit the model $y = \beta_0 + \beta_1 x_2 + \varepsilon$. For each coefficient, test the null hypothesis that it is equal to 0.
c. Fit the model $y = \beta_0 + \beta_1 x_1 + \beta_2 x_2 + \varepsilon$. For each coefficient, test the null hypothesis that it is equal to 0.
d. Which of the models in parts (a) to (c) is the best of the three? Why do you think so?

5. Refer to Exercise 8 in Section 7.4.

a. Divide the data into two groups: points where $R_1 < 4$ in one group, points where $R_1 \geq 4$ in the other. Compute the least-squares line for predicting R_2 from R_1 for each group. (You already did this if you did Exercise 8c in Section 7.4.)
b. For one of the two groups, the relationship is clearly nonlinear. For this group, fit a quadratic model (i.e., using R_1 and R_1^2 as independent variables), a cubic model, and a quartic model. Compute the P-values for each of the coefficients in each of the models.
c. Plot the residuals versus the fitted values for each of the three models in part (b).
d. Compute the correlation coefficient between R_1^3 and R_1^4, and make a scatterplot of the points (R_1^3, R_1^4).
e. On the basis of the correlation coefficient and the scatterplot, explain why the P-values are much different for the quartic model than for the cubic model.
f. Which of the three models in part (b) is most appropriate? Why?

6. The following table lists values for three variables measured for 60 consecutive eruptions of the geyser Old Faithful in Yellowstone National Park. They are the duration of the eruption (x_1), the duration of the dormant period immediately before the eruption (x_2), and the duration of the dormant period immediately after the eruption (y). All the times are in minutes. (Variable x_2 was presented in Supplementary Exercise 3 in Chapter 7.)

x_1	x_2	y	x_1	x_2	y	x_1	x_2	y	x_1	x_2	y
3.5	80	84	1.8	42	91	4.7	88	51	4.1	70	79
4.1	84	50	4.1	91	51	1.8	51	80	3.7	79	60
2.3	50	93	1.8	51	79	4.6	80	49	3.8	60	86
4.7	93	55	3.2	79	53	1.9	49	82	3.4	86	71
1.7	55	76	1.9	53	82	3.5	82	75	4.0	71	67
4.9	76	58	4.6	82	51	4.0	75	73	2.3	67	81
1.7	58	74	2.0	51	76	3.7	73	67	4.4	81	76
4.6	74	75	4.5	76	82	3.7	67	68	4.1	76	83
3.4	75	80	3.9	82	84	4.3	68	86	4.3	83	76
4.3	80	56	4.3	84	53	3.6	86	72	3.3	76	55
1.7	56	80	2.3	53	86	3.8	72	75	2.0	55	73
3.9	80	69	3.8	86	51	3.8	75	75	4.3	73	56
3.7	69	57	1.9	51	85	3.8	75	66	2.9	56	83
3.1	57	90	4.6	85	45	2.5	66	84	4.6	83	57
4.0	90	42	1.8	45	88	4.5	84	70	1.9	57	71

a. Compute the least-squares line for predicting the duration of the dormant period following an eruption (y) from the duration of the eruption (x_1). Is there a linear relationship between the duration of an eruption and the waiting time until the next eruption occurs?

b. Compute the least-squares line for predicting the duration of the dormant period following an eruption (y) from the duration of the dormant period preceding the eruption (x_2). (You already did this if you did Supplementary Exercise 3 in Chapter 7. The results in this problem will be slightly different, since there are 60 points instead of 59.) Is there a linear relationship between the duration of the dormant period preceding an eruption and the waiting time until the next eruption occurs?

c. Fit the multiple regression model that includes both the duration of the eruption x_1 and the duration of the dormant period preceding the eruption x_2 as independent variables.

d. If you could find the value of either x_1 or x_2 but not both, which one would you want to know to predict y? Would it help much to know the other one as well? Explain.

8.3 Model Selection

There are many situations in which a large number of independent variables have been measured, and we need to decide which of them to include in a model. This is the problem of **model selection**, and it is a difficult one. In practice, model selection often proceeds by ad hoc methods, guided by whatever physical intuition may be available. We will not attempt a complete discussion of this extensive and difficult topic. Instead, we will be content to state some basic principles and to present some examples. An advanced reference such as Miller (2002) can be consulted for information on specific methods.

Good model selection rests on a basic principle known as Occam's razor. This principle is stated as follows:

Occam's Razor
The best scientific model is the simplest model that explains the observed facts.

In terms of linear models, Occam's razor implies the **principle of parsimony**:

The Principle of Parsimony
A model should contain the smallest number of variables necessary to fit the data.

There are some exceptions to the principle of parsimony:

1. A linear model should always contain an intercept, unless physical theory dictates otherwise.
2. If a power x^n of a variable is included in a model, all lower powers x, $x^2, \dots, x^{n-1}$ should be included as well, unless physical theory dictates otherwise.
3. If a product $x_i x_j$ of two variables is included in a model, then the variables x_i and x_j should be included separately as well, unless physical theory dictates otherwise.

Models that contain only the variables that are needed to fit the data are called **parsimonious** models. Much of the practical work of multiple regression involves the development of parsimonious models.

We illustrate the principle of parsimony with the following example. The data in Table 8.3 were taken from the article "Capacities and Performance Characteristics of Jaw Crushers" (S. Sastri, *Minerals and Metallurgical Processing*, 1994:80–86). Feed rates and amounts of power drawn were measured for several industrial jaw crushers.

TABLE 8.3 Feed rates and power for industrial jaw crushers

Feed Rate (100 tons/h)	Power (kW)	Feed Rate (100 tons/h)	Power (kW)	Feed Rate (100 tons/h)	Power (kW)	Feed Rate (100 tons/h)	Power (kW)
0.10	11	0.20	15	0.91	45	1.36	58
1.55	60	2.91	84	0.59	12	2.36	45
3.00	40	0.36	30	0.27	24	2.95	75
3.64	150	0.14	16	0.55	49	1.09	44
0.38	69	0.91	30	0.68	45	0.91	58
1.59	77	4.27	150	4.27	150	2.91	149
4.73	83	4.36	144	3.64	100		

The following MINITAB output presents the results for fitting the model

$$\text{Power} = \beta_0 + \beta_1 \text{ FeedRate} + \varepsilon \tag{8.12}$$

```
The regression equation is
Power = 21.0 + 24.6 FeedRate

Predictor        Coef     SE Coef        T         P
Constant       21.028       8.038     2.62     0.015
FeedRate       24.595       3.338     7.37     0.000

S = 26.20         R-Sq = 68.5%     R-Sq(adj) = 67.2%
```

From the output, we see that the fitted model is

$$\text{Power} = 21.028 + 24.595 \text{ FeedRate} \tag{8.13}$$

and that the coefficient for FeedRate is significantly different from 0 ($t = 7.37$, $P \approx 0$). We wonder whether a quadratic model might fit better than this linear one. So we fit

$$\text{Power} = \beta_0 + \beta_1 \text{ FeedRate} + \beta_2 \text{ FeedRate}^2 + \varepsilon \tag{8.14}$$

The results are presented in the following MINITAB output. Note that the values for the intercept and for the coefficient of FeedRate are different than they were in the linear model. This is typical. Adding a new variable to a model can substantially change the coefficients of the variables already in the model.

```
The regression equation is
Power = 19.3 + 27.5 FeedRate -0.64 FeedRate^2

Predictor         Coef     SE Coef        T         P
Constant         19.34       11.56     1.67     0.107
FeedRate         27.47       14.31     1.92     0.067
FeedRate^2     -0.6387       3.090    -0.21     0.838

S = 26.72         R-Sq = 68.5%     R-Sq(adj) = 65.9%
```

The most important point to notice is that the P-value for the coefficient of FeedRate^2 is large (0.838). Recall that this P-value is for the test of the null hypothesis that the coefficient is equal to 0. Thus the data provide no evidence that the coefficient of FeedRate^2 is different from 0. Note also that including FeedRate^2 in the model increases the value of the goodness-of-fit statistic R^2 only slightly, in fact so slightly that the first three digits are unchanged. It follows that there is no evidence that the quadratic model fits the data better than the linear model, so by the principle of parsimony, we should prefer the linear model.

Figure 8.7 on page 586 provides a graphical illustration of the principle of parsimony. The scatterplot of power versus feed rate is presented, and both the least-squares line (8.13) and the quadratic model (8.14) are superimposed. Even though the coefficients of the models are different, we can see that the two curves are almost identical.

There is no reason to include the quadratic term in the model. It makes the model more complicated, without improving the fit.

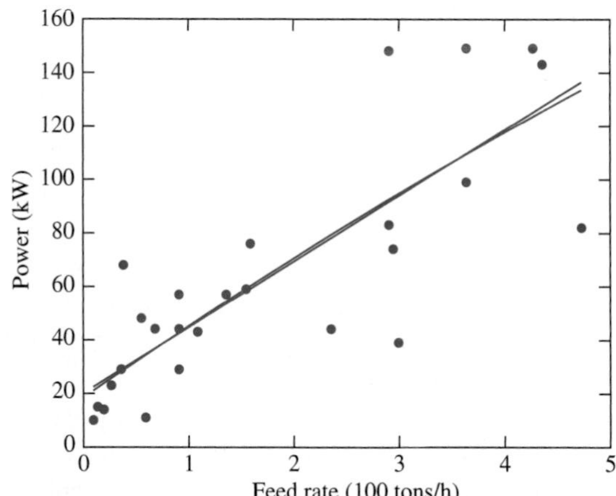

FIGURE 8.7 Scatterplot of power versus feed rate for 27 industrial jaw crushers. The least-squares line and best fitting quadratic model are both superimposed. The two curves are practically identical, which reflects the fact that the coefficient of FeedRate^2 in the quadratic model does not differ significantly from 0.

Determining Whether Variables Can Be Dropped from a Model

It often happens that one has formed a model that contains a large number of independent variables, and one wishes to determine whether a given subset of them may be dropped from the model without significantly reducing the accuracy of the model. To be specific, assume that we know that the model

$$y_i = \beta_0 + \beta_1 x_{1i} + \cdots + \beta_k x_{ki} + \beta_{k+1} x_{k+1\,i} + \cdots + \beta_p x_{pi} + \varepsilon_i \tag{8.15}$$

is correct, in that it represents the true relationship between the x variables and y. We will call this model the "full" model.

We wish to test the null hypothesis

$$H_0\colon \beta_{k+1} = \cdots = \beta_p = 0$$

If H_0 is true, the model will remain correct if we drop the variables $x_{k+1}, \ldots, x_p$, so we can replace the full model with the following reduced model:

$$y_i = \beta_0 + \beta_1 x_{1i} + \cdots + \beta_k x_{ki} + \varepsilon_i \tag{8.16}$$

To develop a test statistic for H_0, we begin by computing the error sum of squares for both the full and the reduced models. We'll call them SSE_{full} and $\text{SSE}_{\text{reduced}}$. The number of degrees of freedom for SSE_{full} is $n - p - 1$, and the number of degrees of freedom for $\text{SSE}_{\text{reduced}}$ is $n - k - 1$.

Now since the full model is correct, we know that the quantity $\text{SSE}_{\text{full}}/(n - p - 1)$ is an estimate of the error variance σ^2; in fact it is just s^2. If H_0 is true, then the reduced model is also correct, so the quantity $\text{SSE}_{\text{reduced}}/(n - k - 1)$ is also an estimate of the error variance. Intuitively, SSE_{full} is close to $(n - p - 1)\sigma^2$, and if H_0 is true, $\text{SSE}_{\text{reduced}}$ is close to $(n-k-1)\sigma^2$. It follows that if H_0 is true, the difference $(\text{SSE}_{\text{reduced}} - \text{SSE}_{\text{full}})$ is close to $(p - k)\sigma^2$, so the quotient $(\text{SSE}_{\text{reduced}} - \text{SSE}_{\text{full}})/(p - k)$ is close to σ^2. The test statistic is

$$f = \frac{(\text{SSE}_{\text{reduced}} - \text{SSE}_{\text{full}})/(p - k)}{\text{SSE}_{\text{full}}/(n - p - 1)} \tag{8.17}$$

Now if H_0 is true, both numerator and denominator of f are estimates of σ^2, so f is likely to be near 1. If H_0 is false, the quantity $\text{SSE}_{\text{reduced}}$ tends to be larger, so the value of f tends to be larger. The statistic f is an F statistic; its null distribution is $F_{p-k,\,n-p-1}$.

The method we have just described is very useful in practice for developing parsimonious models by removing unnecessary variables. However, the conditions under which it is formally valid are seldom met in practice. First, it is rarely the case that the full model is correct; there will be nonrandom quantities that affect the value of the dependent variable y that are not accounted for by the independent variables. Second, for the method to be formally valid, the subset of variables to be dropped must be determined independently of the data. This is usually not the case. More often, a large model is fit, some of the variables are seen to have fairly large P-values, and the F test is used to decide whether to drop them from the model. As we have said, this is a useful technique in practice, but, like most methods of model selection, it should be seen as an informal tool rather than a rigorous theory-based procedure.

We illustrate the method with an example. In mobile ad hoc computer networks, messages must be forwarded from computer to computer until they reach their destinations. The data overhead is the number of bytes of information that must be transmitted along with the messages to get them to the right places. A successful protocol will generally have a low data overhead. Table 8.4 on page 588 presents average speed, pause time, link change rate (LCR), and data overhead for 25 simulated computer networks. The link change rate for a given computer is the rate at which other computers in the network enter and leave the transmission range of the given computer. These data were generated for a study published in the article "Metrics to Enable Adaptive Protocols for Mobile Ad Hoc Networks" (J. Boleng, W. Navidi, and T. Camp, *Proceedings of the 2002 International Conference on Wireless Networks*, 2002:293–298).

We will begin by fitting a fairly large model to these data, namely,

$$\begin{aligned}\text{Overhead} = \beta_0 &+ \beta_1\,\text{LCR} + \beta_2\,\text{Speed} + \beta_3\,\text{Pause} + \beta_4\,\text{Speed} \cdot \text{Pause} + \beta_5\,\text{LCR}^2 \\ &+ \beta_6\,\text{Speed}^2 + \beta_7\,\text{Pause}^2 + \varepsilon\end{aligned}$$

TABLE 8.4 Data overhead, speed, pause time, and link change rate for a mobile computer network

Speed (m/s)	Pause Time (s)	LCR (100/s)	Data Overhead (kB)	Speed (m/s)	Pause Time (s)	LCR (100/s)	Data Overhead (kB)
5	10	9.426	428.90	20	40	12.117	501.48
5	20	8.318	443.68	20	50	10.284	519.20
5	30	7.366	452.38	30	10	33.009	445.45
5	40	6.744	461.24	30	20	22.125	489.02
5	50	6.059	475.07	30	30	16.695	506.23
10	10	16.456	446.06	30	40	13.257	516.27
10	20	13.281	465.89	30	50	11.107	508.18
10	30	11.155	477.07	40	10	37.823	444.41
10	40	9.506	488.73	40	20	24.140	490.58
10	50	8.310	498.77	40	30	17.700	511.35
20	10	26.314	452.24	40	40	14.064	523.12
20	20	19.013	475.97	40	50	11.691	523.36
20	30	14.725	499.67				

The results from fitting this model are as follows.

```
The regression equation is
Overhead = 368 + 3.48 LCR + 3.04 Speed + 2.29 Pause - 0.0122 Speed*Pause
           - 0.1041 LCR^2 - 0.0313 Speed^2 - 0.0132 Pause^2

Predictor        Coef      SE Coef        T        P
Constant       367.96        19.40    18.96    0.000
LCR             3.477        2.129     1.63    0.121
Speed           3.044        1.591     1.91    0.073
Pause          2.2924       0.6984     3.28    0.004
Speed*Pa     -0.01222      0.01534    -0.80    0.437
LCR^2        -0.10412      0.03192    -3.26    0.005
Speed^2      -0.03131      0.01906    -1.64    0.119
Pause^2      -0.01318      0.01045    -1.26    0.224

S = 5.72344        R-Sq = 97.2%        R-Sq(adj) = 96.1%

Analysis of Variance

Source            DF        SS        MS        F        P
Regression         7   19567.5    2795.4    85.33    0.000
Residual Error    17     556.9      32.8
Total             24   20124.3
```

We can see that LCR, Speed · Pause, Speed^2, and Pause^2 have large P-values. We will leave LCR in the model for now, because LCR^2 has a very small P-value, and therefore should stay in the model. We will use the F test to determine whether the reduced model obtained by dropping Speed · Pause, Speed^2, and Pause^2 is a reasonable one. First, from the output for the full model, note that $\text{SSE}_{\text{full}} = 556.9$, and it has 17 degrees of freedom. The number of independent variables in the full model is $p = 7$.

We now fit the reduced model

$$\text{Overhead} = \beta_0 + \beta_1\,\text{LCR} + \beta_2\,\text{Speed} + \beta_3\,\text{Pause} + \beta_5\,\text{LCR}^2 + \varepsilon$$

The results from fitting this model are as follows.

```
The regression equation is
Overhead = 359 + 6.69 LCR + 0.777 Speed + 1.67 Pause - 0.156 LCR^2

Predictor        Coef     SE Coef        T         P
Constant       359.22       13.01    27.61     0.000
LCR             6.695       1.156     5.79     0.000
Speed          0.7766      0.2054     3.78     0.001
Pause          1.6729      0.1826     9.16     0.000
LCR^2         -1.5572     0.02144    -7.26     0.000

S = 6.44304      R-Sq = 95.9%     R-Sq(adj) = 95.0%

Analysis of Variance

Source            DF          SS          MS          F         P
Regression         4     19294.1      4823.5     116.19     0.000
Residual Error    20       830.3        41.5
Total             24     20124.3
```

The P-values for the variables in this model are all quite small. From the output for this reduced model, we note that $\text{SSE}_{\text{reduced}} = 830.3$. The number of variables in this reduced model is $k = 4$.

Now we can compute the F statistic. Using Equation (8.17), we compute

$$f = \frac{(830.3 - 556.9)/(7 - 4)}{556.9/17} = 2.78$$

The null distribution is $F_{3,17}$. From the F table (Table A.7 in Appendix A), we find that $0.05 < P < 0.10$. According to the 5% rule of thumb, since $P > 0.05$, the reduced model is plausible, but only barely so. Rather than settle for a barely plausible model, it is wise to explore further, to look for a slightly less reduced model that has a larger P-value.

To do this, we note that of the three variables we dropped, the variable Speed^2 had the smallest P-value in the full model. We'll take this as an indication that this might be the most important of the variables we dropped, and we'll put it back in the model. We will now fit a second reduced model, which is

$$\text{Overhead} = \beta_0 + \beta_1\,\text{LCR} + \beta_2\,\text{Speed} + \beta_3\,\text{Pause} + \beta_5\,\text{LCR}^2 + \beta_6\,\text{Speed}^2$$

The results from fitting this model are as follows.

```
The regression equation is
Overhead = 373 + 4.80 LCR + 1.99 Speed + 1.45 Pause - 0.123 LCR^2
           -0.0212 Speed^2

Predictor        Coef     SE Coef        T        P
Constant       372.60       16.93    22.00    0.000
LCR             4.799       1.935     2.48    0.023
Speed           1.993       1.023     1.95    0.066
Pause          1.4479      0.2587     5.60    0.000
LCR^2        -0.12345     0.03400    -3.63    0.002
Speed^2      -0.02120     0.01746    -1.21    0.240

S = 6.36809        R-Sq = 96.2%      R-Sq(adj) = 95.2%

Analysis of Variance

Source            DF        SS       MS       F       P
Regression         5   19353.8   3870.8   95.45   0.000
Residual Error    19     770.5     40.6
Total             24   20124.3
```

Note that the P-value for Speed^2 in this model is large (0.240). This is not good. In general we do not want to add a variable whose coefficient might be equal to 0. So we probably won't want to stick with this model. Let's compute the value of the F statistic anyway, just for practice. The value of $\text{SSE}_{\text{reduced}}$ in this model is 770.5. The number of independent variables is $k = 5$. The value of the F statistic, using Equation (8.17), is therefore

$$f = \frac{(770.5 - 556.9)/(7 - 5)}{556.9/17} = 3.26$$

The null distribution is $F_{2,17}$. From the F table (Table A.7), we again find that $0.05 < P < 0.10$, so the reduced model is barely plausible at best.

We chose to put Speed^2 back into the model because it had the smallest P value among the variables we originally dropped. But as we have just seen, this does not guarantee that it will have a small P value when it is put back into the reduced model. Perhaps one of the other variables we dropped will do better. Of the three variables originally dropped, the one with the second smallest P value was Pause^2. We try replacing Speed^2 in the preceding model with Pause^2. So we now fit a third reduced model:

$$\text{Overhead} = \beta_0 + \beta_1\,\text{LCR} + \beta_2\,\text{Speed} + \beta_3\,\text{Pause} + \beta_5\,\text{LCR}^2 + \beta_6\,\text{Pause}^2$$

The results from fitting this model are as follows.

```
The regression equation is
Overhead = 345 + 6.484 LCR + 0.707 Speed + 2.85 Pause - 0.145 LCR^2
          - 0.0183 Pause^2

Predictor          Coef     SE Coef       T       P
Constant         345.42       13.19   26.20   0.000
LCR               6.484       1.050    6.17   0.000
Speed            0.7072      0.1883    3.76   0.001
Pause            2.8537      0.5337    5.35   0.000
LCR^2          -0.14482     0.01996   -7.25   0.000
Pause^2       -0.018334    0.007879   -2.33   0.031

S = 5.83154       R-Sq = 96.8%      R-Sq(adj) = 95.9%

Analysis of Variance

Source            DF        SS        MS         F        P
Regression         5   19478.2    3895.6    114.55    0.000
Residual Error    19     646.1      34.0
Total             24   20124.3
```

This model looks good, at least at first. All the variables have small P values. We'll compute the F statistic to see if this model is plausible. The value of $\text{SSE}_{\text{reduced}}$ in this model is 646.1. The number of independent variables is $k = 5$. The value of the F statistic, using Equation (8.17), is therefore

$$f = \frac{(646.1 - 556.9)/(7 - 5)}{556.9/17} = 1.36$$

The null distribution is $F_{2,17}$. From the F table (Table A.7), we find that the 0.10 point on this F distribution is 2.64. Therefore the P value is much larger than 0.10. This model is clearly plausible.

We have used an informal method to find a good parsimonious model. It is important to realize that this informal procedure could have been carried out somewhat differently, with different choices for variables to drop and to include in the model. We might have come up with a different final model that might have been just as good as the one we actually found. In practice, there are often many models that fit the data about equally well; there is no single "correct" model.

Best Subsets Regression

As we have mentioned, methods of model selection are often rather informal and ad hoc. There are a few tools, however, that can make the process somewhat more systematic. One of them is **best subsets regression**. The concept is quite simple. Assume that there are p independent variables, $x_1, \ldots, x_p$, that are available to be put into the model. Let's assume that we wish to find a good model that contains exactly four independent variables. We can simply fit every possible model containing four of the variables, and rank them in order of their goodness-of-fit, as measured by the coefficient of determination R^2. The subset of four variables that yields the largest value of R^2 is the "best" subset of size 4. One can repeat the process for subsets of other sizes, finding the best subsets of size $1, 2, \ldots, p$. These best subsets can then be examined to see which provides a good fit, while being parsimonious.

The best subsets procedure is computationally intensive. When there are a lot of potential independent variables, there are a lot of models to fit. However, for most data sets, computers today are powerful enough to handle 30 or more independent variables, which is enough to cover many situations in practice. The following MINITAB output is for the best subsets procedure, applied to the data in Table 8.4. There are a total of seven independent variables being considered: Speed, Pause, LCR, Speed · Pause, Speed^2, Pause^2, and LCR^2.

```
Best Subsets Regression

Response is Overhead

                                                         S
                                                 S   P   p
                                                 p   a   e
                                         S   P   e   u   e   L
                                         p   a   e   s   d   C
                                         e   u   L   d   e   *   R
                  Adj.                   e   s   C   ^   ^   P   ^
Vars    R-Sq    R-Sq     C-p          s  d   e   R   2   2   a   2

   1    73.7    72.5   140.6   15.171                        X
   1    54.5    52.6   258.3   19.946        X
   2    82.7    81.2    87.0   12.564        X               X
   2    82.2    80.6    90.3   12.755    X   X
   3    92.9    91.9    26.5   8.2340        X   X               X
   3    89.6    88.1    46.9   9.9870    X       X   X
   4    95.9    95.0    10.3   6.4430    X   X   X               X
   4    95.4    94.5    13.2   6.7991        X   X   X           X
   5    96.8    95.9     6.7   5.8315    X   X   X       X       X
   5    96.7    95.8     7.2   5.9074    X   X       X       X   X
   6    97.1    96.2     6.6   5.6651    X   X   X   X   X       X
   6    97.0    96.0     7.6   5.8164    X   X   X   X       X   X
   7    97.2    96.1     8.0   5.7234    X   X   X   X   X   X   X
```

In this output, both the best and the second-best subset are presented, for sizes 1 through 7. We emphasize that the term *best* means only that the model has the largest value of R^2, and does not guarantee that it is best in any practical sense. We'll explain the output column by column. The first column, labeled "Vars," presents the number of variables in the model. Thus the first row of the table describes the best model that can be made with one independent variable, and the second row describes the second-best such model. The third and fourth rows describe the best and second-best models that can be made with two variables, and so on. The second column presents the coefficient of determination, R^2, for each model. Note that the value of R^2 for the best subset increases as the number of variables increases. It is a mathematical fact that the best subset of $k+1$ variables will always have at least as large an R^2 as the best subset of k variables. We will skip over the next two columns for the moment. The column labeled "s" presents the estimate of the error standard deviation. It is the square root of the estimate s^2 (Equation 8.8 in Section 8.1). Finally, the columns on the right represent the independent variables that are candidates for inclusion into the model. The name of each variable is written vertically above its column. An "X" in the column means that the variable is included in the model. Thus, the best model containing four variables is the one with the variables Speed, Pause, LCR, and LCR^2.

Looking at the best subsets regression output, it is important to note how little difference there is in the fit between the best and second-best models of each size (except for size 1). It is also important to realize that the value of R^2 is a random quantity; it depends on the data. If the process were repeated and new data obtained, the values of R^2 for the various models would be somewhat different, and different models would be "best." For this reason, one should not use this procedure, or any other, to choose a single model. Instead, one should realize that there will be many models that fit the data about equally well.

Nevertheless, methods have been developed to choose a single model, presumably the "best" of the "best." We describe two of them here, with a caution not to take them too seriously. We begin by noting that if we simply choose the model with the highest value of R^2, we will always pick the one that contains all the variables, since the value of R^2 necessarily increases as the number of variables in the model increases. The methods for selecting a model involve statistics that adjust the value of R^2, so as to eliminate this feature.

The first is the **adjusted R^2**. Let n denote the number of observations, and let k denote the number of independent variables in the model. The adjusted R^2 is defined as follows:

$$\text{Adjusted } R^2 = R^2 - \left(\frac{k}{n-k-1}\right)(1 - R^2) \tag{8.18}$$

The adjusted R^2 is always smaller than R^2, since a positive quantity is subtracted from R^2. As the number of variables k increases, R^2 will increase, but the amount subtracted from it will increase as well. The value of k for which the value of adjusted R^2 is a maximum can be used to determine the number of variables in the model, and the best subset of that size can be chosen as the model. In the preceding output, we can see that the adjusted R^2 reaches its maximum (96.2%) at the six-variable model containing the variables Speed, Pause, LCR, Speed^2, Pause^2, and LCR^2.

Another commonly used statistic is **Mallows' C_p**. To compute this quantity, let n be the number of observations, let p be the total number of independent variables under consideration, and let k be the number of independent variables in a subset. As before, let

SSE_{full} denote the error sum of squares for the full model containing all p variables, and let $\text{SSE}_{\text{reduced}}$ denote the error sum of squares for the model containing only the subset of k variables. Mallows' C_p is defined as

$$C_p = \frac{(n - p - 1)\text{SSE}_{\text{reduced}}}{\text{SSE}_{\text{full}}} - (n - 2k - 2) \tag{8.19}$$

For models that contain as many independent variables as necessary, the value of C_p is supposed to be approximately equal to the number of variables, including the intercept, in the model. To choose a single model, one can either choose the model with the minimum value of C_p, or one can choose the model in which the value of C_p is closest to the number of independent variables in the model. In the preceding output, both criteria yield the same six-variable model chosen by the adjusted R^2 criterion. The value of C_p for this model is 6.6.

Finally, we point out that our ad hoc procedure using the F test yielded the five-variable model containing the variables Speed, Pause, LCR, Pause^2, and LCR^2. The output shows that this model is the best five-variable model in terms of R^2. Its adjusted R^2 is 95.9%, and its C_p value is 6.7, both of which are close to their optimum values. In practice, there is no clear reason to prefer the model chosen by adjusted R^2 and Mallows' C_p to this model, or vice versa.

Stepwise Regression

Stepwise regression is perhaps the most widely used model selection technique. Its main advantage over best subsets regression is that it is less computationally intensive, so it can be used in situations where there are a very large number of candidate independent variables and too many possible subsets for every one of them to be examined. The version of stepwise regression that we will describe is based on the P-values of the t statistics for the independent variables. An equivalent version is based on the F statistic (which is the square of the t statistic). Before running the algorithm, the user chooses two threshold P-values, α_{in} and α_{out}, with $\alpha_{\text{in}} \le \alpha_{\text{out}}$. Stepwise regression begins with a step called a **forward selection** step, in which the independent variable with the smallest P-value is selected, provided that it satisfies $P < \alpha_{\text{in}}$. This variable is entered into the model, creating a model with a single independent variable. Call this variable x_1. In the next step, also a forward selection step, the remaining variables are examined one at a time as candidates for the second variable in the model. The one with the smallest P-value is added to the model, again provided that $P < \alpha_{\text{in}}$.

Now it is possible that adding the second variable to the model has increased the P-value of the first variable. In the next step, called a **backward elimination** step, the first variable is dropped from the model if its P-value has grown to exceed the value α_{out}. The algorithm then continues by alternating forward selection steps with backward elimination steps: at each forward selection step adding the variable with the smallest P-value if $P < \alpha_{\text{in}}$, and at each backward elimination step dropping the variable with the largest P-value if $P > \alpha_{\text{out}}$. The algorithm terminates when no variables meet the criteria for being added to or dropped from the model.

The following output is from the MINITAB stepwise regression procedure, applied to the data in Table 8.4. The threshold P-values are $\alpha_{in} = \alpha_{out} = 0.15$. There are a total of seven independent variables being considered: Speed, Pause, LCR, Speed · Pause, Speed^2, Pause^2, and LCR^2.

```
Alpha-to-Enter = 0.15  Alpha-to-Remove = 0.15

Response is Overhead on 7 predictors, with N = 25

Step               1         2         3         4         5
Constant       452.2     437.3     410.7     388.4     338.5

Speed*Pause   0.0470    0.0355    0.0355    0.0304    0.0146
T-Value         8.03      6.00      6.96      5.21      3.52
P-Value        0.000     0.000     0.000     0.000     0.002

Pause                     0.74      3.02      3.75      3.24
T-Value                   3.40      3.78      4.20      6.25
P-Value                  0.003     0.001     0.000     0.000

Pause^2                          -0.0380   -0.0442   -0.0256
T-Value                            -2.94     -3.39     -3.19
P-Value                            0.008     0.003     0.005

LCR                                           0.69      6.97
T-Value                                       1.62      6.95
P-Value                                      0.121     0.000

LCR^2                                                 -0.139
T-Value                                                -6.46
P-Value                                                0.000

S               15.2      12.6      10.8      10.4      5.99
R-Sq           73.70     82.74     87.77     89.19     96.62
R-Sq(adj)      72.55     81.18     86.02     87.02     95.73
Mallows C-p    140.6      87.0      58.1      51.4       7.8
```

In step 1, the variable Speed · Pause had the smallest P-value (0.000) among the seven, so it was the first variable in the model. In step 2, Pause had the smallest P-value (0.003) among the remaining variables, so it was added next. The P-value for Speed · Pause remained at 0.000 after the addition of Pause to the model; since it did not rise to a value greater than $\alpha_{out} = 0.15$, it is not dropped from the model. In steps 3, 4, and 5, the variables Pause^2, LCR, and LCR^2 are added in turn. At no point does the P-value of a variable in the model exceed the threshold $\alpha_{out} = 0.15$, so no variables are dropped. After five steps, none of the variables remaining have P-values less than $\alpha_{in} = 0.15$, so the algorithm terminates. The final model contains the variables Speed · Pause, Pause, Pause^2, LCR, and LCR^2.

The model chosen by stepwise regression is a five-variable model. Comparison with the best subsets output shows that it is not one of the best two five-variable models in terms of R^2. Still, the model fits well, and in terms of fit alone, it is reasonable. We point out that this model has the undesirable feature that it contains the interaction term Speed · Pause without containing the variable Speed by itself. This points out a weakness of all automatic variable selection procedures, including stepwise regression and best subsets. They operate on the basis of goodness-of-fit alone, and are not able to take into account relationships among independent variables that may be important to consider.

Model Selection Procedures Sometimes Find Models When They Shouldn't

When constructing a model to predict the value of a dependent variable, it might seem reasonable to try to start with as many candidate independent variables as possible, so that a model selection procedure has a very large number of models to choose from. Unfortunately, this is not a good idea, as we will now demonstrate.

A correlation coefficient can be computed between any two variables. Sometimes, two variables that have no real relationship will be strongly correlated, just by chance. For example, the statistician George Udny Yule noticed that the annual birthrate in Great Britain was almost perfectly correlated ($r = -0.98$) with the annual production of pig iron in the United States for the years 1875–1920. Yet no one would suggest trying to predict one of these variables from the other. This illustrates a difficulty shared by all model selection procedures. The more candidate independent variables that are provided, the more likely it becomes that some of them will exhibit meaningless correlations with the dependent variable, just by chance.

We illustrate this phenomenon with a simulation. We generated a simple random sample $y_1, \ldots, y_{30}$ of size 30 from a $N(0, 1)$ distribution. We will denote this sample by y. Then we generated 20 more independent samples of size 30 from a $N(0, 1)$ distribution; we will denote these samples by $x_1, \ldots, x_{20}$. To make the notation clear, the sample x_i contains 30 values $x_{i1}, \ldots, x_{i30}$. We then applied both stepwise regression and best subsets regression to these simulated data. None of the x_i are related to y; they were all generated independently. Therefore the ideal output from a model selection procedure would be to produce a model with no dependent variables at all. The actual behavior was quite different. The following two MINITAB outputs are for the stepwise regression and best subsets procedures. The stepwise regression method recommends a model containing six variables, with an adjusted R^2 of 41.89%. The best subsets procedure produces the best-fitting model for each number of variables from 1 to 20. Using the adjusted R^2 criterion, the best subsets procedure recommends a 12-variable model, with an adjusted R^2 of 51.0%. Using the minimum Mallows' C_p criterion, the "best" model is a five-variable model.

Anyone taking this output at face value would believe that some of the independent variables might be useful in predicting the dependent variable. But none of them are. All the apparent relationships are due entirely to chance.

```
Stepwise Regression: Y versus X1, X2, ...

Alpha-to-Enter: 0.15  Alpha-to-Remove: 0.15

Response is Y on 20 predictors, with N = 30

Step                1         2         3         4         5         6
Constant      0.14173   0.11689   0.12016   0.13756   0.09070   0.03589

X15             -0.38     -0.38     -0.28     -0.32     -0.28     -0.30
T-Value         -2.08     -2.19     -1.60     -1.87     -1.69     -1.89
P-Value         0.047     0.037     0.122     0.073     0.105     0.071

X6                         0.39      0.55      0.57      0.57      0.52
T-Value                    2.04      2.76      2.99      3.15      2.87
P-Value                   0.051     0.010     0.006     0.004     0.009

X16                                 -0.43     -0.43     -0.55     -0.73
T-Value                             -1.98     -2.06     -2.60     -3.07
P-Value                             0.058     0.050     0.016     0.005

X12                                            0.33      0.42      0.49
T-Value                                        1.79      2.29      2.66
P-Value                                       0.086     0.031     0.014

X3                                                      -0.42     -0.52
T-Value                                                 -1.83     -2.23
P-Value                                                 0.080     0.035

X17                                                                0.35
T-Value                                                            1.53
P-Value                                                           0.140

S                1.15      1.09      1.04     0.998     0.954     0.928
R-Sq            13.33     24.92     34.75     42.15     49.23     53.91
R-Sq(adj)       10.24     19.36     27.22     32.90     38.66     41.89
Mallows C-p       5.5       3.3       1.7       1.0       0.4       0.7
```

```
Best Subsets Regression: Y versus X1, X2, ...

Response is Y

                                                         X X X X X X X X X X X
                     Mallows           X X X X X X X X X 1 1 1 1 1 1 1 1 1 1 2
Vars  R-Sq  R-Sq(adj)    C-p        S  1 2 3 4 5 6 7 8 9 0 1 2 3 4 5 6 7 8 9 0
   1  13.3       10.2    5.5   1.1539                              X
   2  28.3       23.0    2.0   1.0685            X                   X
   3  34.8       27.2    1.7   1.0390            X                 X X
   4  43.2       34.1    0.6  0.98851      X     X           X       X
   5  49.2       38.7    0.4  0.95391      X     X           X     X X
   6  53.9       41.9    0.7  0.92844      X     X           X     X X X
   7  57.7       44.3    1.3  0.90899  X X       X           X     X X     X
   8  61.2       46.4    2.1  0.89168      X     X           X X   X X X   X
   9  65.0       49.3    2.7  0.86747      X     X X         X X   X X X   X
  10  67.6       50.5    3.8  0.85680      X     X X         X X   X X X X X
  11  69.2       50.4    5.2  0.85803  X X X     X X       X X     X X X   X
  12  71.3       51.0    6.4  0.85267  X X X     X X       X X   X X X X   X
  13  72.4       49.9    8.0  0.86165  X X X     X X     X X X   X X X X   X
  14  73.0       47.8    9.8  0.87965  X X X     X X X   X X X   X X X X   X
  15  74.2       46.5   11.4  0.89122  X X X     X X X   X X X   X X X X X X
  16  74.5       43.1   13.3  0.91886  X X X   X X X X   X X X   X X X X X X
  17  74.8       39.2   15.1  0.94985  X X X   X X X X   X X X X X X X X X X
  18  75.1       34.2   17.1  0.98777  X X X   X X X X   X X X X X X X X X X X
  19  75.1       27.9   19.0   1.0344  X X X   X X X X X X X X X X X X X X X X
  20  75.2       20.1   21.0   1.0886  X X X X X X X X X X X X X X X X X X X X
```

How can one determine which variables, if any, in a selected model are really related to the dependent variable, and which were selected only by chance? Statistical methods are not much help here. The most reliable method is to repeat the experiment, collecting more data on the dependent variable and on the independent variables that were selected for the model. Then the independent variables suggested by the selection procedure can be fit to the dependent variable using the new data. If some of these variables fit well in the new data, the evidence of a real relationship becomes more convincing.

We summarize our discussion of model selection by emphasizing four points.

Summary

When selecting a regression model, keep the following in mind:

- When there is little or no physical theory to rely on, many different models will fit the data about equally well.
- The methods for choosing a model involve statistics (R^2, the F statistic, C_p), whose values depend on the data. Therefore if the experiment is repeated, these statistics will come out differently, and different models may appear to be "best."
- Some or all of the independent variables in a selected model may not really be related to the dependent variable. Whenever possible, experiments should be repeated to test these apparent relationships.
- Model selection is an art, not a science.

Exercises for Section 8.3

1. True or false:

a. For any set of data, there is always one best model.
b. When there is no physical theory to specify a model, there is usually no best model, but many that are about equally good.
c. Model selection methods such as best subsets and stepwise regression, when properly used, are scientifically designed to find the best available model.
d. Model selection methods such as best subsets and stepwise regression, when properly used, can suggest models that fit the data well.

2. The article "Experimental Design Approach for the Optimization of the Separation of Enantiomers in Preparative Liquid Chromatography" (S. Lai and Z. Lin, *Separation Science and Technology*, 2002: 847–875) describes an experiment involving a chemical process designed to separate enantiomers. A model was fit to estimate the cycle time (y) in terms of the flow rate (x_1), sample concentration (x_2), and mobile-phase composition (x_3). The results of a least-squares fit are presented in the following table. (The article did not provide the value of the t statistic for the constant term.)

Predictor	Coefficient	T	P
Constant	1.603		
x_1	−0.619	−22.289	0.000
x_2	0.086	3.084	0.018
x_3	0.306	11.011	0.000
x_1^2	0.272	8.542	0.000
x_2^2	0.057	1.802	0.115
x_3^2	0.105	3.300	0.013
x_1x_2	−0.022	−0.630	0.549
x_1x_3	−0.036	−1.004	0.349
x_2x_3	0.036	1.018	0.343

Of the following, which is the best next step in the analysis?

i. Nothing needs to be done. This model is fine.
ii. Drop x_1^2, x_2^2, and x_3^2 from the model, and then perform an F test.
iii. Drop x_1x_2, x_1x_3, and x_2x_3 from the model, and then perform an F test.
iv. Drop x_1 and x_1^2 from the model, and then perform an F test.
v. Add cubic terms x_1^3, x_2^3, and x_3^3 to the model to try to improve the fit.

3. In the article referred to in Exercise 2, a model was fit to investigate the relationship between the independent variables given in Exercise 2 and the amount of S-isomer collected. The results of a least-squares fit are presented in the following table. (The article did not provide the value of the t statistic for the constant term.)

Predictor	Coefficient	T	P
Constant	3.367		
x_1	−0.018	−1.729	0.127
x_2	1.396	135.987	0.000
x_3	0.104	10.098	0.000
x_1^2	0.017	1.471	0.184
x_2^2	−0.023	−0.909	0.394
x_3^2	−0.030	−2.538	0.039
x_1x_2	−0.006	−0.466	0.655
x_1x_3	0.012	0.943	0.377
x_2x_3	0.055	4.194	0.004

Of the following, which is the best next step in the analysis?

i. Drop x_1^2, x_2^2, and x_3^2 from the model, and then perform an F test.
ii. Nothing needs to be done. This model is fine.
iii. Add cubic terms x_1^3, x_2^3, and x_3^3 to the model to try to improve the fit.
iv Drop x_1x_2, x_1x_3, and x_2x_3 from the model, and then perform an F test.
v. Drop x_2^2, x_1x_2, and x_1x_3 from the model, and then perform an F test.

4. An engineer measures a dependent variable y and independent variables x_1, x_2, and x_3. MINITAB output for the model $y = \beta_0 + \beta_1x_1 + \beta_2x_2 + \beta_3x_3 + \varepsilon$ is presented as follows.

```
The regression equation is
Y= 0.367 + 1.61 X1 - 5.51 X2 + 1.27 X3

Predictor      Coef    SE Coef       T       P
Constant     0.3692     0.9231    0.40   0.698
X1           1.6121     1.3395    1.21   0.254
X2           5.5049     1.4959    3.68   0.004
X3           1.2646     1.9760    0.64   0.537
```

Of the following, which is the best next step in the analysis?

i. Add interaction terms x_1x_2 and x_2x_3 to try to find more variables to put into the model.
ii. Add the interaction term x_1x_3 to try to find another variable to put into the model.
iii. Nothing needs to be done. This model is fine.
iv. Drop x_1 and x_3, and then perform an F test.
v. Drop x_2, and then perform an F test.
vi. Drop the intercept (Constant), since it has the largest P-value.

5. A physiologist is trying to predict people's breathing rates (in m^3/h) during heavy exercise from their breathing rates at rest, while sitting, and during light exercise. The following MINITAB output is from the model

$$\text{Heavy} = \beta_0 + \beta_1 \text{ Light} + \beta_2 \text{ Sit} + \beta_3 \text{ Rest} + \beta_4 \text{ Light} \cdot \text{Sit} + \beta_5 \text{ Light} \cdot \text{Rest} + \beta_6 \text{ Sit} \cdot \text{Rest} + \varepsilon$$

```
The regression equation is
Heavy = 265.6 + 1.48 Light - 10.4 Sit - 1.13 Rest + 0.044 Light*Sit
        + 0.0010 Light*Rest + 0.026 Sit*Rest

Predictor           Coef     SE Coef        T         P
Constant          265.64       10.59    25.08     0.000
Light             1.4755      0.2632     5.61     0.000
Sit              -10.410       1.660    -6.27     0.000
Rest             -1.1338       2.233    -0.51     0.622
Light*Sit       0.043728     0.01513     2.89     0.015
Light*Rest    0.00099612     0.01855     0.05     0.958
Sit*Rest        0.026348     0.02196     1.20     0.255
```

Of the following, which is the best next step in the analysis?

i. Drop Rest, Light · Rest, and Sit · Rest from the model, and then perform an F test.
ii. Drop Light · Sit from the model, and then perform an F test.
iii. Nothing needs to be done. This model is fine.
iv. Add the variables Light^2 and Sit^2 to try to find more variables to put in the model.
v. Add the variable Rest^2 to try to find more variables to put in the model.

6. The following MINITAB output is for a best subsets regression involving five dependent variables $X_1, \ldots, X_5$. The two models of each size with the highest values of R^2 are listed.

```
Best Subsets Regression: Y versus X1, X2, X3, X4, X5

Response is Y

                            Mallows              X  X  X  X  X
Vars  R-Sq   R-Sq(adj)        C-p          S    1  2  3  4  5
   1  77.3        77.1      133.6     1.4051    X
   1  10.2         9.3      811.7     2.7940          X
   2  89.3        89.0       14.6    0.97126    X     X
   2  77.8        77.3      130.5     1.3966    X  X
   3  90.5        90.2        3.6    0.91630    X     X  X
   3  89.4        89.1       14.6    0.96763    X  X  X
   4  90.7        90.3        4.3    0.91446    X  X  X  X
   4  90.6        90.2        5.3    0.91942    X     X  X  X
   5  90.7        90.2        6.0    0.91805    X  X  X  X  X
```

a. Which variables are in the model selected by the minimum C_p criterion?
b. Which variables are in the model selected by the adjusted R^2 criterion?
c. Are there any other good models?

7. The following is supposed to be the result of a best subsets regression involving five independent variables $X_1, \ldots, X_5$. The two models of each size with the highest values of R^2 are listed. Something is wrong. What is it?

```
Best Subsets Regression

Response is Y

                Adj.                          X   X   X   X   X
Vars   R-Sq     R-Sq     C-p        s         1   2   3   4   5

   1   69.1     68.0    101.4    336.79                       X
   1   60.8     59.4    135.4    379.11           X
   2   80.6     79.2     55.9    271.60       X       X
   2   79.5     77.9     60.7    279.59       X   X
   3   93.8     92.8     13.4    184.27       X   X       X
   3   93.7     92.7     18.8    197.88       X   X   X
   4   91.4     90.4      5.5    159.59       X   X       X   X
   4   90.1     88.9      5.6    159.81       X   X   X       X
   5   94.2     93.0      6.0    157.88       X   X   X   X   X
```

8. A certain chemical process is run 12 times, with varying concentrations of three reactants. The concentrations are denoted x_1, x_2, and x_3. The model $y = \beta_0 + \beta_1 x_1 + \beta_2 x_2 + \beta_3 x_3 + \beta_4 x_1 x_2 + \beta_5 x_1 x_3 + \beta_6 x_2 x_3 + \varepsilon$ is fit to the data, and the sum of squares for error is SSE $= 4.6409$. Then the reduced model $y = \beta_0 + \beta_1 x_1 + \beta_2 x_2 + \beta_3 x_3$ is fit, and the sum of squares for error is SSE $= 11.5820$. Is it reasonable to use the reduced model, rather than the model containing all the interactions, to predict yield? Explain.

9. (Continues Exercise 7 in Section 8.1.) To try to improve the prediction of FEV$_1$, additional independent variables are included in the model. These new variables are Weight (in kg), the product (interaction) of Height and Weight, and the ambient temperature (in °C). The following MINITAB output presents results of fitting the model

$$\begin{aligned}\text{FEV}_1 &= \beta_0 + \beta_1 \text{ Last FEV}_1 + \beta_2 \text{ Gender} + \beta_3 \text{ Height} + \beta_4 \text{ Weight} + \beta_5 \text{ Height} \cdot \text{Weight} \\ &\quad + \beta_6 \text{ Temperature} + \beta_7 \text{ Pressure} + \varepsilon\end{aligned}$$

```
The regression equation is
FEV1 = -0.257 + 0.778 Last FEV - 0.105 Gender + 1.213 Height - 0.00624 Weight
+ 0.00386 Height*Weight - 0.00740 Temp - 0.00148 Pressure

Predictor            Coef      SE Coef        T         P
Constant           0.2565       0.7602     0.34     0.736
Last FEV          0.77818      0.05270    14.77     0.000
Gender           -0.10479      0.03647    -2.87     0.005
Height             1.2128       0.4270     2.84     0.005
Weight         -0.0062446      0.01351    -0.46     0.645
Height*Weight   0.0038642     0.008414     0.46     0.647
```

```
Temp                -0.007404    0.009313      -0.79    0.428
Pressure            -0.0014773   0.0005170     -2.86    0.005

S = 0.22189          R-Sq = 93.5%          R-Sq(adj) = 93.2%

Analysis of Variance

Source             DF        SS          MS         F        P
Regression          7    111.35      15.907    323.06    0.000
Residual Error    157    7.7302    0.049237
Total             164    119.08
```

a. The following MINITAB output, reproduced from Exercise 7 in Section 8.1, is for a reduced model in which Weight, Height · Weight, and Temp have been dropped. Compute the F statistic for testing the plausibility of the reduced model.

```
The regression equation is
FEV1 = -0.219 + 0.779 Last FEV - 0.108 Gender + 1.354 Height - 0.00134 Pressure

Predictor            Coef       SE Coef          T          P
Constant         -0.21947        0.4503      -0.49      0.627
Last FEV            0.779       0.04909      15.87      0.000
Gender           -0.10827        0.0352      -3.08      0.002
Height             1.3536        0.2880       4.70      0.000
Pressure       -0.0013431     0.0004722      -2.84      0.005

S = 0.22039          R-Sq = 93.5%          R-Sq(adj) = 93.3%

Analysis of Variance

Source             DF        SS          MS          F          P
Regression          4    111.31      27.826     572.89      0.000
Residual Error    160    7.7716    0.048572
Total             164    119.08
```

b. How many degrees of freedom does the F statistic have?
c. Find the P-value for the F statistic. Is the reduced model plausible?
d. Someone claims that since each of the variables being dropped had large P-values, the reduced model must be plausible, and it was not necessary to perform an F test. Is this correct? Explain why or why not.
e. The total sum of squares is the same in both models, even though the independent variables are different. Is there a mistake? Explain.

10. The article "Optimization of Enterocin P Production by Batch Fermentation of *Enterococcus faecium* P13 at Constant pH" (C. Herran, J. Martinez, et al., *Applied Microbiology and Biotechnology*, 2001:378–383) described a study involving the growth rate of the bacterium *Enterococcus faecium* in media with varying pH. The log of the maximum growth rate for various values of pH are presented in the following table:

ln Growth rate	−2.12	−1.51	−0.89	−0.33	−0.05	−0.11	0.39	−0.25
pH	4.7	5.0	5.3	5.7	6.0	6.2	7.0	8.5

a. Fit the linear model: ln Growth rate $= \beta_0 + \beta_1 \text{ pH} + \varepsilon$. For each coefficient, find the P-value for the null hypothesis that the coefficient is equal to 0. In addition, compute the analysis of variance (ANOVA) table.

b. Fit the quadratic model: ln Growth rate $= \beta_0 + \beta_1 \text{ pH} + \beta_2 \text{ pH}^2 + \varepsilon$. For each coefficient, find the P-value for the null hypothesis that the coefficient is equal to 0. In addition, compute the ANOVA table.
c. Fit the cubic model: ln Growth rate $= \beta_0 + \beta_1 \text{ pH} + \beta_2 \text{ pH}^2 + \beta_3 \text{ pH}^3 + \varepsilon$. For each coefficient, find the P-value for the null hypothesis that the coefficient is equal to 0. In addition, compute the ANOVA table.
d. Which of these models do you prefer, and why?

11. Refer to Exercise 7 in Section 7.4. A windmill is used to generate direct current. Data are collected to determine the relationship between wind speed and current. Let y represent current and let x represent wind speed.

a. Fit the quadratic model $y = \beta_0 + \beta_1 x + \beta_2 x^2 + \varepsilon$. For each coefficient, find the P-value for the null hypothesis that the coefficient is equal to 0.
b. Fit the model $y = \beta_0 + \beta_1 \ln x + \varepsilon$. For each coefficient, find the P-value for the null hypothesis that the coefficient is equal to 0.
c. Make a plot of residuals versus fitted values for each model. Do the assumptions of either model seem to be violated?
d. Using each model, predict the current for wind speeds of 3.0, 5.0, 7.0, and 9.0 mi/h.
e. The quadratic model has two dependent variables (x and x^2), while the log model has only one. Can the F test be used to determine which model is more appropriate? If so, compute the F statistic and the P-value. If not, explain why not.
f. Can you think of a reason to prefer one of these models to the other? Explain.

12. In rock blasting, the peak particle velocity (PPV) depends both on the distance from the blast and on the amount of charge. The article "Prediction of Particle Velocity Caused by Blasting for an Infrastructure Excavation Covering Granite Bedrock" (A. Kahriman, *Mineral Resources Engineering*, 2001:205–218) presents data on PPV, scaled distance (which is equal to the distance divided by the square root of the charge), and the amount of charge. The following table presents the values of PPV, scaled distance, and amount of charge for 15 blasts. (Some of these data were presented in Exercise 4 in Section 7.4.)

PPV (mm/s)	Scaled Distance ($m/kg^{0.5}$)	Amount of Charge (kg)
1.4	47.33	4.2
15.7	9.6	92
2.54	15.8	40
1.14	24.3	48.7
0.889	23	95.7
1.65	12.7	67.7
1.4	39.3	13
26.8	8	70
1.02	29.94	13.5
4.57	10.9	41
6.6	8.63	108.8
1.02	28.64	27.43
3.94	18.21	59.1
1.4	33	11.5
1.4	34	175

a. Fit the model ln PPV $= \beta_0 + \beta_1$ ln Scaled Distance $+ \beta_2$ ln Charge $+ \varepsilon$. Compute the P-value for testing $H_0: \beta_i = 0$ for β_0, β_1, and β_2.
b. The article claims that the model ln PPV $= \beta_0 + \beta_1$ ln Scaled Distance $+ \varepsilon$ is appropriate. Fit this model. Compute the P-value for testing $H_0: \beta_i = 0$ for β_0, β_1, and β_2.
c. Which model do you prefer? Why?

13. The article "Ultimate Load Analysis of Plate Reinforced Concrete Beams" (N. Subedi and P. Baglin, *Engineering Structures*, 2001:1068–1079) presents theoretical and measured ultimate strengths (in kN) for a sample of steel-reinforced concrete beams. The results are presented in the following table (two outliers have been deleted).

Let y denote the measured strength, x the theoretical strength, and t the true strength, which is unknown. Assume that $y = t + \varepsilon$, where ε is the measurement error. It is uncertain whether t is related to x by a linear model $t = \beta_0 + \beta_1 x$ or by a quadratic model $t = \beta_0 + \beta_1 x + \beta_2 x^2$.

Theoretical	Measured	Theoretical	Measured
991	1118	1516	1550
785	902	1071	1167
1195	1373	1480	1609
1021	1196	1622	1756
1285	1609	2032	2119
1167	1413	2032	2237
1519	1668	660	640
1314	1491	565	530
1743	1952	738	893
791	844	682	775

a. Fit the linear model $y = \beta_0 + \beta_1 x + \varepsilon$. For each coefficient, find the P-value for the null hypothesis that the coefficient is equal to 0.
b. Fit the quadratic model $y = \beta_0 + \beta_1 x + \beta_2 x^2 + \varepsilon$. For each coefficient, find the P-value for the null hypothesis that the coefficient is equal to 0.
c. Plot the residuals versus the fitted values for the linear model.
d. Plot the residuals versus the fitted values for the quadratic model.
e. Based on the results in parts (a) through (d), which model seems more appropriate? Explain.
f. Using the more appropriate model, estimate the true strength if the theoretical strength is 1500.
g. Using the more appropriate model, find a 95% confidence interval for the true strength if the theoretical strength is 1500.

14. The article "Permanent Deformation Characterization of Subgrade Soils from RLT Test" (A. Puppala, L. Mohammad, et al., *Journal of Materials in Civil Engineering*, 1999:274–282) presents measurements of plastic strains (in %) on soils at various confining and deviatoric stresses in kPa. The following table presents the plastic strains (y), the confining stress (x_1), and the deviatoric stress (x_2) for tests on a sandy soil.

y	x_1	x_2	y	x_1	x_2
0.01	21	21	0.01	70	140
0.02	21	35	0.07	70	210
0.05	21	52.5	0.002	105	70
0.09	21	70	0.0003	105	105
0.003	35	35	0.0009	105	140
0.006	35	70	0.01	105	210
0.05	35	105	0.001	140	70
0.23	35	140	0.0003	140	105
0.003	70	35	0.0005	140	210
0.0008	70	70	0.03	140	280

a. Fit the model $y = \beta_0 + \beta_1 x_1 + \beta_2 x_2 + \varepsilon$. Plot the residuals versus the fitted values. Does the model seem appropriate?
b. Fit the model $\ln y = \beta_0 + \beta_1 \ln x_1 + \beta_2 \ln x_2 + \varepsilon$. Plot the residuals versus the fitted values. Does the model seem appropriate?
c. Use the more appropriate of the models from (a) and (b) to predict y when $x_1 = 50$ and $x_2 = 100$.
d. Is the model you used in part (c) improved by including an interaction term? Explain.

15. The article "Vehicle-Arrival Characteristics at Urban Uncontrolled Intersections" (V. Rengaraju and V. Rao, *Journal of Transportation Engineering*, 1995:317–323) presents data on traffic characteristics at 10 intersections in Madras, India. The following table provides data on road width in m (x_1), traffic volume in vehicles per lane per hour (x_2), and median speed in km/h (x_3).

y	x_1	x_2	y	x_1	x_2
35.0	76	370	26.5	75	842
37.5	88	475	27.5	92	723
26.5	76	507	28.0	90	923
33.0	80	654	23.5	86	1039
22.5	65	917	24.5	80	1120

a. Fit the model $y = \beta_0 + \beta_1 x_1 + \beta_2 x_2 + \varepsilon$. Find the P-values for testing that the coefficients are equal to 0.
b. Fit the model $y = \beta_0 + \beta_1 x_1 + \varepsilon$. Find the P-values for testing that the coefficients are equal to 0.
c. Fit the model $y = \beta_0 + \beta_1 x_2 + \varepsilon$. Find the P-values for testing that the coefficients are equal to 0.
d. Which of the models (a) through (c) do you think is best? Why?

16. The following table presents measurements of mean noise levels in dBA (y), roadway width in m (x_1), and mean speed in km/h (x_2), for 10 locations in Bangkok, Thailand, as reported in the article "Modeling of Urban Area Stop-and-Go Traffic Noise" (P. Pamanikabud and C. Tharasawatipipat, *Journal of Transportation Engineering* 1999:152–159).

y	x_1	x_2	y	x_1	x_2
78.1	6.0	30.61	78.1	12.0	28.26
78.1	10.0	36.55	78.6	6.5	30.28
79.6	12.0	36.22	78.5	6.5	30.25
81.0	6.0	38.73	78.4	9.0	29.03
78.7	6.5	29.07	79.6	6.5	33.17

Construct a good linear model to predict mean noise levels using roadway width, mean speed, or both, as predictors. Provide the standard deviations of the coefficient estimates and the P-values for testing that they are different from 0. Explain how you chose your model.

17. The article "Modeling Resilient Modulus and Temperature Correction for Saudi Roads" (H. Wahhab, I. Asi, and R. Ramadhan, *Journal of Materials in Civil Engineering*, 2001:298–305) describes a study designed to predict the resilient modulus of pavement from physical properties. The following table presents data for the resilient modulus at 40°C in 10^6 kPa (y), the surface area of the aggregate in m^2/kg (x_1), and the softening point of the asphalt in °C (x_2).

y	x_1	x_2	y	x_1	x_2	y	x_1	x_2
1.48	5.77	60.5	3.06	6.89	65.3	1.88	5.93	63.2
1.70	7.45	74.2	2.44	8.64	66.2	1.90	8.17	62.1
2.03	8.14	67.6	1.29	6.58	64.1	1.76	9.84	68.9
2.86	8.73	70.0	3.53	9.10	68.6	2.82	7.17	72.2
2.43	7.12	64.6	1.04	8.06	58.8	1.00	7.78	54.1

The full quadratic model is $y = \beta_0 + \beta_1 x_1 + \beta_2 x_2 + \beta_3 x_1 x_2 + \beta_4 x_1^2 + \beta_5 x_2^2 + \varepsilon$. Which submodel of this full model do you believe is most appropriate? Justify your answer by fitting two or more models and comparing the results.

18. The article "Models for Assessing Hoisting Times of Tower Cranes" (A. Leung and C. Tam, *Journal of Construction Engineering and Management*, 1999: 385–391) presents a model constructed by a stepwise regression procedure to predict the time needed for a tower crane hoisting operation. Twenty variables were considered, and the stepwise procedure chose a nine-variable model. The adjusted R^2 for the selected model was 0.73. True or false:

a. The value 0.73 is a reliable measure of the goodness of fit of the selected model.
b. The value 0.73 may exaggerate the goodness of fit of the model.
c. A stepwise regression procedure selects only variables that are of some use in predicting the value of the dependent variable.
d. It is possible for a variable that is of no use in predicting the value of a dependent variable to be part of a model selected by a stepwise regression procedure.

Supplementary Exercises for Chapter 8

1. The article "Advances in Oxygen Equivalence Equations for Predicting the Properties of Titanium Welds" (D. Harwig, W. Ittiwattana, and H. Castner, *The Welding Journal*, 2001:126s–136s) reports an experiment to predict various properties of titanium welds. Among other properties, the elongation (in %) was measured, along with the oxygen content and nitrogen content (both in %). The following MINITAB output presents results of fitting the model

$$\text{Elongation} = \beta_0 + \beta_1 \text{ Oxygen} + \beta_2 \text{ Nitrogen} + \beta_3 \text{ Oxygen} \cdot \text{Nitrogen}$$

```
The regression equation is
Elongation = 46.80 - 130.11 Oxygen - 807.1 Nitrogen + 3580.5 Oxy*Nit

Predictor        Coef       SE Coef          T        P
Constant       46.802         3.702      12.64    0.000
Oxygen        -130.11        20.467      -6.36    0.000
Nitrogen      -807.10        158.03     -5.107    0.000
Oxy*Nit        3580.5        958.05      3.737    0.001

S = 2.809            R-Sq = 74.5%         R-Sq(adj) = 72.3%
```

```
Analysis of Variance

Source          DF        SS       MS       F       P
Regression       3    805.43   268.48   34.03   0.000
Residual Error  35    276.11     7.89
Total           38   1081.54
```

a. Predict the elongation for a weld with an oxygen content of 0.15% and a nitrogen content of 0.01%.
b. If two welds both have a nitrogen content of 0.006%, and their oxygen content differs by 0.05%, what would you predict their difference in elongation to be?
c. Two welds have identical oxygen contents, and nitrogen contents that differ by 0.005%. Is this enough information to predict their difference in elongation? If so, predict the elongation. If not, explain what additional information is needed.

2. Refer to Exercise 1.

a. Find a 95% confidence interval for the coefficient of Oxygen.
b. Find a 99% confidence interval for the coefficient of Nitrogen.
c. Find a 98% confidence interval for the coefficient of the interaction term Oxygen · Nitrogen.
d. Can you conclude that $\beta_1 < -75$? Find the P-value.
e. Can you conclude that $\beta_2 > -1000$? Find the P-value.

3. The following MINITAB output is for a multiple regression. Some of the numbers got smudged, becoming illegible. Fill in the missing numbers.

```
Predictor        Coef   SE Coef      T       P
Constant          (a)    0.3501   0.59   0.568
X1             1.8515       (b)   2.31   0.040
X2             2.7241    0.7124    (c)   0.002

S = (d)          R-Sq = 83.4%      R-Sq(adj) = 80.6%

Analysis of Variance

Source          DF       SS     MS     F       P
Regression     (e)      (f)    (g)   (h)   0.000
Residual Error  12    17.28   1.44
Total          (i)   104.09
```

4. An engineer tries three different methods for selecting a linear model. First she uses an informal method based on the F statistic, as described in Section 8.3. Then she runs the best subsets routine, and finds the model with the best adjusted R^2 and the one with the best Mallows C_p. It turns out that all three methods select the same model. The engineer says that since all three methods agree, this model must be the best one. One of her colleagues says that other models might be equally good. Who is right? Explain.

5. In a simulation of 30 mobile computer networks, the average speed, pause time, and number of neighbors were measured. A "neighbor" is a computer within the transmission range of another. The data are presented in the following table.

Neighbors	Speed	Pause	Neighbors	Speed	Pause	Neighbors	Speed	Pause
10.17	5	0	9.36	5	10	8.92	5	20
8.46	5	30	8.30	5	40	8.00	5	50
10.20	10	0	8.86	10	10	8.28	10	20
7.93	10	30	7.73	10	40	7.56	10	50
10.17	20	0	8.24	20	10	7.78	20	20
7.44	20	30	7.30	20	40	7.21	20	50
10.19	30	0	7.91	30	10	7.45	30	20
7.30	30	30	7.14	30	40	7.08	30	50
10.18	40	0	7.72	40	10	7.32	40	20
7.19	40	30	7.05	40	40	6.99	40	50

a. Fit the model with Neighbors as the dependent variable, and independent variables Speed, Pause, Speed · Pause, Speed^2, and Pause^2.

b. Construct a reduced model by dropping any variables whose P-values are large, and test the plausibility of the model with an F test.

c. Plot the residuals versus the fitted values for the reduced model. Are there any indications that the model is inappropriate? If so, what are they?

d. Someone suggests that a model containing Pause and Pause^2 as the only dependent variables is adequate. Do you agree? Why or why not?

e. Using a best subsets software package, find the two models with the highest R^2 value for each model size from one to five variables. Compute C_p and adjusted R^2 for each model.

f. Which model is selected by minimum C_p? By adjusted R^2? Are they the same?

6. The data in Table SE6 on page 610 consist of yield measurements from many runs of a chemical reaction. The quantities varied were the temperature in °C (x_1), the concentration of the primary reactant in % (x_2), and the duration of the reaction in hours (x_3). The dependent variable (y) is the fraction converted to the desired product.

a. Fit the linear model $y = \beta_0 + \beta_1 x_1 + \beta_2 x_2 + \beta_3 x_3 + \varepsilon$.

b. Two of the variables in this model have coefficients significantly different from 0 at the 15% level. Fit a linear regression model containing these two variables.

c. Compute the product (interaction) of the two variables referred to in part (b). Fit the model that contains the two variables along with the interaction term.

d. Based on the results in parts (a) through (c), specify a model that appears to be good for predicting y from x_1, x_2, and x_3.

e. Might it be possible to construct an equally good or better model in another way?

TABLE SE6 Data for Exercise 6

x_1	x_2	x_3	y	x_1	x_2	x_3	y	x_1	x_2	x_3	y
50	19	4.0	27.464	70	27	10.0	38.241	70	31	6.0	35.091
90	38	8.0	49.303	80	32	6.5	34.635	60	23	7.0	34.372
70	28	6.5	37.461	50	26	9.0	44.963	50	19	6.0	26.481
70	25	5.5	36.478	50	22	4.0	30.012	60	22	7.5	36.739
60	26	6.5	33.776	80	34	6.5	41.077	70	30	9.5	36.185
70	29	5.0	35.092	50	21	10.0	41.964	70	25	8.0	38.725
60	23	5.5	31.307	80	34	7.5	44.152	50	17	9.5	32.707
70	28	5.5	37.863	60	22	2.5	29.901	70	28	7.0	32.563
80	34	6.5	41.109	60	24	5.0	26.706	60	25	5.5	36.006
70	26	4.5	28.605	60	23	4.0	28.602	70	25	5.5	33.127
70	26	8.0	35.917	60	29	6.5	33.401	70	29	5.0	32.941
70	26	8.0	33.489	70	27	7.5	41.324	70	29	6.5	33.650
60	30	5.0	31.381	70	32	4.0	24.000	50	19	4.5	34.192
60	26	7.0	38.067	60	25	5.5	38.158	60	24	6.5	24.115
70	25	7.5	31.278	60	26	3.5	25.412	60	26	7.5	37.614
70	31	5.5	32.172	70	28	7.5	37.671	60	28	6.0	29.612
60	27	7.5	36.109	60	22	5.5	27.979	60	22	6.5	39.106
60	23	6.0	31.535	60	22	4.5	31.079	60	28	7.5	36.974
60	23	6.0	33.875	60	27	7.0	30.778	60	25	4.0	28.334
60	24	9.0	37.637	60	25	6.0	28.221	50	20	8.5	33.767
70	31	5.5	40.263	60	23	6.5	30.495	60	26	9.5	38.358
80	32	6.0	36.694	60	27	7.5	38.710	60	25	4.0	33.381
60	26	10.0	45.620	80	31	4.5	27.581	60	29	4.0	37.672
70	28	4.5	38.571	80	36	4.5	38.705	70	30	6.0	36.615
60	24	4.0	19.163	60	22	7.5	40.525	60	26	8.0	39.351
50	21	7.0	31.962	70	28	4.5	29.420	60	24	6.5	38.611
50	17	2.0	23.147	60	26	7.0	37.898	60	25	6.0	36.460
80	34	8.5	40.278	60	25	7.0	40.340	60	24	5.5	23.449
70	27	5.5	32.725	60	24	5.0	27.891	60	24	5.0	23.027
60	24	2.5	28.735	70	32	7.5	38.259	70	26	8.0	31.372

7. In a study to predict temperature from air pressure in a piston–cylinder device, 19 measurements were made of temperature in °F (y) and air pressure in psi (x). Three models were fit: the linear model $y = \beta_0 + \beta_1 x + \varepsilon$, the quadratic model $y = \beta_0 + \beta_1 x + \beta_2 x^2 + \varepsilon$, and the cubic model $y = \beta_0 + \beta_1 x + \beta_2 x^2 + \beta_3 x^3 + \varepsilon$. The residuals and fitted values for each model are presented in the following table. Plot the residuals versus the fitted values for each model. For each model, state whether the model is appropriate, and explain.

Linear Model		Quadratic Model		Cubic Model	
Residual	**Fit**	**Residual**	**Fit**	**Residual**	**Fit**
−56.2	125.6	11.2	58.2	3.3	66.1
−34.0	153.1	−7.4	126.5	−6.7	125.7
8.4	179.8	4.9	183.4	9.2	179.0
21.4	207.2	−3.6	232.2	0.9	227.8
28.6	234.7	−8.2	271.5	−5.9	269.2
46.9	260.9	8.1	299.7	7.3	300.5
47.2	288.1	15.7	319.6	12.1	323.2
8.5	314.4	−7.0	329.8	−11.7	334.6
−7.1	342.0	4.0	330.9	1.0	333.9
−47.1	139.3	−1.3	93.6	−4.1	96.4
−1.6	166.2	9.1	155.5	12.1	152.5
38.0	220.9	5.9	253.0	9.5	249.4
35.7	247.8	−3.2	286.7	−2.4	286.0
34.1	275.1	−2.1	311.3	−4.5	313.7
34.6	301.1	9.9	325.8	5.5	330.2
1.0	328.2	−2.4	331.6	−6.7	335.9
−23.2	355.3	4.1	327.9	3.5	328.5
−50.7	368.4	−5.0	322.7	−2.1	319.9
−72.9	382.1	−5.7	314.9	2.1	307.1

8. The voltage output (y) of a battery was measured over a range of temperatures (x) from 0°C to 50°C. The following figure is a scatterplot of voltage versus temperature, with three fitted curves superimposed. The curves are the linear model $y = \beta_0 + \beta_1 x + \varepsilon$, the quadratic model $y = \beta_0 + \beta_1 x = \beta_2 x^2 + \varepsilon$, and the cubic model $y = \beta_0 + \beta_1 x + \beta_2 x^2 + \beta_3 x^3 + \varepsilon$. Based on the plot, which of the models should be used to describe the data? Explain.

i. The linear model.
ii. The quadratic model.
iii. The cubic model.
iv. All three appear to be about equally good.

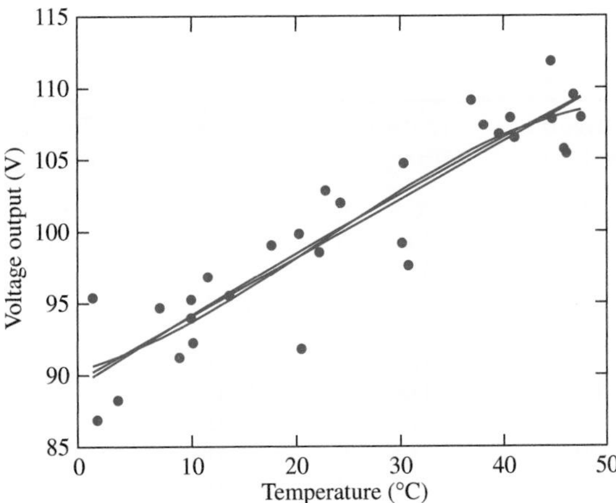

9. Refer to Exercise 2 in Section 8.2.

a. Using each of the three models in turn, predict the NO_x emission when Speed = 1500, Torque = 400, and HP = 150.

b. Using each of the three models in turn, predict the NO_x emission when Speed = 1600, Torque = 300, and HP = 100.

c. Using each of the three models in turn, predict the NO_x emission when Speed = 1400, Torque = 200, and HP = 75.

d. Which model or models appear to be the best? Choose one of the answers, and explain.

i. The model with Speed and Torque as independent variables is the best.

ii. The model with Speed and HP as independent variables is the best.

iii. The model with Speed, Torque, and HP as independent variables is the best.

iv. The model with Speed and Torque and the model with Speed and HP are about equally good; both are better than the model with Speed, Torque, and HP.

v. The model with Speed and Torque and the model with Speed, Torque, and HP are about equally good; both are better than the model with Speed and HP.

vi. The model with Speed and HP and the model with Speed, Torque, and HP are about equally good; both are better than the model with Speed and Torque.

vii. All three models are about equally good.

10. This exercise illustrates a reason for the exceptions to the rule of parsimony (see p. 584).

a. A scientist fits the model $Y = \beta_1 C + \varepsilon$, where C represents temperature in °C and Y can represent any outcome. Note that the model has no intercept. Now convert °C to °F ($C = 0.556F - 17.78$). Does the model have an intercept now?

b. Another scientist fits the model $Y = \beta_0 + \beta_2 C^2$, where C and Y are as in part (a). Note the model has a quadratic term, but no linear term. Now convert °C to °F ($C = 0.556F - 17.78$). Does the model have a linear term now?

c. Assume that x and z are two different units that can be used to measure the same quantity, and that $z = a + bx$, where $a \neq 0$. (°C and °F are an example.) Show that the no-intercept models $y = \beta x$ and $y = \beta z$ cannot both be correct, so that the validity of a no-intercept model depends on the zero point of the units for the independent variable.

d. Let x and z be as in part (c). Show that the models $y = \beta_0 + \beta_2 x^2$ and $y = \beta_0 + \beta_2 z^2$ cannot both be correct, and, thus, that the validity of such a model depends on the zero point of the units for the independent variable.

11. The data presented in the following table give the tensile strength in psi (y) of paper as a function of the percentage of hardwood content (x).

Hardwood Content	Tensile Strength	Hardwood Content	Tensile Strength
1.0	26.8	7.0	52.1
1.5	29.5	8.0	56.1
2.0	36.6	9.0	63.1
3.0	37.8	10.0	62.0
4.0	38.2	11.0	62.5
4.5	41.5	12.0	58.0
5.0	44.8	13.0	52.9
5.5	44.7	14.0	38.2
6.0	48.5	15.0	32.9
6.5	50.1	16.0	21.9

a. Fit polynomial models of degrees 1, 2, and so on to predict tensile strength from hardwood content. For each one, use the F test to compare it with the model of degree one less. Stop when the P-value of the F test is greater than 0.05. What is the degree of the polynomial model chosen by this method?
b. Using the model from part (a), estimate the hardwood concentration that produces the highest tensile strength.

12. The article "Enthalpies and Entropies of Transfer of Electrolytes and Ions from Water to Mixed Aqueous Organic Solvents" (G. Hefter, Y. Marcus, and W. Waghorne, *Chemical Reviews*, 2002:2773–2836) presents measurements of entropy and enthalpy changes for many salts under a variety of conditions. The following table presents the results for entropies of transfer (in J/K · mol) from water to water + methanol of NaCl (table salt) over a range of concentrations of methanol:

Concentration (%)	Entropy
5	1
10	−1
20	−7
30	−17
40	−28
50	−39
60	−52
70	−65
80	−80
90	−98
100	−121

a. Fit polynomial models of degrees 1, 2, and 3 to predict the entropy (y) from the concentration (x).
b. Which degree polynomial is the most appropriate? Explain.
c. Using the most appropriate model, find 99% confidence intervals for the coefficients.

13. A paint company collects data on the durability of its paint and that of its competitors. They measured the lifetimes of three samples of each type of paint in several American cities. The results are given in Table SE13.

TABLE SE13 Data for Exercise 13

	Avg. Temp (°F)		Mean Annual	Lifetime (years)	
City	January	July	Precipitation (in.)	Sponsor's Paint	Competitor's Paint
Atlanta, GA	41.9	78.6	48.6	11.5 10.7 12.3	10.8 11.1 10.2
Boston, MA	29.6	73.5	43.8	11.7 10.1 12.5	10.7 11.6 11.0
Kansas City, KS	28.4	80.9	29.3	12.3 13.4 12.8	11.8 12.2 11.3
Minneapolis, MN	11.2	73.1	26.4	10.5 9.9 11.2	10.4 9.6 9.2
Dallas, TX	45.0	86.3	34.2	11.2 10.6 12.0	10.6 10.1 11.4
Denver, CO	29.5	73.3	15.3	15.2 14.2 13.8	13.4 14.4 13.2
Miami, FL	67.1	82.4	57.5	8.7 7.9 9.4	8.1 8.6 7.6
Phoenix, AZ	52.3	92.3	7.1	11.1 11.8 12.4	10.9 10.1 9.9
San Francisco, CA	48.5	62.2	19.7	16.7 17.2 15.9	15.8 15.4 14.9
Seattle, WA	40.6	65.3	38.9	14.2 14.1 13.6	12.6 13.6 14.1
Washington, DC	35.2	78.9	39.0	12.6 11.5 12.0	11.9 10.9 11.4

a. Prior testing suggests that the most important factors that influence the lifetimes of paint coatings are the minimum temperature (estimated by the average January temperature), the maximum temperature (estimated by the average July temperature), and the annual precipitation. Using these variables, and products and powers of these variables, construct a good model for predicting the lifetime of the sponsor's paint and a good model (perhaps different) for predicting the lifetime of the competitor's paint.
b. Using the models developed in part (a), compute the expected lifetimes for these two paints for someone living in Cheyenne, Wyoming, where the January mean temperature is 26.1°F, the July mean temperature is 68.9°F, and the mean annual precipitation is 13.3 in.

14. The article "Two Different Approaches for RDC Modelling When Simulating a Solvent Deasphalting Plant" (J. Aparicio, M. Heronimo, et al., *Computers and Chemical Engineering*, 2002:1369–1377) reports flow rate (in dm^3/h) and specific gravity measurements for a sample of paraffinic hydrocarbons. The natural logs of the flow rates (y) and the specific gravity measurements (x) are presented in the following table.

y	x
−1.204	0.8139
−0.580	0.8171
0.049	0.8202
0.673	0.8233
1.311	0.8264
1.959	0.8294
2.614	0.8323
3.270	0.8352

a. Fit the linear model $y = \beta_0 + \beta_1 x + \varepsilon$. For each coefficient, test the hypothesis that the coefficient is equal to 0.
b. Fit the quadratic model $y = \beta_0 + \beta_1 x + \beta_2 x^2 + \varepsilon$. For each coefficient, test the hypothesis that the coefficient is equal to 0.
c. Fit the cubic model $y = \beta_0 + \beta_1 x + \beta_2 x^2 + \beta_3 x^3 + \varepsilon$. For each coefficient, test the hypothesis that the coefficient is equal to 0.
d. Which of the models in parts (a) through (c) is the most appropriate? Explain.
e. Using the most appropriate model, estimate the flow rate when the specific gravity is 0.83.

15. The article "Measurements of the Thermal Conductivity and Thermal Diffusivity of Polymer Melts with the Short-Hot-Wire Method" (X. Zhang, W. Hendro, et al., *International Journal of Thermophysics*, 2002:1077–1090) reports measurements of the thermal conductivity (in $W \cdot m^{-1} \cdot K^{-1}$) and diffusivity of several polymers at several temperatures (in 1000°C). The following table presents results for the thermal conductivity of polycarbonate.

Cond.	Temp.	Cond.	Temp.	Cond.	Temp.	Cond.	Temp.
0.236	0.028	0.259	0.107	0.254	0.159	0.249	0.215
0.241	0.038	0.257	0.119	0.256	0.169	0.230	0.225
0.244	0.061	0.257	0.130	0.251	0.181	0.230	0.237
0.251	0.083	0.261	0.146	0.249	0.204	0.228	0.248

a. Denoting conductivity by y and temperature by x, fit the linear model $y = \beta_0 + \beta_1 x + \varepsilon$. For each coefficient, test the hypothesis that the coefficient is equal to 0.
b. Fit the quadratic model $y = \beta_0 + \beta_1 x + \beta_2 x^2 + \varepsilon$. For each coefficient, test the hypothesis that the coefficient is equal to 0.

c. Fit the cubic model $y = \beta_0 + \beta_1 x + \beta_2 x^2 + \beta_3 x^3 + \varepsilon$. For each coefficient, test the hypothesis that the coefficient is equal to 0.
d. Fit the quartic model $y = \beta_0 + \beta_1 x + \beta_2 x^2 + \beta_3 x^3 + \beta_4 x^4 + \varepsilon$. For each coefficient, test the hypothesis that the coefficient is equal to 0.
e. Which of the models in parts (a) through (d) is the most appropriate? Explain.
f. Using the most appropriate model, estimate the conductivity at a temperature of 120°C.

16. The article "Electrical Impedance Variation with Water Saturation in Rock" (Q. Su, Q. Feng, and Z. Shang, *Geophysics*, 2000:68–75) reports measurements of permeabilities (in $10^{-3}\mu m^2$), porosities (in %), and surface area per unit volume of pore space (in 10^4 cm^{-1}) for several rock samples. The results are presented in the following table, denoting ln Permeability by y, porosity by x_1, and surface area per unit volume by x_2.

y	x_1	x_2	y	x_1	x_2
−0.27	19.83	9.55	0.58	10.52	20.03
2.58	17.93	10.97	−0.56	18.92	13.10
3.18	21.27	31.02	−0.49	18.55	12.78
1.70	18.67	28.12	−0.01	13.72	40.28
−1.17	7.98	52.35	−1.71	9.12	53.67
−0.27	10.16	32.82	−0.12	14.39	26.75
−0.53	17.86	57.66	−0.92	11.38	75.62
−0.29	13.48	21.10	2.18	16.59	9.95
4.94	17.49	9.15	4.46	16.77	7.88
1.94	14.18	11.72	2.11	18.55	88.10
3.74	23.88	5.43	−0.04	18.02	10.95

a. Fit the model $y = \beta_0 + \beta_1 x_1 + \beta_2 x_2 + \beta_3 x_1 x_2 + \varepsilon$. Compute the analysis of variance table.
b. Fit the model $y = \beta_0 + \beta_1 x_1 + \beta_2 x_2 + \varepsilon$. Compute the analysis of variance table.
c. Fit the model $y = \beta_0 + \beta_1 x_1 + \varepsilon$. Compute the analysis of variance table.
d. Compute the F statistics for comparing the models in parts (b) and (c) with the model in part (a). Which model do you prefer? Why?

17. The article "Groundwater Electromagnetic Imaging in Complex Geological and Topographical Regions: A Case Study of a Tectonic Boundary in the French Alps" (S. Houtot, P. Tarits, et al., *Geophysics*, 2002:1048–1060) presents measurements of concentrations of several chemicals (in mmol/L) and electrical conductivity (in 10^{-2} S/m) for several water samples in various locations near Gittaz Lake in the French Alps. The results for magnesium and calcium are presented in the following table. Two outliers have been deleted.

Conductivity	Magnesium	Calcium	Conductivity	Magnesium	Calcium
2.77	0.037	1.342	1.10	0.027	0.487
3.03	0.041	1.500	1.11	0.039	0.497
3.09	0.215	1.332	2.57	0.168	1.093
3.29	0.166	1.609	3.27	0.172	1.480
3.37	0.100	1.627	2.28	0.044	1.093
0.88	0.031	0.382	3.32	0.069	1.754
0.77	0.012	0.364	3.93	0.188	1.974
0.97	0.017	0.467	4.26	0.211	2.103

a. To predict conductivity (y) from the concentrations of magnesium (x_1) and calcium (x_2), fit the full quadratic model $y = \beta_0 + \beta_1 x_1 + \beta_2 x_2 + \beta_3 x_1^2 + \beta_4 x_2^2 + \beta_5 x_1 x_2 + \varepsilon$. Compute the analysis of variance table.
b. Use the F test to investigate some submodels of the full quadratic model. State which model you prefer and why.
c. Use a best subsets routine to find the submodels with the maximum adjusted R^2 and the minimum Mallows C_p. Are they the same model? Comment on the appropriateness of this (these) model(s).

18. The article "Low-Temperature Heat Capacity and Thermodynamic Properties of 1,1,1-trifluoro-2, 2-dichloroethane" (R. Varushchenko and A. Druzhinina, *Fluid Phase Equilibria*, 2002:109–119) presents measurements of the molar heat capacity (y) of 1,1,1-trifluoro-2,2-dichloroethane (in $\text{J} \cdot \text{K}^{-1} \cdot \text{mol}^{-1}$) at several temperatures (x) in units of 10 K. The results for every tenth measurement are presented in the following table.

y	x	y	x
5.7037	1.044	60.732	6.765
16.707	1.687	65.042	7.798
29.717	2.531	71.283	9.241
41.005	3.604	75.822	10.214
48.822	4.669	80.029	11.266
55.334	5.722		

a. Fit the simple linear model $y = \beta_0 + \beta_1 x + \varepsilon$. Make a residual plot, and comment on the appropriateness of the model.
b. Fit the simple linear model $y = \beta_0 + \beta_1 \ln x + \varepsilon$. Make a residual plot, and comment on the appropriateness of the model.
c. Compute the coefficients and their standard deviations for polynomials of degrees 2, 3, 4, and 5. Make residual plots for each.
d. The article cited at the beginning of this exercise recommends the quartic model $y = \beta_0 + \beta_1 x + \beta_2 x^2 + \beta_3 x^3 + \beta_4 x^4 + \varepsilon$. Does this seem reasonable? Why or why not?

19. The article "Lead Dissolution from Lead Smelter Slags Using Magnesium Chloride Solutions" (A. Xenidis, T. Lillis, and I. Hallikia) discusses an investigation of leaching rates of lead in solutions of magnesium chloride. The data in the following table (read from a graph) present the percentage of lead that has been extracted at various times (in minutes).

Time (t)	4	8	16	30	60	120
Percent extracted (y)	1.2	1.6	2.3	2.8	3.6	4.4

a. The article suggests fitting a quadratic model $y = \beta_0 + \beta_1 t + \beta_2 t^2 + \varepsilon$ to these data. Fit this model, and compute the standard deviations of the coefficients.
b. The reaction rate at time t is given by the derivative $dy/dt = \beta_1 + 2\beta_2 t$. Estimate the time at which the reaction rate will be equal to 0.05.
c. The reaction rate at $t = 0$ is equal to β_1. Find a 95% confidence interval for the reaction rate at $t = 0$.
d. Can you conclude that the reaction rate is decreasing with time? Explain.

20. The article "The Ball-on-Three-Ball Test for Tensile Strength: Refined Methodology and Results for Three Hohokam Ceramic Types" (M. Beck, *American Antiquity*, 2002:558–569) discusses the strength of ancient ceramics. The following table presents measured weights (in g), thicknesses (in mm), and loads (in kg) required to crack the specimen for a collection of specimens dated between A.D. 1100 and 1300 from the Middle Gila River, in Arizona.

Weight (x_1)	Thickness (x_2)	Load (y)
12.7	5.69	20
12.9	5.05	16
17.8	6.53	20
18.5	6.51	36
13.4	5.92	27
15.2	5.88	35
13.2	4.09	15
18.3	6.14	18
16.2	5.73	24
14.7	5.47	21
18.2	7.32	30
14.8	4.91	20
17.7	6.72	24
16.0	5.85	23
17.2	6.18	21
14.1	5.13	13
16.1	5.71	21

a. Fit the model $y = \beta_0 + \beta_1 x_1 + \beta_2 x_2 + \varepsilon$.
b. Drop the variable whose coefficient has the larger P-value, and refit.
c. Plot the residuals versus the fitted values from the model in part (b). Are there any indications that the model is not appropriate?

21. *Piecewise linear model:* Let $\tilde{x}$ be a known constant, and suppose that a dependent variable y is related to an independent variable x_1 as follows:

$$y = \begin{cases} \beta_0 + \beta_1 x_1 + \varepsilon & \text{if } x_1 \le \tilde{x} \\ \beta_0^* + \beta_1^* x_1 + \varepsilon & \text{if } x_1 > \tilde{x} \end{cases}$$

In other words, y and x_1 are linearly related, but different lines are appropriate depending on whether $x_1 \le \tilde{x}$ or $x_1 > \tilde{x}$. Define a new independent variable x_2 by

$$x_2 = \begin{cases} 0 & \text{if } x_1 \le \tilde{x} \\ 1 & \text{if } x_1 > \tilde{x} \end{cases}$$

Also define $\beta_2 = \beta_0^* - \beta_0$ and $\beta_3 = \beta_1^* - \beta_1$. Find a multiple regression model involving y, x_1, x_2, β_0, β_1, β_2, and β_3 that expresses the relationship described here.

22. The article "Seismic Hazard in Greece Based on Different Strong Ground Motion Parameters" (S. Koutrakis, G. Karakaisis, et al., *Journal of Earthquake Engineering*, 2002:75–109) presents a study of seismic events in Greece during the period 1978–1997. Of interest is the duration of "strong ground motion," which is the length of time that the acceleration of the ground exceeds a specified value. For each event, measurements of the duration of strong ground motion were made at one or more locations. Table SE22 on page 618 presents, for each of 121 such measurements, the data for the duration of time y (in seconds) that the ground acceleration exceeded twice the acceleration due to gravity, the magnitude m of the earthquake, the distance d (in km) of the measurement from the epicenter, and two indicators of the soil type s_1 and s_2, defined as follows: $s_1 = 1$ if the soil consists of soft alluvial deposits, $s_1 = 0$ otherwise, and $s_2 = 1$ if the soil consists of tertiary or older rock, $s_2 = 0$ otherwise. Cases where both $s_1 = 0$ and $s_2 = 0$ correspond to intermediate soil conditions. The article presents repeated measurements at some locations, which we have not included here.

TABLE SE22 Data for Exercise 22

y	m	d	s_1	s_2	y	m	d	s_1	s_2	y	m	d	s_1	s_2
8.82	6.4	30	1	0	4.31	5.3	6	0	0	5.74	5.6	15	0	0
4.08	5.2	7	0	0	28.27	6.6	31	1	0	5.13	6.9	128	1	0
15.90	6.9	105	1	0	17.94	6.9	33	0	0	3.20	5.1	13	0	0
6.04	5.8	15	0	0	3.60	5.4	6	0	0	7.29	5.2	19	1	0
0.15	4.9	16	1	0	7.98	5.3	12	1	0	0.02	6.2	68	1	0
5.06	6.2	75	1	0	16.23	6.2	13	0	0	7.03	5.4	10	0	0
0.01	6.6	119	0	1	3.67	6.6	85	1	0	2.17	5.1	45	0	1
4.13	5.1	10	1	0	6.44	5.2	21	0	0	4.27	5.2	18	1	0
0.02	5.3	22	0	1	10.45	5.3	11	0	1	2.25	4.8	14	0	1
2.14	4.5	12	0	1	8.32	5.5	22	1	0	3.10	5.5	15	0	0
4.41	5.2	17	0	0	5.43	5.2	49	0	1	6.18	5.2	13	0	0
17.19	5.9	9	0	0	4.78	5.5	1	0	0	4.56	5.5	1	0	0
5.14	5.5	10	1	0	2.82	5.5	20	0	1	0.94	5.0	6	0	1
0.05	4.9	14	1	0	3.51	5.7	22	0	0	2.85	4.6	21	1	0
20.00	5.8	16	1	0	13.92	5.8	34	1	0	4.21	4.7	20	1	0
12.04	6.1	31	0	0	3.96	6.1	44	0	0	1.93	5.7	39	1	0
0.87	5.0	65	1	0	6.91	5.4	16	0	0	1.56	5.0	44	1	0
0.62	4.8	11	1	0	5.63	5.3	6	1	0	5.03	5.1	2	1	0
8.10	5.4	12	1	0	0.10	5.2	21	1	0	0.51	4.9	14	1	0
1.30	5.8	34	1	0	5.10	4.8	16	1	0	13.14	5.6	5	1	0
11.92	5.6	5	0	0	16.52	5.5	15	1	0	8.16	5.5	12	1	0
3.93	5.7	65	1	0	19.84	5.7	50	1	0	10.04	5.1	28	1	0
2.00	5.4	27	0	1	1.65	5.4	27	1	0	0.79	5.4	35	0	0
0.43	5.4	31	0	1	1.75	5.4	30	0	1	0.02	5.4	32	1	0
14.22	6.5	20	0	1	6.37	6.5	90	1	0	0.10	6.5	61	0	1
0.06	6.5	72	0	1	2.78	4.9	8	0	0	5.43	5.2	9	0	0
1.48	5.2	27	0	0	2.14	5.2	22	0	0	0.81	4.6	9	0	0
3.27	5.1	12	0	0	0.92	5.2	29	0	0	0.73	5.2	22	0	0
6.36	5.2	14	0	0	3.18	4.8	15	0	0	11.18	5.0	8	0	0
0.18	5.0	19	0	0	1.20	5.0	19	0	0	2.54	4.5	6	0	0
0.31	4.5	12	0	0	4.37	4.7	5	0	0	1.55	4.7	13	0	1
1.90	4.7	12	0	0	1.02	5.0	14	0	0	0.01	4.5	17	0	0
0.29	4.7	5	1	0	0.71	4.8	4	1	0	0.21	4.8	5	0	1
6.26	6.3	9	1	0	4.27	6.3	9	0	1	0.04	4.5	3	1	0
3.44	5.4	4	1	0	3.25	5.4	4	0	1	0.01	4.5	1	1	0
2.32	5.4	5	1	0	0.90	4.7	4	1	0	1.19	4.7	3	1	0
1.49	5.0	4	1	0	0.37	5.0	4	0	1	2.66	5.4	1	1	0
2.85	5.4	1	0	1	21.07	6.4	78	0	1	7.47	6.4	104	0	0
0.01	6.4	86	0	1	0.04	6.4	105	0	1	30.45	6.6	51	1	0
9.34	6.6	116	0	1	15.30	6.6	82	0	1	12.78	6.6	65	1	0
10.47	6.6	117	0	0										

Use the data in Table SE22 to construct a linear model to predict duration y from some or all of the variables m, d, s_1, and s_2. Be sure to consider transformations of the variables, as well as powers of and interactions between the independent variables. Describe the steps taken to construct your model. Plot the residuals versus the fitted values to verify that your model satisfies the necessary assumptions. In addition, note that the data is presented in chronological order, reading down the columns. Make a plot to determine whether time should be included as an independent variable.

23. The article "Estimating Resource Requirements at Conceptual Design Stage Using Neural Networks" (A. Elazouni, I. Nosair, et al., *Journal of Computing in Civil Engineering*, 1997:217–223) suggests that certain resource requirements in the construction of concrete silos can be predicted from a model. These include the quantity of concrete in m^3 (y), the number of crew-days of labor (z), or the number of concrete mixer hours (w) needed for a particular job. Tables SE23A defines 23 potential independent variables that can be used to predict y, z, or w. Values of the dependent and independent variables, collected on 28 construction jobs, are presented in Tables SE23B on page 620 and SE23C on page 621. Unless otherwise stated, lengths are in meters, areas in m^2, and volumes in m^3.

a. Using best subsets regression, find the model that is best for predicting y according to the adjusted R^2 criterion.
b. Using best subsets regression, find the model that is best for predicting y according to the minimum Mallows C_p criterion.
c. Find a model for predicting y using stepwise regression. Explain the criterion you are using to determine which variables to add to or drop from the model.
d. Using best subsets regression, find the model that is best for predicting z according to the adjusted R^2 criterion.
e. Using best subsets regression, find the model that is best for predicting z according to the minimum Mallows C_p criterion.
f. Find a model for predicting z using stepwise regression. Explain the criterion you are using to determine which variables to add to or drop from the model.
g. Using best subsets regression, find the model that is best for predicting w according to the adjusted R^2 criterion.
h. Using best subsets regression, find the model that is best for predicting w according to the minimum Mallows C_p criterion.
i. Find a model for predicting w using stepwise regression. Explain the criterion you are using to determine which variables to add to or drop from the model.

TABLE SE23A Descriptions of Variables for Exercise 23

x_1	Number of bins	x_{13}	Breadth-to-thickness ratio
x_2	Maximum required concrete per hour	x_{14}	Perimeter of complex
x_3	Height	x_{15}	Mixer capacity
x_4	Sliding rate of the slipform (m/day)	x_{16}	Density of stored material
x_5	Number of construction stages	x_{17}	Waste percent in reinforcing steel
x_6	Perimeter of slipform	x_{18}	Waste percent in concrete
x_7	Volume of silo complex	x_{19}	Number of workers in concrete crew
x_8	Surface area of silo walls	x_{20}	Wall thickness (cm)
x_9	Volume of one bin	x_{21}	Number of reinforcing steel crews
x_{10}	Wall-to-floor areas	x_{22}	Number of workers in forms crew
x_{11}	Number of lifting jacks	x_{23}	Length-to-breadth ratio
x_{12}	Length-to-thickness ratio		

TABLE SE23B Data for Exercise 23

y	z	w	x_1	x_2	x_3	x_4	x_5	x_6	x_7	x_8	x_9	x_{10}	x_{11}
1,850	9,520	476	33	4.5	19.8	4.0	4	223	11,072	14,751	335	26.1	72
932	4,272	268	24	3.5	22.3	4.0	2	206	2,615	8,875	109	27.9	64
556	3,296	206	18	2.7	20.3	5.0	2	130	2,500	5,321	139	28.4	48
217	1,088	68	9	3.2	11.0	4.5	1	152	1,270	1,675	141	11.6	40
199	2,587	199	2	1.0	23.8	5.0	1	79	1,370	7,260	685	17.1	21
56	1,560	120	2	0.5	16.6	5.0	1	43	275	1,980	137	22.0	15
64	1,534	118	2	0.5	18.4	5.0	1	43	330	825	165	23.6	12
397	2,660	133	14	3.0	16.0	4.0	1	240	5,200	18,525	371	12.8	74
1,926	11,020	551	42	3.5	16.0	4.0	4	280	15,500	3,821	369	12.8	88
724	3,090	103	15	7.8	15.0	3.5	1	374	4,500	5,600	300	12.2	114
711	2,860	143	25	5.0	16.0	3.5	1	315	2,100	6,851	87	24.8	60
1,818	9,900	396	28	4.8	22.0	4.0	3	230	13,500	13,860	482	17.6	44
619	2,626	202	12	3.0	18.0	5.0	1	163	1,400	2,935	115	26.4	36
375	2,060	103	12	5.8	15.0	3.5	1	316	4,200	4,743	350	11.8	93
214	1,600	80	12	3.5	15.0	4.5	1	193	1,300	2,988	105	20.6	40
300	1,820	140	6	2.1	14.0	5.0	1	118	800	1,657	133	17.0	24
771	3,328	256	30	3.0	14.0	5.0	3	165	2,800	2,318	92	19.9	43
189	1,456	91	12	4.0	17.0	4.5	1	214	2,400	3,644	200	13.6	53
494	4,160	320	27	3.3	20.0	4.5	3	178	6,750	3,568	250	14.0	44
389	1,520	95	6	4.1	19.0	4.0	1	158	2,506	3,011	401	11.8	38
441	1,760	110	6	4.0	22.0	5.0	1	154	2,568	3,396	428	14.1	35
768	3,040	152	12	5.0	24.0	4.0	1	275	5,376	6,619	448	14.5	65
797	3,180	159	9	5.0	25.0	4.0	1	216	4,514	5,400	501	14.8	52
261	1,131	87	3	3.0	17.5	4.0	1	116	1,568	2,030	522	10.5	24
524	1,904	119	6	4.4	18.8	4.0	1	190	3,291	3,572	548	9.8	42
1,262	5,070	169	15	7.0	24.6	3.5	1	385	8,970	9,490	598	12.9	92
839	7,080	354	9	5.2	25.5	4.0	1	249	5,845	6,364	649	13.9	60
1,003	3,500	175	9	5.7	27.7	4.0	1	246	6,095	6,248	677	15.1	60

TABLE SE23C Data for Exercise 23

x_{12}	x_{13}	x_{14}	x_{15}	x_{16}	x_{17}	x_{18}	x_{19}	x_{20}	x_{21}	x_{22}	x_{23}
19.6	17.6	745	0.50	800	6.00	5.50	10	24	7	20	1.12
16.0	16.0	398	0.25	600	7.00	5.00	10	20	6	20	1.00
15.3	13.5	262	0.25	850	7.00	4.50	8	20	5	18	1.13
17.0	13.8	152	0.25	800	5.00	4.00	8	25	6	16	1.23
28.1	27.5	79	0.15	800	7.50	3.50	5	20	4	14	1.02
20.3	20.0	43	0.15	600	5.00	4.00	5	15	1	12	1.02
24.0	18.3	43	0.15	600	5.05	4.25	5	15	2	12	1.31
27.5	23.0	240	0.25	600	6.00	4.00	8	20	7	22	1.20
27.5	23.0	1121	0.25	800	8.00	4.00	10	20	9	24	1.20
21.2	18.4	374	0.75	800	5.00	3.50	10	25	12	24	1.15
10.6	10.0	315	0.50	800	6.00	4.00	10	25	11	20	1.06
20.0	20.0	630	0.50	800	7.00	5.00	10	25	9	18	1.00
13.7	13.9	163	0.25	600	6.00	4.50	8	18	11	18	1.20
20.4	20.4	316	0.50	800	6.50	3.50	10	25	6	14	1.00
13.6	10.2	193	0.50	800	5.00	3.50	10	25	4	14	1.33
13.6	12.8	118	0.25	800	5.00	3.75	8	25	6	14	1.06
13.6	9.6	424	0.25	800	5.00	3.75	8	25	6	14	1.42
18.5	16.0	214	0.50	600	6.00	4.00	8	20	4	14	1.15
19.5	16.0	472	0.25	600	6.50	4.50	10	20	3	14	1.20
21.0	12.8	158	0.50	800	5.50	3.50	6	25	8	14	1.30
20.8	16.0	154	0.50	800	7.00	4.00	8	36	8	14	1.35
23.4	17.3	275	0.50	600	7.50	5.50	8	22	11	16	1.40
16.8	15.4	216	0.50	800	8.00	5.50	8	28	12	16	1.10
26.8	17.8	116	0.25	850	6.50	3.00	6	25	5	14	1.50
23.6	16.1	190	0.50	850	6.50	4.50	5	28	9	16	1.45
23.6	16.6	385	0.75	800	8.00	6.50	15	25	16	20	1.43
25.6	16.0	249	0.50	600	8.00	5.50	12	25	13	16	1.60
22.3	14.3	246	0.50	800	8.50	6.00	8	28	16	16	1.55

24. The article referred to in Exercise 23 presents values for the dependent and independent variables for 10 additional construction jobs. These values are presented in Tables SE24A and SE24B on page 622.

a. Using the equation constructed in part (a) of Exercise 23, predict the concrete quantity (y) for each of these 10 jobs.

b. Denoting the predicted values by $\widehat{y}_1, \ldots, \widehat{y}_{10}$ and the observed values by $y_1, \ldots, y_{10}$, compute the quantities $y_i - \widehat{y}_i$. These are the *prediction errors*.

c. Now compute the fitted values $\widehat{y}_1, \ldots, \widehat{y}_{28}$ from the data in Exercise 23. Using the observed values $y_1, \ldots, y_{28}$ from those data, compute the residuals $y_i - \widehat{y}_i$.

d. On the whole, which are larger, the residuals or the prediction errors? Why will this be true in general?

TABLE SE24A Data for Exercise 24

y	z	w	x_1	x_2	x_3	x_4	x_5	x_6	x_7	x_8	x_9	x_{10}	x_{11}
1,713	3,400	170	6	4.2	27.0	4.0	1	179	4,200	4,980	700.0	15.1	42
344	1,616	101	3	3.4	20.0	5.0	1	133	2,255	2,672	751.5	16.7	30
474	2,240	140	3	3.4	28.0	5.0	1	116	2,396	3,259	798.8	17.0	24
1,336	5,700	190	15	7.0	26.0	3.5	1	344	12,284	9,864	818.9	16.0	86
1,916	9,125	365	18	5.6	26.5	3.5	2	307	15,435	8,140	852.5	12.4	68
1,280	11,980	599	9	2.1	28.3	4.0	1	283	8,064	8,156	896.0	14.0	68
1,683	6,390	213	12	7.9	29.0	3.5	1	361	11,364	10,486	947.0	13.4	87
901	2,656	166	6	5.4	29.5	4.5	1	193	5,592	5,696	932.0	14.8	39
460	2,943	150	3	3.0	30.0	5.0	1	118	2,943	3,540	981.0	17.2	26
826	3,340	167	6	4.9	29.8	4.5	1	211	6,000	6,293	1,000.0	15.1	50

TABLE SE24B Data for Exercise 24

x_{12}	x_{13}	x_{14}	x_{15}	x_{16}	x_{17}	x_{18}	x_{19}	x_{20}	x_{21}	x_{22}	x_{23}
22.5	14.8	179	0.50	850	8.0	5.0	6	28	11	16	1.52
32.0	18.8	133	0.25	800	7.5	3.0	10	25	7	14	1.70
24.6	15.0	116	0.25	800	9.0	4.0	10	28	9	14	1.65
20.2	21.1	344	0.75	850	8.5	6.5	12	28	19	18	1.72
30.0	13.2	540	0.50	600	6.5	7.0	15	25	12	18	1.75
25.3	14.3	283	0.25	800	7.5	6.5	14	30	20	16	1.80
22.7	14.0	361	0.75	800	9.0	7.0	10	30	25	18	1.42
20.5	16.0	193	0.50	850	9.5	5.5	10	30	15	16	1.20
26.0	20.1	118	0.25	600	10.0	4.0	10	25	8	14	1.30
32.0	20.0	211	0.50	600	9.5	5.0	10	25	13	16	1.90

Chapter 9

Factorial Experiments

Introduction

Experiments are essential to the development and improvement of engineering and scientific methods. Only through experimentation can different variants of a method be compared to see which are most effective. To be useful, an experiment must be designed properly, and the data it produces must be analyzed correctly. In this chapter we will discuss the design of and the analysis of data from a class of experiments known as **factorial experiments**.

9.1 One-Factor Experiments

We begin with an example. The article "An Investigation of the $CaCO_3$-CaF_2-K_2SiO_3-SiO_2-Fe Flux System Using the Submerged Arc Welding Process on HSLA-100 and AISI-1081 Steels" (G. Fredrickson, M.S. Thesis, Colorado School of Mines, 1992) describes an experiment in which welding fluxes with differing chemical compositions were prepared. Several welds using each flux were made on AISI-1018 steel base metal. The results of hardness measurements, on the Brinell scale, of five welds using each of four fluxes are presented in Table 9.1.

TABLE 9.1 Brinell hardness of welds using four different fluxes

Flux	Sample Values					Sample Mean	Sample Standard Deviation
A	250	264	256	260	239	253.8	9.7570
B	263	254	267	265	267	263.2	5.4037
C	257	279	269	273	277	271.0	8.7178
D	253	258	262	264	273	262.0	7.4498

Figure 9.1 presents dotplots for the hardnesses using the four fluxes. Each sample mean is marked with an "X." It is clear that the sample means differ. In particular, the welds made using flux C have the largest sample mean and those using flux A have the smallest. Of course, there is uncertainty in the sample means, and the question is whether the sample means differ from each other by a greater amount than could be accounted for by uncertainty alone. Another way to phrase the question is this: Can we conclude that there are differences in the population means among the four flux types?

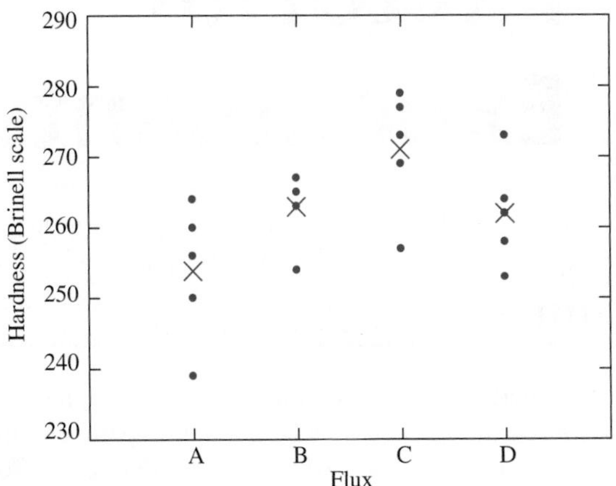

FIGURE 9.1 Dotplots for each sample in Table 9.1. Each sample mean is marked with an "X." The sample means differ somewhat, but the sample values overlap considerably.

This is an example of a factorial experiment. In general a factorial experiment involves several variables. One variable is the **response variable**, which is sometimes called the **outcome variable** or the **dependent variable**. The other variables are called **factors**. The question addressed by a factorial experiment is whether varying the levels of the factors produces a difference in the mean of the response variable. In the experiment described in Table 9.1, the hardness is the response, and there is one factor: flux type. Since there is only one factor, this is a **one-factor experiment**. There are four different values for the flux-type factor in this experiment. These different values are called the **levels** of the factor and can also be called **treatments**. Finally, the objects upon which measurements are made are called **experimental units**. The units assigned to a given treatment are called **replicates**. In the preceding experiment, the welds are the experimental units, and there are five replicates for each treatment.

In this welding experiment, the four particular flux compositions were chosen deliberately by the experimenter, rather than at random from a larger population of fluxes. Such an experiment is said to follow a **fixed effects model**. In some experiments, treatments are chosen at random from a population of possible treatments. In this case the experiment is said to follow a **random effects model**. The methods of analysis for these two models are essentially the same, although the conclusions to be drawn from them differ. We will focus on fixed effects models. Later in this section, we will discuss some of the differences between fixed and random effects models.

Completely Randomized Experiments

In this welding experiment, a total of 20 welds were produced, five with each of the four fluxes. Each weld was produced on a different steel base plate. Therefore, to run the experiment, the experimenter had to choose, from a total of 20 base plates, a group of 5 to be welded with flux A, another group of 5 to be welded with flux B, and so on. The best way to assign the base plates to the fluxes is at random. In this way, the experimental design will not favor any one treatment over another. For example, the experimenter could number the plates from 1 to 20, and then generate a random ordering of the integers from 1 to 20. The plates whose numbers correspond to the first five numbers on the list are assigned to flux A, and so on. This is an example of a **completely randomized experiment**.

Definition

A factorial experiment in which experimental units are assigned to treatments at random, with all possible assignments being equally likely, is called a **completely randomized experiment**.

In many situations, the results of an experiment can be affected by the order in which the observations are taken. For example, the performance of a machine used to make measurements may change over time, due, for example, to calibration drift, or to warm-up effects. In cases like this, the ideal procedure is to take the observations in random order. This requires switching from treatment to treatment as observations are taken, rather than running all the observations that correspond to a given treatment consecutively. In some cases changing treatments involves considerable time or expense, so it is not feasible to switch back and forth. In these cases, the treatments should be run in a random order, with all the observations corresponding to the first randomly chosen treatment being run first, and so on.

In a completely randomized experiment, it is appropriate to think of each treatment as representing a population, and the responses observed for the units assigned to that treatment as a simple random sample from that population. The data from the experiment thus consist of several random samples, each from a different population. The population means are called **treatment means**. The questions of interest concern the treatment means—whether they are all equal, and if not, which ones are different, how big the differences are, and so on.

One-Way Analysis of Variance

To make a formal determination as to whether the treatment means differ, a hypothesis test is needed. We begin by introducing the notation. We have I samples, each from a different treatment. The treatment means are denoted

$$\mu_1, \ldots, \mu_I$$

It is not necessary that the sample sizes be equal, although it is desirable, as we will discuss later in this section. The sample sizes are denoted

$$J_1, \ldots, J_I$$

The total number in all the samples combined is denoted by N.

$$N = J_1 + J_2 + \cdots + J_I$$

The hypotheses we wish to test are

$$H_0\colon \mu_1 = \cdots = \mu_I \quad \text{versus} \quad H_1\colon \text{two or more of the } \mu_i \text{ are different}$$

If there were only two samples, we might use the two-sample t test (Section 6.7) to test the null hypothesis. Since there are more than two samples, we use a method known as **one-way analysis of variance** (ANOVA). To define the test statistic for one-way ANOVA, we first develop the notation for the sample observations. Since there are several samples, we use a double subscript to denote the observations. Specifically, we let X_{ij} denote the jth observation in the ith sample. The sample mean of the ith sample is denoted $\overline{X}_{i.}$.

$$\overline{X}_{i.} = \frac{\sum_{j=1}^{J_i} X_{ij}}{J_i} \tag{9.1}$$

The **sample grand mean**, denoted $\overline{X}_{..}$, is the average of all the sampled items taken together:

$$\overline{X}_{..} = \frac{\sum_{i=1}^{I} \sum_{j=1}^{J_i} X_{ij}}{N} \tag{9.2}$$

With a little algebra, it can be shown that the sample grand mean is also a weighted average of the sample means:

$$\overline{X}_{..} = \frac{\sum_{i=1}^{I} J_i \overline{X}_{i.}}{N} \tag{9.3}$$

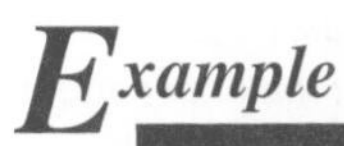

9.1

For the data in Table 9.1, find I, $J_1, \ldots, J_I$, N, X_{23}, $\overline{X}_{3.}$, $\overline{X}_{..}$.

Solution

There are four samples, so $I = 4$. Each sample contains five observations, so $J_1 = J_2 = J_3 = J_4 = 5$. The total number of observations is $N = 20$. The quantity X_{23} is the third observation in the second sample, which is 267. The quantity $\overline{X}_{3.}$ is the sample mean of the third sample. This value is $\overline{X}_{3.} = 271.0$. Finally, we use Equation (9.3) to compute the sample grand mean $\overline{X}_{..}$.

$$\begin{aligned}\overline{X}_{..} &= \frac{(5)(253.8) + (5)(263.2) + (5)(271.0) + (5)(262.0)}{20} \\ &= 262.5\end{aligned}$$

Figure 9.2 presents the idea behind one-way ANOVA. The figure illustrates several hypothetical samples from different treatments, along with their sample means and the sample grand mean. The sample means are spread out around the sample grand mean. One-way ANOVA provides a way to measure this spread. If the sample means are highly spread out, then it is likely that the treatment means are different, and we will reject H_0.

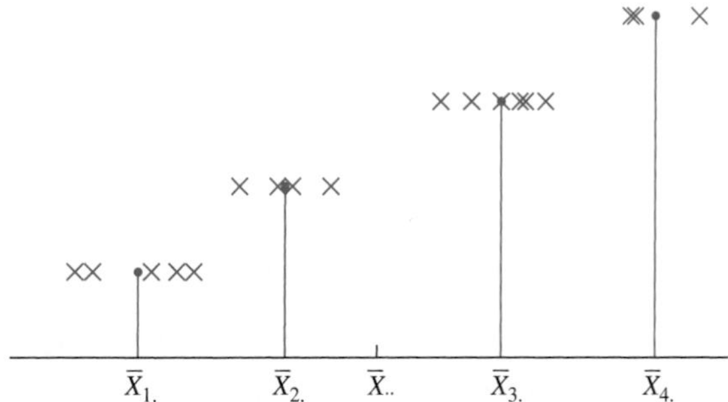

FIGURE 9.2 The variation of the sample means around the sample grand mean can be due both to random uncertainty and to differences among the treatment means. The variation within a given sample around its own sample mean is due only to random uncertainty.

The variation of the sample means around the sample grand mean is measured by a quantity called the **treatment sum of squares** (SSTr for short), which is given by

$$\text{SSTr} = \sum_{i=1}^{I} J_i(\overline{X}_{i.} - \overline{X}_{..})^2 \tag{9.4}$$

Each term in SSTr involves the distance from the sample means to the sample grand mean. Note that each squared distance is multiplied by the sample size corresponding to its sample mean, so that the means for the larger samples count more. SSTr provides an indication of how different the treatment means are from each other. If SSTr is large, then the sample means are spread out widely, and it is reasonable to conclude that the treatment means differ and to reject H_0. If on the other hand SSTr is small, then the sample means are all close to the sample grand mean and therefore to each other, so it is plausible that the treatment means are equal.

An equivalent formula for SSTr, which is a bit easier to compute by hand, is

$$\text{SSTr} = \sum_{i=1}^{I} J_i\overline{X}_{i.}^{\,2} - N\overline{X}_{..}^{\,2} \tag{9.5}$$

In order to determine whether SSTr is large enough to reject H_0, we compare it to another sum of squares, called the **error sum of squares** (SSE for short). SSE measures the variation in the individual sample points around their respective sample means. This variation is measured by summing the squares of the distances from each point to its own sample mean. SSE is given by

$$\text{SSE} = \sum_{i=1}^{I}\sum_{j=1}^{J_i}(X_{ij} - \overline{X}_{i.})^2 \tag{9.6}$$

The quantities $X_{ij} - \overline{X}_{i.}$ are called the **residuals**, so SSE is the sum of the squared residuals. SSE, unlike SSTr, depends only on the distances of the sample points from their own means and is not affected by the location of treatment means relative to one another. SSE therefore measures only the underlying random variation in the process being studied. It is analogous to the error sum of squares in regression.

An equivalent formula for SSE, which is a bit easier to compute by hand, is

$$\text{SSE} = \sum_{i=1}^{I}\sum_{j=1}^{J_i} X_{ij}^2 - \sum_{i=1}^{I} J_i \overline{X}_{i.}^{\,2} \tag{9.7}$$

Another equivalent formula for SSE is based on the sample variances. Let s_i^2 denote the sample variance of the ith sample. Then

$$s_i^2 = \frac{\sum_{j=1}^{J_i}(X_{ij} - \overline{X}_{i.})^2}{J_i - 1} \tag{9.8}$$

It follows from Equation (9.8) that $\sum_{j=1}^{J_i}(X_{ij} - \overline{X}_{i.})^2 = (J_i - 1)s_i^2$. Substituting into Equation (9.6) yields

$$\text{SSE} = \sum_{i=1}^{I}(J_i - 1)s_i^2 \tag{9.9}$$

Example 9.2

For the data in Table 9.1, compute SSTr and SSE.

Solution

The sample means are presented in Table 9.1. They are

$$\overline{X}_{1.} = 253.8 \qquad \overline{X}_{2.} = 263.2 \qquad \overline{X}_{3.} = 271.0 \qquad \overline{X}_{4.} = 262.0$$

The sample grand mean was computed in Example 9.1 to be $\overline{X}_{..} = 262.5$. We now use Equation (9.4) to calculate SSTr.

$$\begin{aligned}\text{SSTr} &= 5(253.8-262.5)^2+5(263.2-262.5)^2+5(271.0-262.5)^2+5(262.0-262.5)^2 \\ &= 743.4\end{aligned}$$

To compute SSE we will use Equation (9.9), since the sample standard deviations s_i have already been presented in Table 9.1.

$$\begin{aligned}\text{SSE} &= (5-1)(9.7570)^2 + (5-1)(5.4037)^2 + (5-1)(8.7178)^2 + (5-1)(7.4498)^2 \\ &= 1023.6\end{aligned}$$

We can use SSTr and SSE to construct a test statistic, provided the following two assumptions are met.

Assumptions for One-Way ANOVA

The standard one-way ANOVA hypothesis tests are valid under the following conditions:

1. The treatment populations must be normal.
2. The treatment populations must all have the same variance, which we will denote by σ^2.

Before presenting the test statistic, we will explain how it works. If the two assumptions for one-way ANOVA are approximately met, we can compute the means of SSE and SSTr. The mean of SSTr depends on whether H_0 is true, because SSTr tends to be smaller when H_0 is true and larger when H_0 is false. The mean of SSTr satisfies the condition

$$\mu_{\text{SSTr}} = (I-1)\sigma^2 \quad \text{when } H_0 \text{ is true} \tag{9.10}$$

$$\mu_{\text{SSTr}} > (I-1)\sigma^2 \quad \text{when } H_0 \text{ is false} \tag{9.11}$$

The likely size of SSE, and thus its mean, does not depend on whether H_0 is true. The mean of SSE is given by

$$\mu_{\text{SSE}} = (N-I)\sigma^2 \quad \text{whether or not } H_0 \text{ is true} \tag{9.12}$$

Derivations of Equations (9.10) and (9.12) are given at the end of this section.

The quantities $I-1$ and $N-I$ are the **degrees of freedom** for SSTr and SSE, respectively. When a sum of squares is divided by its degrees of freedom, the quantity obtained is called a **mean square**. The **treatment mean square** is denoted MSTr, and the **error mean square** is denoted MSE. They are defined by

$$\text{MSTr} = \frac{\text{SSTr}}{I-1} \qquad \text{MSE} = \frac{\text{SSE}}{N-I} \tag{9.13}$$

It follows from Equations (9.10) through (9.13) that

$$\mu_{\text{MSTr}} = \sigma^2 \quad \text{when } H_0 \text{ is true} \tag{9.14}$$

$$\mu_{\text{MSTr}} > \sigma^2 \quad \text{when } H_0 \text{ is false} \tag{9.15}$$

$$\mu_{\text{MSE}} = \sigma^2 \quad \text{whether or not } H_0 \text{ is true} \tag{9.16}$$

Equations (9.14) and (9.16) show that when H_0 is true, MSTr and MSE have the same mean. Therefore, when H_0 is true, we would expect their quotient to be near 1. This quotient is in fact the test statistic. The test statistic for testing $H_0: \mu_1 = \cdots = \mu_I$ is

$$F = \frac{\text{MSTr}}{\text{MSE}} \tag{9.17}$$

When H_0 is true, the numerator and denominator of F are on average the same size, so F tends to be near 1. In fact, when H_0 is true, this test statistic has an F distribution with $I-1$ and $N-I$ degrees of freedom, denoted $F_{I-1,N-I}$. When H_0 is false, MSTr tends to be larger, but MSE does not, so F tends to be greater than 1.

Summary

The *F* test for One-Way ANOVA

To test $H_0: \mu_1 = \cdots = \mu_I$ versus H_1: two or more of the μ_i are different:

1. Compute $\text{SSTr} = \sum_{i=1}^{I} J_i(\overline{X}_{i.} - \overline{X}_{..})^2 = \sum_{i=1}^{I} J_i\overline{X}_{i.}^{\,2} - N\overline{X}_{..}^{\,2}$.
2. Compute $\text{SSE} = \sum_{i=1}^{I}\sum_{j=1}^{J_i}(X_{ij} - \overline{X}_{i.})^2 = \sum_{i=1}^{I}\sum_{j=1}^{J_i} X_{ij}^2 - \sum_{i=1}^{I} J_i\overline{X}_{i.}^{\,2}$
   $= \sum_{i=1}^{I}(J_i - 1)s_i^2$.
3. Compute $\text{MSTr} = \dfrac{\text{SSTr}}{I-1}$ and $\text{MSE} = \dfrac{\text{SSE}}{N-I}$.
4. Compute the test statistic: $F = \dfrac{\text{MSTr}}{\text{MSE}}$.
5. Find the P-value by consulting the F table (Table A.7 in Appendix A) with $I - 1$ and $N - I$ degrees of freedom.

We now apply the method of analysis of variance to the example with which we introduced this section.

Example 9.3

For the data in Table 9.1, compute MSTr, MSE, and F. Find the P-value for testing the null hypothesis that all the means are equal. What do you conclude?

Solution

From Example 9.2, SSTr = 743.4 and SSE = 1023.6. We have $I = 4$ samples and $N = 20$ observations in all the samples taken together. Using Equation (9.13),

$$\text{MSTr} = \frac{743.4}{4-1} = 247.8 \qquad \text{MSE} = \frac{1023.6}{20-4} = 63.975$$

The value of the test statistic F is therefore

$$F = \frac{247.8}{63.975} = 3.8734$$

To find the P-value, we consult the F table (Table A.7). The degrees of freedom are $4 - 1 = 3$ for the numerator and $20 - 4 = 16$ for the denominator. Under H_0, F has an $F_{3,16}$ distribution. Looking at the F table under 3 and 16 degrees of freedom, we find that the upper 5% point is 3.24 and the upper 1% point is 5.29. Therefore the P-value is between 0.01 and 0.05 (see Figure 9.3; a computer software package gives a value of 0.029 accurate to two significant digits). It is reasonable to conclude that the population means are not all equal, and, thus, that flux composition does affect hardness.

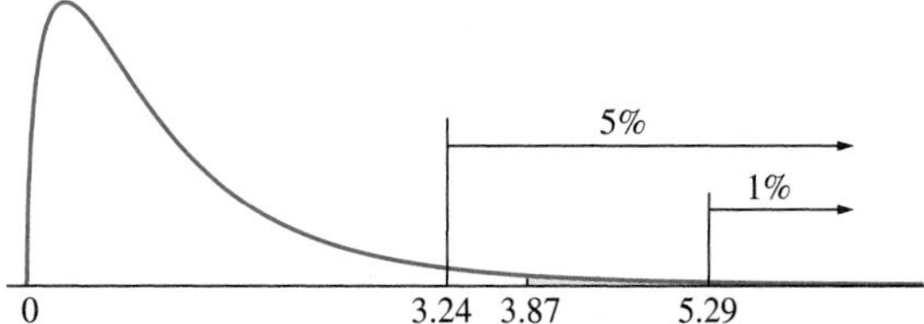

FIGURE 9.3 The observed value of the test statistic is 3.87. The upper 5% point of the $F_{3,16}$ distribution is 3.24. The upper 1% point of the $F_{3,16}$ distribution is 5.29. Therefore the P-value is between 0.01 and 0.05. A computer software package gives a value of 0.029.

Confidence Intervals for the Treatment Means

The observations on the ith treatment are assumed to be a simple random sample from a normal population with mean μ_i and variance σ^2. To construct a confidence interval for μ_i, the first step is to estimate the population variance σ^2. One way to do this would be to use the sample variance s_i^2 of the observations on the ith treatment. However, since we assume that all observations for all treatments have the same variance, it is better to combine all the sample variances into one "pooled" estimate. To do this, note that SSE is a weighted sum of the sample variances (Equation 9.9) and MSE is the weighted average (Equation 9.13). The quantity MSE is therefore the pooled estimate of the variance σ^2. Since $\overline{X}_{i.}$ is the sample mean of J_i observations, the variance of $\overline{X}_{i.}$ is σ^2/J_i, estimated with MSE/J_i. The number of degrees of freedom for MSE is $N - I$. The quantity

$$\frac{\overline{X}_{i.} - \mu_i}{\sqrt{\text{MSE}/J_i}}$$

has a Student's t distribution with $N - I$ degrees of freedom. A confidence interval for μ_i can therefore be constructed by the method described in Section 5.3.

A level $100(1 - \alpha)$ confidence interval for μ_i is given by

$$\overline{X}_{i.} \pm t_{N-I,\,\alpha/2}\sqrt{\frac{\text{MSE}}{J_i}} \qquad (9.18)$$

Example 9.4

Find a 95% confidence interval for the mean hardness of welds produced with flux A.

Solution

From Table 9.1, $\overline{X}_{1.} = 253.8$. The value of MSE was computed in Example 9.3 to be 63.975. There are $I = 4$ treatments, $J_1 = 5$ observations for flux A, and $N = 20$ observations altogether. From the Student's t table we obtain $t_{16,\,.025} = 2.120$. The 95% confidence interval is therefore

$$253.8 \pm 2.120\sqrt{\frac{63.975}{5}} = 253.8 \pm 7.6$$

The ANOVA Table

The results of an analysis of variance are usually summarized in an analysis of variance (ANOVA) table. This table is much like the analysis of variance table produced in multiple regression. The following MINITAB output shows the analysis of variance for the weld data presented in Table 9.1.

```
One-way ANOVA: A, B, C, D

Source    DF       SS       MS     F      P
Factor     3   743.40  247.800  3.87  0.029
Error     16  1023.60   63.975
Total     19  1767.00

S = 7.998   R-Sq = 42.07%   R-Sq(adj) = 31.21%

                             Individual 95% CIs For Mean Based on
                             Pooled StDev
Level  N    Mean  StDev  ----+---------+---------+---------+-----
A      5  253.80   9.76  (-------*------)
B      5  263.20   5.40           (------*-------)
C      5  271.00   8.72                  (-------*-------)
D      5  262.00   7.45         (-------*-------)
                         ----+---------+---------+---------+-----
                           250       260       270       280

Pooled StDev = 8.00
```

In the ANOVA table, the column labeled "DF" presents the number of degrees of freedom for both the treatment ("Factor") and error ("Error") sum of squares. The column labeled "SS" presents SSTr (in the row labeled "Factor") and SSE (in the row labeled "Error"). The row labeled "Total" contains the **total sum of squares**, which is the sum of SSTr and SSE. The column labeled "MS" presents the mean squares MSTr and MSE. The column labeled "F" presents the F statistic for testing the null hypothesis that all the population means are equal. Finally, the column labeled "P" presents the P-value for the F test. Below the ANOVA table, the value "S" is the pooled estimate of the error standard deviation σ, computed by taking the square root of MSE. The quantity "R-sq" is R^2, the coefficient of determination, which is equal to the quotient SSTr/SST. This is analogous to the multiple regression situation (see Equation 8.9 in Section 8.1). The value "R-Sq(adj)" is the adjusted R^2, equal to $R^2 - [(I - 1)/(N - I)](1 - R^2)$, again analogous to multiple regression. The quantities R^2 and adjusted R^2 are not used as much in analysis of variance as they are in multiple regression. Finally, sample means and standard deviations are presented for each treatment group, along with a graphic that illustrates a 95% confidence interval for each treatment mean.

Example 9.5

In the article "Review of Development and Application of CRSTER and MPTER Models" (R. Wilson, *Atmospheric Environment*, 1993:41–57), several measurements of the maximum hourly concentrations (in $\mu g/m^3$) of SO_2 are presented for each of four power plants. The results are as follows (two outliers have been deleted):

Plant 1:	438	619	732	638		
Plant 2:	857	1014	1153	883	1053	
Plant 3:	925	786	1179	786		
Plant 4:	893	891	917	695	675	595

The following MINITAB output presents results for a one-way ANOVA. Can you conclude that the maximum hourly concentrations differ among the plants?

```
One-way ANOVA: Plant 1, Plant 2, Plant 3, Plant 4

Source    DF        SS        MS      F       P
Plant      3    378610    126203   6.21   0.006
Error     15    304838     20323
Total     18    683449

S = 142.6   R-Sq = 55.40%   R-Sq(adj) = 46.48%

                            Individual 95% CIs For Mean Based on
                            Pooled StDev
Level  N   Mean  StDev   -------+---------+---------+---------+--
1      4  606.8  122.9   (------*-------)
2      5  992.0  122.7                       (------*-----)
3      4  919.0  185.3                  (-------*-------)
4      6  777.7  138.8             (-----*-----)
                         -------+---------+---------+---------+--
                              600       800      1000      1200

Pooled StDev = 142.6
```

Solution

In the ANOVA table, the P-value for the null hypothesis that all treatment means are equal is 0.006. Therefore we conclude that not all the treatment means are equal.

Checking the Assumptions

As previously mentioned, the methods of analysis of variance require the assumptions that the observations on each treatment are a sample from a normal population and that the normal populations all have the same variance. A good way to check the normality assumption is with a normal probability plot. If the sample sizes are large enough, one can construct a separate probability plot for each sample. This is rarely the case in practice. When the sample sizes are not large enough for individual probability plots to be informative, the residuals $X_{ij} - \overline{X}_{i.}$ can all be plotted together in a single plot. When the assumptions of normality and constant variance are satisfied, these residuals will be normally distributed with mean zero and should plot approximately on a straight line. Figure 9.4 presents a normal probability plot of the residuals for the weld data of Table 9.1. There is no evidence of a serious violation of the assumption of normality.

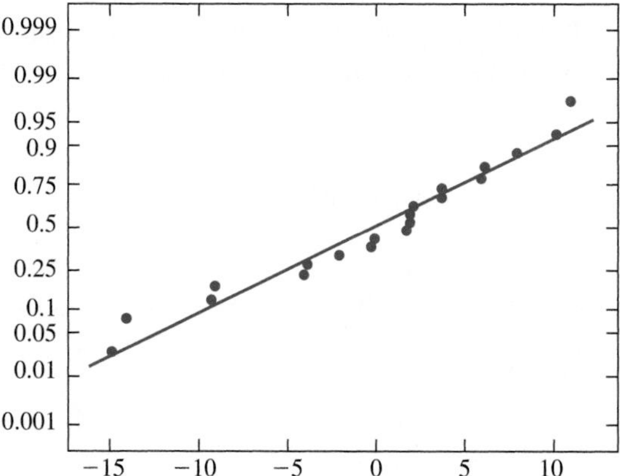

FIGURE 9.4 Probability plot for the residuals from the weld data. There is no evidence of a serious violation of the assumption of normality.

The assumption of equal variances can be difficult to check, because with only a few observations in each sample, the sample standard deviations can differ greatly (by a factor of 2 or more) even when the assumption holds. For the weld data, the sample standard deviations range from 5.4037 to 9.7570. It is reasonable to proceed as though the variances were equal.

The spreads of the observations within the various samples can be checked visually by making a residual plot. This is done by plotting the residuals $X_{ij} - \overline{X}_{i.}$ versus the fitted values, which are the sample means $\overline{X}_{i.}$. If the spreads differ considerably among the samples, the assumption of equal variances is suspect. If one or more of the samples contain outliers, the assumption of normality is suspect as well. Figure 9.5 presents a residual plot for the weld data. There are no serious outliers, and the spreads do not differ greatly among samples.

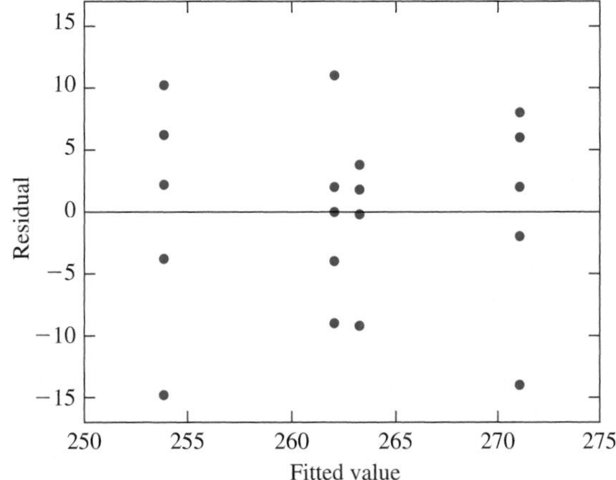

FIGURE 9.5 Residual plot of the values $X_{ij} - \overline{X}_{i.}$ versus $\overline{X}_{i.}$ for the weld data. The spreads do not differ greatly from sample to sample, and there are no serious outliers.

Balanced versus Unbalanced Designs

When equal numbers of units are assigned to each treatment, the design is said to be **balanced**. Although one-way analysis of variance can be used with both balanced and unbalanced designs, balanced designs offer a big advantage. A balanced design is much less sensitive to violations of the assumption of equality of variance than an unbalanced one. Since moderate departures from this assumption can be difficult to detect, it is best to use a balanced design whenever possible, so that undetected violations of the assumption will not seriously compromise the validity of the results. When a balanced design is impossible to achieve, a slightly unbalanced design is preferable to a severely unbalanced one.

Summary

- With a balanced design, the effect of unequal variances is generally not great.
- With an unbalanced design, the effect of unequal variances can be substantial.
- The more unbalanced the design, the greater the effect of unequal variances.

The Analysis of Variance Identity

In both linear regression and analysis of variance, a quantity called the total sum of squares is obtained by subtracting the sample grand mean from each observation, squaring these deviations, and then summing them. An analysis of variance identity is an equation that expresses the total sum of squares as a sum of other sums of squares. We have presented analysis of variance identities for simple linear regression (at the end of Section 7.2) and for multiple regression (Equation 8.7 in Section 8.1).

The total sum of squares for one-way ANOVA is given by

$$\text{SST} = \sum_{i=1}^{I}\sum_{j=1}^{J_i}(X_{ij} - \overline{X}_{..})^2 \tag{9.19}$$

An equivalent formula is given by

$$\text{SST} = \sum_{i=1}^{I}\sum_{j=1}^{J_i} X_{ij}^2 - N\overline{X}_{..}^{\,2} \tag{9.20}$$

Examining Equations (9.5), (9.7), and (9.20) shows that the total sum of squares is equal to the treatment sum of squares plus the error sum of squares. This is the analysis of variance identity for one-way analysis of variance.

The Analysis of Variance Identity

$$\text{SST} = \text{SSTr} + \text{SSE} \tag{9.21}$$

An Alternate Parameterization

Our presentation of one-way analysis of variance, as a method to compare several treatment means by using random samples drawn from each treatment population, is one natural way to view the subject. There is another way to express these same ideas, in somewhat different notation, that is sometimes useful.

For each observation X_{ij}, define $\varepsilon_{ij} = X_{ij} - \mu_i$, the difference between the observation and its mean. By analogy with linear regression, the quantities ε_{ij} are called **errors**. It is clearly true that

$$X_{ij} = \mu_i + \varepsilon_{ij} \tag{9.22}$$

Now since X_{ij} is normally distributed with mean μ_i and variance σ^2, it follows that ε_{ij} is normally distributed with mean 0 and variance σ^2.

In a single-factor experiment, we are interested in determining whether the treatment means are all equal. Given treatment means $\mu_1, \ldots, \mu_I$, the quantity

$$\mu = \frac{1}{I}\sum_{i=1}^{I}\mu_i \tag{9.23}$$

is the average of all the treatment means. The quantity μ is called the **population grand mean**. The ith **treatment effect**, denoted α_i, is the difference between the ith treatment mean and the population grand mean:

$$\alpha_i = \mu_i - \mu \tag{9.24}$$

It follows from the definition of α_i that $\sum_{i=1}^{I} \alpha_i = 0$.

We can now decompose the treatment means as follows:

$$\mu_i = \mu + \alpha_i \tag{9.25}$$

Combining Equations (9.22) and (9.25) yields the **one-way analysis of variance model**:

$$X_{ij} = \mu + \alpha_i + \varepsilon_{ij} \tag{9.26}$$

The null hypothesis $H_0 : \mu_1 = \cdots = \mu_I$ is equivalent to $H_0 : \alpha_1 = \cdots = \alpha_I = 0$.

In one-way ANOVA, it is possible to work with the treatment means μ_i, as we have done, rather than with the treatment effects α_i. In multi-factor experiments, however, the treatment means by themselves are not sufficient and must be decomposed in a manner analogous to the one described here. We will discuss this further in Section 9.3.

Power

When designing a factorial experiment, it is important that the F test have good power, that is, a large probability of rejecting the null hypothesis of equality if in fact the treatment means are not all equal. An experiment with low power is not much use, since it is unlikely to detect a difference in treatments even if one exists. In what follows, we will assume that the experiment is balanced and that the assumptions of normality and equal variance hold. Assume the number of levels is fixed at I.

The power of any test depends first on the rejection criterion: The larger the level at which one is willing to reject, the greater the power. The 5% level is the one most often used in practice. Once the rejection level is set, the power of the F test depends on three quantities: (1) the spread in the true means as measured by the quantity $\sum_i \alpha_i^2$ where α_i is the ith treatment effect, (2) the error standard deviation σ, and (3) the sample size J. Note that if the null hypothesis is true, then $\sum_i \alpha_i^2 = 0$. The larger $\sum_i \alpha_i^2$ is, the farther from the truth is the null hypothesis, and the larger is the power, which is the probability that the null hypothesis is rejected.

A power calculation can serve either of two purposes: to determine the sample size for each treatment necessary to achieve a desired power, or to determine how much power one has with a given sample size. In a traditional power calculation, one specifies the quantity $\sum_i \alpha_i^2$ that one wishes to detect and the value of σ one expects to encounter. Then one can compute the power for a given sample size, or the sample size needed to achieve a given power. In practice, one rarely knows how to specify a value for $\sum_i \alpha_i^2$, but one can often specify the size of a difference between the largest and smallest treatment means that one wishes to detect. For example, in the weld experiment, a metallurgist might be able to specify that a difference of 10 or more between the largest and smallest treatment means is scientifically important, but it is unlikely that she could specify a scientifically important value for $\sum_i \alpha_i^2$.

In MINITAB, one can specify the size of a scientifically important difference between the largest and smallest treatment means and compute the sample size necessary to guarantee that the power to detect that difference will be at least a specified amount. We present an example.

A metallurgist wants to repeat the weld experiment with four different fluxes and wants the design to be sensitive enough so that it is likely to detect a difference of 10 or more in Brinell hardness at the 5% level. He assumes that the error standard deviation will be about the same as the value of 7.998 calculated in the experiment we have been discussing. The following MINITAB output shows the result of a power calculation for an experiment with five observations per treatment. What is the power? What recommendation would you give the metallurgist regarding the usefulness of this proposed experiment?

```
One-way ANOVA

Alpha = 0.05 Assumed standard deviation = 7.998
Number of Levels = 4

   SS     Sample                       Maximum
Means       Size        Power       Difference
   50          5     0.281722               10

The sample size is for each level.
```

Solution

The power is 0.281772. This means that the probability that the proposed experiment will detect a difference of 10 between the largest and smallest treatment means may be no more than about 0.28. The appropriate recommendation is not to run this experiment; it has too little chance of success. Instead, the sample size necessary to provide adequate power should be calculated, and if feasible, an experiment of that size should be run.

The metallurgist in Example 9.6 has taken your advice and has computed the sample size necessary to provide a power of 0.90 to detect a difference of 10 at the 5% level. The MINITAB output follows. What is the power? How many observations will be necessary at each level? How many observations will be necessary in total?

```
One-way ANOVA

Alpha = 0.05 Assumed standard deviation = 7.998
Number of Levels = 4

   SS     Sample     Target                        Maximum
Means       Size      Power   Actual Power      Difference
   50         20        0.9       0.914048              10

The sample size is for each level.
```

Solution

The needed sample size is 20 per level; with four levels there will be 80 observations in total. Note that the actual power of the experiment is approximately 0.914, which is higher than the "target power" of 0.90 that was requested. The reason for this is that the power provided by a sample size of 19 per level could be somewhat less than 0.90; a sample size of 20 is the smallest that is guaranteed to provide a power of 0.90 or more.

Random Effects Models

In many factorial experiments, the treatments are chosen deliberately by the experimenter. These experiments are said to follow a **fixed effects model**. In some cases, the treatments are chosen at random from a population of possible treatments. In these cases the experiments are said to follow a **random effects model**. In a fixed effects model, the interest is on the specific treatments chosen for the experiment. In a random effects model, the interest is in the whole population of possible treatments, and there is no particular interest in the ones that happened to be chosen for the experiment.

The article describing the weld experiment states that the treatments were chosen deliberately and do not represent a random sample from a larger population of flux compositions. This experiment therefore follows a fixed effects model. The four power plants in Example 9.5 are a sample of convenience; they are plants at which measurements were readily available. In some cases it is appropriate to treat a sample of convenience as if it were a simple random sample (see the discussion in Section 1.1). If these conditions hold, then the power plant experiment may be considered to follow a random effects model; otherwise it must be treated as a fixed effects model.

There is an important difference in interpretation between the results of a fixed effects model and those of a random effects model. In a fixed effects model, the only conclusions that can be drawn are conclusions about the treatments actually used in the experiment. In a random effects model, however, since the treatments are a simple random sample from a population of treatments, conclusions can be drawn concerning the whole population, including treatments not actually used in the experiment.

This difference in interpretations results in a difference in the null hypotheses to be tested. In the fixed effects model, the null hypothesis of interest is $H_0: \mu_1 = \cdots = \mu_I$. In the random effects model, the null hypothesis of interest is

$$H_0: \text{the treatment means are equal for every level in the population}$$

In the random effects model, the assumption is made that the population of treatment means is normal.

Interestingly enough, although the null hypothesis for the random effects model differs from that of the fixed effects model, the hypothesis test is exactly the same. The F test previously described is used for the random effects model as well as for the fixed effects model.

Example 9.8

In Example 9.5, assume that it is reasonable to treat the four power plants as a random sample from a large population of power plants, and furthermore, assume that the SO_2 concentrations in the population of plants are normally distributed. Can we conclude that there are differences in SO_2 concentrations among the power plants in the population?

Solution

This is a random effects model, so we can use the F test to test the null hypothesis that all the treatment means in the population are the same. The results of the F test are shown in Example 9.5. The P-value is 0.006. We therefore reject the null hypothesis and conclude that there are differences in mean SO_2 concentrations among the power plants in the population.

Derivations of Equations (9.10) and (9.12)

In what follows it will be easier to use the notation $E(\ )$ to denote the mean of a quantity and $V(\ )$ to denote the variance. So, for example, $E(\text{SSE}) = \mu_{\text{SSE}}$, $E(\text{SSTr}) = \mu_{\text{SSTr}}$, and $V(X_{ij})$ denotes the variance of X_{ij}.

We will show that $E(\text{SSE}) = E[\sum_{i=1}^{I}\sum_{j=1}^{J_i}(X_{ij} - \overline{X}_{i.})^2] = (N-I)\sigma^2$, whether or not the population means are equal. This is Equation (9.12).

We begin by adding and subtracting the treatment mean μ_i from each term in $\sum_{i=1}^{I}\sum_{j=1}^{J_i}(X_{ij} - \overline{X}_{i.})^2$ to obtain

$$\text{SSE} = \sum_{i=1}^{I}\sum_{j=1}^{J_i}[(X_{ij} - \mu_i) - (\overline{X}_{i.} - \mu_i)]^2$$

Multiplying out yields

$$\text{SSE} = \sum_{i=1}^{I}\sum_{j=1}^{J_i}(X_{ij} - \mu_i)^2 - \sum_{i=1}^{I}\sum_{j=1}^{J_i}2(X_{ij} - \mu_i)(\overline{X}_{i.} - \mu_i) + \sum_{i=1}^{I}\sum_{j=1}^{J_i}(\overline{X}_{i.} - \mu_i)^2 \qquad (9.27)$$

Now $\sum_{j=1}^{J_i}(X_{ij} - \mu_i) = J_i(\overline{X}_{i.} - \mu_i)$. Substituting into the middle term of the right-hand side of (9.27) yields

$$\text{SSE} = \sum_{i=1}^{I}\sum_{j=1}^{J_i}(X_{ij} - \mu_i)^2 - 2\sum_{i=1}^{I}J_i(\overline{X}_{i.} - \mu_i)^2 + \sum_{i=1}^{I}\sum_{j=1}^{J_i}(\overline{X}_{i.} - \mu_i)^2$$

Since $\sum_{i=1}^{I}\sum_{j=1}^{J_i}(\overline{X}_{i.} - \mu_i)^2 = \sum_{i=1}^{I}J_i(\overline{X}_{i.} - \mu_i)^2$, this simplifies to

$$\text{SSE} = \sum_{i=1}^{I}\sum_{j=1}^{J_i}(X_{ij} - \mu_i)^2 - \sum_{i=1}^{I} J_i(\overline{X}_{i.} - \mu_i)^2 \tag{9.28}$$

Taking means of both sides of (9.28) yields

$$E(\text{SSE}) = \sum_{i=1}^{I}\sum_{j=1}^{J_i}E(X_{ij} - \mu_i)^2 - \sum_{i=1}^{I} J_i E(\overline{X}_{i.} - \mu_i)^2 \tag{9.29}$$

Now $E(X_{ij}) = E(\overline{X}_{i.}) = \mu_i$. The population variances are all equal; denote their common value by σ^2. It follows that

$$E(X_{ij} - \mu_i)^2 = V(X_{ij}) = \sigma^2$$

$$E(\overline{X}_{i.} - \mu_i)^2 = V(\overline{X}_{i.}) = \frac{\sigma^2}{J_i}$$

Substituting into (9.29) yields

$$E(\text{SSE}) = \sum_{i=1}^{I}\sum_{j=1}^{J_i}\sigma^2 - \sum_{i=1}^{I}\frac{J_i\sigma^2}{J_i} = N\sigma^2 - I\sigma^2 = (N - I)\sigma^2$$

This completes the derivation of $E(\text{SSE})$.

We now show that $E(\text{SSTr}) = E[\sum_{i=1}^{I} J_i(\overline{X}_{i.} - \overline{X}_{..})^2] = (I - 1)\sigma^2$ under the assumption that the treatment means are all equal to a common value denoted by μ. This is Equation (9.10).

We begin by adding and subtracting the common treatment mean μ from each term in $\sum_{i=1}^{I} J_i(\overline{X}_{i.} - \overline{X}_{..})^2$ to obtain

$$\text{SSTr} = \sum_{i=1}^{I} J_i[(\overline{X}_{i.} - \mu) - (\overline{X}_{..} - \mu)]^2$$

Multiplying out, we obtain

$$\text{SSTr} = \sum_{i=1}^{I} J_i(\overline{X}_{i.} - \mu)^2 - 2\sum_{i=1}^{I} J_i(\overline{X}_{i.} - \mu)(\overline{X}_{..} - \mu) + \sum_{i=1}^{I} J_i(\overline{X}_{..} - \mu)^2 \tag{9.30}$$

Now

$$\overline{X}_{..} = \sum_{i=1}^{I}\frac{J_i\overline{X}_{i.}}{N}$$

so

$$\overline{X}_{..} - \mu = \sum_{i=1}^{I}\frac{J_i(\overline{X}_{i.} - \mu)}{N}$$

and

$$\sum_{i=1}^{I} J_i(\overline{X}_{i.} - \mu) = N(\overline{X}_{..} - \mu)$$

Substituting into the middle term of the right-hand side of (9.30), we obtain

$$\text{SSTr} = \sum_{i=1}^{I} J_i(\overline{X}_{i.} - \mu)^2 - 2N(\overline{X}_{..} - \mu)^2 + \sum_{i=1}^{I} J_i(\overline{X}_{..} - \mu)^2$$

Since $\sum_{i=1}^{I} J_i = N$, we obtain

$$\text{SSTr} = \sum_{i=1}^{I} J_i(\overline{X}_{i.} - \mu)^2 - N(\overline{X}_{..} - \mu)^2$$

Taking means of both sides yields

$$E(\text{SSTr}) = \sum_{i=1}^{I} J_i E(\overline{X}_{i.} - \mu)^2 - NE(\overline{X}_{..} - \mu)^2 \qquad (9.31)$$

Now $E(\overline{X}_{i.}) = E(\overline{X}_{..}) = \mu$, so

$$E(\overline{X}_{i.} - \mu)^2 = V(\overline{X}_{i.}) = \frac{\sigma^2}{J_i}$$

$$E(\overline{X}_{..} - \mu)^2 = V(\overline{X}_{..}) = \frac{\sigma^2}{N}$$

Substituting into (9.31) yields

$$E(\text{SSTr}) = \sum_{i=1}^{I} \frac{J_i\sigma^2}{J_i} - \frac{N\sigma^2}{N} = (I-1)\sigma^2$$

Exercises for Section 9.1

1. One of the factors that determines the degree of risk a pesticide poses to human health is the rate at which the pesticide is absorbed into skin after contact. An important question is whether the amount in the skin continues to increase with the length of the contact, or whether it increases for only a short time before leveling off. To investigate this, measured amounts of a certain pesticide were applied to 20 samples of rat skin. Four skins were analyzed at each of the time intervals 1, 2, 4, 10, and 24 hours. The amounts of the chemical (in μg) that were in the skin are given in the following table.

Duration	Amounts Absorbed
1	1.7 1.5 1.2 1.5
2	1.8 1.6 1.8 1.9
4	1.9 1.7 2.1 2.0
10	2.3 1.9 1.7 1.5
24	2.1 2.2 2.5 2.3

a. Construct an ANOVA table. You may give a range for the P-value.
b. Can you conclude that the amount in the skin varies with time?

2. The yield strength of CP titanium welds was measured for welds cooled at rates of 10 °C/s, 15 °C/s, and 28 °C/s. The results are presented in the following table. (Based on the article "Advances in Oxygen Equivalence Equations

for Predicting the Properties of Titanium Welds," D. Harwig, W. Ittiwattana, and H. Castner, *The Welding Journal*, 2001:126s–136s.)

Cooling Rate	Yield Strengths
10	71.00 75.00 79.67 81.00 75.50 72.50 73.50 78.50 78.50
15	63.00 68.00 73.00 76.00 79.67 81.00
28	68.65 73.70 78.40 84.40 91.20 87.15 77.20 80.70 84.85 88.40

a. Construct an ANOVA table. You may give a range for the P-value.
b. Can you conclude that the yield strength of CP titanium welds varies with the cooling rate?

3. The removal of ammoniacal nitrogen is an important aspect of treatment of leachate at landfill sites. The rate of removal (in % per day) is recorded for several days for each of several treatment methods. The results are presented in the following table. (Based on the article "Removal of Ammoniacal Nitrogen from Landfill Leachate by Irrigation onto Vegetated Treatment Planes," S. Tyrrel, P. Leeds-Harrison, and K. Harrison, *Water Research*, 2002:291–299.)

Treatment	Rate of Removal
A	5.21 4.65
B	5.59 2.69 7.57 5.16
C	6.24 5.94 6.41
D	6.85 9.18 4.94
E	4.04 3.29 4.52 3.75

a. Construct an ANOVA table. You may give a range for the P-value.
b. Can you conclude that the treatment methods differ in their rates of removal?

4. In the article "Calibration of an FTIR Spectrometer" (P. Pankratz, *Statistical Case Studies for Industrial and Process Improvement*, SIAM-ASA, 1997:19–38), a spectrometer was used to make five measurements of the carbon content (in ppb) of a certain silicon wafer on four consecutive days. The results are as follows:

Day 1:	358	390	380	372	366
Day 2:	373	376	413	367	368
Day 3:	336	360	370	368	352
Day 4:	368	359	351	349	343

a. Construct an ANOVA table. You may give a range for the P-value.
b. Can you conclude that the calibration of the spectrometer differs among the four days?

5. The article "Quality of the Fire Clay Coal Bed, Southeastern Kentucky" (J. Hower, W. Andrews, et al., *Journal of Coal Quality*, 1994:13–26) contains measurements on samples of coal from several sites in Kentucky. The data on percent TiO_2 ash are as follows (one outlier has been deleted):

Buckeye Hollow:	0.96 0.86 0.94 0.91 0.70 1.28 1.19 1.04 1.42 0.82 0.89 1.45 1.66 1.68 2.10 2.19
Bear Branch:	0.91 1.42 2.54 2.23 2.20 1.44 1.70 1.53 1.84
Defeated Creek:	1.30 1.39 2.58 1.49 1.49 2.07 1.87 1.39 1.02 0.91 0.82 0.67 1.34 1.51
Turkey Creek:	1.20 1.60 1.32 1.24 1.08 2.33 1.81 1.76 1.25 0.81 0.95 1.92

a. Construct an ANOVA table. You may give a range for the P-value.
b. Can you conclude that there are differences in TiO_3 content among these sites?

6. Archaeologists can determine the diets of ancient civilizations by measuring the ratio of carbon-13 to carbon-12 in bones found at burial sites. Large amounts of carbon-13 suggest a diet rich in grasses such as maize, while small amounts suggest a diet based on herbaceous plants. The article "Climate and Diet in Fremont Prehistory: Economic Variability and Abandonment of Maize Agriculture in the Great Salt Lake Basin" (J. Coltrain and S. Leavitt, *American Antiquity*, 2002:453–485) reports ratios, as a difference from a standard in units of parts per thousand, for bones from individuals in several age groups. The data are presented in the following table.

Age Group (years)	Ratio
0–11	17.2 18.4 17.9 16.6 19.0 18.3 13.6 13.5 18.5 19.1 19.1 13.4
12–24	14.8 17.6 18.3 17.2 10.0 11.3 10.2 17.0 18.9 19.2
25–45	18.4 13.0 14.8 18.4 12.8 17.6 18.8 17.9 18.5 17.5 18.3 15.2 10.8 19.8 17.3 19.2 15.4 13.2
46+	15.5 18.2 12.7 15.1 18.2 18.0 14.4 10.2 16.7

a. Construct an ANOVA table. You may give a range for the P-value.
b. Can you conclude that the concentration ratios differ among the age groups?

7. An experiment was set up to measure the yield provided by each of three catalysts in a certain reaction. The experiment was repeated three times for each catalyst. The reactor yields, in grams, are as follows:

Catalyst 1:	84.33	90.25	85.62
Catalyst 2:	88.44	89.81	86.53
Catalyst 3:	94.71	91.19	92.81

a. Construct an ANOVA table. You may give a range for the P-value.
b. Can you conclude that there are differences in the mean yields among the catalysts?

8. An experiment to compare the lifetimes of four different brands of spark plug was carried out. Five plugs of each brand were used, and the number of miles until failure was recorded for each. Following is part of the MINITAB output for a one-way ANOVA.

```
One-way Analysis of Variance

Analysis of Variance
Source      DF        SS       MS      F      P
Brand        3   176.482      (a)    (e)    (f)
Error      (b)       (c)      (d)
Total       19   235.958
```

Fill in the missing numbers for (a) through (f) in the table. You may give a range for the P-value.

9. Refer to Exercise 8. Is it plausible that the brands of spark plug all have the same mean lifetime?

10. Three separation methods were compared in a certain chemical process to study their effects on yield. Three runs were made with each method, and the yields, in percent of a theoretical maximum, are as follows:

Method A:	84.6	83.3	85.1
Method B:	87.3	85.9	88.2
Method C:	87.2	86.0	86.3

a. Construct an ANOVA table. You may give a range for the P-value.
b. Can you conclude that there are differences among the mean yields?

11. An experiment was performed to determine whether the annealing temperature of ductile iron affects its tensile strength. Five specimens were annealed at each of four temperatures. The tensile strength (in ksi) was measured for each. The results are presented in the following table.

Temperature (°C)	Sample Values
750	19.72 20.88 19.63 18.68 17.89
800	16.01 20.04 18.10 20.28 20.53
850	16.66 17.38 14.49 18.21 15.58
900	16.93 14.49 16.15 15.53 13.25

a. Construct an ANOVA table. You may give a range for the P-value.
b. Can you conclude that there are differences among the mean strengths?

12. Refer to Exercise 10.

a. Compute the quantity $s = \sqrt{\text{MSE}}$, the estimate of the error standard deviation σ.
b. Assuming s to be the error standard deviation, find the sample size necessary in each treatment to provide a power of 0.90 to detect a maximum difference of 2 in the treatment means at the 5% level.
c. Using a more conservative estimate of $1.5s$ as the error standard deviation, find the sample size necessary in each treatment to provide a power of 0.90 to detect a maximum difference of 2 in the treatment means at the 5% level.

13. Refer to Exercise 11.

a. Compute the quantity $s = \sqrt{\text{MSE}}$, the estimate of the error standard deviation σ.
b. Assuming s to be the error standard deviation, find the sample size necessary in each treatment to provide a power of 0.90 to detect a maximum difference of 2 in the treatment means at the 5% level.
c. Using a more conservative estimate of $1.5s$ as the error standard deviation, find the sample size necessary in each treatment to provide a power of 0.90 to detect a maximum difference of 2 in the treatment means at the 5% level.

14. The article "The Lubrication of Metal-on-Metal Total Hip Joints: A Slide Down the Stribeck Curve"k (S. Smith, D. Dowson, and A. Goldsmith, *Proceedings of the Institution of Mechanical Engineers*, 2001:483–493) presents results from wear tests done on metal artificial hip joints. Joints with several different diameters were tested. The data presented in the following table on head roughness are consistent with the means and standard deviations reported in the article.

Diameter (mm)	Head Roughness (nm)
16	0.83 2.25 0.20 2.78 3.93
28	2.72 2.48 3.80
36	5.99 5.32 4.59

a. Construct an ANOVA table. You may give a range for the P-value.
b. Can you conclude that mean roughness varies with diameter? Explain.

15. The article "Mechanical Grading of Oak Timbers" (D. Kretschmann and D. Green, *Journal of Materials in Civil Engineering*, 1999:91–97) presents measurements of the modulus of rupture, in MPa, for green mixed oak 7 by 9 timbers from West Virginia and Pennsylvania. The sample means, standard deviations, and sample sizes for four different grades of lumber are presented in the following table.

Grade	Mean	Standard Deviation	Sample Size
Select	45.1	8.52	32
No. 1	42.0	5.50	11
No. 2	33.2	6.71	15
Below grade	38.1	8.04	42

a. Construct an ANOVA table. You may give a range for the P-value.
b. Can you conclude that the mean modulus of rupture differs for different grades of lumber?

16. The article "Withdrawal Strength of Threaded Nails" (D. Rammer, S. Winistorfer, and D. Bender, *Journal of Structural Engineering* 2001:442–449) describes an experiment comparing the withdrawal strengths for several types of nails. The data presented in the following table are consistent with means and standard deviations reported in the article for three types of nails: annularly threaded, helically threaded, and smooth shank. All nails had diameters within 0.1 mm of each other, and all were driven into the same type of lumber.

Nail Type	Withdrawal Strength (N/mm)									
Annularly threaded	36.57	29.67	43.38	26.94	12.03	21.66	41.79	31.50	35.84	40.81
Helically threaded	14.66	24.22	23.83	21.80	27.22	38.25	28.15	36.35	23.89	28.44
Smooth shank	12.61	25.71	17.69	24.69	26.48	19.35	28.60	42.17	25.11	19.98

a. Construct an ANOVA table. You may give a range for the P-value.
b. Can you conclude that the mean withdrawal strength is different for different nail types?

17. The following MINITAB output presents a power calculation.

```
Alpha = 0.05    Assumed standard deviation = 142.6    Number of Levels = 4

           Sample   Target                       Maximum
SS Means     Size    Power    Actual Power    Difference
   20000       14     0.85        0.864138           200

The sample size is for each level.
```

a. What is the power requested by the experimenter?
b. To guarantee a power of 0.864138, how many observations must be taken for all treatments combined?
c. What is the difference between treatment means that can be detected with a power of at least 0.864138?
d. Is the power to detect a maximum difference of 250 greater than 0.864138 or less than 0.864138? Explain.

9.2 Pairwise Comparisons in One-Factor Experiments

In a one-way ANOVA, an F test is used to test the null hypothesis that all the treatment means are equal. If this hypothesis is rejected, we can conclude that the treatment means are not all the same. But the test does not tell us which ones are different from the rest. Sometimes an experimenter has in mind two specific treatments, i and j, and wants to study the difference $\mu_i - \mu_j$. In this case a method known as Fisher's least significant difference (LSD) method is appropriate and can be used to construct confidence intervals for $\mu_i - \mu_j$ or to test the null hypothesis that $\mu_i - \mu_j = 0$. At other times, an experimenter may want to determine all the pairs of means that can be concluded to differ from each other. In this case a type of procedure called a **multiple comparisons method** must be used. We will discuss two methods of multiple comparisons, the Bonferroni method and the Tukey–Kramer method.

Fisher's Least Significant Difference Method

We begin by describing Fisher's LSD method for constructing confidence intervals. The confidence interval for the difference $\mu_i - \mu_j$ is centered at the difference in sample means $\overline{X}_{i.} - \overline{X}_{j.}$. To determine how wide to make the confidence interval, it is necessary to estimate the standard deviation of $\overline{X}_{i.} - \overline{X}_{j.}$. Let J_i and J_j be the sample sizes at levels i and j, respectively. Since by assumption all observations are normally distributed with variance σ^2, it follows that $\overline{X}_{i.} - \overline{X}_{j.}$ is normally distributed with mean $\mu_i - \mu_j$ and variance $\sigma^2(1/J_i + 1/J_j)$. The variance σ^2 is estimated with MSE, for reasons explained previously in the discussion about confidence intervals for the treatment means (Section 9.1). Now the quantity

$$\frac{(\overline{X}_{i.} - \overline{X}_{j.}) - (\mu_i - \mu_j)}{\sqrt{\text{MSE}(1/J_i + 1/J_j)}}$$

has a Student's t distribution with $N - I$ degrees of freedom. (The value $N - I$ is the number of degrees of freedom used in computing MSE; see Equation 9.13.) The quantity $t_{N-I,\alpha/2}\sqrt{\text{MSE}(1/J_i + 1/J_j)}$ is called the least significant difference. This quantity forms the basis for confidence intervals and hypothesis tests.

Fisher's Least Significant Difference Method for Confidence Intervals and Hypothesis Tests

The Fisher's least significant difference confidence interval, at level 100 $(1 - \alpha)$%, for the difference $\mu_i - \mu_j$ is

$$\overline{X}_{i.} - \overline{X}_{j.} \pm t_{N-I,\alpha/2}\sqrt{\text{MSE}\left(\frac{1}{J_i} + \frac{1}{J_j}\right)} \tag{9.32}$$

To test the null hypothesis $H_0: \mu_i - \mu_j = 0$, the test statistic is

$$\frac{\overline{X}_{i.} - \overline{X}_{j.}}{\sqrt{\text{MSE}\left(\frac{1}{J_i} + \frac{1}{J_j}\right)}} \tag{9.33}$$

If H_0 is true, this statistic has a Student's t distribution with $N - I$ degrees of freedom. Specifically, if

$$|\overline{X}_{i.} - \overline{X}_{j.}| > t_{N-I,\alpha/2}\sqrt{\text{MSE}\left(\frac{1}{J_i} + \frac{1}{J_j}\right)} \tag{9.34}$$

then H_0 is rejected at level α.

The reason that the quantity $t_{N-I,\alpha/2}\sqrt{\text{MSE}(1/J_i + 1/J_j)}$ is called the least significant difference is that the null hypothesis of equal means is rejected at level α whenever the difference in sample means $|\overline{X}_{i.} - \overline{X}_{j.}|$ exceeds this value. When the design is balanced, with all sample sizes equal to J, the least significant difference is equal to $t_{N-I,\alpha/2}\sqrt{2\text{MSE}/J}$ for all pairs of means.

Example 9.9

In the weld experiment discussed in Section 9.1, hardness measurements were made for five welds from each of four fluxes A, B, C, and D. The sample mean hardness values were $\overline{X}_{A.} = 253.8$, $\overline{X}_{B.} = 263.2$, $\overline{X}_{C.} = 271.0$, and $\overline{X}_{D.} = 262.0$. The ANOVA table is reproduced in the following box.

```
One-way ANOVA: A, B, C, D

Source    DF        SS        MS      F      P
Factor     3    743.40   247.800   3.87  0.029
Error     16   1023.60    63.975
Total     19   1767.00

S = 7.998    R-Sq = 42.07%    R-Sq(adj) = 31.21%
```

Before the experiment was performed, the carbon contents of the fluxes were measured. Flux B had the lowest carbon content (2.67% by weight), and flux C had the highest (5.05% by weight). The experimenter is therefore particularly interested in comparing the hardnesses obtained with these two fluxes. Find a 95% confidence interval for the difference in mean hardness between welds produced with flux B and those produced with flux C. Can we conclude that the two means differ?

Solution

We use expression (9.32). The sample means are 271.0 for flux C and 263.2 for flux B. The preceding MINITAB output gives the quantity MSE as 63.975. (This value was also computed in Example 9.3 in Section 9.1.) The sample sizes are both equal to 5. There are $I = 4$ levels and $N = 20$ observations in total. For a 95% confidence interval, we consult the t table to find the value $t_{16,.025} = 2.120$. The 95% confidence interval is therefore $271.0 - 263.2 \pm 2.120\sqrt{63.975(1/5 + 1/5)}$ or $(-2.92, 18.52)$.

To perform a test of the null hypothesis that the two treatment means are equal, we compute the value of the test statistic (expression 9.33) and obtain

$$\frac{271.0 - 263.2}{\sqrt{63.975(1/5 + 1/5)}} = 1.54$$

Consulting the t table with $N - I = 16$ degrees of freedom, we find that P is between $2(0.05) = 0.10$ and $2(0.10) = 0.20$ (note that this is a two-tailed test). We cannot conclude that the treatment means differ.

If it is desired to perform a fixed-level test at level $\alpha = 0.05$ as an alternative to computing the P-value, the critical t value is $t_{16,.025} = 2.120$. The left-hand side of the inequality (9.34) is $|271.0-263.2| = 7.8$. The right-hand side is $2.120\sqrt{63.975(1/5+1/5)} = 10.72$. Since 7.8 does not exceed 10.72, we do not reject H_0 at the 5% level.

The following MINITAB output presents 95% Fisher LSD confidence intervals for each difference between treatment means in the weld experiment.

```
Fisher 95% Individual Confidence Intervals
All Pairwise Comparisons

Simultaneous confidence level = 81.11%

A subtracted from:

    Lower   Center    Upper  ------+---------+---------+---------+---
B  -1.324    9.400   20.124              (--------*--------)
C   6.476   17.200   27.924                    (--------*--------)
D  -2.524    8.200   18.924             (--------*--------)
                             ------+---------+---------+---------+---
                                 -12         0        12        24
B subtracted from:

    Lower   Center    Upper  ------+---------+---------+---------+---
C  -2.924    7.800   18.524             (--------*-------)
D -11.924   -1.200    9.524      (--------*-------)
                             ------+---------+---------+---------+---
                                 -12         0        12        24
C subtracted from:

     Lower   Center   Upper  ------+---------+---------+---------+---
D  -19.724   -9.000   1.724  (--------*-------)
                             ------+---------+---------+---------+---
                                 -12         0        12        24
```

The values labeled "Center" are the differences between pairs of treatment means. The quantities labeled "Lower" and "Upper" are the lower and upper bounds, respectively, of the confidence interval. Of particular note is the simultaneous confidence level of 81.11%. This indicates that although we are 95% confident that any given confidence interval contains its true difference in means, we are only 81.11% confident that *all* the confidence intervals contain their true differences.

In Example 9.9, a single test was performed on the difference between two specific means. What if we wanted to test every pair of means, to see which ones we could conclude to be different? It might seem reasonable to perform the LSD test on each pair. However, this is not appropriate, because when several tests are performed, the likelihood of rejecting a true null hypothesis increases. This is the multiple testing problem, which is discussed in some detail in Section 6.14. This problem is revealed in the preceding MINITAB output, which shows that the confidence is only 81.11% that all the 95% confidence intervals contain their true values. When several confidence intervals or hypothesis tests are to be considered simultaneously, the confidence intervals must be wider, and the criterion for rejecting the null hypotheses more strict, than in situations where only a single interval or test is involved. In these situations, multiple comparisons methods are used to produce **simultaneous confidence intervals** and **simultaneous hypothesis tests**. If level $100(1-\alpha)\%$ simultaneous confidence intervals are constructed for differences between every pair of means, then we are confident at the $100(1-\alpha)\%$ level that *every* confidence interval contains the true difference. If simultaneous hypothesis tests are conducted for all null hypotheses of the form $H_0: \mu_i - \mu_j = 0$, then we may reject, at level α, every null hypothesis whose P-value is less than α.

The Bonferroni Method of Multiple Comparisons

The Bonferroni method, discussed in Section 6.14, is a general method, valid anytime that several confidence intervals or tests are considered simultaneously. The method is simple to apply. Let C be the number of pairs of differences to be compared. For example, if there are I treatments, and all pairs of differences are to be compared, then $C = I(I-1)/2$. The Bonferroni method is the same as the LSD method, except that α is replaced with α/C.

The Bonferroni Method for Simultaneous Confidence Intervals and Hypothesis Tests

Assume that C differences of the form $\mu_i - \mu_j$ are to be considered. The Bonferroni simultaneous confidence intervals, at level $100(1-\alpha)\%$, for the C differences $\mu_i - \mu_j$ are

$$\overline{X}_{i.} - \overline{X}_{j.} \pm t_{N-I,\alpha/(2C)}\sqrt{\text{MSE}\left(\frac{1}{J_i}+\frac{1}{J_j}\right)} \tag{9.35}$$

We are $100(1-\alpha)\%$ confident that the Bonferroni confidence intervals contain the true value of the difference $\mu_i - \mu_j$ for all C pairs under consideration.

To test C null hypotheses of the form $H_0: \mu_i - \mu_j = 0$, the test statistics are

$$\frac{\overline{X}_{i.} - \overline{X}_{j.}}{\sqrt{\text{MSE}\left(\frac{1}{J_i}+\frac{1}{J_j}\right)}}$$

To find the P-value for each test, consult the Student's t table with $N - I$ degrees of freedom, and multiply the P-value found there by C.

Specifically, if

$$|\overline{X}_{i.} - \overline{X}_{j.}| > t_{N-I,\alpha/(2C)}\sqrt{\text{MSE}\left(\frac{1}{J_i}+\frac{1}{J_j}\right)}$$

then H_0 is rejected at level α.

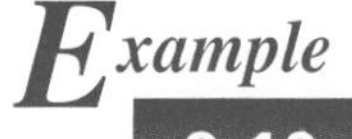

9.10

For the weld data discussed in Example 9.9, use the Bonferroni method to determine which pairs of fluxes, if any, can be concluded, at the 5% level, to differ in their effect on hardness.

Solution

There are $I = 4$ levels, with $J = 5$ observations at each level, for a total of $N = 20$ observations in all. With four levels, there are a total of $C = (4)(3)/2 = 6$ pairs of means to compare.

To test at the $\alpha = 5\%$ level, we compute $\alpha/(2C) = 0.004167$. The critical t value is $t_{16,.004167}$. This value is not in the table; it is between $t_{16,.005} = 2.921$ and $t_{16,.001} = 3.686$. Using computer software, we calculated $t_{16,.004167} = 3.0083$. Without software, one could roughly approximate this value by interpolation. Now MSE $= 63.975$ (see Example 9.9), so $t_{N-I,\,\alpha/(2C)}\sqrt{\text{MSE}(1/J_i + 1/J_j)} = 3.0083\sqrt{63.975(1/5 + 1/5)} = 15.22$. The four sample means are as follows:

Flux	A	B	C	D
Mean hardness	253.8	263.2	271.0	262.0

There is only one pair of sample means, 271.0 and 253.8, whose difference is greater than 15.22. We therefore conclude that welds produced with flux A have different mean hardness than welds produced with flux C. None of the other differences are significant at the 5% level.

Although easy to use, the Bonferroni method has the disadvantage that as C becomes large, the confidence intervals become very wide, and the hypothesis tests have low power. The reason for this is that the Bonferroni method is a general method, not specifically designed for analysis of variance or for normal populations. In many cases C is fairly large, in particular it is often desired to compare all pairs of means. In these cases, a method called the **Tukey–Kramer method** is superior, because it is designed for multiple comparisons of means of normal populations. We now describe this method.

The Tukey–Kramer Method of Multiple Comparisons

The Tukey–Kramer method is based on a distribution called the **Studentized range distribution**, rather than on the Student's t distribution. The Studentized range distribution has two values for degrees of freedom, which for the Tukey–Kramer method are I and $N - I$. (In comparison, the F test uses $I - 1$ and $N - I$ degrees of freedom.) The Tukey–Kramer method uses the $1 - \alpha$ quantile of the Studentized range distribution with I and $N - I$ degrees of freedom; this quantity is denoted $q_{I,N-I,\alpha}$. Table A.8 (in Appendix A) presents values of $q_{I,N-I,\alpha}$ for various values of I, N, and α. The mechanics of the Tukey–Kramer method are the same as those for the LSD method, except that $t_{N-I,\,\alpha/2}\sqrt{\text{MSE}(1/J_i + 1/J_j)}$ is replaced with $q_{I,N-I,\alpha}\sqrt{(\text{MSE}/2)(1/J_i + 1/J_j)}$. The quantity $q_{I,N-I,\alpha}\sqrt{(\text{MSE}/2)(1/J_i + 1/J_j)}$ is sometimes called the **honestly significant difference** (HSD), in contrast to Fisher's least significant difference.

The Tukey–Kramer Method for Simultaneous Confidence Intervals and Hypothesis Tests

The Tukey–Kramer level $100(1 - \alpha)\%$ simultaneous confidence intervals for all differences $\mu_i - \mu_j$ are

$$\overline{X}_{i.} - \overline{X}_{j.} \pm q_{I,N-I,\alpha} \sqrt{\frac{\text{MSE}}{2}\left(\frac{1}{J_i} + \frac{1}{J_j}\right)} \tag{9.36}$$

We are $100(1 - \alpha)\%$ confident that the Tukey–Kramer confidence intervals contain the true value of the difference $\mu_i - \mu_j$ for every i and j.

To test all null hypotheses $H_0: \mu_i - \mu_j = 0$ simultaneously, the test statistics are

$$\frac{\overline{X}_{i.} - \overline{X}_{j.}}{\sqrt{\frac{\text{MSE}}{2}\left(\frac{1}{J_i} + \frac{1}{J_j}\right)}}$$

The P-value for each test is found by consulting the Studentized range table (Table A.8) with I and $N - I$ degrees of freedom.

For every pair of levels i and j for which

$$|\overline{X}_{i.} - \overline{X}_{j.}| > q_{I,N-I,\alpha} \sqrt{\frac{\text{MSE}}{2}\left(\frac{1}{J_i} + \frac{1}{J_j}\right)}$$

the null hypothesis $H_0: \mu_i - \mu_j = 0$ is rejected at level α.

A note on terminology: When the design is balanced, with all sample sizes equal to J, the quantity $\sqrt{(\text{MSE}/2)(1/J_i + 1/J_j)}$ is equal to $\sqrt{\text{MSE}/J}$ for all pairs of levels. In this case, the method is often simply called Tukey's method.

Example 9.11

For the weld data in Table 9.1 (in Section 9.1), which pairs of fluxes, if any, can be concluded, at the 5% level, to differ in their effect on hardness?

Solution

There are $I = 4$ levels, with $J = 5$ observations at each level, for a total of $N = 20$ observations in all. To test at level $\alpha = 0.05$, we consult the Studentized range table (Table A.8) to find $q_{4,16,.05} = 4.05$.

The value of MSE is 63.975 (see Example 9.9). Therefore $q_{I,N-I,\alpha}\sqrt{\text{MSE}/J} = 4.05\sqrt{63.975/5} = 14.49$. The four sample means are as follows:

Flux	A	B	C	D
Mean hardness	253.8	263.2	271.0	262.0

There is only one pair of sample means, 271.0 and 253.8, whose difference is greater than 14.49. We therefore conclude that welds produced with flux A have a different mean hardness than welds produced with flux C. None of the other differences are significant at the 5% level.

Comparing the results of Example 9.11 with those of Example 9.10 shows that in this case the Tukey–Kramer method is slightly more powerful than the Bonferroni method, since its critical value is only 14.49 while that of the Bonferroni method was 15.22. When all possible pairs are compared, as in this example, the Tukey–Kramer method is always more powerful than the Bonferroni method. When only a few of the possible pairs are to be compared, the Bonferroni method is sometimes more powerful.

Sometimes only a single test is performed, but the difference that is tested is chosen by examining the sample means and choosing two whose difference is large. In these cases a multiple comparisons method should be used, even though only one test is being performed. Example 9.12 illustrates the idea.

Example 9.12

An engineer examines the weld data in Table 9.1 and notices that the two treatments with the largest difference in sample means are flux A and flux C. He decides to test the null hypothesis that the mean hardness for welds produced with flux A differs from that for welds produced with flux C. Since he will only perform one test, he uses the Fisher LSD method rather than the Bonferroni or Tukey–Kramer method. Explain why this is wrong.

Solution

The engineer has examined every pair of means and has chosen the two whose difference is largest. Although he is formally performing only one test, he has chosen that test by comparing every pair of sample means. For this reason he should use a multiple comparisons procedure, such as the Bonferroni or Tukey–Kramer method.

The following MINITAB output presents the Tukey–Kramer 95% simultaneous confidence intervals for the weld data.

```
Tukey 95% Simultaneous Confidence Intervals
All Pairwise Comparisons

Individual confidence level = 98.87%

A subtracted from:

    Lower  Center   Upper  ------+---------+---------+---------+---
B  -5.087  9.4000  23.887              (--------*--------)
```

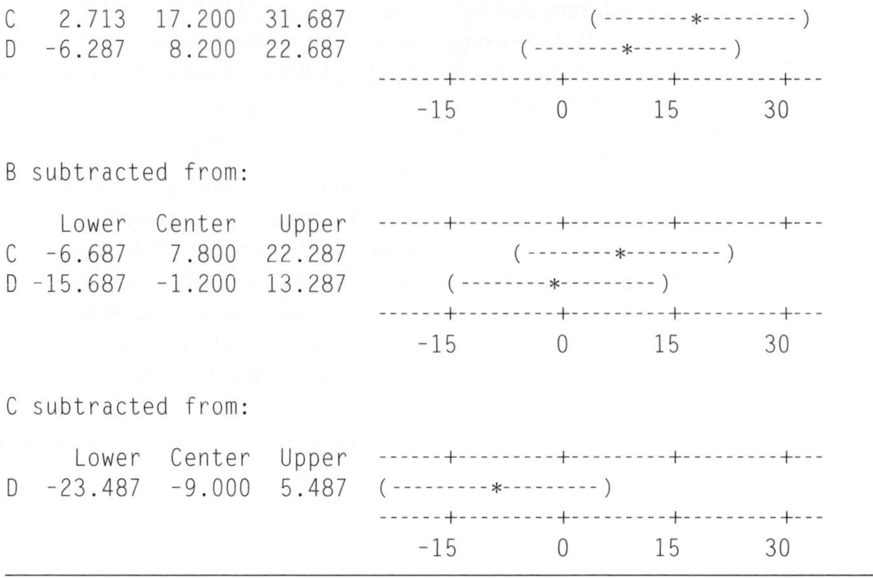

```
C    2.713  17.200  31.687                    (--------*---------)
D   -6.287   8.200  22.687              (--------*---------)
                              ------+---------+---------+---------+---
                                  -15         0        15        30

B subtracted from:

     Lower  Center   Upper    ------+---------+---------+---------+---
C   -6.687   7.800  22.287              (--------*---------)
D  -15.687  -1.200  13.287        (--------*---------)
                              ------+---------+---------+---------+---
                                  -15         0        15        30

C subtracted from:

      Lower  Center  Upper    ------+---------+---------+---------+---
D   -23.487  -9.000  5.487    (---------*--------)
                              ------+---------+---------+---------+---
                                  -15         0        15        30
```

The values labeled "Center" are the differences between pairs of treatment means. The quantities labeled "Lower" and "Upper" are the lower and upper bounds, respectively, of the confidence interval. We are 95% confident that every one of these confidence intervals contains the true difference in treatment means. Note that the "Individual confidence level" is 98.87%. This means that we are 98.87% confident that any one specific confidence interval contains its true value. Finally we point out that because the confidence level for the Tukey–Kramer intervals is higher than that for the Fisher LSD intervals, the Tukey–Kramer intervals are wider.

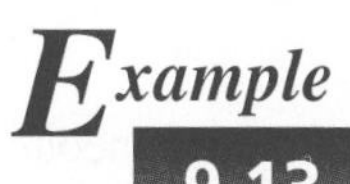

Example 9.13

In Example 9.5 (in Section 9.1), several measurements of the maximum hourly concentrations (in $\mu g/m^3$) of SO_2 were presented for each of four power plants, and it was concluded that the mean concentrations at the four plants were not all the same. The following MINITAB output presents the Tukey–Kramer 95% simultaneous confidence intervals for mean concentrations at the four plants. Which pairs of plants, if any, can you conclude with 95% confidence to have differing means?

```
Tukey 95% Simultaneous Confidence Intervals
All Pairwise Comparisons

Individual confidence level = 98.87%
```

```
1 subtracted from:

   Lower  Center  Upper  -----+---------+---------+---------+---
2  109.4   385.3  661.1                   (--------*--------)
3   21.4   312.3  603.1                (--------*---------)
4  -94.6   170.9  436.4            (--------*--------)
                         -----+---------+---------+---------+---
                           -300         0       300       600

2 subtracted from:

   Lower  Center  Upper  -----+---------+---------+---------+---
3 -348.9   -73.0  202.9    (---------*--------)
4 -463.4  -214.3   34.7  (-------*-------)
                         -----+---------+---------+---------+---
                           -300         0       300       600

3 subtracted from:

   Lower  Center  Upper  -----+---------+---------+---------+---
4 -406.8  -141.3  124.1   (--------*--------)
                         -----+---------+---------+---------+---
                           -300         0       300       600
```

Solution

Among the simultaneous confidence intervals, there are two that do not contain 0. These are the intervals for $\mu_1 - \mu_2$ and for $\mu_1 - \mu_3$. Therefore we conclude that the mean concentrations differ between plants 1 and 2 and between plants 1 and 3.

Exercises for Section 9.2

1. The article "Organic Recycling for Soil Quality Conservation in a Sub-Tropical Plateau Region" (K. Chakrabarti, B. Sarkar, et al., *J. Agronomy and Crop Science*, 2000:137–142) reports an experiment in which soil specimens were treated with six different treatments, with two replicates per treatment, and the acid phosphate activity (in μmol *p*-nitrophenol released per gram of oven-dry soil per hour) was recorded. An ANOVA table for a one-way ANOVA is presented in the following box.

```
One-way ANOVA: Treatments A, B, C, D, E, F

Source      DF       SS         MS        F       P
Treatment    5   1.18547    0.23709   46.64   0.000
Error        6   0.03050    0.00508
Total       11   1.21597
```

The treatment means were

Treatment	A	B	C	D	E	F
Mean	0.99	1.99	1.405	1.63	1.395	1.22

a. Can you conclude that there are differences in acid phosphate activity among the treatments?
b. Use the Tukey–Kramer method to determine which pairs of treatment means, if any, are different at the 5% level.
c. Use the Bonferroni method to determine which pairs of treatment means, if any, are different at the 5% level.
d. Which method is more powerful in this case, the Tukey–Kramer method or the Bonferroni method?
e. The experimenter notices that treatment A had the smallest sample mean, while treatment B had the largest. Of the Fisher LSD method, the Bonferroni method, and the Tukey–Kramer method, which, if any, can be used to test the hypothesis that these two treatment means are equal?

2. The article "Optimum Design of an A-pillar Trim with Rib Structures for Occupant Head Protection" (H. Kim and S. Kang, *Proceedings of the Institution of Mechanical Engineers*, 2001:1161–1169) discusses a study in which several types of A-pillars were compared to determine which provided the greatest protection to occupants of automobiles during a collision. Following is a one-way ANOVA table, where the treatments are three levels of longitudinal spacing of the rib (the article also discussed two insignificant factors, which are omitted here). There were nine replicates at each level. The response is the head injury criterion (HIC), which is a unitless quantity that measures the impact energy absorption of the pillar.

```
One-way ANOVA: Spacing

Source     DF         SS         MS        F        P
Spacing     2    50946.6    25473.3    5.071    0.015
Error      24   120550.9     5023.0
Total      26   171497.4
```

The treatment means were

Treatment	A	B	C
Mean	930.87	873.14	979.41

a. Can you conclude that the longitudinal spacing affects the absorption of impact energy?
b. Use the Tukey–Kramer method to determine which pairs of treatment means, if any, are different at the 5% level.
c. Use the Bonferroni method to determine which pairs of treatment means, if any, are different at the 5% level.
d. Which method is more powerful in this case, the Tukey–Kramer method or the Bonferroni method?

3. Acrylic resins used in the fabrication of dentures should not absorb much water, since water sorption reduces strength. The article "Reinforcement of Acrylic Resin for Provisional Fixed Restorations. Part III: Effects of Addition of Titania and Zirconia Mixtures on Some Mechanical and Physical Properties" (W. Panyayong, Y. Oshida, et al., *Bio-Medical Materials and Engineering*, 2002:353–366) describes a study of the effect on water sorption of adding titanium dioxide (TiO_2) and zirconium dioxide (ZrO_2) to a standard acrylic resin. Twelve specimens from each of several formulations, containing various amounts of TiO_2 and ZrO_2, were immersed in water for one week, and the water sorption (in μg/mm^2) was measured in each. The results are presented in the following table.

Formulation	Volume %		Mean	Standard Deviation
	TIO_2	ZrO_2		
A (control)	0	0	24.03	2.50
B	1	1	14.88	1.55
C	1	2	12.81	1.08
D	1	0.5	11.21	2.98
E	2	2	16.05	1.66
F	2	4	12.87	0.96
G	2	1	15.23	0.97
H	3	3	15.37	0.64

a. Use the Bonferroni method to determine which of the noncontrol formulations (B through H) differ, at the 5% level, in their mean water sorption from the control formulation A.
b. Repeat part (a) using the Tukey–Kramer method.
c. Which method is more powerful for these comparisons? Why?

4. Refer to Exercise 1 in Section 9.1.

a. Use the Bonferroni method to determine which pairs of means, if any, are different at the 5% level.
b. Use the Tukey–Kramer method to determine which pairs of means, if any, are different at the 5% level.
c. Which is the more powerful method to find all the pairs of treatments whose means are different, the Bonferroni method or the Tukey–Kramer method?

5. Refer to Exercise 11 in Section 9.1.

a. Use the Bonferroni method to determine which pairs of means, if any, are different at the 5% level.
b. Use the Tukey–Kramer method to determine which pairs of means, if any, are different at the 5% level.
c. Which is the more powerful method to find all the pairs of treatments whose means are different, the Bonferroni method or the Tukey–Kramer method?

6. Refer to Exercise 1 in Section 9.1. A scientist wants to determine whether the mean amount absorbed in a 24 hour duration differs from the mean amounts absorbed in durations of 1, 2, 4, and 10 hours.

a. Use the Bonferroni method to determine which of the means, if any, for 1, 2, 4, and 10 hours differ from the mean for 24 hours. Use the 5% level.
b. Use the Tukey–Kramer method to determine which of the means, if any, for 1, 2, 4, and 10 hours differ from the mean for 24 hours. Use the 5% level.
c. Which is the more powerful method to find all the treatments whose means differ from that of the 24 hour duration, the Bonferroni method or the Tukey–Kramer method?

7. Refer to Exercise 11 in Section 9.1. A metallurgist wants to determine whether the mean tensile strength for specimens annealed at 900°C differs from the mean strengths for specimens annealed at 750°C, 800°C, and 850°C.

a. Use the Bonferroni method to determine which of the means, if any, for 750°C, 800°C, and 850°C differ from the mean for 900°C.
b. Use the Tukey–Kramer method to determine which of the means, if any, for 750°C, 800°C, and 850°C differ from the mean for 900°C.
c. Which is the more powerful method to find all the pairs of treatments whose means differ from the 900°C mean, the Bonferroni method or the Tukey–Kramer method?

8. Refer to Exercise 3 in Section 9.1.

a. Use the Fisher LSD method to find a 95% confidence interval for the difference between the means for treatments B and D.
b. Use the Tukey–Kramer method to determine which pairs of treatments, if any, differ at the 5% level.

9. Refer to Exercise 5 in Section 9.1.

a. Use the Fisher LSD method to find a 95% confidence interval for the difference between the means for Buckeye Hollow and Bear Branch.
b. Use the Tukey–Kramer method to determine which pairs of treatments, if any, differ at the 5% level.

10. Refer to Exercise 7 in Section 9.1.

a. Use the Fisher LSD method to find a 95% confidence interval for the difference between the means for catalyst 1 and catalyst 3.
b. Use the Tukey–Kramer method to determine which pairs of catalysts, if any, differ at the 5% level.

11. Refer to Exercise 14 in Section 9.1.

a. Use the Fisher LSD method to find a 95% confidence interval for the difference between the means for a diameter of 16 and a diameter of 36.
b. Use the Tukey–Kramer method to determine which pairs of diameters, if any, differ at the 5% level.

12. Refer to Exercise 16 in Section 9.1.

a. Use the Fisher LSD method to find a 95% confidence interval for the difference between the means for annularly threaded and smooth shank nails.
b. Use the Tukey–Kramer method to determine which pairs of nail types, if any, differ at the 5% level.

13. In an experiment to determine the effect of catalyst on the yield of a certain reaction, the mean yields for reactions run with each of four catalysts were $\overline{X}_{1.} = 89.88$, $\overline{X}_{2.} = 89.51$, $\overline{X}_{3.} = 86.98$, and $\overline{X}_{4.} = 85.79$. Assume that five runs were made with each catalyst.

a. If MSE = 3.85, compute the value of the F statistic for testing the null hypothesis that all four catalysts have the same mean yield. Can this null hypothesis be rejected at the 5% level?
b. Use the Tukey–Kramer method to determine which pairs of catalysts, if any, may be concluded to differ at the 5% level.

14. In an experiment to determine the effect of curing time on the compressive strength of a certain type of concrete, the mean strengths, in MPa, for specimens cured for each of four curing times were $\overline{X}_{1.} = 1316$, $\overline{X}_{2.} = 1326$, $\overline{X}_{3.} = 1375$, and $\overline{X}_{4.} = 1389$. Assume that four specimens were cured for each curing time.

a. If MSE = 875.2, compute the value of the F statistic for testing the null hypothesis that all four curing times have the same mean strength. Can this null hypothesis be rejected at the 5% level?
b. Use the Tukey–Kramer method to determine which pairs of curing times, if any, may be concluded to differ at the 5% level.

15. For some data sets, the F statistic will reject the null hypothesis of no difference in mean yields, but the Tukey–Kramer method will not find any pair of means that can be concluded to differ. For the four sample means given in Exercise 13, assuming a sample size of 5 for each treatment, find a value of MSE so that the F statistic rejects the null hypothesis of no difference at the 5% level, while the Tukey–Kramer method does not find any pair of means to differ at the 5% level.

16. For some data sets, the F statistic will reject the null hypothesis of no difference in mean yields, but the Tukey–Kramer method will not find any pair of means that can be concluded to differ. For the four sample means given in Exercise 14, assuming a sample size of 4 for each treatment, find a value of MSE so that the F statistic rejects the null hypothesis of no difference at the 5% level, while the Tukey–Kramer method does not find any pair of means to differ at the 5% level.

9.3 Two-Factor Experiments

In one-factor experiments, discussed in Sections 9.1 and 9.2, the purpose is to determine whether varying the level of a single factor affects the response. Many experiments involve varying several factors, each of which may affect the response. In this section, we will discuss the case in which there are two factors. The experiments, naturally enough, are called **two-factor experiments**. We illustrate with an example.

A chemical engineer is studying the effects of various reagents and catalysts on the yield of a certain process. Yield is expressed as a percentage of a theoretical maximum. Four runs of the process were made for each combination of three reagents and four catalysts. The results are presented in Table 9.2. In this experiment there are two factors, the catalyst and the reagent. The catalyst is called the **row factor**, since its value varies from row to row in the table, while the reagent is called the **column factor**. These designations are arbitrary, in that the table could just as easily have been presented with the rows representing the reagents and the columns representing the catalysts.

TABLE 9.2 Yields for runs of a chemical process with various combinations of reagent and catalyst

	Reagent		
Catalyst	**1**	**2**	**3**
A	86.8 82.4 86.7 83.5	93.4 85.2 94.8 83.1	77.9 89.6 89.9 83.7
B	71.9 72.1 80.0 77.4	74.5 87.1 71.9 84.1	87.5 82.7 78.3 90.1
C	65.5 72.4 76.6 66.7	66.7 77.1 76.7 86.1	72.7 77.8 83.5 78.8
D	63.9 70.4 77.2 81.2	73.7 81.6 84.2 84.9	79.8 75.7 80.5 72.9

In general, there are I levels of the row factor and J levels of the column factor. (In Table 9.2, $I = 4$ and $J = 3$.) There are therefore IJ different combinations of the two factors. The terminology for these factor combinations is not standardized. We will refer to each combination of factors as a **treatment**, but some authors use the term **treatment combination**. Recall that the units assigned to a given treatment are called replicates. When the number of replicates is the same for each treatment, we will denote this number by K. Thus in Table 9.2, $K = 4$.

When observations are taken on every possible treatment, the design is called a **complete design** or a **full factorial design**. Incomplete designs, in which there are no data for one or more treatments, can be difficult to interpret, except for some special cases. When possible, complete designs should be used. When the number of replicates is the same for each treatment, the design is said to be **balanced**. For one-factor experiments, we did not need to assume that the design was balanced. With two-factor experiments, unbalanced designs are much more difficult to analyze than balanced designs. We will restrict our discussion to balanced designs. As with one-factor experiments, the factors may be fixed or random. The methods that we will describe apply to models where both effects are fixed. Later we will briefly describe models where one or both factors are random.

In a completely randomized design, each treatment represents a population, and the observations on that treatment are a simple random sample from that population. We will denote the sample values for the treatment corresponding to the ith level of the row factor and the jth level of the column factor by $X_{ij1}, \ldots, X_{ijK}$. We will denote the population mean outcome for this treatment by μ_{ij}. The values μ_{ij} are often called the **treatment means**. In general, the purpose of a two-factor experiment is to determine whether the treatment means are affected by varying either the row factor, the column factor, or both. The method of analysis appropriate for two-factor experiments is called **two-way analysis of variance**.

Parameterization for Two-Way Analysis of Variance

In a two-way analysis of variance, we wish to determine whether varying the level of the row or column factors changes the value of μ_{ij}. To do this, we must express μ_{ij} in terms of parameters that describe the row and column factors separately. We'll begin this task by describing some notation for the averages of the treatment means for the different levels of the row and column factors.

For any level i of the row factor, the average of all the treatment means μ_{ij} in the ith row is denoted $\overline{\mu}_{i.}$. We express $\overline{\mu}_{i.}$ in terms of the treatment means as follows:

$$\overline{\mu}_{i.} = \frac{1}{J}\sum_{j=1}^{J}\mu_{ij} \tag{9.37}$$

Similarly, for any level j of the column factor, the average of all the treatment means μ_{ij} in the jth column is denoted $\overline{\mu}_{.j}$. We express $\overline{\mu}_{.j}$ in terms of the treatment means as follows:

$$\overline{\mu}_{.j} = \frac{1}{I}\sum_{i=1}^{I}\mu_{ij} \tag{9.38}$$

Finally, we define the **population grand mean**, denoted by μ, which represents the average of all the treatment means μ_{ij}. The population grand mean can also be expressed as the average of the quantities $\overline{\mu}_{i.}$ or of the quantities $\overline{\mu}_{.j}$:

$$\mu = \frac{1}{I}\sum_{i=1}^{I}\overline{\mu}_{i.} = \frac{1}{J}\sum_{j=1}^{J}\overline{\mu}_{.j} = \frac{1}{IJ}\sum_{i=1}^{I}\sum_{j=1}^{J}\mu_{ij} \tag{9.39}$$

Table 9.3 illustrates the relationships among μ_{ij}, $\overline{\mu}_{i.}$, $\overline{\mu}_{.j}$, and μ.

Using the quantities $\overline{\mu}_{i.}$, $\overline{\mu}_{.j}$, and μ, we can decompose the treatment mean μ_{ij} as follows:

$$\mu_{ij} = \mu + (\overline{\mu}_{i.} - \mu) + (\overline{\mu}_{.j} - \mu) + (\mu_{ij} - \overline{\mu}_{i.} - \overline{\mu}_{.j} + \mu) \tag{9.40}$$

Equation (9.40) expresses the treatment mean μ_{ij} as a sum of four terms. In practice, simpler notation is used for the three rightmost terms in Equation (9.40):

$$\alpha_i = \overline{\mu}_{i.} - \mu \tag{9.41}$$

$$\beta_j = \overline{\mu}_{.j} - \mu \tag{9.42}$$

$$\gamma_{ij} = \mu_{ij} - \overline{\mu}_{i.} - \overline{\mu}_{.j} + \mu \tag{9.43}$$

TABLE 9.3 Treatment means and their averages across rows and down columns

Row Level	Column Level 1	2	...	J	Row Mean
1	μ_{11}	μ_{12}	...	μ_{1J}	$\overline{\mu}_{1.}$
2	μ_{21}	μ_{22}	...	μ_{2J}	$\overline{\mu}_{2.}$
⋮	⋮	⋮	...	⋮	⋮
I	μ_{I1}	μ_{I2}	...	μ_{IJ}	$\overline{\mu}_{I.}$
Column Mean	$\overline{\mu}_{.1}$	$\overline{\mu}_{.2}$	...	$\overline{\mu}_{.J}$	μ

Each of quantities μ, α_i, β_j, and γ_{ij} has an important interpretation:

- The quantity μ is the population grand mean, which is the average of all the treatment means.
- The quantity $\alpha_i = \overline{\mu}_{i.} - \mu$ is called the ith **row effect**. It is the difference between the average treatment mean for the ith level of the row factor and the population grand mean. The value of α_i indicates the degree to which the ith level of the row factor tends to produce outcomes that are larger or smaller than the population grand mean.
- The quantity $\beta_j = \overline{\mu}_{.j} - \mu$ is called the jth **column effect**. It is the difference between the average treatment mean for the jth level of the column factor and the population grand mean. The value of β_j indicates the degree to which the jth level of the column factor tends to produce outcomes that are larger or smaller than the population grand mean.
- The quantity $\gamma_{ij} = \mu_{ij} - \overline{\mu}_{i.} - \overline{\mu}_{.j} + \mu$ is called the ij **interaction**. The effect of a level of a row (or column) factor may depend on which level of the column (or row) factor it is paired with. The interaction terms measure the degree to which this occurs. For example, assume that level 1 of the row factor tends to produce a large outcome when paired with column level 1, but a small outcome when paired with column level 2. In this case $\gamma_{1,1}$ would be positive, and $\gamma_{1,2}$ would be negative.

Both row effects and column effects are called **main effects** to distinguish them from the interactions. Note that there are I row effects, one for each level of the row factor, J column effects, one for each level of the column factor, and IJ interactions, one for each treatment. Furthermore, it follows from the definitions of quantities $\overline{\mu}_{i.}$, $\overline{\mu}_{.j}$, and μ in Equations (9.37) through (9.39) that the row effects, column effects, and interactions satisfy the following constraints:

$$\sum_{i=1}^{I} \alpha_i = 0 \qquad \sum_{j=1}^{J} \beta_j = 0 \qquad \sum_{i=1}^{I} \gamma_{ij} = \sum_{j=1}^{J} \gamma_{ij} = 0 \tag{9.44}$$

We now can express the treatment means μ_{ij} in terms of α_i, β_j, and γ_{ij}. From Equation (9.40) it follows that

$$\mu_{ij} = \mu + \alpha_i + \beta_j + \gamma_{ij} \tag{9.45}$$

For each observation X_{ijk}, define $\varepsilon_{ijk} = X_{ijk} - \mu_{ij}$, the difference between the observation and its treatment mean. The quantities ε_{ijk} are called **errors**. It follows that

$$X_{ijk} = \mu_{ij} + \varepsilon_{ijk} \tag{9.46}$$

Combining Equations (9.46) and (9.45) yields the two-way ANOVA model:

$$X_{ijk} = \mu + \alpha_i + \beta_j + \gamma_{ij} + \varepsilon_{ijk} \tag{9.47}$$

When the interactions γ_{ij} are all equal to 0, the **additive model** is said to apply. Under the additive model, Equation (9.45) becomes

$$\mu_{ij} = \mu + \alpha_i + \beta_j \tag{9.48}$$

and Equation (9.47) becomes

$$X_{ijk} = \mu + \alpha_i + \beta_j + \varepsilon_{ijk} \tag{9.49}$$

Under the additive model, the treatment mean μ_{ij} is equal to the population grand mean μ, plus an amount α_i that results from using row level i plus an amount β_j that results from using column level j. In other words, the combined effect of using row level i along with column level j is found by adding the individual main effects of the two levels. When some or all of the interactions are not equal to 0, the additive model does not hold, and the combined effect of a row level and a column level cannot be determined from their individual main effects.

We will now show how to estimate the parameters for the full two-way model (9.47). The procedure for the additive model is exactly the same, except that the interactions γ_{ij} are not estimated. The procedure is straightforward. We first define some notation for various averages of the data X_{ijk}, using the data in Table 9.2 as an example. Table 9.4 presents the average yield for the four runs for each reagent and catalyst in Table 9.2.

TABLE 9.4 Average yields $\overline{X}_{ij.}$ for runs of a chemical process using different combinations of reagent and catalyst

	Reagent			
Catalyst	**1**	**2**	**3**	**Row Mean $\overline{X}_{i..}$**
A	84.85	89.13	85.28	86.42
B	75.35	79.40	84.65	79.80
C	70.30	76.65	78.20	75.05
D	73.18	81.10	77.23	77.17
Column Mean $\overline{X}_{.j.}$	75.92	81.57	81.34	**Sample Grand Mean $\overline{X}_{...}$** = 79.61

Each number in the body of Table 9.4 is the average of the four numbers in the corresponding cell of Table 9.2. These are called the **cell means**. They are denoted $\overline{X}_{ij.}$ and are defined by

$$\overline{X}_{ij.} = \frac{1}{K}\sum_{k=1}^{K} X_{ijk} \tag{9.50}$$

Averaging the cell means across the rows produces the **row means** $\overline{X}_{i..}$:

$$\overline{X}_{i..} = \frac{1}{J}\sum_{j=1}^{J}\overline{X}_{ij.} = \frac{1}{JK}\sum_{j=1}^{J}\sum_{k=1}^{K}X_{ijk} \tag{9.51}$$

Averaging the cell means down the columns produces the **column means** $\overline{X}_{.j.}$:

$$\overline{X}_{.j.} = \frac{1}{I}\sum_{i=1}^{I}\overline{X}_{ij.} = \frac{1}{IK}\sum_{i=1}^{I}\sum_{k=1}^{K}X_{ijk} \tag{9.52}$$

The sample grand mean $\overline{X}_{...}$ can be found by computing the average of the row means, the average of the column means, the average of the cell means, or the average of all the observations:

$$\overline{X}_{...} = \frac{1}{I}\sum_{i=1}^{I}\overline{X}_{i..} = \frac{1}{J}\sum_{j=1}^{J}\overline{X}_{.j.} = \frac{1}{IJ}\sum_{i=1}^{I}\sum_{j=1}^{J}\overline{X}_{ij.} = \frac{1}{IJK}\sum_{i=1}^{I}\sum_{j=1}^{J}\sum_{k=1}^{K}X_{ijk} \tag{9.53}$$

Now we describe how to estimate the parameters in the two-way ANOVA model. The fundamental idea is that the best estimate of the treatment mean μ_{ij} is the cell mean $\overline{X}_{ij.}$, which is the average of the sample observations having that treatment. It follows that the best estimate of the quantity $\overline{\mu}_{i.}$ is the row mean $\overline{X}_{i..}$, the best estimate of the quantity $\overline{\mu}_{.j}$ is the column mean $\overline{X}_{.j.}$, and the best estimate of the population grand mean μ is the sample grand mean $\overline{X}_{...}$. We estimate the row effects α_i, the column effects β_j, and the interactions γ_{ij} by substituting these estimates into Equations (9.41) through (9.43).

$$\widehat{\alpha}_i = \overline{X}_{i..} - \overline{X}_{...} \tag{9.54}$$

$$\widehat{\beta}_j = \overline{X}_{.j.} - \overline{X}_{...} \tag{9.55}$$

$$\widehat{\gamma}_{ij} = \overline{X}_{ij.} - \overline{X}_{i..} - \overline{X}_{.j.} + \overline{X}_{...} \tag{9.56}$$

The row effects, column effects, and interactions satisfy constraints given in Equation (9.44). By performing some algebra, it can be shown that their estimates satisfy the same constraints:

$$\sum_{i=1}^{I}\widehat{\alpha}_i = 0 \qquad \sum_{j=1}^{J}\widehat{\beta}_j = 0 \qquad \sum_{i=1}^{I}\widehat{\gamma}_{ij} = \sum_{j=1}^{J}\widehat{\gamma}_{ij} = 0 \tag{9.57}$$

Example 9.14

Compute the estimated row effects, column effects, and interactions for the data in Table 9.2.

Solution

Using the quantities in Table 9.4 and Equations (9.54) through (9.56), we compute

$\widehat{\alpha}_1 = 86.42 - 79.61 = 6.81$ $\qquad$ $\widehat{\alpha}_2 = 79.80 - 79.61 = 0.19$

$\widehat{\alpha}_3 = 75.05 - 79.61 = -4.56$ $\qquad$ $\widehat{\alpha}_4 = 77.17 - 79.61 = -2.44$

$\widehat{\beta}_1 = 75.92 - 79.61 = -3.69 \qquad \widehat{\beta}_2 = 81.57 - 79.61 = 1.96$

$\widehat{\beta}_3 = 81.34 - 79.61 = 1.73$

$$\begin{array}{lll} \widehat{\gamma}_{11} = 2.12 & \widehat{\gamma}_{12} = 0.75 & \widehat{\gamma}_{13} = -2.87 \\ \widehat{\gamma}_{21} = -0.76 & \widehat{\gamma}_{22} = -2.36 & \widehat{\gamma}_{23} = 3.12 \\ \widehat{\gamma}_{31} = -1.06 & \widehat{\gamma}_{32} = -0.36 & \widehat{\gamma}_{33} = 1.42 \\ \widehat{\gamma}_{41} = -0.30 & \widehat{\gamma}_{42} = 1.97 & \widehat{\gamma}_{43} = -1.67 \end{array}$$

Using Two-Way ANOVA to Test Hypotheses

A two-way analysis of variance is designed to address three main questions:

1. Does the additive model hold?
2. If so, is the mean outcome the same for all levels of the row factor?
3. If so, is the mean outcome the same for all levels of the column factor?

In general, we ask questions 2 and 3 only when we believe that the additive model may hold. We will discuss this further later in this section. The three questions are addressed by performing hypothesis tests. The null hypotheses for these tests are as follows:

1. To test whether the additive model holds, we test the null hypothesis that all the interactions are equal to 0:

$$H_0\colon \gamma_{11} = \gamma_{12} = \cdots = \gamma_{IJ} = 0$$

If this null hypothesis is true, the additive model holds.

2. To test whether the mean outcome is the same for all levels of the row factor, we test the null hypothesis that all the row effects are equal to 0:

$$H_0\colon \alpha_1 = \alpha_2 = \cdots = \alpha_I = 0$$

If this null hypothesis is true, then the mean outcome is the same for all levels of the row factor.

3. To test whether the mean outcome is the same for all levels of the column factor, we test the null hypothesis that all the column effects are equal to 0:

$$H_0\colon \beta_1 = \beta_2 = \cdots = \beta_J = 0$$

If this null hypothesis is true, then the mean outcome is the same for all levels of the column factor.

We now describe the standard tests for these null hypotheses. For the tests to be valid, the following conditions must hold:

Assumptions for Two-Way ANOVA

The standard two-way ANOVA hypothesis tests are valid under the following conditions:

1. The design must be complete.
2. The design must be balanced.
3. The number of replicates per treatment, K, must be at least 2.
4. Within any treatment, the observations $X_{ij1}, \ldots, X_{ijK}$ are a simple random sample from a normal population.
5. The population variance is the same for all treatments. We denote this variance by σ^2.

Just as in one-way ANOVA, the standard tests for these null hypotheses are based on sums of squares. Specifically, they are the row sum of squares (SSA), the column sum of squares (SSB), the interaction sum of squares (SSAB), and the error sum of squares (SSE). Also of interest is the total sum of squares (SST), which is equal to the sum of the others. Formulas for these sums of squares are as follows:

$$\text{SSA} = JK\sum_{i=1}^{I}\widehat{\alpha}_i^2 = JK\sum_{i=1}^{I}(\overline{X}_{i..} - \overline{X}_{...})^2 = JK\sum_{i=1}^{I}\overline{X}_{i..}^2 - IJK\overline{X}_{...}^2 \quad (9.58)$$

$$\text{SSB} = IK\sum_{j=1}^{J}\widehat{\beta}_j^2 = IK\sum_{j=1}^{J}(\overline{X}_{.j.} - \overline{X}_{...})^2 = IK\sum_{j=1}^{J}\overline{X}_{.j.}^2 - IJK\overline{X}_{...}^2 \quad (9.59)$$

$$\begin{aligned}\text{SSAB} &= K\sum_{i=1}^{I}\sum_{j=1}^{J}\widehat{\gamma}_{ij}^2 = K\sum_{i=1}^{I}\sum_{j=1}^{J}(\overline{X}_{ij.} - \overline{X}_{i..} - \overline{X}_{.j.} + \overline{X}_{...})^2 \\ &= K\sum_{i=1}^{I}\sum_{j=1}^{J}\overline{X}_{ij.}^2 - JK\sum_{i=1}^{I}\overline{X}_{i..}^2 - IK\sum_{j=1}^{J}\overline{X}_{.j.}^2 + IJK\overline{X}_{...}^2 \quad (9.60)\end{aligned}$$

$$\text{SSE} = \sum_{i=1}^{I}\sum_{j=1}^{J}\sum_{k=1}^{K}(X_{ijk} - \overline{X}_{ij.})^2 = \sum_{i=1}^{I}\sum_{j=1}^{J}\sum_{k=1}^{K}X_{ijk}^2 - K\sum_{i=1}^{I}\sum_{j=1}^{J}\overline{X}_{ij.}^2 \quad (9.61)$$

$$\text{SST} = \sum_{i=1}^{I}\sum_{j=1}^{J}\sum_{k=1}^{K}(X_{ijk} - \overline{X}_{...})^2 = \sum_{i=1}^{I}\sum_{j=1}^{J}\sum_{k=1}^{K}X_{ijk}^2 - IJK\overline{X}_{...}^2 \quad (9.62)$$

It can be seen from the rightmost expressions in Equations (9.58) through (9.62) that the total sum of squares, SST, is equal to the sum of the others. This is the analysis of variance identity for two-way ANOVA.

The Analysis of Variance Identity

$$\text{SST} = \text{SSA} + \text{SSB} + \text{SSAB} + \text{SSE} \qquad (9.63)$$

Along with each sum of squares is a quantity known as its degrees of freedom. The sums of squares and their degrees of freedom are generally presented in an ANOVA table. Table 9.5 presents the degrees of freedom for each sum of squares, along with the computationally most convenient formula. We point out that the degrees of freedom for SST is the sum of the degrees of freedom for the other sums of squares.

TABLE 9.5 ANOVA table for two-way ANOVA

Source	Degrees of Freedom	Sum of Squares
Rows (SSA)	$I-1$	$JK\sum_{i=1}^{I}\widehat{\alpha}_i^2 = JK\sum_{i=1}^{I}\overline{X}_{i..}^2 - IJK\overline{X}_{...}^2$
Columns (SSB)	$J-1$	$IK\sum_{j=1}^{J}\widehat{\beta}_j^2 = IK\sum_{j=1}^{J}\overline{X}_{.j.}^2 - IJK\overline{X}_{...}^2$
Interactions (SSAB)	$(I-1)(J-1)$	$K\sum_{i=1}^{I}\sum_{j=1}^{J}\widehat{\gamma}_{ij}^2 = K\sum_{i=1}^{I}\sum_{j=1}^{J}\overline{X}_{ij.}^2 - JK\sum_{i=1}^{I}\overline{X}_{i..}^2$ $-IK\sum_{j=1}^{J}\overline{X}_{.j.}^2 + IJK\overline{X}_{...}^2$
Error (SSE)	$IJ(K-1)$	$\sum_{i=1}^{I}\sum_{j=1}^{J}\sum_{k=1}^{K}(X_{ijk}-\overline{X}_{ij.})^2 = \sum_{i=1}^{I}\sum_{j=1}^{J}\sum_{k=1}^{K}X_{ijk}^2 - K\sum_{i=1}^{I}\sum_{j=1}^{J}\overline{X}_{ij.}^2$
Total (SST)	$IJK-1$	$\sum_{i=1}^{I}\sum_{j=1}^{J}\sum_{k=1}^{K}(X_{ijk}-\overline{X}_{...})^2 = \sum_{i=1}^{I}\sum_{j=1}^{J}\sum_{k=1}^{K}X_{ijk}^2 - IJK\overline{X}_{...}^2$

Note that the magnitude of SSA depends on the magnitude of the *estimated* row effects $\widehat{\alpha}_i$. Therefore when the *true* row effects α_i are equal to 0, SSA will tend to be smaller, and when some of the true row effects are not equal to 0, SSA will tend to be larger. We will therefore reject $H_0\colon \alpha_1 = \cdots = \alpha_I = 0$ when SSA is sufficiently large. Similarly, SSB will tend to be smaller when the true column effects β_j are all equal to 0 and larger when some column effects are not zero, and SSAB will tend to be smaller when the true interactions γ_{ij} are all equal to 0 and larger when some interactions are not zero. We will therefore reject $H_0\colon \beta_1 = \cdots = \beta_J = 0$ when SSB is sufficiently large, and we will reject $H_0\colon \gamma_{11} = \cdots = \gamma_{IJ} = 0$ when SSAB is sufficiently large.

We can determine whether SSA, SSB, and SSAB are sufficiently large by comparing them to the error sum of squares, SSE. As in one-way ANOVA (Section 9.1), SSE depends only on the distances between the observations and their own cell means. SSE therefore measures only the random variation inherent in the process and is not affected by the values of the row effects, column effects, or interactions. To compare SSA, SSB, and SSAB with SSE, we first divide each sum of squares by its degrees of freedom, producing quantities known as **mean squares**. The mean squares, denoted MSA, MSB, MSAB, and MSE, are defined as follows:

$$\text{MSA} = \frac{\text{SSA}}{I-1} \qquad \text{MSB} = \frac{\text{SSB}}{J-1} \qquad \text{MSAB} = \frac{\text{SSAB}}{(I-1)(J-1)}$$

$$\text{MSE} = \frac{\text{SSE}}{IJ(K-1)} \tag{9.64}$$

The test statistics for the three null hypotheses are the quotients of MSA, MSB, and MSAB with MSE. The null distributions of these test statistics are F distributions. Specifically,

- Under $H_0\colon \alpha_1 = \cdots = \alpha_I = 0$, the statistic $\dfrac{\text{MSA}}{\text{MSE}}$ has an $F_{I-1,\,IJ(K-1)}$ distribution.
- Under $H_0\colon \beta_1 = \cdots = \beta_J = 0$, the statistic $\dfrac{\text{MSB}}{\text{MSE}}$ has an $F_{J-1,\,IJ(K-1)}$ distribution.
- Under $H_0\colon \gamma_{11} = \cdots = \gamma_{IJ} = 0$, the statistic $\dfrac{\text{MSAB}}{\text{MSE}}$ has an $F_{(I-1)(J-1),\,IJ(K-1)}$ distribution.

In practice, the sums of squares, mean squares, and test statistics are usually calculated with the use of a computer. The following MINITAB output presents the ANOVA table for the data in Table 9.2.

```
Two-way ANOVA: Yield versus Catalyst, Reagent

Source       DF        SS        MS       F       P
Catalyst      3    877.56   292.521    9.36   0.000
Reagent       2    327.14   163.570    5.23   0.010
Interaction   6    156.98    26.164    0.84   0.550
Error        36   1125.33    31.259
Total        47   2487.02

S = 5.591    R-sq = 54.75%    R-Sq(adj) = 40.93%
```

The labels DF, SS, MS, F, and P refer to degrees of freedom, sum of squares, mean square, F statistic, and P-value, respectively. As in one-way ANOVA, the mean square for error (MSE) is an estimate of the error variance σ^2, the quantity labeled "S" is the square root of MSE and is an estimate of the error standard deviation σ. The quantities "R-sq" and "R-sq(adj)" are computed with formulas analogous to those in one-way ANOVA.

Use the preceding ANOVA table to determine whether the additive model is plausible for the yield data. If the additive model is plausible, can we conclude that either the catalyst or the reagent affects the yield?

Solution

We first check to see if the additive model is plausible. The P-value for the interactions is 0.55, which is not small. We therefore do not reject the null hypothesis that all the interactions are equal to 0, and we conclude that the additive model is plausible. Since the additive model is plausible, we now ask whether the row or column factors affect the outcome. We see from the table that the P-value for the row effects (Catalyst) is approximately 0, so we conclude that the catalyst does affect the yield. Similarly, the P-value for the column effects (Reagent) is small (0.010), so we conclude that the reagent affects the yield as well.

The article "Uncertainty in Measurements of Dermal Absorption of Pesticides" (W. Navidi and A. Bunge, *Risk Analysis*, 2002:1175–1182) describes an experiment in which a pesticide was applied to skin at various concentrations and for various lengths of time. The outcome is the amount of the pesticide that was absorbed into the skin. The following MINITAB output presents the ANOVA table. Is the additive model plausible? If so, do either the concentration or the duration affect the amount absorbed?

```
Two-way ANOVA: Absorbed versus Concentration, Duration

Source        DF       SS        MS        F        P
Concent        2   49.991    24.996   107.99    0.000
Duration       2   19.157     9.579    41.38    0.000
Interaction    4    0.337     0.084     0.36    0.832
Error         27    6.250     0.231
Total         35   75.735
```

Solution

The P-value for the interaction is 0.832, so we conclude that the additive model is plausible. The P-values for both concentration and dose are very small. Therefore we can conclude that both concentration and duration affect the amount absorbed.

Checking the Assumptions

A residual plot can be used to check the assumption of equal variances, and a normal probability plot of the residuals can be used to check normality. The residual plot plots the residuals $X_{ijk} - \overline{X}_{ij.}$ versus the fitted values, which are the sample means $\overline{X}_{ij.}$.

Figures 9.6 and 9.7 present both a normal probability plot and a residual plot for the yield data found in Table 9.2. The assumptions appear to be well satisfied.

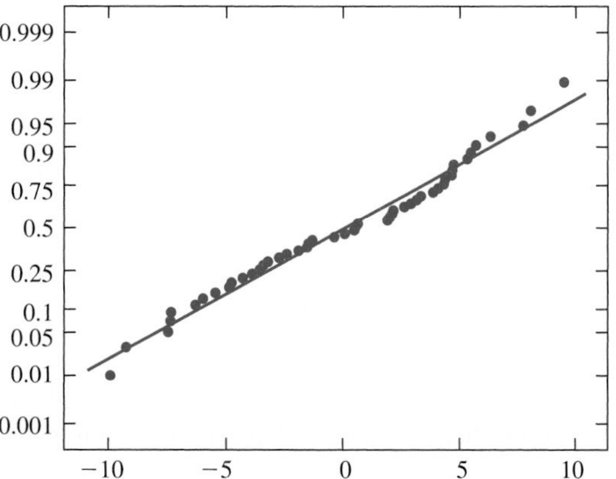

FIGURE 9.6 Normal probability plot for the residuals from the yield data. There is no evidence of a strong departure from normality.

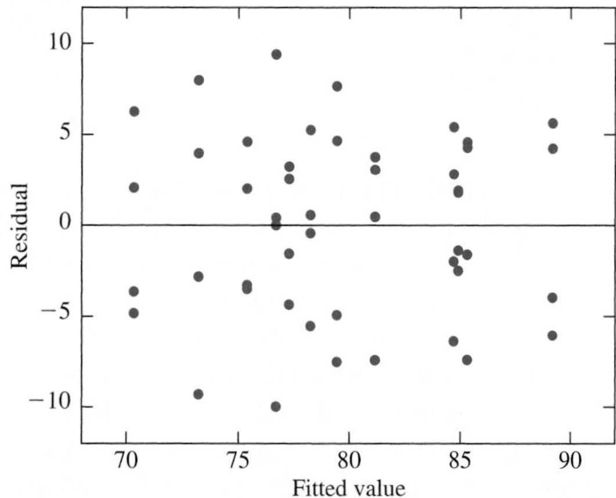

FIGURE 9.7 Residual plot for the yield data. There is no evidence against the assumption of equal variances.

Don't Interpret the Main Effects When the Additive Model Doesn't Hold

When the interactions are small enough so that the additive model is plausible, interpretation of the main effects is fairly straightforward, as shown in Examples 9.15 and 9.16. When the additive model does not hold, however, it is not always easy to interpret the main effects. Here is a hypothetical example to illustrate the point. Assume that a process is run under conditions obtained by varying two factors at two levels each. Two runs are made at each of the four combinations of row and column levels. The yield of the process is measured each time, with the results presented in the following table.

	Column Level	
Row Level	**1**	**2**
1	51, 49	43, 41
2	43, 41	51, 49

Clearly, if it is desired to maximize yield, the row and column factors matter—we want either row level 1 paired with column level 1 or row level 2 paired with column level 2.

Now look at the following ANOVA table.

Source	DF	SS	MS	F	P
Row	1	0.0000	0.0000	0.00	1.000
Column	1	0.0000	0.0000	0.00	1.000
Interaction	1	128.00	128.00	64.00	0.001
Error	4	8.0000	2.0000		
Total	7	136.00			

The main effects sum of squares for both the row and column main effects are equal to 0, and their P-values are equal to 1, which is as large as a P-value can be. If we follow the procedure used in Examples 9.15 and 9.16, we would conclude that neither the row factor nor the column factor affects the yield. But it is clear from the data that the row and column factors do affect the yield. What is happening is that the row and column factors do not matter *on the average*. Level 1 of the row factor is better if level 1 of the column factor is used, and level 2 of the row factor is better if level 2 of the column factor is used. When averaged over the two levels of the column factor, the levels of the row factor have the same mean yield. Similarly, the column levels have the same mean yield when averaged over the levels of the row factor. When the effects of the row levels depend on which column levels they are paired with, and vice versa, the main effects can be misleading.

It is the P-value for the interactions that tells us not to try to interpret the main effects. This P-value is quite small, so we reject the additive model. Then we know that some of the interactions are nonzero, so the effects of the row levels depend on the column levels, and vice versa. For this reason, when the additive model is rejected, we should not try to interpret the main effects. We need to look at the cell means themselves in order to determine how various combinations of row and column levels affect the outcome.

Summary

In a two-way analysis of variance:

- If the additive model *is not* rejected, then hypothesis tests for the main effects can be used to determine whether the row or column factors affect the outcome.
- If the additive model *is* rejected, then hypothesis tests for the main effects should not be used. Instead, the cell means must be examined to determine how various combinations of row and column levels affect the outcome.

Example 9.17

The thickness of the silicon dioxide layer on a semiconductor wafer is crucial to its performance. In the article "Virgin Versus Recycled Wafers for Furnace Qualification: Is the Expense Justified?" (V. Czitrom and J. Reece, *Statistical Case Studies for Process Improvement*, SIAM-ASA, 1997:87–103), oxide layer thicknesses were measured for three types of wafers: virgin wafers, wafers recycled in-house, and wafers recycled by an external supplier. In addition, several furnace locations were used to grow the oxide layer. A two-way ANOVA for three runs at one wafer site for the three types of wafers at three furnace locations was performed. The data are presented in the following table, followed by the MINITAB output.

Furnace Location	Wafer Type	Oxide Layer Thickness (Å)		
1	Virgin	90.1	90.7	89.4
1	In-house	90.4	88.8	90.6
1	External	92.6	90.0	93.3
2	Virgin	91.9	88.6	89.7
2	In-house	90.3	91.9	91.5
2	External	88.3	88.2	89.4
3	Virgin	88.1	90.2	86.6
3	In-house	91.0	90.4	90.2
3	External	91.5	89.8	89.8

```
Two-way ANOVA for Thickness versus Wafer, Location

Source        DF       SS        MS       F        P
Wafer          2   5.8756    2.9378    2.07    0.155
Location       2   4.1089    2.0544    1.45    0.262
Interaction    4   21.349    5.3372    3.76    0.022
Error         18   25.573    1.4207
Total         26   56.907
```

Since recycled wafers are cheaper, the company hopes that there is no difference in the oxide layer thickness among the three types of chips. If possible, determine whether the data are consistent with the hypothesis of no difference. If not possible, explain why not.

Solution

The P-value for the interactions is 0.022, which is small. Therefore the additive model is not plausible, so we cannot interpret the main effects. A good thing to do is to make a table of the cell means. Table 9.6 presents the sample mean for each treatment.

TABLE 9.6 Sample means for each treatment

Furnace	Wafer Type			
Location	Virgin	In-House	External	Row Mean
1	90.067	89.933	91.967	90.656
2	90.067	91.233	88.633	89.978
3	88.300	90.533	90.367	89.733
Column Mean	89.478	90.566	90.322	

From Table 9.6, it can be seen that the thicknesses do vary among wafer types, but no one wafer type consistently produces the thickest, or the thinnest, oxide layer. For example, at furnace location 1 the externally recycled wafers produce the thickest layer while the in-house recycled wafers produce the thinnest. At furnace location 2 the order is reversed: The in-house wafers produce the thickest layer while the external ones produce the thinnest. This is due to the interaction of furnace location and wafer type.

A Two-Way ANOVA Is Not the Same as Two One-Way ANOVAs

Example 9.17 presented a two-way ANOVA with three row levels and three column levels, for a total of nine treatments. If separate one-way ANOVAs were run on the row and column factors separately, there would be only six treatments. This means that in practice, running separate one-way ANOVAs on each factor may be less costly than running a two-way ANOVA. Unfortunately, this "one-at-a-time" design is sometimes used in practice for this reason. It is important to realize that running separate one-way analyses on the individual factors can give results that are misleading when interactions are present. To see this, look at Table 9.6. Assume that an engineer is trying to find the combination of furnace and location that will produce the thinnest oxide layer. He first runs the process once at each furnace location, using in-house recycled wafers, because those wafers are the ones currently being used in production. Furnace location 1 produces the thinnest layer for in-house wafers. Now the engineer runs the process once for each wafer type, all at location 1, which was the best for the in-house wafers. Of the three wafer types, in-house wafers produce the thinnest layer at location 1. So the conclusion drawn from the one-at-a-time analysis is that the thinnest layers are produced by the combination of in-house wafers at furnace location 1. A look at Table 9.6 shows that the conclusion is false. There are two combinations of furnace location and wafer type that produce thinner layers than this.

The one-at-a-time method assumes that the wafer that produces the thinnest layers at one location will produce the thinnest layers at all locations, and that the location that produces the thinnest layers for one wafer type will produce the thinnest layers for all types. This is equivalent to assuming that there are no interactions between the factors, which in the case of the wafers and locations is incorrect. In summary, the one-at-a-time method fails because it cannot detect interactions between the factors.

Summary

- When there are two factors, a two-factor design must be used.
- Examining one factor at a time cannot reveal interactions between the factors.

Interaction Plots

Interaction plots can help to visualize interactions. Figure 9.8 presents an interaction plot for the wafer data. We describe the method by which this plot was constructed. The vertical axis represents the response, which is layer thickness. One factor is chosen to be represented on the horizontal axis. We chose furnace location; it would have been equally acceptable to have chosen wafer type. Now we proceed through the levels of the wafer-type factor. We'll start with external wafers. The three cell means for external wafers, as shown in Table 9.6, are 91.967, 88.633, and 90.367, corresponding to furnace locations 1, 2, and 3, respectively. These values are plotted above their respective furnace locations and are connected with line segments. This procedure is repeated for the other two wafer types to complete the plot.

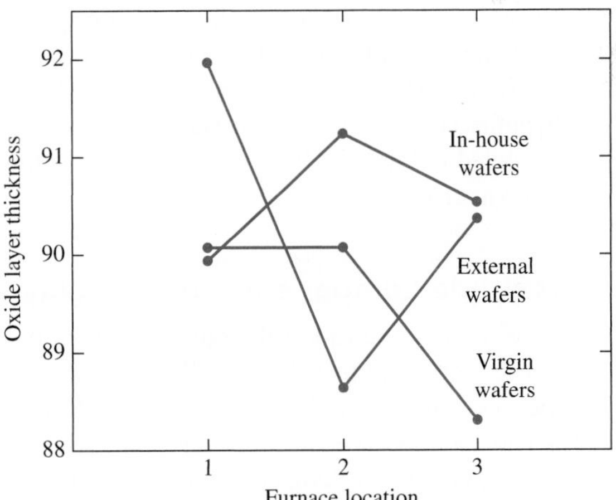

FIGURE 9.8 Interaction plot for the wafer data. The lines are far from parallel, indicating substantial interaction between the factors.

For the wafer data, the means for external wafers follow a substantially different pattern than those for the other two wafer types. This is the source of the significant interaction and is the reason that the main effects of wafer and furnace type cannot be easily interpreted. In comparison, for perfectly additive data, for which the interaction estimates $\widehat{\gamma}_{ij}$ are equal to 0, the line segments in the interaction plot are parallel. Figure 9.9 illustrates this hypothetical case.

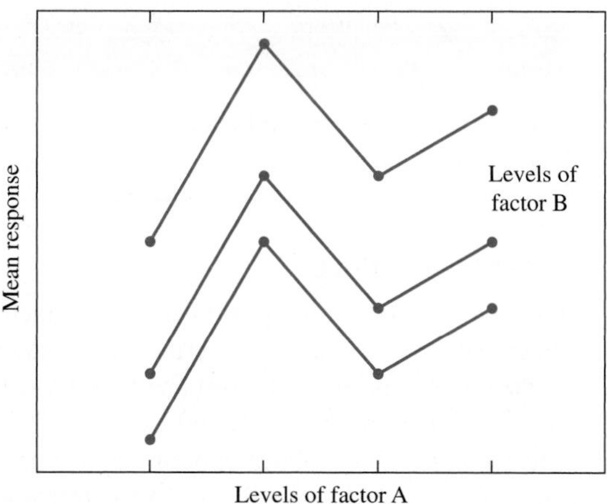

FIGURE 9.9 Interaction plot for hypothetical data with interaction estimates $\widehat{\gamma}_{ij}$ equal to 0. The line segments are parallel.

Figure 9.10 presents an interaction plot for the yield data. The cell means were presented in Table 9.4. The lines are not parallel, but their slopes match better than those for the wafer data. This indicates that the interaction estimates are nonzero, but are smaller than those for the wafer data. In fact, the P-value for the test of the null hypothesis of no interaction was 0.550 (see p. 667). The deviation from parallelism exhibited in Figure 9.10 is therefore small enough to be consistent with the hypothesis of no interaction.

Multiple Comparisons in Two-Way ANOVA

An F test is used to test the null hypothesis that all the row effects (or all the column effects) are equal to 0. If the null hypothesis is rejected, we can conclude that some of the row effects (or column effects) differ from each other. But the hypothesis test does not tell us which ones are different from the rest. If the additive model is plausible, then a method of multiple comparisons known as Tukey's method (related to the Tukey–Kramer method described in Section 9.2) can be applied to determine for which pairs the row effects or column effects can be concluded to differ from one another. The method is described in the following box.

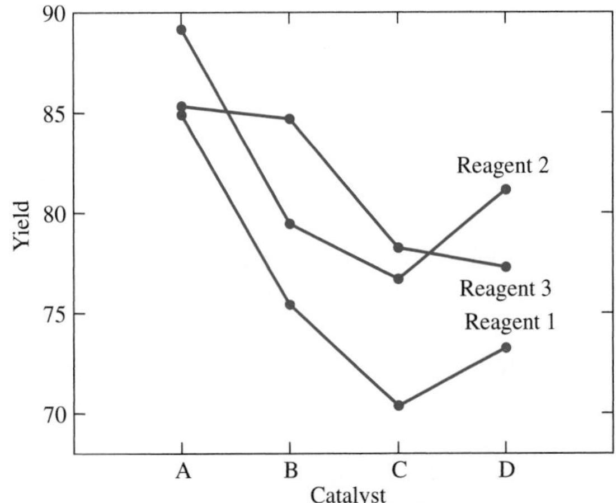

FIGURE 9.10 Interaction plot for yield data.

Tukey's Method for Simultaneous Confidence Intervals and Hypothesis Tests in Two-Way ANOVA

Let I be the number of levels of the row factor, J be the number of levels of the column factor, and K be the sample size for each treatment. **Then, if the additive model is plausible**, the Tukey level $100(1-\alpha)\%$ simultaneous confidence intervals for all differences $\alpha_i - \alpha_j$ (or all differences $\beta_i - \beta_j$) are

$$\widehat{\alpha}_i - \widehat{\alpha}_j \pm q_{I,IJ(K-1),\alpha}\sqrt{\frac{\text{MSE}}{JK}} \qquad \widehat{\beta}_i - \widehat{\beta}_j \pm q_{J,IJ(K-1),\alpha}\sqrt{\frac{\text{MSE}}{IK}}$$

We are $100(1-\alpha)\%$ confident that the Tukey confidence intervals contain the true value of the difference $\alpha_i - \alpha_j$ (or $\beta_i - \beta_j$) for every i and j.

For every pair of levels i and j for which $|\widehat{\alpha}_i - \widehat{\alpha}_j| > q_{I,IJ(K-1),\alpha}\sqrt{\frac{\text{MSE}}{JK}}$, the null hypothesis $H_0: \alpha_i - \alpha_j = 0$ is rejected at level α.

For every pair of levels i and j for which $|\widehat{\beta}_i - \widehat{\beta}_j| > q_{J,IJ(K-1),\alpha}\sqrt{\frac{\text{MSE}}{IK}}$, the null hypothesis $H_0: \beta_i - \beta_j = 0$ is rejected at level α.

Example 9.18

In Example 9.14, the main effects and interactions were computed for the yield data in Table 9.2. An ANOVA table for these data was presented on p. 667. If appropriate, use Tukey's method to determine which pairs of catalysts and which pairs of reagents can be concluded to differ, at the 5% level, in their effect on yield.

Solution

From the ANOVA table, the P-value for interactions is 0.550. Therefore the additive model is plausible, so it is appropriate to use Tukey's method. Catalyst is the row factor and reagent is the column factor, so $I = 4$, $J = 3$, and $K = 4$. From the ANOVA table, MSE = 31.259.

We first find all pairs for which the row effects differ at the 5% level. For the row effects, we should use the value $q_{4,36,.05}$. This value is not found in the Studentized range table (Table A.8 in Appendix A). We will therefore use the value $q_{4,30,.05} = 3.85$, which is close to (just slightly greater than) $q_{4,36,.05}$. We compute $q_{4,30,.05}\sqrt{\text{MSE}/JK} = 3.85\sqrt{31.259/12} = 6.21$.

In Example 9.14, the estimated row effects were computed to be

$$\widehat{\alpha}_1 = 6.81 \qquad \widehat{\alpha}_2 = 0.19 \qquad \widehat{\alpha}_3 = -4.56 \qquad \widehat{\alpha}_4 = -2.44$$

The pairs of row effects whose differences are greater than 6.21 are $\widehat{\alpha}_1$ and $\widehat{\alpha}_2$, $\widehat{\alpha}_1$ and $\widehat{\alpha}_3$, and $\widehat{\alpha}_1$ and $\widehat{\alpha}_4$. We conclude that the mean yield of catalyst A differs from the mean yields of catalysts B, C, and D, but we cannot conclude that the mean yields of catalysts B, C, and D differ from each other.

We now find all pairs for which the column effects differ at the 5% level. For the column effects, we should use the value $q_{3,36,.05}$, but since this value is not found in the Studentized range table, we will use the value $q_{3,30,05} = 3.49$. We compute $q_{3,30,.05}\sqrt{\text{MSE}/IK} = 3.49\sqrt{31.259/16} = 4.88$.

In Example 9.14, the estimated column effects were computed to be

$$\widehat{\beta}_1 = -3.69 \qquad \widehat{\beta}_2 = 1.96 \qquad \widehat{\beta}_3 = 1.73$$

The pairs of column effects whose differences are greater than 4.88 are $\widehat{\beta}_1$ and $\widehat{\beta}_2$ and $\widehat{\beta}_1$ and $\widehat{\beta}_3$. We conclude that the mean yield of reagent 1 differs from the mean yields of reagents 2 and 3, but we cannot conclude that the mean yields of reagents 2 and 3 differ from each other.

Two-Way ANOVA when $K = 1$

The F tests we have presented require the assumption that the sample size K for each treatment be at least 2. The reason for this is that when $K = 1$, the error sum of squares (SSE) will be equal to 0, since $X_{ijk} = \overline{X}_{ij.}$ for each i and j. In addition, the degrees of freedom for SSE, which is $IJ(K-1)$, is equal to 0 when $K = 1$.

When $K = 1$, a two-way ANOVA cannot be performed unless it is certain that the additive model holds. In this case, since the interactions are assumed to be zero, the mean square for interaction (MSAB; see Equation 9.64) and its degrees of freedom can be used in place of MSE to test the main row and column effects.

Random Factors

Our discussion of two-factor experiments has focused on the case where both factors are fixed. Such an experiment is said to follow a **fixed effects model**. Experiments can also be designed in which one or both factors are random. If both factors are random, the experiment is said to follow a **random effects model**. If one factor is fixed and one is random, the experiment is said to follow a **mixed model**.

In the one-factor case, the analysis is the same for both fixed and random effects models, while the null hypothesis being tested differs. In the two-factor case, both the methods of analysis and the null hypotheses differ among fixed effects models, random effects models, and mixed models. Methods for models in which one or more effects are random can be found in more advanced texts, such as Hocking (2003).

Unbalanced Designs

We have assumed that the design is balanced, that is, that the number of replications is the same for each treatment. The methods described here do not apply to unbalanced designs. However, unbalanced designs that are complete may be analyzed with the methods of multiple regression. An advanced text such as Draper and Smith (1998) may be consulted for details.

Exercises for Section 9.3

1. To assess the effect of piston ring type and oil type on piston ring wear, three types of piston ring and four types of oil were studied. Three replications of an experiment, in which the number of milligrams of material lost from the ring in four hours of running was measured, were carried out for each of the 12 combinations of oil type and piston ring type. With oil type as the row effect and piston ring type as the column effect, the following sums of squares were observed: SSA = 1.0926, SSB = 0.9340, SSAB = 0.2485, SSE = 1.7034.

a. How many degrees of freedom are there for the effect of oil type?
b. How many degrees of freedom are there for the effect of piston ring type?
c. How many degrees of freedom are there for interactions?
d. How many degrees of freedom are there for error?
e. Construct an ANOVA table. You may give ranges for the P-values.
f. Is the additive model plausible? Provide the value of the test statistic and the P-value.
g. Is it plausible that the main effects of oil type are all equal to 0? Provide the value of the test statistic and the P-value.
h. Is it plausible that the main effects of piston ring type are all equal to 0? Provide the value of the test statistic and the P-value.

2. A machine shop has three machines used in precision grinding of cam rollers. Three machinists are employed to grind rollers on the machines. In an experiment to determine whether there are differences in output among the machines or their operators, each operator worked on each machine on four different days. The outcome measured was the daily production of parts that met specifications. With the operator as the row effect and the machine as the column effect, the following sums of squares were observed: SSA = 3147.0, SSB = 136.5, SSAB = 411.7, SSE = 1522.0.

a. How many degrees of freedom are there for the operator effect?
b. How many degrees of freedom are there for the machine effect?
c. How many degrees of freedom are there for interactions?
d. How many degrees of freedom are there for error?
e. Construct an ANOVA table. You may give ranges for the P-values.
f. Is the additive model plausible? Provide the value of the test statistic and the P-value.
g. Is it plausible that the main effects of operator are all equal to 0? Provide the value of the test statistic and the P-value.
h. Is it plausible that the main effects of machine are all equal to 0? Provide the value of the test statistic and the P-value.

3. An experiment to determine the effect of mold temperature on tensile strength involved three different alloys and five different mold temperatures. Four specimens of each alloy were cast at each mold temperature. With mold temperature as the row factor and alloy as the column factor, the sums of squares were: SSA = 69,738, SSB = 8958, SSAB = 7275, and SST = 201,816.

a. Construct an ANOVA table. You may give ranges for the P-values.
b. Is the additive model plausible? Explain.
c. Is it plausible that the main effects of mold temperature are all equal to 0? Provide the value of the test statistic and the P-value.
d. Is it plausible that the main effects of alloy are all equal to 0? Provide the value of the test statistic and the P-value.

4. The effect of curing pressure on bond strength was tested for four different adhesives. There were three levels of curing pressure. Five replications were performed for each combination of curing pressure and adhesive. With adhesive as the row factor and curing pressure as the column factor, the sums of squares were: SSA = 155.7, SSB = 287.9, SSAB = 156.7, and SST = 997.3.

a. Construct an ANOVA table. You may give ranges for the P-values.
b. Is the additive model plausible? Explain.
c. Is it plausible that the main effects of curing pressure are all equal to 0? Provide the value of the test statistic and the P-value.
d. Is it plausible that the main effects of adhesive are all equal to 0? Provide the value of the test statistic and the P-value.

5. The article "Change in Creep Behavior of Plexiform Bone with Phosphate Ion Treatment" (R. Regimbal, C. DePaula, and N. Guzelsu, *Bio-Medical Materials and Engineering*, 203:11–25) describes an experiment to study the effects of saline and phosphate ion solutions on mechanical properties of plexiform bone. The following table presents the yield stress measurements for six specimens treated with either saline (NaCl) or phosphate ion (Na_2HPO_4) solution, at a temperature of either 25°C or 37°C. (The article presents means and standard deviations only; the values in the table are consistent with these.)

Solution	Temperature	Yield Stress (MPa)					
NaCl	25°C	138.40	130.89	94.646	96.653	116.90	88.215
NaCl	37°C	92.312	147.28	116.48	88.802	114.37	90.737
Na_2HPO_4	25°C	120.18	129.43	139.76	132.75	137.23	121.73
Na_2HPO_4	37°C	123.50	128.94	102.86	99.941	161.68	136.44

a. Estimate all main effects and interactions.
b. Construct an ANOVA table. You may give ranges for the P-values.
c. Is the additive model plausible? Provide the value of the test statistic and the P-value.
d. Can the effect of solution (NaCl versus Na_2HPO_4) on yield stress be described by interpreting the main effects of solution? If so, interpret the main effects, including the appropriate test statistic and P-value. If not, explain why not.
e. Can the effect of temperature on yield stress be described by interpreting the main effects of temperature? If so, interpret the main effects, including the appropriate test statistic and P-value. If not, explain why not.

6. A study was done to determine the effects of two factors on the latherability of soap. The two factors were type of water (tap versus de-ionized) and glycerol (present or absent). The outcome measured was the amount of foam produced (in mL). The experiment was repeated three times for each combination of the factors. The results are presented in the following table.

Water Type	Glycerol	Foam (mL)
De-ionized	Absent	168 178 168
De-ionized	Present	160 197 200
Tap	Absent	152 142 142
Tap	Present	139 160 160

a. Estimate all main effects and interactions.
b. Construct an ANOVA table. You may give ranges for the P-values.
c. Is the additive model plausible? Provide the value of the test statistic and the P-value.
d. Can the effect of water type on the amount of foam be described by interpreting the main effects of water type? If so, interpret the main effects. If not, explain why not.
e. Can the effect of glycerol on the amount of foam be described by interpreting the main effects of glycerol? If so, interpret the main effects. If not, explain why not.

7. An experiment was performed to measure the effects of two factors on the ability of cleaning solutions to remove oil from a piece of cloth. The factors were the concentration of soap (in % by weight) and the fraction of lauric acid in the solution. The experiment was repeated twice for each combination of factors. The outcome measured was the percentage of oil removed. The results are presented in the following table.

Weight % soap	Fraction Lauric Acid	% Oil Removed
15	10	52.8 54.0
15	30	57.8 53.3
25	10	56.4 58.4
25	30	42.7 45.1

a. Estimate all main effects and interactions.
b. Construct an ANOVA table. You may give ranges for the P-values.
c. Is the additive model plausible? Provide the value of the test statistic and the P-value.
d. Can the effect of soap concentration on the amount of oil removed be described by interpreting the main effects of soap concentration? If so, interpret the main effects. If not, explain why not.
e. Can the effect of lauric acid fraction on the amount of oil removed be described by interpreting the main effects of lauric acid fraction? If so, interpret the main effects. If not, explain why not.

8. The article "A 4-Year Sediment Trap Record of Alkenones from the Filamentous Upwelling Region Off Cape Blanc, NW Africa and a Comparison with Distributions in Underlying Sediments" (P. Müller and G. Fischer, *Deep Sea Research*, 2001:1877–1903) studied sediment trap records to evaluate the transfer of surface water signals into the geological record. The data in the following table are measurements of the total mass flux (in mg/m^2 per day) for traps at two different locations and depths.

Location	Depth	Flux							
A	Upper	109.8	86.5	150.5	69.8	63.2	107.8	72.4	74.4
A	Lower	163.7	139.4	176.9	170.6	123.5	142.9	130.3	111.6
B	Upper	93.2	123.6	143.9	163.2	82.6	63.0	196.7	160.1
B	Lower	137.0	88.3	53.4	104.0	78.0	39.3	117.9	143.0

a. Estimate all main effects and interactions.
b. Construct an ANOVA table. You may give ranges for the P-values.

c. Is the additive model plausible? Provide the value of the test statistic and the P-value.
d. Compute all the cell means. Use them to describe the way in which depth and location affect flux.

9. Artificial joints consist of a ceramic ball mounted on a taper. The article "Friction in Orthopaedic Zirconia Taper Assemblies" (W. Macdonald, A. Aspenberg, et al., *Proceedings of the Institution of Mechanical Engineers*, 2000: 685–692) presents data on the coefficient of friction for a push-on load of 2 kN for taper assemblies made from two zirconium alloys and employing three different neck lengths. Five measurements were made for each combination of material and neck length. The results presented in the following table are consistent with the cell means and standard deviations presented in the article.

Taper Material	Neck Length	Coefficient of Friction
CPTi-ZrO_2	Short	0.254 0.195 0.281 0.289 0.220
CPTi-ZrO_2	Medium	0.196 0.220 0.185 0.259 0.197
CPTi-ZrO_2	Long	0.329 0.481 0.320 0.296 0.178
TiAlloy-ZrO_2	Short	0.150 0.118 0.158 0.175 0.131
TiAlloy-ZrO_2	Medium	0.180 0.184 0.154 0.156 0.177
TiAlloy-ZrO_2	Long	0.178 0.198 0.201 0.199 0.210

a. Compute the main effects and interactions.
b. Construct an ANOVA table. You may give ranges for the P-values.
c. Is the additive model plausible? Provide the value of the test statistic, its null distribution, and the P-value.
d. Can the effect of material on the coefficient of friction be described by interpreting the main effects of material? If so, interpret the main effects. If not, explain why not.
e. Can the effect of neck length on the coefficient of friction be described by interpreting the main effects of neck length? If so, interpret the main effects, using multiple comparisons at the 5% level if necessary. If not, explain why not.

10. The article "Anodic Fenton Treatment of Treflan MTF" (D. Saltmiras and A. Lemley, *Journal of Environmental Science and Health*, 2001:261–274) describes a two-factor experiment designed to study the sorption of the herbicide trifluralin. The factors are the initial trifluralin concentration and the Fe^2:H-2O-2 delivery ratio. There were three replications for each treatment. The results presented in the following table are consistent with the means and standard deviations reported in the article.

Initial Concentration (M)	Delivery Ratio	Sorption (%)
15	1:0	10.90 8.47 12.43
15	1:1	3.33 2.40 2.67
15	1:5	0.79 0.76 0.84
15	1:10	0.54 0.69 0.57
40	1:0	6.84 7.68 6.79
40	1:1	1.72 1.55 1.82
40	1:5	0.68 0.83 0.89
40	1:10	0.58 1.13 1.28
100	1:0	6.61 6.66 7.43
100	1:1	1.25 1.46 1.49
100	1:5	1.17 1.27 1.16
100	1:10	0.93 0.67 0.80

a. Estimate all main effects and interactions.
b. Construct an ANOVA table. You may give ranges for the P-values.
c. Is the additive model plausible? Provide the value of the test statistic, its null distribution, and the P-value.

11. Refer to Exercise 10. The treatments with a delivery ratio of 1:0 were controls, or blanks. It was discovered after the experiment that the high apparent levels of sorption in these controls was largely due to volatility of the trifluralin. Eliminate the control treatments from the data.

a. Estimate all main effects and interactions.
b. Construct an ANOVA table. You may give ranges for the P-values.
c. Is the additive model plausible? Provide the value of the test statistic, its null distribution, and the P-value.
d. Construct an interaction plot. Explain how the plot illustrates the degree to which interactions are present.

12. The article "Use of Taguchi Methods and Multiple Regression Analysis for Optimal Process Development of High Energy Electron Beam Case Hardening of Cast Iron" (M. Jean and Y. Tzeng, *Surface Engineering*, 2003:150–156) describes a factorial experiment designed to determine factors in a high-energy electron beam process that affect hardness in metals. Results for two factors, each with three levels, are presented in the following table. Factor A is the travel speed in mm/s, and factor B is accelerating voltage in volts. The outcome is Vickers hardness. There were six replications for each treatment. In the article, a total of seven factors were studied; the two presented here are those that were found to be the most significant.

A	B	Hardness
10	10	875 896 921 686 642 613
10	25	712 719 698 621 632 645
10	50	568 546 559 757 723 734
20	10	876 835 868 812 796 772
20	25	889 876 849 768 706 615
20	50	756 732 723 681 723 712
30	10	901 926 893 856 832 841
30	25	789 801 776 845 827 831
30	50	792 786 775 706 675 568

a. Estimate all main effects and interactions.
b. Construct an ANOVA table. You may give ranges for the P-values.
c. Is the additive model plausible? Provide the value of the test statistic and the P-value.
d. Can the effect of travel speed on the hardness be described by interpreting the main effects of travel speed? If so, interpret the main effects, using multiple comparisons at the 5% level if necessary. If not, explain why not.
e. Can the effect of accelerating voltage on the hardness be described by interpreting the main effects of accelerating voltage? If so, interpret the main effects, using multiple comparisons at the 5% level if necessary. If not, explain why not.

13. The article "T-Bracing for Stability of Compression Webs in Wood Trusses" (R. Leichti, I. Hofaker, et al., *Journal of Structural Engineering*, 2002:374–381) presents results of experiments in which critical buckling loads (in kN) for T-braced assemblies were estimated by a finite-element method. The following table presents data in which the factors are the length of the side member and its method of attachment. There were 10 replications for each combination of factors. The data are consistent with the means and standard deviations given in the article.

Attachment	Length	Critical Buckling Load									
Adhesive	Quarter	7.90	8.71	7.72	8.88	8.55	6.95	7.07	7.59	7.77	7.86
Adhesive	Half	14.07	13.82	14.77	13.39	11.98	12.72	9.48	13.59	13.09	12.09
Adhesive	Full	26.80	28.57	24.82	23.51	27.57	25.96	24.28	25.68	21.64	28.16
Nail	Quarter	6.92	5.38	5.38	5.89	6.07	6.37	7.14	6.71	4.36	6.78
Nail	Half	9.67	9.17	10.39	10.90	10.06	9.86	10.41	10.24	9.31	11.99
Nail	Full	20.63	21.15	24.75	20.76	21.64	21.47	25.25	22.52	20.45	20.38

a. Compute all main effects and interactions.
b. Construct an ANOVA table. You may give ranges for the P-values.
c. Is the additive model plausible? Provide the value of a test statistic and the P-value.
d. Can the effect of attachment method (nail versus adhesive) on the critical buckling load be described by interpreting the main effects of attachment method? If so, interpret the main effects. If not, explain why not.
e. Can the effect of side member length on the critical buckling load be described by interpreting the main effects of side member length? If so, interpret the main effects, using multiple comparisons at the 5% level if necessary. If not, explain why not.

14. The article referred to in Exercise 13 also presents measurements of Young's modulus for side members of T-braced assemblies. The following table presents data in which the factors are the length of the side member and its method of attachment. There were 10 replications for each combination of factors. The data (in kN/mm^2) are consistent with the means and standard deviations given in the article.

Attachment	Length	Young's Modulus									
Adhesive	Quarter	9.56	10.67	8.82	8.40	9.23	8.20	10.23	9.58	7.57	8.05
Adhesive	Half	8.74	9.24	10.77	9.10	8.08	11.14	10.00	9.17	9.79	8.13
Adhesive	Full	9.84	9.80	8.31	7.37	10.12	9.18	8.93	8.65	7.89	9.07
Nail	Quarter	10.24	9.38	9.38	7.48	9.23	9.64	8.45	8.12	8.86	8.07
Nail	Half	9.84	9.34	9.64	8.21	10.43	9.48	7.46	9.51	10.20	9.66
Nail	Full	7.96	8.32	8.73	9.37	9.12	7.98	9.84	8.89	10.10	8.07

a. Compute all main effects and interactions.
b. Construct an ANOVA table. You may give ranges for the P-values.
c. Is the additive model plausible? Provide the value of a test statistic and the P-value.
d. Can the effect of attachment method (nail versus adhesive) on Young's modulus be described by interpreting the main effects of attachment method? If so, interpret the main effects. If not, explain why not.
e. Can the effect of side member length on Young's modulus be described by interpreting the main effects of side member length? If so, interpret the main effects, using multiple comparisons at the 5% level if necessary. If not, explain why not.

15. Each of three operators made two weighings of several silicon wafers. Results are presented in the following table for three of the wafers. All the wafers had weights very close to 54 g, so the weights are reported in units of μg above 54 g. (Based on the article "Revelation of a Microbalance Warmup Effect," J. Buckner, B. Chin, et al., *Statistical Case Studies for Industrial Process Improvement*, SIAM-ASA, 1997:39–45.)

Wafer	Operator 1		Operator 2		Operator 3	
1	11	15	10	6	14	10
2	210	208	205	201	208	207
3	111	113	102	105	108	111

a. Construct an ANOVA table. You may give ranges for the P-values.
b. Can it be determined from the ANOVA table whether there is a difference in the measured weights among the operators? If so, provide the value of the test statistic and the P-value. If not, explain why not.

16. Refer to Exercise 15. It turns out that the measurements of operator 2 were taken in the morning, shortly after the balance had been powered up. A new policy was instituted to leave the balance powered up continuously. The three operators then made two weighings of three different wafers. The results are presented in the following table.

Wafer	Operator 1	Operator 2	Operator 3
1	152 156	156 155	152 157
2	443 440	442 439	435 439
3	229 227	229 232	225 228

a. Construct an ANOVA table. You may give ranges for the P-values.
b. Compare the ANOVA table in part (a) with the ANOVA table in part (a) of Exercise 15. Would you recommend leaving the balance powered up continuously? Explain your reasoning.

17. The article "Cellulose Acetate Microspheres Prepared by O/W Emulsification and Solvent Evaporation Method" (K. Soppinmath, A Kulkarni, et al., *Journal of Microencapsulation*, 2001:811–817) describes a study of the effects of the concentrations of polyvinyl alcohol (PVAL) and dichloromethane (DCM) on the encapsulation efficiency in a process that produces microspheres containing the drug ibuprofen. There were three concentrations of PVAL (measured in units of % w/v) and three of DCM (in mL). The results presented in the following table are consistent with the means and standard deviations presented in the article.

PVAL	DCM = 50	DCM = 40	DCM = 30
0.5	98.983 99.268 95.149	96.810 94.572 86.718	75.288 74.949 72.363
1.0	89.827 94.136 96.537	82.352 79.156 80.891	76.625 76.941 72.635
2.0	95.095 95.153 92.353	86.153 91.653 87.994	80.059 79.200 77.141

a. Construct an ANOVA table. You may give ranges for the P-values.
b. Discuss the relationships among PVAL concentration, DCM concentration, and encapsulation efficiency.

9.4 Randomized Complete Block Designs

In some experiments, there are factors that vary and have an effect on the response, but whose effects are not of interest to the experimenter. For example, in one commonly occurring situation, it is impossible to complete an experiment in a single day, so the observations have to be spread out over several days. If conditions that can affect the outcome vary from day to day, then the day becomes a factor in the experiment, even though there may be no interest in estimating its effect.

For a more specific example, imagine that three types of fertilizer are to be evaluated for their effect on yield of fruit in an orange grove, and that three replicates will be performed, for a total of nine observations. An area is divided into nine plots, in three rows of three plots each. Now assume there is a water gradient along the plot area, so that the rows receive differing amounts of water. The amount of water is now a factor in the experiment, even though there is no interest in estimating the effect of water amount on the yield of oranges.

If the water factor is ignored, a one-factor experiment could be carried out with fertilizer as the only factor. Each of the three fertilizers would be assigned to three of the plots. In a completely randomized experiment, the treatments would be assigned to the plots at random. Figure 9.11 presents two possible random arrangements. In the arrangement on the left, the plots with fertilizer A get more water than those with the other two fertilizers. In the plot on the right, the plots with fertilizer C get the most water. When the treatments for one factor are assigned completely at random, it is likely that they will not be distributed evenly over the levels of another factor.

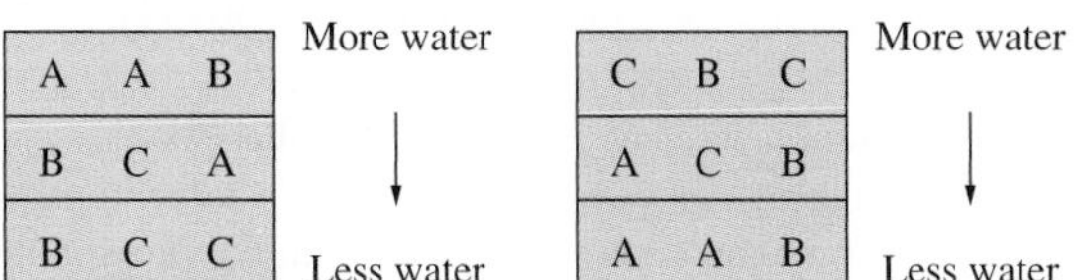

FIGURE 9.11 Two possible arrangements for three fertilizers, A, B, and C, assigned to nine plots completely at random. It is likely that the amounts of water will differ for the different fertilizers.

If the amount of water in fact has a negligible effect on the response, then the completely randomized one-factor design is appropriate. There is no reason to worry about a factor that does not affect the response. But now assume that the water level does have a substantial impact on the response. Then Figure 9.11 shows that in any one experiment, the estimated effects of the treatments are likely to be thrown off the mark, or biased, by the differing levels of water. Different arrangements of the treatments bias the estimates in different directions. If the experiment is repeated several times, the estimates are likely to vary greatly from repetition to repetition. For this reason, the completely randomized one-factor design produces estimated effects that have large uncertainties.

A better design for this experiment is a two-factor design, with water as the second factor. Since the effects of water are not of interest, water is called a **blocking factor**, rather than a treatment factor. In the two-factor experiment, there are nine treatment–block combinations, corresponding to the three fertilizer treatment levels and the three water block levels. With nine experimental units (the nine plots), it is necessary to assign one plot to each combination of fertilizer and water. Figure 9.12 presents two possible arrangements.

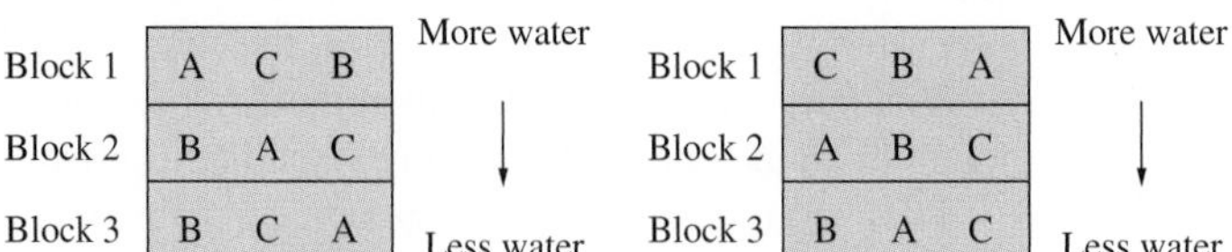

FIGURE 9.12 Two possible arrangements for three fertilizers, A, B, and C, with the restriction that each fertilizer must appear once at each water level (block). The distribution of water levels is always the same for each fertilizer.

In the two-factor design, each treatment appears equally often (once, in this example) in each block. As a result, the effect of the blocking factor does not contribute to uncertainty in the estimate of the main effects of the treatment factor. Because each treatment must appear equally often in each block, the only randomization in the assignment of treatments to experimental units is the order in which the treatments appear in each block. This is not a completely randomized design; it is a design in which the treatments are **randomized within blocks**. Since every possible combination of treatments and blocks is included in the experiment, the design is **complete**. For this reason the design is called a **randomized complete block design**.

Randomized complete block designs can be constructed with several treatment factors and several blocking factors. We will restrict our discussion to the case where there is one treatment factor and one blocking factor. The data from a randomized complete block design are analyzed with a two-way ANOVA, in the same way that data from any complete, balanced two-factor design would be. There is one important consideration, however. The only effects of interest are the main effects of the treatment factor. In order to interpret these main effects, **there must be no interaction between treatment and blocking factors**.

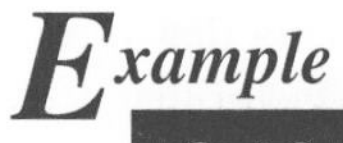

Three fertilizers are studied for their effect on yield in an orange grove. Nine plots of land are used, divided into blocks of three plots each. A randomized complete block design is used, with each fertilizer applied once in each block. The results, in pounds of harvested fruit, are presented in the following table, followed by MINITAB output for the two-way ANOVA. Can we conclude that the mean yields differ among fertilizers? What assumption is made about interactions between fertilizer and plot? How is the sum of squares for error computed?

Fertilizer	Plot 1	Plot 2	Plot 3
A	430	542	287
B	367	463	253
C	320	421	207

```
Two-way ANOVA: Yield versus Block, Fertilizer

Source       DF        SS         MS        F       P
Fertilizer    2   16213.6   8106.778    49.75   0.001
Block         2   77046.9   38523.44    236.4   0.000
Error         4   651.778   162.9444
Total         8   93912.2
```

Solution

The P-value for the fertilizer factor is 0.001, so we conclude that fertilizer does have an effect on yield. The assumption is made that there is no interaction between the fertilizer and the blocking factor (plot), so that the main effects of fertilizer can be interpreted. Since there is only one observation for each treatment–block combination (i.e., $K = 1$), the sum of squares for error (SSE) reported in the MINITAB output is really SSAB, the sum of squares for interaction, and the error mean square (MSE) is actually MSAB. (See the discussion on p. 676.)

A closer look at the ANOVA table in Example 9.19 shows that in this experiment, blocking was necessary to detect the fertilizer effect. To see this, consider the experiment to be a one-factor experiment. The sum of squares for error (SSE) would then be the sum of SSE for the blocked design plus the sum of squares for blocks, or $651.778 + 77{,}046.9 = 77{,}698.7$. The degrees of freedom for error would be equal to the sum of the degrees of freedom for error in the blocked design plus the degrees of freedom for blocks, or $2 + 4 = 6$. The error mean square (MSE) would then be $77{,}698.7/6 \approx 12{,}950$ rather than 162.9444, and the F statistic for the fertilizer effect would be less than 1, which would result in a failure to detect an effect.

In general, using a blocked design reduces the degrees of freedom for error, which by itself tends to reduce the power to detect an effect. However, unless the blocking factor has very little effect on the response, this will usually be more than offset by a reduction in the sum of squares for error. Failing to include a blocking factor that affects the response can reduce the power greatly, while including a blocking factor that does not affect the response reduces the power only modestly in most cases. For this reason it is a good idea to use a blocked design whenever it is thought to be possible that the blocking factor is related to the response.

Summary

- A two-factor randomized complete block design is a complete balanced two-factor design in which the effects of one factor (the treatment factor) are of interest, while the effects of the other factor (the blocking factor) are not of interest. The blocking factor is included to reduce the uncertainty in the main effect estimates of the treatment factor.
- Since the object of a randomized complete block design is to estimate the main effects of the treatment factor, there must be no interaction between the treatment factor and the blocking factor.
- A two-way analysis of variance is used to estimate effects and to perform hypothesis tests on the main effects of the treatment factor.
- A randomized complete block design provides a great advantage over a completely randomized design when the blocking factor strongly affects the response and provides a relatively small disadvantage when the blocking factor has little or no effect. Therefore, when in doubt, it is a good idea to use a blocked design.

9.20

The article "Experimental Design for Process Settings in Aircraft Manufacturing" (R. Sauter and R. Lenth, *Statistical Case Studies: A Collaboration Between Academe and Industry*, SIAM–ASA, 1998:151–157) describes an experiment in which the quality of holes drilled in metal aircraft parts was studied. One important indicator of hole quality is "excess diameter," which is the difference between the diameter of the drill bit and the diameter of the hole. Small excess diameters are better than large ones. Assume we are interested in the effect of the rotational speed of the drill on the excess diameter of the hole. Holes will be drilled in six test pieces (coupons), at three speeds: 6000, 10,000, and 15,000 rpm. The excess diameter can be affected not only by the speed of the drill, but also by the physical properties of the test coupon. Describe an appropriate design for this experiment.

Solution

A randomized complete block design is appropriate, with drill speed as the treatment factor, and test coupon as the blocking factor. Since six observations can be made in each block, each drill speed should be used twice in each block. The order of the speeds within each block should be chosen at random.

9.21

The design suggested in Example 9.20 has been adopted, and the experiment has been carried out. The MINITAB output follows. Does the output indicate any violation of necessary assumptions? What do you conclude regarding the effect of drill speed on excess diameter?

```
Two-way ANOVA: Excess Diameter versus Block, Speed

Source        DF        SS          MS         F        P
Block          5   0.20156   0.0403117      1.08    0.404
Speed          2   0.07835   0.0391750      1.05    0.370
Interaction   10   0.16272   0.0162717      0.44    0.909
Error         18   0.67105   0.0372806
Total         35   1.11368

S = 0.1931      R-Sq = 39.74%      R-Sq(adj) = 0.00%
```

Solution

In a randomized complete block design, there must be no interaction between the treatment factor and the blocking factor, so that the main effect of the treatment factor may be interpreted. The P-value for interactions is 0.909, which is consistent with the hypothesis of no interactions. Therefore there is no indication in the output of any violation of assumptions. The P-value for the main effect of speed is 0.370, which is not small. Therefore we cannot conclude that excess hole diameter is affected by drill speed.

Example 9.22 shows that a paired design (see Section 6.8), in which a t test is used to compare two population means, is a special case of a randomized block design.

A tire manufacturer wants to compare the tread wear of tires made from a new material with that of tires made from a conventional material. There are 10 tires of each type. Each tire will be mounted on the front wheel of a front-wheel drive car and driven for 40,000 miles. Then the tread wear will be measured for each tire. Describe an appropriate design for this experiment.

Solution

The response is the tread wear after 40,000 miles. There is one factor of interest: the type of tire. Since the cars may differ in the amounts of wear they produce, the car is a factor as well, but its effect is not of interest. A randomized complete block design is appropriate, in which one tire of each type is mounted on the front wheels of each car.

You may note that the randomized complete block design in Example 9.22 is the same design that is used when comparing two population means with a paired t test, as discussed in Section 6.8. The paired design described there is a special case of a randomized complete block design, in which the treatment factor has only two levels, and each level appears once in each block. In fact, a two-way analysis of variance applied to data from such a design is equivalent to the paired t test.

Multiple Comparisons in Randomized Complete Block Designs

Once an ANOVA table has been constructed, then if the F test shows that the treatment main effects are not all the same, a method of multiple comparisons may be used to determine which pairs of effects may be concluded to differ. We describe Tukey's method, which is a special case of the Tukey–Kramer method described in Section 9.2. The degrees of freedom, and the mean square used, differ depending on whether each treatment appears only once, or more than once, in each block.

Tukey's Method for Multiple Comparisons in Randomized Complete Block Designs

In a randomized complete block design, with I treatment levels, J block levels, and treatment main effects $\alpha_1, \ldots, \alpha_I$:

- If each treatment appears only once in each block, then the null hypothesis $H_0: \alpha_i - \alpha_j = 0$ is rejected at level α for every pair of treatments i and j for which

$$|\overline{X}_{i.} - \overline{X}_{j.}| > q_{I,(I-1)(J-1),\alpha}\sqrt{\frac{\text{MSAB}}{J}}$$

  where MSAB is the mean square for interaction.
- If each treatment appears $K > 1$ times in each block, then the null hypothesis $H_0: \alpha_i - \alpha_j = 0$ is rejected at level α for every pair of treatments i and j for which

$$|\overline{X}_{i..} - \overline{X}_{j..}| > q_{I,IJ(K-1),\alpha}\sqrt{\frac{\text{MSE}}{JK}}$$

  where MSE is the mean square for error.

For more information on randomized block designs, a text on design of experiments, such as Montgomery (2001a), can be consulted.

Exercises for Section 9.4

1. The article "Methods for Evaluation of Easily-Reducible Iron and Manganese in Paddy Soils" (M. Borges, J. de Mello, et al., *Communication in Soil Science and Plant Analysis*, 2001:3009–3022) describes an experiment in which pH levels of four alluvial soils were measured. Various levels of liming were applied to each soil. The main interest centers on differences among the soils; there is not much interest in the effect of the liming. The results are presented in the following table.

	Liming Level				
Soil	**1**	**2**	**3**	**4**	**5**
A	5.8	5.9	6.1	6.7	7.1
B	5.2	5.7	6.0	6.4	6.8
C	5.5	6.0	6.2	6.7	7.0
D	6.0	6.6	6.7	6.7	7.5

a. Which is the blocking factor, and which is the treatment factor?

b. Construct an ANOVA table. You may give ranges for the P-values.

c. Can you conclude that the soils have differing pH levels?
d. Which pairs of soils, if any, can you conclude to have differing pH levels? Use the 5% level.

2. A study was done to see which of four machines is fastest in performing a certain task. There are three operators; each performs the task twice on each machine. A randomized block design is employed. The MINITAB output follows.

Source	DF	SS	MS	F	P
Machine	(i)	257.678	(ii)	(iii)	0.021
Block	(iv)	592.428	(v)	(vi)	0.000
Interaction	(vii)	(viii)	(ix)	(x)	0.933
Error	(xi)	215.836	17.986		
Total	(xii)	1096.646			

a. Fill in the missing numbers (i) through (xii) in the output.
b. Does the output indicate that the assumptions for the randomized block design are satisfied? Explain.
c. Can you conclude that there are differences among the machines? Explain.

3. An experiment was carried out to determine the effect of broth concentration on the yield of a certain microorganism. There were four different formulations and three different concentrations of broth. There were three replications for each combination of broth formulation and concentration, with the outcome being yield per mL. Broth formulation is a blocking factor; the only effect of interest is the effect of concentration. The following sums of squares were calculated: sum of squares for blocks = 504.7, sum of squares for treatments = 756.7, sum of squares for interactions = 415.3, total sum of squares = 3486.1.

a. Construct an ANOVA table. You may give ranges for the P-values.
b. Are the assumptions for a randomized complete block design satisfied? Explain.
c. Does the ANOVA table provide evidence that broth concentration affects yield? Explain.

4. Three different machines are being considered for use in completing a certain task. Four operators complete the task on each of the three machines, with each operator repeating the task twice. The outcome is the time in minutes needed to complete the task. The only effect of interest is the machine; operator is a blocking factor. The following sums of squares were calculated: sum of squares for blocks = 48.47, sum of squares for treatments = 30.83, sum of squares for interactions = 93.27, total sum of squares = 217.36.

a. Construct an ANOVA table. You may give ranges for the P-values.
b. Are the assumptions for a randomized complete block design satisfied? Explain.
c. Does the ANOVA table provide evidence that mean completion time differs among machines? Explain.

5. The article "Genotype-Environment Interactions and Phenotypic Stability Analyses of Linseed in Ethiopia" (W. Adguna and M. Labuschagne, *Plant Breeding*, 2002:66–71) describes a study in which seed yields of 10 varieties of linseed were compared. Each variety was grown on six different plots. The yields, in kilograms per hectare, are presented in the following table.

Variety	Plot 1	2	3	4	5	6
A	2032	1377	1343	1366	1276	1209
B	1815	1126	1338	1188	1566	1454
C	1739	1311	1340	1250	1473	1617
D	1812	1313	1044	1245	1090	1280
E	1781	1271	1308	1220	1371	1361
F	1703	1089	1256	1385	1079	1318
G	1476	1333	1162	1363	1056	1096
H	1745	1308	1190	1269	1251	1325
I	1679	1216	1326	1271	1506	1368
J	1903	1382	1373	1609	1396	1366

a. Construct an ANOVA table. You may give ranges for the P-values.
b. Can you conclude that the varieties have differing mean yields?

6. The article "Sprinkler Technologies, Soil Infiltration, and Runoff" (D. DeBoer and S. Chu, *Journal of Irrigation and Drainage Engineering*, 2001:234–239) presents a study of the runoff depth (in mm) for various sprinkler types. Each of four sprinklers was tested on each of four days, with two replications per day (there were three replications on a few of the days; these are omitted). It is of interest to determine whether runoff depth varies with sprinkler type; variation from one day to another is not of interest. The data are presented in the following table.

Sprinkler	Day 1	Day 2	Day 3	Day 4
A	8.3 5.5	7.8 4.5	10.7 9.8	10.6 6.6
B	6.5 9.5	3.7 3.6	7.7 10.6	3.6 6.7
C	1.8 1.2	0.5 0.3	1.7 1.9	2.2 2.1
D	0.7 0.8	0.1 0.5	0.1 0.5	0.3 0.5

a. Identify the blocking factor and the treatment factor.
b. Construct an ANOVA table. You may give ranges for the P-values.
c. Are the assumptions of a randomized complete block design met? Explain.
d. Can you conclude that there are differences in mean runoff depth between some pairs of sprinklers? Explain.
e. Which pairs of sprinklers, if any, can you conclude, at the 5% level, to have differing mean runoff depths?

7. The article "Bromate Surveys in French Drinking Waterworks" (B. Legube, B. Parinet, et al., *Ozone Science and Engineering*, 2002:293–304) presents measurements of bromine concentrations (in μg/L) at several waterworks. The measurements made at 15 different times at each of four waterworks are presented in the following table. (The article also presented some additional measurements made at several other waterworks.) It is of interest to determine whether bromine concentrations vary among waterworks; it is not of interest to determine whether concentrations vary over time.

Waterworks	Time 1	2	3	4	5	6	7	8	9	10	11	12	13	14	15
A	29	9	7	35	40	53	38	38	41	34	42	35	38	35	36
B	24	29	21	24	20	25	15	14	8	12	14	35	32	38	33
C	25	17	20	24	19	19	17	23	22	27	17	33	33	39	37
D	31	37	34	30	39	41	34	34	29	33	33	34	16	31	16

a. Construct an ANOVA table. You may give ranges for the P-values.
b. Can you conclude that bromine concentration varies among waterworks?
c. Which pairs of waterworks, if any, can you conclude, at the 5% level, to have differing bromine concentrations?
d. Someone suggests that these data could have been analyzed with a one-way ANOVA, ignoring the time factor, with 15 observations for each of the four waterworks. Does the ANOVA table support this suggestion? Explain.

8. The article "Application of Fluorescence Technique for Rapid Identification of IOM Fractions in Source Waters" (T. Marhaba and R. Lippincott, *Journal of Environmental Engineering*, 2000:1039–1044) presents measurements of concentrations of dissolved organic carbon (in mg/L) at six locations (A, B, C, D, E, F) along the Millstone River in central New Jersey. Measurements were taken at four times of the year: in January, April, July, and October. It is of interest to determine whether the concentrations vary among locations. The data are presented in the following table.

	January	April	July	October
A	3.9	3.7	3.7	4.1
B	4.0	3.5	3.4	5.7
C	4.2	3.4	3.0	4.8
D	4.1	3.3	2.9	4.6
E	4.1	3.4	3.0	3.4
F	4.2	3.5	2.8	4.7

a. Construct an ANOVA table. You may give ranges for the P-values.
b. Can you conclude that the concentration varies among locations?
c. Which pairs of locations, if any, can you conclude, at the 5% level, to have differing concentrations?

9. You have been given the task of designing a study concerning the lifetimes of five different types of electric motor. The initial question to be addressed is whether there are differences in mean lifetime among the five types. There are 20 motors, four of each type, available for testing. A maximum of five motors can be tested each day. The ambient temperature differs from day to day, and this can affect motor lifetime.

a. Describe how you would choose the five motors to test each day. Would you use a completely randomized design? Would you use any randomization at all?
b. If X_{ij} represents the measured lifetime of a motor of type i tested on day j, express the test statistic for testing the null hypothesis of equal lifetimes in terms of the X_{ij}.

10. An engineering professor wants to determine which subject engineering students find most difficult among statistics, physics, and chemistry. She obtains the final exam grades for four students who took all three courses last semester and who were in the same sections of each class. The results are presented in the following table.

	Student			
Course	**1**	**2**	**3**	**4**
Statistics	82	94	78	70
Physics	75	70	81	83
Chemistry	93	82	80	70

a. The professor proposes a randomized complete block design, with the students as the blocks. Give a reason that this is likely not to be appropriate.
b. Describe the features of the data in the preceding table that suggest that the assumptions of the randomized complete block design are violated.

9.5 2^p Factorial Experiments

When an experimenter wants to study several factors simultaneously, the number of different treatments can become quite large. In these cases, preliminary experiments are often performed in which each factor has only two levels. One level is designated as the "high" level, and the other is designated as the "low" level. If there are p factors, there are then 2^p different treatments. Such experiments are called **2^p factorial experiments**. Often, the purpose of a 2^p experiment is to determine which factors have an important effect on the outcome. Once this is determined, more elaborate experiments can be performed, in which the factors previously found to be important are varied over several levels. We will begin by describing 2^3 factorial experiments.

Notation for 2^3 Factorial Experiments

In a 2^3 factorial experiment, there are three factors and $2^3 = 8$ treatments. The **main effect** of a factor is defined to be the difference between the mean response when the factor is at its high level and the mean response when the factor is at its low level. It is common to denote the main effects by A, B, and C. As with any factorial experiment, there can be interactions between the factors. With three factors, there are three two-way interactions, one for each pair of factors, and one three-way interaction. The two-way interactions are denoted by AB, AC, and BC, and the three-way interaction by ABC. The treatments are traditionally denoted with lowercase letters, with a letter indicating that a factor is at its high level. For example, ab denotes the treatment in which the first two factors are at their high level and the third factor is at its low level. The symbol "1" is used to denote the treatment in which all factors are at their low levels.

Estimating Effects in a 2^3 Factorial Experiment

Assume that there are K replicates for each treatment in a 2^3 factorial experiment. For each treatment, the cell mean is the average of the K observations for that treatment. The formulas for the effect estimates can be easily obtained from the 2^3 **sign table**, presented as Table 9.7 on page 694. The signs are placed in the table as follows. For the main effects A, B, C, the sign is $+$ for treatments in which the factor is at its high level, and $-$ for treatments where the factor is at its low level. So for the main effect A, the sign is $+$ for treatments a, ab, ac, and abc, and $-$ for the rest. For the interactions, the signs are computed by taking the product of the signs in the corresponding main effects columns. For example, the signs for the two-way interaction AB are the products of the signs in columns A and B, and the signs for the three-way interaction ABC are the products of the signs in columns A and B and C.

Estimating main effects and interactions is done with the use of the sign table. We illustrate how to estimate the main effect of factor A. Factor A is at its high level in the rows of the table where there is a "+" sign in column A. Each of the cell means $\overline{X}_a$, $\overline{X}_{ab}$,

TABLE 9.7 Sign table for a 2^3 factorial experiment

Treatment	Cell Mean	A	B	C	AB	AC	BC	ABC
1	$\overline{X}_1$	−	−	−	+	+	+	−
a	$\overline{X}_a$	+	−	−	−	−	+	+
b	$\overline{X}_b$	−	+	−	−	+	−	+
ab	$\overline{X}_{ab}$	+	+	−	+	−	−	−
c	$\overline{X}_c$	−	−	+	+	−	−	+
ac	$\overline{X}_{ac}$	+	−	+	−	+	−	−
bc	$\overline{X}_{bc}$	−	+	+	−	−	+	−
abc	$\overline{X}_{abc}$	+	+	+	+	+	+	+

$\overline{X}_{ac}$, and $\overline{X}_{abc}$ is an average response for runs made with factor A at its high level. We estimate the mean response for factor A at its high level to be the average of these cell means.

$$\text{Estimated mean response for } A \text{ at high level} = \frac{1}{4}(\overline{X}_a + \overline{X}_{ab} + \overline{X}_{ac} + \overline{X}_{abc})$$

Similarly, each row with a "−" sign in column A represents a treatment with factor A set to its low level. We estimate the mean response for factor A at its low level to be the average of the cell means in these rows.

$$\text{Estimated mean response for } A \text{ at low level} = \frac{1}{4}(\overline{X}_1 + \overline{X}_b + \overline{X}_c + \overline{X}_{bc})$$

The estimate of the main effect of factor A is the difference in the estimated mean response between its high and low levels.

$$A \text{ effect estimate} = \frac{1}{4}(-\overline{X}_1 + \overline{X}_a - \overline{X}_b + \overline{X}_{ab} - \overline{X}_c + \overline{X}_{ac} - \overline{X}_{bc} + \overline{X}_{abc})$$

The quantity inside the parentheses is called the **contrast** for factor A. It is computed by adding and subtracting the cell means, using the signs in the appropriate column of the sign table. Note that the number of plus signs is the same as the number of minus signs, so the sum of the coefficients is equal to 0. The effect estimate is obtained by dividing the contrast by half the number of treatments, which is $2^3/2$, or 4. Estimates for other main effects and interactions are computed in an analogous manner. To illustrate, we present the effect estimates for the main effect C and for the two-way interaction AB:

$$C \text{ effect estimate} = \frac{1}{4}(-\overline{X}_1 - \overline{X}_a - \overline{X}_b - \overline{X}_{ab} + \overline{X}_c + \overline{X}_{ac} + \overline{X}_{bc} + \overline{X}_{abc})$$

$$AB \text{ interaction estimate} = \frac{1}{4}(\overline{X}_1 - \overline{X}_a - \overline{X}_b + \overline{X}_{ab} + \overline{X}_c - \overline{X}_{ac} - \overline{X}_{bc} + \overline{X}_{abc})$$

Summary

The **contrast** for any main effect or interaction is obtained by adding and subtracting the cell means, using the signs in the appropriate column of the sign table.

For a 2^3 factorial experiment,

$$\text{Effect estimate} = \frac{\text{contrast}}{4} \tag{9.65}$$

A 2^3 factorial experiment was conducted to estimate the effects of three factors on the yield of a chemical reaction. The factors were A: catalyst concentration (low or high), B: reagent (standard formulation or new formulation), and C: stirring rate (slow or fast). Three replicates were obtained for each treatment. The yields, presented in the following table, are measured as a percent of a theoretical maximum. Estimate all effects and interactions.

Treatment	Yield			Cell Mean
1	71.67	70.55	67.40	69.8733
a	78.46	75.42	81.77	78.5500
b	77.14	78.25	78.33	77.9067
ab	79.72	76.17	78.41	78.1000
c	72.65	71.03	73.54	72.4067
ac	80.10	73.91	74.81	76.2733
bc	80.20	73.49	74.86	76.1833
abc	75.58	80.28	71.64	75.8333

Solution

We use the sign table (Table 9.7) to find the appropriate sums and differences of the cell means. We present the calculations for the main effect A, the two-way interaction BC, and the three-way interaction ABC:

$$\begin{aligned} A \text{ effect estimate} &= \frac{1}{4}(-69.8733 + 78.5500 - 77.9067 + 78.1000 \\ &\quad - 72.4067 + 76.2733 - 76.1833 + 75.8333) = 3.0967 \end{aligned}$$

$$\begin{aligned} BC \text{ interaction estimate} &= \frac{1}{4}(69.8733 + 78.5500 - 77.9067 - 78.1000 \\ &\quad - 72.4067 - 76.2733 + 76.1833 + 75.8333) = -1.0617 \end{aligned}$$

$$\begin{aligned} ABC \text{ interaction estimate} &= \frac{1}{4}(-69.8733 + 78.5500 + 77.9067 - 78.1000 \\ &\quad + 72.4067 - 76.2733 - 76.1833 + 75.8333) = 1.0667 \end{aligned}$$

We present all the estimated effects in the following table, rounded off to the same precision as the data:

Variable	Effect
A	3.10
B	2.73
C	−0.93
AB	−3.18
AC	−1.34
BC	−1.06
ABC	1.07

For each effect, we can test the null hypothesis that the effect is equal to 0. When the null hypothesis is rejected, this provides evidence that the factors involved actually affect the outcome. To test these null hypotheses, an ANOVA table is constructed containing the appropriate sums of squares. For the tests we present to be valid, the number of replicates must be the same for each treatment and must be at least 2. In addition, the observations in each treatment must constitute a random sample from a normal population, and the populations must all have the same variance.

We compute the error sum of squares (SSE) by adding the sums of squared deviations from the sample means for all the treatments. To express this in an equation, let $s_1^2, \ldots, s_8^2$ denote the sample variances of the observations in each of the eight treatments, and let K be the number of replicates per treatment. Then

$$\text{SSE} = (K - 1)\sum_{i=1}^{8} s_i^2 \tag{9.66}$$

Each main effect and interaction has its own sum of squares as well. These are easy to compute. The sum of squares for any effect or interaction is computed by squaring its contrast, multiplying by the number of replicates K, and dividing by the total number of treatments, which is $2^3 = 8$.

$$\text{Sum of squares for an effect} = \frac{K(\text{contrast})^2}{8} \tag{9.67}$$

When using Equation (9.67), it is best to keep as many digits in the effect estimates as possible, in order to obtain maximum precision in the sum of squares. For presentation in a table, effect estimates and sums of squares may be rounded to the same precision as the data.

The sums of squares for the effects and interactions have one degree of freedom each. The error sum of squares has $8(K - 1)$ degrees of freedom. The method for computing mean squares and F statistics is the same as the one presented in Section 9.3 for a two-way ANOVA table. Each mean square is equal to its sum of squares divided by its degrees of freedom. The test statistic for testing the null hypothesis that an effect or interaction is equal to 0 is computed by dividing the mean square for the effect estimate by the mean square for error. When the null hypothesis is true, the test statistic has an $F_{1,\,8(K-1)}$ distribution.

Refer to Example 9.23. Construct an ANOVA table. For each effect and interaction, test the null hypothesis that it is equal to 0. Which factors, if any, seem most likely to have an effect on the outcome?

Solution

The ANOVA table follows. The sums of squares for the effects and interactions were computed by using Equation (9.67). The error sum of squares was computed by applying Equation (9.66) to the data in Example 9.23. Each F statistic is the quotient of the mean square with the mean square for error. Each F statistic has 1 and 16 degrees of freedom.

Source	Effect	Sum of Squares	df	Mean Square	F	P
A	3.10	57.54	1	57.54	7.34	0.015
B	2.73	44.72	1	44.72	5.70	0.030
C	−0.93	5.23	1	5.23	0.67	0.426
AB	−3.18	60.48	1	60.48	7.71	0.013
AC	−1.34	10.75	1	10.75	1.37	0.259
BC	−1.06	6.76	1	6.76	0.86	0.367
ABC	1.07	6.83	1	6.83	0.87	0.365
Error		125.48	16	7.84		
Total		317.78	23			

The main effects of factors A and B, as well as the AB interaction, have fairly small P-values. This suggests that these effects are not equal to 0 and that factors A and B do affect the outcome. There is no evidence that the main effect of factor C, or any of its interactions, differ from 0. Further experiments might focus on factors A and B. Perhaps a two-way ANOVA would be conducted, with each of the factors A and B evaluated at several levels, to get more detailed information about their effects on the outcome.

Interpreting Computer Output

In practice, analyses of factorial designs are usually carried out on a computer, using a software package such as MINITAB. The following MINITAB output presents the results of the analysis described in Examples 9.23 and 9.24.

```
Factorial Fit: Yield versus A, B, C

Estimated Effects and Coefficients for Yield (coded units)

Term          Effect      Coef   SE Coef        T       P
Constant                75.641    0.5716   132.33   0.000
A              3.097     1.548    0.5716     2.71   0.015
B              2.730     1.365    0.5716     2.39   0.030
C             -0.933    -0.467    0.5716    -0.82   0.426
A*B           -3.175    -1.587    0.5716    -2.78   0.013
A*C           -1.338    -0.669    0.5716    -1.17   0.259
B*C           -1.062    -0.531    0.5716    -0.93   0.367
A*B*C          1.067     0.533    0.5716     0.93   0.365

S = 2.80040    R-Sq = 60.51%      R-Sq(adj) = 43.24%

Analysis of Variance for Yield (coded units)

Source               DF   Seq SS   Adj SS   Adj MS     F      P
Main Effects          3  107.480  107.480   35.827  4.57  0.017
2-Way Interactions    3   77.993   77.993   25.998  3.32  0.047
3-Way Interactions    1    6.827    6.827    6.827  0.87  0.365
Residual Error       16  125.476  125.476    7.842
  Pure Error         16  125.476  125.476    7.842
Total                23  317.776
```

The table at the top of the MINITAB output presents estimated effects and coefficients. The phrase "coded units" means that the values 1 and -1, rather than the actual values, are used to represent the high and low levels of each factor. The estimated effects are listed in the column labeled "Effect." In the next column are the estimated **coefficients**, each of which is equal to one-half the corresponding effect. While the effect represents the difference in the mean response between the high and low levels of a factor, the coefficient represents the difference between the mean response at the high level and the grand mean response, which is half as much. The coefficient labeled "Constant" is the mean of all the observations, that is, the sample grand mean. Every coefficient estimate has the same standard deviation, which is shown in the column labeled "SE Coef."

MINITAB uses the Student's t statistic, rather than the F statistic, to test the hypotheses that the effects are equal to zero. The column labeled "T" presents the value of the Student's t statistic, which is equal to the quotient of the coefficient estimate (Coef) and its standard deviation. Under the null hypothesis, the t statistic has a Student's t distribution with $2^p(K-1)$ degrees of freedom. The P-values are presented in the column labeled "P." The t test performed by MINITAB is equivalent to the F test described in Example 9.24. The $t_{8(K-1)}$ statistic can be computed by taking the square root of the $F_{1,\,8(K-1)}$ statistic and applying the sign of the effect estimate. The P-values are identical.

We'll discuss the analysis of variance table next. The column labeled "DF" presents degrees of freedom. The columns labeled "Seq SS" (sequential sum of squares) and "Adj SS" (adjusted sum of squares) will be identical in all the examples we will consider and will contain sums of squares. The column labeled "Adj MS" contains mean squares, or sums of squares divided by their degrees of freedom. We will now explain the rows involving error. The row labeled "Pure Error" is concerned with the error sum of squares (SSE) (Equation 9.66). There are $8(K-1) = 16$ degrees of freedom (DF) for pure error. The sum of squares for pure error, found in each of the next two columns is the error sum of squares (SSE). Under the column "Adj MS" is the mean square for error. The row above the pure error row is labeled "Residual Error." The sum of squares for residual error is equal to the sum of squares for pure error, plus the sums of squares for any main effects or interactions that are not included in the model. The degrees of freedom for the residual error sum of squares is equal to the degrees of freedom for pure error, plus the degrees of freedom (one each) for each main effect or interaction not included in the model. Since in this example all main effects and interactions are included in the model, the residual error sum of squares and its degrees of freedom are equal to the corresponding quantities for pure error. The row labeled "Total" contains the total sum of squares (SST). The total sum of squares and its degrees of freedom are equal to the sums of the corresponding quantities for all the effects, interactions, and residual error.

Going back to the top of the table, the first row is labeled "Main Effects." There are three degrees of freedom for main effects, because there are three main effects (A, B, and C), with one degree of freedom each. The sequential sum of squares is the sum of the sums of squares for each of the three main effects. The mean square (Adj MS) is the sum of squares divided by its degrees of freedom. The column labeled "F" presents the F statistic for testing the null hypothesis that all the main effects are equal to zero. The value of the F statistic (4.57) is equal to the quotient of the mean square for main effects (35.827) and the mean square for (pure) error (7.842). The degrees of freedom for the F statistic are 3 and 16, corresponding to the degrees of freedom for the two mean squares. The column labeled "P" presents the P-value for the F test. In this case the P-value is 0.017, which indicates that not all the main effects are zero.

The rows labeled "2-Way Interactions" and "3-Way Interactions" are analogous to the row for main effects. The P-value for two-way interactions is 0.047, which is reasonably strong evidence that at least some of the two-way interactions are not equal to zero. Since there is only one three-way interaction ($A * B * C$), the P-value in the row labeled "3-Way Interactions" is the same (0.365) as the P-value in the table at the top of the MINITAB output for $A * B * C$.

Recall that the hypothesis tests are performed under the assumption that all the observations have the same standard deviation σ. The quantity labeled "S" is the estimate of σ and is equal to the square root of the mean square for error (MSE). The quantities "R-sq" and "R-sq(adj)" are the coefficients of determination R^2 and the adjusted R^2, respectively, and are computed by methods analogous to those in one-way ANOVA.

Estimating Effects in a 2^p Factorial Experiment

A sign table can be used to obtain the formulas for computing effect estimates in any 2^p factorial experiment. The method is analogous to the 2^3 case. The treatments are listed in a column. The sign for any main effect is + in the rows corresponding to treatments where the factor is at its high level, and − in rows corresponding to treatments where the factor is at its low level. Signs for the interactions are found by multiplying the signs corresponding to the factors in the interaction. The estimate for any effect or interaction is found by adding and subtracting the cell means for the treatments, using the signs in the appropriate columns, to compute a contrast. The contrast is then divided by half the number of treatments, or 2^{p-1}, to obtain the effect estimate.

Summary

For a 2^p factorial experiment:

$$\text{Effect estimate} = \frac{\text{contrast}}{2^{p-1}} \tag{9.68}$$

As an example, Table 9.8 presents a sign table for a 2^5 factorial experiment. We list the signs for the main effects and for selected interactions.

Sums of squares are computed by a method analogous to that for a 2^3 experiment. To compute the error sum of squares (SSE), let $s_1, \ldots, s_{2^p}$ be the sample variances of the observations in each of the 2^p treatments. Then

$$\text{SSE} = (K - 1)\sum_{i=1}^{2^p} s_i^2$$

The degrees of freedom for error is $2^p(K - 1)$, where K is the number of replicates per treatment. The sum of squares for each effect and interaction is equal to the square of the contrast, multiplied by the number of replicates K and divided by the number of treatments 2^p. The sums of squares for the effects and interactions have one degree of freedom each.

$$\text{Sum of squares for an effect} = \frac{K(\text{contrast})^2}{2^p} \tag{9.69}$$

F statistics for main effects and interactions are computed by dividing the sum of squares for the effect by the mean square for error. The null distribution of the F statistic is $F_{1,\, 2^p(K-1)}$.

Factorial Experiments without Replication

When the number of factors p is large, it is often not feasible to perform more than one replicate for each treatment. In this case, it is not possible to compute SSE, so the hypothesis tests previously described cannot be performed. If it is reasonable to assume that some of the higher-order interactions are equal to 0, then the sums of squares for those interactions can be added together and treated like an error sum of squares. Then the main effects and lower order interactions can be tested.

TABLE 9.8 Sign table for the main effects and selected interactions for a 2^5 factorial experiment

Treatment	*A*	*B*	*C*	*D*	*E*	*AB*	*CDE*	*ABDE*	*ABCDE*
1	−	−	−	−	−	+	−	+	−
a	+	−	−	−	−	−	−	−	+
b	−	+	−	−	−	−	−	−	+
ab	+	+	−	−	−	+	−	+	−
c	−	−	+	−	−	+	+	+	+
ac	+	−	+	−	−	−	+	−	−
bc	−	+	+	−	−	−	+	−	−
abc	+	+	+	−	−	+	+	+	+
d	−	−	−	+	−	+	+	−	+
ad	+	−	−	+	−	−	+	+	−
bd	−	+	−	+	−	−	+	+	−
abd	+	+	−	+	−	+	+	−	+
cd	−	−	+	+	−	+	−	−	−
acd	+	−	+	+	−	−	−	+	+
bcd	−	+	+	+	−	−	−	+	+
abcd	+	+	+	+	−	+	−	−	−
e	−	−	−	−	+	+	+	−	+
ae	+	−	−	−	+	−	+	+	−
be	−	+	−	−	+	−	+	+	−
abe	+	+	−	−	+	+	+	−	+
ce	−	−	+	−	+	+	−	−	−
ace	+	−	+	−	+	−	−	+	+
bce	−	+	+	−	+	−	−	+	+
abce	+	+	+	−	+	+	−	−	−
de	−	−	−	+	+	+	−	+	−
ade	+	−	−	+	+	−	−	−	+
bde	−	+	−	+	+	−	−	−	+
abde	+	+	−	+	+	+	−	+	−
cde	−	−	+	+	+	+	+	+	+
acde	+	−	+	+	+	−	+	−	−
bcde	−	+	+	+	+	−	+	−	−
abcde	+	+	+	+	+	+	+	+	+

A 2^5 factorial experiment was conducted to estimate the effects of five factors on the quality of lightbulbs manufactured by a certain process. The factors were *A*: plant (1 or 2), *B*: machine type (low or high speed), *C*: shift (day or evening), *D*: lead wire material (standard or new), and *E*: method of loading materials into the assembler (manual or automatic). One replicate was obtained for each treatment. Table 9.9 on page 702 presents the results. Compute estimates of the main effects and interactions, and their sums of squares. Assume that the third-, fourth-, and fifth-order interactions are negligible, and add their sums of squares to use as a substitute for an error sum of squares. Use this substitute to test hypotheses concerning the main effects and second-order interactions.

TABLE 9.9

Treatment	Outcome	Treatment	Outcome	Treatment	Outcome	Treatment	Outcome
1	32.07	*d*	35.64	*e*	25.10	*de*	40.60
a	39.27	*ad*	35.91	*ae*	39.25	*ade*	37.57
b	34.81	*bd*	47.75	*be*	37.77	*bde*	47.22
ab	43.07	*abd*	51.47	*abe*	46.69	*abde*	56.87
c	31.55	*cd*	33.16	*ce*	32.55	*cde*	34.51
ac	36.51	*acd*	35.32	*ace*	32.56	*acde*	36.67
bc	28.80	*bcd*	48.26	*bce*	28.99	*bcde*	45.15
abc	43.05	*abcd*	53.28	*abce*	48.92	*abcde*	48.72

TABLE 9.10

Variable	Effect	Sum of Squares	Variable	Effect	Sum of Squares
A	6.33	320.05	ABD	−0.29	0.67
B	9.54	727.52	ABE	0.76	4.59
C	−2.07	34.16	ACD	0.11	0.088
D	6.70	358.72	ACE	−0.69	3.75
E	0.58	2.66	ADE	−0.45	1.60
AB	2.84	64.52	BCD	0.76	4.67
AC	0.18	0.27	BCE	−0.82	5.43
AD	−3.39	91.67	BDE	−2.17	37.63
AE	0.60	2.83	CDE	−1.25	12.48
BC	−0.49	1.95	$ABCD$	−2.83	63.96
BD	4.13	136.54	$ABCE$	0.39	1.22
BE	0.65	3.42	$ABDE$	0.22	0.37
CD	−0.18	0.26	$ACDE$	0.18	0.24
CE	−0.81	5.23	$BCDE$	−0.25	0.52
DE	0.24	0.46	$ABCDE$	−1.73	23.80
ABC	1.35	14.47			

Solution

We compute the effects, using the rules for adding and subtracting observations given by the sign table, and the sums of squares, using Equation (9.69). See Table 9.10.

Note that none of the three-, four-, or five-way interactions are among the larger effects. If some of them were, it would not be wise to combine their sums of squares. As it is, we add the sums of squares of the three-, four-, and five-way interactions. The results are presented in the following MINITAB output.

```
Factorial Fit: Response versus A, B, C, D, E

Estimated Effects and Coefficients for
Response (coded units)

Term        Effect     Coef   SE Coef       T      P
Constant             39.658    0.5854   67.74  0.000
A            6.325    3.163    0.5854    5.40  0.000
```

```
B              9.536      4.768  0.5854    8.14   0.000
C             -2.066     -1.033  0.5854   -1.76   0.097
D              6.696      3.348  0.5854    5.72   0.000
E              0.576      0.288  0.5854    0.49   0.629
A*B            2.840      1.420  0.5854    2.43   0.027
A*C            0.183      0.091  0.5854    0.16   0.878
A*D           -3.385     -1.693  0.5854   -2.89   0.011
A*E            0.595      0.298  0.5854    0.51   0.618
B*C           -0.494     -0.247  0.5854   -0.42   0.679
B*D            4.131      2.066  0.5854    3.53   0.003
B*E            0.654      0.327  0.5854    0.56   0.584
C*D           -0.179     -0.089  0.5854   -0.15   0.881
C*E           -0.809     -0.404  0.5854   -0.69   0.500
D*E            0.239      0.119  0.5854    0.20   0.841

S = 3.31179       R-Sq = 90.89%      R-Sq(adj) = 82.34%

Analysis of Variance for Response (coded units)

Source               DF   Seq SS   Adj SS   Adj MS       F       P
Main Effects          5   1443.1   1443.1   288.62   26.31   0.000
2-Way Interactions   10    307.1    307.1    30.71    2.80   0.032
Residual Error       16    175.5    175.5    10.97
Total                31   1925.7
```

The estimates have not changed for the main effects or two-way interactions. The residual error sum of squares (175.5) in the analysis of variance table is found by adding the sum of squares for all the higher-order interactions that were dropped from the model. The number of degrees of freedom (16) is equal to the sum of the degrees of freedom (one each) for the 16 higher-order interactions. There is no sum of squares for pure error (SSE), because there is only one replicate per treatment. The residual error sum of squares is used as a substitute for SSE to compute all the quantities that require an error sum of squares.

We conclude from the output that factors A, B, and D are likely to affect the outcome. There seem to be interactions between some pairs of these factors as well. It might be appropriate to plan further experiments to focus on factors A, B, and D.

Using Probability Plots to Detect Large Effects

An informal method that has been suggested to help determine which effects are large is to plot the effect and interaction estimates on a normal probability plot. If in fact none of the factors affect the outcome, then the effect and interaction estimates form a simple random sample from a normal population and should lie approximately on a straight line. In many cases, most of the estimates will fall approximately on a line, and a few will plot far from the line. The main effects and interactions whose estimates plot far from the line are the ones most likely to be important. Figure 9.13 presents a normal probability plot of the main effect and interaction estimates from the data in Example 9.25. It is clear from the plot that the main effects of factors A, B, and D, and the AB and BD interactions, stand out from the rest.

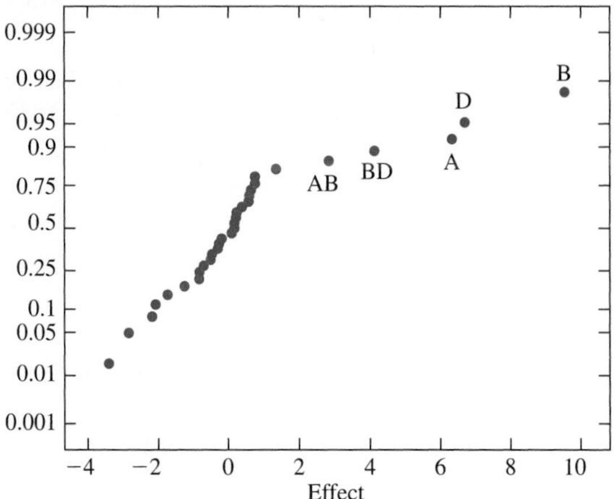

FIGURE 9.13 Normal probability plot of the effect estimates from the data in Example 9.25. The main effects of factors A, B, and D stand out as being larger than the rest.

Fractional Factorial Experiments

When the number of factors is large, it may not be feasible to perform even one replicate for each treatment. In these cases, observations may be taken only for some fraction of the treatments. If these treatments are chosen correctly, it is still possible to obtain information about the factors.

When each factor has two levels, the fraction must always be a power of 2, i.e., one-half, one-quarter, etc. An experiment in which half the treatments are used is called a **half-replicate**; if one-quarter of the treatments are used, it is a **quarter-replicate**, and so on. A half-replicate of a 2^p experiment is often denoted 2^{p-1}, to indicate that while there are p factors, there are only 2^{p-1} treatments being considered. Similarly, a quarter-replicate is often denoted 2^{p-2}. We will focus on half-replicate experiments.

We present a method for choosing a half-replicate of a 2^5 experiment. Such an experiment will have 16 treatments chosen from the 32 in the 2^5 experiment. To choose the 16 treatments, start with a sign table for a 2^4 design that shows the signs for the main effects and the highest-order interaction. This is presented as Table 9.11.

Table 9.11 has the right number of treatments (16), but only four factors. To transform it into a half-replicate for a 2^5 design, we must introduce a fifth factor, E. We do this by replacing the highest-order interaction by E. This establishes the signs for the main effect of E. Then in each row where the sign for E is $+$, we add the letter e to the

TABLE 9.11 Sign table for the main effects and four-way interaction in a 2^4 factorial experiment

Treatment	A	B	C	D	$ABCD$
1	−	−	−	−	+
a	+	−	−	−	−
b	−	+	−	−	−
ab	+	+	−	−	+
c	−	−	+	−	−
ac	+	−	+	−	+
bc	−	+	+	−	+
abc	+	+	+	−	−
d	−	−	−	+	−
ad	+	−	−	+	+
bd	−	+	−	+	+
abd	+	+	−	+	−
cd	−	−	+	+	+
acd	+	−	+	+	−
bcd	−	+	+	+	−
$abcd$	+	+	+	+	+

treatment, indicating that factor E is to be set to its high level for that treatment. Where the sign for E is −, factor E is set to its low level. The resulting design is called the **principal fraction** of the 2^5 design. Table 9.12 presents the signs for the main effects and selected interactions of this design.

TABLE 9.12 Sign table for the main effects and selected interactions for the principal fraction of a 2^5 factorial experiment

Treatment	A	B	C	D	$E = ABCD$	AB	CDE	$ACDE$
e	−	−	−	−	+	+	+	−
a	+	−	−	−	−	−	−	−
b	−	+	−	−	−	−	−	+
abe	+	+	−	−	+	+	+	+
c	−	−	+	−	−	+	+	−
ace	+	−	+	−	+	−	−	−
bce	−	+	+	−	+	−	−	+
abc	+	+	+	−	−	+	+	+
d	−	−	−	+	−	+	+	−
ade	+	−	−	+	+	−	−	−
bde	−	+	−	+	+	−	−	+
abd	+	+	−	+	−	+	+	+
cde	−	−	+	+	+	+	+	−
acd	+	−	+	+	−	−	−	−
bcd	−	+	+	+	−	−	−	+
$abcde$	+	+	+	+	+	+	+	+

There is a price to be paid for using only half of the treatments. To see this, note that in Table 9.12 the AB interaction has the same signs as the CDE interaction, and the $ACDE$ interaction has the same signs as the main effect for B. When two effects have the same signs, they are said to be **aliased**. In fact, the main effects and interactions in a half-fraction form pairs in which each member of the pair is aliased with the other. The alias pairs for this half-fraction of the 2^5 design are

$$\{A, BCDE\} \quad \{B, ACDE\} \quad \{C, ABDE\} \quad \{D, ABCE\} \quad \{E, ABCD\}$$

$$\{AB, CDE\} \quad \{AC, BDE\} \quad \{AD, BCE\} \quad \{AE, BCD\} \quad \{BC, ADE\}$$

$$\{BD, ACE\} \quad \{BE, ACD\} \quad \{CD, ABE\} \quad \{CE, ABD\} \quad \{DE, ABC\}$$

When two effects are aliased, their effect estimates are the same, because they involve the same signs. In fact, when the principal fraction of a design is used, the estimate of any effect actually represents the sum of that effect and its alias. Therefore for the principal fraction of a 2^5 design, each main effect estimate actually represents the sum of the main effect plus its aliased four-way interaction, and each two-way interaction estimate represents the sum of the two-way interaction and its aliased three-way interaction.

In many cases, it is reasonable to assume that the higher-order interactions are small. In the 2^5 half-replicate, if the four-way interactions are negligible, the main effect estimates will be accurate. If in addition the three-way interactions are negligible, the two-way interaction estimates will be accurate as well.

In a fractional design without replication, there is often no good way to compute an error sum of squares, and therefore no rigorous way to test the hypotheses that the effects are equal to 0. In many cases, the purpose of a fractional design is simply to identify a few factors that appear to have the greatest impact on the outcome. This information may then be used to design more elaborate experiments to investigate these factors. For this purpose, it may be enough simply to choose those factors whose effects or two-way interactions are unusually large, without performing hypothesis tests. This can be done by listing the estimates in decreasing order, and then looking to see if there are a few that are noticeably larger than the rest. Another method is to plot the effect and interaction estimates on a normal probability plot, as previously discussed.

Example 9.26

In an emulsion liquid membrane system, an emulsion (internal phase) is dispersed into an external liquid medium containing a contaminant. The contaminant is removed from the external liquid through mass transfer into the emulsion. Internal phase leakage occurs when portions of the extracted material spill into the external liquid. In the article "Leakage and Swell in Emulsion Liquid Membrane Systems: Batch Experiments" (R. Pfeiffer, W. Navidi, and A. Bunge, *Separation Science and Technology*, 2003:519–539), the effects of five factors were studied to determine the effect on leakage in a certain system. The five factors were A: surfactant concentration, B: internal phase lithium hydroxide concentration, C: membrane phase, D: internal phase volume fraction, and E: extraction vessel stirring rate. A half-fraction of a 2^5 design was used. The data are presented in the following table (in the actual experiment, each point actually represented the average of two measurements). Leakage is measured in units of percent. Assume that the third-, fourth-, and fifth-order interactions are negligible. Estimate the main effects and two-way interactions. Which, if any, stand out as being noticeably larger than the rest?

Treatment	Leakage	Treatment	Leakage	Treatment	Leakage	Treatment	Leakage
e	0.61	c	0.35	d	2.03	cde	1.45
a	0.13	ace	0.075	ade	0.64	acd	0.31
b	2.23	bce	7.31	bde	11.72	bcd	1.33
abe	0.095	abc	0.080	abd	0.56	$abcde$	6.24

Solution

Using the sign table (Table 9.12), we compute estimates for the main effects and two-way interactions, shown in the following table.

Variable	Effect	Variable	Effect
A	−2.36	AE	−1.15
B	3.00	BC	0.20
C	−0.11	BD	0.86
D	1.68	BE	2.65
E	2.64	CD	−1.30
AB	−1.54	CE	0.61
AC	1.43	DE	1.32
AD	0.17		

Note that we do not bother to compute sums of squares for the estimates, because we have no SSE to compare them to. To determine informally which effects may be most worthy of further investigation, we rank the estimates in order of their absolute values: B: 3.00, BE: 2.65, E: 2.64, A: −2.36, D: 1.68, and so forth. It seems reasonable to decide that there is a fairly wide gap between the A and D effects, and therefore that factors A, B, and E are most likely to be important.

Exercises for Section 9.5

1. Construct a sign table for the principal fraction for a 2^4 design. Then indicate all the alias pairs.

2. Give an example of a factorial experiment in which failure to randomize can produce incorrect results.

3. A pilot plant study investigated the effects of three factors on the yield (in grams) of monomer for an adhesive formulation. The three factors were A: temperature, B: catalyst level, and C: catalyst type. The levels of the factors were A: 160°C (−1) and 180°C (+1); B: 20% (−1) and 40% (+1); C: vendor E (−1) and vendor J (+1). Two replicates of a 2^3 factorial design were carried out, with the results presented in the following table.

A	*B*	*C*	Yields	Mean Yield
−1	−1	−1	58, 60	59
1	−1	−1	74, 70	72
−1	1	−1	43, 51	47
1	1	−1	75, 73	74
−1	−1	1	56, 60	58
1	−1	1	74, 78	76
−1	1	1	46, 44	45
1	1	1	78, 80	79

a. Compute estimates of the main effects and interactions, and the sum of squares and P-value for each.
b. Which main effects and interactions are important?
c. Other things being equal, will the mean yield be higher when the temperature is 160°C or 180°C? Explain.

4. The article "Efficient Pyruvate Production by a Multi-Vitamin Auxotroph of *Torulopsis glabrata:* Key Role and Optimization of Vitamin Levels" (Y. Li, J. Chen, et al. *Applied Microbiology and Biotechnology*, 2001:680–685) investigates the effects of the levels of several vitamins in a cell culture on the yield (in g/L) of pyruvate, a useful organic acid. The data in the following table are presented as two replicates of a 2^3 design. The factors are A: nicotinic acid, B: thiamine, and C: biotin. (Two statistically insignificant factors have been dropped. In the article, each factor was tested at four levels; we have collapsed these to two.)

A	*B*	*C*	Yields	Mean Yield
−1	−1	−1	0.55, 0.49	0.520
1	−1	−1	0.60, 0.42	0.510
−1	1	−1	0.37, 0.28	0.325
1	1	−1	0.30, 0.28	0.290
−1	−1	1	0.54, 0.54	0.540
1	−1	1	0.54, 0.47	0.505
−1	1	1	0.44, 0.33	0.385
1	1	1	0.36, 0.20	0.280

a. Compute estimates of the main effects and interactions, and the sum of squares and P-value for each.
b. Is the additive model appropriate?
c. What conclusions about the factors can be drawn from these results?

5. The article cited in Exercise 4 also investigated the effects of the factors on glucose consumption (in g/L). A single measurement is provided for each combination of factors (in the article, there was some replication). The results are presented in the following table.

A	*B*	*C*	Glucose Consumption
−1	−1	−1	68.0
1	−1	−1	77.5
−1	1	−1	98.0
1	1	−1	98.0
−1	−1	1	74.0
1	−1	1	77.0
−1	1	1	97.0
1	1	1	98.0

a. Compute estimates of the main effects and the interactions.
b. Is it possible to compute an error sum of squares? Explain.

c. Are any of the interactions among the larger effects? If so, which ones?
d. Assume that it is known from past experience that the additive model holds. Add the sums of squares for the interactions, and use that result in place of an error sum of squares to test the hypotheses that the main effects are equal to 0.

6. A decanter was designed and constructed for the separation of soap solids from a water-based slurry. The purpose of the device is to reduce the water content of the soap slurry so that the soap can be further processed into bars and the liquid (a salt solution) can be reused. A preliminary study was conducted in which three factors were varied. They were *A*: the temperature of the operation (measured in °C), *B*: the flow rate to the device (measured as a residence time, in seconds), and *C*: the holdup of solids in the device (measured as a residence time, in seconds). The outcome is the solid fraction in the slurry. The experiment was performed once for each combination of factors. The results are presented in the following table.

Temperature (*A*)	Flow (*B*)	Solids Holdup (*C*)	Solid Fraction in Slurry
50	170	150	33
70	170	150	44
50	300	150	47
70	300	150	36
50	170	500	51
70	170	500	43
50	300	500	54
70	300	500	56

a. Compute estimates of the main effects and the interactions.
b. Is it possible to compute an error sum of squares? Explain.
c. Are any of the interactions among the larger effects? If so, which ones?
d. Someone claims that the additive model holds. Do the results tend to support this statement? Explain.

7. The measurement of the resistance of a tungsten-coated wafer used in integrated circuit manufacture may be affected by several factors, including *A*: which of two types of cassette was used to hold the wafer, *B*: whether the wafer is loaded on the top or on the bottom of the cassette, and *C*: whether the front or the back cassette station was used. Data from a 2^3 factorial experiment with one replicate are presented in the following table. (Based on the article "Prometrix RS35e Gauge Study in Five Two-Level Factors and One Three-Level Factor," J. Buckner, B. Chin, and J. Henri, *Statistical Case Studies for Industrial Process Improvement*, SIAM-ASA, 1997:9–18.)

A	*B*	*C*	Resistance (mΩ)
−1	−1	−1	85.04
1	−1	−1	84.49
−1	1	−1	82.15
1	1	−1	86.37
−1	−1	1	82.60
1	−1	1	85.14
−1	1	1	82.87
1	1	1	86.44

a. Compute estimates of the main effects and the interactions.
b. Is it possible to compute an error sum of squares? Explain.

c. Plot the estimates on a normal probability plot. Does the plot show that some of the factors influence the resistance? If so, which ones?

8. In a 2^p design with one replicate per treatment, it sometimes happens that the observation for one of the treatments is missing, due to experimental error or to some other cause. When this happens, one approach is to replace the missing value with the value that makes the highest-order interaction equal to 0. Refer to Exercise 7. Assume the observation for the treatment where A, B, and C are all at their low level (-1) is missing.

a. What value for this observation makes the three-way interaction equal to 0?
b. Using this value, compute estimates for the main effects and the interactions.

9. Safety considerations are important in the design of automobiles. The article "An Optimum Design Methodology Development Using a Statistical Technique for Vehicle Occupant Safety" (J. Hong, M. Mun, and S. Song, *Proceedings of the Institution of Mechanical Engineers*, 2001:795–801) presents results from an occupant simulation study. The outcome variable is chest acceleration (in g) 3 ms after impact. Four factors were considered. They were A: the airbag vent characteristic, B: the airbag inflator trigger time, C: the airbag inflator mass flow rate, and D: the stress–strain relationship of knee foam. The results (part of a larger study) are presented in the following table. There is one replicate per treatment.

Treatment	Outcome	Treatment	Outcome	Treatment	Outcome	Treatment	Outcome
1	85.2	*c*	66.0	*d*	85.0	*cd*	62.6
a	79.2	*ac*	69.0	*ad*	82.0	*acd*	65.4
b	84.3	*bc*	68.5	*bd*	84.7	*bcd*	66.3
ab	89.0	*abc*	76.4	*abd*	82.2	*abcd*	69.0

a. Compute estimates of the main effects and the interactions.
b. If you were to design a follow-up study, which factor or factors would you focus on? Explain.

10. The article "Experimental Study of Workpiece-Level Variability in Blind via Electroplating" (G. Poon, J. Chan, and D. Williams, *Proceedings of the Institution of Mechanical Engineers*, 2001:521–530) describes a factorial experiment that was carried out to determine which of several factors influence variability in the thickness of an acid copper electroplate deposit. The results of the experiment are presented here as a complete 2^5 design; two statistically insignificant factors have been omitted. There was one replicate per treatment. The factors are A: Sulphuric acid concentration (g/L), B: Copper sulphate concentration (g/L), C: Average current density (mA/cm^2), D: Electrode separation (cm), and E: Interhole distance (cm). The outcome is the variability in the thickness (in μm), measured by computing the standard deviation of thicknesses measured at several points on a test board, after dropping the largest and smallest measurement. The data are presented in the following table.

Treatment	Outcome	Treatment	Outcome	Treatment	Outcome	Treatment	Outcome
1	1.129	*d*	1.760	*e*	1.224	*de*	1.674
a	0.985	*ad*	1.684	*ae*	1.092	*ade*	1.215
b	1.347	*bd*	1.957	*be*	1.280	*bde*	1.275
ab	1.151	*abd*	1.656	*abe*	1.381	*abde*	1.446
c	2.197	*cd*	2.472	*ce*	1.859	*cde*	2.585
ac	1.838	*acd*	2.147	*ace*	1.865	*acde*	2.587
bc	1.744	*bcd*	2.142	*bce*	1.867	*bcde*	2.339
abc	2.101	*abcd*	2.423	*abce*	2.005	*abcde*	2.629

a. Compute estimates of the main effects and the interactions.
b. If you were to design a follow-up experiment, which factors would you focus on? Why?

11. The article "Factorial Design for Column Flotation of Phosphate Wastes" (N. Abdel-Khalek, *Particulate Science and Technology*, 2000:57–70) describes a 2^3 factorial design to investigate the effect of superficial air velocity (A), frothier concentration (B), and superficial wash water velocity (C) on the percent recovery of P_2O_5. There were two replicates. The data are presented in the following table.

A	B	C	Percent Recovery	
−1	−1	−1	56.30	54.85
1	−1	−1	70.10	72.70
−1	1	−1	65.60	63.60
1	1	−1	80.20	78.80
−1	−1	1	50.30	48.95
1	−1	1	65.30	66.00
−1	1	1	60.53	59.50
1	1	1	70.63	69.86

a. Compute estimates of the main effects and interactions, along with their sums of squares and P-values.
b. Which factors seem to be most important? Do the important factors interact? Explain.

12. The article "An Application of Fractional Factorial Designs" (M. Kilgo, *Quality Engineering*, 1988:19–23) describes a 2^{5-1} design (half-replicate of a 2^5 design) involving the use of carbon dioxide (CO_2) at high pressure to extract oil from peanuts. The outcomes were the solubility of the peanut oil in the CO_2 (in mg oil/liter CO_2), and the yield of peanut oil (in %). The five factors were A: CO_2 pressure, B: CO_2 temperature, C: peanut moisture, D: CO_2 flow rate, and E: peanut particle size. The results are presented in the following table.

Treatment	Solubility	Yield	Treatment	Solubility	Yield
e	29.2	63	*d*	22.4	23
a	23.0	21	*ade*	37.2	74
b	37.0	36	*bde*	31.3	80
abe	139.7	99	*abd*	48.6	33
c	23.3	24	*cde*	22.9	63
ace	38.3	66	*acd*	36.2	21
bce	42.6	71	*bcd*	33.6	44
abc	141.4	54	*abcde*	172.6	96

a. Assuming third- and higher-order interactions to be negligible, compute estimates of the main effects and interactions for the solubility outcome.
b. Plot the estimates on a normal probability plot. Does the plot show that some of the factors influence the solubility? If so, which ones?
c. Assuming third- and higher-order interactions to be negligible, compute estimates of the main effects and interactions for the yield outcome.
d. Plot the estimates on a normal probability plot. Does the plot show that some of the factors influence the yield? If so, which ones?

13. In a 2^{5-1} design (such as the one in Exercise 12) what does the estimate of the main effect of factor A actually represent?

i. The main effect of A.
ii. The sum of the main effect of A and the $BCDE$ interaction.
iii. The difference between the main effect of A and the $BCDE$ interaction.
iv. The interaction between A and $BCDE$.

Supplementary Exercises for Chapter 9

1. The article "Gypsum Effect on the Aggregate Size and Geometry of Three Sodic Soils Under Reclamation" (I. Lebron, D. Suarez, and T. Yoshida, *Journal of the Soil Science Society of America*, 2002:92–98) reports on an experiment in which gypsum was added in various amounts to soil samples before leaching. One of the outcomes of interest was the pH of the soil. Gypsum was added in four different amounts. Three soil samples received each amount added. The pH measurements of the samples are presented in the following table.

Gypsum (g/kg)	pH		
0.00	7.88	7.72	7.68
0.11	7.81	7.64	7.85
0.19	7.84	7.63	7.87
0.38	7.80	7.73	8.00

Can you conclude that the pH differs with the amount of gypsum added? Provide the value of the test statistic and the P-value.

2. The article referred to in Exercise 1 also considered the effect of gypsum on the electric conductivity (in dS m^{-1}) of soil. Two types of soil were each treated with three different amounts of gypsum, with two replicates for each soil–gypsum combination. The data are presented in the following table.

	Soil Type			
Gypsum (g/kg)	las Animas		Madera	
0.00	1.52	1.05	1.01	0.92
0.27	1.49	0.91	1.12	0.92
0.46	0.99	0.92	0.88	0.92

a. Is there convincing evidence of an interaction between the amount of gypsum and soil type?
b. Can you conclude that the conductivity differs among the soil types?
c. Can you conclude that the conductivity differs with the amount of gypsum added?

3. Penicillin is produced by the *Penicillium* fungus, which is grown in a broth whose sugar content must be carefully controlled. Several samples of broth were taken on each of three successive days, and the amount of dissolved sugars (in mg/mL) was measured on each sample. The results were as follows:

Day 1:	4.8	5.1	5.1	4.8	5.2	4.9	5.0	4.9	5.0	4.8	4.8	5.1	5.0
Day 2:	5.4	5.0	5.0	5.1	5.2	5.1	5.3	5.2	5.2	5.1	5.4	5.2	5.4
Day 3:	5.7	5.1	5.3	5.5	5.3	5.5	5.1	5.6	5.3	5.2	5.5	5.3	5.4

Can you conclude that the mean sugar concentration differs among the three days?

4. The following MINITAB output is for a two-way ANOVA. Something went wrong with the printer, and some of the numbers weren't printed.

```
Two-way Analysis of Variance

Analysis of Variance
Source          DF          SS          MS      F       P
Row              3     145.375         (d)    (g)     (j)
Column           2      15.042         (e)    (h)     (k)
Interaction      6         (b)      4.2000    (i)     (l)
Error          (a)         (c)         (f)
Total           23     217.870
```

Fill in the missing numbers in the table for (a) through (l). You may give ranges for the P-values.

5. An experiment was performed to determine whether different types of chocolate take different amounts of time to dissolve. Forty people were divided into five groups. Each group was assigned a certain type of chocolate. Each person dissolved one piece of chocolate, and the dissolve time (in seconds) was recorded. For comparison, each person in each group also dissolved one piece of butterscotch candy; these pieces were identical for all groups. The data, which include the group, the dissolve times for both chocolate and butterscotch, the difference between the dissolve times, and the ratio of the dissolve times, are presented in the following table. Note that the design is slightly unbalanced; group 3 has nine people and group 5 has only seven.

Group	Chocolate	Butterscotch	Difference	Ratio
1	135	60	75	2.25
1	865	635	230	1.36
1	122	63	59	1.94
1	110	75	35	1.47
1	71	37	34	1.92
1	81	58	23	1.40
1	2405	1105	1300	2.18
1	242	135	107	1.79
2	42	38	4	1.11
2	30	30	0	1.00
2	104	110	−6	0.95
2	124	118	6	1.05
2	75	40	35	1.88
2	80	91	−11	0.88
2	255	121	134	2.11
2	71	71	0	1.00
3	51	53	−2	0.96
3	47	40	7	1.18
3	90	155	−65	0.58
3	65	90	−25	0.72
3	27	33	−6	0.82

Continued on page 714

Group	Chocolate	Butterscotch	Difference	Ratio
3	105	68	37	1.54
3	90	72	18	1.25
3	54	52	2	1.04
3	93	77	16	1.21
4	48	30	18	1.60
4	85	55	30	1.55
4	82	50	32	1.64
4	46	22	24	2.09
4	64	46	18	1.39
4	125	45	80	2.78
4	69	30	39	2.30
4	73	44	29	1.66
5	105	45	60	2.33
5	99	58	41	1.71
5	45	23	22	1.96
5	137	64	73	2.14
5	170	105	65	1.62
5	153	93	60	1.65
5	49	28	21	1.75

a. To test whether there are differences in the mean dissolve times for the different types of chocolate, someone suggests performing a one-way ANOVA, using the dissolve times for the chocolate data. Do these data appear to satisfy the assumptions for a one-way ANOVA? Explain.

b. Someone else suggests using the differences (Chocolate − Butterscotch). Do these data appear to satisfy the assumptions for a one-way ANOVA? Explain.

c. Perform a one-way analysis of variance using the ratios. Can you conclude that the mean ratio of dissolve times differs for different types of chocolate?

6. The article "Stability of Silico-Ferrite of Calcium and Aluminum (SFCA) in Air-Solid Solution Limits Between 1240°C and 1390°C and Phase Relationships within the Fe_2O_3-CaO-Al_2O_3-SiO_2 (FCAS) System" (T. Patrick and M. Pownceby, *Metallurgical and Materials Transactions B*, 2002:79–90) investigates properties of silico-ferrites of calcium and aluminum (SFCA). The data in the following table present the ratio of the weights of Fe_2O_3 and CaO for SFCA specimens with several different weight percents of Al_2O_3 and C_4S_3.

Al_2O_3 (%)	C_4S_3	Fe_2O_3/CaO			
1.0	Low (3%–6%)	7.25	6.92	6.60	6.31
1.0	Medium (7%–10%)	6.03	5.78	5.54	5.31
1.0	High (11%–14%)	5.10	4.90	4.71	4.53
5.0	Low (3%–6%)	6.92	6.59	6.29	6.01
5.0	Medium (7%–10%)	5.74	5.26	5.04	4.84
5.0	High (11%–14%)	4.84	4.65	4.47	4.29
10.0	Low (3%–6%)	6.50	6.18	5.89	5.63
10.0	Medium (7%–10%)	5.37	5.14	4.92	4.71
10.0	High (11%–14%)	4.52	4.33	4.16	3.99

a. Estimate all main effects and interactions.

b. Construct an ANOVA table. You may give ranges for the P-values.

c. Do the data indicate that there are any interactions between the weight percent of Al_2O_3 and the weight percent of C_4S_3? Explain.

d. Do the data convincingly demonstrate that the Fe_2O_3/CaO ratio depends on the weight percent of Al_2O_3? Explain.

e. Do the data convincingly demonstrate that the Fe_2O_3/CaO ratio depends on the weight percent of C_4S_3? Explain.

7. A component can be manufactured according to either of two designs and with either a more expensive or a less expensive material. Several components are manufactured with each combination of design and material, and the lifetimes of each are measured (in hours). A two-way analysis of variance was performed to estimate the effects of design and material on component lifetime. The cell means and main effect estimates are presented in the following table.

	Cell Means	
	Design 1	**Design 2**
More expensive	118	120
Less expensive	60	122

Main Effects	
More expensive	14
Less expensive	−14
Design 1	−16
Design 2	16

ANOVA table

Source	DF	SS	MS	F	P
Material	1	2352.0	2352.0	10.45	0.012
Design	1	3072.0	3072.0	13.65	0.006
Interaction	1	2700.0	2700.0	12.00	0.009
Error	8	1800.0	225.00		
Total	11	9924.0			

The process engineer recommends that design 2 should be used along with the more expensive material. He argues that the main effects of both design 2 and the more expensive material are positive, so using this combination will result in the longest component life. Do you agree with the recommendation? Why or why not?

8. The article "Case Study Based Instruction of DOE and SPC" (J. Brady and T. Allen, *The American Statistician*, 2002:312–315) presents the result of a 2^{4-1} factorial experiment to investigate the effects of four factors on the yield of a process that manufactures printed circuit boards. The factors were *A*: transistor power output (upper or lower specification limit), *B*: transistor mounting approach (screwed or soldered), *C*: transistor heat sink type (current or alternative configuration), and *D*: screw position on the frequency adjustor (one-half or two turns). The results are presented in the following table. The yield is a percent of a theoretical maximum.

A	*B*	*C*	*D*	Yield
−1	−1	−1	−1	79.8
1	−1	−1	1	69.0
−1	1	−1	1	72.3
1	1	−1	−1	71.2
−1	−1	1	1	91.3
1	−1	1	−1	95.4
−1	1	1	−1	92.7
1	1	1	1	91.5

a. Estimate the main effects of each of the four factors.
b. Assuming all interactions to be negligible, pool the sums of squares for interaction to use in place of an error sum of squares.
c. Which of the four factors, if any, can you conclude to affect the yield? What is the P-value of the relevant test?

9. The article "Combined Analysis of Real-Time Kinematic GPS Equipment and Its Users for Height Determination" (W. Featherstone and M. Stewart, *Journal of Surveying Engineering*, 2001:31–51) presents a study of the accuracy of global positioning system (GPS) equipment in measuring heights. Three types of equipment were studied, and each was used to make measurements at four different base stations (in the article a fifth station was included, for which the results differed considerably from the other four). There were 60 measurements made with each piece of equipment at each base. The means and standard deviations of the measurement errors (in mm) are presented in the following table for each combination of equipment type and base station.

	Instrument A		Instrument B		Instrument C	
	Mean	Standard Deviation	Mean	Standard Deviation	Mean	Standard Deviation
Base 0	3	15	−24	18	−6	18
Base 1	14	26	−13	13	−2	16
Base 2	1	26	−22	39	4	29
Base 3	8	34	−17	26	15	18

a. Construct an ANOVA table. You may give ranges for the P-values.
b. The question of interest is whether the mean error differs among instruments. It is not of interest to determine whether the error differs among base stations. For this reason, a surveyor suggests treating this as a randomized complete block design, with the base stations as the blocks. Is this appropriate? Explain.

10. Vermont maple sugar producers sponsored a testing program to determine the benefit of a potential new fertilizer regimen. A random sample of 27 maple trees in Vermont were chosen and treated with one of three levels of fertilizer suggested by the chemical producer. In this experimental setup, nine trees (three in each of three climatic zones) were treated with each fertilizer level and the amount of sap produced (in mL) by the trees in the subsequent season was measured. The results are presented in the following table.

	Southern Zone			Central Zone			Northern Zone		
Low fertilizer	76.2	80.4	74.2	79.4	87.9	86.9	84.5	85.2	80.1
Medium fertilizer	87.0	95.1	93.0	98.2	94.7	96.2	88.4	90.4	92.2
High fertilizer	84.2	87.5	83.1	90.3	89.9	93.2	81.4	84.7	82.2

a. Estimate the main effects of fertilizer levels and climatic zone, and their interactions.
b. Construct an ANOVA table. You may give ranges for the P-values.
c. Test the hypothesis that there is no interaction between fertilizer levels and climatic zone.
d. Test the hypothesis that there is no difference in sap production for the three fertilizer levels.

11. A civil engineer is interested in several designs for a drainage canal used to divert floodwaters from around a city. The drainage times of a reservoir attached to each of five different channel designs obtained from a series of experiments using similar initial flow conditions are given in the following table.

Channel Type	Drainage time (min)			
1	41.4	43.4	50.0	41.2
2	37.7	49.3	52.1	37.3
3	32.6	33.7	34.8	22.5
4	27.3	29.9	32.3	24.8
5	44.9	47.2	48.5	37.1

a. Can you conclude that there is a difference in the mean drainage times for the different channel designs?
b. Which pairs of designs, if any, can you conclude to differ in their mean drainage times?

12. A process that manufactures vinyl for automobile seat covers was studied. Three factors were varied: the proportion of a certain plasticizer (A), the rate of extrusion (B), and the temperature of drying (C). The outcome of interest was the thickness of the vinyl (in mils). A 2^3 factorial design with four replicates was employed. The results are presented in the following table. (Based on the article "Split-Plot Designs and Estimation Methods for Mixture Experiments with Process Variables," S. Kowalski, J. Cornell, and G. Vining, *Technometrics*, 2002:72–79.)

A	*B*	*C*	Thickness			
−1	−1	−1	7	5	6	7
1	−1	−1	6	5	5	5
−1	1	−1	8	8	4	6
1	1	−1	9	5	6	9
−1	−1	1	7	6	5	5
1	−1	1	7	7	11	10
−1	1	1	6	4	5	8
1	1	1	8	11	11	9

a. Estimate all main effects and interactions.
b. Construct an ANOVA table. You may give ranges for the P-values.
c. Is the additive model appropriate? Explain.
d. What conclusions about the factors can be drawn from these results?

13. In the article "Occurrence and Distribution of Ammonium in Iowa Groundwater" (K. Schilling, *Water Environment Research*, 2002:177–186), ammonium concentrations (in mg/L) were measured at a large number of wells in the state of Iowa. These included five types of bedrock wells. The number of wells of each type, along with the mean and standard deviation of the concentrations in those wells, is presented in the following table.

Well Type	Sample Size	Mean	Standard Deviation
Cretaceous	53	0.75	0.90
Mississippian	57	0.90	0.92
Devonian	66	0.68	1.03
Silurian	67	0.50	0.97
Cambrian–Ordovician	51	0.82	0.89

Can you conclude that the mean concentration differs among the five types of wells?

14. The article "Enthalpies and Entropies of Transfer of Electrolytes and Ions from Water to Mixed Aqueous Organic Solvents" (G. Hefter, Y. Marcus, and W. Waghorne, *Chemical Reviews*, 2002:2773–2836) presents measurements of entropy and enthalpy changes for many salts under a variety of conditions. The following table presents the results for enthalpy of transfer (in kJ/mol) from water to water + methanol of NaCl (table salt) for several concentrations of methanol. Four independent measurements were made at each concentration.

Concentration (%)	Enthalpy			
5	1.62	1.60	1.62	1.66
10	2.69	2.66	2.72	2.73
20	3.56	3.45	3.65	3.52
30	3.35	3.18	3.40	3.06

a. Is it plausible that the enthalpy is the same at all concentrations? Explain.
b. Which pairs of concentrations, if any, can you conclude to have differing enthalpies?

15. Refer to Exercise 11.

a. Compute the quantity $s = \sqrt{\text{MSE}}$, the estimate of the error standard deviation σ.
b. Assuming s to be the error standard deviation, find the sample size necessary in each treatment to provide a power of 0.90 to detect a maximum difference of 10 in the treatment means at the 5% level.
c. Using a more conservative estimate of $1.5s$ as the error standard deviation, find the sample size necessary in each treatment to provide a power of 0.90 to detect a maximum difference of 10 in the treatment means at the 5% level.

16. Refer to Exercise 14.

a. Compute the quantity $s = \sqrt{\text{MSE}}$, the estimate of the error standard deviation σ.
b. Assuming s to be the error standard deviation, find the sample size necessary in each treatment to provide a power of 0.80 to detect a maximum difference of 0.1 in the treatment means at the 5% level.
c. Using a more conservative estimate of $1.5s$ as the error standard deviation, find the sample size necessary in each treatment to provide a power of 0.80 to detect a maximum difference of 0.1 in the treatment means at the 5% level.

17. The article "Factorial Experiments in the Optimization of Alkaline Wastewater Pretreatment" (M. Prisciandaro, A. Del Borghi, and F. Veglio, *Industrial Engineering and Chemistry Research*, 2002:5034–5041) presents the results of several experiments to investigate methods of treating alkaline wastewater. One experiment was an unreplicated 2^4 design. The four factors were A: concentration of sulfuric acid, B: temperature, C: time, and D: concentration of calcium chloride. The outcome variable is the amount of precipitate in kg/m^3. The results are presented in the following table.

A	B	C	D	Outcome	A	B	C	D	Outcome
−1	−1	−1	−1	6.4	−1	−1	−1	1	11.9
1	−1	−1	−1	12.9	1	−1	−1	1	13.1
−1	1	−1	−1	8.6	−1	1	−1	1	12.1
1	1	−1	−1	12.9	1	1	−1	1	16.0
−1	−1	1	−1	7.4	−1	−1	1	1	12.4
1	−1	1	−1	12.0	1	−1	1	1	16.5
−1	1	1	−1	10.7	−1	1	1	1	15.3
1	1	1	−1	15.0	1	1	1	1	18.3

a. Estimate all main effects and interactions.
b. Which effects seem to be larger than the others?
c. Assume that all third- and higher-order interactions are equal to 0, and add their sums of squares. Use the result in place of an error sum of squares to compute F statistics and P-values for the main effects. Which factors can you conclude to have an effect on the outcome?
d. The article described some replicates of the experiment, in which the error mean square was found to be 1.04, with four degrees of freedom. Using this value, compute F statistics and P-values for all main effects and interactions.
e. Do the results of part (d) help to justify the assumption that the third- and higher-order interactions are equal to 0? Explain.
f. Using the results of part (d), which factors can you conclude to have an effect on the outcome?

18. The Williamsburg Bridge is a suspension bridge that spans the East River, connecting the boroughs of Brooklyn and Manhattan in New York City. An assessment of the strengths of its cables is reported in the article "Estimating Strength of the Williamsburg Bridge Cables" (R. Perry, *The American Statistician*, 2002:211–217). Each suspension cable consists of 7696 wires. From one of the cables, wires were sampled from 128 points. These points came from four locations along the length of the cable (I, II, III, IV). At each location there were eight equally spaced points around the circumference of the cable (*A, B, C, D, E, F, G, H*). At each of the eight points, wires were sampled from four depths: (1) the external surface of the cable, (2) two inches deep, (3) four inches deep, and (4) seven inches deep (the cable is 9.625 inches in radius). Under assumptions made in the article, it is appropriate to consider this as a two-factor experiment with circumferential position and depth as the factors, and with location providing four replicates for each combination of these factors. The minimum breaking strength (in lbf) is presented in the following table for each of the 128 points.

Circumference	Depth	Location			
		I	II	III	IV
A	1	6250	5910	5980	5800
A	2	6650	6690	6780	5540
A	3	5390	6080	6550	5690
A	4	6510	6580	6700	5980
B	1	6200	6240	6180	6740
B	2	6430	6590	6500	6110
B	3	5710	6230	6450	6310
B	4	6510	6600	6250	5660
C	1	5570	5700	6390	6170
C	2	6260	6290	5630	6990
C	3	6050	6120	6290	5800
C	4	6390	6540	6590	6620
D	1	6140	6210	5710	5090
D	2	5090	6000	6020	6480
D	3	5280	5650	5410	5730
D	4	6300	6320	6650	6050
E	1	4890	4830	5000	6490
E	2	5360	5640	5920	6390
E	3	5600	5500	6250	6510
E	4	6640	6810	5760	5200
F	1	5920	5300	5670	6200
F	2	5880	5840	7270	5230

Continued on page 720

Circumference	Depth	Location I	II	III	IV
F	3	6570	6130	5800	6200
F	4	6120	6430	6100	6370
G	1	6070	6980	6570	6980
G	2	6180	6340	6830	6260
G	3	6360	6420	6370	6550
G	4	6340	6380	6480	7020
H	1	5950	5950	6450	5870
H	2	6180	6560	5730	6550
H	3	6560	6560	6450	6790
H	4	6700	6690	6670	6600

a. Construct an ANOVA table. You may give ranges for the P-values.
b. Can you conclude that there are interactions between circumferential position and depth? Explain.
c. Can you conclude that the strength varies with circumferential position? Explain.
d. Can you conclude that the strength varies with depth? Explain.

19. In the article "Nitrate Contamination of Alluvial Groundwaters in the Nakdong River Basin, Korea" (J. Min, S. Yun, et al., *Geosciences Journal*, 2002:35–46), several chemical properties were measured for water samples taken from irrigation wells at three locations. The following table presents the means, standard deviations, and sample sizes for pH measurements.

Location	Mean	SD	Sample Size
Upstream	6.0	0.2	49
Midstream	6.2	0.4	31
Downstream	6.4	0.6	30

Do the data prove conclusively that the pH differs at the different locations?

20. The article cited in Exercise 19 provides measures of electrical conductivity (in μS/cm). The results are presented in the following table.

Location	Mean	SD	Sample Size
Upstream	463	208	49
Midstream	363	98	31
Downstream	647	878	30

Can a one-way analysis of variance be used to determine whether conductivity varies with location? Or is one of the necessary assumptions violated? Explain.

21. The article "Factorial Experiments in the Optimization of Alkaline Wastewater Pretreatment" (M. Prisciandaro, A. Del Borghi, and F. Veglio, *Industrial Engineering and Chemistry Research*, 2002:5034–5041) presents the results of an experiment to investigate the effects of the concentrations of sulfuric acid (H_2SO_4) and calcium chloride

($CaCl_2$) on the amount of black mud precipitate in the treatment of alkaline wastewater. There were three levels of each concentration, and two replicates of the experiment were made at each combination of levels. The results are presented in the following table (all measurements are in units of kg/m^3).

H_2SO_4	$CaCl_2$	Precipitate	
110	15	100.2	98.2
110	30	175.8	176.2
110	45	216.5	206.0
123	15	110.5	105.5
123	30	184.7	189.0
123	45	234.0	222.0
136	15	106.5	107.0
136	30	181.7	189.0
136	45	211.8	201.3

a. Construct an ANOVA table. You may give ranges for the P-values.
b. Is the additive model plausible? Explain.
c. Can you conclude that H_2SO_4 concentration affects the amount of precipitate? Explain.
d. Can you conclude that $CaCl_2$ concentration affects the amount of precipitate? Explain.

22. Fluid inclusions are microscopic volumes of fluid that are trapped in rock during rock formation. The article "Fluid Inclusion Study of Metamorphic Gold-Quartz Veins in Northwestern Nevada, U.S.A.: Characteristics of Tectonically Induced Fluid" (S. Cheong, *Geosciences Journal*, 2002:103–115) describes the geochemical properties of fluid inclusions in several different veins in northwest Nevada. The following table presents data on the maximum salinity (% NaCl by weight) of inclusions in several rock samples from several areas.

Area	Salinity					
Humboldt Range	9.2	10.0	11.2	8.8		
Santa Rosa Range	5.2	6.1	8.3			
Ten Mile	7.9	6.7	9.5	7.3	10.4	7.0
Antelope Range	6.7	8.4	9.9			
Pine Forest Range	10.5	16.7	17.5	15.3	20.0	

Can you conclude that the salinity differs among the areas?

23. The article "Effect of Microstructure and Weathering on the Strength Anisotropy of Porous Rhyolite" (Y. Matsukura, K. Hashizume, and C. Oguchi, *Engineering Geology*, 2002:39–47) investigates the relationship between the angle between cleavage and flow structure and the strength of porous rhyolite. Strengths (in MPa) were measured for a number of specimens cut at various angles. The mean and standard deviation of the strengths for each angle are presented in the following table.

Angle	Mean	Standard Deviation	Sample Size
0°	22.9	2.98	12
15°	22.9	1.16	6
30°	19.7	3.00	4
45°	14.9	2.99	5
60°	13.5	2.33	7
75°	11.9	2.10	6
90°	14.3	3.95	6

Can you conclude that strength varies with the angle?

24. The article "Influence of Supplemental Acetate on Bioremediation for Dissolved Polycyclic Aromatic Hydrocarbons" (T. Ebihara and P. Bishop, *Journal of Environmental Engineering*, 2002:505–513) describes experiments in which water containing dissolved polyaromatic hydrocarbons (PAH) was fed into sand columns. PAH concentrations were measured at various depths after 25, 45, and 90 days. Assume that three independent measurements were made at each depth at each time. The data presented in the following table are naphthalene concentrations (in mg/L) that are consistent with means and standard deviations reported in the article.

Depth	25 days			45 days			90 days		
0	11.15	11.39	11.36	9.28	8.15	8.59	7.68	7.59	7.41
5	14.40	11.78	11.92	9.44	9.34	9.33	7.53	7.92	7.12
15	11.51	11.01	11.09	9.34	9.11	8.94	7.43	7.47	7.53
30	12.77	12.18	11.65	9.37	9.27	9.05	7.60	7.48	7.84
50	11.71	11.29	11.20	9.25	8.97	9.29	7.76	7.84	7.68
75	11.18	11.45	11.27	9.09	8.86	8.78	7.72	7.61	7.74

a. Construct an ANOVA table. You may give ranges for the P-values.

b. Perform a test to determine whether the additive model is plausible. Provide the value of the test statistic and the P-value.

Chapter 10

Statistical Quality Control

Introduction

As the marketplace for industrial goods has become more global, manufacturers have realized that the quality and reliability of their products must be as high as possible for them to be competitive. It is now generally recognized that the most cost-effective way to maintain high quality is through constant monitoring of the production process. This monitoring is often done by sampling units of production and measuring some quality characteristic. Because the units are sampled from some larger population, these methods are inherently statistical in nature.

One of the early pioneers in the area of statistical quality control was Dr. Walter A. Shewart of the Bell Telephone Laboratories. In 1924, he developed the modern control chart, which remains one of the most widely used tools for quality control to this day. After World War II, W. Edwards Deming was instrumental in stimulating interest in quality control; first in Japan, and then in the United States and other countries. The Japanese scientist Genichi Taguchi played a major role as well, developing methods of experimental design with a view toward improving quality. In this chapter, we will focus on the Shewart control charts and on cumulative sum (CUSUM) charts, since these are among the most powerful of the commonly used tools for statistical quality control.

10.1 Basic Ideas

The basic principle of control charts is that in any process there will always be variation in the output. Some of this variation will be due to causes that are inherent in the process and are difficult or impossible to specify. These causes are called **common causes** or **chance causes**. When common causes are the only causes of variation, the process is said to be **in a state of statistical control**, or, more simply, **in control**.

Sometimes special factors are present that produce additional variability. Machines that are malfunctioning, operator error, fluctuations in ambient conditions, and variations in the properties of raw materials are among the most common of these factors. These are called **special causes** or **assignable causes**. Special causes generally produce a higher level of variability than do common causes; this variability is considered to be unacceptable. When a process is operating in the presence of one or more special causes, it is said to be **out of statistical control**.

Control charts enable the quality engineer to decide whether a process appears to be in control, or whether one or more special causes are present. If the process is found to be out of control, the nature of the special cause must be determined and corrected, so as to return the process to a state of statistical control. There are several types of control charts; which ones are used depend on whether the quality characteristic being measured is a **continuous variable**, a **binary variable**, or a **count variable**. For example, when monitoring a process that manufactures aluminum beverage cans, the height of each can in a sample might be measured. Height is a continuous variable. In some situations, it might be sufficient simply to determine whether the height falls within some specification limits. In this case the quality measurement takes on one of only two values: conforming (within the limits) or nonconforming (not within the limits). This measurement is a binary variable, since it has two possible values. Finally, we might be interested in counting the number of flaws on the surface of the can. This is a count variable.

Control charts used for continuous variables are called **variables control charts**. Examples include the $\overline{X}$ chart, the R chart, and the S chart. Control charts used for binary or count variables are called **attribute control charts**. The p chart is most commonly used for binary variables, while the c chart is commonly used for count variables.

Collecting Data—Rational Subgroups

Data to be used in the construction of a control chart are collected in a number of samples, taken over a period of time. These samples are called **rational subgroups**. There are many different strategies for choosing rational subgroups. The basic principle to be followed is that all the variability within the units in a rational subgroup should be due to common causes, and none should be due to special causes. In general, a good way to choose rational subgroups is to decide which special causes are most important to detect, and then choose the rational subgroups to provide the best chance to detect them. The two most commonly used methods are

- Sample at regular time intervals, with all the items in each sample manufactured near the time the sampling is done.
- Sample at regular time intervals, with the items in each sample drawn from all the units produced since the last sample was taken.

For variables data, the number of units in each sample is typically small, often between three and eight. The number of samples should be at least 20. In general, many small samples taken frequently are better than a few large samples taken less frequently. For binary and for count data, samples must in general be larger.

Control versus Capability

It is important to understand the difference between process *control* and process *capability*. A process is in control if there are no special causes operating. The distinguishing feature of a process that is in control is that the values of the quality characteristic vary without any trend or pattern, since the common causes do not change over time. However, it is quite possible for a process to be in control, and yet to be producing output that does not meet a given specification. For example, assume that a process produces steel rods whose lengths vary randomly between 19.9 and 20.1 cm, with no apparent pattern to the fluctuation. This process is in a state of control. However, if the design specification calls for a length between 21 and 21.2 cm, very little of the output would meet the specification. The ability of a process to produce output that meets a given specification is called the **capability** of the process. We will discuss the measurement of process capability in Section 10.5.

Process Control Must Be Done Continually

There are three basic phases to the use of control charts. First, data are collected. Second, these data are plotted to determine whether the process is in control. Third, once the process is brought into control, its capability may be assessed. Of course, a process that is in control and capable at a given time may go out of control at a later time, as special causes re-occur. For this reason processes must be continually monitored.

Exercises for Section 10.1

1. Indicate whether each of the following quality characteristics is a continuous, binary, or count variable.

a. The length of a steel rod.

b. The number of flaws in a section of sheet metal.

c. Whether a concrete specimen meets a strength specification.

d. The length of time taken to perform a final inspection of a finished product.

2. True or false:

a. Control charts are used to determine whether special causes are operating.

b. If no special causes are operating, then most of the output produced will meet specifications.

c. Variability due to common causes does not increase or decrease much over short periods of time.

d. Variability within the items sampled in a rational subgroup is due to special causes.

e. If a process is in a state of statistical control, there will be almost no variation in the output.

3. Fill in the blank. The choices are: *is in control*; *has high capability*.

a. If the variability in a process is approximately constant over time, the process ________.

b. If most units produced conform to specifications, the process ________.

4. Fill in the blank: Once a process has been brought into a state of statistical control, ________

i. It must still be monitored continually.

ii. Monitoring can be stopped for a while, since it is unlikely that the process will go out of control again right away.

iii. The process need not be monitored again, unless it is redesigned.

5. True or false:

a. When a process is in a state of statistical control, then most of the output will meet specifications.

b. When a process is out of control, an unacceptably large proportion of the output will not meet specifications.

c. When a process is in a state of statistical control, all the variation in the process is due to causes that are inherent in the process itself.

d. When a process is out of control, some of the variation in the process is due to causes that are outside of the process.

6. Fill in the blank: When sampling units for rational subgroups, ________

i. it is more important to choose large samples than to sample frequently, since large samples provide more precise information about the process.

ii. it is more important to sample frequently than to choose large samples, so that special causes can be detected more quickly.

10.2 Control Charts for Variables

When a quality measurement is made on a continuous scale, the data are called **variables data**. For these data an R chart or S chart is first used to control the variability in the process, and then an $\overline{X}$-chart is used to control the process mean. The methods described in this section assume that the measurements follow an approximately normal distribution.

We illustrate with an example. The quality engineer in charge of a salt packaging process is concerned about the moisture content in packages of salt. To determine whether the process is in statistical control, it is first necessary to define the rational subgroups, and then to collect some data. Assume that for the salt packaging process, the primary concern is that variation in the ambient humidity in the plant may be causing variation in the mean moisture content in the packages over time. Recall that rational subgroups should be chosen so that the variation within each sample is due only to common causes, not to special causes. Therefore a good choice for the rational subgroups in this case is to draw samples of several packages each at regular time intervals. The packages in each sample will be produced as close to each other in time as possible. In this way, the ambient humidity will be nearly the same for each package in the sample, so the within-group variation will not be affected by this special cause. Assume that five packages of salt are sampled every 15 minutes for eight hours, and that the moisture content in each package is measured as a percentage of total weight. The data are presented in Table 10.1.

Since moisture is measured on a continuous scale, these are variables data. Each row of Table 10.1 presents the five moisture measurements in a given sample, along with their sample mean $\overline{X}$, their sample standard deviation s, and their sample range R (the difference between the largest and smallest value). The last row of the table contains the mean of the sample means ($\overline{\overline{X}}$), the mean of the sample ranges ($\overline{R}$), and the mean of the sample standard deviations ($\overline{s}$).

We assume that each of the 32 samples in Table 10.1 is a sample from a normal population with mean μ and standard deviation σ. The quantity μ is called the **process mean**, and σ is called the **process standard deviation**. The idea behind control charts

TABLE 10.1 Moisture content for salt packages, as a percentage of total weight

Sample	Sample Values	Mean ($\overline{X}$)	Range (R)	SD (s)
1	2.53 2.66 1.88 2.21 2.26	2.308	0.780	0.303
2	2.69 2.38 2.34 2.47 2.61	2.498	0.350	0.149
3	2.67 2.23 2.10 2.43 2.54	2.394	0.570	0.230
4	2.10 2.26 2.51 2.58 2.28	2.346	0.480	0.196
5	2.64 2.42 2.56 2.51 2.36	2.498	0.280	0.111
6	2.64 1.63 2.95 2.12 2.67	2.402	1.320	0.525
7	2.58 2.69 3.01 3.01 2.23	2.704	0.780	0.327
8	2.31 2.39 2.60 2.40 2.46	2.432	0.290	0.108
9	3.03 2.68 2.27 2.54 2.63	2.630	0.760	0.274
10	2.86 3.22 2.72 3.09 2.48	2.874	0.740	0.294
11	2.71 2.80 3.09 2.60 3.39	2.918	0.790	0.320
12	2.95 3.54 2.59 3.31 2.87	3.052	0.950	0.375
13	3.14 2.84 3.77 2.80 3.22	3.154	0.970	0.390
14	2.85 3.29 3.25 3.35 3.59	3.266	0.740	0.267
15	2.82 3.71 3.36 2.95 3.37	3.242	0.890	0.358
16	3.17 3.07 3.14 3.63 3.70	3.342	0.630	0.298
17	2.81 3.21 2.95 3.04 2.85	2.972	0.400	0.160
18	2.99 2.65 2.79 2.80 2.95	2.836	0.340	0.137
19	3.11 2.74 2.59 3.01 3.03	2.896	0.520	0.221
20	2.83 2.74 3.03 2.68 2.49	2.754	0.540	0.198
21	2.76 2.85 2.59 2.23 2.87	2.660	0.640	0.265
22	2.54 2.63 2.32 2.48 2.93	2.580	0.610	0.226
23	2.27 2.54 2.82 2.11 2.69	2.486	0.710	0.293
24	2.40 2.62 2.84 2.50 2.51	2.574	0.440	0.168
25	2.41 2.72 2.29 2.35 2.63	2.480	0.430	0.186
26	2.40 2.33 2.40 2.02 2.43	2.316	0.410	0.169
27	2.56 2.47 2.11 2.43 2.85	2.484	0.740	0.266
28	2.21 2.61 2.59 2.24 2.34	2.398	0.400	0.191
29	2.56 2.26 1.95 2.26 2.40	2.286	0.610	0.225
30	2.42 2.37 2.13 2.09 2.41	2.284	0.330	0.161
31	2.62 2.11 2.47 2.27 2.49	2.392	0.510	0.201
32	2.21 2.15 2.18 2.59 2.61	2.348	0.460	0.231
		$\overline{\overline{X}} = 2.6502$	$\overline{R} = 0.6066$	$\overline{s} = 0.2445$

is that each value of $\overline{X}$ approximates the process mean during the time its sample was taken, while the values of R and s can be used to approximate the process standard deviation. If the process is in control, then the process mean and standard deviation are the same for each sample. If the process is out of control, the process mean μ or the process standard deviation σ, or both, will differ from sample to sample. Therefore the values of $\overline{X}$, R, and s will vary less when the process is in control than when the process is out of control. If the process is in control, the values of $\overline{X}$, R, and s will almost always be contained within computable limits, called **control limits**. If the process is out of control, the values of $\overline{X}$, R, and s will be more likely to exceed these limits. A control chart plots the values of $\overline{X}$, R, or s, along with the control limits, so that it can be easily seen whether the variation is large enough to conclude that the process is out of control.

Now let's see how to determine whether the salt packaging process is in a state of statistical control with respect to moisture content. Since the variation within each sample is supposed to be due only to common causes, this variation should not be too different from one sample to another. Therefore the first thing to do is to check to make sure that the amount of variation within each sample, measured either by the sample range or the sample standard deviation, does not vary too much from sample to sample. For this purpose the R chart can be used to assess variation in the sample range, or the S chart can be used to assess variation in the sample standard deviation. We will discuss the R chart first, since it is the more traditional. We will discuss the S chart at the end of this section.

Figure 10.1 presents the R chart for the moisture data. The horizontal axis represents the samples, numbered from 1 to 32. The sample ranges are plotted on the vertical axis. Most important are the three horizontal lines. The line in the center is plotted at the value $\overline{R}$ and is called the **center line**. The upper and lower lines indicate the 3σ upper and lower **control limits** (UCL and LCL, respectively). The control limits are drawn so that when the process is in control, almost all the points will lie within the limits. A point plotting outside the control limits is evidence that the process is out of control.

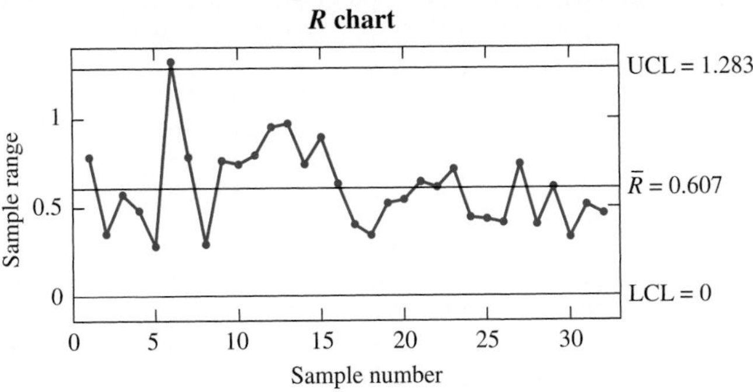

FIGURE 10.1 R chart for the moisture data.

To understand where the control limits are plotted, assume that the 32 sample ranges come from a population with mean μ_R and standard deviation σ_R. The values of μ_R and σ_R will not be known exactly, but it is known that for most populations, it is unusual to observe a value that differs from the mean by more than three standard deviations. For this reason, it is conventional to plot the control limits at points that approximate the values $\mu_R \pm 3\sigma_R$. It can be shown by advanced methods that the quantities $\mu_R \pm 3\sigma_R$ can be estimated with multiples of $\overline{R}$; these multiples are denoted D_3 and D_4. The quantity $\mu_R - 3\sigma_R$ is estimated with $D_3\overline{R}$, and the quantity $\mu_R + 3\sigma_R$ is estimated with $D_4\overline{R}$. The quantities D_3 and D_4 are constants whose values depend on the sample size n. A brief table of values of D_3 and D_4 follows. A more extensive tabulation is provided in Table A.9 (in Appendix A). Note that for sample sizes of 6 or less, the value of D_3 is 0. For these small sample sizes, the quantity $\mu_R - 3\sigma_R$ is negative. In these cases the lower control limit is set to 0, because it is impossible for the range to be negative.

n	2	3	4	5	6	7	8
D_3	0	0	0	0	0	0.076	0.136
D_4	3.267	2.575	2.282	2.114	2.004	1.924	1.864

Compute the 3σ R chart upper and lower control limits for the moisture data in Table 10.1.

Solution

The value of $\overline{R}$ is 0.6066 (Table 10.1). The sample size is $n = 5$. From the table, $D_3 = 0$ and $D_4 = 2.114$. Therefore the upper control limit is $(2.114)(0.6066) = 1.283$, and the lower control limit is $(0)(0.6066) = 0$.

Summary

In an R chart, the center line and the 3σ upper and lower control limits are given by

$$3\sigma \text{ upper limit} = D_4\overline{R}$$
$$\text{Center line} = \overline{R}$$
$$3\sigma \text{ lower limit} = D_3\overline{R}$$

The values D_3 and D_4 depend on the sample size. Values are tabulated in Table A.9.

Once the control limits have been calculated and the points plotted, the R chart can be used to assess whether the process is in control with respect to variation. Figure 10.1 shows that the range for sample number 6 is above the upper control limit, providing evidence that a special cause was operating and that the process variation is not in control. The appropriate action is to determine the nature of the special cause, and then delete the out-of-control sample and recompute the control limits. Assume it is discovered that a technician neglected to close a vent, causing greater than usual variation in moisture content during the time period when the sample was chosen. Retraining the technician will correct that special cause. We delete sample 6 from the data and recompute the R chart. The results are shown in Figure 10.2 (page 730). The process variation is now in control.

Now that the process variation has been brought into control, we can assess whether the process mean is in control by plotting the $\overline{X}$ chart. The $\overline{X}$ chart is presented in Figure 10.3 (page 730). The sample means are plotted on the vertical axis. Note that sample 6 has not been used in this chart since it had to be deleted in order to bring the

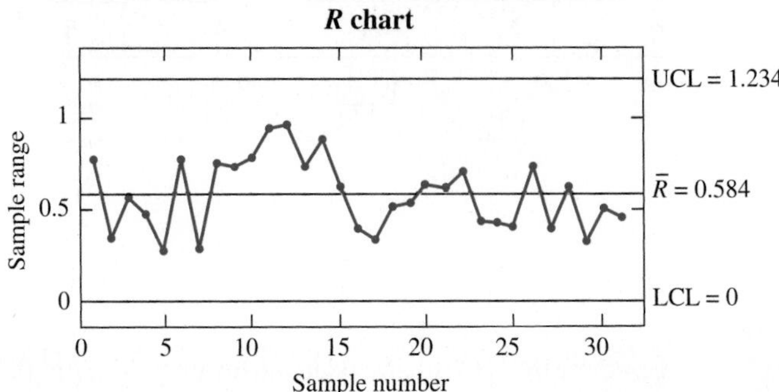

FIGURE 10.2 R chart for the moisture data, after deleting the out-of-control sample.

process variation under control. Like all control charts, the $\overline{X}$ chart has a center line and upper and lower control limits.

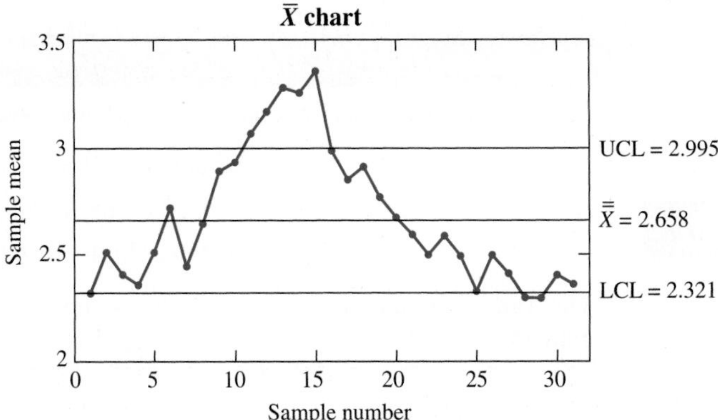

FIGURE 10.3 $\overline{X}$ chart for the moisture data. Sample 6 has been deleted to bring the process variation under control. However, the $\overline{X}$ chart shows that the process mean is out of control.

To compute the center line and the control limits, we can assume that the process standard deviation is the same for all samples, since the R chart has been used to bring the process variation into control. If the process mean μ is in control as well, then it too is the same for all samples. In that case the 32 sample means are drawn from a normal population with mean $\mu_{\overline{X}} = \mu$ and standard deviation $\sigma_{\overline{X}} = \sigma/\sqrt{n}$, where n is the sample size, equal to 5 in this case. Ideally, we would like to plot the center line at μ and the 3σ control limits at $\mu \pm 3\sigma_{\overline{X}}$. However, the values of μ and $\sigma_{\overline{X}}$ are usually unknown and have to be estimated from the data. We estimate μ with $\overline{\overline{X}}$, the average of the sample means. The center line is therefore plotted at $\overline{\overline{X}}$. The quantity $\sigma_{\overline{X}}$ can be estimated by using either the average range $\overline{R}$ or by using the sample standard deviations. We will use $\overline{R}$ here and discuss the methods based on the standard deviation at the end of the section, in conjunction with the discussion of S charts. It can be shown by advanced methods that the quantity $3\sigma_{\overline{X}}$ can be estimated with $A_2\overline{R}$, where A_2 is a constant whose value depends on the sample size. A short table of values of A_2 follows. A more extensive tabulation is provided in Table A.9.

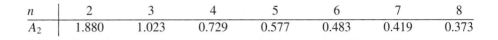

n	2	3	4	5	6	7	8
A_2	1.880	1.023	0.729	0.577	0.483	0.419	0.373

Summary

In an $\overline{X}$ chart, when $\overline{R}$ is used to estimate $\sigma_{\overline{X}}$, the center line and the 3σ upper and lower control limits are given by

$$3\sigma \text{ upper limit} = \overline{\overline{X}} + A_2\overline{R}$$
$$\text{Center line} = \overline{\overline{X}}$$
$$3\sigma \text{ lower limit} = \overline{\overline{X}} - A_2\overline{R}$$

The value A_2 depends on the sample size. Values are tabulated in Table A.9.

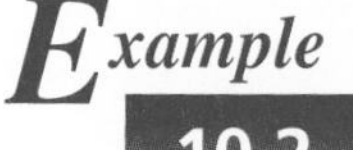

Compute the 3σ $\overline{X}$ chart upper and lower control limits for the moisture data in Table 10.1.

Solution

With sample 6 deleted, the value of $\overline{\overline{X}}$ is 2.658, and the value of $\overline{R}$ is 0.5836. The sample size is $n = 5$. From the table, $A_2 = 0.577$. Therefore the upper control limit is $2.658 + (0.577)(0.5836) = 2.995$, and the lower control limit is $2.658 - (0.577)(0.5836) = 2.321$.

The $\overline{X}$ chart clearly shows that the process mean is not in control, as there are several points plotting outside the control limits. The production manager installs a hygrometer to monitor the ambient humidity and determines that the fluctuations in moisture content are caused by fluctuations in ambient humidity. A dehumidifier is installed to stabilize the ambient humidity. After this special cause is remedied, more data are collected, and a new R chart and $\overline{X}$ chart are constructed. Figure 10.4 (page 732) presents the results. The process is now in a state of statistical control. Of course, the process must be continually monitored, since new special causes are bound to crop up from time to time and will need to be detected and corrected.

Note that while control charts can detect the presence of a special cause, they cannot determine its nature, nor how to correct it. It is necessary for the process engineer to have a good understanding of the process, so that special causes detected by control charts can be diagnosed and corrected.

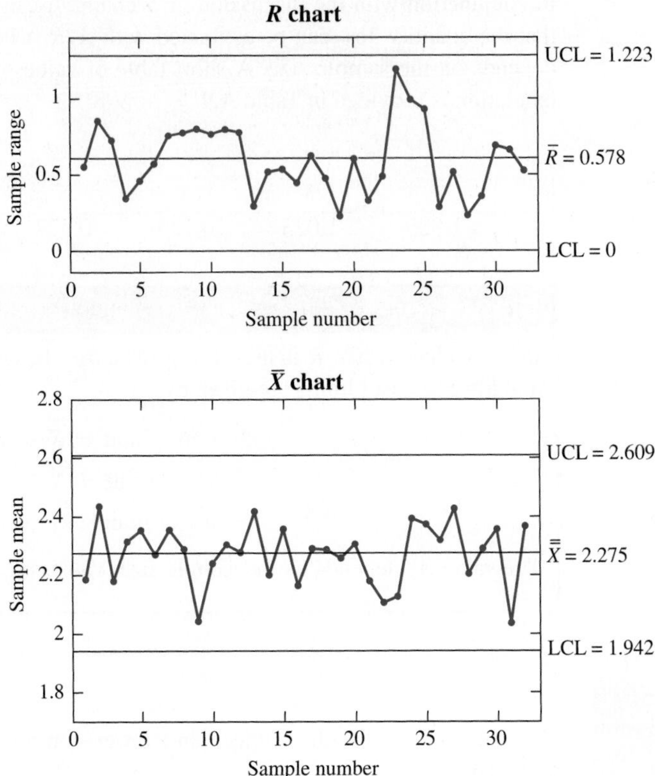

FIGURE 10.4 R chart and $\overline{X}$ chart after special cause is remedied. The process is now in a state of statistical control.

Summary

The steps in using the R chart and $\overline{X}$ chart are

1. Choose rational subgroups.
2. Compute the R chart.
3. Determine the special causes for any out-of-control points.
4. Recompute the R chart, omitting samples that resulted in out-of-control points.
5. Once the R chart indicates a state of control, compute the $\overline{X}$ chart, omitting samples that resulted in out-of-control points on the R chart.
6. If the $\overline{X}$ chart indicates that the process is not in control, identify and correct any special causes.
7. Continue to monitor $\overline{X}$ and R.

Control Chart Performance

There is a close connection between control charts and hypothesis tests. The null hypothesis is that the process is in a state of control. A point plotting outside the 3σ control limits presents evidence against the null hypothesis. As with any hypothesis test, it is possible to make an error. For example, a point will occasionally plot outside the 3σ limits even when the process is in control. This is called a **false alarm**. It can also happen that a process that is not in control may not exhibit any points outside the control limits, especially if it is not observed for a long enough time. This is called a **failure to detect**.

It is desirable for these errors to occur as infrequently as possible. We describe the frequency with which these errors occur with a quantity called the **average run length** (ARL). The ARL is the number of samples that must be observed, on average, before a point plots outside the control limits. We would like the ARL to be large when the process is in control, and small when the process is out of control. We can compute the ARL for an $\overline{X}$ chart if we assume that process mean μ and the process standard deviation σ are known. Then the center line is located at the process mean μ and the control limits are at $\mu \pm 3\sigma_{\overline{X}}$. We must also assume, as is always the case with the $\overline{X}$ chart, that the quantity being measured is approximately normally distributed. Examples 10.3 through 10.6 show how to compute the ARL.

For an $\overline{X}$ chart with control limits at $\mu \pm 3\sigma_{\overline{X}}$, compute the ARL for a process that is in control.

Solution

Let $\overline{X}$ be the mean of a sample. Then $\overline{X} \sim N(\mu, \sigma_{\overline{X}}^2)$. The probability that a point plots outside the control limits is equal to $P(\overline{X} < \mu - 3\sigma_{\overline{X}}) + P(\overline{X} > \mu + 3\sigma_{\overline{X}})$. This probability is equal to $0.00135 + 0.00135 = 0.0027$ (see Figure 10.5). Therefore, on the average, 27 out of every 10,000 points will plot outside the control limits. This is equivalent to 1 every $10{,}000/27 = 370.4$ points. The average run length is therefore equal to 370.4.

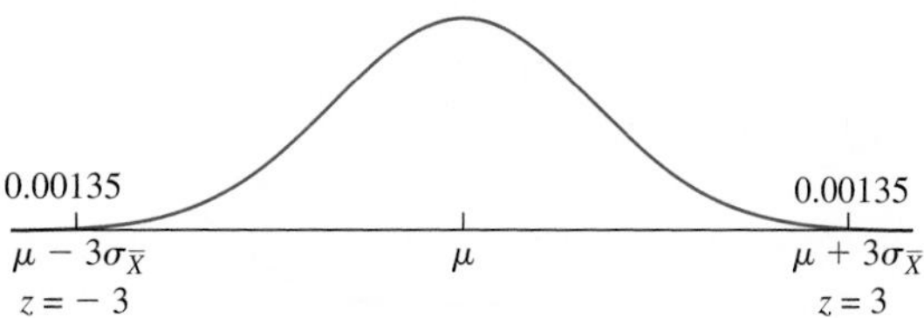

FIGURE 10.5 The probability that a point plots outside the 3σ control limits, when the process is in control, is 0.0027 (0.00135 + 0.00135).

The result of Example 10.3 can be interpreted as follows: If a process is in control, we expect to observe about 370 samples, on the average, before finding one that plots outside the control limits, causing a false alarm. Note also that the ARL in Example 10.3 was 10,000/27, which is equal to 1/0.0027, where 0.0027 is the probability that any given sample plots outside the control limits. This is true in general.

Summary

The average run length (ARL) is the number of samples that will be observed, on the average, before a point plots outside the control limits. If p is the probability that any given point plots outside the control limits, then

$$\text{ARL} = \frac{1}{p} \qquad (10.1)$$

If a process is out of control, the ARL will be less than 370.4. Example 10.4 shows how to compute the ARL for a situation where the process shifts to an out-of-control condition.

A process has mean $\mu = 3$ and standard deviation $\sigma = 1$. Samples of size $n = 4$ are taken. If a special cause shifts the process mean to a value of 3.5, find the ARL.

Solution

We first compute the probability p that a given point plots outside the control limits. Then ARL $= 1/p$. The control limits are plotted on the basis of a process that is in control. Therefore they are at $\mu \pm 3\sigma_{\overline{X}}$, where $\mu = 3$ and $\sigma_{\overline{X}} = \sigma/\sqrt{n} = 1/\sqrt{4} = 0.5$. The lower control limit is thus at 1.5, and the upper control limit is at 4.5. If $\overline{X}$ is the mean of a sample taken after the process mean has shifted, then $\overline{X} \sim N(3.5,\ 0.5^2)$. The probability that $\overline{X}$ plots outside the control limits is equal to $P(\overline{X} < 1.5) + P(\overline{X} > 4.5)$. This probability is 0.0228 (see Figure 10.6). The ARL is therefore equal to $1/0.0228 = 43.9$. We will have to observe about 44 samples, on the average, before detecting this shift.

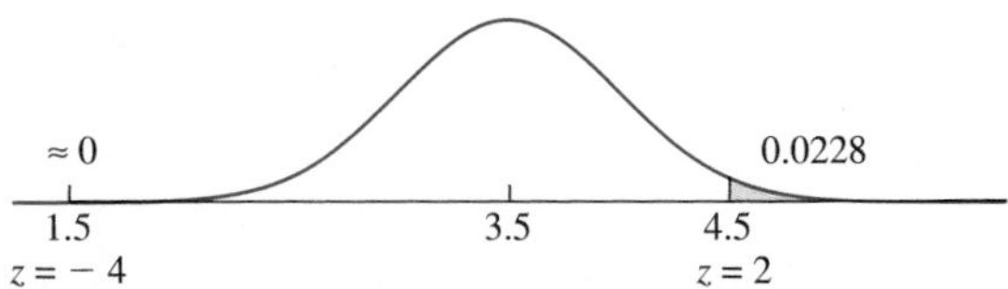

FIGURE 10.6 The process mean has shifted from $\mu = 3$ to $\mu = 3.5$. The upper control limit of 4.5 is now only $2\sigma_{\overline{X}}$ above the mean, indicated by the fact that $z = 2$. The lower limit is now $4\sigma_{\overline{X}}$ below the mean. The probability that a point plots outside the control limits is 0.0228 (0 + 0.0228).

Example 10.5

Refer to Example 10.4. An upward shift to what value can be detected with an ARL of 20?

Solution

Let m be the new mean to which the process has shifted. Since we have specified an upward shift, $m > 3$. In Example 10.4 we computed the control limits to be 1.5 and 4.5. If $\overline{X}$ is the mean of a sample taken after the process mean has shifted, then $\overline{X} \sim N(m, 0.5^2)$. The probability that $\overline{X}$ plots outside the control limits is equal to $P(\overline{X} < 1.5) + P(\overline{X} > 4.5)$ (see Figure 10.7). This probability is equal to $1/\text{ARL} = 1/20 = 0.05$. Since $m > 3$, m is closer to 4.5 than to 1.5. We will begin by assuming that the area to the left of 1.5 is negligible and that the area to the right of 4.5 is equal to 0.05. The z-score of 4.5 is then 1.645, so $(4.5-m)/0.5 = 1.645$. Solving for m, we have $m = 3.68$. We finish by checking our assumption that the area to the left of 1.5 is negligible. With $m = 3.68$, the z-score for 1.5 is $(1.5 - 3.68)/0.5 = -4.36$. The area to the left of 1.5 is indeed negligible.

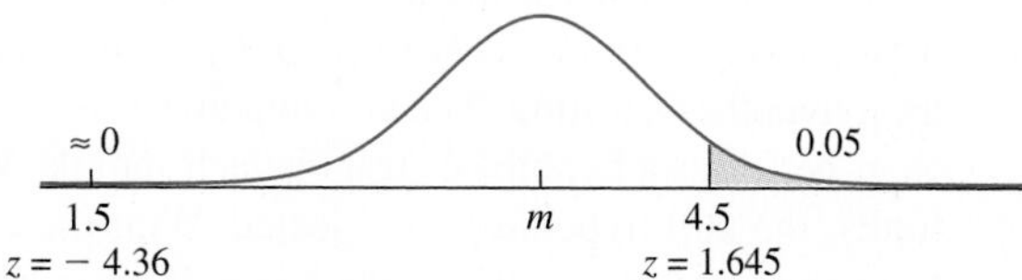

FIGURE 10.7 Solution to Example 10.5.

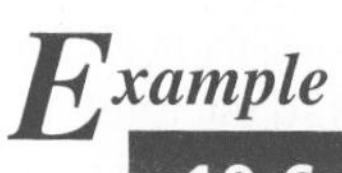

Example 10.6

Refer to Example 10.4. If the sample size remains at $n = 4$, what must the value of the process standard deviation σ be to produce an ARL of 10 when the process mean shifts to 3.5?

Solution

Let σ denote the new process standard deviation. The new control limits are $3 \pm 3\sigma/\sqrt{n}$, or $3 \pm 3\sigma/2$. If the process mean shifts to 3.5, then $\overline{X} \sim N(3.5, \sigma^2/4)$. The probability that $\overline{X}$ plots outside the control limits is equal to $P(\overline{X} < 3-3\sigma/2) + P(\overline{X} > 3+3\sigma/2)$. This probability is equal to $1/\text{ARL} = 1/10 = 0.10$ (see Figure 10.8, page 736) . The process mean, 3.5, is closer to $3 + 3\sigma/2$ than to $3 - 3\sigma/2$. We will assume that the area to the left of $3 - 3\sigma/2$ is negligible and that the area to the right of $3 + 3\sigma/2$ is equal to 0.10. The z-score for $3 + 3\sigma/2$ is then 1.28, so

$$\frac{(3 + 3\sigma/2) - 3.5}{\sigma/2} = 1.28$$

Solving for σ, we obtain $\sigma = 0.58$. We finish by checking that the area to the left of $3 - 3\sigma/2$ is negligible. Substituting $\sigma = 0.58$, we obtain $3 - 3\sigma/2 = 2.13$. The z-score is $(2.13 - 3.5)/(0.58/2) = -4.72$. The area to the left of $3 - 3\sigma/2$ is indeed negligible.

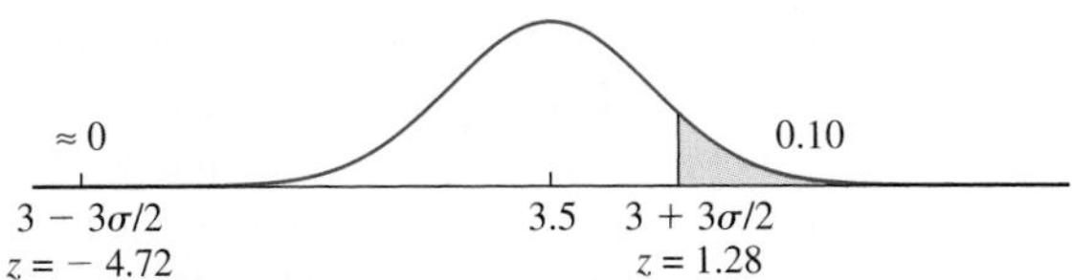

FIGURE 10.8 Solution to Example 10.6.

Examples 10.4 through 10.6 show that $\overline{X}$ charts do not usually detect small shifts quickly. In other words, the ARL is high when shifts in the process mean are small. In principle, one could reduce the ARL by moving the control limits closer to the centerline. This would reduce the size of the shift needed to detect an out-of-control condition, so that changes in the process mean would be detected more quickly. However, there is a trade-off. The false alarm rate would increase as well, because shifts outside the control limits would be more likely to occur by chance. The situation is much like that in fixed-level hypothesis testing. The null hypothesis is that the process is in control. The control chart performs a hypothesis test on each sample. When a point plots outside the control limits, the null hypothesis is rejected. With the control limits at $\pm 3\sigma_{\overline{X}}$, a type I error (rejection of a true null hypothesis) will occur about once in every 370 samples. The price to pay for this low false alarm rate is lack of power to reject the null hypothesis when it is false. Moving the control limits closer together is not the answer. Although it will increase the power, it will also increase the false alarm rate.

Two of the ways in which practitioners have attempted to improve their ability to detect small shifts quickly are by using the **Western Electric rules** to interpret the control chart and by using CUSUM charts. The Western Electric rules are described next. CUSUM charts are described in Section 10.4.

The Western Electric Rules

Figure 10.9 presents an $\overline{X}$ chart. While none of the points fall outside the 3σ control limits, the process is clearly not in a state of control, since there is a nonrandom pattern to the sample means. In recognition of the fact that a process can fail to be in control even when no points plot outside the control limits, engineers at the Western Electric company in 1956 suggested a list of conditions, any one of which could be used as evidence that a process is out of control. The idea behind these conditions is that if a trend or pattern in the control chart persists for long enough, it can indicate the absence of control, even if no point plots outside the 3σ control limits.

To apply the Western Electric rules, it is necessary to compute the 1σ and 2σ control limits. The 1σ control limits are given by $\overline{\overline{X}} \pm A_2\overline{R}/3$, and the 2σ control limits are given by $\overline{\overline{X}} \pm 2A_2\overline{R}/3$.

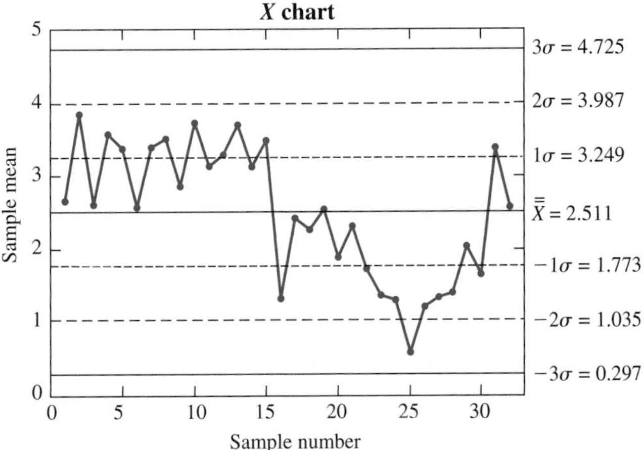

FIGURE 10.9 This $\overline{X}$ chart exhibits nonrandom patterns, indicating a lack of statistical control, even though no points plot outside the 3σ control limits. The 1σ and 2σ control limits are shown on this plot, so that the Western Electric rules can be applied.

The Western Electric Rules

Any one of the following conditions is evidence that a process is out of control:

1. Any point plotting outside the 3σ control limits.
2. Two out of three consecutive points plotting above the upper 2σ limit, or two out of three consecutive points plotting below the lower 2σ limit.
3. Four out of five consecutive points plotting above the upper 1σ limit, or four out of five consecutive points plotting below the lower 1σ limit.
4. Eight consecutive points plotting on the same side of the center line.

In Figure 10.9, the Western Electric rules indicate that the process is out of control at sample number 8, at which time four out of five consecutive points have plotted above the upper 1σ control limit. For more information on using the Western Electric rules to interpret control charts, see Montgomery (2001b).

The *S* chart

The S chart is an alternative to the R chart. Both the S chart and the R chart are used to control the variability in a process. While the R chart assesses variability with the sample range, the S chart uses the sample standard deviation. Figure 10.10 (page 738) presents the S chart for the moisture data in Table 10.1.

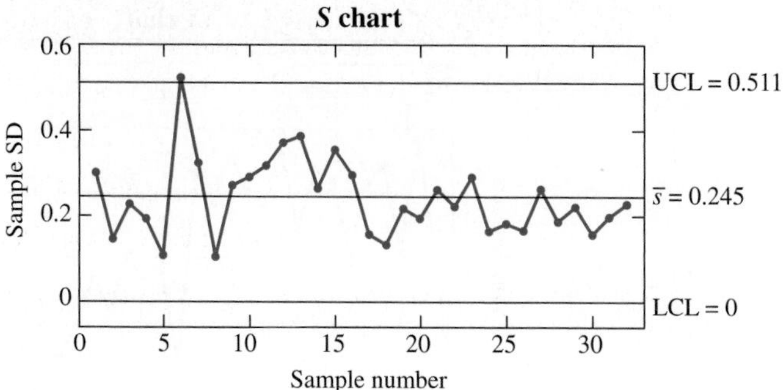

FIGURE 10.10 *S* chart for the moisture data. Compare with Figure 10.1.

Note that the *S* chart for the moisture data is similar in appearance to the *R* chart (Figure 10.1) for the same data. Like the *R* chart, the *S* chart indicates that the variation was out of control in sample 6.

To understand where the control limits are plotted, assume that the 32 sample standard deviations come from a population with mean μ_s and standard deviation σ_s. Ideally we would like to plot the center line at μ_s and the control limits at $\mu_s \pm 3\sigma_s$. These quantities are typically unknown. We approximate μ_s with $\bar{s}$, the average of the sample standard deviations. Thus the center line is plotted at $\bar{s}$. It can be shown by advanced methods that the quantities $\mu_s \pm 3\sigma_s$ can be estimated with multiples of $\bar{s}$; these multiples are denoted B_3 and B_4. The quantity $\mu_s - 3\sigma_s$ is estimated with $B_3\bar{s}$, while the quantity $\mu_s + 3\sigma_s$ is estimated with $B_4\bar{s}$. The quantities B_3 and B_4 are constants whose values depend on the sample size n. A brief table of values of B_3 and B_4 follows. A more extensive tabulation is provided in Table A.9 (Appendix A). Note that for samples of size 5 or less, the value of B_3 is 0. For samples this small, the value of $\mu_s - 3\sigma_s$ is negative. In these cases the lower control limit is set to 0, because it is impossible for a standard deviation to be negative.

n	**2**	**3**	**4**	**5**	**6**	**7**	**8**
B_3	0	0	0	0	0.030	0.118	0.185
B_4	3.267	2.568	2.266	2.089	1.970	1.882	1.815

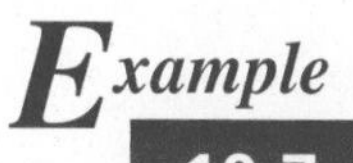

Example 10.7

Compute the center line and the 3σ *S* chart upper and lower control limits for the moisture data in Table 10.1.

Solution

The value of $\bar{s}$ is 0.2445 (Table 10.1). The sample size is $n = 5$. From the table immediately preceding, $B_3 = 0$ and $B_4 = 2.089$. Therefore the upper control limit is $(2.089)(0.2445) = 0.5108$, and the lower control limit is $(0)(0.2445) = 0$.

Summary

In an S chart, the center line and the 3σ upper and lower control limits are given by

$$3\sigma \text{ upper limit} = B_4\bar{s}$$
$$\text{Center line} = \bar{s}$$
$$3\sigma \text{ lower limit} = B_3\bar{s}$$

The values B_3 and B_4 depend on the sample size. Values are tabulated in Table A.9.

The S chart in Figure 10.10 shows that the process variation is out of control in sample 6. We delete this sample and recompute the S chart. Figure 10.11 presents the results. The variation is now in control. Note that this S chart is similar in appearance to the R chart in Figure 10.2.

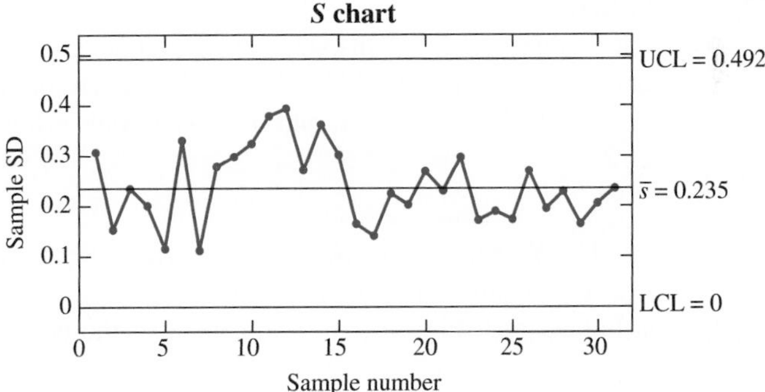

FIGURE 10.11 S chart for the moisture data, after deleting the out-of-control sample. Compare with Figure 10.2.

Once the variation is in control, we compute the $\overline{X}$ chart to assess the process mean. Recall that for the $\overline{X}$ chart, the center line is at $\overline{\overline{X}}$, and the upper and lower control limits would ideally be located a distance $3\sigma_{\overline{X}}$ above and below the center line. Since we used the S chart to assess the process variation, we will estimate the quantity $3\sigma_{\overline{X}}$ with a multiple of $\bar{s}$. Specifically, we estimate $3\sigma_{\overline{X}}$ with $A_3\bar{s}$, where A_3 is a constant whose value depends on the sample size n. A brief table of values of A_3 follows. A more extensive tabulation is provided in Table A.9.

n	2	3	4	5	6	7	8
A_3	2.659	1.954	1.628	1.427	1.287	1.182	1.099

> **Summary**
>
> In an $\overline{X}$ chart, when $\overline{s}$ is used to estimate $\sigma_{\overline{X}}$, the center line and the 3σ upper and lower control limits are given by
>
> $$3\sigma \text{ upper limit} = \overline{\overline{X}} + A_3\overline{s}$$
> $$\text{Center line} = \overline{\overline{X}}$$
> $$3\sigma \text{ lower limit} = \overline{\overline{X}} - A_3\overline{s}$$
>
> The value A_3 depends on the sample size. Values are tabulated in Table A.9.

If Western Electric rules are to be used, 1σ and 2σ control limits must be computed. The 1σ limits are $\overline{\overline{X}} \pm A_3\overline{s}/3$; the 2σ limits are $\overline{\overline{X}} \pm 2A_3\overline{s}/3$.

Example 10.8

Compute the 3σ $\overline{X}$ chart upper and lower control limits for the moisture data in Table 10.1.

Solution

With sample 6 deleted, the value of $\overline{\overline{X}}$ is 2.658, and the value of $\overline{s}$ is 0.2354. The sample size is $n = 5$. From the table, $A_3 = 1.427$. Therefore the upper control limit is $2.658 + (1.427)(0.2354) = 2.994$, and the lower control limit is $2.658 - (1.427)(0.2354) = 2.322$.

The $\overline{X}$ chart for the moisture data with sample 6 deleted is shown in Figure 10.12. The control limits are very similar to those calculated from the sample ranges, as shown in Figure 10.3. Figure 10.12 indicates that the process is out of control. After taking

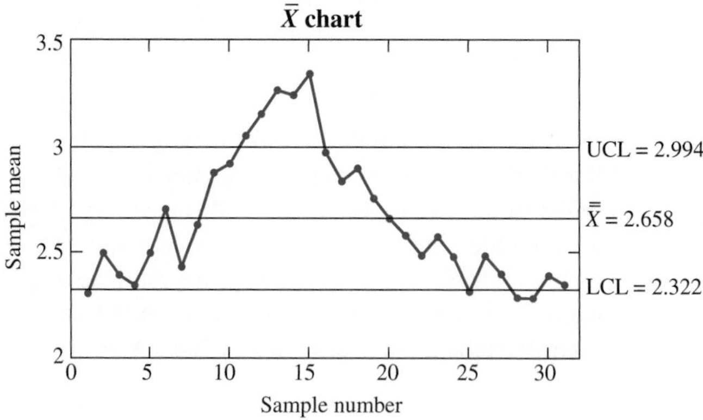

FIGURE 10.12 $\overline{X}$ chart for the moisture data. The control limits are based on the sample standard deviations rather than the sample ranges. Compare with Figure 10.3.

corrective action, a new S chart and $\overline{X}$ chart are constructed. Figure 10.13 presents the results. The process is now in a state of statistical control.

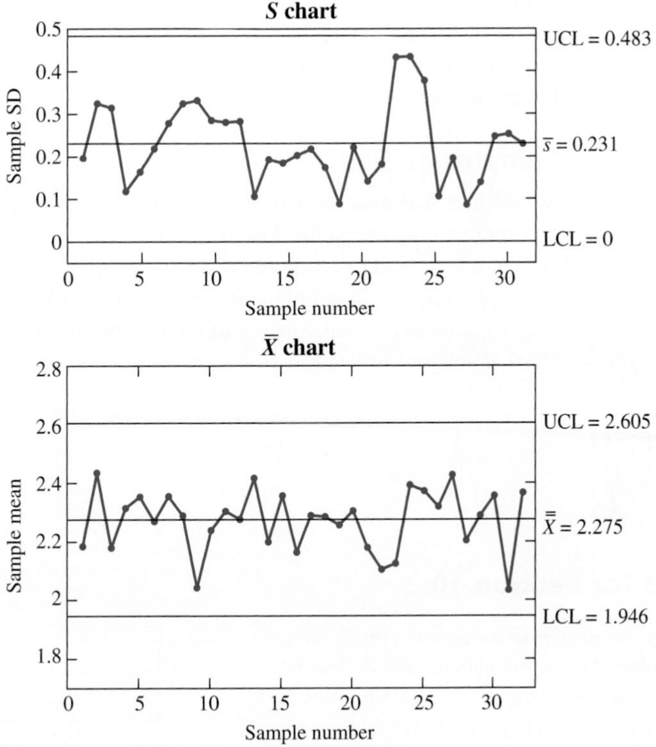

FIGURE 10.13 S chart and $\overline{X}$ chart after special cause is remedied. The process is now in a state of statistical control. Compare with Figure 10.4.

In summary, the S chart is an alternative to the R chart, to be used in combination with the $\overline{X}$ chart. For the moisture data, it turned out that the two charts gave very similar results. This is true in many cases, but it will sometimes happen that the results differ.

Which Is Better, the *S* Chart or the *R* Chart?

Both the R chart and S chart have the same purpose: to estimate the process standard deviation and to determine whether it is in control. It seems more natural to estimate the process standard deviation with the sample standard deviation s than with the range R. In fact, when the population is normal, s is a more precise estimate of the process standard deviation than is R, because it has a smaller uncertainty. To see this intuitively, note that the computation of s involves all the measurements in each sample, while the computation of R involves only two measurements (the largest and the smallest). It turns out that the improvement in precision obtained with s as opposed to R increases as the sample size increases. It follows that the S chart is a better choice, especially for larger sample sizes (greater than 5 or so). The R chart is still widely used, largely through tradition. At one time, the R chart had the advantage that the sample range required less arithmetic to compute than did the sample standard deviation. Now that most calculations are done electronically, this advantage no longer holds. So the S chart is in general the better choice.

Samples of Size 1

Sometimes it is necessary to define rational subgroups in such a way that each sample can contain only one value. For example, if the production rate is very slow, it may not be convenient to wait to accumulate samples larger than $n = 1$. It is impossible to compute a sample range or a sample standard deviation for a sample of size 1, so R charts and S charts cannot be used. Several other methods are available. One method is the CUSUM chart, discussed in Section 10.4.

Exercises for Section 10.2

1. The quality-control plan for a certain production process involves taking samples of size 4. The results from the last 30 samples can be summarized as follows:

$$\sum_{i=1}^{30} \overline{X}_i = 712.5 \quad \sum_{i=1}^{30} R_i = 143.7 \quad \sum_{i=1}^{30} s_i = 62.5$$

a. Compute the 3σ control limits for the R chart.

b. Compute the 3σ control limits for the S chart.

c. Using the sample ranges, compute the 3σ control limits for the $\overline{X}$ chart.

d. Using the sample standard deviations, compute the 3σ control limits for the $\overline{X}$ chart.

2. The following $\overline{X}$ chart depicts the last 50 samples taken from the output of a process. Using the Western Electric rules, is the process detected to be out of control at any time? If so, specify at which sample the process is first detected to be out of control and which rule is violated.

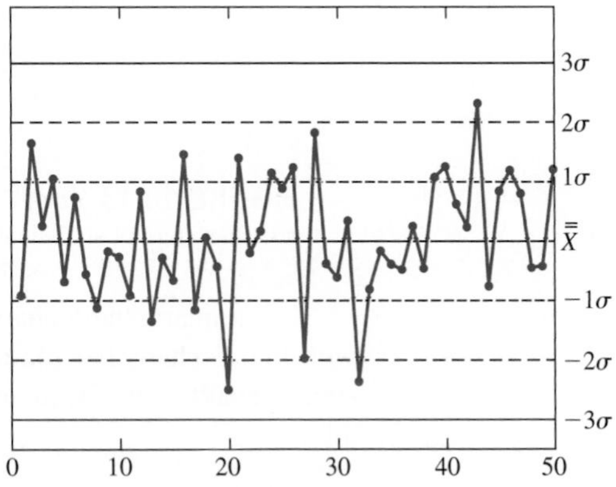

3. The gap (in mm) between the center and side electrodes of spark plugs used in internal combustion engines is measured on samples of size 5. The following table presents the means, ranges, and standard deviations for 20 consecutive samples.

Sample	$\overline{X}$	R	s
1	1.11	0.17	0.07
2	1.09	0.13	0.06
3	1.10	0.13	0.06
4	1.09	0.22	0.09
5	1.11	0.15	0.06
6	1.07	0.13	0.06
7	1.05	0.11	0.05
8	1.04	0.12	0.06
9	1.05	0.12	0.05
10	1.07	0.05	0.02
11	1.11	0.14	0.07
12	1.06	0.10	0.04
13	1.10	0.14	0.07
14	1.14	0.14	0.05
15	1.10	0.19	0.08
16	1.13	0.09	0.04
17	1.19	0.18	0.07
18	1.20	0.06	0.03
19	1.21	0.11	0.05
20	1.18	0.14	0.06

The means are $\overline{\overline{X}} = 1.110$, $\overline{R} = 0.131$, and $\overline{s} = 0.057$.

a. Calculate the 3σ control limits for the R chart. Is the variance under control? If not, delete the samples that are out of control and recompute $\overline{\overline{X}}$ and $\overline{R}$.

b. Based on the sample range R, calculate the 3σ control limits for the $\overline{X}$ chart. Based on the 3σ limits, is the process mean in control? If not, when is it first detected to be out of control?

c. Based on the Western Electric rules, is the process mean in control? If not, when is it first detected to be out of control?

4. Repeat Exercise 3, using the S chart in place of the R chart.

5. A process has mean 10 and standard deviation 2. The process is monitored by taking samples of size 4 at regular intervals. The process is declared to be out of control if a point plots outside the 3σ control limits on an $\overline{X}$ chart.

a. If the process mean shifts to 11, what is the average number of samples that will be drawn before the shift is detected on an $\overline{X}$ chart?

b. An upward shift to what value will be detected with an ARL of 6?

c. If the sample size remains at 4, to what value must the standard deviation be reduced to produce an ARL of 6 when the process mean shifts to 11?

d. If the standard deviation remains at 2, what sample size must be used to produce an ARL no greater than 6 when the process mean shifts to 11?

6. A process has mean 7.2 and standard deviation 1.3. The process is monitored by taking samples of size 6 at regular intervals. The process is declared to be out of control if a point plots outside the 3σ control limits on an $\overline{X}$ chart.

a. If the process mean shifts to 6.7, what is the average number of samples that will be drawn before the shift is detected on an $\overline{X}$ chart?

b. An upward shift to what value will be detected with an ARL of 15?

c. If the sample size remains at 6, to what value must the standard deviation be reduced to produce an ARL of 15 when the process mean shifts to 6.7?

d. If the standard deviation remains at 1.3, what sample size must be used to produce an ARL no greater than 6 when the process mean shifts to 6.7?

7. A process is monitored by taking samples at regular intervals and is declared to be out of control if a point plots outside the 3σ control limits on an $\overline{X}$ chart. Assume the process is in control.

a. What is the probability that a false alarm will occur within the next 50 samples?

b. What is the probability that a false alarm will occur within the next 100 samples?

c. What is the probability that there will be no false alarm within the next 200 samples?

d. Fill in the blank: The probability is 0.5 that there will be a false alarm within the next ________ samples.

8. Samples of six ball bearings are taken periodically, and their diameters (in mm) are measured. The following table presents the means, ranges, and standard deviations for 25 consecutive samples.

Sample	$\overline{X}$	R	s
1	199.5	7.2	2.38
2	200.4	7.9	2.70
3	202.0	3.6	1.35
4	198.9	5.3	1.95
5	199.1	3.4	1.53
6	200.2	4.8	2.19
7	199.5	2.1	0.87
8	198.1	5.5	2.22
9	200.0	5.5	2.09
10	199.0	9.2	3.17
11	199.3	7.8	2.64
12	199.3	5.8	1.98
13	200.5	3.6	1.36
14	200.1	5.7	2.34
15	200.1	3.0	1.11
16	200.4	5.2	1.99
17	200.9	4.1	1.54
18	200.8	5.5	2.01
19	199.3	8.3	2.92
20	199.8	3.8	1.78
21	199.5	3.6	1.60
22	198.9	7.2	2.60
23	199.6	3.8	1.69
24	200.3	3.1	1.15
25	199.9	4.1	1.86

The means are $\overline{\overline{X}} = 199.816$, $\overline{R} = 5.164$, and $\overline{s} = 1.961$.

a. Calculate the 3σ control limits for the R chart. Is the variance under control? If not, delete the samples that are out of control and recompute $\overline{\overline{X}}$ and $\overline{R}$.

b. Based on the sample range R, calculate the 3σ control limits for the $\overline{X}$ chart. Based on the 3σ limits, is the process mean in control? If not, when is it first detected to be out of control?

c. Based on the Western Electric rules, is the process mean in control? If not, when is it first detected to be out of control?

9. Repeat Exercise 8, using the S chart in place of the R chart.

10. A certain type of integrated circuit is connected to its frame by five wires. Thirty samples of five units each were taken, and the pull strength (in grams) of one wire on each unit was measured. The data are presented in Table E10. The means are $\overline{\overline{X}} = 9.81$, $\overline{R} = 1.14$, and $\overline{s} = 0.4647$.

a. Compute the 3σ limits for the R chart. Is the variance out of control at any point? If so, delete the samples that are out of control and recompute $\overline{\overline{X}}$ and $\overline{R}$.

b. Compute the 3σ limits for the $\overline{X}$ chart. On the basis of the 3σ limits, is the process mean in control? If not, at what point is it first detected to be out of control?

c. On the basis of the Western Electric rules, is the process mean in control? If not, when is it first detected to be out of control?

11. Repeat Exercise 10, using the S chart in place of the R chart.

12. Copper wires are coated with a thin plastic coating. Samples of four wires are taken every hour, and the thickness of the coating (in mils) is measured. The data from the last 30 samples are presented in Table E12 on page 746. The means are $\overline{\overline{X}} = 150.075$, $\overline{R} = 6.97$, and $\overline{s} = 3.082$.

a. Compute the 3σ limits for the R chart. Is the variance out of control at any point? If so, delete the samples that are out of control and recompute $\overline{\overline{X}}$ and $\overline{R}$.

b. Compute the 3σ limits for the $\overline{X}$ chart. On the basis of the 3σ limits, is the process mean in control? If not, at what point is it first detected to be out of control?

c. On the basis of the Western Electric rules, is the process mean in control? If not, when is it first detected to be out of control?

13. Repeat Exercise 12, using the S chart in place of the R chart.

TABLE E10 Data for Exercise 10

Sample	Sample Values					$\overline{X}$	R	s
1	10.3	9.8	9.7	9.9	10.2	9.98	0.6	0.26
2	9.9	9.4	10.0	9.4	10.2	9.78	0.8	0.36
3	9.0	9.9	9.6	9.2	10.6	9.66	1.6	0.63
4	10.1	10.6	10.3	9.6	9.7	10.06	1.0	0.42
5	10.8	9.4	9.9	10.1	10.1	10.06	1.4	0.50
6	10.3	10.1	10.0	9.5	9.8	9.94	0.8	0.30
7	8.8	9.3	9.9	8.9	9.3	9.24	1.1	0.43
8	9.4	9.7	9.4	9.9	10.5	9.78	1.1	0.45
9	9.1	8.9	9.8	9.0	9.3	9.22	0.9	0.36
10	8.9	9.4	10.6	9.4	8.7	9.40	1.9	0.74
11	9.0	8.6	9.9	9.6	10.5	9.52	1.9	0.75
12	9.5	9.2	9.4	9.3	9.6	9.40	0.4	0.16
13	9.0	9.4	9.7	9.4	8.6	9.22	1.1	0.43
14	9.4	9.2	9.4	9.3	9.7	9.40	0.5	0.19
15	9.4	10.2	9.0	8.8	10.2	9.52	1.4	0.66
16	9.6	9.5	10.0	9.3	9.4	9.56	0.7	0.27
17	10.2	8.8	10.0	10.1	10.1	9.84	1.4	0.59
18	10.4	9.4	9.9	9.4	9.9	9.80	1.0	0.42
19	11.1	10.5	10.6	9.8	9.4	10.28	1.7	0.68
20	9.3	9.9	10.9	9.5	10.6	10.04	1.6	0.69
21	9.5	10.2	9.7	9.4	10.0	9.76	0.8	0.34
22	10.5	10.5	10.1	9.5	10.3	10.18	1.0	0.41
23	9.8	8.9	9.6	9.8	9.6	9.54	0.9	0.37
24	9.3	9.7	10.3	10.1	9.7	9.82	1.0	0.39
25	10.2	9.6	8.8	9.9	10.2	9.74	1.4	0.58
26	10.8	9.5	10.5	10.5	10.1	10.28	1.3	0.50
27	10.4	9.9	10.1	9.9	10.9	10.24	1.0	0.42
28	11.0	10.8	10.1	9.2	9.9	10.20	1.8	0.72
29	10.3	10.0	10.6	10.0	11.1	10.40	1.1	0.46
30	10.9	10.6	9.9	10.0	10.8	10.44	1.0	0.46

TABLE E12 Data for Exercise 12

Sample	Sample Values	$\overline{X}$	R	s
1	146.0 147.4 151.9 155.2	150.125	9.2	4.22
2	147.1 147.5 151.4 149.4	148.850	4.3	1.97
3	148.7 148.4 149.6 154.1	150.200	5.7	2.65
4	151.3 150.0 152.4 148.2	150.475	4.2	1.81
5	146.4 147.5 152.9 150.3	149.275	6.5	2.92
6	150.2 142.9 152.5 155.5	150.275	12.6	5.37
7	147.8 148.3 145.7 149.7	147.875	4.0	1.66
8	137.1 156.6 147.2 148.9	147.450	19.5	8.02
9	151.1 148.1 145.6 147.6	148.100	5.5	2.27
10	151.3 151.3 142.5 146.2	147.825	8.8	4.29
11	151.3 153.5 150.2 148.7	150.925	4.8	2.02
12	151.9 152.2 149.3 154.2	151.900	4.9	2.01
13	152.8 149.1 148.5 146.9	149.325	5.9	2.50
14	152.9 149.9 151.9 150.4	151.275	3.0	1.38
15	149.0 149.9 153.1 152.8	151.200	4.1	2.06
16	153.9 150.8 153.9 145.0	150.900	8.9	4.20
17	150.4 151.8 151.3 153.0	151.625	2.6	1.08
18	157.2 152.6 148.4 152.6	152.700	8.8	3.59
19	152.7 156.2 146.8 148.7	151.100	9.4	4.20
20	150.2 148.2 149.8 142.1	147.575	8.1	3.75
21	151.0 151.7 148.5 147.0	149.550	4.7	2.19
22	143.8 154.5 154.8 151.6	151.175	11.0	5.12
23	143.0 156.4 149.2 152.2	150.200	13.4	5.64
24	148.8 147.7 147.1 148.2	147.950	1.7	0.72
25	153.8 145.4 149.5 153.4	150.525	8.4	3.93
26	151.6 149.3 155.0 149.0	151.225	6.0	2.77
27	149.4 151.4 154.6 150.0	151.350	5.2	2.32
28	149.8 149.0 146.8 145.7	147.825	4.1	1.90
29	155.8 152.4 150.2 154.8	153.300	5.6	2.51
30	153.9 145.7 150.7 150.4	150.175	8.2	3.38

10.3 Control Charts for Attributes

The *p* Chart

The p chart is used when the quality characteristic being measured on each unit has only two possible values, usually "defective" and "not defective." In each sample, the proportion of defectives is calculated; these sample proportions are then plotted. We will now describe how the center line and control limits are calculated.

Let p be the probability that a given unit is defective. If the process is in control, this probability is constant over time. Let k be the number of samples. We will assume that all samples are the same size, and we will denote this size by n. Let X_i be the number of defective units in the ith sample, and let $\widehat{p}_i = X_i/n$ be the proportion of defective items in the ith sample. Now $X_i \sim \text{Bin}(n,\ p)$, and if $np > 10$, it is approximately true that $\widehat{p}_i \sim N(p,\ p(1-p)/n)$ (see p. 274). Since $\widehat{p}_i$ has mean $\mu = p$ and standard deviation

$\sigma = \sqrt{p(1-p)/n}$, it follows that the center line should be at p, and the 3σ control limits should be at $p \pm 3\sqrt{p(1-p)/n}$. Usually p is not known and is estimated with $\overline{p} = \sum_{i=1}^{k} \widehat{p}_i / k$, the average of the sample proportions $\widehat{p}_i$.

Summary

In a p chart, where the number of items in each sample is n, the center line and the 3σ upper and lower control limits are given by

$$3\sigma \text{ upper limit} = \overline{p} + 3\sqrt{\frac{\overline{p}(1-\overline{p})}{n}}$$

$$\text{Center line} = \overline{p}$$

$$3\sigma \text{ lower limit} = \overline{p} - 3\sqrt{\frac{\overline{p}(1-\overline{p})}{n}}$$

These control limits will be valid if $n\overline{p} > 10$.

We illustrate these ideas with Example 10.9.

In the production of silicon wafers, 30 lots of size 500 are sampled, and the proportion of defective wafers is calculated for each sample. Table 10.2 presents the results. Compute the center line and 3σ control limits for the p chart. Plot the chart. Does the process appear to be in control?

TABLE 10.2 Number and proportion defective, for Example 10.9

Sample	Number Defective	Proportion Defective ($\widehat{p}$)	Sample	Number Defective	Proportion Defective ($\widehat{p}$)
1	17	0.034	16	26	0.052
2	26	0.052	17	19	0.038
3	31	0.062	18	31	0.062
4	25	0.050	19	27	0.054
5	26	0.052	20	24	0.048
6	29	0.058	21	22	0.044
7	36	0.072	22	24	0.048
8	26	0.052	23	30	0.060
9	25	0.050	24	25	0.050
10	21	0.042	25	26	0.052
11	18	0.036	26	28	0.056
12	33	0.066	27	22	0.044
13	29	0.058	28	31	0.062
14	17	0.034	29	18	0.036
15	28	0.056	30	23	0.046

Solution

The average of the 30 sample proportions is $\overline{p} = 0.050867$. The center line is therefore plotted at 0.050867. The control limits are plotted at $0.050867 \pm 3\sqrt{(0.050867)(0.949133)/500}$. The upper control limit is therefore 0.0803, and the lower control limit is 0.0214. Figure 10.14 presents the p chart. The process appears to be in control.

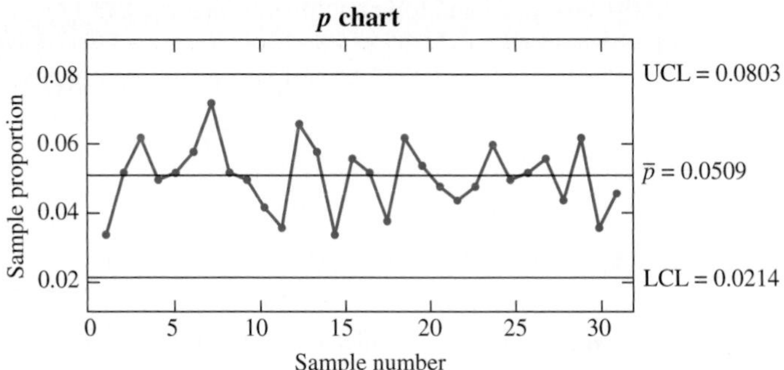

FIGURE 10.14 p chart for the data in Table 10.2

The sample size needed to construct a p chart is usually much larger than that needed for an $\overline{X}$ chart. The reason is that the sample size must be large enough so that there will be several defective items in most of the samples. If defective items are not common, the sample size must be quite large.

Interpreting Out-of-Control Signals in Attribute Charts

When an attribute control chart is used to monitor the frequency of defective units, a point plotting above the upper control limit requires quite a different response than a point plotting below the lower control limit. Both conditions indicate that a special cause has changed the proportion of defective units. A point plotting above the upper control limit indicates that the proportion of defective units has increased, so action must be taken to identify and remove the special cause. A point plotting below the lower control limit, however, indicates that the special cause has *decreased* the proportion of defective units. The special cause still needs to be identified, but in this case, action should be taken to make it continue, so that the proportion of defective items can be decreased permanently.

The *c* Chart

The c chart is used when the quality measurement is a count of the number of defects, or flaws, in a given unit. A *unit* may be a single item, or it may be a group of items large enough so that the expected number of flaws is sufficiently large. Use of the c chart requires that the number of defects follow a Poisson distribution. Assume that k units are sampled, and let c_i denote the total number of defects in the ith unit. Let λ denote the mean total number of flaws per unit. Then $c_i \sim \text{Poisson}(\lambda)$. If the process is in control, the value of λ is constant over time. Now if λ is reasonably large, say $\lambda > 10$, then

$c_i \sim N(\lambda, \lambda)$, approximately (see p. 278). Note that the value of λ can in principle be made large enough by choosing a sufficiently large number of items per unit. The c chart is constructed by plotting the values c_i. Since c_i has mean λ and standard deviation equal to $\sqrt{\lambda}$, the center line should be plotted at λ and the 3σ control limits should be plotted at $\lambda \pm 3\sqrt{\lambda}$. Usually the value of λ is unknown and has to be estimated from the data. The appropriate estimate is $\bar{c} = \sum_{i=1}^{k} c_i/k$, the average number of defects per unit.

Summary

In a c chart, the center line and the 3σ upper and lower control limits are given by

$$3\sigma \text{ upper limit} = \bar{c} + 3\sqrt{\bar{c}}$$

$$\text{Center line} = \bar{c}$$

$$3\sigma \text{ lower limit} = \bar{c} - 3\sqrt{\bar{c}}$$

These control limits will be valid if $\bar{c} > 10$.

Example 10.10 illustrates these ideas.

10.10

Rolls of sheet aluminum, used to manufacture cans, are examined for surface flaws. Table 10.3 presents the numbers of flaws in 40 samples of 100 m^2 each. Compute the center line and 3σ control limits for the c chart. Plot the chart. Does the process appear to be in control?'

Solution

The average of the 40 counts is $\bar{c} = 12.275$. The center line is therefore plotted at 12.275. The 3σ control limits are plotted at $12.275 \pm 3\sqrt{12.275}$. The upper control limit is therefore 22.7857, and the lower control limit is 1.7643. Figure 10.15 on page 750 presents the c chart. The process appears to be in control.

TABLE 10.3 Number of flaws, for Example 10.10

Sample	Number of Flaws (c)	Sample	Number of Flaws (c)	Sample	Number of Flaws (c)	Sample	Number of Flaws (c)
1	16	11	14	21	11	31	10
2	12	12	11	22	16	32	10
3	9	13	10	23	16	33	10
4	13	14	9	24	13	34	12
5	15	15	9	25	12	35	14
6	5	16	14	26	17	36	10
7	13	17	10	27	15	37	15
8	11	18	12	28	13	38	12
9	15	19	8	29	15	39	11
10	12	20	14	30	13	40	14

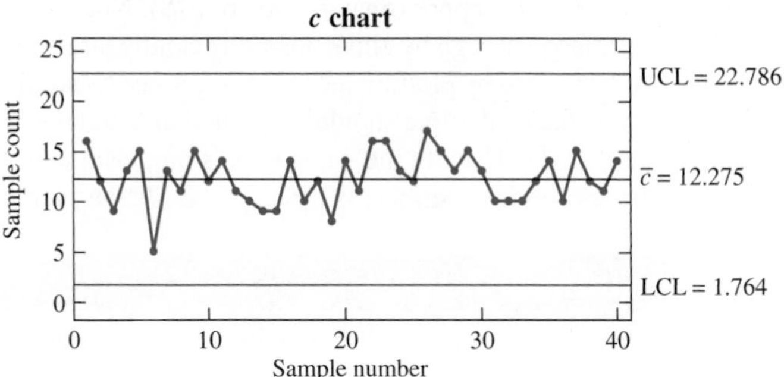

FIGURE 10.15 c chart for the data in Table 10.3

Exercises for Section 10.3

1. A process is monitored for defective items by taking a sample of 200 items each day and calculating the proportion that are defective. Let p_i be the proportion of defective items in the ith sample. For the last 30 samples, the sum of the proportions is $\sum_{i=1}^{30} p_i = 1.64$. Calculate the center line and the 3σ upper and lower control limits for a p chart.

2. The specification for the diameter of a certain rivet head is 13.3–13.5 mm. Each day a sample of 400 rivets is taken, and the number that fail to meet the specification is counted. The numbers of defective rivets for each of the last 20 days are as follows:

   12 25 23 15 20 11 16 20 18 25
   22 16 22 18 37 35 40 40 36 30

   a. Compute the upper and lower 3σ limits for a p chart.
   b. Is the process in control? If not, when is it first detected to be out of control?

3. A process is monitored for defective items by periodically taking a sample of 100 items and counting the number that are defective. In the last 50 samples, there were a total of 622 defective items. Is this enough information to compute the 3σ control limits for a p chart? If so, compute the limits. If not, state what additional information would be required.

4. Refer to Exercise 3. In the last 50 samples, there were a total of 622 defective items. The largest number of defectives in any sample was 24, while the smallest number was 6. Is this enough information to determine whether the process was out of control at any time during the last 50 samples? If so, state whether or not the process was out of control. If not, state what additional information would be required to make the determination.

5. A newly designed quality-control program for a certain process involves sampling 20 items each day and counting the number of defective items. The numbers of defectives in the first 10 samples are 0, 0, 1, 0, 1, 0, 0, 0, 1, 0. A member of the quality-control team asks for advice, expressing concern that the numbers of defectives are too small to construct an accurate p chart. Which of the following is the best advice?

   i. Nothing needs to be changed. An accurate p chart can be constructed when the number of defective items is this small.
   ii. Since the proportion of items that are defective is so small, it isn't necessary to construct a p chart for this process.
   iii. Increase the value of p to increase the number of defectives per sample.
   iv. Increase the sample size to increase the number of defectives per sample.

6. A process that produces aluminum cans is monitored by taking samples of 1000 cans and counting the total number of visual flaws on all the sample cans. Let c_i be the total number of flaws on the cans in the ith sample. For the last 50 samples, the quantity $\sum_{i=1}^{50} c_i = 1476$ has been calculated. Compute the center line and the 3σ upper and lower control limits for a c chart.

7. Refer to Exercise 6. The number of flaws in the 20th sample was 48. Is it possible to determine whether the process was in control at this time? If so, state whether or not the process was in control. If not, state what additional information would be required to make the determination.

8. Each hour, a 10 m^2 section of fabric is inspected for flaws. The numbers of flaws observed for the last 20 hours are as follows:

38	35	35	49	33	48	40	47	45	46
41	53	36	41	51	63	35	58	55	57

a. Compute the upper and lower 3σ limits for a c chart.

b. Is the process in control? If not, when is it first detected to be out of control?

10.4 The CUSUM Chart

One purpose of an $\overline{X}$ chart is to detect a shift in the process mean. Unless a shift is fairly large, however, it may be some time before a point plots outside the 3σ control limits. Example 10.4 (in Section 10.2) showed that when a process mean shifts by an amount equal to $\sigma_{\overline{X}}$, the average run length (ARL) is approximately 44, which means that on the average 44 samples must be observed before the process is judged to be out of control. The Western Electric rules (Section 10.2) provide one method for reducing the ARL. CUSUM charts provide another.

One way that small shifts manifest themselves is with a run of points above or below the center line. The Western Electric rules are designed to respond to runs. Another way to detect smaller shifts is with **cumulative sums**. Imagine that a process mean shifts upward slightly. There will then be a tendency for points to plot above the center line. If we add the deviations from the center line as we go along, and plot the cumulative sums, the points will drift upward and will exceed a control limit much sooner than they would in an $\overline{X}$ chart.

We now describe how to plot the points in a CUSUM chart. We assume that we have m samples of size n, with sample means $\overline{X}_1, \ldots, \overline{X}_m$. To begin, a target value μ must be specified for the process mean. Often μ is taken to be the value $\overline{\overline{X}}$. Then an estimate of $\sigma_{\overline{X}}$, the standard deviation of the sample means, is needed. This can be obtained either with sample ranges, using the estimate $\sigma_{\overline{X}} \approx A_2\overline{R}/3$, or with sample standard deviations, using the estimate $\sigma_{\overline{X}} \approx A_3\overline{s}/3$. If there is only one item per sample ($n = 1$), then an external estimate is needed. Even a rough guess can produce good results, so the CUSUM procedure can be useful when $n = 1$. Finally two constants, usually called k and h, must be specified. Larger values for these constants result in longer average run lengths, and thus fewer false alarms, but also result in longer waiting times to discover that a process is out of control. The values $k = 0.5$ and $h = 4$ or 5 are often used, because they provide a reasonably long ARL when the process is in control but still have fairly good power to detect a shift of magnitude $1\sigma_{\overline{X}}$ or more in the process mean.

For each sample, the quantity $\overline{X}_i - \mu$ is the deviation from the target value. We define two cumulative sums, SH and SL. The sum SH is always either positive or zero and signals that the process mean has become greater than the target value. The sum SL is always either negative or zero and signals that the process mean has become less than the target value. Both these sums are computed recursively: in other words, the current value in the sequence is used to compute the next value. The initial values of SH and SL are

$$\text{SH}_0 = 0 \qquad \text{SL}_0 = 0 \tag{10.2}$$

For $i \geq 1$ the values are

$$\text{SH}_i = \max[0,\ \overline{X}_i - \mu - k\sigma_{\overline{X}} + \text{SH}_{i-1}] \tag{10.3}$$

$$\text{SL}_i = \min[0,\ \overline{X}_i - \mu + k\sigma_{\overline{X}} + \text{SL}_{i-1}] \tag{10.4}$$

If $\text{SH}_i > h\sigma_{\overline{X}}$ for some i, it is concluded that the process mean has become greater than the target value. If $\text{SL}_i < -h\sigma_{\overline{X}}$ for some i, it is concluded that the process mean has become less than the target value.

Figure 10.16 presents a CUSUM chart for the data in Figure 10.9 (in Section 10.2). The values $k = 0.5$ and $h = 4$ were used. The value 2.962 is the quantity $h\sigma_{\overline{X}} = 4(0.738)$. The CUSUM chart indicates an out-of-control condition on the tenth sample. For these data, the CUSUM chart performs about as well as the Western Electric rules, which determined that the process was out of control at the eighth sample (see Figure 10.9).

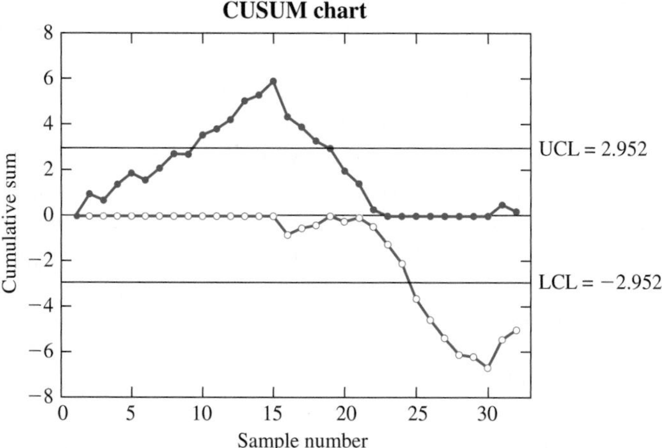

FIGURE 10.16 CUSUM chart for the data in Figure 10.9.

Summary

In a CUSUM chart, two cumulative sums, SH and SL, are plotted.

The initial values are $\text{SH}_0 = \text{SL}_0 = 0$. For $i \geq 1$,

$$\text{SH}_i = \max[0, \overline{X}_i - \mu - k\sigma_{\overline{X}} + \text{SH}_{i-1}]$$

$$\text{SL}_i = \min[0, \overline{X}_i - \mu + k\sigma_{\overline{X}} + \text{SL}_{i-1}]$$

The constants k and h must be specified. Good results are often obtained for the values $k = 0.5$ and $h = 4$ or 5.

If for any i, $\text{SH}_i > h\sigma_{\overline{X}}$ or $\text{SL}_i < -h\sigma_{\overline{X}}$, the process is judged to be out of control.

There are several other methods for constructing CUSUM charts, which are equivalent, or nearly equivalent, to the method presented here. Some people define the deviations to be the z-scores $z_i = (\overline{X}_i - \mu)/\sigma_{\overline{X}}$, and then use z_i in place of $X_i - \mu$, and k in place of $k\sigma_{\overline{X}}$ in the formulas for SH and SL. With this definition, the control limits are plotted at $\pm h$ rather than $\pm h\sigma_{\overline{X}}$.

Other methods for graphing the CUSUM chart are available as well. The most common alternative is the "V-mask" approach. A text on statistical quality control, such as Montgomery (2001b), can be consulted for further information.

Exercises for Section 10.4

1. Refer to Exercise 3 in Section 10.2.

a. Delete any samples necessary to bring the process variation under control. (You did this already if you did Exercise 3 in Section 10.2.)

b. Use $\overline{R}$ to estimate $\sigma_{\overline{X}}$ ($\sigma_{\overline{X}}$ is the difference between $\overline{\overline{X}}$ and the 1σ control limit on an $\overline{X}$ chart).

c. Construct a CUSUM chart, using $\overline{\overline{X}}$ for the target mean μ, and the estimate of $\sigma_{\overline{X}}$ found in part (b) for the standard deviation. Use the values $k = 0.5$ and $h = 4$.

d. Is the process mean in control? If not, when is it first detected to be out of control?

e. Construct an $\overline{X}$ chart, and use the Western Electric rules to determine whether the process mean is in control. (You did this already if you did Exercise 3 in Section 10.2.) Do the Western Electric rules give the same results as the CUSUM chart? If not, how are they different?

2. Refer to Exercise 8 in Section 10.2.

a. Delete any samples necessary to bring the process variation under control. (You did this already if you did Exercise 8 in Section 10.2.)

b. Use $\overline{R}$ to estimate $\sigma_{\overline{X}}$ ($\sigma_{\overline{X}}$ is the difference between $\overline{\overline{X}}$ and the 1σ control limit on an $\overline{X}$ chart).

c. Construct a CUSUM chart, using $\overline{\overline{X}}$ for the target mean μ, and the estimate of $\sigma_{\overline{X}}$ found in part (b) for the standard deviation. Use the values $k = 0.5$ and $h = 4$.

d. Is the process mean in control? If not, when is it first detected to be out of control?

e. Construct an $\overline{X}$ chart, and use the Western Electric rules to determine whether the process mean is in control. (You did this already if you did Exercise 8 in Section 10.2.) Do the Western Electric rules give the same results as the CUSUM chart? If not, how are they different?

3. Refer to Exercise 10 in Section 10.2.

a. Delete any samples necessary to bring the process variation under control. (You did this already if you did Exercise 10 in Section 10.2.)

b. Use $\overline{R}$ to estimate $\sigma_{\overline{X}}$ ($\sigma_{\overline{X}}$ is the difference between $\overline{\overline{X}}$ and the 1σ control limit on an $\overline{X}$ chart).

c. Construct a CUSUM chart, using $\overline{\overline{X}}$ for the target mean μ, and the estimate of $\sigma_{\overline{X}}$ found in part (b) for the standard deviation. Use the values $k = 0.5$ and $h = 4$.

d. Is the process mean in control? If not, when is it first detected to be out of control?

e. Construct an $\overline{X}$ chart, and use the Western Electric rules to determine whether the process mean is in control. (You did this already if you did Exercise 10 in Section 10.2.) Do the Western Electric rules give the same results as the CUSUM chart? If not, how are they different?

4. Refer to Exercise 12 in Section 10.2.

a. Delete any samples necessary to bring the process variation under control. (You did this already if you did Exercise 12 in Section 10.2.)

b. Use $\overline{R}$ to estimate $\sigma_{\overline{X}}$ ($\sigma_{\overline{X}}$ is the difference between $\overline{\overline{X}}$ and the 1σ control limit on an $\overline{X}$ chart).

c. Construct a CUSUM chart, using $\overline{\overline{X}}$ for the target mean μ, and the estimate of $\sigma_{\overline{X}}$ found in part (b) for the standard deviation. Use the values $k = 0.5$ and $h = 4$.

d. Is the process mean in control? If not, when is it first detected to be out of control?

e. Construct an $\overline{X}$ chart, and use the Western Electric rules to determine whether the process mean is in control. (You did this already if you did Exercise 12 in Section 10.2.) Do the Western Electric rules give the same results as the CUSUM chart? If not, how are they different?

5. Concrete blocks to be used in a certain application are supposed to have a mean compressive strength of 1500 MPa. Samples of size 1 are used for quality control. The compressive strengths of the last 40 samples are given in the following table.

Sample	Strength	Sample	Strength
1	1487	21	1507
2	1463	22	1474
3	1499	23	1515
4	1502	24	1533
5	1473	25	1487
6	1520	26	1518
7	1520	27	1526
8	1495	28	1469
9	1503	29	1472
10	1499	30	1512
11	1497	31	1483
12	1516	32	1505
13	1489	33	1507
14	1545	34	1505
15	1498	35	1517
16	1503	36	1504
17	1522	37	1515
18	1502	38	1467
19	1499	39	1491
20	1484	40	1488

Previous results suggest that a value of $\sigma = 15$ is reasonable for this process.

a. Using the value 1500 for the target mean μ, and the values $k = 0.5$ and $h = 4$, construct a CUSUM chart.

b. Is the process mean in control? If not, when is it first detected to be out of control?

6. A quality-control apprentice is preparing a CUSUM chart. The values calculated for SL and SH are presented in the following table. Three of the values have been calculated incorrectly. Which are they?

Sample	SL	SH
1	0	0
2	0	0
3	0	0
4	−1.3280	0
5	−1.4364	0
6	−2.0464	0
7	−1.6370	0
8	−0.8234	0.2767
9	−0.4528	0.1106
10	0	0.7836
11	0.2371	0.0097
12	0.7104	0
13	0	0.2775
14	0	0.5842
15	0	0.3750
16	0	0.4658
17	0	0.1866
18	0	0.3277
19	−0.2036	0
20	0	−0.7345

10.5 Process Capability

Once a process is in a state of statistical control, it is important to evaluate its ability to produce output that conforms to design specifications. We consider variables data, and we assume that the quality characteristic of interest follows a normal distribution.

The first step in assessing process capability is to estimate the process mean and standard deviation. These estimates are denoted $\widehat{\mu}$ and $\widehat{\sigma}$, respectively. The data used to calculate $\widehat{\mu}$ and $\widehat{\sigma}$ are usually taken from control charts at a time when the process is in a state of control. The process mean is estimated with $\widehat{\mu} = \overline{\overline{X}}$. The process standard deviation can be estimated by using either the average sample range $\overline{R}$ or the average sample standard deviation $\overline{s}$. Specifically, it has been shown that $\widehat{\sigma}$ can be computed either by dividing $\overline{R}$ by a constant called d_2, or by dividing $\overline{s}$ by a constant called c_4. The values of the constants d_2 and c_4 depend on the sample size. Values are tabulated in Table A.9 (in Appendix A).

Summary

If a quality characteristic from a process in a state of control is normally distributed, then the process mean $\widehat{\mu}$ and standard deviation $\widehat{\sigma}$ can be estimated from control chart data as follows:

$$\widehat{\mu} = \overline{\overline{X}}$$

$$\widehat{\sigma} = \frac{\overline{R}}{d_2} \quad \text{or} \quad \widehat{\sigma} = \frac{\overline{s}}{c_4}$$

The values of d_2 and c_4 depend on the sample size. Values are tabulated in Table A.9.

Note that the process standard deviation σ is not the same quantity that is used to compute the 3σ control limits on the $\overline{X}$ chart. The control limits are $\mu \pm 3\sigma_{\overline{X}}$, where $\sigma_{\overline{X}}$ is the standard deviation of the sample mean. The process standard deviation σ is the standard deviation of the quality characteristic of individual units. They are related by $\sigma_{\overline{X}} = \sigma/\sqrt{n}$, where n is the sample size.

To be fit for use, a quality characteristic must fall between a lower specification limit (LSL) and an upper specification limit (USL). Sometimes there is only one limit; this situation will be discussed at the end of this section. The specification limits are determined by design requirements. They are *not* the control limits found on control charts. We will assume that the process mean falls between the LSL and the USL.

We will discuss two indices of process capability, C_{pk} and C_p. The index C_{pk} describes the capability of the process as it is, while C_p describes the potential capability of the process. Note that the process capability index C_p has no relation to the quantity called Mallows' C_p that is used for linear model selection (see Chapter 8). It is a coincidence that the two quantities have the same name.

The index C_{pk} is defined to be the distance from $\widehat{\mu}$ to the nearest specification limit, divided by $3\widehat{\sigma}$. Figure 10.17 presents an illustration where $\widehat{\mu}$ is closer to the upper specification limit.

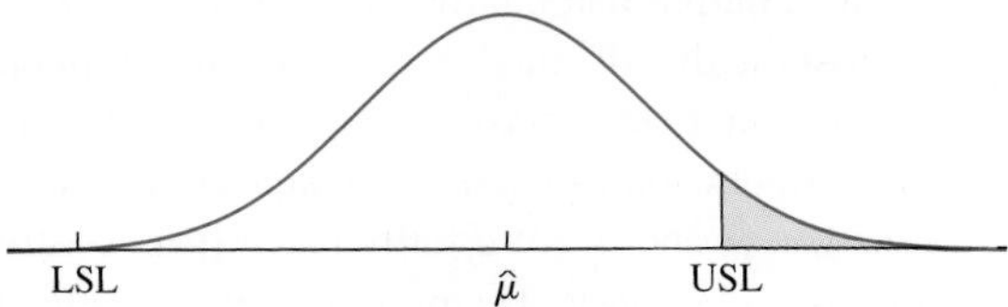

FIGURE 10.17 The normal curve represents the population of units produced by a process. The process mean is closer to the upper specification limit (USL) than to the lower specification limit (LSL). The index C_{pk} is therefore equal to $(\text{USL} - \widehat{\mu})/3\widehat{\sigma}$.

Definition

The index C_{pk} is equal either to

$$\frac{\widehat{\mu} - \text{LSL}}{3\widehat{\sigma}} \quad \text{or} \quad \frac{\text{USL} - \widehat{\mu}}{3\widehat{\sigma}}$$

whichever is less.

By convention, the minimum acceptable value for C_{pk} is 1. That is, a process is considered to be minimally capable if the process mean is three standard deviations from the nearest specification limit. A C_{pk} value of 1.33, indicating that the process mean is four standard deviations from the nearest specification limit, is generally considered good.

Example 10.11

The design specifications for a piston rod used in an automatic transmission call for the rod length to be between 71.4 and 72.8 mm. The process is monitored with an $\overline{X}$ chart and an S chart, using samples of size $n = 5$. These show the process to be in control. The values of $\overline{\overline{X}}$ and $\overline{s}$ are $\overline{\overline{X}} = 71.8$ mm and $\overline{s} = 0.20$ mm. Compute the value of C_{pk}. Is the process capability acceptable?

Solution

We estimate $\widehat{\mu} = \overline{\overline{X}} = 71.8$. To compute $\widehat{\sigma}$, we find, from Table A.9, that $c_4 = 0.9400$ when the sample size is 5. Therefore $\widehat{\sigma} = \overline{s}/c_4 = 0.20/0.9400 = 0.2128$. The specification limits are LSL = 71.4 mm and USL = 72.8 mm. The value $\widehat{\mu}$ is closer to the LSL than to the USL. Therefore

$$C_{pk} = \frac{\widehat{\mu} - \text{LSL}}{3\widehat{\sigma}} = \frac{71.8 - 71.4}{(3)(0.2128)}$$
$$= 0.6266$$

Since $C_{pk} < 1$, the process capability is not acceptable.

Example 10.12

Refer to Example 10.11. Assume that it is possible to adjust the process mean to any desired value. To what value should it be set to maximize the value of C_{pk}? What will the value of C_{pk} be?

Solution

The specification limits are LSL $= 71.4$ and USL $= 72.8$. The value of C_{pk} will be maximized if the process mean is adjusted to the midpoint between the specification limits; that is, if $\mu = 72.1$. The process standard deviation is estimated with $\widehat{\sigma} = 0.2128$. Therefore the maximum value of C_{pk} is $(72.1 - 71.4)/(3)(0.2128) = 1.0965$. The process capability would be acceptable.

The capability that can potentially be achieved by shifting the process mean to the midpoint between the upper and lower specification limits is called the **process capability index**, denoted C_p. If the process mean is at the midpoint between LSL and USL, then the distance from the mean to either specification limit is equal to one-half the distance between the specification limits, that is $\mu - \text{LSL} = \text{USL} - \mu = (\text{USL} - \text{LSL})/2$ (see Figure 10.18). It follows that

$$C_p = \frac{\text{USL} - \text{LSL}}{6\widehat{\sigma}} \tag{10.5}$$

The process capability index C_p measures the potential capability of the process, that is the greatest capability that the process can achieve without reducing the process standard deviation.

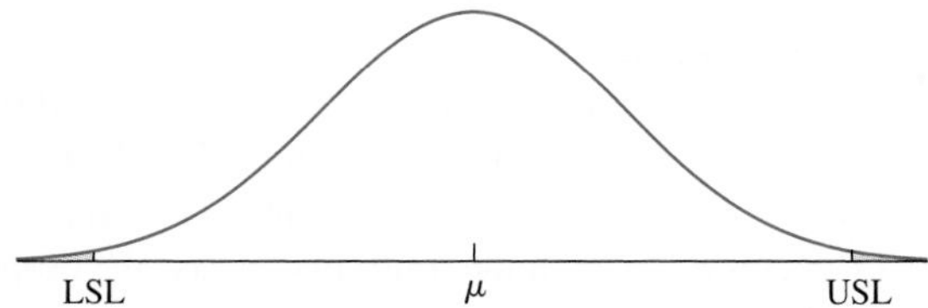

FIGURE 10.18 A process has maximum capability when the process mean is at the midpoint between the specification limits. In this case $\mu - \text{LSL} = \text{USL} - \mu = (\text{USL} - \text{LSL})/2$.

Example 10.13

Specifications for the output voltage of a certain electric circuit are 48 to 52 V. The process is in control with $\widehat{\sigma} = 0.482$ V. Compute the process capability index C_p.

Solution

$$C_p = \frac{\text{USL} - \text{LSL}}{6\widehat{\sigma}} = \frac{52 - 48}{(6)(0.482)}$$
$$= 1.38$$

The process capability is potentially good.

Estimating the Proportion of Nonconforming Units from Process Capability

Many people use the value of C_p to try to estimate the proportion of units that will be nonconforming. For example, if $C_p = 1$, then the specification limits are equal to $\widehat{\mu} - 3\widehat{\sigma}$ and $\widehat{\mu} + 3\widehat{\sigma}$, respectively. Therefore a unit will be nonconforming only if it is more than three standard deviations from the process mean. Now for a normal population, the proportion of items that are more than three standard deviations from the mean is equal to 0.0027. Therefore it is often stated that a process with $C_p = 1$ will produce 27 nonconforming parts per 10,000.

The problem with this is that the normality assumption is only approximate for real processes. The approximation may be very good near the middle of the curve, but it is often not good in the tails. Therefore the true proportion of nonconforming parts may be quite different from that predicted from the normal curve, especially when the proportion is very small. In general, estimates of small probabilities that are based on a normal approximation are extremely crude at best.

Six-Sigma Quality

The phrase "six-sigma quality" has become quite prevalent in discussions of quality control during the last few years. A process is said to have six-sigma quality if the process capability index C_p has a value of 2.0 or greater. Equivalently, a process has six-sigma quality if the difference USL − LSL is at least 12σ. When a process has six-sigma quality, then if the process mean is optimally adjusted, it is six standard deviations from each specification limit. In this case the proportion of nonconforming units will be virtually zero.

An important feature of a six-sigma process is that it can withstand moderate shifts in process mean without significant deterioration in capability. For example, even if the process mean shifts by 3σ in one direction or the other, it is still 3σ from the nearest specification limit, so the capability index will still be acceptable.

Refer to Example 10.13. To what value must the process standard deviation be reduced in order for the process to attain six-sigma quality?

Solution

To attain six-sigma quality, the value of C_p must be at least 2.0. The value of σ for which this occurs is found by setting $C_p = 2.0$ and solving for σ. We obtain

$$2.0 = \frac{52 - 48}{6\sigma}$$

from which $\sigma = 0.33$.

One-Sided Tolerances

Some characteristics have only one specification limit. For example, strengths usually have a lower specification limit but no upper limit, since for most applications a part cannot be too strong. The analog of C_{pk} when there is only a lower specification limit is the **lower capability index** C_{pl}; when there is only an upper limit, it is the **upper capability index** C_{pu}. Each of these quantities is defined to be the difference between the estimated process mean $\widehat{\mu}$ and the specification limit, divided by $3\widehat{\sigma}$.

Summary

If a process has only a lower specification limit (LSL), then the lower capability index is

$$C_{pl} = \frac{\widehat{\mu} - \text{LSL}}{3\widehat{\sigma}}$$

If a process has only an upper specification limit (USL), then the upper capability index is

$$C_{pu} = \frac{\text{USL} - \widehat{\mu}}{3\widehat{\sigma}}$$

There is no analog for C_p for processes with only one specification limit.

Exercises for Section 10.5

1. The design specifications for the intake valve in an internal combustion engine call for the valve clearance to be between 0.18 and 0.22 mm. Data from an $\overline{X}$ chart, based on samples of size 4, that shows that the process is in control, yield values of $\overline{\overline{X}} = 0.205$ mm and $\overline{s} = 0.002$ mm.
   a. Compute the value of C_{pk} for this process.
   b. Is the process capability acceptable? Explain.
2. The specifications for the fill volume of beverage cans is 11.95 to 12.10 oz. Data from an $\overline{X}$ chart, based on samples of size 5, that shows that the process is in control, yield values of $\overline{\overline{X}} = 12.01$ oz and $\overline{R} = 0.124$ oz.
   a. Compute the value of C_{pk} for this process.
   b. Is the process capability acceptable? Explain.
3. Refer to Exercise 1.
   a. To what value should the process mean be set to maximize the process capability?
   b. What will the process capability then be?

4. Refer to Exercise 2.

a. To what value should the process mean be set to maximize the process capability?

b. Is it possible to make the process capability acceptable simply by adjusting the process mean? Explain.

c. When the process mean is adjusted to its optimum value, what value must be attained by the process standard deviation so that the process capability is acceptable?

d. When the process mean is adjusted to its optimum value, what value must be attained by the process standard deviation so that the process has six-sigma quality?

5. A process has a process capability index of $C_p = 1.2$.

a. Assume the process mean is set to its optimal value. Express the upper and lower specification limits in terms of the process mean and standard deviation.

b. Using the normal curve, estimate the proportion of units that will be nonconforming.

c. Is it likely or unlikely that the true proportion of nonconforming units will be quite different from the estimate in part (b)? Explain.

Supplementary Exercises for Chapter 10

1. A process is monitored for defective items by taking a sample of 250 items each day and calculating the proportion that are defective. Let p_i be the proportion of defective items in the ith sample. For the last 50 samples, the sum of the proportions is $\sum_{i=1}^{50} p_i = 2.98$. Calculate the center line and the 3σ upper and lower control limits for a p chart.

2. Someone constructs an $\overline{X}$ chart where the control limits are at $\pm 2.5\sigma_{\overline{X}}$ rather than at $\pm 3\sigma_{\overline{X}}$.

a. If the process is in control, what is the ARL for this chart?

b. If the process mean shifts by $1\sigma_{\overline{X}}$, what is the ARL for this chart?

c. In units of $\sigma_{\overline{X}}$, how large an upward shift can be detected with an ARL of 10?

3. Samples of three resistors are taken periodically, and the resistances, in ohms, are measured. The following table presents the means, ranges, and standard deviations for 30 consecutive samples.

Sample	$\overline{X}$	R	s
1	5.114	0.146	0.077
2	5.144	0.158	0.085
3	5.220	0.057	0.031
4	5.196	0.158	0.081
5	5.176	0.172	0.099
6	5.222	0.030	0.017
7	5.209	0.118	0.059
8	5.212	0.099	0.053
9	5.238	0.157	0.085
10	5.152	0.104	0.054
11	5.163	0.051	0.026
12	5.221	0.105	0.055
13	5.144	0.132	0.071
14	5.098	0.123	0.062
15	5.070	0.083	0.042
16	5.029	0.073	0.038
17	5.045	0.161	0.087
18	5.008	0.138	0.071
19	5.029	0.082	0.042
20	5.038	0.109	0.055
21	4.962	0.066	0.034
22	5.033	0.078	0.041
23	4.993	0.085	0.044
24	4.961	0.126	0.066
25	4.976	0.094	0.047
26	5.005	0.135	0.068
27	5.022	0.120	0.062
28	5.077	0.140	0.074
29	5.033	0.049	0.026
30	5.068	0.146	0.076

The means are $\overline{\overline{X}} = 5.095$, $\overline{R} = 0.110$, and $\overline{s} = 0.058$.

a. Compute the 3σ limits for the R chart. Is the variance out of control at any point? If so, delete the samples that are out of control and recompute $\overline{\overline{X}}$ and $\overline{R}$.

b. Compute the 3σ limits for the $\overline{X}$ chart. On the basis of the 3σ limits, is the process mean in control? If not, at what point is it first detected to be out of control?

c. On the basis of the Western Electric rules, is the process mean in control? If not, when is it first detected to be out of control?

4. Repeat Exercise 3, using the S chart in place of the R chart.

5. Refer to Exercise 3.

a. Delete any samples necessary to bring the process variation under control. (You did this already if you did Exercise 3.)

b. Use $\overline{R}$ to estimate $\sigma_{\overline{X}}$ ($\sigma_{\overline{X}}$ is the difference between $\overline{\overline{X}}$ and the 1σ control limit on an $\overline{X}$ chart).

c. Construct a CUSUM chart, using $\overline{\overline{X}}$ for the target mean μ, and the estimate of $\sigma_{\overline{X}}$ found in part (b) for the standard deviation. Use the values $k = 0.5$ and $h = 4$.

d. Is the process mean in control? If not, when is it first detected to be out of control?

e. Construct an $\overline{X}$ chart, and use the Western Electric rules to determine whether the process mean is in control. (You did this already if you did Exercise 3.) Do the Western Electric rules give the same results as the CUSUM chart? If not, how are they different?

6. A process is monitored for flaws by taking a sample of size 50 each hour and counting the total number of flaws in the sample items. The total number of flaws over the last 30 samples is 658.

a. Compute the center line and upper and lower 3σ control limits.

b. The tenth sample had three flaws. Was the process out of control at that time? Explain.

7. To set up a p chart to monitor a process that produces computer chips, samples of 500 chips are taken daily, and the number of defective chips in each sample is counted. The numbers of defective chips for each of the last 25 days are as follows:

25 22 14 24 18 16 20 27 19 20 22 7 24 26
11 14 18 29 21 32 29 34 34 30 24

a. Compute the upper and lower 3σ limits for a p chart.

b. At which sample is the process first detected to be out of control?

c. Suppose that the special cause that resulted in the out-of-control condition is determined. Should this cause be remedied? Explain.

Appendix 

Tables

- **Table A.1: Cumulative Binomial Distribution**
- **Table A.2: Cumulative Normal Distribution**
- **Table A.3: Upper Percentage Points for the Student's *t* Distribution**
- **Table A.4: Critical Points for the Wilcoxon Signed-Rank Test**
- **Table A.5: Critical Points for the Wilcoxon Rank-Sum Test**
- **Table A.6: Upper Percentage Points for the χ^2 Distribution**
- **Table A.7: Upper Percentage Points for the *F* Distribution**
- **Table A.8: Upper Percentage Points for the Studentized Range q_{ν_1,ν_2}**
- **Table A.9: Control Chart Constants**

TABLE A.1 Cumulative binomial distribution

$$F(x) = P(X \le x) = \sum_{k=0}^{x} \frac{n!}{k!\,(n-k)!} p^k (1-p)^{(n-k)}$$

		p												
n	**x**	**0.05**	**0.10**	**0.20**	**0.25**	**0.30**	**0.40**	**0.50**	**0.60**	**0.70**	**0.75**	**0.80**	**0.90**	**0.95**
2	**0**	0.902	0.810	0.640	0.562	0.490	0.360	0.250	0.160	0.090	0.062	0.040	0.010	0.003
	1	0.997	0.990	0.960	0.938	0.910	0.840	0.750	0.640	0.510	0.438	0.360	0.190	0.098
	2	1.000	1.000	1.000	1.000	1.000	1.000	1.000	1.000	1.000	1.000	1.000	1.000	1.000
3	**0**	0.857	0.729	0.512	0.422	0.343	0.216	0.125	0.064	0.027	0.016	0.008	0.001	0.000
	1	0.993	0.972	0.896	0.844	0.784	0.648	0.500	0.352	0.216	0.156	0.104	0.028	0.007
	2	1.000	0.999	0.992	0.984	0.973	0.936	0.875	0.784	0.657	0.578	0.488	0.271	0.143
	3	1.000	1.000	1.000	1.000	1.000	1.000	1.000	1.000	1.000	1.000	1.000	1.000	1.000
4	**0**	0.815	0.656	0.410	0.316	0.240	0.130	0.062	0.026	0.008	0.004	0.002	0.000	0.000
	1	0.986	0.948	0.819	0.738	0.652	0.475	0.313	0.179	0.084	0.051	0.027	0.004	0.000
	2	1.000	0.996	0.973	0.949	0.916	0.821	0.688	0.525	0.348	0.262	0.181	0.052	0.014
	3	1.000	1.000	0.998	0.996	0.992	0.974	0.938	0.870	0.760	0.684	0.590	0.344	0.185
	4	1.000	1.000	1.000	1.000	1.000	1.000	1.000	1.000	1.000	1.000	1.000	1.000	1.000
5	**0**	0.774	0.590	0.328	0.237	0.168	0.078	0.031	0.010	0.002	0.001	0.000	0.000	0.000
	1	0.977	0.919	0.737	0.633	0.528	0.337	0.187	0.087	0.031	0.016	0.007	0.000	0.000
	2	0.999	0.991	0.942	0.896	0.837	0.683	0.500	0.317	0.163	0.104	0.058	0.009	0.001
	3	1.000	1.000	0.993	0.984	0.969	0.913	0.812	0.663	0.472	0.367	0.263	0.081	0.023
	4	1.000	1.000	1.000	0.999	0.998	0.990	0.969	0.922	0.832	0.763	0.672	0.410	0.226
	5	1.000	1.000	1.000	1.000	1.000	1.000	1.000	1.000	1.000	1.000	1.000	1.000	1.000
6	**0**	0.735	0.531	0.262	0.178	0.118	0.047	0.016	0.004	0.001	0.000	0.000	0.000	0.000
	1	0.967	0.886	0.655	0.534	0.420	0.233	0.109	0.041	0.011	0.005	0.002	0.000	0.000
	2	0.998	0.984	0.901	0.831	0.744	0.544	0.344	0.179	0.070	0.038	0.017	0.001	0.000
	3	1.000	0.999	0.983	0.962	0.930	0.821	0.656	0.456	0.256	0.169	0.099	0.016	0.002
	4	1.000	1.000	0.998	0.995	0.989	0.959	0.891	0.767	0.580	0.466	0.345	0.114	0.033
	5	1.000	1.000	1.000	1.000	0.999	0.996	0.984	0.953	0.882	0.822	0.738	0.469	0.265
	6	1.000	1.000	1.000	1.000	1.000	1.000	1.000	1.000	1.000	1.000	1.000	1.000	1.000
7	**0**	0.698	0.478	0.210	0.133	0.082	0.028	0.008	0.002	0.000	0.000	0.000	0.000	0.000
	1	0.956	0.850	0.577	0.445	0.329	0.159	0.063	0.019	0.004	0.001	0.000	0.000	0.000
	2	0.996	0.974	0.852	0.756	0.647	0.420	0.227	0.096	0.029	0.013	0.005	0.000	0.000
	3	1.000	0.997	0.967	0.929	0.874	0.710	0.500	0.290	0.126	0.071	0.033	0.003	0.000
	4	1.000	1.000	0.995	0.987	0.971	0.904	0.773	0.580	0.353	0.244	0.148	0.026	0.004
	5	1.000	1.000	1.000	0.999	0.996	0.981	0.938	0.841	0.671	0.555	0.423	0.150	0.044
	6	1.000	1.000	1.000	1.000	1.000	0.998	0.992	0.972	0.918	0.867	0.790	0.522	0.302
	7	1.000	1.000	1.000	1.000	1.000	1.000	1.000	1.000	1.000	1.000	1.000	1.000	1.000

Continued on page 765

TABLE A.1 Cumulative binomial distribution (continued)

		p												
n	*x*	0.05	0.10	0.20	0.25	0.30	0.40	0.50	0.60	0.70	0.75	0.80	0.90	0.95
8	0	0.663	0.430	0.168	0.100	0.058	0.017	0.004	0.001	0.000	0.000	0.000	0.000	0.000
	1	0.943	0.813	0.503	0.367	0.255	0.106	0.035	0.009	0.001	0.000	0.000	0.000	0.000
	2	0.994	0.962	0.797	0.679	0.552	0.315	0.145	0.050	0.011	0.004	0.001	0.000	0.000
	3	1.000	0.995	0.944	0.886	0.806	0.594	0.363	0.174	0.058	0.027	0.010	0.000	0.000
	4	1.000	1.000	0.990	0.973	0.942	0.826	0.637	0.406	0.194	0.114	0.056	0.005	0.000
	5	1.000	1.000	0.999	0.996	0.989	0.950	0.855	0.685	0.448	0.321	0.203	0.038	0.006
	6	1.000	1.000	1.000	1.000	0.999	0.991	0.965	0.894	0.745	0.633	0.497	0.187	0.057
	7	1.000	1.000	1.000	1.000	1.000	0.999	0.996	0.983	0.942	0.900	0.832	0.570	0.337
	8	1.000	1.000	1.000	1.000	1.000	1.000	1.000	1.000	1.000	1.000	1.000	1.000	1.000
9	0	0.630	0.387	0.134	0.075	0.040	0.010	0.002	0.000	0.000	0.000	0.000	0.000	0.000
	1	0.929	0.775	0.436	0.300	0.196	0.071	0.020	0.004	0.000	0.000	0.000	0.000	0.000
	2	0.992	0.947	0.738	0.601	0.463	0.232	0.090	0.025	0.004	0.001	0.000	0.000	0.000
	3	0.999	0.992	0.914	0.834	0.730	0.483	0.254	0.099	0.025	0.010	0.003	0.000	0.000
	4	1.000	0.999	0.980	0.951	0.901	0.733	0.500	0.267	0.099	0.049	0.020	0.001	0.000
	5	1.000	1.000	0.997	0.990	0.975	0.901	0.746	0.517	0.270	0.166	0.086	0.008	0.001
	6	1.000	1.000	1.000	0.999	0.996	0.975	0.910	0.768	0.537	0.399	0.262	0.053	0.008
	7	1.000	1.000	1.000	1.000	1.000	0.996	0.980	0.929	0.804	0.700	0.564	0.225	0.071
	8	1.000	1.000	1.000	1.000	1.000	1.000	0.998	0.990	0.960	0.925	0.866	0.613	0.370
	9	1.000	1.000	1.000	1.000	1.000	1.000	1.000	1.000	1.000	1.000	1.000	1.000	1.000
10	0	0.599	0.349	0.107	0.056	0.028	0.006	0.001	0.000	0.000	0.000	0.000	0.000	0.000
	1	0.914	0.736	0.376	0.244	0.149	0.046	0.011	0.002	0.000	0.000	0.000	0.000	0.000
	2	0.988	0.930	0.678	0.526	0.383	0.167	0.055	0.012	0.002	0.000	0.000	0.000	0.000
	3	0.999	0.987	0.879	0.776	0.650	0.382	0.172	0.055	0.011	0.004	0.001	0.000	0.000
	4	1.000	0.998	0.967	0.922	0.850	0.633	0.377	0.166	0.047	0.020	0.006	0.000	0.000
	5	1.000	1.000	0.994	0.980	0.953	0.834	0.623	0.367	0.150	0.078	0.033	0.002	0.000
	6	1.000	1.000	0.999	0.996	0.989	0.945	0.828	0.618	0.350	0.224	0.121	0.013	0.001
	7	1.000	1.000	1.000	1.000	0.998	0.988	0.945	0.833	0.617	0.474	0.322	0.070	0.012
	8	1.000	1.000	1.000	1.000	1.000	0.998	0.989	0.954	0.851	0.756	0.624	0.264	0.086
	9	1.000	1.000	1.000	1.000	1.000	1.000	0.999	0.994	0.972	0.944	0.893	0.651	0.401
	10	1.000	1.000	1.000	1.000	1.000	1.000	1.000	1.000	1.000	1.000	1.000	1.000	1.000
11	0	0.569	0.314	0.086	0.042	0.020	0.004	0.000	0.000	0.000	0.000	0.000	0.000	0.000
	1	0.898	0.697	0.322	0.197	0.113	0.030	0.006	0.001	0.000	0.000	0.000	0.000	0.000
	2	0.985	0.910	0.617	0.455	0.313	0.119	0.033	0.006	0.001	0.000	0.000	0.000	0.000
	3	0.998	0.981	0.839	0.713	0.570	0.296	0.113	0.029	0.004	0.001	0.000	0.000	0.000
	4	1.000	0.997	0.950	0.885	0.790	0.533	0.274	0.099	0.022	0.008	0.002	0.000	0.000
	5	1.000	1.000	0.988	0.966	0.922	0.753	0.500	0.247	0.078	0.034	0.012	0.000	0.000
	6	1.000	1.000	0.998	0.992	0.978	0.901	0.726	0.467	0.210	0.115	0.050	0.003	0.000
	7	1.000	1.000	1.000	0.999	0.996	0.971	0.887	0.704	0.430	0.287	0.161	0.019	0.002
	8	1.000	1.000	1.000	1.000	0.999	0.994	0.967	0.881	0.687	0.545	0.383	0.090	0.015

Continued on page 766

TABLE A.1 Cumulative binomial distribution (continued)

		p												
n	*x*	0.05	0.10	0.20	0.25	0.30	0.40	0.50	0.60	0.70	0.75	0.80	0.90	0.95
11	**9**	1.000	1.000	1.000	1.000	1.000	0.999	0.994	0.970	0.887	0.803	0.678	0.303	0.102
	10	1.000	1.000	1.000	1.000	1.000	1.000	1.000	0.996	0.980	0.958	0.914	0.686	0.431
	11	1.000	1.000	1.000	1.000	1.000	1.000	1.000	1.000	1.000	1.000	1.000	1.000	1.000
12	**0**	0.540	0.282	0.069	0.032	0.014	0.002	0.000	0.000	0.000	0.000	0.000	0.000	0.000
	1	0.882	0.659	0.275	0.158	0.085	0.020	0.003	0.000	0.000	0.000	0.000	0.000	0.000
	2	0.980	0.889	0.558	0.391	0.253	0.083	0.019	0.003	0.000	0.000	0.000	0.000	0.000
	3	0.998	0.974	0.795	0.649	0.493	0.225	0.073	0.015	0.002	0.000	0.000	0.000	0.000
	4	1.000	0.996	0.927	0.842	0.724	0.438	0.194	0.057	0.009	0.003	0.001	0.000	0.000
	5	1.000	0.999	0.981	0.946	0.882	0.665	0.387	0.158	0.039	0.014	0.004	0.000	0.000
	6	1.000	1.000	0.996	0.986	0.961	0.842	0.613	0.335	0.118	0.054	0.019	0.001	0.000
	7	1.000	1.000	0.999	0.997	0.991	0.943	0.806	0.562	0.276	0.158	0.073	0.004	0.000
	8	1.000	1.000	1.000	1.000	0.998	0.985	0.927	0.775	0.507	0.351	0.205	0.026	0.002
	9	1.000	1.000	1.000	1.000	1.000	0.997	0.981	0.917	0.747	0.609	0.442	0.111	0.020
	10	1.000	1.000	1.000	1.000	1.000	1.000	0.997	0.980	0.915	0.842	0.725	0.341	0.118
	11	1.000	1.000	1.000	1.000	1.000	1.000	1.000	0.998	0.986	0.968	0.931	0.718	0.460
	12	1.000	1.000	1.000	1.000	1.000	1.000	1.000	1.000	1.000	1.000	1.000	1.000	1.000
13	**0**	0.513	0.254	0.055	0.024	0.010	0.001	0.000	0.000	0.000	0.000	0.000	0.000	0.000
	1	0.865	0.621	0.234	0.127	0.064	0.013	0.002	0.000	0.000	0.000	0.000	0.000	0.000
	2	0.975	0.866	0.502	0.333	0.202	0.058	0.011	0.001	0.000	0.000	0.000	0.000	0.000
	3	0.997	0.966	0.747	0.584	0.421	0.169	0.046	0.008	0.001	0.000	0.000	0.000	0.000
	4	1.000	0.994	0.901	0.794	0.654	0.353	0.133	0.032	0.004	0.001	0.000	0.000	0.000
	5	1.000	0.999	0.970	0.920	0.835	0.574	0.291	0.098	0.018	0.006	0.001	0.000	0.000
	6	1.000	1.000	0.993	0.976	0.938	0.771	0.500	0.229	0.062	0.024	0.007	0.000	0.000
	7	1.000	1.000	0.999	0.994	0.982	0.902	0.709	0.426	0.165	0.080	0.030	0.001	0.000
	8	1.000	1.000	1.000	0.999	0.996	0.968	0.867	0.647	0.346	0.206	0.099	0.006	0.000
	9	1.000	1.000	1.000	1.000	0.999	0.992	0.954	0.831	0.579	0.416	0.253	0.034	0.003
	10	1.000	1.000	1.000	1.000	1.000	0.999	0.989	0.942	0.798	0.667	0.498	0.134	0.025
	11	1.000	1.000	1.000	1.000	1.000	1.000	0.998	0.987	0.936	0.873	0.766	0.379	0.135
	12	1.000	1.000	1.000	1.000	1.000	1.000	1.000	0.999	0.990	0.976	0.945	0.746	0.487
	13	1.000	1.000	1.000	1.000	1.000	1.000	1.000	1.000	1.000	1.000	1.000	1.000	1.000
14	**0**	0.488	0.229	0.044	0.018	0.007	0.001	0.000	0.000	0.000	0.000	0.000	0.000	0.000
	1	0.847	0.585	0.198	0.101	0.047	0.008	0.001	0.000	0.000	0.000	0.000	0.000	0.000
	2	0.970	0.842	0.448	0.281	0.161	0.040	0.006	0.001	0.000	0.000	0.000	0.000	0.000
	3	0.996	0.956	0.698	0.521	0.355	0.124	0.029	0.004	0.000	0.000	0.000	0.000	0.000
	4	1.000	0.991	0.870	0.742	0.584	0.279	0.090	0.018	0.002	0.000	0.000	0.000	0.000
	5	1.000	0.999	0.956	0.888	0.781	0.486	0.212	0.058	0.008	0.002	0.000	0.000	0.000
	6	1.000	1.000	0.988	0.962	0.907	0.692	0.395	0.150	0.031	0.010	0.002	0.000	0.000
	7	1.000	1.000	0.998	0.990	0.969	0.850	0.605	0.308	0.093	0.038	0.012	0.000	0.000

Continued on page 767

TABLE A.1 Cumulative binomial distribution (continued)

		p												
n	*x*	0.05	0.10	0.20	0.25	0.30	0.40	0.50	0.60	0.70	0.75	0.80	0.90	0.95
14	8	1.000	1.000	1.000	0.998	0.992	0.942	0.788	0.514	0.219	0.112	0.044	0.001	0.000
	9	1.000	1.000	1.000	1.000	0.998	0.982	0.910	0.721	0.416	0.258	0.130	0.009	0.000
	10	1.000	1.000	1.000	1.000	1.000	0.996	0.971	0.876	0.645	0.479	0.302	0.044	0.004
	11	1.000	1.000	1.000	1.000	1.000	0.999	0.994	0.960	0.839	0.719	0.552	0.158	0.030
	12	1.000	1.000	1.000	1.000	1.000	1.000	0.999	0.992	0.953	0.899	0.802	0.415	0.153
	13	1.000	1.000	1.000	1.000	1.000	1.000	1.000	0.999	0.993	0.982	0.956	0.771	0.512
	14	1.000	1.000	1.000	1.000	1.000	1.000	1.000	1.000	1.000	1.000	1.000	1.000	1.000
15	0	0.463	0.206	0.035	0.013	0.005	0.000	0.000	0.000	0.000	0.000	0.000	0.000	0.000
	1	0.829	0.549	0.167	0.080	0.035	0.005	0.000	0.000	0.000	0.000	0.000	0.000	0.000
	2	0.964	0.816	0.398	0.236	0.127	0.027	0.004	0.000	0.000	0.000	0.000	0.000	0.000
	3	0.995	0.944	0.648	0.461	0.297	0.091	0.018	0.002	0.000	0.000	0.000	0.000	0.000
	4	0.999	0.987	0.836	0.686	0.515	0.217	0.059	0.009	0.001	0.000	0.000	0.000	0.000
	5	1.000	0.998	0.939	0.852	0.722	0.403	0.151	0.034	0.004	0.001	0.000	0.000	0.000
	6	1.000	1.000	0.982	0.943	0.869	0.610	0.304	0.095	0.015	0.004	0.001	0.000	0.000
	7	1.000	1.000	0.996	0.983	0.950	0.787	0.500	0.213	0.050	0.017	0.004	0.000	0.000
	8	1.000	1.000	0.999	0.996	0.985	0.905	0.696	0.390	0.131	0.057	0.018	0.000	0.000
	9	1.000	1.000	1.000	0.999	0.996	0.966	0.849	0.597	0.278	0.148	0.061	0.002	0.000
	10	1.000	1.000	1.000	1.000	0.999	0.991	0.941	0.783	0.485	0.314	0.164	0.013	0.001
	11	1.000	1.000	1.000	1.000	1.000	0.998	0.982	0.909	0.703	0.539	0.352	0.056	0.005
	12	1.000	1.000	1.000	1.000	1.000	1.000	0.996	0.973	0.873	0.764	0.602	0.184	0.036
	13	1.000	1.000	1.000	1.000	1.000	1.000	1.000	0.995	0.965	0.920	0.833	0.451	0.171
	14	1.000	1.000	1.000	1.000	1.000	1.000	1.000	1.000	0.995	0.987	0.965	0.794	0.537
	15	1.000	1.000	1.000	1.000	1.000	1.000	1.000	1.000	1.000	1.000	1.000	1.000	1.000
16	0	0.440	0.185	0.028	0.010	0.003	0.000	0.000	0.000	0.000	0.000	0.000	0.000	0.000
	1	0.811	0.515	0.141	0.063	0.026	0.003	0.000	0.000	0.000	0.000	0.000	0.000	0.000
	2	0.957	0.789	0.352	0.197	0.099	0.018	0.002	0.000	0.000	0.000	0.000	0.000	0.000
	3	0.993	0.932	0.598	0.405	0.246	0.065	0.011	0.001	0.000	0.000	0.000	0.000	0.000
	4	0.999	0.983	0.798	0.630	0.450	0.167	0.038	0.005	0.000	0.000	0.000	0.000	0.000
	5	1.000	0.997	0.918	0.810	0.660	0.329	0.105	0.019	0.002	0.000	0.000	0.000	0.000
	6	1.000	0.999	0.973	0.920	0.825	0.527	0.227	0.058	0.007	0.002	0.000	0.000	0.000
	7	1.000	1.000	0.993	0.973	0.926	0.716	0.402	0.142	0.026	0.007	0.001	0.000	0.000
	8	1.000	1.000	0.999	0.993	0.974	0.858	0.598	0.284	0.074	0.027	0.007	0.000	0.000
	9	1.000	1.000	1.000	0.998	0.993	0.942	0.773	0.473	0.175	0.080	0.027	0.001	0.000
	10	1.000	1.000	1.000	1.000	0.998	0.981	0.895	0.671	0.340	0.190	0.082	0.003	0.000
	11	1.000	1.000	1.000	1.000	1.000	0.995	0.962	0.833	0.550	0.370	0.202	0.017	0.001
	12	1.000	1.000	1.000	1.000	1.000	0.999	0.989	0.935	0.754	0.595	0.402	0.068	0.007
	13	1.000	1.000	1.000	1.000	1.000	1.000	0.998	0.982	0.901	0.803	0.648	0.211	0.043
	14	1.000	1.000	1.000	1.000	1.000	1.000	1.000	0.997	0.974	0.937	0.859	0.485	0.189
	15	1.000	1.000	1.000	1.000	1.000	1.000	1.000	1.000	0.997	0.990	0.972	0.815	0.560
	16	1.000	1.000	1.000	1.000	1.000	1.000	1.000	1.000	1.000	1.000	1.000	1.000	1.000

Continued on page 768

TABLE A.1 Cumulative binomial distribution (continued)

		p												
n	*x*	**0.05**	**0.10**	**0.20**	**0.25**	**0.30**	**0.40**	**0.50**	**0.60**	**0.70**	**0.75**	**0.80**	**0.90**	**0.95**
17	**0**	0.418	0.167	0.023	0.008	0.002	0.000	0.000	0.000	0.000	0.000	0.000	0.000	0.000
	1	0.792	0.482	0.118	0.050	0.019	0.002	0.000	0.000	0.000	0.000	0.000	0.000	0.000
	2	0.950	0.762	0.310	0.164	0.077	0.012	0.001	0.000	0.000	0.000	0.000	0.000	0.000
	3	0.991	0.917	0.549	0.353	0.202	0.046	0.006	0.000	0.000	0.000	0.000	0.000	0.000
	4	0.999	0.978	0.758	0.574	0.389	0.126	0.025	0.003	0.000	0.000	0.000	0.000	0.000
	5	1.000	0.995	0.894	0.765	0.597	0.264	0.072	0.011	0.001	0.000	0.000	0.000	0.000
	6	1.000	0.999	0.962	0.893	0.775	0.448	0.166	0.035	0.003	0.001	0.000	0.000	0.000
	7	1.000	1.000	0.989	0.960	0.895	0.641	0.315	0.092	0.013	0.003	0.000	0.000	0.000
	8	1.000	1.000	0.997	0.988	0.960	0.801	0.500	0.199	0.040	0.012	0.003	0.000	0.000
	9	1.000	1.000	1.000	0.997	0.987	0.908	0.685	0.359	0.105	0.040	0.011	0.000	0.000
	10	1.000	1.000	1.000	0.999	0.997	0.965	0.834	0.552	0.225	0.107	0.038	0.001	0.000
	11	1.000	1.000	1.000	1.000	0.999	0.989	0.928	0.736	0.403	0.235	0.106	0.005	0.000
	12	1.000	1.000	1.000	1.000	1.000	0.997	0.975	0.874	0.611	0.426	0.242	0.022	0.001
	13	1.000	1.000	1.000	1.000	1.000	1.000	0.994	0.954	0.798	0.647	0.451	0.083	0.009
	14	1.000	1.000	1.000	1.000	1.000	1.000	0.999	0.988	0.923	0.836	0.690	0.238	0.050
	15	1.000	1.000	1.000	1.000	1.000	1.000	1.000	0.998	0.981	0.950	0.882	0.518	0.208
	16	1.000	1.000	1.000	1.000	1.000	1.000	1.000	1.000	0.998	0.992	0.977	0.833	0.582
	17	1.000	1.000	1.000	1.000	1.000	1.000	1.000	1.000	1.000	1.000	1.000	1.000	1.000
18	**0**	0.397	0.150	0.018	0.006	0.002	0.000	0.000	0.000	0.000	0.000	0.000	0.000	0.000
	1	0.774	0.450	0.099	0.039	0.014	0.001	0.000	0.000	0.000	0.000	0.000	0.000	0.000
	2	0.942	0.734	0.271	0.135	0.060	0.008	0.001	0.000	0.000	0.000	0.000	0.000	0.000
	3	0.989	0.902	0.501	0.306	0.165	0.033	0.004	0.000	0.000	0.000	0.000	0.000	0.000
	4	0.998	0.972	0.716	0.519	0.333	0.094	0.015	0.001	0.000	0.000	0.000	0.000	0.000
	5	1.000	0.994	0.867	0.717	0.534	0.209	0.048	0.006	0.000	0.000	0.000	0.000	0.000
	6	1.000	0.999	0.949	0.861	0.722	0.374	0.119	0.020	0.001	0.000	0.000	0.000	0.000
	7	1.000	1.000	0.984	0.943	0.859	0.563	0.240	0.058	0.006	0.001	0.000	0.000	0.000
	8	1.000	1.000	0.996	0.981	0.940	0.737	0.407	0.135	0.021	0.005	0.001	0.000	0.000
	9	1.000	1.000	0.999	0.995	0.979	0.865	0.593	0.263	0.060	0.019	0.004	0.000	0.000
	10	1.000	1.000	1.000	0.999	0.994	0.942	0.760	0.437	0.141	0.057	0.016	0.000	0.000
	11	1.000	1.000	1.000	1.000	0.999	0.980	0.881	0.626	0.278	0.139	0.051	0.001	0.000
	12	1.000	1.000	1.000	1.000	1.000	0.994	0.952	0.791	0.466	0.283	0.133	0.006	0.000
	13	1.000	1.000	1.000	1.000	1.000	0.999	0.985	0.906	0.667	0.481	0.284	0.028	0.002
	14	1.000	1.000	1.000	1.000	1.000	1.000	0.996	0.967	0.835	0.694	0.499	0.098	0.011
	15	1.000	1.000	1.000	1.000	1.000	1.000	0.999	0.992	0.940	0.865	0.729	0.266	0.058
	16	1.000	1.000	1.000	1.000	1.000	1.000	1.000	0.999	0.986	0.961	0.901	0.550	0.226
	17	1.000	1.000	1.000	1.000	1.000	1.000	1.000	1.000	0.998	0.994	0.982	0.850	0.603
	18	1.000	1.000	1.000	1.000	1.000	1.000	1.000	1.000	1.000	1.000	1.000	1.000	1.000
19	**0**	0.377	0.135	0.014	0.004	0.001	0.000	0.000	0.000	0.000	0.000	0.000	0.000	0.000
	1	0.755	0.420	0.083	0.031	0.010	0.001	0.000	0.000	0.000	0.000	0.000	0.000	0.000

Continued on page 769

TABLE A.1 Cumulative binomial distribution (continued)

		p												
n	***x***	**0.05**	**0.10**	**0.20**	**0.25**	**0.30**	**0.40**	**0.50**	**0.60**	**0.70**	**0.75**	**0.80**	**0.90**	**0.95**
19	**2**	0.933	0.705	0.237	0.111	0.046	0.005	0.000	0.000	0.000	0.000	0.000	0.000	0.000
	3	0.987	0.885	0.455	0.263	0.133	0.023	0.002	0.000	0.000	0.000	0.000	0.000	0.000
	4	0.998	0.965	0.673	0.465	0.282	0.070	0.010	0.001	0.000	0.000	0.000	0.000	0.000
	5	1.000	0.991	0.837	0.668	0.474	0.163	0.032	0.003	0.000	0.000	0.000	0.000	0.000
	6	1.000	0.998	0.932	0.825	0.666	0.308	0.084	0.012	0.001	0.000	0.000	0.000	0.000
	7	1.000	1.000	0.977	0.923	0.818	0.488	0.180	0.035	0.003	0.000	0.000	0.000	0.000
	8	1.000	1.000	0.993	0.971	0.916	0.667	0.324	0.088	0.011	0.002	0.000	0.000	0.000
	9	1.000	1.000	0.998	0.991	0.967	0.814	0.500	0.186	0.033	0.009	0.002	0.000	0.000
	10	1.000	1.000	1.000	0.998	0.989	0.912	0.676	0.333	0.084	0.029	0.007	0.000	0.000
	11	1.000	1.000	1.000	1.000	0.997	0.965	0.820	0.512	0.182	0.077	0.023	0.000	0.000
	12	1.000	1.000	1.000	1.000	0.999	0.988	0.916	0.692	0.334	0.175	0.068	0.002	0.000
	13	1.000	1.000	1.000	1.000	1.000	0.997	0.968	0.837	0.526	0.332	0.163	0.009	0.000
	14	1.000	1.000	1.000	1.000	1.000	0.999	0.990	0.930	0.718	0.535	0.327	0.035	0.002
	15	1.000	1.000	1.000	1.000	1.000	1.000	0.998	0.977	0.867	0.737	0.545	0.115	0.013
	16	1.000	1.000	1.000	1.000	1.000	1.000	1.000	0.995	0.954	0.889	0.763	0.295	0.067
	17	1.000	1.000	1.000	1.000	1.000	1.000	1.000	0.999	0.990	0.969	0.917	0.580	0.245
	18	1.000	1.000	1.000	1.000	1.000	1.000	1.000	1.000	0.999	0.996	0.986	0.865	0.623
	19	1.000	1.000	1.000	1.000	1.000	1.000	1.000	1.000	1.000	1.000	1.000	1.000	1.000
20	**0**	0.358	0.122	0.012	0.003	0.001	0.000	0.000	0.000	0.000	0.000	0.000	0.000	0.000
	1	0.736	0.392	0.069	0.024	0.008	0.001	0.000	0.000	0.000	0.000	0.000	0.000	0.000
	2	0.925	0.677	0.206	0.091	0.035	0.004	0.000	0.000	0.000	0.000	0.000	0.000	0.000
	3	0.984	0.867	0.411	0.225	0.107	0.016	0.001	0.000	0.000	0.000	0.000	0.000	0.000
	4	0.997	0.957	0.630	0.415	0.238	0.051	0.006	0.000	0.000	0.000	0.000	0.000	0.000
	5	1.000	0.989	0.804	0.617	0.416	0.126	0.021	0.002	0.000	0.000	0.000	0.000	0.000
	6	1.000	0.998	0.913	0.786	0.608	0.250	0.058	0.006	0.000	0.000	0.000	0.000	0.000
	7	1.000	1.000	0.968	0.898	0.772	0.416	0.132	0.021	0.001	0.000	0.000	0.000	0.000
	8	1.000	1.000	0.990	0.959	0.887	0.596	0.252	0.057	0.005	0.001	0.000	0.000	0.000
	9	1.000	1.000	0.997	0.986	0.952	0.755	0.412	0.128	0.017	0.004	0.001	0.000	0.000
	10	1.000	1.000	0.999	0.996	0.983	0.872	0.588	0.245	0.048	0.014	0.003	0.000	0.000
	11	1.000	1.000	1.000	0.999	0.995	0.943	0.748	0.404	0.113	0.041	0.010	0.000	0.000
	12	1.000	1.000	1.000	1.000	0.999	0.979	0.868	0.584	0.228	0.102	0.032	0.000	0.000
	13	1.000	1.000	1.000	1.000	1.000	0.994	0.942	0.750	0.392	0.214	0.087	0.002	0.000
	14	1.000	1.000	1.000	1.000	1.000	0.998	0.979	0.874	0.584	0.383	0.196	0.011	0.000
	15	1.000	1.000	1.000	1.000	1.000	1.000	0.994	0.949	0.762	0.585	0.370	0.043	0.003
	16	1.000	1.000	1.000	1.000	1.000	1.000	0.999	0.984	0.893	0.775	0.589	0.133	0.016
	17	1.000	1.000	1.000	1.000	1.000	1.000	1.000	0.996	0.965	0.909	0.794	0.323	0.075
	18	1.000	1.000	1.000	1.000	1.000	1.000	1.000	0.999	0.992	0.976	0.931	0.608	0.264
	19	1.000	1.000	1.000	1.000	1.000	1.000	1.000	1.000	0.999	0.997	0.988	0.878	0.642
	20	1.000	1.000	1.000	1.000	1.000	1.000	1.000	1.000	1.000	1.000	1.000	1.000	1.000

TABLE A.2 Cumulative normal distribution (*z* table)

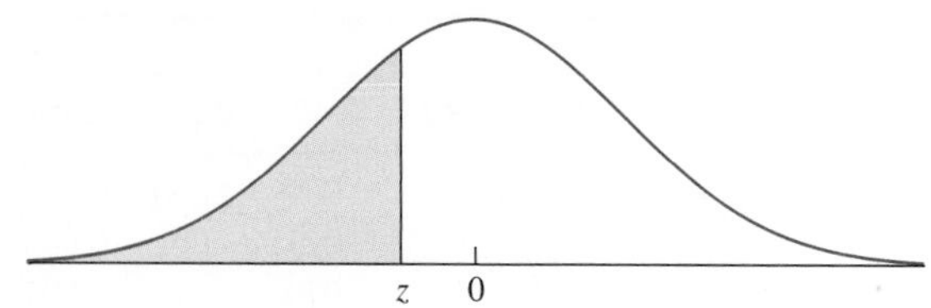

z	0.00	0.01	0.02	0.03	0.04	0.05	0.06	0.07	0.08	0.09
−3.6	.0002	.0002	.0001	.0001	.0001	.0001	.0001	.0001	.0001	.0001
−3.5	.0002	.0002	.0002	.0002	.0002	.0002	.0002	.0002	.0002	.0002
−3.4	.0003	.0003	.0003	.0003	.0003	.0003	.0003	.0003	.0003	.0002
−3.3	.0005	.0005	.0005	.0004	.0004	.0004	.0004	.0004	.0004	.0003
−3.2	.0007	.0007	.0006	.0006	.0006	.0006	.0006	.0005	.0005	.0005
−3.1	.0010	.0009	.0009	.0009	.0008	.0008	.0008	.0008	.0007	.0007
−3.0	.0013	.0013	.0013	.0012	.0012	.0011	.0011	.0011	.0010	.0010
−2.9	.0019	.0018	.0018	.0017	.0016	.0016	.0015	.0015	.0014	.0014
−2.8	.0026	.0025	.0024	.0023	.0023	.0022	.0021	.0021	.0020	.0019
−2.7	.0035	.0034	.0033	.0032	.0031	.0030	.0029	.0028	.0027	.0026
−2.6	.0047	.0045	.0044	.0043	.0041	.0040	.0039	.0038	.0037	.0036
−2.5	.0062	.0060	.0059	.0057	.0055	.0054	.0052	.0051	.0049	.0048
−2.4	.0082	.0080	.0078	.0075	.0073	.0071	.0069	.0068	.0066	.0064
−2.3	.0107	.0104	.0102	.0099	.0096	.0094	.0091	.0089	.0087	.0084
−2.2	.0139	.0136	.0132	.0129	.0125	.0122	.0119	.0116	.0113	.0110
−2.1	.0179	.0174	.0170	.0166	.0162	.0158	.0154	.0150	.0146	.0143
−2.0	.0228	.0222	.0217	.0212	.0207	.0202	.0197	.0192	.0188	.0183
−1.9	.0287	.0281	.0274	.0268	.0262	.0256	.0250	.0244	.0239	.0233
−1.8	.0359	.0351	.0344	.0336	.0329	.0322	.0314	.0307	.0301	.0294
−1.7	.0446	.0436	.0427	.0418	.0409	.0401	.0392	.0384	.0375	.0367
−1.6	.0548	.0537	.0526	.0516	.0505	.0495	.0485	.0475	.0465	.0455
−1.5	.0668	.0655	.0643	.0630	.0618	.0606	.0594	.0582	.0571	.0559
−1.4	.0808	.0793	.0778	.0764	.0749	.0735	.0721	.0708	.0694	.0681
−1.3	.0968	.0951	.0934	.0918	.0901	.0885	.0869	.0853	.0838	.0823
−1.2	.1151	.1131	.1112	.1093	.1075	.1056	.1038	.1020	.1003	.0985
−1.1	.1357	.1335	.1314	.1292	.1271	.1251	.1230	.1210	.1190	.1170
−1.0	.1587	.1562	.1539	.1515	.1492	.1469	.1446	.1423	.1401	.1379
−0.9	.1841	.1814	.1788	.1762	.1736	.1711	.1685	.1660	.1635	.1611
−0.8	.2119	.2090	.2061	.2033	.2005	.1977	.1949	.1922	.1894	.1867
−0.7	.2420	.2389	.2358	.2327	.2296	.2266	.2236	.2206	.2177	.2148
−0.6	.2743	.2709	.2676	.2643	.2611	.2578	.2546	.2514	.2483	.2451
−0.5	.3085	.3050	.3015	.2981	.2946	.2912	.2877	.2843	.2810	.2776
−0.4	.3446	.3409	.3372	.3336	.3300	.3264	.3228	.3192	.3156	.3121
−0.3	.3821	.3783	.3745	.3707	.3669	.3632	.3594	.3557	.3520	.3483
−0.2	.4207	.4168	.4129	.4090	.4052	.4013	.3974	.3936	.3897	.3859
−0.1	.4602	.4562	.4522	.4483	.4443	.4404	.4364	.4325	.4286	.4247
−0.0	.5000	.4960	.4920	.4880	.4840	.4801	.4761	.4721	.4681	.4641

Continued on page 771

TABLE A.2 Cumulative normal distribution (continued)

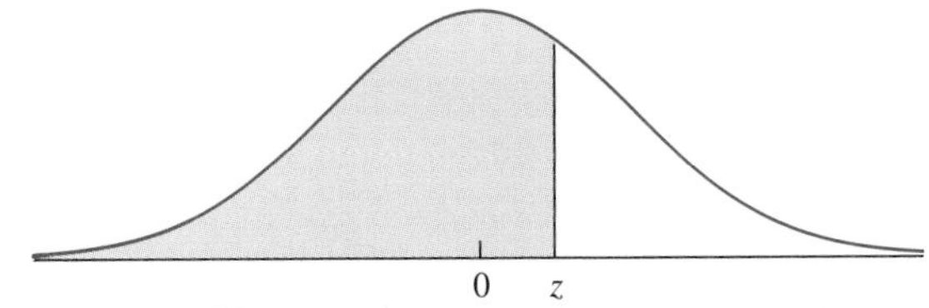

z	0.00	0.01	0.02	0.03	0.04	0.05	0.06	0.07	0.08	0.09
0.0	.5000	.5040	.5080	.5120	.5160	.5199	.5239	.5279	.5319	.5359
0.1	.5398	.5438	.5478	.5517	.5557	.5596	.5636	.5675	.5714	.5753
0.2	.5793	.5832	.5871	.5910	.5948	.5987	.6026	.6064	.6103	.6141
0.3	.6179	.6217	.6255	.6293	.6331	.6368	.6406	.6443	.6480	.6517
0.4	.6554	.6591	.6628	.6664	.6700	.6736	.6772	.6808	.6844	.6879
0.5	.6915	.6950	.6985	.7019	.7054	.7088	.7123	.7157	.7190	.7224
0.6	.7257	.7291	.7324	.7357	.7389	.7422	.7454	.7486	.7517	.7549
0.7	.7580	.7611	.7642	.7673	.7704	.7734	.7764	.7794	.7823	.7852
0.8	.7881	.7910	.7939	.7967	.7995	.8023	.8051	.8078	.8106	.8133
0.9	.8159	.8186	.8212	.8238	.8264	.8289	.8315	.8340	.8365	.8389
1.0	.8413	.8438	.8461	.8485	.8508	.8531	.8554	.8577	.8599	.8621
1.1	.8643	.8665	.8686	.8708	.8729	.8749	.8770	.8790	.8810	.8830
1.2	.8849	.8869	.8888	.8907	.8925	.8944	.8962	.8980	.8997	.9015
1.3	.9032	.9049	.9066	.9082	.9099	.9115	.9131	.9147	.9162	.9177
1.4	.9192	.9207	.9222	.9236	.9251	.9265	.9279	.9292	.9306	.9319
1.5	.9332	.9345	.9357	.9370	.9382	.9394	.9406	.9418	.9429	.9441
1.6	.9452	.9463	.9474	.9484	.9495	.9505	.9515	.9525	.9535	.9545
1.7	.9554	.9564	.9573	.9582	.9591	.9599	.9608	.9616	.9625	.9633
1.8	.9641	.9649	.9656	.9664	.9671	.9678	.9686	.9693	.9699	.9706
1.9	.9713	.9719	.9726	.9732	.9738	.9744	.9750	.9756	.9761	.9767
2.0	.9772	.9778	.9783	.9788	.9793	.9798	.9803	.9808	.9812	.9817
2.1	.9821	.9826	.9830	.9834	.9838	.9842	.9846	.9850	.9854	.9857
2.2	.9861	.9864	.9868	.9871	.9875	.9878	.9881	.9884	.9887	.9890
2.3	.9893	.9896	.9898	.9901	.9904	.9906	.9909	.9911	.9913	.9916
2.4	.9918	.9920	.9922	.9925	.9927	.9929	.9931	.9932	.9934	.9936
2.5	.9938	.9940	.9941	.9943	.9945	.9946	.9948	.9949	.9951	.9952
2.6	.9953	.9955	.9956	.9957	.9959	.9960	.9961	.9962	.9963	.9964
2.7	.9965	.9966	.9967	.9968	.9969	.9970	.9971	.9972	.9973	.9974
2.8	.9974	.9975	.9976	.9977	.9977	.9978	.9979	.9979	.9980	.9981
2.9	.9981	.9982	.9982	.9983	.9984	.9984	.9985	.9985	.9986	.9986
3.0	.9987	.9987	.9987	.9988	.9988	.9989	.9989	.9989	.9990	.9990
3.1	.9990	.9991	.9991	.9991	.9992	.9992	.9992	.9992	.9993	.9993
3.2	.9993	.9993	.9994	.9994	.9994	.9994	.9994	.9995	.9995	.9995
3.3	.9995	.9995	.9995	.9996	.9996	.9996	.9996	.9996	.9996	.9997
3.4	.9997	.9997	.9997	.9997	.9997	.9997	.9997	.9997	.9997	.9998
3.5	.9998	.9998	.9998	.9998	.9998	.9998	.9998	.9998	.9998	.9998
3.6	.9998	.9998	.9999	.9999	.9999	.9999	.9999	.9999	.9999	.9999

TABLE A.3 Upper percentage points for the Student's *t* distribution

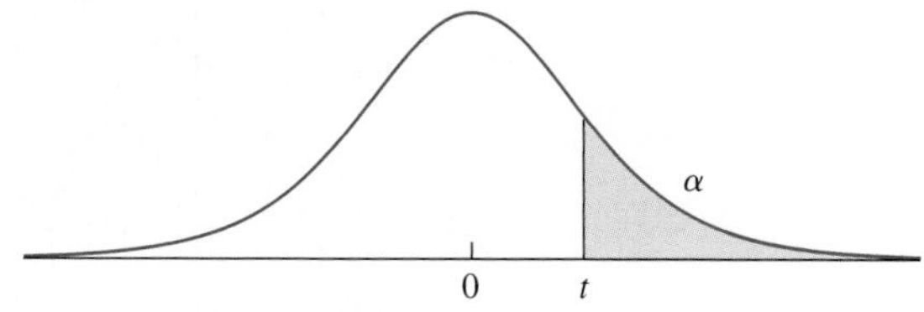

	α								
ν	**0.40**	**0.25**	**0.10**	**0.05**	**0.025**	**0.01**	**0.005**	**0.001**	**0.0005**
1	0.325	1.000	3.078	6.314	12.706	31.821	63.657	318.309	636.619
2	0.289	0.816	1.886	2.920	4.303	6.965	9.925	22.327	31.599
3	0.277	0.765	1.638	2.353	3.182	4.541	5.841	10.215	12.924
4	0.271	0.741	1.533	2.132	2.776	3.747	4.604	7.173	8.610
5	0.267	0.727	1.476	2.015	2.571	3.365	4.032	5.893	6.869
6	0.265	0.718	1.440	1.943	2.447	3.143	3.707	5.208	5.959
7	0.263	0.711	1.415	1.895	2.365	2.998	3.499	4.785	5.408
8	0.262	0.706	1.397	1.860	2.306	2.896	3.355	4.501	5.041
9	0.261	0.703	1.383	1.833	2.262	2.821	3.250	4.297	4.781
10	0.260	0.700	1.372	1.812	2.228	2.764	3.169	4.144	4.587
11	0.260	0.697	1.363	1.796	2.201	2.718	3.106	4.025	4.437
12	0.259	0.695	1.356	1.782	2.179	2.681	3.055	3.930	4.318
13	0.259	0.694	1.350	1.771	2.160	2.650	3.012	3.852	4.221
14	0.258	0.692	1.345	1.761	2.145	2.624	2.977	3.787	4.140
15	0.258	0.691	1.341	1.753	2.131	2.602	2.947	3.733	4.073
16	0.258	0.690	1.337	1.746	2.120	2.583	2.921	3.686	4.015
17	0.257	0.689	1.333	1.740	2.110	2.567	2.898	3.646	3.965
18	0.257	0.688	1.330	1.734	2.101	2.552	2.878	3.610	3.922
19	0.257	0.688	1.328	1.729	2.093	2.539	2.861	3.579	3.883
20	0.257	0.687	1.325	1.725	2.086	2.528	2.845	3.552	3.850
21	0.257	0.686	1.323	1.721	2.080	2.518	2.831	3.527	3.819
22	0.256	0.686	1.321	1.717	2.074	2.508	2.819	3.505	3.792
23	0.256	0.685	1.319	1.714	2.069	2.500	2.807	3.485	3.768
24	0.256	0.685	1.318	1.711	2.064	2.492	2.797	3.467	3.745
25	0.256	0.684	1.316	1.708	2.060	2.485	2.787	3.450	3.725
26	0.256	0.684	1.315	1.706	2.056	2.479	2.779	3.435	3.707
27	0.256	0.684	1.314	1.703	2.052	2.473	2.771	3.421	3.690
28	0.256	0.683	1.313	1.701	2.048	2.467	2.763	3.408	3.674
29	0.256	0.683	1.311	1.699	2.045	2.462	2.756	3.396	3.659
30	0.256	0.683	1.310	1.697	2.042	2.457	2.750	3.385	3.646
35	0.255	0.682	1.306	1.690	2.030	2.438	2.724	3.340	3.591
40	0.255	0.681	1.303	1.684	2.021	2.423	2.704	3.307	3.551
60	0.254	0.679	1.296	1.671	2.000	2.390	2.660	3.232	3.460
120	0.254	0.677	1.289	1.658	1.980	2.358	2.617	3.160	3.373
∞	0.253	0.674	1.282	1.645	1.960	2.326	2.576	3.090	3.291

TABLE A.4 Critical points for the Wilcoxon signed-rank test

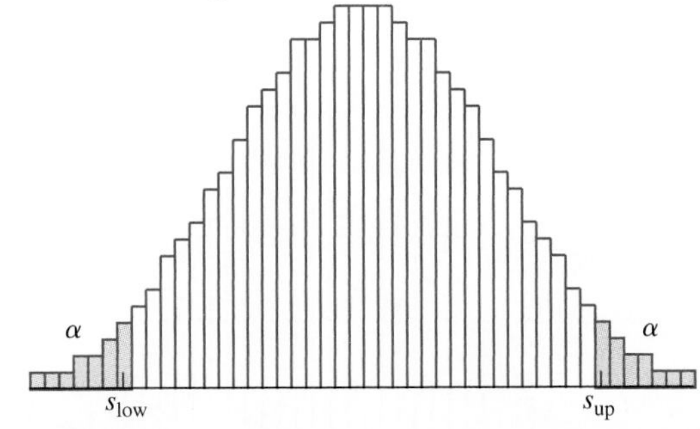

n	s_{low}	s_{up}	α
4	1	9	0.1250
	0	10	0.0625
5	3	12	0.1562
	2	13	0.0938
	1	14	0.0625
	0	15	0.0312
6	4	17	0.1094
	3	18	0.0781
	2	19	0.0469
	1	20	0.0312
	0	21	0.0156
7	6	22	0.1094
	5	23	0.0781
	4	24	0.0547
	3	25	0.0391
	2	26	0.0234
	1	27	0.0156
	0	28	0.0078
8	9	27	0.1250
	8	28	0.0977
	6	30	0.0547
	5	31	0.0391
	4	32	0.0273
	3	33	0.0195
	2	34	0.0117
	1	35	0.0078
	0	36	0.0039
9	11	34	0.1016
	10	35	0.0820
	9	36	0.0645
	8	37	0.0488
	6	39	0.0273
	5	40	0.0195
	4	41	0.0137
	3	42	0.0098
	2	43	0.0059
	1	44	0.0039
10	15	40	0.1162
	14	41	0.0967
	11	44	0.0527
	10	45	0.0420
	9	46	0.0322
	8	47	0.0244
	6	49	0.0137
	5	50	0.0098
	4	51	0.0068
	3	52	0.0049
11	18	48	0.1030
	17	49	0.0874
	14	52	0.0508
	13	53	0.0415
	11	55	0.0269
	10	56	0.0210
	8	58	0.0122
	7	59	0.0093
	6	60	0.0068
	5	61	0.0049
12	22	56	0.1018
	21	57	0.0881
	18	60	0.0549
	17	61	0.0461
	14	64	0.0261
	13	65	0.0212
	10	68	0.0105
	9	69	0.0081
	8	70	0.0061
	7	71	0.0046
13	27	64	0.1082
	26	65	0.0955
	22	69	0.0549
	21	70	0.0471
	18	73	0.0287
	17	74	0.0239
	13	78	0.0107
	12	79	0.0085
	10	81	0.0052
	9	82	0.0040
14	32	73	0.1083
	31	74	0.0969
	26	79	0.0520
	25	80	0.0453
	22	83	0.0290
	21	84	0.0247
	16	89	0.0101
	15	90	0.0083
	13	92	0.0054
	12	93	0.0043
15	37	83	0.1039
	36	84	0.0938
	31	89	0.0535
	30	90	0.0473
	26	94	0.0277
	25	95	0.0240
	20	100	0.0108
	19	101	0.0090
	16	104	0.0051
	15	105	0.0042
16	43	93	0.1057
	42	94	0.0964
	36	100	0.0523
	35	101	0.0467
	30	106	0.0253
	29	107	0.0222
	24	112	0.0107
	23	113	0.0091
	20	116	0.0055
	19	117	0.0046
17	49	104	0.1034
	48	105	0.0950
	42	111	0.0544
	41	112	0.0492
	35	118	0.0253
	34	119	0.0224
	28	125	0.0101
	27	126	0.0087
	24	129	0.0055
	23	130	0.0047
18	56	115	0.1061
	55	116	0.0982
	48	123	0.0542
	47	124	0.0494
	41	130	0.0269
	40	131	0.0241
	33	138	0.0104
	32	139	0.0091
	28	143	0.0052
	27	144	0.0045
19	63	127	0.1051
	62	128	0.0978
	54	136	0.0521
	53	137	0.0478
	47	143	0.0273
	46	144	0.0247
	38	152	0.0102
	37	153	0.0090
	33	157	0.0054
	32	158	0.0047
20	70	140	0.1012
	69	141	0.0947
	61	149	0.0527
	60	150	0.0487
	53	157	0.0266
	52	158	0.0242
	44	166	0.0107
	43	167	0.0096
	38	172	0.0053
	37	173	0.0047

For $n > 20$, compute $z = \dfrac{S_+ - n(n+1)/4}{\sqrt{n(n+1)(2n+1)/24}}$ and use the z table (Table A.2).

TABLE A.5 Critical points for the Wilcoxon rank-sum test

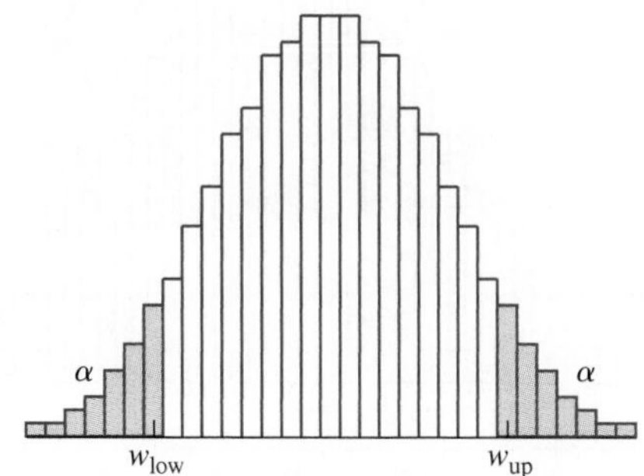

m	n	w_{low}	w_{up}	α
2	5	4	12	0.0952
		3	13	0.0476
	6	4	14	0.0714
		3	15	0.0357
	7	4	16	0.0556
		3	17	0.0278
	8	5	17	0.0889
		4	18	0.0444
		3	19	0.0222
3	4	7	17	0.0571
		6	18	0.0286
	5	8	19	0.0714
		7	20	0.0357
		6	21	0.0179
	6	9	21	0.0833
		8	22	0.0476
		7	23	0.0238
	7	9	24	0.0583
		8	25	0.0333
		7	26	0.0167
		6	27	0.0083
	8	10	26	0.0667
		9	27	0.0424
		8	28	0.0242
		7	29	0.0121
		6	30	0.0061
4	4	12	24	0.0571
		11	25	0.0286
		10	26	0.0143
	5	13	27	0.0556
		12	28	0.0317
		11	29	0.0159
		10	30	0.0079
	6	14	30	0.0571
		13	31	0.0333
		12	32	0.0190
		11	33	0.0095
		10	34	0.0048
	7	15	33	0.0545
		14	34	0.0364
		13	35	0.0212
		12	36	0.0121
		11	37	0.0061
		10	38	0.0030
	8	16	36	0.0545
		15	37	0.0364
		14	38	0.0242
		13	39	0.0141
		12	40	0.0081
		11	41	0.0040
5	5	20	35	0.0754
		19	36	0.0476
		18	37	0.0278
		17	38	0.0159
		16	39	0.0079
		15	40	0.0040
	6	21	39	0.0628
		20	40	0.0411
		19	41	0.0260
		18	42	0.0152
		17	43	0.0087
		16	44	0.0043
	7	22	43	0.0530
		21	44	0.0366
		20	45	0.0240
		19	46	0.0152
		18	47	0.0088
		17	48	0.0051
		16	49	0.0025
	8	24	46	0.0637
		23	47	0.0466
		22	48	0.0326
		21	49	0.0225
		20	50	0.0148
		19	51	0.0093
		18	52	0.0054
		17	53	0.0031
6	6	29	49	0.0660
		28	50	0.0465
		27	51	0.0325
		26	52	0.0206
		25	53	0.0130
		24	54	0.0076
		23	55	0.0043
	7	30	54	0.0507
		29	55	0.0367
		28	56	0.0256
		27	57	0.0175
		26	58	0.0111
		25	59	0.0070
		24	60	0.0041
	8	32	58	0.0539
		31	59	0.0406
		30	60	0.0296
		29	61	0.0213
		28	62	0.0147
		27	63	0.0100
		26	64	0.0063
		25	65	0.0040
7	7	40	65	0.0641
		39	66	0.0487
		37	68	0.0265
		36	69	0.0189
		35	70	0.0131
		34	71	0.0087
		33	72	0.0055
		32	73	0.0035
	8	42	70	0.0603
		41	71	0.0469
		39	73	0.0270
		38	74	0.0200
		36	76	0.0103
		35	77	0.0070
		34	78	0.0047
8	8	52	84	0.0524
		51	85	0.0415
		50	86	0.0325
		49	87	0.0249
		46	90	0.0103
		45	91	0.0074
		44	92	0.0052
		43	93	0.0035

When m and n are both greater than 8, compute $z = \dfrac{W - m(m+n+1)/2}{\sqrt{mn(m+n+1)/12}}$ and use the z table (Table A.2).

TABLE A.6 Upper percentage points for the χ^2 distribution

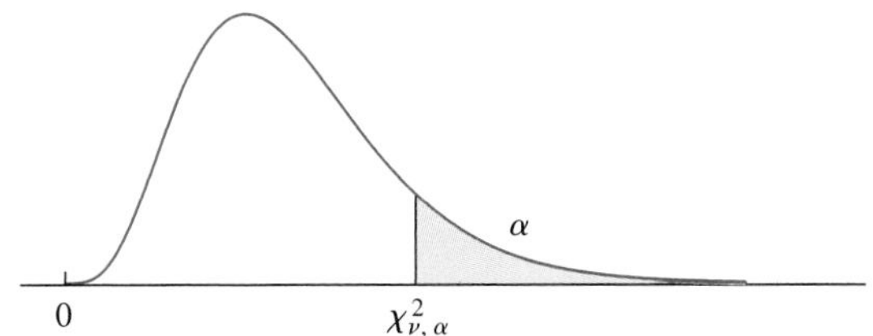

	α									
ν	**0.995**	**0.99**	**0.975**	**0.95**	**0.90**	**0.10**	**0.05**	**0.025**	**0.01**	**0.005**
1	0.000	0.000	0.001	0.004	0.016	2.706	3.841	5.024	6.635	7.879
2	0.010	0.020	0.051	0.103	0.211	4.605	5.991	7.378	9.210	10.597
3	0.072	0.115	0.216	0.352	0.584	6.251	7.815	9.348	11.345	12.838
4	0.207	0.297	0.484	0.711	1.064	7.779	9.488	11.143	13.277	14.860
5	0.412	0.554	0.831	1.145	1.610	9.236	11.070	12.833	15.086	16.750
6	0.676	0.872	1.237	1.635	2.204	10.645	12.592	14.449	16.812	18.548
7	0.989	1.239	1.690	2.167	2.833	12.017	14.067	16.013	18.475	20.278
8	1.344	1.646	2.180	2.733	3.490	13.362	15.507	17.535	20.090	21.955
9	1.735	2.088	2.700	3.325	4.168	14.684	16.919	19.023	21.666	23.589
10	2.156	2.558	3.247	3.940	4.865	15.987	18.307	20.483	23.209	25.188
11	2.603	3.053	3.816	4.575	5.578	17.275	19.675	21.920	24.725	26.757
12	3.074	3.571	4.404	5.226	6.304	18.549	21.026	23.337	26.217	28.300
13	3.565	4.107	5.009	5.892	7.042	19.812	22.362	24.736	27.688	29.819
14	4.075	4.660	5.629	6.571	7.790	21.064	23.685	26.119	29.141	31.319
15	4.601	5.229	6.262	7.261	8.547	22.307	24.996	27.488	30.578	32.801
16	5.142	5.812	6.908	7.962	9.312	23.542	26.296	28.845	32.000	34.267
17	5.697	6.408	7.564	8.672	10.085	24.769	27.587	30.191	33.409	35.718
18	6.265	7.015	8.231	9.390	10.865	25.989	28.869	31.526	34.805	37.156
19	6.844	7.633	8.907	10.117	11.651	27.204	30.144	32.852	36.191	38.582
20	7.434	8.260	9.591	10.851	12.443	28.412	31.410	34.170	37.566	39.997
21	8.034	8.897	10.283	11.591	13.240	29.615	32.671	35.479	38.932	41.401
22	8.643	9.542	10.982	12.338	14.041	30.813	33.924	36.781	40.289	42.796
23	9.260	10.196	11.689	13.091	14.848	32.007	35.172	38.076	41.638	44.181
24	9.886	10.856	12.401	13.848	15.659	33.196	36.415	39.364	42.980	45.559
25	10.520	11.524	13.120	14.611	16.473	34.382	37.652	40.646	44.314	46.928
26	11.160	12.198	13.844	15.379	17.292	35.563	38.885	41.923	45.642	48.290
27	11.808	12.879	14.573	16.151	18.114	36.741	40.113	43.195	46.963	49.645
28	12.461	13.565	15.308	16.928	18.939	37.916	41.337	44.461	48.278	50.993
29	13.121	14.256	16.047	17.708	19.768	39.087	42.557	45.722	49.588	52.336
30	13.787	14.953	16.791	18.493	20.599	40.256	43.773	46.979	50.892	53.672
31	14.458	15.655	17.539	19.281	21.434	41.422	44.985	48.232	52.191	55.003
32	15.134	16.362	18.291	20.072	22.271	42.585	46.194	49.480	53.486	56.328
33	15.815	17.074	19.047	20.867	23.110	43.745	47.400	50.725	54.776	57.648
34	16.501	17.789	19.806	21.664	23.952	44.903	48.602	51.966	56.061	58.964
35	17.192	18.509	20.569	22.465	24.797	46.059	49.802	53.203	57.342	60.275
36	17.887	19.233	21.336	23.269	25.643	47.212	50.998	54.437	58.619	61.581
37	18.586	19.960	22.106	24.075	26.492	48.363	52.192	55.668	59.893	62.883
38	19.289	20.691	22.878	24.884	27.343	49.513	53.384	56.896	61.162	64.181
39	19.996	21.426	23.654	25.695	28.196	50.660	54.572	58.120	62.428	65.476
40	20.707	22.164	24.433	26.509	29.051	51.805	55.758	59.342	63.691	66.766

For $\nu > 40$, $\chi^2_{\nu,\alpha} \approx 0.5(z_\alpha + \sqrt{2\nu - 1})$.

TABLE A.7 Upper percentage points for the F distribution

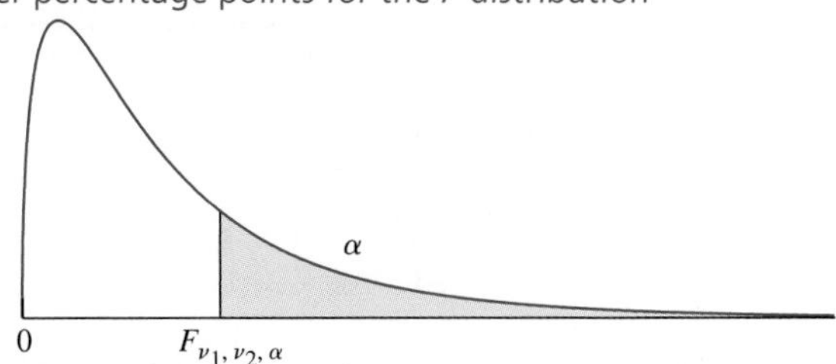

		ν_1								
ν_2	α	**1**	**2**	**3**	**4**	**5**	**6**	**7**	**8**	**9**
1	**0.100**	39.86	49.50	53.59	55.83	57.24	58.20	58.91	59.44	59.86
	0.050	161.45	199.50	215.71	224.58	230.16	233.99	236.77	238.88	240.54
	0.010	4052.18	4999.50	5403.35	5624.58	5763.65	5858.99	5928.36	5981.07	6022.47
	0.001	405284	500012	540382	562501	576405	585938	592874	598144	603040
2	**0.100**	8.53	9.00	9.16	9.24	9.29	9.33	9.35	9.37	9.38
	0.050	18.51	19.00	19.16	19.25	19.30	19.33	19.35	19.37	19.38
	0.010	98.50	99.00	99.17	99.25	99.30	99.33	99.36	99.37	99.39
	0.001	998.50	999.00	999.17	999.25	999.30	999.33	999.36	999.37	999.39
3	**0.100**	5.54	5.46	5.39	5.34	5.31	5.28	5.27	5.25	5.24
	0.050	10.13	9.55	9.28	9.12	9.01	8.94	8.89	8.85	8.81
	0.010	34.12	30.82	29.46	28.71	28.24	27.91	27.67	27.49	27.35
	0.001	167.03	148.50	141.11	137.10	134.58	132.85	131.58	130.62	129.86
4	**0.100**	4.54	4.32	4.19	4.11	4.05	4.01	3.98	3.95	3.94
	0.050	7.71	6.94	6.59	6.39	6.26	6.16	6.09	6.04	6.00
	0.010	21.20	18.00	16.69	15.98	15.52	15.21	14.98	14.80	14.66
	0.001	74.14	61.25	56.18	53.44	51.71	50.53	49.66	49.00	48.47
5	**0.100**	4.06	3.78	3.62	3.52	3.45	3.40	3.37	3.34	3.32
	0.050	6.61	5.79	5.41	5.19	5.05	4.95	4.88	4.82	4.77
	0.010	16.26	13.27	12.06	11.39	10.97	10.67	10.46	10.29	10.16
	0.001	47.18	37.12	33.20	31.09	29.75	28.83	28.16	27.65	27.24
6	**0.100**	3.78	3.46	3.29	3.18	3.11	3.05	3.01	2.98	2.96
	0.050	5.99	5.14	4.76	4.53	4.39	4.28	4.21	4.15	4.10
	0.010	13.75	10.92	9.78	9.15	8.75	8.47	8.26	8.10	7.98
	0.001	35.51	27.00	23.70	21.92	20.80	20.03	19.46	19.03	18.69
7	**0.100**	3.59	3.26	3.07	2.96	2.88	2.83	2.78	2.75	2.72
	0.050	5.59	4.74	4.35	4.12	3.97	3.87	3.79	3.73	3.68
	0.010	12.25	9.55	8.45	7.85	7.46	7.19	6.99	6.84	6.72
	0.001	29.25	21.69	18.77	17.20	16.21	15.52	15.02	14.63	14.33
8	**0.100**	3.46	3.11	2.92	2.81	2.73	2.67	2.62	2.59	2.56
	0.050	5.32	4.46	4.07	3.84	3.69	3.58	3.50	3.44	3.39
	0.010	11.26	8.65	7.59	7.01	6.63	6.37	6.18	6.03	5.91
	0.001	25.41	18.49	15.83	14.39	13.48	12.86	12.40	12.05	11.77
9	**0.100**	3.36	3.01	2.81	2.69	2.61	2.55	2.51	2.47	2.44
	0.050	5.12	4.26	3.86	3.63	3.48	3.37	3.29	3.23	3.18
	0.010	10.56	8.02	6.99	6.42	6.06	5.80	5.61	5.47	5.35
	0.001	22.86	16.39	13.90	12.56	11.71	11.13	10.70	10.37	10.11

Continued on page 777

TABLE A.7 Upper percentage points for the *F* distribution (continued)

		ν_1								
ν_2	α	**10**	**12**	**15**	**20**	**25**	**30**	**40**	**50**	**60**
1	**0.100**	60.19	60.71	61.22	61.74	62.05	62.26	62.53	62.69	62.79
	0.050	241.88	243.91	245.95	248.01	249.26	250.10	251.14	251.77	252.20
	0.010	6055.85	6106.32	6157.29	6208.73	6239.83	6260.65	6286.78	6302.52	6313.03
	0.001	606316	611276	616292	621362	624430	626486	659725	660511	6610390
2	**0.100**	9.39	9.41	9.42	9.44	9.45	9.46	9.47	9.47	9.47
	0.050	19.40	19.41	19.43	19.45	19.46	19.46	19.47	19.48	19.48
	0.010	99.40	99.42	99.43	99.45	99.46	99.47	99.47	99.48	99.48
	0.001	999.40	999.42	999.43	999.45	999.46	999.47	999.47	999.48	999.48
3	**0.100**	5.23	5.22	5.20	5.18	5.17	5.17	5.16	5.15	5.15
	0.050	8.79	8.74	8.70	8.66	8.63	8.62	8.59	8.58	8.57
	0.010	27.23	27.05	26.87	26.69	26.58	26.50	26.41	26.35	26.32
	0.001	129.25	128.32	127.37	126.42	125.84	125.45	124.96	124.66	124.47
4	**0.100**	3.92	3.90	3.87	3.84	3.83	3.82	3.80	3.80	3.79
	0.050	5.96	5.91	5.86	5.80	5.77	5.75	5.72	5.70	5.69
	0.010	14.55	14.37	14.20	14.02	13.91	13.84	13.75	13.69	13.65
	0.001	48.05	47.41	46.76	46.10	45.70	45.43	45.09	44.88	44.75
5	**0.100**	3.30	3.27	3.24	3.21	3.19	3.17	3.16	3.15	3.14
	0.050	4.74	4.68	4.62	4.56	4.52	4.50	4.46	4.44	4.43
	0.010	10.05	9.89	9.72	9.55	9.45	9.38	9.29	9.24	9.20
	0.001	26.92	26.42	25.91	25.39	25.08	24.87	24.60	24.44	24.33
6	**0.100**	2.94	2.90	2.87	2.84	2.81	2.80	2.78	2.77	2.76
	0.050	4.06	4.00	3.94	3.87	3.83	3.81	3.77	3.75	3.74
	0.010	7.87	7.72	7.56	7.40	7.30	7.23	7.14	7.09	7.06
	0.001	18.41	17.99	17.56	17.12	16.85	16.67	16.44	16.31	16.21
7	**0.100**	2.70	2.67	2.63	2.59	2.57	2.56	2.54	2.52	2.51
	0.050	3.64	3.57	3.51	3.44	3.40	3.38	3.34	3.32	3.30
	0.010	6.62	6.47	6.31	6.16	6.06	5.99	5.91	5.86	5.82
	0.001	14.08	13.71	13.32	12.93	12.69	12.53	12.33	12.20	12.12
8	**0.100**	2.54	2.50	2.46	2.42	2.40	2.38	2.36	2.35	2.34
	0.050	3.35	3.28	3.22	3.15	3.11	3.08	3.04	3.02	3.01
	0.010	5.81	5.67	5.52	5.36	5.26	5.20	5.12	5.07	5.03
	0.001	11.54	11.19	10.84	10.48	10.26	10.11	9.92	9.80	9.73
9	**0.100**	2.42	2.38	2.34	2.30	2.27	2.25	2.23	2.22	2.21
	0.050	3.14	3.07	3.01	2.94	2.89	2.86	2.83	2.80	2.79
	0.010	5.26	5.11	4.96	4.81	4.71	4.65	4.57	4.52	4.48
	0.001	9.89	9.57	9.24	8.90	8.69	8.55	8.37	8.26	8.19

Continued on page 778

TABLE A.7 Upper percentage points for the F distribution (continued)

ν_2	α	ν_1 = 1	2	3	4	5	6	7	8	9
10	**0.100**	3.29	2.92	2.73	2.61	2.52	2.46	2.41	2.38	2.35
	0.050	4.96	4.10	3.71	3.48	3.33	3.22	3.14	3.07	3.02
	0.010	10.04	7.56	6.55	5.99	5.64	5.39	5.20	5.06	4.94
	0.001	21.04	14.91	12.55	11.28	10.48	9.93	9.52	9.20	8.96
11	**0.100**	3.23	2.86	2.66	2.54	2.45	2.39	2.34	2.30	2.27
	0.050	4.84	3.98	3.59	3.36	3.20	3.09	3.01	2.95	2.90
	0.010	9.65	7.21	6.22	5.67	5.32	5.07	4.89	4.74	4.63
	0.001	19.69	13.81	11.56	10.35	9.58	9.05	8.66	8.35	8.12
12	**0.100**	3.18	2.81	2.61	2.48	2.39	2.33	2.28	2.24	2.21
	0.050	4.75	3.89	3.49	3.26	3.11	3.00	2.91	2.85	2.80
	0.010	9.33	6.93	5.95	5.41	5.06	4.82	4.64	4.50	4.39
	0.001	18.64	12.97	10.80	9.63	8.89	8.38	8.00	7.71	7.48
13	**0.100**	3.14	2.76	2.56	2.43	2.35	2.28	2.23	2.20	2.16
	0.050	4.67	3.81	3.41	3.18	3.03	2.92	2.83	2.77	2.71
	0.010	9.07	6.70	5.74	5.21	4.86	4.62	4.44	4.30	4.19
	0.001	17.82	12.31	10.21	9.07	8.35	7.86	7.49	7.21	6.98
14	**0.100**	3.10	2.73	2.52	2.39	2.31	2.24	2.19	2.15	2.12
	0.050	4.60	3.74	3.34	3.11	2.96	2.85	2.76	2.70	2.65
	0.010	8.86	6.51	5.56	5.04	4.69	4.46	4.28	4.14	4.03
	0.001	17.14	11.78	9.73	8.62	7.92	7.44	7.08	6.80	6.58
15	**0.100**	3.07	2.70	2.49	2.36	2.27	2.21	2.16	2.12	2.09
	0.050	4.54	3.68	3.29	3.06	2.90	2.79	2.71	2.64	2.59
	0.010	8.68	6.36	5.42	4.89	4.56	4.32	4.14	4.00	3.89
	0.001	16.59	11.34	9.34	8.25	7.57	7.09	6.74	6.47	6.26
16	**0.100**	3.05	2.67	2.46	2.33	2.24	2.18	2.13	2.09	2.06
	0.050	4.49	3.63	3.24	3.01	2.85	2.74	2.66	2.59	2.54
	0.010	8.53	6.23	5.29	4.77	4.44	4.20	4.03	3.89	3.78
	0.001	16.12	10.97	9.01	7.94	7.27	6.80	6.46	6.19	5.98
17	**0.100**	3.03	2.64	2.44	2.31	2.22	2.15	2.10	2.06	2.03
	0.050	4.45	3.59	3.20	2.96	2.81	2.70	2.61	2.55	2.49
	0.010	8.40	6.11	5.18	4.67	4.34	4.10	3.93	3.79	3.68
	0.001	15.72	10.66	8.73	7.68	7.02	6.56	6.22	5.96	5.75
18	**0.100**	3.01	2.62	2.42	2.29	2.20	2.13	2.08	2.04	2.00
	0.050	4.41	3.55	3.16	2.93	2.77	2.66	2.58	2.51	2.46
	0.010	8.29	6.01	5.09	4.58	4.25	4.01	3.84	3.71	3.60
	0.001	15.38	10.39	8.49	7.46	6.81	6.35	6.02	5.76	5.56
19	**0.100**	2.99	2.61	2.40	2.27	2.18	2.11	2.06	2.02	1.98
	0.050	4.38	3.52	3.13	2.90	2.74	2.63	2.54	2.48	2.42
	0.010	8.18	5.93	5.01	4.50	4.17	3.94	3.77	3.63	3.52
	0.001	15.08	10.16	8.28	7.27	6.62	6.18	5.85	5.59	5.39
20	**0.100**	2.97	2.59	2.38	2.25	2.16	2.09	2.04	2.00	1.96
	0.050	4.35	3.49	3.10	2.87	2.71	2.60	2.51	2.45	2.39
	0.010	8.10	5.85	4.94	4.43	4.10	3.87	3.70	3.56	3.46
	0.001	14.82	9.95	8.10	7.10	6.46	6.02	5.69	5.44	5.24

Continued on page 779

TABLE A.7 Upper percentage points for the *F* distribution (continued)

		ν_1								
ν_2	α	**10**	**12**	**15**	**20**	**25**	**30**	**40**	**50**	**60**
10	**0.100**	2.32	2.28	2.24	2.20	2.17	2.16	2.13	2.12	2.11
	0.050	2.98	2.91	2.85	2.77	2.73	2.70	2.66	2.64	2.62
	0.010	4.85	4.71	4.56	4.41	4.31	4.25	4.17	4.12	4.08
	0.001	8.75	8.45	8.13	7.80	7.60	7.47	7.30	7.19	7.12
11	**0.100**	2.25	2.21	2.17	2.12	2.10	2.08	2.05	2.04	2.03
	0.050	2.85	2.79	2.72	2.65	2.60	2.57	2.53	2.51	2.49
	0.010	4.54	4.40	4.25	4.10	4.01	3.94	3.86	3.81	3.78
	0.001	7.92	7.63	7.32	7.01	6.81	6.68	6.52	6.42	6.35
12	**0.100**	2.19	2.15	2.10	2.06	2.03	2.01	1.99	1.97	1.96
	0.050	2.75	2.69	2.62	2.54	2.50	2.47	2.43	2.40	2.38
	0.010	4.30	4.16	4.01	3.86	3.76	3.70	3.62	3.57	3.54
	0.001	7.29	7.00	6.71	6.40	6.22	6.09	5.93	5.83	5.76
13	**0.100**	2.14	2.10	2.05	2.01	1.98	1.96	1.93	1.92	1.90
	0.050	2.67	2.60	2.53	2.46	2.41	2.38	2.34	2.31	2.30
	0.010	4.10	3.96	3.82	3.66	3.57	3.51	3.43	3.38	3.34
	0.001	6.80	6.52	6.23	5.93	5.75	5.63	5.47	5.37	5.30
14	**0.100**	2.10	2.05	2.01	1.96	1.93	1.91	1.89	1.87	1.86
	0.050	2.60	2.53	2.46	2.39	2.34	2.31	2.27	2.24	2.22
	0.010	3.94	3.80	3.66	3.51	3.41	3.35	3.27	3.22	3.18
	0.001	6.40	6.13	5.85	5.56	5.38	5.25	5.10	5.00	4.94
15	**0.100**	2.06	2.02	1.97	1.92	1.89	1.87	1.85	1.83	1.82
	0.050	2.54	2.48	2.40	2.33	2.28	2.25	2.20	2.18	2.16
	0.010	3.80	3.67	3.52	3.37	3.28	3.21	3.13	3.08	3.05
	0.001	6.08	5.81	5.54	5.25	5.07	4.95	4.80	4.70	4.64
16	**0.100**	2.03	1.99	1.94	1.89	1.86	1.84	1.81	1.79	1.78
	0.050	2.49	2.42	2.35	2.28	2.23	2.19	2.15	2.12	2.11
	0.010	3.69	3.55	3.41	3.26	3.16	3.10	3.02	2.97	2.93
	0.001	5.81	5.55	5.27	4.99	4.82	4.70	4.54	4.45	4.39
17	**0.100**	2.00	1.96	1.91	1.86	1.83	1.81	1.78	1.76	1.75
	0.050	2.45	2.38	2.31	2.23	2.18	2.15	2.10	2.08	2.06
	0.010	3.59	3.46	3.31	3.16	3.07	3.00	2.92	2.87	2.83
	0.001	5.58	5.32	5.05	4.78	4.60	4.48	4.33	4.24	4.18
18	**0.100**	1.98	1.93	1.89	1.84	1.80	1.78	1.75	1.74	1.72
	0.050	2.41	2.34	2.27	2.19	2.14	2.11	2.06	2.04	2.02
	0.010	3.51	3.37	3.23	3.08	2.98	2.92	2.84	2.78	2.75
	0.001	5.39	5.13	4.87	4.59	4.42	4.30	4.15	4.06	4.00
19	**0.100**	1.96	1.91	1.86	1.81	1.78	1.76	1.73	1.71	1.70
	0.050	2.38	2.31	2.23	2.16	2.11	2.07	2.03	2.00	1.98
	0.010	3.43	3.30	3.15	3.00	2.91	2.84	2.76	2.71	2.67
	0.001	5.22	4.97	4.70	4.43	4.26	4.14	3.99	3.90	3.84
20	**0.100**	1.94	1.89	1.84	1.79	1.76	1.74	1.71	1.69	1.68
	0.050	2.35	2.28	2.20	2.12	2.07	2.04	1.99	1.97	1.95
	0.010	3.37	3.23	3.09	2.94	2.84	2.78	2.69	2.64	2.61
	0.001	5.08	4.82	4.56	4.29	4.12	4.00	3.86	3.77	3.70

Continued on page 780

TABLE A.7 Upper percentage points for the *F* distribution (continued)

		ν_1								
ν_2	α	1	2	3	4	5	6	7	8	9
21	0.100	2.96	2.57	2.36	2.23	2.14	2.08	2.02	1.98	1.95
	0.050	4.32	3.47	3.07	2.84	2.68	2.57	2.49	2.42	2.37
	0.010	8.02	5.78	4.87	4.37	4.04	3.81	3.64	3.51	3.40
	0.001	14.59	9.77	7.94	6.95	6.32	5.88	5.56	5.31	5.11
22	0.100	2.95	2.56	2.35	2.22	2.13	2.06	2.01	1.97	1.93
	0.050	4.30	3.44	3.05	2.82	2.66	2.55	2.46	2.40	2.34
	0.010	7.95	5.72	4.82	4.31	3.99	3.76	3.59	3.45	3.35
	0.001	14.38	9.61	7.80	6.81	6.19	5.76	5.44	5.19	4.99
23	0.100	2.94	2.55	2.34	2.21	2.11	2.05	1.99	1.95	1.92
	0.050	4.28	3.42	3.03	2.80	2.64	2.53	2.44	2.37	2.32
	0.010	7.88	5.66	4.76	4.26	3.94	3.71	3.54	3.41	3.30
	0.001	14.20	9.47	7.67	6.70	6.08	5.65	5.33	5.09	4.89
24	0.100	2.93	2.54	2.33	2.19	2.10	2.04	1.98	1.94	1.91
	0.050	4.26	3.40	3.01	2.78	2.62	2.51	2.42	2.36	2.30
	0.010	7.82	5.61	4.72	4.22	3.90	3.67	3.50	3.36	3.26
	0.001	14.03	9.34	7.55	6.59	5.98	5.55	5.23	4.99	4.80
25	0.100	2.92	2.53	2.32	2.18	2.09	2.02	1.97	1.93	1.89
	0.050	4.24	3.39	2.99	2.76	2.60	2.49	2.40	2.34	2.28
	0.010	7.77	5.57	4.68	4.18	3.85	3.63	3.46	3.32	3.22
	0.001	13.88	9.22	7.45	6.49	5.89	5.46	5.15	4.91	4.71
26	0.100	2.91	2.52	2.31	2.17	2.08	2.01	1.96	1.92	1.88
	0.050	4.23	3.37	2.98	2.74	2.59	2.47	2.39	2.32	2.27
	0.010	7.72	5.53	4.64	4.14	3.82	3.59	3.42	3.29	3.18
	0.001	13.74	9.12	7.36	6.41	5.80	5.38	5.07	4.83	4.64
27	0.100	2.90	2.51	2.30	2.17	2.07	2.00	1.95	1.91	1.87
	0.050	4.21	3.35	2.96	2.73	2.57	2.46	2.37	2.31	2.25
	0.010	7.68	5.49	4.60	4.11	3.78	3.56	3.39	3.26	3.15
	0.001	13.61	9.02	7.27	6.33	5.73	5.31	5.00	4.76	4.57
28	0.100	2.89	2.50	2.29	2.16	2.06	2.00	1.94	1.90	1.87
	0.050	4.20	3.34	2.95	2.71	2.56	2.45	2.36	2.29	2.24
	0.010	7.64	5.45	4.57	4.07	3.75	3.53	3.36	3.23	3.12
	0.001	13.50	8.93	7.19	6.25	5.66	5.24	4.93	4.69	4.50
29	0.100	2.89	2.50	2.28	2.15	2.06	1.99	1.93	1.89	1.86
	0.050	4.18	3.33	2.93	2.70	2.55	2.43	2.35	2.28	2.22
	0.010	7.60	5.42	4.54	4.04	3.73	3.50	3.33	3.20	3.09
	0.001	13.39	8.85	7.12	6.19	5.59	5.18	4.87	4.64	4.45
30	0.100	2.88	2.49	2.28	2.14	2.05	1.98	1.93	1.88	1.85
	0.050	4.17	3.32	2.92	2.69	2.53	2.42	2.33	2.27	2.21
	0.010	7.56	5.39	4.51	4.02	3.70	3.47	3.30	3.17	3.07
	0.001	13.29	8.77	7.05	6.12	5.53	5.12	4.82	4.58	4.39
31	0.100	2.87	2.48	2.27	2.14	2.04	1.97	1.92	1.88	1.84
	0.050	4.16	3.30	2.91	2.68	2.52	2.41	2.32	2.25	2.20
	0.010	7.53	5.36	4.48	3.99	3.67	3.45	3.28	3.15	3.04
	0.001	13.20	8.70	6.99	6.07	5.48	5.07	4.77	4.53	4.34

Continued on page 781

TABLE A.7 Upper percentage points for the *F* distribution (continued)

		ν_1								
ν_2	α	**10**	**12**	**15**	**20**	**25**	**30**	**40**	**50**	**60**
21	**0.100**	1.92	1.87	1.83	1.78	1.74	1.72	1.69	1.67	1.66
	0.050	2.32	2.25	2.18	2.10	2.05	2.01	1.96	1.94	1.92
	0.010	3.31	3.17	3.03	2.88	2.79	2.72	2.64	2.58	2.55
	0.001	4.95	4.70	4.44	4.17	4.00	3.88	3.74	3.64	3.58
22	**0.100**	1.90	1.86	1.81	1.76	1.73	1.70	1.67	1.65	1.64
	0.050	2.30	2.23	2.15	2.07	2.02	1.98	1.94	1.91	1.89
	0.010	3.26	3.12	2.98	2.83	2.73	2.67	2.58	2.53	2.50
	0.001	4.83	4.58	4.33	4.06	3.89	3.78	3.63	3.54	3.48
23	**0.100**	1.89	1.84	1.80	1.74	1.71	1.69	1.66	1.64	1.62
	0.050	2.27	2.20	2.13	2.05	2.00	1.96	1.91	1.88	1.86
	0.010	3.21	3.07	2.93	2.78	2.69	2.62	2.54	2.48	2.45
	0.001	4.73	4.48	4.23	3.96	3.79	3.68	3.53	3.44	3.38
24	**0.100**	1.88	1.83	1.78	1.73	1.70	1.67	1.64	1.62	1.61
	0.050	2.25	2.18	2.11	2.03	1.97	1.94	1.89	1.86	1.84
	0.010	3.17	3.03	2.89	2.74	2.64	2.58	2.49	2.44	2.40
	0.001	4.64	4.39	4.14	3.87	3.71	3.59	3.45	3.36	3.29
25	**0.100**	1.87	1.82	1.77	1.72	1.68	1.66	1.63	1.61	1.59
	0.050	2.24	2.16	2.09	2.01	1.96	1.92	1.87	1.84	1.82
	0.010	3.13	2.99	2.85	2.70	2.60	2.54	2.45	2.40	2.36
	0.001	4.56	4.31	4.06	3.79	3.63	3.52	3.37	3.28	3.22
26	**0.100**	1.86	1.81	1.76	1.71	1.67	1.65	1.61	1.59	1.58
	0.050	2.22	2.15	2.07	1.99	1.94	1.90	1.85	1.82	1.80
	0.010	3.09	2.96	2.81	2.66	2.57	2.50	2.42	2.36	2.33
	0.001	4.48	4.24	3.99	3.72	3.56	3.44	3.30	3.21	3.15
27	**0.100**	1.85	1.80	1.75	1.70	1.66	1.64	1.60	1.58	1.57
	0.050	2.20	2.13	2.06	1.97	1.92	1.88	1.84	1.81	1.79
	0.010	3.06	2.93	2.78	2.63	2.54	2.47	2.38	2.33	2.29
	0.001	4.41	4.17	3.92	3.66	3.49	3.38	3.23	3.14	3.08
28	**0.100**	1.84	1.79	1.74	1.69	1.65	1.63	1.59	1.57	1.56
	0.050	2.19	2.12	2.04	1.96	1.91	1.87	1.82	1.79	1.77
	0.010	3.03	2.90	2.75	2.60	2.51	2.44	2.35	2.30	2.26
	0.001	4.35	4.11	3.86	3.60	3.43	3.32	3.18	3.09	3.02
29	**0.100**	1.83	1.78	1.73	1.68	1.64	1.62	1.58	1.56	1.55
	0.050	2.18	2.10	2.03	1.94	1.89	1.85	1.81	1.77	1.75
	0.010	3.00	2.87	2.73	2.57	2.48	2.41	2.33	2.27	2.23
	0.001	4.29	4.05	3.80	3.54	3.38	3.27	3.12	3.03	2.97
30	**0.100**	1.82	1.77	1.72	1.67	1.63	1.61	1.57	1.55	1.54
	0.050	2.16	2.09	2.01	1.93	1.88	1.84	1.79	1.76	1.74
	0.010	2.98	2.84	2.70	2.55	2.45	2.39	2.30	2.25	2.21
	0.001	4.24	4.00	3.75	3.49	3.33	3.22	3.07	2.98	2.92
31	**0.100**	1.81	1.77	1.71	1.66	1.62	1.60	1.56	1.54	1.53
	0.050	2.15	2.08	2.00	1.92	1.87	1.83	1.78	1.75	1.73
	0.010	2.96	2.82	2.68	2.52	2.43	2.36	2.27	2.22	2.18
	0.001	4.19	3.95	3.71	3.45	3.28	3.17	3.03	2.94	2.87

Continued on page 782

TABLE A.7 Upper percentage points for the *F* distribution (continued)

		ν_1								
ν_2	α	1	2	3	4	5	6	7	8	9
32	0.100	2.87	2.48	2.26	2.13	2.04	1.97	1.91	1.87	1.83
	0.050	4.15	3.29	2.90	2.67	2.51	2.40	2.31	2.24	2.19
	0.010	7.50	5.34	4.46	3.97	3.65	3.43	3.26	3.13	3.02
	0.001	13.12	8.64	6.94	6.01	5.43	5.02	4.72	4.48	4.30
33	0.100	2.86	2.47	2.26	2.12	2.03	1.96	1.91	1.86	1.83
	0.050	4.14	3.28	2.89	2.66	2.50	2.39	2.30	2.23	2.18
	0.010	7.47	5.31	4.44	3.95	3.63	3.41	3.24	3.11	3.00
	0.001	13.04	8.58	6.88	5.97	5.38	4.98	4.67	4.44	4.26
34	0.100	2.86	2.47	2.25	2.12	2.02	1.96	1.90	1.86	1.82
	0.050	4.13	3.28	2.88	2.65	2.49	2.38	2.29	2.23	2.17
	0.010	7.44	5.29	4.42	3.93	3.61	3.39	3.22	3.09	2.98
	0.001	12.97	8.52	6.83	5.92	5.34	4.93	4.63	4.40	4.22
35	0.100	2.85	2.46	2.25	2.11	2.02	1.95	1.90	1.85	1.82
	0.050	4.12	3.27	2.87	2.64	2.49	2.37	2.29	2.22	2.16
	0.010	7.42	5.27	4.40	3.91	3.59	3.37	3.20	3.07	2.96
	0.001	12.90	8.47	6.79	5.88	5.30	4.89	4.59	4.36	4.18
36	0.100	2.85	2.46	2.24	2.11	2.01	1.94	1.89	1.85	1.81
	0.050	4.11	3.26	2.87	2.63	2.48	2.36	2.28	2.21	2.15
	0.010	7.40	5.25	4.38	3.89	3.57	3.35	3.18	3.05	2.95
	0.001	12.83	8.42	6.74	5.84	5.26	4.86	4.56	4.33	4.14
37	0.100	2.85	2.45	2.24	2.10	2.01	1.94	1.89	1.84	1.81
	0.050	4.11	3.25	2.86	2.63	2.47	2.36	2.27	2.20	2.14
	0.010	7.37	5.23	4.36	3.87	3.56	3.33	3.17	3.04	2.93
	0.001	12.77	8.37	6.70	5.80	5.22	4.82	4.53	4.30	4.11
38	0.100	2.84	2.45	2.23	2.10	2.01	1.94	1.88	1.84	1.80
	0.050	4.10	3.24	2.85	2.62	2.46	2.35	2.26	2.19	2.14
	0.010	7.35	5.21	4.34	3.86	3.54	3.32	3.15	3.02	2.92
	0.001	12.71	8.33	6.66	5.76	5.19	4.79	4.49	4.26	4.08
39	0.100	2.84	2.44	2.23	2.09	2.00	1.93	1.88	1.83	1.80
	0.050	4.09	3.24	2.85	2.61	2.46	2.34	2.26	2.19	2.13
	0.010	7.33	5.19	4.33	3.84	3.53	3.30	3.14	3.01	2.90
	0.001	12.66	8.29	6.63	5.73	5.16	4.76	4.46	4.23	4.05
40	0.100	2.84	2.44	2.23	2.09	2.00	1.93	1.87	1.83	1.79
	0.050	4.08	3.23	2.84	2.61	2.45	2.34	2.25	2.18	2.12
	0.010	7.31	5.18	4.31	3.83	3.51	3.29	3.12	2.99	2.89
	0.001	12.61	8.25	6.59	5.70	5.13	4.73	4.44	4.21	4.02
50	0.100	2.81	2.41	2.20	2.06	1.97	1.90	1.84	1.80	1.76
	0.050	4.03	3.18	2.79	2.56	2.40	2.29	2.20	2.13	2.07
	0.010	7.17	5.06	4.20	3.72	3.41	3.19	3.02	2.89	2.78
	0.001	12.22	7.96	6.34	5.46	4.90	4.51	4.22	4.00	3.82
60	0.100	2.79	2.39	2.18	2.04	1.95	1.87	1.82	1.77	1.74
	0.050	4.00	3.15	2.76	2.53	2.37	2.25	2.17	2.10	2.04
	0.010	7.08	4.98	4.13	3.65	3.34	3.12	2.95	2.82	2.72
	0.001	11.97	7.77	6.17	5.31	4.76	4.37	4.09	3.86	3.69
120	0.100	2.75	2.35	2.13	1.99	1.90	1.82	1.77	1.72	1.68
	0.050	3.92	3.07	2.68	2.45	2.29	2.18	2.09	2.02	1.96
	0.010	6.85	4.79	3.95	3.48	3.17	2.96	2.79	2.66	2.56
	0.001	11.38	7.32	5.78	4.95	4.42	4.04	3.77	3.55	3.38

Continued on page 783

TABLE A.7 Upper percentage points for the *F* distribution (continued)

ν_2	α	ν_1 = 10	12	15	20	25	30	40	50	60
32	**0.100**	1.81	1.76	1.71	1.65	1.62	1.59	1.56	1.53	1.52
	0.050	2.14	2.07	1.99	1.91	1.85	1.82	1.77	1.74	1.71
	0.010	2.93	2.80	2.65	2.50	2.41	2.34	2.25	2.20	2.16
	0.001	4.14	3.91	3.66	3.40	3.24	3.13	2.98	2.89	2.83
33	**0.100**	1.80	1.75	1.70	1.64	1.61	1.58	1.55	1.53	1.51
	0.050	2.13	2.06	1.98	1.90	1.84	1.81	1.76	1.72	1.70
	0.010	2.91	2.78	2.63	2.48	2.39	2.32	2.23	2.18	2.14
	0.001	4.10	3.87	3.62	3.36	3.20	3.09	2.94	2.85	2.79
34	**0.100**	1.79	1.75	1.69	1.64	1.60	1.58	1.54	1.52	1.50
	0.050	2.12	2.05	1.97	1.89	1.83	1.80	1.75	1.71	1.69
	0.010	2.89	2.76	2.61	2.46	2.37	2.30	2.21	2.16	2.12
	0.001	4.06	3.83	3.58	3.33	3.16	3.05	2.91	2.82	2.75
35	**0.100**	1.79	1.74	1.69	1.63	1.60	1.57	1.53	1.51	1.50
	0.050	2.11	2.04	1.96	1.88	1.82	1.79	1.74	1.70	1.68
	0.010	2.88	2.74	2.60	2.44	2.35	2.28	2.19	2.14	2.10
	0.001	4.03	3.79	3.55	3.29	3.13	3.02	2.87	2.78	2.72
36	**0.100**	1.78	1.73	1.68	1.63	1.59	1.56	1.53	1.51	1.49
	0.050	2.11	2.03	1.95	1.87	1.81	1.78	1.73	1.69	1.67
	0.010	2.86	2.72	2.58	2.43	2.33	2.26	2.18	2.12	2.08
	0.001	3.99	3.76	3.51	3.26	3.10	2.98	2.84	2.75	2.69
37	**0.100**	1.78	1.73	1.68	1.62	1.58	1.56	1.52	1.50	1.48
	0.050	2.10	2.02	1.95	1.86	1.81	1.77	1.72	1.68	1.66
	0.010	2.84	2.71	2.56	2.41	2.31	2.25	2.16	2.10	2.06
	0.001	3.96	3.73	3.48	3.23	3.07	2.95	2.81	2.72	2.66
38	**0.100**	1.77	1.72	1.67	1.61	1.58	1.55	1.52	1.49	1.48
	0.050	2.09	2.02	1.94	1.85	1.80	1.76	1.71	1.68	1.65
	0.010	2.83	2.69	2.55	2.40	2.30	2.23	2.14	2.09	2.05
	0.001	3.93	3.70	3.45	3.20	3.04	2.92	2.78	2.69	2.63
39	**0.100**	1.77	1.72	1.67	1.61	1.57	1.55	1.51	1.49	1.47
	0.050	2.08	2.01	1.93	1.85	1.79	1.75	1.70	1.67	1.65
	0.010	2.81	2.68	2.54	2.38	2.29	2.22	2.13	2.07	2.03
	0.001	3.90	3.67	3.43	3.17	3.01	2.90	2.75	2.66	2.60
40	**0.100**	1.76	1.71	1.66	1.61	1.57	1.54	1.51	1.48	1.47
	0.050	2.08	2.00	1.92	1.84	1.78	1.74	1.69	1.66	1.64
	0.010	2.80	2.66	2.52	2.37	2.27	2.20	2.11	2.06	2.02
	0.001	3.87	3.64	3.40	3.14	2.98	2.87	2.73	2.64	2.57
50	**0.100**	1.73	1.68	1.63	1.57	1.53	1.50	1.46	1.44	1.42
	0.050	2.03	1.95	1.87	1.78	1.73	1.69	1.63	1.60	1.58
	0.010	2.70	2.56	2.42	2.27	2.17	2.10	2.01	1.95	1.91
	0.001	3.67	3.44	3.20	2.95	2.79	2.68	2.53	2.44	2.38
60	**0.100**	1.71	1.66	1.60	1.54	1.50	1.48	1.44	1.41	1.40
	0.050	1.99	1.92	1.84	1.75	1.69	1.65	1.59	1.56	1.53
	0.010	2.63	2.50	2.35	2.20	2.10	2.03	1.94	1.88	1.84
	0.001	3.54	3.32	3.08	2.83	2.67	2.55	2.41	2.32	2.25
120	**0.100**	1.65	1.60	1.55	1.48	1.44	1.41	1.37	1.34	1.32
	0.050	1.91	1.83	1.75	1.66	1.60	1.55	1.50	1.46	1.43
	0.010	2.47	2.34	2.19	2.03	1.93	1.86	1.76	1.70	1.66
	0.001	3.24	3.02	2.78	2.53	2.37	2.26	2.11	2.02	1.95

TABLE A.8 Upper percentage points for the Studentized range q_{ν_1, ν_2}

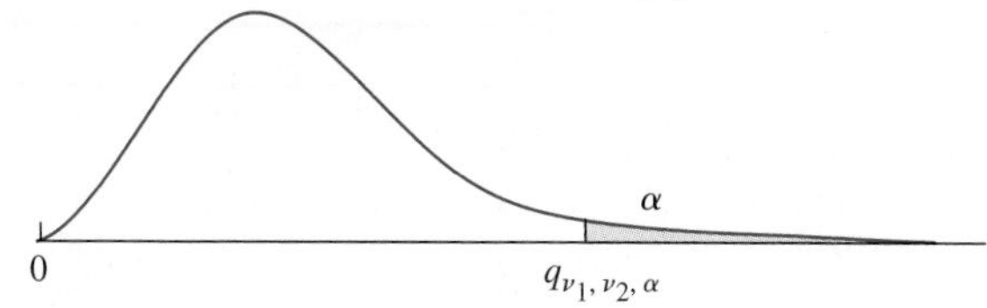

		ν_1													
ν_2	α	**2**	**3**	**4**	**5**	**6**	**7**	**8**	**9**	**10**	**11**	**12**	**13**	**14**	**15**
1	**0.10**	8.93	13.44	16.36	18.49	20.15	21.51	22.64	23.62	24.48	25.24	25.92	26.54	27.10	27.62
	0.05	17.97	26.98	32.82	37.08	40.41	43.12	45.40	47.36	49.07	50.59	51.96	53.20	54.33	55.36
	0.01	90.02	135.0	164.3	185.6	202.2	215.8	227.2	237.0	245.6	253.2	260.0	266.2	271.8	277.0
2	**0.10**	4.13	5.73	6.77	7.54	8.14	8.63	9.05	9.41	9.72	10.01	10.26	10.49	10.70	10.89
	0.05	6.08	8.33	9.80	10.88	11.74	12.44	13.03	13.54	13.99	14.39	14.75	15.08	15.38	15.65
	0.01	14.04	19.02	22.29	24.72	26.63	28.20	29.53	30.68	31.69	32.59	33.40	34.13	34.81	35.43
3	**0.10**	3.33	4.47	5.20	5.74	6.16	6.51	6.81	7.06	7.29	7.49	7.67	7.83	7.98	8.12
	0.05	4.50	5.91	6.82	7.50	8.04	8.48	8.85	9.18	9.46	9.72	9.95	10.15	10.35	10.52
	0.01	8.26	10.62	12.17	13.33	14.24	15.00	15.64	16.20	16.69	17.13	17.53	17.89	18.22	18.52
4	**0.10**	3.01	3.98	4.59	5.04	5.39	5.68	5.93	6.14	6.33	6.49	6.65	6.78	6.91	7.02
	0.05	3.93	5.04	5.76	6.29	6.71	7.05	7.35	7.60	7.83	8.03	8.21	8.37	8.52	8.66
	0.01	6.51	8.12	9.17	9.96	10.58	11.10	11.55	11.93	12.27	12.57	12.84	13.09	13.32	13.53
5	**0.10**	2.85	3.72	4.26	4.66	4.98	5.24	5.46	5.65	5.82	5.97	6.10	6.22	6.34	6.44
	0.05	3.64	4.60	5.22	5.67	6.03	6.33	6.58	6.80	6.99	7.17	7.32	7.47	7.60	7.72
	0.01	5.70	6.98	7.80	8.42	8.91	9.32	9.67	9.97	10.24	10.48	10.70	10.89	11.08	11.24
6	**0.10**	2.75	3.56	4.07	4.44	4.73	4.97	5.17	5.34	5.50	5.64	5.76	5.87	5.98	6.07
	0.05	3.46	4.34	4.90	5.31	5.63	5.90	6.12	6.32	6.49	6.65	6.79	6.92	7.03	7.14
	0.01	5.24	6.33	7.03	7.56	7.97	8.32	8.61	8.87	9.10	9.30	9.49	9.65	9.81	9.95
7	**0.10**	2.68	3.45	3.93	4.28	4.55	4.78	4.97	5.14	5.28	5.41	5.53	5.64	5.74	5.83
	0.05	3.34	4.16	4.68	5.06	5.36	5.61	5.82	6.00	6.16	6.30	6.43	6.55	6.66	6.76
	0.01	4.95	5.92	6.54	7.01	7.37	7.68	7.94	8.17	8.37	8.55	8.71	8.86	9.00	9.12
8	**0.10**	2.63	3.37	3.83	4.17	4.43	4.65	4.83	4.99	5.13	5.25	5.36	5.46	5.56	5.64
	0.05	3.26	4.04	4.53	4.89	5.17	5.40	5.60	5.77	5.92	6.05	6.18	6.29	6.39	6.48
	0.01	4.75	5.64	6.20	6.63	6.96	7.24	7.47	7.68	7.86	8.03	8.18	8.31	8.44	8.55
9	**0.10**	2.59	3.32	3.76	4.08	4.34	4.54	4.72	4.87	5.01	5.13	5.23	5.33	5.42	5.51
	0.05	3.20	3.95	4.42	4.76	5.02	5.24	5.43	5.59	5.74	5.87	5.98	6.09	6.19	6.28
	0.01	4.60	5.43	5.96	6.35	6.66	6.91	7.13	7.33	7.49	7.65	7.78	7.91	8.03	8.13
10	**0.10**	2.56	3.27	3.70	4.02	4.26	4.47	4.64	4.78	4.91	5.03	5.13	5.23	5.32	5.40
	0.05	3.15	3.88	4.33	4.65	4.91	5.12	5.31	5.46	5.60	5.72	5.83	5.93	6.03	6.11
	0.01	4.48	5.27	5.77	6.14	6.43	6.67	6.88	7.05	7.21	7.36	7.49	7.60	7.71	7.81
11	**0.10**	2.54	3.23	3.66	3.96	4.20	4.40	4.57	4.71	4.84	4.95	5.05	5.15	5.23	5.31
	0.05	3.11	3.82	4.26	4.57	4.82	5.03	5.20	5.35	5.49	5.61	5.71	5.81	5.90	5.99
	0.01	4.39	5.15	5.62	5.97	6.25	6.48	6.67	6.84	6.99	7.13	7.25	7.36	7.46	7.56
12	**0.10**	2.52	3.20	3.62	3.92	4.16	4.35	4.51	4.65	4.78	4.89	4.99	5.08	5.16	5.24
	0.05	3.08	3.77	4.20	4.51	4.75	4.95	5.12	5.27	5.40	5.51	5.62	5.71	5.80	5.88
	0.01	4.32	5.05	5.50	5.84	6.10	6.32	6.51	6.67	6.81	6.94	7.06	7.17	7.26	7.36

Continued on page 785

TABLE A.8 Upper percentage points for the Studentized range q_{ν_1, ν_2} (continued)

		ν_1													
ν_2	α	2	3	4	5	6	7	8	9	10	11	12	13	14	15
13	0.10	2.50	3.18	3.59	3.88	4.12	4.30	4.46	4.60	4.72	4.83	4.93	5.02	5.10	5.18
	0.05	3.06	3.73	4.15	4.45	4.69	4.88	5.05	5.19	5.32	5.43	5.53	5.63	5.71	5.79
	0.01	4.26	4.96	5.40	5.73	5.98	6.19	6.37	6.53	6.67	6.79	6.90	7.01	7.10	7.19
14	0.10	2.49	3.16	3.56	3.85	4.08	4.27	4.42	4.56	4.68	4.79	4.88	4.97	5.05	5.12
	0.05	3.03	3.70	4.11	4.41	4.64	4.83	4.99	5.13	5.25	5.36	5.46	5.55	5.64	5.72
	0.01	4.21	4.89	5.32	5.63	5.88	6.08	6.26	6.41	6.54	6.66	6.77	6.87	6.96	7.05
15	0.10	2.48	3.14	3.54	3.83	4.05	4.23	4.39	4.52	4.64	4.75	4.84	4.93	5.01	5.08
	0.05	3.01	3.67	4.08	4.37	4.60	4.78	4.94	5.08	5.20	5.31	5.40	5.49	5.58	5.65
	0.01	4.17	4.84	5.25	5.56	5.80	5.99	6.16	6.31	6.44	6.55	6.66	6.76	6.84	6.93
16	0.10	2.47	3.12	3.52	3.80	4.03	4.21	4.36	4.49	4.61	4.71	4.81	4.89	4.97	5.04
	0.05	3.00	3.65	4.05	4.33	4.56	4.74	4.90	5.03	5.15	5.26	5.35	5.44	5.52	5.59
	0.01	4.13	4.79	5.19	5.49	5.72	5.92	6.08	6.22	6.35	6.46	6.56	6.66	6.74	6.82
17	0.10	2.46	3.11	3.50	3.78	4.00	4.18	4.33	4.46	4.58	4.68	4.77	4.86	4.93	5.01
	0.05	2.98	3.63	4.02	4.30	4.52	4.71	4.86	4.99	5.11	5.21	5.31	5.39	5.47	5.55
	0.01	4.10	4.74	5.14	5.43	5.66	5.85	6.01	6.15	6.27	6.38	6.48	6.57	6.66	6.73
18	0.10	2.45	3.10	3.49	3.77	3.98	4.16	4.31	4.44	4.55	4.65	4.75	4.83	4.90	4.98
	0.05	2.97	3.61	4.00	4.28	4.49	4.67	4.82	4.96	5.07	5.17	5.27	5.35	5.43	5.50
	0.01	4.07	4.70	5.09	5.38	5.60	5.79	5.94	6.08	6.20	6.31	6.41	6.50	6.58	6.65
19	0.10	2.44	3.09	3.47	3.75	3.97	4.14	4.29	4.42	4.53	4.63	4.72	4.80	4.88	4.95
	0.05	2.96	3.59	3.98	4.25	4.47	4.65	4.79	4.92	5.04	5.14	5.23	5.32	5.39	5.46
	0.01	4.05	4.67	5.05	5.33	5.55	5.73	5.89	6.02	6.14	6.25	6.34	6.43	6.51	6.58
20	0.10	2.44	3.08	3.46	3.74	3.95	4.12	4.27	4.40	4.51	4.61	4.70	4.78	4.85	4.92
	0.05	2.95	3.58	3.96	4.23	4.45	4.62	4.77	4.90	5.01	5.11	5.20	5.28	5.36	5.43
	0.01	4.02	4.64	5.02	5.29	5.51	5.69	5.84	5.97	6.09	6.19	6.29	6.37	6.45	6.52
24	0.10	2.42	3.05	3.42	3.69	3.90	4.07	4.21	4.34	4.45	4.54	4.63	4.71	4.78	4.85
	0.05	2.92	3.53	3.90	4.17	4.37	4.54	4.68	4.81	4.92	5.01	5.10	5.18	5.25	5.32
	0.01	3.96	4.55	4.91	5.17	5.37	5.54	5.69	5.81	5.92	6.02	6.11	6.19	6.26	6.33
30	0.10	2.40	3.02	3.39	3.65	3.85	4.02	4.16	4.28	4.38	4.47	4.56	4.64	4.71	4.77
	0.05	2.89	3.49	3.85	4.10	4.30	4.46	4.60	4.72	4.82	4.92	5.00	5.08	5.15	5.21
	0.01	3.89	4.45	4.80	5.05	5.24	5.40	5.54	5.65	5.76	5.85	5.93	6.01	6.08	6.14
40	0.10	2.38	2.99	3.35	3.60	3.80	3.96	4.10	4.21	4.32	4.41	4.49	4.56	4.63	4.69
	0.05	2.86	3.44	3.79	4.04	4.23	4.39	4.52	4.63	4.74	4.82	4.90	4.98	5.05	5.11
	0.01	3.82	4.37	4.70	4.93	5.11	5.27	5.39	5.50	5.60	5.69	5.76	5.83	5.90	5.96
60	0.10	2.36	2.96	3.31	3.56	3.75	3.91	4.04	4.16	4.25	4.34	4.42	4.49	4.56	4.62
	0.05	2.83	3.40	3.74	3.98	4.16	4.31	4.44	4.55	4.65	4.73	4.81	4.88	4.94	5.00
	0.01	3.76	4.28	4.59	4.82	4.99	5.13	5.25	5.36	5.45	5.53	5.60	5.67	5.73	5.79
120	0.10	2.34	2.93	3.28	3.52	3.71	3.86	3.99	4.10	4.19	4.28	4.35	4.42	4.48	4.54
	0.05	2.80	3.36	3.68	3.92	4.10	4.24	4.36	4.47	4.56	4.64	4.71	4.78	4.84	4.90
	0.01	3.70	4.20	4.50	4.71	4.87	5.01	5.12	5.21	5.30	5.38	5.44	5.50	5.56	5.61
∞	0.10	2.33	2.90	3.24	3.48	3.66	3.81	3.93	4.04	4.13	4.21	4.28	4.35	4.41	4.47
	0.05	2.77	3.31	3.63	3.86	4.03	4.17	4.29	4.39	4.47	4.55	4.62	4.68	4.74	4.80
	0.01	3.64	4.12	4.40	4.60	4.76	4.88	4.99	5.08	5.16	5.23	5.29	5.35	5.40	5.45

TABLE A.9 Control chart constants

n	A_2	A_3	B_3	B_4	D_3	D_4	c_4	d_2
2	1.880	2.659	0.000	3.267	0.000	3.267	0.7979	1.128
3	1.023	1.954	0.000	2.568	0.000	2.575	0.8862	1.693
4	0.729	1.628	0.000	2.266	0.000	2.282	0.9213	2.059
5	0.577	1.427	0.000	2.089	0.000	2.114	0.9400	2.326
6	0.483	1.287	0.030	1.970	0.000	2.004	0.9515	2.534
7	0.419	1.182	0.118	1.882	0.076	1.924	0.9594	2.704
8	0.373	1.099	0.185	1.815	0.136	1.864	0.9650	2.847
9	0.337	1.032	0.239	1.761	0.184	1.816	0.9693	2.970
10	0.308	0.975	0.284	1.716	0.223	1.777	0.9727	3.078
11	0.285	0.927	0.321	1.679	0.256	1.744	0.9754	3.173
12	0.266	0.866	0.354	1.646	0.283	1.717	0.9776	3.258
13	0.249	0.850	0.382	1.618	0.307	1.693	0.9794	3.336
14	0.235	0.817	0.406	1.594	0.328	1.672	0.9810	3.407
15	0.223	0.789	0.428	1.572	0.347	1.653	0.9823	3.472
16	0.212	0.763	0.448	1.552	0.363	1.637	0.9835	3.532
17	0.203	0.739	0.466	1.534	0.378	1.622	0.9845	3.588
18	0.194	0.718	0.482	1.518	0.391	1.609	0.9854	3.640
19	0.187	0.698	0.497	1.503	0.403	1.597	0.9862	3.689
20	0.180	0.680	0.510	1.490	0.415	1.585	0.9869	3.735
21	0.173	0.663	0.523	1.477	0.425	1.575	0.9876	3.778
22	0.167	0.647	0.534	1.466	0.434	1.566	0.9882	3.819
23	0.162	0.633	0.545	1.455	0.443	1.557	0.9887	3.858
24	0.157	0.619	0.555	1.445	0.452	1.548	0.9892	3.895
25	0.153	0.606	0.565	1.435	0.459	1.541	0.9896	3.931

For $n > 25$: $A_3 \approx 3/\sqrt{n}$, $B_3 \approx 1 - 3/\sqrt{2n}$, and $B_4 \approx 1 + 3/\sqrt{2n}$.

Appendix B

Partial Derivatives

This appendix presents the mechanics of computing partial derivatives, which are needed in Section 3.4. We begin by recalling that a derivative specifies the rate of change of one variable with respect to another. For example, the volume v of a sphere whose radius is r is given by $v = 4\pi r^3$. If r is allowed to increase (or decrease), the rate at which v increases (or decreases) is given by the derivative of v with respect to r: $dv/dr = 12\pi r^2$.

Partial derivatives are needed when the quantity whose rate of change is to be calculated is a function of more than one variable. Here is an example: The volume v of a cylinder with radius r and height h is given by $v = \pi r^2 h$. If either r or h changes, v will change as well. Now imagine that h is constant, and r is allowed to increase. The rate of increase in v is given by the **partial derivative** of v with respect to r. This derivative is denoted $\partial v/\partial r$, and it is computed exactly like the ordinary derivative of v with respect to r, treating h as a constant: $\partial v/\partial r = 2\pi rh$.

Now assume that r is constant, and h is increasing. The rate of increase in v is the partial derivative of v with respect to h, denoted $\partial v/\partial h$. It is computed exactly like the ordinary derivative of v with respect to h, treating r as a constant: $\partial v/\partial h = \pi r^2$.

> If v is a function of several variables, $v = f(x_1, x_2, \ldots, x_n)$, then the partial derivative of v with respect to any one of the variables $x_1, x_2, \ldots, x_n$ is computed just like the ordinary derivative, holding the other variables constant.

Examples B.1 and B.2 show that computing partial derivatives is no more difficult than computing ordinary derivatives.

Let $v = 12x^2y + 3xy^2$. Find the partial derivatives of v with respect to x and y.

Solution

To compute $\partial v/\partial x$, hold y constant, and compute the derivative with respect to x. The result is

$$\frac{\partial v}{\partial x} = 24xy + 3y^2$$

To compute $\partial v/\partial y$, hold x constant, and compute the derivative with respect to y. The result is

$$\frac{\partial v}{\partial y} = 12x^2 + 6xy$$

Example B.2

Let $v = \dfrac{x^3y + y^3z - xz^3}{x^2 + y^2 + z^2}$. Find the partial derivatives of v with respect to x, y, and z.

Solution

To compute $\partial v/\partial x$, hold both y and z constant, and compute the derivative with respect to x, using the quotient rule:

$$\frac{\partial v}{\partial x} = \frac{(3x^2y - z^3)(x^2 + y^2 + z^2) - (x^3y + y^3z - xz^3)(2x)}{(x^2 + y^2 + z^2)^2}$$

Similarly, we compute the partial derivatives of v with respect to y and z:

$$\frac{\partial v}{\partial y} = \frac{(x^3 + 3y^2z)(x^2 + y^2 + z^2) - (x^3y + y^3z - xz^3)(2y)}{(x^2 + y^2 + z^2)^2}$$

$$\frac{\partial v}{\partial z} = \frac{(y^3 - 3xz^2)(x^2 + y^2 + z^2) - (x^3y + y^3z - xz^3)(2z)}{(x^2 + y^2 + z^2)^2}$$

Exercises for Appendix B

In the following exercises, compute all partial derivatives.

1. $v = 3x + 2xy^4$

2. $w = \dfrac{x^3 + y^3}{x^2 + y^2}$

3. $z = \cos x \sin y^2$

4. $v = e^{xy}$

5. $v = e^x(\cos y + \sin z)$

6. $w = \sqrt{x^2 + 4y^2 + 3z^2}$

7. $z = \ln(x^2 + y^2)$

8. $v = e^{y^2}\cos(xz) + \ln(x^2y + z)$

9. $v = \dfrac{2xy^3 - 3xy^2}{\sqrt{xy}}$

10. $z = \sqrt{\sin(x^2y)}$

Appendix C

Bibliography

Agresti, A. (1996). *An Introduction to Categorical Data Analysis.* John Wiley & Sons, New York. An authoritative and comprehensive treatment of the subject.

Belsey, D., Kuh, E., and Welsch, R. (1980). *Regression Diagnostics: Identifying Influential Data and Sources of Collinearity.* John Wiley & Sons, New York. A presentation of methods for evaluating the reliability of regression estimates.

Bevington, P., and Robinson, D. (2002). *Data Reduction and Error Analysis for the Physical Sciences*, 3rd ed. McGraw-Hill, Boston. An introduction to data analysis, with emphasis on propagation of error and calculation.

Bickel, P., and Doksum, K. (2001). *Mathematical Statistics: Basic Ideas and Selected Topics*, Vol. I, 2nd ed. Prentice-Hall, Upper Saddle River, NJ. A thorough treatment of the mathematical principles of statistics, at a fairly advanced level.

Box, G., and Draper, N. (1987). *Empirical Model-Building and Response Surfaces.* John Wiley & Sons, New York. An excellent practical introduction to the fitting of curves and higher-order surfaces to data.

Box, G., Hunter, W., and Hunter, J. (1978). *Statistics for Experimenters.* John Wiley & Sons, New York. A very intuitive and practical introduction to the basic principles of data analysis and experimental design.

Brockwell, R., and Davis, R. (2002). *Introduction to Time Series and Forecasting*, 2nd ed. Springer-Verlag, New York. An excellent introductory text at the undergraduate level, more rigorous than Chatfield (2003).

Casella, G., and Berger, R. (2002). *Statistical Inference*, 2nd ed. Duxbury, Pacific Grove, CA. A fairly rigorous development of the theory of statistics.

Chatfield, C. (2003). *An Analysis of Time Series: An Introduction*, 6th ed. CRC Press, Boca Raton, FL. An intuitive presentation, at a somewhat less advanced level than Brockwell and Davis (2002).

Chatfield, C. (1983). *Statistics for Technology*, 3rd ed., revised. Chapman and Hall/CRC, Boca Raton, FL. A clear and concise introduction to basic principles of statistics, oriented toward engineers and scientists.

Cochran, W. (1977). *Sampling Techniques*, 3rd ed. John Wiley & Sons, New York. A comprehensive account of sampling theory.

Cook, D., and Weisberg, S. (1994). *An Introduction to Regression Graphics*. John Wiley & Sons, New York. A presentation of interactive methods for analyzing data with linear models.

DeGroot, M., and Schervish, M. (2002). *Probability and Statistics*, 3rd ed. Addison-Wesley, Reading, MA. A very readable introduction at a somewhat higher mathematical level than this book.

Draper, N., and Smith, H. (1998). *Applied Regression Analysis*, 3rd ed. John Wiley & Sons, New York. An extensive and authoritative treatment of linear regression.

Efron, B., and Tibshirani, R. (1993). *An Introduction to the Bootstrap*. Chapman and Hall, New York. A clear and comprehensive introduction to bootstrap methods.

Freedman, D., Pisani, R., and Purves, R. (1998). *Statistics*, 3rd ed. Norton, New York. An excellent intuitive introduction to the fundamental principles of statistics.

Hocking, R. (2003). *Methods and Applications of Linear Models: Regression and the Analysis of Variance*, 2nd ed. John Wiley & Sons, New York. A thorough treatment of the theory and applications of regression and analysis of variance.

Kenett, R., and Zacks, S. (1998). *Modern Industrial Statistics*. Brooks/Cole, Pacific Grove, California. An up-to-date treatment of the subject with emphasis on industrial engineering.

Larsen, R., and Marx, M. (2001). *An Introduction to Mathematical Statistics and Its Applications*, 3rd ed. Prentice-Hall, Upper Saddle River, NJ. An introduction to statistics at a higher mathematical level than that of this book. Contains many good examples.

Lee, P. (1997). *Bayesian Statistics: An Introduction*, 2nd ed. Oxford University Press, New York. A clear and basic introduction to statistical methods that are based on the subjective view of probability.

Lehmann, E. (1998). *Nonparametrics: Statistical Methods Based on Ranks*. Pearson, New York. Thorough presentation of basic distribution-free methods.

Miller, A. (2002). *Subset Selection in Regression*, 2nd ed. Chapman and Hall, London. A strong and concise treatment of the basic principles of model selection.

Miller, R. (1997). *Beyond ANOVA: The Basics of Applied Statistics*. Chapman and Hall/CRC, Boca Raton, FL. A very practical and intuitive treatment of methods useful in analyzing real data, when standard assumptions may not be satisfied.

Montgomery, D. (2001a). *Design and Analysis of Experiments*, 5th ed. John Wiley & Sons, New York. A thorough exposition of the methods of factorial experiments, focusing on engineering applications.

Montgomery, D. (2001b). *Introduction to Statistical Quality Control*, 4th ed. John Wiley & Sons, New York. A comprehensive and readable introduction to the subject.

Mood, A., Graybill, F., and Boes, D. (1974). *Introduction to the Theory of Statistics*, 3rd ed. McGraw-Hill, Boston. A classic introduction to mathematical statistics and an excellent reference.

Mosteller, F., and Tukey, J. (1977). *Data Analysis and Regression*. Addison-Wesley, Reading, MA. An intuitive and philosophical presentation of some very practical ideas.

Rice, J. (1995). *Mathematical Statistics and Data Analysis*, 2nd ed. Wadsworth, Belmont, CA. A good blend of theory and practice, at a somewhat higher level than this book.

Ross, S. (2001). *A First Course in Probability*, 6th ed. Prentice-Hall, Upper Saddle River, NJ. A mathematically sophisticated introduction to probability.

Ross, S. (2004). *Introduction to Probability and Statistics for Engineers and Scientists*, 3rd ed. Harcourt/Academic Press, San Diego. An introduction at a somewhat higher mathematical level than this book.

Salsburg, D. (2001). *The Lady Tasting Tea*. W. H. Freeman and Company, New York. An insightful discussion of the influence of statistics on 20th century science, with many fascinating anecdotes about famous statisticians. The story in Section 4.3 about Student's discovery that the number of particles in a suspension follows a Poisson distribution can be found in this book.

Tanur, J., Pieters, R., and Mosteller, F. (eds.) (1989). *Statistics: A Guide to the Unknown*, 3rd ed. Wadsworth/Brooks-Cole, Pacific Grove, CA. A collection of case studies illustrating a variety of statistical applications.

Taylor, J. (1997). *An Introduction to Error Analysis*, 2nd ed. University Science Books, Sausalito, CA. A thorough treatment of propagation of error, along with a selection of other topics in data analysis.

Tufte, E. (2001). *The Visual Display of Quantitative Information*, 2nd ed. Graphics Press, Cheshire, CT. A clear and compelling demonstration of the principles of effective statistical graphics, containing numerous examples.

Tukey, J. (1977). *Exploratory Data Analysis*. Addison-Wesley, Reading, MA. A wealth of techniques for summarizing and describing data.

Wackerly, D., Mendenhall, W., and Scheaffer, R. (2002). *Mathematical Statistics with Applications*, 6th ed. Duxbury, Pacific Grove, CA. An introduction to statistics at a somewhat higher mathematical level than this book.

Weisberg, S. (1985). *Applied Linear Regression*, 2nd ed. John Wiley & Sons, New York. A concise introduction to the application of linear regression models, including diagnostics, model building, and interpretation of output.

Answers to Odd-Numbered Exercises

Section 1.1

1. (a) The population consists of all the bolts in the shipment. It is tangible.
   (b) The population consists of all measurements that could be made on that resistor with that ohmmeter. It is conceptual.
   (c) The population consists of all residents of the town. It is tangible.
   (d) The population consists of all welds that could be made by that process. It is conceptual.
   (e) The population consists of all parts manufactured that day. It is tangible.

3. (a) False (b) True

5. (a) No. What is important is the population proportion of defectives; the sample percentage is only an approximation. The population proportion for the new process may in fact be greater or less than that of the old process.
   (b) No. The population proportion for the new process may be 10% or more, even though the sample proportion was only 9%.
   (c) Finding two defective bottles in the sample.

7. A good knowledge of the process that generated the data.

Section 1.2

1. False

3. No. In the sample 1, 2, 4 the mean is 7/3, which does not appear at all.

5. The sample size can be any odd number.

7. The mean and standard deviation both increase by 5%.

9. The mean is 2.45, the median is 3, and the standard deviation is 1.2999.

11. (a) Yes, the mean is 2.45. (b) Yes, the median is 3. (c) No, the sample standard deviation depends on the sample size.

13. (a) All would be multiplied by 2.54.
(b) Not exactly the same, since the measurements would be a little different the second time.

15. There is no fourth quartile. There are three quartiles; they are the points that divide the data into four equal groups.

Section 1.3

1. (a)

Stem	Leaf
11	6
12	678
13	13678
14	13368
15	126678899
16	122345556
17	013344467
18	1333558
19	2
20	3

(b) Here is one histogram. Other choices for the endpoints are possible.

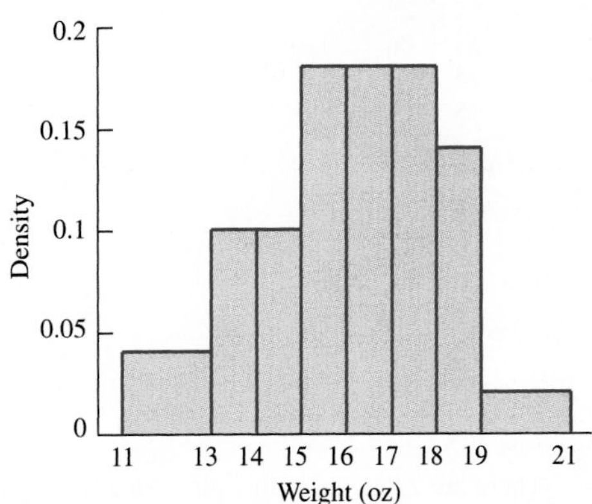

(c)

10 12 14 16 18 20 22 24

Weight (oz)

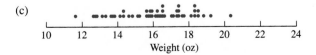

(d)

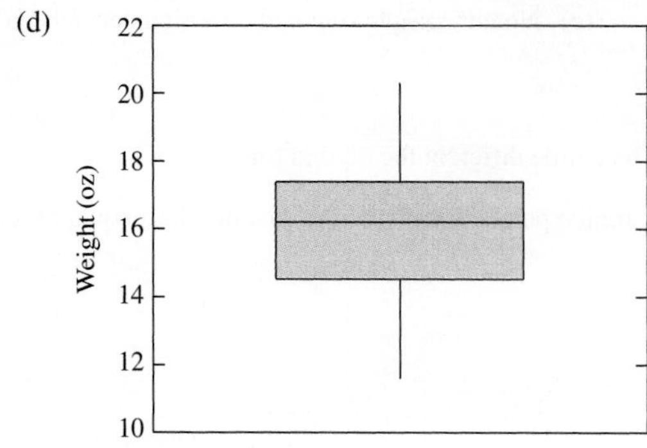

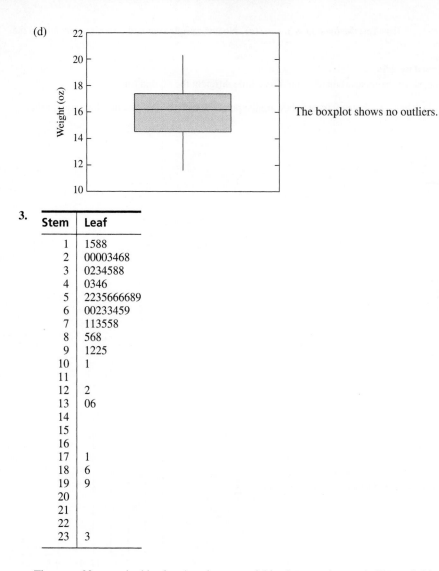

The boxplot shows no outliers.

3.

Stem	Leaf
1	1588
2	00003468
3	0234588
4	0346
5	2235666689
6	00233459
7	113558
8	568
9	1225
10	1
11	
12	2
13	06
14	
15	
16	
17	1
18	6
19	9
20	
21	
22	
23	3

There are 23 stems in this plot. An advantage of this plot over the one in Figure 1.6 is that the values are given to the tenths digit instead of to the ones digit. A disadvantage is that there are too many stems, and many of them are empty.

5. (a) Here are histograms for each group. Other choices for the endpoints are possible.

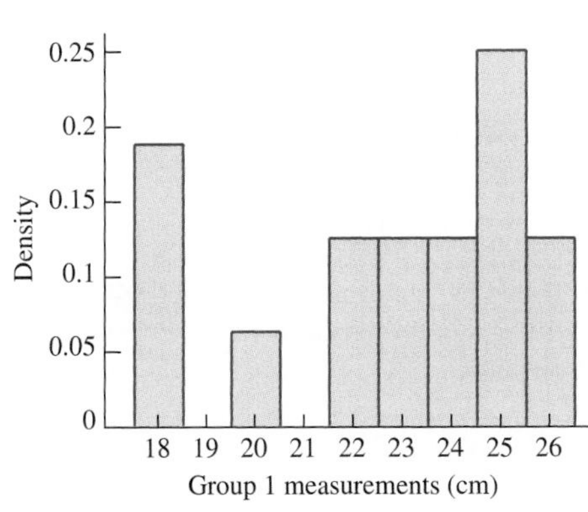

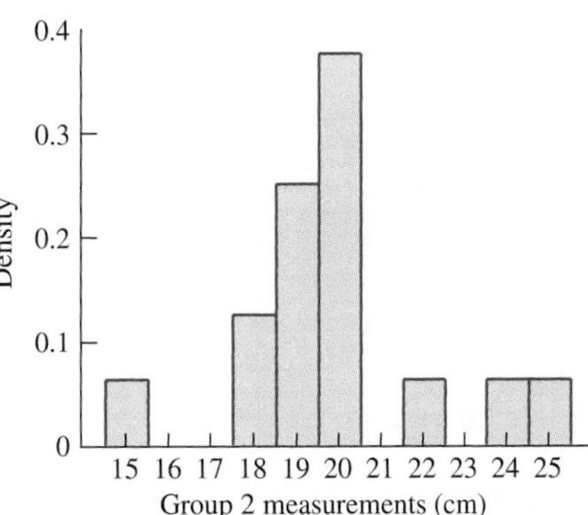

(b)

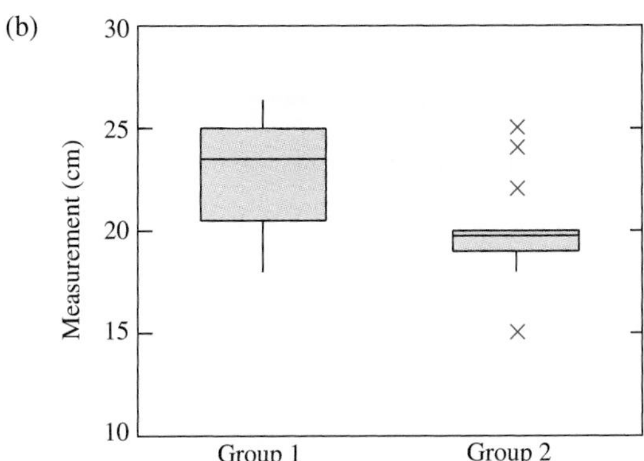

(c) The measurements in Group 1 are generally larger than those in Group 2. The measurements in Group 1 are not far from symmetric, although the boxplot suggests a slight skew to the left since the median is closer to the third quartile than the first. There are no outliers. Most of the measurements for Group 2 are highly concentrated in a narrow range and skewed to the left within that range. The remaining four measurements are outliers.

7. (a) Closest to 25% (b) 130–135 mm (c) 12%

9. (ii)

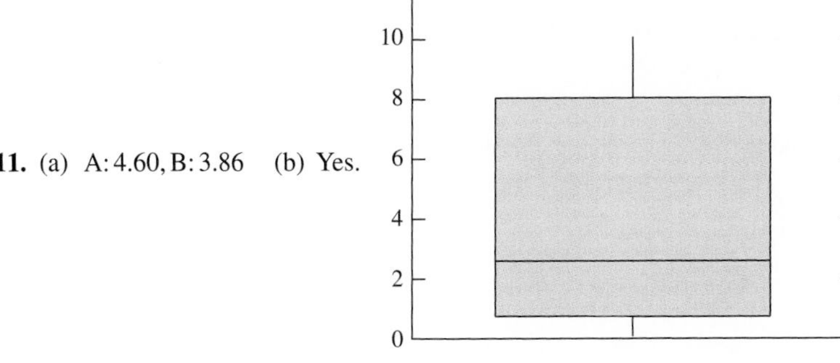

(c) No. The minimum value of −2.235 is an "outlier," since it is more than 1.5 times the interquartile range below the first quartile. The lower whisker should extend to the smallest point that is not an outlier, but the value of this point is not given.

11. (a) A: 4.60, B: 3.86 (b) Yes.

13. (a)

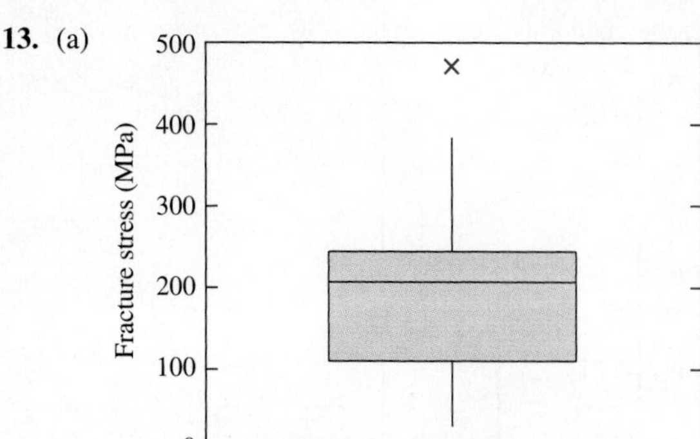

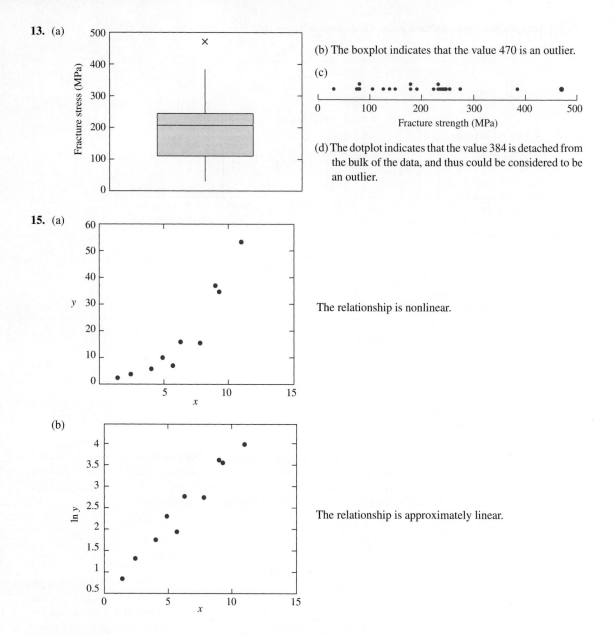

(b) The boxplot indicates that the value 470 is an outlier.

(c)

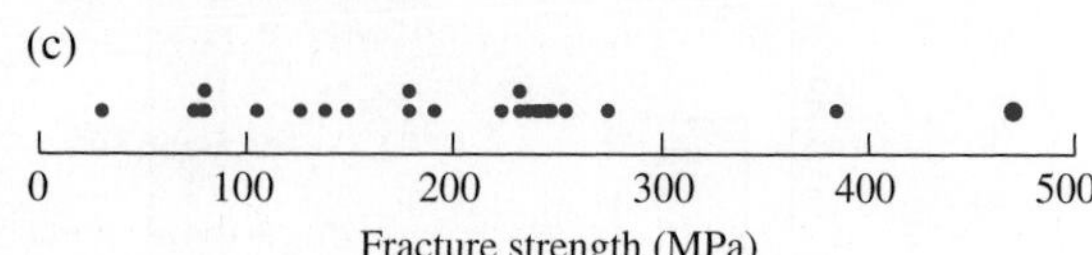

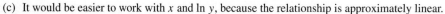

(d) The dotplot indicates that the value 384 is detached from the bulk of the data, and thus could be considered to be an outlier.

15. (a)

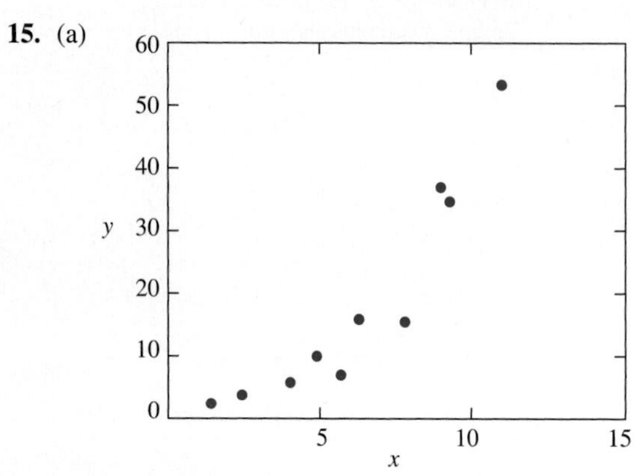

The relationship is nonlinear.

(b)

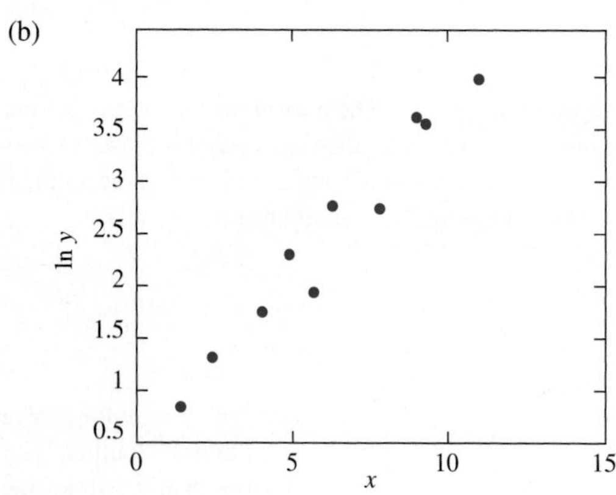

The relationship is approximately linear.

(c) It would be easier to work with x and $\ln y$, because the relationship is approximately linear.

Supplementary Exercises for Chapter 1

1. (a) The mean will be divided by 2.2.
(b) The standard deviation will be divided by 2.2.

3. (a) False (b) True (c) False (d) True

5. (a) It is not possible to tell by how much the mean changes.
(b) If there are more than two numbers on the list, the median is unchanged. If there are only two numbers on the list, the median is changed, but we cannot tell by how much.
(c) It is not possible to tell by how much the standard deviation changes.

7. (a) The mean decreases by 0.774. (b) The mean changes to 24.226.
(c) The median is unchanged.
(d) It is not possible to tell by how much the standard deviation changes.

9. Statement (i) is true.

11. (a) Incorrect (b) Correct (c) Incorrect (d) Correct

13. (a) Skewed to the left. The 85th percentile is much closer to the median (50th percentile) than the 15th percentile is. Therefore the histogram is likely to have a longer left-hand tail than right-hand tail.
(b) Skewed to the right. The 15th percentile is much closer to the median (50th percentile) than the 85th percentile is. Therefore the histogram is likely to have a longer right-hand tail than left-hand tail.

15. (a)

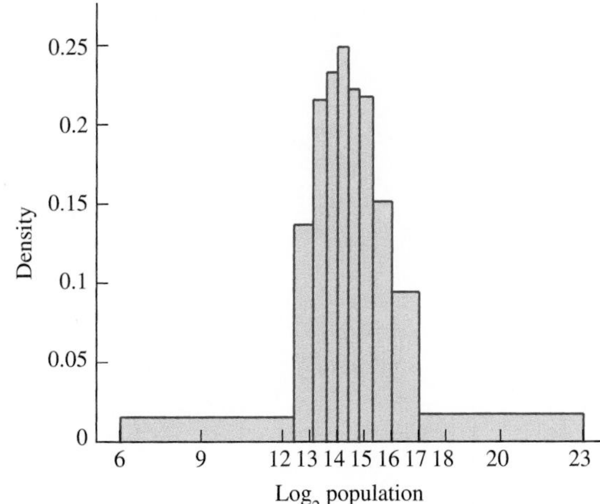

(b) 0.14
(c) Approximately symmetric

(d)

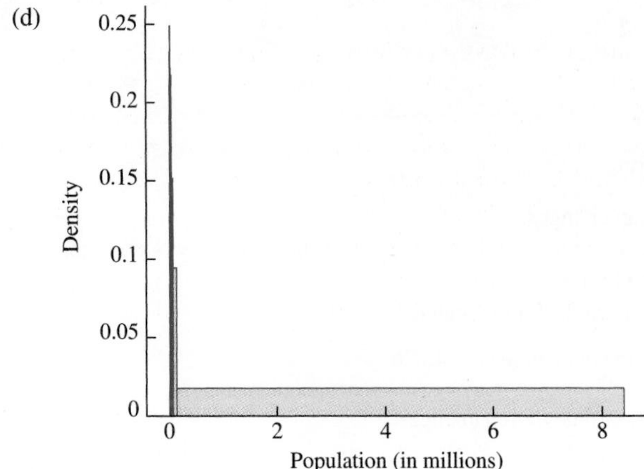

The data on the raw scale are skewed so much to the right that it is impossible to see the features of the histogram.

17. (a)

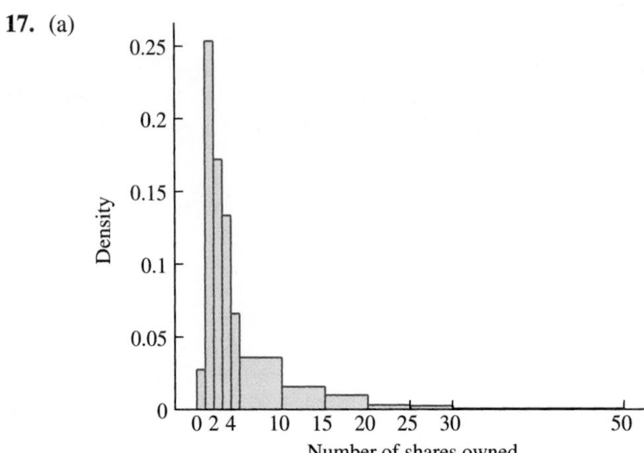

(b) 3.35 (c) 1.88 (d) 7.70 (e) 5.95 (f) 43.34

19. (a)

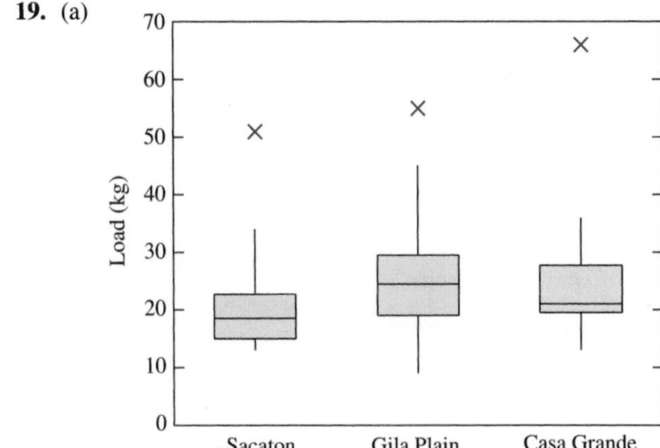

(b) Each sample contains one outlier.

(c) In the Sacaton boxplot, the median is about midway between the first and third quartiles, suggesting that the data between these quartiles are fairly symmetric. The upper whisker of the box is much longer than the lower whisker, and there is an outlier on the upper side. This indicates that the data as a whole are skewed to the right. In the Gila Plain boxplot data, the median is about midway between the first and third quartiles, suggesting that the data between these quartiles are fairly symmetric. The upper whisker is slightly longer than the lower whisker, and there is an outlier on the upper side. This suggests that the data as a whole are somewhat skewed to the right. In the Casa Grande boxplot, the median is very close to the first quartile. This suggests that there are several values very close to each other about one-fourth of the way through the data. The two whiskers are of about equal length, which suggests that the tails are about equal, except for the outlier on the upper side.

Section 2.1

1. 0.92

3. (a) 0.6 (b) 0.9

5. (a) False (b) True

7. 0.94

9. (a) False (b) True (c) False (d) True

Section 2.2

1. (a) 64 (b) 8 (c) 24

3. 210

5. 1,048,576

7. (a) $36^8 = 2.8211 \times 10^{12}$ (b) $36^8 - 26^8 = 2.6123 \times 10^{12}$ (c) 0.9260

9. 0.5238

Section 2.3

1. (a) 2/10 (b) 2/9 (c) 1/9

3. Given that a student is an engineering major, it is almost certain that the student took a calculus course. Therefore $P(B|A)$ is close to 1. Given that a student took a calculus course, it is much less certain that the student is an engineering major, since many nonengineering majors take calculus. Therefore $P(A|B)$ is much less than 1, so $P(B|A) > P(A|B)$.

5. (a) 0.018 (b) 0.728 (c) 0.272

7. (a) 0.8 (b) 0.125 (c) 0.12 (d) 0.167 (e) 0.88 (f) 0.205 (g) 0.795

9. (a) 0.98 (b) 0.02 (c) 0.72 (d) 0.18

11. (a) That the gauges fail independently.
(b) One cause of failure, a fire, will cause both gauges to fail. Therefore, they do not fail independently.

(c) Too low. The correct calculation would use P(second gauge fails|first gauge fails) in place of P(second gauge fails). Because there is a chance that both gauges fail together in a fire, the condition that the first gauge fails makes it more likely that the second gauge fails as well. Therefore P(second gauge fails|first gauge fails) > P(second gauge fails).

13. (a) 3/10 (b) 2/9 (c) 1/15 (d) 7/30 (e) 3/10
(f) No. $P(B) \neq P(B|A)$ [or $P(A \cap B) \neq P(A)P(B)$]

15. $n = 10{,}000$. The two components are a simple random sample from the population. When the population is large, the items in a simple random sample are nearly independent.

17. (a) 0.011 (b) 0.0033

19. (a) 5.08×10^{-5} (b) 0.9801 (c) 0.0001 (d) 0.9801

21. (a) 0.9904 (b) 0.1 (c) 0.2154 (d) 7

23. To show that A^c and B are independent, we show that $P(A^c \cap B) = P(A^c)P(B)$. Now $B = (A^c \cap B) \cup (A \cap B)$, and $(A^c \cap B)$ and $(A \cap B)$ are mutually exclusive. Therefore $P(B) = P(A^c \cap B) + P(A \cap B)$, from which it follows that $P(A^c \cap B) = P(B) - P(A \cap B)$. Since A and B are independent, $P(A \cap B) = P(A)P(B)$. Therefore $P(A^c \cap B) = P(B) - P(A)P(B) = P(B)[1 - P(A)] = P(A^c)P(B)$. To show that A and B^c are independent, it suffices to interchange A and B in the preceding argument. To show that A^c and B^c are independent, replace B with B^c in the preceding argument, and use the fact that A and B^c are independent.

Section 2.4

1. (a) Discrete (b) Continuous (c) Discrete (d) Continuous (e) Discrete

3. (a) 2.3 (b) 1.81 (c) 1.345 (d)

y	10	20	30	40	50
$p(y)$	0.4	0.2	0.2	0.1	0.1

(e) 23 (f) 181 (g) 13.45

5. (a) $c = 0.1$ (b) 0.2 (c) 3 (d) 1 (e) 1

7. (a)

x	$p_1(x)$
0	0.2
1	0.16
2	0.128
3	0.1024
4	0.0819
5	0.0655

(b)

x	$p_2(x)$
0	0.4
1	0.24
2	0.144
3	0.0864
4	0.0518
5	0.0311

(c) $p_2(x)$ appears to be the better model. Its probabilities are all fairly close to the proportions of days observed in the data. In contrast, the probabilities of 0 and 1 for $p_1(x)$ are much smaller than the observed proportions.
(d) No, this is not right. The data are a simple random sample, and the model represents the population. Simple random samples generally do not reflect the population exactly.

9. (a) 2 (b) 0.81 (c) 0.09 (d) 0.9 (e) 0.162

11. (a) 0.444 (b) 50 μg (c) 14.142 μg (d) $F(x) = \begin{cases} 0 & x < 10 \\ (x^2/2 - 10x + 50)/1800 & 10 \le x < 70 \\ 1 & x \ge 70 \end{cases}$ (e) 52.426 μg

13. (a) 5 seconds (b) 5 seconds (c) $F(t) = \begin{cases} 0 & t < 0 \\ 1 - e^{-0.2t} & t \ge 0 \end{cases}$ (d) 0.8647 (e) 3.466 seconds
(f) 11.5129 seconds

15. With this process, the probability that a ring meets the specification is 0.641. With the process in Exercise 14, the probability is 0.568. Therefore this process is better than the one in Exercise 14.

17. (a) 10 (b) 2

Section 2.5

1. (a) $\mu = 21.0, \sigma = 1.0$ (b) $\mu = 4.8, \sigma = 0.583$ (c) $\mu = 42.9, \sigma = 1.62$

3. $\mu = 3500$ hours, $\sigma = 44.7$ hours

5. (a) 0.625 in. (b) 0.0112 in.

7. (a) 0.650 (b) 0.158

9. (a) 150 cm (b) 0.447 cm

11. (a) 0.2993 (b) 0.00288

Section 2.6

1. (a) 0.08 (b) 0.36 (c) 0.86 (d) 0.35 (e) 0.60 (f) 0.65 (g) 0.40

3. (a) $p_{Y|X}(0 \mid 1) = 0.577$, $p_{Y|X}(1 \mid 1) = 0.308$, $p_{Y|X}(2 \mid 1) = 0.115$
(b) $p_{X|Y}(0 \mid 1) = 0.522$, $p_{X|Y}(1 \mid 1) = 0.348$, $p_{X|Y}(2 \mid 1) = 0.130$
(c) 0.538 (d) 0.609

5. (a) $\mu_{X+Y} = 3.90$ (b) $\sigma_{X+Y} = 1.179$ (c) $P(X + Y = 4) = 0.35$

7. (a) $100X + 200Y$ (b) 595 ms (c) 188.35

9. (a) $p_X(0) = 0.10$, $p_X(1) = 0.20$, $p_X(2) = 0.30$, $p_X(3) = 0.25$, $p_X(4) = 0.15$, $p_X(x) = 0$ if $x \ne 0, 1, 2, 3,$ or 4
(b) $p_Y(0) = 0.13$, $p_Y(1) = 0.21$, $p_Y(2) = 0.29$, $p_Y(3) = 0.22$, $p_Y(4) = 0.15$, $p_Y(y) = 0$ if $y \ne 0, 1, 2, 3,$ or 4
(c) No. The joint probability mass function is not equal to the product of the marginals. For example, $p_{X,Y}(0, 0) = 0.05 \ne p_X(0)p_Y(0)$.
(d) $\mu_X = 2.15$, $\mu_Y = 2.05$ (e) $\sigma_X = 1.1948$, $\sigma_Y = 1.2440$
(f) $\text{Cov}(X, Y) = 1.0525$ (g) $\rho_{X,Y} = 0.7081$

11. (a) $p_{Y|X}(0 \mid 3) = 0$, $p_{Y|X}(1 \mid 3) = 0.08$, $p_{Y|X}(2 \mid 3) = 0.32$, $p_{Y|X}(3 \mid 3) = 0.40$, $p_{Y|X}(4 \mid 3) = 0.20$
(b) $p_{X|Y}(0 \mid 4) = 0$, $p_{X|Y}(1 \mid 4) = 0$, $p_{X|Y}(2 \mid 4) = 2/15$, $p_{X|Y}(3 \mid 4) = 1/3$, $p_{X|Y}(4 \mid 4) = 8/15$
(c) $E(Y \mid X = 3) = 2.72$ (d) $E(X \mid Y = 4) = 3.4$

13. (a) $\mu_Z = 2.55$ (b) $\sigma_Z = 1.4309$ (c) $P(Z = 2) = 0.2$

15. (a) $p_{Y|X}(0 \mid 3) = 0.25$, $p_{Y|X}(1 \mid 3) = 0.25$, $p_{Y|X}(2 \mid 3) = 0.5$
(b) $p_{X|Y}(0 \mid 1) = 0.125$, $p_{X|Y}(1 \mid 1) = 0.25$, $p_{X|Y}(2 \mid 1) = 0.5$, $p_{X|Y}(3 \mid 1) = 0.125$
(c) 1.25 (d) 1.625

17. (a) −0.000193 (b) −0.00232

19. (a) $f_X(x) = \begin{cases} x + 1/2 & 0 < x < 1 \\ 0 & \text{for other values of } x \end{cases}$ $f_Y(y) = \begin{cases} y + 1/2 & 0 < y < 1 \\ 0 & \text{for other values of } y \end{cases}$

(b) $f_{Y|X}(y \mid 0.75) = \begin{cases} (4y + 3)/5 & 0 < y < 1 \\ 0 & \text{for other values of } y \end{cases}$

(c) $E(Y \mid X = 0.75) = 0.5667$

21. (a) 0.3 (b) 0.45 (c) 0.135 (d) $\mu_X = 10$ (e) $\mu_Y = 5$ (f) $\mu_A = 50$

23. (a) $f(x, y) = \begin{cases} e^{-x-y} & x > 0 \text{ and } y > 0 \\ 0 & \text{otherwise} \end{cases}$

(b) $P(X \le 1 \text{ and } Y > 1) = e^{-1} - e^{-2} = 0.2325$

(c) 1 (d) 2 (e) $1 - 3e^{-2} = 0.5940$

25. (a) $\mu = 40.25$, $\sigma = 0.11$ (b) $n \approx 52$

27. (a) $0.3X + 0.7Y$ (b) $\mu = \$6$, $\sigma = \$2.52$
(c) $\mu = \$6$, $\sigma = 0.03\sqrt{1.4K^2 - 140K + 10{,}000}$ (d) $K = \$50$
(e) For any correlation ρ, the risk is $0.03\sqrt{K^2 + (100 - K)^2 + 2\rho K(100 - K)}$. If $\rho \ne 1$ this quantity is minimized when $K = 50$.

29. (a) $\text{Cov}(aX, bY) = \mu_{aX \cdot bY} - \mu_{aX}\mu_{bY} = \mu_{abXY} - a\mu_X b\mu_Y = ab\mu_{XY} - ab\mu_X\mu_Y = ab(\mu_{XY} - \mu_X\mu_Y) = ab\,\text{Cov}(X, Y)$.
(b) $\rho_{aX,bY} = \text{Cov}(aX, bY)/(\sigma_{aX}\sigma_{bY}) = ab\,\text{Cov}(X, Y)/(ab\sigma_X\sigma_Y) = \text{Cov}(X, Y)/(\sigma_X\sigma_Y) = \rho_{X,Y}$.

Supplementary Exercises for Chapter 2

1. 0.9997

3. (a) 0.15 (b) 0.6667

5. 0.271

7. 0.82

9. 1/3

11. (a) 0.3996 (b) 0.0821 (c) $f_X(x) = \begin{cases} \frac{1}{2}e^{-x/2} & x > 0 \\ 0 & x \le 0 \end{cases}$ (d) $f_Y(y) = \begin{cases} \frac{1}{3}e^{-y/3} & y > 0 \\ 0 & y \le 0 \end{cases}$

(e) Yes, $f(x, y) = f_X(x) f_Y(y)$.

13. (a) 0.0436 (b) 0.0114 (c) 0.7377

15. 1/3

17. (a) $\mu = 6, \sigma^2 = 9$ (b) $\mu = 4, \sigma^2 = 10$ (c) $\mu = 0, \sigma^2 = 10$ (d) $\mu = 16, \sigma^2 = 328$

19. (a) For additive concentration: $p(0.02) = 0.22$, $p(0.04) = 0.19$, $p(0.06) = 0.29$, $p(0.08) = 0.30$, and $p(x) = 0$ for $x \neq 0.02, 0.04, 0.06$, or 0.08. For tensile strength: $p(100) = 0.14$, $p(150) = 0.36$, $p(200) = 0.50$, and $p(x) = 0$ for $x \neq 100$, 150, or 200.
(b) No, X and Y are not independent. For example, $P(X = 0.02 \cap Y = 100) = 0.05$, but $P(X = 0.02)\ P(Y = 100) = (0.22)(0.14) = 0.0308$. (c) 0.947 (d) 0.867 (e) The concentration should be 0.06.

21. (a) $p_{Y|X}(100 \mid 0.06) = 0.138$, $p_{Y|X}(150 \mid 0.06) = 0.276$, $p_{Y|X}(200 \mid 0.06) = 0.586$
(b) $p_{X|Y}(0.02 \mid 100) = 0.357$, $p_{X|Y}(0.04 \mid 100) = 0.071$, $p_{X|Y}(0.06 \mid 100) = 0.286$, $p_{X|Y}(0.08 | 100) = 0.286$
(c) 172.4 (d) 0.0500

23. (a) $\mu = 3.75, \sigma = 6.68$ (b) $\mu = 2.90, \sigma = 4.91$ (c) $\mu = 1.08, \sigma = 1.81$
(d) Under scenario A, 0.85; under scenario B, 0.89; and under scenario C, 0.99.

25. (a) The joint probability mass function is

	y		
x	0	1	2
0	0.0667	0.2000	0.0667
1	0.2667	0.2667	0
2	0.1333	0	0

(b) 0.8 (c) 0.6 (d) 0.6532 (e) 0.6110 (f) −0.2133 (g) −0.5345

27. (a) $\mu_X = 9/14 = 0.6429$ (b) $\sigma_X^2 = 199/2940 = 0.06769$ (c) $\text{Cov}(X, Y) = -5/588 = -0.008503$
(d) $\rho_{X,Y} = -25/199 = -0.1256$

29. (a) $p_X(0) = 0.6$, $p_X(1) = 0.4$, $p_X(x) = 0$ if $x \neq 0$ or 1. (b) $p_Y(0) = 0.4$, $p_Y(1) = 0.6$, $p_Y(y) = 0$ if $y \neq 0$ or 1.
(c) Yes. It is reasonable to assume that knowledge of the outcome of one coin will not help predict the outcome of the other.
(d) $p(0, 0) = 0.24$, $p(0, 1) = 0.36$, $p(1, 0) = 0.16$, $p(1, 1) = 0.24$, $p(x, y) = 0$ for other values of (x, y).

31. (a) $p_{X,Y}(x, y) = 1/9$ for $x = 1, 2, 3$ and $y = 1, 2, 3$. (b) $p_X(1) = p_X(2) = p_X(3) = 1/3$. p_Y is the same.
(c) $\mu_X = \mu_Y = 2$ (d) $\mu_{XY} = 4$ (e) $\text{Cov}(X, Y) = 0$.

Section 3.1

1. (ii)

3. (a) True (b) False (c) False (d) True

5. (a) No, we cannot determine the standard deviation of the process from a single measurement.
(b) Yes, the bias can be estimated to be 2 lb, because the reading is 2 lb when the true weight is 0.

7. (a) Yes, the uncertainty can be estimated with the standard deviation of the five measurements, which is 21.3 μg.
(b) No, the bias cannot be estimated, since we do not know the true value.

9. We can get a more accurate estimate by subtracting the bias of 26.2 μg, obtaining 100.8 μg above 1 kg.

11. (a) No, they are in increasing order, which would be highly unusual for a simple random sample.
(b) No, since they are not a simple random sample from a population of possible measurements, we cannot estimate the uncertainty.

Section 3.2

1. (a) 0.6 (b) 0.45 (c) 1.3

3. 1.01 ± 0.30 mm

5. 378.0 ± 0.5 in^3

7. 1.04 ± 0.04 cm^2/mol

9. (a) 64.04 ± 0.39°F (b) 64.04 ± 0.11°F

11. (a) The uncertainty in the average of nine measurements is approximately equal to $s/\sqrt{9} = 0.081/3 = 0.027$ cm.
(b) The uncertainty in a single measurement is approximately equal to s, which is 0.081 cm.

13. (a) 87.0 ± 0.7 mL (b) 0.5 mL (c) 25

15. (a) At 65°C, the yield is 70.14 ± 0.28. At 80°C, the yield is 90.50 ± 0.25. (b) 20.36 ± 0.38

17. (a) 0.016 (b) 0.0089
(c) The uncertainty in $\frac{1}{2}\overline{X} + \frac{1}{2}\overline{Y}$ is 0.0091. The uncertainty in $\frac{10}{15}\overline{X} + \frac{5}{15}\overline{Y}$ is 0.011.
(d) $c = 0.24$; the minimum uncertainty is 0.0078.

Section 3.3

1. (a) 3.2 (b) 0.1 (c) 0.025 (d) 0.1 (e) 21.8 (f) 0.26

3. 9.80 ± 0.39 m/s^2

5. (a) 1.7289 ± 0.0058 s (b) 9.79 ± 0.11 m/s^2

7. (a) 0.2555 ± 0.0005 m/s (b) 0.256 ± 0.026 m/s (c) 0.2555 ± 0.0002 m/s

9. (a) 2.3946 ± 0.0011 g/mL

11. (a) 0.27% (b) 0.037% (c) 4.0% (d) 0.5%

13. 9.802 m/s^2 $\pm$ 6.0%

15. (a) 1.856 s $\pm$ 0.29% (b) 9.799 m/s^2 $\pm$ 0.54%

17. (a) 0.2513 m/s $\pm$ 0.33% (b) 0.2513 m/s $\pm$ 2.0% (c) 0.2513 m/s $\pm$ 0.57%

19. 2.484 g/mL $\pm$ 0.19%

Section 3.4

1. (a) 250 ± 16 (b) 125 ± 10 (c) 17.50 ± 0.56

3. (a) 14.25 ± 0.25 MPa (b) Reducing the uncertainty in P_1 to 0.2 MPa.

5. (a) 1.320 ± 0.075 cm (b) Reducing the uncertainty in p to 0.1 cm.

7. 0.259 ± 0.014 m/s

9. (a) 32.6 ± 3.4 MPa (b) Reducing the uncertainty in k to $0.025\ \text{mm}^{-1}$.
(c) Implementing the procedure would reduce the uncertainty in τ only to 3.0 MPa. It is probably not worthwhile to implement the new procedure for a reduction this small.

11. (a) 710.68 ± 0.15 g (b) Reducing the uncertainty in b to 0.1 g.

13. (a) $2264 \pm 608\ \text{N/mm}^2$ (b) R

15. $0.0626 \pm 0.0013\ \text{min}^{-1}$

17. (a) No, they both involve the quantities h and r. (b) $2.68c \pm 0.27c$

19. 283.49 mm/s $\pm$ 2.5%

21. 1.41 cm $\pm$ 6.4%

23. 0.487 $\pm$ 1.7%

25. 3347.9 $\text{N/mm}^2 \pm 29\%$

27. (a) 17.59 μm $\pm$ 18% (b) 5.965 $\mu\text{m}^3 \pm 30\%$ (c) 2.95c $\pm$ 11% (d) No

Supplementary Exercises for Chapter 3

1. (a) 1.8 (b) 4.1 (c) 15 (d) 4.3

3. (a) 0.14 mm (b) 0.035 mm

5. (a) $(1.854 \pm 0.073) \times 10^6$ W (b) 3.9% (c) Reducing the uncertainty in H to 0.05.

7. (a) 6.57 ± 0.17 kcal (b) 2.6% (c) Reducing the uncertainty in the mass to 0.005 g.

9. (a) 26.32 ± 0.33 mm/year (b) 3.799 ± 0.048 years

11. 5.70 ± 0.17 mm

13. (a) 1.4% (b) Reducing the uncertainty in l to 0.5%

15. (a) Yes, the estimated strength is 80,000 lb in both cases.
(b) No, for the ductile wire method the squares of the uncertainties of the 16 wires are added, to obtain $\sigma = \sqrt{16 \times 20^2} = 80$. For the brittle wire method, the uncertainty in the strength of the weakest wire is multiplied by the number of wires, to obtain $\sigma = 16 \times 20 = 320$.

17. (a) $113.1 \pm 6.1\ \text{m}^3/\text{s}$ (b) $100.5 \pm 5.4\ \text{m}^3/\text{s}$ (c) Yes, the relative uncertainty is 5.4%.

19. (a) $10.04 \pm 0.95\ \text{s}^{-1}$ (b) $10.4 \pm 1.2\ \text{s}^{-1}$ (c) 0.78 (d) 0.63

21. (a) $32{,}833 \pm 36\ \text{m}^2$ (b) $12{,}894 \pm 14\ \text{m}^2$
(c) This is not correct. Let s denote the length of a side of the square. Since S and C are both computed in terms of s, they are not independent. In order to compute σ_A correctly, we must express A directly in terms of s: $A = s^2 + 2\pi s^2/8 = s^2(1 + \pi/4)$. So $\sigma_A = (dA/ds)\sigma_s = 2s(1 + \pi/4)\sigma_s = 65\ \text{m}^2$.

23. (a) $P_3 = 11.16871 \pm 0.10$ MPa (b) 11.16916
(c) No. The difference between the two estimates is much less than the uncertainty.

Section 4.1

1. (a) $\mu = 0.55, \sigma^2 = 0.2475$
(b) No. A Bernoulli random variable has possible values 0 and 1. The possible values of Y are 0 and 2.
(c) $\mu = 1.10, \sigma^2 = 0.99$

3. (a) 0.05 (b) 0.20 (c) 0.23 (d) Yes (e) No
(f) No. If the surface has both discoloration and a crack, then $X = 1$, $Y = 1$, and $Z = 1$, but $X + Y = 2$.

5. (a) 1/2 (b) 1/2 (c) 1/4 (d) Yes (e) Yes
(f) Yes. If both coins come up heads, then $X = 1$, $Y = 1$, and $Z = 1$, so $Z = XY$. If not, then $Z = 0$, and either X, Y, or both are equal to 0 as well, so again $Z = XY$.

7. (a) Since the possible values of X and Y are 0 and 1, the possible values of the product $Z = XY$ are also 0 and 1. Therefore Z is a Bernoulli random variable.
(b) $p_Z = P(Z = 1) = P(XY = 1) = P(X = 1 \text{ and } Y = 1) = P(X = 1)P(Y = 1) = p_X p_Y$.

Section 4.2

1. (a) 0.2090 (b) 0.2322 (c) 0.1064 (d) 0.0085 (e) 3.2 (f) 1.92

3. (a) 0.1172 (b) 5 (c) 2.5 (d) 1.58

5. (a) 0.0039 (b) 0.2188 (c) 0.1445 (d) 0.9648

7. (a) 0.120 ± 0.032 (b) 0.050 ± 0.015 (c) 0.070 ± 0.036

9. (a) 0.96 (b) 0.0582

11. (a) 1.346×10^{-4}
(b) Yes, only about 13 or 14 out of every 100,000 samples of size 10 would have seven or more defective items.
(c) Yes, because seven defectives in a sample of size 10 is an unusually large number for a good shipment.
(d) 0.4557 (e) No, in about 45% of the samples of size 10, two or more items would be defective.
(f) No, because two defectives in a sample of size 10 is not an unusually large number for a good shipment.

13. (a) $Y = 7X + 300$ (b) \$930 (c) \$21

15. 0.225 ± 0.064

Section 4.3

1. (a) 0.0733 (b) 0.0183 (c) 0.0916 (d) 0.9084 (e) 4 (f) 2

3. (a) 0.2240 (b) 0.4232 (c) 0.5974 (d) 3 (e) 1.73

5. (a) 0.0916 (b) 0.1048 (c) 0.2381

7. (ii)

9. 78 ± 12

11. (a) 12.5 (b) 7.0 (c) 2.5 (d) 1.9 (e) 5.5 ± 3.1

13. (a) 7.295×10^{-3} (b) Yes. If the mean concentration is 7 particles per mL, then only about 7 in every thousand 1 mL samples will contain 1 or fewer particles.
(c) Yes, because 1 particle in a 1 mL sample is an unusually small number if the mean concentration is 7 particles per mL. (d) 0.4497 (e) No. If the mean concentration is 7 particles per mL, then about 45% of all 1 mL samples will contain 6 or fewer particles.
(f) No, because 6 particles in a 1 mL sample is not an unusually small number if the mean concentration is 7 particles per mL.

15. 0.271 ± 0.019

Section 4.4

1. 0.4196

3. 0.0314

5. (a) 0.1244 (b) 7.5 (c) 11.25

7. (iv)

9. (a) 0.3 (b) $\mu_X = 1.2$ (c) 0.7483

11. (a) 0.00890 (b) 0.1275

Section 4.5

1. (a) 0.8023 (b) 0.2478 (c) 0.4338 (d) 0.7404

3. (a) 0.0073 (b) ≈ 420 (c) 91st percentile (d) 0.4186

5. (a) 0.0764 (b) 9.062 GPa (c) 12.303 GPa

7. (a) 0.0336 (b) Yes, the proportion of days shut down in this case would be only 0.0228.

9. (a) 0.06 cm (b) 0.01458 cm (c) 0.2451 (d) 0.0502 cm (e) 0.7352
(f) The hole diameter should have mean 15.02 cm. The probability of meeting the specification will then be 0.8294.

11. (a) 0.0475 (b) 12.07 oz (c) 0.0215 oz

13. (a) 7.8125 N/m^2 (b) 4.292 N/m^2 (c) 76.65 N/m^2

15. (a) 0.7088 (b) 0.2912 (c) 0.0485

17. (a) The mean is 114.8 J; the standard deviation is 5.006 J.
(b) Yes, only 0.15% of bolts would have breaking torques less than 100 J.
(c) The mean is 117.08 J; the standard deviation is 8.295 J. About 2% of the bolts would have breaking torques less than 100 J, so the shipment would not be accepted.
(d) The bolts in part (c) are stronger.
(e) The method is certainly not valid for the bolts in part (c). This sample contains an outlier (140), so the normal distribution should not be used.

Section 4.6

1. (a) 3.5966 (b) 0.5293 (c) 3.3201 (d) 5.5400

3. (a) 25.212 (b) 3.9828 (c) 24.903 (d) 0.2148 (e) 27.666

5. (a) 46.711 N/mm (b) 33.348 N/mm (c) Annularly threaded nails. The probability is 0.3372 versus 0.0516 for helically threaded nails. (d) 0.0985. (e) A helically threaded nail. Only about 0.01% of annularly threaded nails have strengths this small, while about 4.09% of helically threaded nails do. We can be pretty sure that it was a helically threaded nail.

7. (a) \$1.0565 (b) 0.0934 (c) \$1.0408 (d) 0.2090

Section 4.7

1. (a) 2.2222 (b) 4.9383 (c) 0.2592 (d) 1.5403

3. (a) 4 microns (b) 4 microns (c) 0.5276 (d) 0.0639 (e) 2.7726 microns (f) 5.5452 microns (g) 18.4207 microns

5. No. If the lifetimes were exponentially distributed, the proportion of used components lasting longer than five years would be the same as the proportion of new components lasting longer than five years, because of the lack of memory property.

7. (a) 1/3 year (b) 1/3 year (c) 0.0498 (d) 0.2212 (e) 0.9502

9. (a) 0.6065 (b) 0.0821 (c) The time of the first replacement will be greater than 100 hours if and only if each of the bulbs lasts longer than 100 hours. (d) 0.9179 (e) $P(T \leq t) = 1 - e^{-0.025t}$. (f) Yes, $T \sim \text{Exp}(0.025)$. (g) 40 hours (h) $T \sim \text{Exp}(n\lambda)$

Section 4.8

1. (a) 8 (b) 4 (c) 0.00175 (d) 0.9344

3. (a) 0.6667 (b) 1.4907 (c) 0.8231 (d) 0.0208 (e) 0.0550

5. (a) 0.8490 (b) 0.5410 (c) 1899.2 hours (d) 8.761×10^{-4}

7. (a) 0.3679 (b) 0.2978 (c) 0.4227

9. (a) 0.3679 (b) 0.1353
(c) The lifetime of the system will be greater than five hours if and only if the lifetimes of both components are greater than five hours. (d) 0.8647 (e) $P(T \leq t) = 1 - e^{-0.08t^2}$ (f) Yes, $T \sim \text{Weibull}(2, 0.2828)$.

Section 4.9

1. (a) No (b) No (c) Yes

3.

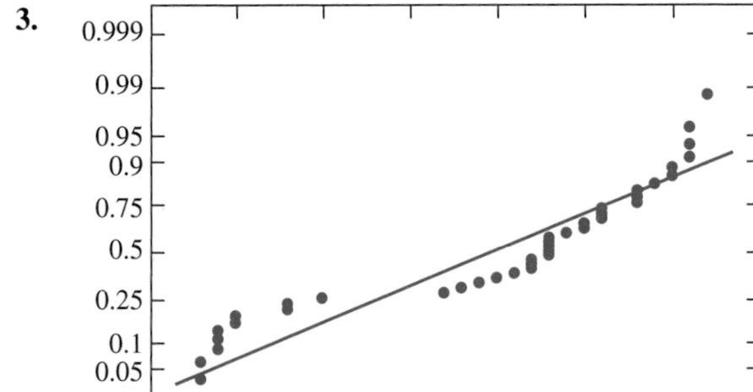

These data do not appear to come from an approximately normal distribution.

5.

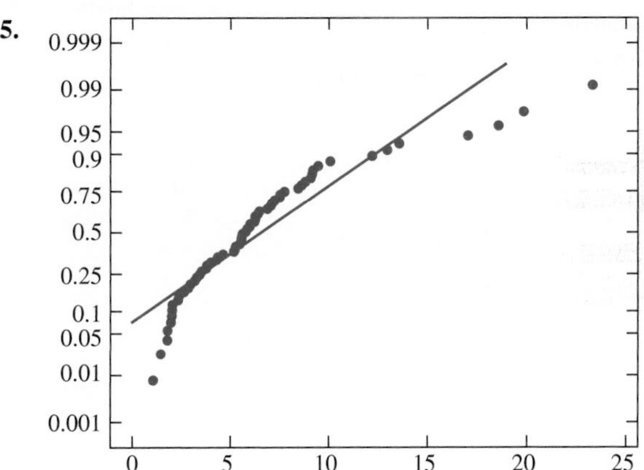

The PM data do not appear to come from an approximately normal distribution.

7. Yes. If the logs of the PM data come from a normal population, then the PM data come from a lognormal population, and vice versa.

Section 4.10

1. (a) 0.0793 (b) 0.0170

3. 0.9772

5. 0.5793

7. (a) 0.0951 (b) 0.0344 (c) 0.181 (0.278 is a spurious root.)

9. (a) 0.6578 (b) 0.4714 (c) 0.6266 (d) 48.02 ml

11. (a) 0.0002 (b) Yes. Only about 2 in 10,000 samples of size 1000 will have 75 or more nonconforming tiles if the goal has been reached. (c) No, because 75 nonconforming tiles in a sample of 1000 is an unusually large number if the goal has been reached.

(d) 0.3594 (e) No. More than one-third of the samples of size 1000 will have 53 or more nonconforming tiles if the goal has been reached. (f) Yes, because 53 nonconforming tiles in a sample of 1000 is not an unusually large number if the goal has been reached.

Section 4.11

1. (a) $X \sim \text{Bin}(100, 0.03)$, $Y \sim \text{Bin}(100, 0.05)$ (b) Answers will vary. (c) ≈ 0.72 (d) ≈ 0.18
(e) The distribution deviates somewhat from the normal.

3. (a) $\mu_A = 6$ exactly (simulation results will be approximate), $\sigma_A^2 \approx 0.25$.
(b) ≈ 0.16 (c) The distribution is approximately normal.

5. (a) ≈ 0.25 (b) ≈ 0.25 (c) ≈ 0.61

7. (a,b,c) Answers will vary. (d) ≈ 0.025

9. (a) Answers will vary. (b) ≈ 2.7 (c) ≈ 0.34 (d) ≈ 1.6
(e) System lifetime is not approximately normally distributed. (f) Skewed to the right.

11. (a) Answers will vary. (b) $\approx 10{,}090$ (c) ≈ 1250 (d) ≈ 0.58 (e) ≈ 0.095
(f) The distribution differs somewhat from normal.

13. (a) $\widehat{\lambda} = 0.25616$ (b,c,d) Answers will vary. (e) Bias ≈ 0.037, $\sigma_{\widehat{\lambda}} \approx 0.12$.

Supplementary Exercises for Chapter 4

1. 0.9744

3. (a) 0.2503 (b) 0.4744 (c) 0.1020 (d) 0.1414 (e) 0.8508

5. (a) 0.9044 (b) 0.00427 (c) 0.00512

7. (a) 0.6826 (b) $z = 1.28$ (c) 0.0010

9. (a) 0.0668 (b) 0.6687 (c) 0.0508

11. (a) 0.8830 (b) 0.4013 (c) 0.0390 (0.1065 is a spurious root.)

13. (a) 28.0 ± 3.7 (b) 28 mL

15. (a) 0.0749 (b) 4.7910 cm (c) 4

17. (a) 0.4889 (b) 0.8679

19. 62

21. (a) 0.4090 (b) No. More than 40% of the samples will have a total weight of 914.8 oz or less if the claim is true. (c) No, because a total weight of 914.8 oz is not unusually small if the claim is true.
(d) ≈ 0 (e) Yes. Almost none of the samples will have a total weight of 910.3 oz or less if the claim is true.
(f) Yes, because a total weight of 910.3 oz is unusually small if the claim is true.

23. (a) $P(X > s) = P(\text{First } s \text{ trials are failures}) = (1-p)^s$.
(b) $P(X > s+t \mid X > s) = P(X > s+t \text{ and } X > s)/P(X > s) = P(X > s+t)/P(X > s) = (1-p)^{s+t}/(1-p)^s = (1-p)^t = P(X > t)$. Note that if $X > s+t$, it must be the case that $X > s$, which is the reason that $P(X > s+t \text{ and } X > s) = P(X > s+t)$.

(c) Let X be the number of tosses of the penny needed to obtain the first head. Then $P(X > 5 \mid X > 3) = P(X > 2) = 1/4$. The probability that the nickel comes up tails twice is also 1/4.

25. (a) $\dfrac{P(X=x)}{P(X=x-1)} = \dfrac{e^{-\lambda}\lambda^x/x!}{e^{-\lambda}\lambda^{x-1}/(x-1)!} = \dfrac{e^{-\lambda}\lambda^x(x-1)!}{e^{-\lambda}\lambda^{x-1}x!} = \dfrac{\lambda}{x}$.

(b) $P(X = x) \geq P(X = x - 1)$ if and only if $\dfrac{\lambda}{x} \geq 1$ if and only if $x \leq \lambda$.

Section 5.1

1. (a) 1.645 (b) 1.37 (c) 2.81 (d) 1.15

3. Up, down

5. (a) (145.10, 154.90) (b) (143.55, 156.45) (c) 76.98% (d) 601 (e) 1041

7. (a) (1.538, 1.582) (b) (1.534, 1.586) (c) 92.66% (d) 385 (e) 543

9. (a) (27.04, 30.96) (b) (26.42, 31.58) (c) 86.64% (d) 312 (e) 540

11. (a) 22.62 (b) 99.38%

13. (a) 70.33 (b) 99.29%

15. 280

17. (83.11, 84.19)

19. (a) False (b) True (c) False

21. The supervisor is underestimating the confidence. The statement that the mean cost is less than \$160 is a one-sided upper confidence bound with confidence level 97.5%.

Section 5.2

1. (a) 0.74 (b) (0.603, 0.842) (c) 74 (d) (0.565, 0.879) (e) 130

3. (a) (0.629, 0.831) (b) (0.645, 0.815) (c) 300 (d) 210 (e) 0.0217

5. 0.339

7. (a) (0.0529, 0.1055) (b) 697 (c) (0.0008, 0.556)

9. (a) (0.490, 0.727) (b) (0.863, 1) (c) (0.572, 0.906)

11. (a) (0.271, 0.382) (b) 658 (c) 748

Section 5.3

1. (a) 1.860 (b) 2.776 (c) 2.763 (d) 12.706

3. (a) 95% (b) 98% (c) 90% (d) 99% (e) 99.9%

5. (5.303, 6.497)

7. Yes, there are no outliers. A 95% confidence interval is (203.81, 206.45).

9. (a)

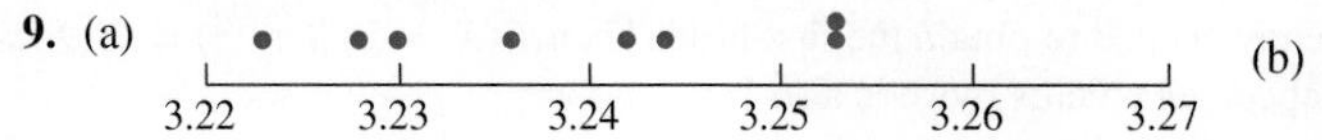

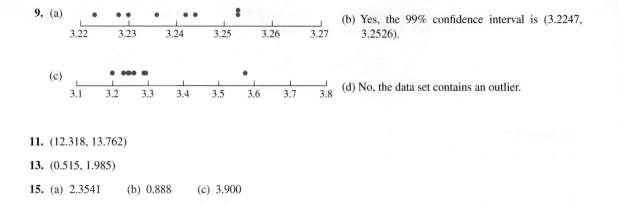

(b) Yes, the 99% confidence interval is (3.2247, 3.2526).

(c)

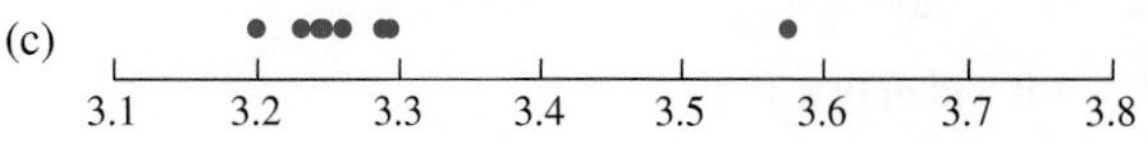

(d) No, the data set contains an outlier.

11. (12.318, 13.762)

13. (0.515, 1.985)

15. (a) 2.3541 (b) 0.888 (c) 3.900

Section 5.4

1. (5.589, 8.211)

3. (9.683, 11.597)

5. (11.018, 32.982)

7. (a) (−1.789, 2.589)
(b) No, since 0 is in the confidence interval, it may be regarded as being a plausible value for the mean difference in hardness.

9. It is not possible. The amounts of time spent in bed and spent asleep in bed are not independent.

Section 5.5

1. (0.0591, 0.208)

3. (−0.232, 0.00148)

5. (a) (−0.0446, 0.103) (b) If 100 additional chips were sampled from the less expensive process, the width of the confidence interval would be approximately ± 0.0721. If 50 additional chips were sampled from the more expensive process, the width of the confidence interval would be approximately ± 0.0638. If 50 additional chips were sampled from the less expensive process and 25 additional chips were sampled from the more expensive process, the width of the confidence interval would be approximately ± 0.0670. Therefore the greatest increase in precision would be achieved by sampling 50 additional chips from the more expensive process.

7. No. The sample proportions come from the same sample rather than from two independent samples.

9. (−0.0481, 0.226)

11. No, these are not simple random samples.

Section 5.6

1. $(-0.0978, 0.00975)$

3. $(7.798, 30.602)$

5. $(20.278, 25.922)$

7. $(0.0447, 0.173)$

9. $(2.500, 33.671)$

11. $(38.931, 132.244)$

13. $(0.614, 11.386)$

Section 5.7

1. $(2.090, 11.384)$

3. $(-0.0456, 0.00558)$

5. $(-30.260, 26.260)$

7. $(9.350, 10.939)$

9. (a) $(0.747, 2.742)$ (b) 80%

Section 5.8

1. (a) $X^* \sim N(8.5, 0.2^2)$, $Y^* \sim N(21.2, 0.3^2)$ (b) Answers will vary. (c) $\sigma_P \approx 0.18429$ (d) Yes, P is approximately normally distributed. (e) $(13.063, 13.785)$ if propagation of error is used to find the standard deviation.

3. (a) Yes, A is approximately normally distributed. (b) $\sigma_A \approx 0.24042$ (c) $(6.140, 7.083)$ if propagation of error is used to find the standard deviation.

5. (a) $N(0.27, 0.40^2/349)$ and $N(1.62, 1.70^2/143)$. Since the values 0.27 and 1.62 are sample means, their variances are equal to the population variances divided by the sample sizes. (b) No, R is not approximately normally distributed. (c) $\sigma_R \approx 0.70966$ (d) It is not appropriate, since R is not approximately normally distributed.

7. (a, b, c) Answers will vary.

9. (a) Coverage probability for Agresti–Coull ≈ 0.98; for traditional interval ≈ 0.89. Mean length for Agresti–Coull ≈ 0.51; for traditional interval ≈ 0.585. (b) Coverage probability for Agresti–Coull ≈ 0.95; for traditional interval ≈ 0.95. Mean length for Agresti–Coull ≈ 0.42; for traditional interval ≈ 0.46. (c) Coverage probability for Agresti–Coull ≈ 0.96; for traditional interval ≈ 0.92. Mean length for Agresti–Coull ≈ 0.29; for traditional interval ≈ 0.305. (d) The traditional method has coverage probability close to 0.95 for $n = 17$, but less than 0.95 for both $n = 10$ and $n = 40$. (e) Agresti–Coull has greater coverage probability for sample sizes 10 and 40, nearly the same for 17. (f) The Agresti–Coull method.

Supplementary Exercises for Chapter 5

1. (1.942, 19.725)

3. (0.0374, 0.0667)

5. (0.084, 0.516)

7. (a) (0.0886, 0.241) (b) 584

9. The narrowest interval, (4.20, 5.83), is the 90% confidence interval, the widest interval, (3.57, 6.46), is the 99% confidence interval, and (4.01, 6.02) is the 95% confidence interval.

11. (−0.420, 0.238)

13. 441

15. (a) False (b) False (c) True (d) False

17. (a) (36.804, 37.196) (b) 68% (c) The measurements come from a normal population.
(d) (36.774, 37.226)

19. (a) Since X is normally distributed with mean $n\lambda$, it follows that for a proportion $1 - \alpha$ of all possible samples, $-z_{\alpha/2}\sigma_X < X - n\lambda < z_{\alpha/2}\sigma_X$. Multiplying by -1 and adding X across the inequality yields $X - z_{\alpha/2}\sigma_X < n\lambda < X + z_{\alpha/2}\sigma_X$, which is the desired result.
(b) Since n is a constant, $\sigma_{X/n} = \sigma_X/n = \sqrt{n\lambda}/n = \sqrt{\lambda/n}$. Therefore $\sigma_{\widehat{\lambda}} = \sigma_X/n$.
(c) Divide the inequality in part (a) by n.
(d) Substitute $\sqrt{\widehat{\lambda}/n}$ for $\sigma_{\widehat{\lambda}}$ in part (c) to show that for a proportion $1 - \alpha$ of all possible samples, $\widehat{\lambda} - z_{\alpha/2}\sqrt{\widehat{\lambda}/n} < \lambda < \widehat{\lambda} + z_{\alpha/2}\sqrt{\widehat{\lambda}/n}$. The interval $\widehat{\lambda} \pm z_{\alpha/2}\sqrt{\widehat{\lambda}/n}$ is therefore a level $1 - \alpha$ confidence interval for λ.
(e) (53.210, 66.790)

21. (a) 234.375 ± 19.639 (b) (195.883, 272.867)
(c) There is some deviation from normality in the tails of the distribution. The middle 95% follows the normal curve closely, so the confidence interval is reasonably good.

23. (a,b,c) Answers will vary.

Section 6.1

1. (a) 0.0014 (b) 0.14%

3. (a) 0.2584 (b) 25.84%

5. (a) ≈ 0 (b) If the mean daily output were 740 tons or more, the chance of observing a sample mean as small as the value of 715 that was actually observed is nearly 0. Therefore we are convinced that the mean daily output is not 740 tons or more, but instead is less than 740 tons.

7. (a) 0.3300 (b) If the mean air velocity is 40 cm/s, there is a 33% chance that a sample will have a mean less than or equal to the observed value of 39.6. Since 33% is not a small probability, it is plausible that the mean air velocity is 40 cm/s.

9. (ii)

11. $P = 0.0456$

13. (a) 0.2153 (b) 2.65 (c) 0.0040

Section 6.2

1. $P = 0.5$

3. (iv)

5. (a) True (b) False (c) True (d) False

7. (a) No. The P-value is 0.196, which is greater than 0.05.
(b) The sample mean is 73.2461. The sample mean is therefore closer to 73 than to 73.5. The P-value for the null hypothesis $\mu = 73$ will therefore be larger than the P-value for the null hypothesis $\mu = 73.5$, which is 0.196. Therefore the null hypothesis $\mu = 73$ cannot be rejected at the 5% level.

9. (a) $H_0 : \mu \leq 8$ (b) $H_0 : \mu \leq 60{,}000$ (c) $H_0 : \mu = 10$

11. (a) (ii) The scale is out of calibration. (b) (iii) The scale might be in calibration. (c) No. The scale is in calibration only if $\mu = 10$. The strongest evidence in favor of this hypothesis would occur if $\overline{X} = 10$. But since there is uncertainty in $\overline{X}$, we cannot be sure even then that $\mu = 10$.

13. No, she cannot conclude that the null hypothesis is true.

15. (i)

17. (a) Yes. Quantities greater than the upper confidence bound will have P-values less than 0.05. Therefore $P < 0.05$.
(b) No, we would need to know the 99% upper confidence bound to determine whether $P < 0.01$.

19. Yes, we can compute the P-value exactly. Since the 95% upper confidence bound is 3.45, we know that $3.40 + 1.645s/\sqrt{n} = 3.45$. Therefore $s/\sqrt{n} = 0.0304$. The z-score is $(3.40 - 3.50)/0.0304 = -3.29$. The P-value is 0.0005, which is less than 0.01.

Section 6.3

1. Yes, $P = 0.0040$.

3. No, $P = 0.1251$.

5. No, $P = 0.2033$.

7. No, $P = 0.1251$.

9. Yes, $P = 0.0011$.

11. (a) 0.69 (b) −0.49 (c) 0.3121

Section 6.4

1. (a) $t_2 = 0.6547$, $0.50 < P < 0.80$. The scale may well be calibrated correctly.
(b) The t test cannot be performed, because the sample standard deviation cannot be computed from a sample of size 1.

3. (a) $H_0: \mu \le 35$ vs. $H_1: \mu > 35$ (b) $t_5 = 2.4495$, $0.025 < P < 0.050$
(c) Yes, the P-value is small, so we conclude that $\mu > 35$.

5. No, $t_6 = 1.7085$, $0.10 < P < 0.20$.

7. Yes, $t_{17} = -2.4244$, $0.01 < P < 0.025$.

9. Yes, $t_3 = -4.0032$, $0.01 < P < 0.025$.

11. (a) 6.0989 (b) 9.190 (c) 17.384 (d) −1.48

Section 6.5

1. Yes, $P = 0.0002$.

3. No, $P = 0.1336$.

5. (a) $H_0: \mu_1 - \mu_2 \le 0$ vs. $H_1: \mu_1 - \mu_2 > 0$, $P = 0.2119$. We cannot conclude that the mean score on one-tailed questions is greater.
(b) $H_0: \mu_1 - \mu_2 = 0$ vs. $H_1: \mu_1 - \mu_2 \ne 0$, $P = 0.4238$. We cannot conclude that the mean score on one-tailed questions differs from the mean score on two-tailed questions.

7. (a) Yes, $P = 0.0217$. (b) No, $P = 0.2514$.

9. Yes, $P = 0.0006$.

11. (a) (i) 11.128, (ii) 0.380484 (b) 0.0424, similar to the P-value computed with the t statistic.
(c) (−0.3967, 5.7367)

Section 6.6

1. (a) $H_0: p_1 - p_2 \ge 0$ vs. $H_1: p_1 - p_2 < 0$ (b) $P = 0.1492$ (c) Machine 1

3. Yes, $P = 0.0018$.

5. Yes, $P = 0.0367$.

7. $P = 0.0643$. The evidence suggests that heavy packaging reduces the proportion of damaged shipments, but may not be conclusive.

9. No, $P = 0.7114$.

11. No, because the two samples are not independent.

13. (a) 0.660131 (b) 49 (c) 1.79 (d) 0.073

Section 6.7

1. (a) Yes, $t_3 = 2.5740$, $0.025 < P < 0.050$. (b) No, $t_3 = 0.5148$, $0.25 < P < 0.40$.

3. $t_{16} = -2.0143$, $0.05 < P < 0.10$. The null hypothesis is suspect.

5. No, $t_7 = 0.3444$, $0.50 < P < 0.80$.

7. No, $t_{15} = 1.0024$, $0.10 < P < 0.25$.

9. Yes, $t_9 = 2.5615$, $0.02 < P < 0.05$.

11. (a) Yes, $t_{16} = 6.1113$, $P < 0.0005$. (b) Yes, $t_{16} = 3.659$, $0.001 < P < 0.005$.

13. No, $t_{14} = 1.0236$, $0.20 < P < 0.50$.

15. (a) 0.197 (b) 0.339 (c) −1.484 (d) −6.805

Section 6.8

1. Yes, $t_6 = 7.0711$, $P < 0.001$.

3. Yes, $t_9 = 2.6434$, $0.02 < P < 0.05$.

5. Yes, $t_4 = 4.7900$, $0.001 < P < 0.005$.

7. Yes, $t_6 = 3.7591$, $0.002 < P < 0.010$.

9. (a) Let μ_R be the mean number of miles per gallon for taxis using radial tires, and let μ_B be the mean number of miles per gallon for taxis using bias tires. The appropriate null and alternate hypotheses are $H_0: \mu_R - \mu_B \leq 0$ vs. $H_1: \mu_R - \mu_B > 0$. The value of the test statistic is $t_9 = 8.9532$, so $P < 0.0005$.
(b) The appropriate null and alternate hypotheses are $H_0: \mu_R - \mu_B \leq 2$ vs. $H_1: \mu_R - \mu_B > 2$. The value of the test statistic is $t_9 = 3.3749$, so $0.001 < P < 0.005$.

11. (a) 1.1050 (b) 2.8479 (c) 4.0665 (d) 3.40

Section 6.9

1. (a) Yes. $S_+ = 25$, $P = 0.0391$. (b) No. $S_+ = 7$, $P > 0.1094$.
(c) No. $S_+ = 23$, $P > 2(0.0781) = 0.1562$.

3. (a) No. $S_+ = 134$, $z = -0.46$, $P = 0.3228$. (b) Yes. $S_+ = 249.5$, $z = 2.84$, $P = 0.0023$.
(c) Yes. $S_+ = 70.5$, $z = -2.27$, $P = 0.0232$.

5.

Difference	0.01	0.01	−0.01	0.03	0.05	−0.05	−0.07	−0.11	−0.13	0.15
Signed rank	2	2	−2	4	5.5	−5.5	−7	−8	−9	10

$S_+ = 2 + 2 + 4 + 5.5 + 10 = 23.5$. From the table, $P > 2(0.1162) = 0.2324$. Do not reject.

7. Yes. $W = 34$, $P = 2(0.0087) = 0.0174$.

9. No. $W = 168$, $z = 0.31$, $P = 0.7566$.

Section 6.10

1. (a) $H_0: p_1 = 0.90,\ p_2 = 0.05,\ p_3 = 0.05$ (b) 900, 50, 50 (c) $\chi^2_2 = 21.7778$
(d) $P < 0.005$. The true percentages differ from 90%, 5%, and 5%.

3. The expected values are

	Net Excess Capacity				
	< 0%	**0–10%**	**11–20%**	**21–30%**	**> 30%**
Small	81.7572	44.7180	14.0026	4.9687	27.5535
Large	99.2428	54.2820	16.9974	6.0313	33.4465

$\chi^2_4 = 12.9451$, $0.01 < P < 0.05$. It is reasonable to conclude that the distributions differ.

5. (a)

10.30	13.35	13.35
6.96	9.02	9.02
9.74	12.62	12.62

(b) $\chi^2_4 = 6.4808$, $P > 0.10$. There is no evidence that the rows and columns are not independent.

7. (iii)

9. $\chi^2_9 = 8.0000$, $P > 0.10$. There is no evidence that the random number generator is not working properly.

11. Yes, $\chi^2_{11} = 41.3289$, $P < 0.005$.

Section 6.11

1. 2.51

3. (a) 0.01 (b) 0.02

5. No. $F_{8,\,13} = 1.0429$, $P > 0.20$.

Section 6.12

1. (a) True (b) False (c) False

3. (a) $H_0: \mu \geq 90$ vs. $H_1: \mu < 90$ (b) $\overline{X} < 89.3284$
(c) This is not an appropriate rejection region. The rejection region should consist of values for $\overline{X}$ that will make the P-value of the test less than a chosen threshold level. This rejection region consists of values for which the P-value will be greater than some level.
(d) This is an appropriate rejection region. The level of the test is 0.0708.
(e) This is not an appropriate rejection region. The rejection region should consist of values for $\overline{X}$ that will make the P-value of the test less than a chosen threshold level. This rejection region contains values of $\overline{X}$ for which the P-value will be large.

5. (a) Type I error (b) Correct decision (c) Correct decision (d) Type II error

7. The 1% level

Section 6.13

1. (a) True (b) True (c) False (d) False

3. Increase

5. (a) $H_0: \mu \geq 50{,}000$ vs. $H_1: \mu < 50{,}000$. H_1 is true. (b) The level is 0.1151; the power is 0.4207.
(c) 0.2578 (d) 0.4364 (e) 618

7. (ii)

9. (a) Two-tailed (b) $p = 0.5$ (c) $p = 0.4$
(d) Less than 0.7. The power for a sample size of 150 is 0.691332, and the power for a smaller sample size of 100 would be less than this.
(e) Greater than 0.6. The power for a sample size of 150 is 0.691332, and the power for a larger sample size of 200 would be greater than this.
(f) Greater than 0.65. The power against the alternative $p = 0.4$ is 0.691332, and the alternative $p = 0.3$ is farther from the null than $p = 0.4$. So the power against the alternative $p = 0.3$ is greater than 0.691332.

(g) It's impossible to tell from the output. The power against the alternative $p = 0.45$ will be less than the power against $p = 0.4$, which is 0.691332. But we cannot tell without calculating whether it will be less than 0.65.

11. (a) Two-tailed (b) Less than 0.9. The sample size of 60 is the smallest that will produce power greater than or equal to the target power of 0.9.
(c) Greater than 0.9. The power is greater than 0.9 against a difference of 3, so it will be greater than 0.9 against any difference greater than 3.

Section 6.14

1. (a) The Bonferroni-adjusted P-value is 0.012. Since this value is small, we can conclude that this setting reduces the proportion of defective parts.
(b) The Bonferroni-adjusted P-value is 0.18. Since this value is not so small, we cannot conclude that this setting reduces the proportion of defective parts.

3. 0.0025

5. (a) No. If the mean burnout amperage is equal to 15 A every day, the probability of rejecting H_0 is 0.05 each day. The number of times in 200 days that H_0 is rejected is then a binomial random variable with $n = 200$, $p = 0.05$. The probability of rejecting H_0 10 or more times in 200 days is then approximately equal to 0.5636. So it would not be unusual to reject H_0 10 times in 200 trials if H_0 is always true.
(b) Yes. If the mean burnout amperage is equal to 15 A every day, the probability of rejecting H_0 is 0.05 each day. The number of times in 200 days that H_0 is rejected is then a binomial random variable with $n = 200$, $p = 0.05$. The probability of rejecting H_0 20 or more times in 200 days is then approximately equal to 0.0010. So it would be quite unusual to reject H_0 20 times in 200 trials if H_0 is always true.

Section 6.15

1. (a) $V = 26.323$, $\sigma_V = 0.3342$ (b) $z = 3.96$, $P \approx 0.$ (c) Yes, V is approximately normally distributed.

3. (a) (ii) and (iv) (b) (i), (ii), and (iv)

5. No, the value 103 is an outlier.

7. (a) $s_A^2 = 200.28$, $s_B^2 = 39.833$, $s_A^2/s_B^2 = 5.02$. (b) No, the F test requires the assumption that the data are normally distributed. These data contain an outlier (103), so the F test should not be used. (c) $P \approx 0.37$.

9. (a) The test statistic is $t = \dfrac{\overline{X} - 7}{s/\sqrt{7}}$. H_0 will be rejected if $|t| > 2.447$. (b) ≈ 0.60.

Supplementary Exercises for Chapter 6

1. This requires a test for the difference between two means. The data are unpaired. Let μ_1 represent the population mean annual cost for cars using regular fuel, and let μ_2 represent the population mean annual cost for cars using premium fuel. Then the appropriate null and alternate hypotheses are $H_0: \mu_1 - \mu_2 \geq 0$ vs. $H_1: \mu_1 - \mu_2 < 0$. The test statistic is the difference in the sample mean costs between the two groups. The z table should be used to find the P-value.

3. This requires a test for a population proportion. Let p represent the population proportion of defective parts under the new program. The appropriate null and alternate hypotheses are $H_0: p \geq 0.10$ vs. $H_1: p < 0.10$. The test statistic is the sample proportion of defective parts. The z table should be used to find the P-value.

5. (a) $H_0: \mu \geq 16$ vs. $H_1: \mu < 16$ (b) $t_9 = -2.7388$ (c) $0.01 < P < 0.025$, reject H_0.

7. (a) $H_0: \mu_1 - \mu_2 = 0$ vs. $H_1: \mu_1 - \mu_2 \neq 0$ (b) $t_6 = 2.1187$ (c) $0.05 < P < 0.10$, H_0 is suspect.

9. Yes. $z = 4.61$, $P \approx 0$.

11. (a) Reject H_0 if $\overline{X} \geq 100.0196$ or if $\overline{X} \leq 99.9804$. (b) Reject H_0 if $\overline{X} \geq 100.01645$ or if $\overline{X} \leq 99.98355$.
(c) Yes (d) No (e) 13.36%

13. (a) 0.05 (b) 0.1094

15. The Bonferroni-adjusted P-value is 0.1228. We cannot conclude that the failure rate on line 3 is less than 0.10.

17. (a) Both samples have a median of 20. (b) $W = 281.5$, $z = 2.03$, $P = 0.0424$. The P-value is fairly small. If the null hypothesis stated that the population medians were equal, this would provide reasonably strong evidence that the population medians were in fact different.
(c) No, the X sample is heavily skewed to the right, while the Y sample is strongly bimodal. It does not seem reasonable to assume that these samples came from populations of the same shape.

19. (a) Let μ_A be the mean thrust/weight ratio for fuel A, and let μ_B be the mean thrust/weight ratio for fuel B. The appropriate null and alternate hypotheses are $H_0: \mu_A - \mu_B \leq 0$ vs. $H_1: \mu_A - \mu_B > 0$.
(b) Yes. $t_{29} = 2.0339$, $0.025 < P < 0.05$.

21. (a) Yes. (b) The conclusion is not justified. The engineer is concluding that H_0 is true because the test failed to reject.

23. No. $\chi^2_2 = 2.1228$, $P > 0.10$.

Section 7.1

1. 0.8214

3. (a) The correlation coefficient is appropriate. The points are approximately clustered around a line.
(b) The correlation coefficient is not appropriate. The relationship is curved, not linear.
(c) The correlation coefficient is not appropriate. The plot contains outliers.

5. More than 0.6

7. (a) Between temperature and yield, $r = 0.7323$; between stirring rate and yield, $r = 0.7513$; between temperature and stirring rate, $r = 0.9064$.
(b) No, the result might be due to confounding, since the correlation between temperature and stirring rate is far from 0.
(c) No, the result might be due to confounding, since the correlation between temperature and stirring rate is far from 0.

9. (a) $(-0.8632, 0.1739)$ (b) $z = -2.31$, $P = 0.0104$, we can conclude that $\rho < 0.3$.
(c) $U = -1.6849$, $0.10 < P < 0.20$, it is plausible that $\rho = 0$.

11. (a) $y = 2$ (b) $y = -3$ (c) $y = -1.1699$ (d) $y = -9.8301$ (e) For the correlation to be equal to -1, the points would have to lie on a straight line with negative slope. There is no value for y for which this is the case.

Section 7.2

1. (a) 319.27 lb (b) 5.65 lb

3. 0.8492

5. (a) 18.869 in. (b) 70.477 in. (c) No, some of the men whose points lie below the least-squares line will have shorter arms.

7. (a)

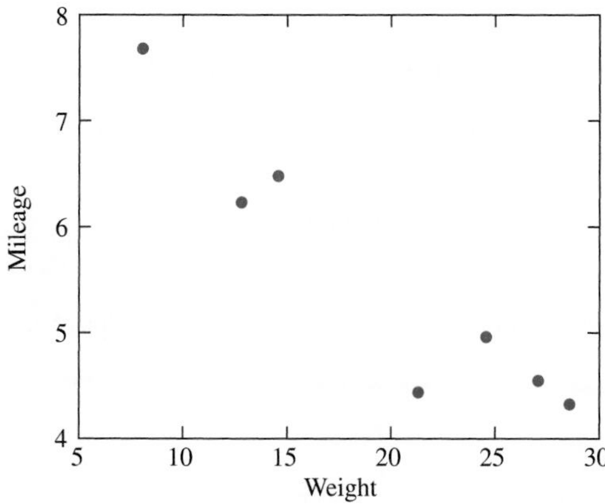

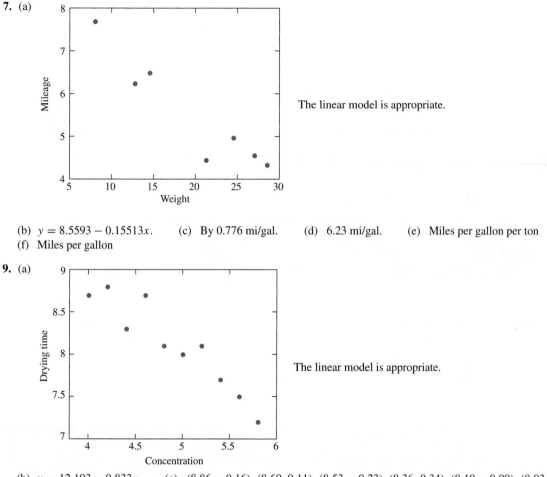

The linear model is appropriate.

(b) $y = 8.5593 - 0.15513x$. (c) By 0.776 mi/gal. (d) 6.23 mi/gal. (e) Miles per gallon per ton (f) Miles per gallon

9. (a)

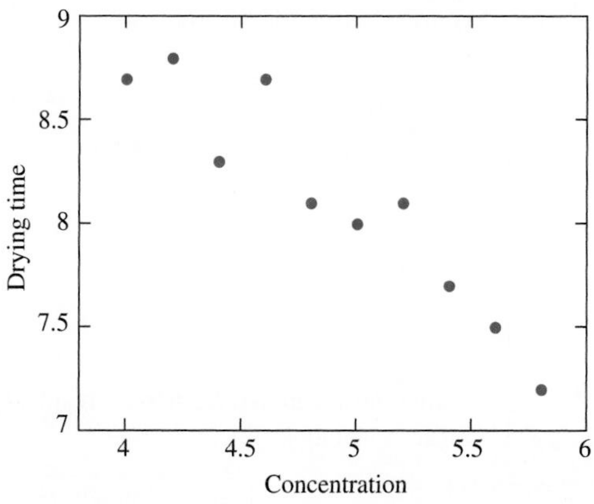

The linear model is appropriate.

(b) $y = 12.193 - 0.833x$ (c) (8.86, −0.16), (8.69, 0.11), (8.53, −0.23), (8.36, 0.34), (8.19, −0.09), (8.03, −0.03), (7.86, 0.24), (7.69, 0.01), (7.53, −0.03), (7.36, −0.16) (d) Decrease by 0.0833 hours. (e) 8.53 hours (f) No, because 7% is outside the range of concentrations present in the data. (g) 4.79% (h) We cannot specify such concentration. According to the least-squares line, a concentration of 7.43% would result in a drying time of 6 hours. However, since 7.43% is outside the range of the data, this prediction is unreliable.

11. $y = 20 + 10x$

13. (iii)

Section 7.3

1. (a) $\widehat{\beta}_0 = 4.7829, \widehat{\beta}_1 = 3.0821$ (b) 0.2169 (c) For β_0: (4.409, 5.157), for β_1: (2.948, 3.216)
(d) No. $t_{23} = 1.271, 0.10 < P < 0.25$. (e) Yes. $t_{23} = -3.965, P < 0.0005$. (f) (9.098, 9.714)
(g) (8.063, 10.749) (h) The confidence interval is more useful, since it is concerned with the true length of the spring, while the prediction interval is concerned with the next measurement of the length.

3. (a) The slope is 0.84451; the intercept is 44.534.
(b) Yes, the P-value for the slope is ≈ 0, so horsepower is related to NO_x.
(c) 52.979 mg/s.
(d) 0.920 (e) (74.6, 82.0) (f) No. A reasonable range of predicted values is given by the 95% prediction interval, which is (29.37, 127.26).

5. (a) $H_0: \beta_C - \beta_E = 0$ (b) Yes. $z = 2.10$, $P = 0.0358$.

7. (a) $y = 2.11 - 0.776x$ (b) For β_0: (2.07, 2.16), for β_1: (−0.829, −0.722). (c) 0.795
(d) (0.742, 0.847) (e) (0.675, 0.914)

9. The confidence interval at 1.5 would be the shortest. The confidence interval at 1.8 would be the longest.

11. 1.388

13. (a) 0.256 (b) 0.80 (c) 1.13448 (d) 0.001

15. (a) 553.71 (b) 162.06 (c) Below
(d) There is a greater amount of vertical spread on the right side of the plot than on the left.

Section 7.4

1. (a) $\ln y = -0.4442 + 0.79833 \ln x$ (b) 330.95 (c) 231.76 (d) (53.19, 1009.89)

3. (a) $y = 20.162 + 1.269x$

(b)

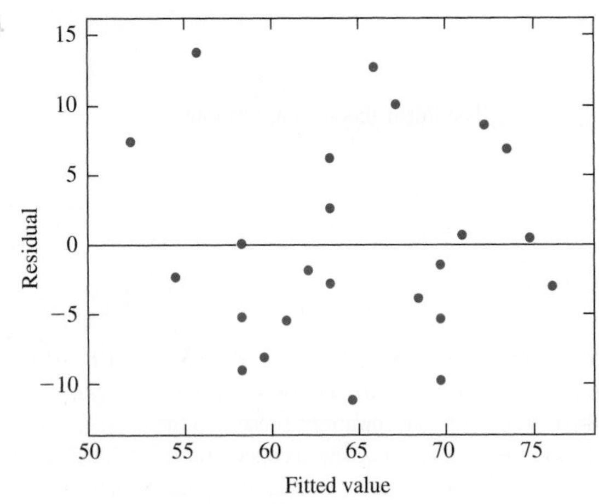

There is no apparent pattern to the residual plot. The linear model looks fine.

(c)

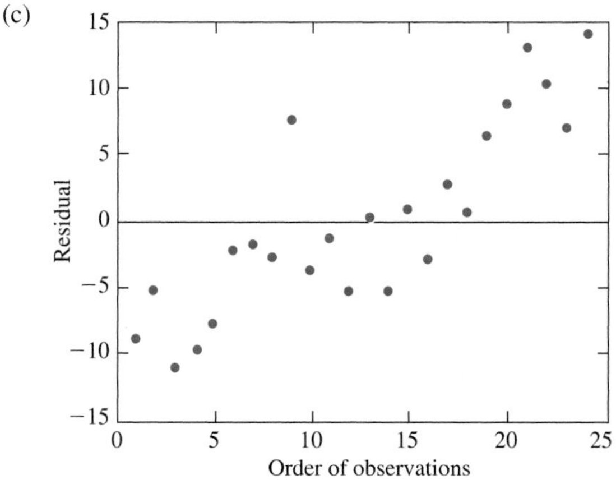

The residuals increase over time. The linear model is not appropriate as is. Time, or other variables related to time, must be included in the model.

5. (a) $y = -235.32 + 0.695x$.

(b)

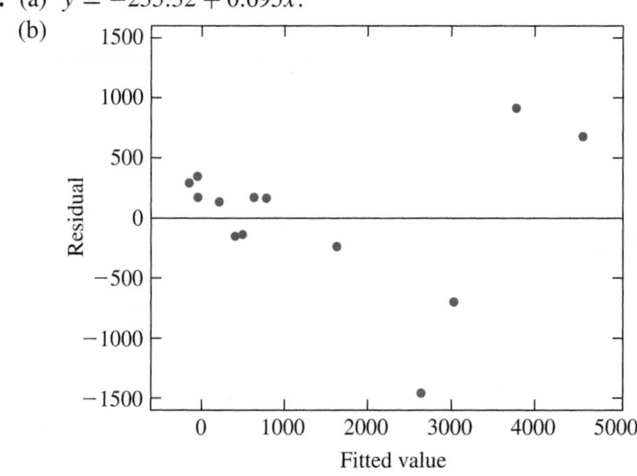

The residual plot shows a pattern, with positive residuals at the higher and lower fitted values, and negative residuals in the middle. The model is not appropriate.

(c) $\ln y = -0.0745 + 0.925 \ln x$.

(d)

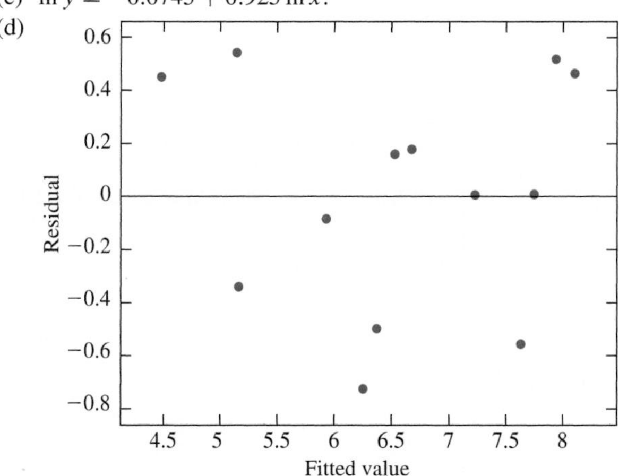

The residual plot shows no obvious pattern. The model is appropriate.

(e) The log model is more appropriate. The 95% prediction interval is (197.26, 1559.76).

7. (a)

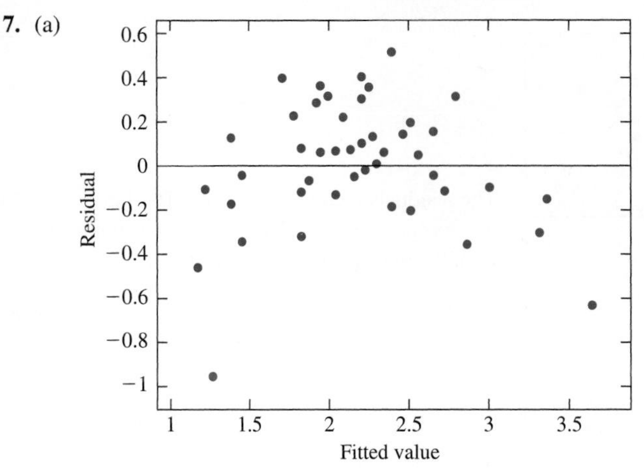

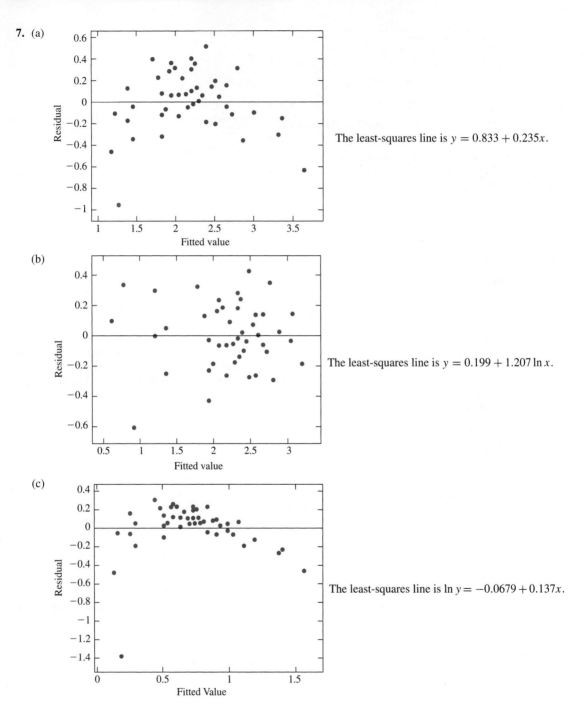

The least-squares line is $y = 0.833 + 0.235x$.

(b)

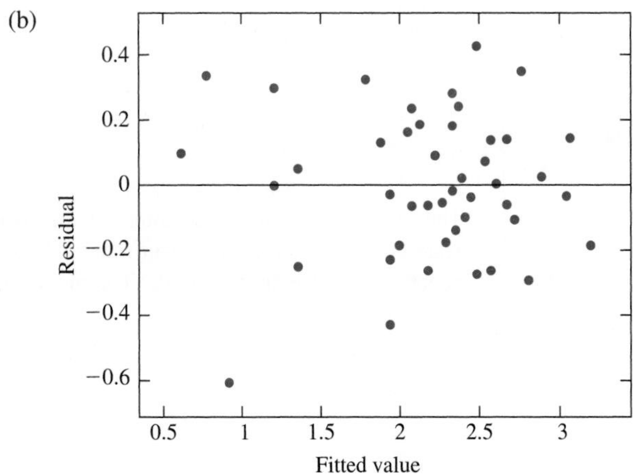

The least-squares line is $y = 0.199 + 1.207 \ln x$.

(c)

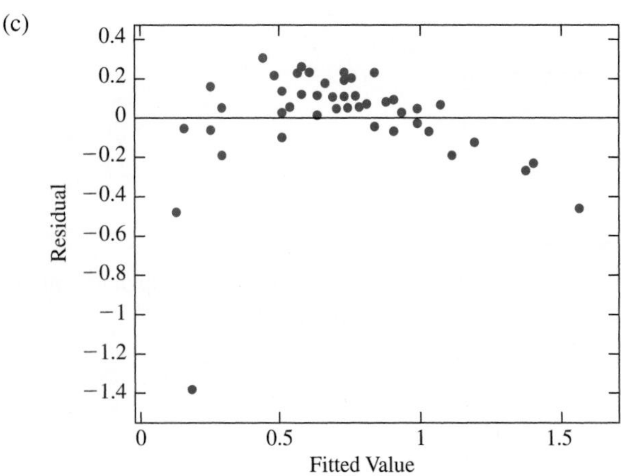

The least-squares line is $\ln y = -0.0679 + 0.137x$.

(d)

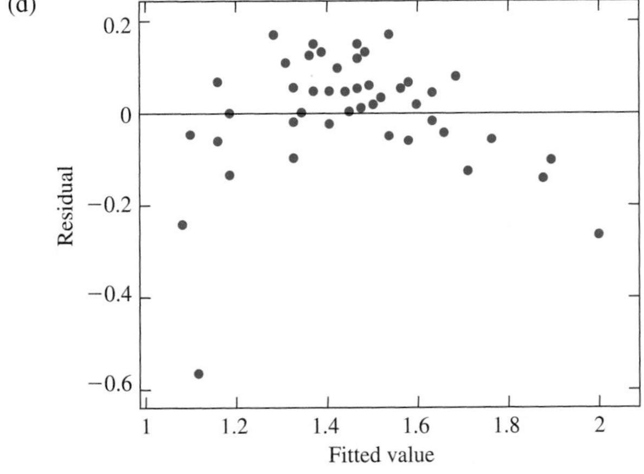

The least-squares line is $\sqrt{y} = 0.956 + 0.0874x$.

(e) The model $y = 0.199 + 1.207 \ln x$ fits best. Its residual plot shows the least pattern.

(f)

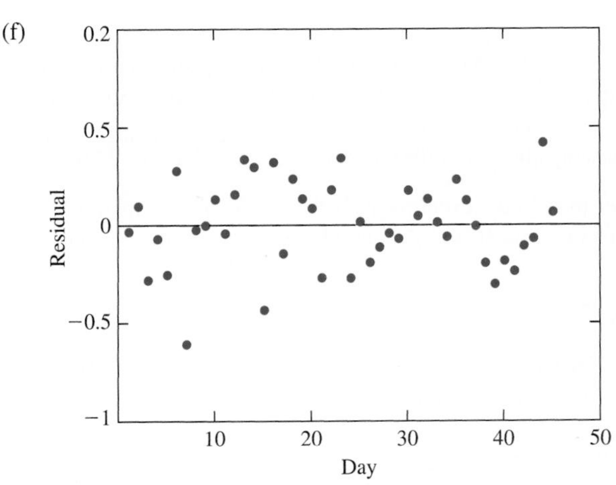

The residuals show no pattern with time.

(g) 2.14 (h) (1.689, 2.594)

9. (a) The model is $\log_{10} y = \beta_0 + \beta_1 \log_{10} x + \varepsilon$. Note that the natural log (ln) could be used in place of $\log_{10}$, but common logs are more convenient since partial pressures are expressed as powers of 10.

(b)

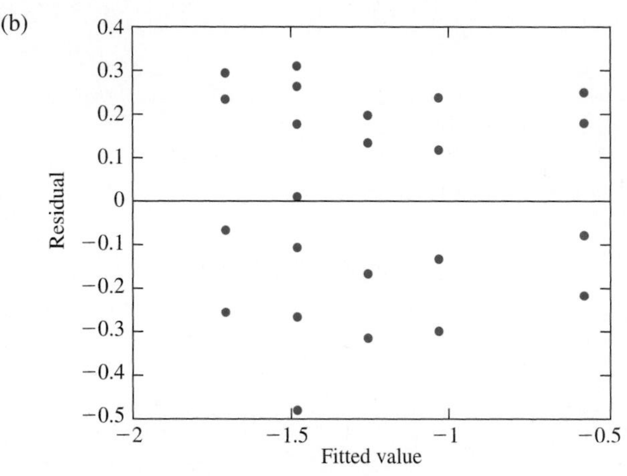

The least-squares line is $\log_{10} y = -3.277 - 0.225 \log_{10} x$. The linear model appears to fit quite well.

(c) The theory says that the coefficient β_1 of $\log_{10} x$ in the linear model is equal to -0.25. The estimated value is $\widehat{\beta} = -0.225$. We determine whether the data are consistent with the theory by testing the hypotheses $H_0 : \beta_1 = -0.25$ vs. $H_1 : \beta_1 \neq -0.25$. The value of the test statistic is $t_{21} = 0.821$, so $0.20 < P < 0.50$. We do not reject H_0, so the data are consistent with the theory.

11. (a) $y = 2049.87 - 4.270x$ (b) (12, 2046) and (13, 1954) are outliers. The least-squares line with (12, 2046) deleted is $y = 2021.85 - 2.861x$. The least-squares line with (13, 1954) deleted is $y = 2069.30 - 5.236x$. The least-squares line with both outliers deleted is $y = 2040.88 - 3.809x$.
(c) The slopes of the least-squares lines are noticeably affected by the outliers. They ranged from -2.861 to -5.236.

13. (a) A physical law. (b) It would be better to redo the experiment. If the results of an experiment violate a physical law, then something was wrong with the experiment, and you can't fix it by transforming variables.

Supplementary Exercises for Chapter 7

1. (a) $\widehat{\beta}_0 = -0.0390, \widehat{\beta}_1 = 1.017$ (b) 0 (c) The molar absorption coefficient M. (d) Testing $H_0 : \beta_0 = 0$ vs. $H_1 : \beta_0 \neq 0$, $t_3 = -0.428$ and $0.50 < P < 0.80$, so the data are consistent with the Beer–Lambert law.

3. (a)

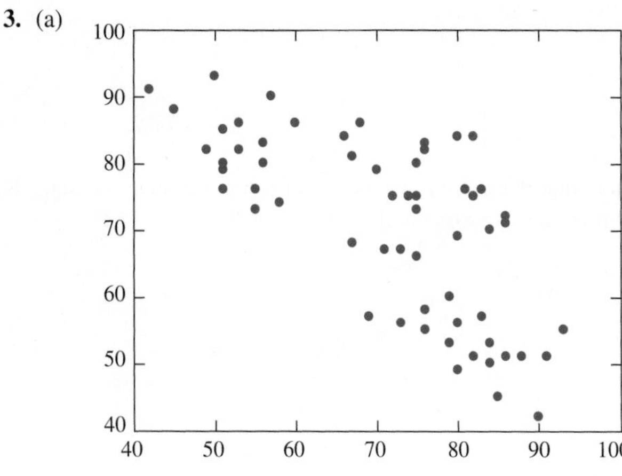

(b) $T_{i+1} = 120.18 - 0.696T_i$. (c) $(-0.888, -0.503)$ (d) 71.48 minutes (e) (68.40, 74.56)
(f) (45.00, 97.95)

5. (a) $\widehat{\beta}_0 = 0.8182, \widehat{\beta}_1 = 0.9418$ (b) No. $t_9 = 1.274, 0.20 < P < 0.50$. (c) Yes. $t_9 = -5.358, P < 0.001$.
(d) Yes, since we can conclude that $\beta_1 \neq 1$, we can conclude that the machine is out of calibration.
(e) (18.58, 20.73) (f) (75.09, 77.23) (g) No, when the true value is 20, the result of part (e) shows that a 95% confidence interval for the mean of the measured values is (18.58, 20.73). Therefore it is plausible that the mean measurement will be 20, so that the machine is in calibration.

7. (a) $y = 0.041496 + 0.0073664x$ (b) $(-0.00018, 0.01492)$ (c) (0.145, 0.232) (d) (0.0576, 0.320)

9. (a) $\ln y = \beta_0 + \beta_1 \ln x$, where $\beta_0 = \ln k$ and $\beta_1 = r$.
(b) The least-squares line is $\ln y = -1.7058 + 0.65033 \ln x$. Therefore $\widehat{r} = 0.65033$ and $\widehat{k} = e^{-1.7058} = 0.18162$.
(c) $t_3 = 4.660, P = 0.019$. No, it is not plausible.

11. (a) $\widehat{\beta}_0 = 60.263, \widehat{\beta}_1 = 0.11653, s^2 = 0.38660$. (b) Yes. $t_{10} = 80.956, P \approx 0$. (c) Yes. $t_{10} = 36.294, P \approx 0$.
(d)

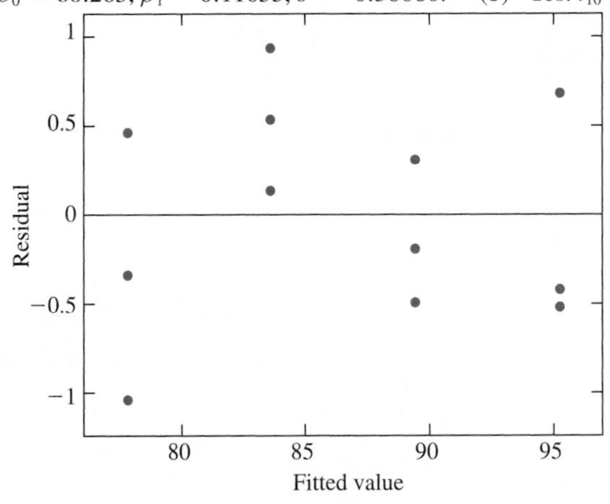

The linear model appears to be appropriate.

(e) (0.10938, 0.12369) (f) (86.083, 86.883) (g) (85.041, 87.925)

13. (ii)

15. (a) 145.63 (b) Yes. $r = -\sqrt{\text{R-Sq}} = -0.988$. Note that r is negative because the slope of the least-squares line is negative. (c) 145.68.

17. (a) We need to minimize the sum of squares $S = \sum(y_i - \widehat{\beta}x_i)^2$. We take the derivative with respect to $\widehat{\beta}$ and set it equal to 0, obtaining $-2\sum x_i(y_i - \widehat{\beta}x_i) = 0$. Then $\sum x_i y_i - \widehat{\beta}\sum x_i^2 = 0$, so $\widehat{\beta} = \sum x_i y_i / \sum x_i^2$.
(b) Let $c_i = x_i / \sum x_i^2$. Then $\widehat{\beta} = \sum c_i y_i$, so $\sigma_{\widehat{\beta}}^2 = \sum c_i^2 \sigma^2 = \sigma^2 \sum x_i^2 / \left(\sum x_i^2\right)^2 = \sigma^2 / \sum x_i^2$.

19. From the answer to Exercise 18, we know that $\sum_{i=1}^n (x_i - \overline{x}) = 0$, $\sum_{i=1}^n \overline{x}(x_i - \overline{x}) = 0$, and $\sum_{i=1}^n x_i(x_i - \overline{x}) = \sum_{i=1}^n (x_i - \overline{x})^2$. Now

$$\mu_{\widehat{\beta}_0} = \sum_{i=1}^n \left[\frac{1}{n} - \frac{\overline{x}(x_i - \overline{x})}{\sum_{i=1}^n (x_i - \overline{x})^2}\right] \mu_{y_i}$$

$$= \sum_{i=1}^n \left[\frac{1}{n} - \frac{\overline{x}(x_i - \overline{x})}{\sum_{i=1}^n (x_i - \overline{x})^2}\right] (\beta_0 + \beta_1 x_i)$$

$$= \beta_0 \sum_{i=1}^{n} \frac{1}{n} + \beta_1 \sum_{i=1}^{n} \frac{x_i}{n} - \beta_0 \frac{\sum_{i=1}^n \overline{x}(x_i - \overline{x})}{\sum_{i=1}^n (x_i - \overline{x})^2} - \beta_1 \frac{\sum_{i=1}^n x_i \overline{x}(x_i - \overline{x})}{\sum_{i=1}^n (x_i - \overline{x})^2}$$

$$= \beta_0 + \beta_1 \overline{x} - 0 - \beta_1 \overline{x} \frac{\sum_{i=1}^n x_i (x_i - \overline{x})}{\sum_{i=1}^n (x_i - \overline{x})^2}$$

$$= \beta_0 + \beta_1 \overline{x} - 0 - \beta_1 \overline{x}$$

$$= \beta_0$$

21.

$$\sigma^2_{\widehat{\beta}_0} = \sum_{i=1}^{n} \left[\frac{1}{n} - \frac{\overline{x}(x_i - \overline{x})}{\sum_{i=1}^n (x_i - \overline{x})^2} \right]^2 \sigma^2$$

$$= \sum_{i=1}^{n} \left[\frac{1}{n^2} - \frac{2\overline{x}}{n} \frac{(x_i - \overline{x})}{\sum_{i=1}^n (x_i - \overline{x})^2} + \overline{x}^2 \frac{\sum_{i=1}^n (x_i - \overline{x})^2}{\left[\sum_{i=1}^n (x_i - \overline{x})^2\right]^2} \right] \sigma^2$$

$$= \left[\sum_{i=1}^{n} \frac{1}{n^2} - 2\frac{\overline{x}}{n} \frac{\sum_{i=1}^n (x_i - \overline{x})}{\sum_{i=1}^n (x_i - \overline{x})^2} + \overline{x}^2 \frac{\sum_{i=1}^n (x_i - \overline{x})^2}{\left[\sum_{i=1}^n (x_i - \overline{x})^2\right]^2} \right] \sigma^2$$

$$= \left[\frac{1}{n} - 2\frac{\overline{x}}{n}(0) + \frac{\overline{x}^2}{\sum_{i=1}^n (x_i - \overline{x})^2} \right] \sigma^2$$

$$= \left[\frac{1}{n} + \frac{\overline{x}^2}{\sum_{i=1}^n (x_i - \overline{x})^2} \right] \sigma^2$$

Section 8.1

1. (a) 6.9619 (b) 0.03925 (c) 0.5413

3.

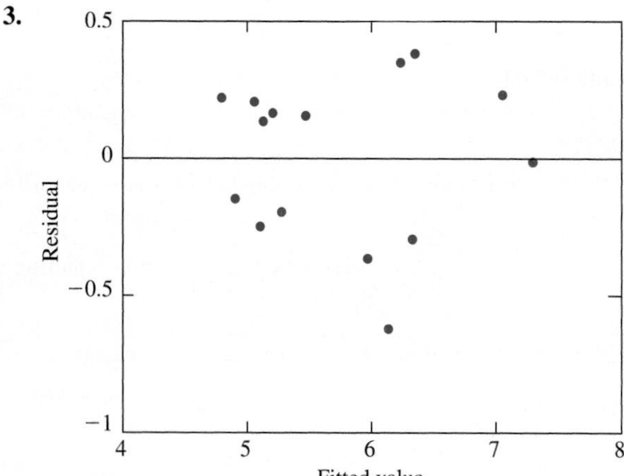

There is no obvious pattern to the residual plot, so the linear model appears to fit well.

5. (a) 25.465 (b) No, the predicted change depends on the values of the other independent variables, because of the interaction terms. (c) 0.9691 (d) $F_{9,17} = 59.204$. Yes, the null hypothesis can be rejected.

7. (a) 2.3411 L (b) 0.06768 L (c) Nothing is wrong. In theory, the constant estimates FEV_1 for an individual whose values for the other variables are all equal to zero. Since these values are outside the range of the data (e.g., no one has zero height), the constant need not represent a realistic value for an actual person. (d) Adjusted FEV_1 = measured $FEV_1 + 0.0013431 \cdot$ pressure.

9. (a) 3.572 (b) 0.098184 (c) Nothing is wrong. The constant estimates the pH for a pulp whose values for the other variables are all equal to zero. Since these values are outside the range of the data (e.g., no pulp has zero density), the constant need not represent a realistic value for an actual pulp. (d) (3.4207, 4.0496) (e) (2.2333, 3.9416) (f) Pulp B. The standard deviation of its predicted pH (SE Fit) is smaller than that of pulp A (0.1351 vs. 0.2510).

11. (a) −2.05 (b) 0.3512 (c) −0.2445 (d) 4.72 (e) 13.92 (f) 18.316 (g) 4.54 (h) 9

13. (a) 135.92°F (b) No. The change in the predicted flash point due to a change in acetic acid concentration depends on the butyric acid concentration as well, because of the interaction between these two variables. (c) Yes. The predicted flash point will change by –13.897°F.

15. (a) −9.17819, −1.42732, 123.26818, −151.35236, 50.12800, −11.43494 (b) SSE = 40,832.432, SST = 1,990,600 (c) $s^2 = 13{,}610.811$ (d) $R^2 = 0.979487$ (e) $F = 71.6257$. There are two and three degrees of freedom. (f) Yes. From the F table, $0.001 < P < 0.01$.

17. (a) 2.0711 (b) 0.17918 (c) PP is more useful, because its P-value is small, while the P-value of CP is fairly large. (d) The percent change in GDP would be expected to be larger in Sweden, because the coefficient of PP is negative.

19. (a) $y = -0.012167 + 0.043258t + 2.9205t^2$ (b) (2.830, 3.011) (c) (5.660, 6.022) (d) $\widehat{\beta}_0$: $t_7 = -1.1766$, $P = 0.278$, $\widehat{\beta}_1$: $t_7 = 1.0017$, $P = 0.350$, $\widehat{\beta}_2$: $t_7 = 76.33$, $P = 0.000$. (e) No, the P-value of 0.278 is not small enough to reject the null hypothesis that $\beta_0 = 0$. (f) No, the P-value of 0.350 is not small enough to reject the null hypothesis that $\beta_1 = 0$.

Section 8.2

1. (a)

Predictor	Coef	StDev	T	P
Constant	6.3347	2.1740	2.9138	0.009
x_1	1.2915	0.1392	9.2776	0.000

β_0 differs from 0 ($P = 0.009$), β_1 differs from 0 ($P = 0.000$).

(b)

Predictor	Coef	StDev	T	P
Constant	53.964	8.7737	6.1506	0.000
x_2	−0.9192	0.2821	−3.2580	0.004

β_0 differs from 0 ($P = 0.000$), β_1 differs from 0 ($P = 0.004$).

(c)

Predictor	Coef	StDev	T	P
Constant	12.844	7.5139	1.7094	0.104
x_1	1.2029	0.1707	7.0479	0.000
x_2	−0.1682	0.1858	−0.90537	0.377

β_0 may not differ from 0 ($P = 0.104$), β_1 differs from 0 ($P = 0.000$), β_2 may not differ from 0 ($P = 0.377$).

(d) The model in part (a) is the best. When both x_1 and x_2 are in the model, only the coefficient of x_1 is significantly different from 0. In addition, the value of R^2 is only slightly greater (0.819 vs. 0.811) for the model containing both x_1 and x_2 than for the model containing x_1 alone.

3. (a) Plot (i) came from engineer B, and plot (ii) came from engineer A. We know this because the variables x_1 and x_2 are both significantly different from 0 for engineer A but not for engineer B. Therefore engineer B is the one who designed the experiment to have the dependent variables nearly collinear.

(b) Engineer A's experiment produced the more reliable results. In engineer B's experiment, the two dependent variables are nearly collinear.

5. (a) For $R_1 < 4$, the least-squares line is $R_2 = 1.23 + 0.264R_1$. For $R_1 \geq 4$, the least-squares line is $R_2 = -0.190 + 0.710R_1$.

(b) The relationship is clearly nonlinear when $R_1 < 4$.

Predictor	Coef	StDev	T	P
Constant	1.2840	0.26454	4.8536	0.000
R_1	0.21661	0.23558	0.91947	0.368
R_1^2	0.0090189	0.044984	0.20049	0.843

Predictor	Coef	StDev	T	P
Constant	−1.8396	0.56292	−3.2680	0.004
R_1	4.4987	0.75218	5.9809	0.000
R_1^2	−1.7709	0.30789	−5.7518	0.000
R_1^3	0.22904	0.039454	5.8053	0.000

Predictor	Coef	StDev	T	P
Constant	−2.6714	2.0117	−1.3279	0.200
R_1	6.0208	3.6106	1.6675	0.112
R_1^2	−2.7520	2.2957	−1.1988	0.245
R_1^3	0.49423	0.61599	0.80234	0.432
R_1^4	−0.02558	0.05930	−0.43143	0.671

(c)

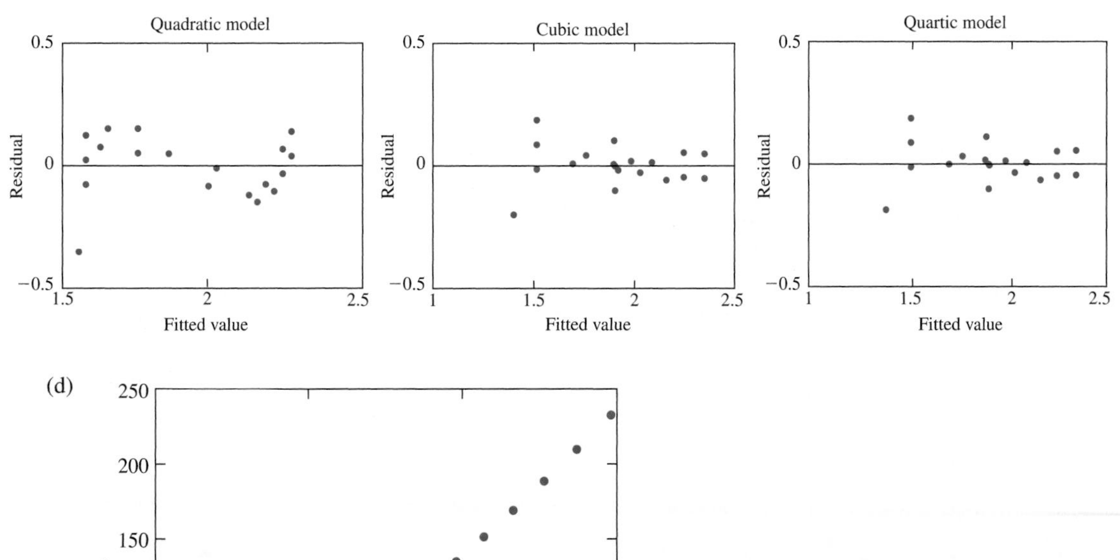

(d)

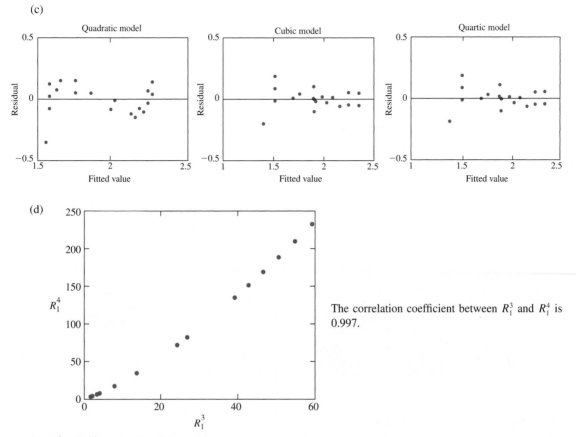

The correlation coefficient between R_1^3 and R_1^4 is 0.997.

(e) R_1^3 and R_1^4 are nearly collinear.

(f) The cubic model is best. The quadratic is inappropriate because the residual plot exhibits a pattern. The residual plots for both the cubic and quartic models look good; however, there is no reason to include R_1^4 in the model since it merely confounds the effect of R_1^3.

Section 8.3

1. (a) False (b) True (c) False (d) True

3. (v)

5. (i)

7. The four-variable model with the highest value of R^2 has a lower R^2 than the three-variable model with the highest value of R^2. This is impossible.

9. (a) 0.2803 (b) Three degrees of freedom in the numerator and 157 in the denominator. (c) $P > 0.10$. The reduced model is plausible. (d) This is not correct. It is possible for a group of variables to be fairly strongly related to an independent variable, even though none of the variables individually is strongly related.
(e) No mistake. If y is the dependent variable, then the total sum of squares is $\sum(y_i - \overline{y})^2$. This quantity does not involve the independent variables.

11. (a)

Predictor	Coef	StDev	T	P
Constant	0.087324	0.16403	0.53237	0.597
x	0.52203	0.055003	9.4909	0.000
x^2	−0.02333	0.0043235	−5.396	0.000

(b)

Predictor	Coef	StDev	T	P
Constant	0.19878	0.11677	1.7023	0.096
$\ln x$	1.2066	0.068272	17.673	0.000

(c)

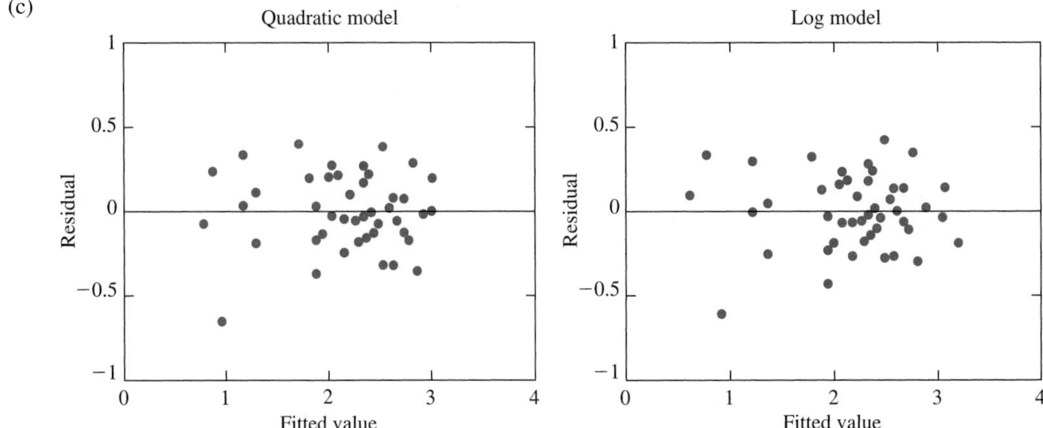

Neither residual plot reveals gross violations of assumptions.

(d)

x	Quadratic Model Prediction	Log Model Prediction
3.0	1.44	1.52
5.0	2.11	2.14
7.0	2.60	2.55
9.0	2.90	2.85

(e) The F test cannot be used. The F test can only be used when one model is formed by dropping one or more independent variables from another model.

(f) The predictions of the two models do not differ much. The log model has only one independent variable instead of two, which is an advantage.

13. (a)

Predictor	Coef	StDev	T	P
Constant	37.989	53.502	0.71004	0.487
x	1.0774	0.041608	25.894	0.000

(b)

Predictor	Coef	StDev	T	P
Constant	−253.45	132.93	−1.9067	0.074
x	1.592	0.22215	7.1665	0.000
x^2	−0.00020052	0.000085328	−2.3499	0.031

(c)

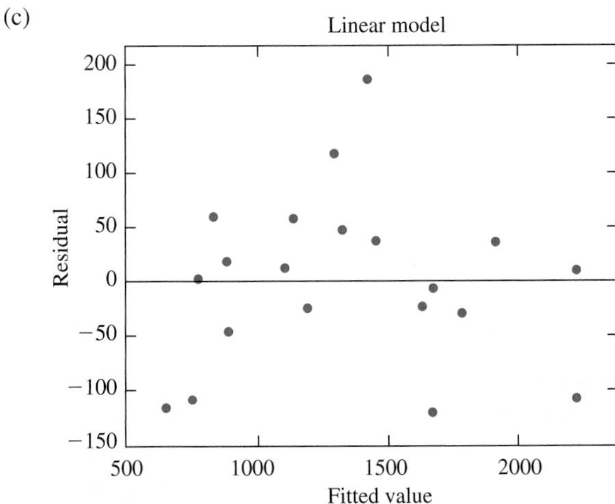

(d)

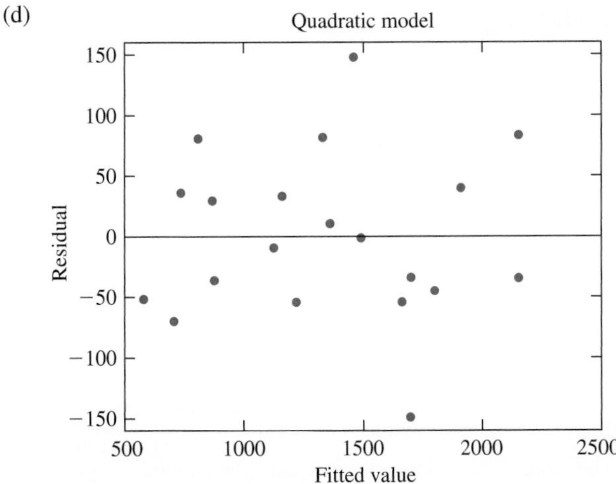

(e) The quadratic model seems more appropriate. The P-value for the quadratic term is fairly small (0.031), and the residual plot for the quadratic model exhibits less of a pattern. (There are a couple of points somewhat detached from the rest of the plot, however.)

(f) 1683.5 (g) (1634.7, 1732.2)

15. (a)

Predictor	Coef	StDev	T	P
Constant	25.613	10.424	2.4572	0.044
x_1	0.18387	0.12353	1.4885	0.180
x_2	−0.015878	0.0040542	−3.9164	0.006

(b)

Predictor	Coef	StDev	T	P
Constant	14.444	16.754	0.86215	0.414
x_1	0.17334	0.20637	0.83993	0.425

(c)

Predictor	Coef	StDev	T	P
Constant	40.370	3.4545	11.686	0.000
x_2	−0.015747	0.0043503	−3.6197	0.007

(d) The model containing x_2 as the only independent variable is best. There is no evidence that the coefficient of x_1 differs from 0.

17. The model $y = \beta_0 + \beta_1 x_2 + \varepsilon$ is a good one. One way to see this is to compare the fit of this model to the full quadratic model. The ANOVA table for the full model is

Source	DF	SS	MS	F	P
Regression	5	4.1007	0.82013	1.881	0.193
Residual error	9	3.9241	0.43601		
Total	14	8.0248			

The ANOVA table for the model $y = \beta_0 + \beta_1 x_2 + \varepsilon$ is

Source	DF	SS	MS	F	P
Regression	1	2.7636	2.7636	6.8285	0.021
Residual error	13	5.2612	0.40471		
Total	14	8.0248			

From these two tables, the F statistic for testing the plausibility of the reduced model is $\dfrac{(5.2612 - 3.9241)/(5-1)}{3.9241/9} = 0.7667$. The null distribution is $F_{4,9}$, $P > 0.10$. The large P-value indicates that the reduced model is plausible.

Supplementary Exercises for Chapter 8

1. (a) 24.6% (b) 5.43% (c) No, we need to know the oxygen content.

3. (a) 0.207 (b) 0.8015 (c) 3.82 (d) 1.200 (e) 2 (f) 86.81 (g) 43.405 (h) 30.14 (i) 14

5. (a)

Predictor	Coef	StDev	T	P
Constant	10.84	0.2749	39.432	0.000
Speed	−0.073851	0.023379	−3.1589	0.004
Pause	−0.12743	0.013934	−9.1456	0.000
Speed²	0.0011098	0.00048887	2.2702	0.032
Pause²	0.0016736	0.00024304	6.8861	0.000
Speed · Pause	−0.00024272	0.00027719	−0.87563	0.390

Analysis of Variance

Source	DF	SS	MS	F	P
Regression	5	31.304	6.2608	56.783	0.000
Residual error	24	2.6462	0.11026		
Total	29	33.95			

(b) We drop the interaction term Speed · Pause.

Predictor	Coef	StDev	T	P
Constant	10.967	0.23213	47.246	0.000
Speed	−0.079919	0.022223	−3.5961	0.001
Pause	−0.13253	0.01260	−10.518	0.000
Speed²	0.0011098	0.00048658	2.2809	0.031
Pause²	0.0016736	0.0002419	6.9185	0.000

Analysis of Variance

Source	DF	SS	MS	F	P
Regression	4	31.22	7.8049	71.454	0.000
Residual error	25	2.7307	0.10923		
Total	29	33.95			

Comparing this model with the one in part (a), $F_{1,24} = 0.77$, $P > 0.10$.

(c)

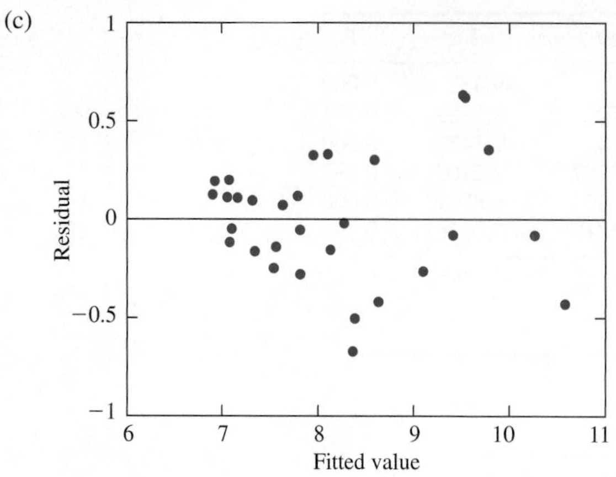

There is some suggestion of heteroscedasticity, but it is hard to be sure without more data.

(d) No, compared with the full model containing Speed, Pause, Speed², and Pause², and Speed · Pause, the F statistic is $F_{3,24} = 15.70$, and $P < 0.001$.

(e)

Vars	R-Sq	R-Sq(adj)	C-p	S	Speed	Pause	Speed2	Pause2	Speed*Pause
1	61.5	60.1	92.5	0.68318		X			
1	60.0	58.6	97.0	0.69600					X
2	76.9	75.2	47.1	0.53888		X		X	
2	74.9	73.0	53.3	0.56198	X	X			
3	90.3	89.2	7.9	0.35621	X	X		X	
3	87.8	86.4	15.5	0.39903		X		X	X
4	92.0	90.7	4.8	0.33050	X	X	X	X	
4	90.5	89.0	9.2	0.35858	X	X		X	X
5	92.2	90.6	6.0	0.33205	X	X	X	X	X

(f) The model containing the dependent variables Speed, Pause, Speed², and Pause² has both the lowest value of C_p and the largest value of adjusted R^2.

7.

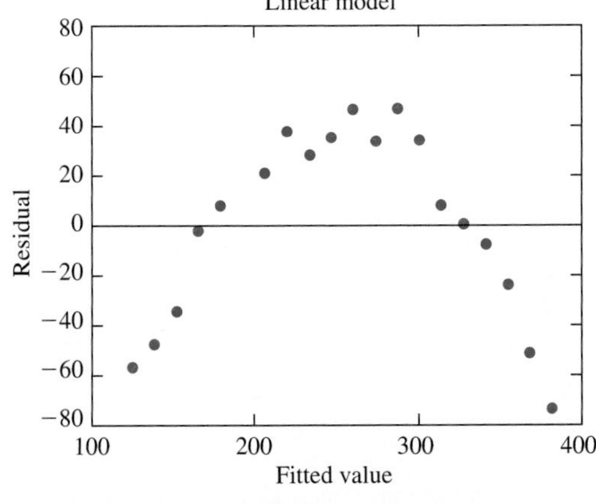

The residual plot shows an obvious curved pattern, so the linear model is not appropriate.

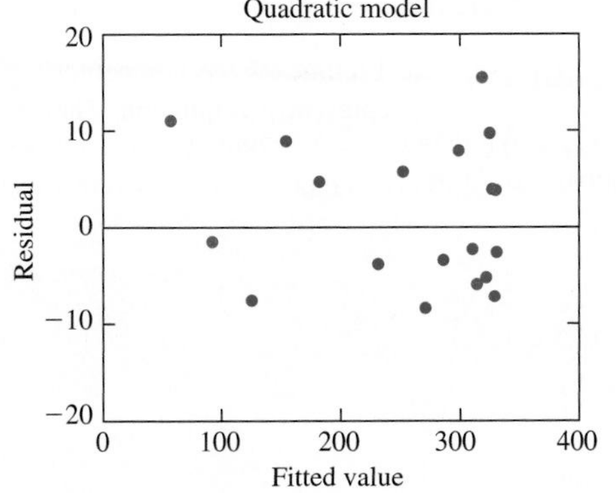

There is no obvious pattern to the residual plot, so the quadratic model appears to fit well.

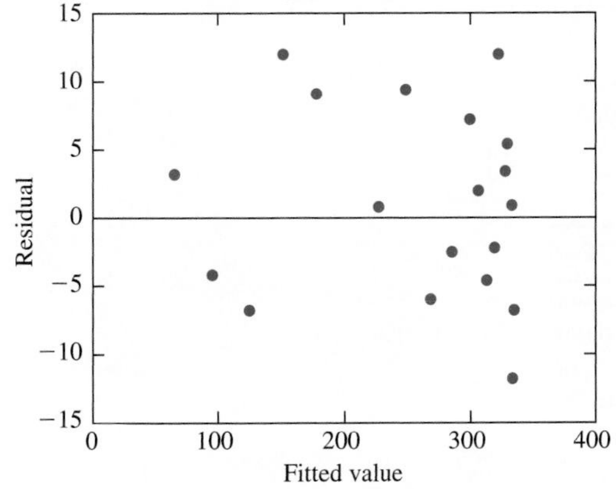

There is no obvious pattern to the residual plot, so the cubic model appears to fit well.

9. (a) 182.52, 166.55, 187.56 (b) 236.39, 234.18, 237.02 (c) 176.80, 163.89, 180.78
(d) (iv). The output does not provide much to choose from between the two-variable models. In the three-variable model, none of the coefficients are significantly different from 0, even though they were significant in the two-variable models. This suggests collinearity.

11. (a) Following are the values of SSE and their degrees of freedom for models of degrees 1, 2, 3, and 4.

Linear	18	2726.55
Quadratic	17	481.90
Cubic	16	115.23
Quartic	15	111.78

To compare quadratic vs. linear, $F_{1,17} = \dfrac{(2726.55 - 481.90)/(18 - 17)}{481.90/17} = 79.185,\ P \approx 0.$

To compare cubic vs. quadratic, $F_{1,16} = \dfrac{(481.90 - 115.23)/(17 - 16)}{115.23/16} = 50.913,\ P \approx 0.$

To compare quartic vs. cubic, $F_{1,15} = \dfrac{(115.23 - 111.78)/(16 - 15)}{111.78/15} = 0.463,\ P > 0.10.$

The cubic model is selected by this procedure.

(b) The cubic model is $y = 27.937 + 0.48749x + 0.85104x^2 - 0.057254x^3$. The estimate y is maximized when $dy/dx = 0$. $dy/dx = 0.48749 + 1.70208x - 0.171762x^2$. Therefore $x = 10.188$.

13. (a) Let y_1 represent the lifetime of the sponsor's paint, y_2 represent the lifetime of the competitor's paint, x_1 represent January temperature, x_2 represent July temperature, and x_3 represent precipitation. Then one good model for y_1 is $y_1 = -4.2342 + 0.79037x_1 + 0.20554x_2 - 0.082363x_3 - 0.0079983x_1x_2 - 0.0018349x_1^2$. A good model for y_2 is $y_2 = 6.8544 + 0.58898x_1 + 0.054759x_2 - 0.15058x_3 - 0.0046519x_1x_2 + 0.0019029x_1x_3 - 0.0035069x_1^2$.

(b) $\widehat{y_1} = 13.83$, $\widehat{y_2} = 13.90$.

15. (a)

Predictor	Coef	StDev	T	P
Constant	0.25317	0.0065217	38.819	0.000
x	–0.041561	0.040281	–1.0318	0.320

(b)

Predictor	Coef	StDev	T	P
Constant	0.21995	0.0038434	57.23	0.000
x	0.58931	0.06146	9.5886	0.000
x^2	–2.2679	0.2155	–10.524	0.000

(c)

Predictor	Coef	StDev	T	P
Constant	0.22514	0.0068959	32.648	0.000
x	0.41105	0.20576	1.9977	0.069
x^2	–0.74651	1.6887	–0.44206	0.666
x^3	–3.6728	4.043	–0.90843	0.382

(d)

Predictor	Coef	StDev	T	P
Constant	0.23152	0.013498	17.152	0.000
x	0.10911	0.58342	0.18702	0.855
x^2	3.4544	7.7602	0.44515	0.665
x^3	−26.022	40.45	−0.64333	0.533
x^4	40.157	72.293	0.55548	0.590

(e) The quadratic model. The coefficient of x^3 in the cubic model is not significantly different from 0. Neither is the coefficient of x^4 in the quartic model.

(f) 0.258

17. (a)

Predictor	Coef	StDev	T	P
Constant	−0.093765	0.092621	−1.0123	0.335
x_1	0.63318	2.2088	0.28666	0.780
x_2	2.5095	0.30151	8.3233	0.000
x_1^2	5.318	8.2231	0.64672	0.532
x_2^2	−0.3214	0.17396	−1.8475	0.094
x_1x_2	0.15209	1.5778	0.09639	0.925

Analysis of Variance

Source	DF	SS	MS	F	P
Regression	5	20.349	4.0698	894.19	0.000
Residual error	10	0.045513	0.0045513		
Total	15	20.394			

(b) The model containing the variables x_1, x_2, and x_2^2 is a good one. Here are the coefficients along with their standard deviations, followed by the analysis of variance table.

Predictor	Coef	StDev	T	P
Constant	−0.088618	0.068181	−1.2997	0.218
x_1	2.1282	0.30057	7.0805	0.000
x_2	2.4079	0.13985	17.218	0.000
x_2^2	−0.27994	0.059211	−4.7279	0.000

Analysis of Variance

Source	DF	SS	MS	F	P
Regression	3	20.346	6.782	1683.9	0.000
Residual error	12	0.048329	0.0040275		
Total	15	20.394			

The F statistic for comparing this model to the full quadratic model is $F_{2,10} = \frac{(0.048329 - 0.045513)/(12-10)}{0.045513/10} = 0.309$, $P > 0.10$, so it is reasonable to drop x_1^2 and x_1x_2 from the full quadratic model. All the remaining coefficients are significantly different from 0, so it would not be reasonable to reduce the model further.

(c) The model with the best adjusted R^2 (0.99716) contains the variables x_2, x_1^2, and x_2^2. This model is also the model with the smallest value of Mallows' C_p (2.2). This is not the best model, since it contains x_1^2 but not x_1. The model containing x_1, x_2, and x_2^2, suggested in the answer to part (b), is better. Note that the adjusted R^2 for the model in part (b) is 0.99704, which differs negligibly from that of the model with the largest adjusted R^2 value.

19. (a)

Predictor	Coef	StDev	T	P
Constant	1.1623	0.17042	6.8201	0.006
t	0.059718	0.0088901	6.7174	0.007
t^2	−0.00027482	0.000069662	−3.9450	0.029

(b) 17.68 minutes (c) (0.0314, 0.0880) (d) The reaction rate is decreasing with time if $\beta_2 < 0$. We therefore test $H_0: \beta_2 \geq 0$ vs. $H_1: \beta_2 < 0$. The test statistic is $t_3 = 3.945$, $P = 0.029/2 = 0.0145$. It is reasonable to conclude that the reaction rate decreases with time.

21. $y = \beta_0 + \beta_1 x_1 + \beta_2 x_2 + \beta_3 x_1 x_2 + \varepsilon$.

23. (a) The 17-variable model containing the independent variables x_1, x_2, x_3, x_6, x_7, x_8, x_9, x_{11}, x_{13}, x_{14}, x_{16}, x_{18}, x_{19}, x_{20}, x_{21}, x_{22}, and x_{23} has adjusted R^2 equal to 0.98446. The fitted model is

$$\begin{aligned} y = &-1569.8 - 24.909x_1 + 196.95x_2 + 8.8669x_3 - 2.2359x_6 - 0.077581x_7 + 0.057329x_8 \\ &- 1.3057x_9 - 12.227x_{11} + 44.143x_{13} + 4.1883x_{14} + 0.97071x_{16} + 74.775x_{18} \\ &+ 21.656x_{19} - 18.253x_{20} + 82.591x_{21} - 37.553x_{22} + 329.8x_{23} \end{aligned}$$

(b) The eight-variable model containing the independent variables x_1, x_2, x_5, x_8, x_{10}, x_{11}, x_{14}, and x_{21} has Mallows' C_p equal to 1.7. The fitted model is

$$y = -665.98 - 24.782x_1 + 76.499x_2 + 121.96x_5 + 0.024247x_8 + 20.4x_{10} - 7.1313x_{11} + 2.4466x_{14} + 47.85x_{21}$$

(c) Using a value of 0.15 for both α-to-enter and α-to-remove, the equation chosen by stepwise regression is $y = -927.72 + 142.40x_5 + 0.081701x_7 + 21.698x_{10} + 0.41270x_{16} + 45.672x_{21}$.

(d) The following 13-variable model has adjusted R^2 equal to 0.95402. (There are also two 12-variable models whose adjusted R^2 is only very slightly lower.)

$$\begin{aligned} z = &\ 8663.2 - 313.31x_3 - 14.46x_6 + 0.358x_7 - 0.078746x_8 + 13.998x_9 + 230.24x_{10} \\ &- 188.16x_{13} + 5.4133x_{14} + 1928.2x_{15} - 8.2533x_{16} + 294.94x_{19} + 129.79x_{22} - 3020.7x_{23} \end{aligned}$$

(e) The two-variable model $z = -1660.9 + 0.67152x_7 + 134.28x_{10}$ has Mallows' C_p equal to −4.0.

(f) Using a value of 0.15 for both α-to-enter and α-to-remove, the equation chosen by stepwise regression is $z = -1660.9 + 0.67152x_7 + 134.28x_{10}$

(g) The following 17-variable model has adjusted R^2 equal to 0.97783.

$$\begin{aligned} w = &\ 700.56 - 21.701x_2 - 20.000x_3 + 21.813x_4 + 62.599x_5 + 0.016156x_7 \\ &- 0.012689x_8 + 1.1315x_9 + 15.245x_{10} + 1.1103x_{11} - 20.523x_{13} - 90.189x_{15} \\ &- 0.77442x_{16} + 7.5559x_{19} + 5.9163x_{20} - 7.5497x_{21} + 12.994x_{22} - 271.32x_{23} \end{aligned}$$

(h) The following 13-variable model has Mallows' C_p equal to 8.0.

$$w = 567.06 - 23.582x_2 - 16.766x_3 + 90.482x_5 + 0.0082274x_7 - 0.011004x_8 + 0.89554x_9 + 12.131x_{10} - 11.984x_{13} - 0.67302x_{16} + 11.097x_{19} + 4.6448x_{20} + 11.108x_{22} - 217.82x_{23}$$

(i) Using a value of 0.15 for both α-to-enter and α-to-remove, the equation chosen by stepwise regression is $w = 130.92 - 28.085x_2 + 113.49x_5 + 0.16802x_9 - 0.20216x_{16} + 11.417x_{19} + 12.068x_{21} - 78.371x_{23}$.

Section 9.1

1. (a)

Source	DF	SS	MS	F	P
Duration	4	1.3280	0.33200	7.1143	0.002
Error	15	0.7000	0.046667		
Total	19	2.0280			

(b) Yes. $F_{4,15} = 7.1143$, $0.001 < P < 0.01$ ($P = 0.002$).

3. (a)

Source	DF	SS	MS	F	P
Treatment	4	19.009	4.7522	2.3604	0.117
Error	11	22.147	2.0133		
Total	15	41.155			

(b) No. $F_{4,11} = 2.3604$, $P > 0.10$ ($P = 0.117$).

5. (a)

Source	DF	SS	MS	F	P
Site	3	1.4498	0.48327	2.1183	0.111
Error	47	10.723	0.22815		
Total	50	12.173			

(b) No. $F_{3,47} = 2.1183$, $P > 0.10$ ($P = 0.111$).

7. (a)

Source	DF	SS	MS	F	P
Catalyst	2	61.960	30.980	5.9926	0.037
Error	6	31.019	5.1698		
Total	8	92.979			

(b) Yes. $F_{2,6} = 5.9926$, $0.01 < P < 0.05$ ($P = 0.037$).

9. No, $F_{3,16} = 15.8255$, $P < 0.001$ ($P \approx 4.8 \times 10^{-5}$).

11. (a)

Source	DF	SS	MS	F	P
Temperature	3	58.650	19.550	8.4914	0.001
Error	16	36.837	2.3023		
Total	19	95.487			

(b) Yes, $F_{3,16} = 8.4914$, $0.001 < P < 0.01$ ($P = 0.0013$).

13. (a) $s = 1.517$ (b) 18 (c) 38

15. (a)

Source	DF	SS	MS	F	P
Grade	3	1721.4	573.81	9.4431	0.000
Error	96	5833.4	60.765		
Total	99	7554.9			

(b) Yes, $F_{3,96} = 9.4431$, $P < 0.001$ ($P \approx 0$).

17. (a) 0.85 (b) 56 (c) 200 (d) Greater than 0.864138. The larger the difference is, the greater the probability is to detect it.

Section 9.2

1. (a) Yes, $F_{5,6} = 46.64$, $P \approx 0$.

(b) $q_{6,6,.05} = 5.63$. The value of MSE is 0.00508. The 5% critical value is therefore $5.63\sqrt{0.00508/2} = 0.284$. Any pair that differs by more than 0.284 can be concluded to be different. The following pairs meet this criterion: A and B, A and C, A and D, A and E, B and C, B and D, B and E, B and F, D and F.

(c) $t_{6,.025/15} = 4.698$ (the value obtained by interpolating is 4.958). The value of MSE is 0.00508. The 5% critical value is therefore $4.698\sqrt{2(0.00508)/2} = 0.335$. Any pair that differs by more than 0.335 may be concluded to be different. The following pairs meet this criterion: A and B, A and C, A and D, A and E, B and C, B and D, B and E, B and F, D and F.

(d) The Tukey–Kramer method is more powerful, since its critical value is smaller (0.284 vs. 0.335).

(e) Either the Bonferroni or the Tukey–Kramer method can be used.

3. (a) MSE = 2.9659, $J_i = 12$ for all i. There are seven comparisons to be made. Now $t_{88,.025/7} = 2.754$, so the 5% critical value is $2.754\sqrt{2.9659(1/12 + 1/12)} = 1.936$. All the sample means of the noncontrol formulations differ from the sample mean of the control formulation by more than this amount. Therefore we conclude at the 5% level that all the noncontrol formulations differ from the control formulation.

(b) There are seven comparisons to be made. We should use the Studentized range value $q_{7,88,.05}$. This value is not in the table, so we will use $q_{7,60,.05} = 4.31$, which is only slightly larger. The 5% critical value is $4.31\sqrt{2.9659/12} = 2.14$. All the noncontrol formulations differ from the sample mean of the control formulation by more than this amount. Therefore we conclude at the 5% level that all the noncontrol formulations differ from the control formulation.

(c) The Bonferroni method is more powerful, because it is based on the actual number of comparisons being made, which is 7. The Tukey–Kramer method is based on the largest number of comparisons that could be made, which is $(7)(8)/2 = 28$.

5. (a) $t_{16,.025/6} = 3.0083$ (the value obtained by interpolating is 3.080). The value of MSE is 2.3023. The 5% critical value is therefore $3.0083\sqrt{2(2.3023)/5} = 2.8869$. We may conclude that the mean for 750°C differs from the means for 850°C and 900°C, and that the mean for 800° differs from the mean for 900°C.

(b) $q_{4,16,.05} = 4.05$. The value of MSE is 2.3023. The 5% critical value is therefore $4.05\sqrt{2.3023/5} = 2.75$. We may conclude that the mean for 750°C differs from the means for 850°C and 900°C, and that the mean for 800° differs from the mean for 900°C.

(c) The Tukey–Kramer method is more powerful, because its critical value is smaller.

7. (a) $t_{16,.025/3} = 2.6730$ (the value obtained by interpolating is 2.696). The value of MSE is 2.3023. The 5% critical value is therefore $2.6730\sqrt{2(2.3023)/5} = 2.5651$. We may conclude that the mean for 900°C differs from the means for 750°C and 800°C.

(b) $q_{4,16,.05} = 4.05$. The value of MSE is 2.3023. The 5% critical value is therefore $4.05\sqrt{2.3023/5} = 2.75$. We may conclude that the mean for 900°C differs from the means for 750°C and 800°C.

(c) The Bonferroni method is more powerful, because its critical value is smaller.

9. (a) $t_{47,.025} = 2.012$, MSE $= 0.22815$, the sample sizes are 16 and 9. The sample means are $\overline{X}_1 = 1.255625$, $\overline{X}_2 = 1.756667$. The 95% confidence interval is $0.501042 \pm 2.012\sqrt{0.22815(1/16 + 1/9)}$, or (0.1006, 0.9015).

(b) The sample sizes are $J_1 = 16$, $J_2 = 9$, $J_3 = 14$, $J_4 = 12$. MSE $= 0.22815$. We should use the Studentized range value $q_{4,47,.05}$. This value is not in the table, so we will use $q_{4,40,.05} = 3.79$, which is only slightly larger. The values of $q_{4,40,.05}\sqrt{(\text{MSE}/2)(1/J_i + 1/J_j)}$ are presented in the following table on the left, and the values of the differences $|\overline{X}_{i.} - \overline{X}_{j.}|$ are presented in the table on the right.

	1	2	3	4
1	—	0.53336	0.46846	0.48884
2	0.53336	—	0.54691	0.56446
3	0.46846	0.54691	—	0.50358
4	0.48884	0.56446	0.50358	

	1	2	3	4
1	0	0.50104	0.16223	0.18354
2	0.50104	0	0.33881	0.3175
3	0.16223	0.33881	0	0.02131
4	0.18354	0.31750	0.02131	0

None of the differences exceeds its critical value, so we cannot conclude at the 5% level that any of the treatment means differ.

11. (a) $t_{8,.025} = 2.306$, MSE $= 1.3718$. The sample means are $\overline{X}_1 = 1.998$ and $\overline{X}_3 = 5.300$. The sample sizes are $J_1 = 5$ and $J_3 = 3$. The 95% confidence interval is therefore $3.302 \pm 2.306\sqrt{1.3718(1/5 + 1/3)}$, or (1.330, 5.274).

(b) The sample means are $\overline{X}_1 = 1.998$, $\overline{X}_2 = 3.0000$, $\overline{X}_3 = 5.300$. The sample sizes are $J_1 = 5$, $J_2 = J_3 = 3$. The upper 5% point of the Studentized range is $q_{3,8,.05} = 4.04$. The 5% critical value for $|\overline{X}_1 - \overline{X}_2|$ and for $|\overline{X}_1 - \overline{X}_3|$ is $4.04\sqrt{(1.3718/2)(1/5 + 1/3)} = 2.44$, and the 5% critical value for $|\overline{X}_2 - \overline{X}_3|$ is $4.04\sqrt{(1.3718/2)(1/3 + 1/3)} = 2.73$. Therefore means 1 and 3 differ at the 5% level.

13. (a) MSTr $= 19.554$ so $F = 19.554/3.85 = 5.08$. There are 3 and 16 degrees of freedom, so $0.01 < P < 0.05$. The null hypothesis of no difference is rejected at the 5% level.

(b) $q_{4,16,.05} = 4.05$, so catalysts whose means differ by more than $4.05\sqrt{3.85/5} = 3.55$ are significantly different at the 5% level. Catalysts 1 and 2 both differ significantly from catalyst 4.

15. Any value of MSE satisfying $5.099 < \text{MSE} < 6.035$.

Section 9.3

1. (a) 3 (b) 2 (c) 6 (d) 24

(e)

Source	DF	SS	MS	F	P
Oil	3	1.0926	0.36420	5.1314	0.007
Ring	2	0.9340	0.46700	6.5798	0.005
Interaction	6	0.2485	0.041417	0.58354	0.740
Error	24	1.7034	0.070975		
Total	35	3.9785			

(f) Yes. $F_{6,24} = 0.58354$, $P > 0.10$ ($P = 0.740$).

(g) No, some of the main effects of oil type are nonzero. $F_{3,24} = 5.1314$, $0.001 < P < 0.01$ ($P = 0.007$).

(h) No, some of the main effects of piston ring type are nonzero. $F_{2,24} = 6.5798$, $0.001 < P < 0.01$ ($P = 0.005$).

3. (a)

Source	DF	SS	MS	F	P
Mold Temp.	4	69,738	17,434.5	6.7724	0.000
Alloy	2	8958	4479.0	1.7399	0.187
Interaction	8	7275	909.38	0.35325	0.939
Error	45	115,845	2574.3		
Total	59	201,816			

(b) Yes. $F_{8,45} = 0.35325$, $P > 0.10$ ($P = 0.939$).
(c) No, some of the main effects of mold temperature are nonzero. $F_{4,45} = 6.7724$, $P < 0.001$ ($P \approx 0$).
(d) Yes. $F_{3,45} = 1.7399$, $P > 0.10$, ($P = 0.187$).

5. (a)

Main Effects of Solution	
NaCl	−9.1148
Na_2HPO_4	9.1148

Main Effects of Temperature	
25°C	1.8101
37°C	−1.8101

Interactions

	Temperature	
Solution	25°C	37°C
NaCl	−0.49983	0.49983
Na_2HPO_4	0.49983	−0.49983

(b)

Source	DF	SS	MS	F	P
Solution	1	1993.9	1993.9	5.1983	0.034
Temperature	1	78.634	78.634	0.20500	0.656
Interaction	1	5.9960	5.9960	0.015632	0.902
Error	20	7671.4	383.57		
Total	23	9750.0			

(c) Yes, $F_{1,20} = 0.015632$, $P > 0.10$ ($P = 0.902$).
(d) Yes, since the additive model is plausible. The mean yield stress differs between Na_2HPO_4 and NaCl: $F_{1,20} = 5.1983$, $0.01 < P < 0.05$ ($P = 0.034$).
(e) There is no evidence that the temperature affects yield stress: $F_{1,20} = 0.20500$, $P > 0.10$ ($P = 0.656$).

7. (a)

Main Effects of Weight	
15	1.9125
25	−1.9125

Main Effects of Lauric Acid	
10	2.8375
30	−2.8375

Interactions

	Fraction = 10	Fraction = 30
Weight = 15	−3.9125	3.9125
Weight = 25	3.9125	−3.9125

(b)

Source	DF	SS	MS	F	P
Weight	1	29.261	29.261	7.4432	0.053
Lauric Acid	1	64.411	64.411	16.384	0.016
Interaction	1	122.46	122.46	31.151	0.005
Error	4	15.725	3.9312		
Total	7	231.86			

(c) No. $F_{1,4} = 31.151$, $0.001 < P < 0.01$ ($P = 0.005$).
(d) No, because the additive model is rejected.
(e) No, because the additive model is rejected.

9. (a)

Main Effects of Material	
$CPTi\text{-}ZrO_2$	0.044367
$TiAlloy\text{-}ZrO_2$	−0.044367

Main Effects of Neck Length	
Short	−0.018533
Medium	−0.024833
Long	0.043367

Interactions

	Neck Length		
	Short	Medium	Long
$CPTi\text{-}ZrO_2$	0.0063333	−0.023767	0.017433
$TiAlloy\text{-}ZrO_2$	0.0063333	0.023767	−0.017433

(b)

Source	DF	SS	MS	F	P
Taper Material	1	0.059052	0.059052	23.630	0.000
Neck Length	2	0.028408	0.014204	5.6840	0.010
Interaction	2	0.0090089	0.0045444	1.8185	0.184
Error	24	0.059976	0.002499		
Total	29	0.15652			

(c) Yes, the interactions may plausibly be equal to 0. The value of the test statistic is 1.8185, its null distribution is $F_{2,24}$, and $P > 0.10$ ($P = 0.184$).

(d) Yes, since the additive model is plausible. The mean coefficient of friction differs between $CPTi\text{-}ZrO_2$ and $TiAlloy\text{-}ZrO_2$: $F_{1,24} = 23.630$, $P < 0.001$.

(e) Yes, since the additive model is plausible. The mean coefficient of friction is not the same for all neck lengths: $F_{2,24} = 5.6840$, $P \approx 0.01$. To determine which pairs of effects differ, we use $q_{3,24,.05} = 3.53$. We compute $3.53\sqrt{0.002499/10} = 0.056$. We conclude that the effect of long neck length differs from both short and medium lengths, but we cannot conclude that the effects of short and medium lengths differ from each other.

11. (a)

Main Effects of Concentration	
15	0.16667
40	−0.067778
100	−0.098889

Main Effects of Delivery Ratio	
1:1	0.73333
1:5	−0.30000
1:10	−0.43333

Interactions

	Delivery Ratio		
Concentration	1:1	1:5	1:10
15	0.66778	−0.30222	−0.36556
40	−0.20111	−0.064444	0.26556
100	−0.46667	0.36667	0.10000

(b)

Source	DF	SS	MS	F	P
Concentration	2	0.37936	0.18968	3.8736	0.040
Delivery Ratio	2	7.34	3.67	74.949	0.000
Interaction	4	3.4447	0.86118	17.587	0.000
Error	18	0.8814	0.048967		
Total	26	12.045			

(c) No. The value of the test statistic is 17.587, its null distribution is $F_{4,18}$, and $P \approx 0$.

(d)

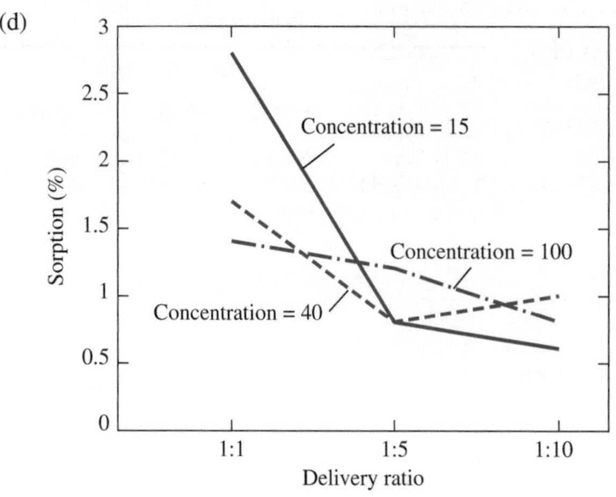

The slopes of the line segments are quite different from one another, indicating a high degree of interaction.

13. (a)

Main Effects of Attachment	
Nail	−1.3832
Adhesive	1.3832

Main Effects of Length	
Quarter	−7.1165
Half	−2.5665
Full	9.683

Interactions

	Length		
Attachment	**Quarter**	**Half**	**Full**
Nail	0.48317	0.33167	−0.51633
Adhesive	−0.48317	−0.33167	0.51633

(b)

Source	DF	SS	MS	F	P
Attachment	1	114.79	114.79	57.773	0.000
Length	2	3019.8	1509.9	759.94	0.000
Interaction	2	10.023	5.0115	2.5223	0.090
Error	54	107.29	1.9869		
Total	59	3251.9			

(c) The additive model is barely plausible: $F_{2,54} = 2.5223$, $0.05 < P < 0.10$ ($P = 0.090$).

(d) Yes, the attachment method does affect the critical buckling load: $F_{1,54} = 57.773$, $P \approx 0$.

(e) Yes, the side member length does affect the critical buckling load: $F_{2,54} = 759.94$, $P \approx 0$. To determine which effects differ at the 5% level, we should use $q_{3,54,.05}$. This value is not found in Table A.8, so we approximate it with $q_{3,40,.05} = 3.44$. We compute $3.44\sqrt{1.9869/20} = 1.08$. We conclude that the effects of quarter, half, and full all differ from each other.

15. (a)

Source	DF	SS	MS	F	P
Wafer	2	114,661.4	57,330.7	11,340.1	0.000
Operator	2	136.78	68.389	13.53	0.002
Interaction	4	6.5556	1.6389	0.32	0.855
Error	9	45.500	5.0556		
Total	17	114,850.3			

(b) There are differences among the operators. $F_{2,9} = 13.53$, $0.01 < P < 0.001$ ($P = 0.002$).

17. (a)

Source	DF	SS	MS	F	P
PVAL	2	125.41	62.704	8.2424	0.003
DCM	2	1647.9	823.94	108.31	0.000
Interaction	4	159.96	39.990	5.2567	0.006
Error	18	136.94	7.6075		
Total	26	2070.2			

(b) Since the interaction terms are not equal to 0 ($F_{4,18} = 5.2567$, $P = 0.006$), we cannot interpret the main effects. Therefore we compute the cell means. These are

	DCM (mL)		
PVAL	50	40	30
0.5	97.8	92.7	74.2
1.0	93.5	80.8	75.4
2.0	94.2	88.6	78.8

We conclude that a DCM level of 50 mL produces greater encapsulation efficiency than either of the other levels. If DCM = 50, the PVAL concentration does not have much effect. Note that for DCM = 50, encapsulation efficiency is maximized at the lowest PVAL concentration, but for DCM = 30 it is maximized at the highest PVAL concentration. This is the source of the significant interaction.

Section 9.4

1. (a) Liming is the blocking factor; soil is the treatment factor.

(b)

Source	DF	SS	MS	F	P
Soil	3	1.178	0.39267	18.335	0.000
Block	4	5.047	1.2617	58.914	0.000
Error	12	0.257	0.021417		
Total	19	6.482			

(c) Yes, $F_{3,12} = 18.335$, $P \approx 0$.

(d) $q_{4,12,.05} = 4.20$, MSAB $= 0.021417$, and $J = 5$. The 5% critical value is therefore $4.20\sqrt{0.021417/5} = 0.275$. The sample means are $\overline{X}_A = 6.32$, $\overline{X}_B = 6.02$, $\overline{X}_C = 6.28$, $\overline{X}_D = 6.70$. We can therefore conclude that D differs from A, B, and C, and that A differs from B.

3. (a)

Source	DF	SS	MS	F	P
Concentration	2	756.7	378.35	5.0185	0.015
Block	3	504.7	168.23	2.2315	0.110
Interaction	6	415.3	69.217	0.91809	0.499
Error	24	1809.4	75.392		
Total	35	3486.1			

(b) Yes. The P-value for interactions is large (0.499).

(c) Yes. The P-value for concentration is small (0.015).

5. (a)

Source	DF	SS	MS	F	P
Variety	9	339,032	37,670	2.5677	0.018
Block	5	1,860,838	372,168	25.367	0.000
Error	45	660,198	14,671		
Total	59	2,860,069			

(b) Yes, $F_{9,45} = 2.5677$, $P = 0.018$.

7. (a)

Source	DF	SS	MS	F	P
Waterworks	3	1253.5	417.84	4.8953	0.005
Block	14	1006.1	71.864	0.84193	0.622
Error	42	3585.0	85.356		
Total	59	5844.6			

(b) Yes, $F_{3,42} = 4.8953$, $P = 0.005$. (c) To determine which effects differ at the 5% level, we should use $q_{4,42,.05}$. This value is not found in Table A.8, so we approximate it with $q_{4,40,.05} = 3.79$. The 5% critical value is $3.79\sqrt{85.356/15} = 9.04$. The sample means are $\overline{X}_A = 34.000$, $\overline{X}_B = 22.933$, $\overline{X}_C = 24.800$, $\overline{X}_D = 31.467$. We can conclude that A differs from both B and C.

(d) The P-value for the blocking factor is large (0.622), suggesting that the blocking factor (time) has only a small effect on the outcome. It might therefore be reasonable to ignore the blocking factor and perform a one-way ANOVA.

9. (a) One motor of each type should be tested on each day. The order in which the motors are tested on any given day should be chosen at random. This is a randomized block design, in which the days are the blocks. It is not a completely randomized design, since randomization occurs only within blocks.

(b) The test statistic is $\dfrac{\sum_{i=1}^{5}(\overline{X}_{i.} - \overline{X}_{..})^2}{\sum_{j=1}^{4}\sum_{i=1}^{5}(X_{ij} - \overline{X}_{i.} - \overline{X}_{.j} - \overline{X}_{..})^2/12}$.

Section 9.5

1.

	A	B	C	D
1	−	−	−	−
ad	+	−	−	+
bd	−	+	−	+
ab	+	+	−	−
cd	−	−	+	+
ac	+	−	+	−
bc	−	+	+	−
abcd	+	+	+	+

The alias pairs are $\{A, BCD\}$, $\{B, ACD\}$, $\{C, ABD\}$, $\{D, ABC\}$, $\{AB, CD\}$, $\{AC, BD\}$, and $\{AD, BC\}$

3. (a)

Variable	Effect	DF	Sum of Squares	Mean Square	F	P
A	23.0000	1	2116.0000	2116.0000	264.5000	0.000
B	–5.0000	1	100.0000	100.0000	12.5000	0.008
C	1.5000	1	9.0000	9.0000	1.1250	0.320
AB	7.5000	1	225.0000	225.0000	28.1250	0.001
AC	3.0000	1	36.0000	36.0000	4.5000	0.067
BC	0.0000	1	0.0000	0.0000	0.0000	1.000
ABC	0.5000	1	1.0000	1.0000	0.1250	0.733
Error		8	64.0000	8.0000		
Total		15	2551.0000			

(b) Main effects A and B, and interaction AB seem most important. The AC interaction is borderline.
(c) The mean yield is higher when the temperature is 180°C.

5. (a)

Variable	Effect
A	3.3750
B	23.625
C	1.1250
AB	–2.8750
AC	–1.3750
BC	–1.6250
ABC	1.8750

(b) No, since the design is unreplicated, there is no error sum of squares.
(c) No, none of the interaction terms are nearly as large as the main effect of factor B.
(d) If the additive model is known to hold, then the following ANOVA table shows that the main effect of B is not equal to 0, while the main effects of A and C may be equal to 0.

Variable	Effect	DF	Sum of Squares	Mean Square	F	P
A	3.3750	1	22.781	22.781	2.7931	0.170
B	23.625	1	1116.3	1116.3	136.86	0.000
C	1.1250	1	2.5312	2.5312	0.31034	0.607
Error		4	32.625	8.1562		
Total		7	1174.2			

7. (a)

Variable	Effect
A	2.445
B	0.140
C	−0.250
AB	1.450
AC	0.610
BC	0.645
ABC	−0.935

(b) No, since the design is unreplicated, there is no error sum of squares.

(c) The estimates lie nearly on a straight line, so none of the factors can clearly be said to influence the resistance.

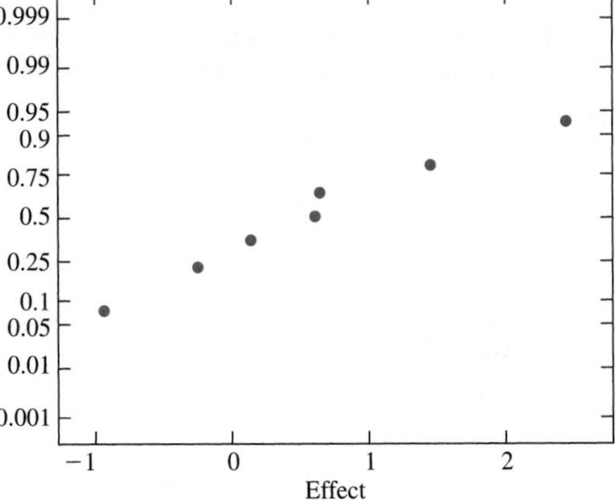

9. (a)

Variable	Effect
A	1.2
B	3.25
C	−16.05
D	−2.55
AB	2
AC	2.9
AD	−1.2
BC	1.05
BD	−1.45
CD	−1.6
ABC	−0.8
ABD	−1.9
ACD	−0.15
BCD	0.8
$ABCD$	0.65

(b) Factor C is the only one that really stands out.

11. (a)

Variable	Effect	DF	Sum of Squares	Mean Square	F	P
A	14.245	1	811.68	811.68	691.2	0.000
B	8.0275	1	257.76	257.76	219.5	0.000
C	−6.385	1	163.07	163.07	138.87	0.000
AB	−1.68	1	11.29	11.29	9.6139	0.015
AC	−1.1175	1	4.9952	4.9952	4.2538	0.073
BC	−0.535	1	1.1449	1.1449	0.97496	0.352
ABC	−1.2175	1	5.9292	5.9292	5.0492	0.055
Error		8	9.3944	1.1743		
Total		15	1265.3			

(b) All main effects are significant, as is the AB interaction. Only the BC interaction has a P-value that is reasonably large. All three factors appear to be important, and they seem to interact considerably with each other.

13. (ii)

Supplementary Exercises for Chapter 9

1.

Source	DF	SS	MS	F	P
Gypsum	3	0.013092	0.0043639	0.28916	0.832
Error	8	0.12073	0.015092		
Total	11	0.13383			

The value of the test statistic is $F_{3,8} = 0.28916$; $P > 0.10$ ($P = 0.832$). There is no evidence that the pH differs with the amount of gypsum added.

3.

Source	DF	SS	MS	F	P
Day	2	1.0908	0.54538	22.35	0.000
Error	36	0.87846	0.024402		
Total	38	1.9692			

We conclude that the mean sugar content differs among the three days ($F_{2,36} = 22.35$, $P \approx 0$).

5. (a) No. The variances are not constant across groups. In particular, there is an outlier in group 1.
(b) No, for the same reasons as in part (a).
(c)

Source	DF	SS	MS	F	P
Group	4	5.2029	1.3007	8.9126	0.000
Error	35	5.1080	0.14594		
Total	39	10.311			

We conclude that the mean dissolve time differs among the groups ($F_{4,35} = 8.9126$, $P \approx 0$).

7. The recommendation is not a good one. The engineer is trying to interpret the main effects without looking at the interactions. The small P-value for the interactions indicates that they must be taken into account. Looking at the cell means, it is clear that if design 2 is used, then the less expensive material performs just as well as the more expensive material. The best recommendation, therefore, is to use design 2 with the less expensive material.

9. (a)

Source	DF	SS	MS	F	P
Base	3	13,495	4498.3	7.5307	0.000
Instrument	2	90,990	45,495	76.164	0.000
Interaction	6	12,050	2008.3	3.3622	0.003
Error	708	422,912	597.33		
Total	719	539,447			

(b) No, it is not appropriate because there are interactions between the row and column effects ($F_{6,708} = 3.3622$, $P = 0.003$).

11. (a) Yes. $F_{4,15} = 8.7139$, $P = 0.001$. (b) $q_{5,20} = 4.23$, MSE $= 29.026$, $J = 4$. The 5% critical value is therefore $4.23\sqrt{29.026/4} = 11.39$. The sample means for the five channels are $\overline{X}_1 = 44.000$, $\overline{X}_2 = 44.100$, $\overline{X}_3 = 30.900$, $\overline{X}_4 = 28.575$, $\overline{X}_5 = 44.425$. We can therefore conclude that channels 3 and 4 differ from channels 1, 2, and 5.

13. No. $F_{4,289} = 1.5974$, $P > 0.10$ ($P = 0.175$).

15. (a) $s = 5.388$ (b) 10 (c) 22

17. (a)

Variable	Effect	Variable	Effect	Variable	Effect	Variable	Effect
A	3.9875	AB	−0.1125	BD	−0.0875	ACD	0.4875
B	2.0375	AC	0.0125	CD	0.6375	BCD	−0.3125
C	1.7125	AD	−0.9375	ABC	−0.2375	$ABCD$	−0.7125
D	3.7125	BC	0.7125	ABD	0.5125		

(b) The main effects are noticeably larger than the interactions, and the main effects for A and D are noticeably larger than those for B and C.

(c)

Variable	Effect	DF	Sum of Squares	Mean Square	F	P
A	3.9875	1	63.601	63.601	68.415	0.000
B	2.0375	1	16.606	16.606	17.863	0.008
C	1.7125	1	11.731	11.731	12.619	0.016
D	3.7125	1	55.131	55.131	59.304	0.001
AB	−0.1125	1	0.050625	0.050625	0.054457	0.825
AC	0.0125	1	0.000625	0.000625	0.00067231	0.980
AD	−0.9375	1	3.5156	3.5156	3.7818	0.109
BC	0.7125	1	2.0306	2.0306	2.1843	0.199
BD	−0.0875	1	0.030625	0.030625	0.032943	0.863
CD	0.6375	1	1.6256	1.6256	1.7487	0.243
Interaction		5	4.6481	0.92963		
Total		15	158.97			

We can conclude that each of the factors A, B, C, and D has an effect on the outcome.

(d) The F statistics are computed by dividing the mean square for each effect (equal to its sum of squares) by the error mean square 1.04. The degrees of freedom for each F statistic are 1 and 4. The results are summarized in the following table.

Variable	Effect	DF	Sum of Squares	Mean Square	F	P
A	3.9875	1	63.601	63.601	61.154	0.001
B	2.0375	1	16.606	16.606	15.967	0.016
C	1.7125	1	11.731	11.731	11.279	0.028
D	3.7125	1	55.131	55.131	53.01	0.002
AB	−0.1125	1	0.050625	0.050625	0.048678	0.836
AC	0.0125	1	0.000625	0.000625	0.00060096	0.982
AD	−0.9375	1	3.5156	3.5156	3.3804	0.140
BC	0.7125	1	2.0306	2.0306	1.9525	0.235
BD	−0.0875	1	0.030625	0.030625	0.029447	0.872
CD	0.6375	1	1.6256	1.6256	1.5631	0.279
ABC	−0.2375	1	0.22563	0.22563	0.21695	0.666
ABD	0.5125	1	1.0506	1.0506	1.0102	0.372
ACD	0.4875	1	0.95063	0.95063	0.91406	0.393
BCD	−0.3125	1	0.39062	0.39062	0.3756	0.573
$ABCD$	−0.7125	1	2.0306	2.0306	1.9525	0.235

(e) Yes. None of the P-values for the third- or higher-order interactions are small.

(f) We can conclude that each of the factors A, B, C, and D has an effect on the outcome.

19. Yes, $F_{2,107} = 9.4427$, $P < 0.001$.

21. (a)

Source	DF	SS	MS	F	P
H_2SO_4	2	457.65	228.83	8.8447	0.008
$CaCl_2$	2	38,783	19,391	749.53	0.000
Interaction	4	279.78	69.946	2.7036	0.099
Error	9	232.85	25.872		
Total	17	39,753			

(b) The P-value for interactions is 0.099. One cannot rule out the additive model.

(c) Yes, $F_{2,9} = 8.8447$, $0.001 < P < 0.01$ ($P = 0.008$).

(d) Yes, $F_{2,9} = 749.53$, $P \approx 0.000$.

23. Yes, $F_{6,39} = 20.302$, $P \approx 0$.

Section 10.1

1. (a) Continuous (b) Count (c) Binary (d) Continuous

3. (a) is in control (b) has high capability

5. (a) False (b) False (c) True (d) True

Section 10.2

1. (a) LCL = 0, UCL = 10.931 (b) LCL = 0, UCL = 4.721 (c) LCL = 20.258, UCL = 27.242
(d) LCL = 20.358, UCL = 27.142

3. (a) LCL = 0, UCL = 0.277. The variance is in control.
(b) LCL = 1.034, UCL = 1.186. The process is out of control for the first time on sample 17.
(c) 1σ limits are 1.085, 1.135; 2σ limits are 1.0596, 1.1604. The process is out of control for the first time on sample 8, where two out of the last three samples are below the lower 2σ control limit.

5. (a) 43.86 (b) 12.03 (c) 0.985 (d) 17

7. (a) 0.126 (b) 0.237 (c) 0.582 (d) 256

9. (a) LCL = 0.0588, UCL = 3.863. The variance is in control.
(b) LCL = 197.292, UCL = 202.340. The process is in control.
(c) 1σ limits are 198.975, 200.657; 2σ limits are 198.133, 201.499. The process is in control.

11. (a) LCL = 0, UCL = 0.971. The variance is in control.
(b) LCL = 9.147, UCL = 10.473. The process is in control.
(c) 1σ limits are 9.589, 10.031; 2σ limits are 9.368, 10.252. The process is out of control for the first time on sample 9, where two of the last three sample means are below the lower 2σ control limit.

13. (a) LCL = 0, UCL = 6.984. The variance is out of control on sample 8. After deleting this sample, $\overline{\overline{X}} = 150.166$, $\overline{R} = 6.538$, $\overline{s} = 2.911$. The new limits for the S chart are 0 and 6.596. The variance is now in control.
(b) LCL = 145.427, UCL = 154.905. The process is in control.
(c) 1σ limits are 148.586, 151.746; 2σ limits are 147.007, 153.325. The process is in control (recall that sample 8 has been deleted).

Section 10.3

1. Center line is 0.0547, LCL is 0.00644, UCL is 0.1029.

3. Yes, the 3σ control limits are 0.0254 and 0.2234.

5. (iv)

7. It was out of control. The UCL is 45.82.

Section 10.4

1. (a) No samples need be deleted. (b) $\sigma_{\overline{X}} = (0.577)(0.131)/3 = 0.0252$

(c)

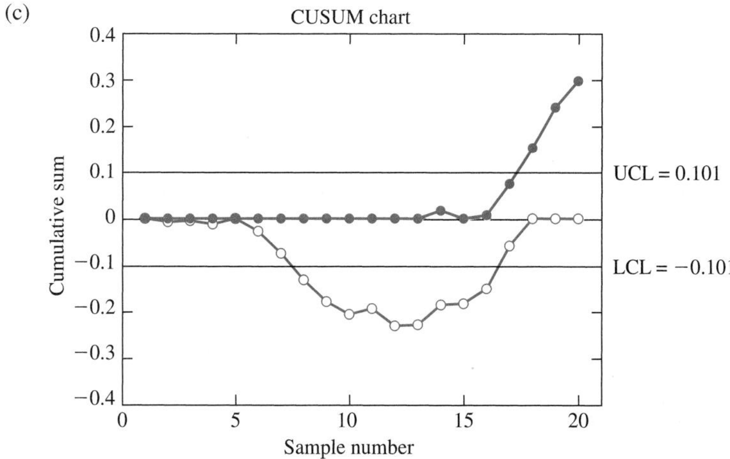

(d) The process is out of control on sample 8. (e) The Western Electric rules specify that the process is out of control on sample 8.

3. (a) No samples need be deleted. (b) $\sigma_{\overline{X}} = (0.577)(1.14)/3 = 0.219$

(c)

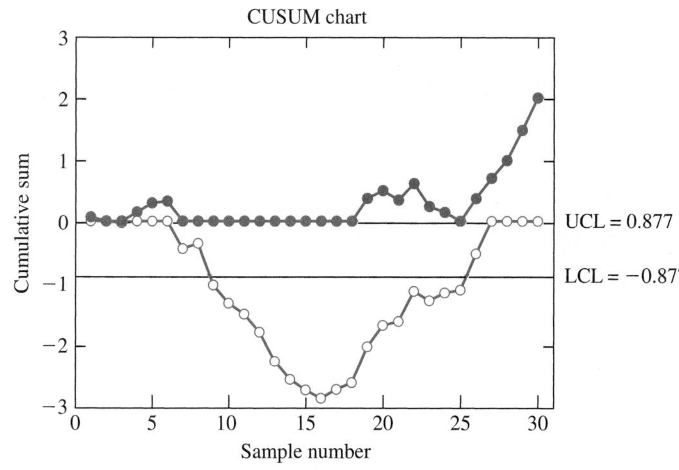

(d) The process is out of control on sample 9. (e) The Western Electric rules specify that the process is out of control on sample 9.

5. (a)

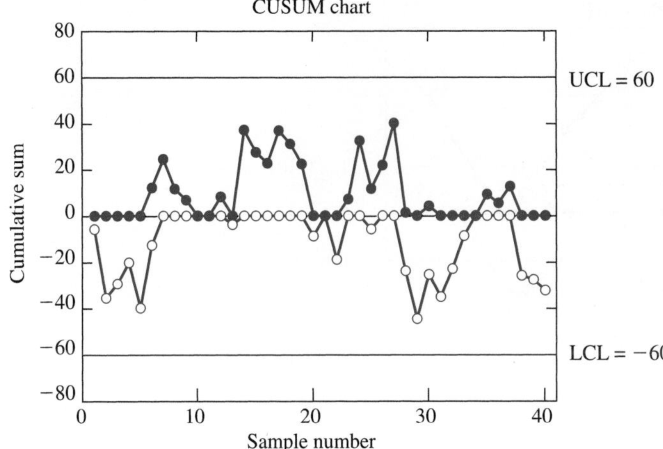

(b) The process is in control.

Section 10.5

1. (a) $C_{pk} = 2.303$ (b) Yes. Since $C_{pk} > 1$, the process capability is acceptable.

3. (a) 0.20 (b) 3.071

5. (a) $\mu \pm 3.6\sigma$ (b) 0.0004 (c) Likely. The normal approximation is likely to be inaccurate in the tails.

Supplementary Exercises for Chapter 10

1. Center line is 0.0596, LCL is 0.0147, UCL is 0.1045.

3. (a) LCL = 0, UCL = 0.283. The variance is in control.
(b) LCL = 4.982, UCL = 5.208. The process is out of control on sample 3.
(c) 1σ limits are 5.057, 5.133; 2σ limits are 5.020, 5.170. The process is out of control for the first time on sample 3, where a sample mean is above the upper 3σ control limit.

5. (a) No samples need be deleted. (b) $\sigma_{\overline{X}} = (1.023)(0.110)/3 = 0.0375$

(c)

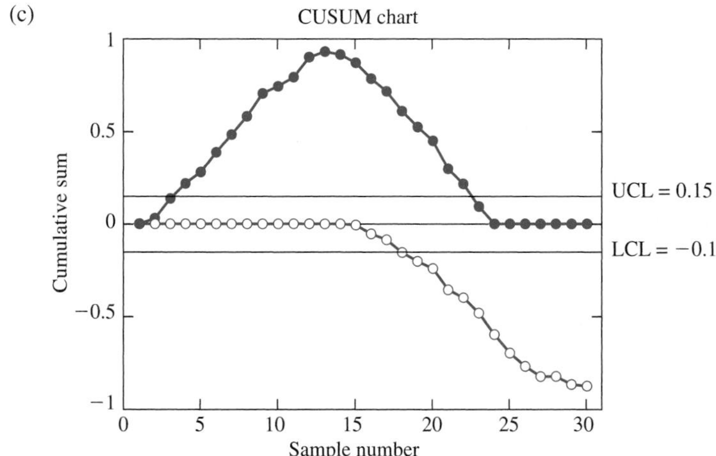

(d) The process is out of control on sample 4. (e) The Western Electric rules specify that the process is out of control on sample 3.

7. (a) LCL = 0.0170, UCL = 0.0726 (b) Sample 12
(c) No, this special cause improves the process. It should be preserved rather than eliminated.

Appendix B

1. $\dfrac{\partial v}{\partial x} = 3 + 2y^4, \quad \dfrac{\partial v}{\partial y} = 8xy^3$

2. $\dfrac{\partial w}{\partial x} = \dfrac{3x^2}{x^2 + y^2} - \dfrac{2x(x^3 + y^3)}{(x^2 + y^2)^2}, \quad \dfrac{\partial w}{\partial y} = \dfrac{3y^2}{x^2 + y^2} - \dfrac{2y(x^3 + y^3)}{(x^2 + y^2)^2}$

3. $\dfrac{\partial z}{\partial x} = -\sin x \sin y^2, \quad \dfrac{\partial z}{\partial y} = 2y \cos x \cos y^2$

4. $\dfrac{\partial v}{\partial x} = ye^{xy}, \quad \dfrac{\partial v}{\partial y} = xe^{xy}$

5. $\dfrac{\partial v}{\partial x} = e^x(\cos y + \sin z), \quad \dfrac{\partial v}{\partial y} = -e^x \sin y, \quad \dfrac{\partial v}{\partial z} = e^x \cos z$

6. $\dfrac{\partial w}{\partial x} = \dfrac{x}{\sqrt{x^2 + 4y^2 + 3z^2}}, \quad \dfrac{\partial w}{\partial y} = \dfrac{4y}{\sqrt{x^2 + 4y^2 + 3z^2}}, \quad \dfrac{\partial w}{\partial z} = \dfrac{3z}{\sqrt{x^2 + 4y^2 + 3z^2}}$

7. $\dfrac{\partial z}{\partial x} = \dfrac{2x}{x^2 + y^2}, \quad \dfrac{\partial z}{\partial y} = \dfrac{2y}{x^2 + y^2}$

8. $\dfrac{\partial v}{\partial x} = \dfrac{2xy}{x^2 y + z} - ze^{y^2} \sin(xz), \quad \dfrac{\partial v}{\partial y} = \dfrac{x^2}{x^2 y + z} + 2ye^{y^2} \cos(xz), \quad \dfrac{\partial v}{\partial z} = \dfrac{1}{x^2 y + z} - xe^{y^2} \sin(xz)$

9. $\dfrac{\partial v}{\partial x} = \sqrt{\dfrac{y^5}{x}} - \dfrac{3}{2}\sqrt{\dfrac{y^3}{x}}, \quad \dfrac{\partial v}{\partial y} = 5\sqrt{xy^3} - \dfrac{9}{2}\sqrt{xy}$

10. $\dfrac{\partial z}{\partial x} = \dfrac{xy \cos(x^2 y)}{\sqrt{\sin(x^2 y)}}, \quad \dfrac{\partial z}{\partial y} = \dfrac{x^2 \cos(x^2 y)}{2\sqrt{\sin(x^2 y)}}$

INDEX

D

E

Q

R

S

T